BIG IDEAS
MATH.

TEACHING EDITION

RED

Ron Larson
Laurie Boswell

Erie, Pennsylvania
BigIdeasLearning.com

Big Ideas Learning, LLC
1762 Norcross Road
Erie, PA 16510-3838
USA

For product information and customer support, contact Big Ideas Learning
at **1-877-552-7766** or visit us at ***BigIdeasLearning.com***.

Printed in the U.S.A.

ISBN 13: 978-1-60840-230-4
ISBN 10: 1-60840-230-4

2 3 4 5 6 7 8 9 10 WEB 15 14 13 12 11

AUTHORS

Ron Larson is a professor of mathematics at Penn State Erie, The Behrend College, where he has taught since receiving his Ph.D. in mathematics from the University of Colorado in 1970. Dr. Larson is well known as the lead author of a comprehensive program for mathematics that spans middle school, high school, and college courses. His high school and Advanced Placement books are published by Holt McDougal. Ron's numerous professional activities keep him in constant touch with the needs of students, teachers, and supervisors. Ron and Laurie Boswell began writing together in 1992. Since that time, they have authored over two dozen textbooks. In their collaboration, Ron is primarily responsible for the pupil edition and Laurie is primarily responsible for the teaching edition of the text.

Laurie Boswell is the Head of School and a mathematics teacher at the Riverside School in Lyndonville, Vermont. Dr. Boswell received her Ed.D. from the University of Vermont in 2010. She is a recipient of the Presidential Award for Excellence in Mathematics Teaching. Laurie has taught math to students at all levels, elementary through college. In addition, Laurie was a Tandy Technology Scholar, and served on the NCTM Board of Directors from 2002 to 2005. She currently serves on the board of NCSM, and is a popular national speaker. Along with Ron, Laurie has co-authored numerous math programs.

ABOUT THE BOOK

The traditional mile-wide and inch-deep programs that have been followed for years have clearly not worked. The Common Core State Standards for Mathematical Practice and Content are the foundation of the Big Ideas Math program. The program has been systematically developed using learning and instructional theory to ensure the quality of instruction. Big Ideas Math provides middle school students a well-articulated curriculum consisting of fewer and more focused standards, conceptual understanding of key ideas, and a continual building on what has been previously taught.

- **DEEPER**   Each section is designed for 2–3 day coverage.
- **DYNAMIC**   Each section begins with a full class period of active learning.
- **DOABLE**   Each section is accompanied by full student and teacher support.
- **DAZZLING**   How else can we say this? This book puts the dazzle back in math!

Ron Larson

Laurie Boswell

TEACHER REVIEWERS

Aaron Eisberg
Napa Valley Unified School District
Napa, CA

Gail Englert
Norfolk Public Schools
Norfolk, VA

Alexis Kaplan
Lindenwold Public Schools
Lindenwold, NJ

Lou Kwiatkowski
Millcreek Township School District
Erie, PA

Marcela Mansur
Broward County Public Schools
Fort Lauderdale, FL

Bonnie Pendergast
Tolleson Union High School District
Tolleson, AZ

Tammy Rush
Hillsborough County Public Schools
Tampa, FL

Patricia D. Seger
Polk County Public Schools
Bartow, FL

Denise Walston
Norfolk Public Schools
Norfolk, VA

STUDENT REVIEWERS

Ashley Benovic

Vanessa Bowser

Sara Chinsky

Kaitlyn Grimm

Lakota Noble

Norhan Omar

Jack Puckett

Abby Quinn

Victoria Royal

Madeline Su

Lance Williams

CONSULTANTS

Patsy Davis
Educational Consultant
Knoxville, Tennessee

Ryan Keating
Special Education Advisor
Gilbert, Arizona

Bob Fulenwider
Mathematics Consultant
Bakersfield, California

Michael McDowell
Project-Based Instruction Specialist
Tahoe City, California

Deb Johnson
Differentiated Instruction Consultant
Missoula, Montana

Sean McKeighan
Interdisciplinary Advisor
Norman, Oklahoma

Mark Johnson
Mathematics Assessment Consultant
Raymond, New Hampshire

Bonnie Spence
Differentiated Instruction Consultant
Missoula, Montana

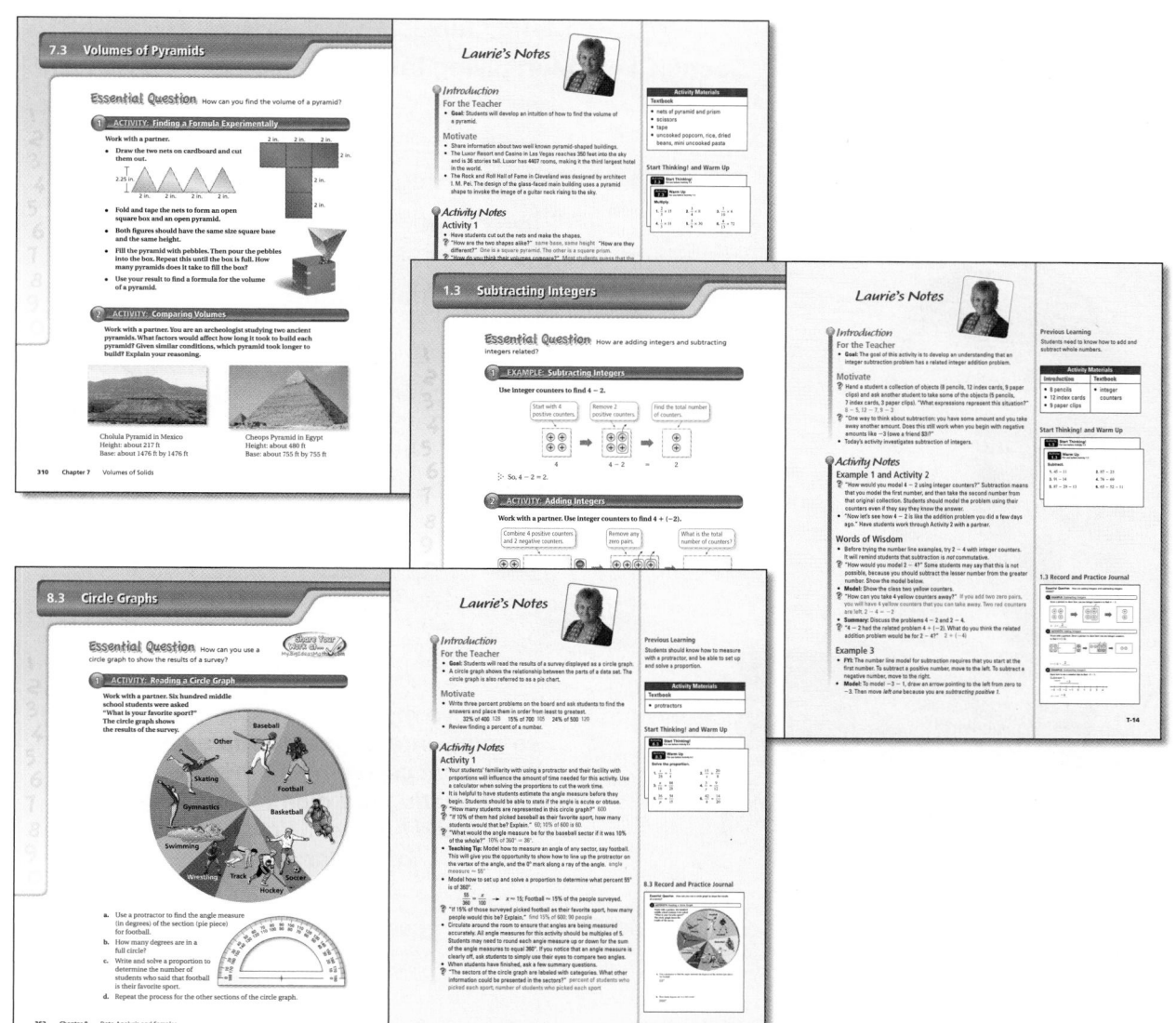

Operations with Integers

"I love my math book. It has so many interesting examples and homework problems. I have always liked math, but I didn't know how it could be used. Now I have lots of ideas."

Rational Numbers and Equations

"I like starting each new lesson with a partner activity. I just moved to this school and the activities helped me make friends."

Proportions and Variation

"I like having the book on the Internet. The online tutorials help me with my homework when I get stuck on a problem."

Percents

"*I love the cartoons. They are funny and they help me remember the math. I want to be a cartoonist some day.*"

Similarity and Transformations

"I like how I can click on the words in the book that is online and hear them read to me. I like to pronouce words correctly, but sometimes I don't know how to do that by just reading the words."

Surface Areas of Solids

"I really liked the projects at the end of the book. The history project on ancient Egypt was my favorite. Someday I would like to visit Egypt and go to the pyramids."

Volumes of Solids

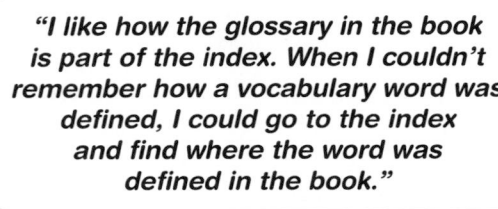

"I like how the glossary in the book is part of the index. When I couldn't remember how a vocabulary word was defined, I could go to the index and find where the word was defined in the book."

Data Analysis and Samples

"I like the practice tests in the book. I get really nervous on tests. So, having a practice test to work on at home helped me to chill out when the real test came."

Probability

"I like the review at the beginning of each chapter. This book has examples to help me remember things from last year. I don't like it when the review is just a list of questions."

Additional Topics

Appendix A:
My Big Ideas Projects

"I like the workbook (Record and Practice Journal). It saved me a lot of work to not have to copy all the questions and graphs."

PROGRAM OVERVIEW
Print
Available in print, online, and in digital format

- **Pupil Edition**
- **Teaching Edition**
- **Record and Practice Journal**
- **Assessment Book**
  - **Pre-Course Test**
  - **Quizzes**
  - **Chapter Tests**
  - **Standardized Test Practice**
  - **Alternative Assessment**
  - **End-of-Course Tests**
- **Resources by Chapter**
  - **Start Thinking! and Warm Up**
  - **Family and Community Involvement: English and Spanish**
  - **School-to-Work**
  - **Graphic Organizers/Study Help**
  - **Financial Literacy**
  - **Technology Connection**
  - **Life Connections**
  - **Stories in History**
  - **Extra Practice**
  - **Enrichment and Extension**
  - **Puzzle Time**
  - **Projects with Rubrics**
  - **Cumulative Practice**

- Differentiating the Lesson
- Skills Review Handbook
- Basic Skills Handbook
- Worked-Out Solutions
- Lesson Plans
- Teacher Tools

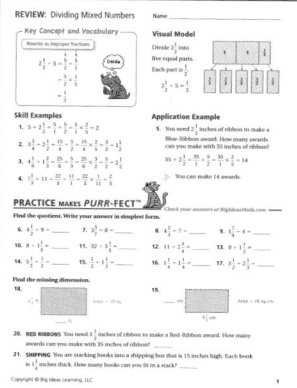

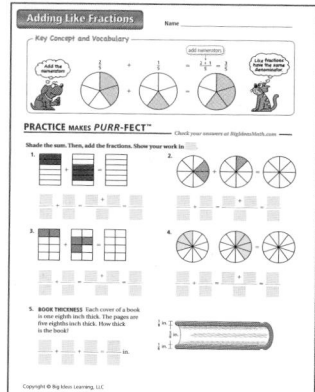

Technology

- Big Ideas **Exam**View® Assessment Suite
  Includes
    - Test Generator
    - Test Player
    - Test Manager

- **Lesson**View® Dynamic Planning Tool

- **Puzzle**View® Vocabulary Puzzle Builder

- **Mind**Point® QuizShow

- Interactive Glossary:
  English and Spanish

- Dynamic Classroom

- Answer Presentation Tool

- *BigIdeasMath.com*
    - Student Companion Website
    - Teacher Companion Website

- Lesson Tutorials

- Online Pupil Edition

COMMON CORE STATE STANDARDS TO BOOK CORRELATION

After a standard is introduced, it is revisited many times in subsequent activities, lessons, and exercises.

Domain: Ratios and Proportional Relationships

Standards

7.RP.1 Compute unit rates associated with ratios of fractions, including ratios of lengths, areas and other quantities measured in like or different units.
- **Section 3.1** Ratios and Rates

7.RP.2 Recognize and represent proportional relationships between quantities.

   a. Decide whether two quantities are in a proportional relationship.
   - **Section 3.3** Proportions
   - **Section 3.4** Writing Proportions
   - **Section 3.7** Direct Variation
   - **Section 3.8** Inverse Variation

   b. Identify the constant of proportionality (unit rate) in tables, graphs, equations, diagrams, and verbal descriptions of proportional relationships.
   - **Section 3.1** Ratios and Rates
   - **Section 3.2** Slope
   - **Section 3.5** Solving Proportions
   - **Section 3.7** Direct Variation
   - **Lesson 3.7b** Proportional Relationships

   c. Represent proportional relationships by equations.
   - **Section 3.7** Direct Variation

   d. Explain what a point (x, y) on the graph of a proportional relationship means in terms of the situation, with special attention to the points $(0, 0)$ and $(1, r)$ where r is the unit rate.
   - **Section 3.5** Solving Proportions
   - **Section 3.7** Direct Variation
   - **Lesson 3.7b** Proportional Relationships

7.RP.3 Use proportional relationships to solve multistep ratio and percent problems.
- **Section 3.6** Converting Measures Between Systems
- **Lesson 3.7b** Proportional Relationships
- **Section 4.1** The Percent Equation
- **Section 4.2** Percents of Increase and Decrease
- **Section 4.3** Discounts and Markups
- **Section 4.4** Simple Interest

Domain: The Number System

Standards

7.NS.1 Apply and extend previous understandings of addition and subtraction to add and subtract rational numbers; represent addition and subtraction on a horizontal or vertical number line diagram.

    a. Describe situations in which opposite quantities combine to make 0.
- **Lesson 2.3b** Number Properties

    b. Understand $p + q$ as the number located a distance $|q|$ from p, in the positive or negative direction depending on whether q is positive or negative. Show that a number and its opposite have a sum of 0 (are additive inverses). Interpret sums of rational numbers by describing real-world contexts.
- **Section 1.2** Adding Integers
- **Section 1.6** The Coordinate Plane
- **Section 2.2** Adding and Subtracting Rational Numbers

    c. Understand subtraction of rational numbers as adding the additive inverse, $p - q = p + (-q)$. Show that the distance between two rational numbers on the number line is the absolute value of their difference, and apply this principle in real-world contexts.
- **Section 1.3** Subtracting Integers
- **Section 2.2** Adding and Subtracting Rational Numbers

    d. Apply properties of operations as strategies to add and subtract rational numbers.
- **Section 1.2** Adding Integers
- **Section 1.3** Subtracting Integers
- **Section 2.2** Adding and Subtracting Rational Numbers
- **Lesson 2.3b** Number Properties

7.NS.2 Apply and extend previous understandings of multiplication and division and of fractions to multiply and divide rational numbers.

    a. Understand that multiplication is extended from fractions to rational numbers by requiring that operations continue to satisfy the properties of operations, particularly the distributive property, leading to products such as $(-1)(-1) = 1$ and the rules for multiplying signed numbers. Interpret products of rational numbers by describing real-world contexts.
- **Section 1.4** Multiplying Integers
- **Section 2.3** Multiplying and Dividing Rational Numbers

    b. Understand that integers can be divided, provided that the divisor is not zero, and every quotient of integers (with non-zero divisor) is a rational number. If p and q are integers, then $-(p/q) = (-p)/q = p/(-q)$. Interpret quotients of rational numbers by describing real-world contexts.
- **Section 1.5** Dividing Integers
- **Section 2.1** Rational Numbers
- **Section 2.3** Multiplying and Dividing Rational Numbers

c. Apply properties of operations as strategies to multiply and divide rational numbers.
  - **Section 1.4** Multiplying Integers
  - **Section 1.5** Dividing Integers
  - **Section 2.3** Multiplying and Dividing Rational Numbers
  - **Lesson 2.3b** Number Properties

d. Convert a rational number to a decimal using long division; know that the decimal form of a rational number terminates in 0s or eventually repeats.
  - **Section 2.1** Rational Numbers

7.NS.3 Solve real-world and mathematical problems involving the four operations with rational numbers.
  - **Section 1.2** Adding Integers
  - **Section 1.3** Subtracting Integers
  - **Section 1.4** Multiplying Integers
  - **Section 1.5** Dividing Integers
  - **Section 2.2** Adding and Subtracting Rational Numbers
  - **Section 2.3** Multiplying and Dividing Rational Numbers

Domain: Expressions and Equations

Standards

7.EE.1 Apply properties of operations as strategies to add, subtract, factor, and expand linear expressions with rational coefficients.
  - **Lesson 2.5b** Algebraic Expressions

7.EE.2 Understand that rewriting an expression in different forms in a problem context can shed light on the problem and how the quantities in it are related.
  - **Lesson 2.5b** Algebraic Expressions
  - **Section 4.3** Discounts and Markups

7.EE.3 Solve multi-step real-life and mathematical problems posed with positive and negative rational numbers in any form (whole numbers, fractions, and decimals), using tools strategically. Apply properties of operations to calculate with numbers in any form; convert between forms as appropriate; and assess the reasonableness of answers using mental computation and estimation strategies.
  - **Section 1.1** Integers and Absolute Value
  - **Section 1.2** Adding Integers
  - **Section 1.3** Subtracting Integers
  - **Section 1.4** Multiplying Integers
  - **Section 1.5** Dividing Integers
  - **Section 2.1** Rational Numbers
  - **Section 2.2** Adding and Subtracting Rational Numbers
  - **Section 2.3** Multiplying and Dividing Rational Numbers
  - **Section 4.1** The Percent Equation
  - **Section 4.2** Percents of Increase and Decrease
  - **Section 4.3** Discounts and Markups
  - **Section 4.4** Simple Interest

7.EE.4 Use variables to represent quantities in a real-world or mathematical problem, and construct simple equations and inequalities to solve problems by reasoning about the quantities.

    a. Solve word problems leading to equations of the form $px + q = r$ and $p(x + q) = r$, where p, q, and r are specific rational numbers. Solve equations of these forms fluently. Compare an algebraic solution to an arithmetic solution, identifying the sequence of the operations used in each approach.

- **Section 2.4** Solving Equations Using Addition or Subtraction
- **Section 2.5** Solving Equations Using Multiplication or Division
- **Section 2.6** Solving Two-Step Equations

    b. Solve word problems leading to inequalities of the form $px + q > r$ or $px + q < r$, where p, q, and r are specific rational numbers. Graph the solution set of the inequality and interpret it in the context of the problem.

- **Lesson 2.6b** Solving Inequalities

Domain: Geometry

Standards

7.G.1 Solve problems involving scale drawings of geometric figures, including computing actual lengths and areas from a scale drawing and reproducing a scale drawing at a different scale.

- **Section 5.1** Identifying Similar Figures
- **Section 5.2** Perimeters and Areas of Similar Figures
- **Section 5.3** Finding Unknown Measures in Similar Figures
- **Section 5.4** Scale Drawings
- **Lesson 5.4b** Scale Drawings

7.G.2 Draw (freehand, with ruler and protractor, and with technology) geometric shapes with given conditions. Focus on constructing triangles from three measures of angles or sides, noticing when the conditions determine a unique triangle, more than one triangle, or no triangle.

- **Section 5.5** Translations
- **Section 5.6** Reflections
- **Section 5.7** Rotations
- **Topic 2** Geometry

7.G.3 Describe the two-dimensional figures that result from slicing three-dimensional figures, as in plane sections of right rectangular prisms and right rectangular pyramids.

- **Section 6.1** Drawing 3-Dimensional Figures
- **Topic 2** Geometry

7.G.4 Know the formulas for the area and circumference of a circle and use them to solve problems; give an informal derivation of the relationship between the circumference and area of a circle.

- **Lesson 6.2b** Circles
- **Section 6.3** Surface Areas of Cylinders
- **Section 6.5** Surface Areas of Cones
- **Section 7.2** Volumes of Cylinders
- **Section 7.4** Volumes of Cones
- **Section 7.6** Surface Areas and Volumes of Similar Solids

7.G.5 Use facts about supplementary, complementary, vertical, and adjacent angles in a multi-step problem to write and solve simple equations for an unknown angle in a figure.

- **Topic 1** Angles

7.G.6 Solve real-world and mathematical problems involving area, volume and surface area of two- and three-dimensional objects composed of triangles, quadrilaterals, polygons, cubes, and right prisms.

- **Section 6.2** Surface Areas of Prisms
- **Section 6.3** Surface Areas of Cylinders
- **Section 6.4** Surface Areas of Pyramids
- **Section 6.5** Surface Areas of Cones
- **Section 6.6** Surface Areas of Composite Solids
- **Section 7.1** Volumes of Prisms
- **Section 7.2** Volumes of Cylinders
- **Section 7.3** Volumes of Pyramids
- **Section 7.4** Volumes of Cones
- **Section 7.5** Volumes of Composite Solids
- **Section 7.6** Surface Areas and Volumes of Similar Solids

Domain: Statistics and Probability

Standards

7.SP.1 Understand that statistics can be used to gain information about a population by examining a sample of the population; generalizations about a population from a sample are valid only if the sample is representative of that population. Understand that random sampling tends to produce representative samples and support valid inferences.

- **Section 8.1** Stem-and-Leaf Plots
- **Section 8.2** Histograms
- **Section 8.3** Circle Graphs
- **Section 8.4** Samples and Populations
- **Lesson 8.4b** Comparing Populations

7.SP.2 Use data from a random sample to draw inferences about a population with an unknown characteristic of interest. Generate multiple samples (or simulated samples) of the same size to gauge the variation in estimates or predictions.

- **Section 8.1** Stem-and-Leaf Plots
- **Section 8.2** Histograms
- **Section 8.3** Circle Graphs
- **Section 8.4** Samples and Populations
- **Lesson 8.4b** Comparing Populations

7.SP.3 Informally assess the degree of visual overlap of two numerical data distributions with similar variabilities, measuring the difference between the centers by expressing it as a multiple of a measure of variability.

- **Lesson 8.4b** Comparing Populations

7.SP.4 Use measures of center and measures of variability for numerical data from random samples to draw informal comparative inferences about two populations.

- **Lesson 8.4b** Comparing Populations

7.SP.5 Understand that the probability of a chance event is a number between 0 and 1 that expresses the likelihood of the event occurring. Larger numbers indicate greater likelihood. A probability near 0 indicates an unlikely event, a probability around 1/2 indicates an event that is neither unlikely nor likely, and a probability near 1 indicates a likely event.

- **Section 9.1** Introduction to Probability

7.SP.6 Approximate the probability of a chance event by collecting data on the chance process that produces it and observing its long-run relative frequency, and predict the approximate relative frequency given the probability.

- **Section 9.3** Experimental Probability

7.SP.7 Develop a probability model and use it to find probabilities of events. Compare probabilities from a model to observed frequencies; if the agreement is not good, explain possible sources of the discrepancy.

a. Develop a uniform probability model by assigning equal probability to all outcomes, and use the model to determine probabilities of events.

- **Section 9.2** Theoretical Probability

b. Develop a probability model (which may not be uniform) by observing frequencies in data generated from a chance process.

- **Section 9.2** Theoretical Probability

7.SP.8 Find probabilities of compound events using organized lists, tables, tree diagrams, and simulation.

    a. Understand that, just as with simple events, the probability of a compound event is the fraction of outcomes in the sample space for which the compound event occurs.
- **Section 9.4** Independent and Dependent Events

    b. Represent sample spaces for compound events using methods such as organized lists, tables and tree diagrams. For an event described in everyday language, identify the outcomes in the sample space which compose the event.
- **Section 9.4** Independent and Dependent Events

    c. Design and use a simulation to generate frequencies for compound events.
- **Section 9.4** Independent and Dependent Events

BOOK TO COMMON CORE STATE STANDARDS CORRELATION

Chapter 1 Operations with Integers

1.1 Integers and Absolute Value

- **7.EE.3** Solve multi-step real-life and mathematical problems posed with positive and negative rational numbers in any form (whole numbers, fractions, and decimals), using tools strategically. Apply properties of operations to calculate with numbers in any form; convert between forms as appropriate; and assess the reasonableness of answers using mental computation and estimation strategies.

1.2 Adding Integers

- **7.NS.1b** Understand $p + q$ as the number located a distance $|q|$ from p, in the positive or negative direction depending on whether q is positive or negative. Show that a number and its opposite have a sum of 0 (are additive inverses). Interpret sums of rational numbers by describing real-world contexts.
- **7.NS.1d** Apply properties of operations as strategies to add and subtract rational numbers.
- **7.NS.3** Solve real-world and mathematical problems involving the four operations with rational numbers.
- **7.EE.3** Solve multi-step real-life and mathematical problems posed with positive and negative rational numbers in any form (whole numbers, fractions, and decimals), using tools strategically. Apply properties of operations to calculate with numbers in any form; convert between forms as appropriate; and assess the reasonableness of answers using mental computation and estimation strategies.

1.3 Subtracting Integers

- **7.NS.1c** Understand subtraction of rational numbers as adding the additive inverse, $p - q = p + (-q)$. Show that the distance between two rational numbers on the number line is the absolute value of their difference, and apply this principle in real-world contexts.
- **7.NS.1d** Apply properties of operations as strategies to add and subtract rational numbers.
- **7.NS.3** Solve real-world and mathematical problems involving the four operations with rational numbers.
- **7.EE.3** Solve multi-step real-life and mathematical problems posed with positive and negative rational numbers in any form (whole numbers, fractions, and decimals), using tools strategically. Apply properties of operations to calculate with numbers in any form; convert between forms as appropriate; and assess the reasonableness of answers using mental computation and estimation strategies.

2.2 Adding and Subtracting Rational Numbers

- **7.NS.1b** Understand $p + q$ as the number located a distance $|q|$ from p, in the positive or negative direction depending on whether q is positive or negative. Show that a number and its opposite have a sum of 0 (are additive inverses). Interpret sums of rational numbers by describing real-world contexts.

- **7.NS.1c** Understand subtraction of rational numbers as adding the additive inverse, $p - q = p + (-q)$. Show that the distance between two rational numbers on the number line is the absolute value of their difference, and apply this principle in real-world contexts.

- **7.NS.1d** Apply properties of operations as strategies to add and subtract rational numbers.

- **7.NS.3** Solve real-world and mathematical problems involving the four operations with rational numbers.

- **7.EE.3** Solve multi-step real-life and mathematical problems posed with positive and negative rational numbers in any form (whole numbers, fractions, and decimals), using tools strategically. Apply properties of operations to calculate with numbers in any form; convert between forms as appropriate; and assess the reasonableness of answers using mental computation and estimation strategies.

2.3 Multiplying and Dividing Rational Numbers

- **7.NS.2a** Understand that multiplication is extended from fractions to rational numbers by requiring that operations continue to satisfy the properties of operations, particularly the distributive property, leading to products such as $(-1)(-1) = 1$ and the rules for multiplying signed numbers. Interpret products of rational numbers by describing real-world contexts.

- **7.NS.2b** Understand that integers can be divided, provided that the divisor is not zero, and every quotient of integers (with non-zero divisor) is a rational number. If p and q are integers, then $-(p/q) = (-p)/q = p/(-q)$. Interpret quotients of rational numbers by describing real-world contexts.

- **7.NS.2c** Apply properties of operations as strategies to multiply and divide rational numbers.

- **7.NS.3** Solve real-world and mathematical problems involving the four operations with rational numbers.

- **7.EE.3** Solve multi-step real-life and mathematical problems posed with positive and negative rational numbers in any form (whole numbers, fractions, and decimals), using tools strategically. Apply properties of operations to calculate with numbers in any form; convert between forms as appropriate; and assess the reasonableness of answers using mental computation and estimation strategies.

2.3b Number Properties

- **7.NS.1a** Describe situations in which opposite quantities combine to make 0.

- **7.NS.1d** Apply properties of operations as strategies to add and subtract rational numbers.

- **7.NS.2c** Apply properties of operations as strategies to multiply and divide rational numbers.

Chapter 7 Volumes of Solids

7.1 Volumes of Prisms

- **7.G.6** Solve real-world and mathematical problems involving area, volume and surface area of two- and three-dimensional objects composed of triangles, quadrilaterals, polygons, cubes, and right prisms.

7.2 Volumes of Cylinders

- **7.G.4** Know the formulas for the area and circumference of a circle and use them to solve problems; give an informal derivation of the relationship between the circumference and area of a circle.
- **7.G.6** Solve real-world and mathematical problems involving area, volume and surface area of two- and three-dimensional objects composed of triangles, quadrilaterals, polygons, cubes, and right prisms.

7.3 Volumes of Pyramids

- **7.G.6** Solve real-world and mathematical problems involving area, volume and surface area of two- and three-dimensional objects composed of triangles, quadrilaterals, polygons, cubes, and right prisms.

7.4 Volumes of Cones

- **7.G.4** Know the formulas for the area and circumference of a circle and use them to solve problems; give an informal derivation of the relationship between the circumference and area of a circle.
- **7.G.6** Solve real-world and mathematical problems involving area, volume and surface area of two- and three-dimensional objects composed of triangles, quadrilaterals, polygons, cubes, and right prisms.

7.5 Volumes of Composite Solids

- **7.G.6** Solve real-world and mathematical problems involving area, volume and surface area of two- and three-dimensional objects composed of triangles, quadrilaterals, polygons, cubes, and right prisms.

7.6 Surface Areas and Volumes of Similar Solids

- **7.G.4** Know the formulas for the area and circumference of a circle and use them to solve problems; give an informal derivation of the relationship between the circumference and area of a circle.
- **7.G.6** Solve real-world and mathematical problems involving area, volume and surface area of two- and three-dimensional objects composed of triangles, quadrilaterals, polygons, cubes, and right prisms.

Chapter 8 Data Analysis and Samples

8.1 Stem-and-Leaf Plots

- **7.SP.1** Understand that statistics can be used to gain information about a population by examining a sample of the population; generalizations about a population from a sample are valid only if the sample is representative of that population. Understand that random sampling tends to produce representative samples and support valid inferences.
- **7.SP.2** Use data from a random sample to draw inferences about a population with an unknown characteristic of interest. Generate multiple samples (or simulated samples) of the same size to gauge the variation in estimates or predictions.

8.2 Histograms

- **7.SP.1** Understand that statistics can be used to gain information about a population by examining a sample of the population; generalizations about a population from a sample are valid only if the sample is representative of that population. Understand that random sampling tends to produce representative samples and support valid inferences.

- **7.SP.2** Use data from a random sample to draw inferences about a population with an unknown characteristic of interest. Generate multiple samples (or simulated samples) of the same size to gauge the variation in estimates or predictions.

8.3 Circle Graphs

- **7.SP.1** Understand that statistics can be used to gain information about a population by examining a sample of the population; generalizations about a population from a sample are valid only if the sample is representative of that population. Understand that random sampling tends to produce representative samples and support valid inferences.

- **7.SP.2** Use data from a random sample to draw inferences about a population with an unknown characteristic of interest. Generate multiple samples (or simulated samples) of the same size to gauge the variation in estimates or predictions.

8.4 Samples and Populations

- **7.SP.1** Understand that statistics can be used to gain information about a population by examining a sample of the population; generalizations about a population from a sample are valid only if the sample is representative of that population. Understand that random sampling tends to produce representative samples and support valid inferences.

- **7.SP.2** Use data from a random sample to draw inferences about a population with an unknown characteristic of interest. Generate multiple samples (or simulated samples) of the same size to gauge the variation in estimates or predictions.

8.4b Comparing Populations

- **7.SP.1** Understand that statistics can be used to gain information about a population by examining a sample of the population; generalizations about a population from a sample are valid only if the sample is representative of that population. Understand that random sampling tends to produce representative samples and support valid inferences.

- **7.SP.2** Use data from a random sample to draw inferences about a population with an unknown characteristic of interest. Generate multiple samples (or simulated samples) of the same size to gauge the variation in estimates or predictions.

- **7.SP.3** Informally assess the degree of visual overlap of two numerical data distributions with similar variabilities, measuring the difference between the centers by expressing it as a multiple of a measure of variability.

- **7.SP.4** Use measures of center and measures of variability for numerical data from random samples to draw informal comparative inferences about two populations.

Chapter 9 Probability

9.1 Introduction to Probability
- **7.SP.5** Understand that the probability of a chance event is a number between 0 and 1 that expresses the likelihood of the event occurring. Larger numbers indicate greater likelihood. A probability near 0 indicates an unlikely event, a probability around 1/2 indicates an event that is neither unlikely nor likely, and a probability near 1 indicates a likely event.

9.2 Theoretical Probability
- **7.SP.7a** Develop a uniform probability model by assigning equal probability to all outcomes, and use the model to determine probabilities of events.
- **7.SP.7b** Develop a probability model (which may not be uniform) by observing frequencies in data generated from a chance process.

9.3 Experimental Probability
- **7.SP.6** Approximate the probability of a chance event by collecting data on the chance process that produces it and observing its long-run relative frequency, and predict the approximate relative frequency given the probability.

9.4 Independent and Dependent Events
- **7.SP.8a** Understand that, just as with simple events, the probability of a compound event is the fraction of outcomes in the sample space for which the compound event occurs.
- **7.SP.8b** Represent sample spaces for compound events using methods such as organized lists, tables and tree diagrams. For an event described in everyday language, identify the outcomes in the sample space which compose the event.
- **7.SP.8c** Design and use a simulation to generate frequencies for compound events.

Additional Topics

Topic 1 Angles
- **7.G.5** Use facts about supplementary, complementary, vertical, and adjacent angles in a multi-step problem to write and solve simple equations for an unknown angle in a figure.

Topic 2 Geometry
- **7.G.2** Draw (freehand, with ruler and protractor, and with technology) geometric shapes with given conditions. Focus on constructing triangles from three measures of angles or sides, noticing when the conditions determine a unique triangle, more than one triangle, or no triangle.
- **7.G.3** Describe the two-dimensional figures that result from slicing three-dimensional figures, as in plane sections of right rectangular prisms and right rectangular pyramids.

NARROWER AND DEEPER™

Middle school students need a new approach to learning mathematics. Big Ideas Math's *Narrower and Deeper*™ program is a revolutionary combination of the discovery and direct instruction approaches. Students gain a deeper understanding of math concepts by narrowing their focus to fewer topics. They master concepts through fun and engaging activities, stepped-out, concise examples, and rich, thought-provoking exercises.

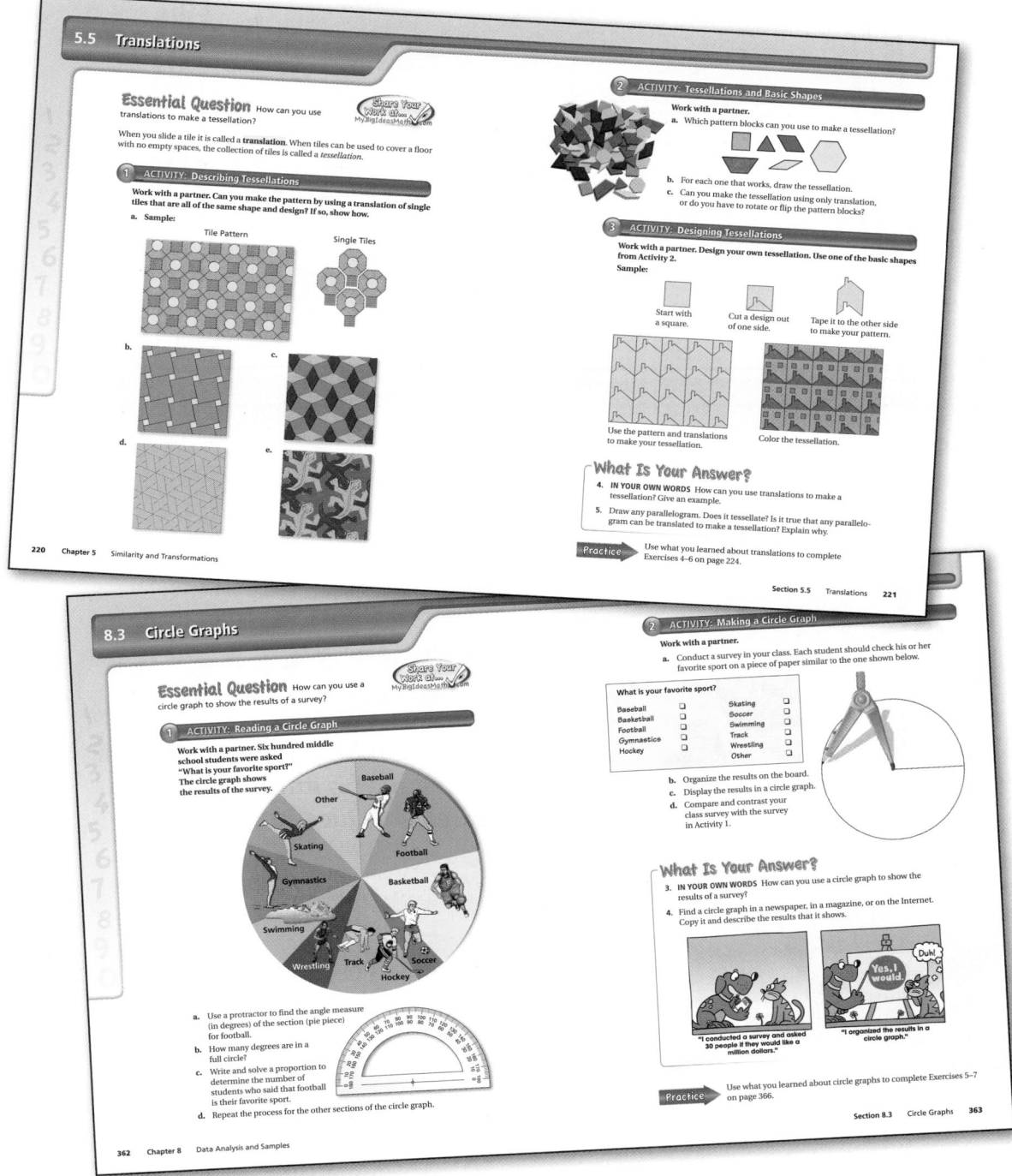

A BALANCED APPROACH

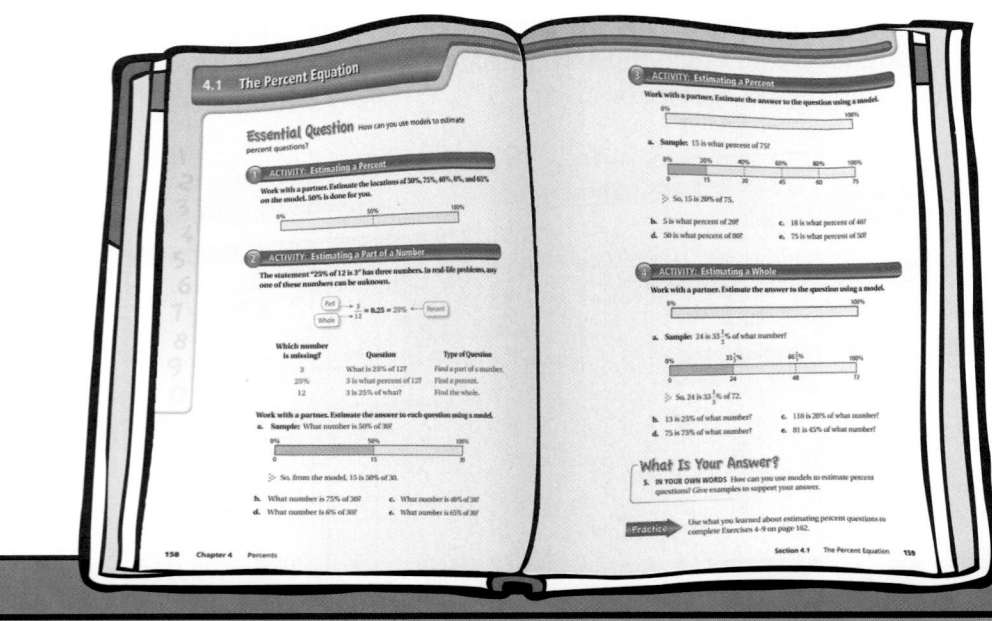

DISCOVERY

Each section begins with a 2-page *Activity* that is introduced by an Essential Question.

- **Deeper**
- **Dynamic**
- **Doable**
- **Dazzling**

TO INSTRUCTION

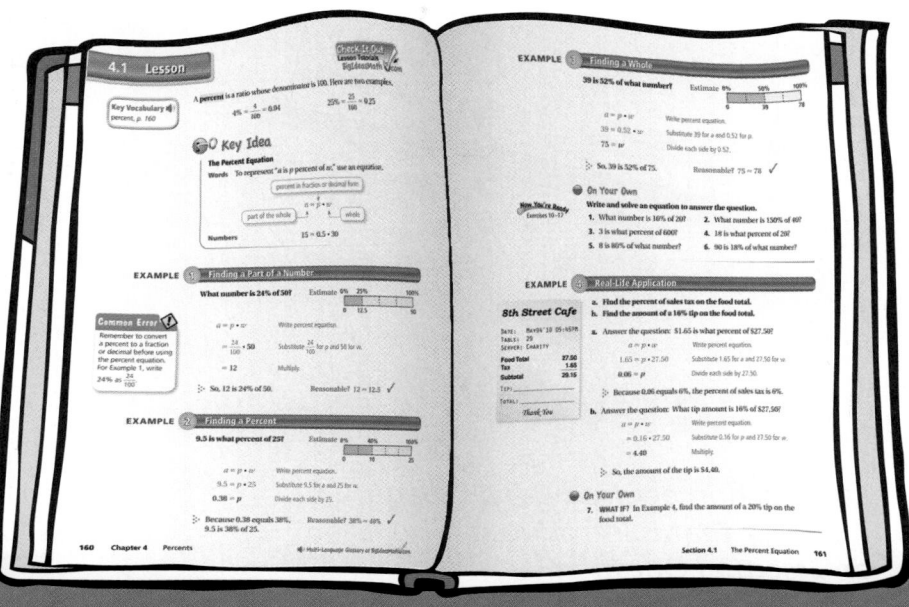

DIRECT INSTRUCTION

After the concept has been introduced with a full-class period **Activity***, it is extended the following day through the* **Lesson***.*

- **Key Ideas**
- **Examples**
- **"On Your Own" Questions**

ENGAGING PUPIL BOOKS

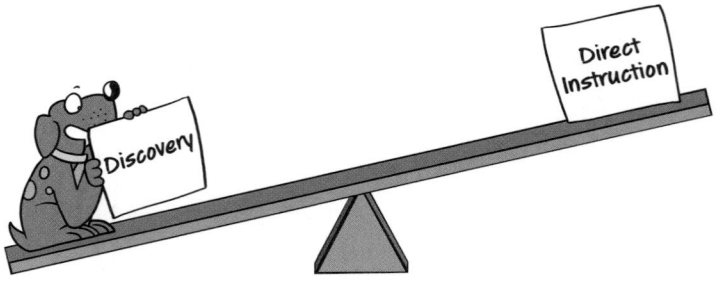

ACTIVITY

Each section begins with a 2-page Activity that is introduced by an Essential Question.

- **Deeper**
- **Dynamic**
- **Doable**
- **Dazzling**

Students gain a deeper understanding of topics through inductive reasoning and exploration.

Students develop communication and problem-solving skills by answering Essential Questions.

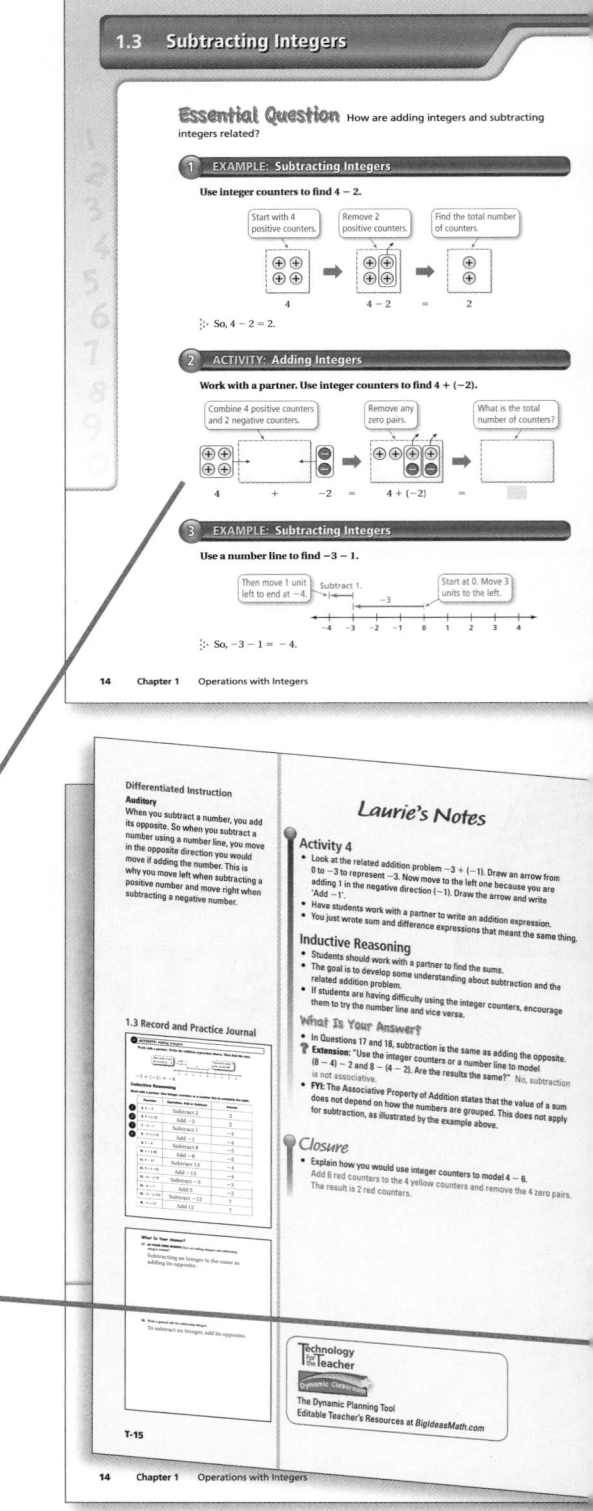

INSPIRING TEACHER BOOKS

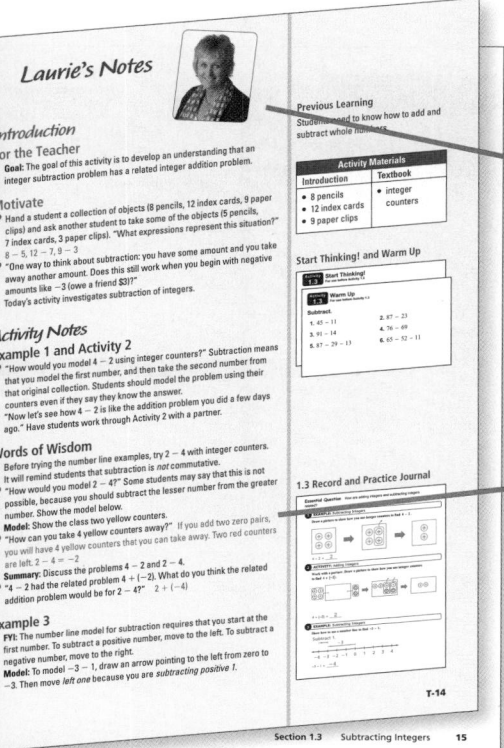

Teachers have the benefit of **Laurie Boswell's** 20-plus years of classroom experience reflected in her lively and informative notes.

Teachers are offered suggestions for questioning that help guide students toward better understanding.

"Laurie's Notes"
COMPLETE ACTIVITY AND TIME MANAGEMENT SUPPORT FROM A MASTER CLASSROOM TEACHER

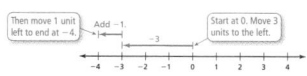

4 ACTIVITY: Adding Integers

Work with a partner. Write the addition expression shown. Then find the sum.

Then move 1 unit left to end at −4. | Add −1. | Start at 0. Move 3 units to the left.

Inductive Reasoning
Work with a partner. Use integer counters or a number line to complete the table.

Exercise	Operation: Add or Subtract	Answer
5. $4 - 2$	Subtract 2	2
6. $4 + (-2)$	Add −2	2
7. $-3 - 1$	Subtract 1	−4
8. $-3 + (-1)$	Add −1	−4
9. $3 - 8$		
10. $3 + (-8)$		
11. $9 - 13$		
12. $9 + (-13)$		
13. $-6 - (-3)$		
14. $-6 + (3)$		
15. $-5 - (-12)$		
16. $-5 + 12$		

What Is Your Answer?

17. IN YOUR OWN WORDS How are adding integers and subtracting integers related?

18. Write a general rule for subtracting integers.

Practice ▸ Use what you learned about subtracting integers to complete Exercises 8–15 on page 18.

Section 1.3   Subtracting Integers   15

CLEAR PUPIL BOOKS

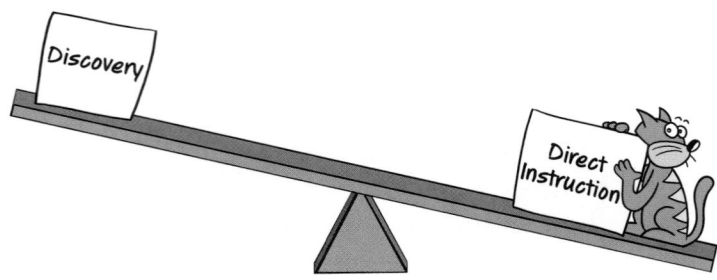

LESSON

After the concept has been introduced with a full-class period **Activity**, *it is extended the following day through the* **Lesson**.

● **Key Ideas**
● **Examples**
● **"On Your Own" Questions**

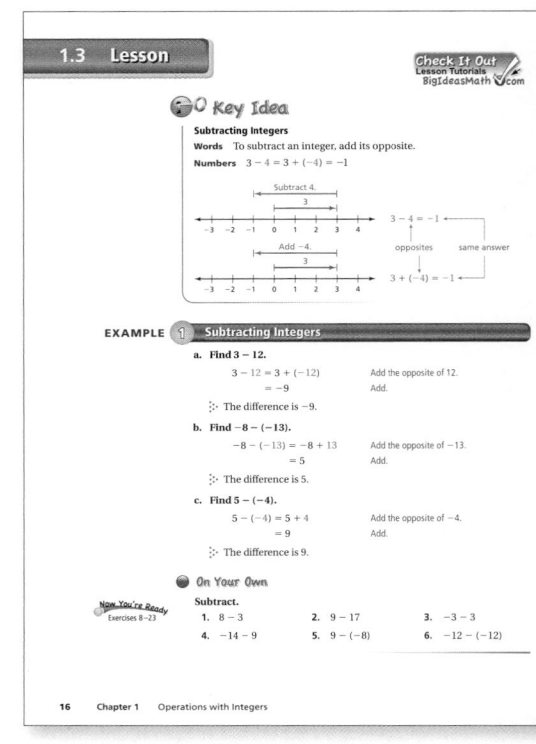

Each concept is accompanied by clear, stepped-out examples.

Each example in the pupil edition is accompanied by teaching suggestions from Laurie.

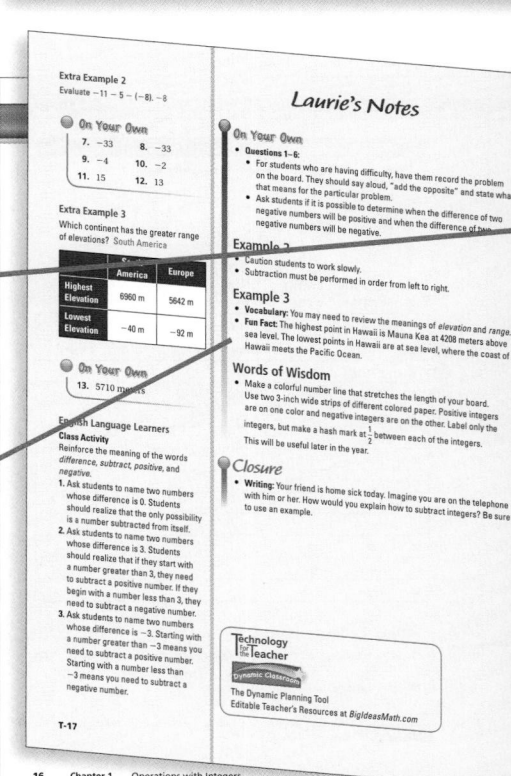

INSIGHTFUL TEACHER BOOKS

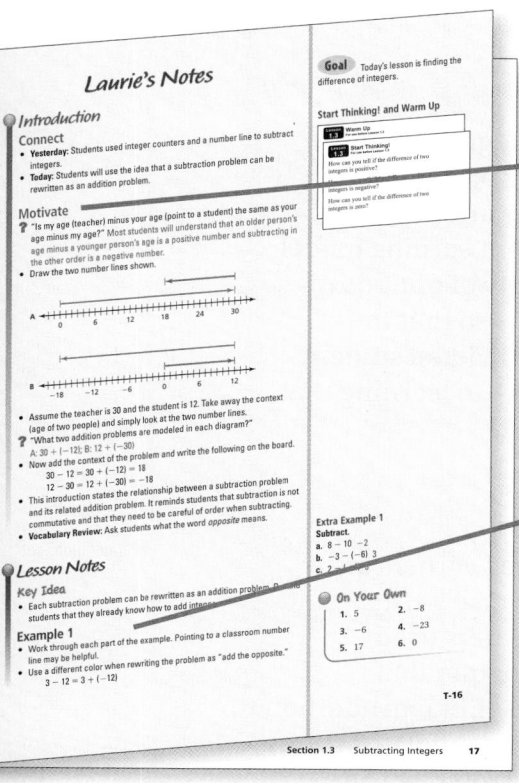

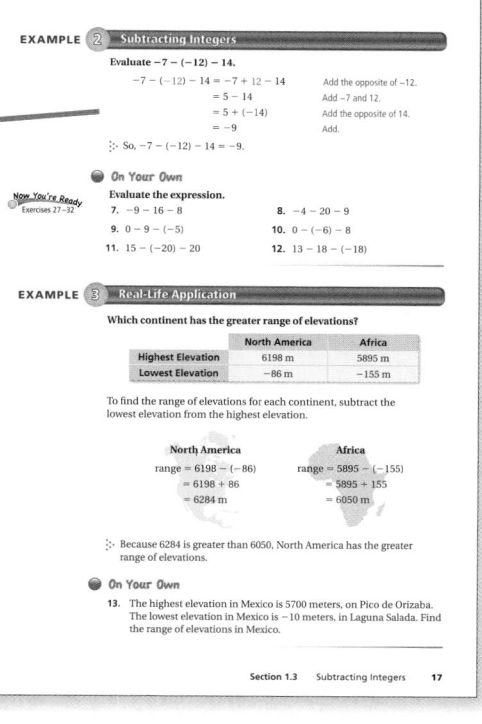

> Student-friendly and teacher-tested motivation activities start each lesson.

> Laurie shares insights she has gained through years of teaching experience.

"Laurie's Notes"

COMPREHENSIVE TEACHING SUPPORT FROM A MASTER CLASSROOM TEACHER

DIFFERENTIATED INSTRUCTION

Opening Doors to Learning

Two primary concerns while developing the Big Ideas Math program were the diversity of the student population and their different learning profiles. The authors developed a curriculum that helps teachers create classrooms that concentrate on learner needs by using the Universal Design for Learning model (UDL). The curriculum is designed to incorporate a wide variety of options to achieve its goals; offering materials, methods, and assessments so that the curriculum in its entirety is flexible and accommodating of individual student needs. By using Differentiated Instruction, teachers open doors to learning that students are unable to open themselves.

English Language Learners

The Big Ideas Math program recognizes that English Language Learners are a highly heterogeneous and complex group of students with diverse gifts, educational needs, backgrounds, languages, and goals. The writers used researched-based recommendations while developing the program that were designed to specifically assist English Language Learners (ELL). In addition to global support, such as curriculum that is organized around Essential Questions that involve both reading and writing, the program includes at-point-of-use ELL notes for the teacher; Home and Community Letters; ebooks with audio (English and Spanish); and a visual glossary.

Differentiating the Lesson Ancillary

Differentiating the Lesson is an online ancillary available at *BigIdeasMath.com*. This ancillary provides complete teaching notes and worksheets that address the needs of the diverse learners in the classroom. The lessons engage students in activities that often incorporate visual learning and kinesthetic learning. Some lessons present an alternative approach to teaching the content while other lessons extend the concepts of the text in a challenging way for advanced students. Each chapter of the *Differentiating the Lesson* ancillary begins with an overview that outlines the differentiated lessons in the chapter, and describes the students who would most benefit from the approach used in each lesson.

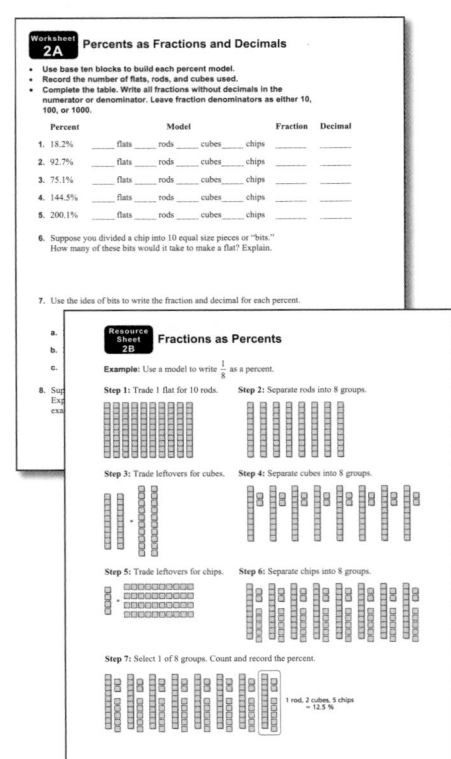

RESPONSE TO INTERVENTION

Through print and digital resources, the Big Ideas Math program completely supports the 3-tier RTI model. Opportunities for daily assessment help identify areas of needs and easy-to-use resources are provided to support the education of all students.

Tier 1: Daily Intervention

The Big Ideas Math program uses research-based instructional strategies to ensure quality instruction. Vocabulary support, cooperative learning opportunities, and graphic organizers are included in the Pupil Edition. Additional strategies can be found throughout the program. Daily student reviews and assessment guarantee that every student is making regular progress. Complete support helps teachers personalize instruction for every student.

- On Your Own
- Mini-Assessment
- Differentiated Instruction
- Skills Review Handbook

Tier 2: Strategic Intervention

The Big Ideas Math program facilitates increased time and focus on instruction for students who are not responding effectively to Tier 1 intervention. Additional support to assist teachers with these struggling learners can be found in the ancillary materials. Extra Examples, Fair Game Reviews, Graphic Organizers, Study Tips, and Real-Life Applications have been specifically written to enhance learning and to engage the diverse students within today's math classrooms. Using the classroom and online resources provided, teachers can reach, challenge, and motivate each student with germane, high-quality instruction targeted to their individual needs.

- Differentiating the Lesson
- Lesson Tutorials
- Basic Skills Handbook
- Record and Practice Journal
- Chapter Resource Books

Tier 3: Customized Learning Intervention

Support for students working below grade level is also available.

Skills Review Handbook

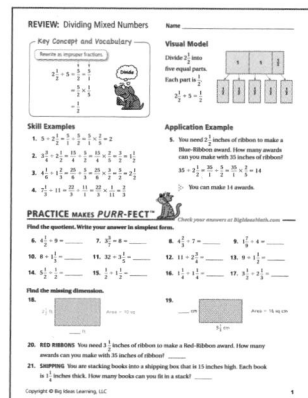

Basic Skills Handbook

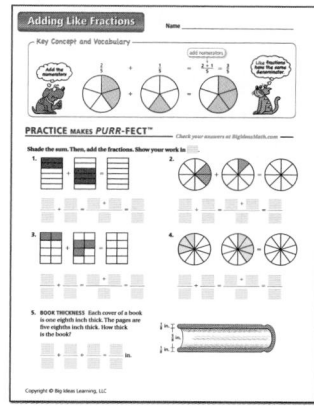

PACING GUIDE

Each page in the book is in the *Pacing Guide*.

Chapters 1–9:   162 days
(Including Additional Topics)

Chapter 1   (17 Days)

Chapter Opener	1 Day
Activity 1.1	1 Day
Lesson 1.1	1 Day
Activity 1.2	1 Day
Lesson 1.2	1 Day
Activity 1.3	1 Day
Lesson 1.3	1 Day
Study Help/Quiz	1 Day
Activity 1.4	1 Day
Lesson 1.4	1 Day
Activity 1.5	1 Day
Lesson 1.5	1 Day
Activity 1.6	1 Day
Lesson 1.6	1 Day
Quiz/Chapter Review	1 Day
Chapter Test	1 Day
Standardized Test Practice	1 Day

Chapter 2   (21 Days)

Chapter Opener	1 Day
Activity 2.1	1 Day
Lesson 2.1	1 Day
Activity 2.2	1 Day
Lesson 2.2	1 Day
Activity 2.3	1 Day
Lesson 2.3	1 Day
Lesson 2.3b	1 Day
Study Help/Quiz	1 Day
Activity 2.4	1 Day
Lesson 2.4	1 Day
Activity 2.5	1 Day
Lesson 2.5	1 Day
Lesson 2.5b	1 Day
Activity 2.6	1 Day
Lesson 2.6	1 Day
Lesson 2.6b	2 Days
Quiz/Chapter Review	1 Day
Chapter Test	1 Day
Standardized Test Practice	1 Day

Chapter 3   (22 Days)

Chapter Opener	1 Day
Activity 3.1	1 Day
Lesson 3.1	1 Day
Activity 3.2	1 Day
Lesson 3.2	1 Day
Activity 3.3	1 Day
Lesson 3.3	1 Day
Activity 3.4	1 Day
Lesson 3.4	1 Day
Activity 3.5	1 Day
Lesson 3.5	1 Day
Study Help/Quiz	1 Day
Activity 3.6	1 Day
Lesson 3.6	1 Day
Activity 3.7	1 Day
Lesson 3.7	1 Day
Lesson 3.7b	1 Day
Activity 3.8	1 Day
Lesson 3.8	1 Day
Quiz/Chapter Review	1 Day
Chapter Test	1 Day
Standardized Test Practice	1 Day

Chapter 4   (13 Days)

Chapter Opener	1 Day
Activity 4.1	1 Day
Lesson 4.1	1 Day
Activity 4.2	1 Day
Lesson 4.2	1 Day
Study Help/Quiz	1 Day
Activity 4.3	1 Day
Lesson 4.3	1 Day
Activity 4.4	1 Day
Lesson 4.4	1 Day
Quiz/Chapter Review	1 Day
Chapter Test	1 Day
Standardized Test Practice	1 Day

Chapter 5 (21 Days)

Chapter Opener	1 Day
Activity 5.1	1 Day
Lesson 5.1	1 Day
Activity 5.2	2 Days
Lesson 5.2	1 Day
Activity 5.3	1 Day
Lesson 5.3	1 Day
Activity 5.4	1 Day
Lesson 5.4	1 Day
Lesson 5.4b	1 Day
Study Help/Quiz	1 Day
Activity 5.5	1 Day
Lesson 5.5	1 Day
Activity 5.6	1 Day
Lesson 5.6	1 Day
Activity 5.7	1 Day
Lesson 5.7	1 Day
Quiz/Chapter Review	1 Day
Chapter Test	1 Day
Standardized Test Practice	1 Day

Chapter 6 (19 Days)

Chapter Opener	1 Day
Activity 6.1	1 Day
Lesson 6.1	1 Day
Activity 6.2	1 Day
Lesson 6.2	1 Day
Lesson 6.2b	1 Day
Activity 6.3	1 Day
Lesson 6.3	1 Day
Study Help/Quiz	1 Day
Activity 6.4	1 Day
Lesson 6.4	1 Day
Activity 6.5	1 Day
Lesson 6.5	1 Day
Activity 6.6	1 Day
Lesson 6.6	2 Days
Quiz/Chapter Review	1 Day
Chapter Test	1 Day
Standardized Test Practice	1 Day

Chapter 7 (18 Days)

Chapter Opener	1 Day
Activity 7.1	1 Day
Lesson 7.1	1 Day
Activity 7.2	1 Day
Lesson 7.2	1 Day
Activity 7.3	1 Day
Lesson 7.3	1 Day
Activity 7.4	1 Day
Lesson 7.4	1 Day
Study Help/Quiz	1 Day
Activity 7.5	1 Day
Lesson 7.5	1 Day
Activity 7.6	1 Day
Lesson 7.6	2 Days
Quiz/Chapter Review	1 Day
Chapter Test	1 Day
Standardized Test Practice	1 Day

Chapter 8 (14 Days)

Chapter Opener	1 Day
Activity 8.1	1 Day
Lesson 8.1	1 Day
Activity 8.2	1 Day
Lesson 8.2	1 Day
Study Help/Quiz	1 Day
Activity 8.3	1 Day
Lesson 8.3	1 Day
Activity 8.4	1 Day
Lesson 8.4	1 Day
Lesson 8.4b	1 Day
Quiz/Chapter Review	1 Day
Chapter Test	1 Day
Standardized Test Practice	1 Day

Chapter 9 (13 Days)

Chapter Opener	1 Day
Activity 9.1	1 Day
Lesson 9.1	1 Day
Activity 9.2	1 Day
Lesson 9.2	1 Day
Study Help/Quiz	1 Day
Activity 9.3	1 Day
Lesson 9.3	1 Day
Activity 9.4	1 Day
Lesson 9.4	1 Day
Quiz/Chapter Review	1 Day
Chapter Test	1 Day
Standardized Test Practice	1 Day

Additional Topics (4 Days)

Opener	1 Day
Topic 1	1 Day
Topic 2	2 Days

PROFESSIONAL DEVELOPMENT

Big Ideas Learning, LLC is a professional development and publishing company founded by Dr. Ron Larson. We are dedicated to providing 21st century teaching and learning in the area of mathematics. We work with middle schools across the country as they implement world-class standards for mathematics.

As teachers and school districts move forward in implementing new Common Core State Standards, the fundamental ideas of rigor and relevance and big ideas that are deep and focused take on new meaning. Big Ideas Learning provides a rich, hands-on experience that allows for astute understanding and practice, not only of the challenging world-class standards, but of the underlying mathematics pedagogy as well.

WORKSHOPS

- Creating Highly Motivating Classrooms
- Implementing the Common Core State Standards

Activities for the Mathematics Classroom:

- Teaching More with Less
- Best Practices in the Mathematics Classroom
- Questioning in the Mathematics Classroom
- The Three R's: Rigor, Relevance and Reality
- Reading in the Content Areas
- Games for Numerical Fluency
- Engaging the Tech Natives

Our professional staff of experienced instructors can also assist you in creating customized training sessions tailored to achieving your desired outcome.

"Name these shapes."

Common Core State Standards for Mathematical Practice

Make sense of problems and persevere in solving them.
- Multiple representations are presented to help students move from concrete to representative and into abstract thinking
- *Essential Questions* help students focus and analyze
- *In Your Own Words* provide opportunities for students to look for meaning and entry points to a problem

Reason abstractly and quantitatively.
- Visual problem solving models help students create a coherent representation of the problem
- Opportunities for students to decontextualize and contextualize problems are presented in every lesson

Construct viable arguments and critique the reasoning of others.
- *Error Analysis*; *Different Words, Same Question*; and *Which One Doesn't Belong* features provide students the opportunity to construct arguments and critique the reasoning of others
- *Inductive Reasoning* activities help students make conjectures and build a logical progression of statements to explore their conjecture

Model with mathematics.
- Real-life situations are translated into diagrams, tables, equations, and graphs to help students analyze relations and to draw conclusions
- Real-life problems are provided to help students learn to apply the mathematics that they are learning to everyday life

Use appropriate tools strategically.
- *Graphic Organizers* support the thought process of what, when, and how to solve problems
- A variety of tool papers, such as graph paper, number lines, and manipulatives, are available as students consider how to approach a problem
- Opportunities to use the web, graphing calculators, and spreadsheets support student learning

Attend to precision.
- *On Your Own* questions encourage students to formulate consistent and appropriate reasoning
- Cooperative learning opportunities support precise communication

Look for and make use of structure.
- *Inductive Reasoning* activities provide students the opportunity to see patterns and structure in mathematics
- Real-world problems help students use the structure of mathematics to break down and solve more difficult problems

Look for and express regularity in repeated reasoning.
- Opportunities are provided to help students make generalizations
- Students are continually encouraged to check for reasonableness in their solutions

Common Core State Standards for Mathematical Content for Grade 7

Chapter Coverage for Standards

Domain — Ratios and Proportional Relationships

- Analyze proportional relationships and use them to solve real-world and mathematical problems.

Domain — The Number System

- Apply and extend previous undertandings of operations with fractions to add, subtract, multiply, and divide rational numbers.

Domain — Expressions and Equations

- Use properties of operations to generate equivalent expressions.
- Solve real-life and mathematical problems using numerical and algebraic expressions and equations.

Domain — Geometry

- Draw, construct, and describe geometrical figures and describe the relationships between them.
- Solve real-life and mathematical problems involving angle measure, area, surface area, and volume.

Domain — Statistics and Probability

- Use random sampling to draw inferences about a population.
- Draw informal comparative inferences about two populations.
- Investigate chance processes and develop, use, and evaluate probability models.

How to Use Your Math Book

● Read the **Essential Question** in the activity.

Work with a partner to decide **What Is Your Answer?**

Now you are ready to do the Practice problems.

● Find the (Key Vocabulary 🔊) words, **highlighted in yellow**.

Read their definitions. Study the concepts in each 🔓 **Key Idea** .
If you forget a definition, you can look it up online in the

🔊 Multi-Language Glossary at BigIdeasMath✓com.

● After you study each **EXAMPLE**, do the exercises in the ● **On Your Own** .

Now You're Ready to do the exercises that correspond to the example.

As you study, look for a  or a  .

● The exercises are divided into 3 parts.

 Vocabulary and Concept Check

 Practice and Problem Solving

 Fair Game Review

If an exercise has a ① next to it, look back at
Example 1 for help with that exercise.

More help is available at  .
Check It Out
Lesson Tutorials
BigIdeasMath✓com

● To help study for your test, use the following.

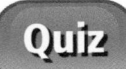 

Quiz   **Study Help**

Chapter Review   **Chapter Test**

SCAVENGER HUNT

Use this *Scavenger Hunt* to find where things are in **Chapter 1**.

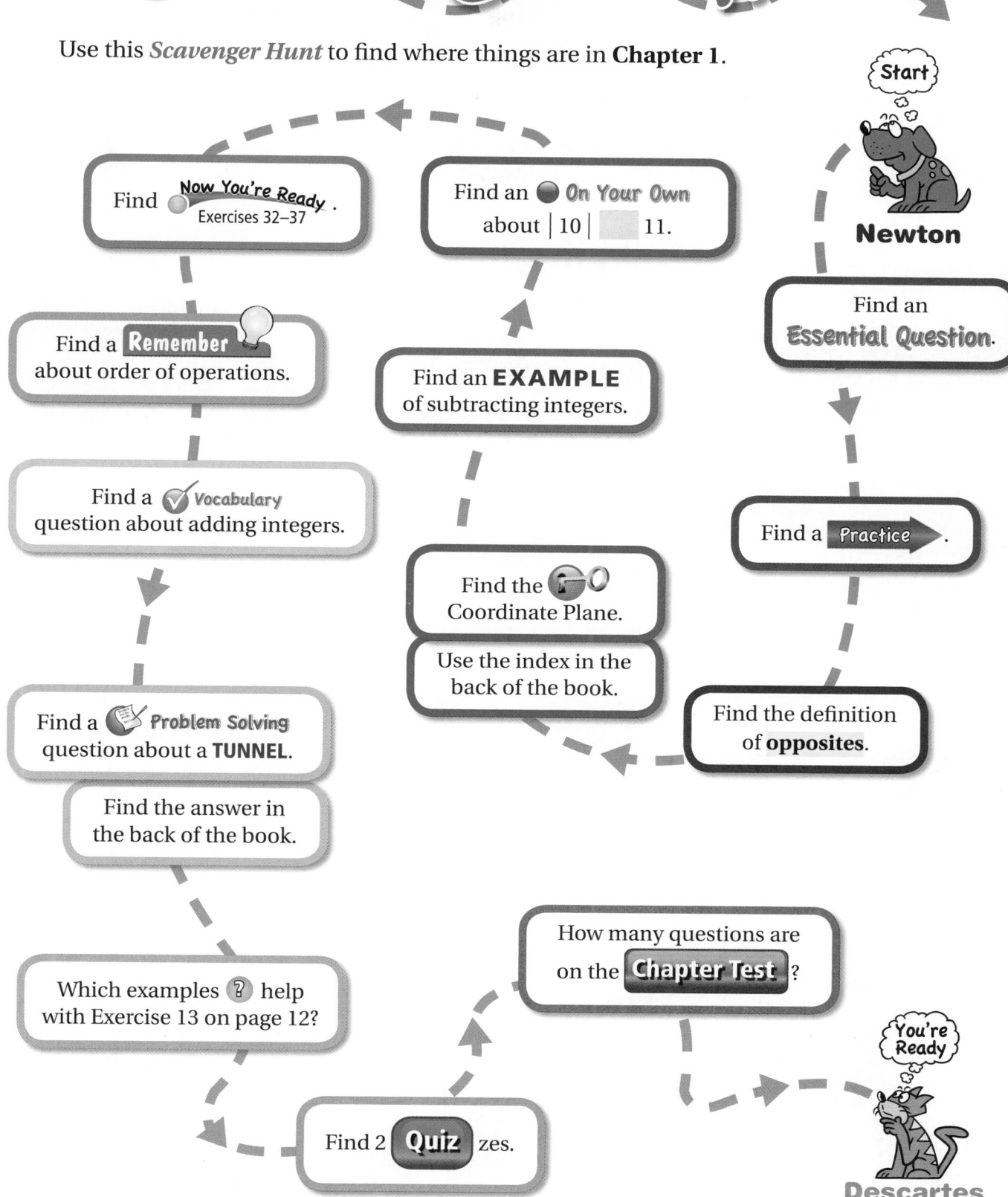

Find **Now You're Ready**. Exercises 32–37.

Find an ● **On Your Own** about |10| ■ 11.

Start

Newton

Find a **Remember** about order of operations.

Find an **EXAMPLE** of subtracting integers.

Find an **Essential Question**.

Find a ✓ **Vocabulary** question about adding integers.

Find a **Practice** ➤.

Find the 🔑 Coordinate Plane.

Use the index in the back of the book.

Find the definition of **opposites**.

Find a 📝 **Problem Solving** question about a **TUNNEL**.

Find the answer in the back of the book.

How many questions are on the **Chapter Test**?

Which examples ❓ help with Exercise 13 on page 12?

You're Ready

Find 2 **Quiz**zes.

Descartes

1 Operations with Integers

"Look, subtraction is not that difficult. Imagine that you have five squeaky mouse toys."

"After your friend Fluffy comes over for a visit, you notice that one of the squeaky toys is missing."

"Now, you go over to Fluffy's and retrieve the missing squeaky mouse toy. It's easy."

"Dear Sir: You asked me to 'find' the opposite of −1."

"I didn't know it was missing."

Connections to Previous Learning

- Describe real-world situations using positive and negative numbers.
- Compare, order, and graph integers.
- Recall multiplication and division facts easily.

- Multiply and divide fractions and decimals efficiently.

- Use and justify rules for addition, subtraction, multiplication, and division of integers.
- Find the absolute value of integers.
- Add, subtract, multiply, and divide integers.

Math in History

As recently as the 18th century, modern mathematicians ignored negative solutions of equations. The concept of a negative number was implied by early cultures.

★ The Chinese book "Nine Chapters on the Mathematical Art" (100 B.C.–50 B.C.) contains methods for finding the areas of regions. Red rods were used to denote positive coefficients. Black rods were used to denote negative coefficients.

★ During the 5th century A.D., negative numbers were in use in India to represent debts.

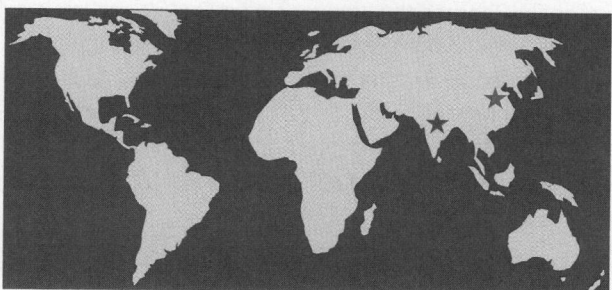

Pacing Guide for Chapter 1

Chapter Opener	1 Day
Section 1 Activity Lesson	1 Day 1 Day
Section 2 Activity Lesson	1 Day 1 Day
Section 3 Activity Lesson	1 Day 1 Day
Study Help / Quiz	1 Day
Section 4 Activity Lesson	1 Day 1 Day
Section 5 Activity Lesson	1 Day 1 Day
Section 6 Activity Lesson	1 Day 1 Day
Quiz / Chapter Review	1 Day
Chapter Test	1 Day
Standardized Test Practice	1 Day
Total Chapter 1	17 Days
Year-to-Date	17 Days

Check Your Resources

- Record and Practice Journal
- Resources by Chapter
- Skills Review Handbook
- Assessment Book
- Worked-Out Solutions

Technology
For the Teacher

The Dynamic Planning Tool
Editable Teacher's Resources at
BigIdeasMath.com

Math Background Notes

Vocabulary Review

- Least and Greatest
- Integer
- Coordinates
- Ordered Pair

Ordering Integers

- Students should know how to order integers and work with the number line.
- Remind students that zero acts as the gate keeper between the negative and positive numbers. Negative numbers appear to the left of zero on the number line, and positive numbers appear to the right.
- **Common Error:** Students will often assume that -6 is greater than -4 because 6 is a greater number than 4. Remind students that the farther left you move from zero, the lesser the values become.

Plotting Points

- Students should know how to plot and identify points in Quadrant I.
- If students struggle with Example 2, review moving in a coordinate plane with your students.
- **Common Error:** Students frequently confuse the *x*- and *y*-coordinates when writing ordered pairs. It may help to point out that the O in "over" appears before the U in "up" in the alphabet. Similarly, because *x* comes before *y* in the alphabet, the *x*-coordinate should come before the *y*-coordinate in the ordered pair. Encourage your students to write their points in alphabetical order!

Try It Yourself

1. $-12, -10, -2, 4, 15$
2. $-5, -3, 1, 3, 7$
3. $(0, 1)$
4. $(1, 3)$
5. $(2, 4)$
6. $(2, 0)$
7. 13
8. 10
9. 48

Record and Practice Journal

1. $-9, -7, 0, 3, 8$
2. $-4, -2, -1, 1, 2$
3. $-11, -8, -6, 5, 9$
4. $-7, -5, 0, 2, 4$
5. $(0, 3)$
6. $(3, 4)$
7. $(4, 1)$
8. $(5, 0)$
9. $-27, -17, -12, 4, 30$
10. 14
11. 3
12. 394
13. 86
14. 76
15. 16
16. a. 386

    b. $4(2) + 2(5^2) + 3^2(6^2) + 2^2 = 386$

Using Order of Operations

- Students should know the order of operations.
- You may want to review the correct order of operations with students. Many students probably learned the pneumonic device *Please Excuse My Dear Aunt Sally*. Ask a volunteer to explain why this phrase is helpful.
- You may want to review exponents with students. Remind students that the exponent tells you how many times the base acts as a factor.
- **Common Error:** Students may misinterpret the exponent in Example 4. Watch for students who incorrectly compute $6^2 = 6 \cdot 2 = 12$.

Reteaching and Enrichment Strategies

If students need help. . .	If students got it. . .
Record and Practice Journal • Fair Game Review Skills Review Handbook Lesson Tutorials	Game Closet at *BigIdeasMath.com* Start the next section

What You Learned Before

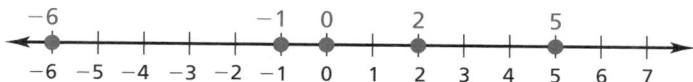

"I liked it because it is the opposite of the freezing point on the Fahrenheit temperature scale."

Ordering Integers

Example 1 Order 0, −1, 2, 5, and −6 from least to greatest.

Try It Yourself

Order the integers from least to greatest.

1. −10, 15, 4, −2, −12

2. 7, −5, 3, −3, 1

Plotting Points

Example 2 Plot the point (2, 3).

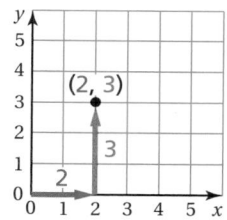

Example 3 Write an ordered pair corresponding to Point Q.

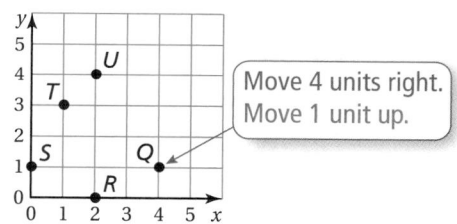

Move 4 units right.
Move 1 unit up.

⋮ The ordered pair (4, 1) corresponds to Point Q.

Try It Yourself

Use the graph in Example 3 to write an ordered pair corresponding to the point.

3. Point S

4. Point T

5. Point U

6. Point R

Using Order of Operations

Example 4 Evaluate $6^2 \div 4 - 2(9 - 5)$.

First:	Parentheses		$6^2 \div 4 - 2(9 - 5) = 6^2 \div 4 - 2 \cdot 4$
Second:	Exponents		$= 36 \div 4 - 2 \cdot 4$
Third:	Multiplication and Division (from left to right)		$= 9 - 8$
Fourth:	Addition and Subtraction (from left to right)		$= 1$

Try It Yourself

Evaluate the expression.

7. $15\left(\dfrac{8}{4}\right) + 2^2 - 3 \cdot 7$

8. $5^2 \cdot 2 \div 10 + 3 \cdot 2 - 1$

9. $3^2 - 1 + 2(4(3 + 2))$

Essential Question How are velocity and speed related?

On these two pages, you will investigate vertical motion (up or down).
- Speed tells how fast an object is moving, but does not tell the direction.
- Velocity tells how fast an object is moving and also tells the direction.

  If velocity is positive, the object is moving up.

  If velocity is negative, the object is moving down.

1 EXAMPLE: Falling Parachute

You are gliding to the ground wearing a parachute. The table shows your height at different times.

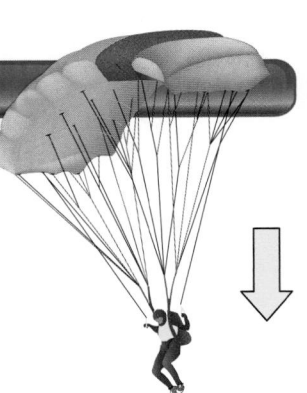

Time (seconds)	0	1	2	3
Height (feet)	45	30	15	0

a. **How many feet do you move each second?**
b. **What is your speed? Give the units.**
c. **Is your velocity positive or negative?**
d. **What is your velocity? Give the units.**

a. For each 1 second of time, your height is 15 feet less.
b. You are moving at 15 feet per second.
c. Because you are moving down, your velocity is negative.
d. Your velocity is −15 feet per second. This can be written as −15 ft/sec.

2 ACTIVITY: Rising Balloons

Work with a partner. The table shows the height of a group of balloons.

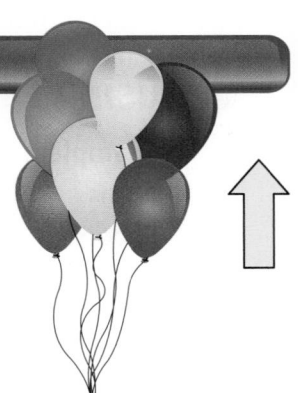

Time (seconds)	0	1	2	3
Height (feet)	0	4	8	12

a. **How many feet do the balloons move each second?**
b. **What is the speed of the balloons? Give the units.**
c. **Is the velocity positive or negative?**
d. **What is the velocity? Give the units.**

Laurie's Notes

Introduction

For the Teacher

- **Goal:** The relationship between velocity and speed is the same as the relationship between an integer and its absolute value.
- *Velocity* has two components—*direction* and *speed.*
  - *Direction* is defined by the person doing the measuring (forward/backward; up/down).
  - *Speed* is the magnitude, or numerical value, without regard to direction.
- **Integers** are the set of whole numbers and their opposites. Two numbers are opposites if they are the same distance from 0 on a number line.

Motivate

- Ask students if they have ever watched the launch of a NASA shuttle. The velocity of the shuttle describes both its speed and its direction. Today's activity looks at the velocities of various objects.
- **Model:** Use a handkerchief to make a simple parachute by stapling or taping a piece of string or yarn to each corner of the handkerchief. Tie a paper clip to the loose ends of the string or yarn.
- Have two students stand on chairs. One student drops a paper clip without the parachute and the other drops the paper clip with the opened parachute.
- Both paper clips have speed, which can be measured in ft/sec. They also have velocity, moving down, so the velocity is negative and the units would still be ft/sec.
- Write the definitions of speed and velocity on the board and ask for (or share) examples of each.
  - **Examples of speed:** a pitcher throws a 94 mi/h fastball; a manatee swims 4 mi/h; a football travels 15 ft/sec
  - **Examples of velocity:** a NASA shuttle has an orbital velocity of 17,500 mi/h; you walk to school at 8 ft/sec; a feather falls at −2 ft/sec

Activity Notes

Example 1 and Activity 2

? "Do you see any pattern(s) in the table? Describe the pattern(s)."
Example 1: The heights are multiples of 15.
Activity 2: The heights are multiples of 4, each height is divisible by 4.

Words of Wisdom

- The number −4 is read as "negative 4" and not "minus 4." Minus is used to describe the operation of subtraction.

Previous Learning

Students should know how to compare, order, and graph integers.

Activity Materials
Introduction
• handkerchief • string or yarn • paper clips

Start Thinking! and Warm Up

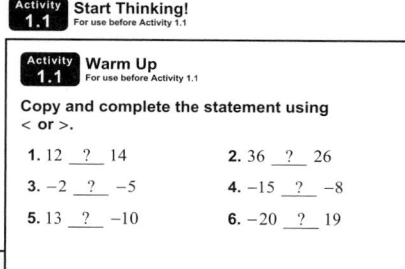

Activity 1.1 Start Thinking! For use before Activity 1.1

Activity 1.1 Warm Up For use before Activity 1.1

Copy and complete the statement using < or >.

1. 12 _?_ 14   2. 36 _?_ 26
3. −2 _?_ −5   4. −15 _?_ −8
5. 13 _?_ −10   6. −20 _?_ 19

1.1 Record and Practice Journal

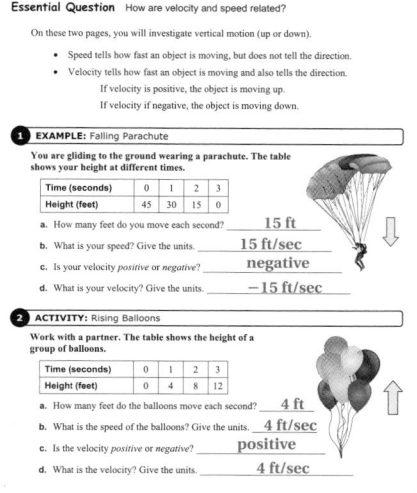

Essential Question How are velocity and speed related?

On these two pages, you will investigate vertical motion (up or down).

- Speed tells how fast an object is moving, but does not tell the direction.
- Velocity tells how fast an object is moving and also tells the direction.
  If velocity is positive, the object is moving up.
  If velocity is negative, the object is moving down.

1 EXAMPLE: Falling Parachute

You are gliding to the ground wearing a parachute. The table shows your height at different times.

Time (seconds)	0	1	2	3
Height (feet)	45	30	15	0

a. How many feet do you move each second? **15 ft**
b. What is your speed? Give the units. **15 ft/sec**
c. Is your velocity *positive* or *negative*? **negative**
d. What is your velocity? Give the units. **−15 ft/sec**

2 ACTIVITY: Rising Balloons

Work with a partner. The table shows the height of a group of balloons.

Time (seconds)	0	1	2	3
Height (feet)	0	4	8	12

a. How many feet do the balloons move each second? **4 ft**
b. What is the speed of the balloons? Give the units. **4 ft/sec**
c. Is the velocity *positive* or *negative*? **positive**
d. What is the velocity? Give the units. **4 ft/sec**

English Language Learners

Vocabulary

On a long strip of paper, mark integers from −10 through 10, with zero in the middle. Hold the strip vertically so that −10 touches the ground. Discuss that when an object is moving from 0 to 10, the *velocity* is positive. Point out the numbers 1 through 10 on the strip are positive integers. Then discuss that an object moving from zero down to the ground has a negative velocity. Again refer to the number strip, pointing out that −1 through −10 are negative integers.

1.1 Record and Practice Journal

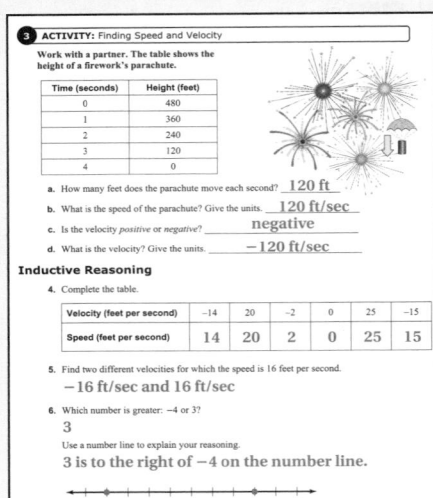

Activity 3

- This activity is modeled after Example 1 and Activity 2.
- **Think-Pair-Share:** Students should read each question independently and then work with a partner to answer the questions. When they have answered the questions, the pair should compare their answers with another group and discuss any discrepancies.

Inductive Reasoning

- In Question 4, if students have connected the relationship between speed and velocity, they will be able to complete the table. To reinforce labeling of answers, have students write *ft/sec* with each entry in the table.
- Questions 5–7 are developing the notion of absolute value—there are two velocities that have a speed of 16 ft/sec, namely 16 ft/sec and −16 ft/sec.
- In Question 6, it is common for students to say −4 > 3. Remind students that a number farther to the right on a number line is greater, so 3 > −4.
- In Question 7, the sign (negative or positive) is not considered because the question concerns speed and the direction does not matter. An object moving 4 ft/sec has a greater speed than an object moving 3 ft/sec.

What Is Your Answer?

- In Questions 8 and 9, speed is the **absolute value** of velocity.

Words of Wisdom

- A formal definition of absolute value will be presented in the lesson.
- Do not let students suggest that **absolute value** simply means to take away the negative sign. This could cause problems in the future when students work with variables or variable expressions within the absolute value symbols. Direct discussions toward the idea that you want to know how far (distance) a number is from zero.

Closure

- **Communication:** Give examples of velocities of two objects (A and B), where Velocity A > Velocity B but Speed A < Speed B.
- Examples are shown below.

Velocity Object A	Velocity Object B	Compare Velocities	Compare Speeds
4 ft/sec	−5 ft/sec	Vel (A) > Vel (B)	Sp (A) < Sp (B)
10 ft/sec	−15 ft/sec	Vel (A) > Vel (B)	Sp (A) < Sp (B)
−5 ft/sec	−15 ft/sec	Vel (A) > Vel (B)	Sp (A) < Sp (B)

Technology
for the **T**eacher

The Dynamic Planning Tool
Editable Teacher's Resources at *BigIdeasMath.com*

3 ACTIVITY: Finding Speed and Velocity

Work with a partner. The table shows the height of a firework's parachute.

Time (seconds)	Height (feet)
0	480
1	360
2	240
3	120
4	0

a. How many feet does the parachute move each second?

b. What is the speed of the parachute? Give the units.

c. Is the velocity positive or negative?

d. What is the velocity? Give the units.

Inductive Reasoning

4. Copy and complete the table.

Velocity (feet per second)	−14	20	−2	0	25	−15
Speed (feet per second)						

5. Find two different velocities for which the speed is 16 feet per second.

6. Which number is greater: −4 or 3? Use a number line to explain your reasoning.

7. One object has a velocity of −4 feet per second. Another object has a velocity of 3 feet per second. Which object has the greater speed? Explain your answer.

What Is Your Answer?

In this lesson, you will study **absolute value**. Here are some examples:

Absolute value of −16 = 16   Absolute value of 16 = 16
Absolute value of 0 = 0   Absolute value of −2 = 2

8. **IN YOUR OWN WORDS** How are velocity and speed related?

9. Which of the following is a true statement? Explain your reasoning.

a. Absolute value of velocity = speed

b. Absolute value of speed = velocity

Practice   Use what you learned about absolute value to complete Exercises 4–11 on page 6.

The following numbers are **integers.**

$$\ldots, -3, -2, -1, 0, 1, 2, 3, \ldots$$

Key Vocabulary
integer, *p. 4*
absolute value, *p. 4*

Key Idea

Absolute Value

Words   The **absolute value** of an integer is the distance between the number and 0 on a number line. The absolute value of a number a is written as $|a|$.

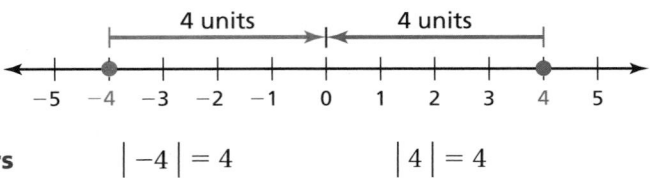

Numbers $|-4| = 4$ $|4| = 4$

EXAMPLE **1** **Finding Absolute Value**

Find the absolute value of 2.

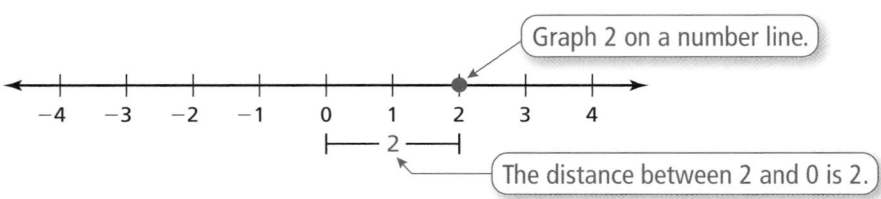

Graph 2 on a number line.

The distance between 2 and 0 is 2.

So, $|2| = 2$.

EXAMPLE **2** **Finding Absolute Value**

Find the absolute value of -3.

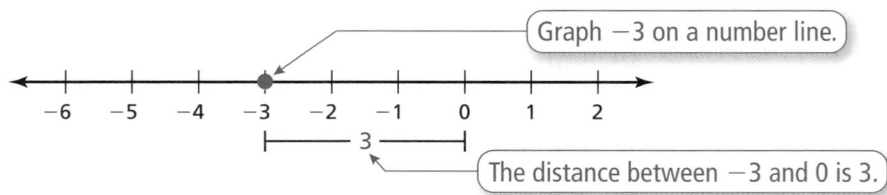

Graph -3 on a number line.

The distance between -3 and 0 is 3.

So, $|-3| = 3$.

On Your Own

Now You're Ready
Exercises 4–19

Find the absolute value of the integer.

1. 7   **2.** -1 **3.** -5 **4.** 14

Laurie's Notes

Introduction

Connect

- **Yesterday:** Students explored speed and velocity in the activity. They used the relationship between speed and velocity to find speed (the absolute value of velocity) when the velocity was known.
- **Today:** Have students think about the distance a number is from 0. This distance is always positive, just like the speed of an object.

Motivate

- Have two students (A and B) stand at the front of the room with a piece of string between them. You hold a piece of paper with the number 0 written on it. State that the distance between A and B is 10 units.
- **?** Position yourself at the midpoint of A and B.
  - "If A is 5, what number does B represent?" —5
  - "Who is closer to me (meaning 0)?" Neither; both are the same distance.
  - "How far away from me is each person?" 5 units
- **?** Move closer to A so that if A is 3, B would be approximately —7.
  - "If A is 3, what number does B represent?" —7
  - "Who is closer to me (meaning 0)?" A
  - "How far away from me is each person?" A is 3 units. B is 7 units.
- **?** Move closer to B so that if B is —2, A would be approximately 8.
  - "If B is —2, what number does A represent?" 8
  - "Who is closer to me (meaning 0)?" B
  - "How far away from me is each person?" A is 8 units. B is 2 units.
- **?** Without the string, ask students "What number or numbers are 6 units from 0?" 6 and —6

Lesson Notes

Key Idea

- When students state that "absolute values are always positive," try to clarify their statement. Students may make incorrect assumptions when there are numeric or variable expressions within absolute value symbols.
- At this stage of development, stress the geometric definition of absolute value. **Absolute value** is the distance a number is from zero.

Example 1 and Example 2

- Work through Examples 1 and 2 and then ask the following:
  - **?** "What is the absolute value of 12?" 12
  - **?** "What is the absolute value of —8?" 8

Words of Wisdom

- Make sure that students understand that when you write the notation for the absolute value, it means *take the absolute value of the number inside the symbols*.

Goal Today's lesson is finding the **absolute value** of an **integer**.

Lesson Materials
Introduction
• string

Start Thinking! and Warm Up

When you go to school in the morning you travel in one direction and then returning home you travel in the other direction. How does this compare to a number line where one direction is positive and the other is negative? Do you ever travel in a negative direction?

Extra Example 1

Find the absolute value of 6. 6

Extra Example 2

Find the absolute value of —11. 11

 On Your Own

1. 7	**2.** 1
3. 5	**4.** 14

Extra Example 3

Compare $|-9|$ and 7. $|-9| > 7$

 On Your Own

5. $|-2| > -1$

6. $-7 < |6|$

7. $|10| < 11$

8. $9 = |-9|$

Extra Example 4

Seawater freezes at $-2°C$. Is the freezing point of honey (from Example 4) or seawater closer to the freezing point of water, $0°C$? seawater

 On Your Own

9. airplane fuel; Because $|-53| < |55|$, the freezing point of airplane fuel is closer to $0°C$, the freezing point of water.

Differentiated Instruction

Kinesthetic

Stand in front of the classroom with two students, one on each side. Tell the students that you represent zero, your left (the students' right) is the positive direction, and your right (the students' left) is the negative direction. Have the student on your left walk three paces away from you. Say, "This student represents $+3$." Have the student on the right walk three paces from you. Say, "This student represents -3." Ask, "How far is each student from me?" (3 paces away). Say, "Positive 3 and negative 3 are the same distance from zero, so they have the same absolute value, which is 3."

Laurie's Notes

Example 3

- Students may mix up or forget the inequality symbols.
- **Common Error:** Students may label the number line left-to-right for both positive and negative integers as shown below. Explain that negative integers are labeled from 0, such that 1 and -1 are both one unit from 0.

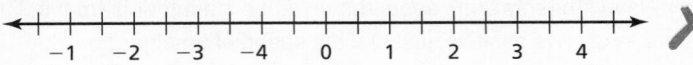

Example 4

- Have students discuss different liquids that they know freeze, such as water, ice cream, and chocolate. Probe to see if students know that water freezes at $0°C$ and $32°F$.
- **FYI:** Citrus fruit trees can sustain damage from low temperatures. Depending upon the temperature, the leaves, wood, or fruit can be damaged, causing economic problems for both the growers and the consumers.
- In part (a), discuss the substances listed in the table.
  - **?** "Why is the point representing -3 closer to 0 than -10?" -3 is 3 units to the left of 0 and -10 is 10 units to the left of 0. Because 3 is less than 10, -3 is closer to 0 than -10.
  - **?** "Why is it important for airplane fuel to have a very low freezing point?" Planes flying in very cold temperatures need fuel to remain in liquid form.
- In part (b), to answer the question "Which substance has a freezing point closer to the freezing point of water?", be sure students understand that the absolute value of the freezing point is how you measure the distance to 0.

Words of Wisdom

- Too often students will say "Because it is," or "It's obvious." Look for a reference to absolute value when students explain their reasoning.

Closure

- **Exit Ticket:** The freezing point of vinegar is $-2°C$. Is the freezing point of vinegar or honey closer to the freezing point of water? Explain your reasoning. vinegar; Because $|-2| < |-3|$, the freezing point of vinegar is closer to $0°C$, the freezing point of water.

Technology For the Teacher

Dynamic Classroom

The Dynamic Planning Tool
Editable Teacher's Resources at *BigIdeasMath.com*

EXAMPLE **3** **Comparing Values**

Remember

A number line can be used to compare and order integers. Numbers to the left are less than numbers to the right. Numbers to the right are greater than numbers to the left.

Compare 1 and $\left|-4\right|$.

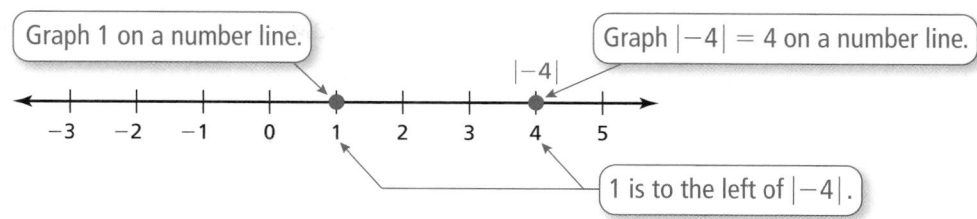

So, $1 < \left|-4\right|$.

On Your Own

Now You're Ready
Exercises 20–25

Copy and complete the statement using <, >, or =.

5. $\left|-2\right|$ ▢ -1

6. -7 ▢ $\left|6\right|$

7. $\left|10\right|$ ▢ 11

8. 9 ▢ $\left|-9\right|$

EXAMPLE **4** **Real-Life Application**

Substance	Freezing Point (°C)
Butter	35
Airplane fuel	−53
Honey	−3
Mercury	−39
Candle wax	55

The *freezing point* is the temperature at which a liquid becomes a solid.

a. Which substance in the table has the lowest freezing point?

b. Is the freezing point of mercury or butter closer to the freezing point of water, 0°C?

a. Graph each freezing point.

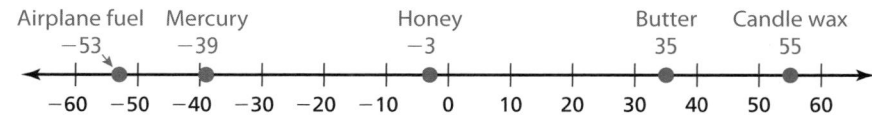

Airplane fuel has the lowest freezing point, −53°C.

b. The freezing point of water is 0°C, so you can use absolute values.

Mercury: $\left|-39\right| = 39$ **Butter:** $\left|35\right| = 35$

Because 35 is less than 39, the freezing point of butter is closer to the freezing point of water.

On Your Own

9. Is the freezing point of airplane fuel or candle wax closer to the freezing point of water? Explain your reasoning.

 Vocabulary and Concept Check

1. **VOCABULARY** Which of the following numbers are integers?
$$9, 3.2, -1, \frac{1}{2}, -0.25, 15$$

2. **VOCABULARY** What is the absolute value of an integer?

3. **WHICH ONE DOESN'T BELONG?** Which expression does *not* belong with the other three? Explain your reasoning.

$$|6| \qquad 6 \qquad -6 \qquad |-6|$$

 **Practice and Problem Solving**

Find the absolute value of the integer.

 **4.** 9   **5.** -6 **6.** -10 **7.** 10

8. -15 **9.** 13   **10.** -7 **11.** -12

12. 5   **13.** -8 **14.** 0   **15.** 18

16. -24 **17.** -45 **18.** 60   **19.** -125

Copy and complete the statement using <, >, or =.

③ 20. $2 \quad \boxed{} \quad |-5|$ **21.** $|-4| \quad \boxed{} \quad 7$ **22.** $-5 \quad \boxed{} \quad |-9|$

23. $|-4| \quad \boxed{} \quad -6$ **24.** $|-1| \quad \boxed{} \quad |-8|$ **25.** $|5| \quad \boxed{} \quad |-5|$

ERROR ANALYSIS Describe and correct the error.

26.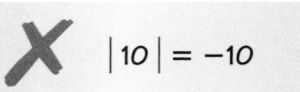

27.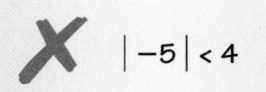

28. **SAVINGS** You deposit $50 in your savings account. One week later, you withdraw $20. Write each amount as an integer.

29. **ELEVATOR** You go down 8 floors in an elevator. Your friend goes up 5 floors in an elevator. Write each amount as an integer.

Order the values from least to greatest.

30. $8, |3|, -5, |-2|, -2$ **31.** $|-6|, -7, 8, |5|, -6$

32. $-12, |-26|, -15, |-12|, |10|$ **33.** $|-34|, 21, -17, |20|, |-11|$

Simplify the expression.

34. $|-30|$ **35.** $-|4|$ **36.** $-|-15|$

Assignment Guide and Homework Check

Level	Day 1 Activity Assignment	Day 2 Lesson Assignment	Homework Check
Basic	4–11, 46–50	1–3, 13–25 odd, 26–29, 31, 33	2, 17, 25, 28, 31
Average	4–11, 46–50	1–3, 17–27 odd, 30–36 even, 37–43 odd	2, 17, 25, 30, 41
Advanced	4–11, 46–50	1–3, 20–26 even, 30–36 even, 37, 38–44 even, 45	24, 30, 38, 40

Common Errors

- **Exercises 4–19** Students may think that the absolute value of a number is its opposite and say $|6| = -6$. Use a number line to show them that the absolute value is a number's distance from 0, so it is always a positive number or zero.
- **Exercises 20–27** When comparing absolute values of negative integers, students may not find the absolute values and instead compare the integers themselves.
- **Exercise 22** A student may find $-5 > |-9|$. The student likely did not find the absolute value of -9 to be 9 and instead compared -5 to -9.
- **Exercise 41** Students may write 14 and 18, rather than -14 and -18, for the diver's positions. These students did not account for the fact that the diver is *below* sea level. Use a vertical scale by turning a number line so that it runs vertically. Point out to students that the exercise defines sea level as 0 on the number line, so the diver's positions are negative.

1.1 Record and Practice Journal

Find the absolute value of the integer.

1. -1	2. -14	3. 0	4. 6
1	14	0	6

Complete the statement using <, >, or =.

5. 6 __ |-2|	6. -7 __ |-8|	7. |-9| __ 5	8. |-2| __ 2
>	<	>	=

Order the values from least to greatest.

9. 4, |7|, -1, |-3|, -4   10. |2|, -3, |-5|, -1, 6   11. |-8|, 0, -9, |-7|, -2

-4, -1, |-3|, 4, |7| -3, -1, |2|, |-5|, 6   -9, -2, 0, |-7|, |-8|

12. You download 12 new songs to your MP3 player. Then you delete 5 old songs. Write each amount as an integer.

12, -5

13. The mantle layer of Earth begins about 70 kilometers underneath the surface. The layer called the outer core begins about 2970 kilometers underneath the surface.

a. Write an integer for the position of each layer relative to the surface.

Mantle layer ___ -70   Outer core ___ -2970

b. Which integer in part (a) is greater?

-70

c. Which integer in part (a) has the greater absolute value? How does this relate to the layer that has the greatest distance from the surface?

-2970; The outer core is the farthest from the surface.

Technology For the Teacher
Answer Presentation Tool
QuizShow

Vocabulary and Concept Check

1. $9, -1, 15$

2. the distance between the integer and zero on a number line

3. -6; All of the other expressions are equal to 6.

Practice and Problem Solving

4. 9   5. 6

6. 10   7. 10

8. 15   9. 13

10. 7   11. 12

12. 5   13. 8

14. 0   15. 18

16. 24   17. 45

18. 60   19. 125

20. $2 < |-5|$

21. $|-4| < 7$

22. $-5 < |-9|$

23. $|-4| > -6$

24. $|-1| < |-8|$

25. $|5| = |-5|$

26. The absolute value of a number cannot be negative. $|10| = 10$

27. Because $|-5| = 5$, the statement is incorrect. $|-5| > 4$

28. $50, -20$ 29. $-8, 5$

30. $-5, -2, |-2|, |3|, 8$

31. $-7, -6, |5|, |-6|, 8$

32. $-15, -12, |10|, |-12|, |-26|$

33. $-17, |-11|, |20|, 21, |-34|$

34. 30   35. -4

36. -15

Practice and Problem Solving

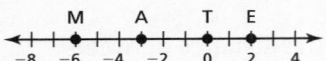

37. a. MATE

M A T E
-8 -6 -4 -2 0 2 4

b. TEAM

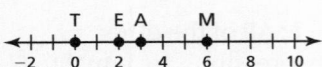

T E A M
-2 0 2 4 6 8 10

38. *Sample answer:* −4

39. $n \geq 0$

40. $n \leq 0$

41. See *Taking Math Deeper*.

42. Loihi

43. a. Player 3

b. Player 2

c. Player 1

44. true; If a number x is negative, then its absolute value is its opposite, $-x$.

45. false; The absolute value of zero is zero, which is neither positive nor negative.

Fair Game Review

46. 51   **47.** 144

48. 398   **49.** 3170

50. A

Mini-Assessment

Find the absolute value of the integer.

1. 6 6   **2.** −13 13

3. −17 17   **4.** 0 0

5. You deposit $125 in your checking account. One month later, you withdraw $65. Write each amount as an integer. 125; −65

Taking Math Deeper

Exercise 41

In this problem, negative numbers are used on a vertical scale to indicate *positions* that are below sea level. The numbers on the number line describe *position* (a negative number), not *depth* below sea level (a positive number).

① **a.** Draw and label a vertical number line.

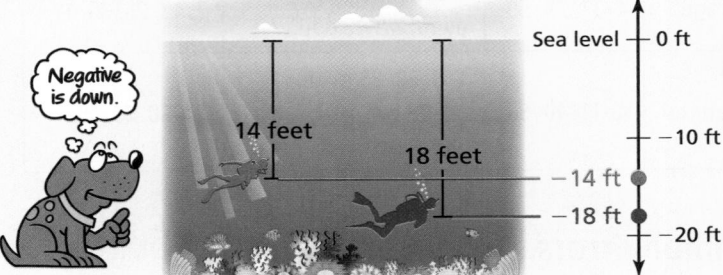

② **b.** Which integer is greater?

Of the two integers −14 and −18, −14 is greater because it is *higher* on a vertical number line.

c. Which integer has the greater absolute value?

$|-18| = 18$ is greater than $|-14| = 14$.

Compare this with the depth of the diver farther from sea level. The numbers are the same. The diver is 18 feet below sea level.

③ In mathematics, many concepts require understanding. Some, however, simply require acceptance. No one really knows why negative means "left" or "down" on a number line. We are not sure of the reason for choosing "right" to be positive, but it could have been something as simple as the fact that René Descartes was right-handed.

Project

Draw a picture that illustrates a real-life use of negative integers. Write a paragraph that explains how negative numbers are used in your picture.

Reteaching and Enrichment Strategies

If students need help...	If students got it...
Resources by Chapter • Practice A and Practice B • Puzzle Time Record and Practice Journal Practice Differentiating the Lesson Lesson Tutorials Skills Review Handbook	Resources by Chapter • Enrichment and Extension Start the next section

37. PUZZLE Use a number line.

    a. Graph and label the following points on a number line: $A = -3, E = 2,$ $M = -6, T = 0.$ What word do the letters spell?

    b. Graph and label the absolute value of each point in part (a). What word do the letters spell now?

38. OPEN-ENDED Write a negative integer whose absolute value is greater than 3.

REASONING Determine whether $n \geq 0$ or $n \leq 0.$

39. $n + \left| -n \right| = 2n$

40. $n + \left| -n \right| = 0$

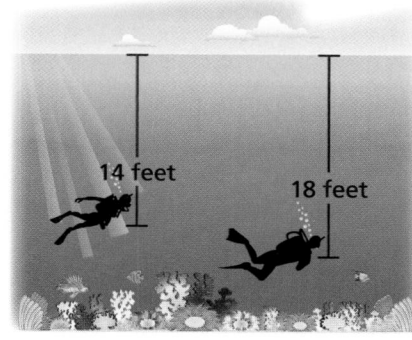

41. CORAL REEF Two scuba divers are exploring a living coral reef.

    a. Write an integer for the position of each diver relative to sea level.

    b. Which integer in part (a) is greater?

    c. Which integer in part (a) has the greater absolute value? Compare this with the position of the diver farther from sea level.

14 feet

18 feet

42. VOLCANOES The *summit elevation* of a volcano is the elevation of the top of the volcano relative to sea level. The summit elevation of the volcano Kilauea in Hawaii is 1277 meters. The summit elevation of the underwater volcano Loihi in the Pacific Ocean is -969 meters. Which summit is closer to sea level?

43. MINIATURE GOLF The table shows golf scores, relative to *par*.

    a. The player with the lowest score wins. Which player wins?

    b. Which player is at par?

    c. Which player is farthest from par?

Player	Score
1	+5
2	0
3	−4
4	−1
5	+2

True or False? **Determine whether the statement is *true* or *false*. Explain your reasoning.**

44. If $x < 0$, then $\left| x \right| = -x.$

45. The absolute value of every integer is positive.

Fair Game Review What you learned in previous grades & lessons

Add. *(Skills Review Handbook)*

46. $19 + 32$ **47.** $50 + 94$ **48.** $181 + 217$ **49.** $1149 + 2021$

50. MULTIPLE CHOICE Which value is *not* a whole number? *(Skills Review Handbook)*

    (A) -5 (B) 0 (C) 4 (D) 113

1.2 Adding Integers

Essential Question Is the sum of two integers *positive*, *negative*, or *zero*? How can you tell?

1 EXAMPLE: Adding Integers with the Same Sign

Use integer counters to find −4 + (−3).

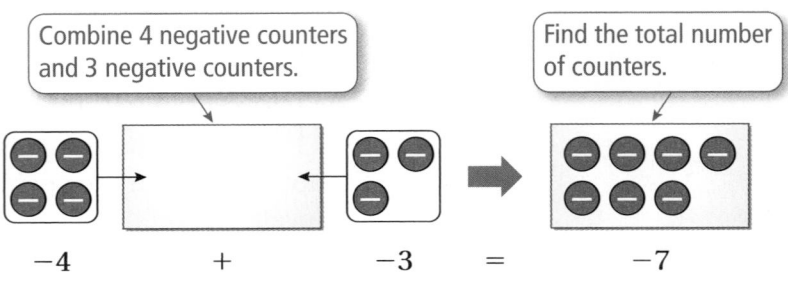

Combine 4 negative counters and 3 negative counters.

Find the total number of counters.

$$-4 \quad + \quad -3 \quad = \quad -7$$

So, −4 + (−3) = −7.

2 ACTIVITY: Adding Integers with Different Signs

Work with a partner. Use integer counters to find −3 + 2.

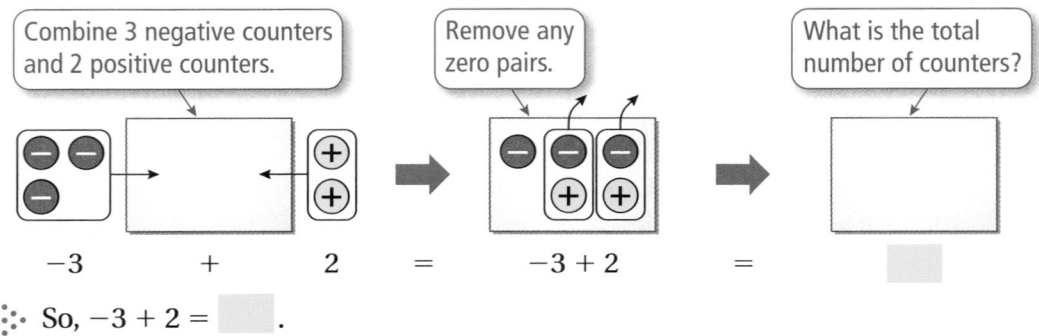

Combine 3 negative counters and 2 positive counters.

Remove any zero pairs.

What is the total number of counters?

$$-3 \quad + \quad 2 \quad = \quad -3 + 2 \quad =$$

So, −3 + 2 = ☐.

3 EXAMPLE: Adding Integers with Different Signs

Use a number line to find 5 + (−3).

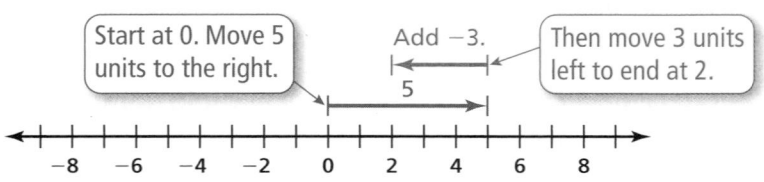

Start at 0. Move 5 units to the right.

Add −3.

5

Then move 3 units left to end at 2.

5

So, 5 + (−3) = 2.

Laurie's Notes

Introduction

For the Teacher

- **Goal:** The goal of this activity is to develop an understanding of the two types of integer addition problems: the signs are the same and the signs are different.
- **Representation:** Two models are introduced in this activity, *integer counters* and a *number line*.

Motivate

- **?** "What is the net result of an 8-yard loss in football followed by a 10-yard gain?" 2-yard gain
- **?** "What is the net result of scoring 25 points in a video game then losing 40 points?" a loss of 15 points
- Today's activity is about how integers are added.

Demonstrate

- If this is the student's first experience with integer counters, define a *yellow* counter as positive 1 (+1) and a *red* counter as negative 1 (−1).
- Counters of opposite color "neutralize" each other, so the net result of such a pair is zero. This is called a *zero pair.*
- **Model:** Show students that there are many ways to model a single integer. For example, the number 2 can be represented by two yellow counters or by three yellow counters with one red counter (2 plus 1 zero pair).

Activity Notes

Example 1 and Activity 2

- **Management Tip:** Store integer counters in self-locking bags. Put 15–20 counters in each bag.
- Students should use counters even if they say they know the answer.
- **Model:** A student volunteer could model Activity 2 at the overhead projector saying aloud what they are doing with the counters.
- **Common Error:** You may hear students say that −3 is greater than 2. You should respond "Gee, 2 is farther to the right on the number line than −3. Are you sure −3 is greater?" Remind students that the number farther to the right on the number line is greater.
- The use of parentheses around the integer −3 is for clarity. Sometimes people write −4 + ⁻3, with the raised negative sign.

Example 3

- Numbers are being represented by *directed line segments*. Positive numbers point to the right and negative numbers point to the left.
- **Connection:** The amount that the two directed line segments overlap is the same as the number of zero pairs that would result if the same problem were modeled using integer counters.

Previous Learning

Students need to know how to add and subtract whole numbers.

Activity Materials
Introduction
• integer counters

Start Thinking! and Warm Up

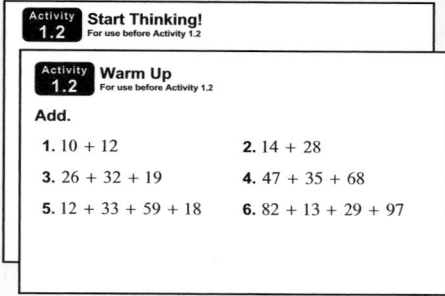

1.2 Record and Practice Journal

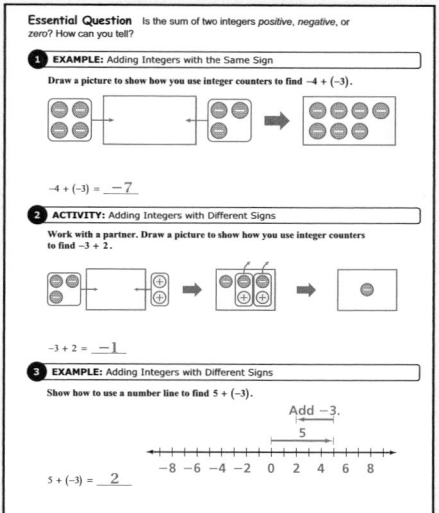

Visual

Use the number line to review and reinforce the Commutative Property of Addition. Draw a number line on the board. Model the expressions $-5 + 2$ and $2 + (-5)$. Students should see from the movement on the number line that the sum (result) is the same. The initial direction of movement does not matter.

1.2 Record and Practice Journal

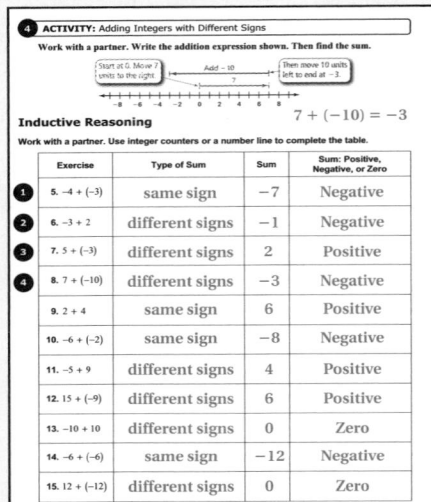

4 **ACTIVITY: Adding Integers with Different Signs**

Work with a partner. Write the addition expression shown. Then find the sum.

$$7 + (-10) = -3$$

Inductive Reasoning

Work with a partner. Use integer counters or a number line to complete the table.

	Exercise	Type of Sum	Sum	Sum: Positive, Negative, or Zero
1	5. $-4 + (-3)$	same sign	-7	Negative
2	6. $-3 + 2$	different signs	-1	Negative
3	7. $5 + (-3)$	different signs	2	Positive
4	8. $7 + (-10)$	different signs	-3	Negative
	9. $2 + 4$	same sign	6	Positive
	10. $-6 + (-2)$	same sign	-8	Negative
	11. $-5 + 9$	different signs	4	Positive
	12. $15 + (-9)$	different signs	6	Positive
	13. $-10 + 10$	different signs	0	Zero
	14. $-6 + (-6)$	same sign	-12	Negative
	15. $12 + (-12)$	different signs	0	Zero

What Is Your Answer?

16. **IN YOUR OWN WORDS** Is the sum of two integers *positive, negative,* or *zero*? How can you tell?

Sample answer: The sum will have the same sign as the integer with the greater absolute value.

17. Write a general rule for adding   Sample answers given.

a. two integers with the same sign.

To add two positive numbers, add normally. To add two negative numbers, ignore the signs and add. Then make the answer negative.

b. two integers with different signs.

Subtract the lesser absolute value from the greater absolute value. Use the sign of the number with the greater absolute value.

c. an integer and its opposite.

The sum is zero.

Laurie's Notes

Activity 4

- **Neighbor Check:** Have students work independently and then have their neighbor check their work. Have students discuss any discrepancies.

? "How would you use the number line to show $-10 + 7$?" Ask a student volunteer to show the solution. Start at zero. Move 10 units to the left to get to -10. Then move 7 units to the right. End at -3.

- Students should recognize that the order in which you add two numbers does not matter (Commutative Property of Addition). You can start at 7 and then move 10 units to the left.

Words of Wisdom

- You can use a standard deck of playing cards to generate random addition problems. Define red = negative, black = positive, Jacks = 11, Queens = 12, Kings = 13, Aces = 1, Jokers = 0.

Inductive Reasoning

- Students should work with a partner to complete the table.
- The goal of Questions 5–15 is to develop some understanding about the two types of addition problems: integers with the same sign and integers with different signs.
- It is important for students to record the *Type of Sum* as shown in Questions 5 and 8.

What Is Your Answer?

- In Questions 16 and 17, the sum of two integers can be positive, negative, or zero. If the two integers have the same sign, the sign of the sum is the same as the integers. If the integers have different signs, then the sum is the sign of the integer with the greater absolute value. (Formal rules will be presented in the lesson.)

? **Extension:** Use the integer counters or a number line to model the sum of three integers.

- "What is the sum of $3 + (-2) + 5$?" 6
- "What is the sum of $(-4) + 2 + (-5)$?" -7

Closure

- "If the sum of two integers is negative, are both integers negative? How do you know?" Both integers could be negative, but they may not be. The sum of 4 and -5 is negative, but both integers are not negative.
- "If the sum of two integers is positive, are both integers positive? How do you know?" Both integers could be positive, but they may not be. The sum of -4 and 5 is positive, but both integers are not positive.

Technology For the Teacher

Dynamic Classroom

The Dynamic Planning Tool
Editable Teacher's Resources at *BigIdeasMath.com*

4 ACTIVITY: Adding Integers with Different Signs

Work with a partner. Write the addition expression shown. Then find the sum.

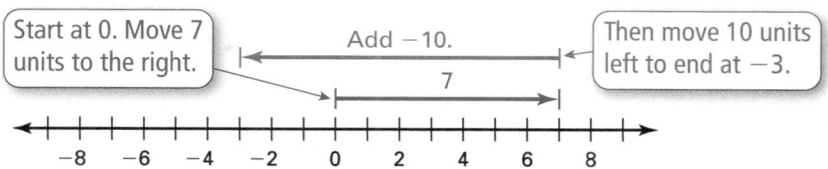

| Start at 0. Move 7 units to the right. | Add −10. | Then move 10 units left to end at −3. |

Inductive Reasoning

Work with a partner. Use integer counters or a number line to complete the table.

	Exercise	Type of Sum	Sum	Sum: Positive, Negative, or Zero
1	**5.** $-4 + (-3)$	Integers with the same sign		
2	**6.** $-3 + 2$			Negative
3	**7.** $5 + (-3)$		2	
4	**8.** $7 + (-10)$	Integers with different signs		
	9. $2 + 4$			
	10. $-6 + (-2)$			
	11. $-5 + 9$			
	12. $15 + (-9)$			
	13. $-10 + 10$			
	14. $-6 + (-6)$			
	15. $12 + (-12)$			

What Is Your Answer?

16. IN YOUR OWN WORDS Is the sum of two integers *positive*, *negative*, or *zero*? How can you tell?

17. Write general rules for adding (a) two integers with the same sign, (b) two integers with different signs, and (c) an integer and its opposite.

Practice

Use what you learned about adding integers to complete Exercises 8–15 on page 12.

Check It Out
Lesson Tutorials
BigIdeasMath com

 Key Idea

Adding Integers with the Same Sign

Words Add the absolute values of the integers. Then use the common sign.

Numbers $2 + 5 = 7$ $\qquad$ $-2 + (-5) = -7$

EXAMPLE ① **Adding Integers with the Same Sign**

Find $-2 + (-4)$. Use a number line to check your answer.

$$-2 + (-4) = -6 \qquad \text{Add } |-2| \text{ and } |-4|.$$

> Use the common sign.

∴ The sum is -6.

Check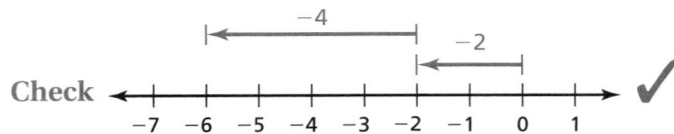

The Meaning of a Word

Opposite

When you sit across from your friend at the lunch table, you sit **opposite** your friend.

● **On Your Own**

Add.

1. $7 + 13$ $\qquad$ **2.** $-8 + (-5)$ $\qquad$ **3.** $-20 + (-15)$

Two numbers that are the same distance from 0, but on opposite sides of 0, are called **opposites.** For example, -3 and 3 are opposites.

 Key Ideas

Adding Integers with Different Signs

Words Subtract the lesser absolute value from the greater absolute value. Then use the sign of the integer with the greater absolute value.

Numbers $8 + (-10) = -2$ $\qquad$ $-13 + 17 = 4$

Additive Inverse Property

Words The sum of an integer and its **additive inverse,** or opposite, is 0.

Numbers $6 + (-6) = 0$ $\qquad$ $-25 + 25 = 0$

Laurie's Notes

Introduction

Connect

- **Yesterday:** Students used integer counters and a number line to add integers of the same sign and of different signs.
- **Today:** Students will add integers without the use of a visual or concrete model.

Motivate

? "Is the sum of 58 and −72 positive or negative? How do you know?"
Negative. *Sample answers:*

Using Counters: Some students may say that there would be more red counters (72) than yellow counters (58), so the sum is negative.

Using a Number Line: Some students may describe a number line: if you go back (left) 72 units and then forward (right) 58 units, you won't get back to 0, so the sum is negative.

Using Definitions: Some students may remember that the sign of the integer with the greater absolute value (72) is negative, so the sum is negative.

Lesson Notes

Example 1

- As you discuss the example, refer to the models from the Activity.
  - When the signs are the *same,* the counters will be the *same color.*
  - When the signs are the *same,* both directed line segments will be going in the *same direction.*

Key Ideas

- Discuss the definition of opposites.
- **?** "When you add two integers with different signs, how do you know if the sum is positive or negative?" Students should be using the concept of absolute value even if they don't use the precise language. You want to hear something about the size of the number, meaning its absolute value.
- **?** **Reasoning:** Write these problems on the board: $14 + (-8) = ?$ and $(-14) + 8 = ?$. Ask "How are the problems alike? How are they different?" *Sample answer:*
  Alike: They each use the numbers 14 and 8. They both consist of two different signs, which are being added together.
  Different: In the first problem, 14 is positive and 8 is negative. In the second problem, 14 is negative and 8 is positive.
- Define how to add integers with different signs.
- Define the Additive Inverse Property. This is a special case of adding integers with different signs.
- **?** "How many zero pairs are there when you add $(-5) + 5$?" 5

Start Thinking! and Warm Up

> Lesson 1.2 **Warm Up** For use before Lesson 1.2
>
> Lesson 1.2 **Start Thinking!** For use before Lesson 1.2
>
> The temperature first rises 10 degrees and then falls 12 degrees. Is the end temperature greater than or less than the starting temperature? How does this compare to adding integers?

Extra Example 1

Find $-3 + (-12)$. -15

On Your Own

1. 20
2. -13
3. -35

Extra Example 2

Add.

a. $-11 + 6$ -5

b. $12 + (-5)$ 7

c. $4 + (-4)$ 0

 **On Your Own**

4. 9	**5.** 5
6. −1	**7.** −4
8. 0	**9.** 0

Extra Example 3

Find the change in the account balance for August.

August Transactions	
Deposit	$35
Deposit	$40
Withdrawal	−$25

increased $50

 On Your Own

10. −5

Differentiated Instruction

Vocabulary

Write a table of opposites on the board. Encourage students to add to the list.

Word	Opposite
little	big
forward	backward

Ask the English learners to write the words in their native language in another column and to share them with the class. Explain to the class that in mathematics, every nonzero number has an opposite. Every pair of opposites consists of a positive number and a negative number. Ask students to name some pairs of opposite numbers. Write opposite numbers in the table and the words that represent them.

Laurie's Notes

Example 2

- For each part of this example, have student volunteers explain how it was computed.
- **Part (a):** Because 5 had the lesser absolute value, you subtract it from the absolute value of −10. In general, use the sign of the number with the greater absolute value. In this case, the answer will be negative.

Words of Wisdom

- Be sure students understand that the subtraction of the two absolute values is connected to the zero pairs that get removed when using integer counters, or it is the overlapping distance when the number line is used.
- If students are not getting correct answers, use integer counters or the number line to model several additional examples.

Example 3

- **Financial Literacy:** Provide a brief description of what *deposit* and *withdrawal* mean in banking. A *deposit* is when you add money to an account, and a *withdrawal* is when you take money out of an account.
- **Model:** Using *play* money and a student volunteer, act out the following.
  Hand $50 to a banker (*deposit*).
  Ask for $40 back (*withdrawal*).
  Hand the banker $75 (*deposit*).
  Ask for $50 back (*withdrawal*).
- "What is the balance in your account?" $35
- "Would the balance be the same if you added the two deposits first, added the two withdrawals next, and then found the sum of the two answers?" yes; $(50 + 75) + (-40 + (-50)) = 125 + (-90) = 35$
- "Is there another way you could add these numbers?"
  Yes, add $50 + (-50)$ first to get 0, then add $-40 + 75$ to get 35. So, $0 + 35 = 35$.
- **FYI:** The Identity Property of Addition states that the sum of any number and zero is that number.

Closure

- What do you know about the sum $A + B$? Explain your reasoning.

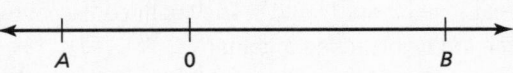

- Two integers have different signs. Their sum is −8. What are possible values for the two integers? *Sample answers:* −9 and 1, −10 and 2, 3 and −11, 4 and −12

Technology
For the Teacher

Dynamic Classroom

The Dynamic Planning Tool
Editable Teacher's Resources at *BigIdeasMath.com*

EXAMPLE 2 **Adding Integers with Different Signs**

a. Find 5 + (−10).

$$5 + (-10) = -5$$

$|-10| > |5|$. So, subtract $|5|$ from $|-10|$.

Use the sign of −10.

∴ The sum is −5.

b. Find −3 + 7.

$$-3 + 7 = 4$$

$|7| > |-3|$. So, subtract $|-3|$ from $|7|$.

Use the sign of 7.

∴ The sum is 4.

c. Find −12 + 12.

$$-12 + 12 = 0$$

The sum is 0 by the Additive Inverse Property.

−12 and 12 are opposites.

∴ The sum is 0.

On Your Own

Exercises 8–23

Add.

4. −2 + 11

5. 13 + (−8)

6. 9 + (−10)

7. −8 + 4

8. 7 + (−7)

9. −31 + 31

EXAMPLE 3 **Adding More than Two Integers**

The list shows four bank account transactions in July. Find the change *C* in the account balance.

JULY TRANSACTIONS	
Deposit	$50
Withdrawal	-$40
Deposit	$75
Withdrawal	-$50

Find the sum of the four transactions.

$$C = 50 + (-40) + 75 + (-50) \quad \text{Write the sum.}$$
$$= 10 + 75 + (-50) \quad \text{Add 50 and } -40.$$
$$= 85 + (-50) \quad \text{Add 10 and 75.}$$
$$= 35 \quad \text{Add 85 and } -50.$$

∴ Because *C* = 35, the account balance increased $35 in July.

On Your Own

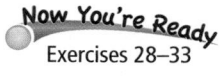
Exercises 28–33

10. WHAT IF? In Example 3, the deposit amounts are $30 and $55. Find the change *C* in the account balance.

 Vocabulary and Concept Check

1. **WRITING** How do you find the additive inverse of an integer?

2. **NUMBER SENSE** Is $3 + (-4)$ the same as $-4 + 3$? Explain.

Tell whether the sum is *positive*, *negative*, or *zero* without adding. Explain your reasoning.

3. $-8 + 20$ 4. $50 + (-50)$ 5. $-10 + (-18)$

Tell whether the statement is *true* or *false*. Explain your reasoning.

6. The sum of two negative integers is always negative.

7. An integer and its absolute value are always opposites.

 **Practice and Problem Solving**

Add.

 8. $6 + 4$ 9. $-4 + (-6)$ 10. $-2 + (-3)$ 11. $-5 + 12$

12. $5 + (-7)$ 13. $8 + (-8)$ 14. $9 + (-11)$ 15. $-3 + 13$

16. $-4 + (-16)$ 17. $-3 + (-4)$ 18. $14 + (-5)$ 19. $0 + (-11)$

20. $-10 + (-15)$ 21. $-13 + 9$ 22. $18 + (-18)$ 23. $-25 + (-9)$

ERROR ANALYSIS Describe and correct the error in finding the sum.

24.
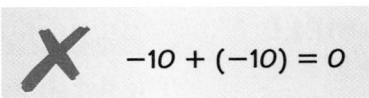
$9 + (-6) = -3$

25.
$-10 + (-10) = 0$

26. **TEMPERATURE** The temperature is $-3°F$ at 7 A.M. During the next four hours, the temperature increases $21°F$. What is the temperature at 11 A.M.?

27. **BANKING** Your bank account has a balance of $-\$12$. You deposit $\$60$. What is your new balance?

Add.

28. $13 + (-21) + 16$ 29. $22 + (-14) + (-35)$ 30. $-13 + 27 + (-18)$

31. $-19 + 26 + 14$ 32. $-32 + (-17) + 42$ 33. $-41 + (-15) + (-29)$

Tell how the Commutative and Associative Properties of Addition can help you find the sum mentally. Then find the sum.

34. $9 + 6 + (-6)$ 35. $-8 + 13 + (-13)$ 36. $9 + (-17) + (-9)$

37. $7 + (-12) + (-7)$ 38. $-12 + 25 + (-15)$ 39. $6 + (-9) + 14$

Assignment Guide and Homework Check

Level	Day 1 Activity Assignment	Day 2 Lesson Assignment	Homework Check
Basic	8–15, 50–54	1–7, 17–25 odd, 26–30, 47	17, 26, 28, 47
Average	8–15, 50–54	1–7, 17–27 odd, 28–32 even, 41, 45, 48	17, 27, 28, 41
Advanced	8–15, 50–54	1–7, 25, 30–36 even, 42–45, 48, 49	30, 36, 43, 45

Common Errors

- **Exercises 8–23, 28–39** Students may try to ignore the signs and just add the integers. Remind them of the meaning of absolute value. Make sure they understand that they should use the sign of the number that is farther from zero. Also remind them of the Key Ideas, and how the signs of the integers determine if they need to add or subtract the integers.
- **Exercise 47** Students may assume that 10 yards to get a first down is a necessary part of the expression and will choose the first expression. Ask how many yards were gained in the three downs. Then ask if the football has traveled far enough for a first down.
- **Exercise 48** Students may not realize that each height measurement is given in reference to the previous point. Tell them to determine the measurement in relation to point A, which would be zero on a number line.

1.2 Record and Practice Journal

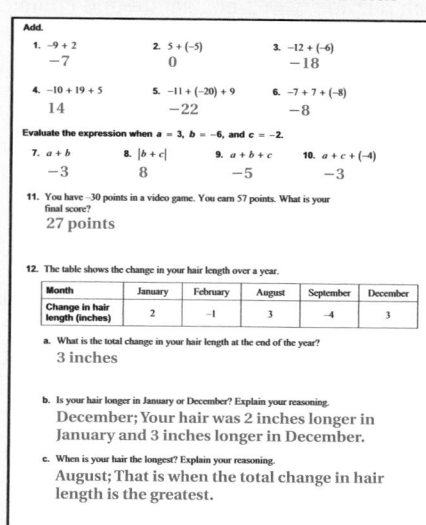

1. Change the sign of the integer.

2. yes; The sums are the same by the Commutative Property of Addition.

3. positive; 20 has the greater absolute value and is positive.

4. zero; 50 and −50 are additive inverses.

5. negative; The common sign is a negative sign.

6. true; To add integers with the same sign, add the absolute values and use the common sign.

7. false; A positive integer and its absolute value are equal, not opposites.

Practice and Problem Solving

8. 10		**9.** −10	
10. −5		**11.** 7	
12. −2		**13.** 0	
14. −2		**15.** 10	
16. −20		**17.** −7	
18. 9		**19.** −11	
20. −25		**21.** −4	
22. 0		**23.** −34	

24. The wrong sign is used. $9 + (-6) = 3$

25. −10 and −10 are not opposites. $-10 + (-10) = -20$

26. 18°F   **27.** $48

28. 8   **29.** −27

30. −4   **31.** 21

32. −7   **33.** −85

Practice and Problem Solving

34. Use the Associative Property to add 6 and −6 first. 9

35. Use the Associative Property to add 13 and −13 first. −8

36. *Sample answer:* Use the Commutative Property to switch the last two terms. −17

37. *Sample answer:* Use the Commutative Property to switch the last two terms. −12

38. *Sample answer:* Use the Commutative Property to switch the last two terms. −2

39. *Sample answer:* Use the Commutative Property to switch the last two terms. 11

40. −1 **41.** −13

42. 9

43–48. See Additional Answers.

49. See *Taking Math Deeper*.

Fair Game Review

50. 31 **51.** 8

52. 114 **53.** 183

54. D

Mini-Assessment

Add.

1. 10 + (−12) −2

2. −7 + (−5) −12

3. −17 + 25 8

4. 65 + (−99) −34

5. The temperature is −2°F at 6 A.M. During the next three hours, the temperature increases to 15°F. What is the temperature at 9 A.M.? 13°F

Taking Math Deeper

Exercise 49

In this puzzle, students get a chance to apply integer addition with *Guess, Check, and Revise.*

 Solve the straightforward part of the puzzle.

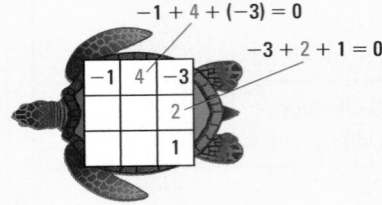

$$-1 + 4 + (-3) = 0$$
$$-3 + 2 + 1 = 0$$

② Make a list of the numbers from −4 to 4. Cross off the numbers you have used.

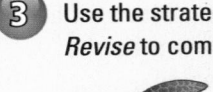

③ Use the strategy *Guess, Check, and Revise* to complete the square.

You can create other magic squares by repeating the same digit in each number in the magic square (−11, 44, −33, etc.).

The emperor Yu-Huang was a legendary figure in China, in the same sense that King Arthur was a legendary figure in Europe. He was called the Jade Emperor and there are many stories about him. In addition to the story of the magic square and the turtle, the Jade Emperor is credited with creating the Chinese Zodiac in which each sequence of 12 years is given the name of an animal, such as the "Year of the Snake" or the "Year of the Rat."

Project

Create a new magic square using integers. Decide which squares will contain numbers and which squares will be blank. Complete your magic square. Then switch puzzles with a classmate and complete the puzzle you receive. Check your answers.

Reteaching and Enrichment Strategies

If students need help...	If students got it...
Resources by Chapter • Practice A and Practice B • Puzzle Time Record and Practice Journal Practice Differentiating the Lesson Lesson Tutorials Skills Review Handbook	Resources by Chapter • Enrichment and Extension Start the next section

ALGEBRA Evaluate the expression when $a = 4$, $b = -5$, and $c = -8$.

40. $a + b$

41. $b + c$

42. $|a + b + c|$

43. OPEN-ENDED Write two integers with different signs that have a sum of -25. Write two integers with the same sign that have a sum of -25.

MENTAL MATH Use mental math to solve the equation.

44. $d + 12 = 2$

45. $b + (-2) = 0$

46. $-8 + m = -15$

47. FIRST DOWN In football, a team must gain 10 yards to get a first down. The team gains 6 yards on the first play, loses 3 yards on the second play, and gains 8 yards on the third play. Which expression can be used to decide whether the team gets a first down?

$$10 + 6 - 3 + 8 \qquad 6 + (-3) + 8 \qquad 6 + (-3) + (-8)$$

48. DOLPHIN Starting at point A, the path of a dolphin jumping out of the water is shown.

    a. Is the dolphin deeper at point C or point E? Explain your reasoning.

    b. Is the dolphin higher at point B or point D? Explain your reasoning.

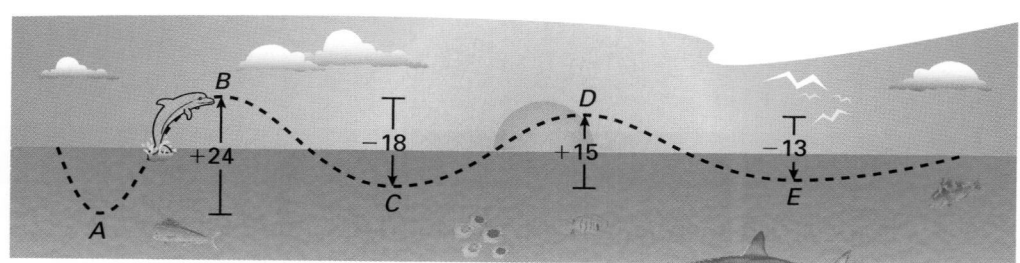

49. _Puzzle_ According to a legend, the Chinese Emperor Yu-Huang saw a magic square on the back of a turtle. In a _magic square_, the numbers in each row and in each column have the same sum. This sum is called the magic sum.

Copy and complete the magic square so that each row and each column has a magic sum of 0. Use each integer from -4 to 4 exactly once.

 **Fair Game Review** What you learned in previous grades & lessons

Subtract. _(Skills Review Handbook)_

50. $69 - 38$

51. $82 - 74$

52. $177 - 63$

53. $451 - 268$

54. MULTIPLE CHOICE What is the range of the numbers below? _(Skills Review Handbook)_

    12, 8, 17, 12, 15, 18, 30

    (A) 12   (B) 15   (C) 18   (D) 22

1.3 Subtracting Integers

Essential Question How are adding integers and subtracting integers related?

1 EXAMPLE: Subtracting Integers

Use integer counters to find 4 − 2.

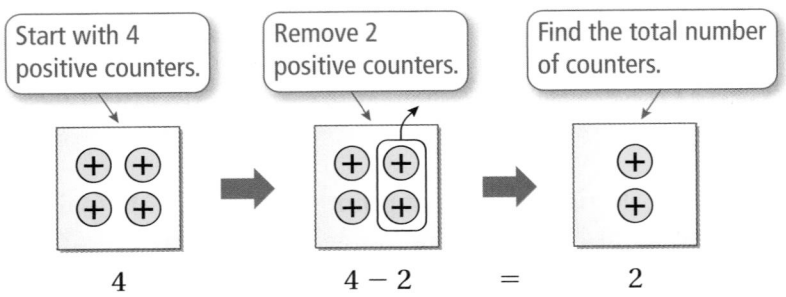

Start with 4 positive counters.

Remove 2 positive counters.

Find the total number of counters.

4   4 − 2   =   2

So, 4 − 2 = 2.

2 ACTIVITY: Adding Integers

Work with a partner. Use integer counters to find 4 + (−2).

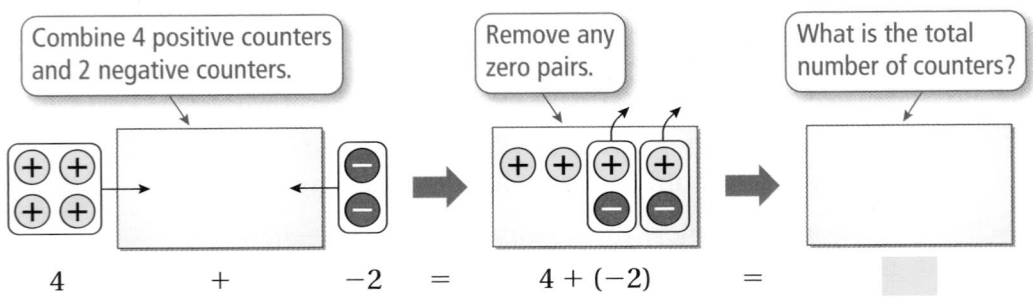

Combine 4 positive counters and 2 negative counters.

Remove any zero pairs.

What is the total number of counters?

4   +   −2   =   4 + (−2)   =

3 EXAMPLE: Subtracting Integers

Use a number line to find −3 − 1.

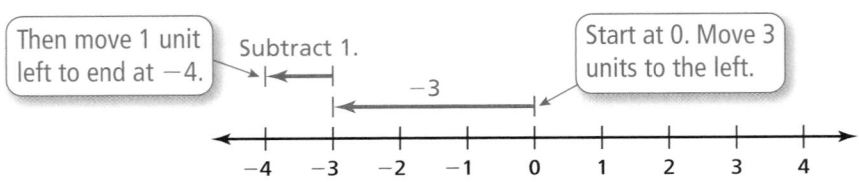

Then move 1 unit left to end at −4.

Subtract 1.

Start at 0. Move 3 units to the left.

−3

So, −3 − 1 = − 4.

Laurie's Notes

Introduction

For the Teacher

- **Goal:** The goal of this activity is to develop an understanding that an integer subtraction problem has a related integer addition problem.

Motivate

- ❓ Hand a student a collection of objects (8 pencils, 12 index cards, 9 paper clips) and ask another student to take some of the objects (5 pencils, 7 index cards, 3 paper clips). "What expressions represent this situation?" $8 - 5, 12 - 7, 9 - 3$
- ❓ "One way to think about subtraction: you have some amount and you take away another amount. Does this still work when you begin with negative amounts like -3 (owe a friend \$3)?"
- Today's activity investigates subtraction of integers.

Activity Notes

Example 1 and Activity 2

- ❓ "How would you model $4 - 2$ using integer counters?" Subtraction means that you model the first number, and then take the second number from that original collection. Students should model the problem using their counters even if they say they know the answer.
- "Now let's see how $4 - 2$ is like the addition problem you did a few days ago." Have students work through Activity 2 with a partner.

Words of Wisdom

- Before trying the number line examples, try $2 - 4$ with integer counters. It will remind students that subtraction is *not* commutative.
- ❓ "How would you model $2 - 4$?" Some students may say that this is not possible, because you should subtract the lesser number from the greater number. Show the model below.
- **Model:** Show the class two yellow counters.
- ❓ "How can you take 4 yellow counters away?" If you add two zero pairs, you will have 4 yellow counters that you can take away. Two red counters are left. $2 - 4 = -2$
- **Summary:** Discuss the problems $4 - 2$ and $2 - 4$.
- ❓ "$4 - 2$ had the related problem $4 + (-2)$. What do you think the related addition problem would be for $2 - 4$?" $2 + (-4)$

Example 3

- **FYI:** The number line model for subtraction requires that you start at the first number. To subtract a positive number, move to the left. To subtract a negative number, move to the right.
- **Model:** To model $-3 - 1$, draw an arrow pointing to the left from zero to -3. Then move *left one* because you are *subtracting positive 1*.

Previous Learning

Students need to know how to add and subtract whole numbers.

Activity Materials	
Introduction	**Textbook**
• 8 pencils	• integer
• 12 index cards	counters
• 9 paper clips	

Start Thinking! and Warm Up

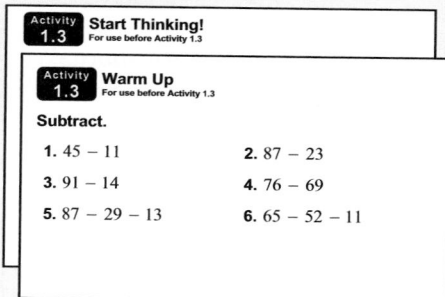

Activity 1.3 Start Thinking!
For use before Activity 1.3

Activity 1.3 Warm Up
For use before Activity 1.3

Subtract.

1. $45 - 11$ 2. $87 - 23$
3. $91 - 14$ 4. $76 - 69$
5. $87 - 29 - 13$ 6. $65 - 52 - 11$

1.3 Record and Practice Journal

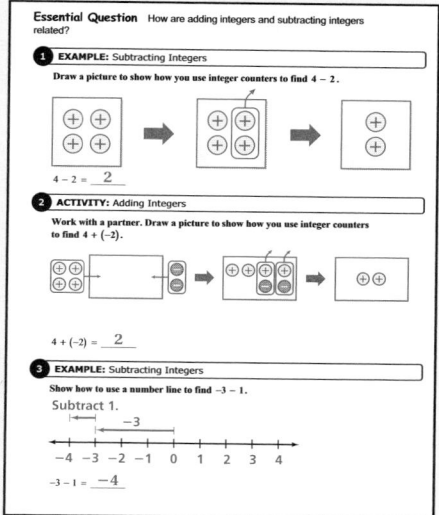

Essential Question How are adding integers and subtracting integers related?

1 EXAMPLE: Subtracting Integers

Draw a picture to show how you use integer counters to find $4 - 2$.

$4 - 2 = \underline{2}$

2 ACTIVITY: Adding Integers

Work with a partner. Draw a picture to show how you use integer counters to find $4 + (-2)$.

$4 + (-2) = \underline{2}$

3 EXAMPLE: Subtracting Integers

Show how to use a number line to find $-3 - 1$.

Subtract 1.

$-3 - 1 = \underline{-4}$

Differentiated Instruction

Auditory

When you subtract a number, you add its opposite. So when you subtract a number using a number line, you move in the opposite direction you would move if adding the number. This is why you move left when subtracting a positive number and move right when subtracting a negative number.

1.3 Record and Practice Journal

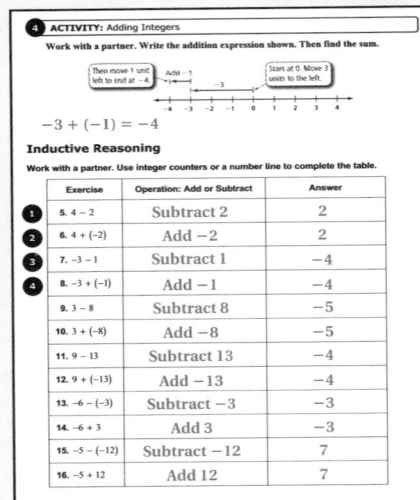

4 ACTIVITY: Adding Integers

Work with a partner. Write the addition expression shown. Then find the sum.

$-3 + (-1) = -4$

Inductive Reasoning

Work with a partner. Use integer counters or a number line to complete the table.

	Exercise	Operation: Add or Subtract	Answer
1	5. $4 - 2$	Subtract 2	2
2	6. $4 + (-2)$	Add -2	2
3	7. $-3 - 1$	Subtract 1	-4
4	8. $-3 + (-1)$	Add -1	-4
	9. $3 - 8$	Subtract 8	-5
	10. $3 + (-8)$	Add -8	-5
	11. $9 - 13$	Subtract 13	-4
	12. $9 + (-13)$	Add -13	-4
	13. $-6 - (-3)$	Subtract -3	-3
	14. $-6 + 3$	Add 3	-3
	15. $-5 - (-12)$	Subtract -12	7
	16. $-5 + 12$	Add 12	7

What Is Your Answer?

17. IN YOUR OWN WORDS How are adding integers and subtracting integers related?

Subtracting an integer is the same as adding its opposite.

18. Write a general rule for subtracting integers.

To subtract an integer, add its opposite.

Laurie's Notes

Activity 4

- Look at the related addition problem $-3 + (-1)$. Draw an arrow from 0 to -3 to represent -3. Now move to the left one because you are adding 1 in the negative direction (-1). Draw the arrow and write 'Add -1'.
- Have students work with a partner to write an addition expression.
- You just wrote sum and difference expressions that meant the same thing.

Inductive Reasoning

- Students should work with a partner to find the sums.
- The goal is to develop some understanding about subtraction and the related addition problem.
- If students are having difficulty using the integer counters, encourage them to try the number line and vice versa.

What Is Your Answer?

- In Questions 17 and 18, subtraction is the same as adding the opposite.
- **Extension:** "Use the integer counters or a number line to model $(8 - 4) - 2$ and $8 - (4 - 2)$. Are the results the same?" No, subtraction is not associative.
- **FYI:** The Associative Property of Addition states that the value of a sum does not depend on how the numbers are grouped. This does not apply for subtraction, as illustrated by the example above.

Closure

- Explain how you would use integer counters to model $4 - 6$.
  Add 6 red counters to the 4 yellow counters and remove the 4 zero pairs. The result is 2 red counters.

Technology For the Teacher

Dynamic Classroom

The Dynamic Planning Tool
Editable Teacher's Resources at *BigIdeasMath.com*

4   ACTIVITY: Adding Integers

Work with a partner. Write the addition expression shown. Then find the sum.

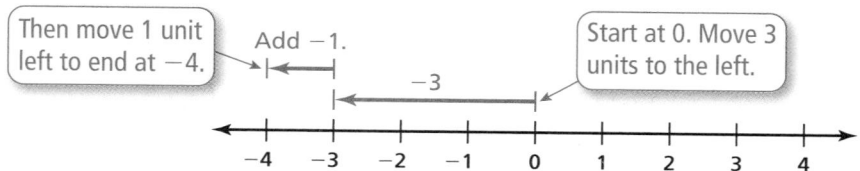

Then move 1 unit left to end at −4.

Add −1.

−3

Start at 0. Move 3 units to the left.

Inductive Reasoning

Work with a partner. Use integer counters or a number line to complete the table.

	Exercise	Operation: Add or Subtract	Answer
1	**5.** $4 - 2$	Subtract 2	2
2	**6.** $4 + (-2)$	Add −2	2
3	**7.** $-3 - 1$	Subtract 1	−4
4	**8.** $-3 + (-1)$	Add −1	−4
	9. $3 - 8$		
	10. $3 + (-8)$		
	11. $9 - 13$		
	12. $9 + (-13)$		
	13. $-6 - (-3)$		
	14. $-6 + (3)$		
	15. $-5 - (-12)$		
	16. $-5 + 12$		

What Is Your Answer?

17. IN YOUR OWN WORDS How are adding integers and subtracting integers related?

18. Write a general rule for subtracting integers.

Practice Use what you learned about subtracting integers to complete Exercises 8–15 on page 18.

 **Key Idea**

Subtracting Integers

Words   To subtract an integer, add its opposite.

Numbers $3 - 4 = 3 + (-4) = -1$

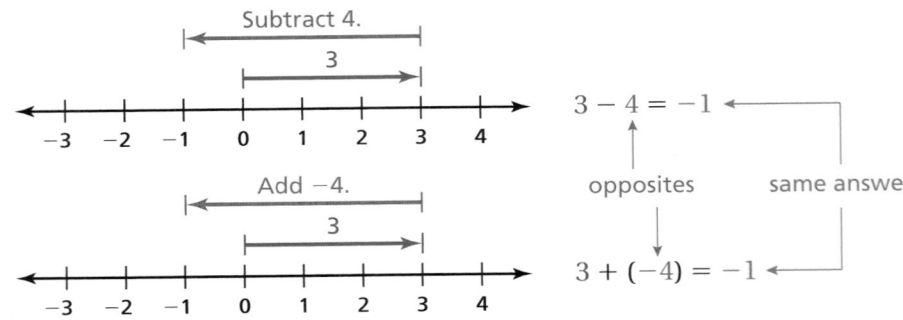

EXAMPLE (1)   **Subtracting Integers**

a. **Find $3 - 12$.**

$$3 - 12 = 3 + (-12) \qquad \text{Add the opposite of 12.}$$
$$= -9 \qquad \text{Add.}$$

⋮⋅ The difference is -9.

b. **Find $-8 - (-13)$.**

$$-8 - (-13) = -8 + 13 \qquad \text{Add the opposite of } -13.$$
$$= 5 \qquad \text{Add.}$$

⋮⋅ The difference is 5.

c. **Find $5 - (-4)$.**

$$5 - (-4) = 5 + 4 \qquad \text{Add the opposite of } -4.$$
$$= 9 \qquad \text{Add.}$$

⋮⋅ The difference is 9.

⬤ **On Your Own**

Now You're Ready
Exercises 8–23

Subtract.

1. $8 - 3$
2. $9 - 17$
3. $-3 - 3$
4. $-14 - 9$
5. $9 - (-8)$
6. $-12 - (-12)$

Laurie's Notes

Introduction

Connect

- **Yesterday:** Students used integer counters and a number line to subtract integers.
- **Today:** Students will use the idea that a subtraction problem can be rewritten as an addition problem.

Motivate

- ❓ **"Is my age (teacher) minus your age (point to a student) the same as your age minus my age?"** Most students will understand that an older person's age minus a younger person's age is a positive number and subtracting in the other order is a negative number.
- Draw the two number lines shown.

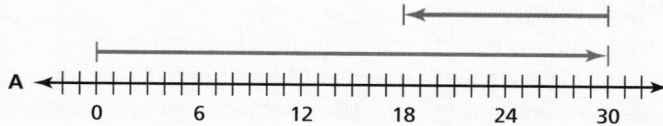

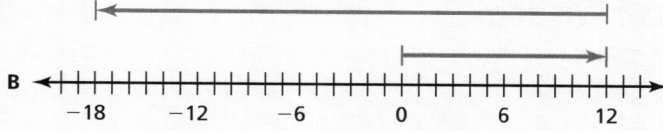

- Assume the teacher is 30 and the student is 12. Take away the context (age of two people) and simply look at the two number lines.
- ❓ **"What two addition problems are modeled in each diagram?"**
  A: $30 + (-12)$; B: $12 + (-30)$
- Now add the context of the problem and write the following on the board.
  $$30 - 12 = 30 + (-12) = 18$$
  $$12 - 30 = 12 + (-30) = -18$$
- This introduction states the relationship between a subtraction problem and its related addition problem. It reminds students that subtraction is not commutative and that they need to be careful of order when subtracting.
- **Vocabulary Review:** Ask students what the word *opposite* means.

Lesson Notes

Key Idea

- Each subtraction problem can be rewritten as an addition problem. Remind students that they already know how to add integers.

Example 1

- Work through each part of the example. Pointing to a classroom number line may be helpful.
- Use a different color when rewriting the problem as "add the opposite."
  $$3 - 12 = 3 + (-12)$$

Start Thinking! and Warm Up

> **Lesson 1.3** Warm Up
> For use before Lesson 1.3
>
> **Lesson 1.3** Start Thinking!
> For use before Lesson 1.3
>
> How can you tell if the difference of two integers is positive?
>
> How can you tell if the difference of two integers is negative?
>
> How can you tell if the difference of two integers is zero?

Extra Example 1
Subtract.
a. $8 - 10$ -2
b. $-3 - (-6)$ 3
c. $2 - (-4)$ 6

On Your Own

1. 5   2. -8

3. -6 4. -23

5. 17   6. 0

Extra Example 2

Evaluate $-11 - 5 - (-8)$. -8

 On Your Own

7. -33	**8.** -33
9. -4	**10.** -2
11. 15	**12.** 13

Extra Example 3

Which continent has the greater range of elevations? South America

	South America	Europe
Highest Elevation	6960 m	5642 m
Lowest Elevation	-40 m	-92 m

 On Your Own

13. 5710 meters

English Language Learners

Class Activity

Reinforce the meaning of the words *difference*, *subtract*, *positive*, and *negative*.

1. Ask students to name two numbers whose difference is 0. Students should realize that the only possibility is a number subtracted from itself.

2. Ask students to name two numbers whose difference is 3. Students should realize that if they start with a number greater than 3, they need to subtract a positive number. If they begin with a number less than 3, they need to subtract a negative number.

3. Ask students to name two numbers whose difference is -3. Starting with a number greater than -3 means you need to subtract a positive number. Starting with a number less than -3 means you need to subtract a negative number.

Laurie's Notes

On Your Own

- **Questions 1–6:**
  - For students who are having difficulty, have them record the problem on the board. They should say aloud, "add the opposite" and state what that means for the particular problem.
  - Ask students if it is possible to determine when the difference of two negative numbers will be positive and when the difference of two negative numbers will be negative.

Example 2

- Caution students to work slowly.
- Subtraction must be performed in order from left to right.

Example 3

- **Vocabulary:** You may need to review the meanings of *elevation* and *range*.
- **Fun Fact:** The highest point in Hawaii is Mauna Kea at 4208 meters above sea level. The lowest points in Hawaii are at sea level, where the coast of Hawaii meets the Pacific Ocean.

Words of Wisdom

- Make a colorful number line that stretches the length of your board. Use two 3-inch wide strips of different colored paper. Positive integers are on one color and negative integers are on the other. Label only the integers, but make a hash mark at $\frac{1}{2}$ between each of the integers. This will be useful later in the year.

Closure

- **Writing:** Your friend is home sick today. Imagine you are on the telephone with him or her. How would you explain how to subtract integers? Be sure to use an example.

Technology
For the Teacher

Dynamic Classroom

The Dynamic Planning Tool
Editable Teacher's Resources at *BigIdeasMath.com*

EXAMPLE 2   Subtracting Integers

Evaluate $-7 - (-12) - 14$.

$$
\begin{aligned}
-7 - (-12) - 14 &= -7 + 12 - 14 &&\text{Add the opposite of } -12. \\
&= 5 - 14 &&\text{Add } -7 \text{ and } 12. \\
&= 5 + (-14) &&\text{Add the opposite of } 14. \\
&= -9 &&\text{Add.}
\end{aligned}
$$

So, $-7 - (-12) - 14 = -9$.

On Your Own

Now You're Ready
Exercises 27–32

Evaluate the expression.

7. $-9 - 16 - 8$

8. $-4 - 20 - 9$

9. $0 - 9 - (-5)$

10. $0 - (-6) - 8$

11. $15 - (-20) - 20$

12. $13 - 18 - (-18)$

EXAMPLE 3   Real-Life Application

Which continent has the greater range of elevations?

	North America	Africa
Highest Elevation	6198 m	5895 m
Lowest Elevation	−86 m	−155 m

To find the range of elevations for each continent, subtract the lowest elevation from the highest elevation.

North America

$$
\begin{aligned}
\text{range} &= 6198 - (-86) \\
&= 6198 + 86 \\
&= 6284 \text{ m}
\end{aligned}
$$

Africa

$$
\begin{aligned}
\text{range} &= 5895 - (-155) \\
&= 5895 + 155 \\
&= 6050 \text{ m}
\end{aligned}
$$

Because 6284 is greater than 6050, North America has the greater range of elevations.

On Your Own

13. The highest elevation in Mexico is 5700 meters, on Pico de Orizaba. The lowest elevation in Mexico is −10 meters, in Laguna Salada. Find the range of elevations in Mexico.

Check It Out
Help with Homework
BigIdeasMath ✓ com

Vocabulary and Concept Check

1. **WRITING** How do you subtract one integer from another?

2. **OPEN-ENDED** Write two integers that are opposites.

3. **DIFFERENT WORDS, SAME QUESTION** Which is different? Find "both" answers.

Find the difference of 3 and −2.	What is 3 less than −2?
How much less is −2 than 3?	Subtract −2 from 3.

MATCHING Match the subtraction expression with the corresponding addition expression.

4. $9 - (-5)$ 5. $-9 - 5$ 6. $-9 - (-5)$ 7. $9 - 5$

   A. $-9 + 5$ B. $9 + (-5)$ C. $-9 + (-5)$ D. $9 + 5$

Practice and Problem Solving

Subtract.

8. $4 - 7$ 9. $8 - (-5)$ 10. $-6 - (-7)$ 11. $-2 - 3$

12. $5 - 8$ 13. $-4 - 6$ 14. $-8 - (-3)$ 15. $10 - 7$

16. $-8 - 13$ 17. $15 - (-2)$ 18. $-9 - (-13)$ 19. $-7 - (-8)$

20. $-6 - (-6)$ 21. $-10 - 12$ 22. $32 - (-6)$ 23. $0 - (20)$

24. **ERROR ANALYSIS** Describe and correct the error in finding the difference $7 - (-12)$.

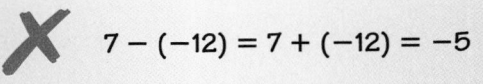

$7 - (-12) = 7 + (-12) = -5$

25. **SWIMMING POOL** The floor of the shallow end of a swimming pool is at −3 feet. The floor of the deep end is 9 feet deeper. Which expression can be used to find the depth of the deep end?

   $-3 + 9$ $-3 - 9$ $9 - 3$

26. **SHARKS** A shark is at −80 feet. It swims up and jumps out of the water to a height of 15 feet. Write a subtraction expression for the vertical distance the shark travels.

Evaluate the expression.

27. $-2 - 7 + 15$ 28. $-9 + 6 - (-2)$ 29. $12 - (-5) - 8$

30. $8 + 14 - (-4)$ 31. $-6 - (-8) + 5$ 32. $-15 - 7 - (-11)$

Assignment Guide and Homework Check

Level	Day 1 Activity Assignment	Day 2 Lesson Assignment	Homework Check
Basic	8–15, 50–56	1–7, 17–23 odd, 24–29	3, 19, 25, 28
Average	8–15, 50–56	1–7, 20–26 even, 27–31 odd, 40, 42, 43, 45	3, 20, 27, 40
Advanced	8–15, 50–56	1–7, 24, 33–41 odd, 42–48 even	33, 37, 41, 44

For Your Information

- **Exercise 3** *Different Words, Same Question* is a new type of exercise. Three of the four choices pose the same question using different words. The remaining choice poses a different question. So there are two answers.

Common Errors

- **Exercises 8–23** Students may change the sign of the first number, or forget to change the problem from subtraction to addition when changing the sign of the second number. Remind them that the first number is a starting point and will never change. Also remind students that the sign of the second number and the operation change.

- **Exercises 27–32** Students may try to do the addition first instead of working left to right. Remind them that the order of operations does not put addition before subtraction, but that addition *and* subtraction is performed from left to right.

- **Exercise 40** Students may try to add $-4 + 11$ instead of subtract $-4 - 11$ because they do not recognize that *change in elevation* means a range (subtraction). Use a number line rotated vertically to help students see the meaning of change in elevation.

1.3 Record and Practice Journal

Subtract.

1. $3 - 8$
-5

2. $6 - (-7)$
13

3. $-10 - 9$
-19

4. $-5 - (-4)$
-1

Evaluate the expression.

5. $11 - (-2) + 14$
27

6. $-16 - (-12) + (-8)$
-12

7. $6 - 17 - 4$
-15

Evaluate the expression when $x = -4$, $y = -8$, and $z = 3$.

8. $6 - x$
10

9. $y - (-10)$
2

10. $-17 + z - x$
-10

11. $|y - x|$
4

12. You begin a hike in Death Valley, California at an elevation of −86 meters. You hike to a point of elevation at 45 meters. What is your change in elevation?
131 meters

13. You sell t-shirts for a fundraiser. It costs $112 to have the t-shirts made. You make $98 in sales. What is your profit?
$-\$14$

14. The table shows the scores of six golfers. Find the range of the scores.

Golfer	1	2	3	4	5	6
Score	−5	−2	3	−3	1	−8

11

 Vocabulary and Concept Check

1. You add the integer's opposite.

2. *Sample answer:* 3, −3

3. What is 3 less than −2?; −5; 5

4. D

5. C

6. A

7. B

 **Practice and Problem Solving**

8. −3	9. 13
10. 1	11. −5
12. −3	13. −10
14. −5	15. 3
16. −21	17. 17
18. 4	19. 1
20. 0	21. −22
22. 38	23. −20

24. The *opposite* of −12 should be added.
$7 - (-12) = 7 + 12 = 19$

25. $-3 - 9$

26. $15 - (-80)$

27. 6	28. −1
29. 9	30. 26
31. 7	32. −11

Practice and Problem Solving

33. $m = 14$ **34.** $w = 4$

35. $c = 15$ **36.** -5

37. 2   **38.** -17

39. 3   **40.** $-15\,m$

41. *Sample answer:* $x = -2$, $y = -1$; $x = -3$, $y = -2$

42. See *Taking Math Deeper*.

43. sometimes; It's positive only if the first integer is greater.

44. sometimes; It's positive only if the first integer is greater.

45. always; It's always positive because the first integer is always greater.

46. never; It's never positive because the first integer is never greater.

47. all values of a and b

48. when a and b both have the same sign, or $a = 0$, or $b = 0$

49. when a and b have the same sign and $|a| > |b|$ or $|a| = |b|$, or $b = 0$

Fair Game Review

50. -20 **51.** -45

52. 40   **53.** 468

54. 1476   **55.** 2378

56. C

Mini-Assessment
Subtract.

1. $6 - 10$ -4

2. $-14 - 16$ -30

3. $-9 - (-4)$ -5

4. $-26 - (-35)$ 9

5. The top of a flag pole is 15 feet high. The base is at -3 feet. Find the length of the flag pole. 18 feet

Taking Math Deeper

Exercise 42

The exercise reviews the concept of the range of a data set. Students should know that the range of a set is the difference between the greatest number and the least number in the set.

 a. Find the range for each month.

	Jan	Feb	Mar	Apr	May	Jun	Jul	Aug	Sep	Oct	Nov	Dec
High (°F)	56	57	56	72	82	92	84	85	73	64	62	53
Low (°F)	-35	-38	-24	-15	1	29	34	31	19	-6	-21	-36
Range (°F)	91	95	80	87	81	63	50	54	54	70	83	89

$56 - (-35) = 91$

 Help me see it.

$$56 - (-35) = 56 + 35$$
$$= 91$$

Add 56 and 35.

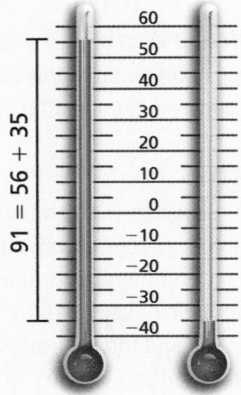

$91 = 56 + 35$

 b. Find all time high and low temperatures.

All time high: 92°F

All time low: -38°F

c. Find the range of the all time high and the all time low.

$$92 - (-38) = 92 + 38$$
$$= 130°F$$

Project

Create a chart showing the high and low temperatures in your town for each month of the year. Give the range of temperatures for each month.

Reteaching and Enrichment Strategies

If students need help. . .	If students got it. . .
Resources by Chapter • Practice A and Practice B • Puzzle Time Record and Practice Journal Practice Differentiating the Lesson Lesson Tutorials Skills Review Handbook	Resources by Chapter • Enrichment and Extension • School-to-Work Start the next section

MENTAL MATH Use mental math to solve the equation.

33. $m - 5 = 9$ **34.** $w - (-3) = 7$ **35.** $6 - c = -9$

ALGEBRA Evaluate the expression when $k = -3$, $m = -6$, and $n = 9$.

36. $4 - n$ **37.** $m - (-8)$

38. $-5 + k - n$ **39.** $|m - k|$

40. PLATFORM DIVING The figure shows a diver diving from a platform. The diver reaches a depth of 4 meters. What is the change in elevation of the dive?

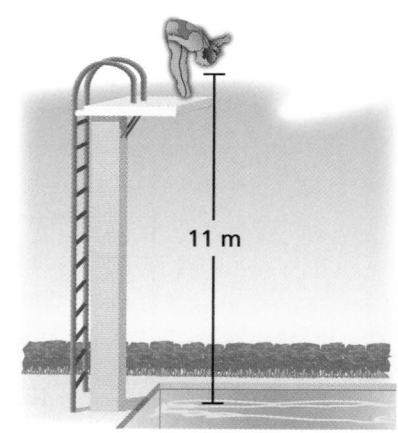

11 m

41. OPEN-ENDED Write two different pairs of negative integers, x and y, that make the statement $x - y = -1$ true.

42. TEMPERATURE The table shows the record monthly high and low temperatures in Anchorage, AK.

	Jan	Feb	Mar	Apr	May	Jun	Jul	Aug	Sep	Oct	Nov	Dec
High (°F)	56	57	56	72	82	92	84	85	73	64	62	53
Low (°F)	−35	−38	−24	−15	1	29	34	31	19	−6	−21	−36

    a. Find the range of temperatures for each month.

    b. What are the all-time high and all-time low temperatures?

    c. What is the range of the temperatures in part (b)?

REASONING Tell whether the difference between the two integers is *always*, *sometimes*, or *never* positive. Explain your reasoning.

43. Two positive integers   **44.** Two negative integers

45. A positive integer and a negative integer   **46.** A negative integer and a positive integer

 For what values of a and b is the statement true?

47. $|a - b| = |b - a|$ **48.** $|a + b| = |a| + |b|$ **49.** $|a - b| = |a| - |b|$

 **Fair Game Review** What you learned in previous grades & lessons

Add. *(Section 1.2)*

50. $-5 + (-5) + (-5) + (-5)$ **51.** $-9 + (-9) + (-9) + (-9) + (-9)$

Multiply. *(Skills Review Handbook)*

52. 8×5 **53.** 6×78 **54.** 36×41 **55.** 82×29

56. MULTIPLE CHOICE Which value of n makes the value of the expression $4n + 3$ a composite number? *(Skills Review Handbook)*

    Ⓐ 1   Ⓑ 2   Ⓒ 3   Ⓓ 4

Check It Out
Graphic Organizer
BigIdeasMath.com

You can use an **idea and examples chart** to organize information about a concept. Here is an example of an idea and examples chart for absolute value.

Absolute Value: **the distance between a number and 0 on the number line**

Example

|3| = 3

Example

|−5| = 5

Example

|0| = 0

On Your Own

Make an idea and examples chart to help you study these topics.

1. integers

2. adding integers

   a. with the same sign

   b. with different signs

3. Additive Inverse Property

4. subtracting integers

After you complete this chapter, make idea and examples charts for the following topics.

5. multiplying integers

   a. with the same sign   **b.** with different signs

6. dividing integers

   a. with the same sign   **b.** with different signs

7. quadrants

8. plotting ordered pairs

"I made an idea and examples chart to give my owner ideas for my birthday next week."

Sample Answers

1.

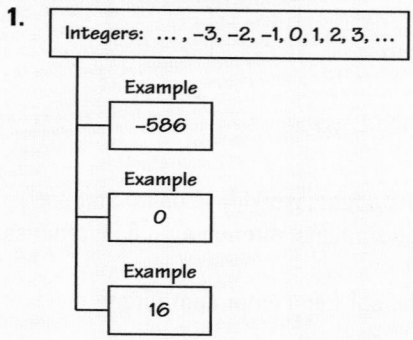

Integers: ... , –3, –2, –1, 0, 1, 2, 3, ...

Example
-586

Example
0

Example
16

2a.

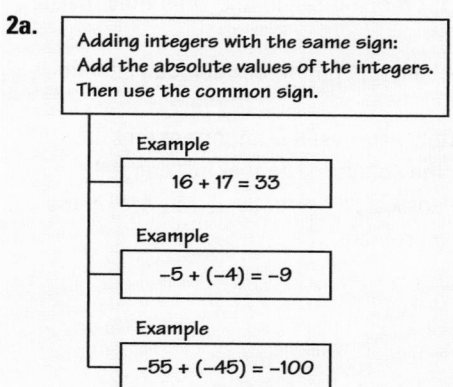

Adding integers with the same sign:
Add the absolute values of the integers.
Then use the common sign.

Example
16 + 17 = 33

Example
–5 + (–4) = –9

Example
–55 + (–45) = –100

2b.

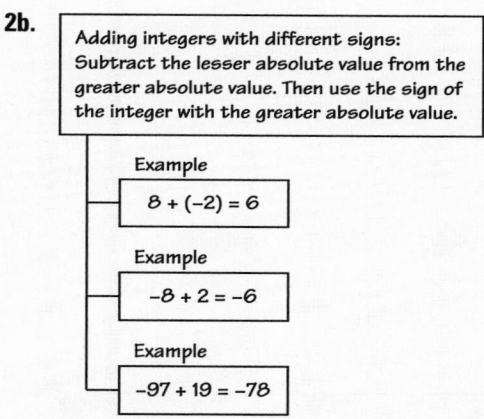

Adding integers with different signs:
Subtract the lesser absolute value from the
greater absolute value. Then use the sign of
the integer with the greater absolute value.

Example
8 + (–2) = 6

Example
–8 + 2 = –6

Example
–97 + 19 = –78

3–4. Available at *BigIdeasMath.com.*

List of Organizers
Available at *BigIdeasMath.com*

Comparison Chart
Concept Circle
Example and Non-Example Chart
Formula Triangle
Four Square
Idea (Definition) and Examples Chart
Information Frame
Information Wheel
Notetaking Organizer
Process Diagram
Summary Triangle
Word Magnet
Y Chart

About this Organizer

An *Idea and Examples Chart* can be used to organize information about a concept. Students fill in the top rectangle with a term and its definition or description. Students fill in the rectangles that follow with examples to illustrate the term. Each sample answer shows 3 examples, but students can show more or fewer examples. Idea and examples charts are useful for concepts that can be illustrated with more than one type of example.

Answers

1. $|-8| > 3$

2. $7 = |-7|$

3. $-6, -4, 3, |-4|, |-5|$

4. $-10, -8, |-9|, 12, |-15|$

5. -11

6. 12

7. -6

8. 0

9. 1

10. 13

11. **a.** $-10, -7$

    b. -7

    c. -10

12. yes; They raised $1129.

13. $130°F$

Assessment Book

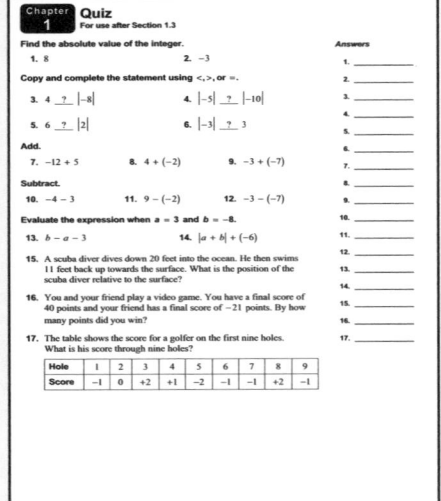

Alternative Quiz Ideas

100% Quiz   Math Log
Error Notebook   Notebook Quiz
Group Quiz   Partner Quiz
Homework Quiz   Pass the Paper

Partner Quiz

- Partner quizzes are to be completed by students working in pairs. Student pairs can be selected by the teacher, by students, through a random process, or any way that works for your class.
- Students are permitted to use their notebooks and other appropriate materials.
- Each pair submits a draft of the quiz for teacher feedback. Then they revise their work and turn it in for a grade.
- When the pair is finished they can submit one paper, or each can submit their own.
- Teachers can give feedback in a variety of ways. It is important that the teacher does not reteach or provide the solution. The teacher can tell students which questions they have answered correctly, if they are on the right track, or if they need to rethink a problem.

Reteaching and Enrichment Strategies

If students need help. . .	If students got it. . .
Resources by Chapter • Study Help • Practice A and Practice B • Puzzle Time Lesson Tutorials *BigIdeasMath.com* Practice Quiz Practice from the Test Generator	Resources by Chapter • Enrichment and Extension • School-to-Work Game Closet at *BigIdeasMath.com* Start the next section

Technology For the Teacher

Answer Presentation Tool
Big Ideas Test Generator

Copy and complete the statement using <, >, or =. *(Section 1.1)*

1. $\left|-8\right|$ ___ 3

2. 7 ___ $\left|-7\right|$

Order the values from least to greatest. *(Section 1.1)*

3. $-4, \left|-5\right|, \left|-4\right|, 3, -6$

4. $12, -8, \left|-15\right|, -10, \left|-9\right|$

Simplify the expression. *(Section 1.2 and Section 1.3)*

5. $-3 + (-8)$

6. $-4 + 16$

7. $3 - 9$

8. $-5 - (-5)$

Evaluate the expression when $a = -2$, $b = -8$, and $c = 5$.
(Section 1.2 and Section 1.3)

9. $4 - a - c$

10. $\left|b - c\right|$

11. EXPLORING Two climbers explore a cave. *(Section 1.1)*

   a. Write an integer for the depth of each climber relative to the surface.

   b. Which integer in part (a) is greater?

   c. Which integer in part (a) has the greater absolute value?

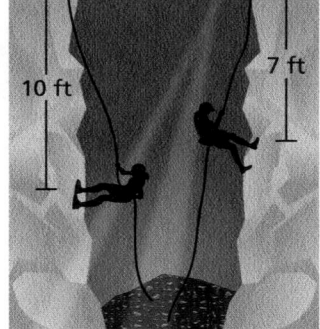

10 ft

7 ft

12. SCHOOL CARNIVAL The table shows the income and expenses for a school carnival. The school's goal was to raise $1100. Did the school reach its goal? Explain. *(Section 1.2)*

Games	Concessions	Donations	Flyers	Decorations
$650	$530	$52	−$28	−$75

13. TEMPERATURE Temperatures in the Gobi Desert reach −40°F in the winter and 90°F in the summer. Find the range of the temperatures. *(Section 1.3)*

1.4 Multiplying Integers

Essential Question Is the product of two integers *positive*, *negative*, or *zero*? How can you tell?

1 EXAMPLE: Multiplying Integers with the Same Sign

Use repeated addition to find 3 • 2.

Recall that multiplication is repeated addition. 3 • 2 means to add 3 groups of 2.

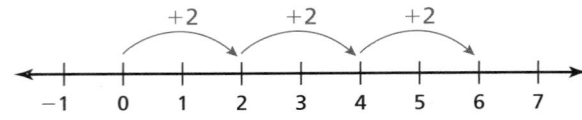

Now you can write
3 • 2 = 2 + 2 + 2 = 6.

So, 3 • 2 = 6.

2 EXAMPLE: Multiplying Integers with Different Signs

Use repeated addition to find 3 • (−2).

3 • (−2) means to add 3 groups of −2.

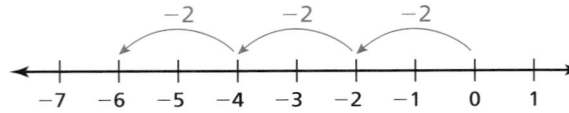

Now you can write
3 • (−2) = (−2) + (−2) + (−2)
= −6.

So, 3 • (−2) = −6.

3 ACTIVITY: Multiplying Integers with Different Signs

Work with a partner. Use a table to find −3 • 2.

Describe the pattern in the table. Use the pattern to complete the table.

2	•	2	=	4
1	•	2	=	2
0	•	2	=	0
−1	•	2	=	
−2	•	2	=	
−3	•	2	=	

Notice the products decrease by 2 in each row.

So, continue the pattern.

−1 • 2: 0 − 2 =

−2 • 2: −2 − 2 =

−3 • 2: −4 − 2 =

So, −3 • 2 =   .

Laurie's Notes

Introduction

For the Teacher

- **Goal:** The goal of the activity is to develop an understanding of when integer multiplication has a positive, negative, or zero product.
- In this activity, students use repeated addition, along with observing patterns, to develop an understanding of how to multiply two integers.

Motivate

- Play *Guess My Rule*. Write the first 4 terms of a sequence on the board. (Use a sequence that involves multiplication.) Ask students to give the next few terms and guess the rule. Here are some possibilities.
  - 2, 4, 8, 16, . . . 32, 64, 128; The rule is to multiply by 2, powers of 2, or doubling (any of those 3 answers are acceptable).
  - 0, 4, 8, 12, . . . 16, 20, 24; The rule is adding 4, multiples of 4, or counting by 4s (any of those 3 answers are acceptable).
  - −6, −3, 0, 3, . . . 6, 9, 12; The rule is adding 3.
- **?** Ask students if any of the patterns seem different than the others. In the sequences above, the third one involves negative numbers.
- Students will have a different way to describe a rule for the third sequence by the end of the class.

Discuss

- **?** "Do you remember *skip counting* in elementary school?"
  If no one remembers, explain to students that skip counting is a fast way to count by a number other than 1.
- Skip counting is one way to show multiplication. For example, skip counting by 5s yields 5, 10, 15, 20, 25, ... You can think of the terms of the sequence formed by skip counting by 5s as $5 \times 1, 5 \times 2, 5 \times 3$, etc.

Activity Notes

Example 1 and Example 2

- This is the *repeated addition model* of multiplication.
- Have students draw a number line to represent 3 groups of 2.
- **?** **Connection:** "I noticed that $3 \times 2 = 2 \times 3$. What property is this?"
  The Commutative Property of Multiplication.
- In Example 2, make sure students understand why the arrows are moving to the left, instead of moving to the right.

Activity 3

- Make sure that students recognize the pattern—the first factor is decreasing by 1, the second factor is constant, and the product is decreasing by 2.

Previous Learning

Students need to know how to multiply whole numbers.

Start Thinking! and Warm Up

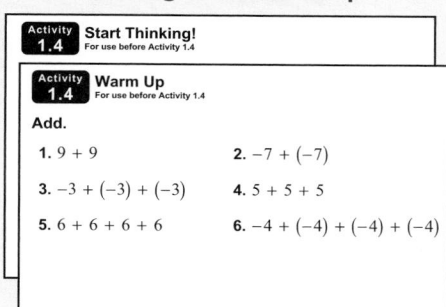

1.4 Record and Practice Journal

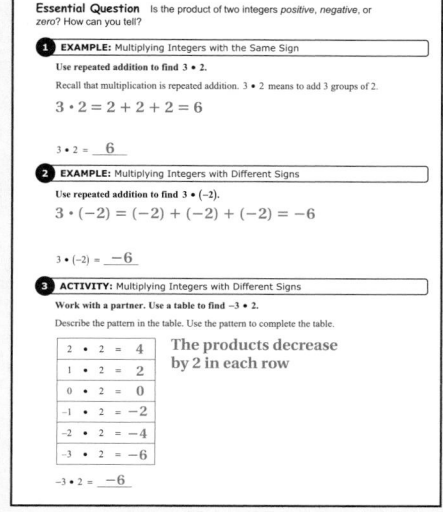

Differentiated Instruction

Visual

Use integer counters to demonstrate that the product of two integers with different signs is negative.

$$3(-4) = (-4) + (-4) + (-4)$$
$$= -12$$

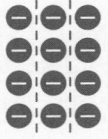

1.4 Record and Practice Journal

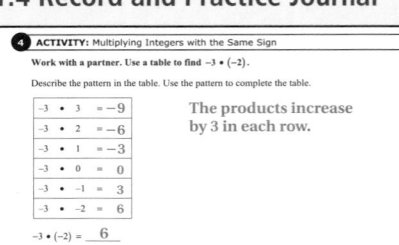

4 ACTIVITY: Multiplying Integers with the Same Sign

Work with a partner. Use a table to find $-3 \cdot (-2)$.

Describe the pattern in the table. Use the pattern to complete the table.

-3 • 3	$= -9$	
-3 • 2	$= -6$	The products increase
-3 • 1	$= -3$	by 3 in each row.
-3 • 0	$= 0$	
-3 • -1	$= 3$	
-3 • -2	$= 6$	

$-3 \cdot (-2) = $ __6__

Inductive Reasoning

Work with a partner. Complete the table.

	Exercise	Type of Product	Product	Product: Positive or Negative
1	5. 3 • 2	same sign	6	Positive
2	6. 3 • (–2)	different signs	–6	Negative
3	7. –3 • 2	different signs	–6	Negative
4	8. –3 • (–2)	same sign	6	Positive
	9. 6 • 3	same sign	18	Positive
	10. 2 • (–5)	different signs	–10	Negative
	11. –6 • 5	different signs	–30	Negative
	12. –5 • (–3)	same sign	15	Positive

13. Write two integers whose product is 0.

Sample answer: 3 and 0

What Is Your Answer?

14. **IN YOUR OWN WORDS** Is the product of two integers *positive, negative, or zero*? How can you tell?

It can be positive, negative, or zero.

15. Write a general rule for multiplying

a. two integers with the same sign.

Multiply the absolute values and make the product positive.

b. two integers with different signs.

Multiply the absolute values and make the product negative.

Laurie's Notes

Activity 4

- **Connection:** Example 1 showed that the product of two positive integers is positive. Example 2 and Activity 3 showed that the product of a positive and a negative (or a negative and a positive) is negative.
- **?** "Are there any other combinations to consider?" the product of two negatives
- Tell students: "Let's look at the product of two negatives." Students should recognize the patterns: the first factor is constant, the second factor is decreasing by 1, and the product is increasing by 3.
- **Extension:** Use the patterns developed to find the product of three numbers, such as $3(-2)(-4)$. 24

Inductive Reasoning

- Students should work with a partner to find the products. The goal is for the students to recognize the bigger pattern. When the factors have the same signs, the product is positive. When the factors have different signs, the product is negative.

Words of Wisdom

- **Common Error:** Students may make mistakes with addition. Review with them that a negative integer added to a negative integer has a negative sum. (Remind students that red counters added to red counters equal red counters.)

What Is Your Answer?

- Students may have a good sense of how to predict the sign of the product; however, they often use language such as: "two positives make a positive and two negatives make a positive." This language should be avoided.

Closure

- "Today we learned that a negative integer multiplied by a negative integer is positive. Be sure to mentally check all of your steps so that you are not confusing anything."

Technology For the Teacher

Dynamic Classroom

The Dynamic Planning Tool
Editable Teacher's Resources at *BigIdeasMath.com*

ACTIVITY: **Multiplying Integers with the Same Sign**

Work with a partner. Use a table to find −3 • (−2).

Describe the pattern in the table. Use the pattern to complete the table.

−3	•	3	=	−9
−3	•	2	=	−6
−3	•	1	=	−3
−3	•	0	=	
−3	•	−1	=	
−3	•	−2	=	

Notice the products increase by 3 in each row.

So, continue the pattern.

−3 • 0: −3 + 3 =

−3 • −1: 0 + 3 =

−3 • −2: 3 + 3 =

So, −3 • (−2) = .

Inductive Reasoning

Work with a partner. Complete the table.

	Exercise	Type of Product	Product	Product: Positive or Negative
1	**5.** 3 • 2	Integers with the same sign		
2	**6.** 3 • (−2)	Integers with different signs		
3	**7.** −3 • 2	Integers with different signs		
4	**8.** −3 • (−2)	Integers with the same sign		
	9. 6 • 3			
	10. 2 • (−5)			
	11. −6 • 5			
	12. −5 • (−3)			

13. Write two integers whose product is 0.

What Is Your Answer?

14. **IN YOUR OWN WORDS** Is the product of two integers *positive*, *negative*, or *zero*? How can you tell?

15. Write general rules for multiplying (a) two integers with the same sign and (b) two integers with different signs.

Practice

Use what you learned about multiplying integers to complete Exercises 8–15 on page 26.

 **Key Ideas**

> **Multiplying Integers with the Same Sign**
>
> **Words** The product of two integers with the same sign is positive.
>
> **Numbers** $2 \cdot 3 = 6$ $-2 \cdot (-3) = 6$
>
> **Multiplying Integers with Different Signs**
>
> **Words** The product of two integers with different signs is negative.
>
> **Numbers** $2 \cdot (-3) = -6$ $-2 \cdot 3 = -6$

EXAMPLE 1 **Multiplying Integers with the Same Sign**

Find $-5 \cdot (-6)$.

The integers have the same sign.

$$-5 \cdot (-6) = 30$$

The product is positive.

:·· The product is 30.

EXAMPLE 2 **Multiplying Integers with Different Signs**

Multiply.

    a. $3(-4)$ **b.** $-7 \cdot 4$

The integers have different signs.

$$3(-4) = -12 \qquad\qquad -7 \cdot 4 = -28$$

The product is negative.

:·· The product is -12.   :·· The product is -28.

On Your Own

Now You're Ready
Exercises 8–23

Multiply.

1. $5 \cdot 5$ **2.** $4(11)$

3. $-1(-9)$ **4.** $-7 \cdot (-8)$

5. $12 \cdot (-2)$ **6.** $4(-6)$

7. $-10(6)$ **8.** $-5 \cdot 7$

Laurie's Notes

Introduction

Connect

- **Yesterday:** Students used repeated addition on a number line to develop a sense of integer multiplication.
- **Today:** Students will use the idea that when the signs of the factors are the same, the product is positive; when the signs of the factors are different, the product is negative.

Motivate

? "Will someone summarize what we learned yesterday?"

- Listen for informal language such as "two negatives make a positive." While it is understood that the remark was made about multiplying two negative numbers, some students may incorrectly remember the comment when they are adding two negatives later on.
- **Vocabulary Review:** Ask students to define *factor, product,* and *Commutative Property of Multiplication.*

Lesson Notes

Key Ideas

- Write the rules for the two cases of multiplying integers. As the examples are written, discuss how multiplication is represented. The multiplication dot is shown in the book and parentheses are used to surround a negative integer. The parentheses are used for clarity so that the negative sign is not confused with the operation of subtraction.

Example 1

- Work through each part of the example.
- Say "You know that 5 times 6 is 30, and because both integers in this example are negative (-5 and -6), the product is 30."

Example 2

- Point out to students how multiplication is represented differently in the two problems. Before doing part (a), you may want to ask if there is another way the problem could be written.
- The goal is for students to be comfortable with all of the ways in which multiplication is represented.

On Your Own

- Students should work independently and check their work.
- Alternately, you could write the problems on index cards. Ask two students to sort the cards into two piles: integers with the same sign, and integers with different signs. Ask "What is true about all of the products (or answers) in this pile?" Point to one of the piles and then repeat the same question for the other pile. Ask for volunteers to do the problems aloud.

Goal Today's lesson is finding the product of integers.

Start Thinking! and Warm Up

 Warm Up
For use before Lesson 1.4

 Start Thinking!
For use before Lesson 1.4

Without using your notes from Activity 1.4, explain to a partner how repeated addition is like multiplication.

Extra Example 1

Find $-8 \cdot (-12)$. 96

Extra Example 2
Multiply.
a. $9(-7)$ -63
b. $-6 \cdot 6$ -36

On Your Own

1. 25	**2.** 44
3. 9	**4.** 56
5. -24	**6.** -24
7. -60	**8.** -35

Extra Example 3
Evaluate the expression.

a. $(-8)^2$ 64

b. -9^2 -81

c. -5^3 -125

On Your Own

9. 9   **10.** -8

11. -49 **12.** -216

Extra Example 4

A football jersey is marked down $10 each week for 3 weeks. Find the total change in the price of the football jersey. $-$30$

On Your Own

13. -45 manatees

English Language Learners
Vocabulary
Make sure that the students understand the mathematical meanings of the words *positive* and *negative*. In math, *positive* means a number greater than zero and *negative* means a number less than zero. Explain that positive and negative do not mean good and bad.

Example 3
- Students should know the meaning of exponents. Write the expression 5^2 on the board and ask students to tell you what it means.
- **Vocabulary Review:** 5 is the *base* and 2 is the *exponent*. The exponent tells you how many factors of the base (how many times you will see the base number) will be multiplied.
- So, $5 \times 5 = 25$. It is read "5 raised to the second power" or "5 squared."
- **Common Error:** When a negative number is raised to a power, the number must be written within parentheses. In part (b), the example is read "the opposite of 5 squared." If you wanted to raise -5 to the second power, it would be written $(-5)^2$. For the given problem, the order of operations says to square the number and then take its opposite. Part (c) shows how to raise a negative integer to a power.
- **Extension:** "When you raise a negative number to a power, is the answer always positive?" No. If the exponent is odd, the answer is negative.

On Your Own
- Students should work with a partner.
- Caution students about Questions 11 and 12.
- **Common Error:** Students sometimes multiply the exponent by the base, particularly if the exponent is greater than 2.

Example 4
- There is no scale written on the vertical axis.
- "Is it possible to determine the number of taxis the company began with?" From the graph, you could estimate 300 taxis to start: each horizontal line is 50 taxis, and the first bar is 6 increments tall.
- Read the verbal model:
  total change = change per year $\times$ number of years.
- **Extension:** "At the same rate, how many years before there are no taxis?" 6

On Your Own
- Note that you do *not* need to know the initial population of manatees in order to answer the question.

Closure
- Write the number -4 on the board. Ask students to write each of the following and then share their responses.
  - "Write a multiplication problem that has -4 as one of the factors, and has a negative product." *Sample answer:* $(-4)(-1)(-1) = -4$
  - "Write a second multiplication problem that has -4 as one of the factors, and has a positive product." *Sample answer:* $(-4)(-1)(1) = 4$

Technology For the Teacher

The Dynamic Planning Tool
Editable Teacher's Resources at *BigIdeasMath.com*

EXAMPLE 3 Using Exponents

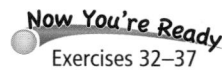

Study Tip

Place parentheses around a negative number to raise it to a power.

a. Evaluate $(-2)^2$.

$(-2)^2 = (-2) \cdot (-2)$ Write $(-2)^2$ as repeated multiplication.

$= 4$ Multiply.

b. Evaluate -5^2.

$-5^2 = -(5 \cdot 5)$ Write 5^2 as repeated multiplication.

$= -25$ Multiply.

c. Evaluate $(-4)^3$.

$(-4)^3 = (-4) \cdot (-4) \cdot (-4)$ Write $(-4)^3$ as repeated multiplication.

$= 16 \cdot (-4)$ Multiply.

$= -64$ Multiply.

On Your Own

Now You're Ready
Exercises 32–37

Evaluate the expression.

9. $(-3)^2$ **10.** $(-2)^3$ **11.** -7^2 **12.** -6^3

EXAMPLE 4 Real-Life Application

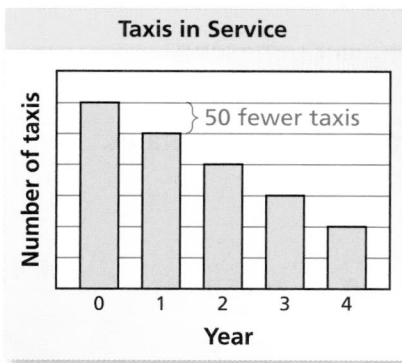

Taxis in Service

50 fewer taxis

Number of taxis

Year

The bar graph shows the number of taxis a company has in service. The number of taxis decreases by the same amount each year for four years. Find the total change in the number of taxis.

The bar graph shows that the number of taxis in service decreases by 50 each year. Use a model to solve the problem.

Total change $=$ Change per year $\cdot$ Number of years

$= -50 \cdot 4$

Use -50 for the change per year because the number *decreases* each year.

$= -200$

∴ The total change in the number of taxis is -200.

On Your Own

13. A manatee population decreases by 15 manatees each year for 3 years. Find the total change in the manatee population.

 Vocabulary and Concept Check

1. **WRITING** What do you know about the signs of two integers whose product is (a) positive and (b) negative?

2. **WRITING** How is $(-2)^2$ different from -2^2?

Tell whether the product is *positive* or *negative* without multiplying. Explain your reasoning.

3. $4(-8)$

4. $-5(-7)$

5. $-3 \cdot (12)$

Tell whether the statement is *true* or *false*. Explain your reasoning.

6. The product of three positive integers is positive.

7. The product of three negative integers is positive.

Practice and Problem Solving

Multiply.

8. $6 \cdot 4$

9. $7(-3)$

10. $-2(8)$

11. $-3(-4)$

12. $-6 \cdot 7$

13. $3 \cdot 9$

14. $8 \cdot (-5)$

15. $-1 \cdot (-12)$

16. $-5(10)$

17. $-13(0)$

18. $-9 \cdot 9$

19. $15(-2)$

20. $-10 \cdot 11$

21. $-6 \cdot (-13)$

22. $7(-14)$

23. $-11 \cdot (-11)$

24. **JOGGING** You burn 10 calories each minute you jog. What integer represents the change in your calories after you jog for 20 minutes?

25. **WETLANDS** About 60,000 acres of wetlands are lost each year in the United States. What integer represents the change in wetlands after 4 years?

Multiply.

26. $3 \cdot (-8) \cdot (-2)$

27. $6(-9)(-1)$

28. $-3(-5)(-4)$

29. $-7(-3)(-5)$

30. $-6 \cdot 3 \cdot (-6)$

31. $3 \cdot (-12) \cdot 0$

Evaluate the expression.

32. $(-4)^2$

33. $(-1)^3$

34. -8^2

35. -6^2

36. $-5^2 \cdot 4$

37. $-2 \cdot (-3)^3$

ERROR ANALYSIS Describe and correct the error in evaluating the expression.

38.
$$\text{✗} \quad -2(-7) = -14$$

39.
$$\text{✗} \quad -10^2 = 100$$

Assignment Guide and Homework Check

Level	Day 1 Activity Assignment	Day 2 Lesson Assignment	Homework Check
Basic	8–15, 49–53	1–7, 17–25 odd, 24, 33–39 odd, 38, 45	2, 21, 24, 33
Average	8–15, 49–53	1–7, 20–23, 30–34 even, 39, 40, 45, 46	2, 21, 32, 45
Advanced	8–15, 49–53	1–7, 30–44 even, 39, 47, 48	7, 34, 40, 47

Common Errors

- **Exercises 8–23** Students may not remember that a negative number multiplied by a negative number is positive. Tell them that it is similar to multiplying by -1, which means to take the opposite. For example, $-6(-13) = (-1 \cdot 6)(-13) = -1[6 \cdot (-13)] = -1(-78) = 78$.
- **Exercises 26–31** Students may multiply all the numbers together ignoring the signs and then place the incorrect sign in front. For example, a student might say $-7(-3)(-5) = 105$. Tell them to multiply only two integers at a time, determine the sign, and then multiply by the last number.
- **Exercises 32–37** Students may erroneously interpret -8^2 as $(-8)(-8)$ instead of $-1(8^2)$. Remind them that the negative sign means multiplication by -1 and that exponents are evaluated before multiplication.

Vocabulary and Concept Check

1. **a.** They are the same.
   b. They are different.

2. $(-2)^2 = 4$ and $-2^2 = -4$; $(-2)^2$ is positive 4 because it is a product of two negative numbers. -2^2 is -4 because it is the opposite of a product of two positive numbers.

3. negative; different signs

4. positive; same signs

5. negative; different signs

6. true; The product of the first two positive integers is positive. The product of the result and the third positive integer is positive.

7. false; The product of the first two negative integers is positive. The product of the positive result and the third negative integer is negative.

1.4 Record and Practice Journal

Multiply.

1. $8 \cdot 9$ 72
2. $7(-7)$ -49
3. $-10 \cdot 4$ -40
4. $-5(-6)$ 30
5. $12 \cdot (-1) \cdot (-2)$ 24
6. $-10(-3)(-7)$ -210
7. $-20 \cdot 0 \cdot (-4)$ 0
8. $-4 \cdot 8 \cdot 3$ -96

Evaluate the expression.

9. $(-8)^2$ 64
10. -11^2 -121
11. $9 \cdot (-5)^2$ 225
12. $(-2)^1 \cdot (-6)$ 48

13. You lose 5 points for every wrong answer in a trivia game. What integer represents the change in your points after answering 8 questions wrong?

   -40

14. A glacier is melting at a rate of 36 cubic miles each year. What integer represents the change in the glacier's volume over 4 years?

   -144

15. The value of a computer is given by the expression $2000 + (-200t)$, where t is the time in years.

   a. Complete the table.

Time	1 year	2 years	3 years	4 years
Value	\$1800	\$1600	\$1400	\$1200

   b. Describe the change in the value of the computer for each year.

   The value of the computer decreases by \$200 each year.

Technology For the Teacher

Answer Presentation Tool
QuizShow

Practice and Problem Solving

8. 24		**9.** -21	
10. -16		**11.** 12	
12. -42		**13.** 27	
14. -40		**15.** 12	
16. -50		**17.** 0	
18. -81		**19.** -30	
20. -110		**21.** 78	
22. -98		**23.** 121	
24. -200		**25.** $-240,000$	
26. 48		**27.** 54	
28. -60		**29.** -105	
30. 108		**31.** 0	
32. 16		**33.** -1	
34. -64		**35.** -36	

Practice and Problem Solving

36. -100 **37.** 54

38. The product should be positive. $-2(-7) = 14$

39. The answer should be negative. $-10^2 = -(10 \cdot 10) = -100$

40. -6 **41.** 32

42. -70

43. $-7500, 37,500$

44. $1792, -7168$

45. -12

46. See *Taking Math Deeper*.

47. a. 153; 141; 129

   b. The price drops $12 every month.

   c. no; yes; In August, you have $135 but the cost is $141. In September, you have $153 and the cost is only $129.

48. -25

Fair Game Review

49. 3   **50.** 8

51. 14   **52.** 17

53. D

Mini-Assessment
Multiply.

1. $-4(-5)$ 20

2. $3(-3)$ -9

3. $-1(-12)$ 12

4. $-2(15)$ -30

5. You have $900 in a checking account. You pay a $60 cell phone bill each month using this account. The account balance is given by $900 + (-60t)$, where t is the time in months. What is the balance of the account after 4 months? $660

Taking Math Deeper

Exercise 46

This is a classic type of problem in mathematics. A real-life measurement (such as height) is modeled by an expression. The height h (in feet) is a function of the time t (in minutes).

$$h = 22,000 + (-480t) = 22,000 - 480t$$

 First, help students understand the model.

$$h = 22,000 - 480t$$

Starting height is 22,000 feet.   Plane descends 480 feet each minute.

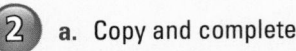

 a. Copy and complete the table.

Time	5 min	10 min	15 min	20 min
Height	19,600	17,200	14,800	12,400

b. When does the plane land?

About 46 minutes

$$\frac{22,000}{480} \approx 45.83 \text{ minutes}$$

The **height** of a plane is called its **altitude**.

Descent rates vary greatly. The rate of 480 feet per minute in this problem is low. Descent rates between 500 and 1500 feet per minute are more common. After take-off, an ascent rate of 1000 to 2000 feet per minute is common.

Project

Draw a graph showing the height of the plane in 5-minute intervals from the time it begins the descent at 22,000 feet until it lands.

Reteaching and Enrichment Strategies

If students need help. . .	If students got it. . .

Resources by Chapter
  • Practice A and Practice B
  • Puzzle Time
Record and Practice Journal Practice
Differentiating the Lesson
Lesson Tutorials
Skills Review Handbook
	Resources by Chapter
	  • Enrichment and Extension
	  • School-to-Work
	Start the next section

ALGEBRA Evaluate the expression when $a = -2$, $b = 3$, and $c = -8$.

40. ab

41. $\left| a^2 c \right|$

42. $ab^3 - ac$

NUMBER SENSE Find the next two numbers in the pattern.

43. $-12, 60, -300, 1500, \ldots$

44. $7, -28, 112, -448, \ldots$

45. GYM CLASS You lose four points each time you attend gym class without sneakers. You forget your sneakers three times. What integer represents the change in your points?

46. AIRPLANE The height of an airplane during a landing is given by $22{,}000 + (-480t)$, where t is the time in minutes.

  a. Copy and complete the table.

  b. Estimate how many minutes it takes the plane to land. Explain your reasoning.

Time	5 min	10 min	15 min	20 min
Height				

47. INLINE SKATES In June, the price of a pair of inline skates is $165. The price changes each of the next three months.

  a. Copy and complete the table.

Month	Price of Skates
June	165 = \$165
July	$165 + \ (-12) = \$$____
August	$165 + 2(-12) = \$$____
September	$165 + 3(-12) = \$$____

  b. Describe the change in the price of the inline skates for each month.

  c. The table at the right shows the amount of money you save each month to buy the inline skates. Do you have enough money saved to buy the inline skates in August? September? Explain your reasoning.

Amount Saved	
June	$35
July	$55
August	$45
September	$18

48. **Reasoning** Two integers, a and b, have a product of 24. What is the least possible sum of a and b?

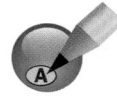 **Fair Game Review** What you learned in previous grades & lessons

Divide. *(Skills Review Handbook)*

49. $27 \div 9$

50. $48 \div 6$

51. $56 \div 4$

52. $153 \div 9$

53. MULTIPLE CHOICE What is the prime factorization of 84? *(Skills Review Handbook)*

  (A) $2^2 \times 3^2$ **(B)** $2^3 \times 7$ **(C)** $3^3 \times 7$ **(D)** $2^2 \times 3 \times 7$

1.5 Dividing Integers

Essential Question Is the quotient of two integers *positive*, *negative*, or *zero*? How can you tell?

1 EXAMPLE: Dividing Integers with Different Signs

Use integer counters to find $-15 \div 3$.

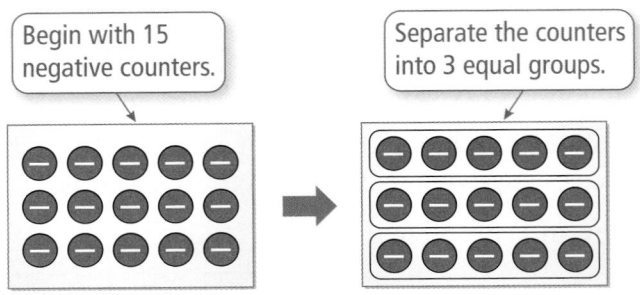

Begin with 15 negative counters.

Separate the counters into 3 equal groups.

Because there are five negative counters in each group, $-15 \div 3 = -5$.

2 ACTIVITY: Rewriting a Product as a Quotient

Work with a partner. Rewrite the product $3 \cdot 4 = 12$ as a quotient in two different ways.

First Way

12 is equal to 3 groups of ⬚.

So, $12 \div 3 = $ ⬚.

Second Way

12 is equal to 4 groups of ⬚.

So, $12 \div 4 = $ ⬚.

3 EXAMPLE: Dividing Integers with Different Signs

Rewrite the product $-3 \cdot (-4) = 12$ as a quotient in two different ways. What can you conclude?

First Way

$12 \div (-3) = -4$

Second Way

$12 \div (-4) = -3$

In each case, when you divide a positive integer by a negative integer, you get a negative integer.

Laurie's Notes

Introduction

For the Teacher

- **Goal:** The goal of the activity is to develop an understanding of when integer division has a positive, negative, or zero quotient.
- In this activity, students write the related multiplication problem for each division problem. Using what they know about integer multiplication, students develop an understanding of the rules for integer division.

Motivate

- **?** "What do you know about football?" Guide students to discuss the length of the field. A football field is 100 yards long, plus two 10-yard end zones, for a total length of 120 yards.
- **?** "If I told you the area of the football field, could you tell me the width of the football field?" The goal is to have students think about the area formula ($A = \ell w$) and realize that if they know the area and one dimension, they can divide to find the other dimension.
- **?** "The area is 6400 yd² and the length is 120 yd. What is the width?" $53\frac{1}{3}$ yd

Discuss

- **?** "What are fact families? Give some examples for multiplication and division." Fact families show the inverse relationship between multiplication and division.

  Sample answers: $2 \times 3 = 6, 3 \times 2 = 6, 6 \div 2 = 3,$ and $6 \div 3 = 2$
  $6 \times 8 = 48, 8 \times 6 = 48, 48 \div 6 = 8,$ and $48 \div 8 = 6$

Activity Notes

Example 1

- **?** Place 15 red integer counters on the overhead, arranged in a 3 × 5 array. "What integer is being modeled?" −15
- **?** Use a ruler to separate the counters into 3 groups of 5. "What division problem does this suggest?" −15 ÷ 3
- There are 5 red counters in every group. The quotient is −5. This is the *grouping model* of division.

Activity 2

- Relate this example back to the football problem and fact families, because length × width = area, area ÷ length = width, and area ÷ width = length.

Example 3

- **?** "What is the product of two negative integers?" a positive integer
- You can rewrite a multiplication problem as a division problem, but be sure to pay attention to the signs.

Previous Learning

Students need to know how to divide whole numbers.

Activity Materials
Textbook
• integer counters

Start Thinking! and Warm Up

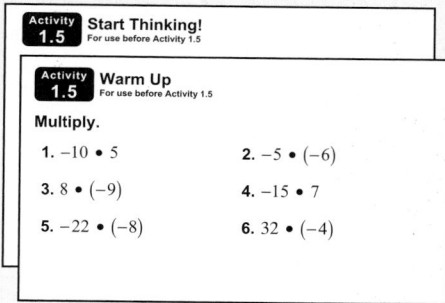

Activity 1.5 Start Thinking!
For use before Activity 1.5

Activity 1.5 Warm Up
For use before Activity 1.5

Multiply.

1. $-10 \cdot 5$ 2. $-5 \cdot (-6)$

3. $8 \cdot (-9)$ 4. $-15 \cdot 7$

5. $-22 \cdot (-8)$ 6. $32 \cdot (-4)$

1.5 Record and Practice Journal

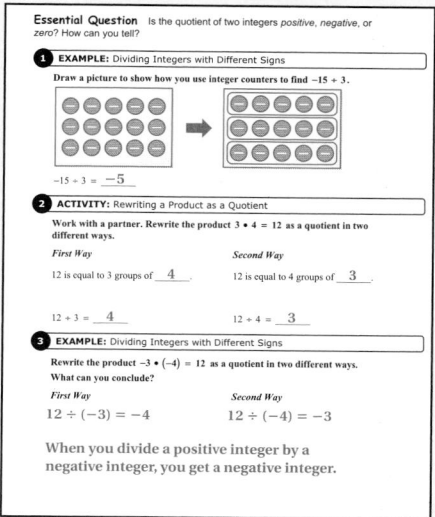

Essential Question Is the quotient of two integers *positive, negative,* or *zero*? How can you tell?

1 EXAMPLE: Dividing Integers with Different Signs

Draw a picture to show how you use integer counters to find −15 ÷ 3.

$-15 \div 3 = -5$

2 ACTIVITY: Rewriting a Product as a Quotient

Work with a partner. Rewrite the product $3 \cdot 4 = 12$ as a quotient in two different ways.

First Way
12 is equal to 3 groups of 4
$12 \div 3 = 4$

Second Way
12 is equal to 4 groups of 3
$12 \div 4 = 3$

3 EXAMPLE: Dividing Integers with Different Signs

Rewrite the product $-3 \cdot (-4) = 12$ as a quotient in two different ways. What can you conclude?

First Way
$12 \div (-3) = -4$

Second Way
$12 \div (-4) = -3$

When you divide a positive integer by a negative integer, you get a negative integer.

Differentiated Instruction

Visual

Have students find the mean of two negative integers. Then tell the students to graph the two numbers and the mean on a number line. Have students share their results with the class. If any student has a mean that is zero or positive, identify the error.

1.5 Record and Practice Journal

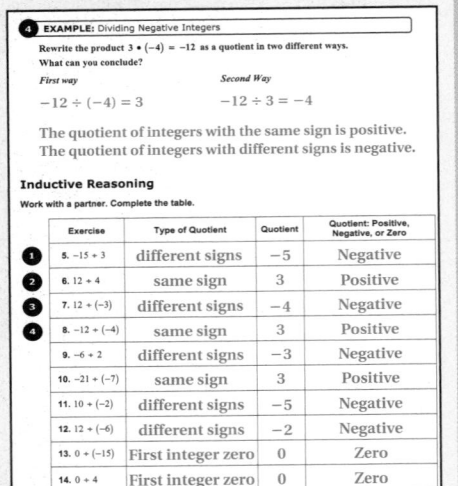

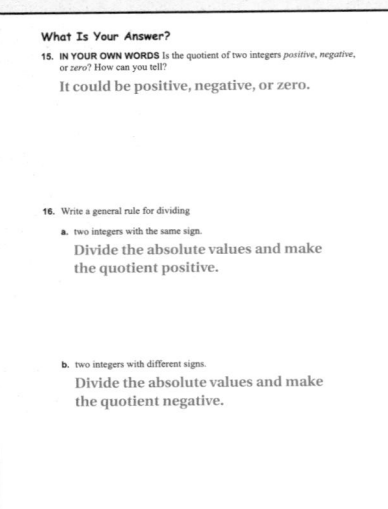

Laurie's Notes

Example 4

● Let's look at one last related problem, $3 \cdot (-4) = -12$. In this example, it is the first way (negative ÷ negative) that is new. The second way (dividing integers with different signs) is similar to Example 3.

Words of Wisdom

? "How are Example 1 and Activity 2 alike? How are Activity 2 and Example 4 alike?" In each problem, you rewrote the integer multiplication problem as an integer division problem.

Inductive Reasoning

● Students should work with a partner to find the quotients. The goal is for the students to recognize the bigger pattern. When the dividend and divisor have the same signs, the quotient is positive. When the dividend and divisor have different signs, the quotient is negative.

● **Extension:**

? "Is division commutative, meaning do 18 ÷ 9 and 9 ÷ 18 have the same quotient?" no

? "What is the relationship between the two solutions?" They are reciprocals.

What Is Your Answer?

● Students may have a good sense of how to predict the sign of the quotient; however, they often use language (as they do with multiplication) such as: two positives make a positive and two negatives make a positive. This language should be avoided.

Closure

● "Today you learned that a negative integer divided by a negative integer is positive. Be sure to mentally check all of your steps so that you are not confusing anything."

Technology For the Teacher

Dynamic Classroom

The Dynamic Planning Tool
Editable Teacher's Resources at *BigIdeasMath.com*

4   EXAMPLE: Dividing Negative Integers

Rewrite the product $3 \cdot (-4) = -12$ as a quotient in two different ways. What can you conclude?

First Way

$-12 \div (-4) = 3$

Second Way

$-12 \div (3) = -4$

When you divide a negative integer by a negative integer, you get a positive integer. When you divide a negative integer by a positive integer, you get a negative integer.

Inductive Reasoning

Work with a partner. Complete the table.

	Exercise	Type of Quotient	Quotient	Quotient: Positive, Negative, or Zero
1	**5.** $-15 \div 3$	Integers with different signs		
2	**6.** $12 \div 4$			Positive
3	**7.** $12 \div (-3)$		-4	
4	**8.** $-12 \div (-4)$	Integers with the same sign		Positive
	9. $-6 \div 2$			
	10. $-21 \div (-7)$			
	11. $10 \div (-2)$			
	12. $12 \div (-6)$			
	13. $0 \div (-15)$			
	14. $0 \div 4$			

What Is Your Answer?

15. IN YOUR OWN WORDS Is the quotient of two integers *positive*, *negative*, or *zero*? How can you tell?

16. Write general rules for dividing (a) two integers with the same sign and (b) two integers with different signs.

Practice Use what you learned about dividing integers to complete Exercises 8–15 on page 32.

 Key Ideas

Dividing Integers with the Same Sign

Words The quotient of two integers with the same sign is positive.

Numbers $8 \div 2 = 4$ $-8 \div (-2) = 4$

Dividing Integers with Different Signs

Words The quotient of two integers with different signs is negative.

Numbers $8 \div (-2) = -4$ $-8 \div 2 = -4$

EXAMPLE 1   Dividing Integers with the Same Sign

Find $-18 \div (-6)$.

The integers have the same sign.

$$-18 \div (-6) = 3$$

The quotient is positive.

∴ The quotient is 3.

EXAMPLE 2   Dividing Integers with Different Signs

Divide.

   a. $75 \div (-25)$ **b.** $\dfrac{-54}{6}$

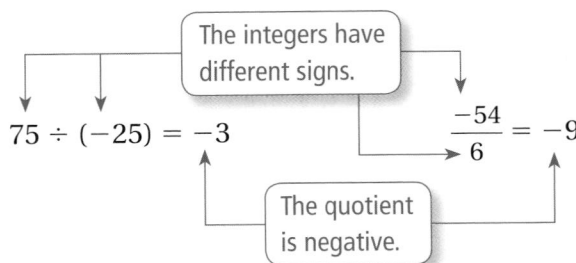

The integers have different signs.

$$75 \div (-25) = -3$$ $$\dfrac{-54}{6} = -9$$

The quotient is negative.

∴ The quotient is -3.   ∴ The quotient is -9.

On Your Own

Now You're Ready
Exercises 8–23

Divide.

1. $14 \div 2$ **2.** $-32 \div (-4)$ **3.** $-40 \div (-8)$

4. $0 \div (-6)$ **5.** $\dfrac{-49}{7}$ **6.** $\dfrac{21}{-3}$

Laurie's Notes

Introduction

Connect

- **Yesterday:** Students used fact families and what they knew about multiplication of integers to develop a sense of integer division.
- **Today:** Students will use the idea that when the signs of the dividend and divisor are the same, the quotient is positive; when the signs of the dividend and divisor are different, the quotient is negative.

Motivate

- **?** "Will someone summarize what we learned yesterday?"
- **Listen:** Again, watch for informal language such as "two negatives make a positive." While it is understood that the remark was made about dividing two negative numbers, some students may incorrectly remember the comment when they are adding or subtracting two negatives later on.
- **Vocabulary Review:** Ask students to define *dividend, divisor, quotient, Commutative Property,* and *division involving zero.*

Lesson Notes

Key Ideas

- Students should know that to check a division problem, you multiply the quotient by the divisor and the answer is the dividend.
- Summary of division involving zero: You can divide 0 by a nonzero number and the answer is 0. You cannot divide a number by 0. Later in this lesson, connect this concept to "0 cannot be in the denominator when division is represented in fraction form."

Example 1

- Work through the example.
- Say "We know that 18 divided by 6 is 3, and because both integers are negative for this example (-18 and -6), the quotient is positive 3."

Example 2

- Point out to students how division is represented differently in the two problems. Before doing part (a), you may want to ask if the problem could be written another way.
- The goal is for students to be comfortable with all of the ways in which division is represented.

On Your Own

- Students should work independently and check their work.

Start Thinking! and Warm Up

> **Lesson 1.5** Warm Up
> For use before Lesson 1.5
>
> **Lesson 1.5** Start Thinking!
> For use before Lesson 1.5
>
> Find the next 2 numbers in the pattern: 12, 6, 24, 12, 48, …. Explain your reasoning.

Extra Example 1

Find $-48 \div (-6)$. 8

Extra Example 2
Divide.

a. $\dfrac{84}{-4}$ -21

b. $-39 \div 3$ -13

On Your Own

1. 7	**2.** 8
3. 5	**4.** 0
5. -7	**6.** -7

Extra Example 3

Evaluate $\dfrac{4-x}{y^2}$ when $x = 8$ and $y = -2$.
-1

 On Your Own

7. 3

8. -4

9. 2

Extra Example 4

The morning high tide at a beach is 57 inches. Six hours later, the afternoon low tide is 12 inches. What is the mean hourly change in the height? -7.5 in.

 On Your Own

10. -6 ft/h

English Language Learners

Visual

Show students that the rules for multiplication can be used to understand the rules for division. For example, to evaluate $16 \div (-8) = ?$, rewrite it using multiplication, $-8 \times ? = 16$. Students should be able to determine that the answer to the multiplication problem is -2 and the answer to the division problem is -2. So, the quotient of two integers with different signs is negative. Use the same approach to demonstrate the other three cases: $-16 \div 8 = ?$, $16 \div 8 = ?$, and $-16 \div (-8) = ?$.

Example 3

- Students should know the order of operations and the meaning of exponents. Students will need to use order of operations. Instead of telling students "remember the order of operations," you want to see what the students remember without prompting.
- **?** Write the problem and ask what it means to "evaluate." To evaluate an expression means to substitute the known values for the variables and perform the operations.
- Rewrite the expression with the values of the variables substituted.
- **Common Error:** If students forget the order of operations, they will perform the operations left to right. Solicit responses as to what operations should be done, in the correct order, and why.

On Your Own

- Students should work with a partner.
- **?** **Extension:** "Can Question 8 be rewritten as $a + 6 \div 3$?" No. There is an implied order of operations by the division bar and, therefore, it would need to be written as $(a + 6) \div 3$.

Example 4

- Discuss how the word *mean* is used in this context. Students often only think of computing a mean by adding values and then dividing by the number of values. So, if students think of *mean* as adding values, it could lead to a problem.
- In this problem, the total change in height is found by finding the difference of the final height and the initial height.

On Your Own

- There is a decrease in the water level, so the hourly change in height is negative. If the tide is coming in, the height would increase, and so the hourly change would be positive.

Closure

- How are the rules for multiplication and division of integers related? Why? The rules are the same because the operations are inverses. You can use fact families to rewrite a division problem as a multiplication problem, and vice versa.

Technology For the Teacher

The Dynamic Planning Tool
Editable Teacher's Resources at *BigIdeasMath.com*

EXAMPLE **Evaluating Expressions**

Evaluate $10 - x^2 \div y$ when $x = 8$ and $y = -4$.

$$10 - x^2 \div y = 10 - 8^2 \div (-4)$$ Substitute 8 for x and -4 for y.

$$= 10 - 8 \cdot 8 \div (-4)$$ Write 8^2 as repeated multiplication.

$$= 10 - 64 \div (-4)$$ Multiply 8 and 8.

$$= 10 - (-16)$$ Divide 64 and -4.

$$= 26$$ Subtract.

Remember

Use order of operations when evaluating an expression.

● **On Your Own**

Now You're Ready
Exercises 28–31

Evaluate the expression when $a = -18$ and $b = -6$.

7. $a \div b$ **8.** $\dfrac{a + 6}{3}$ **9.** $\dfrac{b^2}{a} + 4$

EXAMPLE ④ **Real-Life Application**

You measure the height of the tide using support beams of a pier. Your measurements are shown in the picture. What is the mean hourly change in the height?

59 inches at 2 P.M.→
8 inches at 8 P.M.→

Use a model to solve the problem.

$$\text{Mean hourly change} = \frac{\text{Final height} - \text{Initial height}}{\text{Elapsed Time}}$$

$$= \frac{8 - 59}{6}$$ Substitute. The elapsed time from 2 P.M. to 8 P.M. is 6 hours.

$$= \frac{-51}{6}$$ Subtract.

$$= -8.5$$ Divide.

⋮• The mean change in the height of the tide is -8.5 inches per hour.

● **On Your Own**

10. The height of the tide at the Bay of Fundy in New Brunswick decreases 36 feet in 6 hours. What is the mean hourly change in the height?

 Vocabulary and Concept Check

1. **WRITING** What can you tell about two integers when their quotient is positive? negative? zero?

2. **VOCABULARY** A quotient is undefined. What does this mean?

3. **OPEN-ENDED** Write two integers whose quotient is negative.

4. **WHICH ONE DOESN'T BELONG?** Which expression does *not* belong with the other three? Explain your reasoning.

$$\frac{10}{-5} \qquad \frac{-10}{5} \qquad \frac{-10}{-5} \qquad -\left(\frac{10}{5}\right)$$

Tell whether the quotient is *positive* or *negative* without dividing.

5. $-12 \div 4$

6. $\dfrac{-6}{-2}$

7. $15 \div (-3)$

 **Practice and Problem Solving**

Divide, if possible.

8. $4 \div (-2)$ **9.** $21 \div (-7)$ **10.** $-20 \div 4$ **11.** $-18 \div (-6)$

12. $\dfrac{-14}{7}$ **13.** $\dfrac{0}{6}$ **14.** $\dfrac{-15}{-5}$ **15.** $\dfrac{54}{-9}$

16. $-33 \div 11$ **17.** $-49 \div (-7)$ **18.** $0 \div (-2)$ **19.** $60 \div (-6)$

20. $\dfrac{-56}{14}$ **21.** $\dfrac{18}{0}$ **22.** $\dfrac{65}{-5}$ **23.** $\dfrac{-84}{-7}$

ERROR ANALYSIS Describe and correct the error in finding the quotient.

24.
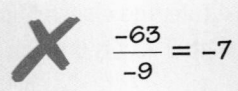
$$\frac{-63}{-9} = -7$$

25.
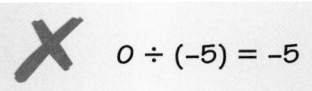
$$0 \div (-5) = -5$$

26. **ALLIGATORS** An alligator population in a nature preserve in the Everglades decreases by 60 alligators over 5 years. What is the mean yearly change in the alligator population?

27. **READING** You read 105 pages of a novel over 7 days. What is the mean number of pages you read each day?

ALGEBRA Evaluate the expression when $x = 10$, $y = -2$, and $z = -5$.

28. $x \div y$ **29.** $\dfrac{10y^2}{z}$ **30.** $\left| \dfrac{xz}{-y} \right|$ **31.** $\dfrac{-x^2 + 6z}{y}$

Assignment Guide and Homework Check

Level	Day 1 Activity Assignment	Day 2 Lesson Assignment	Homework Check
Basic	8–15, 42–45	1–7, 17–31 odd, 26	19, 23, 25, 26, 29
Average	8–15, 42–45	1–7, 20–23, 25, 29–39 odd	22, 25, 29, 37
Advanced	8–15, 42–45	1–7, 25, 28–40 even, 41	25, 28, 36, 38

Common Errors

- **Exercises 8–23** In problems involving zero, students may just say that the quotient is undefined. Remind students that when 0 is the dividend, it means ☐ • −2 = 0, where ☐ = 0. Also, when 0 is the divisor, it means ☐ • 0 = 18, where the answer is undefined.
- **Exercises 28–31 and 34–35** Students may forget to follow the order of operations. Review the order of operations, especially the left-to-right rule in evaluating multiplication/division and addition/subtraction.
- **Exercises 32 and 33** Students may not remember how to find the mean of several numbers. They may get confused by the negative numbers and subtract instead of add. Remind students of the definition of mean.

Vocabulary and Concept Check

1. They have the same sign. They have different signs. The dividend is zero.

2. The divisor is zero.

3. *Sample answer:* −4, 2

4. $\dfrac{-10}{-5}$, which equals 2.

   All the others equal −2.

5. negative

6. positive

7. negative

Practice and Problem Solving

8. −2   9. −3

10. −5   11. 3

12. −2   13. 0

14. 3   15. −6

16. −3   17. 7

18. 0   19. −10

20. −4   21. undefined

22. −13   23. 12

24. The quotient should be positive. $\dfrac{-63}{-9} = 7$

25. The quotient should be 0. $0 \div (-5) = 0$

26. −12 alligators

27. 15 pages

28. −5   29. −8

30. 25   31. 65

1.5 Record and Practice Journal

Divide, if possible.

1. $3 \div (-1)$
   −3

2. $8 \div 2$
   4

3. $-10 \div 5$
   −2

4. $-21 \div (-7)$
   3

5. $\dfrac{48}{-6}$
   −8

6. $\dfrac{-13}{-13}$
   1

7. $\dfrac{0}{3}$
   0

8. $\dfrac{-55}{11}$
   −5

Evaluate the expression.

9. $-63 \div (-7) + 6$
   15

10. $-5 - 12 + 3$
    −9

11. $-8 \cdot 7 + 33 \div (-11)$
    −59

12. An online group loses 20 members over five months. What is the mean monthly change in the group membership?

    −4 members

13. The table shows the number of yards a football player runs in each quarter of a game. Find the mean number of yards the player runs per quarter.

Quarter	1	2	3	4
Yards	−2	14	−18	−6

    −3 yards

14. The far north region of Alaska has an average temperature of −22°F during the month of March. The interior region has an average temperature of −2°F during the month of March. How many times colder is the far north region than the interior region during March?

    11 times colder

Practice and Problem Solving

32. 3   **33.** 5

34. −10   **35.** 4

36. −8, 4; Divide the previous number by −2 to obtain the next number.

37. −400 ft/min

38. See *Taking Math Deeper*.

39. 5

40. 20 people

41. *Sample answer:* −20, −15, −10, −5, 0; Started with −10, then pair −15 with −5 and −20 with 0. The sum of the integers must be $5(-10) = -50$.

Fair Game Review

42–44. See Additional Answers for number lines.

42. $-6, -1, |2|, 4, |-10|$

43. $-8, -3, |0|, 3, |-4|$

44. $-7, -5, -2, |-2|, |5|$

45. B

Mini-Assessment

Divide.

1. $-16 \div (-4)$ 4

2. $-22 \div 11$ −2

3. $35 \div (-5)$ −7

4. $-36 \div (-6)$ 6

5. You play a video game for 15 minutes. You lose 75 points. What integer represents the mean change in points per minute? −5

Taking Math Deeper

Exercise 38

This problem reviews the concept of mean (average) that students learned last year. The difference here is that the data set contains negative numbers. When students studied mean previously, all of the numbers in the data set were positive.

 a. Find the total score.

$$-2 + (-6) + (-7) + (-3) = -18$$

 b. Find the mean score per round.

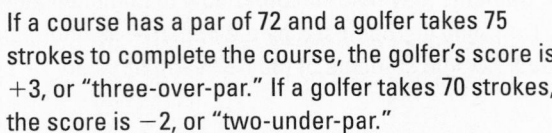

Divide the total score by the number of rounds. $\dfrac{-18}{4} = -4.5$

Scorecard	
Round 1	−2
Round 2	−6
Round 3	−7
Round 4	−3

 In golf, par is the number of strokes it should take to complete a hole. A typical 18-hole championship course may contain four par-3's, ten par-4's, and four par-5's with a total par score of 72.

If a course has a par of 72 and a golfer takes 75 strokes to complete the course, the golfer's score is +3, or "three-over-par." If a golfer takes 70 strokes, the score is −2, or "two-under-par."

Golf tournaments, including the Masters, usually have 4 rounds, for a total par of 4(72) = 288. As an extension, ask students to determine Tiger Woods' total score in 1997. $288 + (-18) = 270$

Project

Make a table showing the winning scores for each Masters Tournament since Tiger Woods won in 1997. Graph the winning scores for each year. Describe any patterns you notice.

Reteaching and Enrichment Strategies

If students need help. . .	If students got it. . .
Resources by Chapter • Practice A and Practice B • Puzzle Time Record and Practice Journal Practice Differentiating the Lesson Lesson Tutorials Skills Review Handbook	Resources by Chapter • Enrichment and Extension • School-to-Work • Financial Literacy Start the next section

Find the mean of the integers.

32. $3, -10, -2, 13, 11$

33. $-26, 39, -10, -16, 12, 31$

Evaluate the expression.

34. $-8 - 14 \div 2 + 5$

35. $24 \div (-4) + (-2) \cdot (-5)$

36. PATTERN Find the next two numbers in the pattern $-128, 64, -32, 16, \ldots$. Explain your reasoning.

37. SNOWBOARDING A snowboarder descends a 1200-foot hill in 3 minutes. What is the mean change in elevation per minute?

38. THE MASTERS In 1997, at the age of 21, Tiger Woods became the youngest golfer to win the Masters Tournament. The table shows his score for each round.

Scorecard	
Round 1	−2
Round 2	−6
Round 3	−7
Round 4	−3

    a. Tiger set the tournament record with the lowest total score. What was his total score?

    b. What was his mean score per round?

39. TUNNEL The Detroit-Windsor Tunnel is an underwater highway that connects the cities of Detroit, Michigan, and Windsor, Ontario. How many times deeper is the roadway than the bottom of the ship?

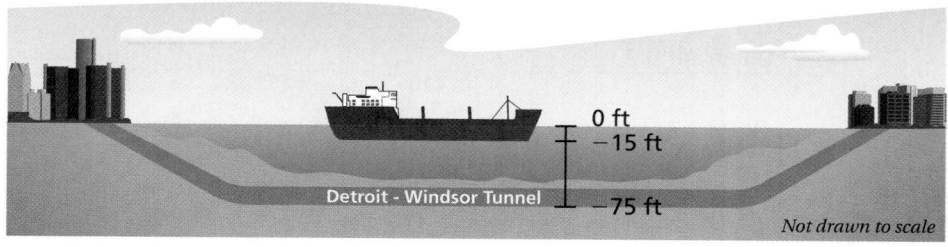

40. AMUSEMENT PARK The regular admission price for an amusement park is $72. For a group of 15 or more, the admission price is reduced by $25. How many people need to be in a group to save $500?

41. 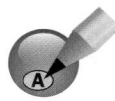 **Number Sense** Write five different integers that have a mean of -10. Explain how you found your answer.

Fair Game Review *What you learned in previous grades & lessons*

Graph the values on a number line. Then order the values from least to greatest. *(Section 1.1)*

42. $-6, 4, |2|, -1, |-10|$

43. $3, |0|, |-4|, -3, -8$

44. $|5|, -2, -5, |-2|, -7$

45. MULTIPLE CHOICE What is the value of $4 \cdot 3 + (12 \div 2)^2$? *(Skills Review Handbook)*

    (A) 15   **(B)** 48   **(C)** 156   **(D)** 324

Essential Question How can you use ordered pairs to locate points in a coordinate plane?

Share Your Work at...
My.BigIdeasMath.com

1 EXAMPLE: Plotting Points in a Coordinate Plane

Plot the ordered pairs. Connect the points to make a picture. Color the picture when you are done.

1 $(4, 12)$ **2** $(9, 9)$ **3** $(12, 4)$ **4** $(12, -3)$ **5** $(10, -9)$

6 $(9, -10)$ **7** $(7, -9)$ **8** $(2, -11)$ **9** $(-1, -11)$ **10** $(-3, -10)$

11 $(-4, -8)$ **12** $(-11, -10)$ **13** $(-12, -9)$ **14** $(-11, -8)$ **15** $(-11, -6)$

16 $(-12, -5)$ **17** $(-11, -4)$ **18** $(-4, -6)$ **19** $(-3, -3)$ **20** $(-4, 0)$

21 $(-8, 2)$ **22** $(-8, 3)$ **23** $(-5, 8)$ **24** $(-1, 11)$

Wildcats

Chiefs

Bulldogs

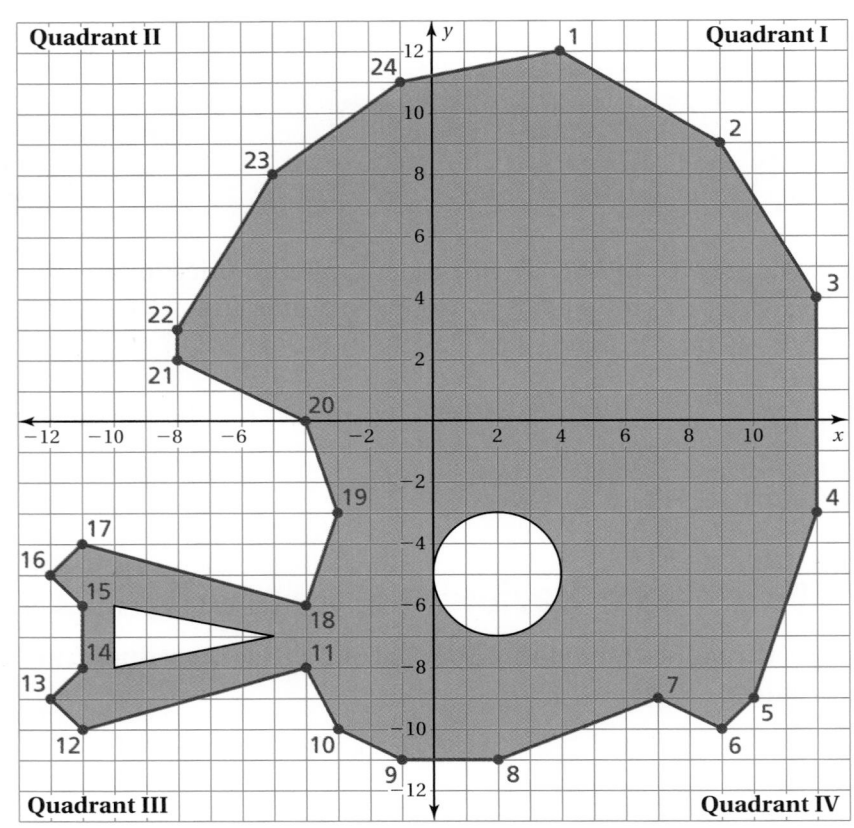

Laurie's Notes

Introduction

For the Teacher

- **Goal:** The goal of the activity is for students to become comfortable with locating and plotting ordered pairs in all four quadrants.
- In this activity, students will plot ordered pairs in all four quadrants. Students who are familiar with games such as *Battleship* or have located points on a map will better understand plotting points.

Motivate

- **Kinesthetic Class Activity:** Create a coordinate grid on your classroom floor or school foyer. Use two strips of masking tape laid perpendicular to one another. Use a thick marker to mark the axes, leaving 1–1.5 feet between each label. Write each of the following points on a piece of paper large enough for the class to see: (3, 2), (−3, 2), (−3, −2), (3, −2).
- Ask four students to volunteer. Have each student start at the origin (the intersection of the masking tape). Have the students face towards the positive *y*-axis. Give the following directions:

    Student 1: Move 3 spaces to the right and forward 2 spaces.
    Student 2: Move 3 spaces to the left and forward 2 spaces.
    Student 3: Move 3 spaces to the left and back 2 spaces.
    Student 4: Move 3 spaces to the right and back 2 spaces.

- ? "What figure would be formed if you connected each student in order with a piece of string?" *a rectangle*
- ? "You have plotted points, or ordered pairs, before. What is the ordered pair for Student 1?" (3, 2) (Hand Student 1 the paper with (3, 2) written on it.)

Discuss

- Ask for suggestions as to what ordered pairs can be given to the other three students. When the correct answer is revealed, have the "plotted students" hold their corresponding ordered pairs.
- Emphasize the positive and negative nature of the *x*- and *y*-coordinates.
- Finish the discussion with how the quadrants are numbered.

Activity Notes

Example 1

- Stress that the points are called ordered pairs and the *order* in which you plot the points is important.
    - The *x*-coordinate is always first. It tells us how far to go horizontally, and in which direction.
    - The *y*-coordinate is always second. It tells us how far to go vertically, and in which direction.
- Leave a summary on the board for students to reference as they work through the example.
- Students should work with a partner and check their work.

Previous Learning

Students should be able to identify and plot ordered pairs in the first quadrant of the coordinate plane.

Activity Materials	
Introduction	**Textbook**
• masking tape • marker • straight edge • colored pencils	• graph paper • colored pencils

Start Thinking! and Warm Up

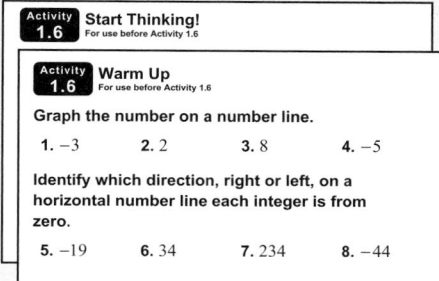

Activity 1.6 Start Thinking!
For use before Activity 1.6

Activity 1.6 Warm Up
For use before Activity 1.6

Graph the number on a number line.

1. −3   2. 2   3. 8   4. −5

Identify which direction, right or left, on a horizontal number line each integer is from zero.

5. −19   6. 34   7. 234   8. −44

1.6 Record and Practice Journal

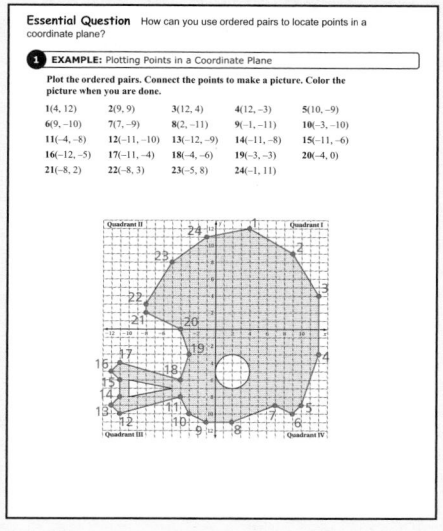

Essential Question How can you use ordered pairs to locate points in a coordinate plane?

1 EXAMPLE: Plotting Points in a Coordinate Plane

Plot the ordered pairs. Connect the points to make a picture. Color the picture when you are done.

1(4, 12)	2(9, 9)	3(12, 4)	4(12, −3)	5(10, −9)
6(9, −10)	7(7, −9)	8(2, −11)	9(−1, −11)	10(−3, −10)
11(−4, −8)	12(−11, −10)	13(−12, −9)	14(−11, −8)	15(−11, −6)
16(−12, −5)	17(−11, −4)	18(−4, −6)	19(−3, −3)	20(−4, 0)
21(−8, 2)	22(−8, 3)	23(−5, 8)	24(−1, 11)	

Differentiated Instruction

Kinesthetic

For kinesthetic learners who have difficulty plotting points in the coordinate plane, suggest they use a finger for tracing. Have students place a finger at the origin and trace left or right along the *x*-axis to the first coordinate, then trace up or down to the second coordinate. Students should also practice writing the coordinates of a plotted point. Guide students with questions such as, "Should you move left or right? How far? Should you move up or down? How far?"

1.6 Record and Practice Journal

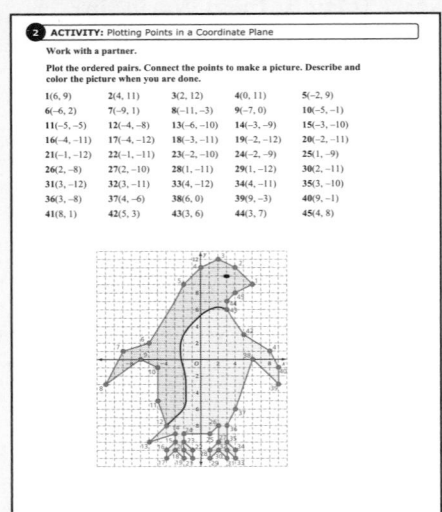

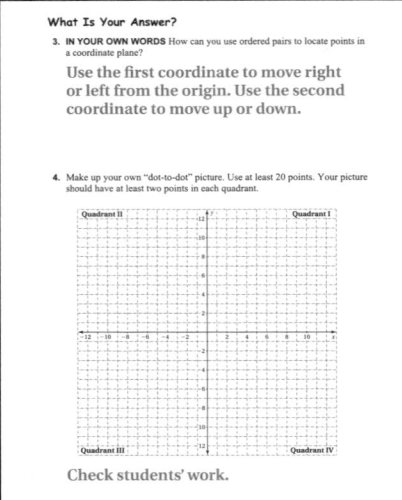

What Is Your Answer?

3. **IN YOUR OWN WORDS** How can you use ordered pairs to locate points in a coordinate plane?

4. Make up your own "dot-to-dot" picture. Use at least 20 points. Your picture should have at least two points in each quadrant.

Check students' work.

Laurie's Notes

Activity 2

- This activity is similar to Example 1.

Words of Wisdom

- Discuss difficulties students had in completing their pictures. Common problems/errors include carelessness, interchanging *x* and *y*, and moving in the opposite direction.
- Another error may occur when students are plotting a point. They may start at the last plotted point, rather than starting at the origin.

What Is Your Answer?

- Students should understand that each point in the coordinate plane has a unique name. The ordered pair specifies an exact spot, similar to latitude and longitude on maps. Students who finish early could start a dot-to-dot design of their own.

- **?** **Extension:** Think of some real-life situations that use coordinates to locate points. "What else besides ordered pairs are used to locate the points?" *Sample answers:* maps often use letters and numbers; stadiums give section-row-seat numbers, games (like Battleship) use numbers and letters; some maps use longitude and latitude

Closure

- "Today you learned about plotting points, or ordered pairs. You did an activity and summary that should help you remember and reinforce these concepts."

Technology For the Teacher

Dynamic Classroom

The Dynamic Planning Tool
Editable Teacher's Resources at *BigIdeasMath.com*

2 ACTIVITY: Plotting Points in a Coordinate Plane

Work with a partner.

**Plot the ordered pairs. Connect the points to make a picture.
Describe and color the picture when you are done.**

1 (6, 9)	**2** (4, 11)	**3** (2, 12)	**4** (0, 11)	**5** (−2, 9)
6 (−6, 2)	**7** (−9, 1)	**8** (−11, −3)	**9** (−7, 0)	**10** (−5, −1)
11 (−5, −5)	**12** (−4, −8)	**13** (−6, −10)	**14** (−3, −9)	**15** (−3, −10)
16 (−4, −11)	**17** (−4, −12)	**18** (−3, −11)	**19** (−2, −12)	**20** (−2, −11)
21 (−1, −12)	**22** (−1, −11)	**23** (−2, −10)	**24** (−2, −9)	**25** (1, −9)
26 (2, −8)	**27** (2, −10)	**28** (1, −11)	**29** (1, −12)	**30** (2, −11)
31 (3, −12)	**32** (3, −11)	**33** (4, −12)	**34** (4, −11)	**35** (3, −10)
36 (3, −8)	**37** (4, −6)	**38** (6, 0)	**39** (9, −3)	**40** (9, −1)
41 (8, 1)	**42** (5, 3)	**43** (3, 6)	**44** (3, 7)	**45** (4, 8)

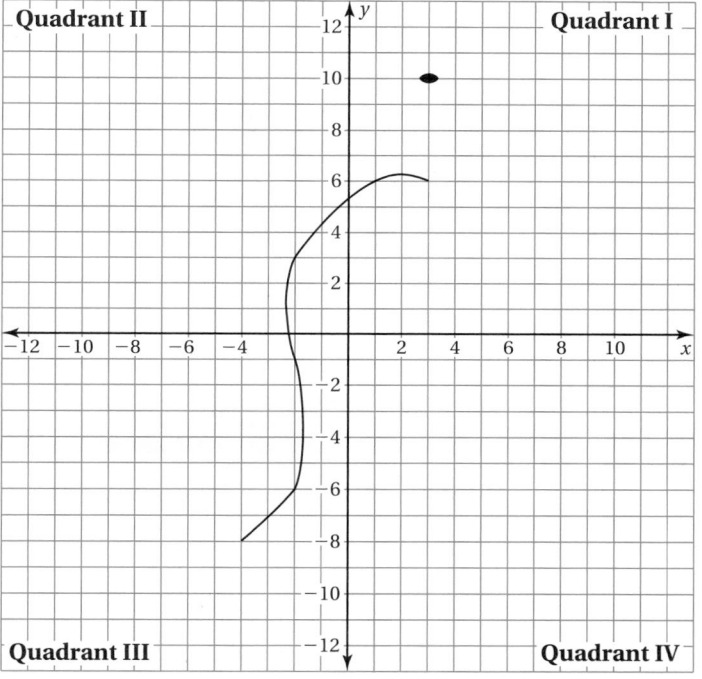

What Is Your Answer?

3. IN YOUR OWN WORDS How can you use ordered pairs to locate points in a coordinate plane?

4. Make up your own "dot-to-dot" picture. Use at least 20 points. Your picture should have at least two points in each quadrant.

Practice Use what you learned about the coordinate plane to complete Exercises 15–18 on page 38.

 **Key Idea**

Key Vocabulary 🔊

coordinate plane, p. 36

origin, p. 36

quadrant, p. 36

x-axis, p. 36

y-axis, p. 36

The Coordinate Plane

A **coordinate plane** is formed by the intersection of a horizontal number line and a vertical number line. The number lines intersect at the **origin** and separate the coordinate plane into four regions called **quadrants.**

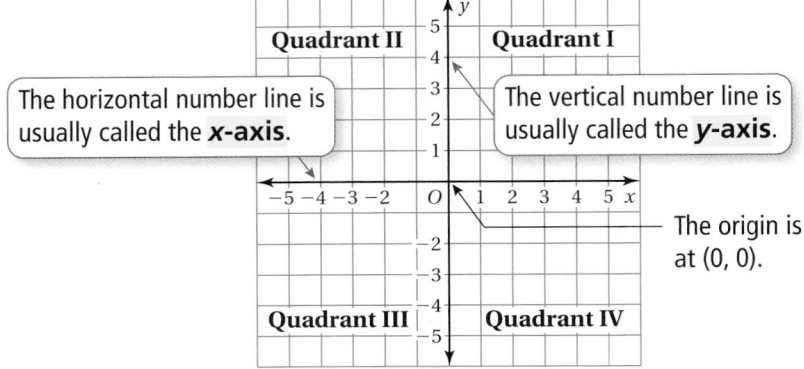

The horizontal number line is usually called the **x-axis.**

The vertical number line is usually called the **y-axis.**

The origin is at (0, 0).

An *ordered pair* is a pair of numbers that is used to locate a point in a coordinate plane.

ordered pair

The *x-coordinate* corresponds to a number on the *x*-axis.

$(4, -2)$

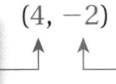

The *y-coordinate* corresponds to a number on the *y*-axis.

EXAMPLE **1** **Standardized Test Practice**

Which ordered pair corresponds to point *T*?

Ⓐ $(-3, -3)$ Ⓑ $(-3, 3)$

Ⓒ $(3, -3)$ Ⓓ $(3, 3)$

Point *T* is 3 units to the right of the origin and 3 units down. So, the *x*-coordinate is 3 and the *y*-coordinate is -3.

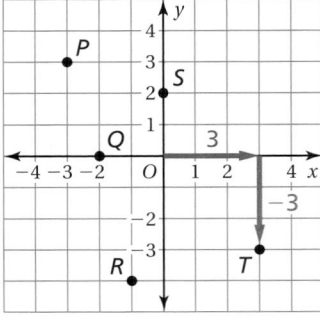

⋮⋗ The ordered pair $(3, -3)$ corresponds to point *T*. The correct answer is Ⓒ.

● **On Your Own**

Now You're Ready
Exercises 5–14

Use the graph in Example 1 to write an ordered pair corresponding to the point.

1. Point *P*   **2.** Point *Q*   **3.** Point *R*   **4.** Point *S*

🔊 Multi-Language Glossary at BigIdeasMath✓com.

Laurie's Notes

Introduction

Connect

- **Yesterday:** Students should be comfortable with plotting ordered pairs from the dot-to-dot activity they completed yesterday.
- **Today:** The vocabulary and concepts are presented more formally.

Motivate

- Share a fictitious story with your students. "Last night when I was out shopping I got hungry, so I decided to get a snack from the vending machine. I decided that I was going to get crackers. To select them, I had to enter the row (a letter from A to E) and a column (a number from 1 to 6). I entered B-4 instead of C-4 and ended up with popcorn instead of crackers. But, I was hungry, so I still ate it!"
- ❓ "How is the vending machine like what you did yesterday?" Students should see that the crackers were in a particular spot associated with the ordered pair (C, 4). When a different ordered pair is entered, you plot a different point. (You get an unexpected item, the popcorn.)

Lesson Notes

Key Idea

- It is important to have a model of the coordinate *grid* versus only a model of scaled axes. The grid is essential in helping the students understand that a point plotted by moving in two directions (horizontal and vertical). You may want to project a coordinate grid, if possible, on the wall or board.
- Use the model of the coordinate grid to identify key vocabulary: coordinate plane, origin, quadrants, *x*-axis, *y*-axis, and ordered pair.
- **Connections:** To help students remember which way is horizontal, hold your arms out horizontally and relate them to the horizon. Origin means "where something starts." If your students have played the game of *four square*, relate the quadrants to the four square court.
- Stress that the ordered pairs (2, 4) and (4, 2) are not the same. The order matters and the ordered pairs are always (*x*, *y*). If it helps, tell students that the order of the coordinates is in alphabetical order, *x* and then *y*.

Example 1

- The colored arrows on the diagram will help students plot in the *x*-direction first, followed by the *y*-direction.

On Your Own

- Ask if there are any questions. Then have students work independently to write the ordered pairs for the remaining points, *P*, *Q*, *R*, and *S*. Ask volunteers to present their answers to the class.

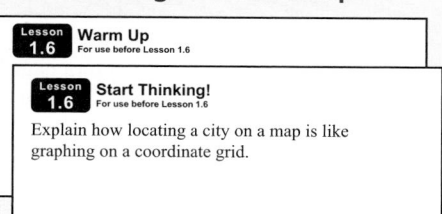
Extra Example 1

Write the ordered pair represented by point *A*, point *B*, point *C*, and point *D*.

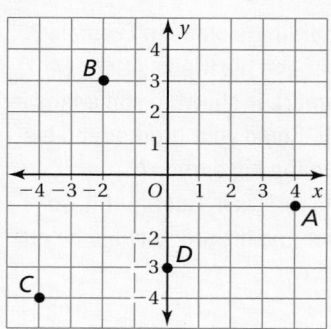

$A(4, -1)$, $B(-2, 3)$, $C(-4, -4)$, $D(0, -3)$

On Your Own

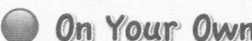

1. $(-3, 3)$ 2. $(-2, 0)$

3. $(-1, -4)$ 4. $(0, 2)$

Extra Example 2

Plot the points in the coordinate plane. Describe the location of each point.

a. $(-4, 0)$

See the graph below. The point is on the *x*-axis.

b. $(1, -3)$

See the graph below. The point is in Quadrant IV.

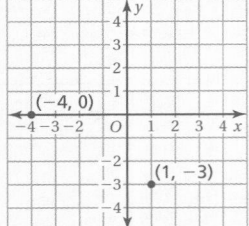

On Your Own

5–8. See Additional Answers for graph.

5. Quadrant I

6. *x*-axis

7. Quadrant III

8. Quadrant IV

Extra Example 3

Use the coordinate plane in Example 3. Your brother and his friend sit at $(-4, 1)$. The fan camera is showing some fans in Quadrant II. Could your brother and his friend be on the big screen?

The point $(-4, 1)$ is in Quadrant II, so your brother and his friend could be on the big screen.

On Your Own

9. your friend

English Language Learners

Visual

English learners may confuse the words *horizontal* and *vertical*. To help students distinguish between the words, draw a picture of the ocean with a ship on the horizon. Connect the word horizontal to the word horizon.

Laurie's Notes

Example 2

- In the previous example, the point was plotted and the student needed to give the ordered pair. Now that is reversed; the ordered pair is given and the student needs to plot the point. Check students' work when they finish.
- **Common Error:** Students often plot points on the axes incorrectly. Be sure the plotted point is $(0, -3)$.

On Your Own

- **Think-Pair-Share:** Students should read each question independently and then work with a partner to answer the questions. When they have answered the questions, the pair should compare their answers with another group and discuss any discrepancies.

Example 3

- Stadium seats are generally located by section, row, and seat number, which is a creative twist on coordinate points!
- Discuss the diagram.
- ❓ Ask "What axis goes through the two end zones, and what axis is parallel to the 50-yard line?" the *x*-axis and the *y*-axis

Closure

- If the *x*-coordinate of an ordered pair is 4, do you know what quadrant the point is in? Explain. The point is in either the first or fourth quadrant, or on the *x*-axis, because those are the only two quadrants with a positive *x*-coordinate.
- If the *y*-coordinate of an ordered pair is negative, do you know what quadrant the point is in? Explain. The point is in either the third quadrant or the fourth quadrant, or on the *y*-axis, because those are the only two quadrants with a negative *y*-coordinate.

Technology For the Teacher

The Dynamic Planning Tool
Editable Teacher's Resources at *BigIdeasMath.com*

EXAMPLE (2) **Plotting Ordered Pairs**

Plot (a) $(-4, 2)$ and (b) $(0, -3)$ in a coordinate plane. Describe the location of each point.

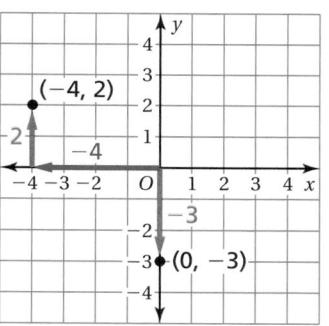

a. Start at the origin. Move 4 units left and 2 units up. Then plot the point.

⋮⋅ The point is in Quadrant II.

b. Start at the origin. Move 3 units down. Then plot the point.

⋮⋅ The point is on the y-axis.

On Your Own

Now You're Ready
Exercises 15–26

Plot the ordered pair in a coordinate plane. Describe the location of the point.

5. $A(2, 3)$

6. $B(-1, 0)$

7. $C(-5, -1)$

8. $D(3, -6)$

EXAMPLE (3) **Real-Life Application**

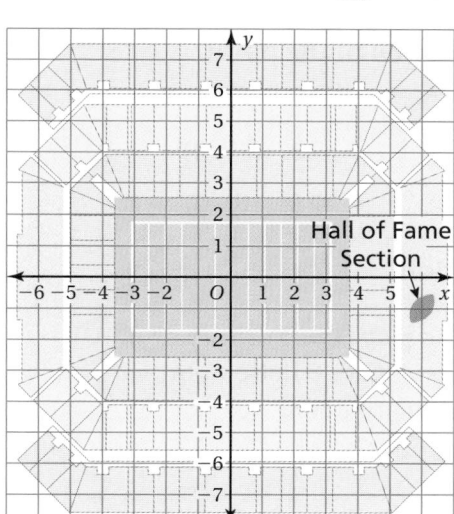

You and a friend have tickets to see a football game. You sit in the Hall of Fame Section and your friend sits at $(-4, -2)$.

a. Write an ordered pair for your location. In which quadrant are you seated?

b. In which quadrant is your friend seated?

c. A fan in Quadrant II is chosen to win a prize. Do you or your friend have a chance to win the prize?

a. The Hall of Fame Section is 6 units to the right of the origin and 1 unit down. So, your seat is located at $(6, -1)$. You are seated in Quadrant IV.

b. Move 4 units to the left of the origin and 2 units down. Your friend is seated in Quadrant III.

c. You are seated in Quadrant IV and your friend is seated in Quadrant III. So, you and your friend do not have a chance to win the prize.

On Your Own

9. **WHAT IF?** In Example 3, a fan sitting in the level closest to the playing field is chosen to win a prize. Do you or your friend have a chance to win the prize?

 Vocabulary and Concept Check

1. **VOCABULARY** How many quadrants are in a coordinate plane?

2. **VOCABULARY** Is the point $(0, -7)$ on the x-axis or the y-axis?

3. **WRITING** How are the locations of the points $(2, -2)$ and $(-2, 2)$ different?

4. **WRITING** Describe the characteristics of ordered pairs in each of the four quadrants.

 Practice and Problem Solving

Write an ordered pair corresponding to the point.

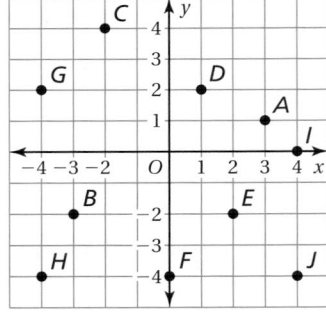

① 5. Point A 6. Point B

7. Point C 8. Point D

9. Point E 10. Point F

11. Point G 12. Point H

13. Point I 14. Point J

Plot the ordered pair in a coordinate plane. Describe the location of the point.

② 15. $K(4, 3)$ 16. $L(-1, 2)$ 17. $M(0, -6)$ 18. $N(3, -2)$

19. $P(2, -4)$ 20. $Q(-2, 4)$ 21. $R(-4, 1)$ 22. $S(7, 0)$

23. $T(-4, -5)$ 24. $U(-2, 5)$ 25. $V(-3, 8)$ 26. $W(-5, -1)$

ERROR ANALYSIS Describe and correct the error in the solution.

27. To plot $(4, 5)$, start at $(0, 0)$ and move 5 units right and 4 units up.

28. To plot $(-6, 3)$, start at $(0, 0)$ and move 6 units right and 3 units down.

29. **REASONING** The coordinates of three vertices of a square are shown in the figure. What are the coordinates of the fourth vertex?

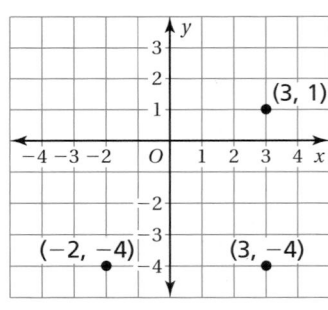

30. **GEOMETRY** The points $D(1, 1)$, $E(1, -2)$, $F(-2, -2)$, and $G(-2, 1)$ are vertices of a figure.

   a. Draw the figure in a coordinate plane.

   b. Find the perimeter of the figure.

   c. Find the area of the figure.

Assignment Guide and Homework Check

Level	Day 1 Activity Assignment	Day 2 Lesson Assignment	Homework Check
Basic	15–18, 45–48	1–4, 5–13 odd, 19–29 odd, 30, 31	7, 13, 23, 30
Average	15–18, 45–48	1–4, 11, 13, 19–29 odd, 33–38	13, 23, 33, 36
Advanced	15–18, 45–48	1–4, 24–30 even, 31–39 odd, 40–44 even	26, 33, 39, 42

Common Errors

- **Exercises 5–14** Students may write the *y*-coordinate first and then the *x*-coordinate for the ordered pair. Tell them that *x* comes before *y* in the alphabet, so the *x*-coordinate must come before the *y*-coordinate in the ordered pair.
- **Exercises 15–26** Students may plot the *x*-coordinate vertically instead of horizontally and the *y*-coordinate horizontally instead of vertically. Remind them that *x* is horizontal and *y* is vertical.
- **Exercise 31** Students may not think about the origin, which is the reason the statement is *sometimes* true. Tell students to think about the *x*-axis as a number line and ask if there is any place where *x* would be 0.

1.6 Record and Practice Journal

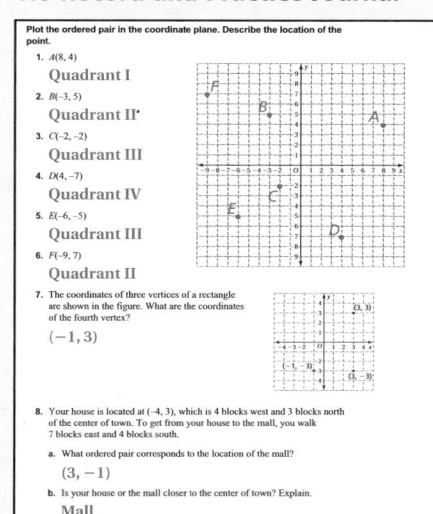

Technology For the Teacher
Answer Presentation Tool
QuizShow

Vocabulary and Concept Check

1. 4   2. *y*-axis

3. $(2, -2)$ is in Quadrant IV, $(-2, 2)$ is in Quadrant II.

4. Quadrant I: (a, b) where $a > 0$ and $b > 0$

   Quadrant II: (a, b) where $a < 0$ and $b > 0$

   Quadrant III: (a, b) where $a < 0$ and $b < 0$

   Quadrant IV: (a, b) where $a > 0$ and $b < 0$

Practice and Problem Solving

5. $(3, 1)$ 6. $(-3, -2)$

7. $(-2, 4)$ 8. $(1, 2)$

9. $(2, -2)$ 10. $(0, -4)$

11. $(-4, 2)$ 12. $(-4, -4)$

13. $(4, 0)$ 14. $(4, -4)$

15–26. See Additional Answers for graph.

15. Quadrant I

16. Quadrant II

17. *y*-axis   18. Quadrant IV

19. Quadrant IV

20. Quadrant II

21. Quadrant II

22. *x*-axis   23. Quadrant III

24. Quadrant II

25. Quadrant II

26. Quadrant III

27. The numbers are reversed. To plot (4, 5), start at (0, 0) and move 4 units right and 5 units up.

28. The directions are reversed. To plot (−6, 3), start at (0, 0) and move 6 units left and 3 units up.

29. $(-2, 1)$

30. a.

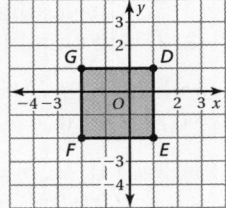

b. 12 units **c.** 9 units^2

31. sometimes; It is true only for $(0, 0)$.

32–33. See Additional Answers.

34–38. See *Taking Math Deeper*.

39–44. See Additional Answers.

Fair Game Review

45. $-\dfrac{16}{2} < -\dfrac{12}{3}$

46. $2\dfrac{2}{5} = \dfrac{24}{10}$

47. $3.45 > 3\dfrac{3}{8}$

48. B

Mini-Assessment

The points $A(-2, 3)$, $B(4, 3)$, $C(-2, -4)$, and $D(4, -4)$ represent the vertices of a garden.

1. Plot the ordered pairs in a coordinate plane.

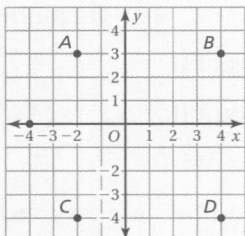

2. What shape does the garden form? a rectangle

3. Describe the location of each point. *A:* Quadrant II, *B:* Quadrant I, *C:* Quadrant III, *D:* Quadrant IV

Taking Math Deeper

Exercises 34–38

This problem shows the common practice of placing a coordinate plane over a real-life diagram or map. The decision for the location of the origin is arbitrary. The scale may also be arbitrary.

 Use the coordinate grid to answer the questions.

(35) Flamingo Cafe is on positive *y*-axis. (34) Reptiles is at (2, 1).

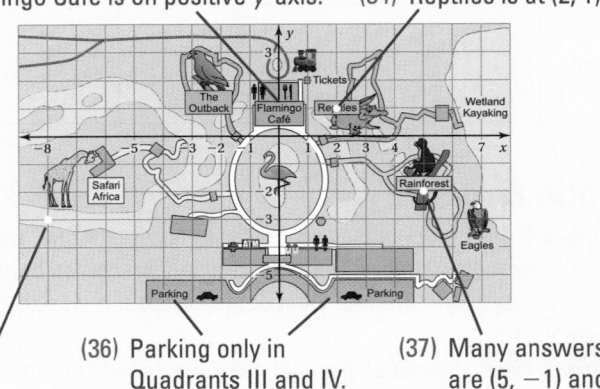

(36) Parking only in Quadrants III and IV. (37) Many answers. Two are (5, −1) and (5, −2).

(38) Safari Africa is closest to (−8, −3).

 The National Zoo in Washington, D.C. is 163 acres (over 7 million square feet). It was created by Congress in 1889. The zoo includes about 400 different species, including giant pandas, which are a symbol of the zoo's conservation efforts.

I love the zoo.

Project

Use a grid to draw a map of your school or town. Include important buildings or rooms, parks, ball fields, playground equipment, and other key landmarks in your school or town.

Reteaching and Enrichment Strategies

If students need help...	If students got it...
Resources by Chapter • Practice A and Practice B • Puzzle Time Record and Practice Journal Practice Differentiating the Lesson Lesson Tutorials Skills Review Handbook	Resources by Chapter • Enrichment and Extension • School-to-Work • Financial Literacy • Technology Connection Start the next section

Tell whether the statement is *sometimes*, *always*, or *never* true. Explain your reasoning.

31. The *x*-coordinate of a point on the *x*-axis is zero.

32. The *y*-coordinate of points in Quadrant III are positive.

33. The *x*-coordinate of a point in Quadrant II has the same sign as the *y*-coordinate of a point in Quadrant IV.

ZOO In Exercises 34–38, use the map of the zoo.

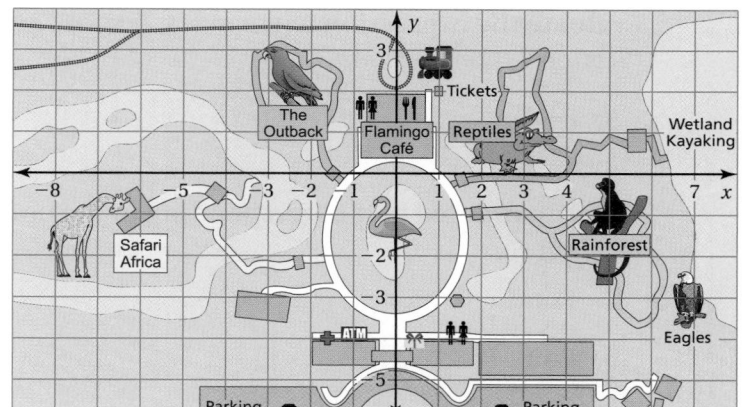

34. Which exhibit is located at (2, 1)?

35. Name an attraction on the positive *y*-axis.

36. Is parking available in Quadrant II? If not, name a quadrant in which you can park.

37. Write two different ordered pairs that represent the location of the Rainforest.

38. Which exhibit is closest to (−8, −3)?

39. **NUMBER SENSE** Name the ordered pair that is 5 units right and 2 units down from (−3, 4).

Plot the ordered pair in a coordinate plane. Describe the location of the point.

40. $A\left(3, -\dfrac{3}{2}\right)$

41. $B\left(-\dfrac{5}{2}, \dfrac{10}{3}\right)$

42. $C(-5.25, -3.5)$

43. $D(-4.75, 0)$

44. **Reasoning** Your school is located at (2, −1), which is 2 blocks east and 1 block south of the center of town. To get from your house to the school, you walk 5 blocks west and 2 blocks north.

   a. What ordered pair corresponds to the location of your house?

   b. Is your house or your school closer to the center of town? Explain.

Fair Game Review *What you learned in previous grades & lessons*

Copy and complete the statement using <, >, or =. *(Section 1.5 and Skills Review Handbook)*

45. $\dfrac{-16}{2} \quad\boxed{}\quad \dfrac{-12}{3}$

46. $2\dfrac{2}{5} \quad\boxed{}\quad \dfrac{24}{10}$

47. $3.45 \quad\boxed{}\quad 3\dfrac{3}{8}$

48. **MULTIPLE CHOICE** What is $\dfrac{1}{3}$ of $3\dfrac{1}{2}$? *(Skills Review Handbook)*

   Ⓐ $\dfrac{1}{2}$ **Ⓑ** $1\dfrac{1}{6}$ **Ⓒ** $1\dfrac{1}{2}$ **Ⓓ** $10\dfrac{1}{2}$

Check It Out
Progress Check
BigIdeasMath.com

Simplify the expression. *(Section 1.4 and Section 1.5)*

1. $-7(6)$

2. $-1(-9)$

3. $\dfrac{-72}{-9}$

4. $-24 \div 3$

Evaluate the expression when $a = 4$, $b = -6$, and $c = -12$.
(Section 1.4 and Section 1.5)

5. c^2

6. $\dfrac{|c - b|}{a}$

Write an ordered pair corresponding to the point.
(Section 1.6)

7. Point A

8. Point B

9. Point C

10. Point D

11. SPEECH In speech class, you lose 3 points for every 30 seconds you go over the time limit. Your speech is 90 seconds over the time limit. What integer represents the change in your points? *(Section 1.4)*

12. MOUNTAIN CLIMBING On a mountain, the temperature decreases by 18°F every 5000 feet. What integer represents the change in temperature at 20,000 feet? *(Section 1.4)*

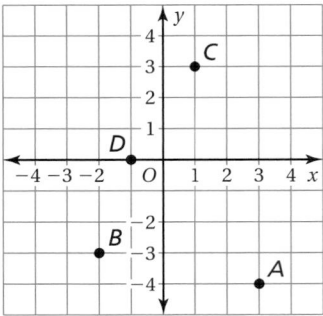

13. GAMING You play a video game for 15 minutes. You lose 165 points. What integer represents the average change in points per minute? *(Section 1.5)*

14. GEOMETRY The points $A(-4, 2)$, $B(1, -1)$, $C(1, 2)$, and $D(-4, -1)$ are the vertices of a figure. *(Section 1.6)*

  a. Draw the figure in a coordinate plane.

  b. Find the perimeter of the figure.

  c. Find the area of the figure.

Alternative Assessment Options

Math Chat   Student Reflective Focus Question
Structured Interview   Writing Prompt

Math Chat

- Have students work in pairs. One student describes the rule for multiplying two integers with the same sign and the rule for multiplying two integers with different signs. The student should include examples. The other student should probe for more information. Students then switch roles and repeat the process for dividing two integers with the same sign and dividing two integers with different signs.
- The teacher should walk around the classroom listening to the pairs and asking questions to ensure understanding.

Study Help Sample Answers

Remind students to complete Graphic Organizers for the rest of the chapter.

5a.

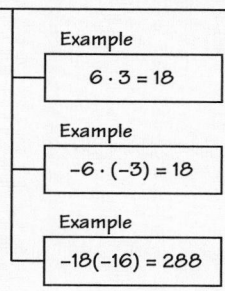

Multiplying integers with the same sign: The product of two integers with the same sign is positive.

Example
$6 \cdot 3 = 18$

Example
$-6 \cdot (-3) = 18$

Example
$-18(-16) = 288$

5b.

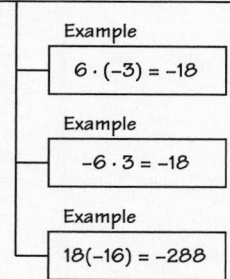

Multiplying integers with different signs: The product of two integers with different signs is negative.

Example
$6 \cdot (-3) = -18$

Example
$-6 \cdot 3 = -18$

Example
$18(-16) = -288$

6–8. Available at *BigIdeasMath.com*.

Reteaching and Enrichment Strategies

If students need help. . .	If students got it. . .

Resources by Chapter
  • Study Help
  • Practice A and Practice B
  • Puzzle Time
Lesson Tutorials
BigIdeasMath.com Practice Quiz
Practice from the Test Generator
	Resources by Chapter
	  • Enrichment and Extension
	  • School-to-Work
	Game Closet at *BigIdeasMath.com*
	Start the Chapter Review

Answers

1. -42 2. 9
3. 8 4. -8
5. 144 6. $\dfrac{3}{2}$
7. $(3, -4)$ 8. $(-2, -3)$
9. $(1, 3)$ 10. $(-1, 0)$
11. -9 12. -72
13. -11

14. a.

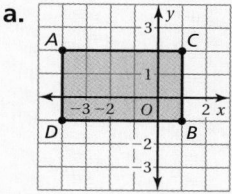

   b. 16 units

   c. 15 square units

Assessment Book

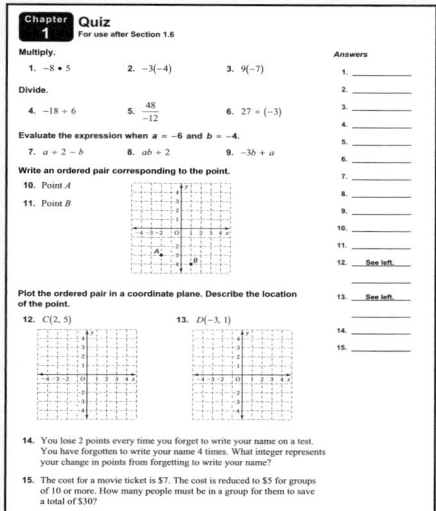

For the Teacher
Additional Review Options
- **Quiz**Show
- Big Ideas Test Generator
- Game Closet at *BigIdeasMath.com*
- Vocabulary Puzzle Builder
- Resources by Chapter
  - Puzzle Time
  - Study Help

Answers

1. 3
2. 9
3. 17
4. 8
5. Mississippi River in Illinois
6. −27
7. −10
8. 25
9. −34

Review of Common Errors

Exercises 1–5
- Students may think they can find the absolute value of a number by changing its sign and incorrectly find $|6| = -6$.

Exercises 6–9
- Students may ignore the signs and just add the integers.
- Remind students of the Key Ideas, and how the signs of the integers determine if the student needs to add or subtract the integers.

Exercises 10–14
- Students may change the sign of the first number, or forget to change the problem from subtraction to addition when changing the sign of the second number.

Exercises 15–22
- Students may not remember that a negative number multiplied or divided by a negative number is positive.

Exercises 23–25
- Remind students of the definition of mean.

Exercises 26–30
- Remind students that x is horizontal and y is vertical.

Review Key Vocabulary

integer, *p. 4*

absolute value, *p. 4*

opposites, *p. 10*

additive inverse, *p. 10*

coordinate plane, *p. 36*

origin, *p. 36*

quadrant, *p. 36*

x-axis, *p. 36*

y-axis, *p. 36*

Review Examples and Exercises

1.1   **Integers and Absolute Value**   *(pp. 2–7)*

Find the absolute value of −2.

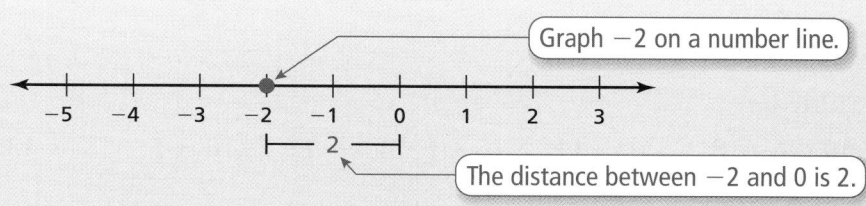

Graph −2 on a number line.

The distance between −2 and 0 is 2.

∴ So, $|-2| = 2$.

Exercises

Find the absolute value of the integer.

  1. 3   **2.** −9   **3.** −17   **4.** 8

  5. ELEVATION The elevation of Death Valley, CA is −282 feet. The Mississippi River in Illinois has an elevation of 279 feet. Which is closer to sea level?

1.2   **Adding Integers**   *(pp. 8–13)*

Find 6 + (−14).

     $6 + (-14) = -8$ $|6|$ is less than $|-14|$. So, subtract $|6|$ from $|-14|$.

Use the sign of −14.

∴ The sum is −8.

Exercises

Add.

  6. −16 + (−11)   **7.** −15 + 5   **8.** 100 + (−75)   **9.** −32 + (−2)

1.3 Subtracting Integers *(pp. 14–19)*

Subtract.

a. $7 - 19 = 7 + (-19)$ Add the opposite of 19.

          $= -12$ Add.

    ∴ The difference is -12.

b. $-6 - (-10) = -6 + 10$ Add the opposite of -10.

             $= 4$ Add.

    ∴ The difference is 4.

Exercises

Subtract.

10. $8 - 18$ **11.** $-16 - (-5)$ **12.** $-18 - 7$ **13.** $-12 - (-27)$

14. GAME SHOW Your score on a game show is -300. You answer the final question incorrectly, so you lose 400 points. What is your final score?

1.4 Multiplying Integers *(pp. 22–27)*

a. Find $-7 \cdot (-9)$.

The integers have the same sign.

$$-7 \cdot (-9) = 63$$

The product is positive.

    ∴ The product is 63.

b. Find $-6(14)$.

The integers have different signs.

$$-6(14) = -84$$

The product is negative.

    ∴ The product is -84.

Exercises

Multiply.

15. $-8 \cdot 6$ **16.** $10(-7)$ **17.** $-3 \cdot (-6)$ **18.** $-12(5)$

Review Game

Integer Operations

Big Ideas
Game Closet

For the Student
Additional Practice
- Lesson Tutorials
- Study Help (textbook)
- Student Website
    Multi-Language Glossary
    Practice Assessments

Materials per Group
- 52 index cards, numbered 1 through 52
- paper for each group member
- pencil for each group member

Directions
This game is like the card game *War*. Divide the class into equally sized groups. One person in each group shuffles and deals the index cards for that group. Eight rounds are played. For each round, each person is dealt two cards. A different operation is used in each round, as follows.

Round 1—addition
Round 2—subtraction
Round 3—multiplication
Round 4—division
Round 5—addition
Round 6—subtraction
Round 7—multiplication
Round 8—division

In Rounds 1 through 4, players evaluate an expression using the number on their first card, the operation for the round, and the number on their second card. In Rounds 5 through 8, players evaluate an expression using the number on their first card, the operation for the round, and *the negative of* the number on their second card.

The person with the greatest result in their group wins the round and takes all the other players' cards. In the event of a tie, the players involved receive one additional card and the operation for the round is performed using *all three* cards. Players continue in this manner until there is one winner.

If a group runs out of cards before the end of Round 8, each player records how many cards they have collected and the cards are collected from the players. The cards are then reshuffled and reused.

Who Wins?
The player with the most cards after Round 8 wins. In the event of a tie, the rounds are repeated, starting with Round 1, until there is a winner.

Answers
10. -10 **11.** -11

12. -25 **13.** 15

14. -700 points

15. -48 **16.** -70

17. 18 **18.** -60

19. -2 **20.** 7

21. -5 **22.** -12

23. -1 **24.** -48

25. $-\$48$

26–29. See graph below.

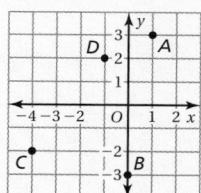

26. Quadrant I

27. y-axis

28. Quadrant III

29. Quadrant II

30.

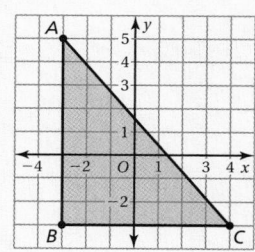

Sample answer: $(0, 0)$, $(0, -1)$, $(-1, 0)$

My Thoughts on the Chapter

What worked. . .

What did not work. . .

What I would do differently. . .

1.5 Dividing Integers *(pp. 28–33)*

Find $30 \div (-10)$.

> The integers have different signs.

$$30 \div (-10) = -3$$

> The quotient is negative.

∴ The quotient is -3.

Exercises

Divide.

19. $-18 \div 9$

20. $\dfrac{-42}{-6}$

21. $\dfrac{-30}{6}$

22. $84 \div (-7)$

Find the mean of the integers.

23. $-3, -8, 12, -15, 9$

24. $-54, -32, -70, -25, -65, -42$

25. PROFITS The table shows the weekly profits of a fruit vendor. What is the mean profit for these weeks?

Week	1	2	3	4
Profit	$-\$125$	$-\$86$	$\$54$	$-\$35$

1.6 The Coordinate Plane *(pp. 34–39)*

Plot (a) $(-3, 0)$ and (b) $(4, -4)$ in a coordinate plane. Describe the location of each point.

a. Start at the origin. Move 3 units left. Then plot the point.

The point is on the x-axis.

b. Start at the origin. Move 4 units right and 4 units down. Then plot the point.

The point is in Quadrant IV.

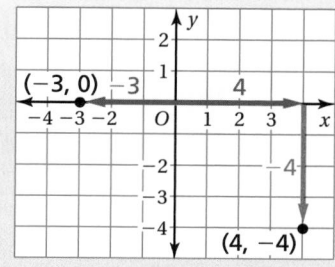

Exercises

Plot the ordered pair in a coordinate plane. Describe the location of the point.

26. $A(1, 3)$

27. $B(0, -3)$

28. $C(-4, -2)$

29. $D(-1, 2)$

30. GEOMETRY The points $A(-3, 5)$, $B(-3, -3)$, and $C(4, -3)$ are vertices of a figure. Draw the figure in a coordinate plane. Name three points that lie inside the figure.

1 Chapter Test

Check It Out
Test Practice
BigIdeasMath.com

Find the absolute value of the integer.

1. -9

2. 64

3. -22

Copy and complete the statement using $<$, $>$, or $=$.

4. $4 \quad \boxed{} \quad \left|-8\right|$

5. $-12 \quad \boxed{} \quad \left|-7\right|$

6. $-7 \quad \boxed{} \quad \left|3\right|$

Simplify the expression.

7. $-6 + (-11)$

8. $2 - (-9)$

9. $-9 \cdot 2$

10. $-72 \div (-3)$

11. $-5 + 17$

12. $-14(21)$

Plot the ordered pair in a coordinate plane. Describe the location of the point.

13. $K(1, 3)$

14. $L(-3, 0)$

15. $M(-4, 5)$

16. $N(2, -1)$

17. BANKING The balance of your checking account is $86. You withdraw $98. What is your new balance?

18. NASCAR A driver receives -25 points for each rule violation. What integer represents the change in points after four rule violations?

19. GOLF The table shows your scores, relative to *par*, for nine holes of golf. What is your total score for the nine holes?

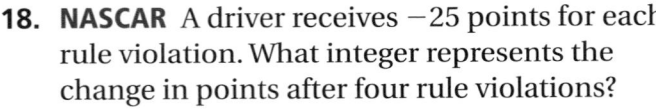

Hole	1	2	3	4	5	6	7	8	9	Total
Score	+1	-2	-1	0	-1	+3	-1	-3	+1	?

20. VISITORS In a recent 10-year period, the change in the number of visitors to U.S. National Parks was about $-11{,}150{,}000$ visitors.

   a. What was the mean yearly change in the number of visitors?

   b. During the seventh year, the change in the number of visitors was about $10{,}800{,}000$. Explain how the change for the 10-year period can be negative.

Test Item References

Chapter Test Questions	Section to Review
1–6, 18	1.1
7–12, 17, 19, 20	1.2, 1.3, 1.4, 1.5
13–16	1.6

Test-Taking Strategies

Remind students to quickly look over the entire test before they start so that they can budget their time. They should not spend too much time on any single problem. Urge students to try to work on a part of each problem because partial credit is better than none. Teach students to use the Stop and Think strategy before answering. **Stop** and carefully read the question, and **Think** about what the answer should look like.

Common Assessment Errors

- **Exercise 2** Students may think that the absolute value of a number is always its opposite and write $|64| = -64$. Use a number line to show students that absolute value is a number's distance from 0, so it is always positive or zero.
- **Exercises 4 and 5** When comparing absolute values of negative integers, students may not find absolute values first and instead just compare the integers. Remind students to find the absolute values first.
- **Exercises 7–12** Students may ignore the signs of the integers when simplifying. Remind students of the Key Ideas for addition, subtraction, multiplication, and division of integers, and that the signs of the integers will affect their answers.
- **Exercises 13–16** Students may plot the x-coordinate vertically instead of horizontally and the y-coordinate horizontally instead of vertically. Remind them that x is horizontal and y is vertical. Also, students may move in the wrong direction when plotting a point. Remind students to follow the directions of the axes.
- **Exercise 19** Students may not know the meaning of the word *par*, so explain it to them.

Reteaching and Enrichment Strategies

If students need help. . .	If students got it. . .
Resources by Chapter • Practice A and Practice B • Puzzle Time Record and Practice Journal Practice Differentiating the Lesson Lesson Tutorials Practice from the Test Generator Skills Review Handbook	Resources by Chapter • Enrichment and Extension • School-to-Work • Financial Literacy Game Closet at *BigIdeasMath.com* Start Standardized Test Practice

Answers

1. 9
2. 64
3. 22
4. $4 < |-8|$
5. $-12 < |-7|$
6. $-7 < |3|$
7. -17
8. 11
9. -18
10. 24
11. 12
12. -294

13–16. See graph below.

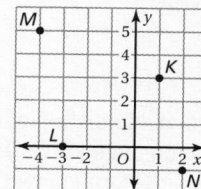

13. Quadrant I
14. x-axis
15. Quadrant II
16. Quadrant IV
17. $-\$12$
18. -100
19. -3
20. a. $-1,115,000$ visitors
    b. As long as the yearly number of visitors at the end of the 10 years was less than the yearly number at the start of the 10 years, the change is negative. During other years, there were more significant changes in visitors in the negative direction.

Assessment Book

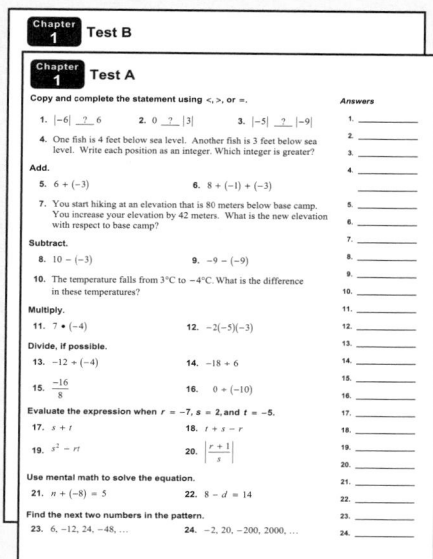

Test-Taking Strategies
Available at *BigIdeasMath.com*

After Answering Easy Questions, Relax
Answer Easy Questions First
Estimate the Answer
Read All Choices before Answering
Read Question before Answering
Solve Directly or Eliminate Choices
Solve Problem before Looking at
   Choices
Use Intelligent Guessing
Work Backwards

About this Strategy

When taking a multiple choice test, be sure to read each question carefully and thoroughly. Before answering a question, determine exactly what is being asked, then eliminate the wrong answers and select the best choice.

Answers

1. C
2. G
3. 25
4. D

Item Analysis

1. **A.** The student treats all numbers as gains and finds their sum.

   B. The student finds the correct difference, but thinks it is a gain instead of a loss.

   C. Correct answer

   D. The student treats all numbers as losses and finds their sum.

2. **F.** The student incorrectly thinks that points in Quadrant II have negative y-coordinates.

   G. Correct answer

   H. The student incorrectly thinks that points in Quadrant II have negative y-coordinates. The student also reverses the x- and y-coordinates.

   I. The student reverses the x- and y-coordinates.

3. **Gridded Response:** Correct answer: 25

   Common Error: The student thinks that $17 - (-8)$ is equivalent to $17 - 8$, getting an answer of 9.

4. **A.** The student chooses an option that does not change the outcome of Sam's incorrect work.

   B. The student thinks that absolute values of numbers are always negative.

   C. The student ignores the absolute value bars.

   D. Correct answer

5. **F.** The student thinks that $27 - 32 = -59$ and then estimates $\frac{5}{9}$ of this difference.

   G. The student finds $\frac{5}{9}$ of 27 and then subtracts 32 from this result.

   H. The student correctly finds the difference of 27 and 32, but does not multiply it by $\frac{5}{9}$.

   I. Correct answer

6. **Gridded Response:** Correct answer: -6

   Common Error: The student incorrectly finds a number halfway between 9 and -21 by adding these two numbers to get -30, half of which is -15.

Standardized Test Practice Icons

 Gridded Response

 Short Response (2-point rubric)

 Extended Response (4-point rubric)

Technology For the Teacher

Big Ideas Test Generator

You ripped out $(-1)^2 + (-2)(-3)$ whiskers. How many did you rip out?

Ⓐ -5   Ⓑ 5   Ⓒ -7   Ⓓ 7

Yeow, why the biggest number?

"You can eliminate A and C. Then, solve directly to determine that the correct answer is D."

1. A football team gains 2 yards on the first play, loses 5 yards on the second play, loses 3 yards on the third play, and gains 4 yards on the fourth play. What is the team's overall gain or loss for all four plays?

   A. a gain of 14 yards   C. a loss of 2 yards

   B. a gain of 2 yards   D. a loss of 14 yards

2. Point P is plotted in the coordinate plane below.

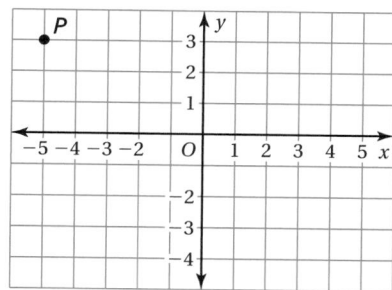

   What are the coordinates of point P?

   F. $(-5, -3)$ H. $(-3, -5)$

   G. $(-5, 3)$ I. $(3, -5)$

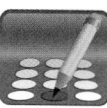

3. What is the value of the expression below?

$$17 - (-8)$$

4. Sam was simplifying an expression in the box below.

$$\left| -8 + 6 + (-3) \right| = \left| -8 \right| + \left| 6 \right| + \left| -3 \right|$$
$$= 8 + 6 + 3$$
$$= 17$$

   What should Sam do to correct the error that he made?

   A. Find the absolute value of the sum of 8, 6, and 3 and make that the final answer.

   B. Find the sum of -8, -6, and -3 and make that the final answer.

   C. Find the sum of -8, 6, and -3 and make that the final answer.

   D. Find the absolute value of the sum of -8, 6, and -3 and make that the final answer.

5. The expression below can be used to find the temperature in degrees Celsius when given F, the temperature in degrees Fahrenheit.

$$\frac{5}{9}(F - 32)$$

What is the temperature in degrees Celsius, to the nearest degree, when the temperature in degrees Fahrenheit is 27°?

F. −33°

H. −5°

G. −17°

I. −3°

6. What is the missing number in the sequence below?

$$39, 24, 9, \underline{\hspace{1em}}, -21$$

7. Which equation is *not* true for all numbers n?

A. $-n + 0 = -n$

C. $n - 0 = -n$

B. $n \cdot (-1) = -n$

D. $-n \cdot 1 = -n$

8. What is the area of the semicircle below? (Use 3.14 for π.)

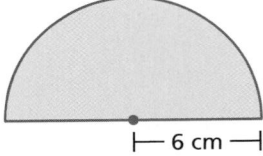

⊢— 6 cm —⊣

F. 18.84 cm^2

H. 56.52 cm^2

G. 37.68 cm^2

I. 226.08 cm^2

9. The campers at a summer camp held a contest in which they had to run across a field carrying buckets of water that were full at the beginning. The team who lost the least water from its bucket was the winner.

- Team A *lost* 40% of the water from its bucket.
- Team B *lost* 0.3 of the water from its bucket.
- Team C *kept* $\frac{5}{8}$ of the water in its bucket.
- Team D *kept* 67% of the water in its bucket.

Which team was the winner?

A. Team A

C. Team C

B. Team B

D. Team D

Item Analysis (continued)

7. **A.** The student does not recognize this equation as an application of the Identity Property of Addition.

   B. The student does not recognize this equation as correctly multiplying n by -1.

   C. Correct answer

   D. The student does not recognize this equation as an application of the Identity Property of Multiplication.

8. **F.** The student multiplied the radius by 2 instead of squaring it.

   G. The student doubled the diameter instead of squaring the radius.

   H. Correct answer

   I. The student squared the diameter instead of the radius.

9. **A.** The student does not realize that losing 40% is a greater loss than losing 0.3, keeping $\frac{5}{8}$, or keeping 67%.

   B. Correct answer

   C. The student does not realize that keeping $\frac{5}{8}$ is a greater loss than losing 0.3 or keeping 67%.

   D. The student does not realize that keeping 67% is a greater loss than losing 0.3.

10. **F.** The student multiplies first instead of subtracting and uses $-5 \cdot 17 - 20$ as an estimate.

    G. The student uses $-5 \cdot (-3)$ as an estimate, but thinks that the product of two negative numbers is negative.

    H. Correct answer

    I. The student multiplies first instead of subtracting and uses $-5 \cdot 17 - 20$ as an estimate. The student then thinks that the sum of two negative numbers is positive.

11. **2 points** The student demonstrates a thorough understanding of plotting points in all four quadrants of the coordinate plane. The student correctly plots and labels points at $(2, -3)$, $(2, -6)$, $(-1, -3)$, $(2, 0)$, and $(5, -3)$.

    1 point The student demonstrates a partial understanding of plotting points in all four quadrants of the coordinate plane. The student shows some knowledge of how to plot points, but does not successfully plot all points.

    0 points The student demonstrates insufficient understanding of plotting points in all four quadrants of the coordinate plane. The student makes many errors in plotting points and/or is unable to correctly draw an x-axis and y-axis.

Answers

5. I

6. -6

7. C

8. H

9. B

Answers

10. H

11. *Part A* and *Part B*

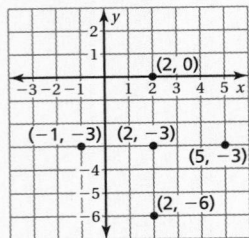

12. D

13. F

Answer for Extra Example

1. A. The student misinterprets the inverse relationship, multiplying instead of dividing. The student also thinks that the product of two negative numbers is negative.

   B. The student thinks that the quotient of two negative numbers is negative.

   C. Correct answer

   D. The student misinterprets the inverse relationship, multiplying instead of dividing.

Item Analysis (continued)

12. A. The student finds the mode instead of the mean.

   B. The student finds the range instead of the mean. The student also makes multiple errors in finding the range, using -8 and 1 instead of -8 and 4, thinking that the range between -8 and 1 is 7, and thinking that the range is negative.

   C. The student finds the median instead of the mean.

   D. Correct answer

13. F. Correct answer

   G. The student thinks that $\frac{1}{2} + \frac{3}{4} = \frac{4}{6} = \frac{2}{3}$ and that $\frac{5}{8} + \frac{1}{4} = \frac{6}{12} = \frac{1}{2}$ (by simply adding the numerators and adding the denominators). The student then finds the difference of $17\frac{2}{3}$ and $17\frac{1}{2}$.

   H. The student finds the total length of Jane's pet rat to be $17\frac{1}{4}$ instead of $18\frac{1}{4}$ (mistake in regrouping).

   I. The student thinks that $10\frac{1}{2} + 7\frac{3}{4} = 17\frac{4}{4} = 18$ and that $9\frac{5}{8} + 8\frac{1}{4} = 17\frac{6}{8} = 17\frac{3}{4}$ (finding common denominators correctly, but not appropriately changing the numerators). The student also chooses Manuel instead of Jane.

Extra Example for Standardized Test Practice

1. What number belongs in the box to make the equation true?

$$-24 = -6 \times \boxed{}$$

   A. -144 **C.** 4

   B. -4 **D.** 144

10. Which integer is closest to the value of the expression below?

$$-5.04 \cdot (16.89 - 20.1)$$

F. -105

H. 15

G. -15

I. 105

11. Answer the following questions in the coordinate plane.

Part A   Draw an x-axis and y-axis in the coordinate plane. Then plot and label the point $(2, -3)$.

Part B   Plot and label *four* points that are 3 units away from $(2, -3)$.

12. What is the mean of the data set in the box below?

$$-8, -6, -2, 0, -6, -8, 4, -7, -8, 1$$

A. -8

C. -6

B. -7

D. -4

13. Jane and Manuel measured the lengths of their pet rats.

- Jane's pet rat
  - Body length: $10\frac{1}{2}$ inches
  - Tail length: $7\frac{3}{4}$ inches

- Manuel's pet rat
  - Body length: $9\frac{5}{8}$ inches
  - Tail length: $8\frac{1}{4}$ inches

The total length of each rat is determined by the sum of its body length and its tail length. Whose rat has the longer total length and by how much?

F. Jane's rat is longer by $\frac{3}{8}$ inch.

H. Manuel's rat is longer by $\frac{5}{8}$ inch.

G. Jane's rat is longer by $\frac{1}{6}$ inch.

I. Manuel's rat is longer by $\frac{1}{4}$ inch.

2 Rational Numbers and Equations

"I can't find my algebra tiles, so I am painting some of my dog biscuits."

"Now I will be able to solve the equation $2x + (-2) = 2$."

"On the count of 5, I'm going to give you half of my dog biscuits."

"1, 2, 3, 4, $4\frac{1}{2}$, $4\frac{3}{4}$, $4\frac{7}{8}$,..."

Connections to Previous Learning

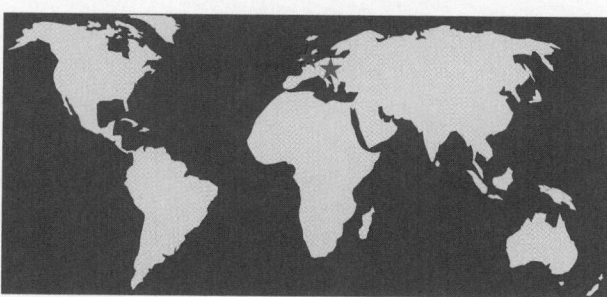

- Use properties of equality to solve numerical and real-world situations.
- Describe mathematical relationships using equations and visual representations.
- Add and subtract fractions and decimals efficiently.

- Write, solve, and graph one-step and two-step linear equations.
- Multiply and divide fractions and decimals efficiently.

- Express rational numbers as terminating or repeating decimals.
- Perform exponential operations with rational bases and whole number exponents.
- Formulate and use different strategies to solve one-step and two-step linear equations, including equations with rational coefficients.
- Use properties of equality to rewrite an equation and to show two equations are equivalent.

Math in History

The symbols $+$, $-$, $\times$, and $\div$ for the four basic operations are relatively recent.

★ The $+$ and $-$ symbols first appeared in print in a German book by Johannes Widmann, published in 1489. However, they referred to surpluses and deficits in business problems, not addition and subtraction.

★ The plus and minus symbols only came into general use in England after they were used by Robert Recorde in 1557 in a book titled *The Whetstone of Whitte*. Recorde wrote, "There be other 2 signes in often use of which the first is made thus $+$ and betokeneth more: the other is thus made $-$ and betokeneth lesse."

Pacing Guide for Chapter 2

Chapter Opener	1 Day
Section 1	
Activity	1 Day
Lesson	1 Day
Section 2	
Activity	1 Day
Lesson	1 Day
Section 3	
Activity	1 Day
Lesson	1 Day
Lesson b	1 Day
Study Help / Quiz	1 Day
Section 4	
Activity	1 Day
Lesson	1 Day
Section 5	
Activity	1 Day
Lesson	1 Day
Lesson b	1 Day
Section 6	
Activity	1 Day
Lesson	1 Day
Lesson b	2 Days
Quiz / Chapter Review	1 Day
Chapter Test	1 Day
Standardized Test Practice	1 Day
Total Chapter 2	21 Days
Year-to-Date	38 Days

Check Your Resources

- Record and Practice Journal
- Resources by Chapter
- Skills Review Handbook
- Assessment Book
- Worked-Out Solutions

The Dynamic Planning Tool
Editable Teacher's Resources at
BigIdeasMath.com

- Place value
- Solving one-step linear equations

Try It Yourself

1. $\frac{51}{100}$ 2. $\frac{731}{1000}$

3. 0.6   4. 0.875

5. $\frac{9}{10}$ 6. $\frac{3}{5}$

7. $\frac{27}{70}$ 8. $\frac{17}{20}$

Record and Practice Journal

1. $\frac{13}{50}$ 2. $\frac{79}{100}$

3. $\frac{571}{1000}$ 4. $\frac{423}{500}$

5. 0.375   6. 0.4

7. 0.6875   8. 0.85

9. $\frac{3}{5}$ 10. $\frac{17}{72}$

11. $\frac{47}{30}$ 12. $\frac{1}{3}$

13. $\frac{2}{35}$ 14. $\frac{5}{27}$

15. $\frac{2}{5}$ 16. $\frac{14}{11}$

17. $\frac{3}{4}$ 18. $7\frac{1}{12}$ cups

Math Background Notes

Vocabulary Review

- Least common multiple
- Common denominator
- Divisor
- Reciprocal

Writing Decimals and Fractions

- Students should know how to convert between decimals and fractions.
- You may need to review place values to the right of the decimal place with students prior to completing Example 1.

Adding and Subtracting Fractions

- Students should know how to add and subtract fractions.
- Remind students that adding and subtracting fractions requires a common denominator.
- You should review the least common multiple with students. This concept will help some students to find a common denominator.
- Using the least common multiple of the denominators will produce the least common denominator. Remind students that there are many common denominators to choose from. Some choices will require students to simplify the fraction at the end.

Multiplying and Dividing Fractions

- Students should know how to multiply and divide fractions.
- Remind students that the rules for multiplying and dividing fractions are different from the rules for adding and subtracting fractions. Multiplying and dividing fractions does not require a common denominator.
- **Teaching Tip:** Most students will remember the process to divide fractions. If your students are comfortable with the process, encourage them to describe it using math vocabulary. Instead of "change the sign and flip the second fraction," encourage "multiply by the reciprocal of the divisor."

Reteaching and Enrichment Strategies

If students need help...	If students got it...
Record and Practice Journal • Fair Game Review Skills Review Handbook Lesson Tutorials	Game Closet at *BigIdeasMath.com* Start the next section

What You Learned Before

"Let's play a game. The goal is to say a positive rational number that is less than the other pet's number... You go first."

This feels like a setup.

● Writing Decimals and Fractions

Example 1 Write 0.37 as a fraction.

$$0.37 = \frac{37}{100}$$

Example 2 Write $\frac{2}{5}$ as a decimal.

$$\frac{2}{5} = \frac{2 \cdot 2}{5 \cdot 2} = \frac{4}{10} = 0.4$$

Try It Yourself
Write the decimal as a fraction or the fraction as a decimal.

1. 0.51   **2.** 0.731   **3.** $\frac{3}{5}$ **4.** $\frac{7}{8}$

● Adding and Subtracting Fractions

Example 3 Find $\frac{1}{3} + \frac{1}{5}$.

$$\frac{1}{3} + \frac{1}{5} = \frac{1 \cdot 5}{3 \cdot 5} + \frac{1 \cdot 3}{5 \cdot 3}$$
$$= \frac{5}{15} + \frac{3}{15}$$
$$= \frac{8}{15}$$

Example 4 Find $\frac{1}{4} - \frac{2}{9}$.

$$\frac{1}{4} - \frac{2}{9} = \frac{1 \cdot 9}{4 \cdot 9} - \frac{2 \cdot 4}{9 \cdot 4}$$
$$= \frac{9}{36} - \frac{8}{36}$$
$$= \frac{1}{36}$$

● Multiplying and Dividing Fractions

Example 5 Find $\frac{5}{6} \cdot \frac{3}{4}$.

$$\frac{5}{6} \cdot \frac{3}{4} = \frac{5 \cdot \overset{1}{\cancel{3}}}{\underset{2}{\cancel{6}} \cdot 4}$$
$$= \frac{5}{8}$$

Example 6 Find $\frac{2}{3} \div \frac{9}{10}$.

$$\frac{2}{3} \div \frac{9}{10} = \frac{2}{3} \cdot \frac{10}{9}$$
$$= \frac{2 \cdot 10}{3 \cdot 9}$$
$$= \frac{20}{27}$$

Multiply by the reciprocal of the divisor.

Try It Yourself
Evaluate the expression.

5. $\frac{1}{4} + \frac{13}{20}$ **6.** $\frac{14}{15} - \frac{1}{3}$ **7.** $\frac{3}{7} \cdot \frac{9}{10}$ **8.** $\frac{4}{5} \div \frac{16}{17}$

Essential Question How can you use a number line to order rational numbers?

The Meaning of a Word ● Rational

The word **rational** comes from the word *ratio*.

If you sleep for 8 hours in a day, then the *ratio* of your sleeping time to the total hours in a day can be written as $\dfrac{8\text{ h}}{24\text{ h}}$.

A **rational number** is a number that can be written as the ratio of two integers.

$$2 = \frac{2}{1} \qquad -3 = \frac{-3}{1} \qquad -\frac{1}{2} = \frac{-1}{2} \qquad 0.25 = \frac{1}{4}$$

1 ACTIVITY: Ordering Rational Numbers

Work in groups of five. Order the numbers from least to greatest.

a. Sample: $-0.5,\ 1.25,\ -\dfrac{1}{3},\ 0.5,\ -\dfrac{5}{3}$

● Make a number line on the floor using masking tape and a marker.

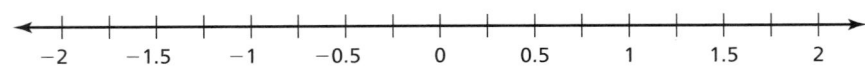

● Write the numbers on pieces of paper. Then each person should choose one.

● Stand on the location of your number on the number line.

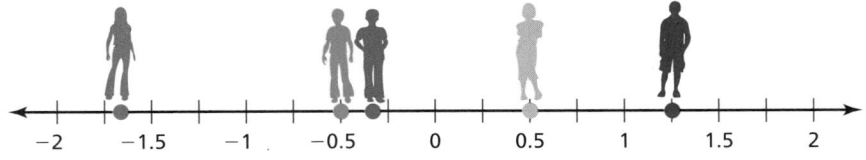

● Use your positions to order the numbers from least to greatest.

∴ So, the numbers from least to greatest are $-\dfrac{5}{3},\ -0.5,\ -\dfrac{1}{3},\ 0.5,$ and 1.25.

b. $-\dfrac{7}{4},\ 1.1,\ \dfrac{1}{2},\ -\dfrac{1}{10},\ -1.3$

c. $-\dfrac{1}{4},\ 2.5,\ \dfrac{3}{4},\ -1.7,\ -0.3$

d. $-1.4,\ -\dfrac{3}{5},\ \dfrac{9}{2},\ \dfrac{1}{4},\ 0.9$

e. $\dfrac{9}{4},\ 0.75,\ -\dfrac{5}{4},\ -0.8,\ -1.1$

Laurie's Notes

Introduction

For the Teacher

- **Goal:** The goal of this activity is for students to practice changing between common fractions and decimals, including their opposites.

  Examples: $\frac{1}{2} = 0.5$ and $-\frac{1}{2} = -0.5$

- The use of money is often effective in helping students think about the decimal form of certain common fractions. One-fourth of a dollar is $0.25; a dime ($0.10) is $\frac{1}{10}$ of a dollar.

Motivate

- A key skill for both activities today will be the ability to compare fractions and decimals. Try a warm up where students need to fill in the following table.

Fraction	$\frac{1}{2}$		$\frac{3}{5}$		$\frac{3}{4}$	
Decimal		0.1		0.8		1.4

- Check for understanding of the process of converting between these two forms of numbers.
- Students have studied operations with whole numbers, fractions, decimals, and integers.
- ❓ "Do you think there is a number halfway between -3 and -4? What is that number?"
- Explain that in this chapter, they will perform operations on numbers such as -3.5. Define rational numbers.

Activity Notes

Activity 1

- In preparing for this activity, be sure to leave sufficient space between the number line marks so that students are able to stand at their location comfortably. If there is enough space in the classroom, make multiple number lines on the floor. If space is limited, pairs of students could do the same problem on the board or on a piece of paper.
- **Discuss:** When the first set of numbers has been located, spend time having students give their reasoning as to why they located the numbers as they did. For instance, how did they know that $-\frac{5}{3}$ was to the left of -0.5 versus to the right of it?
- So that all students have an opportunity to use the number line on the floor, rotate groups at the end of each set of numbers.
- **Extension:** If time permits, ask students to name a decimal between two fractions $\left(-\frac{1}{2} \text{ and } -\frac{3}{4}\right)$ and to name a fraction between two decimals $(-0.6 \text{ and } -0.7)$.

Previous Learning

Students should know how to convert between common fractions (halves, fourths, fifths, and tenths) and decimals. They should also be able to graph common fractions and decimals on a number line.

Activity Materials
Textbook

- masking tape
- marker
- 20 index cards, cut in half, for each group

Start Thinking! and Warm Up

Activity 2.1 Start Thinking! For use before Activity 2.1

Activity 2.1 Warm Up For use before Activity 2.1

Copy and complete the statement using $<, >,$ or $=$.

1. $\frac{2}{3} \underline{\quad?\quad} \frac{4}{6}$ 2. $0.7 \underline{\quad?\quad} \frac{3}{4}$

3. $1.5 \underline{\quad?\quad} \frac{2}{3}$ 4. $-0.6 \underline{\quad?\quad} -1$

5. $3\frac{7}{10} \underline{\quad?\quad} 3\frac{4}{5}$ 6. $-2.4 \underline{\quad?\quad} -2\frac{2}{5}$

2.1 Record and Practice Journal

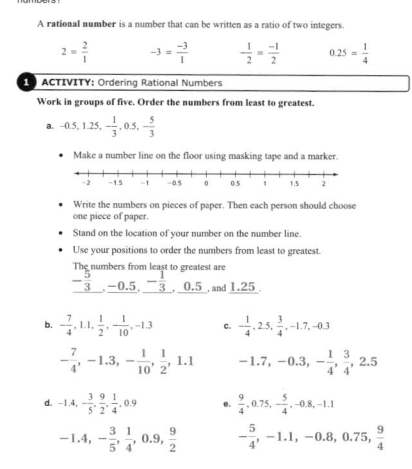

Essential Question How can you use a number line to order rational numbers?

A **rational number** is a number that can be written as a ratio of two integers.

$$2 = \frac{2}{1} \qquad -3 = \frac{-3}{1} \qquad -\frac{1}{2} = \frac{-1}{2} \qquad 0.25 = \frac{1}{4}$$

1 ACTIVITY: Ordering Rational Numbers

Work in groups of five. Order the numbers from least to greatest.

a. $-0.5, 1.25, -\frac{1}{3}, 0.5, -\frac{5}{3}$

- Make a number line on the floor using masking tape and a marker.
- Write the numbers on pieces of paper. Then each person should choose one piece of paper.
- Stand on the location of your number on the number line.
- Use your positions to order the numbers from least to greatest.

The numbers from least to greatest are $-\frac{5}{3}, -0.5, -\frac{1}{3}, 0.5,$ and 1.25.

b. $-\frac{7}{4}, 1.1, \frac{1}{2}, -\frac{1}{10}, -1.3$

$-\frac{7}{4}, -1.3, -\frac{1}{10}, \frac{1}{2}, 1.1$

c. $-\frac{1}{4}, 2.5, \frac{3}{4}, -1.7, -0.3$

$-1.7, -0.3, -\frac{1}{4}, \frac{3}{4}, 2.5$

d. $-1.4, -\frac{3}{5}, \frac{9}{4}, \frac{1}{2}, 0.9$

$-1.4, -\frac{3}{5}, \frac{1}{4}, 0.9, \frac{9}{2}$

e. $\frac{9}{4}, 0.75, -\frac{5}{4}, -0.8, -1.1$

$-\frac{5}{4}, -1.1, -0.8, 0.75, \frac{9}{4}$

English Language Learners

Vocabulary

Let English language learners know that the pronunciation of rational numbers that end in –th such as fourths and fifths is sometimes difficult for native English speakers.

2.1 Record and Practice Journal

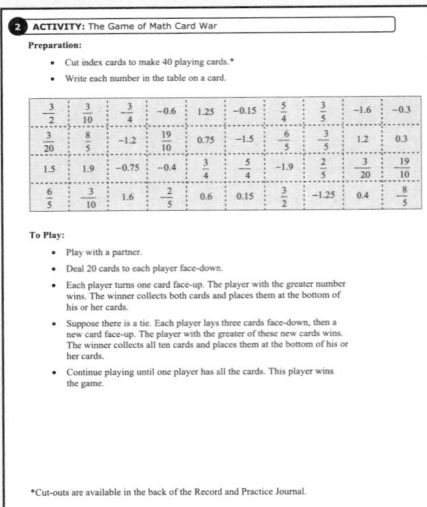

Activity 2

- You may want to make the game cards ahead of time, or have students create the game cards.
- **Management Tip:** To preserve cards for multiple uses, make cards on colored cardstock and store individual sets in sealable plastic bags.
- The card game *War* is familiar to many students. The question asked each play is, "Which number is greater?" The player with the greater value collects both cards. If the cards have an equivalent value, there is a tie. As stated in the text, each player lays 3 cards face down and then 1 card face up. The player with the card of greater value collects all of the cards.
- **Comparing Cards:** The key component of this activity is when students actually compare the two rational numbers. Discuss with students how they will compare the numbers. When both numbers are positive or the signs are different, students will have less difficulty. If both numbers are negative, students need to remember that the farther the number is to the right on the number line, the greater its value.

  For example, $-\frac{3}{5} > -0.75$ because $-\frac{3}{5}$ is to the right of -0.75 on a number line.
- To start play, give students the opportunity to preview the cards. Explain the rules and let students begin. If one group finishes early, have them shuffle the cards and play again.
- **Extension:** The cards can also be used to play the game *Memory*. Put the fraction cards in one group and the decimal cards in another group. Place all cards face down in two grids. Students select one card from each group. If the cards match, (meaning they are equivalent), then the student keeps the cards. If they do not match, the cards are put back face down. A deck of 40 cards is too many! Reduce the deck to 24 (12 in each group). Make sure the equivalent decimals and fractions are in each deck.

What Is Your Answer?

- Listen for the big idea, namely that the farther to the right the number is on the number line, the greater the value of that number.
- For Questions 4–7, students should work with a partner. Have students share their results and their reasoning. Answers will vary, so the explanation is important to hear.

Closure

- Which is greater: *A* or *B*? All have *B* as the greater number.

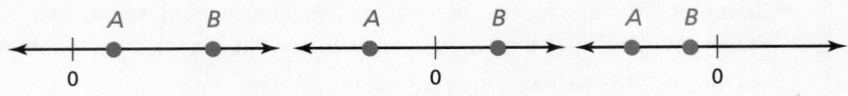

The Dynamic Planning Tool
Editable Teacher's Resources at *BigIdeasMath.com*

2 ACTIVITY: The Game of Math Card War

Preparation:

- Cut index cards to make 40 playing cards.
- Write each number in the table on a card.

To Play:

- Play with a partner.
- Deal 20 cards to each player face-down.
- Each player turns one card face-up. The player with the greater number wins. The winner collects both cards and places them at the bottom of his or her cards.
- Suppose there is a tie. Each player lays three cards face-down, then a new card face-up. The player with the greater of these new cards wins. The winner collects all ten cards and places them at the bottom of his or her cards.
- Continue playing until one player has all the cards. This player wins the game.

$-\dfrac{3}{2}$	$\dfrac{3}{10}$	$-\dfrac{3}{4}$	-0.6	1.25	-0.15	$\dfrac{5}{4}$	$\dfrac{3}{5}$	-1.6	-0.3
$\dfrac{3}{20}$	$\dfrac{8}{5}$	-1.2	$\dfrac{19}{10}$	0.75	-1.5	$-\dfrac{6}{5}$	$-\dfrac{3}{5}$	1.2	0.3
1.5	1.9	-0.75	-0.4	$\dfrac{3}{4}$	$-\dfrac{5}{4}$	-1.9	$\dfrac{2}{5}$	$-\dfrac{3}{20}$	$-\dfrac{19}{10}$
$\dfrac{6}{5}$	$-\dfrac{3}{10}$	1.6	$-\dfrac{2}{5}$	0.6	0.15	$\dfrac{3}{2}$	-1.25	0.4	$-\dfrac{8}{5}$

What Is Your Answer?

3. **IN YOUR OWN WORDS** How can you use a number line to order rational numbers? Give an example.

The numbers are in order from least to greatest. Fill in the blank spaces with rational numbers.

4. $-\dfrac{1}{2},$ ⬜ $, \dfrac{1}{3},$ ⬜ $, \dfrac{7}{5},$ ⬜

5. $-\dfrac{5}{2},$ ⬜ $, -1.9,$ ⬜ $, -\dfrac{2}{3},$ ⬜

6. $-\dfrac{1}{3},$ ⬜ $, -0.1,$ ⬜ $, \dfrac{4}{5},$ ⬜

7. $-3.4,$ ⬜ $, -1.5,$ ⬜ $, 2.2,$ ⬜

Practice

Use what you learned about ordering rational numbers to complete Exercises 28–30 on page 54.

Key Vocabulary 🔊
terminating decimal,
  p. 52
repeating decimal,
  p. 52
rational number,
  p. 52

A **terminating decimal** is a decimal that ends.

> 1.5, –0.25, 10.625

A **repeating decimal** is a decimal that has a pattern that repeats.

$$-1.333\ldots = -1.\overline{3}$$

$$0.151515\ldots = 0.\overline{15}$$

> Use *bar notation* to show which of the digits repeat.

Terminating and repeating decimals are examples of *rational numbers*.

 **Key Idea**

Rational Numbers

A **rational number** is a number that can be written as $\dfrac{a}{b}$ where a and b are integers and $b \neq 0$.

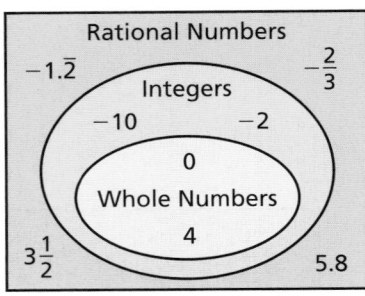

Rational Numbers
$-1.\overline{2}$ $-\dfrac{2}{3}$
Integers
-10 -2
0
Whole Numbers
4
$3\dfrac{1}{2}$ 5.8

EXAMPLE ① **Writing Rational Numbers as Decimals**

a. Write $-2\dfrac{1}{4}$ **as a decimal.**

Notice that $-2\dfrac{1}{4} = -\dfrac{9}{4}$.

> Divide 9 by 4.

$$\begin{array}{r} 2.25 \\ 4\overline{)9.00} \\ -8 \\ \hline 1\,0 \\ -8 \\ \hline 20 \\ -20 \\ \hline 0 \end{array}$$

> The remainder is 0. So, it is a terminating decimal.

∴ So, $-2\dfrac{1}{4} = -2.25$.

b. Write $\dfrac{5}{11}$ **as a decimal.**

> Divide 5 by 11.

$$\begin{array}{r} 0.4545 \\ 11\overline{)5.0000} \\ -4\,4 \\ \hline 60 \\ -55 \\ \hline 50 \\ -44 \\ \hline 60 \\ -55 \\ \hline 5 \end{array}$$

> The remainder repeats. So, it is a repeating decimal.

∴ So, $\dfrac{5}{11} = 0.\overline{45}$.

🔵 **On Your Own**

Now You're Ready
Exercises 11–18

Write the rational number as a decimal.

1. $-\dfrac{6}{5}$ 2. $-7\dfrac{3}{8}$ 3. $-\dfrac{3}{11}$ 4. $1\dfrac{5}{27}$

Laurie's Notes

Introduction

Connect

- **Yesterday:** Students compared fractions and decimals.
- **Today:** Students will extend this knowledge to include repeating decimals.

Motivate

- Ask students to form a "name fraction," where the numerator is the number of letters in their first name and the denominator is the number of letters in their last name.
- ❔ Ask students if their fractions are *close to 0, close to $\frac{1}{2}$, close to 1*, etc.
- Before class, go through your class roster and select two students whose name fractions are nearly equivalent, but one is a terminating decimal and the other is a repeating decimal. Discuss writing the fractions as decimals. Start with the terminating decimal.
- When you look at the repeating decimal, share with students that today's lesson is about writing rational numbers, and some rational numbers may be repeating decimals.

Key Idea

- Define terminating and repeating decimals. Give examples of each.
- **Common Error:** Some students will write $\frac{1}{3}$ as 0.333 and think that is sufficient. They do not realize that the repeat bar represents many more 3s.
- Define rational numbers.
- **Discuss:** Students have worked with fractions and decimals before. Explain that when negative fractions and decimals are included, we refer to these numbers as rational numbers. Also point out that the definition includes the word *can*, meaning that the rational numbers do not have to be written in the form $\frac{a}{b}$, but they *can* be.

Lesson Notes

Example 1

- ❔ "How do you write a fraction as a decimal?" Listen for 3 methods: 1) benchmark fractions you know, 2) write the fraction as an equivalent fraction with a denominator as a power of 10 and use the place value, or 3) divide the numerator by the denominator.
- Complete part (a) as a class. The first step is to write the mixed number as the equivalent improper fraction. Then divide the numerator by the denominator. Point out that the negative sign is simply placed in the answer after the calculations are complete.
- Complete part (b) as a class. Remind students that you always divide the numerator by the denominator, regardless of the size of the numbers!

On Your Own

- **Neighbor Check:** Have students work independently and then have their neighbor check their work. Have students discuss any discrepancies.

Start Thinking! and Warm Up

> **Lesson 2.1 Warm Up** For use before Lesson 2.1
>
> **Lesson 2.1 Start Thinking!** For use before Lesson 2.1
>
> You and two friends are playing basketball. You make 7 out of 15 shots. Your first friend makes 6 out of 10 shots and your second friend makes 5 out of 12 shots. Who is the better shooter?
>
> How would you solve this problem using what you know about rational numbers?

Extra Example 1

Write the rational number as a decimal.

a. $4\frac{3}{16}$ 4.1875

b. $-3\frac{4}{9}$ $-3.\overline{4}$

On Your Own

1. -1.2 2. -7.375

3. $-0.\overline{27}$ 4. $1.\overline{185}$

Laurie's Notes

Extra Example 2

Write -2.625 as a mixed number in simplest form. $-2\frac{5}{8}$

 On Your Own

5. $-\frac{7}{10}$ 6. $\frac{1}{8}$

7. $-3\frac{1}{10}$ 8. $-10\frac{1}{4}$

Extra Example 3

Order the rational numbers $-\frac{5}{9}$, $-1\frac{3}{4}$, $-\frac{13}{8}$, and -0.6 from least to greatest. $-1\frac{3}{4}$, $-\frac{13}{8}$, -0.6, $-\frac{5}{9}$

 On Your Own

9. All of the sea creatures (anglerfish, squid, shark, and whale) are deeper than the dolphin.

Differentiated Instruction

Auditory

Writing terminating decimals as rational numbers is easier if the students read the decimal using place value as opposed to reading the digits.

Terminating decimal: -0.26

Read as place value: *negative twenty-six hundredths*

Read as digits: *negative zero point two six*

Example 2

? "How do you write a decimal as a fraction?" Look at the place value of the last digit in the decimal and that will be the denominator.

• Work through Example 2.

? "How was the fraction simplified?" Both the numerator and the denominator were divided by a common factor of 2.

• **Extension:** Write -0.026 and -2.6 as fractions. This helps students focus on the importance of place value and where the last digit is located.

On Your Own

• **Neighbor Check:** Have students work independently and then have their neighbor check their work. Have students discuss any discrepancies.

• In Questions 7 and 8, the whole number portion of the decimal can be a problem.

Example 3

• Discuss the unit of measure, kilometers.

• Work through the problem. When doing this problem in class, draw the number line vertically and identify sea level.

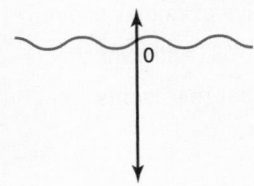

On Your Own

• **Neighbor Check:** Have students work independently and then have their neighbor check their work. Have students discuss any discrepancies.

• **Extension:** If calculators are available to students, explore the repeating patterns for certain sets of fractions (thirds, ninths, elevenths, etc.).

Closure

• **Exit ticket:**
  • Write $-\frac{5}{6}$ as a decimal. $-0.8\overline{3}$
  • Write -0.56 as a fraction. $-\frac{56}{100} = -\frac{14}{25}$

Technology
**For
the**T**eacher**

The Dynamic Planning Tool
Editable Teacher's Resources at *BigIdeasMath.com*

EXAMPLE ② **Writing a Decimal as a Fraction**

Write −0.26 as a fraction in simplest form.

$$-0.26 = -\frac{26}{100}$$

> Write the digits after the decimal point in the numerator.

> The last digit is in the hundredths place. So, use 100 in the denominator.

$$= -\frac{13}{50}$$ Simplify.

On Your Own

Now You're Ready
Exercises 20–27

Write the decimal as a fraction or mixed number in simplest form.

5. −0.7   **6.** 0.125   **7.** −3.1   **8.** −10.25

EXAMPLE ③ **Ordering Rational Numbers**

Creature	Elevations (km)
Anglerfish	$-\dfrac{13}{10}$
Squid	$-2\dfrac{1}{5}$
Shark	$-\dfrac{2}{11}$
Whale	-0.8

The table shows the elevations of four sea creatures relative to sea level. Which of the sea creatures are deeper than the whale? Explain.

Write each rational number as a decimal.

$$-\frac{13}{10} = -1.3$$

$$-2\frac{1}{5} = -2.2$$

$$-\frac{2}{11} = -0.\overline{18}$$

Then graph each decimal on a number line.

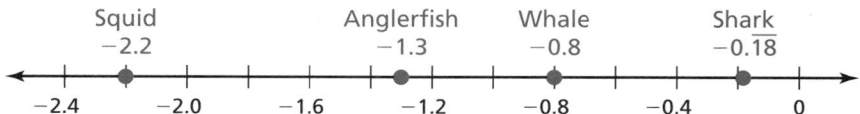

Both −2.2 and −1.3 are less than −0.8. So, the squid and the anglerfish are deeper than the whale.

On Your Own

Now You're Ready
Exercises 28–33

9. WHAT IF? The elevation of a dolphin is $-\dfrac{1}{10}$ kilometer. Which of the sea creatures in Example 3 are deeper than the dolphin? Explain.

Check It Out
Help with Homework
BigIdeasMath com

Vocabulary and Concept Check

1. **VOCABULARY** How can you tell that a number is rational?

2. **WRITING** You have to write 0.63 as a fraction. How do you choose the denominator?

Tell whether the number belongs to each of the following number sets: *rational numbers, integers, whole numbers.*

3. -5 4. $-2.1\overline{6}$ 5. 12 6. 0

Tell whether the decimal is *terminating* or *repeating*.

7. $-0.4848\ldots$ 8. -0.151 9. 72.72 10. $-5.2\overline{36}$

Practice and Problem Solving

Write the rational number as a decimal.

11. $\dfrac{7}{8}$ 12. $\dfrac{5}{11}$ 13. $-\dfrac{7}{9}$ 14. $-\dfrac{17}{40}$

15. $1\dfrac{5}{6}$ 16. $-2\dfrac{17}{18}$ 17. $-5\dfrac{7}{12}$ 18. $8\dfrac{15}{22}$

19. **ERROR ANALYSIS** Describe and correct the error in writing the rational number as a decimal.

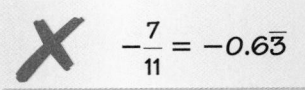

$$-\dfrac{7}{11} = -0.6\overline{3}$$

Write the decimal as a fraction or mixed number in simplest form.

20. -0.9 21. 0.45 22. -0.258 23. -0.312

24. -2.32 25. -1.64 26. 6.012 27. -12.405

Order the numbers from least to greatest.

28. $-\dfrac{3}{4}, 0.5, \dfrac{2}{3}, -\dfrac{7}{3}, 1.2$ 29. $\dfrac{9}{5}, -2.5, -1.1, -\dfrac{4}{5}, 0.8$ 30. $-1.4, -\dfrac{8}{5}, 0.6, -0.9, \dfrac{1}{4}$

31. $2.1, -\dfrac{6}{10}, -\dfrac{9}{4}, -0.75, \dfrac{5}{3}$ 32. $-\dfrac{7}{2}, -2.8, -\dfrac{5}{4}, \dfrac{4}{3}, 1.3$ 33. $-\dfrac{11}{5}, -2.4, 1.6, \dfrac{15}{10}, -2.25$

34. **COINS** You lose one quarter, two dimes and two nickels.

    a. Write the amount as a decimal.

    b. Write the amount as a fraction in simplest form.

35. **HIBERNATION** A box turtle hibernates in sand at $-1\dfrac{5}{8}$ feet. A spotted turtle hibernates at $-1\dfrac{16}{25}$ feet. Which turtle is deeper?

Assignment Guide and Homework Check

Level	Day 1 Activity Assignment	Day 2 Lesson Assignment	Homework Check
Basic	28–30, 48–52	1–10, 15–23 odd, 31–35	15, 21, 31, 34
Average	28–30, 48–52	1–10, 15–23 odd, 31–35 odd, 43–45	15, 21, 31, 44
Advanced	28–30, 48–52	1–10, 19, 28–42 even, 45–47	28, 36, 42, 46

Common Errors

- **Exercises 11–18** Students may forget to carry the negative sign through the division operation. Tell them to create a space for their final answer and to write the sign of the number in the space at the beginning.
- **Exercises 20–27** Students may try to put the decimal number over the denominator. Remind them to remove the decimal before they write it as a fraction. They may also write the whole number in front of the fraction while they are reducing it.
- **Exercises 28–33** Students may just order the fractions or decimals without the negative signs. Remind them that some numbers are negative and will be less than the positive numbers.

2.1 Record and Practice Journal

Write the rational number as a decimal.

1. $-\frac{9}{10}$
   -0.9
2. $-4\frac{2}{3}$
   $-4.\overline{6}$
3. $1\frac{7}{16}$
   1.4375

Write the decimal as a fraction or mixed number in simplest form.

4. -0.84
   $-\frac{21}{25}$
5. 5.22
   $5\frac{11}{50}$
6. -1.716
   $-1\frac{179}{250}$

Order the numbers from least to greatest.

7. $\frac{1}{5}, 0.1, -\frac{1}{2}, -0.25, 0.3$
   $-\frac{1}{2}, -0.25, 0.1, \frac{1}{5}, 0.3$
8. $-1.6, \frac{5}{2}, -\frac{7}{8}, 0.9, -\frac{6}{5}$
   $-1.6, -\frac{6}{5}, -\frac{7}{8}, 0.9, \frac{5}{2}$
9. $-\frac{2}{3}, \frac{5}{9}, 0.5, -1.3, -\frac{10}{3}$
   $-\frac{10}{3}, -1.3, -\frac{2}{3}, 0.5, \frac{5}{9}$

10. Relative to ground level, a black garden ant digs $-20\frac{7}{9}$ feet and a red harvester ant digs $-20\frac{39}{50}$ feet. Which ant is closer to ground level?

    Black garden ant

11. The table shows the position of each runner relative to when the first place finisher crossed the finish line. Who finished in second place? Who finished in fifth place?

Runner	A	B	C	D	E	F
Meters	-1.264	$\frac{5}{4}$	-1.015	-0.480	$\frac{14}{25}$	$\frac{13}{8}$

    Runner D; Runner B

Technology For the Teacher

Answer Presentation Tool
QuizShow

Practice and Problem Solving

11. 0.875
12. $0.\overline{45}$
13. $-0.\overline{7}$
14. -0.425
15. $1.8\overline{3}$
16. $-2.9\overline{4}$
17. $-5.58\overline{3}$
18. $8.68\overline{1}$

19. The bar should be over both digits to the right of the decimal point.

    $-\frac{7}{11} = -0.\overline{63}$

20. $-\frac{9}{10}$
21. $\frac{9}{20}$
22. $-\frac{129}{500}$
23. $-\frac{39}{125}$
24. $-2\frac{8}{25}$
25. $-1\frac{16}{25}$
26. $6\frac{3}{250}$
27. $-12\frac{81}{200}$
28. $-\frac{7}{3}, -\frac{3}{4}, 0.5, \frac{2}{3}, 1.2$

Practice and Problem Solving

29. $-2.5, -1.1, -\frac{4}{5}, 0.8, \frac{9}{5}$

30. $-\frac{8}{5}, -1.4, -0.9, \frac{1}{4}, 0.6$

31. $-\frac{9}{4}, -0.75, -\frac{6}{10}, \frac{5}{3}, 2.1$

32. $-\frac{7}{2}, -2.8, -\frac{5}{4}, 1.3, \frac{4}{3}$

33. $-2.4, -2.25, -\frac{11}{5}, \frac{15}{10}, 1.6$

34. a. -0.55

   b. $-\frac{11}{20}$

35. spotted turtle

36. $-2.2 > -2.42$

37. $-1.82 < -1.81$

38. $\frac{15}{8} = 1\frac{7}{8}$ **39.** $-4\frac{6}{10} > -4.65$

40–45. See Additional Answers.

46. See *Taking Math Deeper*.

47. See Additional Answers.

Fair Game Review

48. $\frac{31}{35}$ **49.** $\frac{7}{30}$

50. 4.72 **51.** 21.15

52. D

Mini-Assessment

Write the rational number as a decimal.

1. $\frac{8}{9}$ $0.\overline{8}$ **2.** $-\frac{11}{12}$ $-0.91\overline{6}$

3. $\frac{4}{11}$ $0.\overline{36}$ **4.** $-\frac{13}{15}$ $-0.8\overline{6}$

5. When your cousin was born, she was $21\frac{4}{5}$ inches long. When your friend was born, he was $21\frac{5}{6}$ inches long. Who was longer at birth?
your friend

Taking Math Deeper

Exercise 46

Students have already learned that it is easier to order numbers in decimal form than in fraction form. This problem gives students practice with this skill using negative numbers. The challenge in this problem is that students need to decide what place values to use for all four decimals.

 Write each number as a decimal.

Week	1	2	3	4
Change (inches)	$-\frac{7}{5}$	$-1\frac{5}{11}$	-1.45	$-1\frac{91}{200}$
Decimal	-1.4000	$-1.45\overline{45}$	-1.4500	-1.4550

 Graph the numbers on a number line.

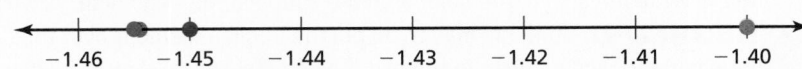

This problem is more difficult than it appears.

Write the numbers in order from least to greatest.

$-1\frac{91}{200}$ $-1\frac{5}{11}$ -1.45 $-\frac{7}{5}$

The U.S. Geological Survey (USGS) records the water levels at various locations in the United States. You can track these measurements by going to www.usgs.org.

Project

Create a chart showing the water levels at various locations in the Great Lakes during the same week. What is the range in water levels? Why do you think the levels vary?

Reteaching and Enrichment Strategies

If students need help. . .	If students got it. . .
Resources by Chapter • Practice A and Practice B • Puzzle Time Record and Practice Journal Practice Differentiating the Lesson Lesson Tutorials Skills Review Handbook	Resources by Chapter • Enrichment and Extension Start the next section

Copy and complete the statement using <, >, or =.

36. -2.2 ▢ -2.42

37. -1.82 ▢ -1.81

38. $\frac{15}{8}$ ▢ $1\frac{7}{8}$

39. $-4\frac{6}{10}$ ▢ -4.65

40. $-5\frac{3}{11}$ ▢ $-5.\overline{2}$

41. $-2\frac{13}{16}$ ▢ $-2\frac{11}{14}$

42. OPEN-ENDED Find one terminating decimal and one repeating decimal between $-\frac{1}{2}$ and $-\frac{1}{3}$.

Player	Hits	At Bats
Eva	42	90
Michelle	38	80

43. SOFTBALL In softball, a batting average is the number of hits divided by the number of times at bat. Does Eva or Michelle have the higher batting average?

44. QUIZ You miss 3 out of 10 questions on a science quiz and 4 out of 15 questions on a math quiz. Which quiz has a higher percent of correct answers?

45. SKATING Is the half pipe deeper than the skating pool? Explain.

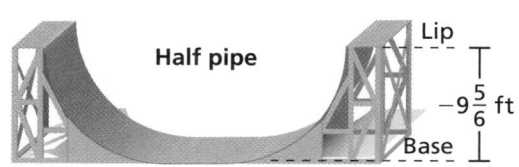

46. ENVIRONMENT The table shows the changes from the average water level of a pond over several weeks. Order the numbers from least to greatest.

Week	1	2	3	4
Change (inches)	$-\frac{7}{5}$	$-1\frac{5}{11}$	-1.45	$-1\frac{91}{200}$

47.  Given: a and b are integers.

 a. When is $-\frac{1}{a}$ positive?

 b. When is $\frac{1}{ab}$ positive?

Fair Game Review What you learned in previous grades & lessons

Add or subtract. *(Skills Review Handbook)*

48. $\frac{3}{5} + \frac{2}{7}$

49. $\frac{9}{10} - \frac{2}{3}$

50. $8.79 - 4.07$

51. $11.81 + 9.34$

52. MULTIPLE CHOICE In one year, a company has a profit of $-\$2$ million. In the next year, the company has a profit of $\$7$ million. How much more money did the company make the second year? *(Section 1.3)*

    Ⓐ $2 million   Ⓑ $5 million   Ⓒ $7 million   Ⓓ $9 million

Adding and Subtracting Rational Numbers

Essential Question How does adding and subtracting rational numbers compare with adding and subtracting integers?

---

1 ACTIVITY: Adding and Subtracting Rational Numbers

Work with a partner. Use a number line to find the sum or difference.

a. **Sample:** $2.7 + (-3.4)$

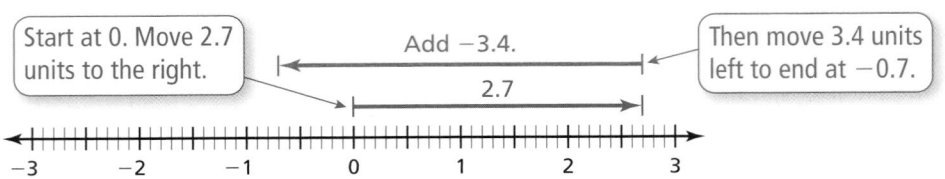

Start at 0. Move 2.7 units to the right.

Add −3.4.

2.7

Then move 3.4 units left to end at −0.7.

∴ So, $2.7 + (-3.4) = -0.7$.

b. $\dfrac{3}{10} + \left(-\dfrac{9}{10}\right)$ c. $-\dfrac{6}{10} - 1\dfrac{3}{10}$

d. $1.3 + (-3.4)$ e. $-1.9 - 0.8$

---

2 ACTIVITY: Adding and Subtracting Rational Numbers

Work with a partner. Write the numerical expression shown on the number line. Then find the sum or difference.

a.

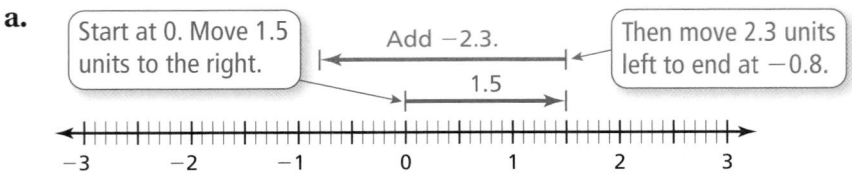

Start at 0. Move 1.5 units to the right.

Add −2.3.

1.5

Then move 2.3 units left to end at −0.8.

b.

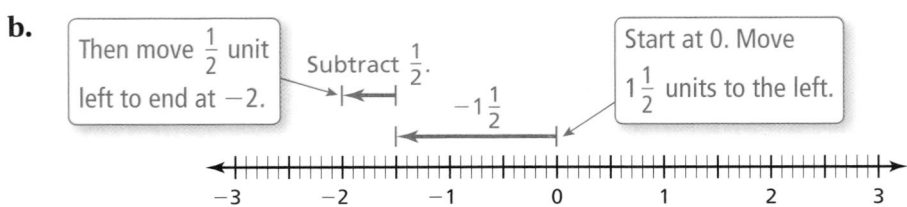

Then move $\dfrac{1}{2}$ unit left to end at −2.

Subtract $\dfrac{1}{2}$.

$-1\dfrac{1}{2}$

Start at 0. Move $1\dfrac{1}{2}$ units to the left.

Laurie's Notes

Introduction

For the Teacher

- **Goal:** The goal of this activity is for students to develop a conceptual understanding of how rational numbers are added and subtracted. The number line model is used to provide a visual and instructional aid.

Motivate

Pose a series of contextual questions that will help students think about negative rational numbers. These questions should suggest why you need to be able to add and subtract rational numbers. Examples:

- "If finding a dollar and a quarter represented $1.25, how would losing a dollar and a quarter be represented?" $-\$1.25$

- "If a half mile above sea level is represented as $\frac{1}{2}$, how would a half mile below sea level be represented?" $-\frac{1}{2}$

- "Represent a loss of $5\frac{1}{2}$ yards on a play in football." $-5\frac{1}{2}$

- "Represent a drop in temperature of 4.2°." -4.2

- Discuss the use of number lines as a model for addition and subtraction. Ask students to describe how addition and subtraction are modeled on a number line. Make the connection between subtraction and adding a negative quantity [e.g., $5 - 2$ and $5 + (-2)$].

Activity Notes

Activity 1

- Student support for part (d): after the arrow is drawn for 1.3, you need to move 3.4 units to the left. This can be done in steps. Move 1 unit left (to 0.3), move a second unit left (to -0.7), move a third unit left (to -1.7), and finally move 0.4 unit left (to -2.1). It is the second move that ends at -0.7 that trips up students. They often think it should end at -0.3.

Activity 2

- Have students write the problem modeled on each number line. Ask what clues helped them figure out the problem. State the solution.
- **Common Error:** Students say that $1.5 + (-2.3) = -1.2$, meaning that students will subtract the lesser digit from the greater digit regardless of how the problem is written. (They subtract 1 from 2 and 0.3 from 0.5.) Take time to look at the number line model.

Previous Learning

Students should be comfortable with decimal addition and subtraction, fraction addition and subtraction, and integer addition and subtraction.

Start Thinking! and Warm Up

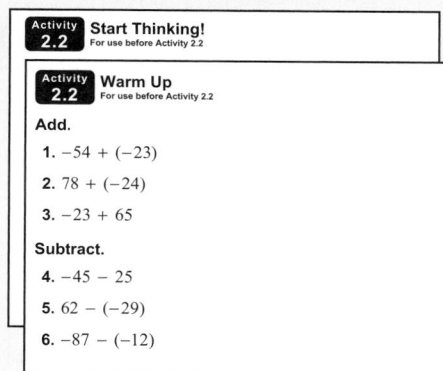

Activity 2.2 Start Thinking! For use before Activity 2.2

Activity 2.2 Warm Up For use before Activity 2.2

Add.
1. $-54 + (-23)$
2. $78 + (-24)$
3. $-23 + 65$

Subtract.
4. $-45 - 25$
5. $62 - (-29)$
6. $-87 - (-12)$

2.2 Record and Practice Journal

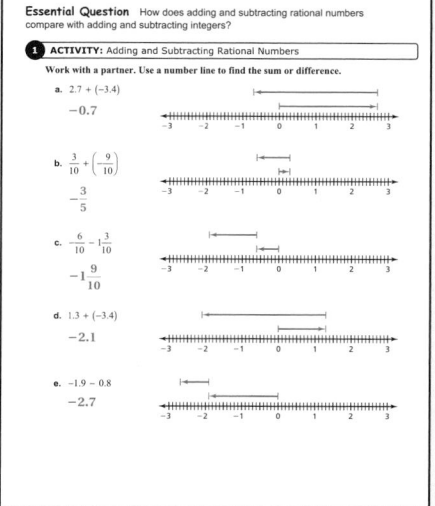

English Language Learners

Vocabulary

Use color-coded examples to help students understand the vocabulary used in this section: denominator, common denominator, least common denominator, improper fraction, and mixed number.

2.2 Record and Practice Journal

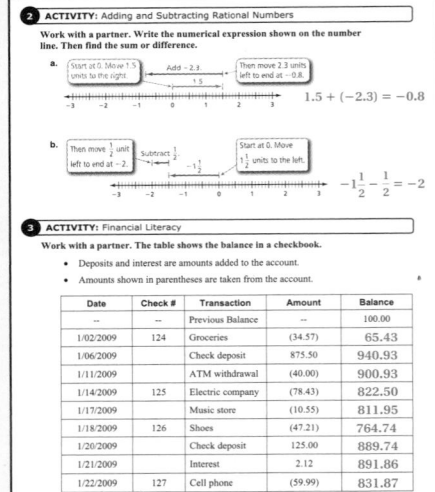

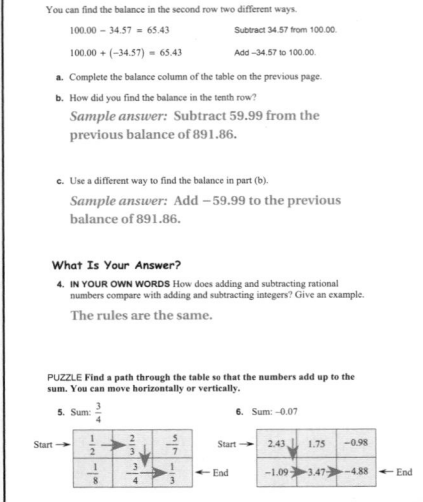

Activity 3

- **Financial Literacy:** Begin with a discussion on how a checkbook and a debit card are used. In order to know your balance at any time, it is necessary to keep a running balance. Checks written must be subtracted from the balance, and deposits are added to the balance. Interest earned is also added to the balance.

- The activity provides additional practice with decimal addition and subtraction. Be sure that students recall the need to align decimal points. Working with a partner, students can check their balances after each transaction.

- Discuss part (c). The check written for $59.99 can be thought of in two ways: balance − $59.99 or balance + (−$59.99).

- Talk about the phrase "in the red," which in accounting means a negative balance. The deposits and interest are "in the black" and are positive. They are being added to the balance and the balance grows (or increases). The checks written are "in the red" and are negative. They are being subtracted from the balance and the balance shrinks (or decreases).

What Is Your Answer?

- Students may choose to add two decimals or two fractions that are opposites for their sum to be 0. The variety of examples written will give you important information as to where your students are in their readiness to add or subtract rational numbers. It is also important for students to see one another's work.

- For Question 5, students should work with a partner. Let students wrestle with the question first, then offer a hint if needed. Five of the six fractions have a common denominator of 24, as does the desired sum of $\frac{3}{4}$. Write each of the fractions as an equivalent fraction with a denominator of 24. The fraction $\left(-\frac{5}{7}\right)$ is not needed to solve the puzzle, so it is not rewritten.

Closure

- **Writing:** Explain how $a - b$ and $a + (-b)$ are equivalent. Create an example to further illustrate what you are explaining.

Technology For the Teacher

Dynamic Classroom

The Dynamic Planning Tool
Editable Teacher's Resources at *BigIdeasMath.com*

3 ACTIVITY: Financial Literacy

Work with a partner. The table shows the balance in a checkbook.

- Black numbers are amounts added to the account.
- Red numbers are amounts taken from the account.

Date	Check #	Transaction	Amount	Balance
––	––	Previous balance	––	100.00
1/02/2009	124	Groceries	34.57	
1/06/2009		Check deposit	875.50	
1/11/2009		ATM withdrawal	40.00	
1/14/2009	125	Electric company	78.43	
1/17/2009		Music store	10.55	
1/18/2009	126	Shoes	47.21	
1/20/2009		Check deposit	125.00	
1/21/2009		Interest	2.12	
1/22/2009	127	Cell phone	59.99	

You can find the balance in the **second row** two different ways.

$$100.00 - 34.57 = 65.43 \qquad \text{Subtract 34.57 from 100.00.}$$
$$100.00 + (-34.57) = 65.43 \qquad \text{Add } -34.57 \text{ to } 100.00.$$

a. Copy the table. Then complete the balance column.

b. How did you find the balance in the **tenth row**?

c. Use a different way to find the balance in part (b).

What Is Your Answer?

4. IN YOUR OWN WORDS How does adding and subtracting rational numbers compare with adding and subtracting integers? Give an example.

PUZZLE **Find a path through the table so that the numbers add up to the sum. You can move horizontally or vertically.**

5. Sum: $\dfrac{3}{4}$

Start →

$\dfrac{1}{2}$	$\dfrac{2}{3}$	$-\dfrac{5}{7}$
$-\dfrac{1}{8}$	$-\dfrac{3}{4}$	$\dfrac{1}{3}$

←End

6. Sum: -0.07

Start →

2.43	1.75	-0.98
-1.09	3.47	-4.88

←End

Practice

Use what you learned about adding and subtracting rational numbers to complete Exercises 7–9 and 16–18 on page 60.

Key Idea

Adding and Subtracting Rational Numbers

Words To add or subtract rational numbers, use the same rules for signs as you used for integers.

Numbers $\dfrac{4}{5} - \dfrac{1}{5} = \dfrac{4-1}{5} = \dfrac{3}{5}$

$-\dfrac{1}{3} + \dfrac{1}{6} = \dfrac{-2}{6} + \dfrac{1}{6} = \dfrac{-2+1}{6} = \dfrac{-1}{6} = -\dfrac{1}{6}$

EXAMPLE **1** **Adding Rational Numbers**

Study Tip

In Example 1, notice how $-\dfrac{8}{3}$ is written as $-\dfrac{8}{3} = \dfrac{-8}{3} = \dfrac{-16}{6}$.

Find $-\dfrac{8}{3} + \dfrac{5}{6}$. **Estimate** $-3 + 1 = -2$

$-\dfrac{8}{3} + \dfrac{5}{6} = \dfrac{-16}{6} + \dfrac{5}{6}$ Rewrite using the LCD (least common denominator).

$= \dfrac{-16 + 5}{6}$ Write the sum of the numerators over the like denominator.

$= \dfrac{-11}{6}$, or $-1\dfrac{5}{6}$ Simplify.

⫶ The sum is $-1\dfrac{5}{6}$.   **Reasonable?** $-1\dfrac{5}{6} \approx -2$ ✓

EXAMPLE **2** **Adding Rational Numbers**

Find $-4.05 + 7.62$.

$-4.05 + 7.62 = 3.57$ $|7.62| > |-4.05|$. So, subtract $|-4.05|$ from $|7.62|$.

Use the sign of 7.62.

⫶ The sum is 3.57.

On Your Own

Now You're Ready
Exercises 4–12

Add.

1. $-\dfrac{7}{8} + \dfrac{1}{4}$

2. $-6\dfrac{1}{3} + \dfrac{20}{3}$

3. $2 + \left(-\dfrac{7}{2}\right)$

4. $-12.5 + 15.3$

5. $-8.15 + (-4.3)$

6. $0.65 + (-2.75)$

Laurie's Notes

Introduction

Connect

- **Yesterday:** Students explored how to add and subtract rational numbers.
- **Today:** Students will formalize the process completed yesterday, and they will add and subtract rational numbers.

Motivate

? Ask students whether the following questions are *true* or *false*.

$\dfrac{\diamond}{8} + \dfrac{\triangle}{8} = \dfrac{\diamond + \triangle}{16}$ false, unless the symbols are opposites

$\dfrac{\diamond}{8} - \dfrac{\triangle}{8} = \dfrac{\diamond - \triangle}{8}$ true

$0.4 + 0.34 = 0.38$ false

$0.4 - 0.34 = 0.14$ false

Lesson Notes

Key Idea

- **Representation:** Take time to talk about how negative fractions are represented, meaning where the negative sign is written. All of the following are equivalent: $-\dfrac{2}{3} = \dfrac{-2}{3} = \dfrac{2}{-3}$.

- **Discuss:** Emphasize the intermediate steps: $\dfrac{4-1}{5}$ and $\dfrac{-2}{6} + \dfrac{1}{6} = \dfrac{-2+1}{6}$. This will help the students a great deal.

Example 1

? "What type of fraction is $-\dfrac{8}{3}$?" improper

? "What would $-\dfrac{8}{3}$ be as a mixed number?" $-2\dfrac{2}{3}$

- Note the *Study Tip*, when working through the example.
- Be sure to tell students to check for reasonableness of their answers.

Example 2

? Write the problem and ask "Should the final answer be *positive* or *negative*? Why?"

On Your Own

- Have three pairs of students complete one of the first three fraction problems at the board, while the other students try the problems at their desks. Have students explain their work at the board.
- ? Ask questions such as "How do you know your answer is reasonable?"
- Have three different pairs of students complete one of the last three problems at the boards, while the other students try the problems at their desks.

Goal Today's lesson is adding and subtracting rational numbers.

Lesson Materials
Textbook
• glass bowl
• toy boat
• water

Start Thinking! and Warm Up

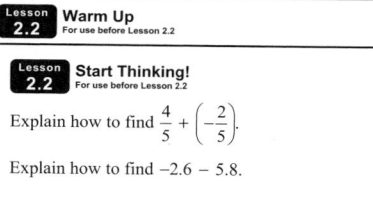

Lesson 2.2 **Warm Up** For use before Lesson 2.2

Lesson 2.2 **Start Thinking!** For use before Lesson 2.2

Explain how to find $\dfrac{4}{5} + \left(-\dfrac{2}{5}\right)$.

Explain how to find $-2.6 - 5.8$.

Extra Example 1

Find $\dfrac{4}{5} + \left(-\dfrac{3}{10}\right)$. $\dfrac{1}{2}$

Extra Example 2

Find $-3.92 + (-6.89)$. -10.81

On Your Own

1. $-\dfrac{5}{8}$
2. $\dfrac{1}{3}$
3. $-1\dfrac{1}{2}$
4. 2.8
5. -12.45
6. -2.1

Extra Example 3

Find $-8\frac{2}{3} - 6\frac{1}{6}$. $-14\frac{5}{6}$

 On Your Own

7. $\frac{2}{3}$

8. $-4\frac{1}{6}$

9. $-\frac{3}{4}$

Extra Example 4

Find $-3.75 - (-0.96)$. -2.79

On Your Own

10. no

Differentiated Instruction

Visual

Some students will incorrectly subtract decimals, especially when the second number has more decimal places than the first. Encourage students to use zeros so that the two numbers have the same number of decimal places. For example, $2.35 - 1.457$ should be written as

$$\begin{array}{r} 2.350 \\ -\ 1.457 \\ \hline 0.893 \end{array}$$ instead of $$\begin{array}{r} 2.35 \\ -\ 1.457 \\ \hline 0.907 \end{array}$$ ✗

T-59

Laurie's Notes

Example 3

❓ **"How do you subtract integers?"** The statement "add the opposite" should be familiar. Once the problem is written as an addition problem, students should recall the rules for integer addition.
- Work through the example by first estimating the answer.
- Be sure the students understand the rule for subtracting rational numbers is the same as the rule for subtracting integers. The challenge will be working with fractions.

Example 4

- To help visualize this problem, fill a glass bowl with water. Float a toy boat in the water so that the distance above the water level is visible.
❓ **"If you know the height of the boat above the water and the depth of the boat below the water, how can you find the total height of the boat?"** Students will probably say to add the two together. It is also acceptable to subtract the lowest point relative to sea level (a negative number) from the highest point (a positive number).

On Your Own

- **Neighbor Check:** Have students work independently and then have their neighbor check their work. Have students discuss any discrepancies.

Words of Wisdom

- Students often think that 2.1 feet is equivalent to 2 feet, 1 inch. Have students explore which is greater: 2.1 feet or 2 feet, 1 inch.

  $0.1 \text{ ft} \times \dfrac{12 \text{ in.}}{1 \text{ ft}} = 1.2 \text{ in.}$, so 2.1 feet is greater than 2 feet, 1 inch.

Closure

- Ask students to explain how addition and subtraction of rational numbers is similar to addition and subtraction of integers. *Sample answer:* The sign of the sum or difference is the sign of the number with the greater absolute value.

EXAMPLE **3** **Subtracting Rational Numbers**

Find $-4\frac{1}{7} - \left(-\frac{6}{7}\right)$.　　　　　Estimate $-4 - (-1) = -3$

$$-4\frac{1}{7} - \left(-\frac{6}{7}\right) = -4\frac{1}{7} + \frac{6}{7}$$　　Add the opposite of $-\frac{6}{7}$.

$$= -\frac{29}{7} + \frac{6}{7}$$　　Write the mixed number as an improper fraction.

$$= \frac{-23}{7}, \text{ or } -3\frac{2}{7}$$　　Simplify.

⋮ The difference is $-3\frac{2}{7}$.　　Reasonable? $-3\frac{2}{7} \approx -3$ ✓

On Your Own

Subtract.

7. $\frac{1}{3} - \left(-\frac{1}{3}\right)$　　　　8. $-3\frac{1}{3} - \frac{5}{6}$　　　　9. $4\frac{1}{2} - 5\frac{1}{4}$

EXAMPLE **4** **Real-Life Application**

Clearance: 11 ft 8 in.

In the water, the bottom of a boat is 2.1 feet below the surface and the top of the boat is 8.7 feet above it. Towed on a trailer, the bottom of the boat is 1.3 feet above the ground. Can the boat and trailer pass under the bridge?

Step 1: Find the height h of the boat.

$$h = 8.7 - (-2.1)$$　　Subtract the lowest point from the highest point.

$$= 8.7 + 2.1$$　　Add the opposite of -2.1.

$$= 10.8$$　　Add.

Step 2: Find the height t of the boat and trailer.

$$t = 10.8 + 1.3$$　　Add the trailer height to the boat height.

$$= 12.1$$　　Add.

⋮ Because 12.1 feet is greater than 11 feet 8 inches, the boat and trailer cannot pass under the bridge.

On Your Own

Now You're Ready
Exercises 13–21

10. **WHAT IF?** In Example 4, the clearance is 12 feet 1 inch. Can the boat and trailer pass under the bridge?

 ## Vocabulary and Concept Check

1. **WRITING** Explain how to find the sum $-8.46 + 5.31$.

2. **OPEN-ENDED** Write an addition expression using fractions that equals $-\frac{1}{2}$.

3. **DIFFERENT WORDS, SAME QUESTION** Which is different? Find "both" answers.

Add -4.8 and 3.9.	What is 3.9 less than -4.8?
What is -4.8 increased by 3.9?	Find the sum of -4.8 and 3.9.

 ## Practice and Problem Solving

Add. Write fractions in simplest form.

1 2   4. $\frac{11}{12} + \left(-\frac{7}{12}\right)$

5. $-\frac{9}{14} + \frac{2}{7}$

6. $\frac{15}{4} + \left(-4\frac{1}{3}\right)$

7. $2\frac{5}{6} + \left(-\frac{8}{15}\right)$

8. $4 + \left(-1\frac{2}{3}\right)$

9. $-4.2 + 3.3$

10. $-3.1 + (-0.35)$

11. $12.48 + (-10.636)$

12. $20.25 + (-15.711)$

Subtract. Write fractions in simplest form.

3 4   13. $\frac{5}{8} - \left(-\frac{7}{8}\right)$

14. $\frac{1}{4} - \frac{11}{16}$

15. $-\frac{1}{2} - \left(-\frac{5}{9}\right)$

16. $-5 - \frac{5}{3}$

17. $-8\frac{3}{8} - 10\frac{1}{6}$

18. $-1 - 2.5$

19. $5.5 - 8.1$

20. $-7.34 - (-5.51)$

21. $6.673 - (-8.29)$

22. **ERROR ANALYSIS** Describe and correct the error in finding the difference.

$$\text{✗} \quad \frac{3}{4} - \frac{9}{2} = \frac{3-9}{4-2} = \frac{-6}{2} = -3$$

23. **SPORTS DRINK** Your sports drink bottle is $\frac{5}{6}$ full. After practice the bottle is $\frac{3}{8}$ full. Write the difference of the amounts after practice and before practice.

24. **BANKING** Your bank account balance is $-\$20.85$. You deposit $\$15.50$. What is your new balance?

Evaluate.

25. $2\frac{1}{6} - \left(-\frac{8}{3}\right) + \left(-4\frac{7}{9}\right)$

26. $6.3 + (-7.8) - (-2.41)$

27. $-\frac{12}{5} + \left|-\frac{13}{6}\right| + \left(-3\frac{2}{3}\right)$

Assignment Guide and Homework Check

Level	Day 1 Activity Assignment	Day 2 Lesson Assignment	Homework Check
Basic	7–9, 16–18, 39–43	1–6, 11–15, 20–25	6, 15, 21, 24, 25
Average	7–9, 16–18, 39–43	1–6, 11–15, 20–22, 25, 29, 30	6, 15, 21, 25, 29
Advanced	7–9, 16–18, 39–43	1–3, 22, 25–28, 30–38	25, 28, 32, 36

Common Errors

- **Exercises 5–8, 13–17** Students may try to identify the sign of the answer before finding a common denominator. Remind them that they need to find the common denominator first.
- **Exercises 9–12, 18–21** Students may forget to line up the decimal points when they add or subtract decimals. Remind them that the decimal points must be lined up before adding or subtracting. Students may want to use half-inch graph paper to help keep the numbers and decimal points aligned.
- **Exercise 15** Students may not know where to put the negative sign in the fraction. Remind them that the negative can go in the numerator or the denominator (although the numerator is usually best when doing calculations), but not both.

2.2 Record and Practice Journal

Add or subtract. Write fractions in simplest form.

1. $\frac{4}{5} + \frac{3}{20}$
$-\frac{13}{20}$

2. $-8 + \left(-\frac{6}{7}\right)$
$-8\frac{6}{7}$

3. $1\frac{2}{15} + \left(-3\frac{1}{2}\right)$
$-2\frac{11}{30}$

4. $\frac{1}{6} - \frac{5}{12}$
$-\frac{7}{12}$

5. $\frac{9}{10} - 3$
$-2\frac{1}{10}$

6. $5\frac{3}{4} - \left(-4\frac{5}{6}\right)$
$10\frac{7}{12}$

7. $0.46 + (-0.642)$
-0.182

8. $0.13 - 5.7$
-5.57

9. $-2.57 - (-3.48)$
0.91

10. Tubs of ice cream are delivered to a store at a temperature of 36.7°F. The ice cream is stored in a -40°F freezer. When a tub is brought out of the freezer, its temperature is 22.2°F. Write the difference between the temperatures of the ice cream after the ice cream is in the freezer and before it is in the freezer.
$-14.5°F$

11. Before a race, you start $4\frac{5}{8}$ feet behind your friend. At the halfway point, you are $3\frac{2}{3}$ feet ahead of your friend. What is the change in distance between you and your friend from the beginning of the race?
$8\frac{7}{24}$ feet

Technology For the Teacher
Answer Presentation Tool
QuizShow

Vocabulary and Concept Check

1. Because $|-8.46| > |5.31|$, subtract $|5.31|$ from $|-8.46|$ and the sign is negative.

2. *Sample answer:* $-\frac{1}{4} + \left(-\frac{1}{4}\right)$

3. What is 3.9 less than −4.8?; −8.7; −0.9

Practice and Problem Solving

4. $\frac{1}{3}$

5. $-\frac{5}{14}$

6. $-\frac{7}{12}$

7. $2\frac{3}{10}$

8. $2\frac{1}{3}$

9. -0.9

10. -3.45

11. 1.844

12. 4.539

13. $1\frac{1}{2}$

14. $-\frac{7}{16}$

15. $\frac{1}{18}$

16. $-6\frac{2}{3}$

17. $-18\frac{13}{24}$

18. -3.5

19. -2.6

20. -1.83

21. 14.963

22. They did not use the least common denominator.

$$\frac{3}{4} - \frac{9}{2} = \frac{3}{4} - \frac{18}{4}$$
$$= \frac{3 - 18}{4}$$
$$= \frac{-15}{4}$$
$$= -3\frac{3}{4}$$

23. $\frac{3}{8} - \frac{5}{6} = -\frac{11}{24}$

24. $-\$5.35$

25. $\frac{1}{18}$

26. 0.91

27. $-3\frac{9}{10}$

28. The difference is an integer when 1.) the decimals have the same sign and the digits to the right of the decimal point are the same, or 2.) the decimals have different signs and the sum of the decimal parts of the numbers add up to 1.

29. No, the cook needs $\frac{1}{12}$ cup more.

30. $-1\frac{7}{8}$ miles

31–33. See *Taking Math Deeper.*

34. $-2x$

35. $-\dfrac{n}{4}$

36. $-\dfrac{13a}{3}$

37. $-\dfrac{b}{24}$

38. $5.24 - (8.85) = -3.61$

Fair Game Review

39. 35.88 40. 3

41. $8\frac{2}{3}$ 42. $2\frac{4}{5}$

43. C

Taking Math Deeper

Exercises 31–33

This problem gives students a chance to find the sum of a long list of signed numbers. To do this efficiently, students can use the Commutative and Associative Properties of Addition.

 Read and interpret the bar graph.

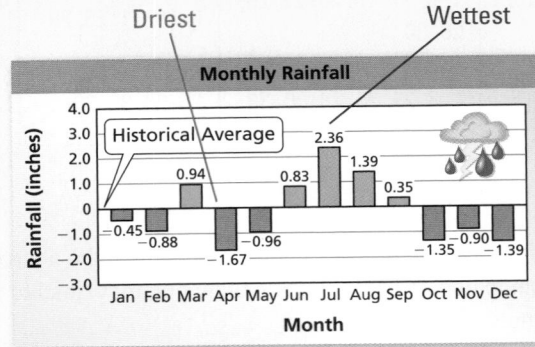

31. Difference $= 2.36 - (-1.67)$
    $= 2.36 + 1.67$
    $= 4.03$ in.

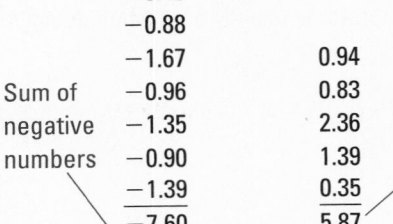 Find the sum of the differences.

Add in 2 groups.

	Sum of negative numbers		Sum of positive numbers
	-0.45		
	-0.88		
	-1.67	0.94	
	-0.96	0.83	
	-1.35	2.36	
	-0.90	1.39	
	-1.39	0.35	
	-7.60	5.87	

32. Total sum: $-7.60 + 5.87 = -1.73$ in.

3 Interpret.

33. The total rainfall for the year was 1.73 inches *less* than the historical average.

Mini-Assessment

Add or subtract. Write fractions in simplest form.

1. $2\frac{4}{5} + \left(-\frac{12}{15}\right)$ 2

2. $-\frac{3}{4} - \left(-\frac{8}{9}\right)$ $\frac{5}{36}$

3. $15.48 + (-17.23)$ -1.75

4. $-3.89 - (-5.34)$ 1.45

5. Your bank account balance is $-\$15.50$. You deposit $\$75$. What is your new balance? $\$59.50$

Reteaching and Enrichment Strategies

If students need help. . .	If students got it. . .
Resources by Chapter • Practice A and Practice B • Puzzle Time Record and Practice Journal Practice Differentiating the Lesson Lesson Tutorials Skills Review Handbook	Resources by Chapter • Enrichment and Extension Start the next section

28. REASONING When is the difference of two decimals an integer? Explain.

29. RECIPE A cook has $2\frac{2}{3}$ cups of flour. A recipe calls for $2\frac{3}{4}$ cups of flour. Does the cook have enough flour? If not, how much more flour is needed?

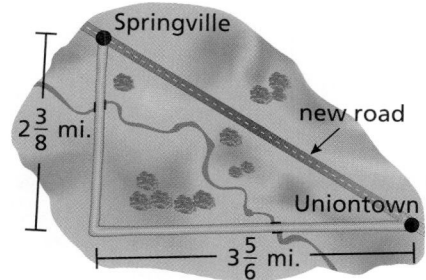
Springville
new road
$2\frac{3}{8}$ mi.
Uniontown
$3\frac{5}{6}$ mi.

30. ROADWAY A new road that connects Uniontown to Springville is $4\frac{1}{3}$ miles long. What is the change in distance when using the new road instead of the dirt roads?

RAINFALL In Exercises 31–33, the bar graph shows the differences in a city's rainfall from the historical average.

31. What is the difference in rainfall between the wettest and driest months?

32. Find the sum of the differences for the year.

33. What does the sum in Exercise 32 tell you about the rainfall for the year?

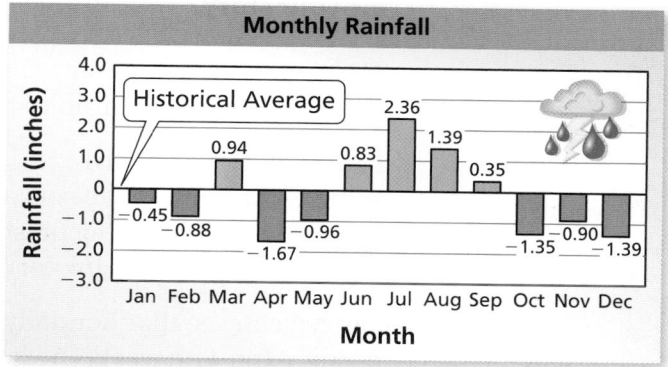

ALGEBRA Add or subtract. Write the answer in simplest form.

34. $-4x + 8x - 6x$

35. $-\frac{3n}{8} + \frac{2n}{8} - \frac{n}{8}$

36. $-4a - \frac{a}{3}$

37. $\frac{5b}{8} + \left(-\frac{2b}{3}\right)$

38. ✏️ **Puzzle** Fill in the blanks to make the solution correct.

$$5.\boxed{}4 - \left(\boxed{}.8\boxed{}\right) = -3.61$$

 **Fair Game Review** *What you learned in previous grades & lessons*

Evaluate. *(Skills Review Handbook)*

39. 5.2×6.9

40. $7.2 \div 2.4$

41. $2\frac{2}{3} \times 3\frac{1}{4}$

42. $9\frac{4}{5} \div 3\frac{1}{2}$

43. MULTIPLE CHOICE A sports store has 116 soccer balls. Over 6 months, it sells eight soccer balls per month. How many soccer balls are in inventory at the end of the 6 months? *(Section 1.3 and Section 1.4)*

  Ⓐ -48 Ⓑ 48 Ⓒ 68 Ⓓ 108

Multiplying and Dividing Rational Numbers

Essential Question How can you use operations with rational numbers in a story?

1 EXAMPLE: Writing a Story

Write a story that uses addition, subtraction multiplication, or division of rational numbers. Draw pictures for your story.

There are many possible stories. Here is an example.

24 Lemons	-$11.75
5 cups sugar	-$1.50
30 plastic glasses	-$1.50
18 sales ($0.50 each)	$9.00
PROFIT	-$5.75

All You Can Drink For 50¢!

Lauryn decides to earn some extra money. She sets up a lemonade stand. To get customers, she uses big plastic glasses and makes a sign saying "All you can drink for 50¢!"

Lauryn can see that her daily profit is negative. But, she decides to keep trying. After one week, she has the same profit each day.

Sunday	Monday	Tuesday	Wednesday	Thursday	Friday	Saturday
-$5.75	-$5.75	-$5.75	-$5.75	-$5.75	-$5.75	-$5.75

Lauryn is frustrated. Her profit for the first week is

$$7(-5.75) = (-5.75) + (-5.75) + (-5.75) + (-5.75) + (-5.75) + (-5.75) + (-5.75)$$
$$= -40.25.$$

She realizes that she has too many customers who are drinking a second and even a third glass of lemonade. So, she decides to try a new strategy. Soon, she has a customer. He buys a glass of lemonade and drinks it.

He hands the empty glass to Lauryn and says "*That was great. I'll have another glass.*" Today, Lauryn says "*That will be 50¢ more, please.*" The man says "*But, you only gave me one glass and the sign says 'All you can drink for 50¢!*'" Lauryn replies, "*One glass IS all you can drink for 50¢.*"

With her new sales strategy, Lauryn starts making a profit of $8.25 per day. Her profit for the second week is

$$7(8.25) = (8.25) + (8.25) + (8.25) + (8.25) + (8.25) + (8.25) + (8.25) = 57.75.$$

Her profit for the two weeks is $-40.25 + 57.75 = \$17.50$. So, Lauryn has made some money. She decides that she is on the right track.

Laurie's Notes

Introduction

For the Teacher

- **Goal:** The goal of this activity is for students to write a story about operations with rational numbers.
- These activities provide a wonderful opportunity for students to be creative, develop their writing skills, and communicate about mathematics. Many students find their first attempt challenging. You may want to give students an opportunity to revise their initial stories.

Motivate

- ? Ask students if they have done any work that required an initial investment of money (such as needing a lawn mower and gas to mow a lawn or needing to purchase supplies to make bracelets for a craft sale). This will help introduce the context of the story in Example 1.
- To make a profit, the initial investment of money must be earned back.

Activity Notes

Example 1

- Ask volunteers to read the story in Example 1. While it is not necessary to read the values in the table, the reader should pause so that students are able to look at the data.
- ? "How does the illustration in Example 1 contribute to the story?" The illustration quickly conveys information, it provides an immediate context, and it draws the reader into the story.
- Discuss each operation and the context for each in the story. See the examples below.
  - addition—the sum of the amounts of money made each week
  - subtraction—the difference of the profits made each week
  - multiplication—the profit of 18 sales at $0.50 each
  - division—the average profit for each sale ($-\$5.75 \div 18$).
- **Extension:**
  - ? "What is the average profit per day for the first two weeks?" $1.25
  - ? "If the third week produces the same profit as the second week, what is the profit for all three weeks?" $75.25

Previous Learning

Students should know integer and decimal multiplication and division.

Start Thinking! and Warm Up

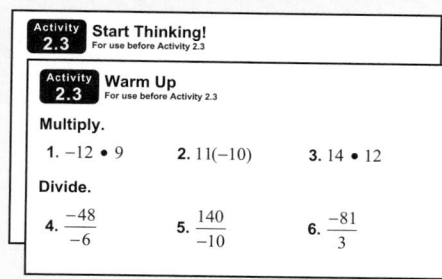

Activity 2.3 Start Thinking!
For use before Activity 2.3

Activity 2.3 Warm Up
For use before Activity 2.3

Multiply.

1. $-12 \cdot 9$ 2. $11(-10)$ 3. $14 \cdot 12$

Divide.

4. $\dfrac{-48}{-6}$ 5. $\dfrac{140}{-10}$ 6. $\dfrac{-81}{3}$

2.3 Record and Practice Journal

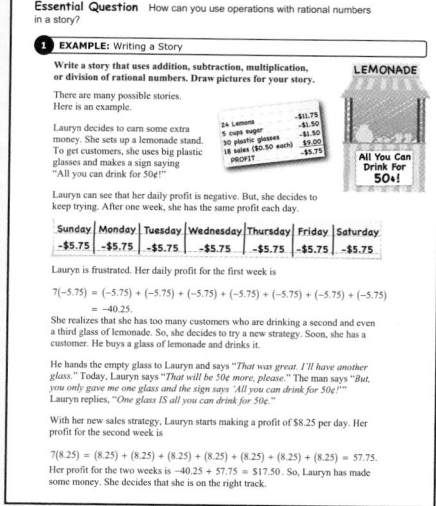

Essential Question How can you use operations with rational numbers in a story?

1 EXAMPLE: Writing a Story

Write a story that uses addition, subtraction, multiplication, or division of rational numbers. Draw pictures for your story.

There are many possible stories. Here is an example.

Lauryn decides to earn some extra money. She sets up a lemonade stand. To get customers, she uses big plastic glasses and makes a sign saying "All you can drink for 50¢!"

Lauryn can see that her daily profit is negative. But, she decides to keep trying. After one week, she has the same profit each day.

Sunday	Monday	Tuesday	Wednesday	Thursday	Friday	Saturday
-$5.75	-$5.75	-$5.75	-$5.75	-$5.75	-$5.75	-$5.75

Lauryn is frustrated. Her daily profit for the first week is

$7(-5.75) = (-5.75) + (-5.75) + (-5.75) + (-5.75) + (-5.75) + (-5.75) + (-5.75)$
$= -40.25$.

She realizes that she has too many customers who are drinking a second and even a third glass of lemonade. So, she decides to try a new strategy. Soon, she has a customer. He buys a glass of lemonade and drinks it.

He hands the empty glass to Lauryn and says "That was great. I'll have another glass." Today, Lauryn says "That will be 50¢ more, please." The man says "But, you only gave me one glass and the sign says 'All you can drink for 50¢!'" Lauryn replies, "One glass IS all you can drink for 50¢."

With her new sales strategy, Lauryn starts making a profit of $8.25 per day. Her profit for the second week is

$7(8.25) = (8.25) + (8.25) + (8.25) + (8.25) + (8.25) + (8.25) + (8.25) = 57.75$.

Her profit for the two weeks is $-40.25 + 57.75 = \$17.50$. So, Lauryn has made some money. She decides that she is on the right track.

English Language Learners

Simplified Language

Writing stories poses a challenge for English learners. You may want to allow students who struggle with language to outline a story or to create a story using pictures. It may also be helpful for students to listen to a translation of the story in the online student textbook.

2.3 Record and Practice Journal

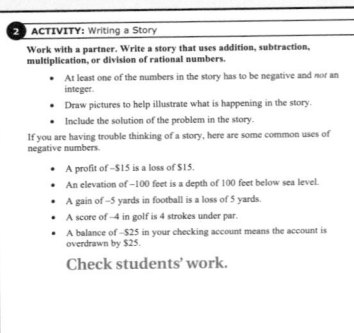

What Is Your Answer?

3. **IN YOUR OWN WORDS** How can you use operations with rational numbers in a story? You already used rational numbers in your story. Describe another use of a negative rational number in a story.

Sample answer: Operations with rational numbers can be used in a story about money, distances, or weights.

PUZZLE Read the cartoon. Fill in the blanks using 4s or 8s to make the equation true.

Words of Wisdom

- Activities 2 and 3 require students to connect prior skills—fluency with fraction, decimal, and integer operations. These activities also involve the literacy skills of reading and writing.

Activity 2

- Read through the directions together as a class.
- Have students work in pairs so that brainstorming can occur. Both students should be actively engaged, with one doing the writing while the other draws a diagram to illustrate the problem.
- The five examples showing where negative numbers are commonly used should help students get started.
- Provide at least 20–25 minutes for the brainstorming and writing process. Students' stories should include computations and a final solution.
- **Discuss:** As time allows, have pairs of students share their stories. To help students see when each operation is used, make a table on the board with four columns, one for each operation. Record the context used for each operation.
- **Interdisciplinary:** Some of your language arts colleagues may want to review the students' stories. Speak with them about different possibilities.

What Is Your Answer?

- **Class Activity:** Answer Question 3 together as a class.
- **Puzzle:** This provides a review of adding, subtracting, multiplying, and dividing fractions and decimals. This review will help prepare students for tomorrow's lesson on multiplying and dividing rational numbers.

Closure

- **Writing:** Have students write brief scenarios for all four operations (addition, subtraction, multiplication, and division). Note: The scenarios *do not* need to be connected to one another. Instead of creating a whole story, students just need to write four sentences.

Technology For the Teacher

Dynamic Classroom

The Dynamic Planning Tool
Editable Teacher's Resources at *BigIdeasMath.com*

2 ACTIVITY: Writing a Story

Work with a partner. Write a story that uses addition, subtraction, multiplication, or division of rational numbers.

- At least one of the numbers in the story has to be negative and *not* an integer.
- Draw pictures to help illustrate what is happening in the story.
- Include the solution of the problem in the story.

If you are having trouble thinking of a story, here are some common uses of negative numbers.

- A profit of $-\$15$ is a loss of $15.
- An elevation of -100 feet is a depth of 100 feet below sea level.
- A gain of -5 yards in football is a loss of 5 yards.
- A score of -4 in golf is 4 strokes under par.
- A balance of $-\$25$ in your checking account means the account is overdrawn by $25.

What Is Your Answer?

3. **IN YOUR OWN WORDS** How can you use operations with rational numbers in a story? You already used rational numbers in your story. Describe another use of a negative rational number in a story.

PUZZLE Read the cartoon. Fill in the blanks using 4s or 8s to make the equation true.

"Dear Mom, I'm in a hurry. To save time I won't be typing any 4's or 8's."

4. $\left(-\dfrac{1}{\boxed{}}\right) + \left(-\dfrac{1}{\boxed{}}\right) = -\dfrac{1}{\boxed{}}$

5. $\left(-\dfrac{1}{\boxed{}}\right) \times \left(-\dfrac{1}{\boxed{}}\right) = \dfrac{1}{6\boxed{}}$

6. $1.\boxed{} \times \left(-0.\boxed{}\right) = -1.\boxed{}\boxed{}$

7. $\left(-\dfrac{3}{\boxed{}}\right) \div \left(\dfrac{3}{\boxed{}}\right) = -\dfrac{1}{2}$

8. $-4.\boxed{} \div 2 = -2.\boxed{}$

Check It Out
Lesson Tutorials
BigIdeasMath ✓com

 **Key Idea**

Multiplying and Dividing Rational Numbers

Words To multiply or divide rational numbers, use the same rules for signs as you used for integers.

Remember

The *reciprocal* of $\frac{a}{b}$ is $\frac{b}{a}$.

Numbers $-\dfrac{2}{7} \cdot \dfrac{1}{3} = \dfrac{-2 \cdot 1}{7 \cdot 3} = \dfrac{-2}{21} = -\dfrac{2}{21}$

$-\dfrac{1}{2} \div \dfrac{4}{9} = \dfrac{-1}{2} \cdot \dfrac{9}{4} = \dfrac{-1 \cdot 9}{2 \cdot 4} = \dfrac{-9}{8} = -\dfrac{9}{8}$

EXAMPLE **1** **Dividing Rational Numbers**

Find $-5\dfrac{1}{5} \div 2\dfrac{1}{3}.$ **Estimate** $-5 \div 2 = -2\dfrac{1}{2}$

$-5\dfrac{1}{5} \div 2\dfrac{1}{3} = -\dfrac{26}{5} \div \dfrac{7}{3}$ Write mixed numbers as improper fractions.

$= \dfrac{-26}{5} \cdot \dfrac{3}{7}$ Multiply by the reciprocal of $\dfrac{7}{3}$.

$= \dfrac{-26 \cdot 3}{5 \cdot 7}$ Multiply the numerators and the denominators.

$= \dfrac{-78}{35}$, or $-2\dfrac{8}{35}$ Simplify.

∴ The quotient is $-2\dfrac{8}{35}$.   **Reasonable?** $-2\dfrac{8}{35} \approx -2\dfrac{1}{2}$ ✓

EXAMPLE **2** **Multiplying Rational Numbers**

Find $-2.5 \cdot 3.6.$

$$
\begin{array}{r}
-2.5 \\
\times\ 3.6 \\
\hline
1\,5\,0 \\
7\,5\,0 \\
\hline
-9.0\,0 \\
\end{array}
$$

The decimals have different signs.

The product is negative.

∴ The product is -9.

Laurie's Notes

Introduction

Connect
- **Yesterday:** Students wrote story problems involving operations with rational numbers.
- **Today:** Students will learn the rules for multiplying and dividing rational numbers.

Discuss
- Before beginning the formal lesson, it would be helpful to review rules for multiplying and dividing integers.
  - same signs → product/quotient is positive
  - different signs → product/quotient is negative

Lesson Notes

Key Idea
- Write the definition for multiplication and division of fractions. Note that the sign of the fraction is written with the numerator when the computation is performed.

Example 1
- **Discuss:** Before starting the first example, take time to discuss estimating products and quotients. This will help students check their answers.
- Work through the problem. Do not skip the initial estimate.
- Remind students that when multiplying or dividing fractions, mixed numbers must be written as improper fractions.
- **Discuss:** There are several important skills involved in this example. Identify each skill with students so that vocabulary is reviewed and each process is made clear.

Example 2
- This example involves multiplying decimals *and* signed numbers.
- Write the example and ask how they might estimate an answer.
- **Extension:** If time permits, repeat this example by converting the decimals to fractions:

$$-2\frac{5}{10} \times 3\frac{6}{10} = -2\frac{1}{2} \times 3\frac{3}{5}$$

$$= \frac{-5}{2} \times \frac{18}{5}$$

$$= \frac{-90}{10}$$

$$= -9$$

Start Thinking! and Warm Up

Lesson 2.3	Warm Up
	For use before Lesson 2.3

Lesson 2.3 **Start Thinking!** For use before Lesson 2.3

A company's profits for a week are as follows: Monday: +$32.65, Tuesday: −$75.32, Wednesday: −$125.75, Thursday: +$100.89, and Friday: +$65.30. Does the company show a gain or loss at the end of the week?

Extra Example 1
Find $3\frac{1}{4} \div \left(-1\frac{1}{8}\right)$. $-2\frac{8}{9}$

Extra Example 2
Find $-4.8(-5.2)$. 24.96

Extra Example 3

Find $\frac{5}{3}\left(-3\frac{3}{5}\right)$. -6

 On Your Own

1. $2\frac{2}{5}$ 2. $-\frac{1}{8}$

3. $-\frac{1}{8}$ 4. -9.18

5. 3.78 6. 1.69

Extra Example 4

Find the mean of -25.63, 37.15, 18.92, and -44.28. -3.46

On Your Own

7. $\$58.65$

Differentiated Instruction

Inclusion

Remind students that one difference between multiplying decimals and dividing decimals is the placement of the decimal point. In multiplication, the decimal point is placed after the decimals are multiplied. In division, the placement of the decimal point is determined before dividing.

Laurie's Notes

Example 3

- This example assesses students' conceptual understanding of fraction multiplication and their ability to use the problem-solving strategy of *Guess, Check, and Revise.*
- A student who first estimates that $-\frac{5}{3}$ is close to -2 will guess that the second factor is near -3. Students may not estimate at all and simply know that the second factor must be negative to give the positive product.

On Your Own

- Have three pairs of students choose one question from 1–3 to complete at the board. Have the other students try these problems at their desks. Have the pairs of students explain their work at the board.

? "How can you check that your answer is reasonable?" use estimation

? **Question 3:** "What does the 3 mean in this problem? What word would you use to describe the 3?" The three means that you multiply $-\frac{1}{2}$ by itself three times. It is an exponent.

- Have three different pairs of students choose one question from 4–6 to complete at the board. Have the other students try these problems at their desks. Have the pairs of students explain their work at the board. Students may need to be reminded that multiplication can be represented using parentheses around one or both of the factors.

Example 4

- **Financial Literacy:** This example uses stock prices to review decimal addition, subtraction, and division. Remind students that the word *mean* is the same as the arithmetic average.
- Explain the stock context and what each column of the table means.

On Your Own

- Predict whether the mean change will be *positive* or *negative*. Explain your reasoning. positive; Students should recognize that the mean of the four stocks is the sum of the change in the first three stocks ($-\$333.63$) and the change in Stock D ($\$568.23$), divided by four. This will be positive because $-\$333.63 + \$568.23$ is positive.
- **Common Error:** When multiplying two mixed numbers, students will often add the product of the whole numbers and the product of the fractions.

Closure

- **Exit Ticket:**

  $-2\frac{1}{3} \times 3\frac{2}{3}$ $-8\frac{5}{9}$ $(-0.5)(-4.2) \div 0.03$ 70

EXAMPLE 3 **Standardized Test Practice**

Which number, when multiplied by $-\dfrac{5}{3}$, gives a product between 5 and 6?

    Ⓐ -6 **Ⓑ** $-3\dfrac{1}{4}$ **Ⓒ** $-\dfrac{1}{4}$ **Ⓓ** 3

Use the guess, check, and revise method.

Guess 1: Because the product is positive and the known factor is negative, choose a number that is negative. Try Choice **Ⓒ**.

$$-\dfrac{1}{4}\left(-\dfrac{5}{3}\right) = \dfrac{-1 \cdot (-5)}{4 \cdot 3} = \dfrac{5}{12}$$

Guess 2: The result of Choice **Ⓒ** is not between 5 and 6. So, choose another number that is negative. Try Choice **Ⓑ**.

$$-3\dfrac{1}{4}\left(-\dfrac{5}{3}\right) = -\dfrac{13}{4}\left(-\dfrac{5}{3}\right) = \dfrac{-13 \cdot (-5)}{4 \cdot 3} = \dfrac{65}{12} = 5\dfrac{5}{12}$$

∴ $5\dfrac{5}{12}$ is between 5 and 6. So, the correct answer is **Ⓑ**.

On Your Own

Now You're Ready
Exercises 10–33

Multiply or divide.

1. $-\dfrac{6}{5} \div \left(-\dfrac{1}{2}\right)$ **2.** $\dfrac{1}{3} \div \left(-2\dfrac{2}{3}\right)$ **3.** $\left(-\dfrac{1}{2}\right)^{3}$

4. $1.8(-5.1)$ **5.** $-6.3(-0.6)$ **6.** $(-1.3)^{2}$

EXAMPLE 4 **Real-Life Application**

Account Positions			
Stock	**Original Value**	**Current Value**	**Change**
A	600.54	420.15	−180.39
B	391.10	518.38	127.28
C	380.22	99.70	−280.52

An investor owns stocks A, B, and C. What is the mean change in value of the stocks?

$$\text{mean} = \dfrac{-180.39 + 127.28 + (-280.52)}{3} = \dfrac{-333.63}{3} = -111.21$$

∴ The mean change in value of the stocks is −$111.21.

On Your Own

7. In Example 4, the change in value of stock D is $568.23. What is the mean change in value of the four stocks?

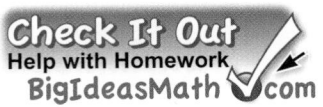
 Vocabulary and Concept Check

1. **WRITING** How is multiplying and dividing rational numbers similar to multiplying and dividing integers?

Find the reciprocal.

2. $-\dfrac{2}{5}$

3. -3

4. $\dfrac{16}{9}$

5. $-2\dfrac{1}{3}$

Tell whether the expression is *positive* or *negative* without evaluating.

6. $-\dfrac{3}{10} \times \left(-\dfrac{8}{15}\right)$

7. $1\dfrac{1}{2} \div \left(-\dfrac{1}{4}\right)$

8. -6.2×8.18

9. $\dfrac{-8.16}{-2.72}$

 Practice and Problem Solving

Divide. Write fractions in simplest form.

① 10. $-\dfrac{7}{10} \div \dfrac{2}{5}$

11. $\dfrac{1}{4} \div \left(-\dfrac{3}{8}\right)$

12. $-\dfrac{8}{9} \div \left(-\dfrac{8}{9}\right)$

13. $-\dfrac{1}{5} \div 20$

14. $-2\dfrac{4}{5} \div (-7)$

15. $-10\dfrac{2}{7} \div \left(-4\dfrac{4}{11}\right)$

16. $-9 \div 7.2$

17. $8 \div 2.2$

18. $-3.45 \div (-15)$

19. $-0.18 \div 0.03$

20. $8.722 \div (-3.56)$

21. $12.42 \div (-4.8)$

Multiply. Write fractions in simplest form.

② ③ 22. $-\dfrac{2}{3} \times \dfrac{2}{9}$

23. $-\dfrac{1}{4} \times \left(-\dfrac{4}{3}\right)$

24. $\dfrac{5}{6}\left(-\dfrac{8}{15}\right)$

25. $-2\left(-1\dfrac{1}{4}\right)$

26. $-3\dfrac{1}{3} \cdot \left(-2\dfrac{7}{10}\right)$

27. $\left(-1\dfrac{2}{3}\right)^3$

28. $0.4 \times (-0.03)$

29. $-0.05 \times (-0.5)$

30. $-8(0.09)$

31. $-9.3 \cdot (-5.1)$

32. $-95.2 \cdot (-0.12)$

33. $(-0.4)^3$

ERROR ANALYSIS Describe and correct the error.

34.

✗ $-2.2 \times 3.7 = 8.14$

35.

✗ $-\dfrac{1}{4} \div \dfrac{3}{2} = -\dfrac{4}{1} \times \dfrac{3}{2} = -\dfrac{12}{2} = -6$

36. **HOUR HAND** The hour hand of a clock moves $-30°$ every hour. How many degrees does it move in $2\dfrac{1}{5}$ hours?

37. **SUNFLOWER SEEDS** How many 0.75-pound packages can be made with 6 pounds of sunflower seeds?

Assignment Guide and Homework Check

Level	Day 1 Activity Assignment	Day 2 Lesson Assignment	Homework Check
Basic	50–54	1–9, 11–33 odd, 35–37	15, 19, 25, 31, 36
Average	50–54	1–9, 15–33 odd, 35, 36, 38, 45, 46	15, 19, 25, 31, 36
Advanced	50–54	1–9, 35, 36, 38–49	36, 40, 44, 48

Common Errors

- **Exercises 10–15** Students may use the reciprocal of the first fraction instead of the second, or they might forget to write a mixed number as an improper fraction before finding the reciprocal. Review multiplying and dividing fractions, and the definition of reciprocal.
- **Exercises 16–21** Students may mix up the dividend and divisor. Remind them that the first number is the dividend and the second is the divisor.
- **Exercises 16–21** Students may forget to shift the decimal point when dividing or they might move the decimal point the wrong number of places. Remind students to use estimation to check their answer and the placement of the decimal.
- **Exercises 38–43** Students may forget to follow the order of operations. Tell them to write parentheses around the multiplication or division parts so that they remember to evaluate them first.

2.3 Record and Practice Journal

Multiply or divide. Write fractions in simplest form.

1. $\frac{8}{9}\left(\frac{18}{25}\right)$

   $\frac{16}{25}$

2. $-4\left(\frac{9}{16}\right)$

   $-2\frac{1}{4}$

3. $-3\frac{3}{7} \times 2\frac{1}{2}$

   $-8\frac{4}{7}$

4. $-\frac{2}{3} \div \frac{5}{9}$

   $-1\frac{1}{5}$

5. $\frac{7}{13} \div (-2)$

   $-\frac{7}{26}$

6. $-5\frac{5}{8} \div \left(-4\frac{7}{12}\right)$

   $1\frac{5}{22}$

7. $-1.39 \times (-6.8)$

   9.452

8. $-10 \div 0.22$

   $-45.\overline{45}$

9. $-12.166 \div (-1.54)$

   7.9

10. In a game of tug of war, your team changes $-1\frac{3}{10}$ feet in position every 10 seconds. What is your change in position after 30 seconds?

    $-3\frac{9}{10}$ ft

11. The table shows the change of gas prices over a month's time. What is the mean change?

    $-\$0.005$

Week	Change
1	−$0.06
2	+$0.10
3	−$0.08
4	+$0.02

Technology For the Teacher
Answer Presentation Tool
QuizShow

Vocabulary and Concept Check

1. The same rules for signs of integers are applied to rational numbers.

2. $-\frac{5}{2}$

3. $-\frac{1}{3}$

4. $\frac{9}{16}$

5. $-\frac{3}{7}$

6. positive

7. negative

8. negative

9. positive

Practice and Problem Solving

10. $-1\frac{3}{4}$

11. $-\frac{2}{3}$

12. 1

13. $-\frac{1}{100}$

14. $\frac{2}{5}$

15. $2\frac{5}{14}$

16. -1.25

17. $3.\overline{63}$

18. 0.23

19. -6

20. -2.45

21. -2.5875

22. $-\frac{4}{27}$

23. $\frac{1}{3}$

24. $-\frac{4}{9}$

25. $2\frac{1}{2}$

26. 9

27. $-4\frac{17}{27}$

28. -0.012

29. 0.025

30. -0.72

31. 47.43

32. 11.424

33. -0.064

34. The answer should be negative. $-2.2 \times 3.7 = -8.14$

35. The wrong fraction was inverted.

$$-\frac{1}{4} \div \frac{3}{2} = -\frac{1}{4} \times \frac{2}{3}$$
$$= -\frac{2}{12}$$
$$= -\frac{1}{6}$$

36. $-66°$

37. 8 packages

38. −19.59   **39.** 1.3

40. −22.667   **41.** $-4\frac{14}{15}$

42. $-5\frac{11}{24}$ **43.** $-1\frac{11}{36}$

44. *Sample answer:* $-\frac{9}{10}, \frac{2}{3}$

45. $191\frac{11}{12}$ yd

46. $3\frac{5}{8}$ gal

47. See *Taking Math Deeper*.

48. −1.28 sec

49. a. −2, 4, −8, 16, −32, 64

   b. When −2 is raised to an odd power, the product is negative. When −2 is raised to an even power, the product is positive.

   c. negative

 Fair Game Review

50. −1.5   **51.** −5.4

52. $4\frac{1}{2}$ **53.** $-8\frac{5}{18}$

54. D

Mini-Assessment

Multiply or divide. Write fractions in simplest form.

1. $-\frac{6}{7}\left(-\frac{5}{2}\right)$ $2\frac{1}{7}$

2. $6\frac{1}{2} \div \left(-2\frac{3}{4}\right)$ $-2\frac{4}{11}$

3. $3.5(-7.65)$ -26.775

4. $-0.25 \div (-0.05)$ 5

5. The cell phone company will add −$2.74 to your next bill for each of the 4 months you were overcharged. How much will be added to your next bill? −$10.96

Taking Math Deeper

Exercise 47

Problems like this one beg for a diagram. It would be easy to misinterpret what "width" is referring to without drawing a diagram and labeling it.

 ① Draw a diagram. Label the known and unknown lengths.

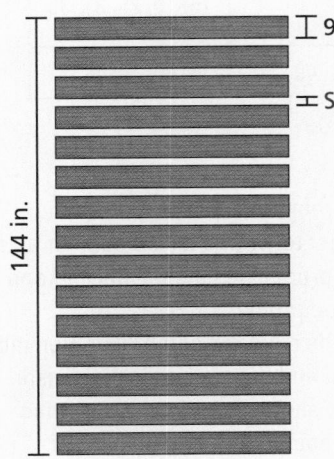

② Find the total width of the boards by multiplying by 15.

$$15\left(9\frac{1}{4}\right) = 15\left(\frac{37}{4}\right) = \frac{555}{4} = 138\frac{3}{4} \text{ in.}$$

 ③ Subtract the width of the boards from 144 inches.

$$144 - 138\frac{3}{4} = 5\frac{1}{4} \text{ in.}$$

There are 14 spaces, so divide $5\frac{1}{4}$ by 14.

$$5\frac{1}{4} \div 14 = \frac{21}{4} \div 14$$
$$= \frac{21}{4} \cdot \frac{1}{14}$$
$$= \frac{3}{8} \text{ in.} \qquad \text{Space}$$

Reteaching and Enrichment Strategies

If students need help...	If students got it...
Resources by Chapter • Practice A and Practice B • Puzzle Time Record and Practice Journal Practice Differentiating the Lesson Lesson Tutorials Skills Review Handbook	Resources by Chapter • Enrichment and Extension • School-to-Work Start the next section

Evaluate.

38. $-4.2 + 8.1 \times (-1.9)$

39. $2.85 - 6.2 \div 2^2$

40. $-3.64 \cdot \left| -5.3 \right| - 1.5^3$

41. $1\frac{5}{9} \div \left(-\frac{2}{3}\right) + \left(-2\frac{3}{5}\right)$

42. $-3\frac{3}{4} \times \frac{5}{6} - 2\frac{1}{3}$

43. $\left(-\frac{2}{3}\right)^2 - \frac{3}{4}\left(2\frac{1}{3}\right)$

44. OPEN-ENDED Write two fractions whose product is $-\frac{3}{5}$.

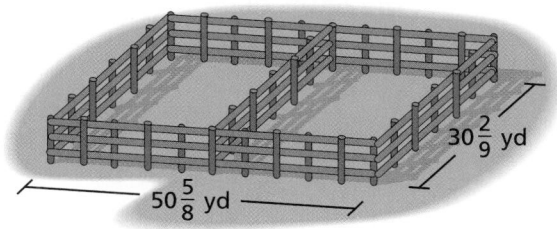

30$\frac{2}{9}$ yd

50$\frac{5}{8}$ yd

45. FENCING A farmer needs to enclose two adjacent rectangular pastures. How much fencing does the farmer need?

46. GASOLINE A 14.5-gallon gasoline tank is $\frac{3}{4}$ full. How many gallons will it take to fill the tank?

47. BOARDWALK A section of a boardwalk is made using 15 boards. Each board is $9\frac{1}{4}$ inches wide. The total width of the section is 144 inches. The spacing between each board is equal. What is the width of the spacing between each board?

48. RUNNING The table shows the changes in the times (in seconds) of four teammates. What is the mean change?

49. *Critical Thinking* Consider $(-2)^1$, $(-2)^2$, $(-2)^3$, $(-2)^4$, $(-2)^5$, and $(-2)^6$.

Teammate	Change
1	−2.43
2	−1.85
3	0.61
4	−1.45

   a. Evaluate each expression.

   b. What pattern do you notice?

   c. What is the sign of $(-2)^{49}$?

Fair Game Review What you learned in previous grades & lessons

Add or subtract. *(Section 2.2)*

50. $-6.2 + 4.7$

51. $-8.1 - (-2.7)$

52. $\frac{9}{5} - \left(-2\frac{7}{10}\right)$

53. $-4\frac{5}{6} + \left(-3\frac{4}{9}\right)$

54. MULTIPLE CHOICE What are the coordinates of the point in quadrant IV? *(Section 1.6)*

   Ⓐ $(-4, 1)$

   Ⓒ $(0, -2)$

   Ⓑ $(-3, -3)$

   Ⓓ $(3, -3)$

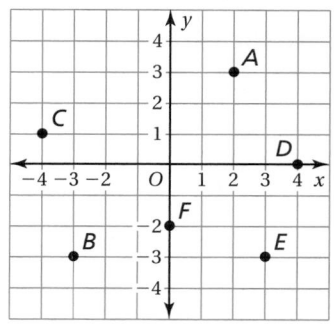

 Key Ideas

Commutative Properties

Words Changing the order of addends or factors does not change the sum or product.

Numbers $-5 + 8 = 8 + (-5)$

$-5 \cdot 8 = 8 \cdot (-5)$

Associative Properties

Words Changing the grouping of addends or factors does not change the sum or product.

Numbers $(7 + 4) + 2 = 7 + (4 + 2)$

$(7 \cdot 4) \cdot 2 = 7 \cdot (4 \cdot 2)$

EXAMPLE 1 **Using Commutative and Associative Properties**

Study Tip

Use number properties to group numbers that are easy to add or multiply.

a. Evaluate 4.7 + 6 + 3.3.

$4.7 + 6 + 3.3 = 6 + 4.7 + 3.3$	Commutative Property of Addition
$= 6 + (4.7 + 3.3)$	Associative Property of Addition
$= 6 + 8$	Add 4.7 and 3.3.
$= 14$	Add 6 and 8.

b. Evaluate $(7 \cdot 4)\frac{1}{4}$.

$(7 \cdot 4)\frac{1}{4} = 7\left(4 \cdot \frac{1}{4}\right)$	Associative Property of Multiplication
$= 7(1)$	Multiply 4 and $\frac{1}{4}$.
$= 7$	Multiplication Property of One

 Practice

Evaluate the expression. Explain each step.

1. $2 + 3 + (-2)$

2. $12 + 6.1 + 5.9$

3. $4 \cdot 19 \cdot \frac{1}{2}$

4. $\frac{1}{3} \cdot 2 \cdot \frac{1}{2}$

5. $5\left(\frac{7}{8} \cdot \frac{2}{5}\right)$

6. $-1.45 + (-8.55 + 2.7)$

Laurie's Notes

Introduction

Connect
- **Yesterday:** Students learned how to multiply and divide rational numbers.
- **Today:** Students will use number properties to evaluate expressions.

Motivate
- **Act It Out:** If there is a bit of an actor in you, start the class by putting your shoes on and then putting your socks over your shoes. This will clearly evoke a few comments about the order in which you performed the two tasks.
- **?** "Oh, does order matter? Hmmm…"
- If wearing a sock over a shoe doesn't work for you, do something else that is obviously in the wrong order. Catch their attention!
- **?** "What are some common things we do where order matters?"

Lesson Notes

Key Ideas
- Write the Commutative and Associative Properties. The key words to remember are *order* for the Commutative Property and *grouping* for the Associative Property.
- Review the use of parentheses in each property. Parentheses are sometimes used to clarify that a number is negative or to show grouping. Parentheses are used in Example 1 part (b) to show multiplication.

Example 1
- Work through the two parts as shown. Because part (a) involves decimals, this is an appropriate time to review previous skills.
- **?** In part (b) ask, "What is the relationship between 4 and $\frac{1}{4}$?" They are reciprocals and their product is 1.

Practice
- **Common Error:** Students may incorrectly label steps, stating the Commutative Property instead of the Associative Property, or vice versa.
- Note that Exercise 5 can involve both properties; the order of factors within the parentheses can be changed, and then the grouping of factors can be changed.

Goal Today's lesson is using number properties to evaluate expressions.

Warm Up

Lesson 2.3b	Warm Up
	For use before Lesson 2.3b

Divide or multiply. Write fractions in simplest form.

1. $-\frac{1}{2} \cdot \frac{3}{2}$ 2. $-\frac{5}{8} \div \left(-\frac{2}{5}\right)$

3. $1.3 \cdot (-5.2)$ 4. $-4 \div 0.4$

5. $-1\frac{3}{4} \times \frac{1}{3}$ 6. $\frac{4}{5} \div \left(-2\frac{2}{5}\right)$

Extra Example 1

a. Evaluate $3.8 + (8.2 + 4)$. 16

b. Evaluate $\frac{1}{3}(8 \cdot 27)$. 72

Practice

1. $2 + 3 + (-2)$
   $= 2 + (-2) + 3$
      Comm. Prop. of Add.
   $= 0 + 3$
      Additive Inverse Property
   $= 3$
      Addition Prop. of Zero

2. $12 + 6.1 + 5.9$
   $= 12 + (6.1 + 5.9)$
      Assoc. Prop. of Add.
   $= 12 + 12$
      Add 6.1 and 5.9.
   $= 24$
      Add 12 and 12.

3–6. See Additional Answers.

Record and Practice Journal Practice
See Additional Answers.

Extra Example 2

A carbon atom has six positively-charged protons and six negatively-charged electrons. The sum of the charges gives the charge of the carbon atom. Find the charge of the atom. 0

Practice

7. 0

8. $30

9. *Sample answer:* Find a map, Lose a compass, Lose a compass

Mini-Assessment

Evaluate the expression. Explain each step.

1. $7\left(\dfrac{2}{7} \cdot \dfrac{1}{4}\right)$ $\dfrac{1}{2}$; Assoc. Prop. of Mult.

2. $4 + (-9) + (-4)$ -9; Comm. Prop. of Add.; Additive Inverse Prop.; Addition Prop. of Zero

4. $-12.3 + (-2.7 + 5.45)$ -9.55; Assoc. Prop. of Add.

4. $9\left(\dfrac{1}{5} \cdot \dfrac{2}{3}\right)$ $1\dfrac{1}{5}$; Comm. Prop. of Mult.; Assoc. Prop. of Mult.

5. The table shows the points gained or lost for each part of a review game. Find the final score.

Points	
Part 1	25 points
Part 2	−10 points
Part 3	−25 points
Part 4	30 points

20 points

Laurie's Notes

Discuss

- Discuss with students the concept of two "actions" balancing one another out. For instance, earning $10 and spending $10 will bring you back to where you started. A football team rushing for 8 yards and then losing 8 yards on a quarterback sack puts the ball back to where it was before the two plays.
- **?** "Can you think of other "actions" or events that balance one another out?" Answers will vary.
- Review the Additive Inverse Property as you work through the example.

Example 2

- **Science Connection:** Atoms are made up of *electrons, protons,* and *neutrons. Electrons* are small, light particles that have a negative electrical charge $(-)$. *Protons* are larger, heavier particles that have a positive charge $(+)$. *Neutrons* are similar in size to protons, but neutrons have no electrical charge.
- Note the use of color to distinguish between positive and negative. Students that have used algebra tiles or two-color counters will be comfortable with this example.

Practice

- These questions will help review addition of rational numbers.

Closure

- On the board, ask volunteers to share examples of a problem that can be simplified using:
  a. the Commutative Property.
  b. the Associative Property.

Technology
For the **T**eacher

The Dynamic Planning Tool
Editable Teacher's Resources at *BigIdeasMath.com*

EXAMPLE 2   Real-Life Application

Helium Atom

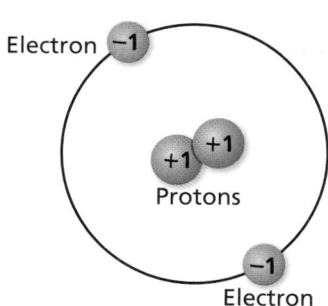

A helium atom has positively-charged protons and negatively-charged electrons, as shown in the diagram. The sum of the charges gives the charge of the helium atom. Find the charge of the atom.

$$\text{Charge of atom} = \text{Charge of protons} + \text{Charge of electrons}$$

$= (+1) + (+1) + (-1) + (-1)$	Substitute.
$= (+2) + (-2)$	Simplify.
$= 0$	Additive Inverse Property

∴ The protons and the electrons are oppositely charged. So, the helium atom has a charge of 0.

Practice

7. **SCIENCE** A lithium atom has positively-charged protons and negatively-charged electrons. The sum of the charges gives the charge of the lithium atom. Find the charge of the atom.

Lithium Atom

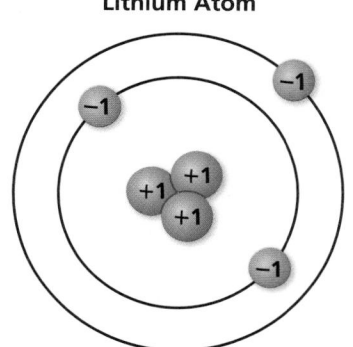

8. **BANK ACCOUNT** The table shows four bank account transactions. Find the change in the account balance.

Transactions	
Deposit	$75
Groceries	−$33.50
Withdrawal	−$75
Deposit	$63.50

9. **VIDEO GAME** The table shows ways to gain and lose points in a video game. List a series of at least three events that results in 0 points.

Event	Points	Event	Points
Find treasure	+100	Find a map	+50
Find a gold coin	+10	Lose a coin	−10
Lose a map	−50	Find a silver coin	+5
Lose a compass	−25	Lose treasure	−100

You can use a **process diagram** to show the steps involved in a procedure. Here is an example of a process diagram for adding rational numbers.

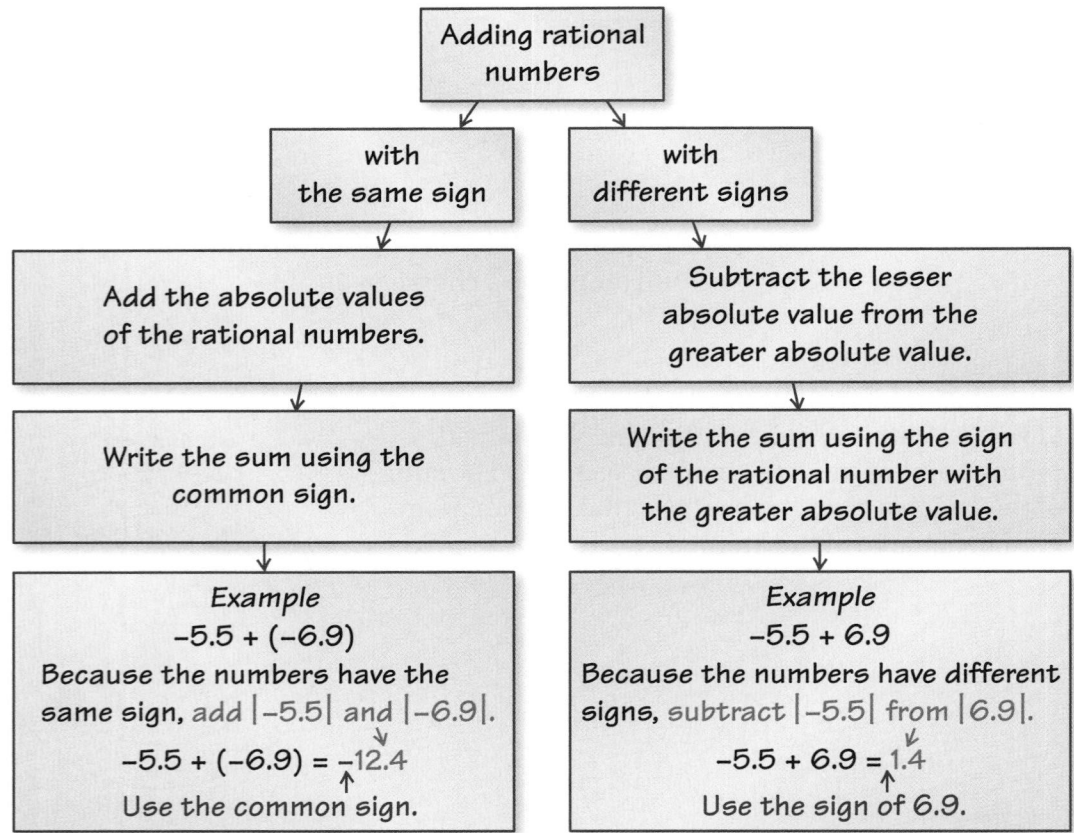

On Your Own

Make a process diagram with examples to help you study these topics. Your process diagram can have one or more branches.

1. writing rational numbers as decimals

2. subtracting rational numbers

3. dividing rational numbers

After you complete this chapter, make process diagrams with examples for the following topics.

4. solving equations using addition or subtraction

5. solving equations using multiplication or division

6. solving two-step equations

"Does this process diagram accurately show how a cat claws furniture?"

Sample Answers

1.

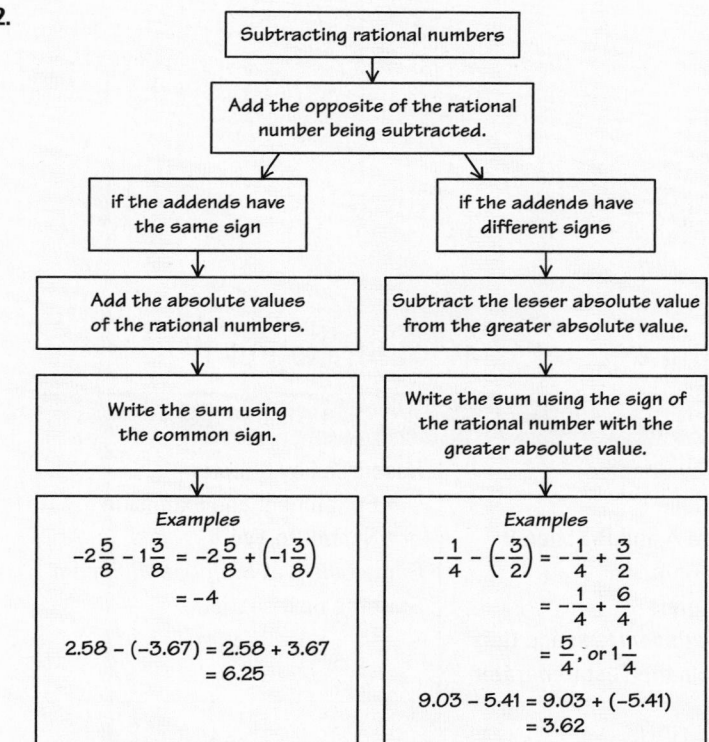

```
┌─────────────────────────────┐
│   Writing rational numbers   │
│        as decimals           │
└─────────────────────────────┘
               ↓
┌─────────────────────────────┐
│   Write the rational number  │
│  as a fraction a/b if necessary. │
└─────────────────────────────┘
               ↓
┌─────────────────────────────┐
│   Use long division to find  │
│      the quotient a ÷ b.     │
└─────────────────────────────┘
```

If the remainder is 0, the rational number is a terminating decimal.

If the remainder repeats, the rational number is a repeating decimal.

Example
Write $-5\frac{1}{5}$ as a decimal.

$$-5\frac{1}{5} = -\frac{26}{5}$$

$$\begin{array}{r} 5.2 \\ 5\overline{)26.0} \\ -25 \\ \hline 1\,0 \\ -1\,0 \\ \hline 0 \end{array}$$

So, $-5\frac{1}{5} = -5.2$.

Example
Write $\frac{2}{15}$ as a decimal.

$$\begin{array}{r} 0.133 \\ 15\overline{)2.000} \\ -15 \\ \hline 50 \\ -45 \\ \hline 50 \\ -45 \\ \hline 5 \end{array}$$

So, $\frac{2}{15} = 0.1\overline{3}$.

2.

```
┌─────────────────────────────┐
│  Subtracting rational numbers │
└─────────────────────────────┘
               ↓
┌─────────────────────────────────┐
│  Add the opposite of the rational │
│   number being subtracted.        │
└─────────────────────────────────┘
```

if the addends have the same sign

if the addends have different signs

Add the absolute values of the rational numbers.

Subtract the lesser absolute value from the greater absolute value.

Write the sum using the common sign.

Write the sum using the sign of the rational number with the greater absolute value.

Examples

$$-2\frac{5}{8} - 1\frac{3}{8} = -2\frac{5}{8} + \left(-1\frac{3}{8}\right)$$
$$= -4$$

$$2.58 - (-3.67) = 2.58 + 3.67$$
$$= 6.25$$

Examples

$$-\frac{1}{4} - \left(-\frac{3}{2}\right) = -\frac{1}{4} + \frac{3}{2}$$
$$= -\frac{1}{4} + \frac{6}{4}$$
$$= \frac{5}{4}, \text{ or } 1\frac{1}{4}$$

$$9.03 - 5.41 = 9.03 + (-5.41)$$
$$= 3.62$$

3. Available at *BigIdeasMath.com*.

List of Organizers
Available at *BigIdeasMath.com*

Comparison Chart
Concept Circle
Example and Non-Example Chart
Formula Triangle
Four Square
Idea (Definition) and Examples Chart
Information Frame
Information Wheel
Notetaking Organizer
Process Diagram
Summary Triangle
Word Magnet
Y Chart

About this Organizer

A **Process Diagram** can be used to show the steps involved in a procedure. Process diagrams are particularly useful for illustrating procedures with two or more steps, and they can have one or more branches. As shown, students' process diagrams can consist of a single flowchart-type diagram, with example(s) included in the last box to illustrate the steps that precede it. Or, the diagram can have two parallel flowcharts, in which the procedure is stepped out in one chart and an example illustrating each step is shown in the other chart.

Technology
For the Teacher

Vocabulary Puzzle Builder

Answers

1. -0.15
2. $-1.8\overline{3}$
3. $-\dfrac{13}{40}$
4. $-1\dfrac{7}{25}$
5. $-1\dfrac{7}{40}$
6. -3.2
7. $1\dfrac{59}{63}$
8. -3.8
9. $-3\dfrac{4}{5}$
10. 44.18
11. $5\dfrac{4}{9}$
12. -4
13. stock B; Because -3.72 is less than -3.68.
14. -79.8 ft
15. $54\dfrac{3}{4}$ yd
16. $-\$13.56$

Alternative Quiz Ideas

100% Quiz   Math Log
Error Notebook   Notebook Quiz
Group Quiz   Partner Quiz
Homework Quiz   Pass the Paper

Pass the Paper

- Work in groups of four. The first student copies the problem and completes the first step, explaining his or her work.
- The paper is passed and the second student works through the next step, also explaining his or her work.
- This process continues until the problem is completed.
- The second member of the group starts the next problem. Students should be allowed to question and debate as they are working through the quiz.
- Student groups can be selected by the teacher, by students, through a random process, or any way that works for your class.
- The teacher walks around the classroom listening to the groups and asks questions to ensure understanding.

Assessment Book

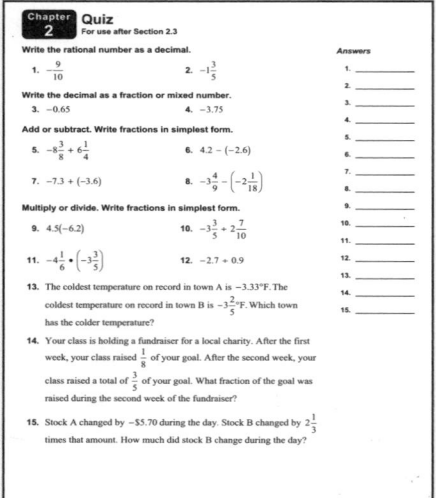

Reteaching and Enrichment Strategies

If students need help. . .	If students got it. . .
Resources by Chapter • Study Help • Practice A and Practice B • Puzzle Time Lesson Tutorials *BigIdeasMath.com* Practice Quiz Practice from the Test Generator	Resources by Chapter • Enrichment and Extension • School-to-Work Game Closet at *BigIdeasMath.com* Start the next section

Technology For the Teacher

Answer Presentation Tool
Big Ideas Test Generator

Check It Out
Progress Check
BigIdeasMath ✓com

Write the rational number as a decimal.   *(Section 2.1)*

1. $-\dfrac{3}{20}$

2. $-\dfrac{11}{6}$

Write the decimal as a fraction or mixed number in simplest form.   *(Section 2.1)*

3. -0.325

4. -1.28

Add or subtract. Write fractions in simplest form.   *(Section 2.2)*

5. $-\dfrac{4}{5} + \left(-\dfrac{3}{8}\right)$

6. $-5.8 + 2.6$

7. $\dfrac{12}{7} - \left(-\dfrac{2}{9}\right)$

8. $9.1 - 12.9$

Multiply or divide. Write fractions in simplest form.   *(Section 2.3)*

9. $-2\dfrac{3}{8} \times \dfrac{8}{5}$

10. $-9.4 \times (-4.7)$

11. $-8\dfrac{5}{9} \div \left(-1\dfrac{4}{7}\right)$

12. $-8.4 \div 2.1$

13. **STOCK** The value of stock A changes $-\$3.68$ and the value of stock B changes $-\$3.72$. Which stock has the greater loss? Explain.   *(Section 2.1)*

14. **PARASAILING** A parasail is at 200.6 feet above the water. After five minutes, the parasail is at 120.8 feet above the water. What is the change in height of the parasail?   *(Section 2.2)*

15. **FOOTBALL** The table shows the statistics of a running back in a football game. How many total yards did he gain?   *(Section 2.2)*

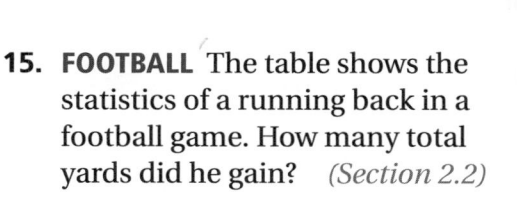

Quarter	1	2	3	4	Total
Yards	$-8\dfrac{1}{2}$	23	$42\dfrac{1}{2}$	$-2\dfrac{1}{4}$	?

16. **LATE FEES** You were overcharged $\$4.52$ on your cell phone bill three months in a row. The cell phone company will add $-\$4.52$ to your next bill for each month you were overcharged. How much will be added to your next bill?   *(Section 2.3)*

Solving Equations Using Addition or Subtraction

Essential Question How can you use inverse operations to solve an equation?

Key: [+] = Variable   [+] = 1   [−] = −1   [+][−] = Zero Pair

1 EXAMPLE: Using Addition to Solve an Equation

Use algebra tiles to model and solve $x - 3 = -4$.

Model the equation $x - 3 = -4$.

To get the green tile by itself, remove the red tiles on the left side by adding three yellow tiles to each side.

Remove the three "zero pairs" from each side.

The remaining tile shows the value of x.

⋮ So, $x = -1$.

2 EXAMPLE: Using Addition to Solve an Equation

Use algebra tiles to model and solve $-5 = n + 2$.

Model the equation $-5 = n + 2$.

Remove the yellow tiles on the right side by adding two red tiles to each side.

Remove the two "zero pairs" from the right side.

The remaining tiles show the value of n.

⋮ So, $-7 = n$ or $n = -7$.

Laurie's Notes

Introduction

For the Teacher

- **Goal:** The goal of this activity is for students to develop a conceptual understanding of how one-step linear equations are solved by adding the same quantity to, or subtracting the same quantity from, each side of an equation.
- This lesson and the next require students to connect two prior skills: solving one-step linear equations and operations with integers.

Motivate

- Show students a collection of algebra tiles and ask them what the collection represents.
- ? "Can the collection be simplified? (Can you remove zero pairs?)"
- ? "What is the expression represented by the collection?"
- **Model:** As a class, model the equations $x + 3 = 7$ and $x + 2 = 5$ using algebra tiles. These do not require a zero pair to solve and will help remind students how to solve equations using algebra tiles.

Activity Notes

Example 1

? There are two points to make at the beginning. First, ask students what it means to solve an equation. Second, mention that students need to think of $x - 3$ as $x + (-3)$ when using algebra tiles. (You can only *add* a positive tile or a negative tile.) To find the value of the variable that makes the equation true.

- Use algebra tiles to model the solution of the equation with the class as the students model the solution at their desks. You may want students to work with a partner.
- **Discuss:** To get one green tile by itself, you need to think of a way to "undo" the three red tiles which represent adding -3. The opposite of adding three red (negative) tiles is to add three yellow (positive) tiles. If you add three yellow tiles to one side of the equation, you need to add three yellow tiles to the other side to keep the equation balanced.
- ? "What does *removing 3 red tiles* mean?" subtract -3

Example 2

- **Representation:** While the equations $-5 = n + 2$ and $n + 2 = -5$ are the same to mathematics teachers, students may see these as very different equations. Students even see $2 + n = -5$ as a different equation. Take time to discuss the equivalence of all three equations.
- Ask students why $x = 4$ is equivalent to $4 = x$. Student conceptions and misconceptions show up when the original equation is modified. Solve equations such as $-2 + n = -5$ and $n - 2 = -5$.
- Model the solution of the equation with the class as the students model the solution at their desks.

Previous Learning

Students have solved equations with whole numbers.

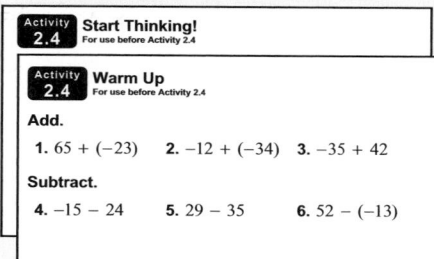

Start Thinking! and Warm Up

Activity 2.4 Start Thinking!
For use before Activity 2.4

Activity 2.4 Warm Up
For use before Activity 2.4

Add.
1. $65 + (-23)$ 2. $-12 + (-34)$ 3. $-35 + 42$

Subtract.
4. $-15 - 24$ 5. $29 - 35$ 6. $52 - (-13)$

2.4 Record and Practice Journal

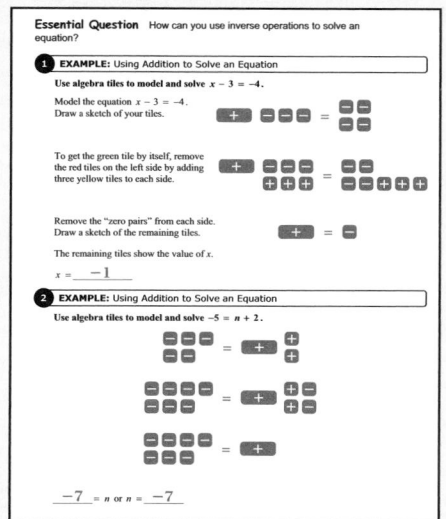

Essential Question How can you use inverse operations to solve an equation?

1 EXAMPLE: Using Addition to Solve an Equation

Use algebra tiles to model and solve $x - 3 = -4$.

Model the equation $x - 3 = -4$.
Draw a sketch of your tiles.

To get the green tile by itself, remove the red tiles on the left side by adding three yellow tiles to each side.

Remove the "zero pairs" from each side.
Draw a sketch of the remaining tiles.

The remaining tiles show the value of x.

$x = \underline{-1}$

2 EXAMPLE: Using Addition to Solve an Equation

Use algebra tiles to model and solve $-5 = n + 2$.

$\underline{-7} = n$ or $n = \underline{-7}$

Differentiated Instruction

Kinesthetic

When working out solutions, ask two students to assist you at the board or overhead. Assign one student to the left side of the equation and the other student to the right side. Each student is responsible for performing the operations on his/her side of the equation. Emphasize that in order for both sides of the equation to remain equal, both students must perform the same operation at the same time to solve the equation.

2.4 Record and Practice Journal

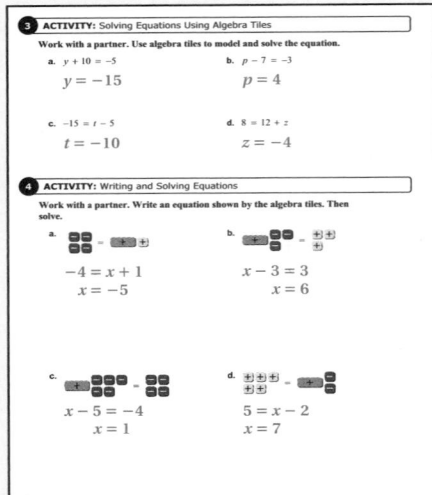

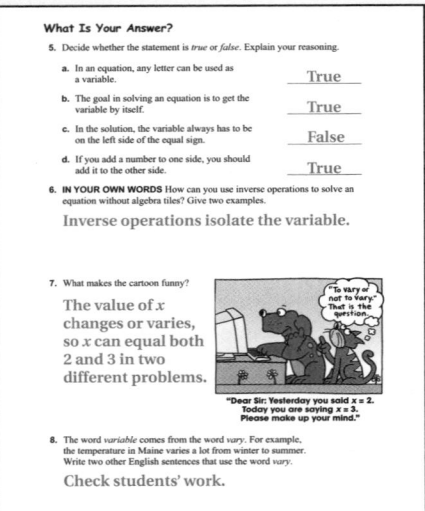

Laurie's Notes

Activity 3

- Have students work with a partner to solve these problems. When they have finished, ask volunteers to solve the problems for the class.

- **?** Some students may have solved part (c) by adding five yellow tiles to each side and some may have subtracted 5 red tiles from each side. Ask students which method they used. Subtracting -5 is equivalent to adding 5, so both methods are acceptable. Numerically, $-(-5) = (+5)$.

Activity 4

- Students are asked to write "an" equation, not "the" equation. This is because there are many correct answers. Students can use any variable and different forms of the equation. For example, part (b) could be written as $n - 3 = 3$ or $n + (-3) = 3$.

- **?** Ask students to share the different equations they wrote.

- **Extension:** Share common tasks that *undo* one another. Examples: tying and untying your shoes, filling and emptying a glass, and opening and closing a door

What Is Your Answer?

- **Think-Pair-Share:** Students should read each question independently and then work with a partner to answer the questions. When they have answered the questions, the pair should compare their answers with another group and discuss any discrepancies.

Closure

- Translate the following model into symbols and explain in words how it could be solved.

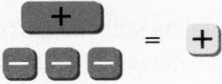

Sample answer: $x - 3 = 1$; Add three yellow tiles to each side. The result will be $x = 4$.

3 ACTIVITY: Solving Equations Using Algebra Tiles

Work with a partner. Use algebra tiles to model and solve the equation.

a. $y + 10 = -5$ **b.** $p - 7 = -3$

c. $-15 = t - 5$ **d.** $8 = 12 + z$

4 ACTIVITY: Writing and Solving Equations

Work with a partner. Write an equation shown by the algebra tiles. Then solve.

a.

b.

c.

d.

What Is Your Answer?

5. Decide whether the statement is *true* or *false*. Explain your reasoning.

    a. In an equation, any letter can be used as a variable.

    b. The goal in solving an equation is to get the variable by itself.

    c. In the solution, the variable always has to be on the left side of the equal sign.

    d. If you add a number to one side, you should add it to the other side.

6. IN YOUR OWN WORDS How can you use inverse operations to solve an equation without algebra tiles? Give two examples.

7. What makes the cartoon funny?

8. The word *variable* comes from the word *vary*. For example, the temperature in Maine varies a lot from winter to summer.

Write two other English sentences that use the word *vary*.

"To vary or not to vary." That is the question.

"Dear Sir: Yesterday you said x = 2. Today you are saying x = 3. Please make up your mind."

Practice

Use what you learned about solving equations using inverse operations to complete Exercises 5–8 on page 74.

Check It Out
Lesson Tutorials
BigIdeasMath com

Key Vocabulary ◀))
equivalent equations,
p. 72

🔑 Key Ideas

Addition Property of Equality

Words Two equations are **equivalent equations** if they have the same solutions. Adding the same number to each side of an equation produces an equivalent equation.

Algebra If $a = b$, then $a + c = b + c$.

Subtraction Property of Equality

Words Subtracting the same number from each side of an equation produces an equivalent equation.

Algebra If $a = b$, then $a - c = b - c$.

EXAMPLE 1 **Solving Equations**

a. Solve $x - 5 = -1$.

$x - 5 = -1$	Write the equation.
$\underline{+5 \quad +5}$	Add 5 to each side.
$x = 4$	Simplify.

∴ So, the solution is $x = 4$.

Check

$$x - 5 = -1$$
$$4 - 5 \stackrel{?}{=} -1$$
$$-1 = -1 \checkmark$$

Remember

To solve equations, use *inverse operations* that "undo" each other. For example, use addition to solve an equation with subtraction.

b. Solve $z + \dfrac{3}{2} = \dfrac{1}{2}$.

$z + \dfrac{3}{2} = \dfrac{1}{2}$	Write the equation.
$\underline{-\dfrac{3}{2} \qquad -\dfrac{3}{2}}$	Subtract $\dfrac{3}{2}$ from each side.
$z = -1$	Simplify.

∴ So, the solution is $z = -1$.

On Your Own

Now You're Ready
Exercises 5–20

Solve the equation. Check your solution.

1. $p - 5 = -2$ 　　2. $w + 13.2 = 10.4$ 　　3. $x - \dfrac{5}{6} = -\dfrac{1}{6}$

Laurie's Notes

Introduction

Connect
- **Yesterday:** Students used algebra tiles to model solving equations.
- **Today:** Students will formalize the process using the Addition and Subtraction Properties of Equality.

Motivate
- Have two students stand at the front of the room and write an "=" on the board between them. Hand each the same number of items (i.e., pencils, paper clips, etc.). The students should verify that they have the same number of items. Then give each two more of the same item. Verify that the number of items they have is equal. Finally, take four of the items from each student. Verify that they have the same amount.
- **Discuss:** This is the essence of the two properties used today—as long as each side of the equation has the same amount added to it or subtracted from it, the two sides of the equation are still equal.

Lesson Notes

Key Ideas
- Write the Properties of Equality on the board.
- Discuss how the activity in the introduction modeled the two properties.

Example 1
- Work through each part as a class. Notice that a vertical format is used. Use color to show the quantity being added to or subtracted from each side.
- **? Discuss:** The equations in parts (a) and (b) have the variable on the left.
  - "Would part (a) have the same solution if it was written as $-1 = x - 5$?" yes
  - "Would part (b) have the same solution if it was written as $\frac{1}{2} = z + \frac{3}{2}$?" yes

On Your Own
- After students have completed Example 1, they should be able to do these questions independently.

Words of Wisdom
- **Struggling Students:** If students have difficulty with the *On Your Own* questions, assess whether it is algebraic (how to solve equations) or computational (how to add or subtract rational numbers). Use this information to guide your instruction. If possible, provide colored pencils so students can record the quantity being added to or subtracted from each side.
- Encourage students to be neat and to keep their equal signs lined up.

Activity Materials
Introduction

- pencils
- paper clips

Start Thinking! and Warm Up

Lesson 2.4 Warm Up For use before Lesson 2.4

Lesson 2.4 Start Thinking! For use before Lesson 2.4

Discuss with a partner, using an example, how inverse operations are used to solve equations.

Extra Example 1
a. Solve $t + 6 = -5$. -11
b. Solve $y - \frac{4}{5} = -\frac{2}{5}$. $\frac{2}{5}$

On Your Own
1. $p = 3$
2. $w = -2.8$
3. $x = \frac{2}{3}$

Laurie's Notes

Extra Example 2

You spent $7.25 this week. This is $3.65 less than you spent last week. Write and solve an equation to find the amount s you spent last week.
$s - 3.65 = 7.25$, $10.90

 On Your Own

    4. $P - 145.25 = 120.50$

Extra Example 3

You have -1 point after Level 2 of a video game. Your score is 24 points less than your friend's score. Write and solve an equation to find your friend's score after Level 2. $-1 = f - 24$, 23 points

 On Your Own

    5. 15 points

English Language Learners

Vocabulary

In this section, students learn to use *inverse* (or *opposite*) operations to solve equations. Students use addition to solve a subtraction equation and use subtraction to solve an addition equation. Review these pairs of words that are essential to understanding mathematics. Give students one word of a pair and ask them to provide the opposite.

odd, even	positive, negative
add, subtract	sum, difference
multiply, divide	product, quotient
plus, minus	

Example 2

- **Financial Literacy:** Discuss the word *profit*, and how it is computed: income − expenses = profit.
- The second sentence contains key information. When translated into symbols, students can tell that "this profit" refers to "the profit this week."
- The color-coding in this text is very helpful in assisting students as they translate from words to symbols. Students may not recognize that "is" translates to "equals," so give a quick example. (Evan is $5\frac{1}{2}$ feet tall means the same as $E = 5.5$.)

On Your Own

- **Neighbor Check:** Have students work independently and then have their neighbor check their work. Have students discuss any discrepancies.

Example 3

- This example includes a line graph as a way to present information about the problem. Take time to have students *read and interpret* the information in the line graph.

? Here are some questions to ask about the graph.
- "What information is displayed on each axis of the line graph?" The horizontal axis shows the level of a video game and the vertical axis shows the number of points scored.
- "Were the scores ever tied?" Yes, at the very start and at some point in Level 3.
- "Who was ahead after Level 2?" your friend
- "What does '33 points' on the line graph mean?" It is the difference of your score and your friend's score after Level 4.
- "Describe each player's performance from start to finish." *Sample answer:* Your friend did better than you at the beginning, but after Level 2 your score increased and your friend's score decreased. You ended up with 33 more points than your friend.

On Your Own

- Encourage students to write the key words and phrases using colored pencils and then translate the words to symbols.

Closure

- **Exit Ticket:**
  $p - 3.5 = -1.3$ 2.2 $-4.2 + m = 8.6$ 12.8

The Dynamic Planning Tool
Editable Teacher's Resources at *BigIdeasMath.com*

EXAMPLE 2 **Standardized Test Practice**

A company has a profit of $750 this week. This profit is $900 more than the profit P last week. Which equation can be used to find P?

(A) $750 = 900 - P$ (B) $750 = P + 900$

(C) $900 = P - 750$ (D) $900 = P + 750$

| Words | The profit this week is $900 more than the profit last week. |

| Equation | 750 | = | P | + | 900 |

The equation is $750 = P + 900$. The correct answer is (B).

On Your Own

Now You're Ready
Exercises 22–25

4. A company has a profit of $120.50 today. This profit is $145.25 less than the profit P yesterday. Write an equation that can be used to find P.

EXAMPLE 3 **Real-Life Application**

The line graph shows the scoring while you and your friend played a video game. Write and solve an equation to find your score after Level 4.

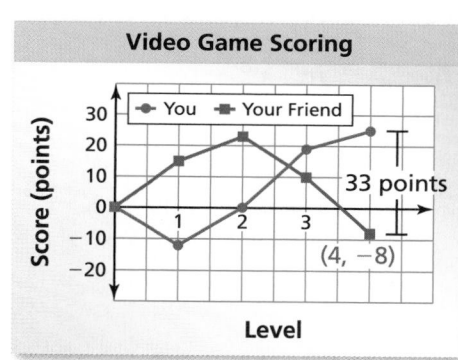

Video Game Scoring

You can determine the following from the graph.

| Words | Your friend's score is 33 points less than your score. |

Variable Let s be your score after Level 4.

| Equation | -8 | = | s | – | 33 |

$-8 = s - 33$ Write equation.

$\underline{+\,33}\quad\underline{+\,33}$ Add 33 to each side.

$25 = s$ Simplify.

Your score after Level 4 is 25 points.

Reasonable? From the graph, your score after Level 4 is between 20 points and 30 points. So, 25 points is a reasonable answer.

On Your Own

5. **WHAT IF?** In Example 3, you have -12 points after Level 1. Your score is 27 points less than your friend's score. What is your friend's score?

Vocabulary and Concept Check

1. **VOCABULARY** What property would you use to solve $m + 6 = -4$?

2. **VOCABULARY** Name two inverse operations.

3. **WRITING** Are the equations $m + 3 = -5$ and $m = -2$ equivalent? Explain.

4. **WHICH ONE DOESN'T BELONG?** Which equation does *not* belong with the other three? Explain your reasoning.

$$x + 3 = -1 \qquad x + 1 = -5 \qquad x - 2 = -6 \qquad x - 9 = -13$$

Practice and Problem Solving

Solve the equation. Check your solution.

5. $a - 6 = 13$

6. $-3 = z - 8$

7. $-14 = k + 6$

8. $x + 4 = -14$

9. $c - 7.6 = -4$

10. $-10.1 = w + 5.3$

11. $\dfrac{1}{2} = q + \dfrac{2}{3}$

12. $p - 3\dfrac{1}{6} = -2\dfrac{1}{2}$

13. $g - 9 = -19$

14. $-9.3 = d - 3.4$

15. $4.58 + y = 2.5$

16. $x - 5.2 = -18.73$

17. $q + \dfrac{5}{9} = \dfrac{1}{6}$

18. $-2\dfrac{1}{4} = r - \dfrac{4}{5}$

19. $w + 3\dfrac{3}{8} = 1\dfrac{5}{6}$

20. $4\dfrac{2}{5} + k = -3\dfrac{2}{11}$

21. **ERROR ANALYSIS** Describe and correct the error in finding the solution.

$$\begin{array}{rcr} x + 8 & = & 10 \\ + 8 & & + 8 \\ \hline x & = & 18 \end{array}$$

Write the verbal sentence as an equation. Then solve.

22. 4 less than a number n is -15.

23. 10 more than a number c is 3.

24. The sum of a number y and -3 is -8.

25. The difference between a number p and 6 is -14.

In Exercises 26–28, write an equation. Then solve.

26. **DRY ICE** The temperature of dry ice is $-109.3°F$. This is $184.9°F$ less than the outside temperature. What is the outside temperature?

27. **PROFIT** A company makes a profit of $1.38 million. This is $2.54 million more than last year. What was the profit last year?

28. **PIER** The difference between the lengths of a paddle boat and a pier is $-7\dfrac{3}{4}$ feet. The pier is $18\dfrac{1}{2}$ feet long. How long is the paddle boat?

Assignment Guide and Homework Check

Level	Day 1 Activity Assignment	Day 2 Lesson Assignment	Homework Check
Basic	5–8, 41–45	1–4, 9–25 odd, 26–29	9, 17, 23, 28
Average	5–8, 41–45	1–4, 13–21 odd, 22–28 even, 31, 33, 34	13, 17, 24, 28
Advanced	5–8, 41–45	1–4, 14–20 even, 21, 30, 32–40	18, 30, 34, 37

Common Errors

- **Exercises 5–20** Students may use the same operation in solving for x instead of the inverse operation. Demonstrate that this will not work to simplify the equation. Students most likely ignored the side with the variable when they made this mistake. Remind them to check their answers in the original equation.

- **Exercises 5–20** Students may add or subtract the number on the side of the equation without the variable. For example, they might write $-14 + 14 = k + 6 + 14$ instead of $-14 - 6 = x + 6 - 6$. Remind students that they are trying to get the variable by itself, so they have to start with the side that the variable is on and use the inverse of that operation.

- **Exercises 29–31** Students may try to use inverse operations to combine like terms. Remind them that inverse operations are used on both sides of the equation.

2.4 Record and Practice Journal

Solve the equation. Check your solution.

1. $y + 12 = -26$
   -38
2. $15 + c = -12$
   -27
3. $-16 = d + 21$
   -37

4. $n + 12.8 = -0.3$
   -13.1
5. $1\frac{1}{8} = g - 4\frac{2}{5}$
   $5\frac{21}{40}$
6. $-5.47 + k = -14.19$
   -8.72

Write the verbal sentence as an equation. Then solve.

7. 42 less than x is -50.
   $x - 42 = -50;\ -8$
8. 32 is the sum of a number z and 9.
   $32 = z + 9;\ 23$

9. A clothing company makes a profit of $2.3 million. This is $4.1 million more than last year. What was the profit last year?
   $-\$1.8$ million

10. A drop on a wooden roller coaster is $-98\frac{1}{2}$ feet. A drop on a steel roller coaster is $100\frac{1}{4}$ feet lower than the drop on the wooden roller coaster. What is the drop on the steel roller coaster?
   $-198\frac{3}{4}$ ft

Technology For the Teacher
Answer Presentation Tool
QuizShow

Practice and Problem Solving

5. $a = 19$
6. $z = 5$
7. $k = -20$
8. $x = -18$
9. $c = 3.6$
10. $w = -15.4$
11. $q = -\dfrac{1}{6}$
12. $p = \dfrac{2}{3}$
13. $g = -10$
14. $d = -5.9$
15. $y = -2.08$
16. $x = -13.53$
17. $q = -\dfrac{7}{18}$
18. $r = -1\dfrac{9}{20}$
19. $w = -1\dfrac{13}{24}$
20. $k = -7\dfrac{32}{55}$

21. The 8 should have been subtracted rather than added.
$$\begin{aligned} x + 8 &= 10 \\ -8 \quad &-8 \\ \hline x &= 2 \end{aligned}$$

22. $n - 4 = -15;\ n = -11$
23. $c + 10 = 3;\ c = -7$
24. $y + (-3) = -8;\ y = -5$
25. $p - 6 = -14;\ p = -8$
26. $T - 184.9 = -109.3;\ 75.6°F$
27. $P + 2.54 = 1.38;$ $-\$1.16$ million
28. $B - 18\dfrac{1}{2} = -7\dfrac{3}{4};\ 10\dfrac{3}{4}$ ft

29. $x + 8 = 12$; 4 cm

30. $x + 20.4 = 24.2$; 3.8 in.

31. $x + 22.7 = 34.6$; 11.9 ft

32. $305 = h + 153$; 152 ft

33. See *Taking Math Deeper*.

34. $d + 24\frac{1}{3} = 65\frac{3}{5}$; $41\frac{4}{15}$ km

35. $m + 30.3 + 40.8 = 180$; 108.9°

36. $p + 63.43 + 87.15 + 81.96 = 311.62$; more than 79.08

37. -9

38. 2, -2

39. 6, -6

40. 13, -13

Fair Game Review

41. -56 **42.** -72

43. -9 **44.** $-6\frac{1}{2}$

45. B

Mini-Assessment

Solve the equation.

1. $x + 3.6 = -4.75$ $x = -8.35$

2. $-15.8 = y - 24.3$ $y = 8.5$

3. $t - 2\frac{2}{3} = -\frac{5}{2}$ $t = \frac{1}{6}$

4. $-\frac{5}{6} = z + \frac{1}{8}$ $z = -\frac{23}{24}$

5. You withdrew $47.25 from your checking account. Now your balance is $-$23.75. Write and solve an equation to find the amount of money in your account before you withdrew the money. $x - 47.25 = -23.75$; $23.50

Taking Math Deeper

Exercise 33

It's surprising how difficult this problem can be for students. There are two reasons for this. One is that you are not given the location of 0 on the vertical number line. The second is that the information is not given in the order it is used.

 Draw a vertical number line. Locate the jumping platform at 0.

 Draw the first jump. Draw the second jump so that the first jump is higher.

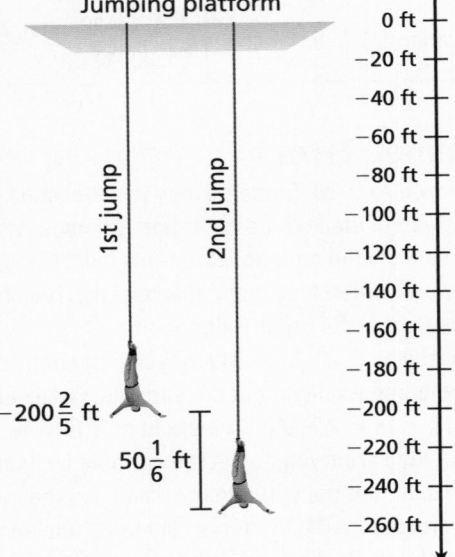

Jumping platform

$-200\frac{2}{5}$ ft

$50\frac{1}{6}$ ft

③ Subtract to find the height of the second jump.

$$-200\frac{2}{5} - 50\frac{1}{6} = -250 - \frac{2}{5} - \frac{1}{6}$$
$$= -250 - \frac{12}{30} - \frac{5}{30}$$
$$= -250\frac{17}{30} \text{ ft}$$

Project

Research bungee jumping. What safety requirements are necessary for a bungee jumping business?

Reteaching and Enrichment Strategies

If students need help. . .	If students got it. . .
Resources by Chapter • Practice A and Practice B • Puzzle Time Record and Practice Journal Practice Differentiating the Lesson Lesson Tutorials Skills Review Handbook	Resources by Chapter • Enrichment and Extension • School-to-Work Start the next section

GEOMETRY Write and solve an equation to find the unknown side length.

29. Perimeter = 12 cm

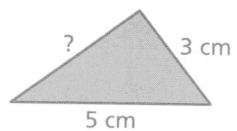

?   3 cm

5 cm

30. Perimeter = 24.2 in.

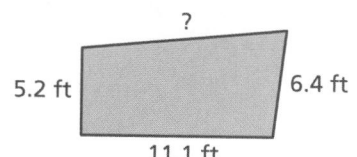

8.3 in.

?   3.8 in.

8.3 in.

31. Perimeter = 34.6 ft

?

5.2 ft   6.4 ft

11.1 ft

In Exercises 32–36, write an equation. Then solve.

305 ft

32. **STATUE OF LIBERTY** The total height of the Statue of Liberty and its pedestal is 153 feet more than the height of the statue. What is the height of the statue?

33. **BUNGEE JUMPING** Your first jump is $50\frac{1}{6}$ feet higher than your second jump. Your first jump reaches $-200\frac{2}{5}$ feet. What is the height of your second jump?

34. **TRAVEL** Boatesville is $65\frac{3}{5}$ kilometers from Stanton. A bus traveling from Stanton is $24\frac{1}{3}$ kilometers from Boatesville. How far has the bus traveled?

35. **GEOMETRY** The sum of the measures of the angles of a triangle equals 180°. What is the measure of the missing angle?

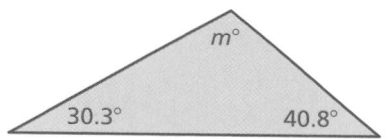

$m°$

30.3°   40.8°

36. **SKATEBOARDING** The table shows your scores in a skateboarding competition. The leader has 311.62 points. What score do you need in the fourth round to win?

Round	1	2	3	4
Points	63.43	87.15	81.96	?

37. **CRITICAL THINKING** Find the value of $2x - 1$ when $x + 6 = 2$.

 **Find the values of x.**

38. $|x| = 2$

39. $|x| - 2 = 4$

40. $|x| + 5 = 18$

 **Fair Game Review** What you learned in previous grades & lessons

Multiply or divide. *(Section 1.4 and Section 1.5)*

41. -7×8

42. $6 \times (-12)$

43. $18 \div (-2)$

44. $-26 \div 4$

45. **MULTIPLE CHOICE** A class of 144 students voted for a class president. Three-fourths of the students voted for you. Of the students who voted for you, $\frac{5}{9}$ are female. How many female students voted for you? *(Section 2.3)*

    (A) 50   (B) 60   (C) 80   (D) 108

Essential Question How can you use multiplication or division to solve an equation?

1 ACTIVITY: Using Division to Solve an Equation

Work with a partner. Use algebra tiles to model and solve the equation.

a. **Sample:** $3x = -12$

Model the equation $3x = -12$.

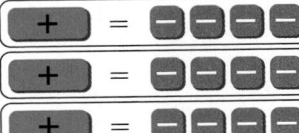

Your goal is to get one green tile by itself. Because there are three green tiles, divide the red tiles into three equal groups.

Keep one of the groups. This shows the value of x.

∴ So, $x = -4$.

b. $2k = -8$

c. $-15 = 3t$

d. $-20 = 5m$

e. $4h = -16$

2 ACTIVITY: Writing and Solving Equations

Work with a partner. Write an equation shown by the algebra tiles. Then solve.

a.

b.

c.

d.

Laurie's Notes

Introduction

For the Teacher

- **Goal:** The goal of this activity is for students to develop conceptual understanding of how to solve a one-step linear equation by multiplying or dividing each side of the equation by the same quantity.

Motivate

- **Model:** Display two green *x*-tiles and four yellow integer-tiles to the class.
- **?** "If two green tiles equal four yellow tiles, what does one green tile equal?" two yellow tiles
- **?** "How did you decide that one green tile equals two yellow tiles?" Divide each side into groups. The number of groups is the number of *x*-tiles.

Activity Notes

Activity 1

- Model the first equation as students model the equation at their desks. Write the corresponding algebraic equation represented by the tiles with the first and last step. Encourage students to do the same.
- **?** **Discuss:** Remind students that the goal is to find the value of just one green tile. "If three green tiles equal 12 red tiles, what is the value of each green tile? How did you find your answer?" 4; To get one *x*-tile, you need three groups. So, divide the 12 red tiles into three equal groups.
- Remind students that variables can be on either side of the equation. If students are more comfortable with variables on the left, they can write part (c) as $3t = -15$.

Activity 2

- **Think-Pair-Share:** After students work on the problems in pairs, ask for volunteers to work the problems for the class. Listen for how students describe the solutions.
- **?** "Why it is difficult to model the equation $\frac{1}{3}x = 6$ with algebra tiles?"

  You can't show $\frac{1}{3}$ of a green *x*-tile, but you can talk about the meaning.

  If $\frac{1}{3}$ of a green *x*-tile is 6, $\frac{2}{3}$ would be 12, and $\frac{3}{3}$ (or a whole green tile) would be 18.

Activity Materials

Introduction	Textbook
• algebra tiles	• algebra tiles • index cards

Start Thinking! and Warm Up

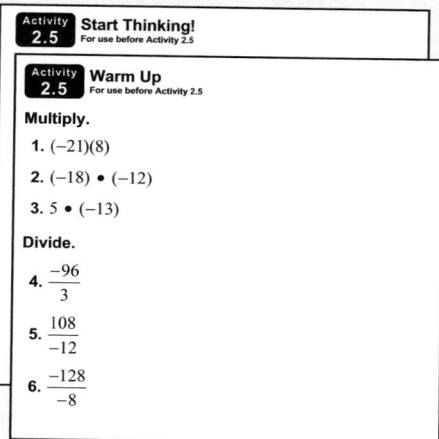

Activity 2.5 Start Thinking! For use before Activity 2.5

Activity 2.5 Warm Up For use before Activity 2.5

Multiply.

1. $(-21)(8)$

2. $(-18) \bullet (-12)$

3. $5 \bullet (-13)$

Divide.

4. $\dfrac{-96}{3}$

5. $\dfrac{108}{-12}$

6. $\dfrac{-128}{-8}$

2.5 Record and Practice Journal

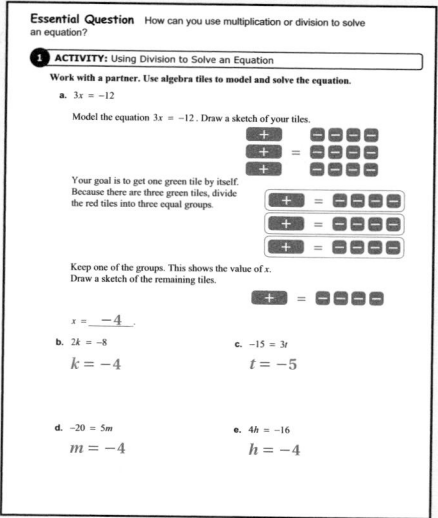

Essential Question How can you use multiplication or division to solve an equation?

1 ACTIVITY: Using Division to Solve an Equation

Work with a partner. Use algebra tiles to model and solve the equation.

a. $3x = -12$

Model the equation $3x = -12$. Draw a sketch of your tiles.

Your goal is to get one green tile by itself. Because there are three green tiles, divide the red tiles into three equal groups.

Keep one of the groups. This shows the value of *x*. Draw a sketch of the remaining tiles.

$x = \underline{-4}$.

b. $2k = -8$

$k = -4$

c. $-15 = 3t$

$t = -5$

d. $-20 = 5m$

$m = -4$

e. $4h = -16$

$h = -4$

Differentiated Instruction

Visual

To model a division equation, such as $\frac{d}{4} = -3$, use the variable tile to represent the fractional part of a variable.

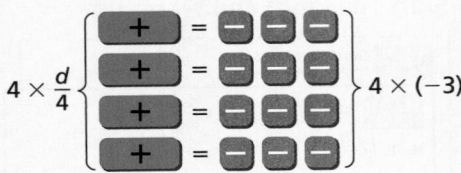

Then, model the solution.

$4 \times \frac{d}{4}$ 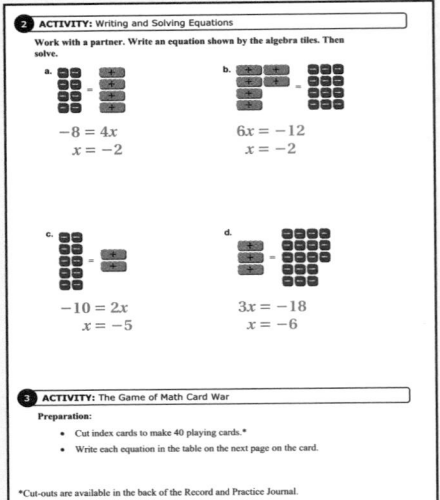 $4 \times (-3)$

So, $d = -12$.

2.5 Record and Practice Journal

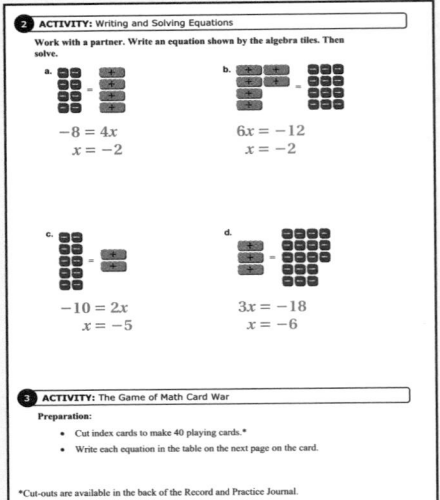

Laurie's Notes

Activity 3

- This activity is based on the card game *War*. In this version, the player with the equation that has the greater solution collects both cards. If the cards have equivalent solutions, there is a tie. As stated in the text, each player lays three cards face down and one card face up. The player that has the card with the greater solution collects all of the cards.
- To begin, give students the opportunity to preview the cards. Explain the rules and let students begin. If one group finishes early, they should shuffle the cards and play again.
- For an added challenge, have three students play the game together, rather than two.
- **Comparing Solutions:** Students will likely use mental math and not the *Guess, Check, and Revise* method to determine the solution.
- **Extension:** You might want to ask some students about the solution of the two equations $\frac{x}{3} = 5$ and $\frac{1}{3}x = 5$. Although the equations are equivalent, the manner in which they are represented often causes students to solve the first equation by multiplying both sides by 3 and the second equation by dividing each side by $\frac{1}{3}$.

Words of Wisdom

- As students play the game in Activity 3, circulate around the room to listen to the conversations.
- Are students only discussing the solution, or are they including the method they used to find the solution, as well? If no discussion is heard, ask the students how they knew their solution was greater.

What Is Your Answer?

- **Neighbor Check:** Have students work independently and then have their neighbor check their work. Have students discuss any discrepancies.

Closure

- **Exit Ticket:** Solve $\frac{x}{2} = -14$ and $2x = -14$. $-28; -7$

Technology For the Teacher

Dynamic Classroom

The Dynamic Planning Tool
Editable Teacher's Resources at *BigIdeasMath.com*

---

3 **ACTIVITY: The Game of Math Card War**

Preparation:

- Cut index cards to make 40 playing cards.
- Write each equation in the table on a card.

To Play:

- Play with a partner. Deal 20 cards to each player face-down.
- Each player turns one card face-up. The player with the greater solution wins. The winner collects both cards and places them at the bottom of his or her cards.
- Suppose there is a tie. Each player lays three cards face-down, then a new card face-up. The player with the greater solution of these new cards wins. The winner collects all ten cards, and places them at the bottom of his or her cards.
- Continue playing until one player has all the cards. This player wins the game.

$-4x = -12$	$x - 1 = 1$	$x - 3 = 1$	$2x = -10$	$-9 = 9x$
$3 + x = -2$	$x = -2$	$-3x = -3$	$\dfrac{x}{-2} = -2$	$x = -6$
$6x = -36$	$-3x = -9$	$-7x = -14$	$x - 2 = 1$	$-1 = x + 5$
$x = -1$	$9x = -27$	$\dfrac{x}{3} = -1$	$-8 = -2x$	$x = 3$
$-7 = -1 + x$	$x = -5$	$-10 = 10x$	$x = -4$	$-2 = -3 + x$
$-20 = 10x$	$x + 9 = 8$	$-16 = 8x$	$x = 2$	$x + 13 = 11$
$x = -3$	$-8 = 2x$	$x = 1$	$\dfrac{x}{2} = -2$	$-4 + x = -2$
$\dfrac{x}{5} = -1$	$-6 = x - 3$	$x = 4$	$x + 6 = 2$	$x - 5 = -4$

What Is Your Answer?

4. IN YOUR OWN WORDS How can you use multiplication or division to solve an equation without using algebra tiles? Give two examples.

Practice

Use what you learned about solving equations to complete Exercises 7–10 on page 80.

Key Ideas

Multiplication Property of Equality

Words Multiplying each side of an equation by the same number produces an equivalent equation.

Algebra If $a = b$, then $a \cdot c = b \cdot c$.

Division Property of Equality

Words Dividing each side of an equation by the same number produces an equivalent equation.

Algebra If $a = b$, then $a \div c = b \div c, c \neq 0$.

EXAMPLE ① **Solving Equations**

a. Solve $\dfrac{x}{3} = -6$.

$\dfrac{x}{3} = -6$ Write the equation.

$3 \cdot \dfrac{x}{3} = 3 \cdot (-6)$ Multiply each side by 3.

$x = -18$ Simplify.

So, the solution is $x = -18$.

b. Solve $18 = -4y$.

$18 = -4y$ Write the equation.

$\dfrac{18}{-4} = \dfrac{-4y}{-4}$ Divide each side by -4.

$-4.5 = y$ Simplify.

So, the solution is $y = -4.5$.

Check

$18 = -4y$

$18 \stackrel{?}{=} -4(-4.5)$

$18 = 18$ ✓

On Your Own

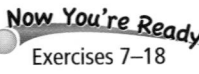
Now You're Ready
Exercises 7–18

Solve the equation. Check your solution.

1. $\dfrac{x}{5} = -2$ 2. $-a = -24$ 3. $3 = -1.5n$

Laurie's Notes

Introduction

Connect
- **Yesterday:** Students used algebra tiles to model solving equations.
- **Today:** Students will formalize the process of solving equations using the Multiplication and Division Properties of Equality.

Motivate
- Have two students stand at the front of the room. Hand a third student an odd number of index cards without telling the student how many cards he or she has been given.
- Ask the student with the index cards to share them equally between the two students. The student may pause when he or she realizes that there is an odd number of cards. Give the student time to realize the remaining card needs to be divided into two pieces and each student will receive one-half of a card.
- Ask the students holding the index cards to verify that they have the same number of cards.

Lesson Notes

Key Ideas
- Write the Properties of Equality on the board.
- ? If you started the class with the index card activity, ask students which property was modeled in the opening activity. Division Property of Equality

Example 1
- Work through each problem. If possible, use colors to show the multiplication or division on each side of the equation.
- **FYI:** Note that the -6 is written in parentheses in the solution of part (a). When you do this step in class, you may want to write both numbers in parentheses $(3)(-6)$ to avoid students thinking that the multiplication dot is a decimal point.
- ? "Could the problem be represented as $(-6) \cdot 3$ instead of $3 \cdot (-6)$? Why or why not?" yes; This is an example of the Commutative Property of Multiplication.

On Your Own
- If students have difficulty as they work these problems, assess whether it is algebraic (how to solve equations) or computational (how to multiply or divide rational numbers). Use this information to guide your instruction.
- You may want to provide colored pencils to students so that they can highlight the quantity being multiplied or divided on each side.
- Encourage students to be neat and to keep the equal signs lined up.
- Students should check their answers. If time permits, have students show their solutions at the board.

Lesson Materials	
Introduction	**Textbook**
• index cards	• colored pencils or highlighters (optional)

Start Thinking! and Warm Up

Lesson 2.5 **Warm Up** For use before Lesson 2.5

Lesson 2.5 **Start Thinking!** For use before Lesson 2.5

With a partner, write and solve a real-life word problem using the equation $13x = 39$.

Then rewrite the word problem using division.

Extra Example 1
a. Solve $\dfrac{c}{8} = -7$. -56

b. Solve $-5p = -32$. 6.4

On Your Own
1. $x = -10$
2. $a = 24$
3. $n = -2$

Extra Example 2

Solve $-\dfrac{5}{9}m = 25$. $\quad -45$

 On Your Own

4. $x = -21$

5. $b = -3\dfrac{1}{8}$

6. $h = -24$

Extra Example 3

The record low temperature in Nevada is $-50°F$. The record low temperature in Montana is 1.4 times the record low temperature in Nevada. What is the record low temperature in Montana? $\quad -70°F$

 On Your Own

7. $-80°F$

English Language Learners

Graphic Organizer

When solving a one-step equation, students must remember to isolate the variable. Encourage students to make a table in their notebooks that will help them remember which operation to use to solve a one-step equation.

Operation on Variable	Operation to Solve	Example
Addition	Subtraction	$a + 3 = -5$
Subtraction	Addition	$b - 4 = 2$
Multiplication	Division	$c \cdot (-2) = 7$
Division	Multiplication	$\dfrac{d}{-4} = -8$

Laurie's Notes

Example 2

Explain that $\dfrac{x}{3}$ and $\dfrac{1}{3}x$ are equivalent. Discuss how to multiply a fraction and a whole number: $\dfrac{1}{3}x = \dfrac{1}{3} \cdot \dfrac{x}{1} = \dfrac{x}{3}$. Repeat to show that $\dfrac{4}{5}x = \dfrac{4x}{5}$.

? "What is x being multiplied by?" $\quad -\dfrac{4}{5}$

? "Can you divide both sides by $-\dfrac{4}{5}$?" yes

- **Discuss:** Dividing by a fraction is equivalent to multiplying by its reciprocal.

? "What is the reciprocal of $-\dfrac{4}{5}$?" $\quad -\dfrac{5}{4}$

- Students may need a quick review of multiplying fractions.
- **FYI:** You may want to emphasize that you are dividing each side by $-\dfrac{4}{5}$.

  This will emphasize the connection to multiplying by the reciprocal $-\dfrac{5}{4}$, and that both of these processes are equivalent.

Words of Wisdom

- When checking a solution, read it out loud. It is helpful for students to hear (as well as to see) what they are reading.

Example 3

- Encourage students to look at the artwork next to the problem. The first sentence contains key information that is translated into the equation.
- The color-coding in the text is very helpful in assisting students as they translate from words to symbols. You may want to use color-coding when you do other examples.

On Your Own

- **Think-Pair-Share:** Students should read each question independently and then work with a partner to answer the questions. When they have answered the questions, the pair should compare their answers with another group and discuss any discrepancies.

Closure

- **Writing:** The variable in a one-step equation is being multiplied by $-\dfrac{3}{4}$. Describe how to solve the equation for x. You divide both sides of the equation by $-\dfrac{3}{4}$, which is the same as multiplying by the reciprocal. So, you multiply both sides of the equation by $-\dfrac{4}{3}$ and simplify.

Technology **F**or **T**he **T**eacher

The Dynamic Planning Tool
Editable Teacher's Resources at *BigIdeasMath.com*

EXAMPLE 2 **Solving an Equation Using a Reciprocal**

Solve $-\dfrac{4}{5}x = -8$.

$$-\dfrac{4}{5}x = -8 \qquad\text{Write the equation.}$$

$$-\dfrac{5}{4} \cdot \left(-\dfrac{4}{5}x\right) = -\dfrac{5}{4} \cdot (-8) \qquad\text{Multiply each side by } -\dfrac{5}{4}, \text{ the reciprocal of } -\dfrac{4}{5}.$$

$$x = 10 \qquad\text{Simplify.}$$

⋮• So, the solution is $x = 10$.

On Your Own

Now You're Ready
Exercises 19–22

Solve the equation. Check your solution.

4. $-14 = \dfrac{2}{3}x$ 5. $-\dfrac{8}{5}b = 5$ 6. $\dfrac{3}{8}h = -9$

EXAMPLE 3 **Real-Life Application**

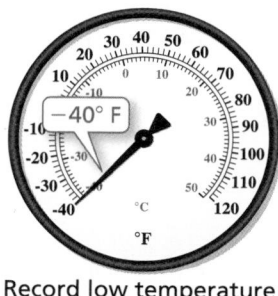

Record low temperature
in Arizona

The record low temperature in Arizona is 1.6 times the record low temperature in Rhode Island. What is the record low temperature in Rhode Island?

Words The record low in Arizona is 1.6 times the record low in Rhode Island.

Variable Let t be the record low in Rhode Island.

Equation $\qquad\qquad -40 \qquad = 1.6 \times \qquad t$

$$-40 = 1.6t \qquad\text{Write equation.}$$

$$-\dfrac{40}{1.6} = \dfrac{1.6t}{1.6} \qquad\text{Divide each side by 1.6.}$$

$$-25 = t \qquad\text{Simplify.}$$

⋮• The record low temperature in Rhode Island is $-25°$F.

On Your Own

7. The record low temperature in Hawaii is –0.15 times the record low temperature in Alaska. The record low temperature in Hawaii is 12°F. What is the record low temperature in Alaska?

 ## Vocabulary and Concept Check

1. **WRITING** Explain why multiplication can be used to solve equations involving division.

2. **OPEN-ENDED** Turning a light on and then turning the light off are considered to be inverse operations. Describe two other real-life situations that can be thought of as inverse operations.

Describe the inverse operation that will undo the given operation.

3. Multiplying by 5   4. Subtracting 12   5. Dividing by -8 6. Adding -6

 ## Practice and Problem Solving

Solve the equation. Check your solution.

7. $3h = 15$

8. $-5t = -45$

9. $\dfrac{n}{2} = -7$

10. $\dfrac{k}{-3} = 9$

11. $5m = -10$

12. $8t = -32$

13. $-0.2x = 1.6$

14. $-10 = -\dfrac{b}{4}$

15. $-6p = 48$

16. $-72 = 8d$

17. $\dfrac{n}{1.6} = 5$

18. $-14.4 = -0.6p$

19. $\dfrac{3}{4}g = -12$

20. $8 = -\dfrac{2}{5}c$

21. $-\dfrac{4}{9}f = -3$

22. $26 = -\dfrac{8}{5}y$

23. **ERROR ANALYSIS** Describe and correct the error in finding the solution.

$$\begin{aligned} -4.2x &= 21 \\ \frac{-4.2x}{4.2} &= \frac{21}{4.2} \\ x &= 5 \end{aligned}$$

Write the verbal sentence as an equation. Then solve.

24. A number divided by -9 is -16.

25. A number multiplied by $\dfrac{2}{5}$ is $\dfrac{3}{20}$.

26. The product of 15 and a number is -75.

27. The quotient of a number and -1.5 is 21.

In Exercises 28 and 29, write an equation. Then solve.

28. **NEWSPAPERS** You make $0.75 for every newspaper you sell. How many newspapers do you have to sell to buy the soccer cleats?

29. **ROCK CLIMBING** A rock climber averages $12\dfrac{3}{5}$ feet per minute. How many feet does the rock climber climb in 30 minutes?

Soccer Cleats $36^{00}

OPEN-ENDED (a) Write a multiplication equation that has the given solution.
(b) Write a division equation that has the same solution.

30. -3 **31.** -2.2 **32.** $-\dfrac{1}{2}$ **33.** $-1\dfrac{1}{4}$

34. REASONING Which of the methods can you use to solve $-\dfrac{2}{3}c = 16$?

Multiply each side by $-\dfrac{2}{3}$.	Multiply each side by $-\dfrac{3}{2}$.
Divide each side by $-\dfrac{2}{3}$.	Multiply each side by 3, then divide each side by -2.

35. STOCK A stock has a return of $-\$1.26$ per day. Write and solve an equation to find the number of days until the total return is $-\$10.08$.

36. ELECTION In a school election, $\dfrac{3}{4}$ of the students vote. There are 1464 ballots. Write and solve an equation to find the number of students.

37. OCEANOGRAPHY Aquarius is an underwater ocean laboratory located in the Florida Keys National Marine Sanctuary. Solve the equation $\dfrac{31}{25}x = -62$ to find the value of x.

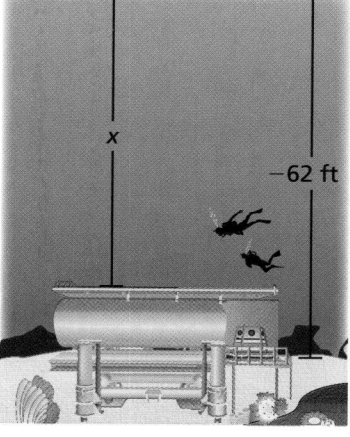
x
-62 ft

38. SHOPPING The price of a bike at store A is $\dfrac{5}{6}$ the price at store B. The price at store A is $\$150.60$. Write and solve an equation to find how much you save by buying the bike at store A.

39. CRITICAL THINKING Solve $-2|m| = -10$.

40. 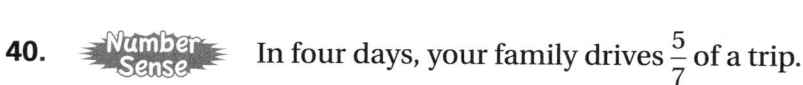 **Number Sense** In four days, your family drives $\dfrac{5}{7}$ of a trip. Your rate of travel is the same throughout the trip. The total trip is 1250 miles. How many more days until you reach your destination?

 **Fair Game Review** What you learned in previous grades & lessons

Subtract. *(Section 1.3)*

41. $5 - 12$ **42.** $-7 - 2$ **43.** $4 - (-8)$ **44.** $-14 - (-5)$

45. MULTIPLE CHOICE Of the 120 apartments in a building, 75 have been scheduled to receive new carpet. What fraction of the apartments have not been scheduled to receive new carpet? *(Section 2.1)*

Ⓐ $\dfrac{1}{4}$ Ⓑ $\dfrac{3}{8}$ Ⓒ $\dfrac{5}{8}$ Ⓓ $\dfrac{3}{4}$

Parts of an algebraic expression are called terms. **Like terms** are terms that have the same variables raised to the same exponents. A term without a variable, such as 4, is called a *constant*. Constant terms are also like terms.

Like Terms	Unlike Terms
3 and -4	x and 5
$-2x$ and $7x$	$2x$ and $-6y$

EXAMPLE **1** **Identifying Terms and Like Terms**

Identify the terms and like terms in each expression.

a. $9x - 2 + 7 - x$

$$9x - 2 + 7 - x$$

Terms: $9x, \ -2, \ 7, \ -x$

Like terms: $9x$ and $-x$, -2 and 7

Same variable raised to same exponent

b. $6 + 5z - 3z + z$

$$6 + 5z - 3z + z$$

Terms: $6, \ 5z, \ -3z, \ z$

Like terms: $5z, \ -3z,$ and z

Remember

The numerical factor of a term that contains a variable is a *coefficient*.

An algebraic expression is in **simplest form** if it has no like terms and no parentheses. To *combine* like terms that have variables, use the Distributive Property to add or subtract the coefficients.

EXAMPLE **2** **Simplifying Algebraic Expressions**

Simplify $\dfrac{3}{4}y + 12 - \dfrac{1}{2}y - 6.$

$\dfrac{3}{4}y$ and $-\dfrac{1}{2}y$ are like terms. 12 and -6 are also like terms.

$$\dfrac{3}{4}y + 12 - \dfrac{1}{2}y - 6 = \dfrac{3}{4}y - \dfrac{1}{2}y + 12 - 6 \qquad \text{Commutative Property of Addition}$$

$$= \left(\dfrac{3}{4} - \dfrac{1}{2}\right)y + 12 - 6 \qquad \text{Distributive Property}$$

$$= \dfrac{1}{4}y + 6 \qquad \text{Simplify.}$$

Practice

Identify the terms and like terms in the expression.

1. $y + 10 - \dfrac{3}{2}y$

2. $2r + 7r - r - 9$

3. $7 + 4p - 5 + p + 2q$

Simplify the expression.

4. $2.5x + 4.3x - 5$

5. $\dfrac{3}{8}b - \dfrac{3}{4}b$

6. $14 - 3z + 8 + z$

◀) Multi-Language Glossary at BigIdeasMath.com.

Laurie's Notes

Introduction

Connect

- **Yesterday:** Students solved equations using the Multiplication and Division Properties of Equality.
- **Today:** Students will simplify algebraic expressions.

Motivate

- Ask students if they have ever heard the phrase "You can't add apples and oranges."
- Some students may have heard this phrase before. Ask them what they think the phrase means. They might talk about needing a common denominator.
- **FYI:** Students often have difficulty simplifying algebraic expressions. They must be comfortable with integer operations. They must also be able to apply the Commutative and Distributive Properties. For example: $5x + 7 - 3x$ can be rewritten as $5x + 7 + (-3x) = 5x + (-3x) + 7$.

Lesson Notes

Discuss

- **Like terms** are also referred to as *similar terms*. Be sure to note that in the definition of like terms, the variables are raised to the same exponents.
- This section only involves *linear* expressions, so the exponent of each variable is 1. In a later course, students will study like terms whose variables have exponents other than 1, such as x^3 and $-2x^3$.

Example 1

- Terms are separated by addition. The expression $x - 3$ can be written as $x + (-3)$, so it has two terms. The expression $3x$ has only one term.
- **Common Error:** When identifying and writing the terms, make sure students include the sign of the term.
- In part (a) note that -2 and 7 are also like terms. More specifically, they are constant terms.

Example 2

- "What do you call the number that is multiplied by the variable?" coefficient
- Discuss what it means to write an algebraic expression in simplest form.
- Ask students to identify the coefficient of each term.
- Remind students about the Commutative and Distributive Properties.
- Have students show the step that uses the Distributive Property until they become proficient.

Practice

- **Common Error:** In Exercise 6, students will often forget the negative sign for the $-3z$ term.

Goal Today's lesson is simplifying algebraic expressions.

Warm Up

Lesson 2.5b Warm Up
For use before Lesson 2.5b

Add or subtract. Write fractions in simplest form.

1. $-\dfrac{1}{2} + \dfrac{1}{4}$ 2. $\dfrac{2}{3} - \dfrac{1}{12}$

3. $-\dfrac{5}{6} + \dfrac{4}{9}$ 4. $\dfrac{3}{4} - \left(-\dfrac{3}{8}\right)$

5. $-\dfrac{5}{8} + \dfrac{11}{12}$ 6. $-\dfrac{1}{9} - \dfrac{5}{27}$

Extra Example 1

Identify the terms and like terms in each expression.

a. $3y - 2 - 4y + 6$
   Terms: $3y$, -2, $-4y$, 6
   Like terms: $3y$ and $-4y$, -2 and 6

b. $1 + 5w + 2w - 7$
   Terms: 1, $5w$, $2w$, -7
   Like terms: 1 and -7, $5w$ and $2w$

Extra Example 2

Simplify $3a + 7.9 - 4.5a - 5.3$.
$-1.5a + 2.6$

Practice

1. Terms: y, 10, $-\dfrac{3}{2}y$

   Like terms: y and $-\dfrac{3}{2}y$

2. Terms: $2r$, $7r$, $-r$, -9
   Like terms: $2r$, $7r$, and $-r$

3. Terms: 7, $4p$, -5, p, $2q$
   Like terms: 7 and -5,
   $4p$ and p

4. $6.8x - 5$

5. $-\dfrac{3}{8}b$

6. $-2z + 22$

Record and Practice Journal Practice

See Additional Answers.

Extra Example 3

Simplify $4(3g + 1) - 5g$. $7g + 4$

Extra Example 4

Each person in Example 4 buys a ticket, a small drink, and a small popcorn. Write an expression in simplest form that represents the amount of money the group spends at the movies. $12.25x$

 ## Practice

7. $3q + 2$

8. $10x - 1$

9. $7g + 3$

10. $14x$;
$7.50x + 3.50x + 3x$
$= (7.50 + 3.50 + 3)x$
$= 14x$

Mini-Assessment

Identify the terms and like terms in the expression.

1. $\frac{5}{8}x + 9 - \frac{1}{4}x$

Terms: $\frac{5}{8}x$, 9, $-\frac{1}{4}x$

Like terms: $\frac{5}{8}x$ and $-\frac{1}{4}x$

2. $3y - 4 - y + 8$

Terms: $3y$, -4, $-y$, 8

Like terms: -4 and 8, $3y$ and $-y$

Simplify the expression.

3. $2.1d + 4 - 3.8d$ $-1.7d + 4$

4. $6\left(\frac{1}{3}g + \frac{5}{6}\right) - 9g$ $-7g + 5$

5. Write an expression in simplest form that represents the perimeter of the polygon. $(3x + 9)$ m

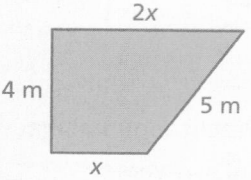

Laurie's Notes

 ## Example 3

? "What are equivalent expressions?" expressions with the same value

- **FYI:** It is not obvious to students why it is okay to rewrite $5n - 40 + 4n$ as $5n + 4n - 40$. The subtraction operation can confuse students. It might be helpful to write the expression as $5n + (-40) + 4n$. (Add the opposite.)
- **Common Error:** Students often distribute the factor of 5 to the first term inside the parentheses and then forget to distribute the 5 to the second term in the parentheses.

Example 4

- Discuss the information provided in the side column. You might ask how these prices compare to the prices at a local movie theater.
- Work through the example with students. Students are writing an algebraic expression, so it is important to identify what the variable represents in the problem. Do not skip this step!
- Take time to translate from the written words of the problem to the algebraic expression.
- When you finish the problem, ask students what the total cost would be for a group of 4 people. $14.25 \times 4 = \$57$

Practice

- **Common Error:** In Exercises 7–9, students may forget to distribute the factor to the second term in the parentheses.

 ## Closure

- Simplify the following algebraic expressions.
  a. $5x - 8 + 2x + 7x$ $14x - 8$
  b. $4n + 6(n - 4)$ $10n - 24$

Technology For the Teacher

The Dynamic Planning Tool
Editable Teacher's Resources at *BigIdeasMath.com*

Which expression is equivalent to $5(n - 8) + 4n$?

Ⓐ $49n$ Ⓑ $9n + 40$ Ⓒ $9n - 40$ Ⓓ $5n - 40$

$$5(n - 8) + 4n = 5(n) - 5(8) + 4n \qquad \text{Distributive Property}$$
$$= 5n - 40 + 4n \qquad \text{Multiply.}$$
$$= 5n + 4n - 40 \qquad \text{Commutative Property of Addition}$$
$$= (5 + 4)n - 40 \qquad \text{Distributive Property}$$
$$= 9n - 40 \qquad \text{Add coefficients.}$$

⋮ The correct answer is Ⓒ.

Evening Tickets $7.50
REFRESHMENTS
Drinks
Small $1.75
Medium $2.75
Large $3.50
Popcorn
Small $3.00
Large $4.00

Each person in a group buys a ticket, a medium drink, and a large popcorn. Write an expression in simplest form that represents the amount of money the group spends at the movies.

Words   Each ticket is $7.50, each medium drink is $2.75, and each large popcorn is $4.

Variable   The same number of each item is purchased. So, x can represent the number of tickets, the number of medium drinks, and the number of large popcorns.

Expression $7.50\,x$ + $2.75\,x$ + $4\,x$

$$7.50x + 2.75x + 4x = (7.50 + 2.75 + 4)x \qquad \text{Distributive Property}$$
$$= 14.25x \qquad \text{Add coefficients.}$$

⋮ The expression $14.25x$ represents the amount of money the group spends at the movies.

Study Tip

In Example 4, rewriting $7.50x + 2.75x + 4x$ as $14.25x$ helps you conclude that the total cost per person is $14.25.

● **Practice**

Simplify the expression.

7. $3(q + 1) - 1$

8. $7x + 4\left(\dfrac{3}{4}x - \dfrac{1}{4}\right)$

9. $2(g + 4) + 5(g - 1)$

10. WHAT IF? In Example 4, each person buys a ticket, a large drink, and a small popcorn. How does the expression change? Explain.

Essential Question In a two-step equation, which step should you do first?

1 EXAMPLE: Solving a Two-Step Equation

Use algebra tiles to model and solve $2x - 3 = -5$.

Model the equation $2x - 3 = -5$.

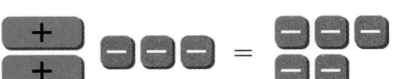

Remove the three red tiles on the left side by adding three yellow tiles to each side.

Remove the three "zero pairs" from each side.

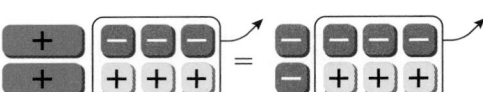

Because there are two green tiles, divide the red tiles into two equal groups.

Keep one of the groups. This shows the value of x.

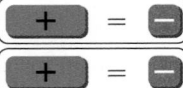

∴ So, $x = -1$.

2 EXAMPLE: The Math Behind the Tiles

Solve $2x - 3 = -5$ without using algebra tiles. Describe each step. Which step is first, adding 3 to each side or dividing each side by 2?

Use the steps in Example 1 as a guide.

$2x - 3 = -5$	Write the equation.
$2x - 3 + 3 = -5 + 3$	Add 3 to each side.
$2x = -2$	Simplify.
$\dfrac{2x}{2} = \dfrac{-2}{2}$	Divide each side by 2.
$x = -1$	Simplify.

∴ So, $x = -1$. Adding 3 to each side is the first step.

Laurie's Notes

Introduction

For the Teacher

- **Goal:** The goal of this activity is to help students develop conceptual understanding of how to solve two-step equations.

Motivate

- Write the number four on a slip of paper and put it in an envelope. Seal the envelope. Write the number 15 on the outside of the envelope.
- Hold the envelope up to your forehead and say "I'm thinking of a number. When I double the number and add 7, I get an answer of 15." Show that the number 15 is written on the outside of the envelope.
- ❓ "What number did I start with?" 4
- ❓ "Can anyone explain how they know what number I was thinking of?" Listen for students to "undo" your process by working backwards: subtract 7 from 15 (to get 8) and then divide by 2 (to get 4).
- Open the envelope and reveal the four on your slip of paper.
- ❓ "Why didn't you divide by 2 first and then subtract 7?" You need to do the steps in the reverse order to undo the calculations.
- Explain that today you will investigate how to solve equations with two operations, like the number puzzle. These are called two-step equations.

Activity Notes

Example 1

- Do the example together as a class using algebra tiles. Write the corresponding algebraic equations that result with each step to connect the model to the algebraic representation.
- **Discuss:** Remind students that the goal is to find the value of just one green tile, so it should seem reasonable to "get rid of" the red integer tiles on the left-hand side. This is referred to as *isolating the variable*.
- ❓ "How can you get the green x-tiles by themselves?" There are two ways to do this. The textbook adds three yellow integer tiles to each side (add the opposite); you can also remove three red integer tiles from each side (do the same thing to both sides of the equation).
- Take time to look at each method discussed in the question above. Adding three yellow tiles is represented by $2x - 3 + 3 = -5 + 3$. Taking three red tiles away from each side results in $2x = -2$.

Example 2

- Take time to point out that adding three is the first step in the solution.
- ❓ After you complete Example 2, ask "What happens if you divide each side by 2 first? Will you get the same answer?"
  $\frac{2x - 3}{2} = -\frac{5}{2}$ simplifies to $x - \frac{3}{2} = -\frac{5}{2}$. This introduces fractions into the problem, but you will still get the same answer.
- ❓ "Which method do you prefer?" Most students will want to avoid fractions.

Previous Learning

Students have solved two-step equations with whole numbers.

Activity Materials	
Introduction	**Textbook**
• envelope	• algebra tiles

Start Thinking! and Warm Up

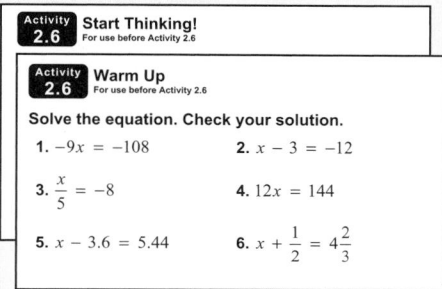

Activity 2.6 Start Thinking! For use before Activity 2.6

Activity 2.6 Warm Up For use before Activity 2.6

Solve the equation. Check your solution.

1. $-9x = -108$ 2. $x - 3 = -12$

3. $\frac{x}{5} = -8$ 4. $12x = 144$

5. $x - 3.6 = 5.44$ 6. $x + \frac{1}{2} = 4\frac{2}{3}$

2.6 Record and Practice Journal

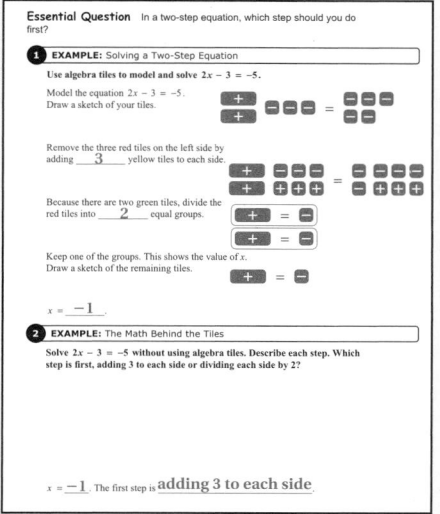

Essential Question In a two-step equation, which step should you do first?

1 EXAMPLE: Solving a Two-Step Equation

Use algebra tiles to model and solve $2x - 3 = -5$.

Model the equation $2x - 3 = -5$. Draw a sketch of your tiles.

Remove the three red tiles on the left side by adding ___3___ yellow tiles to each side.

Because there are two green tiles, divide the red tiles into ___2___ equal groups.

Keep one of the groups. This shows the value of x. Draw a sketch of the remaining tiles.

$x = -1$.

2 EXAMPLE: The Math Behind the Tiles

Solve $2x - 3 = -5$ without using algebra tiles. Describe each step. Which step is first, adding 3 to each side or dividing each side by 2?

$x = -1$. The first step is adding 3 to each side.

Differentiated Instruction

Auditory

Write the equation $5h + 6 = 1$ on the board or overhead. Ask students to tell you the steps needed to solve the equation. Repeat the steps out loud as you solve the problem. Ask students for another way to solve the problem. Which way is more efficient? Why?

2.6 Record and Practice Journal

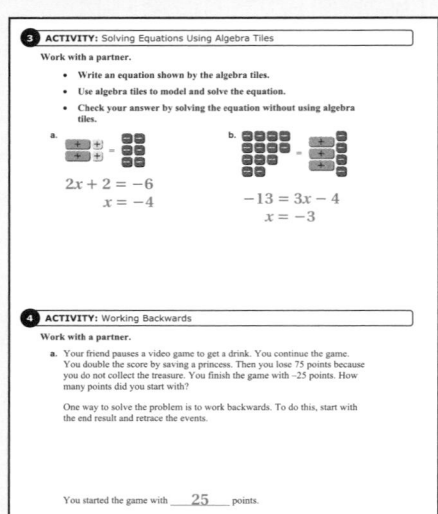

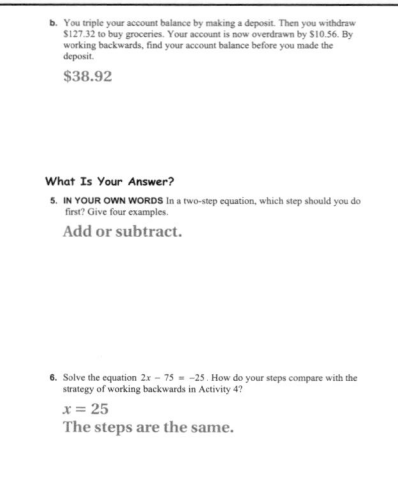

Laurie's Notes

Activity 3

- Ask for volunteers to share their work with the class.
- Listen to the language that students use when they explain their solutions. If they say "I'll put two red tiles on each side," ask them to express their steps mathematically. They *should* say "add 2 red tiles to each side."
- You want students to be able to connect their manipulation of the tiles with operations they'll record symbolically.
- Students may use different methods to solve part (b). Some students may add four yellow tiles to each side (add the opposite), and others may remove four red tiles from each side (subtract −4).
- After one method of solving an equation is described, be sure to ask if anyone approached the problem in another way so students don't think their method is wrong.

Activity 4

- If you did not do the activity in the introduction, you may want to before this activity. Both use the strategy *Working Backwards*.
- Have students make up their own number puzzles and have a partner work backwards to guess their number.

What Is Your Answer?

- Some students may not have the appropriate language to describe the process yet. Have them focus on creating examples.
- For Question 6, listen for the idea that it is the last operation performed that is "undone" first in the solving process.

Closure

- Match the equation with the first step used in solving the equation.

Equation	First Step in Solving
1) $4x - 3 = -7$	A) Divide by -3.
2) $5 = -2x + 4$	B) Subtract 2.
3) $-3x + 2 = 6$	C) Multiply by $\frac{1}{3}$.
4) $-4 = 3x - 2$	D) Add 3.
	E) Add 2.
	F) Subtract 4.

Answers: 1D, 2F, 3B, 4E

Technology For the Teacher

Dynamic Classroom

The Dynamic Planning Tool
Editable Teacher's Resources at *BigIdeasMath.com*

3 ACTIVITY: Solving Equations Using Algebra Tiles

Work with a partner.

- Write an equation shown by the algebra tiles.
- Use algebra tiles to model and solve the equation.
- Check your answer by solving the equation without using algebra tiles.

a.

b.

4 ACTIVITY: Working Backwards

Work with a partner.

a. **Sample:** Your friend pauses a video game to get a drink. You continue the game. You double the score by saving a princess. Then you lose 75 points because you do not collect the treasure. You finish the game with −25 points. How many points did you start with?

One way to solve the problem is to work backwards. To do this, start with the end result and retrace the events.

You have −25 points at the end of the game.	**−25**
You lost 75 points for not collecting the treasure, so add 75 to −25.	**−25 + 75 = 50**
You doubled your score for saving the princess, so find half of 50.	**50 ÷ 2 = 25**

∴ So, you started the game with 25 points.

b. You triple your account balance by making a deposit. Then you withdraw $127.32 to buy groceries. Your account is now overdrawn by $10.56. By working backwards, find your account balance before you made the deposit.

What Is Your Answer?

5. IN YOUR OWN WORDS In a two-step equation, which step should you do first? Give four examples.

6. Solve the equation $2x - 75 = -25$. How do your steps compare with the strategy of working backwards in Activity 4?

Practice

Use what you learned about solving two-step equations to complete Exercises 6–11 on page 86.

2.6 Lesson

EXAMPLE 1 Solving a Two-Step Equation

Solve $-3x + 5 = 2$. Check your solution.

$-3x + 5$	$=$	2	Write the equation.
$\underline{-5}$		$\underline{-5}$	Subtract 5 from each side.
$-3x$	$=$	-3	Simplify.
$\dfrac{-3x}{-3}$	$=$	$\dfrac{-3}{-3}$	Divide each side by -3.
x	$=$	1	Simplify.

Check

$-3x + 5 = 2$

$-3(1) + 5 \overset{?}{=} 2$

$-3 + 5 \overset{?}{=} 2$

$2 = 2$ ✓

⋮ So, the solution is $x = 1$.

On Your Own

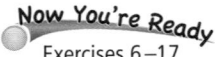
Exercises 6–17

Solve the equation. Check your solution.

1. $2x + 12 = 4$ **2.** $-5c + 9 = -16$ **3.** $3(x - 4) = 9$

EXAMPLE 2 Solving a Two-Step Equation

Solve $\dfrac{x}{8} - \dfrac{1}{2} = -\dfrac{7}{2}$.

$\dfrac{x}{8} - \dfrac{1}{2}$	$=$	$-\dfrac{7}{2}$	Write the equation.
$+\dfrac{1}{2}$		$+\dfrac{1}{2}$	Add $\dfrac{1}{2}$ to each side.
$\dfrac{x}{8}$	$=$	-3	Simplify.
$8 \cdot \dfrac{x}{8}$	$=$	$8 \cdot (-3)$	Multiply each side by 8.
x	$=$	-24	Simplify.

Study Tip

You can simplify the equation in Example 2 before solving. Multiply each side by the LCD of the fractions, 8.

$\dfrac{x}{8} - \dfrac{1}{2} = -\dfrac{7}{2}$

$x - 4 = -28$

$x = -24$

⋮ So, the solution is $x = -24$.

On Your Own

Exercises 20–25

Solve the equation. Check your solution.

4. $\dfrac{m}{2} + 6 = 10$ **5.** $-\dfrac{z}{3} + 5 = 9$ **6.** $\dfrac{2}{5} + 4a = -\dfrac{6}{5}$

Laurie's Notes

Introduction

Connect
- **Yesterday:** Students used algebra tiles to model solving two-step equations.
- **Today:** Students will solve equations by undoing the operations in the reverse order of how the expression would have been evaluated.

Motivate
- ❓ "Four friends each purchase a large beverage and share a $9 pizza. The total bill before tax is $16. What is the cost of each beverage?" $1.75
- Ask students to explain how they solved this problem. Listen for students to mention subtracting the cost of the pizza from the total before dividing by four.

Lesson Notes

Example 1
- ❓ **Vocabulary Review:** "In the expression $-3x + 5$, what is -3 called?" the coefficient
- Work through the example. Before doing each step, ask students what the next step should be.
- Take the time to check the solution so that students see this as important.

On Your Own
- Students may be uncertain of how to solve Question 3 because of the parentheses. Remind students about the Distributive Property.
- Review the methods students use to solve each problem. In Question 3, for example, some students may distribute the three and some may realize they can divide both sides of the equation by three.
- **Challenge:** Ask students to describe two methods for solving Question 1. Some may notice that each number in the equation can be divided by 2.

Example 2
- ❓ "If you knew the value of x, how would you evaluate the expression $\frac{x}{8} - \frac{1}{2}$?" *Sample answer:* Divide the number by 8 and subtract $\frac{1}{2}$; Some students might say that you need to find a common denominator and then subtract the fractions.
- ❓ "What is the first step to solve this equation?" Add $\frac{1}{2}$ to both sides.
- ❓ "What is the second step to solve this equation?" Multiply both sides by 8.

On Your Own
- **Think-Pair-Share:** Students should read each question independently and then work with a partner to answer the questions. When they have answered the questions, the pair should compare their answers with another group and discuss any discrepancies.

Start Thinking! and Warm Up

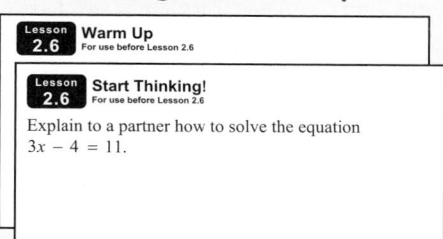

Lesson 2.6 **Warm Up** For use before Lesson 2.6

Lesson 2.6 **Start Thinking!** For use before Lesson 2.6

Explain to a partner how to solve the equation $3x - 4 = 11$.

Extra Example 1
Solve $4t - 7 = -15$. -2

On Your Own
1. $x = -4$
2. $c = 5$
3. $x = 7$

Extra Example 2
Solve $\frac{n}{9} + \frac{2}{3} = -\frac{2}{3}$. -12

On Your Own
4. $m = 8$
5. $z = -12$
6. $a = -\frac{2}{5}$

Extra Example 3

Solve $12.5 = 0.3m - 2.8m$. -5

Extra Example 4

A taxi charges $2.50 plus $2 for every mile traveled. Find the number of miles traveled for a fare of $10.50. 4 miles

 On Your Own

    7. $y = 8$

    8. $x = -5$

    9. $m = 10$

   10. 9.5 ft

English Language Learners

Verbal Clues

English learners should become familiar with words and phrases that give clues to the types of operations required. Word problems calling for two-step equations almost always contain words such as *per*, *each*, and *every*. These are clues to quantities that will appear in the equation, usually as the coefficient of the variable. Point out these words in the exercises and identify the terms associated with them in the equations used to solve the problems.

Laurie's Notes

Example 3

- This problem requires students to *combine like terms* as the first step.
- **?** "What do we call $3y$ and $-8y$?" They are like terms.

Example 4

- Note that the unknown value in this problem is the starting height. Roller coasters do not begin on the ground!
- The table helps students develop their ability to translate from words to symbols.

On Your Own

- **Think-Pair-Share**: Students should read each question independently and then work with a partner to answer the questions. When they have answered the questions, the pair should compare their answers with another group and discuss any discrepancies.

Closure

- What does it mean to *isolate the variable term*? Use inverse operations to get the variable by itself.
- What are like or similar terms? Give examples. Terms that can be combined are like terms. Some examples are $3x$ and $-2x$, $5a$ and $-a$, and 5 and -8.
- Explain how the solutions of the two equations are similar.

$$4x - 5 = 7 \qquad \frac{4}{3}x - \frac{5}{3} = \frac{7}{3}$$

You can multiply each term of the second equation by 3 and then the two equations will be the same. The solution of the two equations is the same.

Technology For the Teacher

Dynamic Classroom

The Dynamic Planning Tool
Editable Teacher's Resources at *BigIdeasMath.com*

EXAMPLE **3** **Combining Like Terms Before Solving**

Solve $3y - 8y = 25$.

$3y - 8y = 25$	Write the equation.
$-5y = 25$	Combine like terms.
$y = -5$	Divide each side by -5.

⋮ So, the solution is $y = -5$.

EXAMPLE **4** **Real-Life Application**

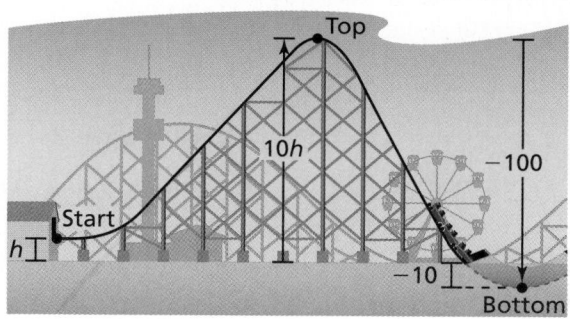

The height at the top of a roller coaster hill is 10 times the height h of the starting point. The height decreases 100 feet from the top to the bottom of the hill. The height at the bottom of the hill is -10 feet. Find h.

Location	Verbal Description	Expression
Start	The height at the start is h.	h
Top of hill	The height at the top of the hill is 10 times the starting height h.	$10h$
Bottom of hill	Height decreases by 100 feet. So, subtract 100.	$10h - 100$

The height at the bottom of the hill is -10 feet. Solve $10h - 100 = -10$ to find h.

$10h - 100 = -10$	Write equation.
$10h = 90$	Add 100 to each side.
$h = 9$	Divide each side by 10.

⋮ The height at the start is 9 feet.

On Your Own

Now You're Ready
Exercises 29–34

Solve the equation. Check your solution.

7. $4 - 2y + 3 = -9$ **8.** $7x - 10x = 15$ **9.** $-8 = 1.3m - 2.1m$

10. WHAT IF? In Example 4, the height at the bottom of the hill is -5 feet. Find the height h.

 Vocabulary and Concept Check

1. **WRITING** How do you solve two-step equations?

Match the equation with the first step to solve it.

2. $4 + 4n = -12$

3. $4n = -12$

4. $\dfrac{n}{4} = -12$

5. $\dfrac{n}{4} - 4 = -12$

A. Add 4.

B. Subtract 4.

C. Multiply by 4.

D. Divide by 4.

 Practice and Problem Solving

Solve the equation. Check your solution.

① 6. $2v + 7 = 3$

7. $4b + 3 = -9$

8. $17 = 5k - 2$

9. $-6t - 7 = 17$

10. $8n + 16.2 = 1.6$

11. $-5g + 2.3 = -18.8$

12. $2t - 5 = -10$

13. $-4p + 9 = -5$

14. $11 = -5x - 2$

15. $4 + 2.2h = -3.7$

16. $-4.8f + 6.4 = -8.48$

17. $7.3y - 5.18 = -51.9$

ERROR ANALYSIS Describe and correct the error in finding the solution.

18.

$$-6 + 2x = -10$$
$$-6 + \dfrac{2x}{2} = -\dfrac{10}{2}$$
$$-6 + x = -5$$
$$x = 1$$

19.

$$-3x + 2 = -7$$
$$-3x = -9$$
$$-\dfrac{3x}{3} = \dfrac{-9}{3}$$
$$x = -3$$

Solve the equation. Check your solution.

② 20. $\dfrac{3}{5}g - \dfrac{1}{3} = -\dfrac{10}{3}$

21. $\dfrac{a}{4} - \dfrac{5}{6} = -\dfrac{1}{2}$

22. $-\dfrac{1}{3} + 2z = -\dfrac{5}{6}$

23. $2 - \dfrac{b}{3} = -\dfrac{5}{2}$

24. $-\dfrac{2}{3}x + \dfrac{3}{7} = \dfrac{1}{2}$

25. $-\dfrac{9}{4}v + \dfrac{4}{5} = \dfrac{7}{8}$

In Exercises 26–28, write an equation. Then solve.

26. **WEATHER** Starting at 1:00 P.M., the temperature changes -4 degrees per hour. How long will it take to reach $-1°$?

27. **BOWLING** It costs $2.50 to rent bowling shoes. Each game costs $2.25. You have $9.25. How many games can you bowl?

28. **CELL PHONES** A cell phone company charges a monthly fee plus $0.25 for each text message. The monthly fee is $30.00 and you owe $59.50. How many text messages did you have?

Temperature at 1:00 P.M.

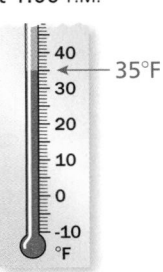

35°F

Assignment Guide and Homework Check

Level	Day 1 Activity Assignment	Day 2 Lesson Assignment	Homework Check
Basic	6–11, 42–46	1–5, 13–25 odd, 26–28, 29–33 odd	13, 21, 28, 29
Average	6–11, 42–46	1–5, 13–25 odd, 28–36 even, 37, 39	13, 21, 28, 30
Advanced	6–11, 42–46	1–5, 18, 19, 24, 25, 30–40 even, 39, 41	24, 30, 36, 40

Common Errors

- **Exercises 6–17** Students may divide the coefficient first instead of adding or subtracting first. Tell them that while this is a valid method, they must remember to divide each part of the equation by the coefficient.
- **Exercises 20–25** Students may immediately multiply each term by one of the denominators without thinking if it will help them solve for the variable. Ask them to check if all the denominators would be eliminated.
- **Exercises 32–34** Students may try to add or subtract without distributing. Remind them that when parentheses are present, they either need to use the Distributive Property or they need to undo the multiplication first. All of the exercises can be solved using either method.

2.6 Record and Practice Journal

Solve the equation. Check your solution.

1. $3a - 5 = -14$
   -3
2. $10 = -2c + 22$
   6
3. $18 = -5b - 17$
   -7

4. $-12 = -8z + 12$
   3
5. $1.3n - 0.03 = -9$
   -6.9
6. $\frac{5}{11}h + \frac{7}{9} = \frac{2}{9}$
   $1\frac{2}{9}$

7. It costs $34.95 to rent a jet ski for four hours plus $15.75 for each additional hour. You have $100. Can you rent the jet ski for 8 hours?
   Explain. **Yes; Solving the equation $34.95 + 15.75h = 100$ gives a solution of $h \approx 4.13$ hours. So you can rent the jet ski for about 8.13 hours. Renting the jet ski for 8 hours costs $97.95 and you have $100.**

8. The length of a rectangle is 3 meters less than twice its width.
   a. Write an equation to find the length of the rectangle.
   $\ell = 2w - 3$
   b. The length of the rectangle is 11 meters. What is the width of the rectangle?
   7 meters

Technology For the Teacher
Answer Presentation Tool
QuizShow

Vocabulary and Concept Check

1. Eliminate the constants on the side with the variable. Then solve for the variable using either division or multiplication.

2. B
3. D
4. C
5. A

Practice and Problem Solving

6. $v = -2$
7. $b = -3$
8. $k = 3\frac{4}{5}$
9. $t = -4$
10. $n = -1.825$
11. $g = 4.22$
12. $t = -2\frac{1}{2}$
13. $p = 3\frac{1}{2}$
14. $x = -2\frac{3}{5}$
15. $h = -3.5$
16. $f = 3.1$
17. $y = -6.4$
18. The steps are out of order.
$$-6 + 2x = -10$$
$$2x = -4$$
$$\frac{2x}{2} = \frac{-4}{2}$$
$$x = -2$$
19. Each side should be divided by -3, not 3.
$$-3x + 2 = -7$$
$$-3x = -9$$
$$\frac{-3x}{-3} = \frac{-9}{-3}$$
$$x = 3$$
20. $g = -5$
21. $a = 1\frac{1}{3}$
22. $z = -\frac{1}{4}$
23. $b = 13\frac{1}{2}$
24. $x = -\frac{3}{28}$
25. $v = -\frac{1}{30}$
26. $-4x + 35 = -1$;
9 hours (10:00 P.M.)

Practice and Problem Solving

27. $2.5 + 2.25x = 9.25$; 3 games

28. $30 + 0.25x = 59.5$; 118 text messages

29. $v = -5$ **30.** $t = -13$

31. $d = -12$ **32.** $x = -1$

33. $m = -9$ **34.** $y = -4$

35. *Sample answer:* You travel halfway up a ladder. Then you climb down two feet and are 8 feet above the ground. How long is the ladder? $x = 20$

36. $12\frac{3}{4}$ ft

37. the initial fee

38. the coldest surface temperature on the moon

39. See *Taking Math Deeper.*

40–41. See Additional Answers.

Fair Game Review

42. -34.72 **43.** $-6\frac{2}{3}$

44. $-3\frac{1}{8}$ **45.** 6.2

46. C

Mini-Assessment

Solve the equation.

1. $4x + 16.4 = -3.6$ $x = -5$

2. $-8.46 = -2.1n - 2.16$ $n = 3$

3. $-\frac{4}{5} + \frac{1}{2}m = -\frac{1}{5}$ $m = 1\frac{1}{5}$

4. $-\frac{5}{9} = \frac{2}{3}\ell - \frac{1}{3}$ $\ell = -\frac{1}{3}$

5. A gym charges $8.75 for each swimming class and a one-time registration fee of $12.50. A student paid a total of $56.25. Write and solve an equation to find the number of swimming classes the student took. $8.75x + 12.5 = 56.25$; 5 classes

Taking Math Deeper

Exercise 39

This problem asks students to answer a question *with* and *without* using algebra.

 Summarize the given information.

   1. You caught x insects on Saturday.

   2. 5 of the insects escaped.

   3. The remaining insects form 3 groups of 9 each.

 a. Work backwards.

   3. There are $3(9) = 27$ insects remaining.

   2. Add the 5 that escaped.

   1. You caught $27 + 5$, or 32, insects on Saturday.

 b. Write and solve an equation.

   1. x insects on Saturday

   2. $(x - 5)$ are remaining.

   3. $\dfrac{(x - 5)}{3} = 9$

$$\frac{x - 5}{3} = 9 \qquad \text{Write the equation.}$$

$$x - 5 = 27 \qquad \text{Multiply each side by 3.}$$

$$x = 32 \qquad \text{Add 5 to each side.}$$

You caught 32 insects on Saturday.

32 is a lot!

Reteaching and Enrichment Strategies

If students need help...	If students got it...

Resources by Chapter
  • Practice A and Practice B
  • Puzzle Time
Record and Practice Journal Practice
Differentiating the Lesson
Lesson Tutorials
Skills Review Handbook
	Resources by Chapter
	  • Enrichment and Extension
	  • School-to-Work
	  • Financial Literacy
	  • Technology Connection
	Start the next section

Solve the equation. Check your solution.

③ 29. $3v - 9v = 30$

30. $12t - 8t = -52$

31. $-8d - 5d + 7d = 72$

32. $6(x - 2) = -18$

33. $-4(m + 3) = 24$

34. $-8(y + 9) = -40$

35. WRITING Write a real-world problem that can be modeled by $\frac{1}{2}x - 2 = 8$. Then solve the equation.

36. GEOMETRY The perimeter of the parallelogram is 102 feet. Find m.

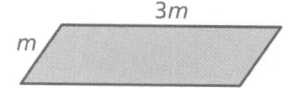

REASONING Exercises 37 and 38 are missing information. Tell what information is needed to solve the problem.

37. TAXI A taxi service charges an initial fee plus \$1.80 per mile. How far can you travel for \$12?

38. EARTH The coldest surface temperature on the moon is 57 degrees colder than twice the coldest surface temperature on Earth. What is the coldest surface temperature on Earth?

39. SCIENCE On Saturday, you catch insects for your science class. Five of the insects escape. The remaining insects are divided into three groups to share in class. Each group has nine insects. How many insects did you catch on Saturday?

   a. Solve the problem by working backwards.

   b. Solve the equation $\dfrac{x - 5}{3} = 9$. How does the answer compare with the answer to part (a)?

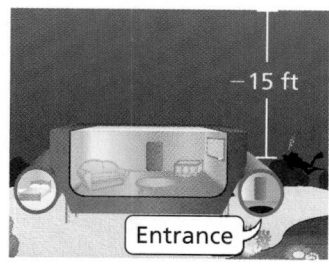

40. UNDERWATER HOTEL You must scuba dive to the entrance of your room at Jule's Undersea Lodge in Key Largo, Florida. The diver is 1 foot deeper than $\frac{2}{3}$ of the elevation of the entrance. What is the elevation of the entrance?

41. ⟐Geometry⟐ How much should you change the length of the rectangle so that the perimeter is 54 centimeters? Write an equation that shows how you found your answer.

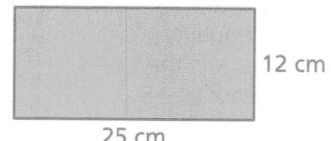

Fair Game Review   *What you learned in previous grades & lessons*

Multiply or divide. *(Section 2.3)*

42. -6.2×5.6

43. $\dfrac{8}{3} \times \left(-2\dfrac{1}{2}\right)$

44. $\dfrac{5}{2} \div \left(-\dfrac{4}{5}\right)$

45. $-18.6 \div (-3)$

46. MULTIPLE CHOICE Which fraction is *not* equivalent to 0.75? *(Skills Review Handbook)*

  Ⓐ $\dfrac{15}{20}$ Ⓑ $\dfrac{9}{12}$ Ⓒ $\dfrac{6}{9}$ Ⓓ $\dfrac{3}{4}$

 **Key Ideas**

Addition Property of Inequality

Words If you add the same number to each side of an inequality, the inequality remains true.

Algebra If $a < b$, then $a + c < b + c$.

Subtraction Property of Inequality

Words If you subtract the same number from each side of an inequality, the inequality remains true.

Algebra If $a < b$, then $a - c < b - c$.

These properties are true for $<$, $>$, $\leq$, and $\geq$.

Study Tip

You can solve inequalities in much the same way you solve equations. Use inverse operations to get the variable by itself.

EXAMPLE 1 **Solving Inequalities Using Addition or Subtraction**

a. Solve $x - 5 < -3$. Graph the solution.

$$x - 5 < -3 \qquad \text{Write the inequality.}$$

Undo the subtraction. ⟶ $\underline{+5 \qquad +5} \qquad \text{Add 5 to each side.}$

$$x < 2 \qquad \text{Simplify.}$$

⋮· The solution is $x < 2$.

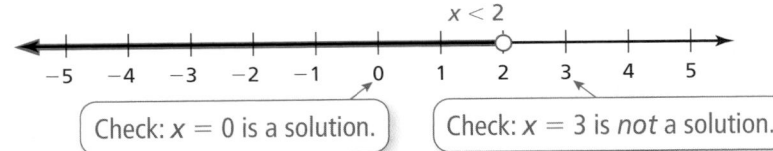

Check: $x = 0$ is a solution.   Check: $x = 3$ is *not* a solution.

Reading

The inequality $-8.3 \leq x$ is the same as $x \geq -8.3$.

b. Solve $-3.5 \leq 4.8 + x$.

$$-3.5 \leq \quad 4.8 + x \qquad \text{Write the inequality.}$$

Undo the addition. ⟶ $\underline{-4.8 \qquad -4.8} \qquad \text{Subtract 4.8 from each side.}$

$$-8.3 \leq x \qquad \text{Simplify.}$$

⋮· The solution is $x \geq -8.3$.

 **Practice**

Solve the inequality. Graph the solution.

1. $x - 2 < 1$

2. $n + 7 \geq -4$

3. $r - 1.2 > -0.5$

4. $2.2 < 4.3 + y$

5. $\dfrac{3}{5} \geq z + \dfrac{2}{5}$

6. $m + \dfrac{1}{2} \leq -\dfrac{1}{2}$

Laurie's Notes

Introduction

Connect
- **Yesterday:** Students solved two-step equations.
- **Today:** Students will use the Addition, Subtraction, Multiplication, and Division Properties of Inequality to solve inequalities.

Motivate
- ❓ "What does TSA stand for?" Transportation Safety Administration
- The TSA has guidelines for the maximum weight of luggage, depending on whether it is carry-on or checked luggage.
- ❓ "If there is a maximum weight restriction of 50 pounds for a checked bag, what inequality does this suggest?" $w \leq 50$
- Suggest different scenarios. "If my bag weighs 40.5 pounds, how much weight can I add? If my bag weighs 56.4 pounds, how much weight must I remove?"
- Today's lesson involves solving inequalities of this type.

Lesson Notes

Key Ideas
- Write the Key Ideas. These properties should look familiar, as they are similar to the Addition and Subtraction Properties of Equality that students have used in solving equations.
- **Teaching Tip:** Summarize these two properties in the following way: George is older than Martha. In two years, George will still be older than Martha.

George's age > Martha's age	If $a > b$,
George's age + 2 > Martha's age + 2	then $a + c > b + c$.

Two years ago, George was older than Martha.

George's age > Martha's age	If $a > b$,
George's age − 2 > Martha's age − 2	then $a - c > b - c$.

Example 1
- ❓ Write the problem in part (a). "How do you isolate the variable, meaning get x by itself?" Add 5 to each side of the inequality.
- Adding 5 is the inverse operation of subtracting 5.
- Solve, graph, and check.
- Part (b) reviews subtraction of rational numbers. The common wrong answer will be -1.3. To help students, write the problem horizontally off to the side: $-3.5 - 4.8 = ?$
- Although $-8.3 \leq x$ is a correct answer, it is standard practice to rewrite inequalities so that the variable is read first, left to right. So, the solution is $x \geq -8.3$. The variable changes sides and the direction of the inequality symbol is reversed.

Practice
- **Note:** These problems integrate review of fraction and decimal operations.

Goal Today's lesson is solving inequalities.

Warm Up

Lesson 2.6b
Warm Up
For use before Lesson 2.6b

Solve the equation. Check your solution.
1. $3x + 1 = 10$
2. $-4x + 15 = 15$
3. $\frac{2}{3}x - \frac{5}{6} = -\frac{1}{3}$
4. $\frac{1}{2} - \frac{2}{3}x = \frac{1}{4}$
5. $0.2 + 0.5x = 0.32$
6. $0.6x + 3 = 30$

Extra Example 1
a. Solve $y - 4 > -7$. Graph the solution.
   $y > -3$
   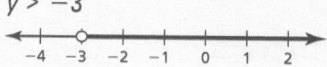

b. Solve $n + 5.2 \leq -1.8$. Graph the solution. $n \leq -7$
   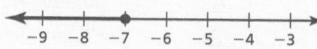

Practice

1. $x < 3$;
   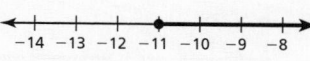

2. $n \geq -11$;

3. $r > 0.7$;

4. $y > -2.1$;
   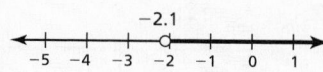

5. $z \leq \frac{1}{5}$;
   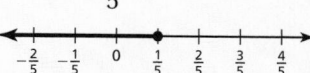

6. $m \leq -1$;

Record and Practice Journal Practice
See Additional Answers.

T-87A

Laurie's Notes

Extra Example 2

a. Solve $\frac{m}{6} \geq -3$. Graph the solution.

$m \geq -18$

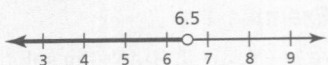

b. Solve $4.2z < 27.3$. Graph the solution.

$z < 6.5$

```
        6.5
  ◄──┼──┼──┼──┼──○──┼──┼──►
     3  4  5  6  7  8  9
```

● Practice

7. $b \geq -40$;

```
  ◄──┼──┼──┼──●━━┿━━┿━━┿━►
    -70 -60 -50 -40 -30 -20 -10
```

8. $g < -6$;

```
  ◄━━┿━━┿━━○──┼──┼──┼──┼──►
     -9  -8  -7  -6  -5  -4  -3
```

9. $m \leq \frac{4}{3}$;

```
               4
               3
  ◄━━┿━━┿━━┿━━●──┼──┼──┼──►
     -2  -1   0   1   2   3   4
```

10. $q > 7$;

```
  ◄──┼──┼──┼──○━━┿━━┿━━┿━►
     4   5   6   7   8   9   10
```

11. $x \geq 25$;

```
  ◄──┼──┼──┼──●━━┿━━┿━━┿━►
    22  23  24  25  26  27  28
```

12. $u > -12$;

```
  ◄──┼──┼──┼──○━━┿━━┿━━┿━►
    -15 -14 -13 -12 -11 -10 -9
```

Key Idea

- Write the Key Idea. These properties should look familiar, as they are similar to the Multiplication and Division Properties of Equality.
- Note that the properties are restricted to multiplying and dividing by a *positive* number. This is very important.

Example 2

- Write the problem in part (a). Read it aloud: "A number x is divided by 10 and the answer is less than or equal to -2."
- **?** "How do you isolate the variable, meaning get x by itself?" Multiply by 10 on each side of the inequality.
- Multiplying by 10 is the inverse operation of dividing by 10.
- **Representation:** Note that multiplication is represented by the dot notation and that -2 is enclosed in parentheses for clarity only; otherwise students might become confused and think 2 is being subtracted.
- Solve, graph, and check.
- Write the problem in part (b). Read it aloud and solve.

Practice

- Remind students that even though the variable may be on the right side of the inequality, the process of solving for the variable remains the same.
- Discuss how to rewrite an answer of $4 < n$ to get the equivalent inequality $n > 4$.
- Exercise 9 involves a fractional coefficient. Students will naturally divide by the coefficient, $\frac{2}{3}$. Discuss the equivalence of multiplying by the reciprocal of the coefficient, $\frac{3}{2}$.

 Key Idea

> **Multiplication and Division Properties of Inequality (Case 1)**
>
> **Words** If you multiply or divide each side of an inequality by the same *positive* number, the inequality remains true.
>
> **Algebra** If $a < b$, then $a \cdot c < b \cdot c$ for a positive number c.
>
> If $a < b$, then $\dfrac{a}{c} < \dfrac{b}{c}$ for a positive number c.

EXAMPLE 2 Solving Inequalities Using Multiplication or Division

a. Solve $\dfrac{x}{10} \le -2$. Graph the solution.

$\dfrac{x}{10} \le -2$	Write the inequality.

Undo the division. ⟶ $10 \cdot \dfrac{x}{10} \le 10 \cdot (-2)$ Multiply each side by 10.

$\qquad\qquad\qquad x \le -20$ Simplify.

:∴: The solution is $x \le -20$.

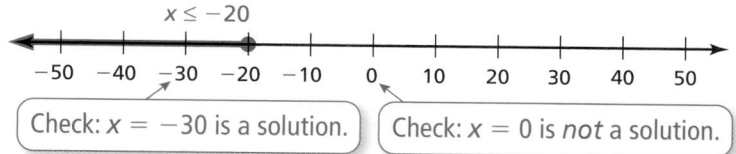

Check: $x = -30$ is a solution. Check: $x = 0$ is *not* a solution.

b. Solve $2.5x > 11.25$. Graph the solution.

$\qquad\qquad 2.5x > 11.25$ Write the inequality.

Undo the multiplication. ⟶ $\dfrac{2.5x}{2.5} > \dfrac{11.25}{2.5}$ Divide each side by 2.5.

$\qquad\qquad\qquad x > 4.5$ Simplify.

:∴: The solution is $x > 4.5$.

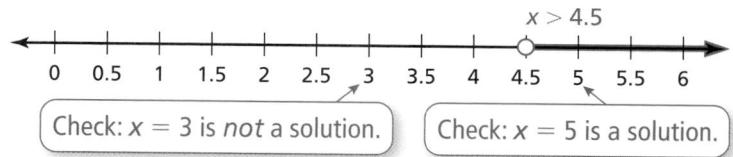

Check: $x = 3$ is *not* a solution. Check: $x = 5$ is a solution.

● **Practice**

Solve the inequality. Graph the solution.

7. $\dfrac{b}{8} \ge -5$

8. $-0.4 > \dfrac{g}{15}$

9. $\dfrac{2}{3}m \le \dfrac{8}{9}$

10. $63 < 9q$

11. $60 \le 2.4x$

12. $1.6u > -19.2$

Key Idea

Multiplication and Division Properties of Inequality (Case 2)

Words If you multiply or divide each side of an inequality by the same *negative* number, the direction of the inequality symbol must be reversed for the inequality to remain true.

Algebra If $a < b$, then $a \cdot c > b \cdot c$ for a negative number c.

If $a < b$, then $\dfrac{a}{c} > \dfrac{b}{c}$ for a negative number c.

EXAMPLE 3 Solving Inequalities Using Multiplication or Division

a. Solve $\dfrac{y}{-4} > 6$. Graph the solution.

$$\dfrac{y}{-4} > 6 \qquad \text{Write the inequality.}$$

Undo the division. → $-4 \cdot \dfrac{y}{-4} < -4 \cdot 6$
Multiply each side by -4.
Reverse the inequality symbol.

$$y < -24 \qquad \text{Simplify.}$$

The solution is $y < -24$.

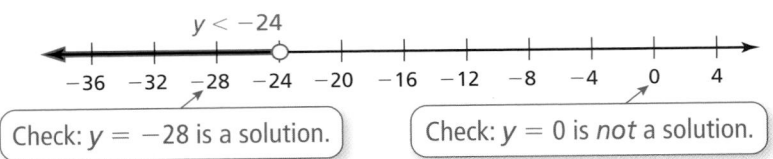

$y < -24$

Check: $y = -28$ is a solution.

Check: $y = 0$ is *not* a solution.

b. Solve $-21 \geq -1.4y$.

$$-21 \geq -1.4y \qquad \text{Write the inequality.}$$

Undo the multiplication. → $\dfrac{-21}{-1.4} \leq \dfrac{-1.4y}{-1.4}$
Divide each side by -1.4.
Reverse the inequality symbol.

$$15 \leq y \qquad \text{Simplify.}$$

The solution is $y \geq 15$.

● Practice

Solve the inequality. Graph the solution.

13. $7 > \dfrac{j}{-1.5}$

14. $\dfrac{a}{-3} \leq -2$

15. $-2.5 < k \div (-4.8)$

16. $-2s < 24$

17. $-3.1z \geq 62$

18. $-3.9 \geq -0.6d$

Laurie's Notes

Key Idea

- This Key Idea addresses the negative coefficient.
- Write the Key Idea. These properties look similar to what students have been using in the lesson, *except* now the direction of the inequality symbol must be reversed for the inequality to remain true because they are multiplying or dividing by a *negative* quantity.
- The short version of the property: When you multiply or divide by a negative quantity, reverse the direction of the inequality symbol.
- Use the following example to help students understand why you reverse the inequality symbol.

$$-x < 4$$
$$0 < x + 4 \qquad \text{Add } x \text{ to each side}$$
$$-4 < x \qquad \text{Subtract 4 from each side.}$$
$$x > -4 \qquad \text{Rewrite the inequality.}$$

- Explain that by using the rule in the Key Idea, students will get the same answer but use fewer steps.
- Draw the connection between reversing the inequality symbol and the example above, where you move the variable term with a negative coefficient to the other side of the inequality. The negative sign is eliminated, but the inequality symbol is different when you rewrite the inequality. You can show this with more difficult inequalities, such as $-2x < 4$ or $7 \geq -3x + 10$.
- **Common Error:** When students solve $2x < -4$, they sometimes reverse the inequality symbol because there's a negative number in the problem. The inequality symbol is reversed *only* when both sides of the inequality are multiplied or divided by a negative number.

Example 3

- Write the problem in part (a).
- **?** "What operation is being performed?" Divide by -4.
- **?** "How do you undo dividing by -4?" Multiply by -4.
- Solve as usual, reversing the direction of the inequality symbol.
- When graphing, the endpoint is an open circle because the inequality is strictly less than.
- Write the problem in part (b).
- **?** "What operation is being performed?" Multiply by -1.4.
- **?** "How do you undo multiplying by -1.4?" Divide by -1.4.
- Solve as usual, reversing the direction of the inequality symbol. Also, remember that the quotient of two negatives is positive.
- If you graph the solution, remember the endpoint is a closed circle because the inequality is greater than or equal to.

Practice

- **Note:** These problems integrate review of decimal operations.
- **Common Error:** Students may forget to reverse the inequality symbol.

Extra Example 3

a. Solve $\dfrac{k}{-2} \leq 12$. Graph the solution.

$k \geq -24$

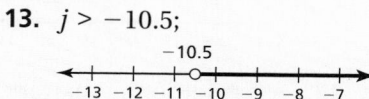

b. Solve $-2.1d > -16.8$. Graph the solution. $d < 8$

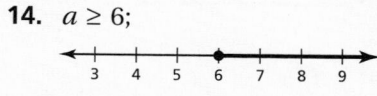

Practice

13. $j > -10.5;$

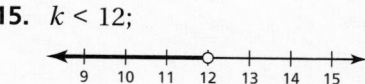

14. $a \geq 6;$

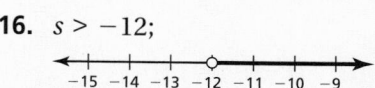

15. $k < 12;$

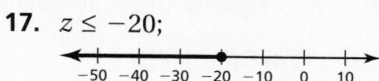

16. $s > -12;$

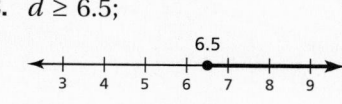

17. $z \leq -20;$

18. $d \geq 6.5;$

Laurie's Notes

Extra Example 4

Solve $-\frac{1}{2}y + 4 > 1$. Graph the solution. $y < 6$

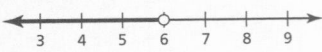

Extra Example 5

You have 74 points in a video game. For each jewel lost, you lose 2 points. You need at least 50 points to advance to the next level. Write and solve an inequality to find the number of jewels you can lose and still advance to the next level. $-2x + 74 \geq 50;\ x \leq 12$

● Practice

19. $n < 3$;

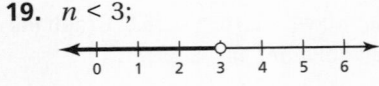

20. $w < 1$;

21. $c \geq 18$;

22. See Additional Answers.

● Example 4

- Discuss with students that you solve two-step inequalities in much the same way you solve two-step equations. The first step is to isolate the variable term. The second step is to solve for the variable.
- **?** "What is the first step in solving the equation $-3x + 2 = 11$?" Subtract 2 from each side of the equation.
- Point out that the answer to the question above is also the first step in solving the inequality $-3x + 2 > 11$.

● Example 5

- Before beginning to solve the example, ask students to give an example of a weight that would satisfy the goal of "at least 30 pounds."
- **?** "Would 28 pounds be a solution?" no "Would 35 pounds be a solution?" yes "Would 30 pounds be a solution?" yes
- Ask students to interpret the information in the Progress Report. They should recognize that 14 pounds have already been lost.
- You may want to show students a verbal model for the inequality.

Practice

- There are many skills reviewed in these problems. Review the Distributive Property. Remind students to change the direction of the inequality symbol when multiplying or dividing both sides of an inequality by a negative number.
- **Common Error:** In Exercises 20 and 21, students may forget to reverse the inequality symbol.

● Closure

- **Exit Ticket:** Solve the inequality and graph the solution.

  $\dfrac{x}{-3} \leq -9$ $x \geq 27$

  $-8 > 4x$ $x < -2$

  $5x - 4.5 > -2$ $x > \dfrac{1}{2}$

Mini-Assessment

Solve the inequality. Graph the solution.

1. $x - 3.2 > 5.4$ $x > 8.6$

2. $\dfrac{m}{-4} \leq 2$ $m \geq -8$

3. $6y + 7.3 \leq 25.3$ $y \leq 3$

Technology For the Teacher

The Dynamic Planning Tool
Editable Teacher's Resources at *BigIdeasMath.com*

T-87D

EXAMPLE 4 **Solving a Two-Step Inequality**

Solve $-3x + 2 > 11$.

$$-3x + 2 > 11$$ Write the inequality.

Step 1: Undo the addition. ⟶ $\underline{-2 \quad -2}$ Subtract 2 from each side.

$$-3x > 9$$ Simplify.

Step 2: Undo the multiplication. ⟶ $\dfrac{-3x}{-3} < \dfrac{9}{-3}$ Divide each side by -3. Reverse the inequality symbol.

$$x < -3$$ Simplify.

The solution is $x < -3$.

EXAMPLE 5 **Real-Life Application**

Progress Report	
Month	Pounds Lost
1	9
2	5
3	x
4	x

A contestant in a weight loss competition wants to lose at least 30 pounds in 4 months. Write and solve an inequality to find the average number x of pounds the contestant must lose in each of the last 2 months to meet the goal.

Use the progress report to write an expression for the number of pounds lost.

Pounds lost: $9 + 5 + x + x = 14 + 2x$

Because the contestant wants to lose *at least* 30 pounds, use the symbol $\geq$.

$$14 + 2x \geq 30$$ Write an inequality.

$$\underline{-14 \qquad\quad -14}$$ Subtract 14 from each side.

$$2x \geq 16$$ Simplify.

$$\dfrac{2x}{2} \geq \dfrac{16}{2}$$ Divide each side by 2.

$$x \geq 8$$ Simplify.

The contestant must lose an average of at least 8 pounds in each of the last 2 months to meet the goal.

● **Practice**

Solve the inequality. Graph the solution.

19. $5n - 3 < 12$ **20.** $-3(w - 10) > 27$ **21.** $-7 \geq \dfrac{c}{-2} + 2$

22. BICYCLE You want to purchase a bicycle that costs $265. So far, you have saved $128 and you plan to save an additional $20 per week.

   a. Write and solve an inequality to find the number of weeks it will take to save at least $265.

   b. Graph the solution in part (a). Will you have saved enough money after 6 weeks? 8 weeks? Explain.

Solve the equation. Check your solution. *(Section 2.4 and Section 2.5)*

1. $-6.5 + x = -4.12$

2. $4\dfrac{1}{2} + p = -5\dfrac{3}{4}$

3. $-\dfrac{b}{7} = 4$

4. $2h = -57$

Write the verbal sentence as an equation. Then solve. *(Section 2.4 and Section 2.5)*

5. The difference between a number b and 7.4 is -6.8.

6. $5\dfrac{2}{5}$ more than a number a is $7\dfrac{1}{2}$.

7. A number x multiplied by $\dfrac{3}{8}$ is $-\dfrac{15}{32}$.

8. The quotient of two times a number k and -2.6 is 12.

Write and solve an equation to find the value of x. *(Section 2.4 and Section 2.6)*

9. Perimeter = 26

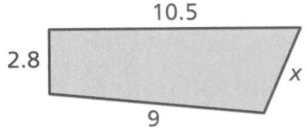

10. Perimeter = 23.59

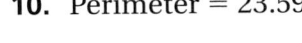
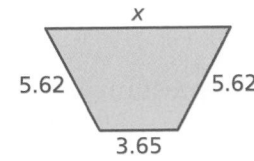

11. Perimeter = 33

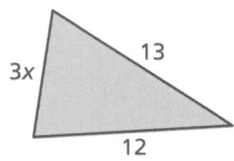

12. BANKING You withdraw $29.79 from your bank account. Now, your balance is $-$20.51. Write and solve an equation to find the amount of money in your bank account before you withdrew the money. *(Section 2.4)*

13. WATER LEVEL During a drought, the water level of a lake changes $-3\dfrac{1}{5}$ feet per day. Write and solve an equation to find how long it takes for the water level to change -16 feet. *(Section 2.5)*

14. BASKETBALL A basketball game has four quarters. The length of a game is 32 minutes. You play the entire game except $4\dfrac{1}{2}$ minutes. Write and solve an equation to find the mean time you play per quarter. *(Section 2.6)*

15. SCRAPBOOKING The mat needs to be cut to have a 0.5-inch border on all four sides. *(Section 2.6)*

  a. How much should you cut from the left and right sides?

  b. How much should you cut from the top and bottom?

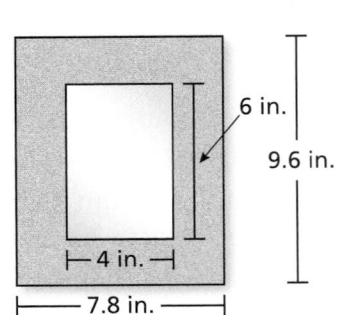

Alternative Assessment Options

Math Chat   Student Reflective Focus Question
Structured Interview   Writing Prompt

Math Chat

- Work in groups of four. Discuss the similarities and differences of solving equations with integers compared to solving equations with rational numbers. When they are finished, each group explains their findings to the other groups in the class.
- The teacher should walk around the classroom listening to the groups and ask questions to ensure understanding.

Study Help Sample Answers

Remind students to complete Graphic Organizers for the rest of the chapter.

4.

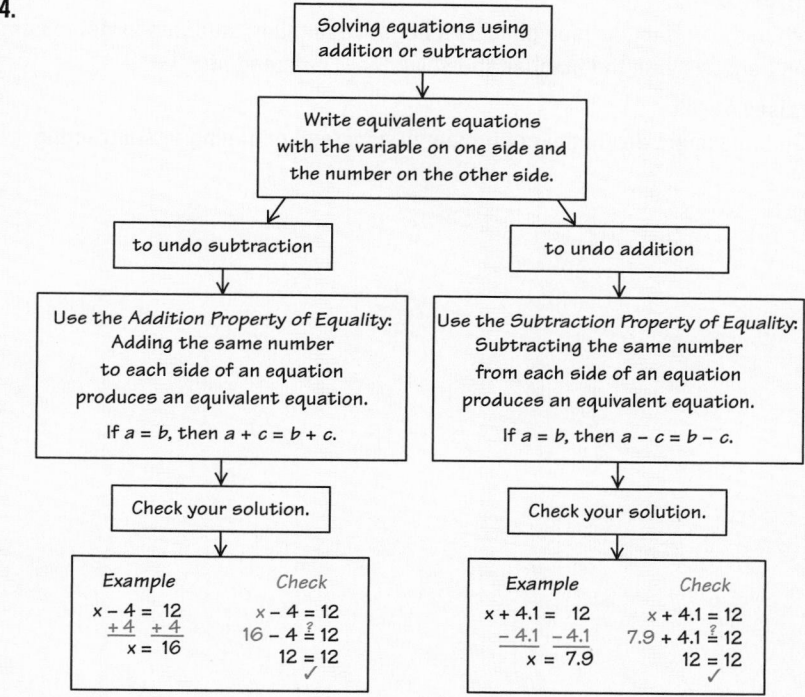

5–6. Available at *BigIdeasMath.com*.

Reteaching and Enrichment Strategies

If students need help. . .	If students got it. . .
Resources by Chapter • Study Help • Practice A and Practice B • Puzzle Time Lesson Tutorials *BigIdeasMath.com* Practice Quiz Practice from the Test Generator	Resources by Chapter • Enrichment and Extension • School-to-Work Game Closet at *BigIdeasMath.com* Start the Chapter Review

Technology
For the Teacher
Answer Presentation Tool

Assessment Book

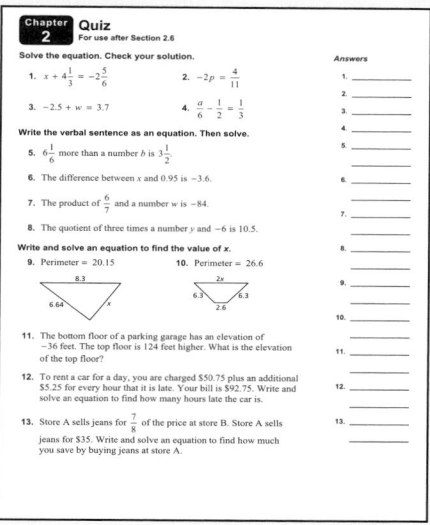

Review of Common Errors

Exercises 1–4
- Students may forget to carry the negative through the division operation.

Exercises 5–8
- Students may use the wrong numerator.

Exercises 9–13
- When adding and subtracting decimals, students may forget to line up the decimal points.

Exercises 14–22
- Students may use the reciprocal of the first fraction instead of the second.

Exercises 23–31
- Students may use the same operation in solving for x instead of the inverse operation.

Exercises 32–40
- When the variable is multiplied by a negative number, students forget to keep the negative with the number and solve for $-x$ instead of x.

Exercises 41–45
- Students might divide the coefficient first instead of adding or subtracting.

Answers

1. $-0.5\overline{3}$ 2. 0.625

3. $-2.1\overline{6}$ 4. 1.4375

5. $-\dfrac{3}{5}$ 6. $-\dfrac{7}{20}$

7. $-5\dfrac{4}{5}$ 8. $24\dfrac{23}{100}$

9. $-3\dfrac{2}{3}$ 10. $-\dfrac{43}{60}$

11. 2.22 12. 11.25

13. $-11\dfrac{5}{12}$ inches

Review Key Vocabulary

terminating decimal, *p. 52*   rational number, *p. 52*
repeating decimal, *p. 52*   equivalent equations, *p. 72*

Review Examples and Exercises

2.1 Rational Numbers *(pp. 50–55)*

Write −0.14 as a fraction in simplest form.

$$-0.14 = -\frac{14}{100}$$

> Write the digits after the decimal point in the numerator.

> The last digit is in the hundredths place. So, use 100 in the denominator.

$$= -\frac{7}{50}$$ Simplify.

Exercises

Write the rational number as a decimal.

1. $-\dfrac{8}{15}$ **2.** $\dfrac{5}{8}$ **3.** $-\dfrac{13}{6}$ **4.** $1\dfrac{7}{16}$

Write the decimal as a fraction or mixed number in simplest form.

5. -0.6 **6.** -0.35 **7.** -5.8 **8.** 24.23

2.2 Adding and Subtracting Rational Numbers *(pp. 56–61)*

Find −8.18 + 3.64.

$$-8.18 + 3.64 = -4.54 \quad |-8.18| > |3.64|. \text{ So, subtract } |3.64| \text{ from } |-8.18|.$$

> Use the sign of −8.18.

Exercises

Add or subtract. Write fractions in simplest form.

9. $-4\dfrac{5}{9} + \dfrac{8}{9}$ **10.** $-\dfrac{5}{12} - \dfrac{3}{10}$ **11.** $-2.53 + 4.75$ **12.** $3.8 - (-7.45)$

13. TURTLES A turtle is $20\dfrac{5}{6}$ inches below the surface of a pond. It dives to a depth of $32\dfrac{1}{4}$ inches. How far did it dive?

2.3 Multiplying and Dividing Rational Numbers *(pp. 62–67)*

Find $-4\frac{1}{6} \div 1\frac{1}{3}$.

$$-4\frac{1}{6} \div 1\frac{1}{3} = -\frac{25}{6} \div \frac{4}{3}$$ Write mixed numbers as improper fractions.

$$= \frac{-25}{6} \cdot \frac{3}{4}$$ Multiply by the reciprocal of $\frac{4}{3}$.

$$= \frac{-25 \cdot 3}{6 \cdot 4}$$ Multiply the numerators and the denominators.

$$= \frac{-25}{8}, \text{ or } -3\frac{1}{8}$$ Simplify.

Exercises

Multiply or divide. Write fractions in simplest form.

14. $-\frac{4}{9}\left(-\frac{7}{9}\right)$ **15.** $\frac{9}{10} \div \left(-\frac{6}{5}\right)$ **16.** $\frac{8}{15}\left(-\frac{2}{3}\right)$ **17.** $-\frac{4}{11} \div \frac{2}{7}$

18. $-5.9(-9.7)$ **19.** $6.4 \div (-3.2)$ **20.** $4.5(-5.26)$ **21.** $-15.4 \div (-2.5)$

22. SUNKEN SHIP The elevation of a sunken ship is -120 feet. Your elevation is $\frac{5}{8}$ of the ship's elevation. What is your elevation?

2.4 Solving Equations Using Addition or Subtraction *(pp. 70–75)*

Solve $x - 9 = -6$.

$$x - 9 = -6$$ Write the equation.

$$\underline{+9 \quad +9}$$ Add 9 to each side.

$$x = 3$$ Simplify.

Exercises

Solve the equation. Check your solution.

23. $p - 3 = -4$ **24.** $6 + q = 1$ **25.** $-2 + j = -22$ **26.** $b - 19 = -11$

27. $n + \frac{3}{4} = \frac{1}{4}$ **28.** $v - \frac{5}{6} = -\frac{7}{8}$ **29.** $t - 3.7 = 1.2$ **30.** $\ell + 15.2 = -4.5$

31. GIFT CARD A shirt costs $24.99. After using a gift card as a partial payment, you still owe $9.99. What is the value of the gift card?

Review Game

Rational Numbers and Equations

Big Ideas
Game Closet

For the Student
Additional Practice
- Lesson Tutorials
- Study Help (textbook)
- Student Website
    Multi-Language Glossary
    Practice Assessments

Materials
- questions from the chapter's homework, quizzes, examples, or tests
- 5 index cards, each with a letter in the word HORSE written on it, for each group

Directions
Divide the class into groups. Ask a group one of the questions. If the group is correct, the game continues to the next group. If they answer incorrectly, they receive an H. Each wrong answer in a group will result in that group receiving the next letter in the word HORSE. When a group has all 5 letters, they are out of the game. Choose questions with a wide range of difficulty to control how long the game takes.

Who Wins?
The last group with 4 or fewer letters wins.

Answers

14. $\dfrac{28}{81}$ **15.** $-\dfrac{3}{4}$

16. $-\dfrac{16}{45}$ **17.** $-1\dfrac{3}{11}$

18. 57.23 **19.** -2

20. -23.67 **21.** 6.16

22. -75 ft   **23.** $p = -1$

24. $q = -5$ **25.** $j = -20$

26. $b = 8$ **27.** $n = -\dfrac{1}{2}$

28. $v = -\dfrac{1}{24}$ **29.** $t = 4.9$

30. $\ell = -19.7$ **31.** $\$15$

32. $x = -24$ **33.** $y = -49$

34. $z = 3$ **35.** $w = 50$

36. $x = -2$ **37.** $y = -5$

38. $z = 6$ **39.** $w = -0.5$

40. $-16°\text{F}$ **41.** $c = 7$

42. $w = -\dfrac{8}{9}$

43. $w = -12$

44. $x = -3.5$ **45.** 11 years

My Thoughts on the Chapter

What worked. . .

What did not work. . .

What I would do differently. . .

2.5 Solving Equations Using Multiplication or Division (pp. 76–81)

Solve $\dfrac{x}{5} = -7$.

$$\dfrac{x}{5} = -7 \qquad \text{Write the equation.}$$

$$5 \cdot \dfrac{x}{5} = 5 \cdot (-7) \qquad \text{Multiply each side by 5.}$$

$$x = -35 \qquad \text{Simplify.}$$

Exercises

Solve the equation. Check your solution.

32. $\dfrac{x}{3} = -8$ **33.** $-7 = \dfrac{y}{7}$ **34.** $-\dfrac{z}{4} = -\dfrac{3}{4}$ **35.** $-\dfrac{w}{20} = -2.5$

36. $4x = -8$ **37.** $-10 = 2y$ **38.** $-5.4z = -32.4$ **39.** $-6.8w = 3.4$

40. TEMPERATURE The mean temperature change is $-3.2°F$ per day for five days. What is the total change over the five-day period?

2.6 Solving Two-Step Equations (pp. 82–87)

Solve $\dfrac{x}{5} + \dfrac{7}{10} = -\dfrac{3}{10}$.

$$\dfrac{x}{5} + \dfrac{7}{10} = -\dfrac{3}{10} \qquad \text{Write the equation.}$$

$$\dfrac{x}{5} = -1 \qquad \text{Subtract } \dfrac{7}{10} \text{ from each side.}$$

$$x = -5 \qquad \text{Multiply each side by 5.}$$

Exercises

Solve the equation. Check your solution.

41. $-2c + 6 = -8$ **42.** $3(3w - 4) = -20$

43. $\dfrac{w}{6} + \dfrac{5}{8} = -1\dfrac{3}{8}$ **44.** $-3x - 4.6 = 5.9$

45. EROSION The floor of a canyon has an elevation of -14.5 feet. Erosion causes the elevation to change by -1.5 feet per year. How many years will it take for the canyon floor to have an elevation of -31 feet?

Check It Out
Test Practice
BigIdeasMath.com

Write the rational number as a decimal.

1. $\dfrac{7}{40}$

2. $-\dfrac{1}{9}$

3. $-\dfrac{21}{16}$

4. $\dfrac{36}{5}$

Write the decimal as a fraction or mixed number in simplest form.

5. -0.122

6. 0.33

7. -4.45

8. -7.09

Add or subtract. Write fractions in simplest form.

9. $-\dfrac{4}{9} + \left(-\dfrac{23}{18}\right)$

10. $\dfrac{17}{12} - \left(-\dfrac{1}{8}\right)$

11. $9.2 + (-2.8)$

12. $2.86 - 12.1$

Multiply or divide. Write fractions in simplest form.

13. $3\dfrac{9}{10} \times \left(-\dfrac{8}{3}\right)$

14. $-1\dfrac{5}{6} \div 4\dfrac{1}{6}$

15. $-4.4 \times (-6.02)$

16. $-5 \div 1.5$

Solve the equation. Check your solution.

17. $7x = -3$

18. $2(x + 1) = -2$

19. $\dfrac{2}{9}g = -8$

20. $z + 14.5 = 5.4$

21. $-14 = 6c$

22. $\dfrac{2}{7}k - \dfrac{3}{8} = -\dfrac{19}{8}$

23. MARATHON A marathon is a 26.2-mile race. You run three marathons in one year. How many miles do you run?

24. RECORD A runner is compared with the world record holder during a race. A negative number means the runner is ahead of the time of the world record holder, and a positive number means that the runner is behind the time of the world record holder. The table shows the time difference between the runner and the world record holder for each lap. What time difference does the runner need for the fourth lap to match the world record?

Lap	Time Difference
1	-1.23
2	0.45
3	0.18
4	?

25. GYMNASTICS You lose 0.3 point for stepping out of bounds during a floor routine. Your final score is 9.124. Write and solve an equation to find your score before the penalty.

26. PERIMETER The perimeter of the triangle is 45. Find the value of x.

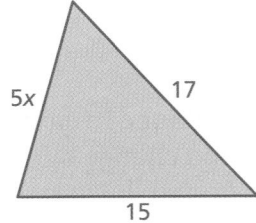

$5x$ 17   15

Test Item References

Chapter Test Questions	Section to Review
1–8	2.1
9–12, 23, 24, 25, 26	2.2
13–16, 23	2.3
17–22, 24, 25, 26	2.4
17–22	2.5
17–22, 24, 26	2.6

Test-Taking Strategies

Remind students to quickly look over the entire test before they start so that they can budget their time. On tests, it is really important for students to **Stop** and **Think.** When students hurry on a test dealing with signed numbers, they often make 'sign' errors. Sometimes it helps to represent each problem with a number line to ensure that they are thinking through the process.

Common Assessment Errors

- **Exercises 9 and 10** Students may forget to find a common denominator. Remind students that adding and subtracting fractions always requires a common denominator.
- **Exercises 11 and 12** Students may forget to line up the decimal points when they add or subtract decimals. Remind students that the decimal points must be lined up before adding or subtracting.
- **Exercises 15 and 16** Students may place the decimal point incorrectly in their answers. Remind students of the rules for multiplying and dividing decimals. Also, remind students to use estimation to check their answers.
- **Exercises 17–22** Students may use the same operations, instead of inverse operations, when solving for the variable. Demonstrate to students that this will not give the correct solution. Also, students may use the properties of equality improperly by adding, subtracting, multiplying, or dividing on one side of the equation only, or by using inverse operations on opposite sides of the equation. Remind students of the properties and to check their answers in the original equation.

Reteaching and Enrichment Strategies

If students need help. . .	If students got it. . .
Resources by Chapter • Practice A and Practice B • Puzzle Time Record and Practice Journal Practice Differentiating the Lesson Lesson Tutorials Practice from the Test Generator Skills Review Handbook	Resources by Chapter • Enrichment and Extension • School-to-Work • Financial Literacy Game Closet at *BigIdeasMath.com* Start Standardized Test Practice

Answers

1. 0.175
2. $-0.\overline{1}$
3. -1.3125
4. 7.2
5. $-\dfrac{61}{500}$
6. $\dfrac{33}{100}$
7. $-4\dfrac{9}{20}$
8. $-7\dfrac{9}{100}$
9. $-1\dfrac{13}{18}$
10. $1\dfrac{13}{24}$
11. 6.4
12. -9.24
13. $-10\dfrac{2}{5}$
14. $-\dfrac{11}{25}$
15. 26.488
16. $-3.\overline{3}$
17. $x = -\dfrac{3}{7}$
18. $x = -2$
19. $g = -36$
20. $z = -9.1$
21. $c = -2\dfrac{1}{3}$
22. $k = -7$
23. 78.6 miles
24. 0.6
25. $x - 0.3 = 9.124;\ 9.424$
26. $2\dfrac{3}{5}$

Assessment Book

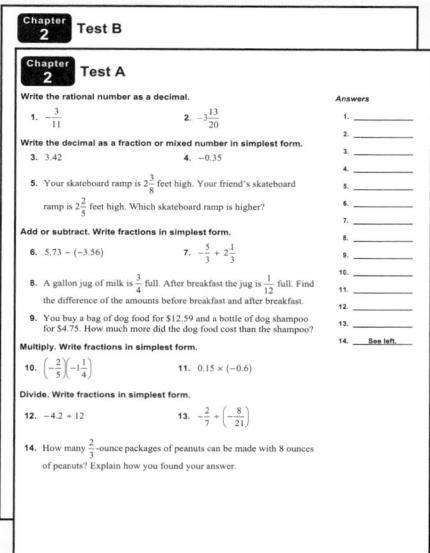

Chapter 2 Test B

Chapter 2 Test A

Write the rational number as a decimal.

1. $-\dfrac{3}{11}$
2. $-3\dfrac{13}{20}$

Write the decimal as a fraction or mixed number in simplest form.

3. 3.42
4. -0.35

5. Your skateboard ramp is $2\dfrac{3}{8}$ feet high. Your friend's skateboard ramp is $2\dfrac{2}{5}$ feet high. Which skateboard ramp is higher?

Add or subtract. Write fractions in simplest form.

6. $5.73 - (-3.56)$
7. $-\dfrac{5}{3} + 2\dfrac{1}{3}$

8. A gallon jug of milk is $\dfrac{3}{4}$ full. After breakfast the jug is $\dfrac{1}{12}$ full. Find the difference of the amounts before breakfast and after breakfast.

9. You buy a bag of dog food for $12.59 and a bottle of dog shampoo for $4.75. How much more did the dog food cost than the shampoo?

Multiply. Write fractions in simplest form.

10. $\left(-\dfrac{2}{5}\right)\left(-1\dfrac{1}{4}\right)$
11. $0.15 \times (-0.6)$

Divide. Write fractions in simplest form.

12. $-4.2 \div 12$
13. $-\dfrac{2}{7} \div \left(-\dfrac{8}{21}\right)$

14. How many $\dfrac{2}{3}$-ounce packages of peanuts can be made with 8 ounces of peanuts? Explain how you found your answer.

Answers

1. _____
2. _____
3. _____
4. _____
5. _____
6. _____
7. _____
8. _____
9. _____
10. _____
11. _____
12. _____
13. _____
14. See left.

Test-Taking Strategies

Available at *BigIdeasMath.com*

After Answering Easy Questions, Relax
Answer Easy Questions First
Estimate the Answer
Read All Choices before Answering
Read Question before Answering
Solve Directly or Eliminate Choices
Solve Problem before Looking at
   Choices
Use Intelligent Guessing
Work Backwards

About this Strategy

When taking a multiple choice test, be sure to read each question carefully and thoroughly. After reading the question, estimate the answer before trying to solve.

Answers

1. A
2. F
3. −18
4. C

Item Analysis

1. **A.** Correct answer
   B. The student correctly finds José's height at 5 years old, which was 41 inches, but then reverses the relationship between José and Sean.
   C. When multiplying the rate of growth by the number of elapsed years, the student multiplies only the whole number parts to get $16\frac{3}{4}$.
   D. When multiplying the rate of growth by the number of elapsed years, the student multiplies only the whole number parts to get $16\frac{3}{4}$. After using this to find José's height at 5 years old, the student also reverses the relationship between José and Sean.

2. **F.** Correct answer
   G. The student confuses the ordered pairs of $(-3, 0)$ and $(0, -3)$. The students does not recognize that $(0, -3)$ lies on the line, whereas $(-3, 0)$ does not.
   H. The student does not recognize that $(3, -1)$ lies on the line, whereas $(-3, 0)$ does not.
   I. The student does not recognize that $(6, 1)$ lies on the line, whereas $(-3, 0)$ does not.

3. **Gridded Response:** Correct answer: -18

   Common Error: The student thinks that the common ratio is 2 rather than -2, and gets an answer of 18.

4. **A.** The student thinks that the absolute value of each individual number is negative and finds the sum of -2 and -2.5.
   B. The student correctly simplifies the expression inside the absolute value bars, but then thinks that the absolute value means to take the opposite.
   C. Correct answer
   D. The student takes the absolute value of each individual number and finds the sum of 2 and 2.5.

5. **F.** Correct answer
   G. The student adds $\frac{1}{8}$ to both sides instead of subtracting.
   H. The student creates an equation that is not equivalent to the given equation, because if the student multiplies both sides of the equation by $-\frac{4}{3}$, this would require distributing the multiplication on the left side of the equation, thereby also multiplying $\frac{1}{8}$ by $-\frac{4}{3}$.
   I. The student creates an equation that is not equivalent to the given equation, because if the student multiplies both sides of the equation by $-\frac{3}{4}$, this would require distributing the multiplication on the left side of the equation, thereby also multiplying $\frac{1}{8}$ by $-\frac{3}{4}$. Furthermore, multiplying $-\frac{3}{4}x$ by $-\frac{3}{4}$ does not result in a product of x.

1. When José and Sean were each 5 years old, José was $1\frac{1}{2}$ inches taller than Sean. José grew at an average rate of $2\frac{3}{4}$ inches per year from the time that he was 5 years old until the time he was 13 years old. José was 63 inches tall when he was 13 years old. How tall was Sean when he was 5 years old?

   A. $39\frac{1}{2}$ in.

   B. $42\frac{1}{2}$ in.

   C. $44\frac{3}{4}$ in.

   D. $47\frac{3}{4}$ in.

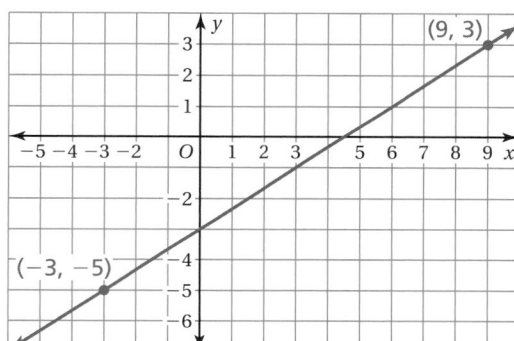

One-fourth of the 36 cats in our town are tabbies. How many are not tabbies?

   Ⓐ 9   Ⓑ 18   Ⓒ 27   Ⓓ 36

IC.

"Using estimation you can see that there are about 10 tabbies. So about 30 are not tabbies."

2. A line is graphed in the coordinate plane below.

   Which point is *not* on the line?

   F. $(-3, 0)$

   G. $(0, -3)$

   H. $(3, -1)$

   I. $(6, 1)$

3. What is the missing number in the sequence below?

   $$\frac{9}{16},\ -\frac{9}{8},\ \frac{9}{4},\ -\frac{9}{2},\ 9,\ \underline{\quad}$$

4. What is the value of the expression below?

   $$\left| -2 - (-2.5) \right|$$

   A. -4.5

   B. -0.5

   C. 0.5

   D. 4.5

5. Which equation is equivalent to the equation shown below?

$$-\frac{3}{4}x + \frac{1}{8} = -\frac{3}{8}$$

F. $-\frac{3}{4}x = -\frac{3}{8} - \frac{1}{8}$

G. $-\frac{3}{4}x = -\frac{3}{8} + \frac{1}{8}$

H. $x + \frac{1}{8} = -\frac{3}{8} \cdot \left(-\frac{4}{3}\right)$

I. $x + \frac{1}{8} = -\frac{3}{8} \cdot \left(-\frac{3}{4}\right)$

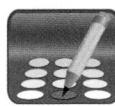

6. What is the value of the expression below?

$$-5 \div 20$$

7. Karina was solving the equation in the box below.

$$-96 = -6(15 - 2x)$$
$$-96 = -90 - 12x$$
$$-96 + 90 = -90 + 90 - 12x$$
$$-6 = -12x$$
$$\frac{-6}{-12} = \frac{-12x}{-12}$$
$$\frac{1}{2} = x$$

What should Karina do to correct the error that she made?

A. First add 6 to both sides of the equation.

B. First add $2x$ to both sides of the equation.

C. Distribute the -6 to get $90 - 12x$.

D. Distribute the -6 to get $-90 + 12x$.

8. Current, voltage, and resistance are related according to the formula below, where I represents the current, in amperes, V represents the voltage, in volts, and R represents the resistance, in ohms.

$$I = \frac{V}{R}$$

What is the voltage when the current is 0.5 ampere and the resistance is 0.8 ohm?

F. 4.0 volts

G. 1.3 volts

H. 0.4 volt

I. 0.3 volt

Item Analysis (continued)

6. **Gridded Response:** Correct answer: -0.25 or $-\dfrac{1}{4}$

   Common Error: The student divides the greater number by the lesser number, getting -4.

7. **A.** The student thinks that the inverse operation of multiplying by -6 is adding 6.

   B. The student disregards that the -6 first must be distributed and violates the order of operations.

   C. The student does not distribute the negative sign to the two terms inside the parentheses.

   D. Correct answer

8. **F.** The student divides 0.8 by 0.5 to get 4.0, or the student makes a place value error when multiplying 0.5 by 0.8.

   G. The student adds 0.8 to both sides of the equation instead of multiplying by 0.8.

   H. Correct answer

   I. The student subtracts 0.5 from 0.8 instead of multiplying both sides of the equation by 0.8.

9. **A.** The student finds $\dfrac{1}{2}$ of the sum of the base and the height.

   B. Correct answer

   C. The student finds the sum of the base and the height.

   D. The student finds the product of the base and the height.

10. **F.** Correct answer

    G. The student only multiplies the whole number parts together and the decimal parts together and then combines these two products to write the answer.

    H. The student does not double the radius.

    I. The student does not double the radius. The student then only multiplies the whole number parts together and the decimal parts together and then combines these two products to write the answer.

11. **4 points** The student demonstrates a thorough understanding of interpreting rational numbers on a number line and a thorough conceptual understanding of the four operations using rational numbers. In Part A, the student correctly recognizes that the two greatest values, T and U, have the greatest sum, which is approximately 1.7. In Part B, the student correctly recognizes that the two values that are the farthest apart, U and R, have the greatest difference, which is approximately 4. In Part C, the student correctly recognizes that the two values that have the same sign and also the greatest magnitude, R and S, have the greatest product, which is approximately 0.87. In Part D, the student correctly recognizes that the two values that have the same sign and also the greatest ratio, R and S, have the greatest quotient, which is approximately 10. The student provides clear and complete explanations of the reasoning used.

Answers

5. F

6. -0.25 or $-\dfrac{1}{4}$

7. D

8. H

Answers

9. B

10. F

11. *Part A* 1.7
    Part B 4
    Part C 0.87
    Part D 10

12. B

Answer for Extra Example

1. **A.** The student combines two terms that are not like terms.

   B. Correct answer

   C. The student distributes the 1.5 by adding instead of multiplying.

   D. The student makes a sign error in distributing the 1.5 to the second term inside parentheses.

Item Analysis (continued)

11. (continued)

    3 points The student demonstrates an understanding of interpreting rational numbers on a number line and a good conceptual understanding of the four operations using rational numbers, but the student's work and explanations demonstrate an essential but less than thorough understanding.

    2 points The student demonstrates an understanding of interpreting rational numbers on a number line and a partial conceptual understanding of the four operations using rational numbers. The student's work and explanations demonstrate a lack of essential understanding.

    1 point The student demonstrates a partial understanding of interpreting rational numbers on a number line and a limited conceptual understanding of the four operations using rational numbers. The student's response is incomplete and exhibits many flaws.

    0 points The student provided no response, a completely incorrect or incomprehensible response, or a response that demonstrates insufficient understanding of interpreting rational numbers on a number line and an insufficient conceptual understanding of the four operations using rational numbers.

12. **A.** The student thinks that $-0.4 + 0.8 = 1.2$.

    B. Correct answer

    C. The student thinks that $\dfrac{-0.4}{-0.2} = -2$.

    D. The student thinks that $\dfrac{-0.4}{-1} = -0.4$ and that $-0.4 + 0.8 = 1.2$.

Extra Example for Standardized Test Practice

1. Raj was solving an equation in the box below.

$$57 = 1.5(8 - 2w)$$
$$57 = 12 - 3w$$
$$57 - 12 = 12 - 12 - 3w$$
$$45 = 3w$$
$$\frac{45}{3} = \frac{3w}{3}$$
$$15 = w$$

   What should Raj do to correct the error that he made?

   A. add $3w$ to both sides to get $60w = 12$

   B. bring down $-3w$ instead of $3w$

   C. distribute the 1.5 to get $9.5 - 3.5x$

   D. distribute the 1.5 to get $12 + 3w$

T-95

9. What is the area of a triangle with a base length of $2\frac{1}{2}$ inches and a height of 3 inches?

A. $2\frac{3}{4}$ in.^{2}

B. $3\frac{3}{4}$ in.^{2}

C. $5\frac{1}{2}$ in.^{2}

D. $7\frac{1}{2}$ in.^{2}

10. What is the circumference of the circle below? (Use 3.14 for π.)

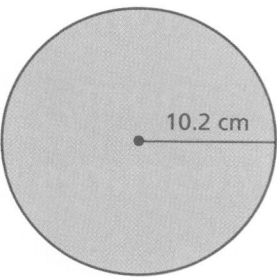

10.2 cm

F. 64.056 cm

G. 60.028 cm

H. 32.028 cm

I. 30.028 cm

11. Four points are graphed on the number line below.

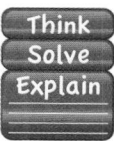

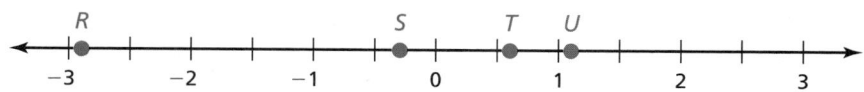

Part A   Choose the two points whose values have the greatest sum. Approximate this sum. Explain your reasoning.

Part B   Choose the two points whose values have the greatest difference. Approximate this difference. Explain your reasoning.

Part C   Choose the two points whose values have the greatest product. Approximate this product. Explain your reasoning.

Part D   Choose the two points whose values have the greatest quotient. Approximate this quotient. Explain your reasoning.

12. What number belongs in the box to make the equation true?

$$\frac{-0.4}{\square} + 0.8 = -1.2$$

A. 1

B. 0.2

C. -0.2

D. -1

3 Proportions and Variation

"I am doing an experiment with slope. I want you to run up and down the board 10 times."

"Now with 2 more dog biscuits, do it again and we'll compare your rates."

"Dear Sir: I counted the number of bacon, cheese, and chicken dog biscuits in the box I bought."

"There were 16 bacon, 12 cheese, and only 8 chicken. That's a ratio of 4:3:2. Please go back to the original ratio of 1:1:1."

Connections to Previous Learning

- Compare, contrast, and convert units within the same dimension.
- Use appropriate measures and tools to estimate and solve real-world area problems.

- Interpret and compare ratio and rates.
- Use reasoning about multiplication and division to solve ratio and rate problems.

- Distinguish between proportional and non-proportional situations.
- Use proportions to solve problems.
- Compare, contrast, and convert between measurement systems.
- Graph proportional relationships and identify the unit rate as the slope of the related linear function.
- Distinguish direct variation from other relationships, including inverse variation.

Math in History

All ancient cultures had methods of counting time. There were many different versions of calendars.

★ Our modern calendar is called the Gregorian Calendar, named after Pope Gregory XIII. This calendar was suggested by a German mathematician named Christoph Clavius. It was put into use in the year 1582. When people went to bed on October 4, 1582, they woke up the next day on October 15, 1582.

★ The loss of 10 days was due to the use of the Julian calendar from the years 46 A.D. to 1582 A.D. The Julian calendar was named after Julius Caesar, the first emperor of Rome. A year in the Julian calendar is $365\frac{1}{4}$ days on average, which is about 11 minutes longer than a true solar year.

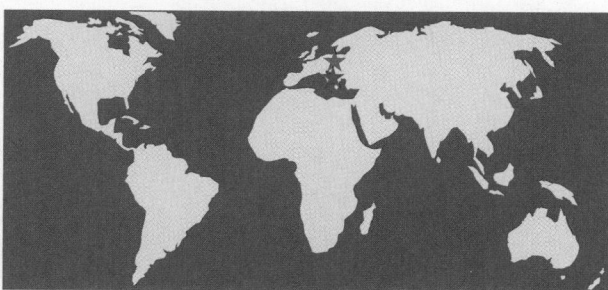

Pacing Guide for Chapter 3

Chapter Opener	1 Day
Section 1	
Activity	1 Day
Lesson	1 Day
Section 2	
Activity	1 Day
Lesson	1 Day
Section 3	
Activity	1 Day
Lesson	1 Day
Section 4	
Activity	1 Day
Lesson	1 Day
Section 5	
Activity	1 Day
Lesson	1 Day
Study Help / Quiz	1 Day
Section 6	
Activity	1 Day
Lesson	1 Day
Section 7	
Activity	1 Day
Lesson	1 Day
Lesson b	1 Day
Section 8	
Activity	1 Day
Lesson	1 Day
Quiz / Chapter Review	1 Day
Chapter Test	1 Day
Standardized Test Practice	1 Day
Total Chapter 3	22 Days
Year-to-Date	60 Days

Check Your Resources

- Record and Practice Journal
- Resources by Chapter
- Skills Review Handbook
- Assessment Book
- Worked-Out Solutions

Technology For the Teacher

The Dynamic Planning Tool
Editable Teacher's Resources at
BigIdeasMath.com

Math Background Notes

Additional Topics for Review

- Standard conversions
- Identifying patterns in tables

Try It Yourself

1. $\frac{3}{4}$ 2. $\frac{2}{3}$

3. $\frac{1}{12}$ 4. $\frac{1}{3}$

5. no   6. no

7. no   8. yes

9. 5   10. 10

11. 3

Record and Practice Journal

1. $\frac{1}{6}$ 2. $\frac{2}{3}$

3. $\frac{1}{5}$ 4. $\frac{1}{2}$

5. $\frac{4}{9}$ 6. $\frac{4}{5}$

7. no   8. yes

9. yes   10. no

11. $\frac{6}{29}$ 12. 4 yards

13. 7 gallons   14. 4 feet

15. 5 tons   16. 18 cups

17. 1280 ounces

18. 180 inches

19. 1.75 pounds

20. 48 cups

Vocabulary Review

- Greatest Common Factor
- Magic One

Simplifying Fractions

- Students should know how to simplify fractions.
- Some students may have learned simplifying fractions as reducing fractions.
- Remind students that you must divide the numerator and the denominator by the same factor. This is equivalent to dividing by one which does not change the value of the fraction but does change the form.

Comparing Equivalent Fractions

- Students should know how to compare equivalent fractions.
- Encourage students to simplify the fraction with the greater numbers. If the fraction simplifies to the second fraction, students can conclude the fractions are equivalent.
- **Teaching Tip:** Some students may have difficulty simplifying fractions. This makes the search for equivalent fractions difficult. Encourage these students to start with the fraction with lesser numbers and multiply the numerator and denominator by the same factor to see if they can produce the second fraction.

Converting Measures

- Students should know how to convert measures.
- Remind students that they must multiply by a conversion factor equal to one so that the value of the original quantity does not change.
- Remind students that converting measures only changes the units of the quantity and not the value.
- **Common Error:** Some students will use the conversion factors upside-down. Remind students to always start with the given quantity. This will help students to use the conversion factor correctly.

Reteaching and Enrichment Strategies

If students need help. . .	If students got it. . .
Record and Practice Journal • Fair Game Review Skills Review Handbook Lesson Tutorials	Game Closet at *BigIdeasMath.com* Start the next section

What You Learned Before

"I wonder if our rate is proportional to the slope of the hill."

...or possibly proportional to our stupidity!

Simplifying Fractions

Example 1 Simplify $\frac{4}{8}$.

$$\frac{4 \div 4}{8 \div 4} = \frac{1}{2}$$

> Simplify fractions by using the Greatest Common Factor.

Example 2 Simplify $\frac{10}{15}$.

$$\frac{10 \div 5}{15 \div 5} = \frac{2}{3}$$

Try It Yourself
Simplify.

1. $\frac{75}{100}$

2. $\frac{16}{24}$

3. $\frac{12}{144}$

4. $\frac{15}{45}$

Comparing Equivalent Fractions

Example 3 Is $\frac{1}{4}$ equivalent to $\frac{13}{52}$?

$$\frac{13 \div 13}{52 \div 13} = \frac{1}{4}$$

∴ $\frac{1}{4}$ is equivalent to $\frac{13}{52}$.

Example 4 Is $\frac{30}{64}$ equivalent to $\frac{5}{8}$?

$$\frac{30 \div 2}{64 \div 2} = \frac{15}{32}$$

∴ $\frac{30}{64}$ is *not* equivalent to $\frac{5}{8}$.

Try It Yourself
Are the fractions equivalent? Explain.

5. $\frac{15}{60} \overset{?}{=} \frac{3}{4}$

6. $\frac{2}{5} \overset{?}{=} \frac{24}{144}$

7. $\frac{15}{20} \overset{?}{=} \frac{3}{5}$

8. $\frac{2}{8} \overset{?}{=} \frac{16}{64}$

Converting Measures

Example 5 A person must be at least 56 inches tall to drive a race car at an amusement park. Gina is 4 feet 11 inches tall. Is she tall enough to drive?

$$4 \text{ feet} \times \frac{12 \text{ inches}}{1 \text{ foot}} + 11 \text{ inches} = 48 \text{ inches} + 11 \text{ inches} = 59 \text{ inches}$$

∴ Because 59 inches is greater than 56 inches, Gina is tall enough to drive.

Try It Yourself
Convert.

9. 15 feet = ▢ yards

10. 5 quarts = ▢ pints

11. 6000 pounds = ▢ tons

Essential Question How do rates help you describe real-life problems?

The Meaning of a Word ● Rate

When you rent snorkel gear at the beach, you should pay attention to the rental **rate**. The rental rate is in dollars per hour.

Snorkel Rentals $8.75 per hour

Snorkel Rentals $7.25 per hour

1 ACTIVITY: Finding Reasonable Rates

Work with a partner.

a. Match each description with a verbal rate.

b. Match each verbal rate with a numerical rate.

c. Give a reasonable numerical rate for each description. Then give an unreasonable rate.

Description	*Verbal Rate*	*Numerical Rate*
Your pay rate for washing cars	inches per month	$\dfrac{\text{m}}{\text{sec}}$
The average rainfall rate in a rain forest	pounds per acre	$\dfrac{\text{people}}{\text{yr}}$
Your average driving rate along an interstate	meters per second	$\dfrac{\text{lb}}{\text{acre}}$
The growth rate for the length of a baby alligator	people per year	$\dfrac{\text{mi}}{\text{h}}$
Your running rate in a 100-meter dash	dollars per hour	$\dfrac{\text{in.}}{\text{yr}}$
The population growth rate of a large city	dollars per year	$\dfrac{\text{in.}}{\text{mo}}$
The average pay rate for a professional athlete	miles per hour	$\dfrac{\$}{\text{h}}$
The fertilization rate for an apple orchard	inches per year	$\dfrac{\$}{\text{yr}}$

Laurie's Notes

Introduction

For the Teacher

- **Goal:** Students will explore and use a variety of rates that describe common real-life situations.

Motivate

- **Model:** In an area visible to students, set a wind-up toy in motion. If a toy is not available, (quietly) ask a student to walk across the room at a constant speed.
- **?** "How fast is the toy or student moving?" There will be no exact answer.
- **?** "How do you measure the *rate* that the toy or student is moving? Which two pieces of information do you need to find the *rate*?"
- Provide measuring tape and a stop watch. Ask volunteers to compute the rate.
- Discuss why you might use a convenient unit of time (i.e., 5 seconds) versus trying to use a convenient unit of distance (i.e., 10 feet).
- Write the information measured on the board in words, and also as a numerical rate.
- Explain that today they will investigate how rates are written and how they can be rewritten with different units.

Discuss

- Begin with a general discussion of **rates** that students should understand and ask for reasonable values for each: speed limit (65 miles per hour), heart rate (70 beats per minute), gas mileage (35 miles per gallon), and homework time (2 hours per night).
- Also ask for unreasonable values for each rate.

Activity Notes

Activity 1

- Have students work with a partner to complete the activity.
- Share and discuss sample answers when students have finished.
- Students might choose "inches per year" to measure the growth rate of a baby alligator. Discuss with students that young animals usually grow at faster rates than when they are older, so "inches per month" is a better choice.
- **?** "How would you describe what a rate is to someone who doesn't know?" Listen for a comparison of two quantities where the units are different. The word "per" is used when comparing the two quantities.
- *Note:* All of the rates in this activity are **unit rates**. The denominator references a single unit (i.e., inches per one month, miles per one hour).
- Share with students that rates do not have to be unit rates. Example: $1.89 per 12 ounces and 100 shared minutes per 4 people.
- Unit rates are presented formally in the lesson that follows.

Previous Learning

Students have used reasoning about multiplication and division to solve ratio and rate problems. Knowing measurement abbreviations is helpful.

Activity Materials
Introduction
• stop watch
• measuring tape
• self-propelled wind-up toy

Start Thinking! and Warm Up

Activity 3.1 Start Thinking! For use before Activity 3.1

Activity 3.1 Warm Up For use before Activity 3.1

Convert the measurement.

1. 30 min = __?__ h
2. 4 h = __?__ min
3. 15 sec = __?__ min
4. 60 h = __?__ days
5. 3 days = __?__ h
6. 1 wk = __?__ h

3.1 Record and Practice Journal

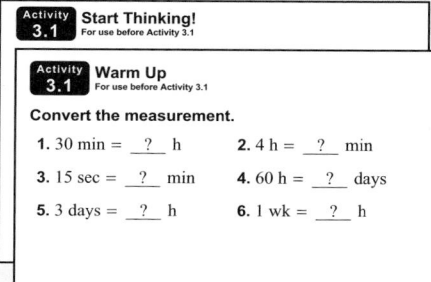

Essential Question How do rates help you describe real-life problems?

1 ACTIVITY: Finding Reasonable Rates

Work with a partner. See Additional Answers.

a. Match each description with a verbal rate.

b. Match each verbal rate with a numerical rate.

c. Give a reasonable numerical rate for each description. Then give an unreasonable rate.

Description	Verbal Rate	Numerical Rate
Your pay rate for washing cars	inches per month	□ m/sec : □ m/sec
The average rainfall in a rain forest	pounds per acre	□ people/yr : □ people/yr
Your average driving rate along an interstate	meters per second	□ lb/acre : □ lb/acre
The growth rate for the length of a baby alligator	people per year	□ mi/h : □ mi/h
Your running rate in a 100-meter dash	dollars per hour	□ in./yr : □ in./yr
The population growth rate of a large city	dollars per year	□ in./mo : □ in./mo
The average pay rate for a professional athlete	miles per hour	$□/h : $□/h
The fertilization rate for an apple orchard	inches per year	$□/yr : $□/yr

English Language Learners

Have students add a glossary to their math notebook. Key vocabulary words should be added as they are introduced. Illustrations next to the vocabulary words will help in understanding and reinforcing the concept.

3.1 Record and Practice Journal

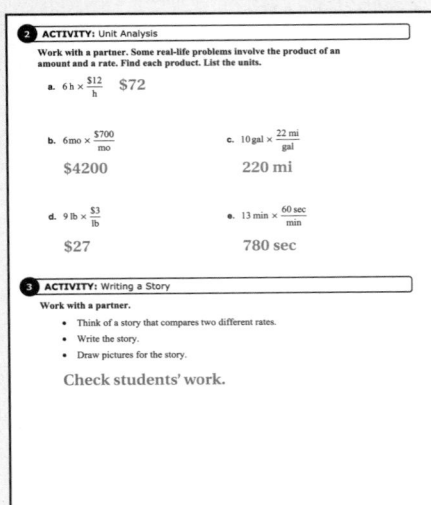

Activity 2

- **FYI:** Unit analysis is also called dimensional analysis, or the factor-label method. It uses the fact that multiplying a number or expression by one does not change its value.
- In these problems, a quantity is multiplied by a unit rate whose denominator is the same unit as the initial quantity. The common units divide out. Example:

$$3 \text{ weeks} \times \frac{40 \text{ hours}}{\text{week}} = 120 \text{ hours}$$

- **Common Error:** Students will often try to write the unit rate first. Encourage students to always write the given information (3 weeks) first and then multiply by the unit rate.

Activity 3

- To get students started, have them look at the sample rates on the previous page. Pick one and use it as a starting point.
- The second rate may or may not use the exact same units (inches per month and inches per year).
- This writing activity could become part of the homework assignment.

What Is Your Answer?

- The research assignment in Question 4 could become part of a bulletin board display.
- The estimation rule in Question 6 is very interesting.
- Students will need to try a few examples to see why it works. One example to try: the 40-hour work week.

Closure

- Describe three common rates and give a numerical example of each. Listen to be sure that students are comparing two units using the word "per."

Technology For the Teacher

Dynamic Classroom

The Dynamic Planning Tool
Editable Teacher's Resources at *BigIdeasMath.com*

2 ACTIVITY: Unit Analysis

Work with a partner. Some real-life problems involve the product of an amount and a rate. Find each product. List the units.

a. **Sample:** $6 \text{ h} \times \dfrac{\$12}{\text{h}} = 6\,\cancel{\text{h}} \times \dfrac{\$12}{\cancel{\text{h}}}$ Divide out "hours."

$= \$72$ Multiply. Answer is in dollars.

b. $6 \text{ mo} \times \dfrac{\$700}{\text{mo}}$

c. $10 \text{ gal} \times \dfrac{22 \text{ mi}}{\text{gal}}$

d. $9 \text{ lb} \times \dfrac{\$3}{\text{lb}}$

e. $13 \text{ min} \times \dfrac{60 \text{ sec}}{\text{min}}$

3 ACTIVITY: Writing a Story

Work with a partner.

- **Think of a story that compares two different rates.**
- **Write the story.**
- **Draw pictures for the story.**

What Is Your Answer?

4. **RESEARCH** Use newspapers, the Internet, or magazines to find examples of salaries. Try to find examples of each of the following ways to write salaries.

    a. dollars per hour   b. dollars per month   c. dollars per year

5. **IN YOUR OWN WORDS** How do rates help you describe real-life problems? Give two examples.

6. To estimate the annual salary for a given hourly pay rate, multiply by 2 and insert "000" at the end.

    Sample: $10 per hour is about $20,000 per year.

    a. Explain why this works. Assume the person is working 40 hours a week.
    b. Estimate the annual salary for an hourly pay rate of $8 per hour.
    c. You earn $1 million per month. What is your annual salary?
    d. Why is the cartoon funny?

"We had someone apply for the job. He says he would like $1 million a month, but will settle for $8 an hour."

Practice Use what you discovered about ratios and rates to complete Exercises 7–10 on page 102.

Check It Out
Lesson Tutorials
BigIdeasMath com

Key Vocabulary 🔊
ratio, p. 100
rate, p. 100
unit rate, p. 100

A **ratio** is a comparison of two quantities using division.

$$\frac{3}{4}, 3 \text{ to } 4, 3:4$$

A **rate** is a ratio of two quantities with different units.

$$\frac{60 \text{ miles}}{2 \text{ hours}}$$

A rate with a denominator of 1 is called a **unit rate**.

$$\frac{30 \text{ miles}}{1 \text{ hour}}$$

EXAMPLE ① **Finding Ratios and Rates**

There are 45 males and 60 females in a subway car. The subway car travels 2.5 miles in 5 minutes.

a. Find the ratio of males to females.

b. Find the speed of the subway car.

a. $\dfrac{\text{males}}{\text{females}} = \dfrac{45}{60} = \dfrac{3}{4}$

∴ The ratio of males to females is $\dfrac{3}{4}$.

b. $2.5 \text{ miles in } 5 \text{ minutes} = \dfrac{2.5 \text{ mi}}{5 \text{ min}} = \dfrac{2.5 \text{ mi} \div 5}{5 \text{ min} \div 5} = \dfrac{0.5 \text{ mi}}{1 \text{ min}}$

∴ The speed is 0.5 mile per minute.

EXAMPLE ② **Finding a Rate from a Table**

The table shows the amount of money you can raise by walking for a charity. Find your unit rate in dollars per mile.

	+2	+2	+2	
Distance (miles)	2	4	6	8
Money (dollars)	24	48	72	96
	+24	+24	+24	

Use the table to find the unit rate.

$\dfrac{\text{change in money}}{\text{change in distance}} = \dfrac{\$24}{2 \text{ mi}}$ The money raised increases by $24 every 2 miles.

$= \dfrac{\$12}{1 \text{ mi}}$ Simplify.

∴ Your unit rate is $12 per mile.

🔊 Multi-Language Glossary at BigIdeasMath✓com.

Laurie's Notes

Introduction

Connect

- **Yesterday:** Students explored many common unit rates and gave reasonable numeric values for each.
- **Today:** Students determine rates from words, tables, and graphs.

Motivate

- ❓ "A pitcher for a baseball team is able to throw a fastball approximately 132 feet in 1 second. How fast would this be in miles per hour?" 90 mi/h
- Explain that in this lesson they will be working with equivalent rates written in different units.

Write

- Remind students of the definitions for *ratio, rate,* and *unit rate.*
- ❓ Ask students to give their own examples of each.

Lesson Notes

Example 1

- **Common Error:** Be sure students read the question carefully. Order matters when writing a ratio. Writing $\frac{60}{45}$ would be incorrect.
- In part (b), 2.5 miles per 5 minutes is a rate. 0.5 mile per 1 minute is a unit rate.
- ❓ **Extension:** "How would you change 2.5 miles per 5 minutes into a unit rate in miles per hour?" To answer this question, use dimensional analysis introduced in the investigation.

$$\frac{2.5 \text{ miles}}{5 \text{ minutes}} \times \frac{60 \text{ minutes}}{1 \text{ hour}} = 30 \text{ miles per hour}$$

  - Besides dividing out a common unit of minutes, divide out a common factor of 5.

Example 2

- Copy the table, explaining that it shows the amounts of money you can raise walking for a charity.
- To assess students' understanding of the table, ask how much is raised when you walk 6 miles.
- ❓ "What is the pattern in the distance walked?" Students might say even numbers, but help them think about the pattern. Students should realize that the numbers are increasing by 2.
- ❓ "What is the pattern in money earned?" increasing by 24
- ❓ "Do you think the unit rate should be money raised per mile walked, or miles walked per dollar raised?" Although it could be either, it's more natural to think of how much you raise for each mile you walk.
- Check to see that students can distinguish between rate and unit rate.

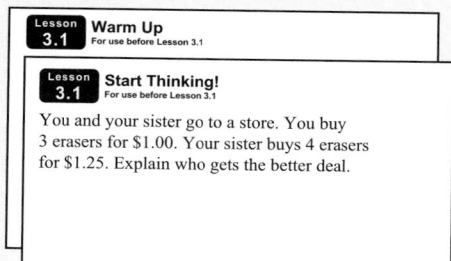
Extra Example 1

a. There are 12 dogs and 15 cats at the pet store. Find the ratio of cats to dogs. $\frac{5}{4}$

b. You bicycle 30 blocks in 20 minutes. Find your speed. 1.5 blocks per minute

Extra Example 2

The table shows the amount of money you can raise by dancing for a charity. Find your unit rate in dollars per hour.

Time (hours)	6	12	18	24
Money (dollars)	$90	$180	$270	$360

$15 per hour

On Your Own

1. $\frac{4}{3}$ 2. $\frac{4}{7}$

3. 4.8 mi per sec

Extra Example 3

The graph shows the distance that you walk. Find your rate in feet per second.

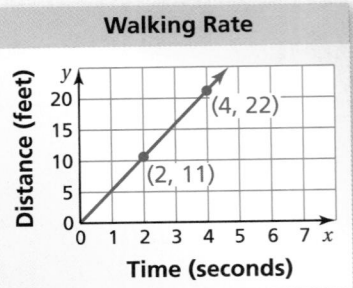

Walking Rate

5.5 feet per second

On Your Own

4. No, because the rate is the same along the line.

5. 0.35 km per sec

6. water; $\dfrac{1.5 \text{ km}}{\text{sec}} > \dfrac{0.35 \text{ km}}{\text{sec}}$

Differentiated Instruction

Visual

Students may see little difference between fractions and ratios. Fractions are one type of ratio, part-to-whole. Ratios include part-to-part and whole-to-whole.

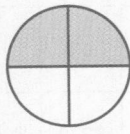
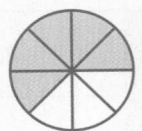

Circle 1   Circle 2

Part-to-whole of Circle 1:

$\dfrac{\text{unshaded parts}}{\text{whole number of parts}} = \dfrac{2}{4}$

Part-to-part of Circle 2:

$\dfrac{\text{shaded parts}}{\text{unshaded parts}} = \dfrac{5}{3}$

Whole-to-whole:

$\dfrac{\text{parts of Circle 1}}{\text{parts of Circle 2}} = \dfrac{4}{8}$

Laurie's Notes

On Your Own

? "How do you calculate a speed?" distance ÷ time

? "Speed is a rate. Why?" You are comparing two quantities with different units.

• Students should work with a partner on these three problems.

• **Extension:** Have students estimate the speed in miles per hour.

$$\dfrac{14.4 \text{ miles}}{3 \text{ seconds}} \times \dfrac{3600 \text{ seconds}}{1 \text{ hour}} = \dfrac{17,280 \text{ miles}}{\text{hour}}$$

Example 3

• Give the students time to read the graph.

• Work through the problem. To help students see the connection between this lesson and the next, use language such as "When x increases by 2, y increases by 3."

• **Common Error:** Students often write $\frac{2}{3}$ as the speed, because the x-coordinate is first.

• Remind students that speed is distance traveled (y) ÷ time (x).

• **Misconception:** The scale in the x- and y- directions does *not* need to be equal. In this problem, the scale is different.

• **Multiple Representations:** Three points are plotted: (0, 0), (2, 3), and (4, 6).

• This same data could be displayed in a table. If time permits, have students represent it in tabular format.

? "Using the pattern, what would the y-coordinate be for x = 8?"
12; (8, 12) is the point.

On Your Own

• **Connection:** Question 4 connects the concept of slope in the next lesson. The change (in each variable) can be measured from (0, 0).

• Watch for common errors like time ÷ distance.

• Discuss the answer(s) to Question 6.

Closure

• **Exit Ticket:** Write 4.8 meters per 3 seconds as a unit rate.
1.6 meters per second

Technology For the Teacher

Dynamic Classroom

The Dynamic Planning Tool
Editable Teacher's Resources at *BigIdeasMath.com*

On Your Own

Now You're Ready
Exercises 11–26

1. In Example 1, find the ratio of females to males.

2. In Example 1, find the ratio of females to total passengers.

3. The table shows the distance that the International Space Station travels while orbiting Earth. Find the speed in miles per second.

Time (seconds)	3	6	9	12
Distance (miles)	14.4	28.8	43.2	57.6

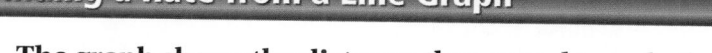

EXAMPLE 3 **Finding a Rate from a Line Graph**

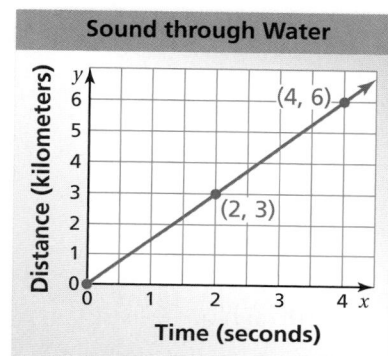

Sound through Water

The graph shows the distance that sound travels through water. Find the speed of sound in kilometers per second.

Step 1: Choose a point on the line.

The point $(2, 3)$ shows you that sound travels 3 kilometers in 2 seconds.

Step 2: Find the speed.

$$\frac{\text{distance traveled}}{\text{elapsed time}} = \frac{3}{2} \quad \begin{matrix} \leftarrow \text{kilometers} \\ \leftarrow \text{seconds} \end{matrix}$$

$$= \frac{1.5 \text{ km}}{1 \text{ sec}} \quad \text{Simplify.}$$

∴ The speed is 1.5 kilometers per second.

On Your Own

4. **WHAT IF?** In Example 3, you use the point $(4, 6)$ to find the speed. Does your answer change? Why or why not?

5. The graph shows the distance that sound travels through air. Find the speed of sound in kilometers per second.

6. Does sound travel faster in water or in air? Explain.

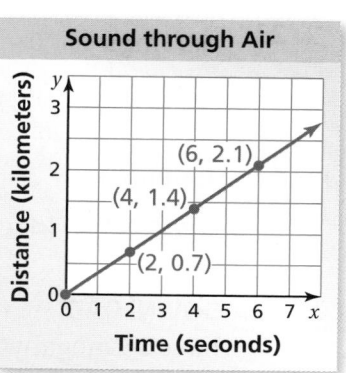

Sound through Air

Section 3.1 Ratios and Rates **101**

Check It Out
Help with Homework
BigIdeasMath ✓.com

 ## Vocabulary and Concept Check

1. **VOCABULARY** How can you tell when a rate is a unit rate?

2. **WRITING** Why do you think rates are usually written as unit rates?

3. **OPEN-ENDED** Write a real-life rate that applies to you.

Estimate the unit rate.

4. $74.75

Gloss White PAINT 5 gal

5. $1.19

GRAPE JUICE 12 fl oz

6. $2.35

12 Grade AA Eggs

 ## Practice and Problem Solving

Find the product. List the units.

7. $8 \text{ h} \times \dfrac{\$9}{\text{h}}$

8. $8 \text{ lb} \times \dfrac{\$3.50}{\text{lb}}$

9. $14 \text{ sec} \times \dfrac{60 \text{ MB}}{\text{sec}}$

10. $6 \text{ h} \times \dfrac{19 \text{ mi}}{\text{h}}$

Write the ratio as a fraction in simplest form.

11. 25 to 45

12. $63 : 28$

13. 35 girls : 15 boys

14. 2 feet : 8 feet

15. 16 dogs to 12 cats

16. 51 correct : 9 incorrect

Find the unit rate.

17. 180 miles in 3 hours

18. 256 miles per 8 gallons

19. $9.60 for 4 pounds

20. $4.80 for 6 cans

21. 297 words in 5.5 minutes

22. 54 meters in 2.5 hours

Use the table to find the rate.

23.

Servings	0	1	2	3
Calories	0	90	180	270

24.

Days	0	1	2	3
Liters	0	1.6	3.2	4.8

25.

Packages	3	6	9	12
Servings	13.5	27	40.5	54

26.

Years	2	6	10	14
Feet	7.2	21.6	36	50.4

27. **DOWNLOAD** At 1 P.M., you have 24 megabytes of a movie. At 1:15 P.M., you have 96 megabytes. What is the download rate in megabytes per minute?

28. **POPULATION** In 2000, the U.S. population was 281 million people. In 2008, it was 305 million. What was the rate of population change per year?

Assignment Guide and Homework Check

Level	Day 1 Activity Assignment	Day 2 Lesson Assignment	Homework Check
Basic	7–10, 34–38	1–6, 11–25 odd, 27–29	2, 13, 17, 23, 28
Average	7–10, 34–38	1–6, 11–21 odd, 24–28 even, 29, 31	2, 13, 17, 24, 28
Advanced	7–10, 34–38	1–6, 18–26 even, 30–33	18, 24, 30, 31

Common Errors

- **Exercises 11–13** Students may put the wrong number in the numerator. Remind them that the first number or object is the numerator and the second is the denominator.
- **Exercises 17–22** Students may find the unit rate, but forget to include the units. Remind them that the units are necessary for understanding a unit rate, or any rate.
- **Exercises 23–26** When finding the rate from the table, students may put the wrong unit on top. Tell them that the unit on the bottom of the table will be the unit in the numerator of the rate.

3.1 Record and Practice Journal

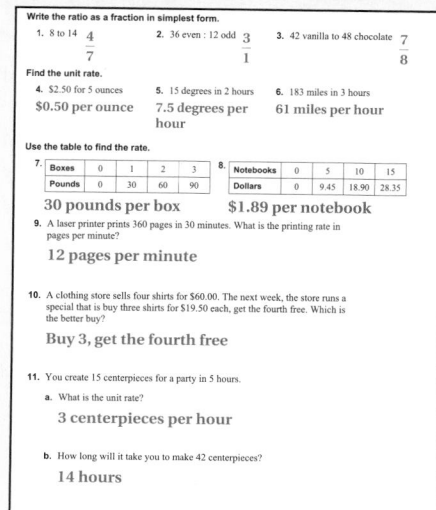

Technology For the Teacher
Answer Presentation Tool
QuizShow

Vocabulary and Concept Check

1. It has a denominator of 1.

2. Unit rates are easier to compare.

3. *Sample answer:* A basketball player runs 10 feet down the court in 2 seconds.

4. $15 per gal

5. $0.10 per fl oz

6. $0.20 per egg

Practice and Problem Solving

7. $72   8. $28

9. 840 MB   10. 114 mi

11. $\frac{5}{9}$ 12. $\frac{9}{4}$

13. $\frac{7}{3}$ 14. $\frac{1}{4}$

15. $\frac{4}{3}$ 16. $\frac{17}{3}$

17. 60 mi/h

18. 32 mi per gal

19. $2.40 per lb

20. $0.80 per can

21. 54 words per min

22. 21.6 m per h

23. 90 calories per serving

24. 1.6 L per day

25. 4.5 servings per package

26. 3.6 ft per yr

27. 4.8 MB per min

28. 3,000,000 people per yr

29. a. It costs $122 for 4 tickets.

    b. $30.50 per ticket

    c. $305

Practice and Problem Solving

30. No; Although the relative number of boys and girls are the same, the two ratios are inverses.

31. The 9-pack is the best buy at $2.55 per container.

32. a. whole milk

b. orange juice

33. See *Taking Math Deeper*.

Fair Game Review

34–37.

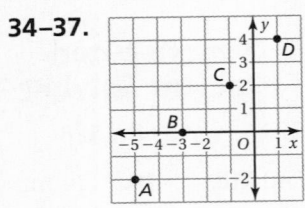

38. B

Mini-Assessment

Write the ratio as a fraction in simplest form.

1. 30 to 50 $\frac{3}{5}$ **2.** 3 : 12 $\frac{1}{4}$

Find the unit rate.

3. 165 miles in 3 hours 55 mi/h

4. $9.60 for 8 cans $1.20 per can

5. The graph shows the cost of buying movie tickets.

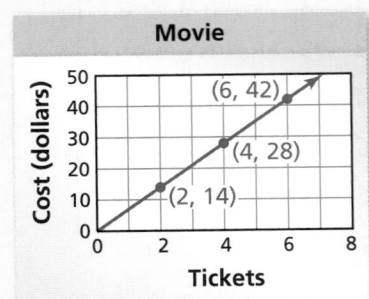

a. What does the point (4, 28) represent? $28 for 4 tickets

b. What is the unit rate? $7 per ticket

c. What is the cost of buying 9 tickets? $63

Taking Math Deeper

Exercise 33

This is a nice real-life problem that deals with rates. If students search the Internet for "fire hydrant colors" they can get the information about the rates in gallons per minute (GPM).

 Perform an Internet search and make a table from the results.

Blue	1500 + GPM	Very good
Green	1000–1499 GPM	Good
Yellow	500–999 GPM	Adequate
Red	Below 500 GPM	Inadequate

 A number line helps display the information.

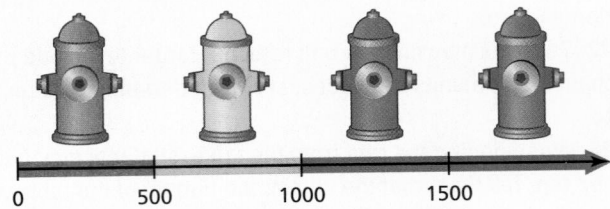

Rates in gallons per minute (GPM)

 Knowing the rate at which the water comes out of the fire hydrant is critical. If a fire fighter pumps water out at too high a rate, the system of water pipes in the ground could be stressed and burst.

Reteaching and Enrichment Strategies

If students need help. . .	If students got it. . .
Resources by Chapter • Practice A and Practice B • Puzzle Time Record and Practice Journal Practice Differentiating the Lesson Lesson Tutorials Skills Review Handbook	Resources by Chapter • Enrichment and Extension Start the next section

29. TICKETS The graph shows the cost of buying tickets to a concert.

   a. What does the point (4, 122) represent?

   b. What is the unit rate?

   c. What is the cost of buying 10 tickets?

30. CRITICAL THINKING Are the two statements equivalent? Explain your reasoning.

   ● The ratio of boys to girls is 2 to 3.

   ● The ratio of girls to boys is 3 to 2.

31. TENNIS A sports store sells three different packs of tennis balls. Which pack is the best buy? Explain.

 $11.49

 $16.79

 $22.99

Beverage	Serving Size	Calories	Sodium
Whole milk	1 cup	146	98 mg
Orange juice	1 pt	210	10 mg
Apple juice	24 fl oz	351	21 mg

32. NUTRITION The table shows nutritional information for three beverages.

   a. Which has the most calories per fluid ounce?

   b. Which has the least sodium per fluid ounce?

33. Open-Ended Fire hydrants are painted four different colors to indicate the rate at which water comes from the hydrant.

   a. RESEARCH Use the Internet to find the ranges of the rates for each color.

   b. Research why a firefighter needs to know the rate at which water comes out of the hydrant.

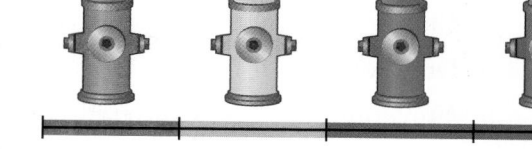

Fair Game Review   What you learned in previous grades & lessons

Plot the ordered pair in a coordinate plane.   *(Section 1.6)*

34. $A(-5, -2)$ **35.** $B(-3, 0)$ **36.** $C(-1, 2)$ **37.** $D(1, 4)$

38. MULTIPLE CHOICE Which fraction is greater than $-\frac{2}{3}$ and less than $-\frac{1}{2}$? *(Section 2.1)*

   Ⓐ $-\frac{3}{4}$ Ⓑ $-\frac{7}{12}$ Ⓒ $-\frac{5}{12}$ Ⓓ $-\frac{3}{8}$

3.2 Slope

Essential Question How can you compare two rates graphically?

1 ACTIVITY: Comparing Unit Rates

Work with a partner. The table shows the maximum speeds of several animals.

a. Find the missing speeds. Round your answers to the nearest tenth.

b. Which animal is fastest? Which animal is slowest?

c. Explain how you convert between the two units of speed.

Animal	Speed (miles per hour)	Speed (feet per second)
Antelope	61.0	
Black Mamba Snake		29.3
Cheetah		102.6
Chicken		13.2
Coyote	43.0	
Domestic Pig		16.0
Elephant		36.6
Elk		66.0
Giant Tortoise	0.2	
Giraffe	32.0	
Gray Fox		61.6
Greyhound	39.4	
Grizzly Bear		44.0
Human		41.0
Hyena	40.0	
Jackal	35.0	
Lion		73.3
Peregrine Falcon	200.0	
Quarter Horse	47.5	
Spider		1.76
Squirrel	12.0	
Thomson's Gazelle	50.0	
Three-Toed Sloth		0.2
Tuna	47.0	

Laurie's Notes

Introduction

For the Teacher

- **Goal:** Students will explore the maximum speeds of various animals and then use information about two particular animals to explore what it means to run "at a constant speed."

Motivate

- If students are not comfortable with dimensional analysis from the previous lesson, you will need to help them get started with this activity.
- **Big Idea:** You begin with a speed in certain units (i.e., miles per hour) and you want a speed with different units (i.e., feet per second).
- You need to set up the factors so that the unwanted units divide out. The units you want to divide out should always appear diagonally from one another.
- Example: $\dfrac{61 \cancel{\text{ miles}}}{\cancel{\text{hour}}} \times \dfrac{1 \cancel{\text{ hour}}}{3600 \text{ seconds}} \times \dfrac{5280 \text{ feet}}{1 \cancel{\text{ mile}}}$
- Because 1 hour = 3600 seconds, multiplying by $\dfrac{1 \text{ hour}}{3600 \text{ seconds}}$ or $\dfrac{3600 \text{ seconds}}{1 \text{ hour}}$ is equivalent to multiplying by 1.
- Divide out by a common factor of 240, and the answer is $89\dfrac{7}{15}$.
- Converting from feet per second to miles per hour will require students to multiply by the reciprocal of each of the conversion factors.

Activity Notes

Activity 1

- It would be appropriate for students to work with a partner and use a calculator to complete the activity.
- ❓ "Look through the list and predict the fastest animal. Mark it with the letter F. Predict the slowest animal. Mark it with the letter S.
- Did students select the Peregrine falcon or cheetah as the fastest?
- Did students select the giant tortoise or three-toed sloth as the slowest?
- You may want to help students see how to convert from miles per hour to feet per second and vice versa.

$$\dfrac{\text{miles}}{\cancel{\text{hour}}} \times \underset{\uparrow \text{ equals } 1}{\dfrac{1 \cancel{\text{ hour}}}{3600 \text{ seconds}}} \times \underset{\uparrow \text{ equals } 1}{\dfrac{5280 \text{ feet}}{1 \cancel{\text{ mile}}}}$$

- When students have finished, discuss the results, the answer to part (c), and their predictions.

Previous Learning

In this activity, students will find the product of three fractions. Remember, $\dfrac{a}{b} \cdot \dfrac{c}{d} \cdot \dfrac{e}{f} = \dfrac{ace}{bdf}$. Divide out common factors, if possible.

Activity Materials
Textbook
• straightedge or ruler • calculators

Start Thinking! and Warm Up

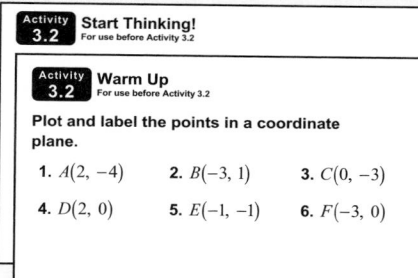

Activity 3.2 Start Thinking! For use before Activity 3.2

Activity 3.2 Warm Up For use before Activity 3.2

Plot and label the points in a coordinate plane.

1. $A(2, -4)$ 2. $B(-3, 1)$ 3. $C(0, -3)$
4. $D(2, 0)$ 5. $E(-1, -1)$ 6. $F(-3, 0)$

3.2 Record and Practice Journal

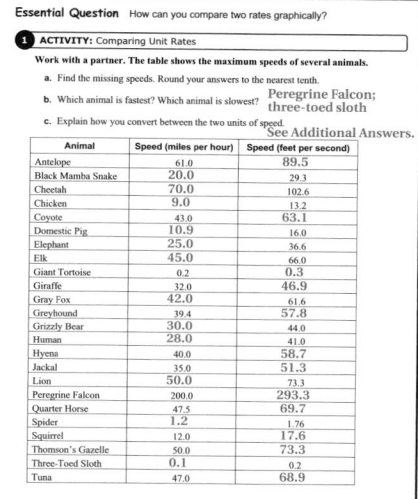

Essential Question How can you compare two rates graphically?

1 ACTIVITY: Comparing Unit Rates

Work with a partner. The table shows the maximum speeds of several animals.

a. Find the missing speeds. Round your answers to the nearest tenth.

b. Which animal is fastest? Which animal is slowest? Peregrine Falcon; three-toed sloth

c. Explain how you convert between the two units of speed. See Additional Answers.

Animal	Speed (miles per hour)	Speed (feet per second)
Antelope	61.0	89.5
Black Mamba Snake	20.0	29.3
Cheetah	70.0	102.6
Chicken	9.0	13.2
Coyote	43.0	63.1
Domestic Pig	10.9	16.0
Elephant	25.0	36.6
Elk	45.0	66.0
Giant Tortoise	0.2	0.3
Giraffe	32.0	46.9
Gray Fox	42.0	61.6
Greyhound	39.4	57.8
Grizzly Bear	30.0	44.0
Human	28.0	41.0
Hyena	40.0	58.7
Jackal	35.0	51.3
Lion	50.0	73.3
Peregrine Falcon	200.0	293.3
Quarter Horse	47.5	69.7
Spider	1.2	1.76
Squirrel	12.0	17.6
Thomson's Gazelle	50.0	73.3
Three-Toed Sloth	0.1	0.2
Tuna	47.0	68.9

English Language Learners

Labels

English learners may recognize the fraction bar as division, but may not be familiar with its use in the concept of rate. The following unit rates are equivalent.

$\frac{3\text{ m}}{1\text{ h}}$, 3 m/h, 3 meters per hour

Each of these rates can be read as "three meters *for every* hour."

3.2 Record and Practice Journal

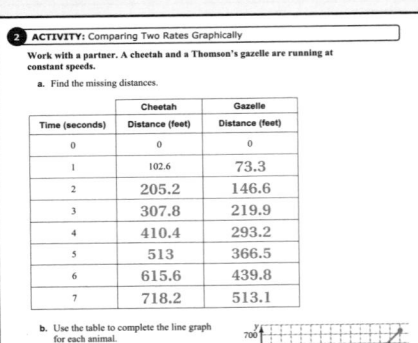

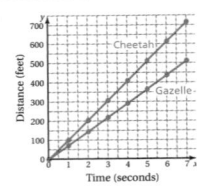

Activity 2

- *Note:* The formal definition of **slope** will come in the lesson. The idea in this activity is to get students to understand that a steeper line translates to a greater speed.
- Discuss the concept of a constant speed. Talk about a car on cruise control or items traveling on an assembly line in a factory. Over short periods of time it is possible for animals to run at a constant speed.
- ❓ "If a cheetah is running at a constant speed and has traveled a distance of 102.6 feet in 1 second, how far will it run in two seconds?" 205.2 ft
- Have students complete the table. This is a good review of decimal multiplication or decimal addition (depending on how the student completes the work). You may find it helpful to review the rules for multiplying and adding decimals beforehand.
- ❓ **Discuss:** "After 3 seconds, how far has each animal run? after 7 seconds?"
- ❓ "Is there ever a time when the gazelle has run farther than the cheetah?"
- ❓ "What does a line graph show?" the trend in data over time
- Students might find it helpful to use a straightedge or ruler to connect their points for a line graph.
- ❓ "If the cheetah was not running at a constant speed, would your graph look different? Explain." Yes. The points would not be in a line. The graph would go up and down accordingly.
- Discuss the relationship between the speed of animals and the steepness of the two graphs.

What Is Your Answer?

- Question 4 can be done as homework.

Closure

- An airplane is traveling at a constant speed of 6 miles per minute. Make a table to show the distance traveled each minute for 8 minutes. Make a line plot of your data and describe the graph.

Minutes, x	0	1	2	3	4	5	6	7	8
Miles, y	0	6	12	18	24	30	36	42	48

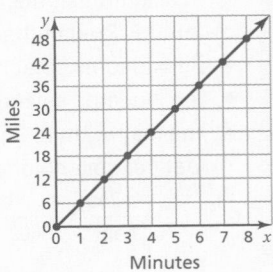

Technology For the Teacher

The Dynamic Planning Tool
Editable Teacher's Resources at *BigIdeasMath.com*

ACTIVITY: Comparing Two Rates Graphically

Work with a partner. A cheetah and a Thomson's gazelle are running at constant speeds.

a. Find the missing distances.

	Cheetah	Gazelle
Time (seconds)	Distance (feet)	Distance (feet)
0	0	0
1	102.6	
2		
3		
4		
5		
6		
7		

b. Use the table to complete the line graph for each animal.

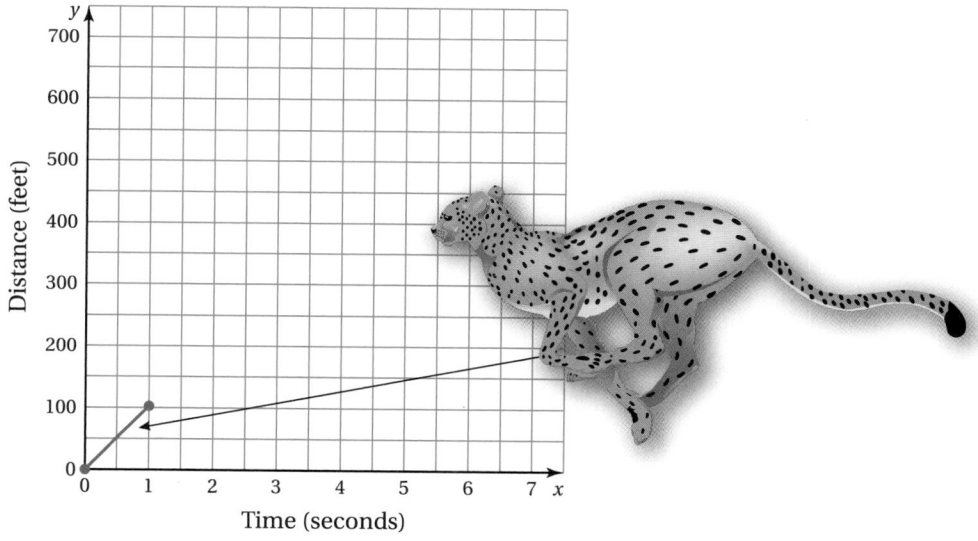

c. Which graph is steeper? The speed of which animal is greater?

What Is Your Answer?

3. **IN YOUR OWN WORDS** How can you compare two rates graphically? Explain your reasoning. Give some examples with your answer.

4. Choose 10 animals from Activity 1.

   a. Make a table for each animal similar to the table in Activity 2.

   b. Sketch a graph of the distances for each animal.

   c. Compare the steepness of the 10 graphs. What can you conclude?

Check It Out
Lesson Tutorials
BigIdeasMath ✓com

Key Vocabulary 🔊
slope, p. 106

🔑 Key Idea

Slope

Slope is the rate of change between any two points on a line. It is a measure of the *steepness* of a line.

To find the slope of a line, find the ratio of the change in y (vertical change) to the change in x (horizontal change).

$$\text{slope} = \frac{\text{change in } y}{\text{change in } x}$$

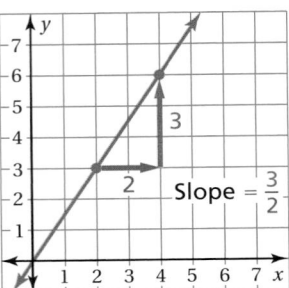

Slope $= \frac{3}{2}$

EXAMPLE 1   Finding Slopes

Find the slope of each line.

a.

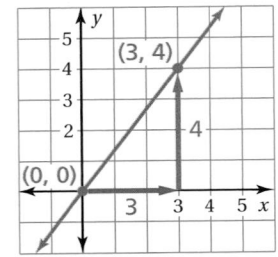

$$\text{slope} = \frac{\text{change in } y}{\text{change in } x}$$

$$= \frac{4}{3}$$

⋮ The slope of the line is $\frac{4}{3}$.

b.

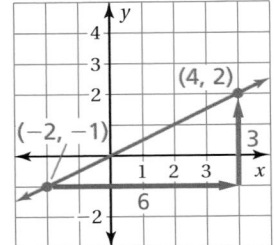

$$\text{slope} = \frac{\text{change in } y}{\text{change in } x}$$

$$= \frac{3}{6} = \frac{1}{2}$$

⋮ The slope of the line is $\frac{1}{2}$.

🔵 On Your Own

Now You're Ready
Exercises 4–9

Find the slope of the line.

1.

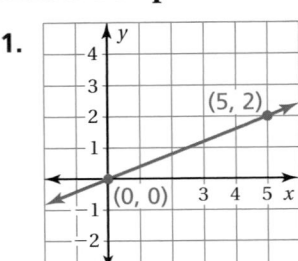

2.

🔊 Multi-Language Glossary at BigIdeasMath ✓com.

Laurie's Notes

Introduction

Connect

- **Yesterday:** Students explored constant rates of speed in a table and on a graph.
- **Today:** Students will define slope and determine the slope of a line from its graph.

Motivate

- Ask students about the slope of a half-pipe at a skateboard park, the slope of a local hill that they are familiar with, or the slope of a wheelchair ramp at school. (Chances are they will all have different slopes.)
- ❓ "What does the word slope mean when talking about a wheelchair ramp?" Listen for words such as "steepness" or "incline." Students often use their hands to demonstrate slope.

Lesson Notes

Key Idea

- Write the definition of slope on the board.
- Remind students that a rate is a ratio. Slopes are often thought of as rates.
- **FYI:** You may remember learning that the formula for slope is $\dfrac{y_2 - y_1}{x_2 - x_1}$.

  Students are not formally introduced to this formula until future courses. Try to enforce the concept rather than teach the formula.

Example 1

- The arrows on the graphs are good visual aids to help students think about how much each variable changes.
- Writing the amount of change on the graph is good reinforcement.
- At this stage, have them think about reading the graph left-to-right.
- ❓ "For the second graph, what would the slope be if you used the points (0, 0) and (4, 2)?" the same, $\dfrac{2}{4} = \dfrac{1}{2}$
- **Common Error:** A very common error is for students to find slope by writing the change in x over the change in y.

On Your Own

- **Neighbor Check:** Have students work independently and then have their neighbor check their work. Have students discuss any discrepancies.

Start Thinking! and Warm Up

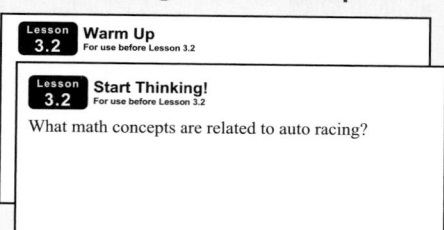

Lesson 3.2 **Warm Up** For use before Lesson 3.2

Lesson 3.2 **Start Thinking!** For use before Lesson 3.2

What math concepts are related to auto racing?

Extra Example 1

Find the slope of each line.

a.
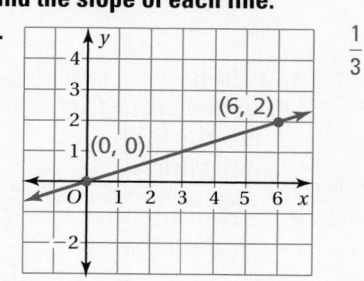
$\dfrac{1}{3}$

b.
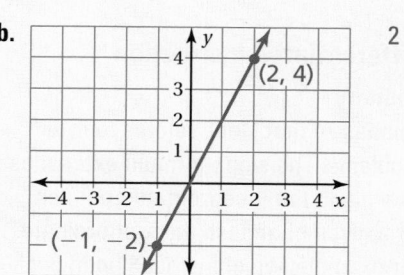
2

On Your Own

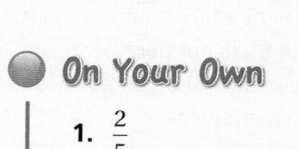

1. $\dfrac{2}{5}$

2. 3

Extra Example 2

The table shows your earnings for mowing lawns.

Lawns Mowed, x	3	6	9	12
Earnings, y (dollars)	45	90	135	180

a. Graph the data.

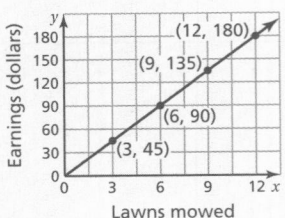

b. Find and interpret the slope of the line through the points. 15; You earn $15 per lawn mowed.

On Your Own

3. 5; No

4. a. Your friend's line is steeper. Your friend's pay rate is greater than yours.

    b. 7; Your friend earns $7 per hour babysitting.

Differentiated Instruction

Auditory

Emphasize that slope relates to rate problems. The slope formula expresses how much the y-coordinate changes for a given change in the x-coordinate. For example, weighing out 3 pounds of apples and paying $4.05 is equivalent to the unit rate of $1.35 per pound. So, the change in the cost is $1.35 for every one pound increase of apples.

Laurie's Notes

Example 2

- Students must first read and understand the data (table of ordered pairs), and then plot the data.
- Start to use language that will prepare students for future courses.
- The number of hours worked is x and the amount of dollars earned is y. "The amount of money earned depends upon how many hours you worked."
- Slopes are often rates. Here, the rate is *dollars per hour*. Be sure to have students explain the meaning of this concept.
- **Big Idea:** Students will often be asked to interpret a slope. This means to look at the context of the problem and decide what a slope means with regard to the two variables.

On Your Own

- Share answers as a class.
- **Extension:** "Read information from the graph. How many hours did you work to earn $30?" 6

Closure

- The cost of admission to the local museum is given in the table. Graph the data and determine the slope. Interpret the slope in the context of the problem.

Number of People, x	2	3	4
Total Cost of Admission, y	$4.50	$6.75	$9.00

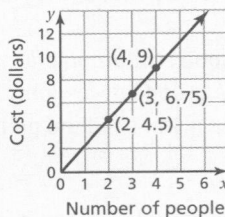

slope: 2.25; The slope represents the cost per person.

EXAMPLE ② **Finding a Slope**

The table shows your earnings for babysitting.

a. Graph the data.

b. Find and interpret the slope of the line through the points.

Hours, x	0	2	4	6	8	10
Earnings, y (dollars)	0	10	20	30	40	50

a. Graph the data. Draw a line through the points.

b. Choose any two points to find the slope of the line.

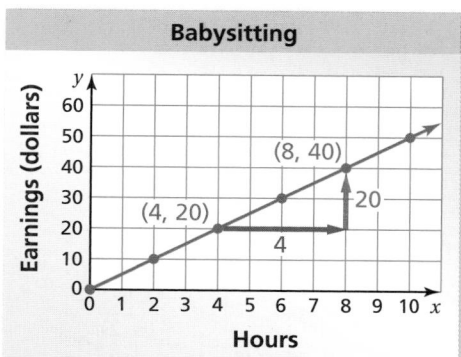

$$\text{slope} = \frac{\text{change in } y}{\text{change in } x}$$

$$= \frac{20}{4} \quad \leftarrow \text{dollars} \\ \leftarrow \text{hours}$$

$$= 5$$

∴ The slope of the line is 5. So, you earn $5 per hour babysitting.

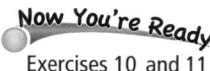 **On Your Own**

Now You're Ready
Exercises 10 and 11

3. In Example 2, use two other points to find the slope. Does the slope change?

4. The graph shows the earnings of you and your friend for babysitting.

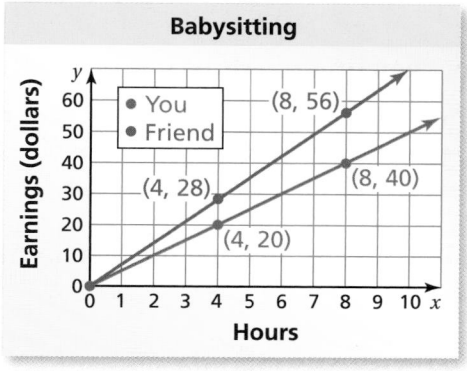

a. Compare the steepness of the lines. What does this mean in the context of the problem?

b. Find and interpret the slope of the blue line.

Section 3.2 Slope **107**

Check It Out
Help with Homework
BigIdeasMath ✓com

Vocabulary and Concept Check

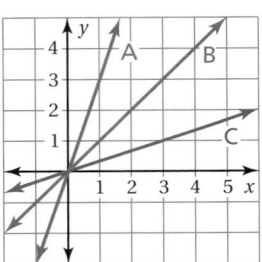

1. **VOCABULARY** Is there a connection between rate and slope? Explain.

2. **REASONING** Which line has the greatest slope?

3. **REASONING** Is it more difficult to run up a ramp with a slope of $\frac{1}{5}$ or a ramp with a slope of 5? Explain.

Practice and Problem Solving

Find the slope of the line.

4.

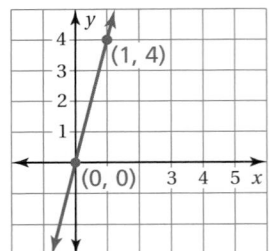

5.

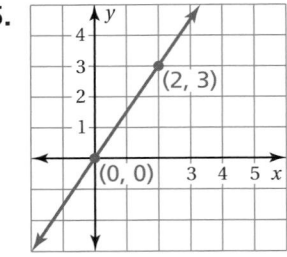

6.

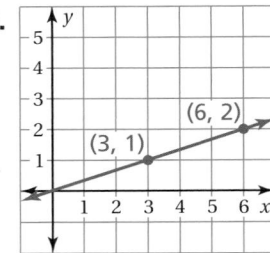

7.

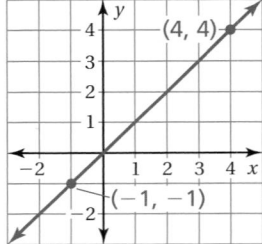

8.

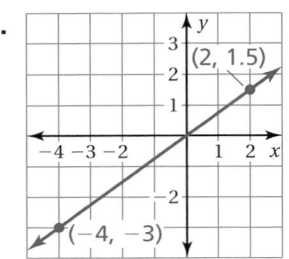

9.

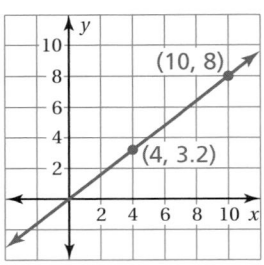

Graph the data. Then find the slope of the line through the points.

10.

Minutes, x	3	5	7	9
Words, y	135	225	315	405

11.

Gallons, x	5	10	15	20
Miles, y	162.5	325	487.5	650

Graph the line that passes through the two points. Then find the slope of the line.

12. (0, 0), (5, 8)

13. (−2, −2), (2, 2)

14. (10, 4), (−5, −2)

15. ERROR ANALYSIS Describe and correct the error in finding the slope of the line passing through (0, 0) and (4, 5).

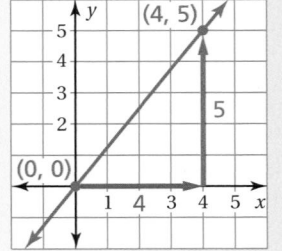

$$\text{slope} = \frac{4}{5}$$

Assignment Guide and Homework Check

Level	Day 1 Activity Assignment	Day 2 Lesson Assignment	Homework Check
Basic	20–26	1–3, 5–15 odd, 16	2, 5, 11, 16
Average	20–26	1–3, 5–11 odd, 12, 14–17	2, 5, 11, 16
Advanced	20–26	1–3, 4–14 even, 15, 17–19	4, 10, 12, 18

Common Errors

- **Exercises 4–11** Students may put the change in *x* over the change in *y*. Remind them that the vertical change is written over the horizontal change and that slope is a rate. Tell the students to label the axes with units that represent a common rate to help them remember which change goes on top. For example, label the *y*-axis "miles" and the *x*-axis "gallons." This should help students identify that the change in *y* goes on top because miles is first in the rate.
- **Exercises 12–14** Students may not remember how to plot ordered pairs with negative numbers. Remind students of Quadrants II, III, and IV, and which coordinate is negative in each quadrant.

3.2 Record and Practice Journal

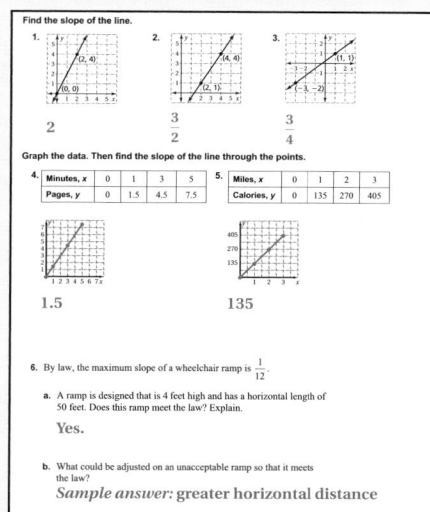

Find the slope of the line.

1. (2, 4) (0, 0) — 2

2. (4, 4) (2, 1) — $\frac{3}{2}$

3. (1, 1) (-3, -2) — $\frac{3}{4}$

Graph the data. Then find the slope of the line through the points.

4.
Minutes, *x*	0	1	3	5
Pages, *y*	0	1.5	4.5	7.5

1.5

5.
Miles, *x*	0	1	2	3
Calories, *y*	0	135	270	405

135

6. By law, the maximum slope of a wheelchair ramp is $\frac{1}{12}$.

a. A ramp is designed that is 4 feet high and has a horizontal length of 50 feet. Does this ramp meet the law? Explain.
Yes.

b. What could be adjusted on an unacceptable ramp so that it meets the law?
Sample answer: greater horizontal distance

Technology For the Teacher
Answer Presentation Tool
QuizShow

Vocabulary and Concept Check

1. yes; Slope is the rate of change of a line.

2. A

3. 5; A ramp with a slope of 5 increases 5 units vertically for every 1 unit horizontally. A ramp with a slope of $\frac{1}{5}$ increases 1 unit vertically for every 5 units horizontally.

Practice and Problem Solving

4. 4

5. $\frac{3}{2}$

6. $\frac{1}{3}$

7. 1

8. $\frac{3}{4}$

9. $\frac{4}{5}$

10.

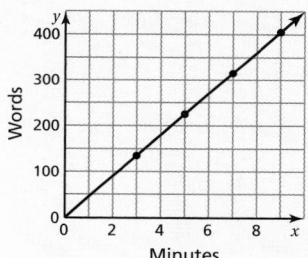

slope = 45

11.

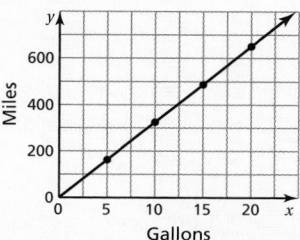

slope = 32.5

12–14. See Additional Answers.

15. The change in *y* should be in the numerator. The change in *x* should be in the denominator.

Slope = $\frac{5}{4}$

Practice and Problem Solving

16. See *Taking Math Deeper*.

17. See Additional Answers.

18. 0; The change in y is 0 because the y-values do not change. So, the slope is 0.

19. $y = 6$

Fair Game Review

20. $>$ 21. $<$

22. $=$ 23. $-\dfrac{4}{5}$

24. $-\dfrac{3}{5}$ 25. 3

26. C

Mini-Assessment

Find the slope of the line that passes through the two points.

1. $(0, 0)$, $(3, 2)$ $\dfrac{2}{3}$

2. $(-2, -2)$, $(5, 5)$ 1

3. $(-3, -4)$, $(6, 8)$ $\dfrac{4}{3}$

4. $(-4, -2)$, $(-2, -1)$ $\dfrac{1}{2}$

5. The graph shows the amount of money you are saving for a computer.

   a. Find the slope of the line.   40

   b. Interpret the slope of the line in the context of the problem.   You are saving $40 per week.

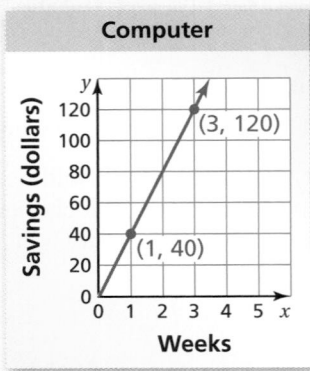

Taking Math Deeper

Exercise 16

This is a classic type of problem that uses linear models to predict future events. Each person is saving money at a constant rate (constant slope). The fact that the rate is constant is what makes the graph a line. The prediction of when $165 will be saved assumes that the constant rate continues into the future.

 Interpret the slope in context.

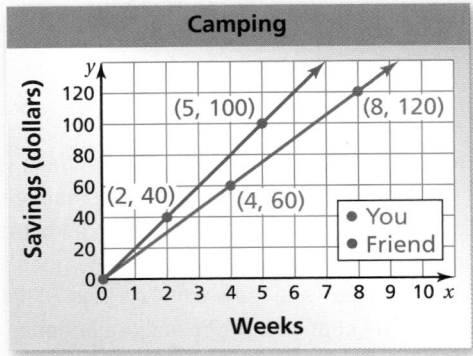

a. Your friend's saving rate (in dollars per week) is greater than yours.

 Find the slope of each line.

b. Slopes   Rates

You: $\dfrac{120 - 60}{8 - 4} = \dfrac{60}{4} = \$15$ per week

Friend: $\dfrac{100 - 40}{5 - 2} = \dfrac{60}{3} = \$20$ per week

c. Your friend saves $20 - 15 = \$5$ more per week.

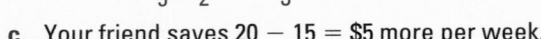 How long will it take for you to save $165?

d. At $15 per week, it will take $\dfrac{165}{15} = 11$ weeks.

Reteaching and Enrichment Strategies

If students need help. . .	If students got it. . .
Resources by Chapter • Practice A and Practice B • Puzzle Time Record and Practice Journal Practice Differentiating the Lesson Lesson Tutorials Skills Review Handbook	Resources by Chapter • Enrichment and Extension Start the next section

16. **CAMPING** The graph shows the amount of money you and a friend are saving for a camping trip.

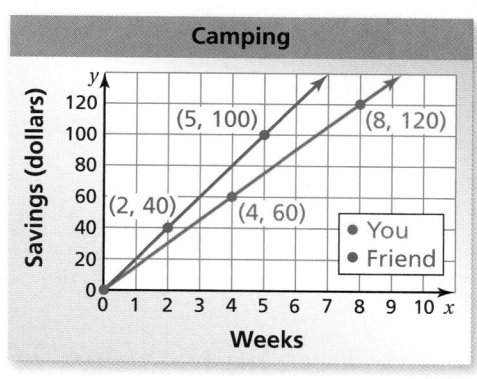

a. Compare the steepness of the lines. What does this mean in the context of the problem?

b. Find the slope of each line.

c. How much more money does your friend save each week than you?

d. The camping trip costs $165. How long will it take you to save enough money?

17. **MAPS** The table shows data from a key to a map of Ohio.

Distance on Map (mm), x	10	20	30	40
Actual Distance (mi), y	25	50	75	100

a. Graph the data.

b. Find the slope of the line. What does this mean in the context of the problem?

c. The map distance between Toledo and Columbus is 48 millimeters. What is the actual distance?

d. Cincinnati is about 225 miles from Cleveland. What is the distance between these cities on the map?

18. **CRITICAL THINKING** What is the slope of a line that passes through the points (2, 0) and (5, 0)? Explain.

19. **Number Sense** A line has a slope of 2. It passes through the points (1, 2) and (3, y). What is the value of y?

Fair Game Review What you learned in previous grades & lessons

Copy and complete the statement using <, >, or =. *(Section 2.1)*

20. $\dfrac{9}{2}$ ▨ $\dfrac{8}{3}$

21. $-\dfrac{8}{15}$ ▨ $\dfrac{10}{18}$

22. $\dfrac{-6}{24}$ ▨ $\dfrac{-2}{8}$

Multiply. *(Section 2.3)*

23. $-\dfrac{3}{5} \times \dfrac{8}{6}$

24. $1\dfrac{1}{2} \times \left(-\dfrac{6}{15}\right)$

25. $-2\dfrac{1}{4} \times -1\dfrac{1}{3}$

26. **MULTIPLE CHOICE** You have 18 stamps from Mexico in your stamp collection. These stamps are $\dfrac{3}{8}$ of your collection. The rest of the stamps are from the United States. How many stamps are from the United States? *(Section 2.5)*

Ⓐ 12　　　　Ⓑ 24　　　　Ⓒ 30　　　　Ⓓ 48

Essential Question How can proportions help you decide when things are "fair?"

The Meaning of a Word ● Proportional

When you work toward a goal, your success is usually **proportional** to the amount of work you put in.

An equation stating that two ratios are equal is a **proportion**.

1 ACTIVITY: Determining Proportions

Work with a partner. Tell whether the two ratios are equivalent. If they are not equivalent, change the second day to make the ratios equivalent. Explain your reasoning.

a. On the first day, you pay $5 for 2 boxes of popcorn. The next day, you pay $7.50 for 3 boxes.

First Day

$$\frac{\$5.00}{\$7.50} \overset{?}{=} \frac{2 \text{ boxes}}{3 \text{ boxes}}$$

Next Day

b. On the first day, it takes you 3 hours to drive 135 miles. The next day, it takes you 5 hours to drive 200 miles.

First Day

$$\frac{3 \text{ h}}{5 \text{ h}} \overset{?}{=} \frac{135 \text{ mi}}{200 \text{ mi}}$$

Next Day

c. On the first day, you walk 4 miles and burn 300 calories. The next day, you walk 3 miles and burn 225 calories.

First Day

$$\frac{4 \text{ mi}}{3 \text{ mi}} \overset{?}{=} \frac{300 \text{ cal}}{225 \text{ cal}}$$

Next Day

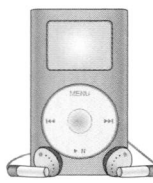

d. On the first day, you download 5 songs and pay $2.25. The next day, you download 4 songs and pay $2.00.

First Day

$$\frac{5 \text{ songs}}{4 \text{ songs}} \overset{?}{=} \frac{\$2.25}{\$2.00}$$

Next Day

Laurie's Notes

Introduction

For the Teacher

- **Goal:** Students will explore two ratios and try to reason whether they are equivalent or not.

Motivate

- Ask for two volunteers. Hand student A 8 square tiles and student B 4 square tiles.
- Make up a story as to why student A starts off with more than student B.
- ? "What is the ratio of student A's tiles to student B's tiles?" 2 : 1
- ? "What is the ratio of student B's tiles to student A's tiles?" 1 : 2
- ? "If I give student A 2 more tiles, how many should I give student B so that they still have the same ratio? Explain your reasoning." 1; Student A has twice as many as student B, so you need to give him/her twice as many tiles each time.
- Hand each student 2 more tiles.
- ? Is the ratio of student A's tiles to student B's tiles still 2 : 1? Explain." No, the ratio is 10 : 6 = 5 : 3 ≠ 2 : 1.
- Ask additional questions if time permits.

Discuss

- Explain the meaning of the word proportional. Reference the activity used to motivate today's lesson for examples.

Activity Notes

Activity 1

- The color coding will help students focus on the writing of the proportion.
- Note that the numerators contain information from the first day and the denominators contain information from the second day. In doing it this way, like units are contained within the same ratio.
- *When the activity is finished,* you should point out to students that the ratios could have been written so that the ratio on the left contains information from the first day and the ratio on the right contains information from the second day.
- Example: $\dfrac{\$5.00}{2\text{ boxes}} = \dfrac{\$7.50}{3\text{ boxes}}$. In writing them this way, they look like rates.
- **Management Tip:** For this activity, students will work in pairs. To allow the activity to run smoothly and to save time in class, plan a partner for each student before class.
- Discuss student explanations.
- Encourage students to share their strategies. How did they decide whether the ratios were equal or not? There will be different strategies, and it is important to hear a variety.

Previous Learning

Students have written and simplified ratios.

Activity Materials
Introduction
• any small manipulatives (wooden cubes, square tiles, counters, etc.)

Start Thinking! and Warm Up

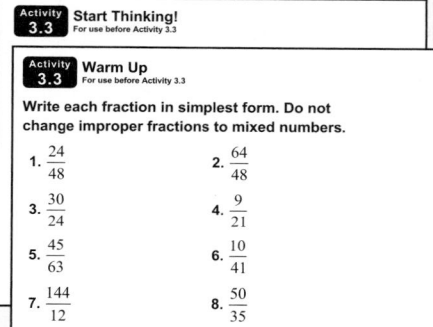

Activity 3.3 Start Thinking!
For use before Activity 3.3

Activity 3.3 Warm Up
For use before Activity 3.3

Write each fraction in simplest form. Do not change improper fractions to mixed numbers.

1. $\dfrac{24}{48}$ 2. $\dfrac{64}{48}$

3. $\dfrac{30}{24}$ 4. $\dfrac{9}{21}$

5. $\dfrac{45}{63}$ 6. $\dfrac{10}{41}$

7. $\dfrac{144}{12}$ 8. $\dfrac{50}{35}$

3.3 Record and Practice Journal

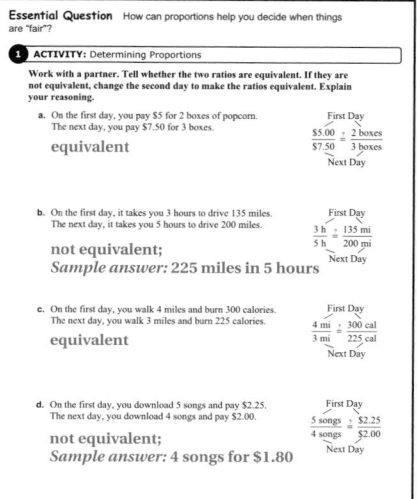

Essential Question How can proportions help you decide when things are "fair"?

1 ACTIVITY: Determining Proportions

Work with a partner. Tell whether the two ratios are equivalent. If they are not equivalent, change the second day to make the ratios equivalent. Explain your reasoning.

a. On the first day, you pay $5 for 2 boxes of popcorn. The next day, you pay $7.50 for 3 boxes.
equivalent

b. On the first day, it takes you 3 hours to drive 135 miles. The next day, it takes you 5 hours to drive 200 miles.
not equivalent;
Sample answer: **225 miles in 5 hours**

c. On the first day, you walk 4 miles and burn 300 calories. The next day, you walk 3 miles and burn 225 calories.
equivalent

d. On the first day, you download 5 songs and pay $2.25. The next day, you download 4 songs and pay $2.00.
not equivalent;
Sample answer: **4 songs for $1.80**

Word Problems

Most word problems follow a standard format that allows English learners to recognize key words that are integral to writing a mathematical statement of the problem. Most numbers given in a word problem are used. Analyzing the units in the mathematical statement and determining the units of the solution give students confidence that they are on the right path for solving the problem.

3.3 Record and Practice Journal

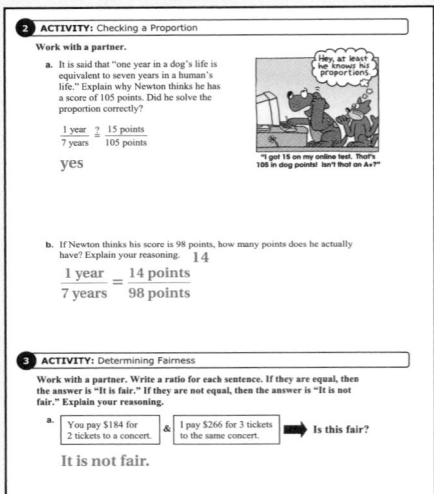

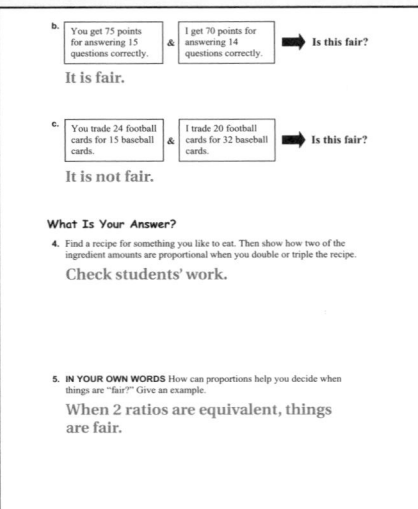

Laurie's Notes

Activity 2

- Help students think through this problem.
- The ratio often heard is 1 dog year : 7 human years. Years may be in the ratio of 1 : 7, but there is no reason why test scores need to be.
- The proportion that is being used is $\dfrac{1 \text{ dog year}}{7 \text{ human years}} = \dfrac{15 \text{ test points}}{105 \text{ dog points}}$.
- While the numeric proportion is a true proportion, the units of $\dfrac{1}{7} = \dfrac{15}{105}$ make no sense.

Activity 3

- Students like to think about fairness when deciding if two ratios are equal. When they finish, have students share their reasoning.
- One common strategy is to compare unit rates.
- *Note:* Part (b) is the common misconception referenced earlier.

Words of Wisdom

- All of the work today is to build an understanding of what it means for two ratios to be equal and to develop different strategies for deciding if the ratios are equal.
- Student discussion can often be quite interesting as they explain their reasoning.

What Is Your Answer?

- Question 4 can be assigned for homework.

Closure

- **Writing Prompt:** One way to decide if 6 hours for $4.80 is the same rate as 10 hours for $8 is . . . *Sample answer:* to write each comparison as a ratio and then find the unit rates.

$$\dfrac{\$4.80}{6 \text{ hours}} = \$0.80 \text{ per hour}$$

$$\dfrac{\$8.00}{10 \text{ hours}} = \$0.80 \text{ per hour}$$

Technology **F**or the **T**eacher

The Dynamic Planning Tool
Editable Teacher's Resources at *BigIdeasMath.com*

2 ACTIVITY: Checking a Proportion

Work with a partner.

a. It is said that "one year in a dog's life is equivalent to seven years in a human's life." Explain why Newton thinks he has a score of 105 points. Did he solve the proportion correctly?

$$\frac{1 \text{ year}}{7 \text{ years}} \stackrel{?}{=} \frac{15 \text{ points}}{105 \text{ points}}$$

b. If Newton thinks his score is 98 points, how many points does he actually have? Explain your reasoning.

Hey, at least he knows his proportions.

"I got 15 on my online test. That's 105 in dog points! Isn't that an A+?"

3 ACTIVITY: Determining Fairness

Work with a partner. Write a ratio for each sentence. If they are equal, then the answer is "It is fair." If they are not equal, then the answer is "It is not fair." Explain your reasoning.

a.

You pay $184 for 2 tickets to a concert. & I pay $266 for 3 tickets to the same concert. ➤ **Is this fair?**

b.

You get 75 points for answering 15 questions correctly. & I get 70 points for answering 14 questions correctly. ➤ **Is this fair?**

c.

You trade 24 football cards for 15 baseball cards. & I trade 20 football cards for 32 baseball cards. ➤ **Is this fair?**

What Is Your Answer?

4. Find a recipe for something you like to eat. Then show how two of the ingredient amounts are proportional when you double or triple the recipe.

5. IN YOUR OWN WORDS How can proportions help you decide when things are "fair?" Give an example.

Practice ➤ Use what you discovered about proportions to complete Exercises 17–22 on page 114.

 Key Idea

Key Vocabulary 🔊

proportion, *p. 112*
proportional, *p. 112*
cross products, *p. 113*

Proportions

Words   A **proportion** is an equation stating that two ratios are
equivalent. Two quantities that form a proportion are
proportional.

Numbers $\dfrac{2}{3} = \dfrac{4}{6}$ The proportion is read "2 is to 3 as 4 is to 6."

EXAMPLE   1   Determining Whether Ratios Form a Proportion

Tell whether the ratios form a proportion.

a. $\dfrac{4}{10}$ and $\dfrac{10}{25}$

Compare the ratios in simplest form.

$$\frac{4}{10} = \frac{4 \div 2}{10 \div 2} = \frac{2}{5}$$

The ratios are equivalent.

$$\frac{10}{25} = \frac{10 \div 5}{25 \div 5} = \frac{2}{5}$$

∴ So, $\dfrac{4}{10}$ and $\dfrac{10}{25}$ form a proportion.

b. $\dfrac{6}{4}$ and $\dfrac{8}{12}$

Compare the ratios in simplest form.

$$\frac{6}{4} = \frac{6 \div 2}{4 \div 2} = \frac{3}{2}$$

The ratios are not equivalent.

$$\frac{8}{12} = \frac{8 \div 4}{12 \div 4} = \frac{2}{3}$$

∴ So, $\dfrac{6}{4}$ and $\dfrac{8}{12}$ do not form a proportion.

⬤ **On Your Own**

Now You're Ready
Exercises 5–16

Tell whether the ratios form a proportion.

1. $\dfrac{1}{2}, \dfrac{5}{10}$ **2.** $\dfrac{4}{6}, \dfrac{18}{24}$ **3.** $\dfrac{10}{3}, \dfrac{5}{6}$ **4.** $\dfrac{25}{20}, \dfrac{15}{12}$

🔊 Multi-Language Glossary at BigIdeasMath ✓com.

Laurie's Notes

Introduction

Connect
- **Yesterday:** Students explored pairs of rates and decided if they were equivalent, or fair.
- **Today:** Students will use multiplication and division, and the Cross Products Property to decide if two ratios are equal.

Words of Wisdom
- This lesson is about deciding if you have a proportion. It is *not* about writing or solving proportions.
- Students are likely developing some mental math strategies. Encourage students to share their strategies with the class.

Motivate
- Draw the following on the board: $\dfrac{\square}{\square} = \dfrac{\square}{\square}$
- Ask students to use the numbers 2, 3, 4, and 6 placing one number in each square to make two ratios. They should list all combinations that are different $\left(\text{i.e., } \dfrac{2}{4} = \dfrac{3}{6} \text{ is not different from } \dfrac{3}{6} = \dfrac{2}{4}\right)$.
- "How did you decide where to place the numbers?" Listen for ideas related to fractions (reducing, using the Cross Products Property, etc.).
- Record student solutions to reference later.

Lesson Notes

Key Idea
- Write the definition of proportion on the board.
- **FYI:** Without units associated with the numeric values, students think of proportions as fractions.
- If students are comfortable with writing equivalent fractions and simplifying fractions, they will generally have a good sense about working with proportions.

Example 1
- The strategy used for each problem is to write the ratios in simplest form.
- "What is the relationship between $\dfrac{2}{3}$ and $\dfrac{3}{2}$?" They are reciprocals.

On Your Own
- **Common Error:** When students work quickly, and without thinking, they often answer Question 3 as being equal ratios because the numbers look right. Ten is divided by 2, while three is multiplied by 2. The ratios are not equal!

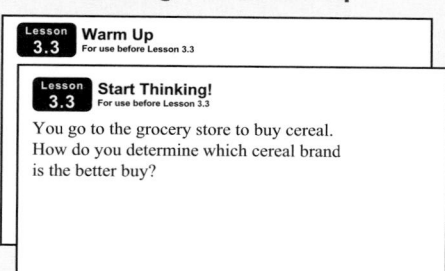
Extra Example 1

Tell whether the ratios form a proportion.

a. $\dfrac{10}{18}$ and $\dfrac{45}{81}$ form a proportion

b. $\dfrac{12}{9}$ and $\dfrac{15}{12}$ do not form a proportion

On Your Own
1. yes
2. no
3. no
4. yes

Extra Example 2

You run the first 3 laps around the gym in 1.5 minutes. You complete 24 laps in 12 minutes. Is the number of laps proportional to your time? *The number of laps is proportional to the time.*

On Your Own

5. yes

Differentiated Instruction

Visual

Students should understand that the Cross Products Property is an application of the Multiplication Property of Equality. Show the following steps on the board or overhead.

$$\frac{a}{b} = \frac{c}{d}$$ Proportion

$$\frac{a}{b} \cdot b \cdot d = \frac{c}{d} \cdot b \cdot d$$ Multiply each side by $b \cdot d$.

$$a \cdot d = c \cdot b$$ Simplify.

$$ad = bc$$ Commutative Property of Multiplication

Laurie's Notes

Key Ideas

- Write the Key Ideas on the board.
- ❓ "Why can't *b* or *d* equal zero?" or "Which number should *b* or *d* not be equal to? Why not?" *Zero, because it would lead to division by zero.*
- Use the Cross Products Property to verify that each of the solutions written at the beginning of class is a proportion.

Example 2

- Ask the students to read the problem.
- Work through each method of the solution.
- ❓ **Connection:** When you have finished each method, tie this lesson to the slope by asking "If you are swimming at a constant rate and you swam 4 laps in 2.4 minutes, how long should it take you to complete 16 laps?" *4 times as long, or 9.6 min*
- ❓ "What does this mean in the context of the problem?" *You slowed down after 4 laps.*

Closure

- Write an example of two ratios that are equal. Explain how you know they are equal.
- Write an example of two ratios that are not equal. Explain how you know they are not equal.

Technology For the Teacher

Dynamic Classroom

The Dynamic Planning Tool
Editable Teacher's Resources at *BigIdeasMath.com*

 **Key Ideas**

Cross Products

In the proportion $\frac{a}{b} = \frac{c}{d}$, the products $a \cdot d$ and $b \cdot c$ are called **cross products**.

Cross Products Property

Words The cross products of a proportion are equal.

Numbers

$$\frac{2}{3} \bowtie \frac{4}{6}$$

$$2 \cdot 6 = 3 \cdot 4$$

Algebra

$$\frac{a}{b} \bowtie \frac{c}{d}$$

$$ad = bc,$$
where $b \neq 0$ and $d \neq 0$

EXAMPLE (2) **Identifying Proportional Relationships**

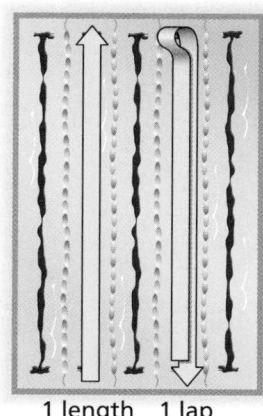

1 length   1 lap

You swim your first 4 laps in 2.4 minutes. You complete 16 laps in 12 minutes. Is the number of laps proportional to your time?

Method 1: Compare unit rates.

$$\overset{\div\,4}{\frac{2.4 \text{ min}}{4 \text{ laps}}} = \frac{0.6 \text{ min}}{1 \text{ lap}} \qquad \overset{\div\,16}{\frac{12 \text{ min}}{16 \text{ laps}}} = \frac{0.75 \text{ min}}{1 \text{ lap}}$$

$\div\,4$ $\div\,16$

The unit rates are not equivalent.

∴ So, the number of laps is not proportional to the time.

Method 2: Use the Cross Products Property.

$$\frac{2.4 \text{ min}}{4 \text{ laps}} \overset{?}{=} \frac{12 \text{ min}}{16 \text{ laps}} \qquad \text{Test to see if the rates are equivalent.}$$

$$2.4 \cdot 16 \overset{?}{=} 4 \cdot 12 \qquad \text{Find the cross products.}$$

$$38.4 \neq 48 \qquad \text{The cross products are not equal.}$$

∴ So, the number of laps is not proportional to the time.

● **On Your Own**

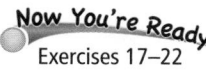
Exercises 17–22

5. You read the first 20 pages of a book in 25 minutes. You read 36 pages in 45 minutes. Is the number of pages read proportional to your time?

 Vocabulary and Concept Check

1. **VOCABULARY** What does it mean for two ratios to form a proportion?

2. **VOCABULARY** What are two ways you can tell that two ratios form a proportion?

3. **OPEN-ENDED** Write two ratios that are equivalent to $\frac{3}{5}$.

4. **WHICH ONE DOESN'T BELONG?** Which ratio does *not* belong with the other three? Explain your reasoning.

$$\frac{4}{10} \qquad \frac{2}{5} \qquad \frac{3}{5} \qquad \frac{6}{15}$$

 Practice and Problem Solving

Tell whether the ratios form a proportion.

① 5. $\frac{1}{3}, \frac{7}{21}$ 6. $\frac{1}{5}, \frac{6}{30}$ 7. $\frac{3}{4}, \frac{24}{18}$

8. $\frac{2}{5}, \frac{40}{16}$ 9. $\frac{48}{9}, \frac{16}{3}$ 10. $\frac{18}{27}, \frac{33}{44}$

11. $\frac{7}{2}, \frac{16}{6}$ 12. $\frac{12}{10}, \frac{14}{12}$ 13. $\frac{27}{15}, \frac{18}{10}$

14. $\frac{4}{15}, \frac{15}{42}$ 15. $\frac{76}{36}, \frac{19}{9}$ 16. $\frac{49}{77}, \frac{38}{57}$

Tell whether the two rates form a proportion.

② 17. 7 inches in 9 hours; 42 inches in 54 hours

18. 12 players from 21 teams; 15 players from 24 teams

19. 440 calories in 4 servings; 300 calories in 3 servings

20. 120 units made in 5 days; 88 units made in 4 days

21. 66 wins in 82 games; 99 wins in 123 games

22. 68 hits in 172 at bats; 43 hits in 123 at bats

23. **FITNESS** You can do 90 sit-ups in 2 minutes. Your friend can do 135 sit-ups in 3 minutes. Are these rates proportional? Explain.

24. **HEARTBEAT** Find the heartbeat rates of you and your friend. Do these rates form a proportion? Explain.

	Heartbeats	Seconds
You	22	20
Friend	18	15

Assignment Guide and Homework Check

Level	Day 1 Activity Assignment	Day 2 Lesson Assignment	Homework Check
Basic	17–22, 35–39	1–4, 5–15 odd, 23, 24, 29	5, 11, 24, 29
Average	17–22, 35–39	1–4, 5–15 odd, 25, 28–31	5, 11, 28, 29
Advanced	17–22, 35–39	1–4, 11–16, 25–27, 31–34	12, 13, 26, 32

Common Errors

- **Exercises 5–16** Students may have difficulty understanding why you can simplify the ratio to simplest form. Tell students to compare the ratio to a fraction. Simplifying ratios is the same as writing equivalent fractions.

- **Exercises 17–22** Students may mix up the rates and incorrectly find that they are not proportional. For example, they might write $\dfrac{7 \text{ inches}}{9 \text{ hours}} \overset{?}{=} \dfrac{54 \text{ hours}}{42 \text{ inches}}$. Remind students about writing a rate and help them to identify which unit goes in the numerator.

3.3 Record and Practice Journal

Tell whether the ratios form a proportion.

1. $\dfrac{1}{5}, \dfrac{5}{15}$ no
2. $\dfrac{2}{3}, \dfrac{12}{18}$ yes
3. $\dfrac{15}{2}, \dfrac{4}{30}$ no
4. $\dfrac{56}{21}, \dfrac{8}{3}$ yes

5. $\dfrac{5}{8}, \dfrac{62.5}{100}$ yes
6. $\dfrac{17}{20}, \dfrac{90.1}{106}$ yes
7. $\dfrac{3.2}{4}, \dfrac{16}{24}$ no
8. $\dfrac{34}{50}, \dfrac{6.8}{10}$ yes

Tell whether the two rates form a proportion.

9. 28 points in 3 games; 112 points in 12 games yes
10. 32 notes in 4 measures; 12 notes in 2 measures no

11. You can type 105 words in two minutes. Your friend can type 210 words in four minutes. Are these rates proportional? Explain. yes

12. You make punch for a party. The ratio of ginger ale to fruit juice is 8 cups to 3 cups. You decide to add 4 more cups of ginger ale. How many more cups of fruit juice do you need to add to keep the correct ratio? Explain.
1.5 cups of fruit juice; *Sample answer:* You add 1.5 cups because $8 \cdot 1.5 = 3 \cdot 4$.

Technology
For the **Teacher**
Answer Presentation Tool
QuizShow

✓ Vocabulary and Concept Check

1. Both ratios are equal.

2. Compare the ratios in simplest form and compare the cross products.

3. *Sample answer:* $\dfrac{6}{10}, \dfrac{12}{20}$

4. $\dfrac{3}{5}$; The others are equal to $\dfrac{2}{5}$.

Practice and Problem Solving

5. yes		**6.** yes	
7. no		**8.** no	
9. yes		**10.** no	
11. no		**12.** no	
13. yes		**14.** no	
15. yes		**16.** no	
17. yes		**18.** no	
19. no		**20.** no	
21. yes		**22.** no	

23. yes; Both can do 45 sit-ups per minute.

24. you: 1.1 beats per second friend: 1.2 beats per second No, the rates are not equivalent.

25. yes

26. no

27. yes

28. **a.** $7 per hour

b. $9 per hour

c. no; Your friend earns more money per hour.

Practice and Problem Solving

29. yes; They are both $\frac{4}{5}$.

30. no; $17.40 per CD $\neq$ $12.49 per CD

31. **a.** Pitcher 3

  b. Pitcher 2 and Pitcher 4

32. See *Taking Math Deeper*.

33. **a.** no

  b. *Sample answer:* If the collection has 50 quarters and 30 dimes, when 10 of each coin are added, the new ratio of quarters to dimes is 3 : 2.

34. yes; Because ratio *A* is equivalent to ratio *B*, ratios *A* and *B* simplify to the same ratio. Because ratio *B* is equivalent to ratio *C*, ratios *B* and *C* simplify to the same ratio. Ratios *A* and *C* simplify to the same ratio, so they are equivalent.

Fair Game Review

35. -13 **36.** -17

37. -18 **38.** -3

39. D

Mini-Assessment

Tell whether the ratios form a proportion.

1. $\frac{4}{12}, \frac{5}{15}$ yes **2.** $\frac{8}{4}, \frac{12}{8}$ no

3. $\frac{14}{17}, \frac{42}{51}$ yes **4.** $\frac{16}{12}, \frac{26}{22}$ no

5. You can do 40 push-ups in 2 minutes. Your friend can do 57 push-ups in 3 minutes. Are these rates proportional? no

Taking Math Deeper

Exercise 32

In this problem, students are given a mixture of red and yellow pigment and are asked to decide whether they should add more red or more yellow to get a desired shade.

Add red or yellow.

① Summarize given information.

  Ratio for desired shade: $\dfrac{7 \text{ parts red}}{2 \text{ parts yellow}} = 3.5$

  Given mixture: $\dfrac{35 \text{ parts red}}{8 \text{ parts yellow}} = 4.375$

② Interpret the given information.

  The ratio for the given mixture needs more yellow to take it down to the desired ratio of 3.5 (red to yellow).

③ Use a table with *Guess, Check, and Revise* to find how much yellow to add to the given mixture.

Red	Yellow	Ratio
35 qt	8 qt	$\frac{35}{8} = 4.375$
35 qt	9 qt	$\frac{35}{9} = 3.\overline{8}$
35 qt	10 qt	$\frac{35}{10} = 3.5$

Add 2 quarts of yellow.

0% magenta   50% magenta   100% magenta
100% yellow   50% yellow   0% yellow

Here are some mixtures of yellow and magenta (red).

Reteaching and Enrichment Strategies

If students need help. . .	If students got it. . .
Resources by Chapter • Practice A and Practice B • Puzzle Time Record and Practice Journal Practice Differentiating the Lesson Lesson Tutorials Skills Review Handbook	Resources by Chapter • Enrichment and Extension • School-to-Work Start the next section

Tell whether the ratios form a proportion.

25. $\dfrac{3}{8}, \dfrac{31.5}{84}$

26. $\dfrac{14}{30}, \dfrac{75.6}{180}$

27. $\dfrac{2.5}{4}, \dfrac{7}{11.2}$

28. PAY RATE You earn $56 walking your neighbor's dog for 8 hours. Your friend earns $36 painting your neighbor's fence for 4 hours.

   a. What is your pay rate?

   b. What is your friend's pay rate?

   c. Are the pay rates equivalent? Explain.

29. GEOMETRY Are the ratios of h to b in the two triangles proportional? Explain.

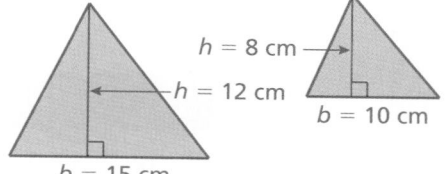

$h = 8$ cm

$h = 12$ cm

$b = 10$ cm

$b = 15$ cm

30. MUSIC You can buy 3 CDs for $52.20 or 5 CDs for $62.45. Are the rates proportional? Explain.

31. BASEBALL The table shows pitching statistics for four pitchers during the 2008 season.

   a. Which pitcher has the highest ratio of strikeouts to walks?

   b. Which of the pitchers have equivalent strikeout to walk ratios?

2008 Season		
Pitcher	**Strikeouts**	**Walks**
Pitcher 1	6	8
Pitcher 2	8	4
Pitcher 3	10	1
Pitcher 4	10	5

32. NAIL POLISH A specific shade of red nail polish requires 7 parts red to 2 parts yellow. A mixture contains 35 quarts of red and 8 quarts of yellow. How can you fix the mixture to make the correct shade of red?

33. COIN COLLECTION The ratio of quarters to dimes in a coin collection is $5:3$. The same number of new quarters and dimes are added to the collection.

   a. Is the ratio of quarters to dimes still $5:3$?

   b. If so, illustrate your answer with an example. If not, show why with a "counterexample."

34. *Critical Thinking* Ratio A is equivalent to ratio B. Ratio B is equivalent to ratio C. Is ratio A equivalent to ratio C? Explain.

Fair Game Review What you learned in previous grades & lessons

Add or subtract. *(Sections 1.2 and 1.3)*

35. $-28 + 15$

36. $-6 + (-11)$

37. $-10 - 8$

38. $-17 - (-14)$

39. MULTIPLE CHOICE Which fraction is not equivalent to $\dfrac{2}{6}$? *(Skills Review Handbook)*

   Ⓐ $\dfrac{1}{3}$ Ⓑ $\dfrac{12}{36}$ Ⓒ $\dfrac{4}{12}$ Ⓓ $\dfrac{6}{9}$

Essential Question How can you write a proportion that solves a problem in real life?

1 ACTIVITY: Writing Proportions

Work with a partner. A rough rule for finding the correct bat length is "The bat length should be half of the batter's height." So, a 62-inch-tall batter uses a bat that is 31 inches long. Write a proportion to find the bat length for each given batter height.

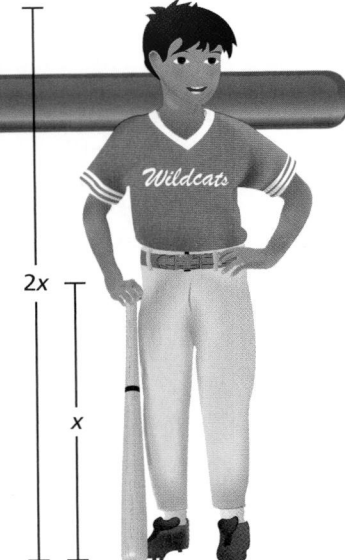

2x

x

a. 58 inches

b. 60 inches

c. 64 inches

2 ACTIVITY: Bat Lengths

Work with a partner. Here is a more accurate table for determining the bat length for a batter. Find all of the batter heights for which the rough rule in Activity 1 is exact.

	Height of Batter (inches)							
Weight of Batter (pounds)	45–48	49–52	53–56	57–60	61–64	65–68	69–72	Over 72
Under 61	28	29	29					
61–70	28	29	30	30				
71–80	28	29	30	30	31			
81–90	29	29	30	30	31	32		
91–100	29	30	30	31	31	32		
101–110	29	30	30	31	31	32		
111–120	29	30	30	31	31	32		
121–130	29	30	30	31	32	33	33	
131–140	30	30	31	31	32	33	33	
141–150	30	30	31	31	32	33	33	
151–160	30	31	31	32	32	33	33	33
161–170		31	31	32	32	33	33	34
171–180				32	33	33	34	34
Over 180					33	33	34	34

Bat length

Laurie's Notes

Introduction

For the Teacher

- **Goal:** Students will write and solve proportions using mental math.
- This activity has a baseball theme. If appropriate in your setting, wearing a baseball team shirt or a baseball hat would add interest for the day.
- You may choose to allow students to use a page with common conversions on it for reference when doing homework, quizzes, and even tests.

Motivate

- **Management Tip:** You may want to pre-cut several lengths of string prior to this activity so students can join in.
- Ask for a volunteer. Say "I can estimate the distance around your neck without actually measuring your neck!"
- Use the string to measure the distance around the student's wrist.
- Double this length and it will be approximately the distance around his/her neck. Have the student verify this length by measuring the distance around his/her own neck.
- **Write:** $\dfrac{\text{distance around wrist}}{\text{distance around neck}} = \dfrac{1}{2}$
- Write a new proportion substituting the length around the wrist:

  $\dfrac{8.5 \text{ inches}}{x \text{ inches}} = \dfrac{1}{2}$

- **Solve:** The distance around the neck is two times 8.5 inches, or 17 inches.

Activity Notes

Activity 1

- Borrow a baseball bat for this activity to use as a prop.
- Help students translate the words in the activity into a proportion: $\dfrac{\text{length of bat}}{\text{height of batter}} = \dfrac{1}{2}$. Say "The ratio of the length of the bat to the height of the batter is 1 : 2. A proportion to determine the bat length for a player 58 inches tall is $\dfrac{1}{2} = \dfrac{b}{58}$."
- You want students to understand that two measures are being compared and the order in which you write (and say) them does matter.

Activity 2

- This activity promotes reading information from a table.
- The batter's height *and* weight factor into selecting the correct bat length.
- Practice reading information from the table.
- **?** "A 50-inch tall batter weighing 95 pounds should use what length bat?" 30"
- **?** "A 5-foot, 5-inch tall batter weighing 95 pounds should use what length bat?" 32"

Previous Learning

Students have written and simplified ratios and determined if two ratios are equal.

Activity Materials
Introduction
• string or yarn • baseball bat (optional)

Start Thinking! and Warm Up

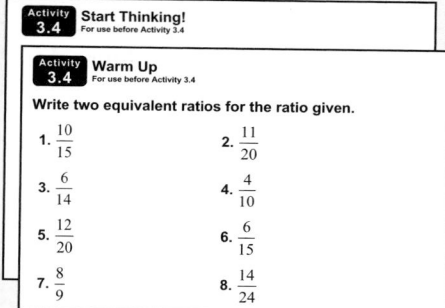

Activity 3.4 Start Thinking! For use before Activity 3.4

Activity 3.4 Warm Up For use before Activity 3.4

Write two equivalent ratios for the ratio given.

1. $\dfrac{10}{15}$ 2. $\dfrac{11}{20}$

3. $\dfrac{6}{14}$ 4. $\dfrac{4}{10}$

5. $\dfrac{12}{20}$ 6. $\dfrac{6}{15}$

7. $\dfrac{8}{9}$ 8. $\dfrac{14}{24}$

3.4 Record and Practice Journal

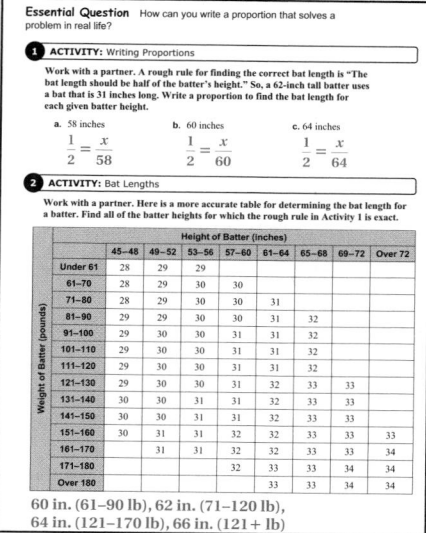

Essential Question How can you write a proportion that solves a problem in real life?

1 ACTIVITY: Writing Proportions

Work with a partner. A rough rule for finding the correct bat length is "The bat length should be half of the batter's height." So, a 62-inch tall batter uses a bat that is 31 inches long. Write a proportion to find the bat length for each given batter height.

a. 58 inches $\dfrac{1}{2} = \dfrac{x}{58}$

b. 60 inches $\dfrac{1}{2} = \dfrac{x}{60}$

c. 64 inches $\dfrac{1}{2} = \dfrac{x}{64}$

2 ACTIVITY: Bat Lengths

Work with a partner. Here is a more accurate table for determining the bat length for a batter. Find all of the batter heights for which the rough rule in Activity 1 is exact.

	Height of Batter (inches)							
Weight of Batter (pounds)		45–48	49–52	53–56	57–60	61–64	65–68	69–72	Over 72
Under 61	28	29	29					
61–70	28	29	30	30				
71–80	28	29	30	30	31			
81–90	29	29	30	30	31	32		
91–100	29	30	30	31	31	32		
101–110	29	30	30	31	31	32		
111–120	29	30	30	31	31	32		
121–130	29	30	30	31	32	33	33	
131–140	30	30	31	31	32	33	33	
141–150	30	30	31	31	32	33	33	
151–160	30	31	31	32	32	33	33	33
161–170		31	31	32	32	33	33	34
171–180				32	33	33	34	34
Over 180					33	33	34	34

60 in. (61–90 lb), 62 in. (71–120 lb),
64 in. (121–170 lb), 66 in. (121+ lb)

English Language Learners

Visual Aids

English learners might find it useful to use a general template when writing a proportion problem.

$$\frac{\text{part}}{\text{whole}} = \frac{\text{part}}{\text{whole}}$$

Activity 3

? "Does anyone know what the term *batting average* means and how it is computed?" It is actually more involved than explained in the text. For instance, if a batter walks, has a sacrifice fly, or is hit by a pitch, it is not considered an "at bat."

- After the discussion of batting average, write the formula followed by the example. To have a proportion, the decimal form of the fraction is written.
- Remind students of how ratios are simplified (by dividing out common factors).
- Example: "Determine how many hits a batter has if he has 100 at bats and his batting average is 0.300.

$$\frac{H}{100} = \frac{300}{1000} \rightarrow \text{dividing out a common factor of 10} \rightarrow \frac{H}{100} = \frac{30}{100}$$

- **Common Error:** Students have divided out common factors when simplifying a simple fraction or when multiplying two fractions. A proportion is neither of these! A common error at this point is for students to divide out a factor from the numerator on one side of the equal sign with a factor in the denominator on the other side of the equal sign.

- For instance, in this example a student would incorrectly write $\frac{H}{1} = \frac{3}{1000}$ and get an answer of 0.003 hit.

- Have pairs of students show their work at the board.

What Is Your Answer?

- To answer Question 5, students may need to be reminded of how to write a fraction as a decimal. A calculator will be helpful.

3.4 Record and Practice Journal

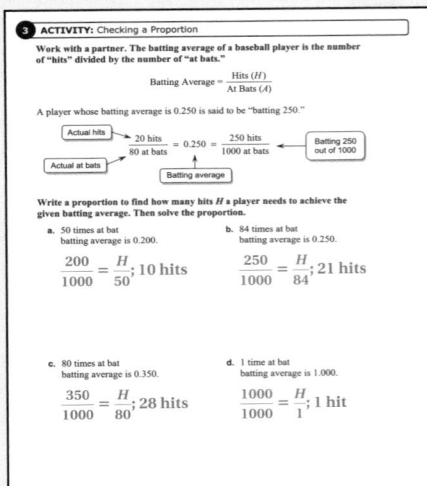

Closure

- Write and solve a proportion: A stadium holds approximately 45,000 people during a baseball game. If the ratio of season ticket holders to all tickets is 1 : 3, approximately how many season ticket holders are there? 15,000 season ticket holders

Technology For the Teacher

The Dynamic Planning Tool
Editable Teacher's Resources at *BigIdeasMath.com*

3 **ACTIVITY: Writing Proportions**

Work with a partner. The batting average of a baseball player is the number of "hits" divided by the number of "at bats."

$$\text{Batting average} = \frac{\text{Hits } (H)}{\text{At bats } (A)}$$

A player whose batting average is 0.250 is said to be "batting 250."

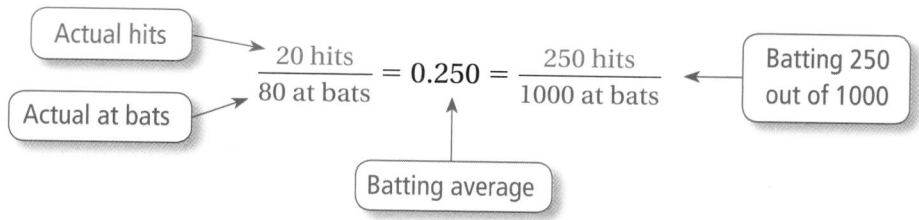

Write a proportion to find how many hits H a player needs to achieve the given batting average. Then solve the proportion.

   a. 50 times at bat; batting average is 0.200.

   b. 84 times at bat; batting average is 0.250.

   c. 80 times at bat; batting average is 0.350.

   d. 1 time at bat; batting average is 1.000.

What Is Your Answer?

4. **IN YOUR OWN WORDS** How can you write a proportion that solves a problem in real life?

5. Two players have the same batting average.

	At Bats	Hits	Batting Average
Player 1	132	45	
Player 2	132	45	

Player 1 gets four hits in the next five at bats. Player 2 gets three hits in the next three at bats.

   a. Who has the higher batting average?

   b. Does this seem fair? Explain your reasoning.

Practice

Use what you discovered about proportions to complete Exercises 4–7 on page 120.

One way to write a proportion is to use a table.

	Last Month	**This Month**
Purchase	2 ringtones	3 ringtones
Total Cost	6 dollars	x dollars

Use the columns or the rows to write a proportion.

Use columns:

$$\frac{2 \text{ ringtones}}{6 \text{ dollars}} = \frac{3 \text{ ringtones}}{x \text{ dollars}}$$

> Numerators have the same units.
>
> Denominators have the same units.

Use rows:

$$\frac{2 \text{ ringtones}}{3 \text{ ringtones}} = \frac{6 \text{ dollars}}{x \text{ dollars}}$$

> The units are the same on each side of the proportion.

EXAMPLE **1** **Writing a Proportion**

Black Bean Soup

1.5 cups black beans
0.5 cup salsa
2 cups water
1 tomato
2 teaspoons seasoning

A chef increases the amounts of ingredients in a recipe to make a proportional recipe. The new recipe has 6 cups of black beans. Write a proportion that gives the number x of tomatoes in the new recipe.

Organize the information in a table.

	Original Recipe	**New Recipe**
Black Beans	1.5 cups	6 cups
Tomatoes	1 tomato	x tomatoes

⋮⋮ One proportion is $\dfrac{1.5 \text{ cups beans}}{1 \text{ tomato}} = \dfrac{6 \text{ cups beans}}{x \text{ tomatoes}}$.

On Your Own

Now You're Ready
Exercises 8–11

1. In Example 1, write a different proportion that gives the number x of tomatoes in the new recipe.

2. In Example 1, write a proportion that gives the amount y of water in the new recipe.

Laurie's Notes

Introduction

Connect

- **Yesterday:** Students wrote and solved proportions related to baseball that used simple mental math.
- **Today:** Students will write and solve a proportion using mental math.

Motivate

? "A student is reading a 280-page book. On average he/she reads 8 pages in 3 minutes. How long will it take to read the book?"

- If students can offer strategies, work on this problem now. If not, wait until later in the class.
- There are two equivalent proportions that can be set up. Be sure to use labels.

$$\frac{8 \text{ pages}}{3 \text{ minutes}} = \frac{280 \text{ pages}}{x \text{ minutes}} \text{ or } \frac{8 \text{ pages}}{280 \text{ pages}} = \frac{3 \text{ minutes}}{x \text{ minutes}}$$

- The Cross Products Property is not needed. Because $280 \div 8 = 35$, mental math strategies are sufficient.

Discuss

? "What is the information in the table saying?" You purchased 2 ringtones last month for $6. This month, you purchase 3 ringtones and the cost is unknown.

- Work through the example showing how the labels (ringtones and dollars) are used to identify the numbers.
- **Big Idea:** As identified in the text, when you use the rows or columns from the table, the proportion will be set up correctly. If the table had *not* been provided, it is possible for students to incorrectly set up the proportion:

$$\frac{2 \text{ ringtones}}{x \text{ dollars}} = \frac{3 \text{ ringtones}}{6 \text{ dollars}} \text{ or } \frac{2 \text{ ringtones}}{3 \text{ ringtones}} = \frac{x \text{ dollars}}{6 \text{ dollars}}$$

- What students must also see besides the labels is that the information from one month (i.e., 2 ringtones, $6) must be in one ratio *or* in the numerators, and the information from the second month (i.e., 3 ringtones, $x) must be in a second ratio *or* in the denominators.
- The column and the rows from a correct table ensure this happening.

Lesson Notes

Example 1

- Organizing the information in a table helps to write the proportion correctly.

? "Can the rows and columns be interchanged?" yes

On Your Own

- **Think-Pair-Share:** Students should read each question independently and then work with a partner to answer the questions. When they have answered the questions, the pair should compare their answers with another group and discuss any discrepancies.

Goal Today's lesson is writing and solving a proportion using mental math.

Start Thinking! and Warm Up

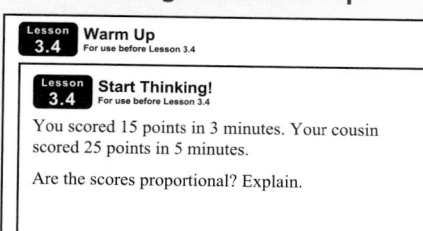

Lesson 3.4 Warm Up
For use before Lesson 3.4

Lesson 3.4 Start Thinking!
For use before Lesson 3.4

You scored 15 points in 3 minutes. Your cousin scored 25 points in 5 minutes.

Are the scores proportional? Explain.

Extra Example 1

The chef increases the amounts of ingredients in the recipe in Example 1 to make a proportional recipe. The new recipe has 3 cups of salsa. Write a proportion that gives the amount w of water in the new recipe.

Sample answer:

$$\frac{0.5 \text{ cup salsa}}{3 \text{ cups salsa}} = \frac{2 \text{ cups water}}{w \text{ cups water}}$$

On Your Own

1. $\dfrac{1.5 \text{ cups beans}}{6 \text{ cups beans}} = \dfrac{1 \text{ tomato}}{x \text{ tomatoes}}$

2. *Sample answer:*

$$\frac{1.5 \text{ cups beans}}{2 \text{ cups water}} = \frac{6 \text{ cups beans}}{y \text{ cups water}}$$

Extra Example 2

Solve $\frac{8}{5} = \frac{n}{15}$. 24

Extra Example 3

In Extra Example 1, how much water is in the new recipe? 12 cups water

 On Your Own

3. $d = 32$

4. $z = 5$

5. $x = 7$

6. $\frac{48}{95} = \frac{f}{950}$;
   480 female students

Differentiated Instruction

Kinesthetic

Provide students with counters and two pieces of blank paper. Draw fraction bar lines on each of the papers. Place counters on the papers to represent the two ratios. For each ratio, rearrange the counters in the numerator and denominator in stacks of equal number. If the stacks in each individual ratio are the same size and there are the same number of stacks in each ratio, then the ratios are equal and form a proportion. For instance, in Example 2, the first ratio has 3 stacks of 1 in the numerator and 2 stacks of 1 in the denominator. The second ratio has 3 stacks of 4 in the numerator and 2 stacks of 4 in the denominator. Because there are 3 stacks over 2 stacks in each of the ratios, the ratios form a proportion.

Laurie's Notes

Example 2

- This example has no context. The focus is on the process, and how mental math is used in solving the proportion.
- When finished, present the following problems to assess if students can distinguish when mental math is a reasonable approach.

? Tell whether you can easily use mental math to solve these problems:

a. $\frac{3}{7} = \frac{x}{27}$ **b.** $\frac{3}{7} = \frac{27}{x}$

Answers:

a. 7 is not a factor of 27, so mental math is not an easy approach.

b. 3 is a factor of 27, so mental math can be used.

Example 3

- Work through this example.
- Not all students will know that $4 \times 1.5 = 6$. If students are still struggling with multiplying decimals, you may want to review these rules prior to Example 3.
- Two ideas to help develop fluency:
  $4 \times 15 = 60$, so $4 \times 1.5 = 6$.
  $1 \times 4 = 4$, and half (0.5) of 4 is 2, so add $2 + 4$ to get $4 \times 1.5 = 6$.

On Your Own

- **Neighbor Check:** Have students work independently and then have their neighbor check their work. Have students discuss any discrepancies.

Closure

- **Exit Ticket:** The ratio of quarts to gallons is 4 : 1. If a recipe calls for 14 quarts, how many gallons would be needed? 3.5

Technology For the Teacher
Dynamic Classroom
The Dynamic Planning Tool
Editable Teacher's Resources at *BigIdeasMath.com*

EXAMPLE **2** **Solving Proportions Using Mental Math**

Solve $\dfrac{3}{2} = \dfrac{x}{8}$.

Step 1: Think: The product of 2 and what number is 8?

$$\dfrac{3}{2} = \dfrac{x}{8}$$

$2 \times ? = 8$

Step 2: Because the product of 2 and 4 is 8, multiply the numerator by 4 to find x.

$3 \times 4 = 12$

$$\dfrac{3}{2} = \dfrac{x}{8}$$

$2 \times 4 = 8$

∴ The solution is $x = 12$.

EXAMPLE **3** **Solving Proportions Using Mental Math**

In Example 1, how many tomatoes are in the new recipe?

Solve the proportion $\dfrac{1.5}{1} = \dfrac{6}{x}$. ← cups black beans
← tomatoes

Step 1: Think: The product of 1.5 and what number is 6?

$1.5 \times ? = 6$

$$\dfrac{1.5}{1} = \dfrac{6}{x}$$

Step 2: Because the product of 1.5 and 4 is 6, multiply the denominator by 4 to find x.

$1.5 \times 4 = 6$

$$\dfrac{1.5}{1} = \dfrac{6}{x}$$

$1 \times 4 = 4$

∴ So, there are 4 tomatoes in the new recipe.

On Your Own

Now You're Ready
Exercises 16–21

Solve the proportion.

3. $\dfrac{5}{8} = \dfrac{20}{d}$

4. $\dfrac{7}{z} = \dfrac{14}{10}$

5. $\dfrac{21}{24} = \dfrac{x}{8}$

6. A school has 950 students. The ratio of female students to all students is $\dfrac{48}{95}$. Write and solve a proportion to find the number f of students that are female.

Check It Out
Help with Homework
BigIdeasMath.com

Vocabulary and Concept Check

1. **WRITING** Describe two ways you can use a table to write a proportion.

2. **WRITING** What is your first step when solving $\frac{x}{15} = \frac{3}{5}$? Explain.

3. **OPEN-ENDED** Write a proportion using an unknown value x and the ratio $5:6$. Then solve it.

Practice and Problem Solving

Write a proportion to find how many points a student needs to score on the test to get the given score.

4. Test worth 50 points; test score of 40%

5. Test worth 50 points; test score of 78%

6. Test worth 80 points; test score of 80%

7. Test worth 150 points; test score of 96%

Use the table to write a proportion.

8.

	Game 1	Game 2
Points	12	18
Shots	14	w

9.

	May	June
Winners	n	34
Entries	85	170

10.

	Today	Yesterday
Miles	15	m
Hours	2.5	4

11.

	Race 1	Race 2
Meters	100	200
Seconds	x	22.4

12. **ERROR ANALYSIS** Describe and correct the error in writing the proportion.

	Monday	Tuesday
Dollars	2.08	d
Ounces	8	16

$$\frac{2.08}{16} = \frac{d}{8}$$

13. **T-SHIRTS** You can buy three T-shirts for $24. Write a proportion that gives the cost c of buying seven T-shirts.

14. **COMPUTERS** A school requires two computers for every five students. Write a proportion that gives the number c of computers needed for 145 students.

15. **SWIM TEAM** The school team has 80 swimmers. The ratio of 6th grade swimmers to all swimmers is $5:16$. Write a proportion that gives the number s of 6th grade swimmers.

Assignment Guide and Homework Check

Level	Day 1 Activity Assignment	Day 2 Lesson Assignment	Homework Check
Basic	4–7, 26–30	1–3, 9, 11–15, 17–21 odd	9, 12, 14, 17
Average	4–7, 26–30	1–3, 9, 11, 12, 15, 19–21, 22, 23	9, 12, 20, 22
Advanced	4–7, 26–30	1–3, 8–12 even, 16–20 even, 23–25	8, 12, 18, 24

Common Errors

- **Exercises 8–11** Students may write half of the proportion using rows and the other half using columns. They will have forgotten to include one of the values. Remind students that they need to pick a method for writing proportions with tables and be consistent throughout the problem.
- **Exercises 16–21** Students may get confused with using mental math to find the value of the variable and try to multiply the numerator and denominator by different numbers. Tell students that they are finding an equivalent ratio, or an equivalent fraction. They are multiplying the original fraction by 1 $\left(\text{or } \frac{4}{4}, \text{for example}\right)$ to find the equivalent fraction, so they must multiply the numerator and denominator by the same number.

3.4 Record and Practice Journal

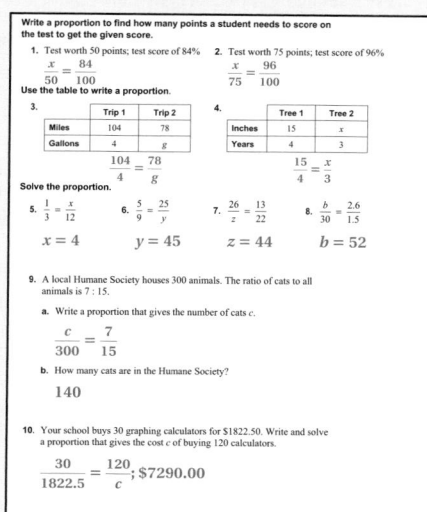

1. You can use the columns or the rows of the table to write a proportion.

2. Find the number that when multiplied by 5 is 15.

3. *Sample answer:* $\frac{x}{12} = \frac{5}{6}$; $x = 10$

 Practice and Problem Solving

4. $\dfrac{x}{50} = \dfrac{40}{100}$

5. $\dfrac{x}{50} = \dfrac{78}{100}$

6. $\dfrac{x}{80} = \dfrac{80}{100}$

7. $\dfrac{x}{150} = \dfrac{96}{100}$

8. $\dfrac{12 \text{ points}}{14 \text{ shots}} = \dfrac{18 \text{ points}}{w \text{ shots}}$

9. $\dfrac{n \text{ winners}}{85 \text{ entries}} = \dfrac{34 \text{ winners}}{170 \text{ entries}}$

10. $\dfrac{15 \text{ miles}}{2.5 \text{ hours}} = \dfrac{m \text{ miles}}{4 \text{ hours}}$

11. $\dfrac{100 \text{ meters}}{x \text{ seconds}} = \dfrac{200 \text{ meters}}{22.4 \text{ seconds}}$

12. The proportion cannot be written using diagonals of the table. $\dfrac{2.08}{8} = \dfrac{d}{16}$

13. $\dfrac{\$24}{3 \text{ shirts}} = \dfrac{c}{7 \text{ shirts}}$

14. $\dfrac{2 \text{ computers}}{5 \text{ students}} = \dfrac{c \text{ computers}}{145 \text{ students}}$

15. $\dfrac{5 \text{ 6th grade swimmers}}{16 \text{ swimmers}} = \dfrac{s \text{ 6th grade swimmers}}{80 \text{ swimmers}}$

Practice and Problem Solving

16. $z = 5$ **17.** $y = 16$

18. $k = 15$ **19.** $c = 24$

20. $b = 20$ **21.** $g = 14$

22. a. $\dfrac{1 \text{ trombone}}{3 \text{ violas}} = \dfrac{t \text{ trombones}}{9 \text{ violas}}$

  b. 3 trombones

23. $\dfrac{1}{200} = \dfrac{19.5}{x}$; Dimensions for the model are in the numerators and the corresponding dimensions for the actual space shuttle are in the denominators.

24. no; The solution of that equation is $x = 1.5$, but using mental math, you can see that the solution of the proportion is $x = 24$.

25. See *Taking Math Deeper.*

Fair Game Review

26. $x = 150$ **27.** $x = 9$

28. $x = 75$ **29.** $x = 140$

30. D

Mini-Assessment

Write a proportion to find how many points a student needs to score on the test to get the given score.

1. Test worth 60 points; test score of 60% $\dfrac{x}{60} = \dfrac{60}{100}$

2. Test worth 50 points; test score of 70% $\dfrac{x}{50} = \dfrac{70}{100}$

3. Test worth 100 points; test score of 85% $\dfrac{x}{100} = \dfrac{85}{100}$

4. Test worth 120 points; test score of 88% $\dfrac{x}{120} = \dfrac{88}{100}$

5. You can buy four DVDs for $48. Write a proportion that gives the cost c of buying six DVDs. $\dfrac{4}{48} = \dfrac{6}{c}$

Taking Math Deeper

Exercise 25

Although this problem does not have difficult or messy mathematics, it is still difficult for many students to know how to start. Emphasize that it is good to just "write things down" and organize the given facts. In this problem, it might help the visual learner to sketch 3 white lockers for every 5 blue lockers.

Draw a diagram.

 Draw a diagram that shows the given values and the unknown values.

3 white lockers for every 5 blue lockers

180 white lockers

 Write and solve a proportion.

$$\dfrac{x \text{ blue lockers}}{180 \text{ white lockers}} = \dfrac{5}{3} \qquad \text{Write a proportion.}$$

$$x = 300 \qquad \text{Use mental math.}$$

 Answer the question.

There are 180 white lockers and 300 blue lockers. So, there are a total of 480 lockers in the school.

Reteaching and Enrichment Strategies

If students need help. . .	If students got it. . .
Resources by Chapter • Practice A and Practice B • Puzzle Time Record and Practice Journal Practice Differentiating the Lesson Lesson Tutorials Skills Review Handbook	Resources by Chapter • Enrichment and Extension • School-to-Work Start the next section

Solve the proportion.

②③ 16. $\dfrac{1}{4} = \dfrac{z}{20}$

17. $\dfrac{3}{4} = \dfrac{12}{y}$

18. $\dfrac{35}{k} = \dfrac{7}{3}$

19. $\dfrac{15}{8} = \dfrac{45}{c}$

20. $\dfrac{b}{36} = \dfrac{5}{9}$

21. $\dfrac{1.4}{2.5} = \dfrac{g}{25}$

22. ORCHESTRA In an orchestra, the ratio of trombones to violas is 1 to 3.

    a. There are nine violas. Write a proportion that gives the number t of trombones in the orchestra.

    b. How many trombones are in the orchestra?

23. ATLANTIS Your science teacher has a 1 : 200 scale model of the Space Shuttle Atlantis. Which of the proportions can be used to find the actual length x of Atlantis? Explain.

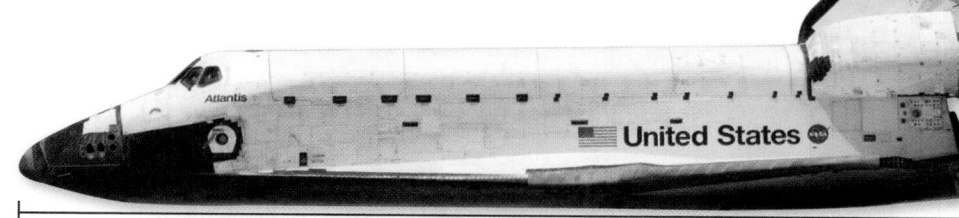

$$\dfrac{1}{200} = \dfrac{19.5}{x} \qquad \dfrac{1}{200} = \dfrac{x}{19.5} \qquad \dfrac{200}{19.5} = \dfrac{x}{1} \qquad \dfrac{x}{200} = \dfrac{1}{19.5}$$

19.5 cm

24. YOU BE THE TEACHER Your friend says "$48x = 6 \cdot 12$." Is your friend right? Explain.

    Solve $\dfrac{6}{x} = \dfrac{12}{48}$.

25. Reasoning There are 180 white lockers in the school. There are 3 white lockers for every 5 blue lockers. How many lockers are in the school?

Fair Game Review What you learned in previous grades & lessons

Solve the equation. *(Section 2.5)*

26. $\dfrac{x}{6} = 25$

27. $8x = 72$

28. $150 = 2x$

29. $35 = \dfrac{x}{4}$

30. MULTIPLE CHOICE Which is the slope of a line? *(Section 3.2)*

    Ⓐ $\dfrac{\text{change in } y}{1}$
    Ⓑ $\dfrac{\text{change in } x}{1}$
    Ⓒ $\dfrac{\text{change in } x}{\text{change in } y}$
    Ⓓ $\dfrac{\text{change in } y}{\text{change in } x}$

3.5 Solving Proportions

Essential Question How can you use ratio tables and cross products to solve proportions in science?

1 ACTIVITY: Solving a Proportion in Science

SCIENCE Scientists use *ratio tables* to determine the amount of a compound (like salt) that is dissolved in a solution. Work with a partner to show how scientists use cross products to determine the unknown quantity in a ratio.

a.   Sample: Salt Water

Salt Water	1 L	3 L
Salt	250 g	x g

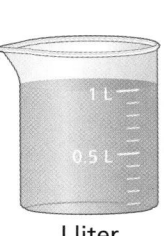 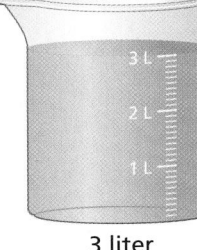

I liter   3 liter

$$\frac{3\,\cancel{L}}{1\,\cancel{L}} = \frac{x\,g}{250\,\cancel{g}}$$ Write proportion.

$$3 \cdot 250 = 1 \cdot x$$ Set cross products equal.

$$750 = x$$ Simplify.

⋮• So, there are 750 grams of salt in the 3-liter solution.

b.   White Glue Solution

Water	½ cup	1 cup
White Glue	½ cup	x cups

c.   Borax Solution

Borax	1 tsp	2 tsp
Water	1 cup	x cups

d.   Slime (see recipe)

Borax Solution	½ cup	1 cup
White Glue Solution	y cups	x cups

Recipe for SLIME

1. Add ½ cup of water and ½ cup white glue. Mix thoroughly. This is your white glue solution.

2. Add a couple drops of food coloring to the glue solution. Mix thoroughly.

3. Add 1 teaspoon of borax to 1 cup of water. Mix thoroughly. This is your borax solution (about 1 cup).

4. Pour the borax solution and the glue solution into a separate bowl.

5. Place the slime that forms in a plastic bag and squeeze the mixture repeatedly to mix it up.

Laurie's Notes

Introduction

For the Teacher

- **Goal:** Students will solve proportions using ratio tables and cross products.
- **Interdisciplinary:** This activity has a recipe for Slime. Ask a science colleague for assistance, and make a batch of Slime.

Motivate

- **Model:** Display two containers, each filled with water. The ratio of their volumes should be 2 : 1. Let students watch you put 4 drops of food coloring in the smaller one. Stir the water.

❓ "How many drops do I need to put in the larger container so that the water is the same darkness as the smaller vessel?" Students should note the ratio of the volumes of the two vessels. It may be helpful to label the volume of each container. Students should have no difficulty in understanding that 8 drops are needed.

❓ "Suppose that the two volumes are not in the ratio 1 : 2. The smaller container has a volume of 400 milliliters and the larger vessel has a volume of 600 milliliters. If I add 4 drops of food coloring to the smaller vessel, how much should I add to the larger vessel?" 6 drops

- This should get students thinking about a proportion.
- **Discuss:** Ask for volunteers to share their thinking.

Activity Notes

Activity 1

- Write the ratio table.

❓ "If there are 250 grams of salt in 1 liter of salt water, how many grams of salt is in 3 liters?" Students should quickly answer 750 grams, not just 750. The units are extremely important!

- Set up the proportion from the ratio table and discuss how common units divide out.
- Then use the Cross Products Property to verify the answer.

❓ "Could the proportion $\frac{1 \text{ L}}{250 \text{ g}} = \frac{3 \text{ L}}{x \text{ g}}$ be used to solve for x?" yes; Listen for an explanation that 1 liter has 250 grams of salt and you're solving for the number of grams of salt in 3 liters.

❓ "Is there another way you might solve this proportion without using the Cross Products Property?" mental math: $1 \times 250 = 250$, so $3 \times 250 = 750$

- Ask students to solve for x in each of the three parts—white glue solution, borax solution, and Slime.

❓ **Extension:** Ask the following:

- "If you only had $\frac{1}{4}$ cup of white glue, how much water should you mix with it to make the white glue solution?" $\frac{1}{4}$ cup

- "If you have 1 tablespoon of borax, how much water do you need to make the borax solution?" 3 cups

Start Thinking! and Warm Up

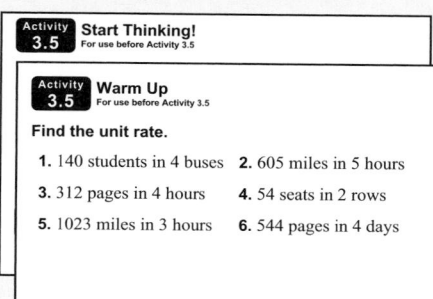

Activity 3.5 **Start Thinking!**
For use before Activity 3.5

Activity 3.5 **Warm Up**
For use before Activity 3.5

Find the unit rate.

1. 140 students in 4 buses
2. 605 miles in 5 hours
3. 312 pages in 4 hours
4. 54 seats in 2 rows
5. 1023 miles in 3 hours
6. 544 pages in 4 days

3.5 Record and Practice Journal

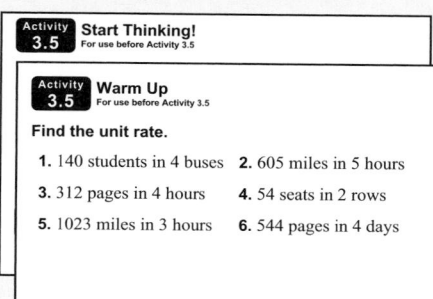

Essential Question How can you use ratio tables and cross products to solve proportions in science?

1 ACTIVITY: Solving a Proportion in Science

SCIENCE Scientists use *ratio tables* to determine the amount of a compound (like salt) that is dissolved in a solution. Work with a partner to show how scientists use cross products to determine the unknown quantity in a ratio.

a. **Sample:** Salt Water

Salt Water	1 L	3 L
Salt	250 g	x g

There are ___750___ grams of salt in the 3-liter solution.

b. **White Glue Solution**

Water	$\frac{1}{2}$ cup	1 cup
White Glue	$\frac{1}{2}$ cup	x cups

1 cup

c. **Borax Solution**

Borax	1 tsp	2 tsp
Water	1 cup	x cups

2 cups

English Language Learners

Visual

When solving a proportion using the Cross Products Property, use a visual X through the equal sign.

$$\frac{4}{3} \diagup\!\!\!\!\diagdown \frac{f}{27}$$ Write the proportion.

$4 \cdot 27 = 3 \cdot f$ Cross Products Property

$108 = 3f$ Multiply.

$36 = f$ Divide.

3.5 Record and Practice Journal

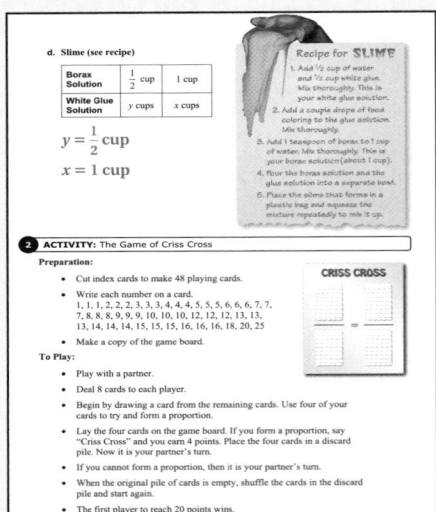

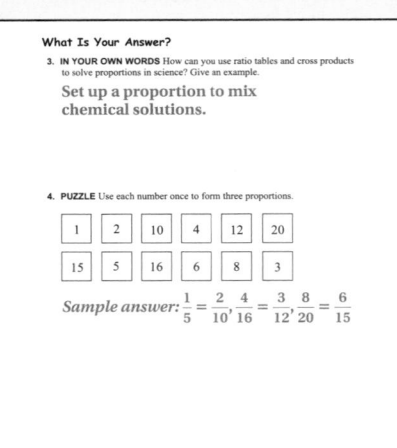

Activity 2

- **Management Tip:** The cards can be photocopied onto heavier weight paper, laminated, cut apart, and stored in locking plastic bags for easy distribution *and* for use next year.
- **Game Notes:**
  - The number of cards in your hand will vary depending upon whether you have been able to make a proportion.
  - The game moves along fairly quickly.
  - If a pair of students finishes early, they can play again.
- After students have played for a period of time, ask what strategies they used to decide if they have a proportion or not. Make the decision on whether or not they had a proportion.
- Some students may only think about equivalent fractions. Others may think about the Cross Products Property.

What Is Your Answer?

? Question 4 is a nice recap of the lesson. When students have finished, ask: "Do you think everyone has the same 3 proportions?" Students will likely say yes, forgetting that there are different arrangements of the 4 numbers in a proportion that will form another proportion.

Closure

- Use the information in the table to solve for *x*.

# of bracelets	3	x
Yellow twine	48 in.	80 in.

$$\frac{3}{48} = \frac{x}{80}$$

$$x = 5$$

five bracelets

The Dynamic Planning Tool
Editable Teacher's Resources at *BigIdeasMath.com*

2 ACTIVITY: The Game of Criss Cross

Preparation:

- Cut index cards to make 48 playing cards.

- Write each number on a card.

  1, 1, 1, 2, 2, 2, 3, 3, 3, 4, 4, 4, 5, 5, 5, 6, 6, 6, 7, 7,

  7, 8, 8, 8, 9, 9, 9, 10, 10, 10, 12, 12, 12, 13, 13,

  13, 14, 14, 14, 15, 15, 15, 16, 16, 16, 18, 20, 25

- Make a copy of the game board.

CRISS CROSS

To Play:

- Play with a partner.

- Deal 8 cards to each player.

- Begin by drawing a card from the remaining cards. Use four of your cards to try to form a proportion.

- Lay the four cards on the game board. If you form a proportion, say "Criss Cross" and you earn 4 points. Place the four cards in a discard pile. Now it is your partner's turn.

- If you cannot form a proportion, then it is your partner's turn.

- When the original pile of cards is empty, shuffle the cards in the discard pile and start again.

- The first player to reach 20 points wins.

What Is Your Answer?

3. **IN YOUR OWN WORDS** How can you use ratio tables and cross products to solve proportions in science? Give an example.

4. **PUZZLE** Use each number once to form three proportions.

| 1 | 2 | 10 | 4 | 12 | 20 |

| 15 | 5 | 16 | 6 | 8 | 3 |

Practice

Use what you discovered about solving proportions to complete Exercises 10–13 on page 126.

Key Idea

Solving Proportions

Method 1 Use mental math. *(Section 3.4)*

Method 2 Use the Multiplication Property of Equality. *(Section 3.5)*

Method 3 Use the Cross Products Property. *(Section 3.5)*

EXAMPLE 1 Solving Proportions Using Multiplication

Solve $\dfrac{5}{7} = \dfrac{x}{21}$.

$$\dfrac{5}{7} = \dfrac{x}{21} \qquad \text{Write the proportion.}$$

$$21 \cdot \dfrac{5}{7} = 21 \cdot \dfrac{x}{21} \qquad \text{Multiply each side by 21.}$$

$$15 = x \qquad \text{Simplify.}$$

The solution is 15.

On Your Own

Now You're Ready
Exercises 4−9

Solve the proportion using multiplication.

1. $\dfrac{w}{6} = \dfrac{6}{9}$ **2.** $\dfrac{12}{10} = \dfrac{a}{15}$ **3.** $\dfrac{y}{6} = \dfrac{2}{4}$

EXAMPLE 2 Solving Proportions Using the Cross Products Property

Solve each proportion.

a. $\dfrac{x}{8} = \dfrac{7}{10}$

$$x \cdot 10 = 8 \cdot 7 \qquad \text{Use the Cross Products Property.}$$

$$10x = 56 \qquad \text{Multiply.}$$

$$x = 5.6 \qquad \text{Divide.}$$

The solution is 5.6.

b. $\dfrac{9}{y} = \dfrac{3}{17}$

$$9 \cdot 17 = y \cdot 3$$

$$153 = 3y$$

$$51 = y$$

The solution is 51.

Laurie's Notes

Introduction

Connect
- **Yesterday:** Students solved proportions using the Cross Products Property.
- **Today:** Students will solve proportions using different strategies.

Motivate
- The Cross Products Property can be used to solve any proportion, but you want students to recognize when it is more efficient to use simple mental math or the Multiplication Property of Equality.
- As you work through problems with students, share with them the wisdom of analyzing the problem first to decide what method makes the most sense.
- **Common Error:** Students sometimes confuse multiplication of fractions and the Cross Products Property.

Lesson Notes

Example 1
- **?** If you wanted to put a context around this problem: "There are 5 weekdays and 7 entire days in a week. How many weekdays are there in 3 weeks?" 15 weekdays
- The property works because the variable is in the numerator.
- If this same problem had been $\frac{7}{5} = \frac{21}{x}$, you could not solve by multiplying both sides of the equation by $\frac{1}{21}$ because that would simplify to $\frac{1}{15} = \frac{1}{x}$ and you still haven't solved for x.
- Be sure to check for understanding with this idea.
- **?** "Could you use another strategy such as mental math to solve this problem?" Yes. Listen for the idea of equivalent fractions.

On Your Own
- **Think-Pair-Share:** Students should read each question independently and then work with a partner to answer the questions. When they have answered the questions, the pair should compare their answers with another group and discuss any discrepancies.
- **?** Ask students to share their strategies.
- At least one pair of students should solve Question 3 by simplifying $\frac{2}{4} = \frac{1}{2}$, and using mental math to finish.

Example 2
- **?** "How are the two examples different?" In part (a), the variable is in the numerator and in part (b), the variable is in the denominator. Part (b) involves one numerator that is a factor of the other numerator.
- **?** "Can you easily use the Multiplication Property of Equality to solve both examples?" Using the Multiplication Property of Equality would be difficult in part (b) because the variable is in the denominator.

Goal Today's lesson is solving proportions using a variety of strategies.

Start Thinking! and Warm Up

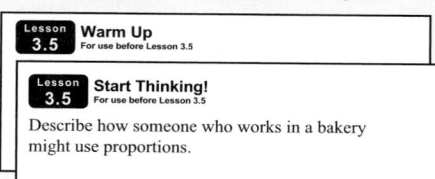

Lesson 3.5 **Warm Up** For use before Lesson 3.5

Lesson 3.5 **Start Thinking!** For use before Lesson 3.5

Describe how someone who works in a bakery might use proportions.

Extra Example 1

Solve $\frac{c}{12} = \frac{5}{3}$. 20

On Your Own
1. $w = 4$
2. $a = 18$
3. $y = 3$

Extra Example 2

Solve each proportion.

a. $\frac{3}{4} = \frac{u}{6}$ 4.5

b. $\frac{4}{13} = \frac{12}{h}$ 39

On Your Own

4. $x = 8$

5. $y = 2.5$

6. $z = 15$

Extra Example 3

The toll due on the Pennsylvania Turnpike is proportional to the number of miles driven. How much does it cost to drive 100 miles? $6

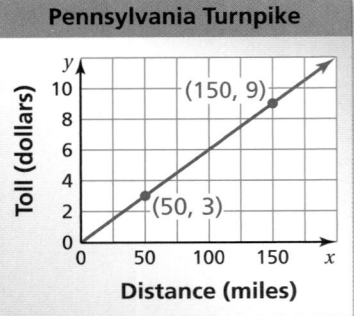

Pennsylvania Turnpike

On Your Own

7. $5.63

Differentiated Instruction

Visual

To reinforce the Cross Products Property, write each number and variable of a proportion on a card. Give the set of cards to a student. Have the student set up the proportion.

$$\frac{\boxed{3}}{\boxed{8}} = \frac{\boxed{k}}{\boxed{4}}$$

Then have the student find the cross products with the cards.

$$\boxed{3} \cdot \boxed{4} = \boxed{8} \cdot \boxed{k}$$

Finish by having the student complete the solution using paper and pencil.

Laurie's Notes

On Your Own

- **Think-Pair-Share:** Students should read each question independently and then work with a partner to answer the questions. When they have answered the questions, the pair should compare their answers with another group and discuss any discrepancies.
- ? Ask students to share their strategies.
- Although the directions say to solve using the Cross Products Property, one pair of students might solve Questions 4 and 5 by using mental math and recognize equivalent fractions.

Example 3

- Ask a student to read the problem out loud.
- ? "What do the ordered pairs (100, 7.5) and (200, 15) mean on the graph?" If you drive 100 miles the toll is $7.50, and if you drive 200 miles the toll is $15.
- ? "What does the ordered pair (0, 0) mean on the graph?" If you don't drive on the turnpike, you don't have to pay a toll.
- **Connection:** This is a connection to an earlier section in the chapter. Students should be comfortable solving this problem by looking at the graph or by using proportions.
- ? "In Method 1, how was the change in y and change in x found?" Using (0, 0) and (100, 7.5) *or* (100, 7.5) and (200, 15). Subtract the y-values and subtract the x-values.
- ? "What does a slope of 0.075 mean in the context of this problem?" The unit rate is $0.075 per mile. Each mile driven costs about 8 cents in tolls.
- ? "In Method 2, could the ordered pair (200, 15) be used to write the proportion?" yes
- ? "How about (0, 0)? Explain." (0, 0) could not be used because there would be a 0 in the denominator. Division by zero is not possible.

On Your Own

- **Neighbor Check:** Have students work independently and then have their neighbor check their work. Have students discuss any discrepancies.

Closure

- Write and solve 3 proportions. One should use mental math to solve, one should use the Multiplication Property of Equality, and one should use the Cross Products Property.

Technology For the Teacher

Dynamic Classroom

The Dynamic Planning Tool
Editable Teacher's Resources at *BigIdeasMath.com*

On Your Own

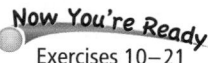
Now You're Ready
Exercises 10–21

Solve the proportion using the Cross Products Property.

4. $\dfrac{2}{7} = \dfrac{x}{28}$

5. $\dfrac{12}{5} = \dfrac{6}{y}$

6. $\dfrac{40}{z+1} = \dfrac{15}{6}$

EXAMPLE ③ **Real-Life Application**

The toll due on a turnpike is proportional to the number of miles driven. How much does it cost to drive 150 miles?

> **TOLL PLAZA**
> **¹/₂ MILE**
> **REDUCE SPEED**

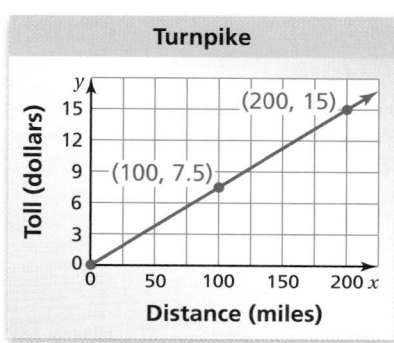

Method 1: Interpret the slope as a unit rate.

$$\text{slope} = \frac{\text{change in } y}{\text{change in } x}$$

$$= \frac{7.5}{100} \qquad \text{Substitute.}$$

$$= 0.075 \qquad \text{Divide.}$$

The unit rate is $0.075 per mile. Multiply to find the total cost.

$$150 \text{ mi} \cdot \frac{\$0.075}{1 \text{ mi}} = \$11.25$$

∵ It costs $11.25 to drive 150 miles on the turnpike.

Method 2: Write and solve a proportion.

$$\frac{7.5}{100} = \frac{x}{150} \quad \leftarrow \boxed{\text{dollars}} \qquad \text{Use (100, 7.5) to write a proportion.}$$
$$\leftarrow \boxed{\text{miles}}$$

$$150 \cdot \frac{7.5}{100} = 150 \cdot \frac{x}{150} \qquad \text{Multiply each side by 150.}$$

$$11.25 = x \qquad \text{Simplify.}$$

∵ It costs $11.25 to drive 150 miles on the turnpike.

On Your Own

7. **WHAT IF?** In Example 3, how much does it cost to drive 75 miles on the turnpike?

 Vocabulary and Concept Check

1. **WRITING** What are three ways you can solve a proportion?

2. **OPEN-ENDED** Which way would you choose to solve $\dfrac{3}{x} = \dfrac{6}{14}$? Explain your reasoning.

3. **NUMBER SENSE** Does $\dfrac{x}{4} = \dfrac{15}{3}$ have the same solution as $\dfrac{x}{15} = \dfrac{4}{3}$? Use the Cross Products Property to explain your answer.

 **Practice and Problem Solving**

Solve the proportion using multiplication.

1 **4.** $\dfrac{9}{5} = \dfrac{z}{20}$ **5.** $\dfrac{h}{15} = \dfrac{16}{3}$ **6.** $\dfrac{w}{4} = \dfrac{42}{24}$

7. $\dfrac{35}{28} = \dfrac{n}{12}$ **8.** $\dfrac{7}{16} = \dfrac{x}{4}$ **9.** $\dfrac{y}{9} = \dfrac{44}{54}$

Solve the proportion using the Cross Products Property.

2 **10.** $\dfrac{a}{6} = \dfrac{15}{2}$ **11.** $\dfrac{10}{7} = \dfrac{8}{k}$ **12.** $\dfrac{3}{4} = \dfrac{v}{14}$ **13.** $\dfrac{5}{n} = \dfrac{16}{32}$

14. $\dfrac{36}{42} = \dfrac{24}{r}$ **15.** $\dfrac{9}{10} = \dfrac{d}{6.4}$ **16.** $\dfrac{x}{8} = \dfrac{3}{12}$ **17.** $\dfrac{8}{m} = \dfrac{6}{15}$

18. $\dfrac{4}{24} = \dfrac{c}{36}$ **19.** $\dfrac{20}{16} = \dfrac{d}{12}$ **20.** $\dfrac{30}{20} = \dfrac{w}{14}$ **21.** $\dfrac{2.4}{1.8} = \dfrac{7.2}{k}$

22. **ERROR ANALYSIS** Describe and correct the error in solving the proportion $\dfrac{m}{8} = \dfrac{15}{24}$.

$$\dfrac{m}{8} = \dfrac{15}{24}$$
$$8 \cdot m = 24 \cdot 15$$
$$m = 45$$

23. **PENS** Forty-eight pens are packaged in four boxes. How many pens are packaged in nine boxes?

24. **PIZZA PARTY** How much does it cost to buy 10 medium pizzas?

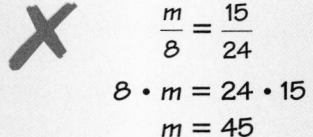

3 Medium Pizzas for $10.50

Solve the proportion.

25. $\dfrac{2x}{5} = \dfrac{9}{15}$ **26.** $\dfrac{5}{2} = \dfrac{d-2}{4}$ **27.** $\dfrac{4}{k+3} = \dfrac{8}{14}$

Assignment Guide and Homework Check

Level	Day 1 Activity Assignment	Day 2 Lesson Assignment	Homework Check
Basic	10–13, 36–40	1–3, 5–9 odd, 15–21 odd, 22–24, 29, 30	2, 5, 17, 24
Average	10–13, 36–40	1–3, 5–9 odd, 15–21 odd, 22, 29–32	2, 5, 17, 30
Advanced	10–13, 36–40	1–3, 18–22, 25–28, 32–35	18, 25, 28, 34

Common Errors

- **Exercises 4–9** Some students may multiply by the denominator of the fraction without the variable. Remind them that they are trying to get the variable alone, so they want to multiply both sides by the denominator of the fraction with the variable. Give students an example without a fraction on the other side of the equation to remind them of the process.
- **Exercises 10–21** Students may divide instead of multiply when finding the cross products, or they may multiply across the numerators and the denominators as if they were multiplying fractions. Remind students that the proportions have an equal sign between them, not a multiplication sign. Also tell them that when they use the Cross Products Property, it produces an "X" which means multiplication.

3.5 Record and Practice Journal

Solve the proportion using multiplication.

1. $\frac{a}{40} = \frac{3}{10}$ $a = 12$
2. $\frac{6}{11} = \frac{c}{77}$ $c = 42$
3. $\frac{b}{65} = \frac{7}{13}$ $b = 35$

Solve the proportion using the Cross Products Property.

4. $\frac{k}{6} = \frac{8}{16}$ $k = 3$
5. $\frac{5.4}{7} = \frac{27}{h}$ $h = 35$
6. $\frac{15}{n} = \frac{20}{8}$ $n = 6$

Solve the proportion.

7. $\frac{5}{2} = \frac{4x}{8}$ $x = 5$
8. $\frac{8}{11} = \frac{4}{y+2}$ $y = 3.5$
9. $\frac{3}{z-1} = \frac{9}{15}$ $z = 6$

10. A cell phone company charges $5 for 250 text messages. How much does the company charge for 300 text messages?
$6

11. There are 84 players on a football team. The ratio of offensive players to defensive players is 4 to 3. How many offensive players are on the team?
48

 Vocabulary and Concept Check

1. mental math; Multiplication Property of Equality; Cross Products Property

2. *Sample answer:* mental math; Because $3 \cdot 2 = 6$, the product of x and 2 is 14. So, $x = 7$.

3. yes; Both cross products give the equation $3x = 60$.

 Practice and Problem Solving

4. $z = 36$
5. $h = 80$
6. $w = 7$
7. $n = 15$
8. $x = 1\frac{3}{4}$
9. $y = 7\frac{1}{3}$
10. $a = 45$
11. $k = 5.6$
12. $v = 10.5$
13. $n = 10$
14. $r = 28$
15. $d = 5.76$
16. $x = 2$
17. $m = 20$
18. $c = 6$
19. $d = 15$
20. $w = 21$
21. $k = 5.4$

22. They did not perform the cross multiplication properly.
$$\frac{m}{8} = \frac{15}{24}$$
$$m \cdot 24 = 8 \cdot 15$$
$$m = 5$$

23. 108 pens

24. $35

25. $x = 1.5$

26. $d = 12$

27. $k = 4$

28. true; Both cross products give the equation $3a = 2b$.

29. $769.50

30. 15.5 lb

31. a. 16 mo

   b. 40 mo

32. a. about 14 min 34 sec

   b. *Sample answer:* 5.84 sec

33. See *Taking Math Deeper.*

34. 4 bags

35. $2; \dfrac{\frac{1}{2}}{\frac{1}{4}} = \dfrac{1}{2} \times \dfrac{4}{1} = 2$

Fair Game Review

36. 5.3   **37.** 6400

38. 3.5   **39.** 7920

40. C

Mini-Assessment
Solve the proportion.

1. $\dfrac{x}{12} = \dfrac{3}{8}$ $x = 4.5$

2. $\dfrac{6}{11} = \dfrac{9}{m}$ $m = 16.5$

3. $\dfrac{6}{12} = \dfrac{c}{36}$ $c = 18$

4. $\dfrac{18}{3} = \dfrac{24}{b}$ $b = 4$

5. Thirty-six pencils are packed in three boxes. How many pencils are packed in five boxes? 60 pencils

Taking Math Deeper

Exercise 33

Sometimes it is a good suggestion to "forget about algebra and just answer the question." After answering the question, we might "relax" and try to look for an efficient or clever way to answer the question.

 Start with a table. Adults : children is 5 : 3.

Adults	Children	Total
5	3	8
10	6	16
15	9	24

 I can see this will take a while. I'm going to jump ahead in the table.

Adults	Children	Total
100	60	160 too many
80	48	128 too few
90	54	144 just right

 The answer is 90 adults. I wonder if I can get the answer algebraically.

   Adults $= 5x$ Children $= 3x$

$$5x + 3x = 144$$
$$8x = 144$$
$$x = 18$$

Adults $= 5 \cdot 18 = 90$

Reteaching and Enrichment Strategies

If students need help. . .	If students got it. . .
Resources by Chapter • Practice A and Practice B • Puzzle Time Record and Practice Journal Practice Differentiating the Lesson Lesson Tutorials Skills Review Handbook	Resources by Chapter • Enrichment and Extension • School-to-Work • Financial Literacy Start the next section

28. **TRUE OR FALSE?** Tell whether the statement is *true* or *false*. Explain.

If $\dfrac{a}{b} = \dfrac{2}{3}$, then $\dfrac{3}{2} = \dfrac{b}{a}$.

29. **CLASS TRIP** It costs $95 for 20 students to visit an aquarium. How much does it cost for 162 students?

30. **GRAVITY** A person who weighs 120 pounds on Earth weighs 20 pounds on the moon. How much does a 93-pound person weigh on the moon?

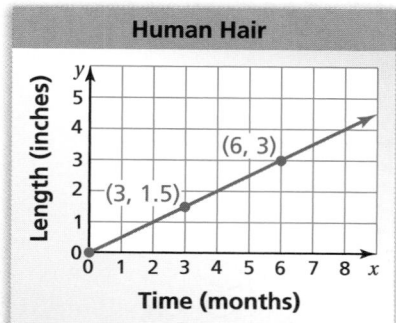

31. **HAIR** The length of human hair is proportional to the number of months it has grown.

 a. How long does it take hair to grow 8 inches?

 b. Use a different method than the one in part (a) to find how long it takes hair to grow 20 inches.

32. **CHEETAH** Cheetahs are the fastest mammals in the world. They can reach speeds of 70 miles per hour.

 a. At this speed, how long would it take a cheetah to run 17 miles?

 b. **RESEARCH** Use the Internet or library to find how long a cheetah can maintain a speed of 70 miles per hour.

33. **AUDIENCE** There are 144 people in an audience. The ratio of adults to children is 5 to 3. How many are adults?

34. **LAWN SEED** Three pounds of lawn seed covers 1800 square feet. How many bags are needed to cover 8400 square feet?

35. **Critical Thinking** Consider the proportions $m = \dfrac{1}{2}$ and $k = \dfrac{1}{4}$.

What is the ratio $\dfrac{m}{k}$? Explain your reasoning.

Fair Game Review What you learned in previous grades & lessons

Copy and complete. *(Skills Review Handbook)*

36. $530 \text{ cm} = \boxed{} \text{ m}$

37. $6.4 \text{ kg} = \boxed{} \text{ g}$

38. $56 \text{ oz} = \boxed{} \text{ lb}$

39. $1\dfrac{1}{2} \text{ mi} = \boxed{} \text{ ft}$

40. **MULTIPLE CHOICE** How many cups of milk are shown? *(Skills Review Handbook)*

 Ⓐ $\dfrac{7}{10}$ c

 Ⓑ $\dfrac{7}{8}$ c

 Ⓒ $1\dfrac{3}{4}$ c

 Ⓓ 14 c

3 Study Help

You can use an **information wheel** to organize information about a concept. Here is an example of an information wheel for slope.

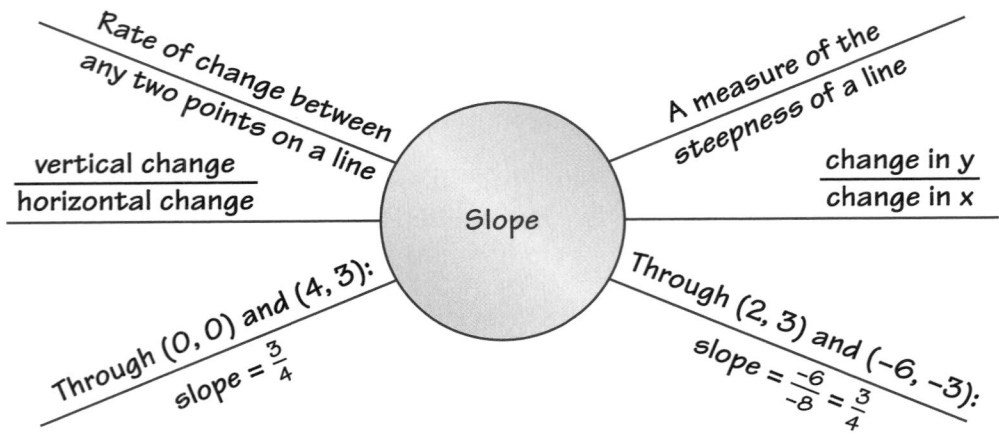

On Your Own

Make an information wheel to help you study these topics.

1. ratio

2. rate

3. unit rate

4. proportion

5. cross products

6. solving proportions

After you complete this chapter, make information wheels for the following topics.

7. U.S. customary system

8. metric system

9. converting units

10. direct variation

11. inverse variation

"My **information wheel** summarizes how cats act when they get baths."

Sample Answers

1.

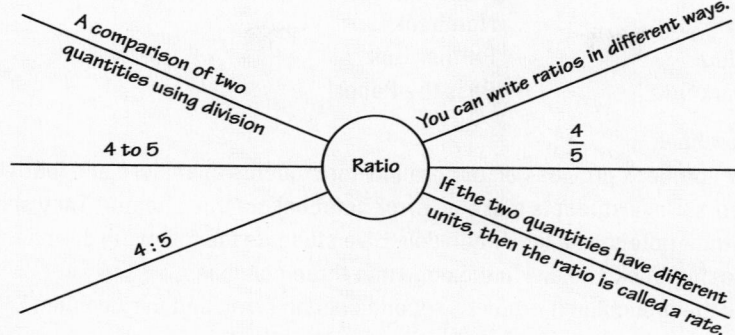

A comparison of two quantities using division

You can write ratios in different ways.

Ratio

4 to 5

$\frac{4}{5}$

4 : 5

If the two quantities have different units, then the ratio is called a rate.

2.

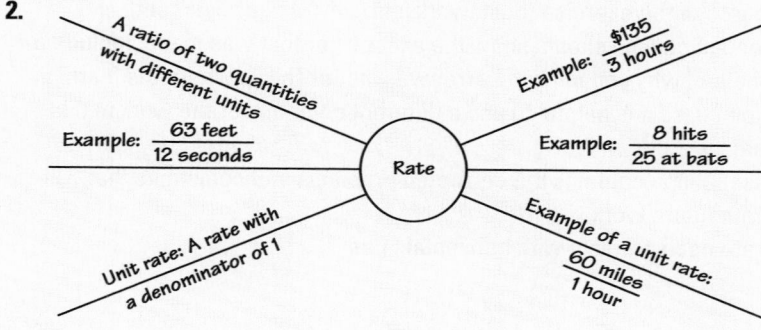

A ratio of two quantities with different units

Example: $\frac{\$135}{3\ hours}$

Example: $\frac{63\ feet}{12\ seconds}$

Rate

Example: $\frac{8\ hits}{25\ at\ bats}$

Unit rate: A rate with a denominator of 1

Example of a unit rate: $\frac{60\ miles}{1\ hour}$

3.

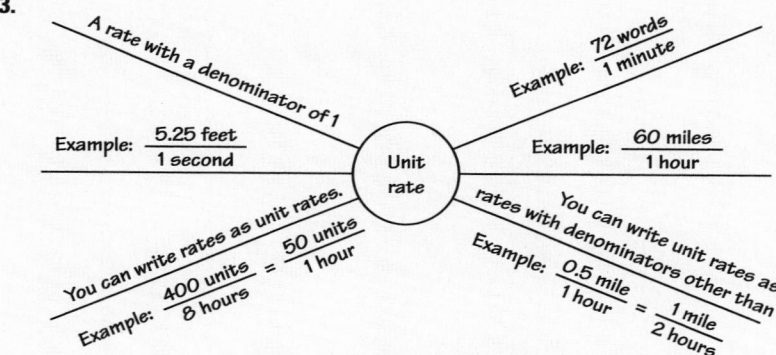

A rate with a denominator of 1

Example: $\frac{72\ words}{1\ minute}$

Example: $\frac{5.25\ feet}{1\ second}$

Unit rate

Example: $\frac{60\ miles}{1\ hour}$

You can write rates as unit rates. Example: $\frac{400\ units}{8\ hours} = \frac{50\ units}{1\ hour}$

You can write unit rates as rates with denominators other than 1. Example: $\frac{0.5\ mile}{1\ hour} = \frac{1\ mile}{2\ hours}$

4.

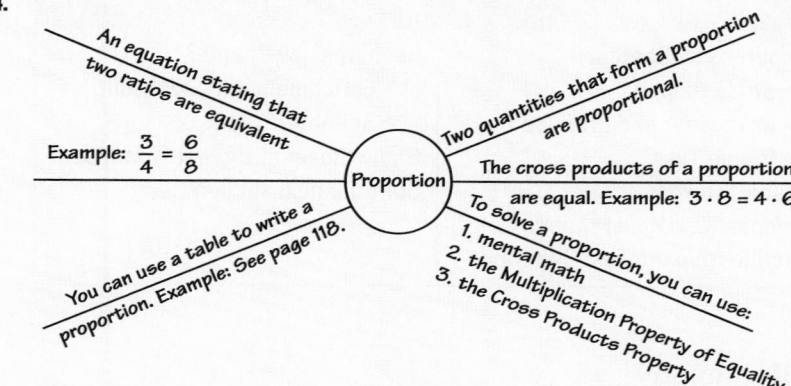

An equation stating that two ratios are equivalent

Two quantities that form a proportion are proportional.

Example: $\frac{3}{4} = \frac{6}{8}$

Proportion

The cross products of a proportion are equal. Example: $3 \cdot 8 = 4 \cdot 6$

To solve a proportion, you can use:
1. mental math
2. the Multiplication Property of Equality
3. the Cross Products Property

You can use a table to write a proportion. Example: See page 118.

5–6. Available at *BigIdeasMath.com.*

Technology For the Teacher
Vocabulary Puzzle Builder

Answers

1. $\dfrac{3}{2}$　　2. $\dfrac{10}{1}$

3. 36 feet per second

4. 30 miles per gallon

5. $\dfrac{1}{2}$　　6. $\dfrac{2}{5}$

7. yes　　8. no

9. *Sample answer:* $\dfrac{\$56}{\$42} = \dfrac{h \text{ hours}}{6 \text{ hours}}$

10. *Sample answer:*

$$\dfrac{g \text{ games}}{4 \text{ wins}} = \dfrac{6 \text{ games}}{3 \text{ wins}}$$

11. $x = 5$　　12. $z = 99$

13. $\dfrac{1}{3}$ megabyte per second

14. no; Your rate is 5 minutes per level and your friend's rate is 4 minutes per level.

15. $\dfrac{150 \text{ minutes}}{3 \text{ classes}} = \dfrac{x \text{ minutes}}{5 \text{ classes}}$;

    250 minutes

16. $10\dfrac{2}{3}$ hours or

    10 hours 40 minutes

Assessment Book

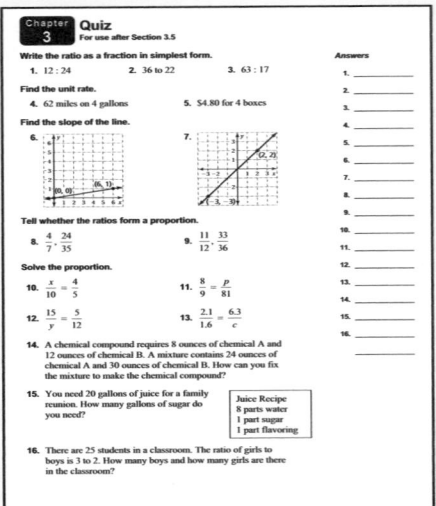

Alternative Quiz Ideas

100% Quiz　　　　　Math Log
Error Notebook　　Notebook Quiz
Group Quiz　　　　Partner Quiz
Homework Quiz　　Pass the Paper

Error Notebook

An error notebook provides an opportunity for students to analyze and learn from their errors. Have students make an error notebook for this chapter. They should work in their notebook a little each day. Give students the following directions.

- Use a notebook and divide the page into three columns.
- Label the first column *problem*, second column *error*, and third column *correction*.
- In the first column, write the exercise in which the errors were made. Record the source of the exercise (homework, quiz, in-class assignment).
- The second column should show the exact error that was made. Include a statement of why you think the error was made. This is where the learning takes place, so it is helpful to use a different color ink for the work in this column.
- The last column contains the corrected problems and comments that will help with future work.
- Separate each problem with horizontal lines.

Reteaching and Enrichment Strategies

If students need help. . .	If students got it. . .
Resources by Chapter • Study Help • Practice A and Practice B • Puzzle Time Lesson Tutorials *BigIdeasMath.com* Practice Quiz Practice from the Test Generator	Resources by Chapter • Enrichment and Extension • School-to-Work Game Closet at *BigIdeasMath.com* Start the next section

Technology For the Teacher

Answer Presentation Tool
Big Ideas Test Generator

Write the ratio as a fraction in simplest form.   *(Section 3.1)*

1. 18 red buttons : 12 blue buttons

2. 30 inches to 3 inches

Find the unit rate.   *(Section 3.1)*

3. 108 feet per 3 seconds

4. 360 miles per 12 gallons

Find the slope of the line.   *(Section 3.2)*

5.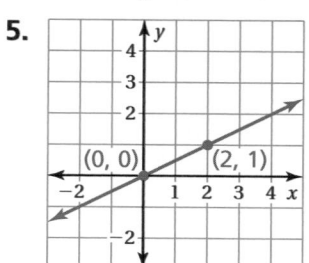

6.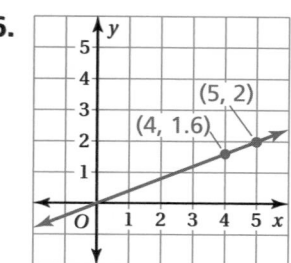

Tell whether the ratios form a proportion.   *(Section 3.3)*

7. $\dfrac{1}{8}, \dfrac{4}{32}$

8. $\dfrac{2}{3}, \dfrac{10}{30}$

Use the table to write a proportion.   *(Section 3.4)*

9.

	Monday	Tuesday
Dollars	42	56
Hours	6	h

10.

	Series 1	Series 2
Games	g	6
Wins	4	3

Solve the proportion.   *(Section 3.5)*

11. $\dfrac{x}{2} = \dfrac{40}{16}$

12. $\dfrac{3}{11} = \dfrac{27}{z}$

13. **MUSIC DOWNLOAD** The amount of time needed to download music is shown in the table. Find the rate in megabytes per second.   *(Section 3.1)*

Megabytes	Seconds
2	6
4	12
6	18
8	24

14. **GAMING** You advance 3 levels in 15 minutes. Your friend advances 5 levels in 20 minutes. Are these rates proportional? Explain.   *(Section 3.3)*

15. **CLASS TIME** You spend 150 minutes in three classes. Write and solve a proportion to find how many minutes you spend in five classes.   *(Section 3.4)*

16. **CONCERT** A benefit concert with three performers lasts 8 hours. At this rate, how many hours is a concert with four performers?   *(Section 3.5)*

Essential Question How can you compare lengths between the customary and metric systems?

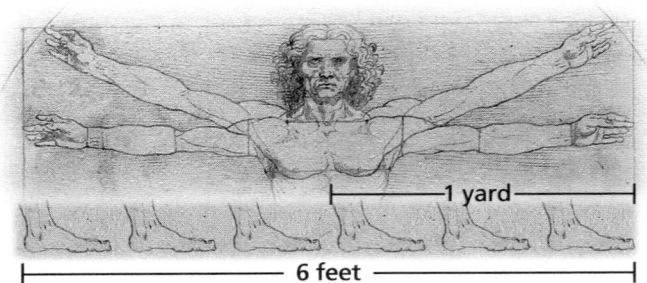

---1 yard---

----- 6 feet -----

1 ACTIVITY: Customary Measure History

Work with a partner.

a. Match the measure of length with its historical beginning.

Length	*Historical Beginning*
Inch	The length of a human foot.
Foot	The width of a human thumb.
Yard	The distance a human can walk in 1000 paces (two steps).
Mile	The distance from a human nose to the end of an outstretched human arm.

b. Use a ruler to measure your thumb, arm, and foot. How do your measurements compare to your answers from part (a)? Are they close to the historical measures?

You know how to convert measures within the customary and metric systems.

Equivalent Customary Lengths

1 ft = 12 in.   1 yd = 3 ft   1 mi = 5280 ft

Equivalent Metric Lengths

1 m = 1000 mm   1 m = 100 cm   1 km = 1000 m

You will learn how to convert between the two systems.

Converting Between Systems

1 in. ≈ 2.54 cm

1 mi ≈ 1.6 km

2.54 cm

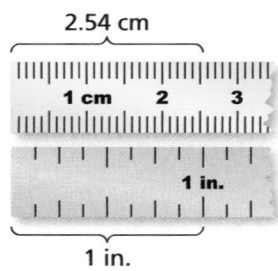

1 in.

Laurie's Notes

Introduction

For the Teacher

- **Goal:** Compare lengths between the customary and metric systems.
- Today's activity focuses solely on linear measurement. Tomorrow's lesson expands measurement to capacity and weight. Begin to collect materials that can be used for both lessons.

Discuss

- Discuss what linear measurement means—the distance between two points. Draw the following examples:

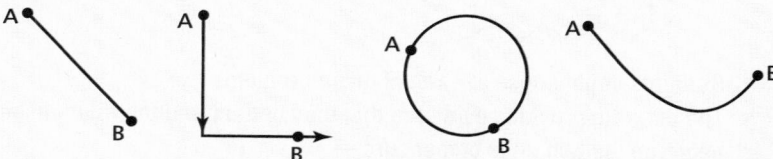

- The idea is that linear measurement doesn't mean it has to involve a straight line. Techniques can be used to find the lengths of curves, rivers, arteries, and so on.
- The first activity looks at the history of customary measure. Someone had to define how long an inch would be, and it wasn't a math teacher!

Activity Notes

Activity 1

- A transparent ruler for the overhead is helpful. Lay small objects on the overhead and have a student come to the overhead to measure it.
- **Common Error:** Not all rulers begin scaling with 0 at the left edge. Check rulers that will be used by students and if possible, make an overhead version for students to see the difference.
- Have students work with a partner to answer part (a), and then measure and record the length in part (b).
- If comfortable, students should remove their shoe to measure the length of their foot. They do not need to step directly onto the ruler.
- **Safety:** It is a wise idea to use a disinfectant on all measuring devices on a regular basis.
- Discuss the need for *standard* units of measure that are not related to body parts or other distances that are subject to change.
- Review customary lengths. In addition to the listed facts, 1 yd = 36 in. is considered a benchmark fact.
- **FYI:** We generally do not convert between systems of measurement; however, there are a few conversions that are considered *common knowledge* that people encounter in daily life. These common conversions are the focus of this investigation and the lesson.
- Write the two common conversions relating inches to centimeters, and miles to kilometers.
- Note that the conversions are *from* customary *to* metric.

Previous Learning

Students have studied measurement systems. They should be familiar with metric prefixes such as kilo, centi, and milli.

Start Thinking! and Warm Up

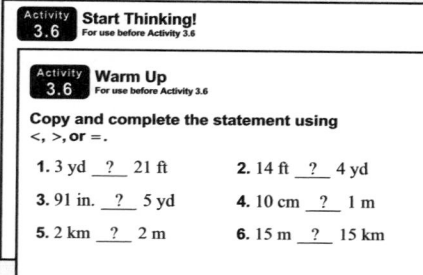

3.6 Record and Practice Journal

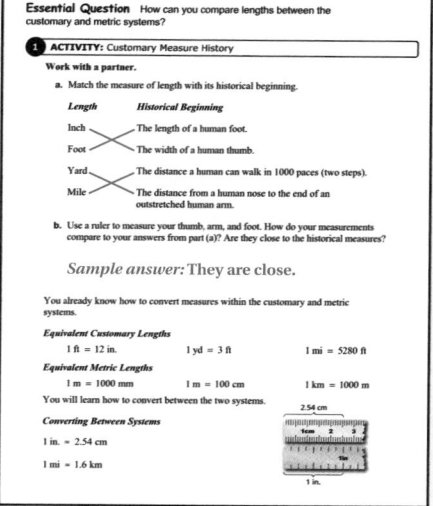

English Language Learners

Words and Abbreviations

Make a poster to display both customary and metric units of measure. Show the full word, both singular and plural, and the abbreviation.

Word		
Singular	**Plural**	**Abbreviation**
inch	inches	in.
foot	feet	ft
meter	meters	m

3.6 Record and Practice Journal

2 ACTIVITY: Comparing Measures

Work with a partner. Answer each question. Explain your answer. Use a diagram in your explanation.

	Metric	Customary
a. Car Speed: Which is faster?	80 km/h	60 mi/h
60 mi/h		
b. Trip Distance: Which is farther?	200 km	200 mi
200 mi		
c. Human Height: Who is taller?	180 cm	5 ft 8 in.
180 cm		
d. Wrench Width: Which is wider?	8 mm	5/16 in.
8 mm		
e. Swimming Pool Depth: Which is deeper?	1.4 m	4 ft
1.4 m		
f. Mountain Elevation: Which is higher?	2000 m	7000 ft
7000 ft		
g. Room Width: Which is wider?	3.5 m	12 ft
12 ft		

What Is Your Answer?

3. IN YOUR OWN WORDS How can you compare lengths between the customary and metric systems? Give examples with your description.

Check students' work.

4. HISTORY The meter and the metric system originated in France. In 1791, the French Academy of Sciences was instructed to create a new system of measurement. This new system would be based on powers of 10.

The fundamental units of this system would be based on natural values that were unchanging. The French Academy of Sciences decided to find the length of an imaginary arc that began at the North Pole and ended at the equator.

They would then divide this line into exactly ten million identical pieces. The length of one of these pieces would be the base unit of length for the new system of measurement.

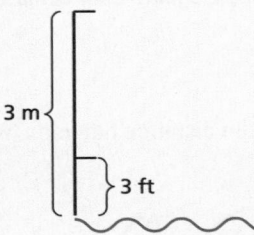

North Pole
Meter = 1 ten-millionth of this distance
Equator

a. Find the distance around Earth in meters.

40,000,000 m

b. Find the distance around Earth in kilometers.

40,000 km

5. Find the distance around Earth in miles.

about 25,000 miles

Laurie's Notes

Activity 2

? "Which diving board is higher, a 3-meter diving board or a 3-foot diving board?" Listen for reasoning that 1 meter is a little longer than 1 yard and because 1 yard equals 3 feet, the 3-meter diving board is about 3 times higher.

• A diagram would be helpful. It could look something like the following:

3 m
3 ft

• Students could guess at each of these problems.
• The diagram provides evidence that they understand the relationship between the two units of measure.

What Is Your Answer?

• **Whole Class Activity:** Students should work with a partner. When students are finished, have a class discussion.

Closure

• By eyesight, estimate the length and width of the classroom in feet and then in meters. Answers will vary depending upon classroom size.

Technology For the Teacher

Dynamic Classroom

The Dynamic Planning Tool
Editable Teacher's Resources at *BigIdeasMath.com*

ACTIVITY: Comparing Measures

Work with a partner. Answer each question. Explain your answer. Use a diagram in your explanation.

		Metric	*Customary*
a.	Car Speed: Which is faster?	80 km/h	60 mi/h
b.	Trip Distance: Which is farther?	200 km	200 mi
c.	Human Height: Who is taller?	180 cm	5 ft 8 in.
d.	Wrench Width: Which is wider?	8 mm	5/16 in.
e.	Swimming Pool Depth: Which is deeper?	1.4 m	4 ft
f.	Mountain Elevation: Which is higher?	2000 m	7000 ft
g.	Room Width: Which is wider?	3.5 m	12 ft

What Is Your Answer?

3. **IN YOUR OWN WORDS** How can you compare lengths between the customary and metric systems? Give examples with your description.

4. **HISTORY** The meter and the metric system originated in France. In 1791, the French Academy of Sciences was instructed to create a new system of measurement. This new system would be based on powers of 10.

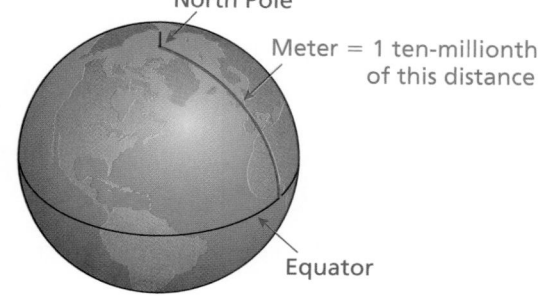

North Pole

Meter = 1 ten-millionth of this distance

Equator

The fundamental units of this system would be based on natural values that were unchanging. The French Academy of Sciences decided to find the length of an imaginary arc that began at the North Pole and ended at the equator.

They would then divide this arc into exactly ten million identical pieces. The length of one of these pieces would be the base unit of length for the new system of measurement.

   a. Find the distance around Earth in meters.

   b. Find the distance around Earth in kilometers.

5. Find the distance around Earth in miles.

Practice

Use what you learned about converting measures between systems to complete Exercises 4–9 on page 134.

Check It Out
Lesson Tutorials
BigIdeasMath com

Key Vocabulary 🔊
U.S. customary
system, *p. 132*
metric system, *p. 132*

The **U.S. customary system** is a system of measurement that contains units for length, capacity, and weight. The **metric system** is a decimal system of measurement, based on powers of 10, that contains units for length, capacity, and mass.

Use the relationships below to convert units *between* systems.

Length	Capacity	Weight and Mass
1 in. ≈ 2.54 cm	1 qt ≈ 0.95 L	1 lb ≈ 0.45 kg
1 mi ≈ 1.6 km		

EXAMPLE **1** **Converting Units**

Convert 5 liters to quarts.

Method 1: Convert using a ratio.

$$1 \text{ qt} \approx 0.95 \text{ L}$$

$$5 \,\cancel{L} \times \frac{1 \text{ qt}}{0.95 \,\cancel{L}} \approx 5.26 \text{ qt}$$

∴ So, 5 liters is about 5.26 quarts.

Method 2: Convert using a proportion.

Let x be the number of quarts equivalent to 5 liters.

quarts → $\dfrac{1}{0.95} = \dfrac{x}{5}$ ← quarts

liters → $\phantom{\dfrac{1}{0.95}}$ ← liters

Write a proportion.

$$5 = 0.95x$$ Use the Cross Products Property.

$$5.26 \approx x$$ Divide each side by 0.95.

∴ So, 5 liters is about 5.26 quarts.

● **On Your Own**

Now You're Ready
Exercises 10–22

Copy and complete the statement. Round to the nearest hundredth, if necessary.

1. 7 mi ≈ ▢ km

2. 12 qt ≈ ▢ L

3. 25 kg ≈ ▢ lb

4. 8 cm ≈ ▢ in.

🔊 Multi-Language Glossary at BigIdeasMath✓com.

Laurie's Notes

Introduction

Connect

- **Yesterday:** Students explored linear measurement in customary and metric units, and converted between the two systems.
- **Today:** Students will convert units of length, capacity, and weight/mass between systems (customary and metric).
- **Interdisciplinary:** Science standards address the concept of *mass*. In everyday use, especially in common conversation, weight is usually used as a synonym for mass.

Motivate

- Hold up empty containers of a quart of milk and a liter of iced tea. "I was really thirsty and drank both."
- ❓ Did I drink the same amount of each beverage? If students are not familiar with the containers, you may need to use the words "quart" and "liter." A quart is actually slightly less than a liter.
- Pretend to hold up containers (boxes) and say, "This box held a pound of chocolates and the other held a kilogram of chocolates."
- ❓ Did I eat the same amount of chocolates out of each box? One student may know that a kilogram is more than twice a pound. If not, it helps transition to today's lesson.
- **FYI:** If your students know that you have a fondness for a particular item, such as coffee, replace chocolates with the item.
- ❓ "Could someone give an example of a common item that would be measured in inches? in pounds? in gallons?" Answers will vary. *Sample answer:* inches of rain; pound of coffee; gallon of milk

Lesson Notes

Example 1

- ❓ You know 1 quart is approximately equal to 0.95 liter. So, should 5 liters be more than 5 quarts or less than 5 quarts? Listen for student reasoning. A quart is less than a liter, so it takes more than 5 quarts to equal 5 liters.
- The two methods used in this example help to review skills from this chapter—unit analysis and solving proportions using the Cross Products Property.
- **Common Student Questions:** A common question when doing conversions is, "Do I divide or multiply?" When unit analysis is used, the operation becomes evident. If the original question had been to convert 5 quarts to liters, notice how unit analysis is still used. Compare this to the problem in the text.

$$5 \text{ qt} \times \frac{0.95 \text{ L}}{1 \text{ qt}} = 4.75 \text{ L}$$

On Your Own

- Students should work with a partner. Encourage them to use each method (convert using a ratio and convert using a proportion) at least once.

Goal Today's lesson is converting units of measure between systems.

Lesson Materials
Introduction

- empty 1 quart milk container
- empty 1 liter tea bottle
- empty boxes

Start Thinking! and Warm Up

Lesson 3.6 Warm Up For use before Lesson 3.6

Lesson 3.6 Start Thinking! For use before Lesson 3.6

Can you fit an entire gallon of milk into a 2-liter bottle? Explain.

Extra Example 1

Convert 16 kilometers to miles.
about 10 miles

On Your Own

1. 11.2
2. 11.4
3. 55.56
4. 3.15

Laurie's Notes

Extra Example 2

Extra Example 2

Copy and complete the statement using
< or > : 2 ft __?__ 55 cm. 2 ft > 55 cm

On Your Own

5. <

6. <

7. >

Extra Example 3

A kitchen faucet has a maximum flow rate of 8.3 liters per minute and a bathroom faucet has a maximum flow rate of 24 cups per minute. Which faucet has a faster flow rate? kitchen faucet

On Your Own

8. 0.2 km/min, 70 km/h, 50 mi/h

Differentiated Instruction

Visual

Create a poster to display the units of 1 (or ratios) that are used in converting units of measure between the customary and metric systems.

Length

$\dfrac{1 \text{ in.}}{2.54 \text{ cm}}, \dfrac{2.54 \text{ cm}}{1 \text{ in.}}, \dfrac{1 \text{ mi}}{1.6 \text{ km}}, \dfrac{1.6 \text{ km}}{1 \text{ mi}}$

Capacity

$\dfrac{1 \text{ qt}}{0.95 \text{ L}}, \dfrac{0.95 \text{ L}}{1 \text{ qt}}$

Weight and Mass

$\dfrac{1 \text{ lb}}{0.45 \text{ kg}}, \dfrac{0.45 \text{ kg}}{1 \text{ lb}}$

Example 2

- These are multi-step problems that require students to possess knowledge of common equivalences, such as 1 pound equals 16 ounces. Before looking at these problems, it would be helpful to make a list of measurement facts. Ask students for examples.
- Work through Example 2. Unit analysis is used to keep track of the units in each ratio.
- **Common Student Questions:** A common question is how to recognize when to use $\dfrac{1 \text{ pound}}{16 \text{ ounces}}$ and when to use $\dfrac{16 \text{ ounces}}{1 \text{ pound}}$. Because you are beginning with 25 ounces, the unit of ounces must be in the denominator of the ratio in order for the units to divide out. Then, because pounds are in the numerator of that ratio, pounds must be in the denominator of the next ratio in order for the units of pounds to divide out. In other words, start with the quantity you're trying to convert, then decide on a conversion factor based on the units.

On Your Own

- These questions will take some time. Students should work together to set up an appropriate expression that uses unit analysis. Use of calculators would be appropriate.

Example 3

- This problem involves rates, so there are units associated with the numerator and denominator. For this reason, many students tend to have less difficulty with rate problems.
- The check is important. It focuses attention on the ability to write the relationship between miles and kilometers in two ways that are reciprocals of one another: $\dfrac{1.6 \text{ kilometers}}{1 \text{ mile}}$ or $\dfrac{1 \text{ mile}}{1.6 \text{ kilometers}}$.

On Your Own

- **Neighbor Check:** Have students work independently and then have their neighbor check their work. Have students discuss any discrepancies.

Closure

- Explain how you would convert 40 miles to kilometers. Start by writing 40 miles. Then multiply by the conversion factor so that 1.6 kilometers is in the numerator and 1 mile is in the denominator. Now, multiply to get 64 kilometers.

Technology
For the **T**eacher

Dynamic Classroom

The Dynamic Planning Tool
Editable Teacher's Resources at *BigIdeasMath.com*

EXAMPLE 2 **Comparing Units**

Copy and complete the statement using < or >: 25 oz ▢ 2 kg.

Convert 25 ounces to kilograms.

| 1 lb = 16 oz | | 1 lb ≈ 0.45 kg |

$$25 \text{ oz} \times \frac{1 \text{ lb}}{16 \text{ oz}} \times \frac{0.45 \text{ kg}}{1 \text{ lb}} = \frac{25 \cdot 1 \cdot 0.45 \text{ kg}}{16 \cdot 1} \approx 0.70 \text{ kg}$$

∴ Because 0.70 kilogram is less than 2 kilograms, 25 oz < 2 kg.

On Your Own

Now You're Ready
Exercises 25–30

Copy and complete the statement using < or >.

5. 7 cm ▢ 3 in. **6.** 8 c ▢ 2 L **7.** 3 oz ▢ 70 g

EXAMPLE 3 **Converting a Rate**

Which of the two remote controlled planes is faster?

Convert 50 miles per hour to kilometers per hour.

$$\frac{50 \text{ mi}}{1 \text{ h}} \times \frac{1.6 \text{ km}}{1 \text{ mi}} = \frac{80 \text{ km}}{1 \text{ h}}$$

Biplane
70 kilometers per hour

Monoplane
50 miles per hour

The speed of the monoplane is 80 kilometers per hour. The speed of the biplane is 70 kilometers per hour.

∴ So, the monoplane is faster.

Check Convert 70 kilometers per hour to miles per hour.

$$\frac{70 \text{ km}}{1 \text{ h}} \times \frac{1 \text{ mi}}{1.6 \text{ km}} = \frac{44 \text{ mi}}{1 \text{ h}}$$

Monoplane   *Biplane*

$$\frac{50 \text{ mi}}{1 \text{ h}} > \frac{44 \text{ mi}}{1 \text{ h}} \checkmark$$

On Your Own

Now You're Ready
Exercises 31–34

8. The speed of a remote controlled car is 0.2 kilometer per minute. Order the speeds of the car and the two planes in Example 3 from least to greatest.

 Vocabulary and Concept Check

1. **WRITING** Describe two methods you can use to convert measurements.

2. **OPEN-ENDED** Which method would you use to convert 10 miles to kilometers? Explain your reasoning.

3. **DIFFERENT WORDS, SAME QUESTION** Which is different? Find "both" answers.

> Convert 5 inches to centimeters.

> Find the number of inches in 5 centimeters.

> How many centimeters are in 5 inches?

> Five inches equals how many centimeters?

 **Practice and Problem Solving**

Copy and complete the statement using < or >.

4. 1 ft ▢ 1 cm

5. 450 yd ▢ 450 cm

6. 30 in. ▢ 30 mm

7. 125 in. ▢ 125 cm

8. 100 ft/h ▢ 100 km/h

9. 10 L ▢ 10 gal

Copy and complete the statement using a ratio. Round to the nearest hundredth, if necessary.

① 10. 3 mi ≈ ▢ km

11. 10 qt ≈ ▢ L

12. 68 kg ≈ ▢ lb

13. 8.3 in. ≈ ▢ cm

14. 25.5 lb ≈ ▢ kg

15. 5 km ≈ ▢ mi

16. **ERROR ANALYSIS** Describe and correct the error in using a ratio to convert 12 kilometers to miles.

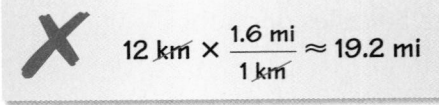

$$12 \text{ km} \times \frac{1.6 \text{ mi}}{1 \text{ km}} \approx 19.2 \text{ mi}$$

Copy and complete the statement using a proportion. Round to the nearest hundredth, if necessary.

17. 48 in. ≈ ▢ cm

18. 2 km ≈ ▢ mi

19. 165 cm ≈ ▢ in.

20. 85 lb ≈ ▢ kg

21. 2.5 qt ≈ ▢ L

22. 14.2 L ≈ ▢ qt

23. **CAVES** Mammoth Cave is the longest cave system in the world. So far, 365 miles of the cave have been explored. What is this distance in kilometers?

24. **IGUANA** How long is the iguana in inches?

Length: 24.7 cm

Assignment Guide and Homework Check

Level	Day 1 Activity Assignment	Day 2 Lesson Assignment	Homework Check
Basic	4–9, 41–43	1–3, 11–33 odd, 16, 24	11, 17, 24, 25, 31
Average	4–9, 41–43	1–3, 11–21 odd, 16, 29–33, 35–37	11, 17, 30, 32, 36
Advanced	4–9, 41–43	1–3, 14–18 even, 26–36 even, 37, 39, 40	16, 26, 32, 37

Common Errors

- **Exercises 4–9, 25–30** Students may try to compare the measurements without converting the units. Remind them that they cannot compare quantities unless the units are the same, so they must choose one unit to convert.
- **Exercises 10–15** When using a ratio, students may put the unit that they are solving for in the denominator. This is an incorrect conversion. Demonstrate to students visually that the original units will not cancel.

  For example, $3 \text{ mi} \times \dfrac{1 \text{ mi}}{1.6 \text{ km}} = \dfrac{1.875 \text{ mi}^2}{\text{km}}$.

- **Exercises 31–34** Students may forget to carry the unchanged unit into the final answer. Remind them that rates have two units, and both must be in the answer.

3.6 Record and Practice Journal

Complete the statement using a ratio. Round to the nearest hundredth, if necessary.

1. 10 mi ≈ __16__ km   2. 15 kg ≈ 33.33 lb   3. 6 qt ≈ __5.7__ L

Complete the statement using < or >.

4. 11 in. __>__ 22 cm   5. 12 kg __<__ 500 oz

6. 8 gal __>__ 25 L   7. 10 m __>__ 30 ft

Complete the statement. Round to the nearest hundredth, if necessary.

8. 60 mi/h ≈ __96__ km/h   9. 6 ft/sec ≈ 182.88 cm/sec

10. 52 gal/min ≈ 197.6 L/min   11. 5 kg/day ≈ 177.78 oz/day

12. One lap around a high school track is 400 meters. How many laps do you run around the track if you run 2 miles?
about 8 laps

13. A doctor prescribes 200 milligrams of medicine for a patient. How many ounces of medicine is the patient taking?
about 0.007 ounce

14. The lightest weight class for young men competing in freestyle wrestling is from 29 kilograms to 32 kilograms. What is the range of the weight class in pounds?
about 64.4 pounds to about 71.1 pounds

Vocabulary and Concept Check

1. To convert between measurements, multiply by the ratio of the given relationship such that the desired unit is in the numerator, or set up and solve a proportion using the given relationship as one of the ratios.

2. *Sample answer:* ratio; There are fewer steps needed.

3. Find the number of inches in 5 cm; 5 cm ≈ 1.97 in.; 5 in. ≈ 12.7 cm

Practice and Problem Solving

4. >   5. >

6. >   7. >

8. <   9. <

10. 4.8   11. 9.5

12. 151.11   13. 21.08

14. 11.48   15. 3.13

16. There are 1.6 kilometers in a mile, not the other way around.
$$\dfrac{12 \ \cancel{\text{km}}}{1} \times \dfrac{1 \text{ mi}}{1.6 \ \cancel{\text{km}}} = 7.5 \text{ mi}$$

17. 121.92   18. 1.25

19. 64.96   20. 38.25

21. 2.38   22. 14.95

23. about 584 km

24. about 9.72 in.

25. >   26. <

27. >   28. >

29. <   30. >

31. 72   32. 19

33. 4.72   34. 2

35. about 77,400 kg

36. no; 2 L > 2 qt

37. See *Taking Math Deeper.*

38. about 45.74 m wide and about 73.18 m long

39. about 3.7 gal

40. about 675,000,000 mi/h

Fair Game Review

41.

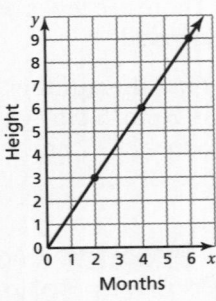

slope $= \dfrac{3}{2}$

42.

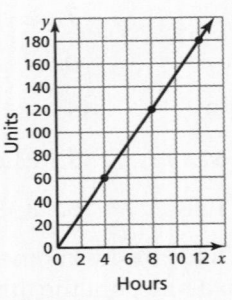

slope $= 15$

43. C

Mini-Assessment

Complete the statement.

1. 5 mi ≈ [ ] km 8

2. 16 qt ≈ [ ] L 15.2

3. 3 in. ≈ [ ] cm 7.62

4. 5 lb ≈ [ ] kg 2.25

5. A new born baby weighs 8 pounds. How much does the baby weigh in kilograms? about 3.6 kg

Taking Math Deeper

Exercise 37

This is a practical problem that students need as a real-life skill—changing between miles per hour and kilometers per hour. Most countries in the world, including our neighbors Canada and Mexico, measure driving distance in kilometers.

 Help me see the conversion factors.

$$\frac{5}{8} \text{ mi} \approx 1 \text{ km} \qquad 1 \text{ mi} \approx \frac{8}{5} \text{ km}$$

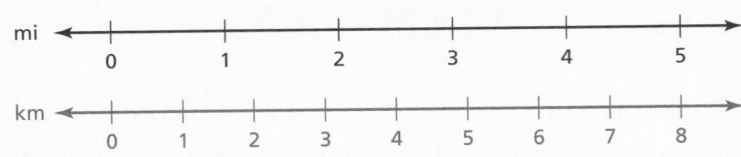

A kilometer is shorter than a mile.

 Convert 110 miles per hour to kilometers per hour.

$$\frac{110 \text{ mi}}{1 \text{ h}} \cdot \frac{8 \text{ km}}{5 \text{ mi}} = \frac{176 \text{ km}}{1 \text{ h}}$$

a. 176 kilometers per hour is above the speed limit of 130 kilometers per hour.

③ Convert 80 miles per hour to kilometers per hour.

$$\frac{80 \text{ mi}}{1 \text{ h}} \cdot \frac{8 \text{ km}}{5 \text{ mi}} = \frac{128 \text{ km}}{1 \text{ h}}$$

b. 128 kilometers per hour is just below the speed limit of 130 kilometers per hour.

Project

Use a map to plan a family trip to a place you would like to visit outside of your state. If the car travels at an average speed of 90 kilometers per hour, how long will your round trip take?

Reteaching and Enrichment Strategies

If students need help. . .	If students got it. . .
Resources by Chapter • Practice A and Practice B • Puzzle Time Record and Practice Journal Practice Differentiating the Lesson Lesson Tutorials Skills Review Handbook	Resources by Chapter • Enrichment and Extension • School-to-Work • Financial Literacy • Technology Connection Start the next section

Copy and complete the statement using < or >.

② **25.** 8 kg ▢ 30 oz

26. 6 ft ▢ 300 cm

27. 3 gal ▢ 6 L

28. 10 in. ▢ 200 mm

29. 1200 g ▢ 5 lb

30. 1500 m ▢ 3000 ft

Copy and complete the statement. Round to the nearest hundredth, if necessary.

③ **31.** 45 mi/h ≈ ▢ km/h

32. 5 gal/min ≈ ▢ L/min

33. 120 mm/sec ≈ ▢ in./sec

34. 900 g/day ≈ ▢ lb/day

35. BRACHIOSAURUS One of the largest dinosaurs was the brachiosaurus. How much did it weigh in kilograms?

36. BOTTLE Can you pour the water from a full 2-liter bottle into a 2-quart pitcher without spilling any? Explain.

37. AUTOBAHN Germany suggests a speed limit of 130 kilometers per hour on highways.

Weight: 172,000 lb

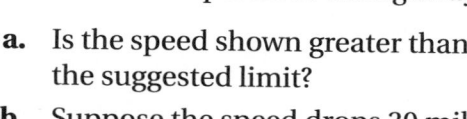

    a. Is the speed shown greater than the suggested limit?

    b. Suppose the speed drops 30 miles per hour. Is the new speed below the suggested limit?

38. SOCCER The size of a soccer field is 50 yards wide by 80 yards long. What is the size in meters?

39. PAINT One liter of paint covers 100 square feet. How many gallons does it take to cover 1400 square feet?

40. *Critical Thinking* The speed of light is about 300,000 kilometers per second. Convert the speed to miles per hour.

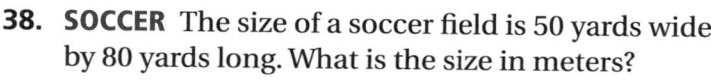

Fair Game Review What you learned in previous grades & lessons

Graph the data. Then find the slope of the line through the points. *(Section 3.2)*

41.

Months, x	Height, y
2	3
4	6
6	9

42.

Hours, x	Units, y
4	60
8	120
12	180

43. MULTIPLE CHOICE Which equation has a solution of 4? *(Section 2.6)*

    Ⓐ $2x + 7 = -1$ Ⓑ $-3 + 2x = -11$ Ⓒ $2x - 11 = -3$ Ⓓ $11 + 2x = 3$

Essential Question How can you use a graph to show the relationship between two variables that vary directly? How can you use an equation?

1 ACTIVITY: Math in Literature

Gulliver's Travels was written by Jonathan Swift and published in 1725. Gulliver was shipwrecked on the island Lilliput, where the people were only 6 inches tall. When the Lilliputians decided to make a shirt for Gulliver, a Lilliputian tailor stated that he could determine Gulliver's measurements by simply measuring the distance around Gulliver's thumb. He said "Twice around the thumb equals once around the wrist. Twice around the wrist is once around the neck. Twice around the neck is once around the waist."

Work with a partner. Use the tailor's statement to complete the table.

Thumb, *t*	Wrist, *w*	Neck, *n*	Waist, *x*
0 in.	0 in.		
1 in.	2 in.		
2 in.	4 in.		
3 in.	6 in.		
4 in.	8 in.		
5 in.	10 in.		

Laurie's Notes

Introduction

For the Teacher

- **Goal:** Students investigate what it means for two quantities to vary directly.

Motivate

- Write the following table.

	Paper	Pencil	$1 Bill	Stick of Gum
Length (inches)	11	8	6	?
Length (centimeters)	27.94	20.32	?	6

- ❓ "Does anyone have an idea of how to find the missing values in the table without measuring?" number of inches × 2.54 = number of centimeters
- This question is checking to see if students remember the relationship between inches and centimeters. It is also a perfect example for direct variation.

Discuss

- **Big Idea:** Direct variation means that the ratio of one quantity to another is a *constant*.
- The ordered pair (0, 0) is always a solution of a direct variation equation.
- When solutions of a direct variation equation are plotted, the *constant ratio* is the slope of the line.
- When two variables vary directly, we also say they are directly proportional.

Activity Notes

Activity 1

- Ask a student to read the introduction.
- ❓ "Have any of you read *Gulliver's Travels*?" Give students an opportunity to share information about the story if they have read it.
- Students should work with a partner to complete the table.
- ❓ "Are there any patterns in the table? Describe the patterns." There are many patterns! Listen for patterns in a single column, between adjacent columns, and between non-adjacent columns. Students might also mention the only odd numbers in the table appear in the first column.

Previous Learning

Students have used a variable to write an expression and have plotted points in the coordinate plane. Students have represented functions as equations, writing an equation in two variables.

Activity Materials
Textbook
• string

Start Thinking! and Warm Up

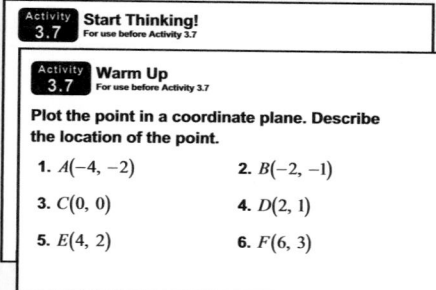

Activity 3.7 **Start Thinking!** For use before Activity 3.7

Activity 3.7 **Warm Up** For use before Activity 3.7

Plot the point in a coordinate plane. Describe the location of the point.

1. $A(-4, -2)$ 2. $B(-2, -1)$

3. $C(0, 0)$ 4. $D(2, 1)$

5. $E(4, 2)$ 6. $F(6, 3)$

3.7 Record and Practice Journal

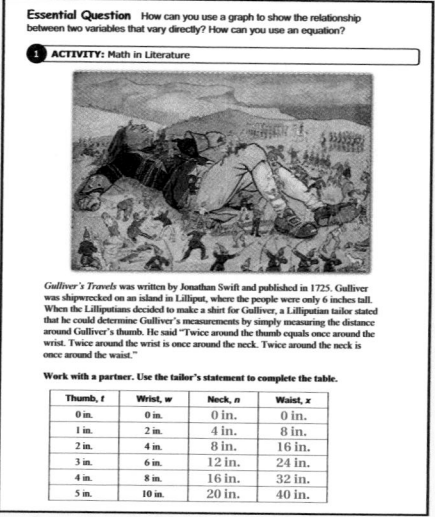

Essential Question How can you use a graph to show the relationship between two variables that vary directly? How can you use an equation?

1 ACTIVITY: Math in Literature

Gulliver's Travels was written by Jonathan Swift and published in 1725. Gulliver was shipwrecked on an island in Lilliput, where the people were only 6 inches tall. When the Lilliputians decided to make a shirt for Gulliver, a Lilliputian tailor stated that he could determine Gulliver's measurements by simply measuring the distance around Gulliver's thumb. He said "Twice around the thumb equals once around the wrist. Twice around the wrist is once around the neck. Twice around the neck is once around the waist."

Work with a partner. Use the tailor's statement to complete the table.

Thumb, t	Wrist, w	Neck, n	Waist, x
0 in.	0 in.	0 in.	0 in.
1 in.	2 in.	4 in.	8 in.
2 in.	4 in.	8 in.	16 in.
3 in.	6 in.	12 in.	24 in.
4 in.	8 in.	16 in.	32 in.
5 in.	10 in.	20 in.	40 in.

Differentiated Instruction

Auditory

A common mistake when plotting points is to confuse the order of the coordinates in the ordered pair. The simple phrase "over and up" will assist the students in moving over the x-axis with the first number, and then moving up with the second number.

3.7 Record and Practice Journal

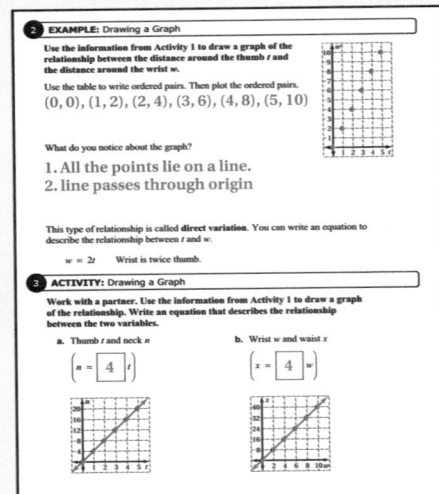

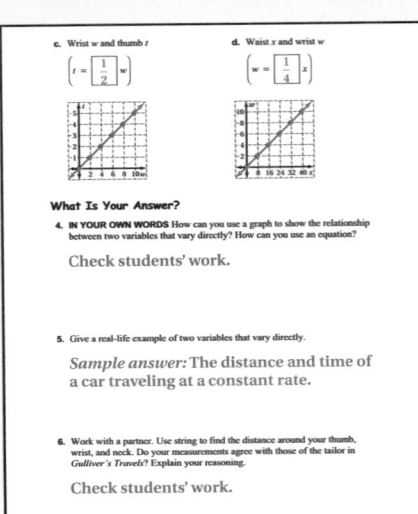

Laurie's Notes

Example 2

- Introduce the next activity.
- ❓ **"How do you plot ordered pairs?"** The first coordinate tells the distance and direction you move horizontally. The second coordinate is the distance and direction you move vertically.
- If the ordered pairs have a context, it is helpful to write the words (thumb length, wrist length, etc.). *Note:* Although the word "length" is used, it is referring to the *distance* around the wrist and the *distance* around the thumb.
- Discuss student observations about the graph. The points lie on a line and the line passes through the origin (0, 0). Students might make additional observations.
- ❓ **Connection: "What is the slope of the line?"** slope = 2; For every 1 inch that the distance around the thumb increases, the distance around the wrist increases by 2 inches.
- Note how the equation ($w = 2t$) can be translated into words (the distance around the wrist is twice the distance around the thumb).

Activity 3

- **Scaffolding:** If time is short, you may need to scaffold this activity. Have different groups do a different problem and record a sample of each on the board.
- **Common Error:** Students may reverse the ordered pairs (wrist length, waist length) versus (waist length, wrist length).
- Remind students that as the problems are written, the first variable mentioned is the first coordinate.
- They should do a test point to see if it satisfies the equation they wrote.
- **Connection:** There are two pairs of equations that share a pattern:
  $w = 2t$ and $t = \frac{1}{2}w$; $x = 4w$ and $w = \frac{1}{4}x$. The slopes are reciprocals.

What Is Your Answer?

- Question 5: Students are sometimes stumped by this question, yet it is very common. For instance, the cost of an item varies directly with how many items are purchased (1 newspaper, $1.25), (2 newspapers, $2.50), and so on.
- Question 6: Leave sufficient time for this problem.

Closure

- **Exit Ticket:** Write an equation that describes the relationship between the length of objects measured in inches x and measured in centimeters y.
  $y = 2.54x$

Technology For the Teacher

The Dynamic Planning Tool
Editable Teacher's Resources at *BigIdeasMath.com*

2 EXAMPLE: Drawing a Graph

Use the information from Activity 1 to draw a graph of the relationship between the distance around the thumb t and the distance around the wrist w.

Use the table to write ordered pairs. Then plot the ordered pairs.

$(0, 0)$, $(1, 2)$, $(2, 4)$, $(3, 6)$, $(4, 8)$, $(5, 10)$

Notice the following about the graph:

1. All the points lie on a line.

2. The line passes through the origin.

This type of relationship is called **direct variation**. You can write an equation to describe the relationship between t and w.

$$w = 2t \qquad \text{Wrist is twice thumb.}$$

3 ACTIVITY: Drawing a Graph

Work with a partner. Use the information from Activity 1 to draw a graph of the relationship. Write an equation that describes the relationship between the two variables.

a. Thumb t and neck n ($n = \boxed{} t$)

b. Wrist w and waist x ($x = \boxed{} w$)

c. Wrist w and thumb t ($t = \boxed{} w$)

d. Waist x and wrist w ($w = \boxed{} x$)

What Is Your Answer?

4. **IN YOUR OWN WORDS** How can you use a graph to show the relationship between two variables that vary directly? How can you use an equation?

5. Give a real-life example of two variables that vary directly.

6. Work with a partner. Use string to find the distance around your thumb, wrist, and neck. Do your measurements agree with those of the tailor in *Gulliver's Travels*? Explain your reasoning.

Practice → Use what you learned about direct variation to complete Exercises 4–7 on page 140.

Check It Out
Lesson Tutorials
BigIdeasMath.com

Key Vocabulary ◀))
direct variation,
  p. 138

 **Key Idea**

Direct Variation

Words Two quantities x and y show **direct variation** when $y = kx$, where k is a number and $k \neq 0$.

Graph The graph of $y = kx$ is a line that passes through the origin.

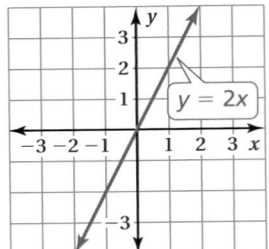

EXAMPLE **1** **Identifying Direct Variation**

Tell whether x and y show direct variation. Explain your reasoning.

a.

x	1	2	3	4
y	−2	0	2	4

Plot the points. Draw a line through the points.

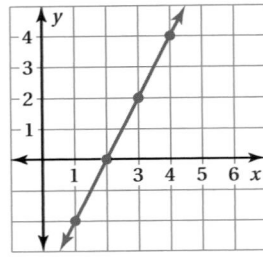

b.

x	0	2	4	6
y	0	2	4	6

Plot the points. Draw a line through the points.

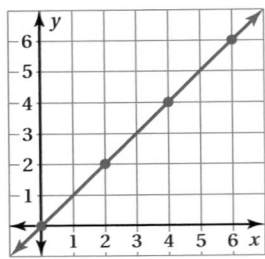

Study Tip

Other ways to say that x and y show direct variation are "y varies directly with x" and "x and y are directly proportional."

∴ The line does not pass through the origin. So, x and y do *not* show direct variation.

∴ The line passes through the origin. So, x and y show direct variation.

EXAMPLE **2** **Identifying Direct Variation**

Tell whether x and y show direct variation. Explain your reasoning.

a. $y + 1 = 2x$

  $y = 2x - 1$ Solve for y.

∴ The equation *cannot* be written as $y = kx$. So, x and y do *not* show direct variation.

b. $\frac{1}{2}y = x$

  $y = 2x$ Solve for y.

∴ The equation can be written as $y = kx$. So, x and y show direct variation.

◀)) Multi-Language Glossary at BigIdeasMath✓com.

Laurie's Notes

Introduction

Connect
- **Yesterday:** Students completed a table of values, plotted ordered pairs, and wrote equations as they developed an understanding of direct variation.
- **Today:** Students will use a formal definition of direct variation.

Motivate
- Some states have returnable bottle laws.
- **?** "If you receive $0.05 for each bottle, how much money do you receive for 4 bottles? 10 bottles?" $0.20; $0.50
- Have students make a table to show the relationship between the number x of bottles collected and the amount y of money received. Let $x = 0, 1, 2, 4, 6,$ and 10.
- Then have students make a quick sketch of the ordered pairs.
- Observe that (0, 0) is on the graph and the ordered pairs lie on a line.

Lesson Notes

Key Idea
- Write the Key Idea on the board.
- The equation $y = kx$ can be confusing to students. They see three variables. Remind them that this is the *general form*. The variables y and x will remain in the final equation but k will be replaced by a number.
- Examples of equations in general form include $y = 2x$ and $y = 1.25x$. Point out to students that k has been replaced with a number.
- **?** "Why should $k \neq 0$?" If k did equal 0, the resulting equation would be $y = 0 \cdot x$ or $y = 0$. While k has been replaced with a number and y still appears in the final equation, x no longer does. Because there is no x, the equation is not in general form and does not show direct variation.
- Mention a key feature of this graph: it passes through the origin.

Example 1
- Work through each example. This is good practice in plotting ordered pairs. Some students may still be reversing the coordinates.
- **Common Error:** Students may look at the table of values and believe that the first example is not direct variation simply because x and y are increasing by different rates in the table. x is increasing by 1 and y is increasing by 2.
- **Connection:** When two variables vary directly, we say they vary proportionally.

Example 2
- This example requires students to recall equations in two variables.
- Students need to think about how they solved equations using the Properties of Equality. The difference is now the equation contains two variables.

Goal Today's lesson is using a formal definition of **direct variation**.

Start Thinking! and Warm Up

> **Lesson 3.7** Warm Up
> For use before Lesson 3.7
>
> **Lesson 3.7** Start Thinking!
> For use before Lesson 3.7
>
> You receive $20 every time you mow your neighbor's lawn. How many times do you need to mow the lawn so that you can buy a digital camera that costs $189? Write an equation to represent the situation. Does the equation show direct variation? Why or why not?

Extra Example 1

Tell whether x and y show direct variation. Explain your reasoning.

a.

x	1	2	3	4
y	-1	0	1	2

The line does not pass through the origin. So, x and y do not show direct variation.

b.

x	0	3	6	9
y	0	1	2	3

The line passes through the origin. So, x and y show direct variation.

Extra Example 2

Tell whether x and y show direct variation. Explain your reasoning.

a. $y - 6 = 3x$

The equation cannot be written as $y = kx$. So, x and y do not show direct variation.

b. $x = 4y$

The equation can be written as $y = kx$. So, x and y show direct variation.

Laurie's Notes

On Your Own

1. no; The line does not pass through the origin.

2. yes; The line passes through the origin.

3. no; The points do not lie on a line.

4. no; The equation cannot be written as $y = kx$.

5. yes; The equation can be written as $y = kx$.

6. no; The equation cannot be written as $y = kx$.

Extra Example 3

The height y of a movie screen varies directly with its width x. An equation that models this relationship is $y = \dfrac{20}{47}x$.

a. Find the height of the movie screen when the width is 11.75 meters.
   5 meters

b. Sketch a graph of the equation.

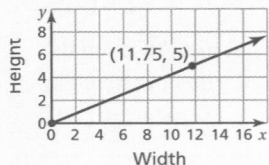

On Your Own

7. $37.50

English Language Learners

Vocabulary

Students may confuse the words *variation* and *variable*. A variable is a number that changes and is represented by a letter. Stress that variation refers to how the variable *y varies* in relation to the variable *x*.

On Your Own

- **Think-Pair-Share:** Students should read each question independently and then work with a partner to answer the questions. When they have answered the questions, the pair should compare their answers with another group and discuss any discrepancies.
- If students have difficulty with Questions 4–6, they could make a quick table of values and plot the ordered pairs.

Example 3

- Read the information from the illustration.
- Write the direct variation equation $y = \dfrac{9}{16}x$.
- ❓ "What meaning does $\dfrac{9}{16}$ have in the equation?" $\dfrac{9}{16}$ represents the ratio of the height to the width. Companies usually advertise television ratios as width to height.
- Substitute the width, 48 inches, into the equation to find the height.
- **FYI:** Standard television sets have a width to height ratio of $4:3$, whereas High Definition televisions display with a ratio of $16:9$ as in this example. Ask if students have heard of the term "aspect ratio" for televisions. These ratios are called aspect ratios.
- **Connection:** Note that the slope of the line $\left(\dfrac{9}{16}\right)$ is k.

On Your Own

- **Neighbor Check:** Have students work independently and then have their neighbor check their work. Have students discuss any discrepancies.

Closure

- Tell whether x and y show direct variation. Explain your reasoning.

x	y
1	0
3	2
5	4
7	6

no; The line passes through $(1, 0)$, so it does not pass through the origin.

Technology For the Teacher

Dynamic Classroom

The Dynamic Planning Tool
Editable Teacher's Resources at *BigIdeasMath.com*

Tell whether *x* and *y* show direct variation. Explain your reasoning.

1.

x	y
0	−2
1	1
2	4
3	7

2.

x	y
1	4
2	8
3	12
4	16

3.

x	y
−2	4
−1	2
0	0
1	2

4. $xy = 3$

5. $x = \frac{1}{3}y$

6. $y + 1 = x$

EXAMPLE **3** **Using a Direct Variation Model**

The height *y* of a television screen varies directly with its width *x*.

$$y = \frac{9}{16}x$$

a. Find the height when the width is 48 inches.

b. Sketch the graph of the equation.

a. Use the equation to find the height when *x* = 48 inches.

$$y = \frac{9}{16}(48) \qquad \text{Substitute 48 for } x.$$

$$= 27 \qquad \text{Simplify.}$$

   So, when the width is 48 inches, the height is 27 inches.

b. To sketch a graph, plot the point (48, 27). Then draw the line that passes through this point and the origin.

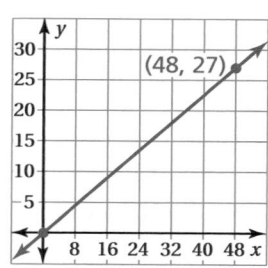

On Your Own

Now You're Ready
Exercises 24–29

7. Your earnings *y* (in dollars) vary directly with the number *x* of lawns you mow. Use the equation $y = 7.5x$ to find how much you earn when you mow 5 lawns.

Vocabulary and Concept Check

1. **VOCABULARY** What does it mean for x and y to vary directly?

2. **WRITING** What point is on the graph of every direct variation equation?

3. **WHICH ONE DOESN'T BELONG?** Which graph does *not* belong with the other three? Explain your reasoning.

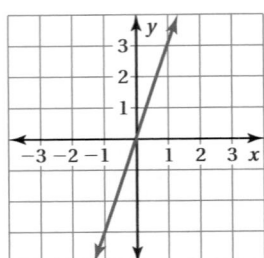 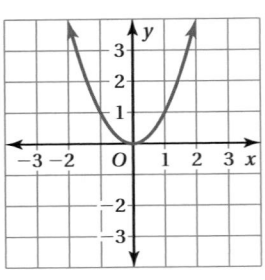 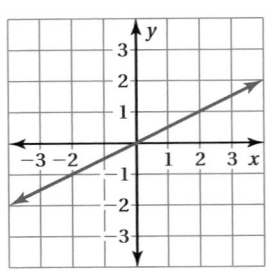 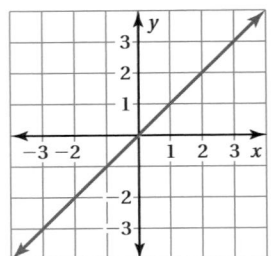

Practice and Problem Solving

Tell whether x and y show direct variation. Explain your reasoning.

4. $(-1, -1), (0, 0), (1, 1), (2, 2)$

5. $(-4, -2), (-2, 0), (0, 2), (2, 4)$

6. $(1, 2), (1, 4), (1, 6), (1, 8)$

7. $(2, 1), (6, 3), (10, 5)\ (14, 7)$

① 8.

x	1	2	3	4
y	2	4	6	8

9.

x	−2	−1	0	1
y	0	2	4	6

10.

x	−1	0	1	2
y	−2	−1	0	1

11.

x	4	8	12	16
y	1	2	3	4

12.

x	−1	0	1	2
y	1	0	1	2

13.

x	3	6	9	12
y	2	4	6	8

② 14. $y - x = 4$

15. $x = \dfrac{2}{5}y$

16. $y + 3 = x + 6$

17. $y - 5 = 2x$

18. $x - y = 0$

19. $\dfrac{x}{y} = 2$

20. $8 = xy$

21. $x^2 = y$

22. **ERROR ANALYSIS** Describe and correct the error in telling whether x and y show direct variation.

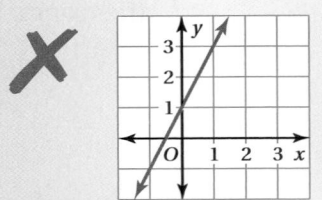

The graph is a line, so it shows direct variation.

23. **RECYCLING** The table shows the profit y for recycling x pounds of aluminum. Tell whether x and y show direct variation.

Aluminum, x	10	20	30	40
Profit, y	$4.50	$9.00	$13.50	$18.00

Assignment Guide and Homework Check

Common Errors

- **Exercises 8–13** Students may immediately state that the table does not show direct variation because (0, 0) is not listed. Encourage them to find the change in x and change in y. Then use that knowledge to go back to $x = 0$ and determine if the table satisfies both requirements for direct variation.
- **Exercises 14–21** Students may try to identify the direct variation equations without solving for y. Remind them that just like when they first began comparing ratios and rates to determine proportionality, they need to simplify the equation.
- **Exercise 21** Students may say that it shows direct variation because the graph goes through (0, 0). Ask them if the graph will be linear.
- **Exercises 24–29** Students may not grasp how to write the equation. Remind them of slope to determine the coefficient of x.

Technology For the Teacher
Answer Presentation Tool
QuizShow

3.7 Record and Practice Journal

Tell whether x and y show direct variation. Explain your reasoning.

1.
x	1	2	3	4
y	3	6	9	12

yes

2.
x	-1	0	1	2
y	1	3	7	13

no

3.
x	0	2	4	6
y	8	5	2	-1

no

4. $y + 2 = x$

no

5. $3y = x$

yes

6. $\frac{y}{x} = 4$

yes

The variables x and y vary directly. Use the values to write an equation that relates x and y.

7. $y = 8; x = 2$

$y = 4x$

8. $y = 14, x = 16$

$y = \frac{7}{8}x$

9. $y = 25, x = 35$

$y = \frac{5}{7}x$

10. The table shows the cups c of dog food needed to feed a dog that weighs p pounds. Tell whether p and c show direct variation.

Pounds, p	10	20	40	70
Food, c	$\frac{3}{4}$	$1\frac{1}{4}$	2	$2\frac{3}{4}$

No, p and c do not show direct variation.

11. Write a direct variation equation that relates x tires to y cars.

$y = \frac{1}{4}x$

12. Tell whether h and m show direct variation. If so, write an equation of direct variation.

Hours, h	1	2	4	5
Miles, m	60.5	121	242	302.5

Yes, $m = 60.5h$

19. yes; The equation can be written as $y = kx$.

20. no; The equation cannot be written as $y = kx$.

21. no; The equation cannot be written as $y = kx$.

22. The line does not pass through the origin, so x and y do not show direct variation.

23. yes

24. $y = 2x$ 25. $y = 5x$

26. $y = 4x$ 27. $y = 24x$

28. $y = \dfrac{5}{3}x$ 29. $y = \dfrac{9}{8}x$

30. $y = 2.54x$

31. See *Taking Math Deeper*.

32. See Additional Answers.

33. no   34. 76,000 mg

35. See Additional Answers.

36. $x = -9$

37. $y = -60$

38. $m = -32$

39. $d = -59\dfrac{1}{2}$

40. D

Mini-Assessment

Tell whether x and y show direct variation.

1. $y - 3 = 4x$ no

2. $\dfrac{1}{4}y = x$ yes

3. $x - y = 2$ no

4. $6y + 3 = 12x + 3$ yes

5. One mile is approximately equal to 1.6 kilometers. Write a direct variation equation that relates x miles to y kilometers. $y = 1.6x$

Taking Math Deeper

Exercise 31

The problem shows students how the coordinate plane can help draw a blueprint.

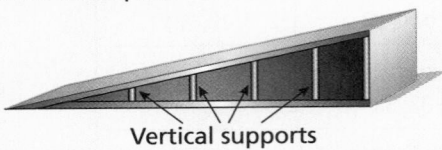

Vertical supports

To design the jet ski ramp, locate the beginning of the ramp at the origin. The horizontal distances are the x-values. The vertical distances are the y-values.

① Design the ramp.

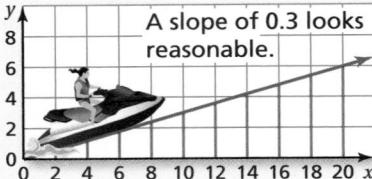

A slope of 0.3 looks reasonable.

The jet ski leaves this ramp at a height of 6 feet. If this seems too high, redesign the ramp with a slope of 0.2. Then, the final height will be 4 feet.

② Write a direct variation equation.

For the slope in the graph, $y = 0.3x$.

③ Plan 10 vertical support heights.

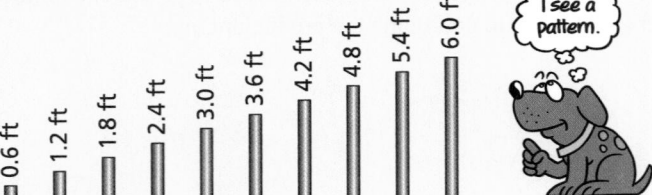

0.6 ft 1.2 ft 1.8 ft 2.4 ft 3.0 ft 3.6 ft 4.2 ft 4.8 ft 5.4 ft 6.0 ft

I see a pattern.

Project

Write a report comparing jumping ramps that are used in a variety of sports. Include your opinion as to why some ramps are higher and others are lower.

Reteaching and Enrichment Strategies

If students need help. . .	If students got it. . .
Resources by Chapter • Practice A and Practice B • Puzzle Time Record and Practice Journal Practice Differentiating the Lesson Lesson Tutorials Skills Review Handbook	Resources by Chapter • Enrichment and Extension • School-to-Work • Financial Literacy • Technology Connection • Life Connections Start the next section

The variables x and y vary directly. Use the values to write an equation that relates x and y.

③ **24.** $y = 4; x = 2$

25. $y = 25; x = 5$

26. $y = 60; x = 15$

27. $y = 72; x = 3$

28. $y = 20; x = 12$

29. $y = 45; x = 40$

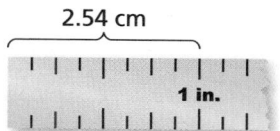
2.54 cm
1 in.

30. MEASUREMENT Write a direct variation equation that relates x inches to y centimeters.

31. JET SKI RAMP Design a jet ski ramp. Show how you can use direct variation to plan the heights of the vertical supports.

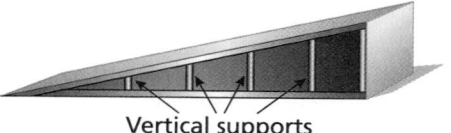
Vertical supports

32. JUPITER The weight of an object in our solar system varies directly with the weight of the object on Earth.

   a. Copy and complete the table.

   b. RESEARCH Why does weight vary throughout our solar system?

Location	Earth	Jupiter	Moon
Weight (lb)	100	214	
Weight (lb)	120		20

Minutes, x	500	700	900	1200
Cost, y	$40	$50	$60	$75

33. CELL PHONE PLANS Tell whether x and y show direct variation. If so, write an equation of direct variation.

34. CHLORINE The amount of chlorine in a swimming pool varies directly with the volume of water. The pool has 2.5 milligrams of chlorine per liter of water. How much chlorine is in the pool?

8000 gallons

35. *Critical Thinking* Is the graph of every direct variation equation a line? Does the graph of every line represent a direct variation equation? Explain your reasoning.

Fair Game Review *What you learned in previous grades & lessons*

Solve the equation. *(Section 2.5)*

36. $-4x = 36$

37. $\dfrac{y}{6} = -10$

38. $-\dfrac{3}{4}m = 24$

39. $-17 = \dfrac{2}{7}d$

40. MULTIPLE CHOICE Which rate is *not* equivalent to 180 feet per 8 seconds? *(Section 3.1)*

  Ⓐ $\dfrac{225 \text{ ft}}{10 \text{ sec}}$

  Ⓑ $\dfrac{45 \text{ ft}}{2 \text{ sec}}$

  Ⓒ $\dfrac{135 \text{ ft}}{6 \text{ sec}}$

  Ⓓ $\dfrac{180 \text{ ft}}{1 \text{ sec}}$

EXAMPLE ❶ **Interpreting a Proportional Relationship**

The distance traveled by a high speed train is proportional to the number of hours traveled. Interpret each plotted point in the graph.

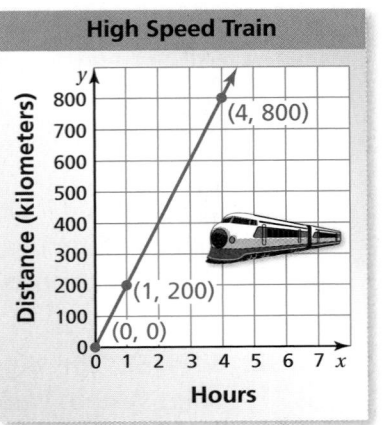

High Speed Train

(0, 0): The train travels 0 kilometers in 0 hours.

Study Tip

In the graph of a proportional relationship, you can find the unit rate from the point (1, y).

(1, 200): The train travels 200 kilometers in 1 hour.

This point represents the unit rate, $\dfrac{200 \text{ km}}{1 \text{ h}}$, or 200 kilometers per hour.

(4, 800): The train travels 800 kilometers in 4 hours.

Because the relationship is proportional, you can also use this point to find the unit rate.

$\dfrac{800 \text{ km}}{4 \text{ h}} = \dfrac{200 \text{ km}}{1 \text{ h}}$, or 200 kilometers per hour

Practice

Interpret each plotted point in the graph of the proportional relationship.

1.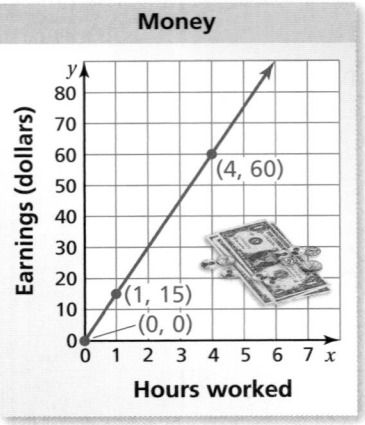

Money

2.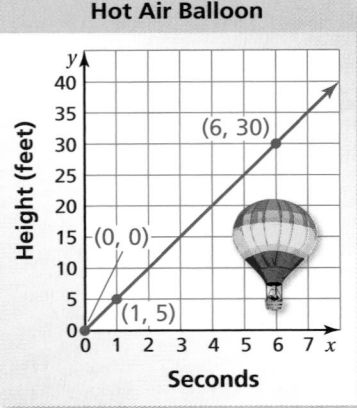

Hot Air Balloon

3. **REASONING** The graph of a proportional relationship passes through (3, 4.5) and (1, y). Find y.

Laurie's Notes

Introduction

Connect
- **Yesterday:** Students studied proportions and direct variation.
- **Today:** Students will extend their work with direct variation by interpreting and comparing graphs of proportional relationships.

Motivate
- Share a story with students about purchasing a new showerhead that has a flow rate of 2.5 gallons per minute.
- **?** "How much water is used during a 6-minute shower? Explain." 15 gallons; One minute uses 2.5 gallons, so six minutes uses six times as much; $6 \times 2.5 = 15$
- **?** "How long was a shower that used 22.5 gallons of water? Explain." 9 minutes; Divide 22.5 by 2.5 to find the number of minutes.
- Write proportions for the problems: $\dfrac{2.5}{1} = \dfrac{15}{6}$ and $\dfrac{2.5}{1} = \dfrac{22.5}{9}$.
- **?** "How do you know these are proportions?" The cross products are equal.
- Draw a table on the board for the shower problem and ask students to complete the table. Discuss the proportional relationship shown.

Minutes	0	1	2	3	4
Gallons	0	2.5	5	7.5	10

- Explain to students that today they will take another look at graphs of proportional relationships.

Lesson Notes

Example 1
- Discuss why the relationship is proportional (line through origin). Write the proportion $\dfrac{800 \text{ km}}{4 \text{ h}} = \dfrac{200 \text{ km}}{1 \text{ h}}$.
- **?** "What is a unit rate?" a rate with a denominator of 1
- Point out that the unit rate for this problem can be found by solving a proportion: $\dfrac{800 \text{ km}}{4 \text{ h}} = \dfrac{x \text{ km}}{1 \text{ h}}$, or by using the graph: starting at (0, 0), travel right one unit and up 200 units.
- **Connection:** The unit rate is the slope of the line, 200 km/h. The slope between (0, 0) and (4, 800) is also 200 km/h.

Practice
- Students should be thinking about the labels on the axes as they interpret the ordered pairs.
- **Connection:** For Exercise 3, students can draw a line through (3, 4.5) and (0, 0), and then think about where the ordered pair (1, y) would be if it were on the same line.

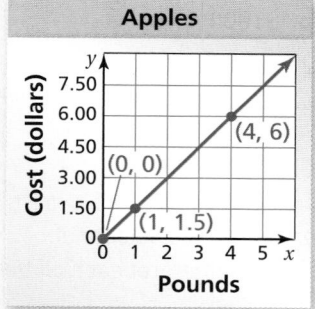

T-141A

Extra Example 2

The state sales taxes in New Jersey and Colorado are proportional to the price of a purchase, as shown in the graph.

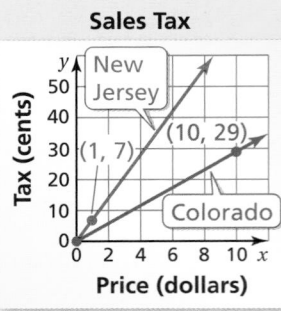

Sales Tax

a. Express the tax rate for each state as a percent.
New Jersey: 7%; Colorado: 2.9%

b. What is the sales tax on a $30 purchase in Colorado? $0.87

Practice

4. **a.** Salesman A: 10%;
Salesman B: 3.75%

   b. $1000

   c. $625

Mini-Assessment

1. The distance traveled by a bicyclist is proportional to the number of hours traveled. Interpret each plotted point in the graph.

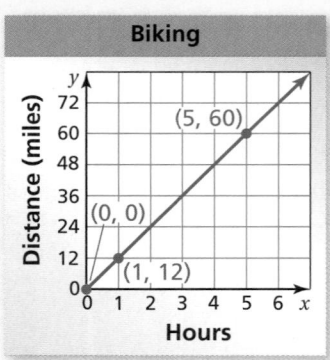

Biking

(0, 0): The bicyclist travels 0 miles in 0 hours.

(1, 12): The bicyclist travels 12 miles in 1 hour; unit rate: $\dfrac{12 \text{ mi}}{1 \text{ h}}$

(5, 60): The bicyclist travels 60 miles in 5 hours; unit rate: $\dfrac{60 \text{ mi}}{5 \text{ h}} = \dfrac{12 \text{ mi}}{1 \text{ h}}$

Laurie's Notes

Example 2

- Discuss why the relationship is proportional.
- ❓ "What is the meaning of the ordered pair (1, 5)?" There is a tax of 5¢ for every dollar of the purchase price in Maine.
- **Language:** It is common to hear a sales tax of 5% referred to as "5¢ on the dollar."
- Remind students that a percent is the number of parts per 100. This example reviews the percent skills learned in Grade 6.
- Work through both parts of the example as shown.
- ❓ "How can you compare the tax rates in Maine and New York from looking at the graph?" The slope of the line representing Maine is steeper so the sales tax rate is greater.
- ❓ Connect this lesson to the previous lesson by asking "What is the direct variation equation for each line?" Maine: $y = 0.05x$; New York: $y = 0.04x$
- Note that both graphs pass through (0, 0). There is no sales tax when you do not purchase anything.

Practice

- Define a commission for students unfamiliar with the term.
- Note that again both graphs pass through (0, 0). There is no commission when there is no sale.
- **Common Error:** Students may forget to divide $150 by $4000 to find the commission rate of Salesman B.

Closure

- Have students graph the line through (0, 0) and (4, 12). If these points represent the gallons y of water used for a shower of x minutes, what is the flow rate of the showerhead? 3 gallons per minute

Technology For the Teacher

Dynamic Classroom

The Dynamic Planning Tool
Editable Teacher's Resources at *BigIdeasMath.com*

EXAMPLE **2** **Comparing Graphs of Proportional Relationships**

Sales Tax

The state sales taxes in Maine and New York are proportional to the price of a purchase, as shown in the graph.

a. Express the sales tax rate for each state as a percent.

> *Maine:* (1, 5) indicates that the tax is 5 cents per dollar.
>
> $$\frac{5 \text{ cents}}{1 \text{ dollar}} = \frac{5 \text{ cents}}{100 \text{ cents}} = 5\%$$
>
> *New York:* (5, 20) indicates that the tax is 20 cents per 5 dollars.
>
> $$\frac{20 \text{ cents}}{5 \text{ dollars}} = \frac{4 \text{ cents}}{1 \text{ dollar}} = \frac{4 \text{ cents}}{100 \text{ cents}} = 4\%$$

⋮∙ Maine has a 5% sales tax and New York has a 4% sales tax.

b. What is the sales tax on a $12 purchase in New York?

Method 1: Write and solve a proportion to find the sales tax.

$$\frac{20}{5} = \frac{y}{12}$$ ← cents ← dollars   Use (5, 20) to write a proportion.

$$48 = y$$ Multiply each side by 12.

⋮∙ The sales tax on a $12 purchase in New York is 48 cents.

Method 2: Find 4% of $12.

$$4\% \text{ of } \$12 = 0.04 \cdot 12$$ Write 4% as a decimal.

$$= 0.48$$ Multiply.

⋮∙ The sales tax on a $12 purchase in New York is $0.48, or 48 cents.

Practice

4. **COMMISSION** The graph shows that the commissions of two salesmen are proportional to the amounts of sales.

   a. Express the commission rate for each salesman as a percent.

   b. What commission does Salesman A receive for a $10,000 sale?

   c. How much more commission does Salesman A receive than Salesman B for a $10,000 sale?

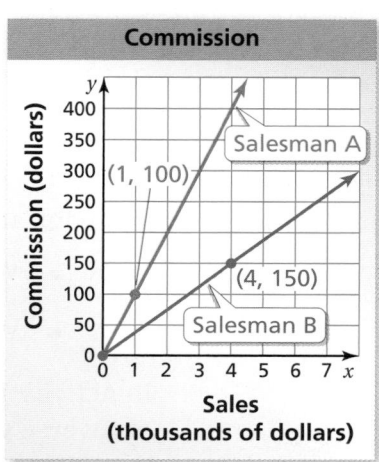

Essential Question How can you recognize when two variables are inversely proportional?

1 ACTIVITY: Comparing the Height and the Base

Work with a partner.

a. There are nine ways to arrange 36 square blocks to form a rectangle. Here are two ways. Find the other seven ways.

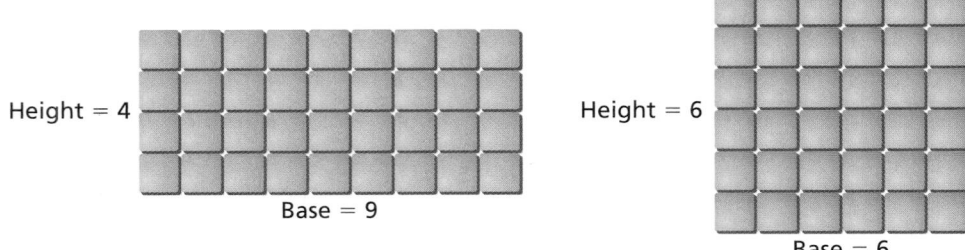

Height = 4   Base = 9

Height = 6   Base = 6

b. Order the nine ways according to height. Record your results in a table.

Height, h	Base, b	Area, A
4	9	$A = 9 \cdot 4 = 36$
6	6	$A = 6 \cdot 6 = 36$

c. Look at the first and second columns. Complete each sentence.

- When the height increases, the base   .

- When the height decreases, the base   .

In Activity 1, the relationship between the height and the base is an example of **inverse variation**. You can describe the relationship with an equation.

$$h = \frac{36}{b}$$ h and b are inversely proportional.

Laurie's Notes

Introduction

For the Teacher

- **Goal:** Students investigate what it means for two quantities to vary inversely, or to be inversely proportional.

Motivate

- Demonstrate how one variable increases while the other decreases.
- Pour the water from the pitcher into 4 cups.
- ❓ "How full is each cup?" Each cup is full.
- Pour the water back into the pitcher. Then pour the water into 6 cups.
- ❓ "How full is each cup?" Each cup is two-thirds full.
- Pour the water back into the pitcher. Then pour the water into 8 cups.
- ❓ "How full is each cup?" Each cup is one-half full.
- ❓ "As I continue to increase the number of cups, what happens to the amount in each cup?" The amount decreases proportionally.
- **Big Idea: Inverse variation** means that the product of one quantity and a second quantity is a *constant*. As one variable increases, the other decreases proportionally.

Activity Notes

Activity 1

- ❓ "How do you find the area of a rectangle?" length × width *or* base × height
- ❓ "Can the dimensions be the same?" Yes, it would be a square.
- Provide square tiles for each group.
- Check to see that the results are written in order according to height. If this is not done, students may not observe a pattern.
- **Common Misconception:** Students sometimes believe that the height of a rectangle must be the vertical dimension and the base is the horizontal dimension. These are arbitrary terms.
- ❓ "What patterns do you observe in the table?" As the height increases, the base decreases and vice versa. The product is always 36. Students might also mention that at least one of the dimensions is always even.
- ❓ **Extension:** "Could this rectangle have a dimension that is a fraction?" yes; $\frac{1}{2} \times 72 = 36$, but because whole blocks are being used, the dimensions modeled here are to be whole numbers.
- Discuss the idea of *inverse variation* in this activity.
- ❓ "How does inverse variation differ from direct variation?" Listen for direct variation as the variables having a constant ratio and inverse variation as the variables having a constant product.

Previous Learning

Students know how to find the area of a rectangle. Students have represented functions as equations, writing equations in two variables.

Activity Materials	
Introduction	**Textbook**
• a one quart pitcher of water • eight 8 fl oz plastic cups	• square tiles

Start Thinking! and Warm Up

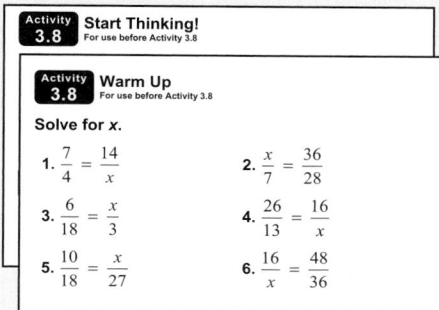

Activity 3.8 Start Thinking! For use before Activity 3.8

Activity 3.8 Warm Up For use before Activity 3.8

Solve for *x*.

1. $\dfrac{7}{4} = \dfrac{14}{x}$ 2. $\dfrac{x}{7} = \dfrac{36}{28}$

3. $\dfrac{6}{18} = \dfrac{x}{3}$ 4. $\dfrac{26}{13} = \dfrac{16}{x}$

5. $\dfrac{10}{18} = \dfrac{x}{27}$ 6. $\dfrac{16}{x} = \dfrac{48}{36}$

3.8 Record and Practice Journal

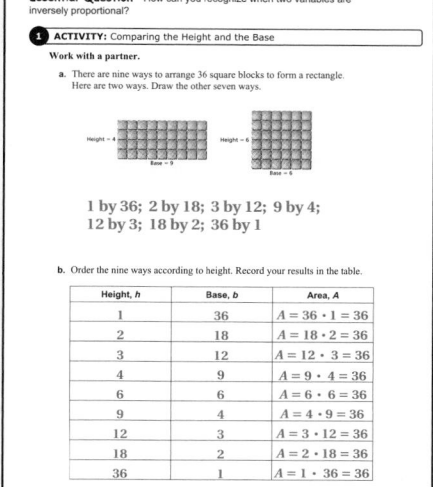

Essential Question How can you recognize when two variables are inversely proportional?

1 ACTIVITY: Comparing the Height and the Base

Work with a partner.

a. There are nine ways to arrange 36 square blocks to form a rectangle. Here are two ways. Draw the other seven ways.

1 by 36; 2 by 18; 3 by 12; 9 by 4; 12 by 3; 18 by 2; 36 by 1

b. Order the nine ways according to height. Record your results in the table.

Height, h	Base, b	Area, A
1	36	$A = 36 \cdot 1 = 36$
2	18	$A = 18 \cdot 2 = 36$
3	12	$A = 12 \cdot 3 = 36$
4	9	$A = 9 \cdot 4 = 36$
6	6	$A = 6 \cdot 6 = 36$
9	4	$A = 4 \cdot 9 = 36$
12	3	$A = 3 \cdot 12 = 36$
18	2	$A = 2 \cdot 18 = 36$
36	1	$A = 1 \cdot 36 = 36$

Differentiated Instruction

Visual

On a long strip of paper, mark integers from −10 through 10, with zero in the middle. Hold the strip vertically so that −10 touches the ground. Discuss that when an object is moving from 0 to 10, the *velocity* is positive. Point out that numbers 1 through 10 on the strip are positive integers. Then discuss that an object moving from zero down to the ground has a negative velocity. Again refer to the number strip, pointing out that −1 through −10 are negative integers.

3.8 Record and Practice Journal

---
c. Look at the first and second columns of the table. Complete each sentence.
- When the height increases, the base **decreases** .
- When the height decreases, the base **increases** .

In Activity 1, the relationship between the height and base is an example of **inverse variation**. You can describe the relationship with an equation.

$h = \dfrac{36}{b}$ h and b are inversely proportional.

② ACTIVITY: Comparing Direct and Inverse Variation

Work with a partner. Discuss each description. Tell whether the two variables are examples of *direct variation* or *inverse variation*. Use a table to explain your reasoning. Write an equation that relates the variables.

a. You bring 200 cookies to a party. Let n represent the number of people at the party and c represent the number of cookies each person receives.

inverse variation; $c = \dfrac{200}{n}$

b. You work at a restaurant for 20 hours. Let r represent your hourly pay rate and p represent the total amount you earn.

direct variation; $p = 20r$

c. You are going on a 240-mile trip. Let t represent the number of hours driving and s represent the speed of the car.

inverse variation; $s = \dfrac{240}{t}$

What Is Your Answer?

3. **IN YOUR OWN WORDS** How can you recognize when two variables are inversely proportional? Explain how a table can help you recognize inverse variation.

When one variable increases the other variable decreases at a proportional rate.

4. **SCIENCE** The *wing beat frequency* of a bird is the number of times per second the bird flaps its wings.

Hummingbird   **Mallard Duck**

Canada Goose   **Albatross**

Which of the following seems true? Explain your reasoning.
- Wing length and wing beat frequency are directly proportional.
- Wing length and wing beat frequency are inversely proportional.
- Wing length and wing beat frequency are unrelated.

5. **SCIENCE** Think of an example in science where two variables are inversely proportional.

Check students' work.

Laurie's Notes

Activity 2

- Introduce the next activity.
- Have students work with a partner.
- You may wish to offer additional help. In part (a), each person receives the same number of cookies.
- If students are stuck, suggest that they make a table of values to begin with and then decide if it is direct or inverse variation.
- The first step is deciding what the ordered pairs will look like.
  a. (number of people at the party, number of cookies each receives)
  b. (hourly pay rate, total amount earned)
  c. (number of hours driving, speed of the car)
- Students should ask themselves, "As the first variable increases, what will happen to the other variable?"
- Share the results as a whole class.

What Is Your Answer?

- **FYI:** The average hummingbird flaps its wings 50 times per second!

Closure

- How would you describe the difference between direct variation and inverse variation?

Technology For the Teacher

Dynamic Classroom

The Dynamic Planning Tool
Editable Teacher's Resources at *BigIdeasMath.com*

2 ACTIVITY: Comparing Direct and Inverse Variation

Work with a partner. Discuss each description. Tell whether the two variables are examples of *direct variation* or *inverse variation*. Use a table to explain your reasoning. Write an equation that relates the variables.

a. You bring 200 cookies to a party. Let *n* represent the number of people at the party and *c* represent the number of cookies each person receives.

b. You work at a restaurant for 20 hours. Let *r* represent your hourly pay rate and *p* represent the total amount you earn.

c. You are going on a 240-mile trip. Let *t* represent the number of hours driving and *s* represent the speed of the car.

What Is Your Answer?

3. IN YOUR OWN WORDS How can you recognize when two variables are inversely proportional? Explain how a table can help you recognize inverse variation.

4. SCIENCE The *wing beat frequency* of a bird is the number of times per second the bird flaps its wings.

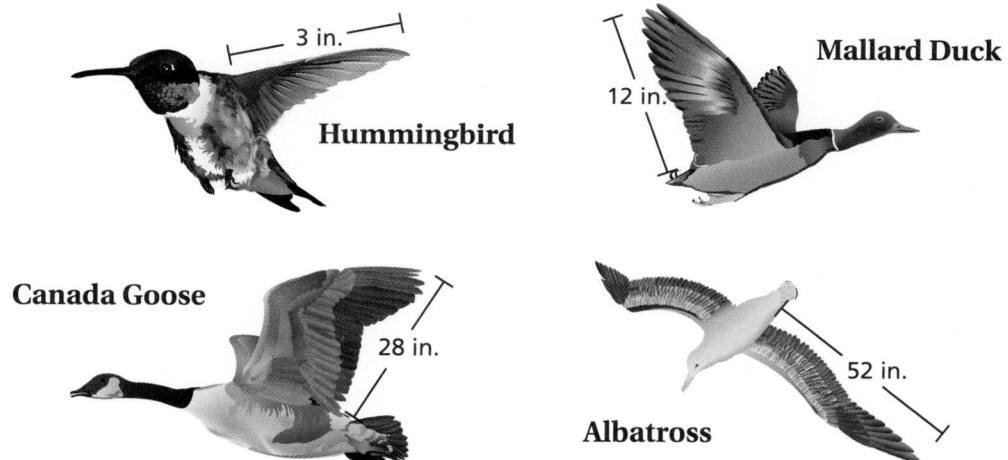

3 in.

Hummingbird

Mallard Duck

12 in.

Canada Goose

28 in.

52 in.

Albatross

Which of the following seems true? Explain your reasoning.

- Wing length and wing beat frequency are directly proportional.
- Wing length and wing beat frequency are inversely proportional.
- Wing length and wing beat frequency are unrelated.

5. SCIENCE Think of an example in science where two variables are inversely proportional.

Practice

Use what you learned about inverse variation to complete Exercises 4–7 on page 146.

Check It Out
Lesson Tutorials
BigIdeasMath.com

Key Vocabulary
inverse variation,
  p. 144

Key Idea

Inverse Variation

Words

Two quantities x and y show **inverse variation** when $y = \dfrac{k}{x}$, where k is a number and $k \neq 0$.

Graph

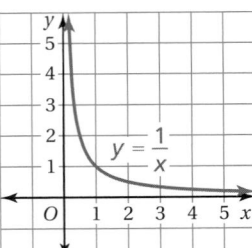

EXAMPLE **1** **Identifying Direct and Inverse Variation**

Tell whether x and y show *direct variation, inverse variation,* or *neither.* Explain your reasoning.

Study Tip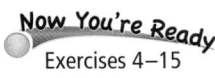

Other ways to say that x and y show inverse variation are "y varies inversely with x" and "x and y are inversely proportional."

a. $5y = x$

$\quad y = \dfrac{1}{5}x$ ⠀⠀⠀⠀Solve for y.

∴ The equation can be written as $y = kx$. So, x and y show direct variation.

b. $\dfrac{1}{3}y = \dfrac{1}{x}$

$\quad y = \dfrac{3}{x}$ ⠀⠀⠀⠀Solve for y.

∴ The equation can be written as $y = \dfrac{k}{x}$. So, x and y show inverse variation.

c. ⠀$-x = y + 7$

$\quad -x - 7 = y$ ⠀⠀⠀Solve for y.

∴ The equation cannot be written as $y = kx \; or \; y = \dfrac{k}{x}$. So, x and y do *not* show direct or inverse variation.

On Your Own

Now You're Ready
Exercises 4–15

Tell whether x and y show *direct variation, inverse variation,* or *neither.* Explain your reasoning.

1. $y - 1 = 2x$ ⠀⠀⠀⠀**2.** $\dfrac{1}{5}y = x$ ⠀⠀⠀⠀**3.** $2y = \dfrac{1}{x}$

⠀⠀🔊 Multi-Language Glossary at BigIdeasMath✓com.

Laurie's Notes

Introduction

Connect

- **Yesterday:** Students completed an activity with the dimensions of a rectangle that had a fixed area of 36 square units and began to develop an understanding of inverse variation.
- **Today:** Students will use a formal definition of inverse variation.

Motivate

- **Tell a Story:** "You plan to purchase one large pizza for dinner and invite some friends over. One thing to consider is how many pieces each person will get. The pizza has 12 slices." Ask students to help you fill out this table.

# of people	1	2	3	4	6	12
# of slices for each person						

- ? "What pattern do you observe in the table?" The number of slices decreases as the number of people increases.
- ? "So if I'm really hungry, how many friends should I invite over?" about 2

Lesson Notes

Key Idea

- The equation $y = \dfrac{k}{x}$ can be confusing to students. They see three variables. Remind them that this is the *general form*.

Example 1

- This example requires students to recall equations in two variables.
- Students should think about how they solved equations previously. The difference here is that the equation contains two variables.
- ? "What does the direct variation equation look like? The inverse variation equation?" $y = kx$; $y = \dfrac{k}{x}$
- The strategy for these examples is to solve the equation for y and decide if the equation shows direct or inverse variation. Students will need help with the equation solving.
- **Part (a):** To solve for y, multiply both sides by $\dfrac{1}{5}$ as shown, or divide both sides by 5. Dividing by 5 would give $y = \dfrac{x}{5}$, which is the same as $y = \dfrac{1}{5}x$.
- **Part (b):** To solve for y, multiply both sides of the equation by 3.
- **Part (c):** To solve for y, subtract 7 from both sides of the equation.

On Your Own

- Students should work with a partner.
- **Common Question:** "What is the first step?"

Goal Today's lesson is distinguishing, using, and solving equations involving **inverse variation**.

Start Thinking! and Warm Up

Lesson 3.8 Warm Up
For use before Lesson 3.8

Lesson 3.8 Start Thinking!
For use before Lesson 3.8

Does the distance formula, $d = rt$, show *direct variation* or *inverse variation*? Explain your reasoning.

Extra Example 1

Tell whether x and y show *direct variation*, *inverse variation*, or *neither*. Explain your reasoning.

a. $x = \dfrac{1}{7}y$ The equation can be written as $y = kx$. So, x and y show direct variation.

b. $\dfrac{1}{x} = 4y$ The equation can be written as $y = \dfrac{k}{x}$. So, x and y show inverse variation.

c. $y + 3 = x$ The equation cannot be written as $y = kx$ or $y = \dfrac{k}{x}$. So, x and y do not show direct or inverse variation.

On Your Own

1. neither; The equation cannot be written as $y = kx$ or $y = \dfrac{k}{x}$.

2. direct; The equation can be written as $y = kx$.

3. inverse; The equation can be written as $y = \dfrac{k}{x}$.

Laurie's Notes

Extra Example 2

Suppose y varies inversely with x and $y = 5$ when $x = 1$. Write an equation that relates x and y. $y = \dfrac{5}{x}$

Extra Example 3

The equation $y = \dfrac{12}{x}$ shows the number y of hours it takes x workers to mow and edge the city park.

a. Do x and y show direct variation or inverse variation? inverse variation

b. How many hours does it take two workers to mow and edge the city park? 6 hours

On Your Own

4. $y = \dfrac{3}{x}$

5. $2\dfrac{2}{3}$ hours

English Language Learners

Vocabulary

Begin by discussing words used in everyday language that are examples of opposites. Create a table and write each pair of English words in the first column. Start with *little and big, easy and hard, down and up,* and so on. Encourage students to expand the list. Ask the English learners to write the words in their native language in a second column and to share the words with the class. Explain to the class that in mathematics, *opposites* has a specific meaning. Every nonzero number has an opposite. Every pair of opposites consists of a positive number and a negative number. Ask students to name some pairs of opposite numbers. Write them on the board in numeric and word form.

Example 2

- One strategy for solving this problem is shown.
- Work through the solution.
- An alternative strategy for this multiple choice question is to eliminate choice C and choice D because they are *not* in the form $y = \dfrac{k}{x}$.
- Then, substitute 2 for x into the equations in choice A and choice B and solve for y.
- The only equation that gives the y-value of 1 is choice B.

Example 3

- Take time to "read" the graph. In words it says, "as the number of workers increases, the amount of time it takes to paint a room decreases."
- This concept should make sense to students.
- Remind them of a famous quote by John Heywood, "Many hands make light work."
- **FYI:** This is actually an example of a discrete graph. There can only be whole numbers of people, not 1.5.
- ? "Is it possible to have 16 people? If so, how much time would it take to paint the room?" Yes, with 16 people it would take $\dfrac{1}{2}$ hour. From a practical standpoint, there is a point at which there wouldn't be enough space for everyone to fit in the room.

On Your Own

- **Neighbor Check:** Students should work independently and then have their neighbor check their work. Have students discuss any discrepancies.

Closure

- Match the equation with a description.

1. $y = \dfrac{12}{x}$ A. direct variation

2. $y = 12x$ B. inverse variation

3. $y = x + 12$ C. neither

1. B   **2.** A   **3.** C

Technology For the Teacher

The Dynamic Planning Tool
Editable Teacher's Resources at *BigIdeasMath.com*

EXAMPLE **2** **Standardized Test Practice**

In the graph, x and y show inverse variation. Which equation relates x and y?

A $y = -\dfrac{2}{x}$ **B** $y = \dfrac{2}{x}$

C $y = -2x$ **D** $y = 2x$

The graph passes through (2, 1). Substitute to find k.

$y = \dfrac{k}{x}$ Write inverse variation equation.

$1 = \dfrac{k}{2}$ Substitute 2 for x and 1 for y.

$2 = k$ Solve for k.

So, the equation $y = \dfrac{2}{x}$ relates x and y. The correct answer is **B**.

EXAMPLE **3** **Real-Life Application**

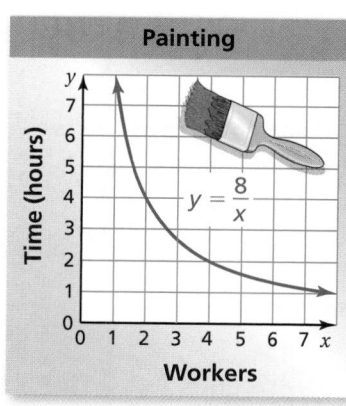

Painting

Time (hours)

$y = \dfrac{8}{x}$

Workers

The graph shows the number of hours y it takes x workers to paint a room. (a) How does y change as x increases? (b) Do x and y show direct or inverse variation? (c) How many hours does it take five workers to paint the room?

a. From the graph, you can see that y decreases as x increases. So, as the number of workers increases, the time to paint the room decreases.

b. The equation is written as $y = \dfrac{k}{x}$. So, x and y show inverse variation.

c. Use the equation to find y when $x = 5$.

$y = \dfrac{8}{x}$ Write equation.

$= \dfrac{8}{5} = 1.6$ Substitute. Then simplify.

It takes 1.6 hours for five workers to paint the room.

On Your Own

Now You're Ready
Exercises 23 and 24

4. Suppose y varies inversely with x and $y = 3$ when $x = 1$. Write an equation that relates x and y.

5. **WHAT IF?** In Example 3, how many hours does it take three workers to paint the room?

 Vocabulary and Concept Check

1. **WRITING** What does it mean for x and y to vary inversely?

2. **NUMBER SENSE** When x increases from 1 to 10, does $\frac{1}{x}$ increase or decrease?

3. **OPEN-ENDED** Describe a real-life situation that shows inverse variation.

 Practice and Problem Solving

Tell whether x and y show *direct variation, inverse variation,* or *neither.* **Explain your reasoning.**

4. $y = \dfrac{1}{x}$

5. $xy = 8$

6. $y - x = 0$

7. $\dfrac{1}{2}y = 2x$

8. $\dfrac{y}{3} = \dfrac{2}{x}$

9. $y - 2 = \dfrac{7}{x}$

10. $x = y + 9$

11. $x = 4y$

12. $y = \dfrac{5}{2x}$

13. $2y = \dfrac{6}{x}$

14. $\dfrac{5x}{3} = \dfrac{y}{4}$

15. $x = \dfrac{7 + y}{2}$

16. **ERROR ANALYSIS** Describe and correct the error in telling whether x and y show inverse variation.

$$\frac{y}{2} = \frac{8}{x}$$

The equation does not show inverse variation because it is not of the form $y = \dfrac{k}{x}$.

Graph the data. Tell whether x and y show *direct variation* or *inverse variation.*

17.
x	−2	2	4	6
y	−1	1	2	3

18.
x	0.5	1	3	6
y	6	3	1	0.5

19.
x	2	5	8	20
y	10	4	2.5	1

20.
x	2	4	8	11
y	1.5	3	6	8.25

Tell whether x and y show *direct variation* or *inverse variation.* **Explain.**

21. **STADIUM** The time y it takes to empty a stadium and the number x of open exits are related by the equation $y = \dfrac{0.8}{x}$.

22. **TRAVEL** The number y of miles driven and the number x of gallons of gas used are related by the equation $y = 28.5x$.

Assignment Guide and Homework Check

Level	Day 1 Activity Assignment	Day 2 Lesson Assignment	Homework Check
Basic	4–7, 29–37	1–3, 9–23 odd, 16, 22	9, 13, 16, 17, 23
Average	4–7, 29–37	1–3, 12–16, 17–25 odd	12, 16, 17, 23, 25
Advanced	4–7, 29–37	1–3, 12–20 even, 24–28 even, 27	12, 16, 18, 24, 26

Common Errors

- **Exercises 4–15** Students may struggle with multiplying or dividing by x or y when solving for y. Remind them of the Multiplication and Division Properties of Equality, and let them know that these properties apply to variables as well as numbers.
- **Exercises 23 and 24** Students may substitute the wrong values for x and y. Remind them that x is the first coordinate and y is the second.
- **Exercises 23 and 24** Students may also substitute the x-coordinate for k. Tell them to be careful when substituting and that the variable k should still be in the numerator after substituting in the coordinates.

3.8 Record and Practice Journal

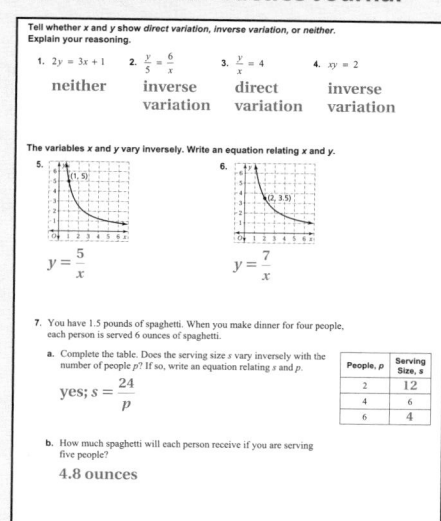

Tell whether x and y show *direct variation, inverse variation,* or *neither.* Explain your reasoning.

1. $2y = 3x + 1$ 2. $\frac{y}{5} = \frac{6}{x}$ 3. $\frac{y}{x} = 4$ 4. $xy = 2$

   neither   inverse variation   direct variation   inverse variation

The variables x and y vary inversely. Write an equation relating x and y.

5. $y = \frac{5}{x}$ 6. $y = \frac{7}{x}$

7. You have 1.5 pounds of spaghetti. When you make dinner for four people, each person is served 6 ounces of spaghetti.

   a. Complete the table. Does the serving size s vary inversely with the number of people p? If so, write an equation relating s and p.

   yes; $s = \frac{24}{p}$

People, p	Serving Size, s
2	12
4	6
6	4

   b. How much spaghetti will each person receive if you are serving five people?

   4.8 ounces

1. As x increases, y decreases.

2. decrease

3. *Sample answer:* The wingspan of a bird varies inversely with its wing beat frequency.

Practice and Problem Solving

4. inverse variation; The equation can be written as $y = \frac{k}{x}$.

5. inverse variation; The equation can be written as $y = \frac{k}{x}$.

6. direct variation; The equation can be written as $y = kx$.

7. direct variation; The equation can be written as $y = kx$.

8. inverse variation; The equation can be written as $y = \frac{k}{x}$.

9. neither; The equation cannot be written as $y = kx$ or $y = \frac{k}{x}$.

10. neither; The equation cannot be written as $y = kx$ or $y = \frac{k}{x}$.

11. direct variation; The equation can be written as $y = kx$.

12. inverse variation; The equation can be written as $y = \frac{k}{x}$.

13. inverse variation; The equation can be written as $y = \frac{k}{x}$.

14. direct variation; The equation can be written as $y = kx$.

15. neither; The equation cannot be written as $y = kx$ or $y = \dfrac{k}{x}$.

16. The equation does show inverse variation because it can be written as $y = \dfrac{16}{x}$.

17–22. See Additional Answers.

23. $y = \dfrac{4}{x}$ 24. $y = \dfrac{12}{x}$

25. a. yes; $t = \dfrac{12}{s}$

    b. 3 h

26. See *Taking Math Deeper.*

27. decreases

28. 8 people

29.	88	30.	88
31.	63	32.	63
33.	yes	34.	no
35.	yes	36.	yes
37.	B		

Mini-Assessment

Tell whether *x* and *y* show *direct variation*, *inverse variation*, or *neither*.

1. $y = x + 9$ neither

2. $y = \dfrac{1}{6}x$ direct variation

3. $-x - y = 10$ neither

4. $\dfrac{y}{4} = \dfrac{3}{x}$ inverse variation

5. The number *y* of miles driven and the number *x* of gallons of gas used are related by the equation $y = 25x$. Tell whether *x* and *y* show direct variation or inverse variation.
   direct variation

Taking Math Deeper

Exercise 26

This problem asks students to use their physical intuition to decide whether direct or inverse variation apply.

 1   Help me see it.

The closer the chairs are together, the more weight the spaghetti can support without breaking.

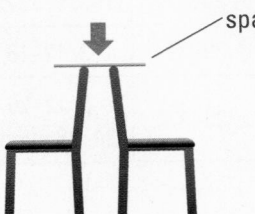

spaghetti

2   Make a conclusion.

a. I can see that more force is needed to break a short (brittle) board. Less force is needed to break a long (brittle) board. So, inverse variation applies.

I see that.

3   Find the force.

Let $y =$ force. Let $x =$ length.

$y = \dfrac{k}{x}$	Write inverse variation equation.
$3.6 = \dfrac{k}{5}$	Substitute given values.
$18 = k$	Solve for k.
$y = \dfrac{18}{x}$	Write inverse variation equation.
$y = \dfrac{18}{3}$	Substitute 3 for x.
b. $y = 6$ lb	Solve for force.

Project

What is the world record for the number of boards one person has broken? When and where was the record set? How much training and practice would it take to master some type of martial art?

Reteaching and Enrichment Strategies

If students need help. . .	If students got it. . .
Resources by Chapter • Practice A and Practice B • Puzzle Time Record and Practice Journal Practice Differentiating the Lesson Lesson Tutorials Skills Review Handbook	Resources by Chapter • Enrichment and Extension • School-to-Work • Financial Literacy • Technology Connection • Life Connections • Stories in History Start the next section

The variables x and y vary inversely. Write an equation relating x and y.

23.

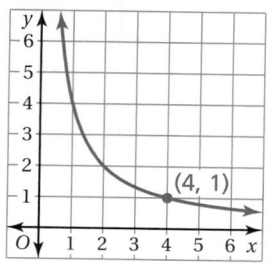

(4, 1)

24.

(2, 6)

25. BICYCLING The table shows the times it takes to bicycle 12 miles at various speeds.

Speed (mi/h)	12	6	3	2
Time (h)	1	2	4	6

   a. Does the time t vary inversely with the speed s? If so, write an equation relating t and s.

   b. What time corresponds to a speed of 4 miles per hour?

3 ft

26. MARTIAL ARTS It takes 3.6 pounds of force to break a 5-foot board.

   a. You remember from science that force and board length vary directly or inversely, but you've forgotten which. How can you use reason to remember?

   b. How much force does it take to break the board shown?

27. SALARY A salesperson has a fixed weekly salary. The person works twice as many hours this week as last week. What happens to the person's hourly rate?

28. **Reasoning** The price per person to rent a limousine varies inversely with the number of passengers. It costs $90 each for five people. How many people are renting the limousine when the cost per person is $56.25?

Fair Game Review *What you learned in previous grades & lessons*

Find the percent of the number. *(Skills Review Handbook)*

29. 40% of 220   **30.** 32% of 275   **31.** 84% of 75   **32.** 21% of 300

Tell whether the ratios form a proportion. *(Section 3.3)*

33. $\dfrac{9}{15}, \dfrac{18}{30}$ **34.** $\dfrac{21}{9}, \dfrac{18}{8}$ **35.** $\dfrac{42}{91}, \dfrac{24}{52}$ **36.** $\dfrac{24}{38}, \dfrac{36}{57}$

37. MULTIPLE CHOICE A gumball machine contains 1000 gumballs. The ratio of red gumballs to the total number of gumballs is 1 : 4. How many red gumballs are in the machine? *(Section 3.5)*

   Ⓐ 150   **Ⓑ** 250   **Ⓒ** 400   **Ⓓ** 750

Copy and complete the statement. Round to the nearest hundredth, if necessary.
(Section 3.6)

1. $10 \text{ mi} \approx$ ___ km

2. $3 \text{ qt} \approx$ ___ L

3. $29 \text{ kg} \approx$ ___ lb

4. $6.8 \text{ in.} \approx$ ___ cm

Tell whether x and y show direct variation. Explain your reasoning. *(Section 3.7)*

5.

x	y
−3	0
−1	1
1	2
3	3

6.

x	y
−1	−2
0	0
1	2
2	4

7. $y - 9 = 6 + x$

8. $x = \dfrac{5}{8}y$

Tell whether x and y show *direct variation*, *inverse variation*, or *neither*. Explain your reasoning. *(Section 3.8)*

9. $y = \dfrac{12}{x}$

10. $y - x = 9$

11.

x	y
−3	−2
−1	0
1	0
3	2

12.

x	y
1	2
2	1
4	0.5
8	0.25

13. HEIGHT The tallest player in Euroleague Basketball is 229 centimeters. The tallest player in the National Basketball Association is 90 inches. Which league has the tallest player? *(Section 3.6)*

14. PIE SALE The table shows the profit of a pie sale. Tell whether there is direct variation between the two data sets. If so, write the equation of direct variation. *(Section 3.7)*

Pies Sold	10	12	14	16
Profit	$79.50	$95.40	$111.30	$127.20

15. JEWELRY The number of beads on a bracelet varies inversely with the length of the beads. You use 8-millimeter beads to make a bracelet with 25 beads. How many 10-millimeter beads would you need to make a bracelet? *(Section 3.8)*

Alternative Assessment Options

Math Chat   Student Reflective Focus Question
Structured Interview   Writing Prompt

Structured Interview

Interviews can occur formally or informally. Ask a student to perform a task and to explain what and why as they work. Have them describe their thought process. Probe the student for more information. Do not ask leading questions. Keep a rubric or notes.

Teacher Prompts	Student Answers	Teacher Notes
Tell me a story about taking a vacation. Include this sentence. 4 miles is approximately equal to ? kilometers.	I took a four mile bike ride. Four miles is approximately equal to 6.4 kilometers.	Student can convert measures between systems.
Add to your story using this sentence. 18 pounds is approximately equal to ? kilograms.	My bicycle weighs about 18 pounds. 18 pounds is approximately equal to 8.1 kilograms.	Student can convert measures between systems.

Study Help

Remind students to complete Graphic Organizers for the rest of the chapter.

7.

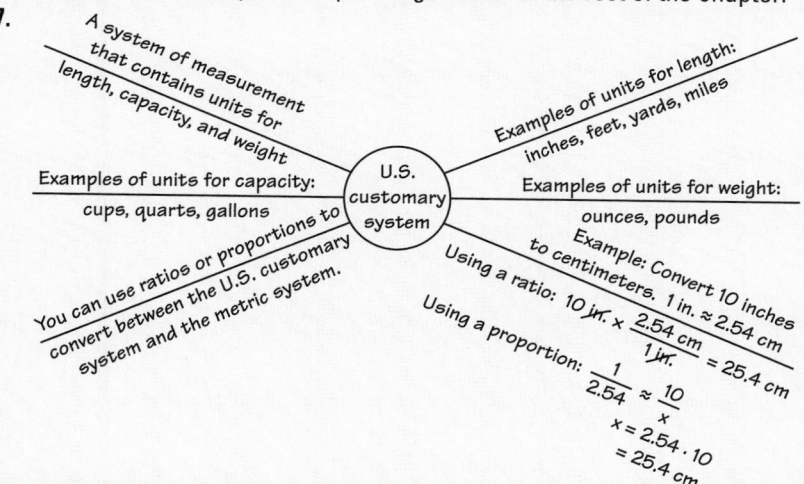

8–11. Available at *BigIdeasMath.com*

Reteaching and Enrichment Strategies

If students need help. . .	If students got it. . .
Resources by Chapter • Study Help • Practice A and Practice B • Puzzle Time Lesson Tutorials *BigIdeasMath.com* Practice Quiz Practice from the Test Generator	Resources by Chapter • Enrichment and Extension • School-to-Work Game Closet at *BigIdeasMath.com* Start the Chapter Review

Answers

1. 16
2. 2.85
3. 64.44
4. 17.27
5. no; The line does not pass through the origin.
6. yes; The line passes through the origin.
7. no; The equation cannot be written as $y = kx$.
8. yes; The equation can be written as $y = kx$.
9. inverse variation; The equation can be written as $y = \dfrac{k}{x}$.
10. neither; The equation cannot be written as $y = kx$ or $y = \dfrac{k}{x}$.
11. neither; The equation cannot be written as $y = kx$ or $y = \dfrac{k}{x}$.
12. inverse variation; The equation can be written as $y = \dfrac{k}{x}$.
13. Euroleague Basketball
14. direct variation; $y = 7.95x$
15. 20 beads

Technology For the Teacher
Answer Presentation Tool

Assessment Book

Chapter 3 Quiz For use after Section 3.8

Copy and complete the statement using < or >.

1. 1 cm ? 1 in.
2. 47 kg ? 47 lb
3. 63 mi ? 63 km
4. 15 L ? 15 qt

Copy and complete the statement using a ratio. Round to the nearest hundredth, if necessary.

5. 13 L ≈ ? qt
6. 10 lb ≈ ? kg
7. 30 km/h ≈ ? mi/h
8. 26 in. ≈ ? m

Tell whether x and y show direct variation. Explain your reasoning.

9.
x	y
1	5
2	8
3	11
4	14

10.
x	y
−1	−2.5
1	0.5
3	3.5
5	6.5

Tell whether x and y show direct variation, inverse variation, or neither. Explain your reasoning.

11. $y = \dfrac{5}{x}$
12. $x = y$

13.
x	y
−3	7
0	7
2	7
4	7

14.
x	y
2	8
4	4
8	2
16	1

15. Would it be possible to stretch a string that is 10 kilometers long between two cities that are 7 miles apart? If not, how many kilometers short is the string?

Answers
1. _____
2. _____
3. _____
4. _____
5. _____
6. _____
7. _____
8. _____
9. _____
10. _____
11. _____
12. _____
13. _____
14. _____
15. _____

For the Teacher
Additional Review Options

- **Quiz**Show
- Big Ideas Test Generator
- Game Closet at *BigIdeasMath.com*
- Vocabulary Puzzle Builder
- Resources by Chapter
  Puzzle Time
  Study Help

Answers

1. 28.9 miles per gallon

2. 325 revolutions per minute

3.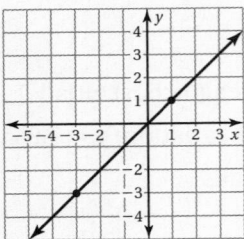

slope $= 1$

4.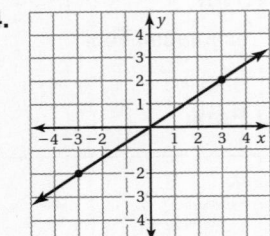

slope $= \dfrac{2}{3}$

5.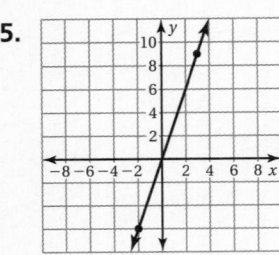

slope $= 3$

Review of Common Errors

Exercises 1 and 2
- Students may find the unit rate but forget to include the units.

Exercises 3–5
- Students may put the change in *x* over the change in *y*.

Exercises 6–9
- Remind students that equivalent ratios are the same as equivalent fractions.

Exercises 10 and 11
- Students may write one half of the proportion using rows and the other half using columns.

Exercises 12–15
- Students may multiply across the numerator and the denominator as if they were multiplying fractions.

Exercises 16–18
- Students may put the unit that they are solving for in the denominator.

Exercises 19–22
- Students may try to identify the direct variation equations without solving for *y*.

Exercises 23–26
- Students may struggle with multiplying or dividing by *x* or *y* when solving for *y*.

Check It Out
Vocabulary Help
BigIdeasMath ✓com

Review Key Vocabulary

ratio, *p. 100*
rate, *p. 100*
unit rate, *p. 100*
slope, *p. 106*

proportion, *p. 112*
proportional, *p. 112*
cross products, *p. 113*
U.S. customary system,
    p. 132

metric system, *p. 132*
direct variation, *p. 138*
inverse variation, *p. 144*

Review Examples and Exercises

3.1 Ratios and Rates (pp. 98–103)

Find the unit rate of calories per serving.

Servings	2	4	6	8
Calories	240	480	720	960

The calories increase by 240 for every 2 servings.

$$\frac{\text{change in calories}}{\text{change in servings}} = \frac{240}{2} = \frac{120}{1}$$

∴ The rate is 120 calories per serving.

Exercises

Find the unit rate.

1. 289 miles on 10 gallons

2. 975 revolutions in 3 minutes

3.2 Slope (pp. 104–109)

Find the slope of the line.

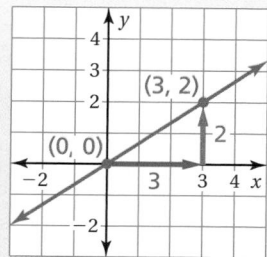

$$\text{slope} = \frac{\text{change in } y}{\text{change in } x}$$

$$= \frac{2}{3}$$

∴ The slope of the line is $\frac{2}{3}$.

Exercises

Graph the line that passes through the two points. Then find the slope of the line.

3. $(-3, -3), (1, 1)$

4. $(-3, -2), (3, 2)$

5. $(3, 9), (-2, -6)$

3.3 Proportions (pp. 110–115)

Tell whether the ratios $\dfrac{9}{12}$ and $\dfrac{6}{8}$ form a proportion.

$$\frac{9}{12} = \frac{9 \div 3}{12 \div 3} = \frac{3}{4} \qquad \frac{6}{8} = \frac{6 \div 2}{8 \div 2} = \frac{3}{4}$$

∴∴ The ratios are equivalent. So, the ratios form a proportion.

Exercises

Tell whether the ratios form a proportion.

6. $\dfrac{4}{9}, \dfrac{2}{3}$

7. $\dfrac{12}{22}, \dfrac{18}{33}$

8. $\dfrac{8}{50}, \dfrac{4}{10}$

9. $\dfrac{32}{40}, \dfrac{12}{15}$

3.4 Writing Proportions (pp. 116–121)

Write a proportion that gives the number r of returns on Saturday.

	Friday	Saturday
Sales	40	85
Returns	32	r

$$\frac{40 \text{ sales}}{32 \text{ returns}} = \frac{85 \text{ sales}}{r \text{ returns}}$$

Exercises

Use the table to write a proportion.

10.

	Game 1	Game 2
Penalties	6	8
Minutes	16	m

11.

	Concert 1	Concert 2
Songs	15	18
Hours	2.5	h

3.5 Solving Proportions (pp. 122–127)

Solve $\dfrac{x}{9} = \dfrac{4}{5}$.

$x \cdot 5 = 9 \cdot 4$ Use the Cross Products Property.

$5x = 36$ Multiply.

$x = 7.2$ Divide.

Exercises

Solve the proportion.

12. $\dfrac{x}{4} = \dfrac{2}{5}$

13. $\dfrac{5}{12} = \dfrac{y}{15}$

14. $\dfrac{z}{7} = \dfrac{3}{16}$

15. $\dfrac{8}{20} = \dfrac{6}{w}$

Review Game
Proportions

Big Ideas
Game Closet

Materials
- 1 deck of cards for each group
- paper for each student
- pencil for each student

Directions
Play in groups of 4 to 6 people. Each player is dealt four cards. Two are dealt face up. The other two are dealt face down and are held in the player's hand. The remainder of the deck is placed face down between the players.

The face up cards represent the denominators of two fractions. The object is to use the face down cards and other cards to form a proportion. Several cards can be added together in the numerator and denominator of both fractions until a proportion is obtained.

When it is a player's turn, he or she must do one of three things: lay a card down in either fraction's numerator or denominator, ask another player for a specific card, or draw from the pile. A student who draws from the pile or obtains a card from another student must wait until their next turn to lay a card down.

Card values are as follows:
> 2 through 10: face value
> Jack: -1
> Queen: -2
> King: -3
> Ace: -4

Points are awarded as follows:
> First person to form a proportion: 10 points
> Second: 9 points
> Third: 8 points
> Fourth: 7 points, and so on.

Who Wins?
After a set amount of time, the player with the most points wins.

For the Student
Additional Practice
- Lesson Tutorials
- Study Help (textbook)
- Student Website
    Multi-Language Glossary
    Practice Assessments

Answers

6. no   7. yes

8. no   9. yes

10. *Sample answer:*
$$\frac{8 \text{ penalties}}{6 \text{ penalties}} = \frac{m \text{ minutes}}{16 \text{ minutes}}$$

11. *Sample answer:*
$$\frac{15 \text{ songs}}{2.5 \text{ hours}} = \frac{18 \text{ songs}}{h \text{ hours}}$$

12. $x = 1.6$ 13. $y = 6.25$

14. $z = 1.3125$ 15. $w = 15$

16. 3.16   17. 22.86

18. 6.75

19. no; The equation cannot be written as $y = kx$.

20. yes; The equation can be written as $y = kx$.

21. yes; The equation can be written as $y = kx$.

22. no; The equation cannot be written as $y = kx$.

23. direct variation; The equation can be written as $y = kx$.

24. neither; The equation cannot be written as $y = kx$ or $y = \frac{k}{x}$.

25. direct variation; The equation can be written as $y = kx$.

26. inverse variation; The equation can be written as $y = \frac{k}{x}$.

My Thoughts on the Chapter

What worked. . .

What did not work. . .

What I would do differently. . .

3.6 Converting Measures Between Systems (pp. 130–135)

Convert 8 kilometers to miles.

$$8 \text{ km} \times \frac{1 \text{ mi}}{1.6 \text{ km}} \approx 5 \text{ mi}$$

1 mi ≈ 1.6 km, so use the ratio $\frac{1 \text{ mi}}{1.6 \text{ km}}$.

Exercises

Copy and complete the statement. Round to the nearest hundredth, if necessary.

16. 3 L ≈ ▢ qt

17. 9 in. ≈ ▢ cm

18. 15 lb ≈ ▢ kg

3.7 Direct Variation (pp. 136–141)

Tell whether x and y show direct variation. Explain your reasoning.

a. $x + y - 1 = 3$

$y = 4 - x$ Solve for y.

∴ The equation *cannot* be written as $y = kx$. So, x and y do *not* show direct variation.

b. $x = 8y$

$\frac{1}{8}x = y$ Solve for y.

∴ The equation can be written as $y = kx$. So, x and y show direct variation.

Exercises

Tell whether x and y show direct variation. Explain your reasoning.

19. $x + y = 6$

20. $y - x = 0$

21. $\frac{x}{y} = 20$

22. $x = y + 2$

3.8 Inverse Variation (pp. 142–147)

Tell whether x and y show inverse variation. Explain your reasoning.

$$xy = 5$$

$$y = \frac{5}{x}$$ Solve for y.

∴ The equation can be written as $y = \frac{k}{x}$. So, x and y show inverse variation.

Exercises

Tell whether x and y show *direct variation*, *inverse variation*, or *neither*. Explain your reasoning.

23. $\frac{x}{y} = 6$

24. $3x + y = 7$

25. $8y = 4x$

26. $xy = 12$

Write the ratio as a fraction in simplest form.

1. 34 cars : 26 trucks

2. 3 feet to 9 feet

Find the unit rate.

3. 84 miles in 12 days

4. $3.20 for 8 ounces

Graph the line that passes through the two points. Then find the slope of the line.

5. $(15, 9)$, $(-5, -3)$

6. $(2, 9)$, $(4, 18)$

Tell whether the ratios form a proportion.

7. $\dfrac{1}{9}, \dfrac{6}{54}$

8. $\dfrac{9}{12}, \dfrac{8}{72}$

Use the table to write a proportion.

9.

	Monday	Tuesday
Gallons	6	8
Miles	180	m

10.

	Thursday	Friday
Classes	6	c
Hours	8	4

Solve the proportion.

11. $\dfrac{x}{8} = \dfrac{9}{4}$

12. $\dfrac{17}{3} = \dfrac{y}{6}$

Copy and complete the statement. Round to the nearest hundredth, if necessary.

13. 5 L ≈ ▢ qt

14. 56 lb ≈ ▢ kg

Tell whether x and y show *direct variation*, *inverse variation*, or *neither*. Explain your reasoning.

15. $xy - 11 = 5$

16. $x = \dfrac{3}{y}$

17. $\dfrac{y}{x} = 8$

18. MOVIE TICKETS Five movie tickets cost $36.25. What is the cost of eight movie tickets?

19. VOLLEYBALL COURT Find the dimensions of the volleyball court in feet.

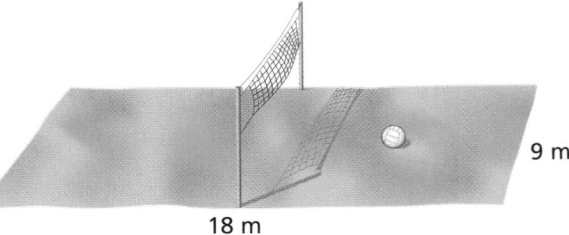

9 m

18 m

20. GLAZE A specific shade of green glaze requires 5 parts blue to 3 parts yellow. A glaze mixture contains 25 quarts of blue and 9 quarts of yellow. How can you fix the mixture to make the specific shade of green glaze?

Test Item References

Chapter Test Questions	Section to Review
1–4	3.1
5, 6	3.2
7, 8	3.3
9, 10	3.4
11, 12, 18, 20	3.5
13, 14, 19	3.6
15–17	3.7
15–17	3.8

Test-Taking Strategies

Remind students to quickly look over the entire test before they start so that they can budget their time. There are conversions on the test, so remind students to jot down conversions on the back of their test before they start. Students should use **Stop** and **Think** strategies to ensure that they understand what is being asked before they write an answer.

Common Assessment Errors

- **Exercises 3 and 4** Students may find the unit rate correctly, but forget to include the units. Remind students that units are necessary for understanding a rate.
- **Exercises 5 and 6** Students may find the ratio of horizontal change to vertical change, not vertical change to horizontal change, when finding slope. Remind students that slope is the ratio of the change in y to the change in x.
- **Exercises 11 and 12** Students might cross multiply incorrectly by multiplying the numerators and multiplying the denominators.
- **Exercises 13 and 14** When converting units, students may use a ratio that is the reciprocal of what it should be. Remind students that the ratio should be such that the original units divide out and the new units remain.
- **Exercises 15–17** Students may have difficulty solving for y. Remind students of the properties of equality and let them know that these properties can apply to variables as well as to numbers.

Reteaching and Enrichment Strategies

If students need help. . .	If students got it. . .
Resources by Chapter • Practice A and Practice B • Puzzle Time Record and Practice Journal Practice Differentiating the Lesson Lesson Tutorials Practice from the Test Generator Skills Review Handbook	Resources by Chapter • Enrichment and Extension • School-to-Work • Financial Literacy • Life Connections • Stories in History Game Closet at *BigIdeasMath.com* Start Standardized Test Practice

Answers

1. $\dfrac{17}{13}$ 2. $\dfrac{1}{3}$

3. 7 miles per day

4. \$0.40 per ounce

5–6. See Additional Answers.

7. yes   8. no

9. *Sample answer:*
$$\frac{8 \text{ gallons}}{6 \text{ gallons}} = \frac{m \text{ miles}}{180 \text{ miles}}$$

10. *Sample answer:*
$$\frac{6 \text{ classes}}{8 \text{ hours}} = \frac{c \text{ classes}}{4 \text{ hours}}$$

11. $x = 18$ 12. $y = 34$

13. 5.26   14. 25.2

15. inverse variation; The equation can be written as $y = \dfrac{k}{x}$.

16. inverse variation; The equation can be written as $y = \dfrac{k}{x}$.

17. direct variation; The equation can be written as $y = kx$.

18. \$58

19. about 29.53 feet by about 59.06 feet

20. Add 6 quarts of yellow

Assessment Book

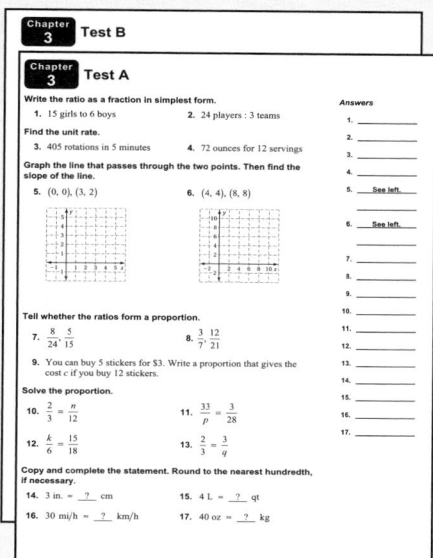

T-152

After Answering Easy Questions, Relax
Answer Easy Questions First
Estimate the Answer
Read All Choices before Answering
Read Question before Answering
Solve Directly or Eliminate Choices
Solve Problem before Looking at Choices
Use Intelligent Guessing
Work Backwards

About this Strategy

When taking a multiple choice test, be sure to read each question carefully and thoroughly. It is also very important to read each answer choice carefully. Do not pick the first answer that you think is correct!

Answers

1. B
2. I
3. 0.25 or $\frac{1}{4}$
4. D

Item Analysis

1. **A.** The cost for 4 pencils is $0.10 more than $0.10 times 4. So, the student thinks that the cost of 10 pencils is $0.10 more than $0.10 times 10.

   B. Correct answer

   C. The student multiplies 4 times $0.50.

   D. The student multiplies 10 times $0.50.

2. **F.** The student thinks that absolute value is always negative.

   G. The student thinks that absolute value is always negative. The student also thinks that the negative sign in III is irrelevant.

   H. The student ignores the absolute values bars as well as the negative sign outside the absolute value bars in IV.

   I. Correct answer

3. **Correct answer:** 0.25 or $\frac{1}{4}$

   Common Error: The student divides the greater number by the lesser number, getting an answer of 4.

4. **A.** The student does not apply the concept of slope and instead picks (0, −5) because the coordinates most resemble the coordinates in the given points.

   B. The student does not apply the concept of slope. The student sketches the line containing (0, 5) and (5, 0), and then picks (3, 3) because this point appears to be close to the line.

   C. The student does not apply the concept of slope. The student instead makes an ordered pair using the *x*-coordinate from (5, 0) and the *y*-coordinate from (0, 5).

   D. Correct answer

5. **F.** Correct answer

   G. The student finds the mode score.

   H. The student finds the greatest number of **X**s.

   I. The student finds the range of the scores.

6. **A.** Correct answer

   B. The student ignores the negative sign when distributing $-\frac{1}{2}$.

   C. The student thinks that the quotient of two negative numbers is negative.

   D. The student thinks that the inverse operation of multiplication by a negative number is addition of the opposite of the number.

7. **F.** Correct answer

   G. The student adds −4 instead of subtracting −4 from 72.

   H. The student thinks that the product of −9 and −8 is −72.

   I. The student thinks that the product of −9 and −8 is −72. The student also adds −4 instead of subtracting −4 from −72.

1. The school store sells 4 pencils for $0.50. At that rate, what would be the cost of 10 pencils?

   A. $1.10

   B. $1.25

   C. $2.00

   D. $5.00

2. Which expressions do *not* have a value of 3?

   I. $|3|$ II. $|-3|$

   III. $-|3|$ IV. $-|-3|$

   F. I and II

   G. I and III

   H. II and IV

   I. III and IV

Test-Taking Strategy
Read Question Before Answering

"Be sure to read the question before choosing your answer. You may find a word that changes the meaning."

3. What is the value of y in the equation below when $x = 12$ and $k = 3$?

   $$xy = k$$

4. Use the coordinate plane to answer the question below.

   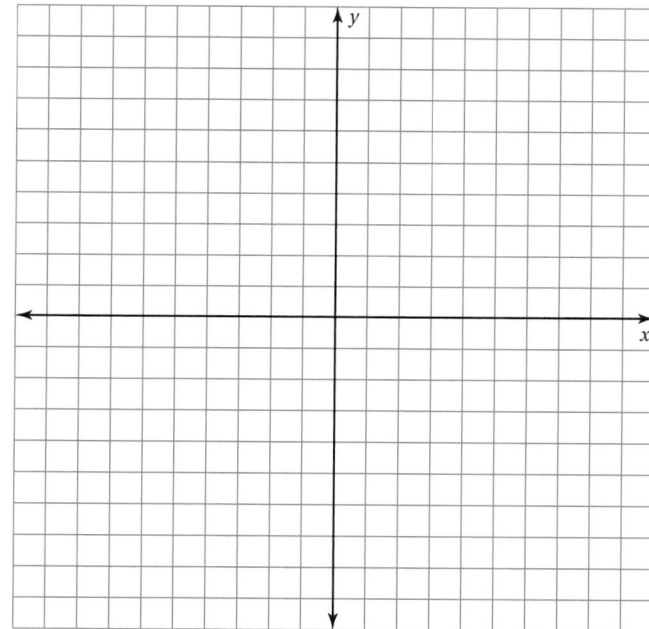

   A line contains both the point (0, 5) and the point (5, 0). Which of the following points is also on this line?

   A. (0, −5)

   B. (3, 3)

   C. (5, 5)

   D. (7, −2)

5. The scores from a diving competition are shown in the line plot below.

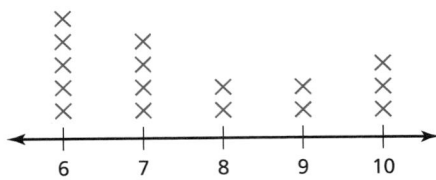

What is the median score?

F. 7

H. 5

G. 6

I. 4

6. Meli was solving the equation in the box below.

$$-\frac{1}{2}(4x - 10) = -16$$

$$-2x - 5 = -16$$

$$-2x - 5 + 5 = -16 + 5$$

$$-2x = -11$$

$$\frac{-2x}{-2} = \frac{-11}{-2}$$

$$x = \frac{11}{2}$$

What should Meli do to correct the error that she made?

A. Distribute the $-\frac{1}{2}$ to get $-2x + 5$.

B. Distribute the $-\frac{1}{2}$ to get $2x - 5$.

C. Divide -11 by -2 to get $-\frac{11}{2}$.

D. Add 2 to -11 to get -9.

7. What is the value of the expression below when $n = -8$ and $p = -4$?

$$-9n - p$$

F. 76

H. -68

G. 68

I. -76

Item Analysis (continued)

8. **Correct answer:** 500 mm

   Common Errors: The student reverses the conversion factor and gets an answer of 0.8 millimeter. The student either does not know or does not apply the fact that a millimeter is less than an inch.

9. **A.** The student divides 24 by 8.

   B. Correct answer

   C. The student thinks that because there are 3 more dogs, each dog gets 3 fewer treats.

   D. The student uses direct variation instead of inverse variation (answer rounded to nearest whole number).

10. **F.** The student simply adds the four numbers in the figure.

    G. The student does not multiply by $\frac{1}{2}$ when finding the area of the triangle, gets 16 for the area of the triangle, and subtracts the result from 56.

    H. Correct answer

    I. The student thinks that the composite figure can be divided and recomposed as two rectangles, one 7×4 and one 8×4.

11. **2 points** The student demonstrates a thorough understanding of writing and solving proportions. For Part A, the student correctly writes a proportion such as $\frac{800}{15} = \frac{6000}{m}$ and provides an appropriate explanation. For Part B, the student correctly gets a value of 112.5 for m, shows appropriate work, and states that it would take 112.5 minutes.

    1 point The student demonstrates a partial understanding of writing and solving proportions. The student writes a correct proportion but does not solve it successfully, or the student does not write a correct proportion but demonstrates the ability to solve a proportion.

    0 points The student demonstrates insufficient understanding of writing and solving proportions. The student does not write a correct proportion and shows little or no evidence of being able to solve proportions.

Answers

5. F

6. A

7. F

Answers

8. 500 mm

9. B

10. H

11. *Part A Sample answer:*

$$\frac{800}{15} = \frac{6000}{m}$$

Part B 112.5 min

12. A

Answer for Extra Example

1. **A.** The student incorrectly thinks that proportions involve addition.

   B. The student incorrectly thinks that proportions involve subtraction.

   C. Correct answer

   D. The student switches the 40 and the 27 in the proportion, resulting in a proportion that is not equivalent to the original proportion.

Item Analysis (continued)

12. **A.** Correct answer

   B. The student multiplies both sides of the equation by $-\frac{1}{4}$ instead of dividing.

   C. The student adds $-\frac{1}{4}$ to both sides of the equation instead of dividing both sides by $-\frac{1}{4}$.

   D. The student thinks that the quotient of $5\frac{7}{8}$ and $-\frac{1}{4}$ is positive because $5\frac{7}{8}$ has the greater absolute value.

Extra Example For Standardized Test Practice

1. Nathaniel was solving a proportion in the box below.

$$\frac{16}{40} = \frac{p}{27}$$
$$16 \cdot p = 40 \cdot 27$$
$$16p = 1080$$
$$\frac{16p}{16} = \frac{1080}{16}$$
$$p = 67.5$$

What should Nathaniel do to correct the error that he made?

   A. Add 40 to 16 and 27 to *p*.

   B. Subtract 16 from 40 and 27 from *p*.

   C. Multiply 16 by 27 and *p* by 40.

   D. Divide 16 by 27 and *p* by 40.

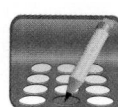

8. How many millimeters are equivalent to 20 inches?
   (Use 1 millimeter ≈ 0.04 inch.)

9. If 5 dogs share equally a bag of dog treats, each dog gets 24 treats. Suppose 8 dogs share equally the bag of treats. How many treats does each dog get?

   A. 3

   B. 15

   C. 21

   D. 38

10. The figure below consists of a rectangle and a right triangle.

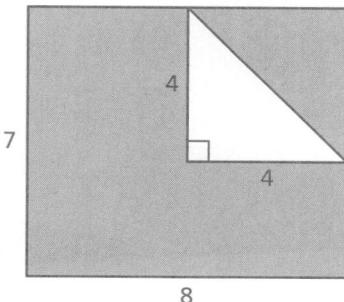

    What is the area of the shaded region?

    F. 23 units^2

    G. 40 units^2

    H. 48 units^2

    I. 60 units^2

11. You can mow 800 square feet of lawn in 15 minutes. At this rate, how many minutes will you take to mow a lawn that measures 6000 square feet?

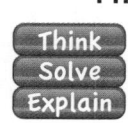

    Part A   Write a proportion to represent the problem. Use m to represent the number of minutes. Explain your reasoning.

    Part B   Solve the proportion you wrote in Part A and use it to answer the problem. Show your work.

12. What number belongs in the box to make the equation true?

$$5\frac{7}{8} = \boxed{} \cdot \left(-\frac{1}{4}\right)$$

    A. $-23\frac{1}{2}$

    B. $-1\frac{15}{32}$

    C. $6\frac{1}{8}$

    D. $23\frac{1}{2}$

4 Percents

"Here's my sales strategy. I buy each dog bone for $0.05."

"Then I mark each one up to $1. Then, I have a 75% off sale. Cool, huh?"

"Dear Vet: I have this strange feeling that I am wagging my tail 15% fewer times than I used to wag it."

"Oh look. He already answered me."

"Dear Newton, I only practice general vet work. I need to refer you to a dog tail specialist."

Connections to Previous Learning

- Write common fractions as percents, only include halves, fourths, tenths, and hundredths.

- Use equivalent forms of fractions, decimals, and percents.
- Estimate the results of computations with percents, and verify reasonableness.

- Solve percent problems involving percents of increase and decrease.
- Solve percent problems including discounts, markups, simple interest, taxes, and tips.

Math in History

The concept of percent goes back to Roman times. However, the percent symbol is more recent.

★ Percent has been used since the end of the fifteenth century in business problems such as computing interest, profit, and taxes. However, the idea had its origin much earlier. When the Roman emperor Augustus levied a tax on all goods sold at auction, the rate was $\frac{1}{100}$.

★ In the Middle Ages, as large denominations of money came to be used, 100 became a common base for computation. Italian manuscripts of the fifteenth century contained such expressions as "20 p 100" to indicate 20%. The percent sign, %, evolved from a symbol introduced in an anonymous Italian manuscript from 1425. Instead of "per 100" or "P cento," which were common at that time, this author used the symbol ͦ⁄ₒ. The current symbol, using a slanted line, is relatively modern.

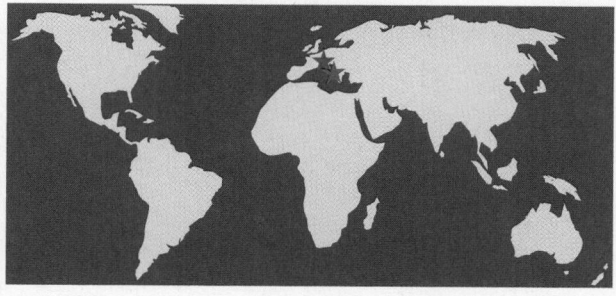

Pacing Guide for Chapter 4

Chapter Opener	1 Day
Section 1 Activity Lesson	 1 Day 1 Day
Section 2 Activity Lesson	 1 Day 1 Day
Study Help / Quiz	1 Day
Section 3 Activity Lesson	 1 Day 1 Day
Section 4 Activity Lesson	 1 Day 1 Day
Quiz / Chapter Review	1 Day
Chapter Test	1 Day
Standardized Test Practice	1 Day
Total Chapter 4	13 Days
Year-to-Date	73 Days

Check Your Resources

- Record and Practice Journal
- Resources by Chapter
- Skills Review Handbook
- Assessment Book
- Worked-Out Solutions

The Dynamic Planning Tool
Editable Teacher's Resources at
BigIdeasMath.com

Math Background Notes

Additional Topics for Review

- The division algorithm
- Greatest common factor
- Simplifying fractions
- Place value

Try It Yourself

1. 6%
2. 100%
3. 80%
4. $0.35, \dfrac{7}{20}$
5. $60\%, \dfrac{3}{5}$
6. 52%, 0.52
7. $0.1, \dfrac{1}{10}$
8. $85\%, \dfrac{17}{20}$
9. 20%, 0.2

Record and Practice Journal

1. 18%
2. 10%
3. 58%
4. 93%
5. 0.625
6. 0.525
7. $\dfrac{13}{50}$
8. $\dfrac{79}{100}$
9. $\dfrac{13}{20}$
10. 65%
11. 94%
12. $\dfrac{13}{25}$
13. $\dfrac{31}{100}$
14. 6%
15. 84%
16. 0.22
17. 1.91

18–20.

Percent	Decimal	Fraction
45%	0.45	$\dfrac{9}{20}$
73%	0.73	$\dfrac{73}{100}$
30%	0.3	$\dfrac{3}{10}$

Vocabulary Review

- Numerator
- Denominator
- Equivalent fraction
- Percent

Writing Percents Using Models

- Students should know how to work with models and percents.
- Remind students that the word percent comes from per cent, or per one hundred.
- To express what percent of the model is shaded, students should count the number of blocks per hundred that are shaded.
- If time permits, you may want to provide examples of percents greater than 100%.

Writing Percents, Decimals, and Fractions

- Students know how to convert between percents, decimals, and fractions. Students may require additional practice to achieve mastery.
- **Multiple Representations:** To convert from fractions to decimals, suggest that students convert the given fraction to an equivalent fraction with a denominator that is a power of ten. In Example 3, students can rewrite $\dfrac{3}{5}$ as $\dfrac{6}{10}$. Students that have mastered place value will realize that six-tenths can be expressed as the decimal 0.6 without having to employ the division algorithm. Alternatively, students could simply use the division algorithm to complete Example 3.

Reteaching and Enrichment Strategies

If students need help. . .	If students got it. . .
Record and Practice Journal • Fair Game Review Skills Review Handbook Lesson Tutorials	Game Closet at *BigIdeasMath.com* Start the next section

What You Learned Before

Writing Percents Using Models

What percent of the model is shaded?

Example 1

$$= \frac{70}{100}$$

$$= 70\%$$

Example 2

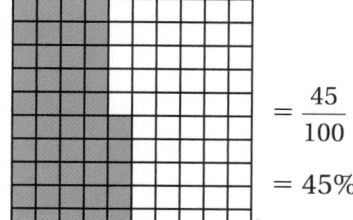

$$= \frac{45}{100}$$

$$= 45\%$$

Try It Yourself
What percent of the model is shaded?

1.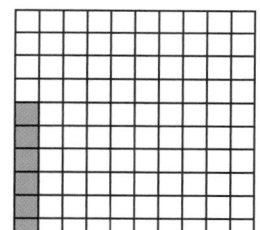

2.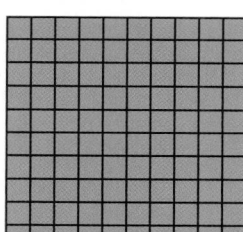

3.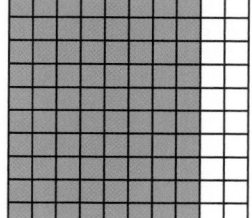

Writing Decimals, Percents, and Fractions

Example 3 Write $\frac{3}{5}$ as a decimal.

$$\frac{3}{5} = \frac{3 \cdot 2}{5 \cdot 2} = \frac{6}{10} = 0.6$$

Example 4 Write $\frac{3}{5}$ as a percent.

$$\frac{3}{5} = \frac{3 \cdot 20}{5 \cdot 20} = \frac{60}{100} = 60\%$$

> Multiply to make the denominator 100.

Try It Yourself
Copy and complete the table.

	Percent	Decimal	Fraction
4.	35%		
5.		0.6	
6.			$\frac{13}{25}$

	Percent	Decimal	Fraction
7.	10%		
8.		0.85	
9.			$\frac{1}{5}$

Essential Question How can you use models to estimate percent questions?

1 ACTIVITY: Estimating a Percent

Work with a partner. Estimate the locations of 50%, 75%, 40%, 6%, and 65% on the model. 50% is done for you.

0%   50%   100%

2 ACTIVITY: Estimating a Part of a Number

The statement "25% of 12 is 3" has three numbers. In real-life problems, any one of these numbers can be unknown.

Part → $\dfrac{3}{12}$ = 0.25 = 25% ← Percent
Whole →

Which number is missing?	Question	Type of Question
3	What is 25% of 12?	Find a part of a number.
25%	3 is what percent of 12?	Find a percent.
12	3 is 25% of what?	Find the whole.

Work with a partner. Estimate the answer to each question using a model.

a. Sample: What number is 50% of 30?

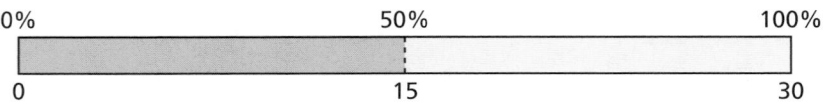

0%   50%   100%

0   15   30

So, from the model, 15 is 50% of 30.

b. What number is 75% of 30?   **c.** What number is 40% of 30?

d. What number is 6% of 30?   **e.** What number is 65% of 30?

Laurie's Notes

Introduction

For the Teacher

- **Goal:** Students will use the percent bar model to help solve three types of percent problems.

Motivate

- Share with students that sometimes their thinking can get *scrambled up* while solving percent problems, so an egg model would be a good way to introduce the chapter!
- Use an egg carton to ask and to help visualize a few simple percent problems.
  - "What is 75% of 12?" 9
  - "3 is what percent of 12?" 25%
  - "12 is 50% of what number?" 24

Activity Notes

Activity 1

- **FYI:** You may want to begin with a quick review of fractional equivalents of the following common percents:
  10%, 20%, 30%, 40%, 60%, 70%, 80%, 90%, 25%, 50%, 75%, $33\frac{1}{3}$%, $66\frac{2}{3}$%

- **Representation:** The percent bar model is an effective tool for estimating an answer, or judging the reasonableness of an answer if students have an understanding of fractional parts of a whole.
- The length of the bar is 100%, the whole. Percents near 50% are about $\frac{1}{2}$ of the whole.
- Students should be able to judge percents near 25% $\left(\frac{1}{4}\right)$ and 75% $\left(\frac{3}{4}\right)$.

- Students should locate the percents on the same model.
- When students have finished, draw a percent bar model on the board. Have volunteers share their answers.
- Remind students that these are approximations. Check for reasonableness in their approximations. For example, 40% is closer to 50% than 25%.

Activity 2

- Use an egg carton as a visual model when discussing the three numbers in the statement "25% of 12 is 3."
- Students work with a partner to answer the questions.
- **FYI:** Some students may find it helpful to use a long strip of paper that they can fold or write on when answering questions.
- **Summarize:** In each of the questions, the whole (30) was known and a part (percent) of it was found. Because all of the percents are less than 100%, all of the parts are less than 30 (the whole).
- **Extension:** "What number is 150% of 30?" 45

Previous Learning

Students should know how to solve simple percent problems.

Activity Materials
Introduction
• egg carton (one dozen)

Start Thinking! and Warm Up

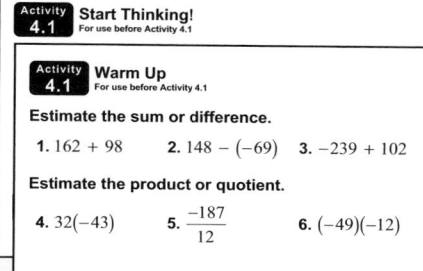

Activity 4.1 Start Thinking!
For use before Activity 4.1

Activity 4.1 Warm Up
For use before Activity 4.1

Estimate the sum or difference.

1. $162 + 98$ **2.** $148 - (-69)$ **3.** $-239 + 102$

Estimate the product or quotient.

4. $32(-43)$ **5.** $\dfrac{-187}{12}$ **6.** $(-49)(-12)$

4.1 Record and Practice Journal

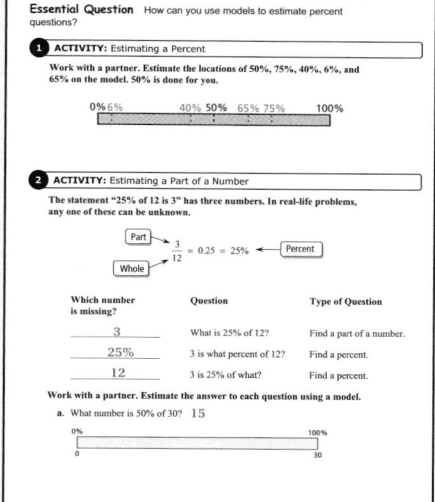

Differentiated Instruction

Visual

Some students will benefit from seeing how fractions and percents relate. Draw a circle on the board. Write 100% on top of the circle and 1 underneath. Explain that both of these values describe the area of the circle. Draw one-half of a circle and one-fourth of a circle and ask students to give you two representations.

$\frac{1}{2}$ and 50%, $\frac{1}{4}$ and 25%

4.1 Record and Practice Journal

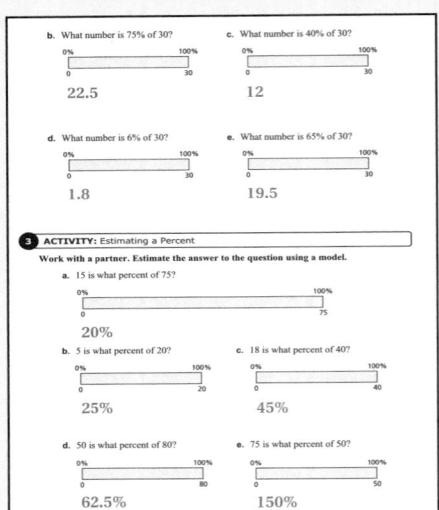

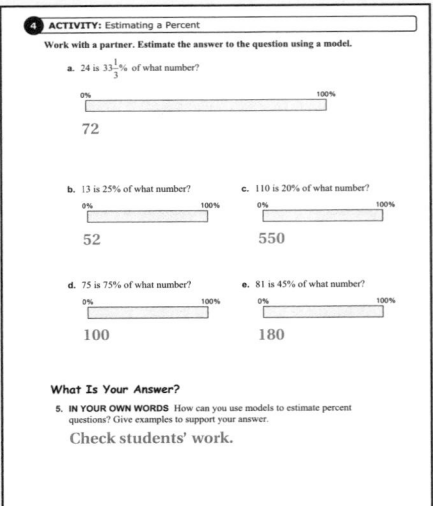

Laurie's Notes

Activity 3

- Encourage students to draw a percent bar model for each problem. You could also provide strips of paper that students can fold or write on.
- It is important that students be able to approximate the part in the whole model. Ask questions such as, "Is it greater than or less than one-half? Is it greater than or less than one-quarter?"
- **Common Error:** In part (e), students may misread 50 as the part of 75. Help students with this by drawing a bar model to represent the whole (50).
- ❓ "How much more is needed to make 75?" 25
- ❓ "How much do I need to extend the bar to show 25?"

  a length $\frac{1}{2}$ as long as 50
- ❓ "So, 75 is what percent of 50?" 150%

Activity 4

- For most students, this is the most challenging of the 3 types of percent problems.
- Talk about the model shown for the sample. The 24 is drawn first and you know it is $\frac{1}{3}$ of the whole $\left(33\frac{1}{3}\%\right)$. Next, draw two more thirds. Each one represents 24.
- **Strategy:** You are given the part of the whole. Add enough additional parts until you have a whole. For part (d), you know 3 equal parts would make 75, so each part must be 25.
- You may need to offer a hint for part (e). Ask what 81 and 45 have as a common factor. 9 From that information, you can conclude that 5% must be 9.

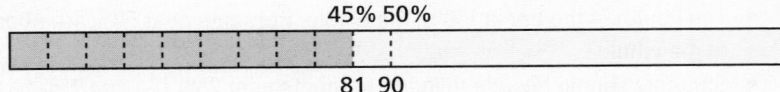

- When students have finished, have them describe their strategy to their classmates. Hearing others' thinking is helpful.

Closure

- Use the model shown. What 3 questions could be asked? "24 is what percent of 60?"; "What is 40% of 60?"; "24 is 40% of what number?"

0%	20%	40%	60%	80%	100%
0	12	24	36	48	60

The Dynamic Planning Tool
Editable Teacher's Resources at *BigIdeasMath.com*

3 ACTIVITY: Estimating a Percent

Work with a partner. Estimate the answer to the question using a model.

0%   100%

a. **Sample:** 15 is what percent of 75?

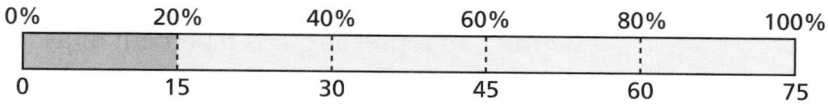

0%   20%   40%   60%   80%   100%

0   15   30   45   60   75

⋮⋅ So, 15 is 20% of 75.

b. 5 is what percent of 20?   c. 18 is what percent of 40?

d. 50 is what percent of 80?   e. 75 is what percent of 50?

4 ACTIVITY: Estimating a Whole

Work with a partner. Estimate the answer to the question using a model.

0%   100%

a. **Sample:** 24 is $33\frac{1}{3}$% of what number?

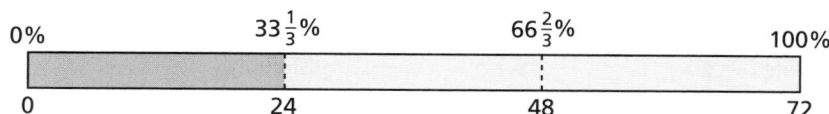

0% $33\frac{1}{3}$% $66\frac{2}{3}$%   100%

0   24   48   72

⋮⋅ So, 24 is $33\frac{1}{3}$% of 72.

b. 13 is 25% of what number?   c. 110 is 20% of what number?

d. 75 is 75% of what number?   e. 81 is 45% of what number?

What Is Your Answer?

5. **IN YOUR OWN WORDS** How can you use models to estimate percent questions? Give examples to support your answer.

Practice

Use what you learned about estimating percent questions to complete Exercises 4–9 on page 162.

A **percent** is a ratio whose denominator is 100. Here are two examples.

$$4\% = \frac{4}{100} = 0.04 \qquad\qquad 25\% = \frac{25}{100} = 0.25$$

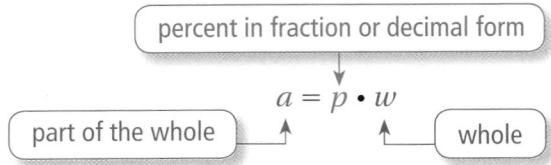 **Key Idea**

The Percent Equation

Words To represent "*a* is *p* percent of *w*," use an equation.

percent in fraction or decimal form

$$a = p \cdot w$$

part of the whole ↑   ↑ whole

Numbers $15 = 0.5 \cdot 30$

EXAMPLE 1   Finding a Part of a Number

What number is 24% of 50?   **Estimate** 0% 25% 100%

0   12.5   50

 **Common Error**

Remember to convert a percent to a fraction or decimal before using the percent equation. For Example 1, write 24% as $\frac{24}{100}$.

$a = p \cdot w$	Write percent equation.
$= \dfrac{24}{100} \cdot 50$	Substitute $\dfrac{24}{100}$ for *p* and 50 for *w*.
$= 12$	Multiply.

∴ So, 12 is 24% of 50.   **Reasonable?** $12 \approx 12.5$ ✓

EXAMPLE 2   Finding a Percent

9.5 is what percent of 25?   **Estimate** 0% 40% 100%

0   10   25

$a = p \cdot w$	Write percent equation.
$9.5 = p \cdot 25$	Substitute 9.5 for *a* and 25 for *w*.
$0.38 = p$	Divide each side by 25.

∴ Because 0.38 equals 38%,   **Reasonable?** $38\% \approx 40\%$ ✓
9.5 is 38% of 25.

 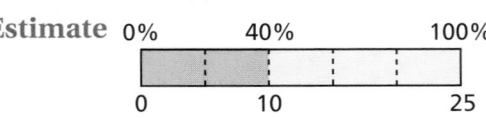

Laurie's Notes

Introduction

Connect
- **Yesterday:** Students used the percent bar model to explore three types of percent problems.
- **Today:** Students will use the percent equation to solve three types of percent problems.

Motivate
- The 2007 population of the United States was approximately 300 million (*Source:* U.S. Census Bureau) with about 25% being under 18 years old. About how many people in the U.S. are under the age of 18?
  about 75,000,000

Lesson Notes

Key Idea
- Write the Key Idea.
- **Connection:** Students should know how to find a percent of a number by multiplying. The percent equation builds upon this idea to find the missing percent or the unknown whole. When you know two of the three quantities in this equation, you can solve for the third.
- To help students think through how the equation will be used, use a numeric example ($2 \times 4 = 8$) or a variable example ($a \cdot b = c$).
  - "If you know *a* and *b*, how do you solve for *c*?" Multiply *a* and *b*.
  - "If you know *a* and *c*, how do you solve for *b*?" Divide *c* by *a*.
  - "If you know *b* and *c*, how do you solve for *a*?" Divide *c* by *b*.
- **FYI:** Students often get lost in the language of these problems. It is important to help students translate the problems and make sense of the information that is given.

Example 1
- Another way to phrase this question is "24% of 50 is what number?"
- **Estimate:** 24% is close to 25%, and 25% is $\frac{1}{4}$.
- "What is $\frac{1}{4}$ of 50?" 12.5
- If time permits, write 24% as a decimal and work the problem again.

Example 2
- Read the example as "9.5 is a part of 25."
- "Is 9.5 more or less than half of 25?" less
- Draw the percent bar model explaining that it represents 25. Draw the half mark (50%) and ask how much that would represent. 12.5
- Now draw the quarter mark (25%) and ask how much that would represent. 6.25 Through this process, students should recognize that 9.5 is between 25% and 50% of 25.
- **Common Error:** Students may forget that the decimal answer to the division problem needs to be rewritten as a percent.
- Note that the percent bar model is divided into 5 equal parts instead of 4.

Goal Today's lesson is finding **percents** using the percent equation.

Start Thinking! and Warm Up

Lesson 4.1 Warm Up
For use before Lesson 4.1

Lesson 4.1 Start Thinking!
For use before Lesson 4.1

At a soccer game, your team scored 6 goals during regular play and 4 goals on penalty kicks. Use a model to show the percent of goals that were not scored by penalty kicks.

Extra Example 1
What number is 73% of 200? 146

Extra Example 2
36.4 is what percent of 40? 91%

Laurie's Notes

Extra Example 3

18 is 15% of what number? 120

 On Your Own

1. $a = 0.1 \cdot 20$; 2
2. $a = 1.5 \cdot 40$; 60
3. $3 = p \cdot 600$; 0.5%
4. $18 = p \cdot 20$; 90%
5. $8 = 0.8 \cdot w$; 10
6. $90 = 0.18 \cdot w$; 500

Extra Example 4

Your total cost for lunch is $18.50 for food and $1.48 for tax.

a. Find the percent of sales tax on the food total. 8%
b. Find the amount of an 18% tip on the food total. $3.33

 On Your Own

7. $5.50

English Language Learners

Vocabulary

English learners may have trouble identifying which is the *whole* and which is the *part of the whole* in a percent equation. Have students write percent equations for the statements "20% of 300 is 60" and "125% of 50 is 62.5." Suggest that they start by substituting the percent *p* into the equation. Next, substitute the whole *w*. In most cases, this is the number after the word *of*. The remaining number is the part of the whole *a*.

Example 3

- This type of problem, finding a whole, is a bit harder. Knowing fractional equivalents is extremely helpful in developing a sense about the size of the answer.
- **?** "What is the part?" 39 "So, 39 is a part of something."
- **?** "How big of a part is it, approximately?" 52%, about half
- Help students reason that if 39 is half of something, the whole must be about 80. Only at this point does it make sense to translate what is known into an equation. 39 is 52% of some number.
- **Common Error:** Students may divide 39 by 52 and ignore the decimals completely.

On Your Own

- Have students work with a partner on these problems. Encourage students to sketch the percent bar model and record the information they know. Then write the percent equation.
- Have students put their work on the board.

Example 4

- **?** "In addition to paying for what you ordered (food and drink), what other costs are there when you eat at a restaurant?" sales tax and tip
- Review decimal operations as you work through each part.

On Your Own

- Model finding 10%, and then double for 20%.

Closure

- **Exit Ticket:** Use the percent equation to answer the question, 12 is what percent of 48? 25%

The Dynamic Planning Tool
Editable Teacher's Resources at *BigIdeasMath.com*

EXAMPLE **3** **Finding a Whole**

39 is 52% of what number? Estimate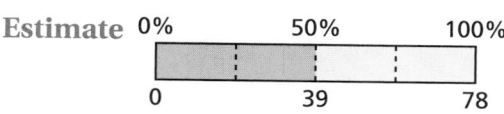

$a = p \cdot w$ Write percent equation.

$39 = 0.52 \cdot w$ Substitute 39 for a and 0.52 for p.

$75 = w$ Divide each side by 0.52.

⋮ So, 39 is 52% of 75.   **Reasonable?** $75 \approx 78$ ✓

On Your Own

Now You're Ready
Exercises 10–17

Write and solve an equation to answer the question.

1. What number is 10% of 20?

2. What number is 150% of 40?

3. 3 is what percent of 600?

4. 18 is what percent of 20?

5. 8 is 80% of what number?

6. 90 is 18% of what number?

EXAMPLE **4** **Real-Life Application**

8th Street Cafe

DATE: MAY04'10 05:45PM
TABLE: 29
SERVER: CHARITY

Food Total	**27.50**
Tax	**1.65**
Subtotal	**29.15**

TIP: _____

TOTAL: _____

Thank You

a. **Find the percent of sales tax on the food total.**

b. **Find the amount of a 16% tip on the food total.**

a. Answer the question: $1.65 is what percent of $27.50?

$a = p \cdot w$ Write percent equation.

$1.65 = p \cdot 27.50$ Substitute 1.65 for a and 27.50 for w.

$0.06 = p$ Divide each side by 27.50.

⋮ Because 0.06 equals 6%, the percent of sales tax is 6%.

b. Answer the question: What tip amount is 16% of $27.50?

$a = p \cdot w$ Write percent equation.

$= 0.16 \cdot 27.50$ Substitute 0.16 for p and 27.50 for w.

$= 4.40$ Multiply.

⋮ So, the amount of the tip is $4.40.

On Your Own

7. **WHAT IF?** In Example 4, find the amount of a 20% tip on the food total.

 Vocabulary and Concept Check

1. **VOCABULARY** Write the percent equation in words.

2. **REASONING** A number *n* is 150% of number *m*. Is *n* *greater than*, *less than*, or *equal to m*? Explain your reasoning.

3. **DIFFERENT WORDS, SAME QUESTION** Which is different? Find "both" answers.

What number is 20% of 55?	55 is 20% of what number?
20% of 55 is what number?	0.2 • 55 is what number?

 **Practice and Problem Solving**

Estimate the answer to the question using a model.

4. What number is 24% of 80?
5. 15 is what percent of 40?
6. 15 is 30% of what number?
7. What number is 120% of 70?
8. 20 is what percent of 52?
9. 48 is 75% of what number?

Write and solve an equation to answer the question.

① 10. 20% of 150 is what number?
   11. 45 is what percent of 60?

② 12. 35% of what number is 35?
   13. 32% of 25 is what number?

③ 14. 29 is what percent of 20?
   15. 0.5% of what number is 12?

16. What percent of 300 is 51?
   17. 120% of what number is 102?

ERROR ANALYSIS Describe and correct the error in using the percent equation.

18. What number is 35% of 20?

✗ $a = p \cdot w$
$= 35 \cdot 20$
$= 700$

19. 30 is 60% of what number?

✗ $a = p \cdot w$
$= 0.6 \cdot 30$
$= 18$

20. **BASEBALL** A pitcher throws 75 pitches. Of these, 72% were strikes. How many strikes did the pitcher throw?

21. **FUNDRAISING** Your school raised 125% of its fundraising goal. The school raised $6750. What was the goal?

22. **SURFBOARD** The sales tax on a surfboard is $12. What is the percent of sales tax?

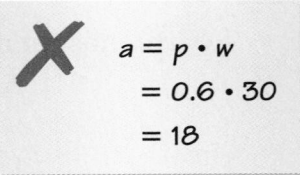

SALE
$240

Assignment Guide and Homework Check

Level	Day 1 Activity Assignment	Day 2 Lesson Assignment	Homework Check
Basic	4–9, 32–36	1–3, 11–17 odd, 18–22, 28	11, 15, 20, 28
Average	4–9, 32–36	1–3, 11–19 odd, 22–26, 28, 29	11, 15, 23, 28
Advanced	4–9, 32–36	1–3, 18, 19, 22–27, 29–31	24, 26, 27, 30

Common Errors

- **Exercises 4–17** Students may not know what number to substitute for each variable. Walk through each type of question with the students. Emphasize that the word *is* means *equals*, and *of* means *to multiply*. Tell students to write the question and then write the meaning of each word or group of words underneath.
- **Exercises 20–22** Students will mix up the whole and the part when trying to write the percent equation for the word problems. Ask them to identify each part of the equation before writing it in the equation format. For example, in Exercise 20, ask "How many total pitches did the pitcher throw?" 75 "Which variable in the percent equation does this number represent?" The whole Continue to ask questions for each of the variables.
- **Exercise 28** Students may not realize that the sum of the parts of a circle graph equals 100%.

4.1 Record and Practice Journal

Write and solve an equation to answer the question.

1. 40% of 60 is what number?

   $a = 0.4 \cdot 60;\ 24$

2. 17 is what percent of 50?

   $17 = p \cdot 50;\ 34\%$

3. 38% of what number is 57?

   $57 = 0.38 \cdot w;\ 150$

4. 44% of 25 is what number?

   $a = 0.44 \cdot 25;\ 11$

5. 52 is what percent of 50?

   $52 = p \cdot 50;\ 104\%$

6. 150% of what number is 18?

   $18 = 1.5 \cdot w;\ 12$

7. You put 60% of your paycheck into your savings account. Your paycheck is $235. How much money do you put in your savings account?

   $141

8. You made lemonade and iced tea for a school fair. You made 15 gallons of lemonade and 60% is gone. About 52% of the iced tea is gone. The ratio of gallons of lemonade to gallons of iced tea was 3 : 2.

   a. How many gallons of lemonade are left?

   6 gallons

   b. How many gallons of iced tea did you make?

   10 gallons

   c. About how many gallons of iced tea are left?

   4.8 gallons

1. A part of the whole is equal to a percent times the whole.
2. greater than; Because $150\% = 1.5$, $n = 1.5 \cdot m$.
3. 55 is 20% of what number?; 275; 11

 Practice and Problem Solving

4. 20
5. 37.5%
6. 50
7. 84
8. about 38%
9. 64
10. $a = 0.2 \cdot 150;\ 30$
11. $45 = p \cdot 60;\ 75\%$
12. $35 = 0.35 \cdot w;\ 100$
13. $a = 0.32 \cdot 25;\ 8$
14. $29 = p \cdot 20;\ 145\%$
15. $12 = 0.005 \cdot w;\ 2400$
16. $51 = p \cdot 300;\ 17\%$
17. $102 = 1.2 \cdot w;\ 85$
18. The percent was not converted to a decimal or fraction.

    $a = p \cdot w$
    $= 0.35 \cdot 20$
    $= 7$
19. 30 represents the part of the whole.

    $30 = 0.6 \cdot w$
    $50 = w$
20. 54 strikes
21. $5400
22. 5%

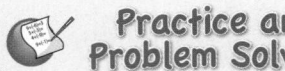
Practice and Problem Solving

23. 26 years old

24. 70 years old

25. 56 signers

26. 70%

27. If the percent is less than 100%, the percent of a number is less than the number. If the percent is equal to 100%, the percent of a number will equal the number. If the percent is greater than 100%, the percent of a number is greater than the number.

28. a. 80 students

b. 30 students

29. See *Taking Math Deeper.*

30. false; If W is 25% of Z, then $Z : W$ is $100 : 25$, because Z represents the whole.

31. 92%

Fair Game Review

32. 0.6   **33.** 0.88

34. 0.25   **35.** 0.36

36. A

Mini-Assessment

Write and solve an equation to answer the question.

1. 52 is what percent of 80? 65%

2. 28 is 35% of what number? 80

3. What number is 25% of 92? 23

4. What percent of 250 is 60? 24%

5. A new laptop computer costs $800. The sales tax on the computer is $48. What is the percent of sales tax? 6%

Taking Math Deeper

Exercise 29

Any problem that has this much given information is difficult for students. Encourage students to begin by organizing the information with a table or a diagram. When organizing the information, it is a good idea to add as much other information as you can find... *before looking at the questions.*

① Organize the given information.

② Add other information.

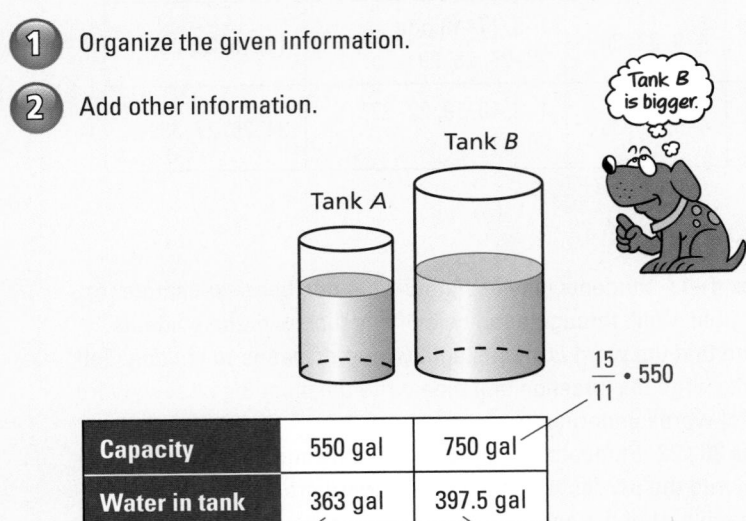

Capacity	550 gal	750 gal
Water in tank	363 gal	397.5 gal

0.66 • 550   0.53 • 750

$\frac{15}{11}$ • 550

③ Now the questions are easy.
  a. Tank *A* has 363 gallons of water.
  b. The capacity of tank *B* is 750 gallons.
  c. Tank *B* has 397.5 gallons of water.

Project

Use your school library or the Internet to research how a water tower works. How does the water get into the tower? How long does it take for the water to drain out? How often is the water completely exchanged; in other words, if a gallon goes in today when will that gallon be draining out? What other interesting things did you discover?

Reteaching and Enrichment Strategies

If students need help...	If students got it...

Resources by Chapter
  • Practice A and Practice B
  • Puzzle Time
Record and Practice Journal Practice
Differentiating the Lesson
Lesson Tutorials
Skills Review Handbook
	Resources by Chapter
	  • Enrichment and Extension
	Start the next section

PUZZLE There were *w* signers of the Declaration of Independence. The youngest was Edward Rutledge, who was *x* years old. The oldest was Benjamin Franklin, who was *y* years old.

23. *x* is 25% of 104. What was Rutledge's age?

24. 7 is 10% of *y*. What was Franklin's age?

25. *w* is 80% of *y*. How many signers were there?

26. *y* is what percent of $(w + y - x)$?

Favorite Sport

Other

40.0%

37.5%

27. REASONING How can you tell whether the percent of a number will be *greater than*, *less than*, or *equal to* the number?

28. SURVEY In a survey, a group of students were asked their favorite sport. "Other" sports were chosen by 18 people.

  a. How many students participated?

  b. How many chose football?

29. WATER TANK Water tank *A* has a capacity of 550 gallons and is 66% full. Water tank *B* is 53% full. The ratio of the capacity of tank *A* to tank *B* is $11 : 15$.

  a. How much water is in tank *A*?

  b. What is the capacity of tank *B*?

  c. How much water is in tank *B*?

30. TRUE OR FALSE? Tell whether the statement is *true* or *false*. Explain your reasoning.

If *W* is 25% of *Z*, then $Z : W$ is $75 : 25$.

31. *Reasoning* The table shows your test results for math class. What test score is needed on the last exam to earn 90% of the total points?

Test Score	Point Value
83%	100
91.6%	250
88%	150
?	300

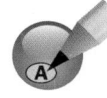

Fair Game Review *What you learned in previous grades & lessons*

Simplify. Write as a decimal. *(Skills Review Handbook)*

32. $\dfrac{10 - 4}{10}$

33. $\dfrac{25 - 3}{25}$

34. $\dfrac{105 - 84}{84}$

35. $\dfrac{170 - 125}{125}$

36. MULTIPLE CHOICE There are 160 people in a grade. The ratio of boys to girls is 3 to 5. Which proportion can you use to find the number *x* of boys? *(Section 3.4)*

Ⓐ $\dfrac{3}{8} = \dfrac{x}{160}$

Ⓑ $\dfrac{3}{5} = \dfrac{x}{160}$

Ⓒ $\dfrac{5}{8} = \dfrac{x}{160}$

Ⓓ $\dfrac{3}{5} = \dfrac{160}{x}$

4.2 Percents of Increase and Decrease

Essential Question What is a percent of decrease? What is a percent of increase?

1 ACTIVITY: Percent of Decrease

Each year in the Columbia River Basin, adult salmon swim up river to streams to lay eggs and hatch their young.

To go up the river, the adult salmon use fish ladders. But, to go down the river, the young salmon must pass through several dams.

There are electric turbines at each of the eight dams on the main stem of the Columbia and Snake Rivers. About 88% of the young salmon pass through these turbines unharmed.

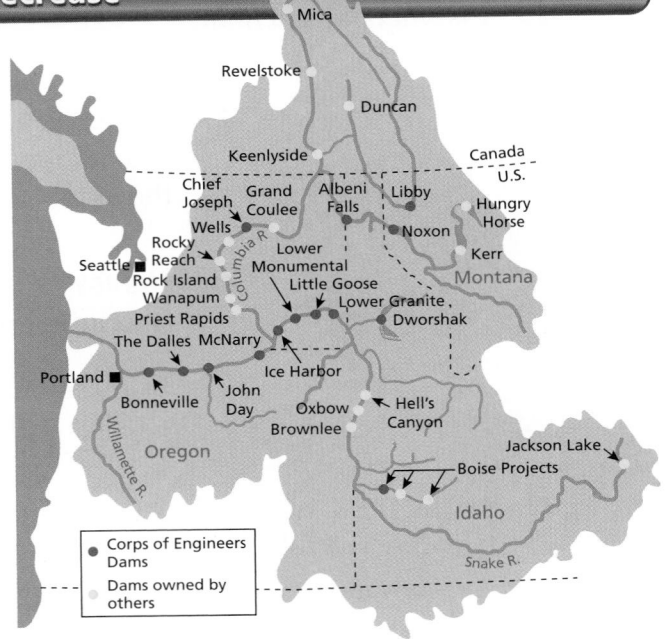

Copy and complete the table and the bar graph to show the number of young salmon that make it through the dams.

Dam	0	1	2	3	4	5	6	7	8
Salmon	1000	880	774						

88% of 1000 = 0.88 • 1000
= 880

88% of 880 = 0.88 • 880
= 774.4 ≈ 774

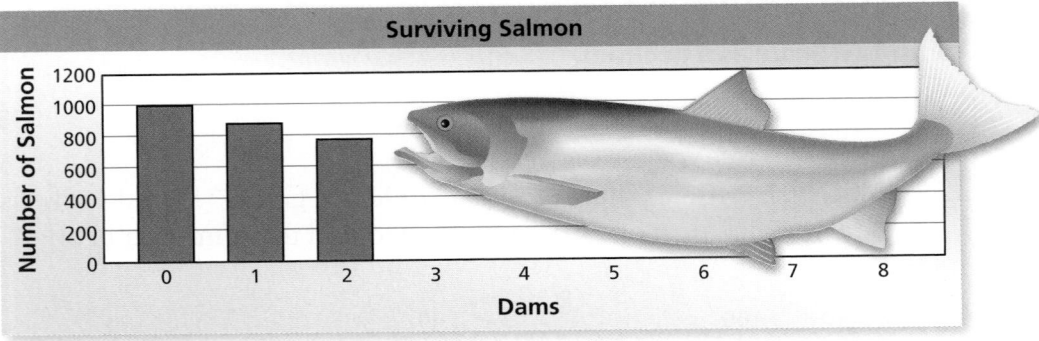

Laurie's Notes

Introduction

For the Teacher

- **Goal:** Students will explore percent decrease and increase by working through real-life problems.
- Use of a calculator will help facilitate the computation so that students can focus on how the numbers are changing.

Motivate

- Talk about the new compact fluorescent light bulbs. Fluorescent light bulbs use 75% less energy than incandescent light bulbs (percent decrease) and last up to 900% longer (percent increase).
- If your school has replaced incandescent light bulbs with fluorescent light bulbs, discuss the potential savings.

Activity Notes

Activity 1

- **Representation:** Although the difference between the decimal point and the multiplication dot are clear in the textbook, it may not be as clear when you write it on the board. You may consider using parentheses to show the multiplication: (0.88)(1000).
- Have a student read the story information.
- **?** "What percent of salmon makes it through each dam?" 88%
- **?** "What percent of salmon does not make it through each dam?" 12%
- Discuss the general concept of fewer salmon at dam 2 than dam 1.
- **FYI:** Electric turbines in the dams generate electricity. These turbines are what affect the survival rate of the young salmon.
- Students should follow the two calculations shown. Remind students to round their answers to a whole number of salmon at each dam.
- Check students' results before completing the graph.
- **Big Idea:** Each entry is 12% less than the previous entry. The *amount* of decrease is changing, but the *percent* is not.
- Have students describe patterns they observe in the numbers and the bar graph.

Previous Learning

Students should be able to find a percent of a number, round decimal values, and convert between fractions, decimals, and percents.

Activity Materials
Textbook
• calculators

Start Thinking! and Warm Up

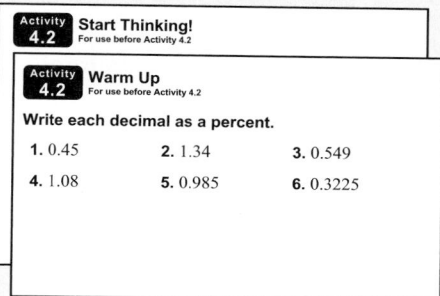

Activity 4.2 Start Thinking!
For use before Activity 4.2

Activity 4.2 Warm Up
For use before Activity 4.2

Write each decimal as a percent.

1. 0.45   2. 1.34   3. 0.549
4. 1.08   5. 0.985   6. 0.3225

4.2 Record and Practice Journal

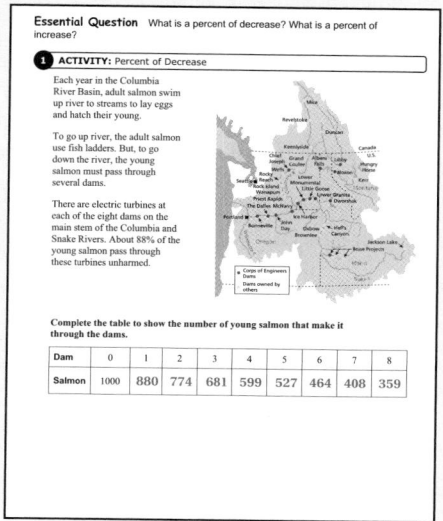

Essential Question What is a percent of decrease? What is a percent of increase?

1 ACTIVITY: Percent of Decrease

Each year in the Columbia River Basin, adult salmon swim up river to streams to lay eggs and hatch their young.

To go up river, the adult salmon use fish ladders. But, to go down the river, the young salmon must pass through several dams.

There are electric turbines at each of the eight dams on the main stem of the Columbia and Snake Rivers. About 88% of the young salmon pass through these turbines unharmed.

Complete the table to show the number of young salmon that make it through the dams.

Dam	0	1	2	3	4	5	6	7	8
Salmon	1000	880	774	681	599	527	464	408	359

English Language Learners

Visual Aid

Demonstrate writing a percent as a decimal. Locate the decimal point in the 7%.

   7.%

Draw two arrows to show that the decimal point moves two places *left*.

   .7.%

Write zeros to the left of the number if needed.

   007.%

Rewrite as a decimal with the decimal point two places to the left and without the percent sign.

   0.07

4.2 Record and Practice Journal

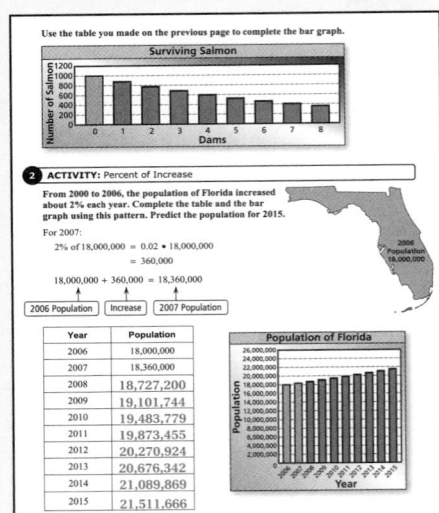

What Is Your Answer?

3. In Activity 1, by what percent does the number of young salmon decrease with each dam?

   12%

4. Describe real-life examples of a percent of decrease and a percent of increase.

   Check students' work.

5. IN YOUR OWN WORDS What is a percent of decrease? What is a percent of increase?

   percent of change when the original amount decreases or increases

Laurie's Notes

Activity 2

- Read the information. Look at the map and the information on current population.
- Work through the first year with the students. There are two steps involved: 1) find the amount the population has increased, and 2) add this amount to the current population.
- ? "How much did the population increase in 2007?" 360,000
- ? "What percent did the population increase in 2007?" 2%
- Remind students to round their answers to a whole number and add this number to the current population.
- Check students' results before completing the graph.
- **Big Idea:** Each entry is 2% more than the previous entry. The *amount* of increase is changing, but the *percent* is not.
- **Extension:** Discuss how projections are made based upon current trends.

What Is Your Answer?

- For Question 4, students could discuss this at home and bring ideas to class.

Closure

- "You scored 80 points on your first test. If your score increased 10% on the next test, what is your score?" 88 points

Technology For the Teacher

Dynamic Classroom

The Dynamic Planning Tool
Editable Teacher's Resources at *BigIdeasMath.com*

From 2000 to 2006, the population of Florida increased about 2% each year. Copy and complete the table and the bar graph using this pattern. Predict the population in 2015.

For 2007:

$$2\% \text{ of } 18{,}000{,}000 = 0.02 \cdot 18{,}000{,}000$$

$$= 360{,}000$$

$$18{,}000{,}000 + 360{,}000 = 18{,}360{,}000$$

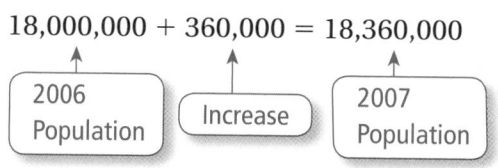

2006 Population   Increase   2007 Population

2006 Population 18,000,000

Year	Population
2006	18,000,000
2007	18,360,000
2008	
2009	
2010	
2011	
2012	
2013	
2014	
2015	

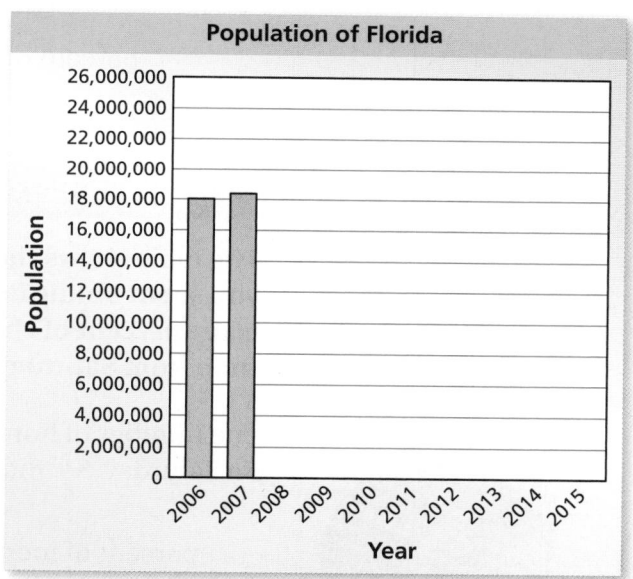

Population of Florida

What Is Your Answer?

3. In Activity 1, by what percent does the number of young salmon decrease with each dam?

4. Describe real-life examples of a percent of decrease and a percent of increase.

5. **IN YOUR OWN WORDS** What is a percent of decrease? What is a percent of increase?

Practice Use what you learned about percent of increase and percent of decrease to complete Exercises 13–18 on page 168.

4.2 Lesson

Check It Out
Lesson Tutorials
BigIdeasMath ✓com

Key Vocabulary
percent of change,
  p. 166
percent of increase,
  p. 166
percent of decrease,
  p. 166

A **percent of change** is the percent that a quantity changes from the original amount.

$$\text{percent of change} = \frac{\text{amount of change}}{\text{original amount}}$$

Key Idea

Percents of Increase and Decrease

When the original amount increases, the percent of change is called a **percent of increase**.

$$\text{percent of increase} = \frac{\text{new amount} - \text{original amount}}{\text{original amount}}$$

When the original amount decreases, the percent of change is called a **percent of decrease**.

$$\text{percent of decrease} = \frac{\text{original amount} - \text{new amount}}{\text{original amount}}$$

EXAMPLE 1 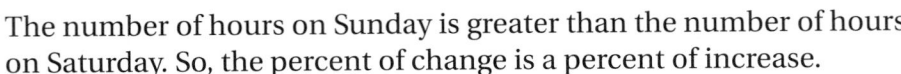 **Finding a Percent of Increase**

The table shows the number of hours you spent online last weekend. What is the percent of change in your online time from Saturday to Sunday?

Day	Hours Online
Saturday	2
Sunday	4.5

The number of hours on Sunday is greater than the number of hours on Saturday. So, the percent of change is a percent of increase.

$$\text{percent of increase} = \frac{\text{new amount} - \text{original amount}}{\text{original amount}}$$

$$= \frac{4.5 - 2}{2} \qquad \text{Substitute.}$$

$$= \frac{2.5}{2} \qquad \text{Subtract.}$$

$$= 1.25, \text{ or } 125\% \qquad \text{Write as a percent.}$$

∴ Your online time increased 125% from Saturday to Sunday.

On Your Own

Find the percent of change. Round to the nearest tenth of a percent, if necessary.

1. 10 inches to 25 inches

2. 57 people to 65 people

◀) Multi-Language Glossary at BigIdeasMath✓com.

Laurie's Notes

Introduction

Connect

- **Yesterday:** Students explored two real-life problems with quantities that decreased or increased by a percent.
- **Today:** Students will use a percent change formula to solve problems.

Motivate

- Cell phone ownership is increasing each year. Pose a question such as: "If 400 people in your neighborhood had a cell phone last year and one year later 500 people had a cell phone, what percent has cell phone ownership increased?"

Lesson Notes

Key Idea

- Explain the difference between *amount* of change and *percent* of change. Refer to the salmon and population activities.
- Use the cell phone example to help identify vocabulary:
  original amount $= 400$,
  amount of change $= 100$,
  percent of change $= \dfrac{100}{400} = 25\%$.

Example 1

- Read the information in the table to your students.
- **?** "Did the online use increase or decrease from Saturday to Sunday?" increase
- **?** "How much did the online use increase from Saturday to Sunday?" 2.5 h
- Have students write the equation, substitute the values, and then simplify. The original amount is 2. The new amount is 4.5. Because the number of hours increased, you are finding a **percent of increase**. Percent of increase $= \dfrac{4.5 - 2}{2} = 1.25 = 125\%$.
- **Common Error:** Students think the answer is 1.25. This decimal must still be converted to a percent. This often happens when the percent answer is greater than 100%.
- **Connection:** Draw a percent bar model of this problem.

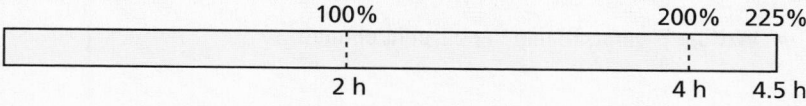

The percent of increase is 125% beyond the 100%.

On Your Own

- In Question 1, the length has more than doubled, so the percent of increase is greater than 100%.
- In Question 2, the number of people has not doubled, so the percent of increase is less than 100%.

Goal

Today's lesson is using a **percent change** formula to solve problems.

Start Thinking! and Warm Up

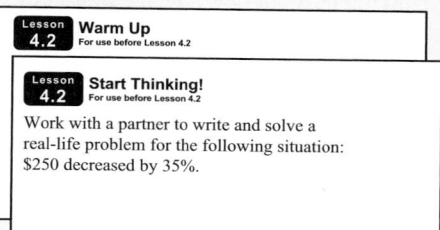

| Lesson 4.2 | **Warm Up** For use before Lesson 4.2 |

| Lesson 4.2 | **Start Thinking!** For use before Lesson 4.2 |

Work with a partner to write and solve a real-life problem for the following situation: $250 decreased by 35%.

Extra Example 1

Find the percent of change from 40 hours to 50 hours. increase of 25%

On Your Own

1. 150% increase
2. about 14.0% increase

Extra Example 2

Find the percent of change from 20 days to 12 days. decrease of 40%

 On Your Own

　　3. about 44.44% decrease

Extra Example 3

Your email account contains 135 messages. You delete 60% of the messages. How many messages are left? 54 messages

 On Your Own

　　4. 220 songs

Differentiated Instruction

Vocabulary

Make sure students understand the difference between *increased by 150%* and *increased to 150%*. The percent of increase in the first case is 150%. The percent of increase in the second case is 50%, because the increase does not include the original amount. By the same token, there is a difference between *decreased by 25%* and *decreased to 25%*. In the first case, it means taking 25% and leaving 75%. In the second case, it means to leave 25% and take away 75%.

Laurie's Notes

Example 2

- "How much did the number of home runs change each year?" decrease of 8, increase of 18, decrease of 8
- ❓ "What is the original amount?" 28
- ❓ "What is the new amount?" 20
- Students should now use the **percent of decrease** formula.
- This problem involves a number of skills: reading a bar graph, using the percent of decrease formula, converting a fraction to a decimal, and converting a decimal to a percent.
- The answer is rounded to the nearest tenth of a percent.
- ❓ "Is the percent change from 2006 to 2007 more or less than 100%? How do you know?" The number of home runs more than doubled, so the increase is greater than 100%.

On Your Own

- If you have small white boards available, have each student solve Question 3 on their white board. Have students hold up their white boards and then have students determine their mistakes.

Example 3

- Have students identify what information is given in the problem.
- ❓ What is the amount of increase or decrease? not given
- What percent are you given? 20% decrease
- Because you are given the percent of decrease (percent) and the original amount (whole), you can use the percent equation to find the amount of decrease (part).
- **Review:** The *amount* of change is 50 songs. The *percent* of change is 20%.
- **Common Error:** If students do not read carefully, they may answer 50 songs. 50 is how many songs were deleted. The question is asking how many songs are left.
- **Connection:** Find 10% using mental math and double to find 20%.
- **Alternative Method:** If you delete 20%, then you have 80% left on your MP3 player.

$$80\% \text{ of } 250 = 0.8 \cdot 250$$
$$= 200$$

 Closure

- **Writing Prompt:** To find the percent change…

Technology For the **Teacher**

The Dynamic Planning Tool
Editable Teacher's Resources at *BigIdeasMath.com*

EXAMPLE 2 · Finding a Percent of Decrease

The bar graph shows a softball player's home run totals. What was the percent of change from 2007 to 2008?

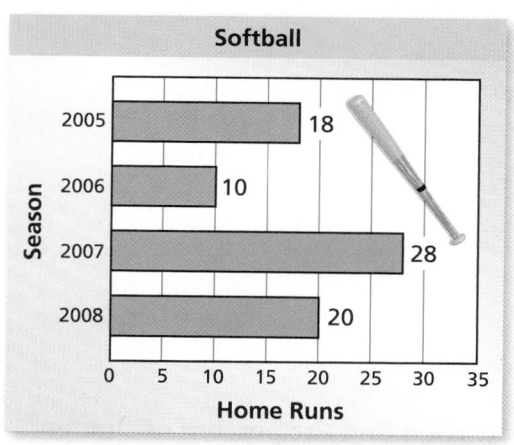

The number of home runs decreased from 2007 to 2008. So, the percent of change is a percent of decrease.

$$\text{percent of decrease} = \frac{\text{original amount} - \text{new amount}}{\text{original amount}}$$

$$= \frac{28 - 20}{28} \qquad \text{Substitute.}$$

$$= \frac{8}{28} \qquad \text{Subtract.}$$

$$\approx 0.286, \text{ or } 28.6\% \qquad \text{Write as a percent.}$$

⁘ The number of home runs decreased about 28.6%.

On Your Own

Now You're Ready
Exercises 4–11

3. What was the percent of change from 2005 to 2006?

EXAMPLE 3 · Standardized Test Practice

You have 250 songs on your MP3 player. You delete 20% of the songs. How many songs are left?

Ⓐ 50　　　　Ⓑ 150　　　　Ⓒ 200　　　　Ⓓ 300

Find the amount of decrease.

$$20\% \text{ of } 250 = 0.2 \cdot 250 \qquad \text{Write as multiplication.}$$

$$= 50 \qquad \text{Multiply.}$$

The decrease is 50 songs. So, there are $250 - 50 = 200$ songs left.

⁘ The correct answer is Ⓒ.

On Your Own

Now You're Ready
Exercises 13–22

4. WHAT IF? After deleting the 50 songs in Example 3, you add 10% more songs. How many songs are on the MP3 player?

 Vocabulary and Concept Check

1. **VOCABULARY** How do you know whether a percent of change is a *percent of increase* or a *percent of decrease*?

2. **NUMBER SENSE** Without calculating, which has a greater percent of increase?
   - 5 bonus points on a 50-point exam
   - 5 bonus points on a 100-point exam

3. **WRITING** What does it mean to have a 100% decrease?

 Practice and Problem Solving

Identify the percent of change as an *increase* or *decrease*. Then find the percent of change. Round to the nearest tenth of a percent, if necessary.

① ②   4. 12 inches to 36 inches

5. 75 people to 25 people

6. 50 pounds to 35 pounds

7. 24 songs to 78 songs

8. 10 gallons to 24 gallons

9. 72 paper clips to 63 paper clips

10. 16 centimeters to 44.2 centimeters

11. 68 miles to 42.5 miles

12. **ERROR ANALYSIS** Describe and correct the error in finding the percent increase from 18 to 26.

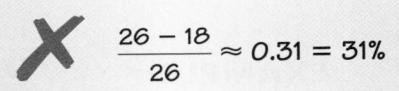
$$\frac{26 - 18}{26} \approx 0.31 = 31\%$$

Find the new amount.

③ 13. 8 meters increased by 25%

14. 15 liters increased by 60%

15. 50 points decreased by 26%

16. 25 penalties decreased by 32%

17. 68 students increased by 125%

18. 1000 grams decreased by 94%

19. 62 kilograms decreased by 32%

20. 124 ounces decreased by 67%

21. **ERROR ANALYSIS** Describe and correct the error in using the percent of change to find a new amount.

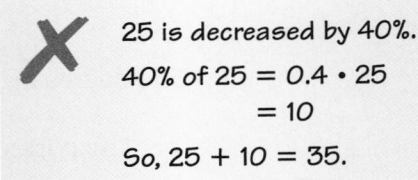
25 is decreased by 40%.
40% of 25 = 0.4 • 25
= 10
So, 25 + 10 = 35.

22. **VIDEO GAME** Last week, you finished Level 2 of a video game in 32 minutes. Today, you finish Level 2 in 28 minutes. What is your percent of change?

Assignment Guide and Homework Check

Level	Day 1 Activity Assignment	Day 2 Lesson Assignment	Homework Check
Basic	13–18, 34–38	1–3, 5–11 odd, 12, 19–22, 28	9, 12, 20, 22
Average	13–18, 34–38	1–3, 8–12, 19–21, 23, 25, 28, 29	9, 12, 20, 23, 28
Advanced	13–18, 34–38	1–3, 12, 21, 23–27, 30–33	12, 24, 27, 30

Common Errors

- **Exercises 4–11** Students may mix up where to place the numbers in the equation to find percent of change. When they do not put the numbers in the right place, they might find a negative number in the numerator. First, emphasize that students must know if it is increasing or decreasing before they start the problem. Next, tell students that the number in the denominator is going to be the original or starting number given for both increasing and decreasing percents of change. Finally, the numerator should never have a negative answer. If students get a negative number, it is because they found the wrong difference. The numerator is always the greater number minus the lesser number.

- **Exercises 13–20** Students may find the percent of the number and forget to add or subtract from the original amount. Remind them that these are two-step problems. Before evaluating, tell students to write down what needs to be done for each step.

4.2 Record and Practice Journal

Identify the percent of change as an *increase* or *decrease*. Then find the percent of change. Round to the nearest tenth of a percent, if necessary.

1. 25 points to 50 points
increase; 100%

2. 125 invitations to 75 invitations
decrease; 40%

3. 32 pages to 28 pages
decrease; 12.5%

4. 7 players to 10 players
increase; 42.9%

Find the new amount.

5. 120 books increased by 55%
186 books

6. 80 members decreased by 65%
28 members

7. One week, 72 people got a speeding ticket. The next week, only 36 people got a speeding ticket. What is the percent of change in speeding tickets?
50% decrease

8. The number of athletes participating in the Paralympics rose from 130 athletes in 1952 to 3806 athletes in 2004. What is the percent of change? Round your answer to the nearest tenth of a percent.
2827.7% increase

Technology For the Teacher
Answer Presentation Tool
QuizShow

Vocabulary and Concept Check

1. If the original amount decreases, the percent of change is a percent of decrease. If the original amount increases, the percent of change is a percent of increase.

2. 5 bonus points on a 50-point exam

3. The new amount is now 0.

Practice and Problem Solving

4. increase; 200%

5. decrease; 66.7%

6. decrease; 30%

7. increase; 225%

8. increase; 140%

9. decrease; 12.5%

10. increase; 176.3%

11. decrease; 37.5%

12. The denominator should be 18, which is the original amount.
$$\frac{26 - 18}{18} \approx 0.44 = 44\%$$

13. 10 m   **14.** 24 L

15. 37 points   **16.** 17 penalties

17. 153 students

18. 60 g

19. 42.16 kg   **20.** 40.92 oz

21. They should have subtracted 10 in the last step because 25 is decreased by 40%.
$$40\% \text{ of } 25 = 0.4 \cdot 25$$
$$= 10$$
So, $25 - 10 = 15$.

22. 12.5% decrease

23. increase; 100%

24. decrease; 25%

Practice and Problem Solving

25. increase; 133.3%

26. decrease; 70%

27. Increasing 20 to 40 is the same as increasing 20 by 20. So, it is a 100% increase. Decreasing 40 to 20 is the same as decreasing 40 by one-half of 40. So, it is a 50% decrease.

28. a. about 16.95% increase

   b. 161,391 people

29. a. 100% increase

   b. 300% increase

30. about 24.52% decrease

31. See Additional Answers.

32. See *Taking Math Deeper*.

33. 10 girls

Fair Game Review

34. 16   **35.** 35%

36. 100   **37.** 56.25

38. B

Mini-Assessment

Identify the percent of change as an *increase* or *decrease*. Then find the percent of change.

1. 15 meters to 36 meters increase; 140%

2. 20 songs to 70 songs increase; 250%

3. 90 people to 45 people decrease; 50%

4. 65 pounds to 40 pounds decrease; 38.5%

5. Yesterday, it took 40 minutes to drive to school. Today, it took 32 minutes to drive to school. What is your percent of change? The number of minutes it took to get to school decreased by 20%.

Taking Math Deeper

Exercise 32

This exercise is difficult because the percent of increase is given backwards. A good way to start is to use a table to organize the given information.

(1) Organize given information.

	Donation	Increase over previous year
This year	$10,120	15%
1 year ago	x	10%
2 years ago	y	

(2) Find last year's donation.

$x + 0.15x = 10{,}120$	Write the equation.
$1.15x = 10{,}120$	Combine like terms.
$x = \$8800$	Divide each side by 1.15.

(3) Find donation from 2 years ago.

$y + 0.1y = 8800$	Write the equation.
$1.1y = 8800$	Combine like terms.
$y = \$8000$	Divide each side by 1.1.

Project

Plan a fund raiser for your school. Write a proposal that includes the purpose of the fund raiser, the type of activity, the length of time, and the amount of money you would like to raise. Be prepared to present your proposal to the class.

Reteaching and Enrichment Strategies

If students need help. . .	If students got it. . .
Resources by Chapter • Practice A and Practice B • Puzzle Time Record and Practice Journal Practice Differentiating the Lesson Lesson Tutorials Skills Review Handbook	Resources by Chapter • Enrichment and Extension Start the next section

Identify the percent of change as an *increase* or *decrease*. Then find the percent of change. Round to the nearest tenth of a percent, if necessary.

23. $\frac{1}{4}$ to $\frac{1}{2}$ **24.** $\frac{4}{5}$ to $\frac{3}{5}$ **25.** $\frac{3}{8}$ to $\frac{7}{8}$ **26.** $\frac{5}{4}$ to $\frac{3}{8}$

27. CRITICAL THINKING Explain why a change from 20 to 40 is a 100% increase, but a change from 40 to 20 is a 50% decrease.

28. POPULATION The table shows population data for a community.

Year	Population
2000	118,000
2006	138,000

   a. What is the percent of change from 2000 to 2006?

   b. Use this percent of change to predict the population in 2012.

29. GEOMETRY Suppose the length and width of the sandbox are doubled.

   a. Find the percent of change in the perimeter.

   b. Find the percent of change in the area.

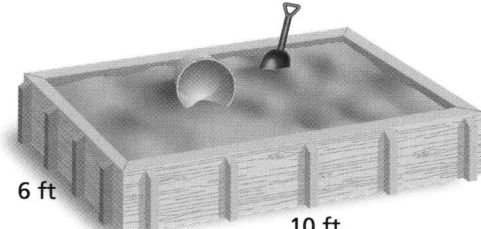

6 ft

10 ft

June   September

30. RUNNING Find the percent of change in the time to run a mile from June to September.

31. CRITICAL THINKING A number increases by 10% and then decreases by 10%. Will the result be *greater than*, *less than*, or *equal to* the original number? Explain.

32. DONATIONS Donations to an annual fundraiser are 15% greater this year than last year. Last year, donations were 10% greater than the year before. The amount raised this year is $10,120. How much was raised 2 years ago?

33. Reasoning Forty students are in the science club. Of those, 45% are girls. This percent increases to 56% after new girls join the club. How many new girls join?

Fair Game Review *What you learned in previous grades & lessons*

Write and solve an equation to answer the question. *(Section 4.1)*

34. What number is 25% of 64?

35. 39.2 is what percent of 112?

36. 5 is 5% of what number?

37. 18 is 32% of what number?

38. MULTIPLE CHOICE Which equation shows direct variation? *(Section 3.7)*

   (A) $y - x = 1$ **(B)** $\frac{y}{x} = 10$ **(C)** $y = \frac{4}{x}$ **(D)** $xy = 5$

You can use a **summary triangle** to explain a concept. Here is an example of a summary triangle for finding a percent of a number.

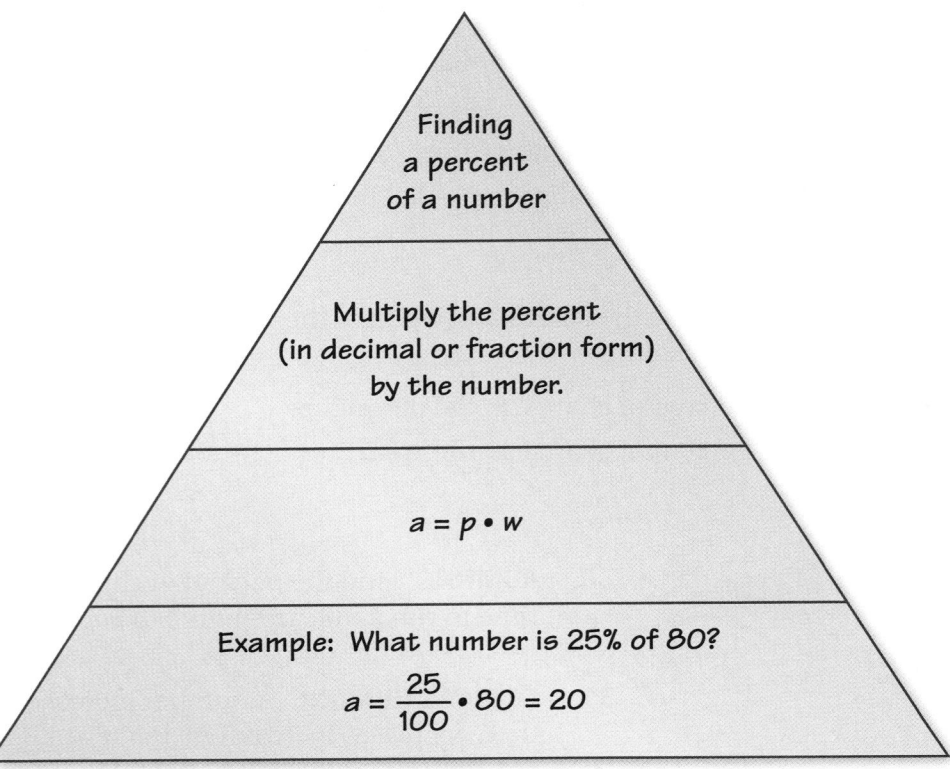

Finding
a percent
of a number

Multiply the percent
(in decimal or fraction form)
by the number.

$a = p \cdot w$

Example: What number is 25% of 80?

$$a = \frac{25}{100} \cdot 80 = 20$$

On Your Own

Make a summary triangle to help you study these topics.

1. finding the percent given a number and a part of the number

2. finding the number given a part of the number and a percent

3. percent of increase

4. percent of decrease

After you complete this chapter, make summary triangles for the following topics.

5. discount

6. markup

7. simple interest

"I hope my owner sees my summary triangle. I just can't seem to learn 'roll over'."

Sample Answers

1.

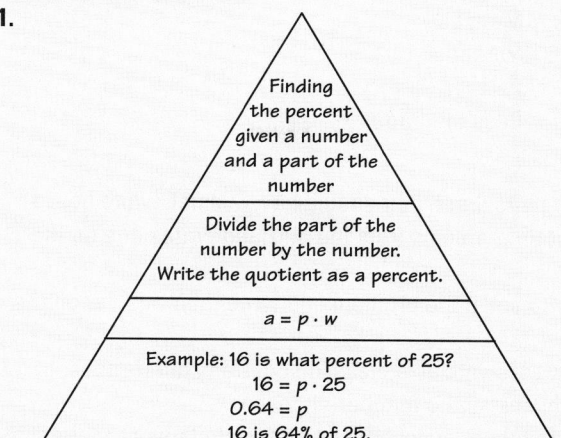

Finding the percent given a number and a part of the number

Divide the part of the number by the number. Write the quotient as a percent.

$a = p \cdot w$

Example: 16 is what percent of 25?
$16 = p \cdot 25$
$0.64 = p$
16 is 64% of 25.

2.

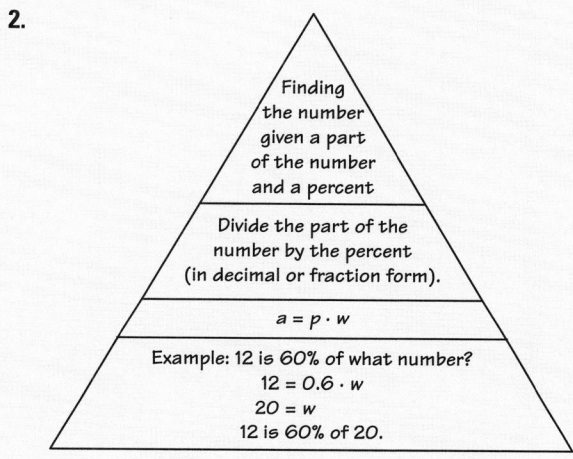

Finding the number given a part of the number and a percent

Divide the part of the number by the percent (in decimal or fraction form).

$a = p \cdot w$

Example: 12 is 60% of what number?
$12 = 0.6 \cdot w$
$20 = w$
12 is 60% of 20.

3.

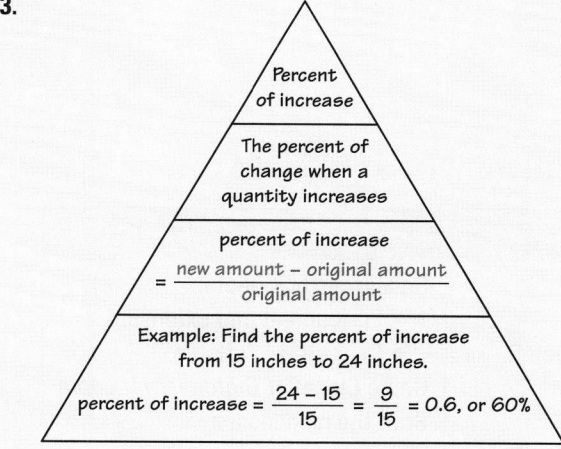

Percent of increase

The percent of change when a quantity increases

$\text{percent of increase} = \dfrac{\text{new amount} - \text{original amount}}{\text{original amount}}$

Example: Find the percent of increase from 15 inches to 24 inches.

$\text{percent of increase} = \dfrac{24 - 15}{15} = \dfrac{9}{15} = 0.6, \text{ or } 60\%$

4. Available at *BigIdeasMath.com*.

List of Organizers
Available at *BigIdeasMath.com*

Comparison Chart
Concept Circle
Example and Non-Example Chart
Formula Triangle
Four Square
Idea (Definition) and Examples Chart
Information Frame
Information Wheel
Notetaking Organizer
Process Diagram
Summary Triangle
Word Magnet
Y Chart

About this Organizer

A **Summary Triangle** can be used to explain a concept. Typically, the summary triangle is divided into 3 or 4 parts. In the top part, students write the concept being explained. In the middle part(s), students write any procedure, explanation, description, definition, theorem, and/or formula(s). In the bottom part, students write an example to illustrate the concept. A summary triangle can be used as an assessment tool, in which blanks are left for students to complete. Also, students can place their summary triangles on note cards to use as a quick study reference.

Technology For the Teacher
Vocabulary Puzzle Builder

Answers

1. $a = 0.28 \cdot 75$; 21

2. $42 = 0.21 \cdot w$; 200

3. $36 = p \cdot 45$; 80%

4. $a = 0.68 \cdot 12$; 8.16

5. $66 = p \cdot 55$; 120%

6. increase; 200%

7. decrease; 30%

8. decrease; 61.9%

9. increase; 43.8%

10. decrease; 17.3%

11. increase; 100%

12. 50 text messages

13. 17 passes

14. 93.33%

15. a. 2.5% increase

    b. 7.5% increase

16. $16,000

Alternative Quiz Ideas

100% Quiz
Error Notebook
Group Quiz
Homework Quiz

Math Log
Notebook Quiz
Partner Quiz
Pass the Paper

Partner Quiz

- Students should work in pairs. Each pair should have a small white board.
- The teacher selects certain problems from the quiz and writes one on the board.
- The pairs work together to solve the problem and write their answer on the white board.
- Students show their answers and, as a class, discuss any differences.
- Repeat for as many problems as the teacher chooses.
- For the word problems, teachers may choose to have students read them out of the book.

Assessment Book

Reteaching and Enrichment Strategies

If students need help. . .	If students got it. . .
Resources by Chapter • Study Help • Practice A and Practice B • Puzzle Time Lesson Tutorials *BigIdeasMath.com* Practice Quiz Practice from the Test Generator	Resources by Chapter • Enrichment and Extension • School-to-Work Game Closet at *BigIdeasMath.com* Start the next lesson

Technology For the Teacher

Answer Presentation Tool
Big Ideas Test Generator

Write and solve an equation to answer the question.   *(Section 4.1)*

1. What number is 28% of 75?

2. 42 is 21% of what number?

3. 36 is what percent of 45?

4. What number is 68% of 12?

5. 66 is what percent of 55?

Identify the percent of change as an *increase* or *decrease*. Then find the percent of change. Round to the nearest tenth of a percent, if necessary.   *(Section 4.2)*

6. 8 inches to 24 inches

7. 300 miles to 210 miles

8. $42.00 to $16.00

9. 32 points to 46 points

10. 185 pounds to 153 pounds

11. 35 people to 70 people

12. **TEXT MESSAGES** You have 44 text messages in your inbox. How many messages can your cell phone hold?   *(Section 4.1)*

13. **COMPLETIONS** A quarterback completed 68% of his passes in a game. He threw 25 passes. How many passes did the quarterback complete?   *(Section 4.1)*

14. **QUIZ** You answered 14 questions correctly on a 15-question quiz. What percent did you receive on the quiz? Round to the nearest hundredth.   *(Section 4.1)*

15. **FRUIT JUICE** The graph shows the amount of fruit juice available per person in the United States during a six-year period.   *(Section 4.2)*

    a. What is the percent of change from 2002 to 2005?

    b. What is the percent of change from 2002 to 2003?

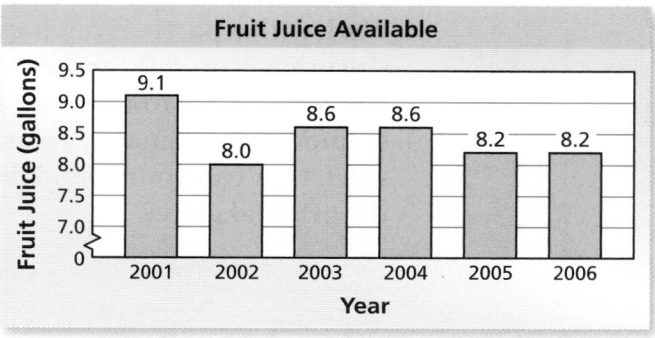

16. **CAR** A car loses 15% of its original value each year. After one year, a car has a value of $13,600. What is the original value of the car?   *(Section 4.2)*

Essential Question How can you find discounts and markups efficiently?

1 ACTIVITY: Comparing Discounts

Work with a partner. The same pair of sneakers is on sale at three stores. Which one is the best buy?

a. Regular Price: $45   **b.** Regular Price: $49   **c.** Regular Price: $39

a.
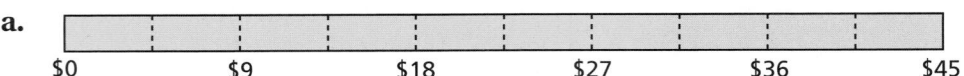

b.
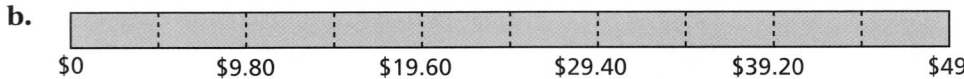

c.
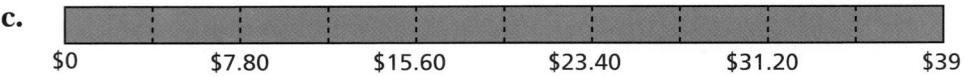

2 ACTIVITY: Finding the Original Price

Work with a partner. You buy a shirt that is on sale for 30% off. You pay $22.40. Your friend wants to know the original price of the shirt. How can your friend find the original price?

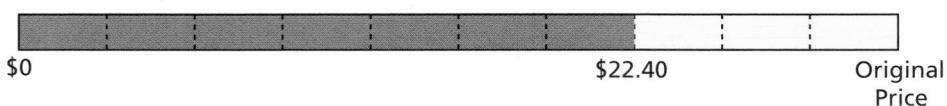

Laurie's Notes

Introduction

For the Teacher

- **Goal:** Students will use a percent bar model to visualize discounts and markups.

Motivate

- Show a newspaper circular that advertises a discount (sale).

Activity Notes

Activity 1

- Explain that sale items involve a *percent* discount and the *amount* of discount. If possible, use the newspaper circular to make this distinction.
- The percent bar models are divided into 10 equal parts. Dollar amounts for items are shown on the bars.
- Discuss how the dollar amounts can be computed. Students can use mental math to find 10% and multiply by the correct amount.
- **Big Idea:** When you *save* 40% ($18), you *pay* 60% ($27). Starting at $45, move to the left 40%, that is the savings.
- **Extension:** Determine the amount you save *and* the price paid. This will not be possible for the last example because the amount of discount varies.
- **?** "How do you decide the best buy?" Listen for the lowest final price instead of the greatest savings because the original prices may vary.
- **?** "What does the phrase, "up to 70% off" mean?" Percent off will vary from 0% up to 70%.

Activity 2

- **Connection:** Finding the original price is the same as finding the whole. $22.40 is the part.
- **?** "What percent does $22.40 represent of the original price?" 70%
- **?** "How does the percent bar model help you think about the original price?"

  Students might describe the $22.40 as 70% or about $\frac{2}{3}$ of the original price.

  So, another $\frac{1}{3}$ has to be added on to find the original price.
- Use the percent equation: $22.40 is 70% of what number?
- **?** "Why is 70% used instead of 30%?" Because $22.40 is the part and it is 70% of the whole, or original price.
- **Struggling Students:** Students sometimes struggle with this concept. Reinforce by constantly telling students "30% off the original price is the same as paying 70% of the original price."

Previous Learning

Students should be able to find a percent of a number, round decimal values, and convert between fractions, decimals, and percents.

Activity Materials	
Introduction	**Textbook**
• newspaper circular	• calculator

Start Thinking! and Warm Up

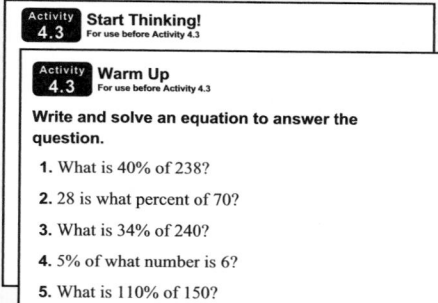

4.3 Record and Practice Journal

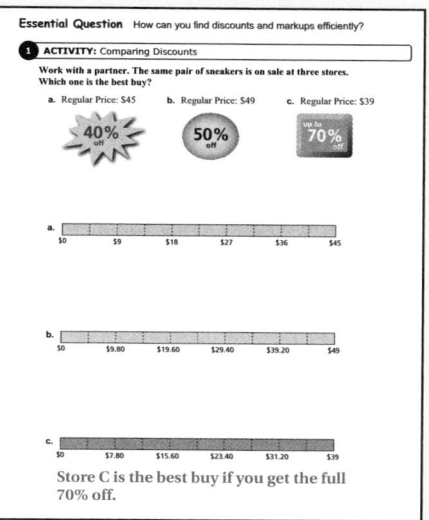

English Language Learners

Vocabulary

English learners may not be familiar with terms used in business, such as *discount, markup, purchase price,* and *selling price*. Take time to explain these terms.

Laurie's Notes

Activity 3

- **Discuss:** A store purchases an item for *x* dollars. The store needs to sell this item for more than *x* dollars (markup) to cover operating costs and to make a profit.
- **Explain:** A store purchases an item for $2 and sells it for $4. This represents a 100% markup ($2 + 100% of $2).
- **Tip:** Have students put 100% above the $250 store cost in part (a).
- **?** "How does the selling price compare to the price the store paid for the item?" 125% greater, or 225% of the store's cost
- **Reasoning:** If a $10 item sells for $25, the $10 item was *marked up 150%* and the selling price is *250% of $10*. One way to show this to students is to write: (100% of $10) + (150% of $10) = (250% of $10) = $25.

Words of Wisdom

- Be careful with language. This is not an obvious concept for students. Try to use consistent language with every example; store purchase price, store selling price, original price, markup amount, discount amount, and sale price.

What Is Your Answer?

- You want students to discover that they can find the selling price after a 25% discount by multiplying by 0.75 (one step) *or* by multiplying by 0.25 and then subtracting the result from the original price (two steps).
- Similarly, for markups, you can multiply the cost by 1.75 (one step) for a 75% markup *or* multiply by 0.75 and then add to the cost (two steps).

Closure

- You purchased an item marked 25% off. What percent of the original price did you pay? 75%

4.3 Record and Practice Journal

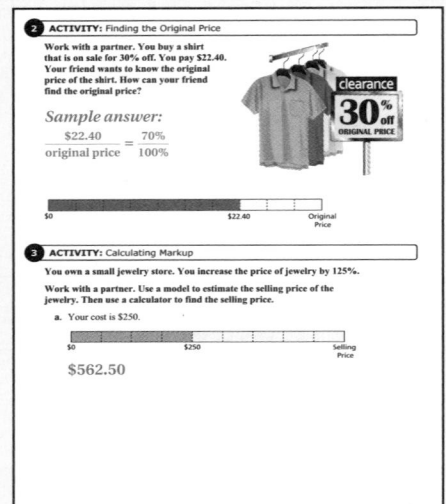

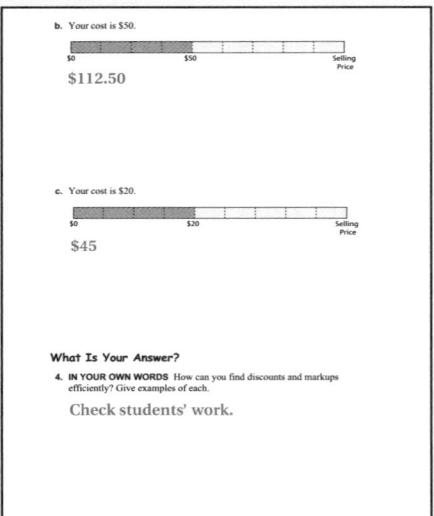

You own a small jewelry store. You increase the price of the jewelry by 125%.

Work with a partner. Use a model to estimate the selling price of the jewelry. Then use a calculator to find the selling price.

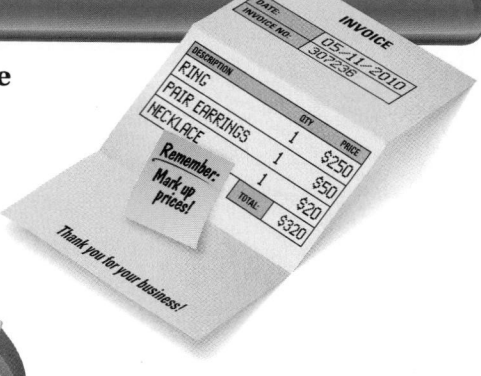

a. Your cost is $250.

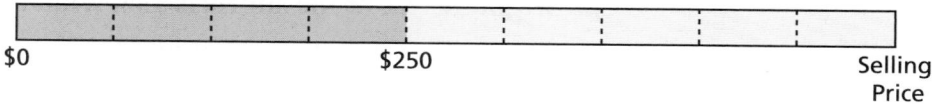

$0 $250   Selling Price

b. Your cost is $50.

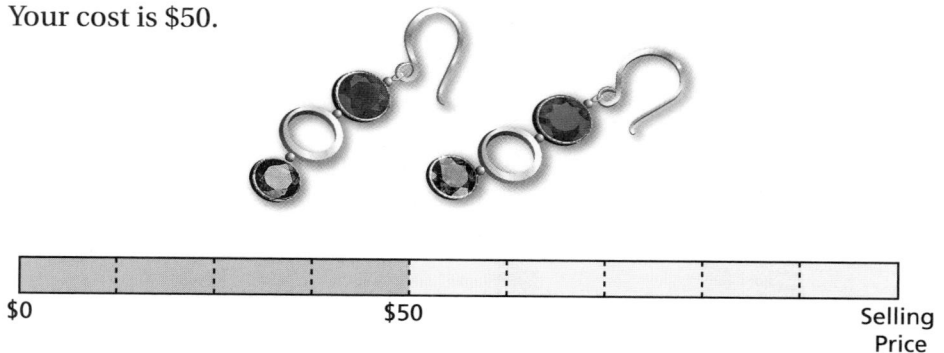

$0 $50   Selling Price

c. Your cost is $20.

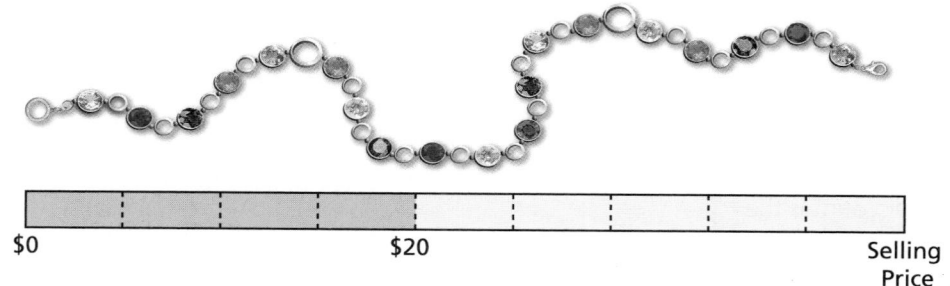

$0 $20   Selling Price

What Is Your Answer?

4. IN YOUR OWN WORDS How can you find discounts and markups efficiently? Give examples of each.

Practice

Use what you learned about discounts and markups to complete Exercises 4, 9, 14, and 18–20 on pages 176 and 177.

Check It Out
Lesson Tutorials
BigIdeasMath com

Key Vocabulary
discount, *p. 174*
markup, *p. 174*

 **Key Ideas**

Discounts

A **discount** is a decrease in the original price of an item.

Markups

To make a profit, stores charge more than what they pay. The increase from what the store pays to the selling price is called a **markup**.

EXAMPLE (1) **Finding a Sale Price**

The original price of the shorts is $35. What is the sale price?

Method 1: First, find the discount. The discount is 25% of $35.

$a = p \cdot w$		Write percent equation.
$= 0.25 \cdot 35$		Subsitute 0.25 for *p* and 35 for *w*.
$= 8.75$		Multiply.

Next, find the sale price.

sale price	=	original price	−	discount
	=	35	−	8.75
	=	26.25		

∴ The sale price is $26.25.

Method 2: First, find the percent of the original price.

$$100\% - 25\% = 75\%$$

Next, find the sale price.

$$\text{sale price} = 75\% \text{ of } \$35$$
$$= 0.75 \cdot 35$$
$$= 26.25$$

Study Tip

A 25% discount is the same as paying 75% of the original price.

∴ The sale price is $26.25.   **Check**

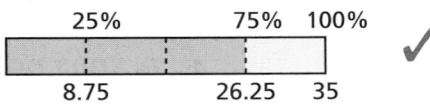

● **On Your Own**

Now You're Ready
Exercises 4–8

1. The original price of a skateboard is $50. The sale price includes a 20% discount. What is the sale price?

◀ Multi-Language Glossary at BigIdeasMath com.

Laurie's Notes

Introduction

Connect
- **Yesterday:** Students explored discounts and markups using a percent bar model.
- **Today:** Students will use the percent equation to find discounts and markups of items.

Motivate
? **Story Time:** "A store buys an MP3 player for $100 and marks it up 50%. The store has a 50% off sale. You purchase the MP3 player. What do you pay?" $75 "Did the store lose money?" yes

Lesson Notes

Key Ideas
- Discuss each concept using examples from the previous day's activity.
- Use the following to help students understand the vocabulary.

wholesale price + markup = retail price
(or selling price)

what a store pays   increase in price   price you pay

Example 1
- Two methods are shown. Both methods require two steps. In the first method, you multiply to find the amount of discount, then you subtract to find the sale price. In the second method, you subtract first to find the percent of the original price you will pay, then you use the percent equation to find the sale price.
- Work through each method.
- **Connection:** The amount of discount is a *part* of the *whole* original price. The percent equation is used to find the amount of the discount.
- **Common Error:** Students find the discount or the amount saved ($8.75) instead of the sale price ($26.25).
- Discuss the *Study Tip.* Try other discounts (i.e., 30%) and ask what percent you are paying (70%).
- **?** "Why is the percent bar model divided into 4 parts?"

Because the discount is 25% or $\frac{1}{4}$.

On Your Own
? "How should the percent bar model be divided and why?" 5 parts

because 20% = $\frac{1}{5}$

Start Thinking! and Warm Up

Lesson 4.3 **Warm Up** For use before Lesson 4.3

Lesson 4.3 **Start Thinking!** For use before Lesson 4.3

You go to a store to buy a new pair of jeans. You find 2 pairs of jeans each on sale for a different price. Explain which is a better bargain.

Regular Price: $35; Discount: 30%

Regular Price: $40; Discount: 35%

Extra Example 1
The original price of a T-shirt is $15. The sale price includes a 35% discount. What is the sale price? $9.75

On Your Own

1. $40

Laurie's Notes

Extra Example 2

The discount on a package of athletic socks is 15%. It is on sale for $17. What is the original price of the package of athletic socks? $20

Extra Example 3

A store pays $15 for a baseball cap. The percent markup is 60%. What is the selling price? $24

 On Your Own

2. $20

3. $90

4. The selling price is 120% of $70. So, the selling price is 1.2(70) = $84.

Differentiated Instruction

Visual

Some students may have a hard time remembering the relationships between *sale price, selling price, discount,* and *markup*. Have them copy the verbal models into their notebooks.

Discount

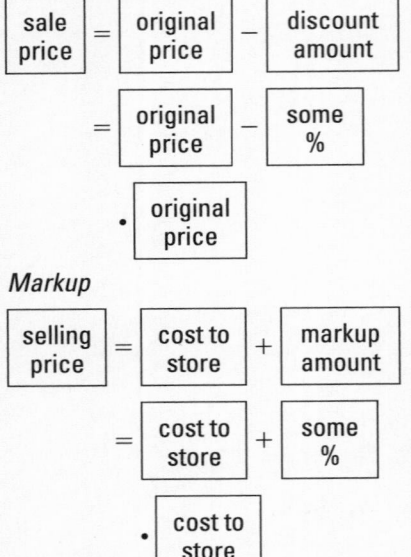

$$\boxed{\begin{array}{c}\text{sale}\\\text{price}\end{array}} = \boxed{\begin{array}{c}\text{original}\\\text{price}\end{array}} - \boxed{\begin{array}{c}\text{discount}\\\text{amount}\end{array}}$$

$$= \boxed{\begin{array}{c}\text{original}\\\text{price}\end{array}} - \boxed{\begin{array}{c}\text{some}\\\%\end{array}}$$

$$\cdot \boxed{\begin{array}{c}\text{original}\\\text{price}\end{array}}$$

Markup

$$\boxed{\begin{array}{c}\text{selling}\\\text{price}\end{array}} = \boxed{\begin{array}{c}\text{cost to}\\\text{store}\end{array}} + \boxed{\begin{array}{c}\text{markup}\\\text{amount}\end{array}}$$

$$= \boxed{\begin{array}{c}\text{cost to}\\\text{store}\end{array}} + \boxed{\begin{array}{c}\text{some}\\\%\end{array}}$$

$$\cdot \boxed{\begin{array}{c}\text{cost to}\\\text{store}\end{array}}$$

Example 2

- "What is the percent equation?" $a = p \cdot w$
- ? "What do you know in this problem?" 33 is the part and 60% is the percent.
- **Common Error:** Students multiply 33 by 60% (or 40%). Students need to remember that 33 is a *part* of the original price, it's not the *whole*.

Example 3

- Work through the problem as shown. Encourage students to use mental math to find 20% of $70. 10% of 70 is 7. So, 20% of 70 is 2(7), or 14.
- Two steps were used to answer the question: 1) Find 20% of $70 and 2) add this amount to the original amount of $70.
- ? "Could this problem be done in one step? Explain." yes; 120% of $70 = $84
- ? "Explain why the 120% makes sense." You pay 100% of the store's cost plus an additional 20% markup for a total of 120%.
- **Common Error:** Students find the markup ($14) instead of the selling price ($84).
- **Extension:** Have students draw a percent bar model for this problem. The model will be divided into fifths.

Closure

- **Writing Prompt:** Explain two ways to find the sale price for an item marked 30% off. 1) Find the amount of discount and subtract from the original price. 2) Find the percent of the original price and multiply the percent by the original price.
- **Extension:** "If an item is marked up and then discounted the same percent, will the store make a profit? Explain." no; the *amount of markup* will be less than the *amount of discount*, so the store will sell the item for less than what it paid.
- **Extension:** "Is a 25% discount followed by a 10% discount the same as a 35% discount? Explain." no; The sale price for an item discounted 25% followed by a 10% discount would be 0.75(0.9) = 0.675, or 67.5% of the original price. The sale price for an item discounted 35% would be 65% of the original price.

Technology For the Teacher

The Dynamic Planning Tool
Editable Teacher's Resources at *BigIdeasMath.com*

EXAMPLE **2** **Finding an Original Price**

What is the original price of the shoes?

The sale price is
$100\% - 40\% = 60\%$
of the original price.

Answer the question: 33 is 60% of what number?

$a = p \cdot w$	Write percent equation.
$33 = 0.6 \cdot w$	Substitute 33 for a and 0.6 for p.
$55 = w$	Divide each side by 0.6.

∴ The original price of the shoes is $55.

Check

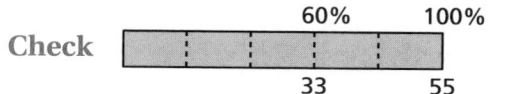

EXAMPLE **3** **Finding a Selling Price**

A store pays $70 for a bicycle. The percent of markup is 20%. What is the selling price?

First, find the markup. The markup is 20% of $70.

$a = p \cdot w$	Write percent equation.
$= 0.20 \cdot 70$	Substitute 0.20 for p and 70 for w.
$= 14$	Multiply.

Next, find the selling price.

selling price	=	cost to store	+	markup
	=	70	+	14
	= 84			

∴ The selling price is $84.

On Your Own

Exercises 9–20

2. The discount on a DVD is 50%. It is on sale for $10. What is the original price of the DVD?

3. A store pays $75 for an aquarium. The markup is 20%. What is the selling price?

4. Solve Example 3 using a different method.

Vocabulary and Concept Check

1. **WRITING** Describe how to find the sale price of an item that has been discounted 25%.

2. **WRITING** Describe how to find the selling price of an item that has been marked up 110%.

3. **REASONING** Which would you rather pay? Explain your reasoning.

   a. 6% tax on a discounted price   or   6% tax on the original price

   b. 30% markup on a $30 shirt   or $30 markup on a $30 shirt

Practice and Problem Solving

Copy and complete the table.

	Original Price	Percent of Discount	Sale Price
4.	$80	20%	
5.	$42	15%	
6.	$120	80%	
7.	$112	32%	
8.	$69.80	60%	
9.		25%	$40
10.		5%	$57
11.		80%	$90
12.		64%	$72
13.		15%	$146.54
14.	$60		$45
15.	$82		$65.60
16.	$95		$61.75

17. **YOU BE THE TEACHER** The cost to a store for an MP3 player is $60. The selling price is $105. A classmate says that the markup is 175% because $\frac{\$105}{\$60} = 1.75$. Is your classmate correct? If not, explain how to find the correct percent of markup.

Assignment Guide and Homework Check

Level	Day 1 Activity Assignment	Day 2 Lesson Assignment	Homework Check
Basic	4, 9, 14, 18–20, 26–29	1–3, 5, 7, 11–17 odd, 21	5, 11, 15, 17
Average	4, 9, 14, 18–20, 26–29	1–3, 5, 7, 11–17 odd, 21, 22	5, 11, 15, 17
Advanced	4, 9, 14, 18–20, 26–29	1–3, 8, 12, 16, 17, 21–25	8, 12, 16, 17, 24

Common Errors

- **Exercises 4–8** Students may write the discount amount as the sale price instead of subtracting it from the original amount. When students copy the table, ask them to add another column titled "Discount Amount." Remind them to subtract the discount amount from the original price.
- **Exercises 9–16** Remind students that there is an extra step in the problem. They should subtract the percent of discount from 100% to find the percent of the original price of the item.
- **Exercises 18–20** Students may find the markup and not the selling price. Remind them that they must add the markup to the cost to obtain the selling price.

4.3 Record and Practice Journal

Complete the table.

	Original Price	Percent of Discount	Sale Price
1.	$20	20%	$16
2.	$95	35%	$61.75
3.	$222	75%	$55.50
4.	$130	40%	$78

Find the cost to store, percent of markup, or selling price.

5. Cost to store: $20
Markup: 15%
Selling price: ?
$23

6. Cost to store: ?
Markup: 80%
Selling Price: $100.80
$56

7. Cost to store: $110
Markup: ?
Selling price: $264
140%

8. A store buys an item for $10. To earn a profit of $25, what percent does the store need to markup the item?
250%

9. Your dinner at a restaurant costs $13.65 after you use a coupon for a 25% discount. You leave a tip for $3.00.
a. How much was your dinner before the discount?
$18.20
b. You tip your server based on the price before the discount. What percent tip did you leave? Round your answer to the nearest tenth of a percent.
16.5%

Technology For the Teacher
Answer Presentation Tool
QuizShow

Vocabulary and Concept Check

1. *Sample answer:* Multiply the original price by $100\% - 25\% = 75\%$ to find the sale price.

2. Find the markup by taking 110% of the amount. Then add the amount and the markup to find the selling price.

3. a. 6% tax on a discounted price; The discounted price is less, so the tax is less.
   b. 30% markup on a $30 shirt; 30% of $30 is less than $30.

Practice and Problem Solving

4. $64
5. $35.70
6. $24
7. $76.16
8. $27.92
9. $53.33
10. $60
11. $450
12. $200
13. $172.40
14. 25%
15. 20%
16. 35%
17. no; Only the amount of markup should be in the numerator, $\frac{105 - 60}{60} = 0.75$. So, the percent of markup is 75%.

18. $77   **19.** $36

20. 140%

21. "Multiply $45.85 by 0.1" and "Multiply $45.85 by 0.9, then subtract from $45.85." Both will give the sale price of $4.59. The first method is easier because it is only one step.

22. a. store C

b. The markup percent of store A may decrease so it may be cheaper there.

23. no; $31.08

24. See *Taking Math Deeper.*

25. $30

Fair Game Review

26. 170   **27.** 180

28. 1152   **29.** C

Mini-Assessment

Find the price, discount, markup, or cost to store.

1. Original price: $50
Discount: 15%
Sale price: ? $42.50

2. Original price: $35
Discount: ?
Sale price: $31.50 10%

3. Cost to store: $75
Markup: ?
Selling price: $112.50 50%

4. Cost to store: ?
Markup: 15%
Selling price: $85.10 $74

5. The sale price for a bicycle is $89.90. The sale price includes a discount of 20%. What is the original price of the bicycle? $112.38

Taking Math Deeper

Exercise 24

A good way to approach this problem is to take things one step at a time. Also, in problems like this, it is much easier to round up to $40 and $30 for easier calculations.

① Find the percent of discount.

a. 10 is 25% of 40.

It is easier to round $39.99 to $40 before doing the calculations.

Jeans	$40	39.99
Discount		-10.00
Subtotal		29.99
Sales Tax		1.95
Total		31.94

② Find the percent of sales tax.

1.95 is what % of 30?

$$1.95 = p \cdot 30$$
$$0.065 = p$$

b. Sales tax = 6.5%.

Jeans		39.99
Discount		-10.00
Subtotal	$30	29.99
Sales Tax		1.95
Total		31.94

③ Find the actual markup.

$$x + 0.6x = 40$$
$$1.6x = 40$$
$$x = \$25 \qquad \text{Wholesale}$$

$5 markup

The $40 jeans cost the store $25. After the discount of $10, the markup is $5. Find the percent of markup by answering "5 is what % of 25?"

c. $5 is a 20% markup on $25.

Project

Check the newspaper or local advertisements for a store near you. Select five items that are on sale. Prepare a chart that shows the original price, the percent of discount, and the sale price. How much would you save if you purchased all five items at the sale price?

Reteaching and Enrichment Strategies

If students need help. . .	If students got it. . .

Resources by Chapter
  • Practice A and Practice B
  • Puzzle Time
Record and Practice Journal Practice
Differentiating the Lesson
Lesson Tutorials
Skills Review Handbook
	Resources by Chapter
	  • Enrichment and Extension
	  • School-to-Work
	Start the next section

Find the cost to store, percent of markup, or selling price.

③ **18.** Cost to store: $70
Markup: 10%
Selling price:

19. Cost to store:
Markup: 75%
Selling price: $63

20. Cost to store: $75
Markup:
Selling price: $180

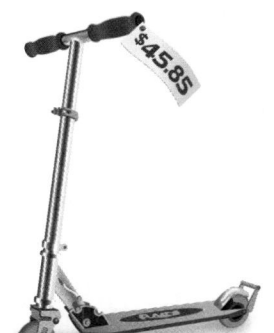

21. SCOOTER The scooter is on sale for 90% off the original price. Which of the methods can you use to find the sale price? Which method do you prefer? Explain.

Multiply $45.85 by 0.9.

Multiply $45.85 by 0.1.

Multiply $45.85 by 0.9, then add to $45.85.

Multiply $45.85 by 0.9, then subtract from $45.85.

22. GAMING You are shopping for a video game system.

   a. At which store should you buy the system?

   b. Store A has a weekend sale. How can this change your decision in part (a)?

Store	Cost to Store	Markup
A	$162	40%
B	$155	30%
C	$160	25%

23. STEREO A $129.50 stereo is discounted 40%. The next month, the sale price is discounted 60%. Is the stereo now "free"? If not, what is the sale price?

24. CLOTHING You buy a pair of jeans at a department store.

   a. What is the percent of discount to the nearest percent?

   b. What is the percent of sales tax to the nearest tenth of a percent?

   c. The price of the jeans includes a 60% markup. After the discount, what is the percent of markup to the nearest percent?

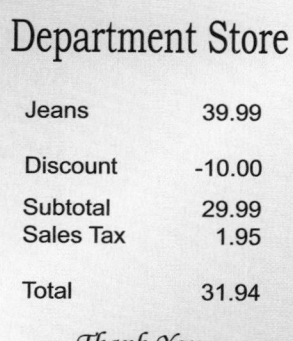

Department Store

Jeans	39.99
Discount	-10.00
Subtotal	29.99
Sales Tax	1.95
Total	31.94

Thank You

25. *Critical Thinking* You buy a bicycle helmet for $22.26, which includes 6% sales tax. The helmet is discounted 30% off the selling price. What is the original price?

Fair Game Review *What you learned in previous grades & lessons*

Evaluate. *(Skills Review Handbook)*

26. $2000(0.085)$

27. $1500(0.04)(3)$

28. $3200(0.045)(8)$

29. MULTIPLE CHOICE Which measurement is greater than 1 meter? *(Section 3.6)*

   Ⓐ 38 inches   Ⓑ 1 yard   Ⓒ 3.4 feet   Ⓓ 98 centimeters

4.4 Simple Interest

Essential Question How can you find the amount of simple interest earned on a savings account? How can you find the amount of interest owed on a loan?

Simple interest is money earned on a savings account or an investment. It can also be money you pay for borrowing money.

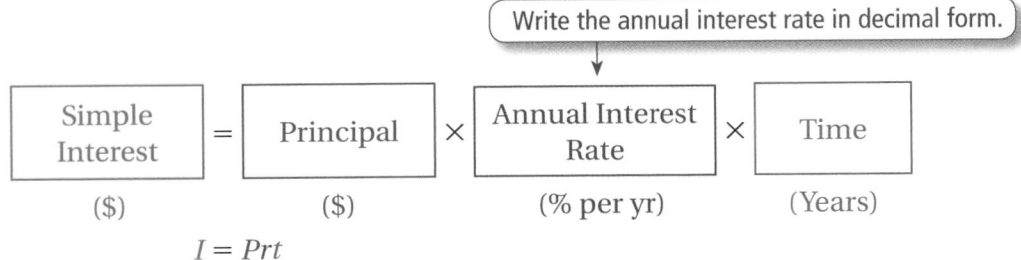

Write the annual interest rate in decimal form.

Simple Interest	=	Principal	×	Annual Interest Rate	×	Time
($)		($)		(% per yr)		(Years)

$$I = Prt$$

1 ACTIVITY: Finding Simple Interest

Work with a partner. You put $100 in a savings account. The account earns 6% simple interest per year. (a) Find the interest earned and the balance at the end of 6 months. (b) Copy and complete the table. Then make a bar graph that shows how the balance grows in 6 months.

a. $I = Prt$ Write simple interest formula

$= 100(0.06)\left(\dfrac{6}{12}\right)$ Substitute values.

$= 3$ Multiply.

⋮ At the end of 6 months, you earn $3 in interest. So, your balance is $100 + $3 = $103.

b.

Time	Interest	Balance
0 month	$0	$100
1 month		
2 months		
3 months		
4 months		
5 months		
6 months	$3	$103

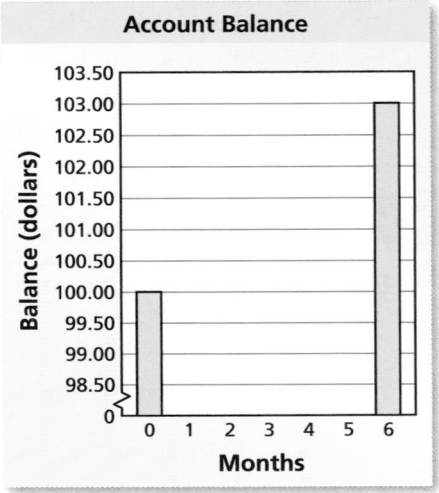

Account Balance

Laurie's Notes

Introduction

For the Teacher
- **Goal:** Students will use the simple interest formula to determine the amount of interest earned in a savings account.
- Students should assume that deposits are made at the beginning of the interest period in all banking problems, unless otherwise stated.

Motivate
- **Tell the Story:** Baseball legend Ken Griffey Jr. played a prank on (former) teammate Josh Fogg one season. He owed Fogg some money and paid him back in pennies. Griffey stacked 60 cartons, each holding $25 worth of pennies, in Fogg's locker. Not only did Fogg not get paid interest, he had to haul all the pennies to the bank!

Activity Notes

Discuss
- Today's investigation involves three activities. Given time constraints and your own students, you may not complete all three.
- **Financial Literacy:** You want students to have some understanding of the cost of borrowing money or the ability to earn money when it is deposited in a bank, not to become trained loan officers.
- **Discuss:** When you *deposit* money, you should *earn* money. When you *borrow* money, you should *pay* money.
- Define *simple interest formula*.
- **Discuss:** Interest earned/owed is influenced by how much money is involved (principal), the rate you pay/earn, and the amount of time.
- Make clear that it is an *annual* interest rate and the time is in *years*.

Activity 1
- This activity uses the simple interest formula. The principal stays the same for each month's calculation.
- **Demonstrate:** After one month, you earn $100(0.06)\left(\frac{1}{12}\right) = \$0.50$. This $0.50 is added to the principal.
- Get students started on month 2. Students should use $100 for the principal and $\frac{2}{12}$ for the time. Interest earned = $100(0.06)\left(\frac{2}{12}\right) = \$1.00$.
- Students should work with a partner to complete the table and the graph.

4.4 Record and Practice Journal

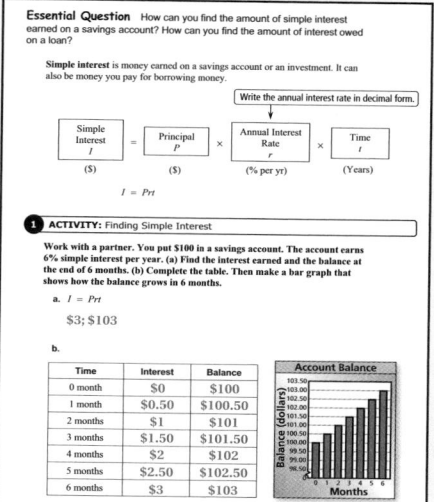

Essential Question How can you find the amount of simple interest earned on a savings account? How can you find the amount of interest owed on a loan?

Simple interest is money earned on a savings account or an investment. It can also be money you pay for borrowing money.

Write the annual interest rate in decimal form.

$$\underset{(\$)}{\text{Simple Interest } I} = \underset{(\$)}{\text{Principal } P} \times \underset{(\% \text{ per yr})}{\text{Annual Interest Rate } r} \times \underset{(\text{Years})}{\text{Time } t}$$

$$I = Prt$$

1 ACTIVITY: Finding Simple Interest

Work with a partner. You put $100 in a savings account. The account earns 6% simple interest per year. (a) Find the interest earned and the balance at the end of 6 months. (b) Complete the table. Then make a bar graph that shows how the balance grows in 6 months.

a. $I = Prt$

   $3; \$103$

b.

Time	Interest	Balance
0 month	$0	$100
1 month	$0.50	$100.50
2 months	$1	$101
3 months	$1.50	$101.50
4 months	$2	$102
5 months	$2.50	$102.50
6 months	$3	$103

Discuss the meaning of the word *interest*. An interest rate is often expressed as an annual percentage of the principal.

Laurie's Notes

Activity 2

- You may wish to use the information to demonstrate the impact of carrying a large credit card debt.
- **Discuss:** How a credit card operates, how you can get one, and how it works (consumer, store, bank).
- **Community:** If your local bank has an education or outreach coordinator, consider having them come in as a guest speaker.
- Read through the information given. Calculate the interest owed for one month, $5000(0.18)\left(\dfrac{1}{12}\right) = \$75$, or at the higher interest rate, $5000(0.20)\left(\dfrac{1}{12}\right) \approx \$83.33$.
- **Discuss:** Why should you shop around for the lowest interest rates? Why should you keep your principal as small as possible?
- Remind the students that the consumer needs to pay the interest ($75 to $83.34) *plus* they need to be paying off the principal.

Activity 3

- The national debt is a complicated concept. The intent of this activity is to raise awareness, and to use the simple interest formula with a really large number.
- **Caution:** If you use a calculator for this problem, the debt and simple interest will appear in scientific notation.
- **Representation:** It will be helpful to write out the simple interest formula using the decimal numbers so that students can see all of the zeros: $I = (10,000,000,000,000)(0.03)(1)$. This has a greater impact than scientific notation.
- **Extension:** Many local newspapers print the national debt and the approximate per person debt each day. Record this information once a week for about 2 months to get a sense for how the numbers are changing. You can even do this for the entire school year.

4.4 Record and Practice Journal

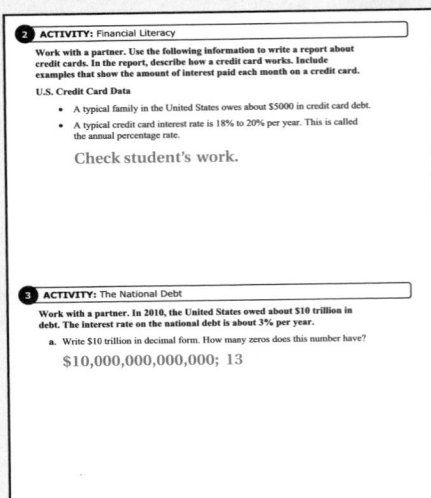

Closure

- **Exit Ticket:** What do you need to know in order to compute simple interest? Principal, annual interest rate, and time

Technology For the Teacher

Dynamic Classroom

The Dynamic Planning Tool
Editable Teacher's Resources at *BigIdeasMath.com*

2 ACTIVITY: Financial Literacy

Work with a partner. Use the following information to write a report about credit cards. In the report, describe how a credit card works. Include examples that show the amount of interest paid each month on a credit card.

U.S. Credit Card Data

• A typical family in the United States owes about $5000 in credit card debt.

• A typical credit card interest rate is 18% to 20% per year. This is called the annual percentage rate.

3 ACTIVITY: The National Debt

Work with a partner. In 2010, the United States owed about $10 trillion in debt. The interest rate on the national debt is about 3% per year.

a. Write $10 trillion in decimal form. How many zeros does this number have?

b. How much interest does the United States pay each year on its national debt?

c. How much interest does the United States pay each day on its national debt?

$10 Trillion in Debt

d. The United States has a population of about 300 million people. Estimate the amount of interest that each person pays per year toward interest on the national debt.

What Is Your Answer?

4. IN YOUR OWN WORDS How can you find the amount of simple interest earned on a savings account? How can you find the amount of interest owed on a loan? Give examples with your answer.

Practice

Use what you learned about simple interest to complete Exercises 4–7 on page 182.

Check It Out
Lesson Tutorials
BigIdeasMath ✓com

Interest is money paid or earned for the use of money. The **principal** is the amount of money borrowed or deposited.

Key Vocabulary 🔊
interest, *p. 180*
principal, *p. 180*
simple interest,
   p. 180

 **Key Idea**

Simple Interest

Words **Simple interest** is money paid or earned only on the principal.

Algebra

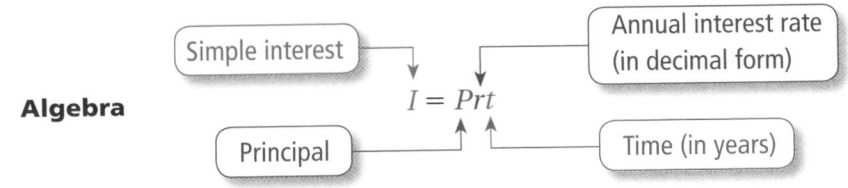

$$I = Prt$$

where: Simple interest, Principal, Annual interest rate (in decimal form), Time (in years)

EXAMPLE 1 Finding Interest Earned

You put $500 in a savings account. The account earns 3% simple interest per year. **(a)** What is the interest earned after 3 years? **(b)** What is the balance after 3 years?

a. $I = Prt$ Write simple interest formula.

$\quad\ = 500(0.03)(3)$ Substitute 500 for *P*, 0.03 for *r*, and 3 for *t*.

$\quad\ = 45$ Multiply.

∴ The interest earned is $45 after 3 years.

b. To find the balance, add the interest to the principal.

∴ So, the balance is $500 + $45 = $545 after 3 years.

EXAMPLE 2 Finding an Annual Interest Rate

You put $1000 in an account. The account earns $100 simple interest in 4 years. What is the annual interest rate?

$\qquad\quad I = Prt$ Write simple interest formula.

$\quad 100 = 1000(r)(4)$ Substitute 100 for *I*, 1000 for *P*, and 4 for *t*.

$\quad 100 = 4000r$ Simplify.

$\ 0.025 = r$ Divide each side by 4000.

∴ The annual interest rate of the account is 0.025, or 2.5%.

🔊 Multi-Language Glossary at BigIdeasMath✓com.

Laurie's Notes

Introduction

Connect
- **Yesterday:** Students explored the simple interest formula, applying it to several consumer applications.
- **Today:** Students will use the simple interest formula and knowledge of equation solving to solve for different variables in the formula.

Motivate
- Just imagine that when you are older, you win a $5 million lottery. If you deposit that money for 10 years at 6% simple interest, how much will you have at the end of 10 years? $8,000,000

Lesson Notes

Key Idea
- **Vocabulary:** interest, money paid or earned, principal, amount of money borrowed or deposited, balance
- **Representation:** Write the formula in words first.
  Simple Interest = (Principal)(Annual interest rate)(Time)
- **Explain:** Simple interest is only one type of interest. There are also compound and exponential interest calculations. The interest rate is written as a decimal. Time is written in terms of years. When time is given in months, remember to express it as a fraction of a year or as a decimal. For example, 9 months $= \frac{9}{12}$ or 0.75 year.
- **Connection:** This formula is similar to the volume formula for a rectangular prism; three variables are multiplied together. Knowing 3 of the 4 variables, you can solve for the fourth.

Example 1
- Read the information given.
- There are two parts to the problem: Calculate the interest earned and then determine the amount (balance) in the account.
- **?** "What operation is performed in writing *Prt*?" multiplication
- **?** In calculating 500(0.03)(3), what order is the multiplication performed? Order doesn't matter, multiplication is commutative.
- **Explain:** Your balance is the original principal *plus* the interest earned.
- **Extension:** If time permits, "What would your balance be if the interest rate had been 6% instead of 3%?" $590

Example 2
- This example uses the Division Property of Equality to solve for the interest rate.
- **?** "Why does 1000(*r*)(4) = 4000*r*?" Commutative Property of Multiplication
- **Common Error:** Students divide 4000 by 100 instead of 100 by 4000.
- **?** "How do you write a decimal as a percent?" Move the decimal point two places to the right. (Multiply by 100.) Then add a percent symbol.

Goal
Today's lesson is using the simple interest formula.

Start Thinking! and Warm Up

> **Lesson 4.4** Warm Up
> For use before Lesson 4.4

> **Lesson 4.4** Start Thinking!
> For use before Lesson 4.4
>
> You earned $150 babysitting. You want to open a savings account. What factors must you consider before opening an account?

Extra Example 1

You put $200 in a savings account. The account earns 2% simple interest per year.
a. What is the interest earned after 5 years? $20
b. What is the balance after 5 years? $220

Extra Example 2

You put $700 in an account. The account earns $224 simple interest in 8 years. What is the annual interest rate? 4%

On Your Own

1. $511.25

2. 2%

Extra Example 3

Using the pictograph in Example 3, how long does it take an account with a principal of $400 to earn $36 interest? 6 years

Extra Example 4

You borrow $300 to buy a guitar. The simple interest rate is 12%. You pay off the loan after 4 years. How much do you pay for the loan? $444

On Your Own

3. 2.5 yr

4. $270

English Language Learners

Vocabulary

Review with English learners the mathematical meanings of principal, interest, and balance because these words have multiple meanings in the English language. They should understand that interest is paid to customers when they deposit money into an account. When a person borrows money, the person pays interest to the bank.

Laurie's Notes

On Your Own

- **Neighbor Check**: Have students work independently and then have their neighbor check their work. Have students discuss any discrepancies.
- Check accuracy of decimals in these problems.

Example 3

- Discuss the diagram.
- **?** "Why would a bank offer different interest rates for different principals?" Students may not understand that banks are using deposited money to loan to other people.
- Work through the problem.
- **?** "What is 6.25 as a mixed number?" $6\frac{1}{4}$
- **Connection:** Students may wonder why anyone would want to know how long it takes to earn $100 in interest. Use an example of depositing money for a future purchase (car, house, college education).

Example 4

- Remind students that the simple interest formula is used to calculate interest *earned* when you *deposit* money and to calculate interest *owed* when you *borrow* money.
- **Discuss:** There are two parts to the problem: 1) Calculate the interest owed and 2) determine the total cost you must pay back for the loan.
- **Extension:** Have students find the monthly payment. $1050 ÷ 60 = $17.50

On Your Own

- **Neighbor Check:** Have students work independently and then have their neighbor check their work. Have students discuss any discrepancies.

Closure

- **Exit Ticket:** Assume $1000 was deposited at 5% simple interest when you were born. Approximately how much is the account worth today? age 11: $1550, age 12: $1600, age 13: $1650, age 14: $1700

Technology For the Teacher

Dynamic Classroom

The Dynamic Planning Tool
Editable Teacher's Resources at *BigIdeasMath.com*

On Your Own

Now You're Ready
Exercises 4–16

1. In Example 1, what is the balance of the account after 9 months?

2. You put $350 in an account. The account earns $17.50 simple interest in 2.5 years. What is the annual interest rate?

EXAMPLE ③ Finding an Amount of Time

A bank offers three savings accounts. The simple interest rate is determined by the principal. How long does it take an account with a principal of $800 to earn $100 interest?

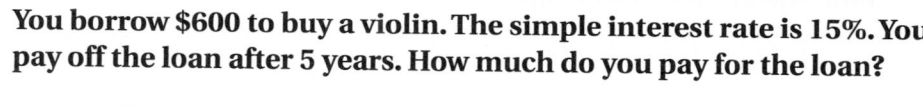

The pictogram shows that the interest rate for a principal of $800 is 2%.

$$I = Prt$$ Write simple interest formula.

$$100 = 800(0.02)(t)$$ Substitute 100 for I, 800 for P, and 0.02 for r.

$$100 = 16t$$ Simplify.

$$6.25 = t$$ Divide each side by 16.

⋮⋮ The account earns $100 in interest in 6.25 years.

EXAMPLE ④ Finding Amount Paid on a Loan

You borrow $600 to buy a violin. The simple interest rate is 15%. You pay off the loan after 5 years. How much do you pay for the loan?

$$I = Prt$$ Write simple interest formula.

$$= 600(0.15)(5)$$ Substitute 600 for P, 0.15 for r, and 5 for t.

$$= 450$$ Multiply.

To find the amount you pay, add the interest to the loan amount.

⋮⋮ So, you pay $600 + $450 = $1050 for the loan.

On Your Own

Now You're Ready
Exercises 17–27

3. In Example 3, how long does it take an account with a principal of $10,000 to earn $750 interest?

4. WHAT IF? In Example 4, you pay off the loan after 2 years. How much money do you save?

Check It Out
Help with Homework
BigIdeasMath.com

Vocabulary and Concept Check

1. **VOCABULARY** Define each variable in $I = Prt$.

2. **WRITING** In each situation, tell whether you would want a *higher* or *lower* interest rate. Explain your reasoning.

    a. You borrow money   **b.** You open a savings account

3. **REASONING** An account earns 6% simple interest. You want to find the interest earned on $200 after 8 months. What conversions do you need to make before you can use the formula $I = Prt$?

Practice and Problem Solving

An account earns simple interest. (a) Find the interest earned. (b) Find the balance of the account.

  ①   **4.** $600 at 5% for 2 years   **5.** $1500 at 4% for 5 years

       6. $350 at 3% for 10 years   **7.** $1800 at 6.5% for 30 months

       8. $700 at 8% for 6 years   **9.** $1675 at 4.6% for 4 years

     10. $925 at 2% for 2.4 years   **11.** $5200 at 7.36% for 54 months

12. ERROR ANALYSIS Describe and correct the error in finding the simple interest earned on $500 at 6% for 18 months.

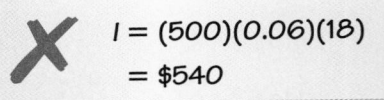

$$I = (500)(0.06)(18)$$
$$= \$540$$

Find the annual simple interest rate.

  ②   **13.** $I = \$24$, $P = \$400$, $t = 2$ years   **14.** $I = \$562.50$, $P = \$1500$, $t = 5$ years

     15. $I = \$54$, $P = \$900$, $t = 18$ months   **16.** $I = \$160.67$, $P = \$2000$, $t = 8$ months

Find the amount of time.

  ③   **17.** $I = \$30$, $P = \$500$, $r = 3\%$ **18.** $I = \$720$, $P = \$1000$, $r = 9\%$

     19. $I = \$54$, $P = \$800$, $r = 4.5\%$ **20.** $I = \$450$, $P = \$2400$, $r = 7.5\%$

21. BANKING A savings account earns 5% annual simple interest. The principal is $1200. What is the balance after 4 years?

22. SAVINGS You put $400 in an account. The account earns $18 simple interest in 9 months. What is the annual interest rate?

23. CD You put $3000 in a CD (certificate of deposit) at the promotional rate. How long will it take to earn $336 in interest?

Certificate of Deposit

This certificate is the original Specimen and valid document from the treasury and Security department of this new trust financial group & associates. The agreement herein construed are thorough, correct and binding on the parties. Alterations made on this after it has been legally issued are prohibited

Promotional Rate 5.6% Simple Interest

DIRECTOR'S SIGNATURE

Assignment Guide and Homework Check

Level	Day 1 Activity Assignment	Day 2 Lesson Assignment	Homework Check
Basic	4–7, 38–42	1–3, 9–27 odd, 12, 22	9, 13, 17, 22, 25
Average	4–7, 38–42	1–3, 8–12 even, 13–19 odd, 23, 25, 27, 32–34	8, 13, 17, 25, 32
Advanced	4–7, 38–42	1–3, 12, 23, 24, 26, 28–32, 34–37	24, 30, 34, 36

Common Errors

- **Exercises 4–11** Students may forget to change the percent to a decimal. Remind them that before they can put the percent into the equation, they must change the percent to a fraction or a decimal.
- **Exercises 7 and 11** Students may not change months into years and calculate a much greater interest amount. Remind them that the simple interest formula is for *years* and that the time must be changed to years.
- **Exercises 15 and 16** Students may not change the time from months to years. Remind them that the time is in years.
- **Exercises 24–27** Students may only find the amount of interest paid for the loan. Remind them that the total amount paid on a loan is the original principal plus the interest.

4.4 Record and Practice Journal

An account earns simple interest. (a) Find the interest earned. (b) Find the balance of the account.

1. $400 at 7% for 3 years
   a. $84
   b. $484

2. $1200 at 5.6% for 4 years
   a. $268.80
   b. $1468.80

Find the annual simple interest rate.

3. $I = \$18$, $P = \$200$, $t = 18$ months
   6%

4. $I = \$310$, $P = \$1000$, $t = 5$ years
   6.2%

Find the amount of time.

5. $I = \$60$, $P = \$750$, $r = 4\%$
   2 years

6. $I = \$825$, $P = \$2500$, $r = 5.5\%$
   6 years

7. You put $500 in a savings account. The account earns $15.75 simple interest in 6 months. What is the annual interest rate?
   6.3%

8. You put $1000 in an account. The simple interest rate is 4.5%. After a year, you put in another $550. What is your total interest after 2 years from the time you opened the account?
   $114.75

Technology For the Teacher
Answer Presentation Tool
Quiz*Show*

Vocabulary and Concept Check

1. I = simple interest,
   P = principal,
   r = annual interest rate (in decimal form),
   t = time (in years)

2. **a.** lower interest rate because you would pay less

   b. higher interest rate because you would receive more

3. You have to change 6% to a fraction or decimal and 8 months to years.

Practice and Problem Solving

4. **a.** $60   **b.** $660
5. **a.** $300   **b.** $1800
6. **a.** $105   **b.** $455
7. **a.** $292.50   **b.** $2092.50
8. **a.** $336   **b.** $1036
9. **a.** $308.20   **b.** $1983.20
10. **a.** $44.40   **b.** $969.40
11. **a.** $1722.24   **b.** $6922.24
12. They didn't convert 18 months to years.
    $$I = 500(0.06)\left(\frac{18}{12}\right)$$
    $$= \$45$$
13. 3%   14. 7.5%
15. 4%   16. 12.05%
17. 2 yr   18. 8 yr
19. 1.5 yr   20. 2.5 yr
21. $1440
22. 6%
23. 2 yr

Practice and Problem Solving

24. $1770   25. $2720

26. $3660   27. $6700.80

28. $2550   29. $8500

30. 4 yr   31. 5.25%

32. See *Taking Math Deeper*.

33. 4 yr

34. $77.25

35. 12.5 yr; Substitute $2000 for *P* and *I*, 0.08 for *r*, and solve for *t*.

36. $300

37. Year 1 = $520
    Year 2 = $540.80
    Year 3 = $562.43

 ## Fair Game Review

38. $x = 27$

39. $n = 5$

40. $m = 3.5$

41. $z = 9$

42. A

Mini-Assessment

Find the annual simple interest rate.

1. $I = \$60$, $P = \$500$, $t = 3$ years 4%

2. $I = \$45$, $P = \$600$, $t = 2$ years 3.75%

Find the amount of time.

3. $I = \$117$, $P = \$1300$, $r = 3\%$ 3 yr

4. $I = \$71.50$, $P = \$1100$, $r = 3.25\%$ 2 yr

5. A savings account earns 4.5% annual simple interest. The principal is $1300. What is the balance after 3 years? $1475.50

Taking Math Deeper

Exercise 32

This problem isn't particularly difficult. However, it is a good opportunity for students to pick up some financial literacy. That is, when you pay for items with a credit card, you almost always have to pay interest. In other words, you are taking out a loan.

 Find the amount spent.

Total = $175.54

Zoo Trip	
Tickets	67.70
Food	62.34
Gas	45.50
Total Cost	175.54

② Find the interest paid.

$I = Prt$ Write the formula.

$= 175.54 \cdot 0.12 \cdot \dfrac{3}{12}$ Substitute amounts.

$\approx \$5.27$ Simplify.

③ Find the total cost of the trip.

Total = 175.54 + 5.27
      = $180.81

How much interest would I pay if I didn't pay the charge for 1 year? for 2 years?

Project

Many credit cards charge different rates of interest. Use the school library or the internet to research the amount of interest charged by three different credit card companies. Compare the cost of the trip to the zoo based on the different interest rates. Why should you be careful when selecting a credit card and charging items to the card?

Reteaching and Enrichment Strategies

If students need help. . .	If students got it. . .
Resources by Chapter • Practice A and Practice B • Puzzle Time Record and Practice Journal Practice Differentiating the Lesson Lesson Tutorials Skills Review Handbook	Resources by Chapter • Enrichment and Extension • School-to-Work Start the next section

Find the amount paid for the loan.

④ **24.** $1500 at 9% for 2 years

25. $2000 at 12% for 3 years

26. $2400 at 10.5% for 5 years

27. $4800 at 9.9% for 4 years

Copy and complete the table.

	Principal	Interest Rate	Time	Simple Interest
28.	$12,000	4.25%	5 years	
29.		6.5%	18 months	$828.75
30.	$15,500	8.75%		$5425.00
31.	$18,000		54 months	$4252.50

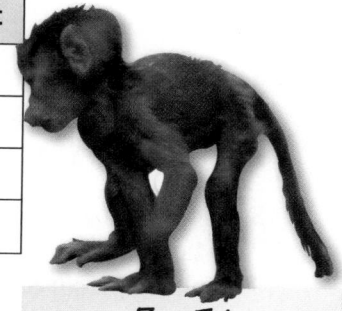

32. ZOO A family charges a trip to the zoo on a credit card. The simple interest rate is 12%. The charges are paid after 3 months. What is the total amount paid for the trip?

33. MONEY MARKET You deposit $5000 in an account earning 7.5% simple interest. How long will it take for the balance of the account to be $6500?

Zoo Trip

Tickets   67.70
Food   62.34
Gas   45.50
Total Cost   ?

11.8% Simple Interest
Equal monthly
payments for 2 years.

34. LOANS A music company offers a loan to buy a drum set for $1500. What is the monthly payment?

35. REASONING How many years will it take for $2000 to double at a simple interest rate of 8%? Explain how you found your answer.

36. LOANS You have two loans, for 2 years each. The total interest for the two loans is $138. On the first loan, you pay 7.5% simple interest on a principal of $800. On the second loan, you pay 3% simple interest. What is the principal for the second loan?

37. Critical Thinking You put $500 in an account that earns 4% annual interest. The interest earned each year is added to the principal to create a new principal. Find the total amount in your account after each year for 3 years.

 **Fair Game Review** What you learned in previous grades & lessons

Solve the proportion. *(Section 3.5)*

38. $\dfrac{4}{9} = \dfrac{12}{x}$

39. $\dfrac{15}{36} = \dfrac{n}{12}$

40. $\dfrac{m}{6.5} = \dfrac{14}{26}$

41. $\dfrac{2.4}{z} = \dfrac{3}{11.25}$

42. MULTIPLE CHOICE What is the solution of $4x + 5 = -11$? *(Section 2.6)*

    Ⓐ -4 Ⓑ -1.5 Ⓒ 1.5 Ⓓ 4

Check It Out
Progress Check
BigIdeasMath ✔com

Find the price, discount, markup, or cost to store.   *(Section 4.3)*

1. Original price: $30
Discount: 10%
Sale price: ?

2. Original price: $55
Discount: ?
Sale price: $46.75

3. Original price: ?
Discount: 75%
Sale price: $74.75

4. Cost to store: $152
Markup: 50%
Selling price: ?

5. Cost to store: $20
Markup: ?
Selling price: $32

6. Cost to store: ?
Markup: 80%
Selling price: $21.60

An account earns simple interest. Find the interest earned, principal, interest rate, or time.   *(Section 4.4)*

7. Interest earned: ?
Principal: $1200
Interest rate: 2%
Time: 5 years

8. Interest earned: $25
Principal: $500
Interest rate: 5%
Time: ?

9. Interest earned: $76
Principal: $800
Interest rate: ?
Time: 2 years

10. Interest earned: $119.88
Principal: ?
Interest rate: 3.6%
Time: 3 years

11. DIGITAL CAMERA A digital camera costs $229. The camera is on sale for 30% off and you have a coupon for an additional 15% off the original price. What is the final price?   *(Section 4.3)*

12. WATER SKIS The original price of the water skis was $200. What is the percent of discount?   *(Section 4.3)*

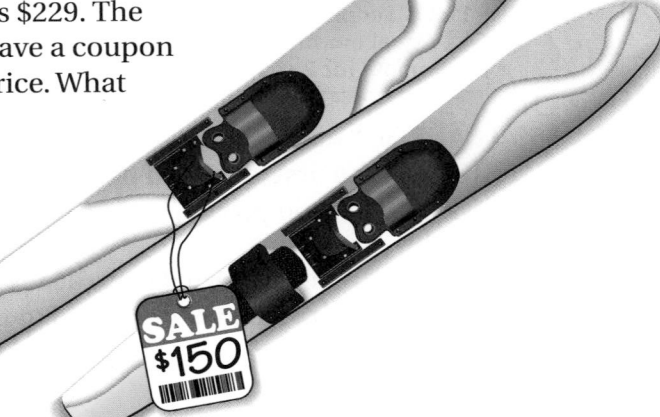

2 Ways to Own:
1. $75 cash back with 3.5% simple interest
2. No interest for 2 years

13. SAXOPHONE A saxophone costs $1200. A store offers two loan options. Which option saves more money if you pay the loan in 2 years?   *(Section 4.4)*

14. LOAN You borrow $200. The simple interest rate is 12%. You pay off the loan after 2 years. How much do you pay for the loan?   *(Section 4.4)*

Alternative Assessment Options

Math Chat   Student Reflective Focus Question
Structured Interview   **Writing Prompt**

Writing Prompt
Ask students to write a story about making purchases and saving money. The students should include discounts and markups in the story. If they have money left over from their purchases, they should place it in a savings account. The students should include simple interest in the story. Then have students share their stories with the class.

Study Help Sample Answers

Remind students to complete Graphic Organizers for the rest of the chapter.

5.

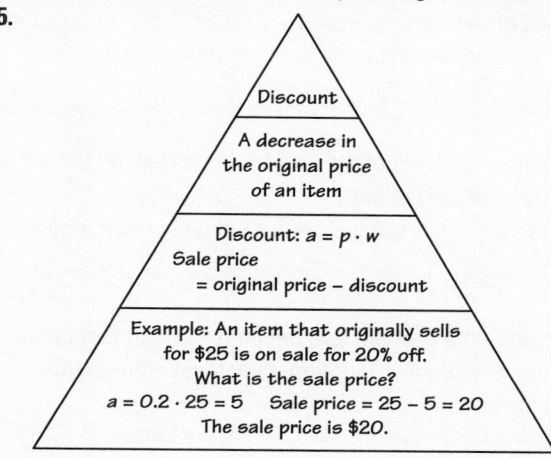

Discount

A decrease in the original price of an item

Discount: $a = p \cdot w$
Sale price = original price – discount

Example: An item that originally sells for $25 is on sale for 20% off. What is the sale price?
$a = 0.2 \cdot 25 = 5$ Sale price = 25 – 5 = 20
The sale price is $20.

6.

Markup

To make a profit, stores charge more than what they pay. The increase from what the store pays to the selling price is called a markup.

Markup: $a = p \cdot w$
Selling price = cost to store + markup

Example: The wholesale price of an item is $15. The percent of markup is 12%. What is the selling price?
$a = 0.12 \cdot 15 = 1.8$ Selling price = 15 + 1.8 = 16.8
The selling price is $16.80.

7. Available at *BigIdeasMath.com*

Reteaching and Enrichment Strategies

If students need help. . .	If students got it. . .
Resources by Chapter • Study Help • Practice A and Practice B • Puzzle Time Lesson Tutorials *BigIdeasMath.com* Practice Quiz Practice from the Test Generator	Resources by Chapter • Enrichment and Extension • School-to-Work Game Closet at *BigIdeasMath.com* Start the Chapter Review

Technology For the Teacher

Answer Presentation Tool

Assessment Book

Chapter 4 Quiz
For use after Section 4.4

Find the price, discount, markup, or cost to store.

1. Original price: $130
   Discount: 60%
   Sale price: ?

2. Original price: $32
   Discount: ?
   Sale price: $8

3. Original price: ?
   Discount: 20%
   Sale price: $14.40

4. Cost to store: $45
   Markup: 35%
   Selling price: ?

5. Cost to store: $50
   Markup: ?
   Selling price: $60

6. Cost to store: ?
   Markup: 110%
   Selling price: $105

An account earns simple interest. Find the interest earned, principal, interest rate, or time.

7. Interest earned: ?
   Principal: $1450
   Interest rate: 9%
   Time: 5 years

8. Interest earned: $10
   Principal: $250
   Interest rate: 4%
   Time: ?

9. Interest earned: $40
   Principal: $400
   Interest rate: ?
   Time: 2 years

10. Interest earned: $45
    Principal: ?
    Interest rate: 3%
    Time: 2 years

11. Interest earned: $750
    Principal: ?
    Interest rate: 12.5%
    Time: 4 years

12. Interest earned: $120
    Principal: $640
    Interest rate: 6.25%
    Time: ?

13. Store A sells a watch for $50 and offers a 5% discount. Store B sells the same watch for $60 and offers a 20% discount. From which store should you buy?

14. A store sells a television for $1000. Customers can choose to receive a 10% discount and pay it off with a loan at a simple interest rate of 4%, or they can choose to pay the full price and pay it off in 3 years with no interest. If the customer plans to pay it off in 3 years, which option is better?

15. A store offers a loan for $900 to buy a computer. The terms of the loan are for 9% simple interest and equal monthly payments for three years. What is the monthly payment?

Answers

1. _____
2. _____
3. _____
4. _____
5. _____
6. _____
7. _____
8. _____
9. _____
10. _____
11. _____
12. _____
13. _____
14. _____
15. _____

Answers

1. $a = 0.24 \cdot 25$; 6
2. $9 = p \cdot 20$; 45%
3. $10.2 = 0.85 \cdot w$; 12
4. $a = 0.83 \cdot 20$; 16.6
5. 120 parking spaces

Review of Common Errors

Exercises 1–5
- Students may not know what number to substitute for each variable. Walk through each type of question with the students. Emphasize that the word "is" means "equals," and "of" means "multiplied by."
- Students may mix up the whole and the part when trying to write the percent equation for the word problems. Ask students to identify each part of the equation before writing it in the equation format.

Exercises 6–9
- Students may mix up where to place the numbers in the equation to find percent of change. When students do not put the numbers in the right place, they might find a negative number in the numerator. Emphasize that students must know if it is increasing or decreasing before they can do anything else. The numerator should never have a negative answer. If students get a negative number, then they need to switch the order of the numbers in the problem and then subtract.

Exercises 10 and 11
- Students may just find the markup and not the selling price. Remind them that they must add the markup to the cost to store.
- Remind students that the sale price is not the percent of discount multiplied by the original price.

Exercises 12–18
- Students may forget to change the percent to a decimal. Remind them that before they can put the percent into the equation, they must change the percent to a fraction or a decimal.

Review Key Vocabulary

percent, *p. 160*
percent of change, *p. 166*
percent of increase, *p. 166*

percent of decrease, *p. 166*
discount, *p. 174*
markup, *p. 174*

interest, *p. 180*
principal, *p. 180*
simple interest, *p. 180*

Review Examples and Exercises

4.1 The Percent Equation (pp. 158–163)

What number is 72% of 25?

$a = p \cdot w$	Write percent equation.
$= 0.72 \cdot 25$	Substitute 0.72 for *p* and 25 for *w*.
$= 18$	Multiply.

So, 72% of 25 is 18.

28 is what percent of 70?

$a = p \cdot w$	Write percent equation.
$28 = p \cdot 70$	Substitute 28 for *a* and 70 for *w*.
$0.4 = p$	Divide each side by 70.

Because 0.4 = 40%, 28 is 40% of 70.

22.1 is 26% of what number?

$a = p \cdot w$	Write percent equation.
$22.1 = 0.26 \cdot w$	Substitute 22.1 for *a* and 0.26 for *p*.
$85 = w$	Divide each side by 0.26.

So, 22.1 is 26% of 85.

Exercises

Write and solve an equation to answer the question.

1. What number is 24% of 25?
2. 9 is what percent of 20?
3. 85% of what number is 10.2?
4. 83% of 20 is what number?

5. **PARKING** 15% of the school parking spaces are handicap spaces. The school has 18 handicap spaces. How many parking spaces are there?

4.2 Percents of Increase and Decrease *(pp. 164–169)*

The table shows the number of skim boarders at a beach on Saturday and Sunday. What was the percent of change in boarders from Saturday to Sunday?

The number of skim boarders on Sunday is less than the number of skim boarders on Saturday. So, the percent of change is a percent of decrease.

Day	Number of Skim Boarders
Saturday	12
Sunday	9

$$\text{percent of decrease} = \frac{\text{original amount} - \text{new amount}}{\text{original amount}}$$

$$= \frac{12 - 9}{12} \qquad \text{Substitute.}$$

$$= \frac{3}{12} \qquad \text{Subtract.}$$

$$= 0.25 = 25\% \qquad \text{Write as a percent.}$$

❖ The number of skim boarders decreased by 25% from Saturday to Sunday.

Exercises

Identify the percent of change as an *increase* or *decrease*. Then find the percent of change. Round to the nearest tenth of a percent, if necessary.

6. 6 yards to 36 yards

7. 6 hits to 3 hits

8. 120 meals to 52 meals

9. 35 words to 115 words

4.3 Discounts and Markups *(pp. 172–177)*

What is the original price of the tennis racquet?

The sale price is 100% − 30% = 70% of the original price.

Answer the question: 21 is 70% of what number?

$$a = p \cdot w \qquad \text{Write percent equation.}$$

$$21 = 0.7 \cdot w \qquad \text{Substitute 21 for } a \text{ and 0.7 for } p.$$

$$30 = w \qquad \text{Divide each side by 0.7.}$$

❖ The original price of the tennis racquet is $30.

SALE 30% off Now $21

Exercises

Find the price.

10. Original price: $50
Discount: 15%
Sale price: ?

11. Original price: ?
Discount: 20%
Sale price: $75

Review Game

Percents of Increase and Decrease

Big Ideas
Game Closet

Materials per Group
- 1 deck of cards with the jacks, queens, kings, and aces removed
- paper
- pencil
- calculator

Directions
Each group starts with 108 points. The cards are placed face down in the middle of the group. One member of the group turns a card over. If the card is red, the face value of the card is subtracted from the number of points. If the card is black, the face value of the card is added to the number of points. Group members take turns calculating the percent increase or decrease and turning cards over. The starting number of points at each player's turn is the same as the ending number of points at the previous player's turn. The group should be back to 108 points after going through all of the cards.

Who Wins?
The group with the highest mean percent increase wins. To find the mean percent increase, add the percent increases and divide the sum by 18.

For the Student
Additional Practice
- Lesson Tutorials
- Study Help (textbook)
- Student Website
    Multi-Language Glossary
    Practice Assessments

Answers
6. increase; 500%
7. decrease; 50%
8. decrease; 56.7%
9. increase; 228.6%
10. $42.50
11. $93.75
12. **a.** $36
    b. $336
13. **a.** $280
    b. $2280
14. 1.7%
15. 7.1%
16. 3 years
17. 6 years
18. 4%

My Thoughts on the Chapter

What worked. . .

What did not work. . .

What I would do differently. . .

4.4 Simple Interest (pp. 178–183)

You put $200 in a savings account. The account earns 2% simple interest per year.

a. What is the interest after 4 years?

b. What is the balance after 4 years?

a. $I = Prt$ Write simple interest formula.

     $= 200(0.02)(4)$ Substitute 200 for P, 0.02 for r, and 4 for t.

     $= 16$ Multiply.

     The interest earned is $16 after 4 years.

b. The balance is the principal plus the interest.

     So, the balance is $200 + $16 = $216 after 4 years.

You put $500 in an account. The account earns $55 simple interest in 5 years. What is the annual interest rate?

        $I = Prt$ Write simple interest formula.

      $55 = 500(r)(5)$ Substitute 55 for I, 500 for P, and 5 for t.

      $55 = 2500r$ Simplify.

   $0.022 = r$ Divide each side by 2500.

     The annual interest rate of the account is 0.022, or 2.2%.

Exercises

An account earns simple interest.

a. Find the interest earned.

b. Find the balance of the account.

12. $300 at 4% for 3 years   **13.** $2000 at 3.5% for 4 years

Find the annual simple interest rate.

14. $I = \$17$, $P = \$500$, $t = 2$ years   **15.** $I = \$426$, $P = \$1200$, $t = 5$ years

Find the amount of time.

16. $I = \$60$, $P = \$400$, $r = 5\%$ **17.** $I = \$237.90$, $P = \$1525$, $r = 2.6\%$

18. SAVINGS You put $100 in an account. The account earns $2 simple interest in 6 months. What is the annual interest rate?

Write and solve an equation to answer the question.

1. 16% of 150 is what number?

2. 10 is 40% of what number?

3. 27 is what percent of 75?

4. What number is 35% of 56?

Identify the percent of change as an *increase* or *decrease*. Then find the percent of change. Round to the nearest tenth of a percent, if necessary.

5. 4 strikeouts to 10 strikeouts

6. $24.00 to $18.00

Find the price, discount, or markup.

7. Original price: $15
   Discount: 5%
   Sale price: ?

8. Original price: $189
   Discount: ?
   Sale price: $75.60

9. Cost to store: $15
   Markup: ?
   Selling price: $24.75

10. Cost to store: $5.50
    Markup: 75%
    Selling price: ?

An account earns simple interest. Find the interest earned, principal, interest rate, or time.

11. Interest earned: ?
    Principal: $450
    Interest rate: 6%
    Time: 8 years

12. Interest earned: $27
    Principal: ?
    Interest rate: 1.5%
    Time: 2 years

13. Interest earned: $116.25
    Principal: $1550
    Interest rate: ?
    Time: 9 months

14. Interest earned: $45.60
    Principal: $2400
    Interest rate: 3.8%
    Time: ?

15. **MOVIE PREVIEWS** There are eight previews before a movie. Seventy-five percent of the previews are for comedies. How many previews are for comedies?

16. **BOOK** What was the original price of the book?

The WORLD around us

A pictu 20% off

Only $7.00

17. **TEXT MESSAGES** The cost of a text message increases from $0.10 per message to $0.25 per message. What is the percent increase in the cost of sending a text message?

18. **INVESTMENT** You put $800 in an account that earns 4% simple interest. Find the total amount in your account after each year for 3 years.

Test Item References

Chapter Test Questions	Section to Review
1–4, 15	4.1
5, 6, 16	4.2
7–10, 16, 17	4.3
11–14, 18	4.4

Test-Taking Strategies

Remind students to quickly look over the entire test before they start so that they can budget their time. Students should estimate and check for reasonableness as they work through the test. Some students will benefit from putting essential information on the back of their test before they begin.

Common Assessment Errors

- **Exercises 1–4** Students may not know what numbers to substitute for the variables. Review each type of question with students. Emphasize that the word "is" means "equals" and "of" means "multiplied by." Ask students to identify the whole, the part of the whole, and the percent.
- **Exercises 5 and 6** Students might place the numbers in the percent of change formulas incorrectly. Remind them that they should have the difference between the greater amount and the lesser amount in the numerator, so the numerator should never be negative. Also point out that the original amount should always be in the denominator.
- **Exercises 7 and 10** Students may write the discount or markup amount as the new price instead of subtracting it from or adding it to the original price. Remind them to subtract or add as appropriate to find the sale or selling price.
- **Exercises 8 and 9** Students may treat the difference in the prices as the percent of discount or markup. Remind students that the discount or markup should be a *percent*, and that this percent is found by using the original price and the difference in prices in the percent equation.
- **Exercises 11–14** Students may forget to write the percent as a decimal, forget to convert time to years (if necessary), or use the wrong inverse operation to solve for the unknown value. Review the simple interest formula and the Division Property of Equality.

Reteaching and Enrichment Strategies

If students need help. . .	If students got it. . .
Resources by Chapter • Practice A and Practice B • Puzzle Time Record and Practice Journal Practice Differentiating the Lesson Lesson Tutorials Practice from the Test Generator Skills Review Handbook	Resources by Chapter • Enrichment and Extension • School-to-Work Game Closet at *BigIdeasMath.com* Start the next chapter

Answers

1. $a = 0.16 \cdot 150$; 24
2. $10 = 0.4 \cdot w$; 25
3. $27 = p \cdot 75$; 36%
4. $a = 0.35 \cdot 56$; 19.6
5. increase; 150%
6. decrease; 25%
7. $14.25
8. 60%
9. 65%
10. $9.63
11. $216
12. $900
13. 10%
14. 6 months
15. 6 previews
16. $8.75
17. 150%
18. Year 1: $832
    Year 2: $864
    Year 3: $896

Assessment Book

Chapter 4 **Test B**

Chapter 4 **Test A**

Write and solve an equation to answer the question.

1. 17 is what percent of 68? 2. What number is 16% of 80?
3. 35% of what number is 21? 4. 70 is what percent of 56?

Identify the percent of change as an *increase* or *decrease*. Then find the percent of change. Round to the nearest tenth of a percent, if necessary.

5. 15 books to 21 books 6. 60 cars to 24 cars
7. 12 calculators to 3 calculators 8. 100 pennies to 101 pennies

Use the percent of change to find the new amount.

9. 40 employees increased by 15% 10. 120 pounds decreased by 30%
11. $84 increased by 12% 12. 820 brushes decreased by 25%

Find the price, discount, or markup.

13. Original price: $82 14. Original price: $125
    Discount: 10% Discount: ?
    Sale price: ? Sale price: $81.25
15. Original price: ? 16. Original price: $148
    Discount: 36% Discount: ?
    Sale price: $32 Sale price: $125.80
17. Cost to store: $32 18. Cost to store: $3
    Markup: 16% Markup: ?
    Selling price: ? Selling price: $5.70

Answers

1. _____
2. _____
3. _____
4. _____
5. _____
6. _____
7. _____
8. _____
9. _____
10. _____
11. _____
12. _____
13. _____
14. _____
15. _____
16. _____
17. _____
18. _____

After Answering Easy Questions, Relax
Answer Easy Questions First
Estimate the Answer
Read All Choices before Answering
Read Question before Answering
Solve Directly or Eliminate Choices
Solve Problem before Looking at Choices
Use Intelligent Guessing
Work Backwards

About this Strategy

When taking a multiple choice test, be sure to read each question carefully and thoroughly. It is also very important to read each answer choice carefully. Do not pick the first answer you think is correct. If two answer choices are the same, eliminate them both. There can only be one correct answer.

Answers

1. C
2. G
3. 152 lb
4. D

Item Analysis

1. **A.** The student finds 30% of $8.50 but does not subtract this amount from $8.50.

   B. The student thinks that 30% is equivalent to $3.00 and subtracts this amount from $8.50.

   C. Correct answer

   D. The student thinks that 30% is equivalent to $0.30 and subtracts this amount from $8.50.

2. **F.** The student divides incorrectly or converts measures incorrectly to choose an incorrect box.

   G. Correct answer

   H. The student divides incorrectly or converts measures incorrectly to choose an incorrect box.

   I. The student divides incorrectly or converts measures incorrectly to choose an incorrect box.

3. **Gridded Response:** Correct answer: 152 lb

   Common Error: The student finds only the loss, getting an answer of 8.

4. **A.** The student chooses a proportion that will find what percent 17 is of 43.

   B. The student chooses a proportion that will find 43% of 17.

   C. The student chooses a proportion that will find 17% of 43.

   D. Correct answer

5. **F.** Correct answer

   G. The student incorrectly thinks that $|21| = -21$, so the opposite of $|21|$ is 21.

   H. The student incorrectly thinks that the absolute value of any number is its opposite.

   I. The student makes an order of operations error and does not first find the sum within the absolute value bars.

6. **A.** The student finds the minimum number of hours.

   B. The student finds the mode of the numbers of hours.

   C. Correct answer

   D. The student finds the maximum number of hours or the middle value in the list as it is given.

Technology
For the Teacher

Big Ideas Test Generator

1. A movie theatre offers 30% off the price of a movie ticket to students from your school. The regular price of a movie ticket is $8.50. What is the discounted price that you would pay for a ticket?

   A. $2.55   C. $5.95

   B. $5.50   D. $8.20

Test-Taking Strategy

Read Question Before Answering

About 0.4 of cats are polydactyl. Of 80 cats, how many have 5 toes per paw?
Ⓐ 32   Ⓑ 30%   Ⓒ 48   Ⓓ 58

Not fair. I'm a cartoon character and I have only 4 toes per paw.

"Keep on your toes and read the questions before choosing your answer."

2. You are comparing the prices of four boxes of cereal. Two of the boxes contain free extra cereal.

   - Box F costs $3.59 and contains 16 ounces.

   - Box G costs $3.79 and contains 16 ounces, plus an additional 10% for free.

   - Box H costs $4.00 and contains 500 grams.

   - Box I costs $4.69 and contains 500 grams, plus an additional 20% for free.

   Which box has the least unit cost? (1 ounce = 28.35 grams)

   F. Box F   H. Box H

   G. Box G   I. Box I

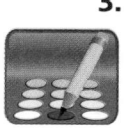

3. James is getting ready for wrestling season. As part of his preparation, he plans to lose 5% of his body weight. James currently weighs 160 pounds. How much will he weigh, in pounds, after he loses 5% of his weight?

4. Which proportion represents the problem below?

   "17% of a number is 43. What is the number?"

   A. $\dfrac{17}{43} = \dfrac{n}{100}$

   C. $\dfrac{n}{43} = \dfrac{17}{100}$

   B. $\dfrac{n}{17} = \dfrac{43}{100}$

   D. $\dfrac{43}{n} = \dfrac{17}{100}$

5. Betty was simplifying the expression in the box below.

$$-|8 + (-13)| = -(|8| + |-13|)$$
$$= -|8 + 13|$$
$$= -|21|$$
$$= -21$$

What should Betty do to correct the error that she made?

F. Simplify $-|8 + (-13)|$ to get $-|-5|$.

G. Find the opposite of $|21|$, which is 21.

H. Find the absolute value of 8, which is -8.

I. Distribute the negative sign to get $|-8 + (-13)|$.

6. The students from the Math Club participated in a long-distance walk as a fundraiser. The number of hours each club member took to complete the walk is shown in the bar graph below.

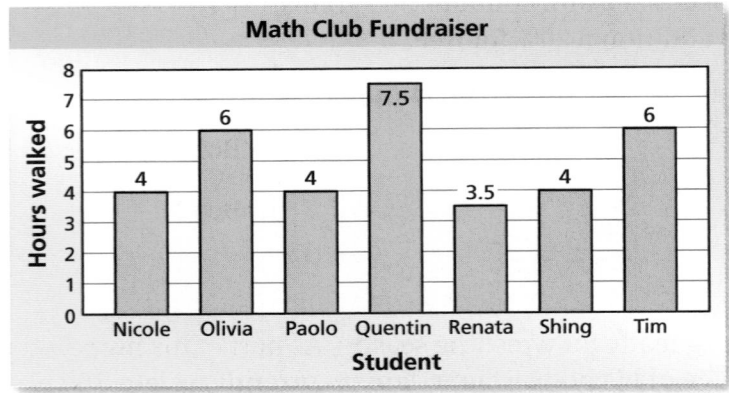

What is the mean number of hours the club members took to complete the walk?

A. 3.5 h

B. 4 h

C. 5 h

D. 7.5 h

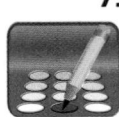

7. A lighting store is holding a clearance sale. The store is offering discounts on all the lamps it sells. As the sale progresses, the store will increase the percent of discount it is offering.

You want to buy a lamp that has an original price of $40. You will buy the lamp when its price is marked down to $10. What percent discount will you have received?

Item Analysis (continued)

7. **Gridded Response:** Correct answer: 75%

   Common Error: The student finds what percent 10 is of 40, and gets an answer of 25.

8. **F.** The student finds what percent 60 is of 660.

   G. Correct answer

   H. The student finds 60% of 660 and misplaces the decimal point.

   I. The student thinks that the difference of the scores is equivalent to the percent.

9. **A.** The student chooses an inequality that matches a ray that points to the left.

   B. The student chooses an inequality that matches a ray that points to the left and a circle that is solid.

   C. Correct answer

   D. The student chooses an inequality that matches a circle that is solid.

10. **4 points** The student demonstrates a thorough understanding of interpreting a problem involving simple interest. In Part A, the student correctly determines that it would take 2.5 years. In Part B, the student correctly determines that an initial deposit of $4000 is not large enough. The student provides clear and complete work and explanations.

    3 points The student demonstrates an understanding of interpreting a problem involving simple interest, but the student's work and explanations demonstrate an essential but less than thorough understanding.

    2 points The student demonstrates a partial understanding of interpreting a problem involving simple interest. The student's work and explanations demonstrate a lack of essential understanding.

    1 point The student demonstrates a limited understanding of interpreting a problem involving simple interest. The student's response is incomplete and exhibits many flaws.

    0 points The student provides no response, a completely incorrect or incomprehensible response, or a response that demonstrates insufficient understanding of interpreting a problem involving simple interest.

Answers

5. F

6. C

7. 75%

Answers

8. G

9. C

10. *Part A* 2.5 years

   Part B no

11. I

12. D

Answers for Extra Examples

1. **A.** Correct answer

   B. The student aligns the decimal point of the estimated product with the longer of the two estimated factors.

   C. The student thinks that the product is positive because the greater factor is positive.

   D. The student aligns the decimal point of the estimated product with the longer of the two estimated factors. The student also thinks that the product is positive because the greater factor is positive.

2. **F.** The student finds the value of $2 - 6 - 9$.

   G. The student finds the value of $-2 + 6 - 9$.

   H. Correct answer

   I. The student finds the value of $-2 + 6 - (-9)$.

Item Analysis (continued)

11. **F.** The student does not recognize that -3 was already distributed correctly.

   G. The student does not recognize that -3 was already distributed correctly.

   H. The student makes the right correction but finds the incorrect sum of -45 and 6.

   I. Correct answer

12. **A.** The student thinks that the product of two negative numbers is negative.

   B. The student thinks that the product of two negative numbers is negative, and the student thinks that $-9\frac{7}{8}$ is closer to -9.

   C. The student thinks that $-9\frac{7}{8}$ is closer to -9.

   D. Correct answer

Extra Examples for Standardized Test Practice

1. Which of the following is closest to the value of the expression below?

   $$0.041 \cdot (-0.0038)$$

   A. -0.00016 **C.** 0.00016

   B. -0.016 **D.** 0.016

2. What is the value of the expression below?

   $$2 - 6 - (-9)$$

   F. -13 **H.** 5

   G. -5 **I.** 13

8. A student scored 600 the first time she took the mathematics portion of her college entrance exam. The next time she took the exam, she scored 660. Her second score represents what percent increase over her first score?

   F. 9.1%   **H.** 39.6%

   G. 10%   **I.** 60%

9. Which inequality is represented by the graph on the number line below?

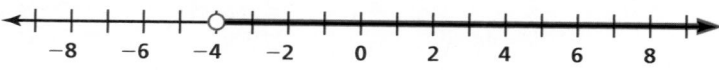

   A. $x < -4$ **C.** $x > -4$

   B. $x \le -4$ **D.** $x \ge -4$

10. You are planning to deposit $4000 into an account that earns 5% simple interest per year. You will not make any other deposits or withdrawals.

   Part A   How long would it take for your account to contain $4500? Show your work and explain your reasoning.

   Part B   You would like the account to contain $5100 after 4 years. Would your initial $4000 deposit be large enough? Show your work and explain your reasoning.

11. Brad was solving the equation in the box shown.

What should Brad do to correct the error that he made?

   F. Distribute -3 to get $6 - 15w$.

   G. Distribute -3 to get $-6 - 15w$.

   H. Add 6 to both sides to get $15w = -51$.

   I. Add 6 to both sides to get $15w = -39$.

$$-3(2 - 5w) = -45$$
$$-6 + 15w = -45$$
$$9w = -45$$
$$\frac{9w}{9} = \frac{-45}{9}$$
$$w = -5$$

12. Which integer is closest to the value of the expression below?

$$-7\frac{1}{4} \cdot \left(-9\frac{7}{8}\right)$$

   A. -70 **C.** 63

   B. -63 **D.** 70

5 Similarity and Transformations

"Just 2 more minutes. I'm almost done with my 'cat tessellation' painting."

"If you hold perfectly still..."

"...each frame becomes a horizontal..."

"...translation of the previous frame..."

Connections to Previous Learning

- Identify and describe basic transformations including figures with line and rotational symmetry.
- Use a formula to find the areas of parallelograms, triangles, and trapezoids.

- Find the area of a plane figure.
- Use given information to find a missing dimension of a plane figure.

- Solve problems involving similar figures.
- Apply proportionality to measurement, including scale drawings and constant speed.
- Determine how changes in dimensions affect the perimeter and area of similar geometric figures, and apply these relationships to solve problems.
- Predict the results of transformations and draw transformed figures, with and without the coordinate plane.

Math in History

Many ancient cultures used symmetry in art and design.

★ The Navajo and Pueblo people of the southwestern United States often used rotational symmetry in weaving designs for blankets and rugs. Many of the rugs have the property that when the rug is rotated 180 degrees, the pattern is unchanged.

★ The Arabic tile patterns called Zellige have rotational symmetry. These patterns were used to create ceramic mosaics for decorations on walls, ceilings, fountains, floors, pools, and tables.

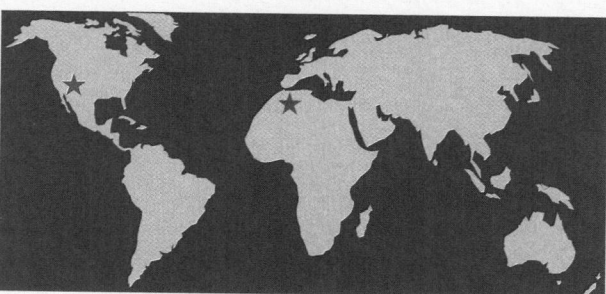

Pacing Guide for Chapter 5

Chapter Opener	1 Day
Section 1 Activity Lesson	1 Day 1 Day
Section 2 Activity Lesson	2 Days 1 Day
Section 3 Activity Lesson	1 Day 1 Day
Section 4 Activity Lesson Lesson b	1 Day 1 Day 1 Day
Study Help / Quiz	1 Day
Section 5 Activity Lesson	1 Day 1 Day
Section 6 Activity Lesson	1 Day 1 Day
Section 7 Activity Lesson	1 Day 1 Day
Quiz / Chapter Review	1 Day
Chapter Test	1 Day
Standardized Test Practice	1 Day
Total Chapter 5	21 Days
Year-to-Date	94 Days

Check Your Resources

- Record and Practice Journal
- Resources by Chapter
- Skills Review Handbook
- Assessment Book
- Worked-Out Solutions

Technology For the **Teacher**

Dynamic Classroom

The Dynamic Planning Tool
Editable Teacher's Resources at
BigIdeasMath.com

Additional Topics for Review

- Identifying polygons
- Solving one-step equations
- Review angles (acute, right, obtuse)

Try It Yourself

1. 40 mm
2. 20 in.
3. about 33.41 cm
4. 6
5. 2
6. 42
7. 4

Record and Practice Journal

1. 58 ft
2. 40 in.
3. about 69.08 cm
4. 74 in.
5. about 30.84 mm
6. 57 in.
7. 40 m
8. $x = 16$
9. $x = 8$
10. $x = 11.25$
11. $x = 3$
12. $x = 6$
13. $x = 12$
14. $x = \$37.50$

Math Background Notes

Vocabulary Review

- Perimeter
- Circumference
- Proportion
- Cross Products Property

Finding Perimeter

- Students should be familiar with the concept of perimeter.
- Remind students that perimeter is a measure of the distance around the outside of an object.
- Remind students that the perimeter of a circle is referred to as the circumference.
- **Teaching Tip:** There are many ways to present the idea of perimeter. These varied methods will help all types of learners. For example, verbally describing perimeter to an auditory learner will be sufficient. Consider allowing kinesthetic learners to measure the perimeter of the classroom. Encourage visual learners to draw a scenario in which they need to know the perimeter (placing a fence around a dog house). You might ask tactile learners to use craft sticks to construct polygons with specified perimeters.
- You will want to be sure that students have mastered the concept of perimeter (both what it is and how to calculate it) before moving forward. In this chapter, students will use perimeter with similar figures. Being confident in an old skill will make learning a new one easier.

Solving Proportions

- Students know how to solve proportions in Chapter 3. Because this skill is fairly recent, students may need more time and practice to master the material.
- You may wish to review the Cross Products Property. Remind students that this property is unique as it can only be used in proportions.
- **Teaching Tip:** Help students to visualize the Cross Products Property by drawing the X as you work through the problem. Example:

$$\frac{2}{5} \diagdown\!\!\!\!\diagup \frac{x}{15}$$

$$30 = 5x$$

Reteaching and Enrichment Strategies

If students need help. . .	If students got it. . .
Record and Practice Journal • Fair Game Review Skills Review Handbook Lesson Tutorials	Game Closet at *BigIdeasMath.com* Start the next section

What You Learned Before

"These clouds are making me hungry."

● Finding Perimeter

Example 1 Find the perimeter.

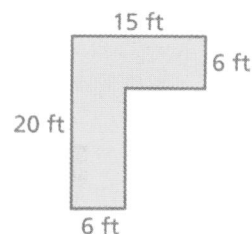

15 ft
6 ft
20 ft
6 ft

$P = 15 + 6 + 9 + 14 + 6 + 20$

$= 70$ ft

Example 2 Find the circumference.

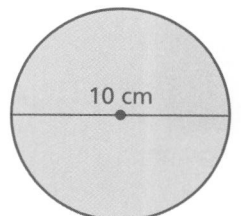

10 cm

$C = 2\pi r$

$= 2 \cdot 3.14 \cdot 5$ ⟵ $r = \dfrac{d}{2} = \dfrac{10}{2} = 5$

$= 31.4$ cm

Try It Yourself
Find the perimeter.

1.
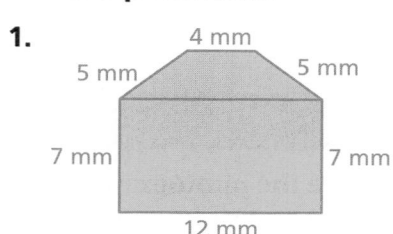
4 mm
5 mm   5 mm
7 mm   7 mm
12 mm

2.
4 in.
4 in.   5 in.
7 in.

3.
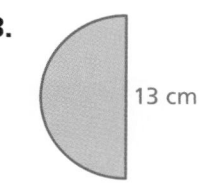
13 cm

● Solving Proportions

Example 3 Solve the proportion.

a. $\dfrac{x}{32} = \dfrac{3}{4}$

$\dfrac{x}{32} = \dfrac{3}{4}$ Write the proportion.

$4x = 96$ Use the Cross Products Property.

$x = 24$ Solve for x.

b. $\dfrac{3x}{20} = \dfrac{3}{5}$

$\dfrac{3x}{20} = \dfrac{3}{5}$

$15x = 60$

$x = 4$

Try It Yourself
Solve the proportion.

4. $\dfrac{2}{7} = \dfrac{x}{21}$

5. $\dfrac{3}{4} = \dfrac{3y}{8}$

6. $\dfrac{3}{14} = \dfrac{9}{y}$

7. $\dfrac{8}{9x} = \dfrac{2}{9}$

5.1 Identifying Similar Figures

Essential Question How can you use proportions to help make decisions in art, design, and magazine layouts?

In a computer art program, when you click and drag on a side of a photograph, you distort it.

But when you click and drag on a corner of the photograph, it remains proportional to the original.

Original Photograph

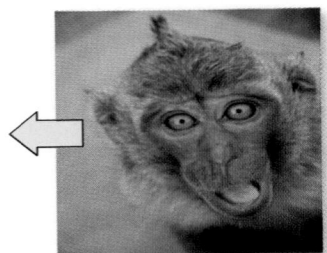

Distorted

Distorted

Proportional

1 ACTIVITY: Reducing Photographs

Work with a partner. You are trying to reduce the photograph to the indicated size for a nature magazine. Can you reduce the photograph to the indicated size without distorting or cropping? Explain your reasoning.

a.

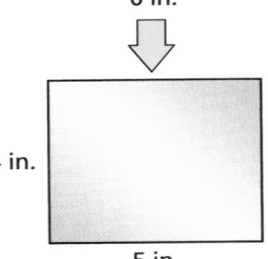

5 in.

6 in.

4 in.

5 in.

b.

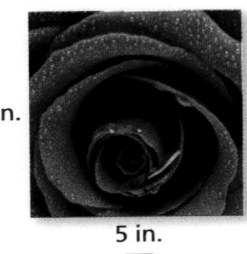

5 in.

5 in.

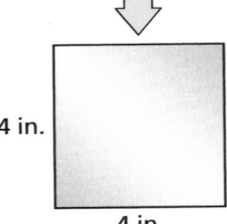

4 in.

4 in.

c.

6 in.

8 in.

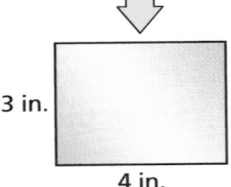

3 in.

4 in.

Laurie's Notes

Introduction

For the Teacher

- **Goal:** Students will decide if two polygons are proportional or if they are distortions of the original polygon.

Motivate

- Draw a simple stick figure or other image on a stretchable surface, such as a balloon, physical therapy elastic, or play putty.
- **?** Ask students what they think will happen to the figure when you pull the picture to the right. Students should recognize that the image will be distorted. Pull one of the sides of the picture to confirm.
- Pull the top of the picture so that students see this result as the same.
- **?** Ask students what they think will happen if the stretchable surface is pulled in both directions (right and up). Students should recognize that the image will enlarge proportionally.
- **Alternative:** If you can display computer images to the class, you can distort actual images on-screen by dragging the corners of the image.

Activity Notes

Activity 1

- Photography is a good context to use to examine similarity. A common misconception is that standard photo sizes are proportional and, in fact, most are not. A 5" × 7" photo and a 4" × 6" photo are not proportional.
- **Common Misconception:** Students often believe that if you subtract the same amount from each dimension, the resulting ratio will be proportional to the first. For example, $\frac{5}{7} \neq \frac{5-1}{7-1} = \frac{4}{6}$.
- **?** Ask questions about proportions and ratios:
  - "What is a proportion?" two equal ratios
  - "Are the two ratios 2:3 and 4:6 equal?" yes
  - "Are the ratios 2:3 and 8:9 equal? Explain." No, listen for students to get at the idea that $2 \times 4 = 8$, but 3×4 is 12, not 9.
- Remember, students have *not* learned a formal definition for similar figures. Remind students that the task is to decide if the photograph can be reduced to the new dimensions without distorting it. Therefore, students must use the information about keeping the side lengths proportional.

Words of Wisdom

- Listen to how students describe their proportions. There are many correct ways to set up a proportion and some students might hear one way and incorrectly think their way is wrong.
- **?** Ask students "Did anyone set up their proportions differently?" Here are two possibilities. The key is to make sure *like things* are being compared.

$$\frac{\text{length (original)}}{\text{width (original)}} = \frac{\text{length (new size)}}{\text{width (new size)}} \, ; \, \frac{\text{length (original)}}{\text{length (new size)}} = \frac{\text{width (original)}}{\text{width (new size)}}$$

Previous Learning

Students should know how to write ratios and have a basic understanding of proportions.

Activity Materials
Introduction

- balloon or physical therapy elastic
- overhead marker
- computer art program

Start Thinking! and Warm Up

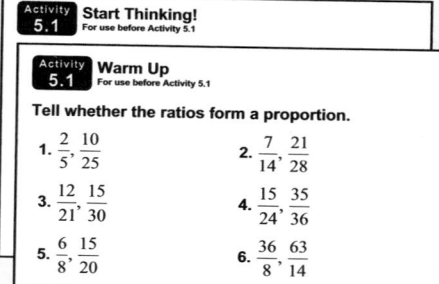

Activity 5.1	Start Thinking! For use before Activity 5.1

Activity 5.1	Warm Up For use before Activity 5.1

Tell whether the ratios form a proportion.

1. $\frac{2}{5}, \frac{10}{25}$ 2. $\frac{7}{14}, \frac{21}{28}$

3. $\frac{12}{21}, \frac{15}{30}$ 4. $\frac{15}{24}, \frac{35}{36}$

5. $\frac{6}{8}, \frac{15}{20}$ 6. $\frac{36}{8}, \frac{63}{14}$

5.1 Record and Practice Journal

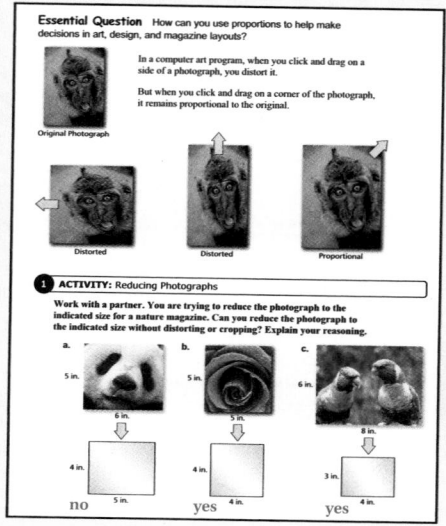

Essential Question How can you use proportions to help make decisions in art, design, and magazine layouts?

In a computer art program, when you click and drag on a side of a photograph, you distort it.

But when you click and drag on a corner of the photograph, it remains proportional to the original.

Original Photograph

Distorted   Distorted   Proportional

1 ACTIVITY: Reducing Photographs

Work with a partner. You are trying to reduce the photograph to the indicated size for a nature magazine. Can you reduce the photograph to the indicated size without distorting or cropping? Explain your reasoning.

a.   b.   c.

no   yes   yes

English Language Learners

Vocabulary

Ask students what *similar* means. Ask them if *similar* things are exactly alike. Explain that *similar figures* are not exactly alike. Similar figures have the same shape, but not necessarily the same size.

5.1 Record and Practice Journal

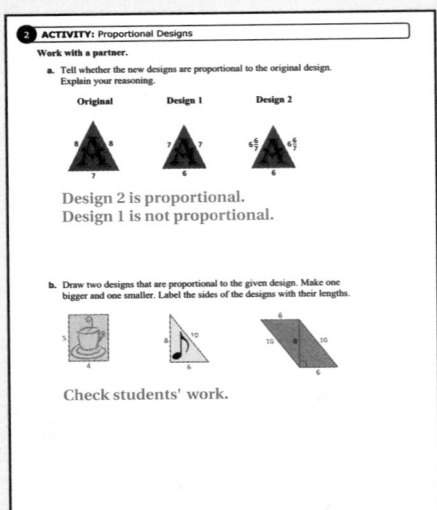

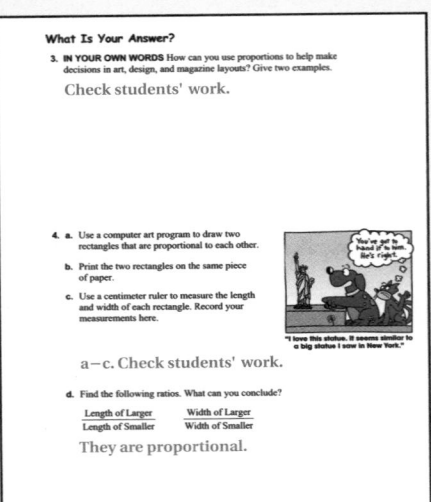

Activity 2

- This activity is similar to the first activity. In Activity 1, students were asked if the original figure would become distorted. In this activity, students are asked if the designs are proportional.

? "What type of triangle is the original design?" Isosceles

? "Triangles don't have a *length* and *width* as rectangles do. What dimensions will you compare to decide if the triangular designs are proportional?" Listen for language such as base and sides or base and legs.

- **Common Error:** Students think that subtracting 1 from each side length of the triangle produces a new triangle proportional to the original triangle.

- In completing part (b), students are not expected to make a scale drawing. They are drawing a figure that should look similar to the original. The rectangle should not look like a square. The right scalene triangle should not look equilateral.

- Have students share the dimensions of their new figures. They should explain how they came up with the new dimensions. Listen for methods that use multiplication, not addition.

What Is Your Answer?

- **Technology:** Question 4 provides a great opportunity to have students work with a computer art program to enhance their understanding of similar figures.

Closure

- Are all three of these triangles proportional? Yes.

Technology For the Teacher

Dynamic Classroom

The Dynamic Planning Tool
Editable Teacher's Resources at *BigIdeasMath.com*

Work with a partner.

a. Tell whether the new designs are proportional to the original design. Explain your reasoning.

Original

8 8

7

Design 1

7 7

6

Design 2

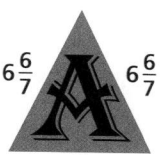

$6\frac{6}{7}$ $6\frac{6}{7}$

6

b. Draw two designs that are proportional to the given design. Make one bigger and one smaller. Label the sides of the designs with their lengths.

5

4

8 10

6

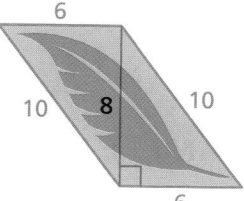

6

10 8 10

6

What Is Your Answer?

3. IN YOUR OWN WORDS How can you use proportions to help make decisions in art, design, and magazine layouts? Give two examples.

4. a. Use a computer art program to draw two rectangles that are proportional to each other.

"I love this statue. It seems similar to a big statue I saw in New York."

b. Print the two rectangles on the same piece of paper.

c. Use a centimeter ruler to measure the length and width of each rectangle.

d. Find the following ratios. What can you conclude?

$$\frac{\text{Length of Larger}}{\text{Length of Smaller}} \qquad \frac{\text{Width of Larger}}{\text{Width of Smaller}}$$

Practice Use what you learned about similar figures to complete Exercises 9 and 10 on page 198.

Check It Out
Lesson Tutorials
BigIdeasMath⟍com

 Key Idea

Similar Figures

Figures that have the same shape but not necessarily the same size are called **similar figures**. The triangles below are similar.

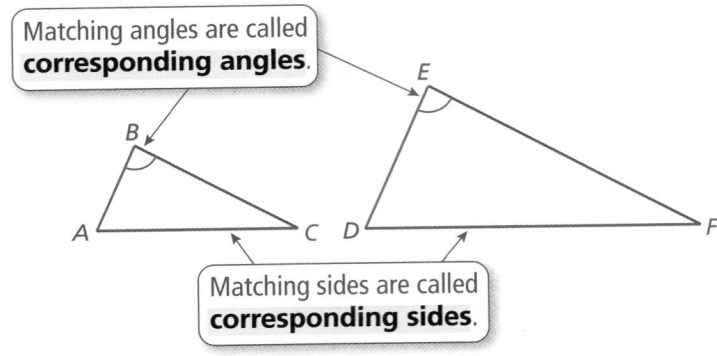

Matching angles are called **corresponding angles**.

Matching sides are called **corresponding sides**.

EXAMPLE **1** **Naming Corresponding Parts**

The trapezoids are similar. (a) Name the corresponding angles.
(b) Name the corresponding sides.

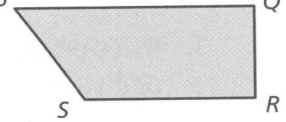

a. Corresponding angles:

∠A and ∠P

∠B and ∠Q

∠C and ∠R

∠D and ∠S

b. Corresponding sides:

Side AB and Side PQ

Side BC and Side QR

Side CD and Side RS

Side AD and Side PS

● On Your Own

Now You're Ready
Exercises 5 and 6

1. The figures are similar.

   a. Name the corresponding angles.

   b. Name the corresponding sides.

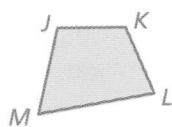
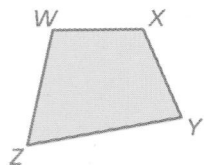

◀) Multi-Language Glossary at BigIdeasMath⟍com.

Laurie's Notes

Introduction

Connect

- **Yesterday:** Students developed an intuitive understanding about proportional polygons.
- **Today:** Students will use the formal definition of similar figures.

Motivate

- Place an item on an overhead projector, such as an index card, school ID, or other rectangular item. Ask questions about the actual item and its projected image.
- **?** "How does the actual item compare to its projection?" Listen for: "they look alike," "they have the same shape," or "they're similar," because it is unlikely that they know the mathematical definition of similar.
- Place a different-shaped item on the overhead.
- **?** "There are two items and two projected images. Which projection goes with which item? How do you know?" Listen for students to say the items are the same shape but different sizes.

Lesson Notes

Key Idea

- Students often think about corresponding parts in terms of the longest/shortest sides and greatest/least angle measures. Students are influenced by the orientation of the shapes.
- All of the shapes below are similar. Students have difficulty working with the third shape because its orientation is different from the first two.

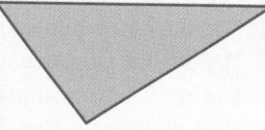

- In the lesson, the similar shapes are presented in the same orientation. In the practice exercises, some figures are *not* in the same orientation.

Example 1

- **Representation:** Have students use color-coding to show corresponding parts.
- **Notation:** Remind students that the angle symbol (∠) is needed when talking about corresponding angles; they should not write the letter only. In this textbook, *AB* means segment *AB* or the length of segment *AB*.
- **?** "Is side *AB* the same as side *BA*?" Yes.

On Your Own

- When students have finished, ask a volunteer to share their answers. Students should use proper terminology such as "angle *K*" and "side *JK*."

Goal Today's lesson is using proportions to determine if two figures are similar.

Lesson Materials
Introduction
• index cards
• school ID

Start Thinking! and Warm Up

Lesson 5.1	**Warm Up** For use before Lesson 5.1

Lesson 5.1 **Start Thinking!** For use before Lesson 5.1

Explain how to determine if two figures are similar.

Extra Example 1

The triangles are similar.

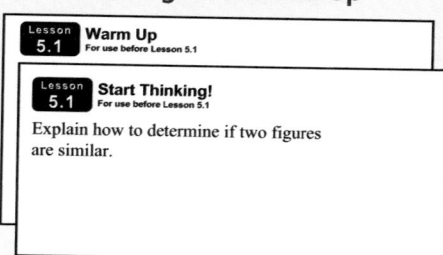

a. Name the corresponding angles. ∠E and ∠L, ∠F and ∠M, ∠G and ∠N

b. Name the corresponding sides. Side *EF* and Side *LM*, Side *FG* and Side *MN*, and Side *GE* and Side *NL*

On Your Own

1. **a.** ∠*J* and ∠*W*, ∠*M* and ∠*Z*, ∠*K* and ∠*X*, ∠*L* and ∠*Y*

   b. Side *JK* and Side *WX*, Side *KL* and Side *XY*, Side *LM* and Side *YZ*, Side *JM* and Side *WZ*

Differentiated Instruction

Visual

Bring in examples of figures that are similar and figures that have the same size and shape. Ask students to identify the figures that have the same size and shape (congruent). Then ask students to identify the figures that have the same shape (similar).

Extra Example 2

Which parallelogram is similar to Parallelogram A?

Parallelogram A

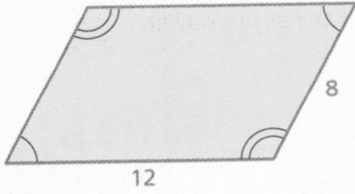

Parallelogram B

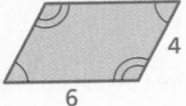

Parallelogram C

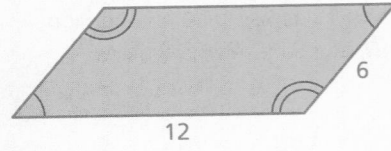

Parallelogram B

🔵 On Your Own

2. Rectangle B

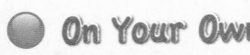

Laurie's Notes

🔘 Words of Wisdom

- Draw two rectangles on the chalkboard that are *not* similar.

- **?** "Do you think these two rectangles are similar? Explain." No, you want students to recognize that similarity is more than just being the same shape. You need to consider the size and how the corresponding sides compare.
- **?** **Big Idea:** "How will you decide if two figures are similar?" You want students to consider this question before going on to the next Key Idea. Similar is more than just looking at corresponding sides and it is more than just looking at corresponding angles—it is both!

Key Idea

- Discuss the tilde symbol ~ that denotes similarity. Explain that the order in which the vertices of the triangle are written identifies how the sides and angles correspond.
- **Big Idea:** Discuss the need for two conditions to be met for two figures to be similar: **corresponding side** lengths are proportional *and* **corresponding angles** have the same measure.
- **Representation:** Point out the color-coding, which should help students see the corresponding parts.
- Take your time in this section. There is a great deal of vocabulary, symbols, representations, *and* the fundamental concept of similarity. Give students time to ask questions and think about all that is being presented.

Example 2

- **?** "What do you know about the angles of a rectangle?" 4 right angles
- **?** "Do the corresponding angles have the same measure?" Yes.
- **?** "What else must you check to know that the rectangles are similar?" corresponding side lengths are proportional
- Note that the problem has students focus on the dimensions of the rectangles, using the words *length* and *width*, without using the side names that can confuse students.

🔘 Closure

- Sketch two figures that look similar. Describe how you would determine if your figures are actually similar.

 Key Idea

Reading

Red arcs are used to indicate angles that have the same measure. The symbol ~ means "is similar to."

Identifying Similar Figures

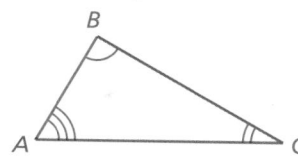
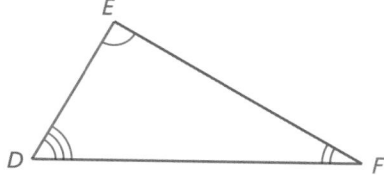

Triangle *ABC* is similar to triangle *DEF*: $\triangle ABC \sim \triangle DEF$

Words Two figures are similar if

- corresponding side lengths are proportional, and
- corresponding angles have the same measure.

Common Error

When writing a similarity statement, make sure to list the vertices of the figures in the same order.

Symbols | ***Side Lengths*** | | ***Angles*** |

$$\frac{AB}{DE} = \frac{BC}{EF} = \frac{AC}{DF}$$

$\angle A$ has the same measure as $\angle D$.

$\angle B$ has the same measure as $\angle E$.

$\angle C$ has the same measure as $\angle F$.

EXAMPLE **2** **Identifying Similar Figures**

Which rectangle is similar to Rectangle A?

Rectangle A
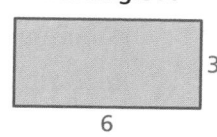
6

Rectangle B
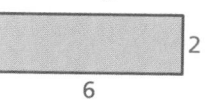
6

Rectangle C
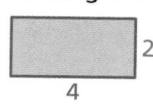
4

Each figure is a rectangle. So, corresponding angles have the same measure. Check to see if corresponding side lengths are proportional.

Rectangle A and Rectangle B

$\dfrac{\text{Length of A}}{\text{Length of B}} = \dfrac{6}{6} = 1$ $\qquad \dfrac{\text{Width of A}}{\text{Width of B}} = \dfrac{3}{2}$ Not proportional

Rectangle A and Rectangle C

$\dfrac{\text{Length of A}}{\text{Length of C}} = \dfrac{6}{4} = \dfrac{3}{2}$ $\qquad \dfrac{\text{Width of A}}{\text{Width of C}} = \dfrac{3}{2}$ Proportional

So, Rectangle C is similar to Rectangle A.

On Your Own

Now You're Ready
Exercises 7–12

2. Rectangle D is 3 units long and 1 unit wide. Which rectangle in Example 2 is similar to Rectangle D?

Vocabulary and Concept Check

1. **VOCABULARY** How are corresponding angles of two similar figures related?

2. **VOCABULARY** How are corresponding side lengths of two similar figures related?

3. **OPEN-ENDED** Give examples of two real-world objects whose shapes are similar.

4. **CRITICAL THINKING** Are two figures that have the same size and shape similar? Explain.

Practice and Problem Solving

Name the corresponding angles and the corresponding sides of the similar figures.

5.

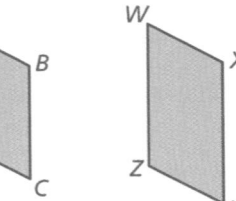

6.

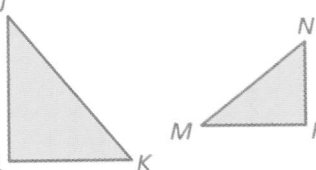

In a coordinate plane, draw the figures with the given vertices. Which figures are similar? Explain your reasoning.

7. Triangle A: (0, 0), (3, 0), (0, 3)
   Triangle B: (0, 0), (5, 0), (0, 5)
   Triangle C: (0, 0), (3, 0), (0, 6)

8. Rectangle A: (0, 0), (4, 0), (4, 2), (0, 2)
   Rectangle B: (0, 0), (−6, 0), (−6, 3), (0, 3)
   Rectangle C: (0, 0), (4, 0), (4, 2), (0, 2)

Tell whether the two figures are similar. Explain your reasoning.

9.

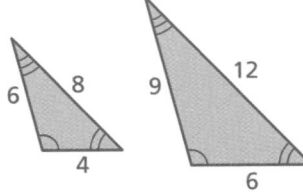

10.

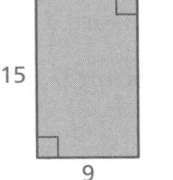
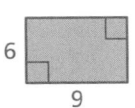

11. **MEXICO** A Mexican flag is 63 inches long and 36 inches high. Is the drawing at the right similar to the Mexican flag?

12. **DESKS** A student's rectangular desk is 30 inches long and 18 inches wide. The teacher's rectangular desk is 60 inches long and 36 inches wide. Are the desks similar?

8.5 in.

11 in.

Assignment Guide and Homework Check

Level	Day 1 Activity Assignment	Day 2 Lesson Assignment	Homework Check
Basic	9, 10, 23–27	1–8, 11–13, 15	2, 6, 8, 12, 13
Average	9, 10, 23–27	1–5, 7, 13–17, 20	2, 5, 7, 13, 16
Advanced	9, 10, 23–27	1–4, 8, 14–22 even, 19, 21	2, 8, 14, 16, 20

Common Errors

- **Exercises 5 and 6** Students may forget to write the angle symbol with the angle name. Remind them that A is a point and $\angle A$ is the angle.
- **Exercises 7 and 8** Students may think that because the side lengths have changed by the same amount, the triangles or rectangles are similar. Remind them that the *ratios* of the corresponding side lengths must be the same for the two figures to be similar.
- **Exercises 13–15** Students may have difficulty remembering which angles are corresponding. Tell them to write out the corresponding angles before writing the angle measures.

5.1 Record and Practice Journal

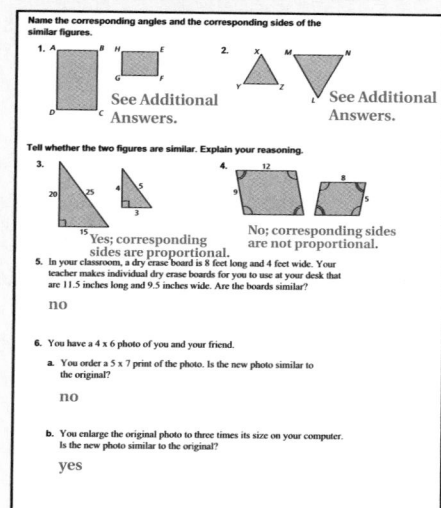

1. They have the same measure.

2. They are proportional.

3. *Sample answer:* A photograph of size 3" × 5" and another photograph of size 6" × 10".

4. Yes, because the angles have the same measure and the side lengths are proportional.

 Practice and Problem Solving

5. $\angle A$ and $\angle W$, $\angle B$ and $\angle X$, $\angle C$ and $\angle Y$, $\angle D$ and $\angle Z$; Side AB and Side WX, Side BC and Side XY, Side CD and Side YZ, Side AD and Side WZ

6. $\angle J$ and $\angle M$, $\angle L$ and $\angle P$, $\angle K$ and $\angle N$; Side JK and Side MN, Side KL and Side NP, Side JL and Side MP

7–8. See Additional Answers.

9. similar; Corresponding angles have the same measure.
Because $\dfrac{4}{6} = \dfrac{6}{9} = \dfrac{8}{12}$, the corresponding side lengths are proportional.

10. not similar; Corresponding side lengths are not proportional.

11. no

12. yes

Practice and Problem Solving

13. 48°   **14.** 90°

15. 42°

16. See Additional Answers.

17. See *Taking Math Deeper*.

18–22. See Additional Answers.

Fair Game Review

23. $\frac{16}{81}$ **24.** $\frac{9}{64}$

25. $\frac{49}{16}$ **26.** $\frac{169}{16}$

27. B

Mini-Assessment

Tell whether the rectangles are similar. Explain your reasoning.

1.

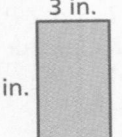
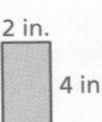

3 in.

6 in.

2 in.

4 in.

yes; corresponding side lengths are proportional and corresponding angles have the same measure

2.

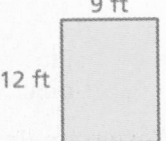
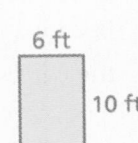

9 ft

6 ft

12 ft

10 ft

no; corresponding side lengths are not proportional

3. Are the two triangular stickers similar? Explain your reasoning.

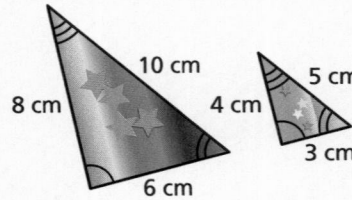

10 cm

8 cm

6 cm

5 cm

4 cm

3 cm

yes; corresponding side lengths are proportional and corresponding angles have the same measure

T-199

Taking Math Deeper

Exercise 17

If you are able to show students the same photo at different sizes, they can see that the photos need to be cropped in different ways. This is true because only two of the five sizes are similar. The 8 × 12 and 18 × 27 both have a 2 : 3 ratio.

① Help me see it.

4 × 5

5 × 7

8 × 12   11 × 14   18 × 27

② Write ratios as simplified fractions.

$\frac{4}{5}$ $\frac{5}{7}$ $\frac{8}{12} = \frac{2}{3}$ $\frac{11}{14}$ $\frac{18}{27} = \frac{2}{3}$

Only these 2 are similar.

I see it.

③ Use a computer drawing program. Scan a 4 × 5 photo and a 5 × 7 photo. Try to enlarge the 4 × 5 photo to make it fit exactly on top of the 5 × 7 photo.

Project

Draw a picture that measures 5" × 7". Draw a similar picture that measures 7.5" on one side. What are the dimensions of the new picture? How did you draw the new picture so that everything was similar to the first?

Reteaching and Enrichment Strategies

If students need help. . .	If students got it. . .
Resources by Chapter • Practice A and Practice B • Puzzle Time Record and Practice Journal Practice Differentiating the Lesson Lesson Tutorials Skills Review Handbook	Resources by Chapter • Enrichment and Extension Start the next section

The two triangles are similar. Find the measure of the angle.

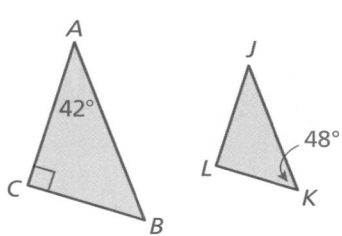

13. $\angle B$ **14.** $\angle L$ **15.** $\angle J$

16. REASONING Given $\triangle FGH \sim \triangle QRT$, name the corresponding angles and the corresponding sides.

17. PHOTOS You want to buy only photos that are similar rectangles. Which of the photo sizes should you buy?

18. CRITICAL THINKING Are the following figures *always, sometimes,* or *never* similar? Explain.

    a. Two triangles   **b.** Two squares

    c. Two rectangles   **d.** A square and a triangle

Photo Size
4 in. × 5 in.
5 in. × 7 in.
8 in. × 12 in.
11 in. × 14 in.
18 in. × 27 in.

19. CRITICAL THINKING Can you draw two quadrilaterals each having two 130° angles and two 50° angles that are *not* similar? Justify your answer.

20. SIGN All of the angle measures in the sign are 90°.

    a. Each side length is increased by 20%. Is the new sign similar to the original?

    b. Each side length is increased by 6 inches. Is the new sign similar to the original?

21. GEOMETRY Use a ruler to draw two different isosceles triangles similar to the one shown. Measure the heights of each triangle to the nearest centimeter.

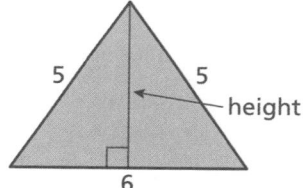

    a. Is the ratio of the corresponding heights proportional to the ratio of the corresponding side lengths?

    b. Do you think this is true for all similar triangles? Explain.

22. **Critical Thinking** Given $\triangle ABC \sim \triangle DEF$ and $\triangle DEF \sim \triangle JKL$, is $\triangle ABC \sim \triangle JKL$? Give an example or non-example.

Fair Game Review *What you learned in previous grades & lessons*

Simplify. *(Skills Review Handbook)*

23. $\left(\dfrac{4}{9}\right)^2$ **24.** $\left(\dfrac{3}{8}\right)^2$ **25.** $\left(\dfrac{7}{4}\right)^2$ **26.** $\left(\dfrac{6.5}{2}\right)^2$

27. MULTIPLE CHOICE Which equation shows inverse variation? *(Section 3.8)*

    Ⓐ $3y = 8x$ **Ⓑ** $y = \dfrac{8}{3x}$ **Ⓒ** $\dfrac{y}{3} = \dfrac{x}{8}$ **Ⓓ** $y = 8x - 3$

Essential Question How do changes in dimensions of similar geometric figures affect the perimeters and areas of the figures?

> **1 ACTIVITY: Comparing Perimeters and Areas**

Work with a partner. Use pattern blocks to make a figure whose dimensions are 2, 3, and 4 times greater than those of the original figure. Find the perimeter P and area A of each larger figure.

a. Sample: Square

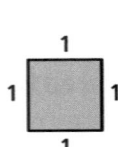 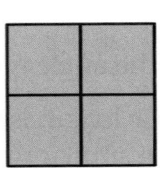 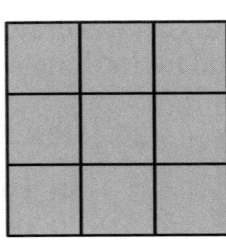 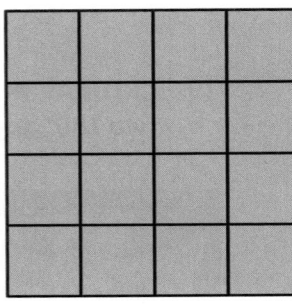

$P = 4$ $P = 8$ $P = 12$ $P = 16$

$A = 1$ $A = 4$ $A = 9$ $A = 16$

b. Triangle

 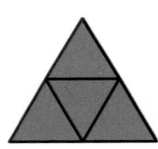  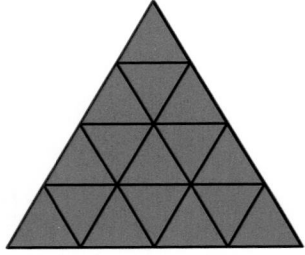

$P = 3$ $P = 6$ $P = \boxed{}$ $P = \boxed{}$

$A = B$ $A = 4B$ $A = \boxed{}$ $A = \boxed{}$

c. Rectangle

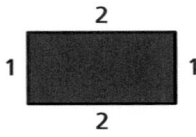

$P = 6$

$A = 2$

d. Parallelogram

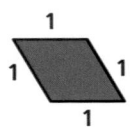

$P = 4$

$A = C$

Laurie's Notes

Introduction

For the Teacher

- **Goal:** Students will develop intuitive understanding about what happens to the perimeter and area of a figure when it is enlarged (or reduced) proportionally.
- Today's investigation explores an important concept that will extend naturally into the study of volume. This concept is used in many real-life applications, and is much easier to understand if presented in a visual fashion.

Motivate

- Show examples of the fractal known as The Sierpinski Triangle shown at different stages. Ask students how many triangles (of various sizes) they see in each stage.

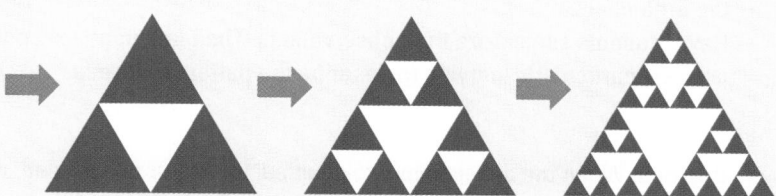

Activity Notes

Activity 1

- This activity is most effective if you use pattern blocks. The visual and tactile experience helps students learn and retain the information.
- ❓ Ask questions to review perimeter and area.
    - **"What does perimeter mean?"** the distance around a figure (you don't want a formula here, you want an understanding of what perimeter means)
    - **"How is it found?"** Add the lengths of the sides of the figure.
    - **"What does area mean?"** the amount of surface that a figure covers (you don't want a formula here, you want an understanding of what area means)
    - **"How is it found?"** The type of figure determines which area formula you use.
- Students may struggle using a variable to represent the area of the triangle and the parallelogram. Explain that B is simply the area of a single green triangle. Because there are 4 green triangles in the second grouping, the area is 4 times the area of a single green triangle and can be written as $4B$. Similarly, the area of four blue parallelograms is $4C$.
- As you circulate around the room, observe how students are building each successive model. Do they add on to the previous model? If so, the area has to increase by the number of new pieces that are added on.
- ❓ **"When you make a figure whose dimensions are twice the original dimensions, is the new figure similar to the original? Explain."** Yes, the corresponding sides are proportional and the corresponding angles have the same measure.

Start Thinking! and Warm Up

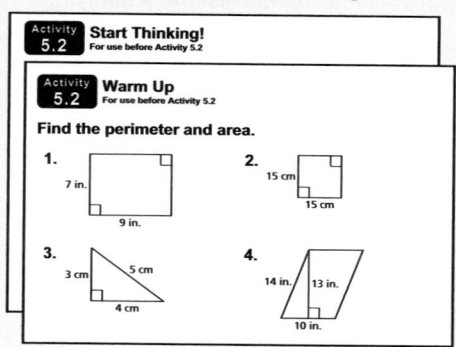

5.2 Record and Practice Journal

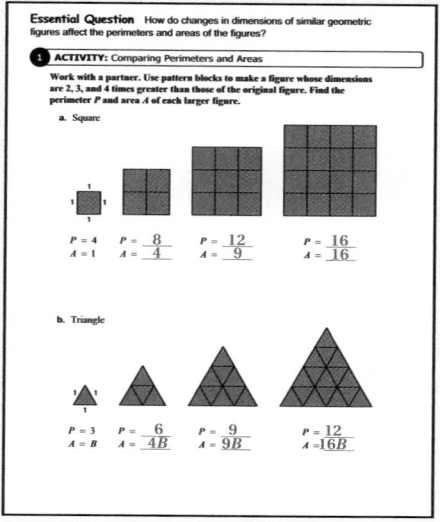

English Language Learners

Vocabulary

Have students select objects around the classroom and use tape to mark the perimeters of the objects. Label the objects "perimeter." Have students select other objects in the classroom and cover them with square sheets of paper. Label these objects "area." Have students identify which units are best for measuring perimeter and area. Add this information to the labels. Keep the objects in the classroom until the students understand the concepts of perimeter and area.

5.2 Record and Practice Journal

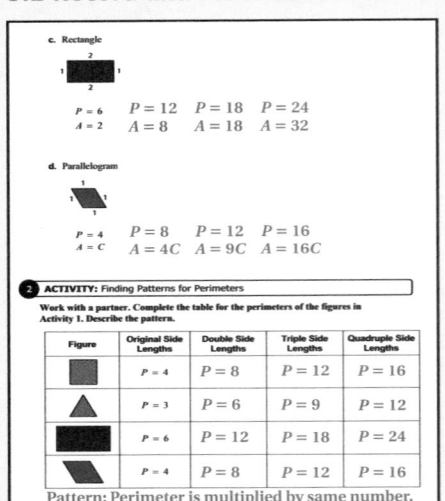

c. **Rectangle**

$P = 6$	$P = 12$	$P = 18$	$P = 24$
$A = 2$	$A = 8$	$A = 18$	$A = 32$

d. **Parallelogram**

$P = 4$	$P = 8$	$P = 12$	$P = 16$
$A = C$	$A = 4C$	$A = 9C$	$A = 16C$

2 ACTIVITY: Finding Patterns for Perimeters

Work with a partner. Complete the table for the perimeters of the figures in Activity 1. Describe the pattern.

Figure	Original Side Lengths	Double Side Lengths	Triple Side Lengths	Quadruple Side Lengths
	$P = 4$	$P = 8$	$P = 12$	$P = 16$
	$P = 3$	$P = 6$	$P = 9$	$P = 12$
	$P = 6$	$P = 12$	$P = 18$	$P = 24$
	$P = 4$	$P = 8$	$P = 12$	$P = 16$

Pattern: Perimeter is multiplied by same number.

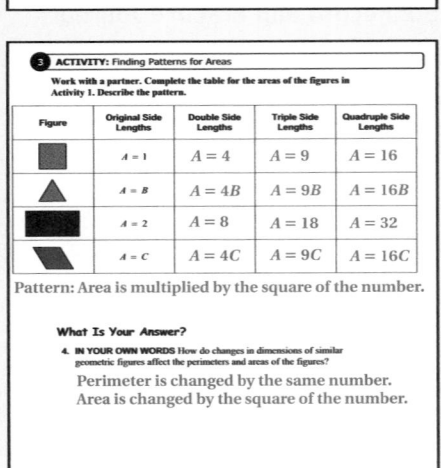

3 ACTIVITY: Finding Patterns for Areas

Work with a partner. Complete the table for the areas of the figures in Activity 1. Describe the pattern.

Figure	Original Side Lengths	Double Side Lengths	Triple Side Lengths	Quadruple Side Lengths
	$A = 1$	$A = 4$	$A = 9$	$A = 16$
	$A = B$	$A = 4B$	$A = 9B$	$A = 16B$
	$A = 2$	$A = 8$	$A = 18$	$A = 32$
	$A = C$	$A = 4C$	$A = 9C$	$A = 16C$

Pattern: Area is multiplied by the square of the number.

What Is Your Answer?

4. **IN YOUR OWN WORDS** How do changes in dimensions of similar geometric figures affect the perimeters and areas of the figures?

Perimeter is changed by the same number.
Area is changed by the square of the number.

Laurie's Notes

Activity 2 and Activity 3

- Students should work with a partner to complete both activities *before* you discuss the answers. Students should be able to find many patterns to describe. The patterns can be found in the rows and the chart as a whole.
- **Common Error:** Students may only look for numeric patterns and forget how the numbers in the chart were found, meaning the dimensions were doubled, tripled, or quadrupled. For example, students read across the first row for perimeter and say, "The perimeter is increasing by 4." While that is a true statement, it does not connect to what is happening with the dimensions of the figure.
- **Suggestion:** When students look at any entry in the table, they need to consider the following statements:
  "I doubled the original side lengths and the perimeter _____." doubled
  "I doubled the original side lengths and the area is _____." 4 times greater
- This strategy should be repeated for each figure in the perimeter chart and the area chart.
- Have students summarize their observations. Their language may not be precise, particularly in trying to describe the pattern for areas.

What Is Your Answer?

- **Big Idea:** When the dimensions are doubled, tripled, or quadrupled, the resulting figure is similar to the original figure. Students should see a pattern in the perimeters and in the areas. The Key Ideas in the lesson will define these patterns further.

Closure

- **Exit Ticket:** Find the perimeter and area of a rectangle with dimensions 3 inches by 4 inches. If you double the dimensions, what will be the new perimeter and area? New dimensions: perimeter is $2(14) = 28$ in., area is $4(12) = 48$ in.^{2}

Technology For the Teacher

The Dynamic Planning Tool
Editable Teacher's Resources at *BigIdeasMath.com*

2 ACTIVITY: Finding Patterns for Perimeters

Work with a partner. Copy and complete the table for the perimeters of the figures in Activity 1. Describe the pattern.

Figure	Original Side Lengths	Double Side Lengths	Triple Side Lengths	Quadruple Side Lengths
 ▢	$P = 4$	$P = 8$	$P = 12$	$P = 16$
 △	$P = 3$	$P = 6$		
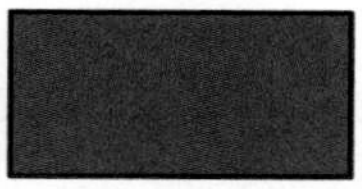 ▬	$P = 6$			
 ▱	$P = 4$			

Perimeters

3 ACTIVITY: Finding Patterns for Areas

Work with a partner. Copy and complete the table for the areas of the figures in Activity 1. Describe the pattern.

Figure	Original Side Lengths	Double Side Lengths	Triple Side Lengths	Quadruple Side Lengths
▢	$A = 1$	$A = 4$	$A = 9$	$A = 16$
 △	$A = B$	$A = 4B$		
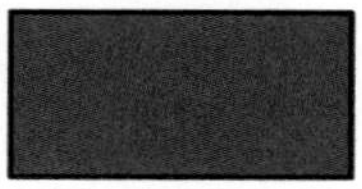 ▬	$A = 2$			
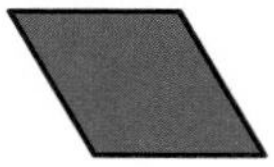 ▱	$A = C$			

Areas

What Is Your Answer?

4. **IN YOUR OWN WORDS** How do changes in dimensions of similar geometric figures affect the perimeters and areas of the figures?

Practice Use what you learned about perimeters and areas of similar figures to complete Exercises 8–11 on page 204.

Key Idea

Perimeters of Similar Figures

If two figures are similar, then the ratio of their perimeters is equal to the ratio of their corresponding side lengths.

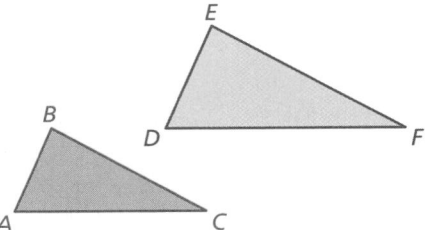

$$\frac{\text{Perimeter of } \triangle ABC}{\text{Perimeter of } \triangle DEF} = \frac{AB}{DE} = \frac{BC}{EF} = \frac{AC}{DF}$$

EXAMPLE 1 Finding Ratios of Perimeters

Find the ratio (red to blue) of the perimeters of the similar rectangles.

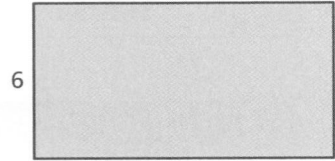

$$\frac{\text{Perimeter of red rectangle}}{\text{Perimeter of blue rectangle}} = \frac{4}{6} = \frac{2}{3}$$

⋮• The ratio of the perimeters is $\frac{2}{3}$.

On Your Own

1. The height of Figure A is 9 feet. The height of a similar Figure B is 15 feet. What is the ratio of the perimeter of A to the perimeter of B?

Key Idea

Areas of Similar Figures

If two figures are similar, then the ratio of their areas is equal to the *square* of the ratio of their corresponding side lengths.

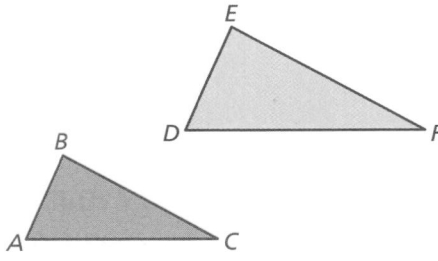

$$\frac{\text{Area of } \triangle ABC}{\text{Area of } \triangle DEF} = \left(\frac{AB}{DE}\right)^2 = \left(\frac{BC}{EF}\right)^2 = \left(\frac{AC}{DF}\right)^2$$

Laurie's Notes

Introduction

Connect

- **Yesterday:** Students used pattern blocks to investigate how changes in the dimensions of similar figures affect the perimeter and area of the figures.
- **Today:** Students will use the stated relationships to solve problems.

Motivate

- **Story Time:** Tell students that your neighbor's lawn is twice the size of your lawn. In other words, it is twice as long and twice as wide. If it takes you one-half hour to mow your lawn, about how long does it take your neighbor to mow their lawn? We will answer this in the Closure activity.

Lesson Notes

Key Idea

- ❓ "How do you identify similar triangles?" corresponding sides are proportional and corresponding angles have the same measure
- ❓ "If the corresponding sides are proportional, does it make sense that the perimeters have the same ratio?" Yes.
- Write some side lengths on the two triangles, such as 3-4-5 and 6-8-10. Then find the two perimeters, 12 and 24, to show the same ratio of 1 : 2.

Example 1

- Only the values of the corresponding sides are given. Use the relationship stated in the Key Idea.
- ❓ "What is the ratio of the corresponding sides?" 4 : 6 or 2 : 3 The ratio of the perimeters will also be 2 : 3.

On Your Own

- **Think-Pair-Share:** Students should read the question independently and then work with a partner to answer the question. When they have answered the question, the pair should compare their answer with another group and discuss any discrepancies.

Key Idea

- **Representation:** Draw two triangles so they look like they have the ratio of 1 : 2.

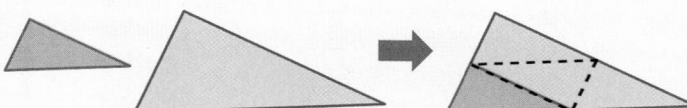

- ❓ "If the corresponding sides have a ratio of 1 : 2, what will the ratio of the areas be?" 1 : 4; You can also use pattern blocks to show this relationship.
- There are 4 copies of the smaller triangle inside the larger. Another way to state this relationship is that the larger triangle has an area 4 times greater than the area of the smaller triangle.

Goal Today's lesson is finding ratios of perimeters and areas of similar figures.

Start Thinking! and Warm Up

| Lesson 5.2 | **Warm Up** For use before Lesson 5.2 |

| Lesson 5.2 | **Start Thinking!** For use before Lesson 5.2 |

Your neighbor wants to replace his rectangular deck with one that is double the side lengths. Use what you have learned in Activity 5.2 to explain to your neighbor what will happen to the perimeter of the deck.

Extra Example 1

Find the ratio (red to blue) of the perimeters of the similar trapezoids. $\frac{7}{5}$

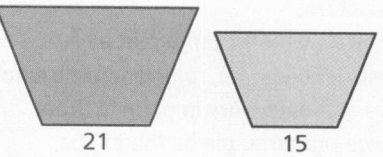

21   15

On Your Own

1. $\frac{3}{5}$

Extra Example 2

Find the ratio (red to blue) of the areas of the similar parallelograms. $\dfrac{81}{36} = \dfrac{9}{4}$

9   6

Extra Example 3

You place a photo with a width of 15 centimeters on a scrapbook page with a width of 30 centimeters. The page and the photo are similar rectangles.

a. How many times greater is the area of the page than the area of the photo?   4 times greater

b. The area of the photo is 165 square centimeters. What is the area of the page?   660 cm^2

 On Your Own

2. $\dfrac{64}{49}$ 3. 36 in.

Differentiated Instruction

Kinesthetic

Materials needed: large piece of construction paper, ruler, and protractor. Have students work in pairs to draw a large right triangle on the paper. Record the lengths of the sides and the measures of the angles in a table. Connect the midpoints of each side of the large triangle and record the side lengths and angle measures of the second triangle in the table. Connect the midpoints of the sides of the second triangle to form a third triangle. Record the side lengths and angle measures of the third triangle in the table. Have students determine if the triangles are similar. If they are similar, find the ratios of the perimeters and areas of each pair of triangles.

Laurie's Notes

Example 2

- As with Example 1, only the ratio of the corresponding sides is given. The actual areas cannot be computed. Students must use the relationship stated in the Key Idea.
- Remind students that $\left(\dfrac{3}{5}\right)^2$ means $\left(\dfrac{3}{5}\right) \cdot \left(\dfrac{3}{5}\right) = \dfrac{9}{25}$.

Example 3

- **Common Difficulty:** When students read "how many times greater" they generally think of a whole number answer. They do not see the ratio $\dfrac{16}{9}$ as being the answer to the question. Refer back to the triangle problem in the Key Idea (larger triangle is 4 times greater in area).
- **Estimation:** If the ratio of the areas was $\dfrac{18}{9}$, the area of the page would be twice the area of the photo. Because the ratio is $\dfrac{16}{9}$, it is less than twice the area of the photo.

On Your Own

- Have students work with a partner on Questions 2 and 3.

Closure

- **Exit Ticket:** Return to the question used to motivate at the beginning of the lesson. Your neighbor's lawn is twice the size of your lawn, meaning twice as long and twice as wide. If it takes you one-half hour to mow your lawn, about how long does it take your neighbor to mow their lawn? Explain your reasoning. (Assume that both people mow their lawns at the same rate.)

  2 hours; because the dimensions of your neighbor's lawn are double the dimensions of your lawn, you have the ratio neighbor : me = 2 : 1. The area of your neighbor's lawn is $\left(\dfrac{2}{1}\right)^2 = \dfrac{4}{1}$ times the area of your lawn, so it should take your neighbor 4 times longer to mow his or her lawn, $4\left(\dfrac{1}{2}\right) = 2$ hours.

Technology For the **Teacher**

Dynamic Classroom

The Dynamic Planning Tool
Editable Teacher's Resources at *BigIdeasMath.com*

EXAMPLE (2) **Finding Ratios of Areas**

Find the ratio (red to blue) of the areas of the similar triangles.

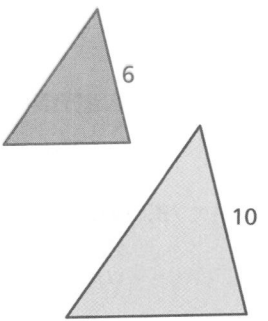

$$\frac{\text{Area of red triangle}}{\text{Area of blue triangle}} = \left(\frac{6}{10}\right)^2$$

$$= \left(\frac{3}{5}\right)^2 = \frac{9}{25}$$

∴ The ratio of the areas is $\frac{9}{25}$.

EXAMPLE (3) **Real-Life Application**

├── 6 in. ──┤

├──── 8 in. ────┤

You place a picture on a page of a photo album. The page and the picture are similar rectangles.

a. How many times greater is the area of the page than the area of the picture?

b. The area of the picture is 45 square inches. What is the area of the page?

a. Find the ratio of the area of the page to the area of the picture.

$$\frac{\text{Area of page}}{\text{Area of picture}} = \left(\frac{\text{length of page}}{\text{length of picture}}\right)^2$$

$$= \left(\frac{8}{6}\right)^2 = \left(\frac{4}{3}\right)^2 = \frac{16}{9}$$

∴ The area of the page is $\frac{16}{9}$ times greater than the area of the picture.

b. Multiply the area of the picture by $\frac{16}{9}$.

$$45 \cdot \frac{16}{9} = 80$$

∴ The area of the page is 80 square inches.

● **On Your Own**

Now You're Ready
Exercises 4–13

2. The base of Triangle P is 8 meters. The base of a similar Triangle Q is 7 meters. What is the ratio of the area of P to the area of Q?

3. In Example 3, the perimeter of the picture is 27 inches. What is the perimeter of the page?

Check It Out
Help with Homework
BigIdeasMath.com

 Vocabulary and Concept Check

1. **WRITING** How are the perimeters of two similar figures related?

2. **WRITING** How are the areas of two similar figures related?

3. **VOCABULARY** Rectangle *ABCD* is similar to Rectangle *WXYZ*. The area of *ABCD* is 30 square inches. What is the area of *WXYZ*? Explain.

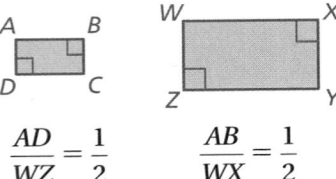

$$\frac{AD}{WZ} = \frac{1}{2} \qquad \frac{AB}{WX} = \frac{1}{2}$$

 **Practice and Problem Solving**

The two figures are similar. Find the ratios (red to blue) of the perimeters and of the areas.

 4.

 11    6

5.

 5   8

6.

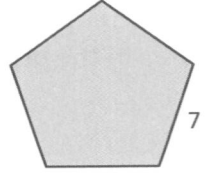 7    4

7.

9 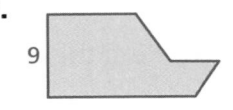   14

8. How does doubling the side lengths of a triangle affect its perimeter?

9. How does tripling the side lengths of a triangle affect its perimeter?

10. How does doubling the side lengths of a rectangle affect its area?

11. How does quadrupling the side lengths of a rectangle affect its area?

12. **FOOSBALL** The playing surfaces of two foosball tables are similar. The ratio of the corresponding side lengths is 10 : 7. What is the ratio of the areas?

13. **LAPTOP** The ratio of the corresponding side lengths of two similar computer screens is 13 : 15. The perimeter of the smaller screen is 39 inches. What is the perimeter of the larger screen?

Triangle *ABC* is similar to Triangle *DEF*. Tell whether the statement is *true* or *false*. Explain your reasoning.

14. $\dfrac{\text{Perimeter of } \triangle ABC}{\text{Perimeter of } \triangle DEF} = \dfrac{AB}{DE}$

15. $\dfrac{\text{Area of } \triangle ABC}{\text{Area of } \triangle DEF} = \dfrac{AB}{DE}$

Assignment Guide and Homework Check

Level	Day 1 Activity Assignment	Day 2 Lesson Assignment	Homework Check
Basic	8–11, 21–24	1–7, 12–15	2, 4, 12, 14
Average	8–11, 21–24	1–7, 14–17	2, 4, 14, 16
Advanced	8–11, 21–24	1–4, 6, 14, 15, 17–20	4, 14, 18, 19

Common Errors

- **Exercises 4–7** Students may find the reciprocal of the ratio. For example, they may find the ratio of blue to red instead of red to blue. Remind students to read the directions carefully.
- **Exercise 17** Students may only find the area of the larger merry-go-round and forget to find the percent of increase between the areas of the bases. Remind them to read the problem carefully and answer the question.

5.2 Record and Practice Journal

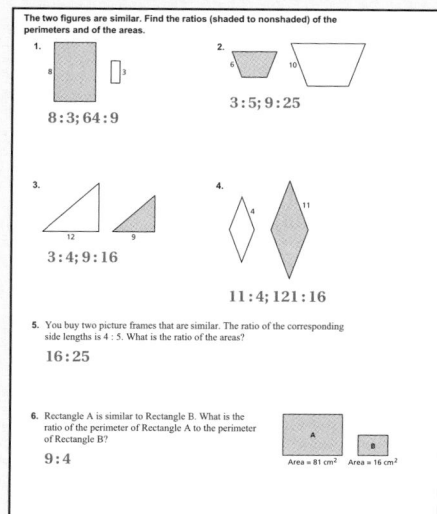

1. The ratio of the perimeters is equal to the ratio of the corresponding side lengths.

2. The ratio of the areas is equal to the square of the ratio of the corresponding side lengths.

3. 120 in.^{2}; Because the ratio of the corresponding side lengths is $\frac{1}{2}$, the ratio of the areas is equal to $\left(\frac{1}{2}\right)^2$. To find the area, solve the proportion $\frac{30}{x} = \frac{1}{4}$.

Practice and Problem Solving

4. $\frac{11}{6}$; $\frac{121}{36}$

5. $\frac{5}{8}$; $\frac{25}{64}$

6. $\frac{4}{7}$; $\frac{16}{49}$

7. $\frac{14}{9}$; $\frac{196}{81}$

8. perimeter doubles

9. perimeter triples

10. area quadruples

11. area is 16 times larger

12. 100 : 49

13. 45 in.

14. true; Because the triangles are similar, the ratio of the perimeters is equal to the ratio of the corresponding side lengths.

15. false; $\dfrac{\text{Area of } \triangle ABC}{\text{Area of } \triangle DEF} = \left(\dfrac{AB}{DE}\right)^2$

Practice and Problem Solving

16. See *Taking Math Deeper*.

17. 39,900%; The ratio of the corresponding lengths is $\dfrac{6 \text{ in.}}{120 \text{ in.}} = \dfrac{1}{20}$. So, the ratio of the areas is $\dfrac{1}{400}$ and the area of the actual merry-go-round is 180,000 square inches. The percent of increase is $\dfrac{180,000 - 450}{450} = 399 = 39,900\%$.

18. See Additional Answers.

19. $\dfrac{3}{4}$ **20.** 15 m

Fair Game Review

21. 25% increase

22. 30% decrease

23. 42.7% decrease

24. C

Mini-Assessment

The two figures are similar. Find the ratio (red to blue) of the perimeters and of the areas.

1.

$\dfrac{3}{2}, \dfrac{9}{4}$

2.

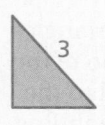 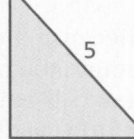

$\dfrac{3}{5}, \dfrac{9}{25}$

3. The ratio of the corresponding side lengths of two similar cellular phones is 3 : 4. The perimeter of the smaller phone is 9 inches. What is the perimeter of the larger phone? 12 in.

Taking Math Deeper

Exercise 16

There are several very different ways to solve this problem. This would be a good problem to encourage students to "think outside the box." You might have students work in pairs to see how many different ways they can solve the problem.

① Recognize that the smaller piece is one-fourth the size of the larger piece.

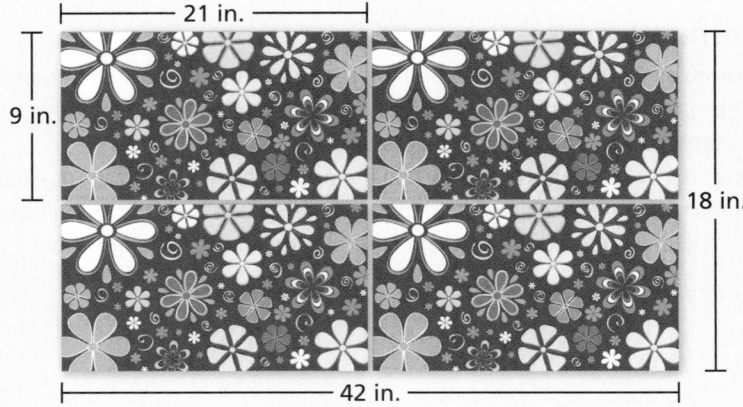

② Use unit prices.
$$\frac{\text{Cost}}{\text{Area}} = \frac{1.31}{9 \times 21} \approx \$0.007 \text{ per in.}^2$$

Area of the 18×42 piece $= 18 \cdot 42 = 756$ in.^{2}
Cost of the 18×42 piece $\approx 756 \cdot 0.007 \approx \$5.29$

③ Use a proportion.
$$\frac{\text{Area}}{\text{Area}} = \frac{\text{Cost}}{\text{Cost}}$$
$$\frac{756}{189} = \frac{x}{1.31}$$
$$\$5.24 = x$$

4 times more

Reteaching and Enrichment Strategies

If students need help. . .	If students got it. . .
Resources by Chapter • Practice A and Practice B • Puzzle Time Record and Practice Journal Practice Differentiating the Lesson Lesson Tutorials Skills Review Handbook	Resources by Chapter • Enrichment and Extension Start the next section

21 in.

9 in.

16. FABRIC The cost of the fabric is $1.31. What would you expect to pay for a similar piece of fabric that is 18 inches by 42 inches?

6 in.

17. AMUSEMENT PARK A model of a merry-go-round has a base area of about 450 square inches. What is the percent of increase of the base area from the model to the actual merry-go-round? Explain.

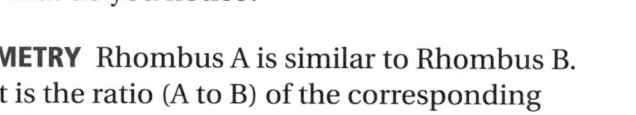

Model 450 in.²

10 ft

18. CRITICAL THINKING The circumference of Circle K is π. The circumference of Circle L is 4π.

a. What is the ratio of their circumferences? of their radii? of their areas?

b. What do you notice?

Circle K

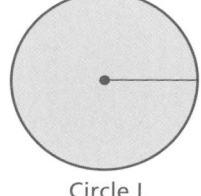

Circle L

19. GEOMETRY Rhombus A is similar to Rhombus B. What is the ratio (A to B) of the corresponding side lengths?

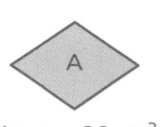

A

B

Area = 36 cm²

Area = 64 cm²

20. ⭐**Geometry** A triangle with an area of 10 square meters has a base of 4 meters. A similar triangle has an area of 90 square meters. What is the *height* of the larger triangle?

 **Fair Game Review** What you learned in previous grades & lessons

Find the percent of change. Round to the nearest tenth of a percent, if necessary. *(Section 4.2)*

21. 24 feet to 30 feet   **22.** 90 miles to 63 miles   **23.** 150 liters to 86 liters

24. MULTIPLE CHOICE A runner completes an 800-meter race in 2 minutes 40 seconds. What is the runner's speed? *(Section 3.1)*

Ⓐ $\dfrac{3 \text{ sec}}{10 \text{ m}}$ Ⓑ $\dfrac{160 \text{ sec}}{1 \text{ m}}$ Ⓒ $\dfrac{5 \text{ m}}{1 \text{ sec}}$ Ⓓ $\dfrac{10 \text{ m}}{3 \text{ sec}}$

Essential Question What information do you need to know to find the dimensions of a figure that is similar to another figure?

1 ACTIVITY: Drawing and Labeling Similar Figures

Work with a partner. You are given the red rectangle. Find a blue rectangle that is similar and has one side from $(-1, -6)$ to $(5, -6)$. Label the vertices.

a. Sample:

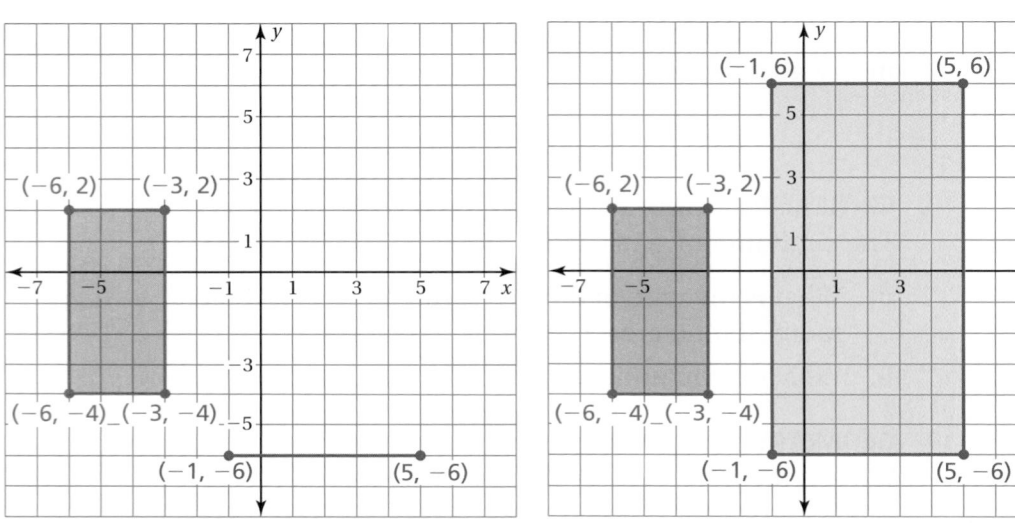

You can see that the two rectangles are similar by showing that ratios of corresponding sides are equal.

$$\frac{\text{Red Length}}{\text{Blue Length}} \overset{?}{=} \frac{\text{Red Width}}{\text{Blue Width}}$$

$$\frac{\text{change in } y}{\text{change in } y} \overset{?}{=} \frac{\text{change in } x}{\text{change in } x}$$

$$\frac{6}{12} \overset{?}{=} \frac{3}{6}$$

$$\frac{1}{2} = \frac{1}{2}$$

⋮ The ratios are equal. So, the rectangles are similar.

b. There are three other blue rectangles that are similar to the red rectangle and have the given side.

• Draw each one. Label the vertices of each.

• Show that each is similar to the original red rectangle.

Laurie's Notes

Introduction

For the Teacher

- **Goal:** Students will construct similar rectangles in the coordinate plane to find a missing measure.
- The typical approach to finding an unknown measure in similar figures is to set up a proportion. This approach depends upon the fact that students have a grasp of proportional reasoning, but it lacks a visual connection for students. In the first activity, the unknown measure is found through construction instead of computation. This spatial reasoning skill is often underdeveloped in students and is one on which you need to focus.

Motivate

- Using 3 student volunteers and yarn, construct a pair of similar triangles. Student A stays fixed; Student B walks 10 paces from Student A (parallel to a wall); Student C walks 6 paces from Student A and perpendicular to the segment *AB*. String yarn between the 3 students to form a triangle.
- Have 3 new students construct a triangle similar to the first triangle. Student D stays fixed; Student E walks 5 paces from Student D (maybe parallel to the same wall as *AB*).
- **?** "If Student F walks perpendicular to *DE*, how many steps should this student take so the two triangles are similar?" *3 steps*

Activity Notes

Activity 1

- **?** "What are the dimensions of the red rectangle?" *3 units by 6 units*
- It is okay for students to put their fingers on the sides and count units.
- **Common Error:** Students count the lattice points beginning with the vertex and end up with dimensions 4 by 7 instead of 3 by 6.
- **Connection:** Notice that in the sample we are using language that is also used for slope. To compute the "change in *y*," students should just look at the diagram and count.
- Do not skip the last step of showing that the rectangle is similar to the original red rectangle. Students need the practice of writing the proportion. Note that while the terms length and width are interchangeable, the length here is the longer of the two sides. The color reference is also easier for students to understand rather than saying "corresponding side in the left rectangle to the corresponding side in the right rectangle."
- Keep the language simple so students focus on the concept.
- Have students share their three solutions. Check to see that the coordinates are correctly labeled, in other words (5, −6) and not (−6, 5).
- **?** "What was the ratio of corresponding sides when the rectangles were the same size?" *1 : 1*

Previous Learning

Students should know how to plot ordered pairs. Students also need to remember how to solve a proportion.

Activity Materials	
Introduction	**Textbook**
• yarn or string	• rulers

Start Thinking! and Warm Up

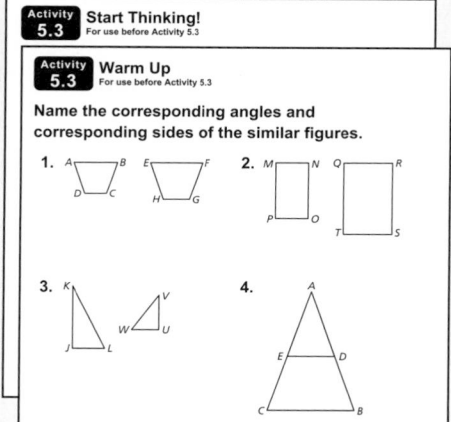

5.3 Record and Practice Journal

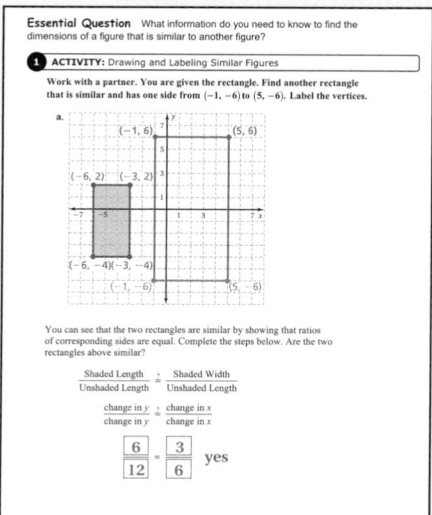

Differentiated Instruction

Auditory

In Activity 2, have students verbalize the process of solving a proportion to find the length of the blue rectangle. The techniques used to solve the problem have been learned in previous sections. Students should set up the proportion, apply the Cross Products Property, and then divide.

5.3 Record and Practice Journal

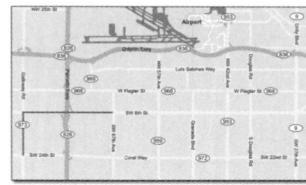

Laurie's Notes

Activity 2

• Part (a) connects directly to Activity 1. The coordinate grid is removed but the visual model is the same. Students should be comfortable setting up a proportion using the same language.

$$\frac{\text{red length}}{\text{blue length}} = \frac{\text{red width}}{\text{blue width}}$$

? "What observations can be made about the two rectangles?" *Sample answers:* ratio of length to width is 2 : 1; red rectangle is measured in inches and the blue rectangle is measured in miles.

• Part (b) is setting the stage for scale drawings.

• You may want to provide rulers, although they are not necessary. Students can use a piece of paper as a measuring device. The edge of the paper can be used to mark a length equal to 1 mile. Next, turn the paper, so it is horizontal. The horizontal distance is twice as long.

• Another approach is to observe that the dimensions of the red rectangle in part (a) are the same as the red segments in part (b).

• **Communication:** Students may have discovered other approaches. Ask them to explain their approach at the end of the activity.

• **Extension:** *As the crow flies* - approximate the distance across the diagonal.

What Is Your Answer?

• **Neighbor Check:** Have students work independently and then have their neighbor check their work. Have students discuss any discrepancies.

Closure

• Revisit Activity 2, part (a). The red and green rectangles are similar. Find the length of the green rectangle. Explain your reasoning.

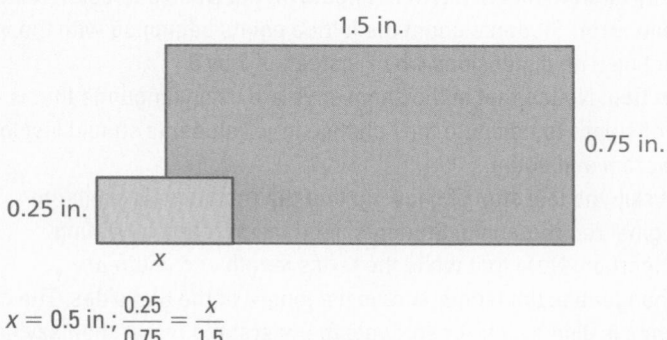

1.5 in.

0.75 in.

0.25 in.

x

$$x = 0.5 \text{ in.}; \quad \frac{0.25}{0.75} = \frac{x}{1.5}$$

Technology For the Teacher

Dynamic Classroom

The Dynamic Planning Tool
Editable Teacher's Resources at *BigIdeasMath.com*

Work with a partner.

a. The red and blue rectangles are similar. Find the length of the blue rectangle. Explain your reasoning.

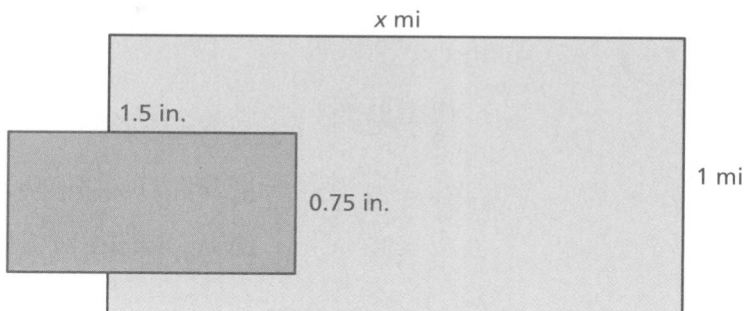

b. The distance marked by the vertical red line on the map is 1 mile. Find the distance marked by the horizontal red line. Explain your reasoning.

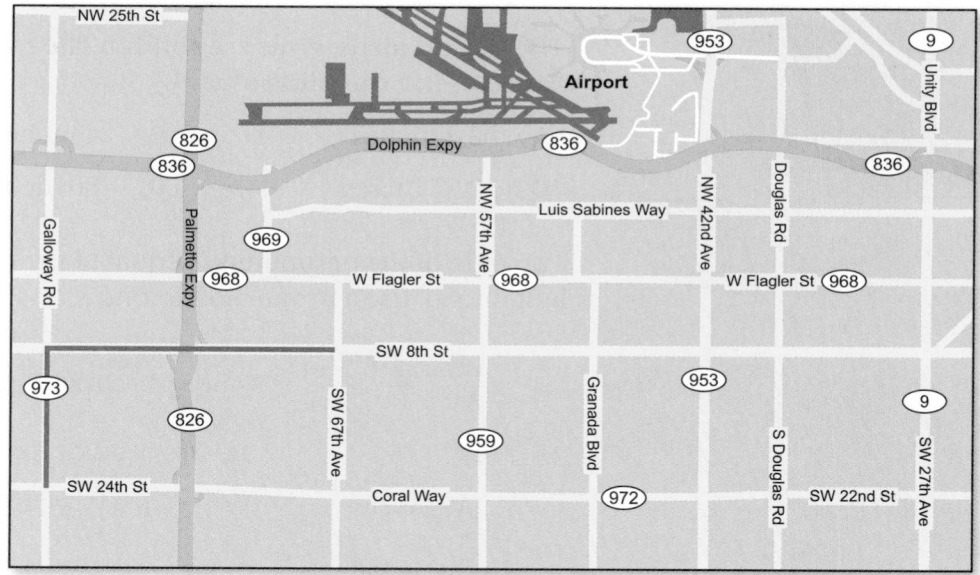

What Is Your Answer?

3. **IN YOUR OWN WORDS** What information do you need to know to find the dimensions of a figure that is similar to another figure? Give some examples using two rectangles.

4. When you know the length and width of one rectangle and the length of a similar rectangle, can you always find the missing width? Why or why not?

Practice

Use what you learned about finding unknown measures in similar figures to complete Exercises 3 and 4 on page 210.

5.3 Lesson

Check It Out
Lesson Tutorials
BigIdeasMath.com

EXAMPLE 1 Finding an Unknown Measure

The two triangles are similar. Find the value of x.

Key Vocabulary
indirect measurement,
p. 209

Corresponding side lengths are proportional. So, use a proportion to find x.

$$\frac{6}{9} = \frac{8}{x}$$ Write a proportion.

$6x = 72$ Use Cross Products Property.

$x = 12$ Divide each side by 6.

∴ So, x is 12 meters.

EXAMPLE 2 Standardized Test Practice

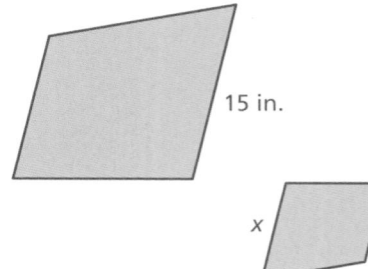

15 in.

x

The two quadrilaterals are similar. The ratio of their perimeters is 12 : 5. Find the value of x.

Ⓐ 2.4 inches   Ⓑ 4 inches

Ⓒ 6.25 inches   Ⓓ 36 inches

The ratio of the perimeters is equal to the ratio of corresponding side lengths. So, use a proportion to find x.

$$\frac{12}{5} = \frac{15}{x}$$ Write a proportion.

$12x = 75$ Use Cross Products Property.

$x = 6.25$ Divide each side by 12.

∴ So, x is 6.25 inches. The correct answer is Ⓒ.

On Your Own

Now You're Ready
Exercises 3–8

1. The two quadrilaterals are similar. The ratio of the perimeters is 3 : 4. Find the value of x.

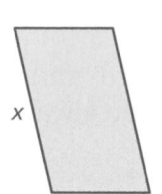

x

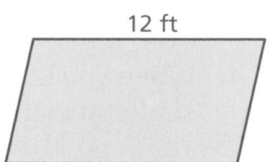

12 ft

Laurie's Notes

Introduction

Connect

- **Yesterday:** Students found unknown measures in similar figures by drawing the figure in a coordinate plane.
- **Today:** Students will find unknown measures in similar figures by setting up a proportion.

Motivate

- *Easy as 1, 2, 3!* Solve the proportions:

  $$\frac{1}{2} = \frac{3}{?} \quad \frac{1}{3} = \frac{2}{?} \quad \frac{1}{2} = \frac{?}{3} \quad \frac{1}{3} = \frac{?}{2}$$

- Where you place the numbers makes a difference. The first two problems have whole number solutions. 6 and 6 The last two problems have fractional solutions. $\frac{3}{2}$ and $\frac{2}{3}$

- This example draws students' attention to the fact that they need to be careful when they set up their proportions.

Lesson Notes

Example 1

- Note that the vertices of the triangles are not named. Instead, there is a simple statement that the two triangles are similar. The triangles in the diagram are in the same orientation. So, the corresponding sides should be obvious to students.
- ❓ "Which side length in the smaller triangle corresponds to the side labeled *x* in the larger triangle?" 8 m
- ❓ "Is there enough information to solve for *x*? Explain." Yes; the other corresponding sides are labeled 6 m and 9 m.

Example 2

- ❓ "The quadrilaterals are similar. What do you know about the perimeter of the quadrilaterals?" The ratio of the perimeters is the same as the ratio of corresponding sides.
- There is only one side measured in each quadrilateral. Students should think "larger to smaller" when setting up the proportion.
- If students set up the proportion incorrectly, the result is 36 inches, which is not reasonable for the problem.

On Your Own

- Note that the orientation of the second figure is different from the first figure. Students should think, longer side to longer side, or shorter side to shorter side.

Goal Today's lesson is finding unknown measures in similar figures.

Start Thinking and Warm-up

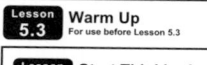

Lesson 5.3	Warm Up
	For use before Lesson 5.3

Lesson 5.3	Start Thinking!
	For use before Lesson 5.3

Two rectangles are similar. Explain to a partner how to find the missing length of the first rectangle, knowing the width of the rectangle. The length and width are given in the second rectangle.

Extra Example 1

A rectangle has a width of 3 feet and a length of 10 feet. A similar rectangle has a width of 12 feet and a length of ℓ feet. Find the value of ℓ. 40 ft

Extra Example 2

Two regular pentagons are similar. The ratio of their perimeters is 4 : 7. The smaller pentagon has a side length of 18 meters. Find the side length of the larger pentagon. 31.5 m

On Your Own

1. 9 ft

Laurie's Notes

Discuss

- Define **indirect measurement** and give examples: finding heights of trees without climbing the trees, distance across a lake without paddling across the lake, and so on.

Example 3

- Discuss the diagram and relate it to what students have noticed about their own shadows and the shadows of trees and buildings.
- Discuss how shadow lengths vary depending upon the time of day.
- At a given time of day, the taller the object, the longer the shadow. Two objects in the same vicinity will have shadows proportional to their heights.
- **Note:** You do not need to use the Cross Products Property because the variable is in the numerator.

On Your Own

- **Neighbor Check:** Have students work independently and then have their neighbor check their work. Have students discuss any discrepancies.

Example 4

- ❓ "What do you know about the areas of the similar quadrilaterals?" The ratio of the areas is the square of the ratio of corresponding sides.
- ❓ "Do you know the area of each rectangle?" No; we do not know the area of the swimming pool.
- ❓ "Can you find the area of the swimming pool? Explain." Yes; because the pool is similar to the volleyball court and you know the lengths of corresponding sides, the area can be found indirectly.
- **Common Error:** When squaring a number, students may multiply the number by 2 instead of multiplying the number by itself.

On Your Own

- Determine how students found the perimeter of the pool. There are several ways. Have students share their answers and methods.

Closure

- **Exit Ticket:** The two triangles are similar. Find the value of *x*.

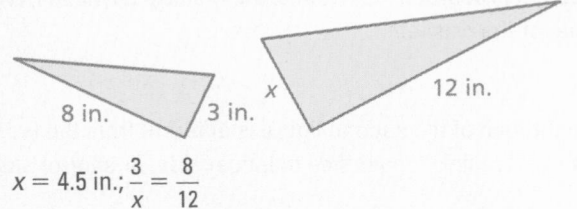

8 in.   3 in.   *x*   12 in.

$x = 4.5$ in.; $\dfrac{3}{x} = \dfrac{8}{12}$

Extra Example 3

A person that is 5 feet tall casts a 2-foot-long shadow. A nearby statue casts a 4-foot-long shadow. What is the height *h* of the statue? Assume the triangles are similar.   10 ft

 On Your Own

   2. 5 ft

Extra Example 4

The hypotenuse of a right triangle is 5 meters and its area is 6 square meters. The hypotenuse of a similar larger right triangle is 25 meters. What is the area *A* of the larger triangle? 150 m^2

 On Your Own

   3. 108 yd

English Language Learners

Vocabulary

Point out to English language learners that there are several prefixes in the English language that mean "not." One of these is the prefix *in-* as in *indirect*. An indirect measurement does *not* involve the direct reading of a measuring tool.

Indirect measurement uses similar figures to find a missing measure that is difficult to find directly.

EXAMPLE 3 Using Indirect Measurement

A person that is 6 feet tall casts a 3-foot-long shadow. A nearby palm tree casts a 15-foot-long shadow. What is the height h of the palm tree? Assume the triangles are similar.

h ft

 6 ft

15 ft   3 ft

Corresponding side lengths are proportional.

$$\frac{h}{6} = \frac{15}{3} \qquad \text{Write a proportion.}$$

$$6 \cdot \frac{h}{6} = \frac{15}{3} \cdot 6 \qquad \text{Multiply each side by 6.}$$

$$h = 30 \qquad \text{Simplify.}$$

∴ The palm tree is 30 feet tall.

On Your Own

Now You're Ready
Exercise 9

2. **WHAT IF?** Later in the day, the palm tree in Example 3 casts a 25-foot-long shadow. How long is the shadow of the person?

EXAMPLE 4 Using Proportions to Find Area

A swimming pool is similar in shape to a volleyball court. What is the area A of the pool?

$$\frac{\text{Area of court}}{\text{Area of pool}} = \left(\frac{\text{width of court}}{\text{width of pool}}\right)^2$$

$$\frac{200}{A} = \left(\frac{10}{18}\right)^2 \qquad \text{Substitute.}$$

$$\frac{200}{A} = \frac{100}{324} \qquad \text{Simplify.}$$

$$A = 648 \qquad \text{Solve the proportion.}$$

∴ The area of the pool is 648 square yards.

18 yd

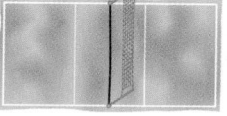

10 yd

Area = 200 yd²

On Your Own

3. The length of the volleyball court in Example 4 is 20 yards. What is the perimeter of the pool?

 Vocabulary and Concept Check

1. **REASONING** How can you use corresponding side lengths to find unknown measures in similar figures?

2. **CRITICAL THINKING** In which of the situations would you likely use indirect measurement? Explain your reasoning.

Finding the height of a statue   Finding the width of a doorway

Finding the width of a river   Finding the length of a lake

 **Practice and Problem Solving**

The polygons are similar. Find the value of x.

 3.
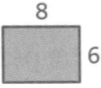
20
8
6
x

4.
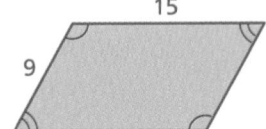
15
9
x
4

5.
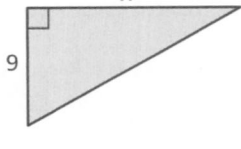
x
9

8
5

6.

21
9
6
x

7. The ratio of the perimeters is 7 : 10.

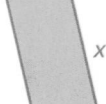 
x
12

8. The ratio of the perimeters is 8 : 5.

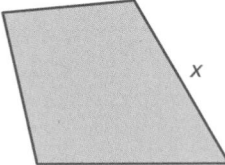 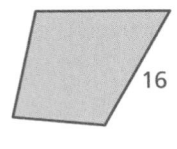
x
16

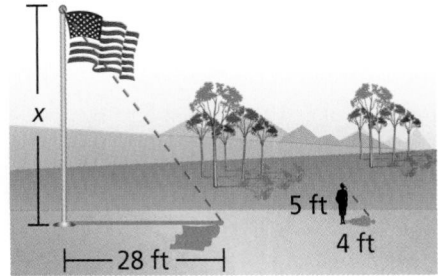

9. **FLAGPOLE** What is the height x of the flagpole? Assume the triangles are similar.

10. **CHEERLEADING** A rectangular school banner has a length of 44 inches and a perimeter of 156 inches. The cheerleaders make signs similar to the banner. The length of a sign is 11 inches. What is its perimeter?

x
5 ft
4 ft
28 ft

Assignment Guide and Homework Check

Level	Day 1 Activity Assignment	Day 2 Lesson Assignment	Homework Check
Basic	3, 4, 16–22	1, 2, 5–11	2, 6, 8, 10
Average	3, 4, 16–22	1, 2, 5–8, 10–13	2, 6, 8, 10
Advanced	3, 4, 16–22	1, 2, 6, 8, 10, 12–15	6, 8, 12, 14

Common Errors

- **Exercises 3–6** Students may write the proportion incorrectly. For example, they may write $\frac{8}{20} = \frac{x}{6}$ instead of $\frac{8}{20} = \frac{6}{x}$. Remind them that the corresponding sides should both be in one ratio. Suggest students always set up their proportions as either $\frac{\text{left figure}}{\text{right figure}}$ or $\frac{\text{smaller figure}}{\text{larger figure}}$.

- **Exercise 13** Students may forget to add the length of the person's shadow onto the distance between the person and the street light. Tell them to draw a picture of the problem and label the information that is given in the picture, as well as the information in the problem, before writing a proportion.

- **Exercise 14** Students may forget to square the ratio of the side lengths when finding the area. Remind them that for area, the ratio of the side lengths must be squared.

Vocabulary and Concept Check

1. You can set up a proportion and solve for the unknown measure.

2. All of them except the doorway because the others are large measurements and would be hard to measure directly.

Practice and Problem Solving

3. 15

4. $6\frac{2}{3}$

5. 14.4

6. 14

7. 8.4

8. 25.6

9. 35 ft

10. 39 in.

5.3 Record and Practice Journal

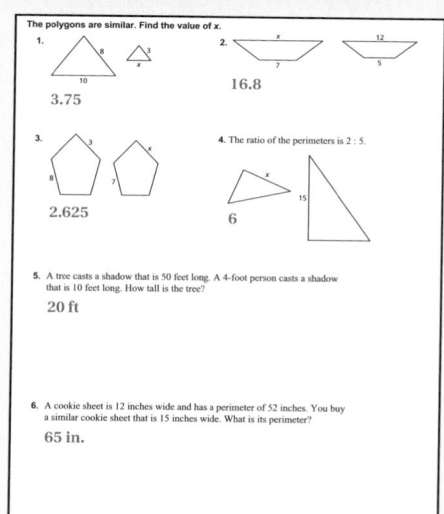

The polygons are similar. Find the value of x.

1. 3.75

2. 16.8

3. 2.625

4. The ratio of the perimeters is 2 : 5. 6

5. A tree casts a shadow that is 50 feet long. A 4-foot person casts a shadow that is 10 feet long. How tall is the tree? 20 ft

6. A cookie sheet is 12 inches wide and has a perimeter of 52 inches. You buy a similar cookie sheet that is 15 inches wide. What is its perimeter? 65 in.

Practice and Problem Solving

11. 108 yd

12. See *Taking Math Deeper*.

13. 3 times

14. 60 min

15. 12.5 bottles

Fair Game Review

16. 6.4   **17.** 31.75

18. 244.44   **19.** 3.88

20. 3.94   **21.** 41.63

22. B

Mini-Assessment

The polygons are similar. Find the value of *x*.

1.

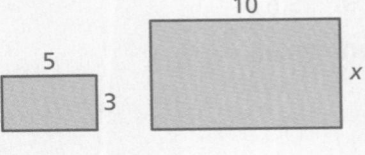

$x = 6$

2.

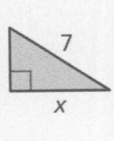 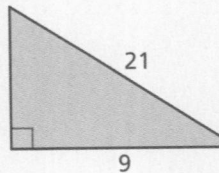

$x = 3$

3. A rectangular picture has a side length of 6 inches and a perimeter of 20 inches. You make an enlarged copy of the picture. The length of the new picture is 12 inches. What is its perimeter?   40 in.

Taking Math Deeper

Exercise 12

There are many different ways that similar triangles can be used to measure the distance indirectly. Right triangles are often used. However, this is a nice example of how other types of triangles can also be used.

 Label the unknown distance *x*.

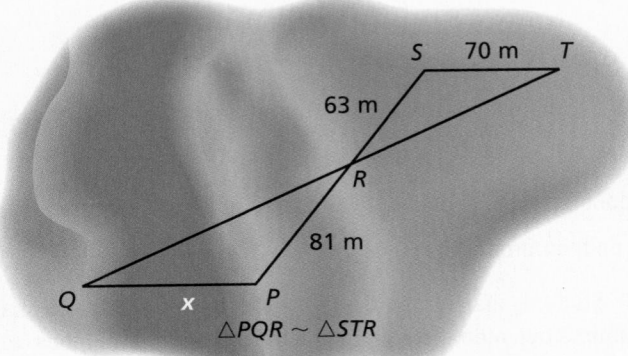

$\triangle PQR \sim \triangle STR$

 Write and solve a proportion.

$$\frac{x}{70} = \frac{81}{63}$$

$$70 \cdot \frac{x}{70} = 70 \cdot \frac{81}{63}$$

$$x = 90 \text{ meters}$$

Put *x* in numerator.

 Answer the question.
No, the width of the river is less than 100 meters.

Reteaching and Enrichment Strategies

If students need help. . .	If students got it. . .
Resources by Chapter • Practice A and Practice B • Puzzle Time Record and Practice Journal Practice Differentiating the Lesson Lesson Tutorials Skills Review Handbook	Resources by Chapter • Enrichment and Extension • School-to-Work Start the next section

11. **SQUARE** The ratio of the side length of Square A to the side length of Square B is 4 : 9. The side length of Square A is 12 yards. What is the perimeter of Square B?

12. **RIVER** Is the distance QP across the river greater than 100 meters? Explain.

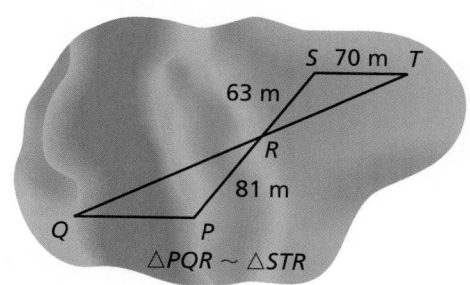

△PQR ~ △STR

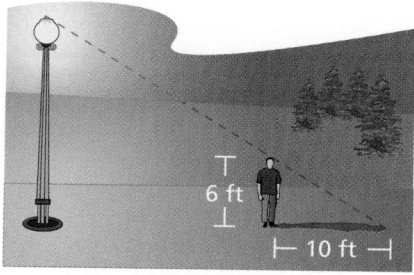

13. **STREET LIGHT** A person standing 20 feet from a street light casts a shadow as shown. How many times taller is the street light than the person? Assume the triangles are similar.

14. **AREA** A school playground is similar in shape to the community park. You can mow 250 square yards of grass in 15 minutes. How long would it take you to mow the grass on the playground?

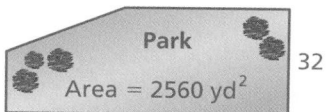

15. **Critical Thinking** Two bottles of fertilizer are needed to treat the flower garden shown. How many bottles are needed to treat a similar garden with a perimeter of 105 feet?

Fair Game Review What you learned in previous grades & lessons

Copy and complete the statement using a ratio. Round to the nearest hundredth, if necessary. *(Section 3.6)*

16. 4 mi ≈ ___ km

17. 12.5 in. ≈ ___ cm

18. 110 kg ≈ ___ lb

19. 6.2 km ≈ ___ mi

20. 10 cm ≈ ___ in.

21. 92.5 lb ≈ ___ kg

22. **MULTIPLE CHOICE** A recipe that makes 8 pints of salsa uses 22 tomatoes. Which proportion can you use to find the number n of tomatoes needed to make 12 pints of salsa? *(Section 3.4)*

Ⓐ $\dfrac{n}{8} = \dfrac{22}{12}$

Ⓑ $\dfrac{8}{22} = \dfrac{12}{n}$

Ⓒ $\dfrac{22}{n} = \dfrac{12}{8}$

Ⓓ $\dfrac{8}{22} = \dfrac{n}{12}$

5.4 Scale Drawings

Essential Question How can you use a scale drawing to estimate the cost of painting a room?

1 ACTIVITY: Making Scale Drawings

Work with a partner. You have decided that your classroom needs to be painted. Start by making a **scale drawing** of each of the four walls.

- Measure each of the walls.
- Measure the locations and dimensions of parts that will *not* be painted.
- Decide on a **scale** for your drawings.
- Make a scale drawing of each of the walls.

Sample: Wall #1

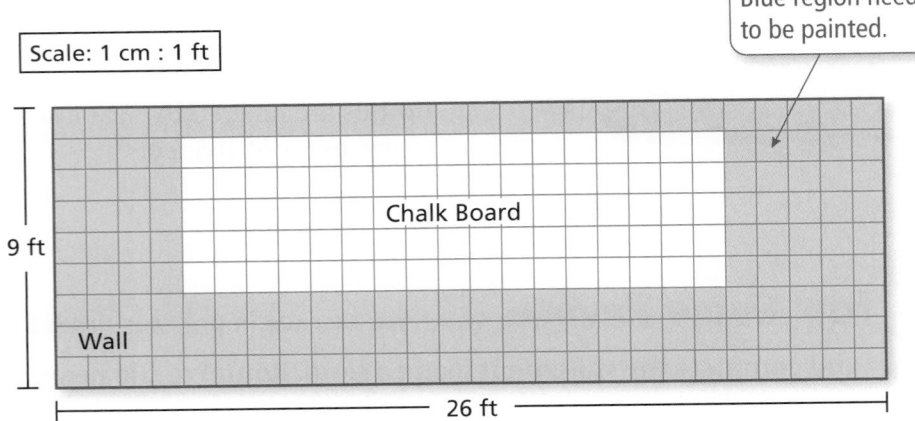

Scale: 1 cm : 1 ft

Blue region needs to be painted.

Chalk Board

9 ft

Wall

26 ft

- **For each wall, find the area of the part that needs to be painted.**

	Dimensions	*Area*
Dimensions of the wall	9 ft by 26 ft	$9 \times 26 = 234$ sq ft
Dimensions of the part that will *not* be painted	5 ft by 17 ft	$5 \times 17 = 85$ sq ft
Area of painted part		149 sq ft

Laurie's Notes

Introduction

For the Teacher

- **Goal:** Students will draw scale models of the walls of their classroom in order to decide how much paint is needed to apply two coats of paint.

Motivate

- If available, hold up a can of paint or a paintbrush. The building maintenance people may be able to help.
- **?** To get students interested and excited about the first activity, ask "Do you think a gallon of paint (or this can of paint) will be enough to paint this classroom once? How can we decide?"

Activity Notes

Activity 1

- Review with students what the word scale means in this context. Students should be familiar with maps and seeing a scale written on the map. A formal definition for scale is given in the lesson.
- **Management Tip:** You will need to have an organized plan as to how students can spread out to do the measuring. You do not want all of the students trying to measure the same dimension simultaneously.
- **Whole Class:** Discuss what needs to be measured. Make a list on the board. Remember to list the items that will not be painted such as bulletin boards, chalkboard, windows, and doors.
- Because the formal definition of scale is presented in the lesson, a great deal of potential frustration and inaccurate drawings can be avoided if a convenient size grid paper is used where a possible scale is obvious, such as 1 cm : 1 ft.
- Student pairs should be self-directed today once they get started. Continue to circulate as students are measuring, sketching, and calculating the area.
- **Big Idea:** Students need to determine how much paint is going to be needed to paint the classroom. The important part of the lesson is deciding on the scale and making the scale drawing. Students have been working with similar figures in this chapter. The drawing is similar to the actual room.
- **Connection:** Students could actually *count* the area of each wall from their scale drawing. For instance, if the scale is 1 cm : 1 ft, how many squares are shaded (to be painted)? Counting should give an area equal to the computation.
- **Extension:** Have students determine what percent of the total wall area is being painted.

Previous Learning

Students should know common units of length, both U.S. customary and metric systems. Students should also know how to solve proportions.

Activity Materials	
Introduction	**Textbook**
• can of paint	• grid paper

Start Thinking! and Warm Up

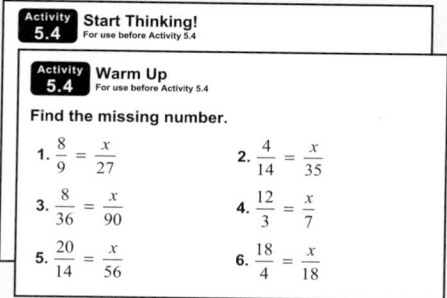

Activity 5.4 Start Thinking!
For use before Activity 5.4

Activity 5.4 Warm Up
For use before Activity 5.4

Find the missing number.

1. $\dfrac{8}{9} = \dfrac{x}{27}$ 2. $\dfrac{4}{14} = \dfrac{x}{35}$

3. $\dfrac{8}{36} = \dfrac{x}{90}$ 4. $\dfrac{12}{3} = \dfrac{x}{7}$

5. $\dfrac{20}{14} = \dfrac{x}{56}$ 6. $\dfrac{18}{4} = \dfrac{x}{18}$

5.4 Record and Practice Journal

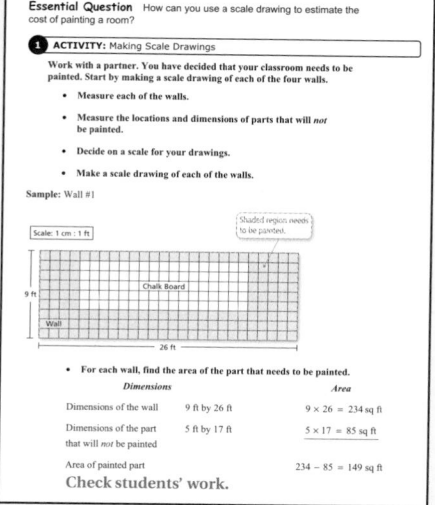

Essential Question How can you use a scale drawing to estimate the cost of painting a room?

1 ACTIVITY: Making Scale Drawings

Work with a partner. You have decided that your classroom needs to be painted. Start by making a scale drawing of each of the four walls.

- Measure each of the walls.
- Measure the locations and dimensions of parts that will *not* be painted.
- Decide on a scale for your drawings.
- Make a scale drawing of each of the walls.

Sample: Wall #1

- For each wall, find the area of the part that needs to be painted.

	Dimensions	*Area*
Dimensions of the wall	9 ft by 26 ft	$9 \times 26 = 234$ sq ft
Dimensions of the part that will *not* be painted	5 ft by 17 ft	$5 \times 17 = 85$ sq ft
Area of painted part		$234 - 85 = 149$ sq ft

Check students' work.

Differentiated Instruction

Kinesthetic

Materials needed: poster board, ruler, colored markers or pencils, copies of the design below.

Students are to recreate the design in a larger size. They will need to choose a ratio to enlarge the design, measure the design, and set up proportions to find the new dimensions.

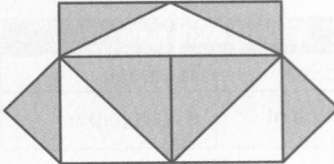

5.4 Record and Practice Journal

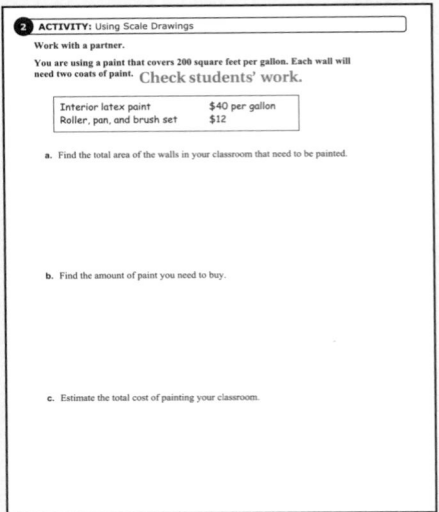

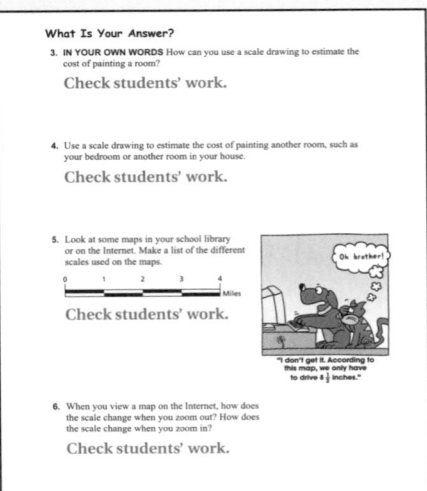

Laurie's Notes

Activity 2

- Students will use the information from Activity 1 to find the total area to be painted. Remind students that each wall will have two coats of paint.
- As you circulate, notice if students are significantly off in their calculations. Try to assist and correct the error before they find the total area of the walls. It may be a mistake in multiplication, measurement, or a procedural one.
- **Connection:** This is a context where rounding up is needed even if the answer is less than a half-gallon, because you can only purchase full gallons of paint.

What Is Your Answer?

- Students can work in small groups for Question 5. Each group can then present their examples of scales used on maps to the class.

Closure

- **Writing Prompt:** To make a scale drawing you . . .

Technology **F**or the **T**eacher

Dynamic Classroom

The Dynamic Planning Tool
Editable Teacher's Resources at *BigIdeasMath.com*

2 ACTIVITY: Using Scale Drawings

Work with a partner.

You are using a paint that covers 200 square feet per gallon. Each wall will need two coats of paint.

a. Find the total area of the walls from Activity 1 that needs to be painted.

b. Find the amount of paint you need to buy.

c. Estimate the total cost of painting your classroom.

| Interior latex paint | $40 per gallon |
| Roller, pan, and brush set | $12 |

What Is Your Answer?

3. **IN YOUR OWN WORDS** How can you use a scale drawing to estimate the cost of painting a room?

4. Use a scale drawing to estimate the cost of painting another room, such as your bedroom or another room in your house.

5. Look at some maps in your school library or on the Internet. Make a list of the different scales used on the maps.

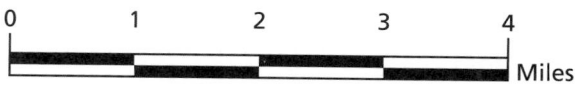

6. When you view a map on the Internet, how does the scale change when you zoom out? How does the scale change when you zoom in?

"I don't get it. According to this map, we only have to drive $8\frac{1}{2}$ inches."

Practice Use what you learned about scale drawings to complete Exercises 4–7 on page 216.

Check It Out
Lesson Tutorials
BigIdeasMath ✓com

Key Vocabulary 🔊
scale drawing, *p. 214*
scale model, *p. 214*
scale, *p. 214*
scale factor, *p. 215*

Key Ideas

Scale Drawings and Models

A **scale drawing** is a proportional two-dimensional drawing of an object.
A **scale model** is a proportional three-dimensional model of an object.

Scale

Measurements in scale drawings and models are proportional to the measurements of the actual object. The **scale** gives the ratio that compares the measurements of the drawing or model with the actual measurements.

$$\frac{1 \text{ in.}}{10 \text{ mi}} \xleftarrow{\text{drawing distance}}_{\text{actual distance}}$$

$1 \text{ in.} : 10 \text{ mi}$

drawing   actual

Study Tip ✏️

Scales are written so that the drawing distance comes first in the ratio.

EXAMPLE ① **Finding an Actual Distance**

What is the actual distance *d* between Cadillac and Detroit?

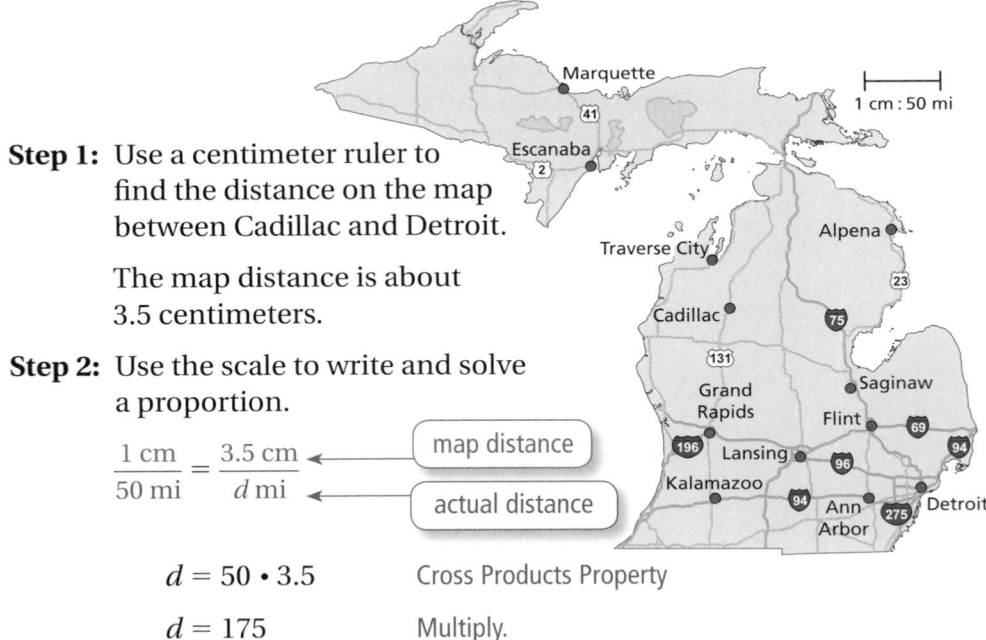

Step 1: Use a centimeter ruler to find the distance on the map between Cadillac and Detroit.

The map distance is about 3.5 centimeters.

Step 2: Use the scale to write and solve a proportion.

$$\frac{1 \text{ cm}}{50 \text{ mi}} = \frac{3.5 \text{ cm}}{d \text{ mi}} \xleftarrow{\text{map distance}}_{\text{actual distance}}$$

$d = 50 \cdot 3.5$ Cross Products Property

$d = 175$ Multiply.

∴ The distance between Cadillac and Detroit is about 175 miles.

On Your Own

Now You're Ready
Exercises 8–11

1. What is the actual distance between Traverse City and Marquette?

🔊 **Multi-Language Glossary at BigIdeasMath✓com.**

Laurie's Notes

Introduction

Connect
- **Yesterday:** Students created (sketched) a scale drawing of the classroom.
- **Today:** Students will use a scale drawing to find a missing measure.

Motivate
- Show the class some items that have scales written on them: map (print one from the internet), matchbook car, blueprint, or floor plan. Ask about the meaning of the scale for each item.
- **Trivia:** Share a fact about scale models:
  The world's biggest baseball bat is 120 feet and leans against the Louisville Slugger Museum & Factory in Kentucky. It is an exact-scale replica of Babe Ruth's 34-inch Louisville Slugger bat.

Lesson Notes

Key Ideas
- Be careful and consistent with your language today. Continually refer to the ratio of the scale drawing (or scale model) to the actual object.
- Review the *Study Tip*.

Example 1
- Have students explore the map. Ask if they see the scale at the bottom. You can use the width of your baby finger to approximate 1 centimeter. Ask students to use their baby fingers to approximate the distance across the bottom of Michigan. about 160 mi
- Use centimeter rulers to measure the distance from Cadillac to Detroit.
- **Common Error:** Students might measure in inches instead of in centimeters, or they may confuse centimeters with millimeters.
- **?** "What is the map distance from Cadillac to Detroit?" 3.5 cm
- Set up the proportion using the language, "1 centimeter is to 50 miles as 3.5 centimeters is to what?"
- **Common Question:** Students will often ask how you can use the Cross Products Property because you are multiplying two different units together (50 mi and 3.5 cm). Because both numerators are the same unit (cm) and both denominators are the same unit (mi), it is okay to multiply. If the units were not the same in the numerators and were the same in the denominators, this could not be done. A quick way to explain this to students is to use a simple problem: If the scale is 1 cm : 5 ft, then 2 cm would be what actual distance? 10 ft
- Encourage students to write the units when they write the initial proportion. This ensures that the proportion has been set up correctly. When the numeric answer has been found, label with the correct units of measure.
- **Check for Reasonableness:** If the map distance is 3.5 centimeters, the actual distance should be 3.5 times the scale distance of 50 miles. Three times 50 miles is 150 miles, so an answer of 175 miles seems reasonable.

Goal Today's lesson is using scale drawings to find missing measurements.

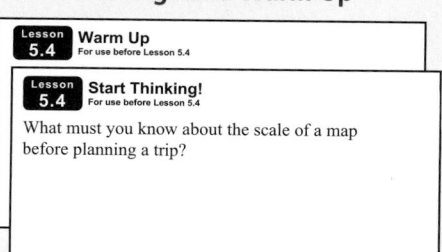

Lesson Materials	
Introduction	**Textbook**
• map • blueprint	• rulers

Start Thinking! and Warm Up

Lesson 5.4	**Warm Up** For use before Lesson 5.4

Lesson 5.4	**Start Thinking!** For use before Lesson 5.4

What must you know about the scale of a map before planning a trip?

Extra Example 1

Using the map from Example 1, what is the distance *d* between Detroit and Marquette? about 350 mi

On Your Own
1. about 150 mi

Laurie's Notes

Extra Example 2

The Earth's crust has a thickness of 80 kilometers on some of the continents. Using the scale model from Example 2, how thick is the crust of the model? 0.16 in.

 On Your Own

> **2.** 5.8 in.

Extra Example 3

A sketch of a fashion designer's shirt is 9 centimeters long. The actual shirt is 1 meter long.

a. What is the scale of the drawing? 9 cm : 1 m

b. What is the scale factor of the drawing? 9 : 100

 On Your Own

> **3.** 1 : 200

English Language Learners

Illustrate

Students have seen maps in classrooms and perhaps on road trips with their families. Hand out road maps to students in small groups. Have students find distances between cities on the map. Ask students how they can find the distances between cities using proportions.

Example 2

- This question is looking for a dimension of the scale model, not for an actual distance as in Example 1. Be careful with the language: scale to actual.
- Explain that this is an example of a scale model (3-Dimensional), not a scale drawing (2-Dimensional).
- Note that the units are written in the original proportion.
- **Check for Reasonableness:** Because the scale is 1 in. : 500 km, 2 inches would represent 1000 kilometers and 4 inches would represent 2000 kilometers. An answer of 4.6 inches is reasonable.

On Your Own

- **Think-Pair-Share:** Students should read each question independently and then work with a partner to answer the questions. When they have answered the questions, the pair should compare their answers with another group and discuss any discrepancies.

Example 3

? "What measurements are given?" drawing length is 5 cm, actual length is 10 mm; Make sure students reference the units.

? "Is the scale drawing larger or smaller than the actual spider? Explain." The scale drawing is larger because 5 centimeters is longer than 10 millimeters.

- Distinguish between the two questions—find the scale of the drawing and then find the scale factor. The units must be the same for a scale factor.

On Your Own

? "Is the scale model larger or smaller than the actual item? Explain." The scale model is smaller because 1 millimeter is shorter than 20 centimeters.

Closure

- **Exit Ticket:** A common model train scale is called the HO Scale, where the scale factor is 1 : 87. If the diameter of a wheel on a model train is 1 inch, what is the diameter of the actual wheel? 87 in.

Technology For the Teacher

Dynamic Classroom

The Dynamic Planning Tool
Editable Teacher's Resources at *BigIdeasMath.com*

EXAMPLE 2 **Standardized Test Practice**

The liquid outer core of Earth is 2300 kilometers thick. A scale model of the layers of Earth has a scale of 1 in. : 500 km. How thick is the liquid outer core of the model?

Ⓐ 0.2 in.   Ⓑ 4.6 in.   Ⓒ 0.2 km   Ⓓ 4.6 km

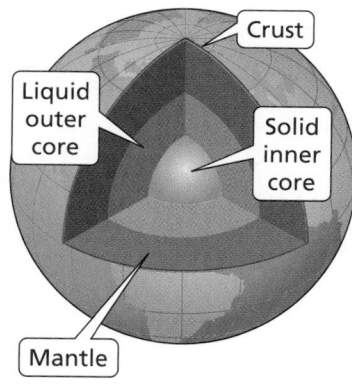

$$\frac{1 \text{ in.}}{500 \text{ km}} = \frac{x \text{ in.}}{2300 \text{ km}}$$ ← model thickness
← actual thickness

$$\frac{1 \text{ in.}}{500 \text{ km}} \cdot 2300 \text{ km} = \frac{x \text{ in.}}{2300 \text{ km}} \cdot 2300 \text{ km}$$ Multiply each side by 2300 km.

$$4.6 = x$$ Simplify.

⋮• The liquid outer core of the model is 4.6 inches thick. The correct answer is Ⓑ.

On Your Own

2. The mantle of Earth is 2900 kilometers thick. How thick is the mantle of the model?

A scale can be written without units when the units are the same. A scale without units is called a **scale factor**.

EXAMPLE 3 **Finding a Scale Factor**

A scale drawing of a spider is 5 centimeters long. The actual spider is 10 millimeters long. (a) What is the scale of the drawing? (b) What is the scale factor of the drawing?

5 cm

a. $\dfrac{\text{drawing length}}{\text{actual length}} = \dfrac{5 \text{ cm}}{10 \text{ mm}} = \dfrac{1 \text{ cm}}{2 \text{ mm}}$

⋮• The scale is 1 cm : 2 mm.

b. Write the scale with the same units. Use the fact that 1 cm = 10 mm.

$$\text{scale factor} = \frac{1 \text{ cm}}{2 \text{ mm}} = \frac{10 \text{ mm}}{2 \text{ mm}} = \frac{5}{1}$$

⋮• The scale factor is 5 : 1.

On Your Own

Now You're Ready
Exercises 12–16

3. A model has a scale of 1 mm : 20 cm. What is the scale factor of the model?

Vocabulary and Concept Check

1. **VOCABULARY** Compare and contrast the terms *scale* and *scale factor*.

2. **CRITICAL THINKING** The scale of a drawing is 2 cm : 1 mm. Is the scale drawing *larger* or *smaller* than the actual object? Explain.

3. **REASONING** How would you find a scale factor of a drawing that shows a length of 4 inches when the actual object is 8 feet long?

 Practice and Problem Solving

Use the drawing and a centimeter ruler.

4. What is the actual length of the flower garden?

5. What are the actual dimensions of the rose bed?

6. What are the actual perimeters of the perennial beds?

7. The area of the tulip bed is what percent of the area of the rose bed?

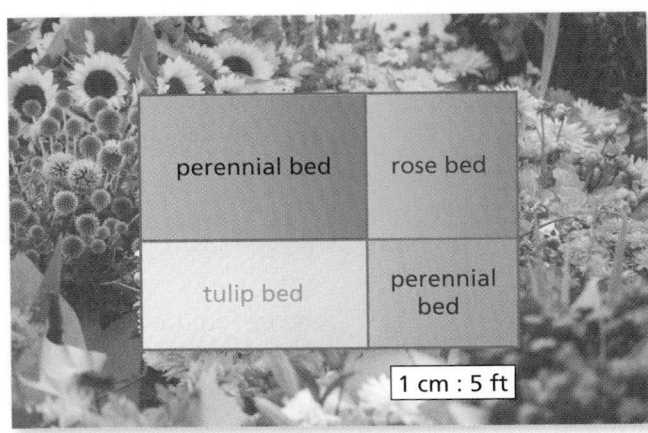

Use the map in Example 1 to find the actual distance between the cities.

8. Kalamazoo and Ann Arbor

9. Lansing and Flint

10. Grand Rapids and Escanaba

11. Saginaw and Alpena

Find the missing dimension. Use the scale factor 1 : 12.

Item	Model	Actual
Mattress	Length: 6.25 in.	Length: ___ in.
Corvette	Length: ___ in.	Length: 15 ft
Water Tower	Depth: 32 cm	Depth: ___ m
Wingspan	Width: 5.4 ft	Width: ___ yd
Football Helmet	Diameter: ___ mm	Diameter: 21 cm

17. **ERROR ANALYSIS** A scale is 1 cm : 20 m. Describe and correct the error in finding the actual distance that corresponds to 5 cm.

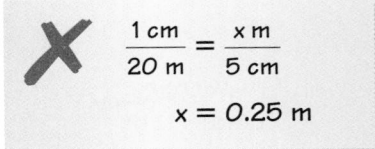

$$\frac{1 \text{ cm}}{20 \text{ m}} = \frac{x \text{ m}}{5 \text{ cm}}$$

$$x = 0.25 \text{ m}$$

Assignment Guide and Homework Check

Level	Day 1 Activity Assignment	Day 2 Lesson Assignment	Homework Check
Basic	4–7, 25–29	1–3, 9, 11, 13–19	2, 9, 14, 18
Average	4–7, 25–29	1–3, 9–17 odd, 18–22	2, 9, 13, 18
Advanced	4–7, 25–29	1–3, 10–18 even, 17, 22–24	10, 14, 18, 22

Common Errors

- **Exercises 8–11** When measuring with a centimeter ruler, students may not start at zero on the ruler. As a result, they will get a much greater number. Ask students to estimate the scale distance before measuring so they can check the reasonableness of their measurement.
- **Exercises 12–16** Students may mix up the proportion values when solving for the missing dimension. Remind them that the model is in the numerator and the actual size is in the denominator in both ratios.
- **Exercise 22** Students may count the squares in the blueprint of the bathroom and use that as the area of the bathroom. Remind them that they need to find the actual length and width of each room and then find the area.

1. A scale is the ratio that compares the measurements of the drawing or model with the actual measurements. A scale factor is a scale without any units.

2. larger; because 2 cm > 1 mm

3. Convert one of the lengths into the same units as the other length. Then, form the scale and simplify.

 Practice and Problem Solving

4. 25 ft

5. 10 ft by 10 ft

6. 50 ft; 35 ft

7. 112.5%

8. 100 mi

9. 50 mi

10. 200 mi

11. 110 mi

12. 75 in.

13. 15 in.

14. 3.84 m

15. 21.6 yd

16. 17.5 mm

17. The "5 cm" should be in the numerator.
$$\frac{1\ cm}{20\ m} = \frac{5\ cm}{x\ m}$$
$$x = 100\ m$$

5.4 Record and Practice Journal

Find the missing dimension. Use the scale factor 1 : 8.

Item	Model	Actual	
1. Statue	Height: 168 in.	Height: __112__ ft	
2. Painting	Width: __2500__ cm	Width: 200 m	
3. Alligator	Height: __9.6__ in.	Height: 6.4 ft	
4. Train	Length: 36.5 in.	Length: _____ ft	$24\frac{1}{3}$

5. The diameter of the moon is 2160 miles. A model has a scale of 1 in. : 150 mi. What is the diameter of the model?

14.4 in.

6. A map has a scale of 1 in. : 4 mi.

   a. You measure 3 inches between your house and the movie theater. How many miles is it from your house to the movie theater?

   12 mi

   b. It is 17 miles to the mall. How many inches is that on the map?

   4.25 in.

Technology For the Teacher

Answer Presentation Tool
QuizShow

18. 4 cm; 1 cm : 30 m

19. 2.4 cm; 1 cm : 10 mm

20. The length or width of the actual item

21. a. *Answer should include, but is not limited to:* Make sure words and picture match the product.

　b. Answers will vary.

22. a. $480

　b. $1536

　c. tile; Because $5 per square foot is greater than $2 per square foot, the tile has a higher unit cost.

23. See *Taking Math Deeper.*

24. 1 in. : 25 mi

Fair Game Review

25–28.

29. C

Mini-Assessment

Find the missing dimension. Use the scale factor 1 : 6.

	Model	Actual
1.	12 in.	72 in.
2.	3 ft	18 ft
3.	20 cm	120 cm
4.	2 yd	12 yd

5. A fish in an aquarium is 4 feet long. A scale model of the fish is 2 inches long. What is the scale factor?
1 : 24

Taking Math Deeper

Exercise 23

This is a short, but nice problem. It requires that students make a reasonable estimate for the diameter of a baseball. Then it requires that students use this information to determine the reasonability of making a scale model.

1 Draw and label a diagram.

Model: 3 in.

Actual: 6378 km

Model: *x* in.

Actual 695,500 km

2 Write and solve a proportion.

$$\frac{x}{695{,}500} = \frac{3}{6378}$$

$$695{,}000 \cdot \frac{x}{695{,}500} = 695{,}500 \cdot \frac{3}{6378}$$

$$x \approx 327 \text{ in.}$$

3 Answer the question.

327 inches is equal to 27.25 feet. The model for the Sun would have to be about the width of a classroom. So, it is not reasonable to use a baseball for Earth. A reasonable model for Earth would have a diameter of one quarter inch and the diameter for the model of the Sun would be 27.26 inches.

Project

Make a scale drawing of the solar system. Make sure students include the scale they used.

Reteaching and Enrichment Strategies

If students need help. . .	If students got it. . .
Resources by Chapter • Practice A and Practice B • Puzzle Time Record and Practice Journal Practice Differentiating the Lesson Lesson Tutorials Skills Review Handbook	Resources by Chapter • Enrichment and Extension • School-to-Work Start the next section

Use a centimeter ruler to measure the segment shown. Find the scale of the drawing.

18.
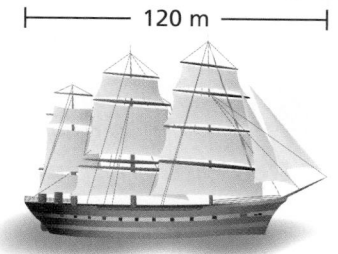
├── 120 m ──┤

19.
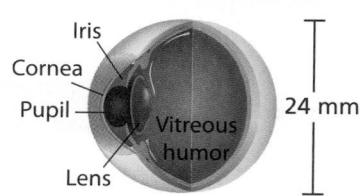
Iris
Cornea
Pupil
Vitreous humor
Lens
24 mm

20. **REASONING** You know the length and width of a scale model. What additional information do you need to know to find the scale of the model?

21. **OPEN-ENDED** You are in charge of creating a billboard advertisement with the dimensions shown.

   a. Choose a product. Then design the billboard using words and a picture.

   b. What is the scale factor of your design?

16 ft
8 ft
YOUR AD HERE

Reduced drawing of blueprint

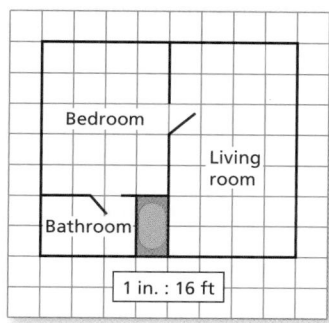
Bedroom
Living room
Bathroom
1 in. : 16 ft

22. **BLUEPRINT** In a blueprint, each square has a side length of $\frac{1}{4}$ inch.

   a. Ceramic tile costs $5 per square foot. How much would it cost to tile the bathroom?

   b. Carpet costs $18 per square yard. How much would it cost to carpet the bedroom and living room?

   c. Which has a higher unit cost, the tile or the carpet? Explain.

23. **REASONING** You are making a scale model of the solar system. The radius of Earth is 6378 kilometers. The radius of the Sun is 695,500 kilometers. Is it reasonable to choose a baseball as a model of Earth? Explain your reasoning.

24.  *Critical Thinking* A map on the Internet has a scale of 1 in. : 10 mi. You zoom out one level. The map has been reduced so that 2.5 inches on the old map appears as 1 inch on the new map. What is the scale of the new map?

🖊️ **Fair Game Review** *What you learned in previous grades & lessons*

Plot and label the ordered pair in a coordinate plane. *(Section 1.6)*

25. $A(-4, 3)$ 26. $B(2, -6)$ 27. $C(5, 1)$ 28. $D(-3, -7)$

29. **MULTIPLE CHOICE** A backpack is on sale for 15% off the original price. The original price is $68. What is the sale price? *(Section 4.3)*

   Ⓐ $10.20   Ⓑ $53   Ⓒ $57.80   Ⓓ $78.20

5.4b Scale Drawings

Check It Out
Lesson Tutorials
BigIdeasMath.com

EXAMPLE 1 Finding an Actual Area

Central Park is a rectangular park in New York City. Find the actual area of the park.

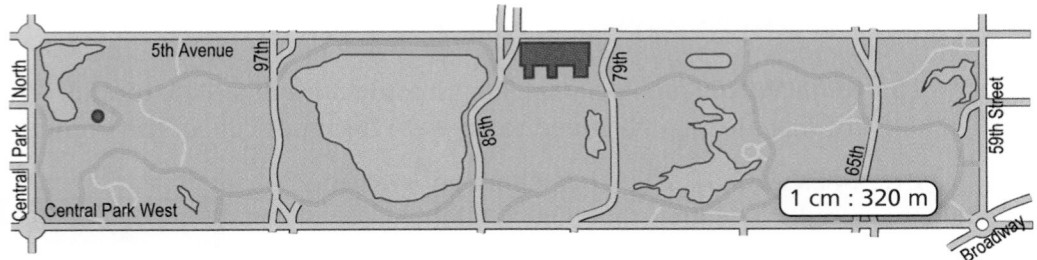

1 cm : 320 m

Step 1: Use a centimeter ruler to find the length and width of the park in the scale drawing.

The scale drawing of the park is 12.5 centimeters long and 2.5 centimeters wide.

Step 2: Use the scale to write and solve proportions to find the actual length and width of the park. Let ℓ be the actual length and let w be the actual width.

$$\frac{1 \text{ cm}}{320 \text{ m}} = \frac{12.5 \text{ cm}}{\ell \text{ m}} \quad \leftarrow \text{drawing distance} \quad \frac{1 \text{ cm}}{320 \text{ m}} = \frac{2.5 \text{ cm}}{w \text{ m}}$$
$$\leftarrow \text{actual distance}$$

$$\ell = 320 \cdot 12.5 \qquad\qquad w = 320 \cdot 2.5$$

$$\ell = 4000 \qquad\qquad\qquad w = 800$$

Step 3: Use a formula to find the area.

$$A = \ell w \qquad\qquad \text{Write formula.}$$
$$= 4000(800) \qquad \text{Substitute 4000 for } \ell \text{ and 800 for } w.$$
$$= 3,200,000 \qquad \text{Multiply.}$$

∴ The actual area of Central Park is 3,200,000 square meters.

Practice

The shuffleboard diagram has a scale of 1 cm : 1 ft. Find the actual area of the region.

1. red region

2. blue region

3. green region

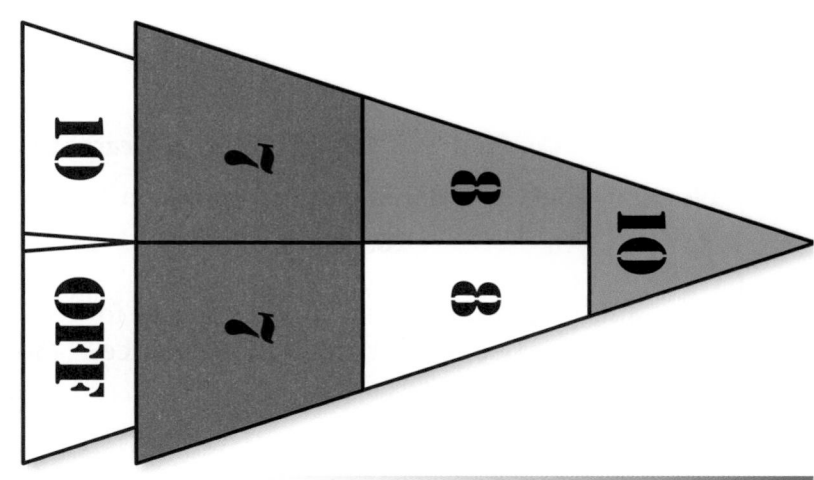

Laurie's Notes

Introduction

Connect

- **Yesterday:** Students used scaled drawings to find missing measurements.
- **Today:** Students will find areas using scale drawings and recreate scale drawings at different scales.

Motivate

- **?** Ask students, "Have any of you visited Central Park in New York?" Have them share their thoughts about the relative size of the park.
- Share some information about the park. Central Park is approximately 843 acres, a rectangle about 2.5 miles long and 0.5 mile wide. There are playgrounds, running trails, ice-skating rinks, and a zoo within the park.
- Explain that in today's lesson they will find the area of the park, but not in acres or square miles.

Lesson Notes

Example 1

- **FYI:** A centimeter is approximately the width of your pinky finger.
- Have students measure the artwork to find the drawing dimensions of Central Park. Because the scale for the drawing is in centimeters, it should make sense to students that their measurements will be in centimeters.
- **Common Error:** Students may set up the proportion incorrectly. It is very important to label the units when writing the proportion. Remind students that the measurements in the numerators are the drawing distances and the measurements in the denominators are the actual distances.
- Work through the three steps of the example as shown.
- **?** "What does a square meter look like? Is it bigger than the top of your desk?" yes; A square meter is probably bigger than the tops of their desks.
- If time permits, have students estimate the area of the classroom floor in square meters.

Practice

- These problems review measuring with a centimeter ruler and area formulas for triangles and trapezoids.
- **Common Error:** Students may forget to multiply by $\frac{1}{2}$ in the area formulas.

Goal Today's lesson is finding areas using scale drawings and recreating scale drawings.

Lesson Materials
Textbook
• centimeter ruler • inch ruler

Warm Up

Extra Example 1

The drawing of a triangular garden has a scale of 1 in. : 4 ft. Find the actual area of the garden. 24 ft²

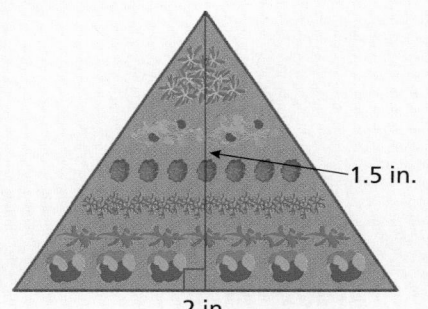

1.5 in.

2 in.

Practice

1. 15 ft^2
2. 4.5 ft^2
3. 3 ft^2

Record and Practice Journal Practice

See Additional Answers.

Laurie's Notes

Extra Example 2

Recreate the scale drawing of a Madagascan flag to have a scale of 1 cm : 8 ft.

1 cm : 4 ft

1 cm : 8 ft

Practice

4. See Additional Answers.

5. See Additional Answers.

Mini-Assessment

The scale drawing has a scale of 1 cm : 2 m. Find the actual area of the region.

1.

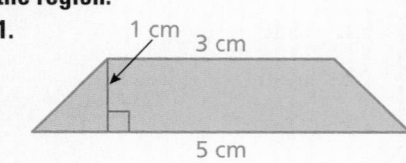

1 cm   3 cm

5 cm

16 m²

2.

3 cm

2 cm

12 m²

Example 2

- If possible, find two maps of the same state or two flags of the same country that have different scales to display for this example.
- Students are making a new scale drawing of a flag. The new scale drawing is twice the size of the original.
- Check to see that students are measuring correctly with their inch rulers.
- **? Extension:** "How do the areas of the two flags compare?" The area of the new scale drawing is 4 times greater than the area of the original scale drawing.
- **Extension:** Have students find the ratio of the new drawing's scale to the original drawing's scale. Help students see that this ratio indicates that measurements in the new scale drawing are two times the measurements in the original scale drawing.

Practice

- **Note:** Exercise 4 is an enlargement and Exercise 5 is a reduction. You may want to help students compare the measurements in Exercise 5:

Original drawing $\vdash\!\dfrac{1\text{ cm}}{2\text{ m}}\!\mid\!\dfrac{1\text{ cm}}{2\text{ m}}\!\dashv$

New drawing $\vdash\!\dfrac{1\text{ cm}}{4\text{ m}}\!\dashv$

- **FYI:** Exercise 4 shows the flag of the Bahamas.
- **Common Error:** Students may draw a reduction in Exercise 4 and an enlargement in Exercise 5 because they misunderstand how the scale changes the drawing.

Closure

- **Exit Ticket:** The scale drawing of a rectangular garden is 3 centimeters by 4 centimeters. If the scale is 1 cm : 2.5 m, what is the area of the actual garden? 75 square meters

EXAMPLE 2 — Recreating a Scale Drawing

1 in. : 8 ft

Recreate the scale drawing of a Romanian flag so that it has a scale of 1 in. : 4 ft.

Step 1: Compare measurements in the original scale drawing to measurements in the new scale drawing.

Original drawing
$$\text{1 in.} \mid \text{8 ft}$$

New drawing
$$\text{1 in.} \mid \text{4 ft} \quad \text{1 in.} \mid \text{4 ft}$$

Measurements in the new scale drawing will be 2 times longer than measurements in the original scale drawing.

Step 2: Use an inch ruler to measure the original scale drawing. Multiply the measurements by 2 and create the new scale drawing.

Original scale drawing	*New scale drawing*
Length: 1.5 in.	Length: $1.5 \cdot 2 = 3$ in.
Width: 1 in.	Width: $1 \cdot 2 = 2$ in.
Blue bar width: 0.5 in.	Blue bar width: $0.5 \cdot 2 = 1$ in.
Yellow bar width: 0.5 in.	Yellow bar width: $0.5 \cdot 2 = 1$ in.
Red bar width: 0.5 in.	Red bar width: $0.5 \cdot 2 = 1$ in.

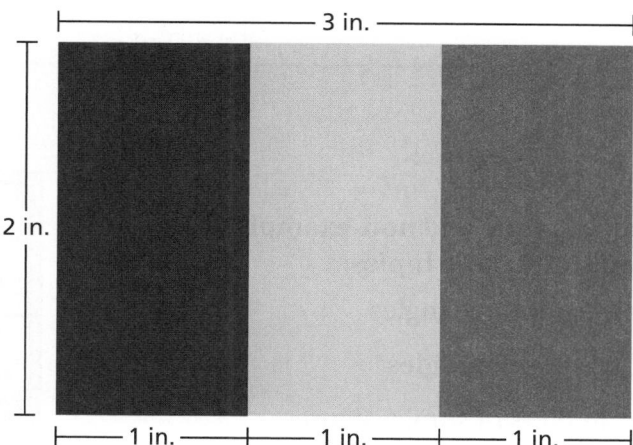

3 in.

2 in.

1 in.   1 in.   1 in.

Practice

Recreate the scale drawing so that it has a scale of 1 cm : 4 m.

4.

1 cm : 8 m

5.

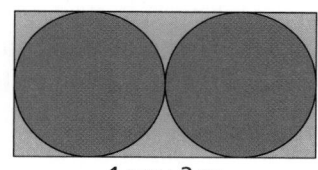

1 cm : 2 m

You can use an **example and non-example chart** to list examples and non-examples of a vocabulary word or term. Here is an example and non-example chart for similar figures.

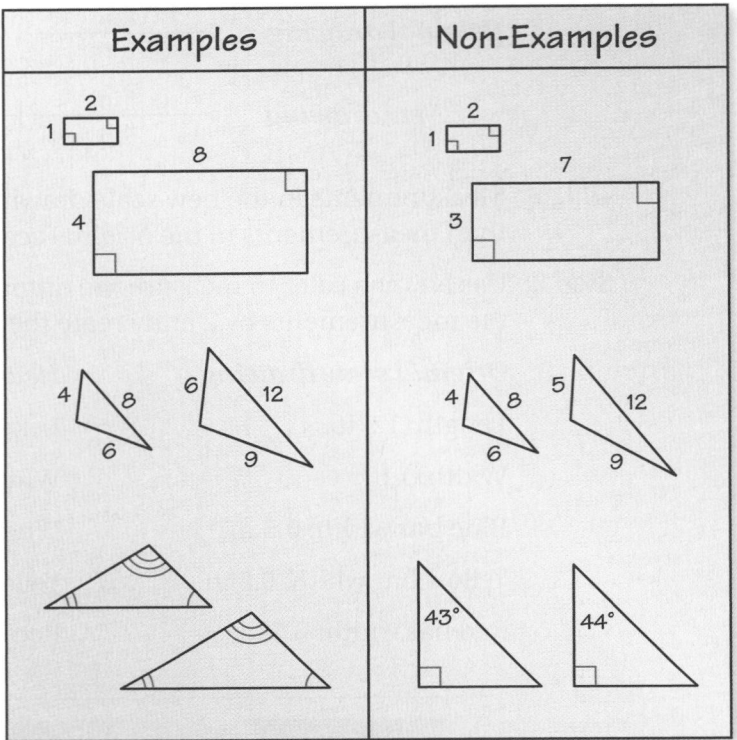

On Your Own

Make an example and non-example chart to help you study these topics.

1. corresponding angles

2. corresponding sides

3. perimeters of similar figures

4. areas of similar figures

5. indirect measurements

6. scale drawings

After you complete this chapter, make example and non-example charts for the following topics.

7. transformations

    a. translations   **b.** reflections   **c.** rotations

"I'm using an example and non-example chart for a talk on cat hygiene."

Sample Answers

1. Corresponding Angles

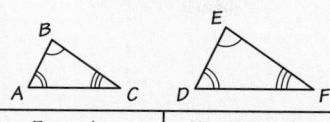

Examples	Non-Examples
∠A and ∠D	∠A and ∠E
∠B and ∠E	∠A and ∠F
∠C and ∠F	∠B and ∠D
	∠B and ∠F
	∠C and ∠D
	∠C and ∠E

2. Corresponding Sides

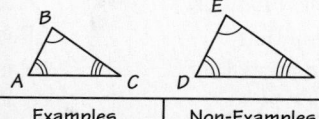

Examples	Non-Examples
Side AB and Side DE	Side AB and Side EF
Side BC and Side EF	Side AB and Side DF
Side AC and Side DF	Side BC and Side DE
	Side BC and Side DF
	Side AC and Side DE
	Side AC and Side EF

3. Perimeters of Similar Figures

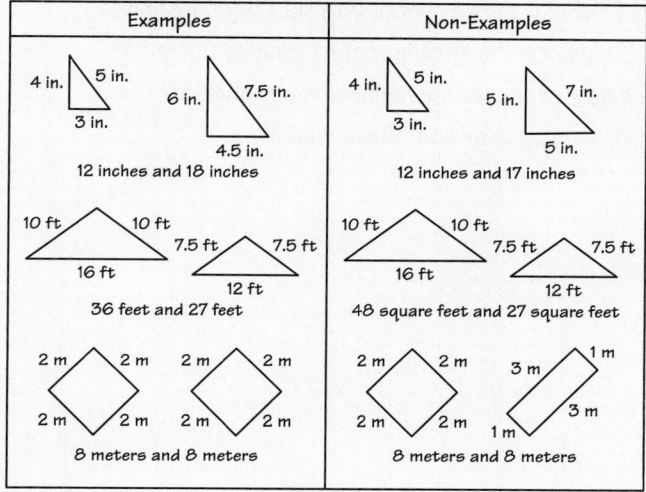

4–6. Available at *BigIdeasMath.com*.

List of Organizers
Available at *BigIdeasMath.com*

Comparison Chart
Concept Circle
Example and Non-Example Chart
Formula Triangle
Four Square
Idea (Definition) and Examples Chart
Information Frame
Information Wheel
Notetaking Organizer
Process Diagram
Summary Triangle
Word Magnet
Y Chart

About this Organizer

An **Example and Non-Example Chart** can be used to list examples and non-examples of a vocabulary word or term. Students write examples of the word or term in the left column and non-examples in the right column. This type of organizer serves as a good tool for assessing students' knowledge of pairs of topics that have subtle but important differences, such as translations and rotations. Blank example and non-example charts can be included on tests or quizzes for this purpose.

Technology
For
the Teacher
Vocabulary Puzzle Builder

Answers

1. yes; Corresponding angles have the same measure.

   Because $\dfrac{4}{10} = \dfrac{8}{20}$, the corresponding side lengths are proportional.

2. $\dfrac{3}{2}; \dfrac{9}{4}$

3. $\dfrac{4}{15}; \dfrac{16}{225}$

4. $18\dfrac{2}{3}$

5. 16.5

6. 4.5 in.

7. 75 ft

8. 7 in. : 20 in.

9. 7 : 288

10. no; Corresponding side lengths are not proportional.

Assessment Book

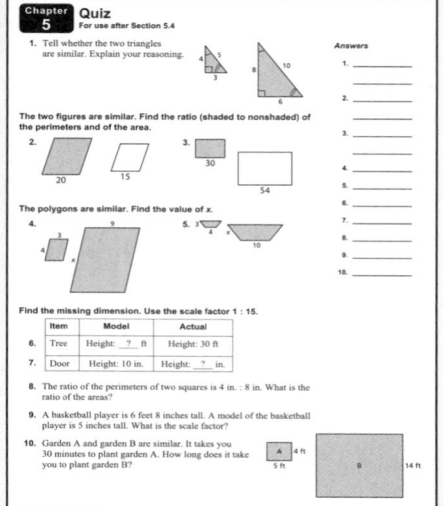

Alternative Quiz Ideas

100% Quiz	Math Log
Error Notebook	Notebook Quiz
Group Quiz	Partner Quiz
Homework Quiz	Pass the Paper

Notebook Quiz

A notebook quiz is used to check students' notebooks. Students should be told at the beginning of the course what the expectations are for their notebooks: notes, class work, homework, date, problem number, goals, definitions, or anything else that you feel is important for your class. They also need to know that it is their responsibility to obtain the notes when they miss class.

1. On a certain day, what was the answer to the Warm Up question?
2. On a certain day, how was this vocabulary term defined?
3. In Lesson 5.2, what is the answer to On Your Own Question 1?
4. In Section 5.3, what is the answer to the Essential Question?
5. On a certain day, what was the homework assignment?

Give the students 5 minutes to answer these questions.

Reteaching and Enrichment Strategies

If students need help. . .	If students got it. . .
Resources by Chapter • Study Help • Practice A and Practice B • Puzzle Time Lesson Tutorials *BigIdeasMath.com* Practice Quiz Practice from the Test Generator	Resources by Chapter • Enrichment and Extension • School-to-Work Game Closet at *BigIdeasMath.com* Start the next section

Technology For the Teacher

Answer Presentation Tool
Big Ideas Test Generator

1. Tell whether the two rectangles are similar. Explain your reasoning. *(Section 5.1)*

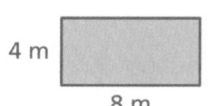

The two figures are similar. Find the ratios (red to blue) of the perimeters and of the areas. *(Section 5.2)*

2.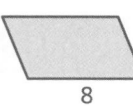

3.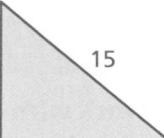

The polygons are similar. Find the value of *x*. *(Section 5.3)*

4.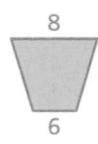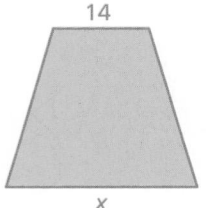

5.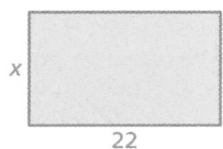

Find the missing dimension. Use the scale factor 1 : 20. *(Section 5.4)*

	Item	Model	Actual
6.	Basketball Player	Height: in.	Height: 90 in.
7.	Dinosaur	Length: 3.75 ft	Length: ft

8. **POSTERS** The ratio of the corresponding side lengths of two similar posters is 7 in. : 20 in. What is the ratio of the perimeters? *(Section 5.2)*

9. **DOLPHIN** A dolphin in an aquarium is 12 feet long. A scale model of the dolphin is $3\frac{1}{2}$ inches long. What is the scale factor of the model? *(Section 5.4)*

10. **TENNIS COURT** The tennis courts for singles and doubles matches are different sizes. Are the courts similar? Explain. *(Section 5.1)*

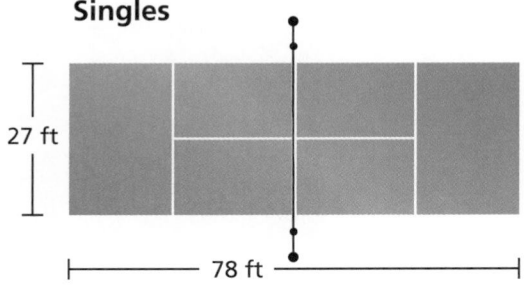

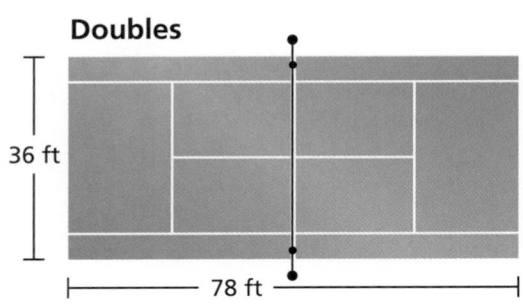

Essential Question How can you use translations to make a tessellation?

When you slide a tile it is called a **translation**. When tiles can be used to cover a floor with no empty spaces, the collection of tiles is called a *tessellation*.

1 ACTIVITY: Describing Tessellations

Work with a partner. Can you make the pattern by using a translation of single tiles that are all of the same shape and design? If so, show how.

a. Sample:

Tile Pattern   Single Tiles

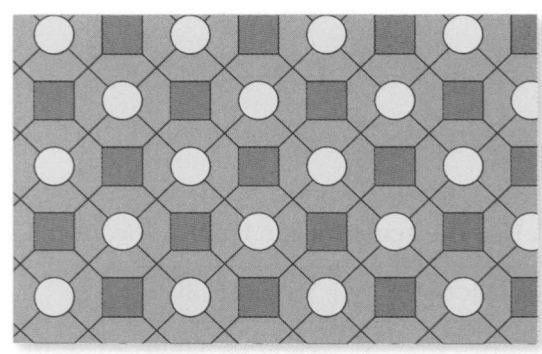   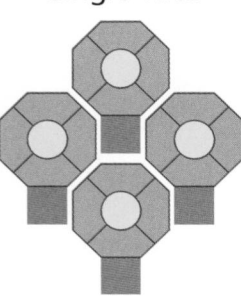

b.

c.

d.

e.
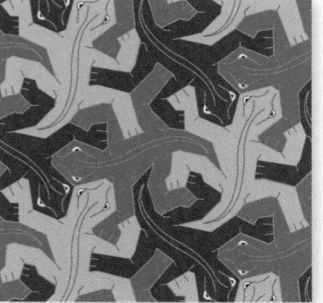

Laurie's Notes

Introduction

For the Teacher

- **Goal:** Students will explore translations, or slides, in visual patterns using tracing paper and pattern blocks.
- Students need to develop spatial reasoning skills. When teaching transformations, it is important to have materials available for students to explore the concepts. Tracing paper or transparencies to overlay on designs, pattern blocks that can be manipulated, and lots of visual samples will help students develop this skill.

Motivate

- **Whole Class Activity:** You can model translations by having all students stand at the front of the room facing in the same direction. Give single directions such as: two steps forward; three steps right; one step at a 45° angle with the right foot.
- You can also model a translation at the overhead. When you move a transparency in a certain direction, it is a translation or slide.
- **Big Idea: Translations** slide a figure to a new location in the plane. The figure does not change in size. All points in the figure move the same distance and direction. This key idea is presented formally in the lesson.

Activity Notes

Discuss

- The tiles on the floor or on the ceiling are common examples of tessellations in your classroom. Perhaps there is special tiling in the bathrooms at your school. Wallpaper books are also very useful. You can also find examples on the Internet.
- **FYI:** The word tessellation comes from the Latin word tessellae, which was the name given by the Romans to the small tiles used for pavements and walls in ancient Rome.

Activity 1

- Provide tracing paper (the local deli or doughnut shop may donate a box) so that students can trace over the single tile that they believe can be translated to cover a floor.
- As you circulate, remind students that they need to find a single tile that can be traced repeatedly to cover the floor. This is equivalent to purchasing a box of tiles to put down a new floor. All of the tiles in the box are the same size and shape.
- Make overhead transparencies of the designs so that when students share their solutions, they can trace their single tile on a clear transparency and slide it over the page.
- **Note:** All of the given tessellations are translations of a single tile.

Previous Learning

Students should know how to plot points in the coordinate plane.

Activity Materials
Textbook
- tracing paper - pattern blocks

Start Thinking! and Warm Up

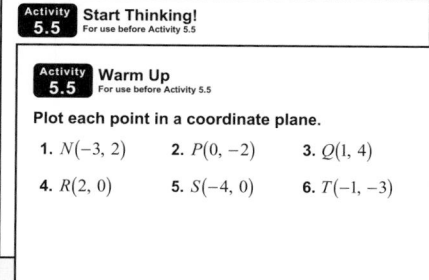

5.5 Record and Practice Journal

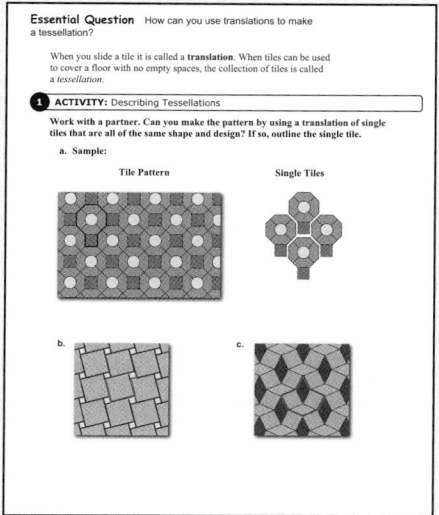

Auditory

To help students remember what each type of transformation is, use the words *slide* for translation, *flip* for reflection, and *turn* for rotation.

5.5 Record and Practice Journal

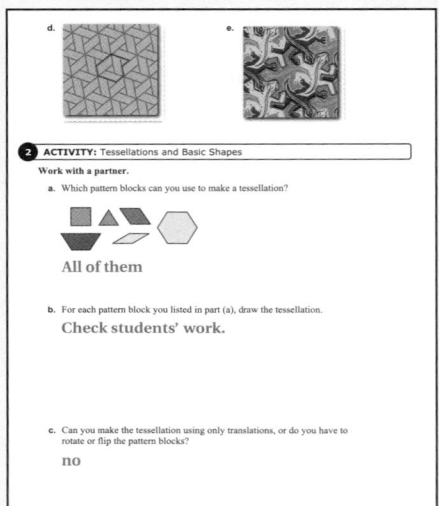

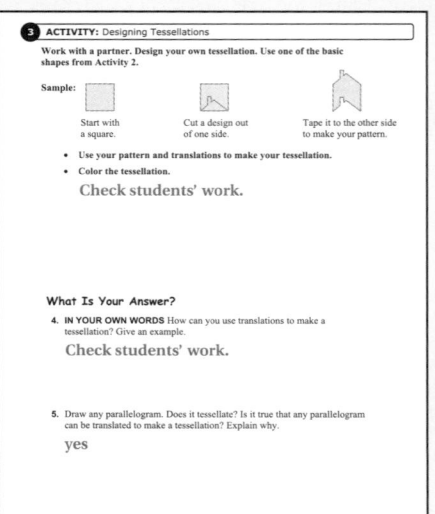

Laurie's Notes

Activity 2

• **Management Tip:** Place sufficient quantities of pattern blocks in a zipped plastic bag for each pair of students. Each bag should have a minimum of 10 triangles and 6 each of the remaining shapes. This will help with distribution and pick-up of materials.

• **Review:** "What are the names of each pattern block shape?" square, equilateral triangle, rhombus or parallelogram (tan and blue), hexagon, trapezoid

• **Whole Class Discussion:** Discuss the findings of Activity 2. Four of the six pieces can be used alone to make a tessellation by only translating the figure. You need 2 of the triangular pieces, with one rotated, to make a tessellation using translations. This is also true for the trapezoidal piece.

Activity 3

• This activity could be started today and the students could finish their designs for homework.

• Explain to the students that they need to start with one of the basic shapes from Activity 2.

• Use the *cut and bump* method to create the tessellation. Whatever shape is *cut* from one edge of the shape must *bump out* on the edge to where the shape slides. After the initial shape is created, additional artwork (such as windows and roof coloring) can be added to provide additional details.

• **Management Tip:** Some students find it helpful to draw their design on grid paper. The grid helps the students be more accurate with their design.

What Is Your Answer?

• **Think-Pair-Share:** Students should read each question independently and then work with a partner to answer the questions. When they have answered the questions, the pair should compare their answers with another group and discuss any discrepancies.

Closure

• Would *any* rectangle cover or tessellate a flat surface? Explain. Yes, because the sides are parallel.

Technology
For
the **T**eacher

Dynamic Classroom

The Dynamic Planning Tool
Editable Teacher's Resources at *BigIdeasMath.com*

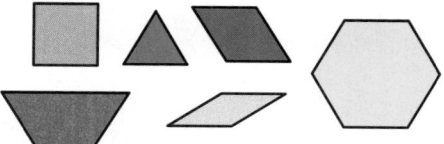

ACTIVITY: Tessellations and Basic Shapes

Work with a partner.

a. Which pattern blocks can you use to make a tessellation?

b. For each one that works, draw the tessellation.

c. Can you make the tessellation using only translation, or do you have to rotate or flip the pattern blocks?

ACTIVITY: Designing Tessellations

Work with a partner. Design your own tessellation. Use one of the basic shapes from Activity 2.

Sample:

Start with a square.

Cut a design out of one side.

Tape it to the other side to make your pattern.

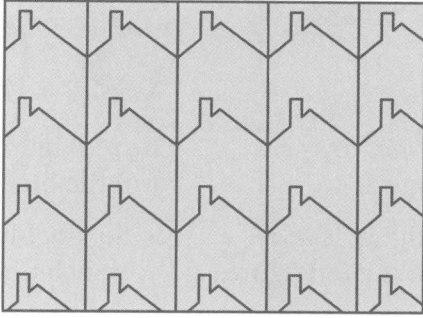

Use the pattern and translations to make your tessellation.

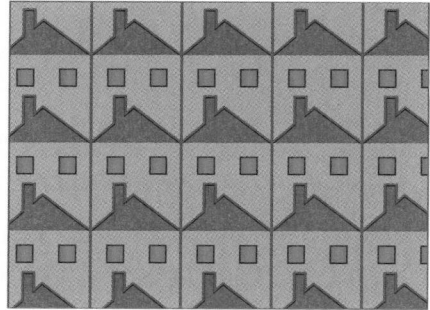

Color the tessellation.

What Is Your Answer?

4. IN YOUR OWN WORDS How can you use translations to make a tessellation? Give an example.

5. Draw any parallelogram. Does it tessellate? Is it true that any parallelogram can be translated to make a tessellation? Explain why.

Use what you learned about translations to complete Exercises 4–6 on page 224.

Check It Out
Lesson Tutorials
BigIdeasMath⟋com

A **transformation** changes a figure into another figure. The new figure is called the **image**.

Key Vocabulary 🔊

transformation, *p. 222*

image, *p. 222*

translation, *p. 222*

 **Key Idea**

Translations

A **translation** is a transformation in which a figure *slides* but does not turn. Every point of the figure moves the same distance and in the same direction.

Slide

Slide

The original figure and its image have the same size and shape.

EXAMPLE **1** **Identifying a Translation**

Tell whether the blue figure is a translation of the red figure.

a.

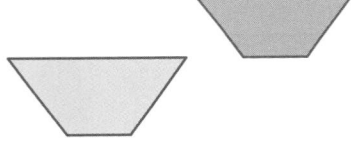

The red figure *slides* to form the blue figure.

∴ So, the blue figure is a translation of the red figure.

b.

3 ɜ

The red figure *turns* to form the blue figure.

∴ So, the blue figure is *not* a translation of the red figure.

🔵 **On Your Own**

Now You're Ready
Exercises 4–9

Tell whether the blue figure is a translation of the red figure. Explain.

1.

2.

3.

4.

🔊 Multi-Language Glossary at BigIdeasMath⟋com.

Laurie's Notes

Introduction

Connect

- **Yesterday:** Students explored translations by manipulating pattern blocks and sketching translations.
- **Today:** Students will use their visual skills to draw translations in the coordinate plane.

Motivate

- Share a quick story about movie animation. Perhaps some of your students have made flipbooks where images are drawn at a slightly different location on each card, so that as you flip through the cards the image appears to move. If the image were a baseball, it would be translated to a new location on each card and appear to be moving as you flip the cards.
- Make a flipbook in advance to share with the class.

Lesson Notes

Key Idea

- **Representation:** An arrow is often used to represent the direction of the translation (slide). In a formal geometry course, this is called the *translation vector*.
- The example of the bicycle is used in each transformation section to describe each type of transformation (slide, flip, turn). Have students focus on the *entire* bicycle and not just the tires.
- **Management Tip:** Make 2 copies of a transparency with one bicycle on each. To demonstrate the translation, move the top transparency leaving the bottom transparency stationary. After you slide the transparency to a new location, discuss the vocabulary and definition. The new figure is the image. All parts of the bicycle moved the same distance and in the same direction.

Example 1

- **Common Misconception:** The translation does not need to be in a horizontal or vertical direction. It can also be in a diagonal direction.
- Students generally have little difficulty identifying translations.

On Your Own

- **Think-Pair-Share:** Students should read each question independently and then work with a partner to answer the questions. When they have answered the questions, the pair should compare their answers with another group and discuss any discrepancies.
- For the questions that are translations, describe the direction of the translation.

Goal Today's lesson is identifying translations.

Lesson Materials
Introduction
• flipbook

Start Thinking! and Warm Up

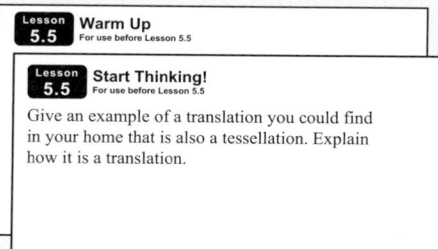

Lesson 5.5 Warm Up
For use before Lesson 5.5

Lesson 5.5 Start Thinking!
For use before Lesson 5.5

Give an example of a translation you could find in your home that is also a tessellation. Explain how it is a translation.

Extra Example 1

Tell whether the blue figure is a translation of the red figure.

a.

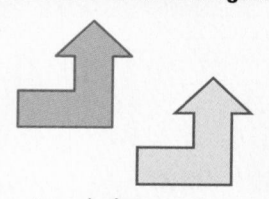

a translation

b.

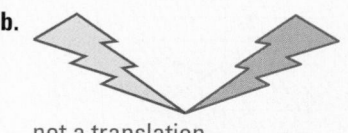

not a translation

On Your Own

1. no; The blue figure is larger than the red figure.

2. yes; The red figure slides to form the blue figure.

3. yes; The red figure slides to form the blue figure.

4. no; The red figure flips to form the blue figure.

Extra Example 2

The vertices of a triangle are $A(-2, -1)$, $B(0, 2)$, and $C(3, 0)$. The triangle is translated 2 units right and 5 units up. What are the coordinates of the image? $A'(0, 4)$, $B'(2, 7)$, $C'(5, 5)$

 On Your Own

    5. $A'(-6, 3)$, $B'(-2, 7)$,
       $C'(-3, 4)$

Extra Example 3

The vertices of a rectangle are $A(1, 4)$, $B(3, 4)$, $C(3, 1)$ and $D(1, 1)$. Draw the figure and its image after a translation 3 units left and 4 units down.

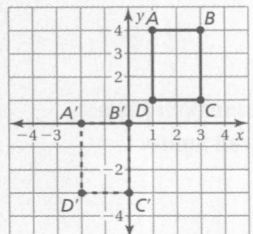

 On Your Own

    6.

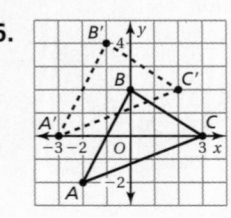

English Language Learners

Visual

Make a poster in the classroom to illustrate the movement of a point in a coordinate plane based on the coordinate notation.

$x + h$	$\rightarrow$
$x - h$	$\leftarrow$
$y + k$	$\uparrow$
$y - k$	$\downarrow$

Example 2

- Draw $\triangle ABC$ and label the vertices on a transparency. Slide the transparency 3 units to the right and down 3 units.
- **Representation:** The **image** of a transformation (translation, reflection, or rotation) is written with the prime symbol. This helps to distinguish the image from the original figure, often referred to as the pre-image.
- After the result of the translation has been drawn, you can draw an arrow from A to A', B to B', and C to C'. The resulting figure appears to be a 3-D diagram of a triangular prism.
- Explain that translating the triangle on a diagonal is equivalent to translating the triangle horizontally and then vertically. The two steps focus on what happens to each of the coordinates in an ordered pair.
- **?** "Is the blue triangle the same size and shape as the red triangle?" yes
- Reinforce the concept of same size and shape by talking about the lengths of corresponding sides, the measures of the corresponding angles, and the perimeters and areas of the two triangles.

Example 3

- Plot the four ordered pairs.
- **Common Error:** Students may interchange x- and y-directions in plotting the ordered pairs.
- **?** Ask questions about the translation.
  - "In what quadrant is the original square?" IV
  - "If a figure is translated in the coordinate plane 4 units left, what will change, the x-coordinate or the y-coordinate?" x-coordinate
  - "If a figure is translated in the coordinate plane 6 units up, what will change, the x-coordinate or the y-coordinate?" y-coordinate
- Explain the notation in the table. Use an alternate color to draw attention to the repeated pattern (subtracting 4 and adding 6) that occurs with each ordered pair.
- Draw the new image.
- **?** "In what quadrant is the image?" II
- **?** "Is the blue square the same size and shape as the red square?" yes

On Your Own

- **Neighbor Check:** Have students work independently and then have their neighbor check their work. Have students discuss any discrepancies.

Closure

- Draw a right triangle in Quadrant II. Translate the triangle so that the image is in Quadrant IV. Describe the translation.

The Dynamic Planning Tool
Editable Teacher's Resources at *BigIdeasMath.com*

EXAMPLE 2 **Translating a Figure**

Translate the red triangle 3 units right and 3 units down. What are the coordinates of the image?

Reading

A′ is read "*A* prime." Use *prime* symbols when naming an image.

$A \longrightarrow A'$

$B \longrightarrow B'$

$C \longrightarrow C'$

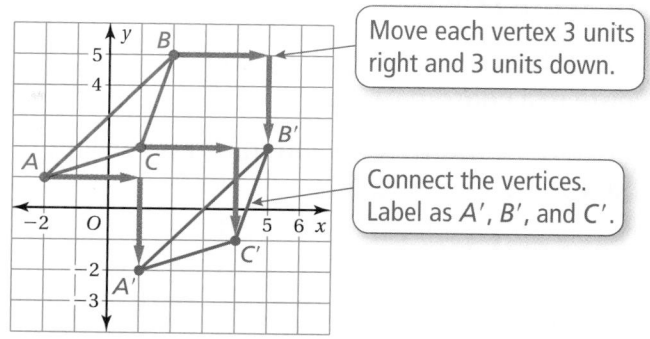

Move each vertex 3 units right and 3 units down.

Connect the vertices. Label as *A′*, *B′*, and *C′*.

∴ The coordinates of the image are $A'(1, -2)$, $B'(5, 2)$, and $C'(4, -1)$.

On Your Own

Now You're Ready
Exercises 10 and 11

5. The red triangle is translated 4 units left and 2 units up. What are the coordinates of the image?

EXAMPLE 3 **Translating a Figure**

The vertices of a square are $A(1, -2)$, $B(3, -2)$, $C(3, -4)$, and $D(1, -4)$. Draw the figure and its image after a translation 4 units left and 6 units up.

Subtract 4 from each *x*-coordinate.

Add 6 to each *y*-coordinate.

Vertices of *ABCD*	$(x - 4, y + 6)$	Vertices of *A′B′C′D′*
$A(1, -2)$	$(1 - 4, -2 + 6)$	$A'(-3, 4)$
$B(3, -2)$	$(3 - 4, -2 + 6)$	$B'(-1, 4)$
$C(3, -4)$	$(3 - 4, -4 + 6)$	$C'(-1, 2)$
$D(1, -4)$	$(1 - 4, -4 + 6)$	$D'(-3, 2)$

∴ The figure and its image are shown at the right.

On Your Own

Now You're Ready
Exercises 12–15

6. The vertices of a triangle are $A(-2, -2)$, $B(0, 2)$, and $C(3, 0)$. Draw the figure and its image after a translation 1 unit left and 2 units up.

 Vocabulary and Concept Check

1. **VOCABULARY** Which figure is the image?

2. **VOCABULARY** How do you translate a figure in a coordinate plane?

3. **CRITICAL THINKING** Can you translate the letters in the word TOKYO to form the word KYOTO? Explain.

Slide

 **Practice and Problem Solving**

Tell whether the blue figure is a translation of the red figure.

4.

5.

6.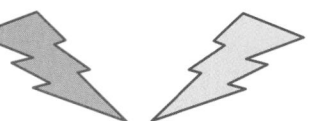

7.

8.

9.

10. Translate the triangle 4 units right and 3 units down. What are the coordinates of the image?

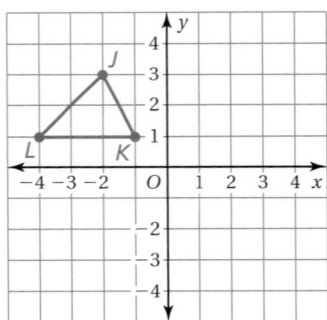

11. Translate the figure 2 units left and 4 units down. What are the coordinates of the image?

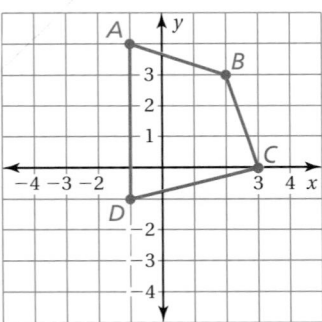

The vertices of a triangle are $L(0, 1)$, $M(1, -2)$, and $N(-2, 1)$. Draw the figure and its image after the translation.

12. 1 unit left and 6 units up

13. 5 units right

14. 2 units right and 3 units up

15. 3 units left and 4 units down

16. **ICONS** You can click and drag an icon on a computer screen. Is this an example of a translation? Explain.

Assignment Guide and Homework Check

Level	Day 1 Activity Assignment	Day 2 Lesson Assignment	Homework Check
Basic	4–6, 24–28	1–3, 7–19 odd, 16	7, 11, 13, 17
Average	4–6, 24–28	1–3, 8–16 even, 17–21 odd	8, 10, 12, 17
Advanced	4–6, 24–28	1–3, 12–20 even, 21–23	12, 18, 20, 22

For Your Information

- **Exercise 3** The Japanese language is composed of symbols, not letters. KYO means *capitol* and TO means *new*. So, KYO TO was the ancient capitol of Japan and TO KYO is the modern day capitol of Japan.

Common Errors

- **Exercises 4–9** Students may forget that the objects must be the same size to be a translation. Remind them that the size stays the same. Tell students that when the size is different, it is a scale drawing.
- **Exercises 10–15** Students may translate the shape the wrong direction or mix up the units for the translation. Tell them to redraw the original on graph paper. Also, tell students to write the direction of the translation using arrows to show the movement left, right, up, or down.
- **Exercises 17 and 18** Students may struggle finding the translation. Encourage students to plot the points in a coordinate plane and count the change left, right, up, or down.
- **Exercises 19 and 20** Students may count the translation to the wrong point. Ask them to label the red figure with points *A*, *B*, etc. and the corresponding points on the blue figure as *A'*, *B'*, etc.

5.5 Record and Practice Journal

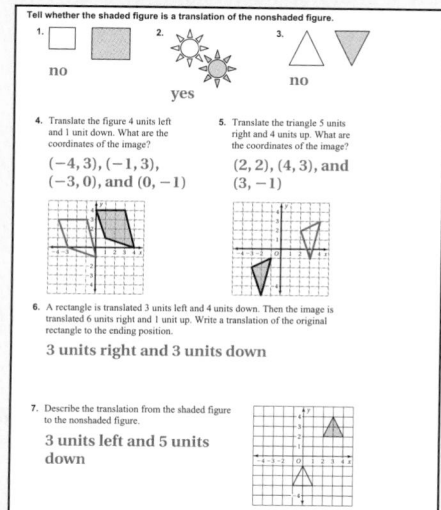

Vocabulary and Concept Check

1. A

2. Move each vertex according to the translation.

3. yes; Translate the letters T and O to the end.

Practice and Problem Solving

4. yes
5. no
6. no
7. yes
8. yes
9. no
10. $J'(2, 0)$, $K'(3, -2)$, $L'(0, -2)$
11. $A'(-3, 0)$, $B'(0, -1)$, $C'(1, -4)$, $D'(-3, -5)$

12.

13.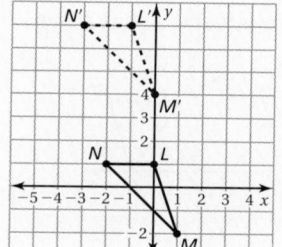

14.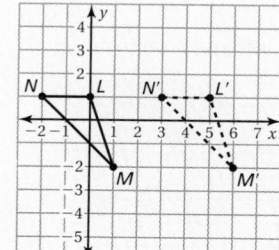

15. See Additional Answers.

16. Yes, because the figure slides.

17. 2 units left and 2 units up

18. 5 units right and 9 units up

19. 6 units right and 3 units down

Practice and Problem Solving

20. 5 units left and 2 units down

21–22. See Additional Answers.

23. See *Taking Math Deeper*.

Fair Game Review

24. yes   **25.** no

26. no   **27.** yes

28. B

Mini-Assessment

The vertices of a triangle are *A* (1, 3), *B* (4, 3), and *C* (3, 0). Draw the figure and its image after the translation.

1. 2 units left and 3 units down

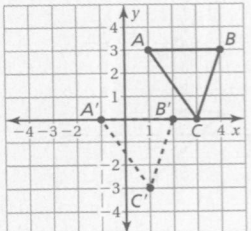

2. 1 unit left and 4 units down

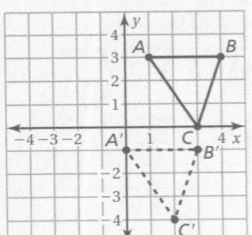

3. Describe a translation of the helicopter from point *A* to point *B*.

5 units right and 7 units up

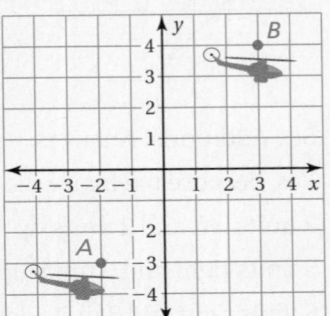

Taking Math Deeper

Exercise 23

There are thousands of correct answers to this question. This could be a nice discussion question for pairs or groups of students.

 Here is one translation that takes 5 moves.

 Here is one translation that takes only 3 moves.

③ It is not possible to move from g8 to g5 in less than 3 moves.

Project

Create a board game similar to chess. Write the rules for your game and play your game with another student.

Reteaching and Enrichment Strategies

If students need help. . .	If students got it. . .

Resources by Chapter
  • Practice A and Practice B
  • Puzzle Time
Record and Practice Journal Practice
Differentiating the Lesson
Lesson Tutorials
Skills Review Handbook
	Resources by Chapter
	  • Enrichment and Extension
	  • School-to-Work
	  • Financial Literacy
	Start the next section

Describe the translation of the point to its image.

17. $(3, -2) \rightarrow (1, 0)$

18. $(-8, -4) \rightarrow (-3, 5)$

Describe the translation from the red figure to the blue figure.

19.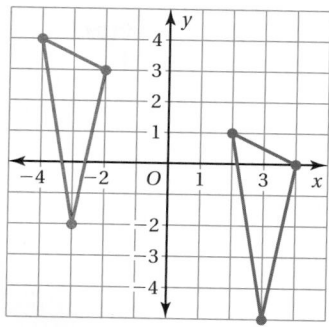

20.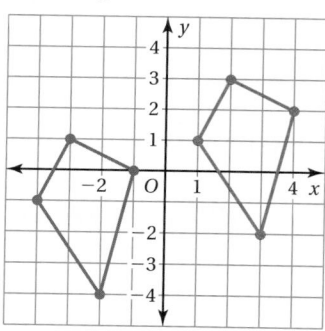

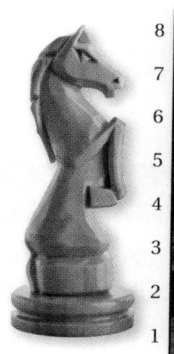

21. **FISHING** A school of fish translates from point F to point D.

    a. Describe the translation of the school of fish.

    b. Can the fishing boat make a similar translation? Explain.

    c. Describe a translation the fishing boat could make to get to point D.

22. **REASONING** A triangle is translated 5 units right and 2 units up. Then the image is translated 3 units left and 8 units down. Write a translation of the original triangle to the ending position.

23. **Critical Thinking** In chess, a knight can move only in an L-shape pattern:

- *two* vertical squares then *one* horizontal square;
- *two* horizontal squares then *one* vertical square;
- *one* vertical square then *two* horizontal squares; or
- *one* horizontal square then *two* vertical squares.

Write a series of translations to move the knight from g8 to g5.

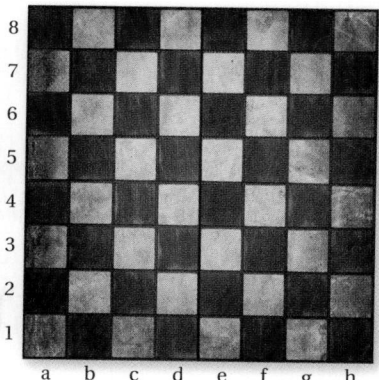

 **Fair Game Review** *What you learned in previous grades & lessons*

Tell whether each figure can be folded in half so that one side matches the other.
(Skills Review Handbook)

24.

25.

26.

27.

28. **MULTIPLE CHOICE** You put \$550 in an account that earns 4.4% simple interest per year. How much interest do you earn in 6 months? *(Section 4.4)*

    A \$1.21   **B** \$12.10   **C** \$121.00   **D** \$145.20

Essential Question How can you use reflections to classify a frieze pattern?

The Meaning of a Word ● Reflection

When you look at a mountain by a lake, you can see the **reflection**, or mirror image, of the mountain in the lake.

If you fold the photo on its axis, the mountain and its reflection will align.

Actual mountain

Axis

Reflection of mountain

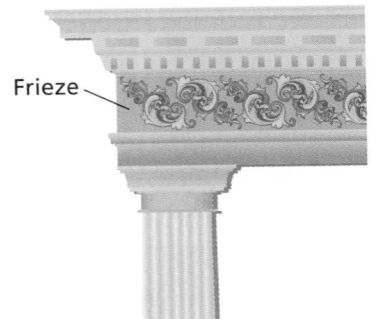

Frieze

A *frieze* is a horizontal band that runs at the top of a building. A frieze is often decorated with a design that repeats.

- All frieze patterns are translations of themselves.
- Some frieze patterns are reflections of themselves.

1 EXAMPLE: Frieze Patterns

Is the frieze pattern a reflection of itself when folded horizontally? vertically?

- Fold (reflect) on horizontal axis. The pattern coincides.

- Fold (reflect) on vertical axis. The pattern coincides.

∴ This frieze pattern is a reflection of itself when folded horizontally *and* vertically.

Laurie's Notes

Introduction

For the Teacher

- **Goal:** Students will explore reflections in frieze patterns.
- Today's investigation involves working with a pattern known as a frieze. The text shows an example of an architectural frieze on a building. Friezes also occur on wallpaper borders, designs on pottery, ironwork railings, and the headbands and belts of the indigenous people of North America, to name a few.
- A frieze is a pattern which repeats in one direction and can always be translated onto itself. Friezes may also contain reflections, and that is the focus of this investigation.
- Use the Internet to find additional examples of frieze patterns. There are many examples that can be shared with students.

Motivate

- Before class, practice folding a long strip of scrap paper. Cut it to make a frieze pattern. A common design is the stick figure. Practice various folds so you can create reflections.

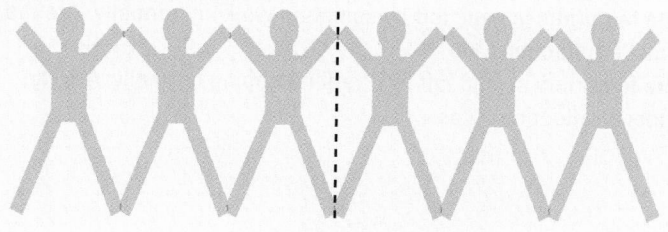

Vertical line of symmetry

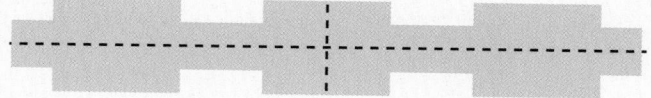

Horizontal and vertical lines of symmetry

- **Teaching Tip:** Lay the cut-out, still folded, on the overhead projector. Ask students to visualize and then describe what the figure will look like when it is opened up.

Activity Notes

Example 1

- **Management Tip:** When I have students cutting paper, I tape plastic bags in many locations around the room. Having numerous places where scraps of paper can be thrown away is easier than having only one wastebasket.
- Holding the folded paper up to the light or a window will help students *see* if the pattern is folding onto itself.
- **Extension:** Ask students to identify the smallest design region that can be translated left and right to continue the pattern.

Previous Learning

Students should know how to plot points in the coordinate plane.

Activity Materials	
Introduction	**Textbook**
• paper cut-out	• tracing paper

Start Thinking! and Warm Up

Activity 5.6 Start Thinking!
For use before Activity 5.6

Activity 5.6 Warm Up
For use before Activity 5.6

The vertices of a triangle are $A(-4, 4)$, $B(-4, 1)$, and $C(-1, 1)$. Draw the figure and its image after the translation.

1. 6 units right
2. 1 unit left and 3 units down

5.6 Record and Practice Journal

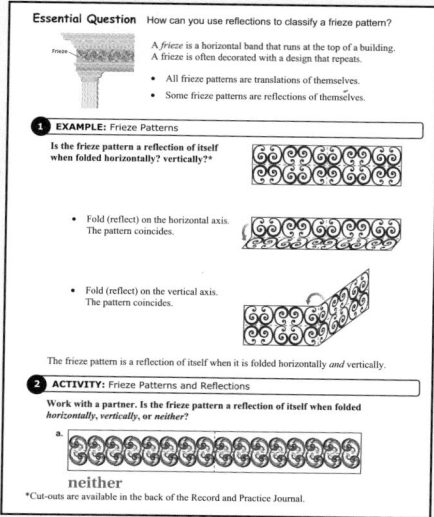

Differentiated Instruction

Kinesthetic

Have students fold paper in half or in quarters and use scissors to cut out various shapes. Open the paper and find the lines of symmetry. Depending on the cut out, there may be more than one line of symmetry.

Activity 2

- Students should work with a partner.
- Students may wish to have tracing paper to test their thinking about the patterns shown.
- Make an overhead transparency of the designs to help facilitate discussion. Have clear transparencies available for students to trace their answers.
- Remind students that a reflection that folds onto itself in a frieze must be horizontal or vertical.
- **Common Error:** Students will see the rotation in the pattern and identify it as a reflection.

What Is Your Answer?

- To facilitate Questions 3–5, provide students with strips of grid paper, perhaps 2 to 3 inches wide. Colored pencils are helpful for students to use to decorate their pattern. One or more of these questions may be completed for homework.

5.6 Record and Practice Journal

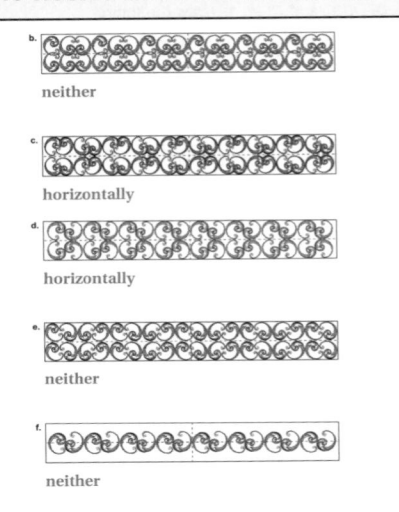

b.
neither

c.
horizontally

d.
horizontally

e.
neither

f.
neither

Closure

- Imagine footprints in sand left by someone walking normally. Are the footprints a reflection? no
- Imagine footprints in mud left by a rabbit hopping normally. Are the footprints a reflection? yes

What Is Your Answer?

3. Draw a frieze pattern that is a reflection of itself when folded horizontally.

Check students' work.

4. Draw a frieze pattern that is a reflection of itself when folded vertically.

Check students' work.

5. Draw a frieze pattern that is not a reflection of itself when folded horizontally or vertically.

Check students' work.

6. **IN YOUR OWN WORDS** How can you use reflections to classify a frieze pattern?

Fold a frieze pattern horizontally or vertically. If it coincides, then it is a reflection of itself.

Technology For the Teacher

Dynamic Classroom

The Dynamic Planning Tool
Editable Teacher's Resources at *BigIdeasMath.com*

ACTIVITY: Frieze Patterns and Reflections

Work with a partner. Is the frieze pattern a reflection of itself when folded *horizontally, vertically,* or *neither?*

a.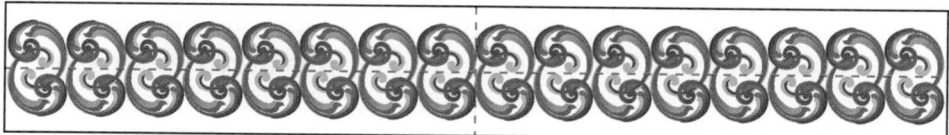

b.

c.

d.

e.

f.

What Is Your Answer?

3. Draw a frieze pattern that is a reflection of itself when folded horizontally.

4. Draw a frieze pattern that is a reflection of itself when folded vertically.

5. Draw a frieze pattern that is not a reflection of itself when folded horizontally or vertically.

6. IN YOUR OWN WORDS How can you use reflections to classify a frieze pattern?

Practice Use what you learned about reflections to complete Exercises 4–6 on page 230.

Check It Out
Lesson Tutorials
BigIdeasMath ✓com

Key Vocabulary 🔊
reflection, *p. 228*
line of reflection,
p. 228

 **Key Idea**

Reflections

A **reflection**, or flip, is a transformation in which a figure is reflected in a line called the **line of reflection**. A reflection creates a mirror image of the original figure.

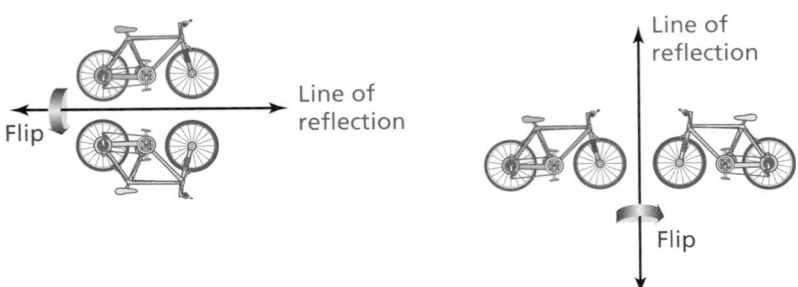

The original figure and its image have the same size and shape.

EXAMPLE **1** **Identifying a Reflection**

Tell whether the blue figure is a reflection of the red figure.

a.

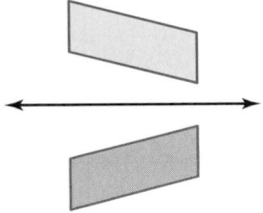

The red figure can be *flipped* to form the blue figure.

∴ So, the blue figure is a reflection of the red figure.

b.

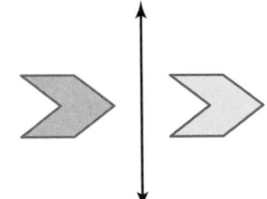

If the red figure were *flipped*, it would point to the left.

∴ So, the blue figure is *not* a reflection of the red figure.

On Your Own

Now You're Ready
Exercises 4–9

Tell whether the blue figure is a reflection of the red figure. Explain.

1.

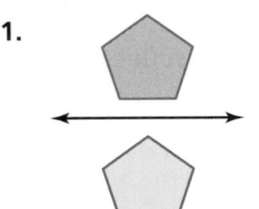

2.

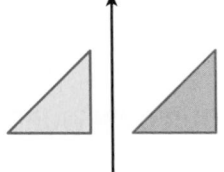

3.

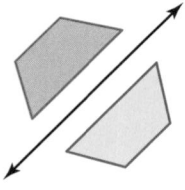

🔊 Multi-Language Glossary at BigIdeasMath ✓com.

Laurie's Notes

Introduction

Connect
- **Yesterday:** Students explored reflections in frieze patterns.
- **Today:** Students will use their visual skills to draw reflections in the coordinate plane.

Motivate
❓ Write the word **MOM** on a transparency, and ask a few questions.
- "What is special about this word?" Listen for ideas about reflection.
- "Describe the result when the word is reflected over the red line." MOM
- "Describe the result when the word is reflected over the green line." WOW

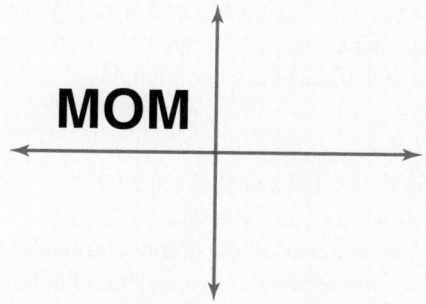

❓ "Can you think of other words that behave in a similar fashion?"

Lesson Notes

Key Idea
- Reflections are informally called flips.
- Use the bicycle transparencies made for the translation lesson. Overlay the two transparencies. Pick up the top transparency and reflect it vertically, then horizontally. You can use a piece of spaghetti to show the line of reflection.
- Discuss the difference between **line of reflection** and line symmetry. We think of the heart shape as having line symmetry—the left and right halves are the same shape and size. A line of reflection means that the *entire* shape is being reflected over the line.
- The word **MOM** can be used to show both line symmetry and line of reflection. This lesson is about lines of reflection.

Example 1
- **Common Error:** Students may call part (b) a reflection because the shapes remain the same size and the orientation is the same. It is actually a translation.
- Offer tracing paper to students who struggle with spatial reasoning.

On Your Own
- **Neighbor Check:** Have students work independently and then have their neighbor check their work. Have students discuss any discrepancies.

Goal Today's lesson is identifying and drawing **reflections**.

Lesson Materials
Textbook
• tracing paper

Start Thinking! and Warm Up

Lesson 5.6 Warm Up For use before Lesson 5.6

Lesson 5.6 Start Thinking! For use before Lesson 5.6

Words like "racecar" and "deed" are known as palindromes. What are some other examples of palindromes? Are palindromes reflections? Explain.

Extra Example 1

Tell whether the blue figure is a reflection of the red figure.

a.

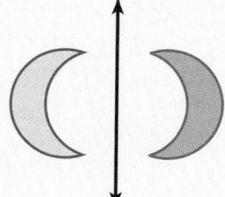

a reflection

b.

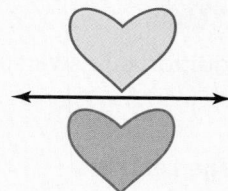

not a reflection

On Your Own

1. no; it is a translation

2. no; it is a translation

3. yes; The red figure can be flipped to form the blue figure.

Extra Example 2

The vertices of a parallelogram are $A(-1, -1)$, $B(2, -1)$, $C(4, -3)$ and $D(1, -3)$. Draw this parallelogram and its reflection in the x-axis. What are the coordinates of the image?

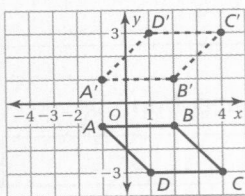

$A'(-1, 1)$, $B'(2, 1)$, $C'(4, 3)$, $D'(1, 3)$

Extra Example 3

The vertices of a triangle are $A(1, -2)$, $B(4, -1)$, and $C(2, 4)$. Draw this triangle and its reflection in the y-axis. What are the coordinates of the image?

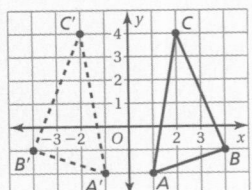

$A'(-1, -2)$, $B'(-4, -1)$, $C'(-2, 4)$

On Your Own

4. See Additional Answers.

English Language Learners

Vocabulary

Tell students that different words may be used to describe a reflection in a coordinate plane. For example, a figure is a reflection *in* the x-axis, *about* the x-axis, *across* the x-axis, or *over* the x-axis. The same words can be used to describe a reflection in the y-axis as well.

Laurie's Notes

Example 2

- Draw $\triangle ABC$ and label the vertices.
- ? "Which is the x-axis?" the horizontal axis
- We want to reflect the triangle from above the x-axis to below the x-axis.
- Note the suggestion boxes on the graph. Start with point A. Say, "Because A is 1 unit above the x-axis, it will be reflected to 1 unit below the x-axis." Repeat using similar language for points B and C. When students don't use this approach, they can easily translate the triangle instead of reflecting it.
- **Common Error:** The numbers written horizontally along the x-axis may cause students to be off by one number when they find the coordinates of each point in the blue triangle.
- ? "Is the blue triangle the same size and shape as the red triangle?" yes
- Reinforce the concept of same size and shape by talking about the lengths of corresponding sides, the measures of the corresponding angles, and the perimeters and areas of the two triangles.

Example 3

- This problem is similar to Example 2 except it is reflected in the y-axis, and the original figure is a quadrilateral.
- Note the suggestion boxes on the graph. Start with point P. Say, "Because P is 2 units to the left of the y-axis, it will be reflected to 2 units to the right of the y-axis." Repeat using similar language for points Q, R, and S.
- ? "Is the blue quadrilateral the same size and shape as the red quadrilateral?" yes
- **Extension:** List the coordinates of each quadrilateral side-by-side. Ask students to identify patterns in the coordinates. The y-coordinates stay the same and the x-coordinates are opposites of one another when you reflect in the y-axis. List the coordinates of each triangle from Example 2. Ask students to identify patterns in the coordinates. The x-coordinates stay the same and the y-coordinates are opposites of one another when you reflect in the x-axis.

On Your Own

- **Think-Pair-Share:** Students should read each question independently and then work with a partner to answer the questions. When they have answered the questions, the pair should compare their answers with another group and discuss any discrepancies.

Closure

- Draw a right triangle in Quadrant II. Reflect the triangle in the x-axis. Reflect the original triangle in the y-axis.

Technology
For the Teacher

The Dynamic Planning Tool
Editable Teacher's Resources at *BigIdeasMath.com*

EXAMPLE **2**   **Reflecting a Figure in the x-axis**

The vertices of a triangle are $A(-1, 1)$, $B(2, 3)$, and $C(6, 3)$. Draw this triangle and its reflection in the x-axis. What are the coordinates of the image?

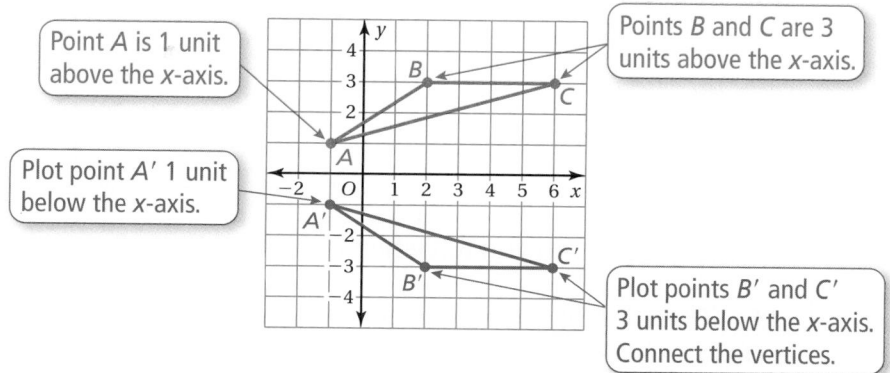

Point A is 1 unit above the x-axis.

Points B and C are 3 units above the x-axis.

Plot point A' 1 unit below the x-axis.

Plot points B' and C' 3 units below the x-axis. Connect the vertices.

∴ The coordinates of the image are $A'(-1, -1)$, $B'(2, -3)$, and $C'(6, -3)$.

EXAMPLE **3**   **Reflecting a Figure in the y-axis**

The vertices of a quadrilateral are $P(-2, 5)$, $Q(-1, -1)$, $R(-4, 2)$, and $S(-4, 4)$. Draw this quadrilateral and its reflection in the y-axis. What are the coordinates of the image?

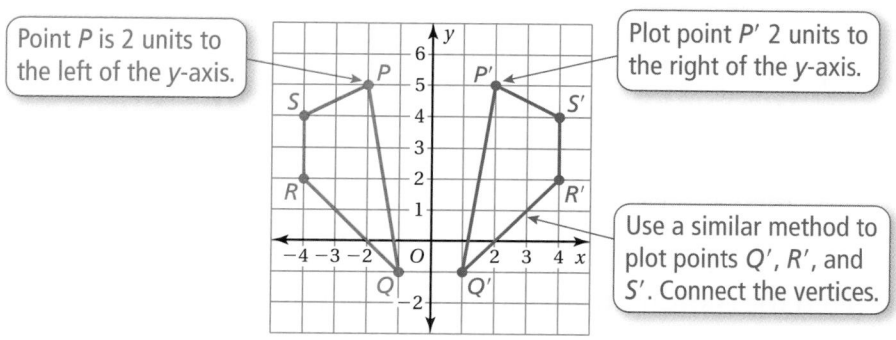

Point P is 2 units to the left of the y-axis.

Plot point P' 2 units to the right of the y-axis.

Use a similar method to plot points Q', R', and S'. Connect the vertices.

∴ The coordinates of the image are $P'(2, 5)$, $Q'(1, -1)$, $R'(4, 2)$, and $S'(4, 4)$.

On Your Own

Exercises 10–17

4. The vertices of a rectangle are $A(-4, -3)$, $B(-4, -1)$, $C(-1, -1)$, and $D(-1, -3)$.

   a. Draw the rectangle and its reflection in the x-axis.

   b. Draw the rectangle and its reflection in the y-axis.

   c. Are the images in parts (a) and (b) the same size and shape? Explain.

Vocabulary and Concept Check

1. **WHICH ONE DOESN'T BELONG?** Which transformation does *not* belong with the other three? Explain your reasoning.

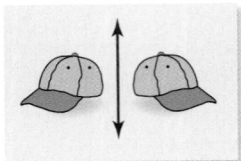 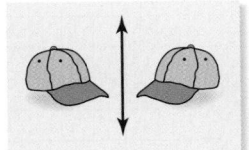 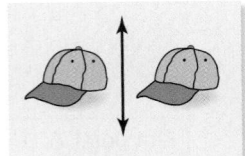 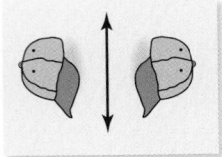

2. **WRITING** How can you tell when one figure is a reflection of another figure?

3. **REASONING** A figure lies entirely in Quadrant I. The figure is reflected in the *x*-axis. In which quadrant is the image?

Practice and Problem Solving

Tell whether the blue figure is a reflection of the red figure.

① **4.**

5.

6.

7.
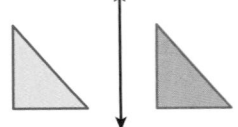

8.
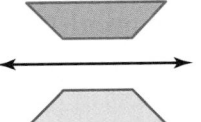

9.
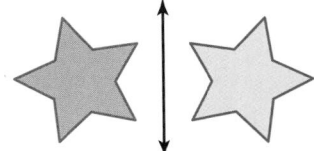

Find the coordinates of the figure after reflecting in the *x*-axis.

② **10.** $A(3, 2), B(4, 4), C(1, 3)$

11. $M(-2, 1), N(0, 3), P(2, 2)$

12. $H(2, -2), J(4, -1), K(6, -3), L(5, -4)$

13. $D(-2, -1), E(0, -2), F(1, -5), G(-1, -4)$

Find the coordinates of the figure after reflecting in the *y*-axis.

③ **14.** $Q(-4, 2), R(-2, 4), S(-1, 1)$

15. $T(1, -1), U(4, 2), V(6, -2)$

16. $W(2, -1), X(5, -2), Y(5, -5), Z(2, -4)$

17. $J(2, 2), K(7, 4), L(9, -2), M(3, -1)$

18. **ALPHABET** Which letters look the same when reflected in the line ?

A B C D E F G H I J K L M N O P Q R S T U V W X Y Z

Assignment Guide and Homework Check

Level	Day 1 Activity Assignment	Day 2 Lesson Assignment	Homework Check
Basic	4–6, 28–32	1–3, 7–21 odd, 18	7, 11, 15, 19
Average	4–6, 28–32	1–3, 8–18 even, 19–25 odd, 24	8, 10, 14, 19
Advanced	4–6, 28–32	1–3, 10–22 even, 24–27	10, 14, 20, 26

Common Errors

- **Exercises 4–9** Some students may struggle with the visual and think that a translation is actually a reflection. Give students tracing paper to trace the objects, and then fold the paper to see if the vertices line up.
- **Exercises 10–17** Students may reflect in the incorrect axis. Refer them back to Examples 2 and 3. Ask students to make a chart that describes which direction to move when reflecting over each axis.
- **Exercise 18** Students may need to copy the alphabet and fold their paper on the line to see which letters look the same.

A B C D E F G H I J K L M N O P Q R S T U V W X Y Z
⟵⟶
Ɐ B C D E Ⅎ Ɠ H I ſ K Ⅎ W N O Ԁ Ò Я Ƨ ⊥ ∩ ∧ M X ⅄ Z

- **Exercise 25** Students may not be able to see which face is a reflection. Give them a small mirror that they can place in the middle of each face to see if it is the same on both sides.

5.6 Record and Practice Journal

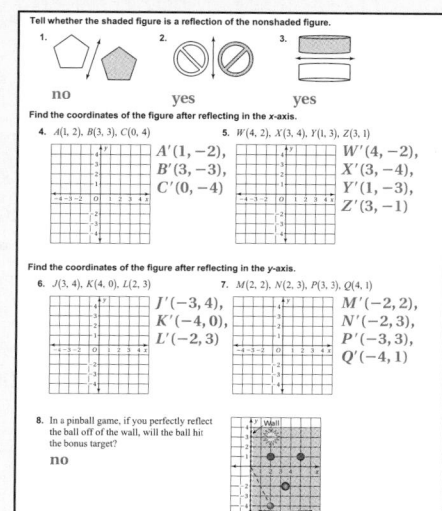

Technology For the Teacher
Answer Presentation Tool
QuizShow

 Vocabulary and Concept Check

1. The third one because it is not a reflection.

2. A figure is a reflection of another figure if one is a mirror image of the other.

3. Quadrant IV

Practice and Problem Solving

4. no   5. yes

6. yes   7. no

8. yes   9. no

10. $A'(3, -2), B'(4, -4), C'(1, -3)$

11. $M'(-2, -1), N'(0, -3), P'(2, -2)$

12. $H'(2, 2), J'(4, 1), K'(6, 3), L'(5, 4)$

13. $D'(-2, 1), E'(0, 2), F'(1, 5), G'(-1, 4)$

14. $Q'(4, 2), R'(2, 4), S'(1, 1)$

15. $T'(-1, -1), U'(-4, 2), V'(-6, -2)$

16. $W'(-2, -1), X'(-5, -2), Y'(-5, -5), Z'(-2, -4)$

17. $J'(-2, 2), K'(-7, 4), L'(-9, -2), M'(-3, -1)$

18. B, C, D, E, H, I, K, O, X

19. x-axis   20. y-axis

21. y-axis   22. x-axis

23.
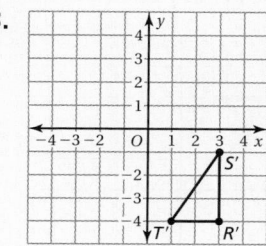

T-230

24. very different; The left side and right side of the faces are not mirror images.

25. the first one; The left side of the face is a mirror image of the right side.

26. See *Taking Math Deeper*.

27. See Additional Answers.

 Fair Game Review

28. obtuse   **29.** straight

30. right   **31.** acute

32. B

Mini-Assessment

Find the coordinates of the figure after reflecting in the *y*-axis.

1. $A(-2, 4)$, $B(-4, 2)$, $C(-1, -1)$
$A'(2, 4)$, $B'(4, 2)$, $C'(1, -1)$

2. $A(-2, 5)$, $B(-5, 1)$, $C(-3, -4)$
$A'(2, 5)$, $B'(5, 1)$, $C'(3, -4)$

Find the coordinates of the figure after reflecting in the *x*-axis.

3. $A(-4, -2)$, $B(4, -1)$, $C(1, -6)$
$A'(-4, 2)$, $B'(4, 1)$, $C'(1, 6)$

4. $A(-2, 5)$, $B(4, 8)$, $C(5, 1)$
$A'(-2, -5)$, $B'(4, -8)$, $C'(5, -1)$

5. Will the letter E look the same when reflected in the *y*-axis? no

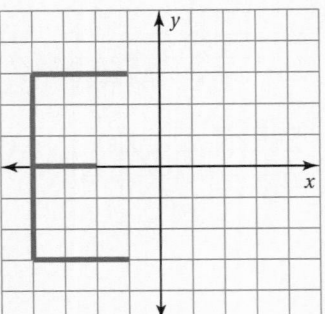

Taking Math Deeper

Exercise 26

Students need a mirror to see this one.

① Looking straight on, this is the ambulance.

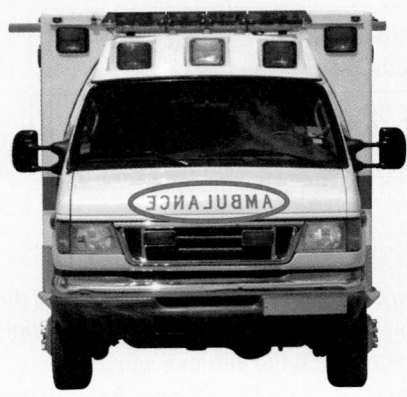

② Looking in a mirror, this is what you see.

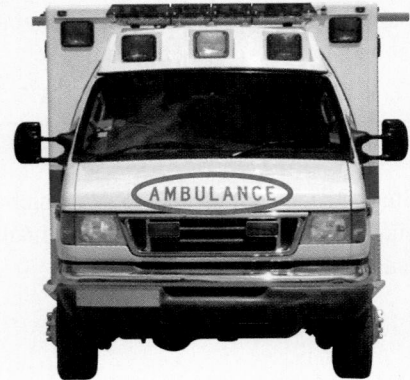

I see it.

③ The word "AMBULANCE" is printed backwards so that when people see the ambulance coming up behind their car, the word will look correct in the rear-view mirror.

Reteaching and Enrichment Strategies

If students need help. . .	If students got it. . .

Resources by Chapter
  • Practice A and Practice B
  • Puzzle Time
Record and Practice Journal Practice
Differentiating the Lesson
Lesson Tutorials
Skills Review Handbook
	Resources by Chapter
	  • Enrichment and Extension
	  • School-to-Work
	  • Financial Literacy
	  • Technology Connections
	Start the next section

The coordinates of a point and its image are given. Is the reflection in the x-axis or y-axis?

19. $(2, -2) \longrightarrow (2, 2)$

20. $(-4, 1) \longrightarrow (4, 1)$

21. $(-2, -5) \longrightarrow (2, -5)$

22. $(-3, -4) \longrightarrow (-3, 4)$

23. Translate the triangle 1 unit right and 5 units down. Then reflect the image in the y-axis.

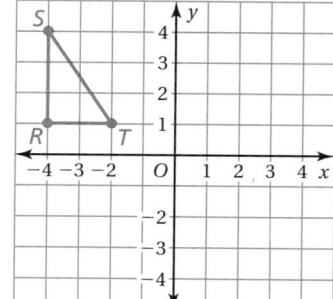

24. PROJECT Use a computer drawing program to create photographs of people by copying one side of the person's face and reflecting it in a vertical line. Does the person look normal or very different?

25. MIRROR IMAGE One of the faces shown is an exact reflection of itself. Which one is it? How can you tell?

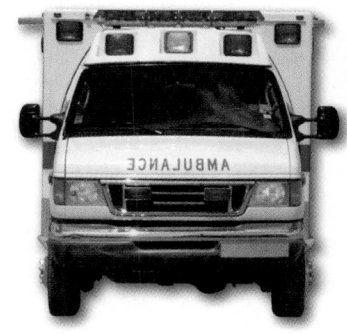

26. EMERGENCY VEHICLE Hold a mirror to the left side of the photo of the vehicle.

    a. What word do you see in the mirror?

    b. Why do you think it is written that way on the front of the vehicle?

27. *Critical Thinking* Reflect the triangle in the line $y = x$. How are the x- and y-coordinates of the image related to the x- and y-coordinates of the original triangle?

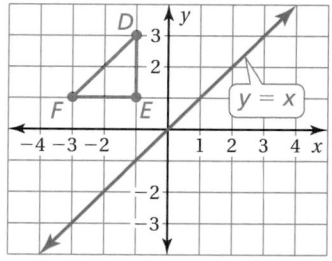

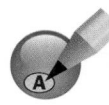 **Fair Game Review** *What you learned in previous grades & lessons*

Classify the angle as *acute*, *right*, *obtuse*, or *straight*. *(Skills Review Handbook)*

28. **29.** **30.** **31.**

32. MULTIPLE CHOICE 36 is 75% of what number? *(Section 4.1)*

    A 27   **B** 48   **C** 54   **D** 63

Essential Question What are the three basic ways to move an object in a plane?

The Meaning of a Word ● Rotate

A bicycle wheel

can **rotate** clockwise

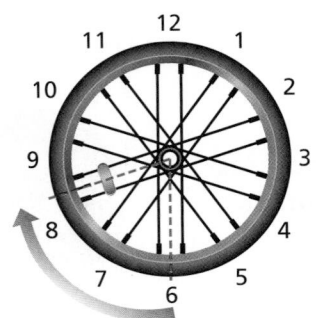

or counterclockwise.

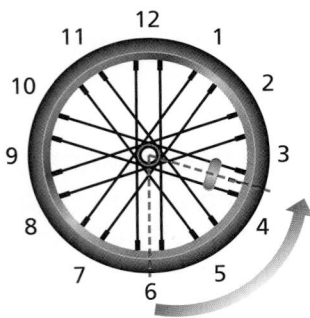

---

1 **ACTIVITY: Three Basic Ways to Move Things**

There are three basic ways to move objects on a flat surface.

1. Translate the object.

2. Reflect the object.

3. Rotate the object.

Work with a partner.

- Cut out a paper triangle that is the same size as the blue triangle shown.
- Decide how you can move the blue triangle to make each red triangle.
- Is each move a *translation*, a *reflection*, or a *rotation*?
- Draw four other red triangles in a coordinate plane. Describe how you can move the blue triangle to make each red triangle.

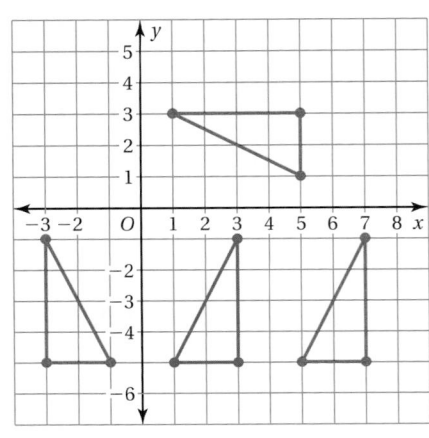

Laurie's Notes

Introduction

For the Teacher

- **Goal:** Students will explore rotations in the coordinate plane by using tracing paper and cut-out polygons.
- The rotation is the most challenging for students to visualize.
- Review the vocabulary, *clockwise* and *counterclockwise*.

Motivate

- **Time to Play:** *Name Five Twice*. In this game, students will name things that rotate: the first five objects rotate about a point in the center of the object (like a wheel) and the next five objects rotate about a point not in the center of the object (like a windshield wiper). Give students time to work with a partner to generate two lists of five.

  Example 1: car tire, Ferris wheel, merry-go-round, dial on a combination lock

  Example 2: windshield wiper, lever—as on a mechanical arm or wrench

Activity Notes

Activity 1

- The introduction reviews two transformations, translations and reflections, and introduces the third, rotations. Hold an object in your hand, like a small flag, to demonstrate all three transformations.
- Use scissors and scrap paper to cut out the blue triangle.
- **?** When students have finished the first portion of the activity, identifying the transformation for each red triangle, ask the following questions.
  - "What is the line of reflection?" *y*-axis
  - "Describe the translation." 4 units to the right
  - "Is the rotation clockwise or counterclockwise?" counterclockwise
- **Extension:** How many degrees is the triangle rotated? Explain.
  90°; It turns one-quarter of the way about the origin.
- **Alternate Approach:** Try this approach for the last part of the activity. One student slides, flips, or turns the blue triangle and traces the new image without the other watching. The other student then tries to decide how to move the blue triangle onto the new image. When finished, switch roles and try again.

Previous Learning

Students should know how to plot points in the coordinate plane.

Activity Materials
Textbook
• scissors • cardstock paper

Start Thinking! and Warm Up

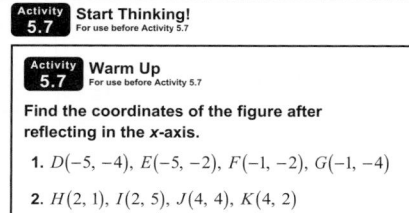

Activity 5.7 Start Thinking!
For use before Activity 5.7

Activity 5.7 Warm Up
For use before Activity 5.7

Find the coordinates of the figure after reflecting in the *x*-axis.

1. $D(-5, -4)$, $E(-5, -2)$, $F(-1, -2)$, $G(-1, -4)$

2. $H(2, 1)$, $I(2, 5)$, $J(4, 4)$, $K(4, 2)$

5.7 Record and Practice Journal

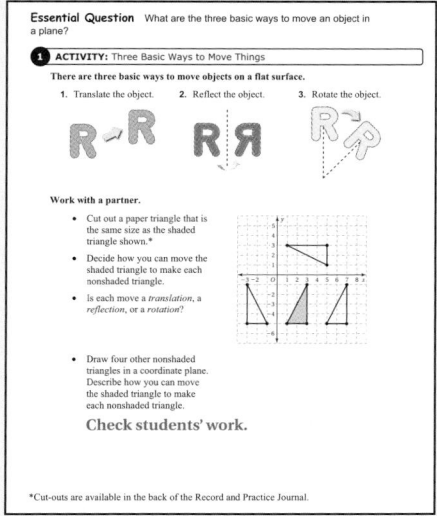

Differentiated Instruction

Kinesthetic

Project a coordinate plane using an overhead projector. Give one student a geometric shape to place on one quadrant of the coordinate plane. Give another student a duplicate shape and have them place it in another quadrant of the coordinate plane so that the shape is a rotation of the first shape. Discuss whether the duplicate shape could also be a translation or reflection of the first shape.

5.7 Record and Practice Journal

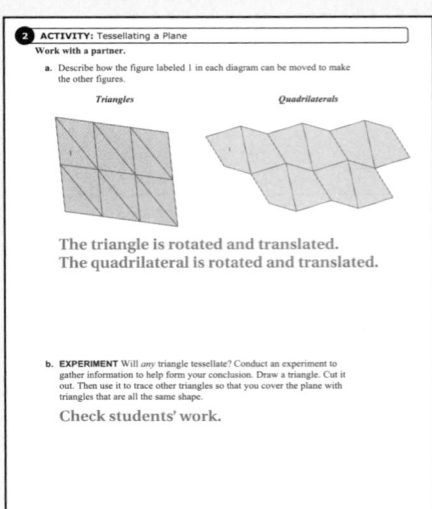

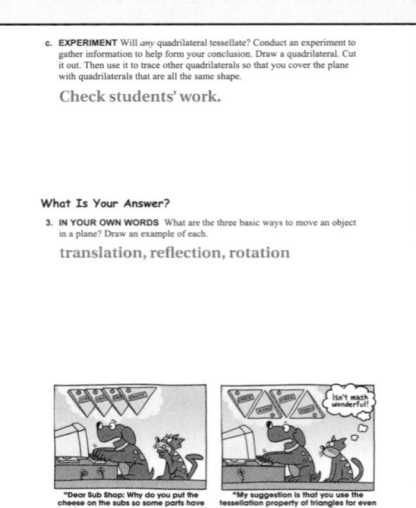

Activity 2

- This is a very open-ended activity. Some students may quickly see how the plane is tessellated, while other students will need to explore by manipulating a cut-out of the polygon or by using tracing paper.
- Discuss the two tessellations shown to make sure that students know what basic shape is used to create each. The tessellation on the left is composed of scalene triangles, while the one on the right is a scalene quadrilateral. Give students time to explore each pattern in order to describe how the tessellation is being created.
- Ask for volunteers to share their observations.
- If students are uncertain as to what is meant by the question, pose the following scenario: You have an unlimited number of purple triangular tiles like the triangle labeled 1.
- **?** "How would you place a tile to the left of triangle 1 so that the tessellation pattern would be continued?" Rotate the triangle and slide it next to triangle 1 so that sides of the same length touch.
- **?** "How would you place a tile to the left, top, or bottom of quadrilateral 1 so that the tessellation pattern would be continued?" Rotate the quadrilateral and slide it next to quadrilateral 1 so that sides of the same length touch.
- Students will need time to explore and experiment with parts (b) and (c).
- Give each pair of students some heavier weight paper, such as cardstock or old file folders. They can use this to cut out *any* triangle to see if it will tessellate. They can also cut out *any* quadrilateral and see if it will tessellate. Students should trace around their cut-outs to explore different possibilities.

What Is Your Answer?

- Students should work independently on Question 3 and then share their results with the class.

Closure

- Draw a right triangle in Quadrant I with the right angle at (0, 0). Rotate the triangle 90° clockwise about the origin.

Technology **For** **the** **Teacher**

Dynamic Classroom

The Dynamic Planning Tool
Editable Teacher's Resources at *BigIdeasMath.com*

Work with a partner.

a. Describe how the figure labeled 1 in each diagram can be moved to make the other figures.

Triangles

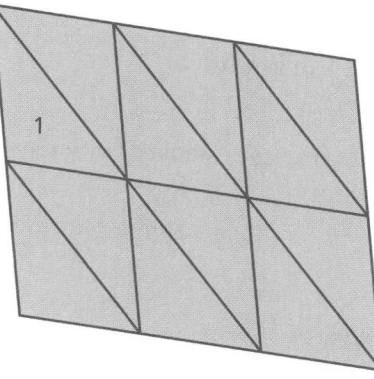

Quadrilaterals

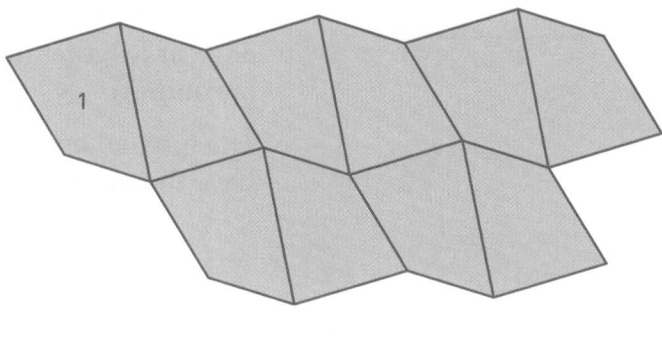

b. **EXPERIMENT** Will *any* triangle tessellate? Conduct an experiment to gather information to help form your conclusion. Draw a triangle. Cut it out. Then use it to trace other triangles so that you cover the plane with triangles that are all the same shape.

c. **EXPERIMENT** Will *any* quadrilateral tessellate? Conduct an experiment to gather information to help form your conclusion. Draw a quadrilateral. Cut it out. Then use it to trace other quadrilaterals so that you cover the plane with quadrilaterals that are all the same shape.

What Is Your Answer?

3. **IN YOUR OWN WORDS** What are the three basic ways to move an object in a plane? Draw an example of each.

"Dear Sub Shop: Why do you put the cheese on the subs so some parts have double coverage and some have none?"

"My suggestion is that you use the tessellation property of triangles for even cheese coverage."

Practice

Use what you learned about rotations to complete Exercises 7–9 on page 236.

Check It Out
Lesson Tutorials
BigIdeasMath.com

Key Vocabulary 🔊
rotation, *p. 234*
center of rotation,
  p. 234
angle of rotation,
  p. 234

🔑 Key Idea

Rotations

A **rotation**, or *turn*, is a transformation in which a figure is rotated about a point called the **center of rotation**. The number of degrees a figure rotates is the **angle of rotation**.

The original figure and its image have the same size and shape.

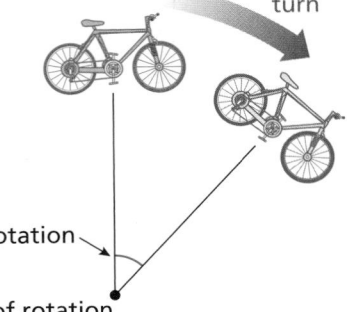

turn

angle of rotation

center of rotation

EXAMPLE 1   Standardized Test Practice

You must rotate the puzzle piece 270° clockwise about point *P* to fit it into a puzzle. Which piece fits in the puzzle as shown?

• *P*

Ⓐ    Ⓑ    Ⓒ    Ⓓ

Rotate the puzzle piece 270° clockwise about point *P*.

Study Tip

When rotating figures, it may help to sketch the rotation in several steps, as shown in Example 1.

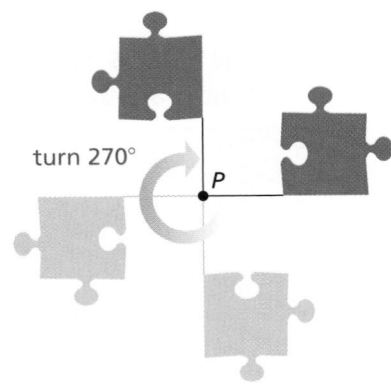

turn 270°

P

:⁘ The correct answer is Ⓒ.

On Your Own

Now You're Ready
Exercises 7–12

1. Which piece is a 90° counterclockwise rotation about point *P*?

2. Is choice D a rotation of the original puzzle piece? If not, what kind of transformation does the image show?

Laurie's Notes

Introduction

Connect

- **Yesterday:** Students explored rotations by manipulating cut-out polygons and sketching rotations.
- **Today:** Students will use their visual skills to draw rotations in the coordinate plane.

Motivate

- Use a marker to make two sizeable dots, one at the tip of your middle finger and one at the base of your palm. Anchor your elbow on a level surface. Wave at the class so that your elbow is the pivot and your forearm does the moving.
- **?** Do a "wave" of 90°, meaning start in the horizontal position and "wave" to the vertical position. Ask the following questions.
  - "Through how many degrees did I wave my hand?" 90°
  - "Did my elbow move?" no
  - "Did the two points move the same distance?" no "If not, which point moved farther?" The point on the tip of the middle finger moved farther.
- Relate this motion to that of a windshield wiper. There is a fixed point that rotates, and the farther out on the wiper blade you go, the farther the point travels.

Lesson Notes

Key Idea

- Rotations are informally called turns.
- The **rotation** is hard to visualize because the **center of rotation** is generally not attached to the shape being rotated. My hand is connected to my forearm, which is connected to my elbow, so the "wave" is easier to see as a rotation. When a diagram only shows the original figure and the image, it is harder to see the **angle of rotation**.

Example 1

- Model a rotation of 270° clockwise using a transparency with an arrow. Lightly place your finger on the middle to act as the center of rotation. Turn the transparency 90°, 3 times, stopping each time for students to see where the arrow is pointing. After 270° the arrow will be pointing up.

On Your Own

- If you have a puzzle piece that you can use to model these two questions, it would help those students who have difficulty visualizing the movement of the pieces.

Goal Today's lesson is identifying and drawing **rotations**.

Lesson Materials
Textbook
• tracing paper
• puzzle piece

Start Thinking! and Warm Up

Lesson 5.7 | **Warm Up** For use before Lesson 5.7

Lesson 5.7 | **Start Thinking!** For use before Lesson 5.7

Give an example of a translation, reflection, and rotation in a basketball game.

Extra Example 1

Tell whether the blue figure is a 180° clockwise rotation of the red figure.

a.

Is a 180° clockwise rotation.

b.

Is not a 180° clockwise rotation.

On Your Own

1. C
2. no; reflection

Extra Example 2

The vertices of a trapezoid are $A(1, 2)$, $B(4, 4)$, $C(4, 6)$, and $D(1, 6)$. Rotate the trapezoid 270° counterclockwise about the origin. What are the coordinates of the image?

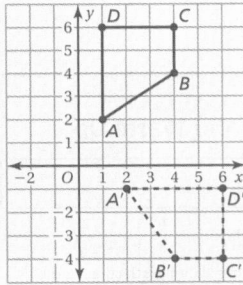

$A'(2, -1)$, $B'(4, -4)$, $C'(6, -4)$, $D'(6, -1)$

Extra Example 3

The vertices of a triangle are $A(-4, 1)$, $B(-1, 6)$, and $C(-1, 1)$. Rotate the triangle 90° clockwise about vertex C. What are the coordinates of the image?

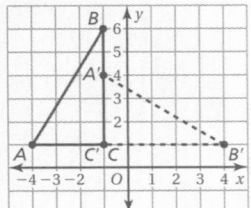

$A'(-1, 4)$, $B'(4, 1)$, $C'(-1, 1)$

On Your Own

3. See Additional Answers.

English Language Learners

Vocabulary

Discuss the meanings of the words *translation*, *reflection*, and *rotation*. Students may think of translation as a process of writing text in another (parallel) language. Mathematically, a translation is when all points of a figure move along parallel lines. Have students visualize a sun setting on the horizon of the ocean. At the point where half of the sun has set, its reflection in the water gives the appearance of a full-circled sun. A rotation about a point in the plane is similar to a nail rotating around the wheel of a tire.

Laurie's Notes

Example 2

- Tracing paper or a transparency will be needed by many students for this example. Students need to *see* where the trapezoid rotates, before they can plot the ordered pairs.
- Draw trapezoid *WXYZ* and label the vertices.
- **Common Error:** Students will rotate the trapezoid about vertex *Z* instead of rotating about the origin.
- **Teaching Strategy:** Remind students that when a figure is rotated 180°, what was on the top will rotate to the bottom, and vice versa. Model this by holding a sheet of paper and rotating it 180°.
- **Extension:** List the coordinates of the original trapezoid and the image.

$$
\begin{array}{ccc}
\underline{WXYZ} & \longrightarrow & \underline{W'\,X'\,Y'\,Z'} \\
W(-4, 2) & \longrightarrow & W'(4, -2) \\
X(-3, 4) & \longrightarrow & X'(3, -4) \\
Y(-1, 4) & \longrightarrow & Y'(1, -4) \\
Z(-1, 2) & \longrightarrow & Z'(1, -2)
\end{array}
$$

? "Do you notice any patterns in the ordered pairs?" The x- and y-coordinates are opposites of one another.

Example 3

- This problem is similar to Example 2, except the triangle is rotated 90° counterclockwise. The center of rotation is a vertex of the triangle, which is easier for students to visualize than the origin.
- ? Hold a sheet of paper facing the students. "If the paper is rotated 90° counterclockwise, to where will the top of the paper rotate?" left side
- **Big Idea:** The lengths of the sides of the triangle will not change when rotated. Find the length of the vertical and horizontal legs. The sides of the image triangle will have the same lengths.
- LJ is 5 units in length and vertical. Draw segment $L'J'$ 5 units in length and horizontal. Once J' is located, draw segment $J'K'$ 3 units in length and vertical.

On Your Own

- **Think-Pair-Share:** Students should read each question independently and then work with a partner to answer the questions. When they have answered the questions, the pair should compare their answers with another group and discuss any discrepancies.

Closure

- Draw a right triangle in Quadrant II. Reflect the triangle in the x-axis. Rotate the original triangle about the origin 90° clockwise.

Technology For the Teacher

The Dynamic Planning Tool
Editable Teacher's Resources at *BigIdeasMath.com*

EXAMPLE 2 Rotating a Figure

The vertices of a trapezoid are $W(-4, 2)$, $X(-3, 4)$, $Y(-1, 4)$, and $Z(-1, 2)$. Rotate the trapezoid 180° clockwise about the origin. What are the coordinates of the image?

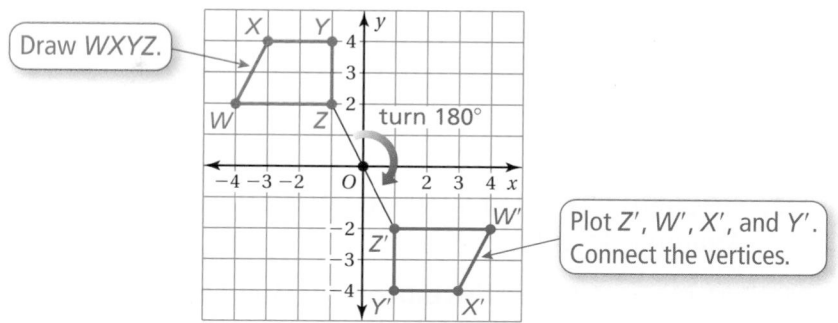

Draw WXYZ.

turn 180°

Plot Z', W', X', and Y'. Connect the vertices.

The coordinates of the image are $W'(4, -2)$, $X'(3, -4)$, $Y'(1, -4)$, and $Z'(1, -2)$.

EXAMPLE 3 Rotating a Figure

The vertices of a triangle are $J(1, 2)$, $K(4, 2)$, and $L(1, -3)$. Rotate the triangle 90° counterclockwise about vertex L. What are the coordinates of the image?

Common Error

Be sure to pay attention to whether a rotation is clockwise or counterclockwise.

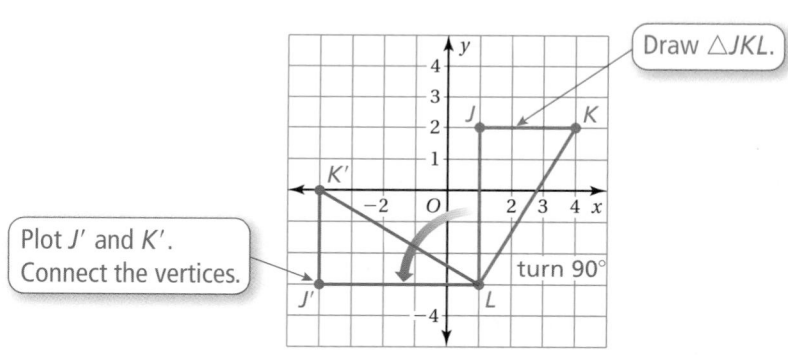

Draw △JKL.

Plot J' and K'. Connect the vertices.

turn 90°

The coordinates of the image are $J'(-4, -3)$, $K'(-4, 0)$, and $L'(1, -3)$.

On Your Own

Now You're Ready
Exercises 13–16

3. A triangle has vertices $Q(4, 5)$, $R(4, 0)$, and $S(1, 0)$.

 a. Rotate the triangle 90° counterclockwise about the origin.

 b. Rotate the triangle 180° about vertex S.

 c. Are the images in parts (a) and (b) the same size and shape? Explain.

 ## Vocabulary and Concept Check

1. **VOCABULARY** Identify the transformation shown.

   a.

   b.

   c.

2. **VOCABULARY** What are the coordinates of the center of rotation in Example 2? Example 3?

MENTAL MATH A figure lies entirely in Quadrant II. In which quadrant will the figure lie after the given clockwise rotation about the origin?

3. 90°   **4.** 180°   **5.** 270°   **6.** 360°

 ## Practice and Problem Solving

Tell whether the blue figure is a rotation of the red figure about the origin. If so, give the angle and direction of rotation.

7.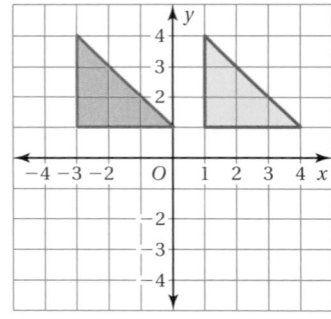

8.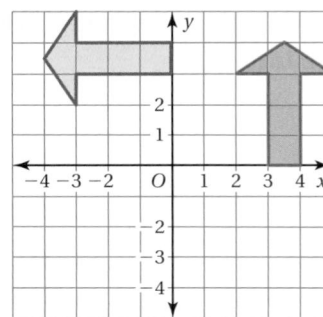

9.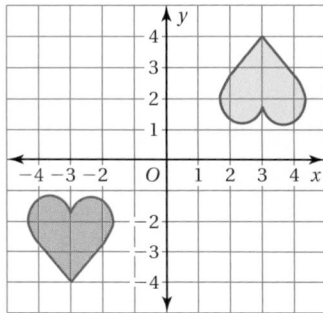

A figure has *rotational symmetry* if a rotation of 180° or less produces an image that fits exactly on the original figure. Explain why the figure has rotational symmetry.

10.

11.

12.

Assignment Guide and Homework Check

Level	Day 1 Activity Assignment	Day 2 Lesson Assignment	Homework Check
Basic	7–9, 21–23	1–6, 10–12, 13–17 odd	4, 10, 13, 17
Average	7–9, 21–23	1–6, 11–19 odd, 18	4, 11, 13, 18
Advanced	7–9, 21–23	1–6, 14, 16–20	4, 14, 17, 18

For Your Information

- **Exercise 18** A dilation is a transformation in which a figure is enlarged or reduced. In this exercise, the figure is enlarged by a scale factor of 2. You may want to repeat this exercise with a scale factor between 0 and 1 to show a reduction.

Common Errors

- **Exercises 7–9** Students with minimal spatial skills may not be able to tell whether or not the figure is rotated. Give them tracing paper and have them copy the original. They can rotate it to compare with the second figure in the book.
- **Exercises 13–16** Students may rotate the parallelogram the wrong direction. Remind them what clockwise and counterclockwise mean. Before rotating the figure, it may be helpful for students to draw an arrow for the direction that the parallelogram will rotate.

5.7 Record and Practice Journal

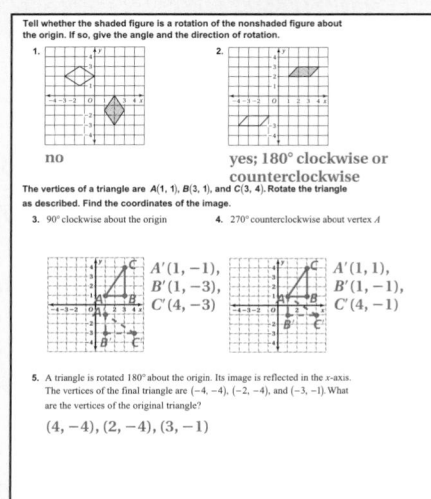

1. **a.** reflection

    b. rotation

    c. translation

2. $(0, 0)$; $(1, -3)$

3. Quadrant I

4. Quadrant IV

5. Quadrant III

6. Quadrant II

Practice and Problem Solving

7. No

8. yes; 90° counterclockwise

9. yes; 180° clockwise or counterclockwise

10. It only needs to rotate 120° to produce an identical image.

11. It only needs to rotate 90° to produce an identical image.

12. It only needs to rotate 180° to produce an identical image.

13.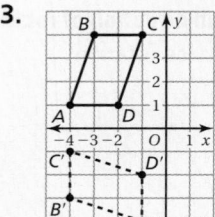

    $A'(-1, -4)$, $B'(-4, -3)$,
    $C'(-4, -1)$, $D'(-1, -2)$

14.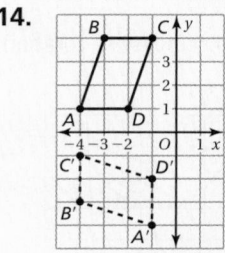

    $A'(-1, -4)$, $B'(-4, -3)$,
    $C'(-4, -1)$, $D'(-1, -2)$

15–16. See Additional Answers.

Practice and Problem Solving

17. because both ways will produce the same image

18. See Additional Answers.

19. See *Taking Math Deeper*.

20. (2, 4), (4, 1), (1, 1)

Fair Game Review

21. triangular prism

22. cylinder

23. C

Mini-Assessment

Tell whether the blue figure is a rotation of the red figure about the origin. If so, give the angle and direction of rotation.

1. yes; 90° clockwise rotation

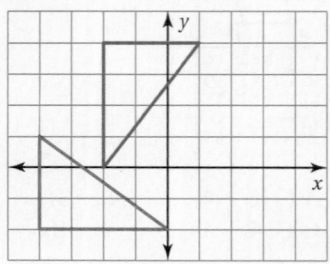

2. yes; 90° counterclockwise rotation

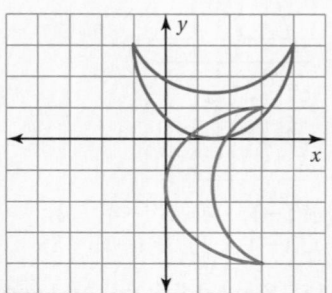

Taking Math Deeper

Exercise 19

Students can use *Guess, Check, and Revise* to find a correct sequence of transformations. This is a good question for students to discuss in pairs or groups.

 Do the rotations.

Original position

Rotate 180° about the origin.

Rotate 90° counterclockwise about the origin.

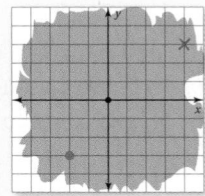
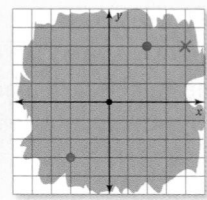
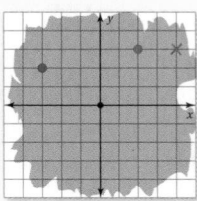

2 Do the reflection.

Position after rotations.

Reflect in *y*-axis.

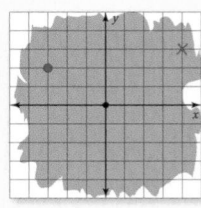
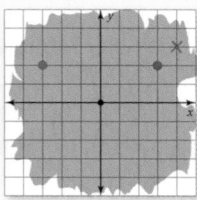

3 Do the translation.

Position after rotations and reflection.

Translate 1 unit right and 1 unit up.

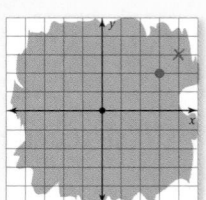
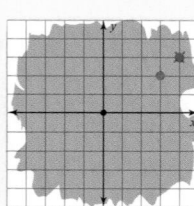

That's hard.

Reteaching and Enrichment Strategies

If students need help. . .	If students got it. . .
Resources by Chapter • Practice A and Practice B • Puzzle Time Record and Practice Journal Practice Differentiating the Lesson Lesson Tutorials Skills Review Handbook	Resources by Chapter • Enrichment and Extension • School-to-Work • Financial Literacy • Technology Connection • Life Connections Start the next section

The vertices of a parallelogram are $A(-4, 1)$, $B(-3, 4)$, $C(-1, 4)$, and $D(-2, 1)$. Rotate the parallelogram as described. Find the coordinates of the image.

②③ **13.** 90° counterclockwise about the origin

14. 270° clockwise about the origin

15. 180° clockwise about vertex D

16. 90° counterclockwise about vertex B

17. **WRITING** Why is it *not* necessary to use the words *clockwise* and *counterclockwise* when describing a rotation of 180°?

18. **DILATIONS** A *dilation* is a transformation in which a figure is enlarged or reduced.

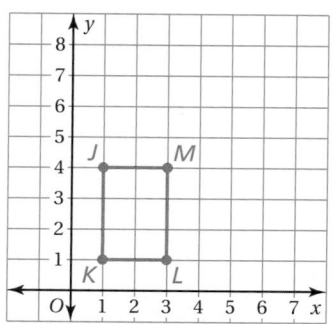

   a. Dilate Rectangle *JKLM* by multiplying the x- and y-coordinates of each vertex by 2. Compare the original figure and its image.

   b. Are the rectangles identical? Are they similar? Explain.

   c. How do dilations differ from translations, reflections, and rotations?

19. **TREASURE MAP** You want to find the treasure located on the map at ✕. You are located at ●. The following transformations will lead you to the treasure, but they are not in the correct order. Find the correct order. Use each transformation exactly once.

   ● Rotate 180° about the origin.

   ● Reflect in the y-axis.

   ● Rotate 90° counterclockwise about the origin.

   ● Translate 1 unit right and 1 unit up.

20. **Reasoning** A triangle is rotated 90° counterclockwise about the origin. Its image is translated 1 unit left and 2 units down. The vertices of the final triangle are $(-5, 0)$, $(-2, 2)$, and $(-2, -1)$. What are the vertices of the original triangle?

 **Fair Game Review** What you learned in previous grades & lessons

Identify the solid. *(Skills Review Handbook)*

21.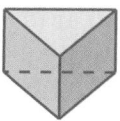

22.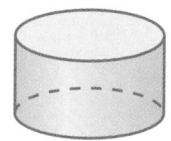

23. **MULTIPLE CHOICE** What is the value of $x - y$ when $x = -5$ and $y = -8$? *(Section 1.3)*

   Ⓐ -13 Ⓑ -3 Ⓒ 3 Ⓓ 13

1. Translate the triangle 2 units right and 3 units down. What are the coordinates of the image? *(Section 5.5)*

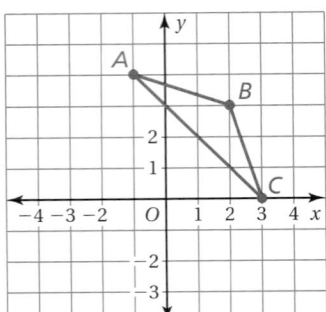

2. Translate the figure 2 units left and 4 units down. What are the coordinates of the image? *(Section 5.5)*

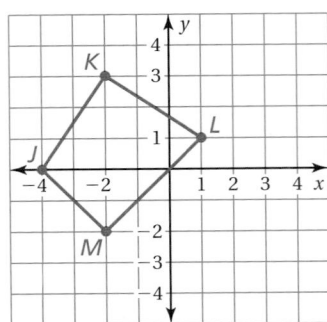

Find the coordinates of the figure after reflecting in the (a) *x*-axis and (b) *y*-axis. *(Section 5.6)*

3. $A(2, 0)$, $B(1, 5)$, $C(4, 3)$

4. $D(-2, -5)$, $E(-2, -2)$, $F(1, -2)$, $G(2, -5)$

Tell whether the blue figure is a rotation of the red figure about the origin. If so, give the angle and direction of rotation. *(Section 5.7)*

5.

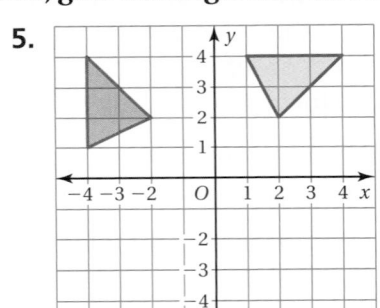

6.

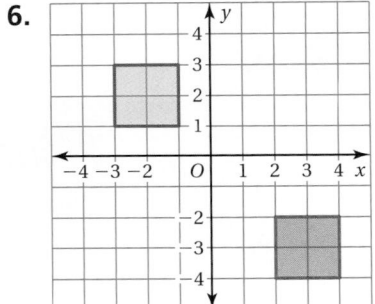

7. **AIRPLANE** Describe a translation of the airplane from point *A* to point *B*. *(Section 5.5)*

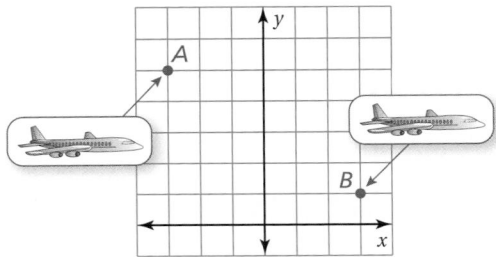

8. **MINI GOLF** You plan to hit the golf ball along the red path so that its image will be a reflection in the *y*-axis. Does the golf ball land in the hole? Explain. *(Section 5.6)*

9. **GEOMETRY** The pivot point of a compass is at the origin. A circle is drawn starting at (3, 6). What point is the compass pencil on when the compass has rotated 270° counterclockwise? *(Section 5.7)*

Alternative Assessment Options

Math Chat   Student Reflective Focus Question

Structured Interview   Writing Prompt

Math Chat

- Put students in pairs to complete and discuss the exercises from the quiz. The discussion of transformations should include comparing and contrasting translations, reflections, and rotations.
- The teacher should walk around the classroom listening to the pairs and ask questions to ensure understanding.

Study Help Sample Answers

Remind students to complete Graphic Organizers for the rest of the chapter.

7a.

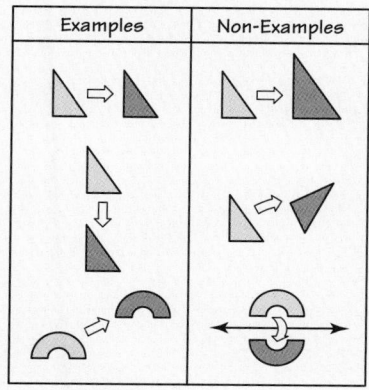

Translations

7b, c. Available at *BigIdeasMath.com*.

Reteaching and Enrichment Strategies

If students need help. . .	If students got it. . .

Resources by Chapter
  • Study Help
  • Practice A and Practice B
  • Puzzle Time
Lesson Tutorials
BigIdeasMath.com Practice Quiz
Practice from the Test Generator
	Resources by Chapter
	  • Enrichment and Extension
	  • School-to-Work
	Game Closet at *BigIdeasMath.com*
	Start the Chapter Review

Technology
For the Teacher

Answer Presentation Tool

Answers

1.

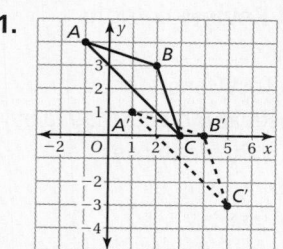

$A'(1, 1)$, $B'(4, 0)$, $C'(5, -3)$

2.

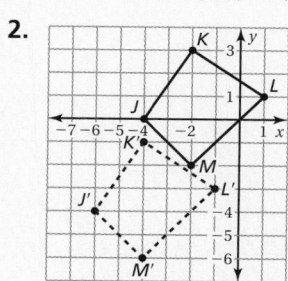

$J'(-6, -4)$, $K'(-4, -1)$,
$L'(-1, -3)$, $M'(-4, -6)$

3. a. $A'(2, 0)$, $B'(1, -5)$, $C'(4, -3)$

   b. $A'(-2, 0)$, $B'(-1, 5)$,
      $C'(-4, 3)$

4. a. $D'(-2, 5)$, $E'(-2, 2)$,
      $F'(1, 2)$, $G'(2, 5)$

   b. $D'(2, -5)$, $E'(2, -2)$,
      $F'(-1, -2)$, $G'(-2, -5)$

5. yes; 90° clockwise

6. no

7. 6 units right and 4 units down

8. no; It will be 1 unit to the right of the hole.

9. $(6, -3)$

Assessment Book

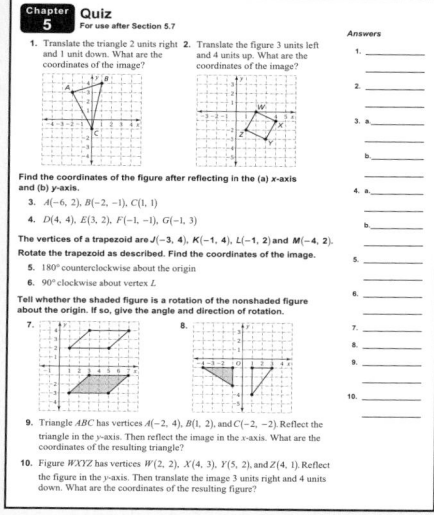

T-238

Review of Common Errors

Exercises 1–4

• Students may think that because the side lengths have changed by the same amount, the figures are similar. Remind them that the *ratios* of corresponding side lengths must be the same for the two objects to be similar.

Answers

1. no; The lengths of corresponding sides are not proportional.

2. yes; The lengths of corresponding sides are proportional and corresponding angles have the same measure.

3. yes; The lengths of corresponding sides are proportional and corresponding angles have the same measure.

4. no; The lengths of corresponding sides are not proportional.

Check It Out
Vocabulary Help
BigIdeasMath ✓com

Review Key Vocabulary

similar figures, *p. 196*
corresponding angles, *p. 196*
corresponding sides, *p. 196*
indirect measurement, *p. 209*
scale drawing, *p. 214*
scale model, *p. 214*
scale, *p. 214*
scale factor, *p. 215*

transformation, *p. 222*
image, *p. 222*
translation, *p. 222*
reflection, *p. 228*
line of reflection, *p. 228*
rotation, *p. 234*
center of rotation, *p. 234*
angle of rotation, *p. 234*

Review Examples and Exercises

5.1 Identifying Similar Figures *(pp. 194–199)*

Is Rectangle A similar to Rectangle B?

Each figure is a rectangle. So, corresponding angles have the same measure. Check to see if corresponding side lengths are proportional.

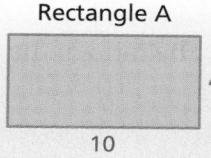
Rectangle A
Rectangle B

$$\frac{\text{Length of A}}{\text{Length of B}} = \frac{10}{5} = 2 \qquad \frac{\text{Width of A}}{\text{Width of B}} = \frac{4}{2} = 2 \qquad \text{Proportional}$$

∴ So, Rectangle A is similar to Rectangle B.

Exercises

Tell whether the two figures are similar. Explain your reasoning.

1.

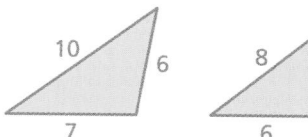

2.

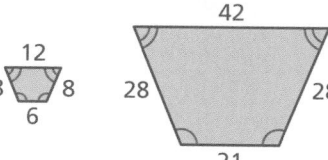

3.

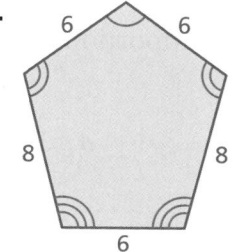

4.

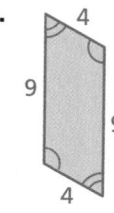
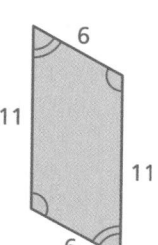

5.2 Perimeters and Areas of Similar Figures (pp. 200–205)

Find the ratio (red to blue) of the perimeters of the similar parallelograms.

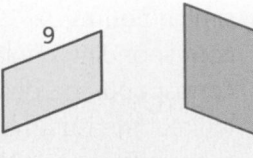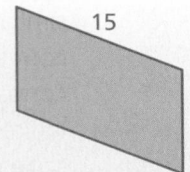

$$\frac{\text{Perimeter of red parallelogram}}{\text{Perimeter of blue parallelogram}} = \frac{15}{9}$$

$$= \frac{5}{3}$$

∴ The ratio of the perimeters is $\frac{5}{3}$.

Find the ratio (red to blue) of the areas of the similar figures.

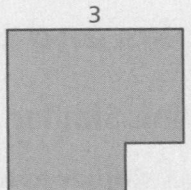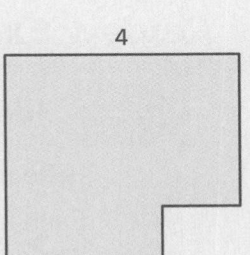

$$\frac{\text{Area of red figure}}{\text{Area of blue figure}} = \left(\frac{3}{4}\right)^2$$

$$= \frac{9}{16}$$

∴ The ratio of the areas is $\frac{9}{16}$.

Exercises

The two figures are similar. Find the ratios (red to blue) of the perimeters and of the areas.

5.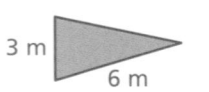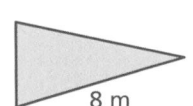

6.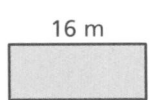

7. PHOTOS Two photos are similar. The ratio of the corresponding side lengths is $3:4$. What is the ratio of their areas?

5.3 Finding Unknown Measures in Similar Figures (pp. 206–211)

The two rectangles are similar. Find the value of x.

Corresponding side lengths of similar figures are proportional.
So, use a proportion to find x.

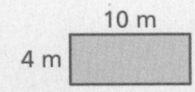

$$\frac{10}{24} = \frac{4}{x} \qquad \text{Write a proportion.}$$

$$10x = 96 \qquad \text{Use Cross Products Property.}$$

$$x = 9.6 \qquad \text{Divide each side by 10.}$$

∴ So, x is 9.6 meters.

Review of Common Errors (continued)

Exercises 5–7

- Students may find the reciprocal of the ratio. For example, they may find the ratio of blue to red instead of red to blue.
- When finding the ratio of the areas, students may forget to square the ratio. Remind them of the Key Idea.

8. 10 in.

9. 9 cm

10. 6

11. 8.32

12. 4 cm; 1 cm : 48 ft

13. 6 cm; 1 cm : 5 in.

Review of Common Errors (continued)

Exercises 8–11, 12, and 13

- Students may write the proportion incorrectly. For example, for Exercise 8 they may write $\frac{14}{20} = \frac{x}{7}$ instead of $\frac{14}{20} = \frac{7}{x}$. Remind them that the corresponding sides should both be in the numerator or the denominator OR the side lengths of the larger shape or smaller shape should be in the numerators, and the corresponding sides of the other shape should be in the denominators.

Exercises 14 and 15

- Students may translate the shape in the wrong direction or mix up the units for the translation.

Exercises 16 and 17

- Some students may struggle with the visual and think that a translation is actually a reflection. Give them tracing paper to trace the objects and then fold the paper to see if the vertices line up.

Exercises 18–21

- Students with lower spatial skills may not be able to tell whether a figure is rotated or how to rotate a figure. Give them tracing paper and have them copy the red figure and rotate it.

Exercises

The polygons are similar. Find the value of _x_.

8.

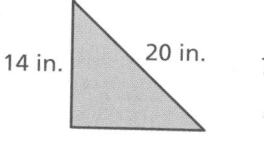 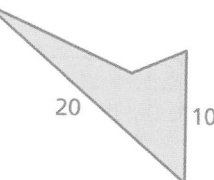

14 in. 20 in. 7 in. _x_

9.

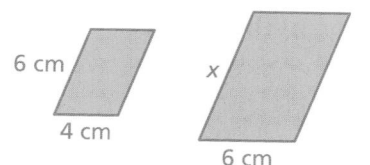

6 cm 4 cm _x_ 6 cm

10.

20 10 12 _x_

11.

26 8 _x_ 25

5.4   **Scale Drawings**   *(pp. 212–217)*

A lighthouse is 160 feet tall. A scale model of the lighthouse has a scale of 1 in. : 8 ft. How tall is the model of the lighthouse?

$$\frac{1 \text{ in.}}{8 \text{ ft}} = \frac{x \text{ in.}}{160 \text{ ft}}$$

model height ←
actual height ←

$$\frac{1 \text{ in.}}{8 \text{ ft}} \cdot 160 \text{ ft} = \frac{x \text{ in.}}{160 \text{ ft}} \cdot 160 \text{ ft} \qquad \text{Multiply each side by 160 ft.}$$

$$20 = x \qquad\qquad\qquad \text{Simplify.}$$

∴ The model of the lighthouse is 20 inches tall.

Exercises

Use a centimeter ruler to measure the segment shown. Find the scale of the drawing.

12.

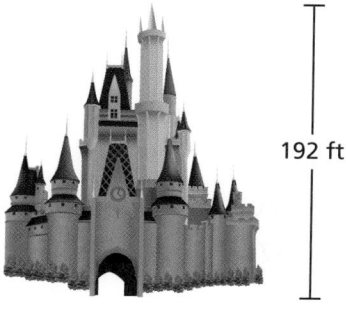

192 ft

13.

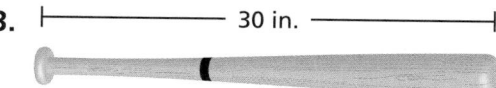

30 in.

5.5 Translations (pp. 220–225)

Translate the red triangle 4 units left and 1 unit down. What are the coordinates of the image?

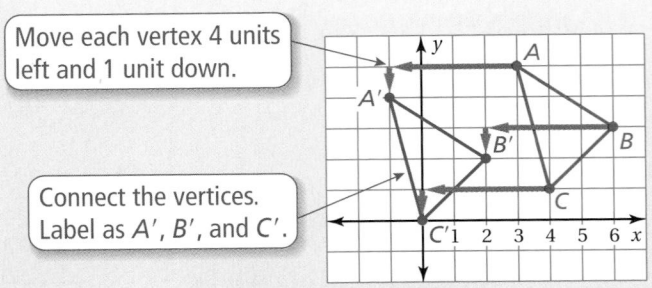

Move each vertex 4 units left and 1 unit down.

Connect the vertices. Label as A', B', and C'.

The coordinates of the image are $A'(-1, 4)$, $B'(2, 2)$, and $C'(0, 0)$.

Exercises

Translate the figure as described. What are the coordinates of the image?

14. 3 units left and 2 units down

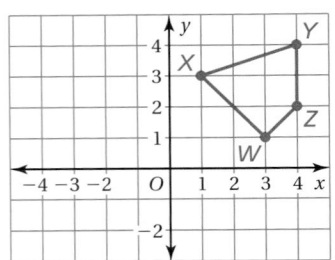

15. 5 units right and 4 units up

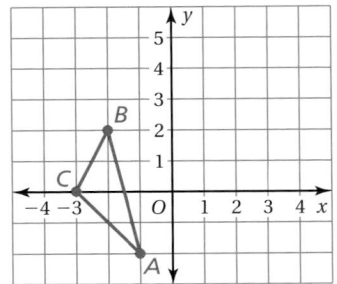

5.6 Reflections (pp. 226–231)

Tell whether the blue figure is a reflection of the red figure.

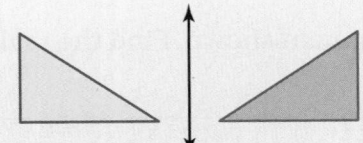

The red figure can be *flipped* to form the blue figure.

So, the blue figure is a reflection of the red figure.

Exercises

Tell whether the blue figure is a reflection of the red figure.

16.

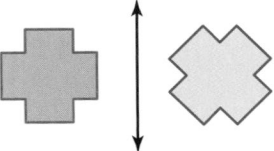

17.

Review Game

Similar Shapes

Big Ideas
Game Closet

For the Student
Additional Practice
- Lesson Tutorials
- Study Help (textbook)
- Student Website
    Multi-Language Glossary
    Practice Assessments

Materials per Group

- a variety of measuring devices (ruler, yardstick, tape measure, etc.)

Directions

Groups can consist of as many students as desired. Each group chooses two similar-looking objects in the classroom, announcing the objects as they are chosen. Nothing is off limits except objects that have been chosen by other groups. The groups measure their objects and decide whether they are similar. Groups can change objects in the middle of the game if necessary. Groups report their findings to the class with proof.

If the game is repeated, objects cannot be reused. The idea is to make the game progressively more difficult. If necessary, objects from outside the classroom (such as in the gymnasium, cafeteria, playground, etc.) can be used.

Who Wins?

Each time a group finds a pair of similar objects, they receive a point. The group with the most points at the end of the game wins.

Answers

14. $W'(0, -1)$, $X'(-2, 1)$, $Y'(1, 2)$, $Z'(1, 0)$

15. $A'(4, 2)$, $B'(3, 6)$, $C'(2, 4)$

16. no

17. yes

18. no

19. yes; 180° counterclockwise or clockwise

20. $A'(1, 2)$, $B'(3, 1)$, $C'(4, 3)$, $D'(2, 4)$

21. $L'(2, 3)$, $M'(0, 2)$, $N'(2, 5)$

My Thoughts on the Chapter

What worked. . .

Teacher Tip

Not allowed to write in your teaching edition? Use sticky notes to record your thoughts.

What did not work. . .

What I would do differently. . .

5.7 **Rotations** *(pp. 232–237)*

Tell whether the blue figure is a rotation of the red figure about a vertex. If so, give the angle and direction of rotation.

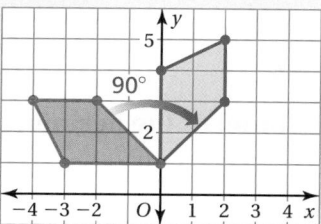

The red figure can be turned 90° clockwise about (0, 1) to form the blue figure.

So, the blue figure is a 90° clockwise rotation of the red figure.

Rotate the red triangle 90° counterclockwise about the origin. What are the coordinates of the image?

Plot *A′*, *B′*, and *C′*. Connect the vertices.

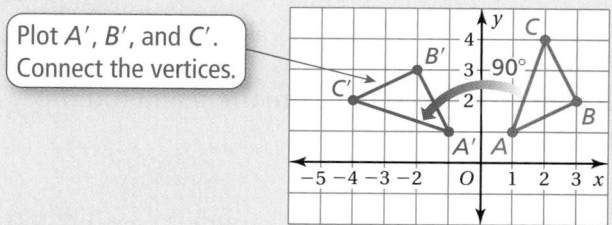

The coordinates of the image are *A′*(−1, 1), *B′*(−2, 3), and *C′*(−4, 2).

Exercises

Tell whether the blue figure is a rotation of the red figure about the origin. If so, give the angle and direction of rotation.

18.

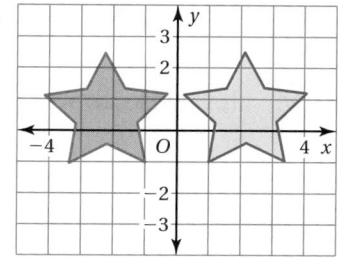

19.

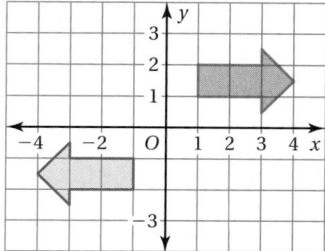

Rotate the figure as described. What are the coordinates of the image?

20. 270° counterclockwise about the origin

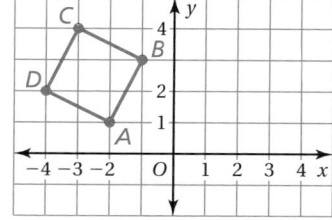

21. 180° clockwise about vertex *M*.

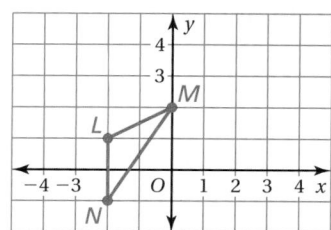

Check It Out
Test Practice
BigIdeasMath.com

1. Tell whether the parallelograms are similar. Explain your reasoning.

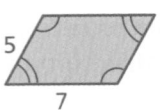
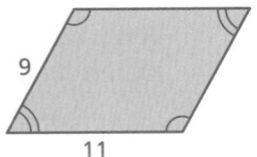

The two figures are similar. Find the ratios (red to blue) of the perimeters and of the areas.

2.

3.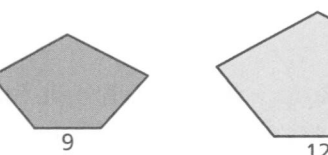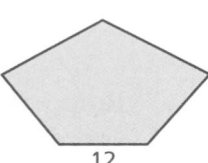

4. Use a centimeter ruler to measure the fish. Find the scale factor of the drawing.

15 mm

5. The vertices of a triangle are $A(2, 4)$, $B(2, 1)$, and $C(5, 1)$. Draw the triangle and its image after a translation of 1 unit left and 3 units down.

6. Find the coordinates of the triangle whose vertices are $A(2, 5)$, $B(1, 2)$, and $C(3, 1)$ after reflecting in (a) the x-axis and (b) the y-axis.

The vertices of a triangle are $D(-2, -2)$, $E(-1, 1)$, and $F(1, -1)$. Rotate the triangle as described. Find the coordinates of the image.

7. 180° counterclockwise about the origin

8. 90° clockwise about the vertex D

9. **SCREENS** A wide screen television measures 36 inches by 54 inches. A movie theater screen measures 42 feet by 63 feet. Are the screens similar? Explain.

10. **HOCKEY** An air hockey table and an ice hockey rink are similar. The ratio of their corresponding side lengths is 1 inch : 2 feet. What is the ratio of their areas?

11. **HEIGHT** You are five feet tall and cast a seven-foot eight-inch shadow. At the same time, a basketball hoop casts a 19-foot shadow. How tall is the basketball hoop? Assume the triangles are similar.

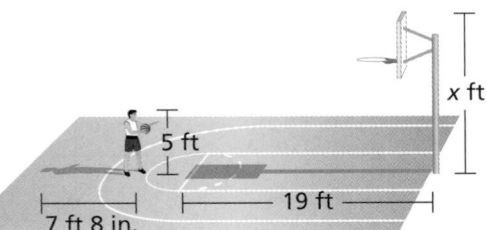

x ft

5 ft

19 ft

7 ft 8 in.

Test Item References

Chapter Test Questions	Section to Review
1, 9	5.1
2, 3, 10	5.2
11	5.3
4	5.4
5	5.5
6	5.6
7, 8	5.7

Test-Taking Strategies

Remind students to quickly look over the entire test before they start so that they can budget their time. Students need to **Stop** and **Think** as they work through the test. Remind them to be careful and read slowly so that they will set up the proportions correctly.

Common Assessment Errors

- **Exercise 1** Students may think that because the side lengths change by the same amount, the objects are similar. Remind them that the *ratios* of corresponding side lengths must be the same for the objects to be similar.
- **Exercises 2, 3, and 10** When finding the ratio of the areas, students may forget to square the ratio.
- **Exercise 6** Students may reflect in the incorrect axis. Ask them to make a chart that describes which direction to move when reflecting in each axis.

Reteaching and Enrichment Strategies

If students need help. . .	If students got it. . .

Resources by Chapter
  • Practice A and Practice B
  • Puzzle Time
Record and Practice Journal Practice
Differentiating the Lesson
Lesson Tutorials
Practice from the Test Generator
Skills Review Handbook
	Resources by Chapter
	  • Enrichment and Extension
	  • School-to-Work
	  • Financial Literacy
	  • Life Connections
	Game Closet at *BigIdeasMath.com*
	Start Standardized Test Practice

Answers

1. no; The lengths of corresponding sides are not proportional.

2. $\dfrac{7}{4}$; $\dfrac{49}{16}$ 3. $\dfrac{3}{4}$; $\dfrac{9}{16}$

4. 5 cm; 10 : 3

5.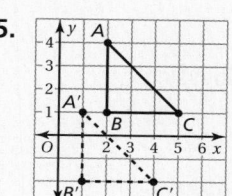

6. a. $A'(2, -5)$, $B'(1, -2)$, $C'(3, -1)$

   b. $A'(-2, 5)$, $B'(-1, 2)$, $C'(-3, 1)$

7–8. See Additional Answers.

9. yes; Because both screens are rectangles, the corresponding angle measures are the same. Corresponding side lengths are proportional.

10. 1 in.^{2} : 4 ft^2

11. 12.39 ft

Assessment Book

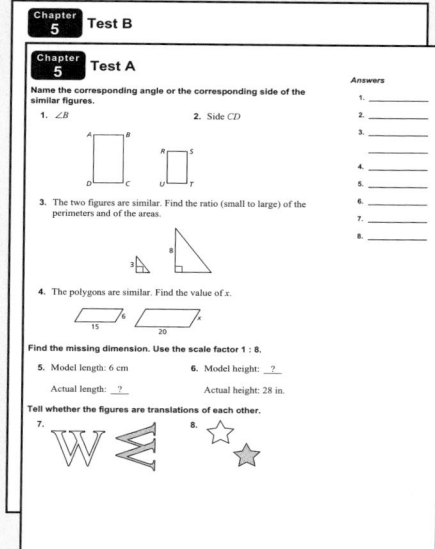

Available at *BigIdeasMath.com*

After Answering Easy Questions, Relax
Answer Easy Questions First
Estimate the Answer
Read All Choices before Answering
Read Question before Answering
Solve Directly or Eliminate Choices
Solve Problem before Looking at Choices
Use Intelligent Guessing
Work Backwards

About this Strategy

When taking a multiple choice test, be sure to read each question carefully and thoroughly. Sometimes it is easier to solve the problem and then look for the answer among the choices.

Answers

1. B
2. G
3. 270°
4. A

Item Analysis

1. **A.** The student would get the correct mean but thinks that one of the data values has to be 7 for the median to be 7.

   B. Correct answer

   C. The student would get the correct mean but would get a median of 6.

   D. The student would get the correct median but would get a mean of 6.4.

2. **F.** The student uses the diameter instead of the radius to find the area of the circle. The student then subtracts $\frac{1}{4}$ of the incorrect area of the circle from the area of the square.

   G. Correct answer

   H. The student subtracts $\frac{1}{4}$ of the area of the circle from the area of the square.

   I. The student doubles the radius instead of squaring it to find the area of the circle.

3. **Gridded Response:** Correct answer: 270°

   Common Error: The student confuses a rotation of 90° with a rotation of 180° and thinks that a 90° clockwise rotation has the same result as a 90° counterclockwise rotation and gets an answer of 90°.

4. **A.** Correct answer

   B. The student computes the area in square meters but then uses the given scale factor, which is for length, not area.

   C. The student reverses the relationship between the actual park and the scale model. The student also computes the area in square meters but then uses the given scale factor, which is for length, not area.

   D. The student reverses the relationship between the actual park and the scale model.

Technology For the Teacher
Big Ideas Test Generator

1. A set of data is shown below. Two of the data are missing.

    8, 2, 10, 4, 8, 4, 8, 8, ____ , ____

    The mean of the complete set of data is 6, and the median is 7. What are the two missing data?

    A. 1 and 7

    B. 2 and 6

    C. 4 and 4

    D. 6 and 6

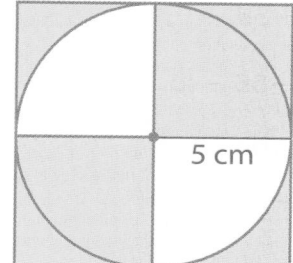

2. What is the area of the shaded region in the figure below? (Use 3.14 for π.)

    5 cm

    F. 21.5 cm²

    G. 60.75 cm²

    H. 80.375 cm²

    I. 84.3 cm²

3. A clockwise rotation of 90° is equivalent to a counterclockwise rotation of how many degrees?

4. You are building a scale model of a park that is planned for a city. The model uses the scale below.

    1 centimeter = 2 meters

    The park will have a rectangular reflecting pool with a length of 20 meters and a width of 12 meters. In your scale model, what will be the area of the reflecting pool?

    A. 60 cm²

    B. 120 cm²

    C. 480 cm²

    D. 960 cm²

5. In the figure, $\triangle EFG \sim \triangle HIJ$.

   Which proportion is *not* necessarily correct for $\triangle EFG$ and $\triangle HIJ$?

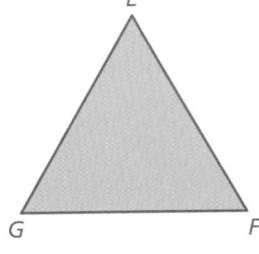

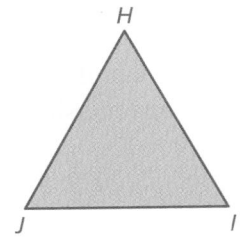

   F. $\dfrac{EF}{FG} = \dfrac{HI}{IJ}$

   H. $\dfrac{GE}{EF} = \dfrac{JH}{HI}$

   G. $\dfrac{EG}{HI} = \dfrac{FG}{IJ}$

   I. $\dfrac{EF}{HI} = \dfrac{GE}{JH}$

6. Brett was solving the equation in the box below.

   $$\frac{c}{5} - (-15) = -35$$

   $$\frac{c}{5} + 15 = -35$$

   $$\frac{c}{5} + 15 - 15 = -35 - 15$$

   $$\frac{c}{5} = -50$$

   $$\frac{c}{5} = \frac{-50}{5}$$

   $$c = -10$$

   What should Brett do to correct the error that he made?

   A. Subtract 15 from -35 to get -20.

   B. Rewrite $\dfrac{c}{5} - (-15)$ as $\dfrac{c}{5} - 15$.

   C. Multiply both sides of the equation by 5 to get $c = -250$.

   D. Multiply both sides of the equation by -5 to get $c = 250$.

7. In the figure below, $\triangle ABC \sim \triangle DEF$.

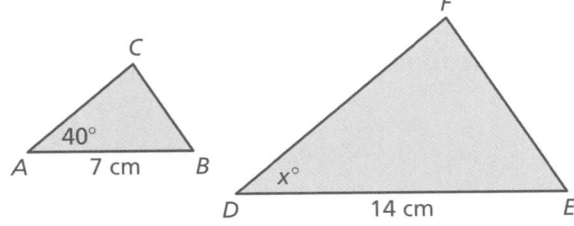

   What is the value of x?

Item Analysis (continued)

5. **F.** The student chooses a proportion that correctly represents a relationship between pairs of corresponding sides of the triangles.

   G. Correct answer

   H. The student chooses a proportion that correctly represents a relationship between pairs of corresponding sides of the triangles.

   I. The student chooses a proportion that correctly represents a relationship between pairs of corresponding sides of the triangles.

6. **A.** The student subtracts incorrectly. The original difference of -50 is correct.

   B. The student subtracts incorrectly. The original expression, $\frac{c}{5} + 15$, is correct.

   C. Correct answer

   D. The student identifies the operation error but then thinks that division by a number is undone by multiplication by the opposite of the number.

7. **Gridded Response:** Correct answer: 40

   Common Error: The student applies the ratio of the corresponding sides to the corresponding angles, getting an answer of 80.

8. **F.** Correct answer

   G. The student chooses an answer based only on visual approximation.

   H. The student thinks that because 12 is 4 more than 8, x should be 4 more than 12.

   I. The student thinks that because 8 is 4 less than 12, x should be 4 less than 21.

9. **A.** The student finds only the number of smaller cubes along one edge of the larger cube.

   B. The student finds only the number of smaller cubes along one face of the larger cube.

   C. Correct answer

   D. The student finds the number of smaller cubes along all six faces of the larger cube, without regard to the cubes that would be counted multiple times.

Answers

8. F

9. C

10. *Part A* 90 miles

   Part B $3\frac{1}{4}$ inches

Answers for Extra Examples

1. **A.** Correct answer

   B. The student uses the numerator and denominator as digits in the decimal part.

   C. The student finds the reciprocal of a fraction that is equivalent to the given mixed number.

   D. The student improperly applies a percent sign to the decimal equivalent of the mixed number.

2. **F.** Correct answer

   G. The student chooses the familiar 3–4–5 triangle.

   H. The student subtracts 1 from the length of each side of the given triangle.

   I. The student adds 1 to the length of each side of the given triangle.

Item Analysis (continued)

10. **2 points** The student demonstrates a thorough understanding of working with scale drawings. In Part A, the student correctly determines that the actual distance is 90 miles. In Part B, the student correctly determines that the distance on the map should be $3\frac{1}{4}$ inches. The student provides clear and complete work and explanations.

   1 point The student demonstrates a partial understanding of working with scale drawings. The student provides some correct work and explanation.

   0 points The student demonstrates insufficient understanding of working with scale drawings. The student is unable to make any meaningful progress toward a correct answer.

Extra Examples for Standardized Test Practice

1. Which of the following is equivalent to $1\frac{2}{5}$?

   A. 1.4

   B. 1.25

   C. $\frac{5}{7}$

   D. 1.4%

2. Which of the following triangles is similar to the triangle below?

   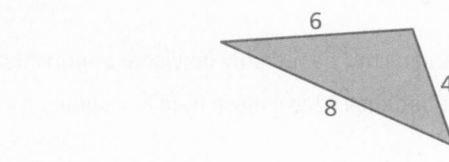

   F.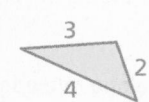

   G.

   H.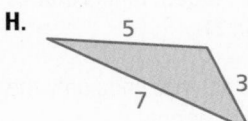

   I.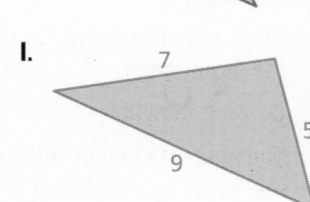

8. In the figure below, rectangle *EFGH* ~ rectangle *IJKL*.

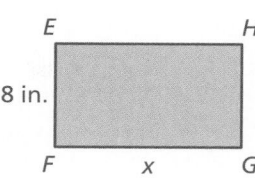
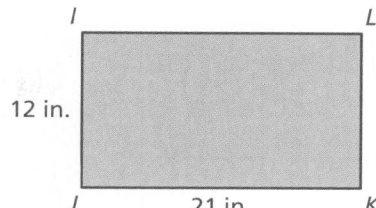

What is the value of *x*?

F. 14 in.

G. 15 in.

H. 16 in.

I. 17 in.

9. Two cubes are shown below.

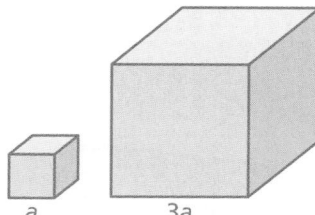

How many of the smaller cubes can be stacked to completely fill the larger cube?

A. 3

B. 9

C. 27

D. 54

10. A map of Donna's state has the following scale:

$$\frac{1}{2} \text{ inch} = 10 \text{ miles}$$

Part A   Donna measured the distances between her town and the state capitol on the map. Her measurement was $4\frac{1}{2}$ inches. Based on Donna's measurement, what is the actual distance, in miles, between her town and the state capitol? Show your work and explain your reasoning.

Part B   Donna wants to mark her favorite campsite on the map. She knows that the campsite is 65 miles north of her town. What distance, in inches, on the map represents an actual distance of 65 miles? Show your work and explain your reasoning.

6 Surface Areas of Solids

"I want to paint my dog house. To make sure I buy the correct amount of paint, I want to calculate the lateral surface area."

"Then, because I want to paint the inside and the outside, I will multiply by 2. Does this seem right to you?"

"Dear Sir: Why do you sell dog food in tall cans and sell cat food in short cans?"

"Neither of these shapes is the optimal use of surface area when compared to volume."

Connections to Previous Learning

- Distinguish between two-dimensional figures and three-dimensional solids, including the number of edges, faces, and vertices.
- Identify and build three dimensional solids from two dimensional figures and vice versa.
- Find the surface area of prisms without formulas.

- Find the area of a plane figure.
- Find perimeters and areas of composite two-dimensional figures, including semi-circles.

- Justify and apply formulas for the surface area of pyramids, prisms, cylinders, and cones.
- Use formulas to find the surface area of three dimensional composite solids.

Math in History

There are infinitely many regular polygons. However, there are only five possible regular polyhedrons (a convex polyhedron whose vertices are all congruent and whose sides are congruent regular polygons).

★ Although named after the Greek mathematician Plato, the Platonic solids have been known since antiquity. Models of them can be found among the carved stone balls created by the late Neolithic people of Scotland at least 1000 years before Plato.

★ The ancient Greeks studied the Platonic solids extensively. Some sources credit Pythagoras with their discovery. Other evidence suggests he may have only been familiar with the tetrahedron, cube, and dodecahedron, and that the discovery of the octahedron and icosahedron belong to Theaetetus, a contemporary of Plato.

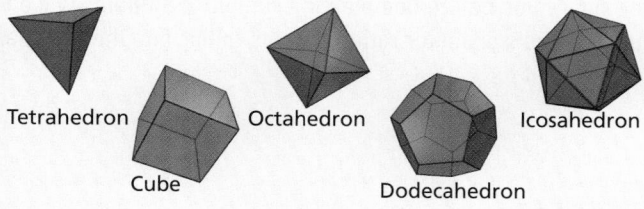

Tetrahedron   Octahedron   Icosahedron

Cube   Dodecahedron

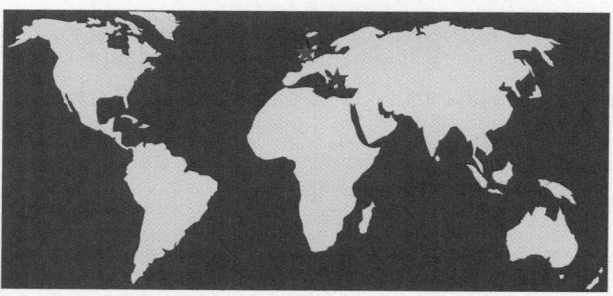

Pacing Guide for Chapter 6

Chapter Opener	1 Day
Section 1 Activity Lesson	 1 Day 1 Day
Section 2 Activity Lesson Lesson b	 1 Day 1 Day 1 Day
Section 3 Activity Lesson	 1 Day 1 Day
Study Help / Quiz	1 Day
Section 4 Activity Lesson	 1 Day 1 Day
Section 5 Activity Lesson	 1 Day 1 Day
Section 6 Activity Lesson	 1 Day 2 Days
Quiz / Chapter Review	1 Day
Chapter Test	1 Day
Standardized Test Practice	1 Day
Total Chapter 6	19 Days
Year-to-Date	113 Days

Check Your Resources

- Record and Practice Journal
- Resources by Chapter
- Skills Review Handbook
- Assessment Book
- Worked-Out Solutions

The Dynamic Planning Tool
Editable Teacher's Resources at
BigIdeasMath.com

Additional Topics for Review

- Identifying polygons
- Basic area formulas (square, rectangle, triangle, parallelogram, trapezoid, etc.)
- Exponents
- Finding surface area of prisms without formulas
- Faces, edges, and vertices

Try It Yourself

1. about 145.12 m^2
2. 86 cm^2
3. about 78.5 ft^2
4. about 530.66 in.^{2}
5. about 38.465 cm^2

Record and Practice Journal

1. 51 m^2
2. about 146.93 m^2
3. 74 in.^{2}
4. 171 in.^{2}
5. 81 ft^2
6. 88 in.^{2}
7. $444
8. about 314 in.^{2}
9. about 113.04 m^2
10. about 452.16 cm^2
11. about 153.86 ft^2
12. about 490.625 yd^2
13. about 706.5 mm^2
14. about 502.4 cm^2

Vocabulary Review

- Area
- Composite figures
- Pi

Finding the Area of a Composite Figure

- Students should be able to compute areas of composite figures.
- Remind students to identify the basic figures contained in the composite figure before they consider the area.
- Remind students that to find the area of a composite figure, all they need do is sum the areas of the basic figures together.
- **Teaching Tip:** Sometimes students find it helpful to "break up" the composite figure. For instance, rather than working with the composite figure in Example 1, have students draw the triangle separately from the square and mark the dimensions on each figure. Ask students to find the area of each figure and then sum these quantities to determine the area of the composite figure.
- **Common Error:** Students will often think that the problem does not provide enough information to be solved. In Example 1, some students may think they have not been given the base of the triangle. Try to help students see that the basic shapes contained in the figure are just as important as how the shapes fit together. Because the base of the triangle stretches the same length as the top of the square, the base must measure 10 inches.

Finding the Area of Circles

- Students should be able to compute areas of circles.
- You may wish to review the concept of pi with students. Pi is the ratio of a circle's circumference (perimeter) to its diameter. This ratio is constant regardless of the size of the circle. As a result of its frequent appearance in mathematics, the symbol π is used to represent the ratio. Students should be familiar with using 3.14 as an approximate value of pi.
- **Common Error:** You may want to review the relationship between a circle's diameter and radius before completing Example 3. Students will often substitute a circle's diameter rather than its radius into the formula.

Reteaching and Enrichment Strategies

If students need help. . .	If students got it. . .
Record and Practice Journal • Fair Game Review Skills Review Handbook Lesson Tutorials	Game Closet at *BigIdeasMath.com* Start the next section

What You Learned Before

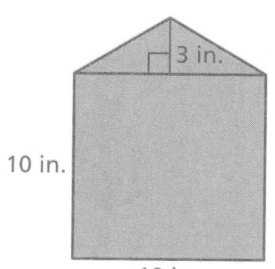

"Name these shapes."

Polly Prism, Prissy Pyramid, Cici Cylinder, and Connie Cone

Finding the Area of a Composite Figure

Example 1 Find the area.

Area = Area of square + Area of triangle

$$A = s^2 + \frac{1}{2}bh$$

$$= 10^2 + \left(\frac{1}{2} \cdot 10 \cdot 3\right)$$

$$= 100 + 15$$

$$= 115 \text{ in.}^2$$

Try It Yourself

Find the area.

1.

8 m
15 m

2.

9 cm
4 cm
14 cm
5 cm

Finding the Area of Circles

Example 2 Find the area.

7 mm

$$A = \pi r^2$$

$$\approx 3.14(7)^2$$

$$= 3.14 \cdot 49$$

$$= 153.86 \text{ mm}^2$$

Example 3 Find the area.

24 yd

$$A = \pi r^2$$

$$\approx 3.14(12)^2$$

$$= 3.14 \cdot 144$$

$$= 452.16 \text{ yd}^2$$

Try It Yourself

Find the area.

3.

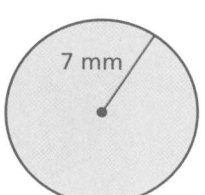

5 ft

4.

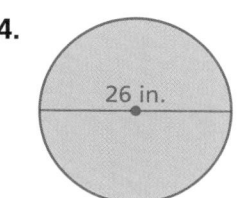

26 in.

5.

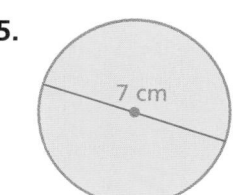

7 cm

Essential Question How can you draw three-dimensional figures?

Dot paper can help you draw three-dimensional figures, or solids. Shading parallel sides the same color helps create a three-dimensional illusion.

Square Dot Paper   **Isometric Dot Paper**

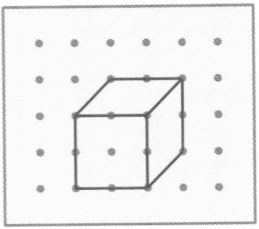   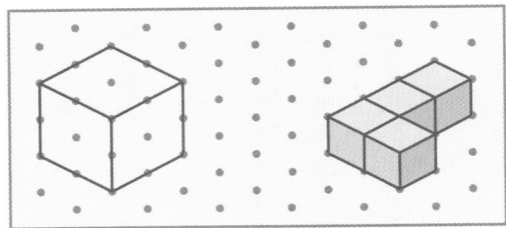

Face-On View   **Corner View**

1 ACTIVITY: Finding Surface Areas and Volumes

Work with a partner.

Draw the front, side, and top views of each stack of cubes. Then find the surface area and volume. Each small cube has side lengths of 1 unit.

a. Sample:

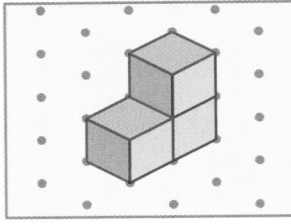   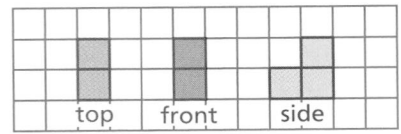

Volume: 3 cubic units

Surface Area: 14 square units

b.   **c.**   **d.**

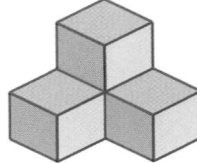      

e.   **f.**   **g.**

Laurie's Notes

Introduction

For the Teacher

- **Goal:** Students will use cubes and isometric dot paper to investigate the surface area and volume of solids.

Motivate

- Place a cube-shaped tissue box on a desk. Ask students to describe what they see when standing directly in front of the cube (front view), to the side of the cube (side view), and looking down on the cube (top view).
- Now place a rectangular prism (a shoe box) on the same surface. Describe all three views.
- Place a cube on top of a rectangular prism creating a solid that might be similar to the one shown in part (a) of Activity 1.
- Describe all three views. Students are challenged to ignore the difference in depth when describing a solid from one of the view points. They should focus on the surface that they see, not the depth of the solid. This can often be confusing to students.

Activity Notes

Activity 1

- If enough cubes are available, give each pair of students 10 cubes to create the models on their desk or table. To see each view, the students need to be at "eye level" with the solid.
- **Teaching Tip:** Another way to help students think about the top view is to ask them, "If you want to paint the top, what shapes would you paint?" Two squares would be painted. Repeat this strategy for each view.
- Discuss surface area. The surface area is the painting concept. How many square faces would be painted if the entire figure was painted?
- **Connection:** Although only 3 views are drawn, a solid has 6 views. Students may discover through this activity that the surface area can be found by sketching the 3 views, counting the total square units of the 3 views, and then doubling this number to find the surface area of the solid.
- Discuss volume. How many cubes are necessary to build the figure? Students are often confused by the cubes that are not visible in the picture. Explain that parts (b), (d), (f), and (g) would fit into the corner of a room. There are cubes behind cubes that are supporting the top cubes.
- **Scaffolding:** You may want half the class to work on parts (b), (d), and (f), while the other half works on parts (c), (e), and (g).
- Generally, it is the side view that is often challenging to draw.
- Check to see if students notice a connection between the squares drawn in the three views and the surface area.
- Check to see if students understand that the volume is the same number as the number of cubes used to build the figure.

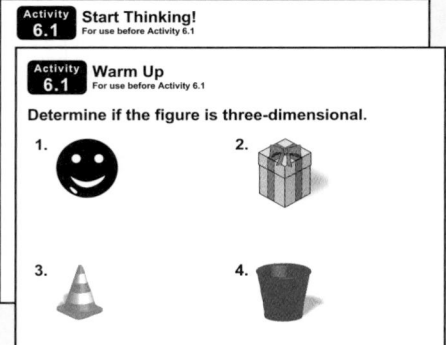
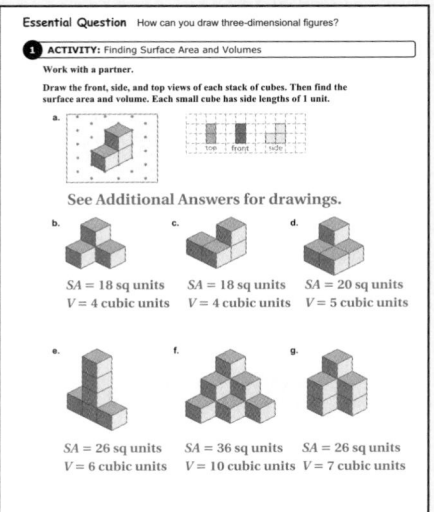

Differentiated Instruction

Kinesthetic

Provide building blocks or cubes for students to use. Students may work in pairs. Each student builds a solid out of sight of his/her partner and draws front, top, and side views. Students meet at a neutral site and trade drawings. At the building site, the student builds the solid shown in the drawing. Students get together to compare their drawings and models.

6.1 Record and Practice Journal

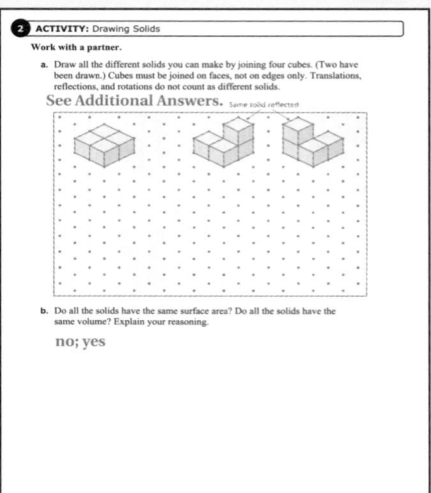

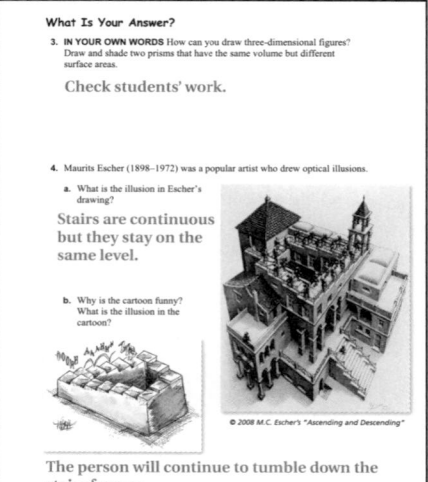

Activity 2

- If enough cubes are available, give each pair of students 4 cubes.
- If needed, suggest to students that they draw all of the possibilities where the cubes are in one layer (no stacking). There are 5 arrangements possible. Then try to stack just one cube, resulting in a solid different from the previous 5 solids.
- **Common Error:** Students forget that a solid can be turned and rotated, giving a new appearance, but it is still congruent to the original view.
- Each time a new solid is made, have students record the surface area and volume next to the sketch in order to answer part (b).
- Check to see if students recognized that all solids had a volume of 4.
- Check to see if students found a smaller surface area for the *compact* view of the $2 \times 2 \times 1$ prism shown.
- **Connection:** Students are often surprised that two figures can have the same volume, but different surface areas. This is similar to the concept of two rectangles having the same area but different perimeters.

What Is Your Answer?

- For Question 3, students should refer back to their sketches from the first activity.
- Ask if any students have heard of M.C. Escher. Even if they don't think they have heard of him, they have probably seen his work. Discuss the picture and give some background information on Escher and his artwork.

Closure

- How did the three views of a solid help you determine the surface area of the figure? Answers will vary.

Technology
For the
Teacher

Dynamic Classroom

The Dynamic Planning Tool
Editable Teacher's Resources at *BigIdeasMath.com*

Work with a partner.

a. Draw all the different solids you can make by joining four cubes. (Two have been drawn.) Cubes must be joined on faces, not on edges only. Translations, reflections, and rotations do not count as different solids.

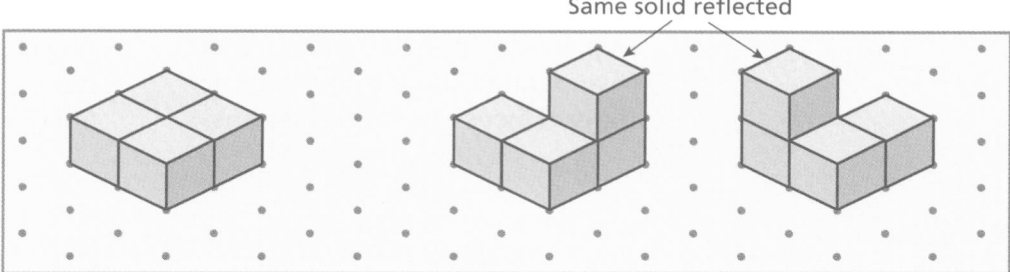

Same solid reflected

b. Do all the solids have the same surface area? Do all the solids have the same volume? Explain your reasoning.

What Is Your Answer?

3. **IN YOUR OWN WORDS** How can you draw three-dimensional figures? Draw and shade two prisms that have the same volume but different surface areas.

4. Maurits Escher (1898–1972) was a popular artist who drew optical illusions.

   a. What is the illusion in Escher's drawing?

   b. Why is the cartoon funny? What is the illusion in the cartoon?

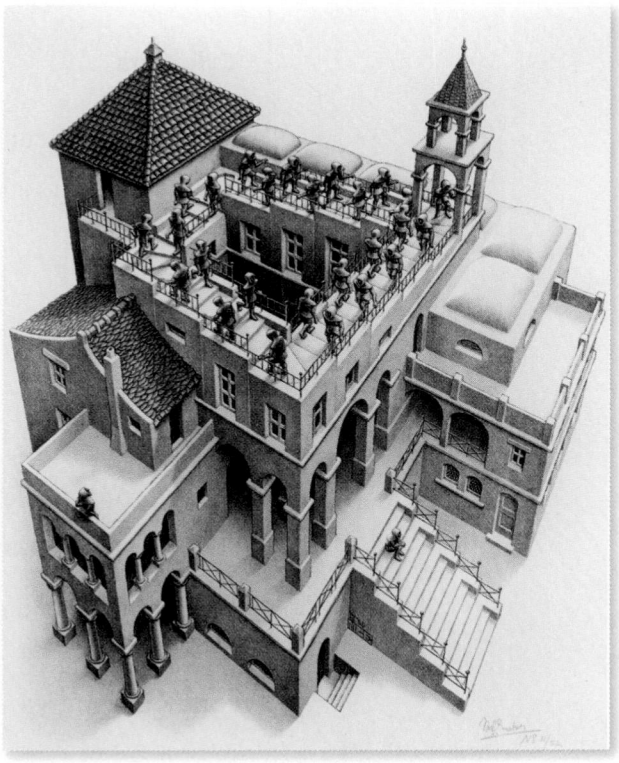

©2010 M.C. Escher's "Ascending and Descending"

Practice → Use what you learned about three-dimensional figures to complete Exercises 7–9 on page 254.

6.1 Lesson

Key Vocabulary 🔊
three-dimensional
   figure, *p. 252*
polyhedron, *p. 252*
lateral face, *p. 252*

A **three-dimensional figure**, or *solid*, has length, width, and depth.
A **polyhedron** is a three-dimensional figure whose faces are all polygons.

🔑 Key Ideas

Prisms

A prism is a polyhedron that has
two parallel, identical bases. The
lateral faces are parallelograms.

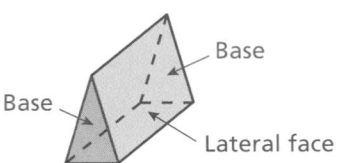

Base
Base
Lateral face

Triangular Prism

Pyramids

A pyramid is a polyhedron that
has one base. The lateral faces
are triangles.

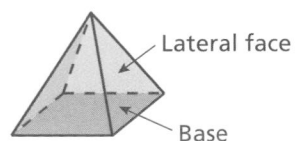

Lateral face
Base

Rectangular Pyramid

The shape of the base tells the
name of the prism or the pyramid.

Cylinders

A cylinder is a solid that has two
parallel, identical circular bases.

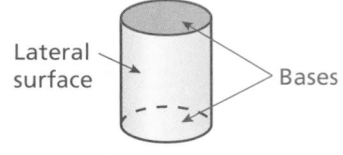

Lateral
surface
Bases

Cones

A cone is a solid that has one
circular base and one vertex.

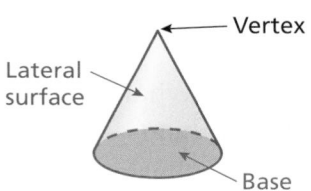

Vertex
Lateral
surface
Base

EXAMPLE ① **Drawing a Prism**

Draw a rectangular prism.

Step 1

Draw identical
rectangular bases.

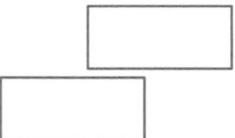

Step 2

Connect corresponding
vertices.

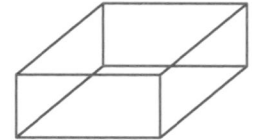

Step 3

Change any *hidden*
lines to dashed lines.

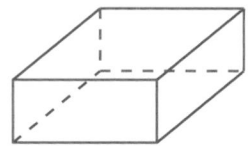

Laurie's Notes

Introduction

Connect

- **Yesterday:** Students explored surface area and volume by using cubes and isometric dot paper.
- **Today:** Students will be introduced to the vocabulary of common solids and asked to sketch several solids.

Motivate

- Collect and display solids from school or home that are examples of the four solids introduced in today's lesson: prisms, pyramids, cylinders, and cones.
- Ask a volunteer to select one solid and describe it using mathematical vocabulary. You want students to say "cylinder" instead of "can of peas." Besides naming the object, they should identify its features or attributes. They may say that the cylinder has two circles, top and bottom (circular bases), and then a round side (lateral portion).
- **?** "Are there any other solids in the collection that share the same features or attributes?" Students may pick up another cylinder, or they may pick up a cone because it has one circular face, or a prism because it has two congruent bases and a lateral portion. There is no one correct answer. Listen to students' reasoning as to what attribute(s) the second solid shares with the first.
- Repeat this process for several solids.

Lesson Notes

Key Ideas

- The point of the vocabulary is not to memorize definitions, but to have a sense as to the attributes of the solid. This will help in generalizing surface area and volume formulas later.
- Mention to students that prisms and pyramids have a qualifying name, given the type of base. A triangular prism has two bases that are triangles.
- **Common Error:** Students often think that the face that is "on the bottom" is the base. The solid does not need to be oriented so that it is resting on a base. Demonstrate this with several solids.
- It may be helpful to use vocabulary, such as vertex (vertices) and edge.

Example 1

- Explain that hidden edges are those that are not visible when looking at the solid. You can demonstrate this by holding a prism and facing it toward the students.
- **?** "What is the maximum number of faces that any student can see?" 3
- **?** "What edges are not visible?" Listen for students' descriptions. Generally speaking, it is the edges in the back of the prism.

Goal Today's lesson is sketching **three-dimensional solids** accurately.

Lesson Materials
Introduction
• real-life prisms, pyramids, cylinders, and cones

Start Thinking! and Warm Up

| Lesson 6.1 | Warm Up |
| For use before Lesson 6.1 | |

Lesson 6.1 **Start Thinking!** For use before Lesson 6.1

Explain how to find the surface area of a three-dimensional figure that is made up of cubes.

English Language Learners

Vocabulary

By this time, students should be able to tell the difference between a prism and a pyramid. Discuss the terms *base* and *face*. Explain that all the surfaces of prisms and pyramids are *faces*. Prisms have two bases that are congruent polygons and parallel to each other. Pyramids have only one face that is called the base.

Extra Example 1

Draw a pentagonal prism.

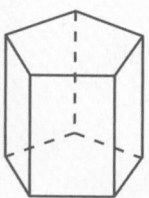

Extra Example 2

Draw a rectangular pyramid.

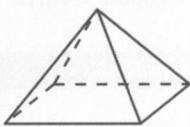

On Your Own

1.

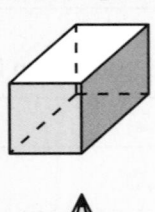

2.

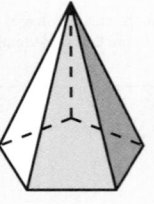

Extra Example 3

Draw the front, side, and top views of the solid.

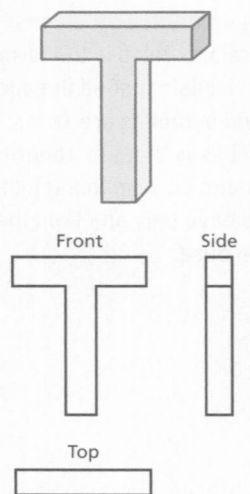

Front   Side

Top

On Your Own

3–5. See Additional Answers.

Laurie's Notes

Example 2

• Demonstrate how to sketch a triangular pyramid.
• Drawings of objects may differ based on perspective. For instance, the point in step 1 could be placed below the triangle, or different lines in step 3 could be dashed. Each would offer a different perspective of the same object.

On Your Own

• Ask volunteers to draw their sketches at the board.

Example 3

• This example connects to yesterday's activity.
• Remind students that they need to take a *bird's eye view* in each of the three directions. It is hard for students to see that the depth is not what is being drawn when the cone and pyramid slope away from the base. Students may try to draw a triangle that is sloping backward.

On Your Own

• **Think-Pair-Share:** Students should view each solid independently and then work with a partner to draw the views of the solids. When they have drawn the views, the pair should compare their drawings with another group and discuss any discrepancies.
• Have volunteers share their drawings.

Closure

• Sketch a rectangular prism. Draw the front, side, and top views.

Technology For the Teacher

Dynamic Classroom

The Dynamic Planning Tool
Editable Teacher's Resources at *BigIdeasMath.com*

EXAMPLE 2 **Drawing a Pyramid**

Draw a triangular pyramid.

Step 1

Draw a triangular base and a point.

Step 2

Connect the vertices of the triangle to the point.

Step 3

Change any *hidden* lines to dashed lines.

On Your Own

Now You're Ready
Exercises 10–15

Draw the solid.

1. Square prism

2. Pentagonal pyramid

EXAMPLE 3 **Drawing Views of a Solid**

Draw the front, side, and top views of the paper cup.

The front view is a triangle.

The side view is a triangle.

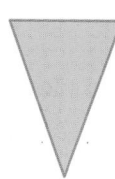

The top view is a circle.

On Your Own

Now You're Ready
Exercises 16–21

Draw the front, side, and top views of the solid.

3.

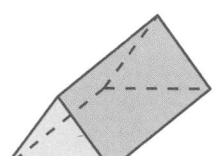

4.

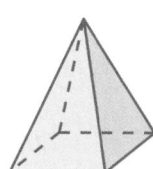

5.

6.1 Exercises

Check It Out
Help with Homework
BigIdeasMath.com

✓ Vocabulary and Concept Check

1. **VOCABULARY** Compare and contrast prisms and cylinders.

2. **VOCABULARY** Compare and contrast pyramids and cones.

3. **WRITING** Give examples of prisms, pyramids, cylinders, and cones in real life.

Identify the shape of the base. Then name the solid.

4.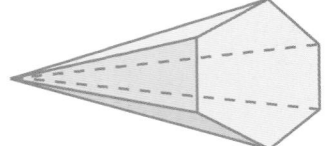

5.

6.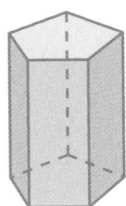

Practice and Problem Solving

Draw the front, side, and top views of the stack of cubes. Then find the surface area and volume.

7.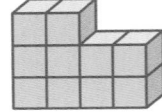

8.

9.

Draw the solid.

10. Triangular prism

11. Pentagonal prism

12. Rectangular pyramid

13. Hexagonal pyramid

14. Cone

15. Cylinder

Draw the front, side, and top views of the solid.

16.

17.

18.

19.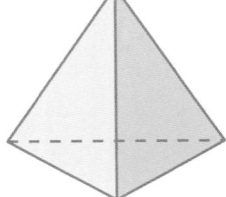

20.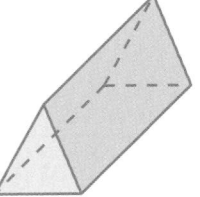

21.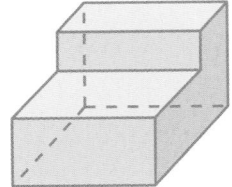

Assignment Guide and Homework Check

Level	Day 1 Activity Assignment	Day 2 Lesson Assignment	Homework Check
Basic	7–9, 29–32	1–6, 11–25 odd, 22, 24	4, 11, 19, 24
Average	7–9, 29–32	1–6, 11–25 odd, 24, 26	4, 11, 19, 24
Advanced	7–9, 29–32	1–6, 11–13, 20, 21, 24–28	4, 11, 21, 24

For Your Information

• **Exercise 22** The pyramid was built 18 B.C.–12 B.C. as a tomb for a Roman magistrate. It measures 100 Roman feet (22 meters) square at the base and is 125 Roman feet (27 meters) high.

Common Errors

• **Exercises 10–15** Students may mix up the different types of solids. Remind them of the definition of each solid and give a few real-life examples of each solid.
• **Exercises 16–21** Students may have difficulty visualizing the front, side, and top views of the solid. Create paper objects for those who are struggling to draw the different sides of the solid.
• **Exercises 24 and 25** Students may not be able to see how the shapes go together. Have them cut out pieces of paper or use blocks to model the solid.

6.1 Record and Practice Journal

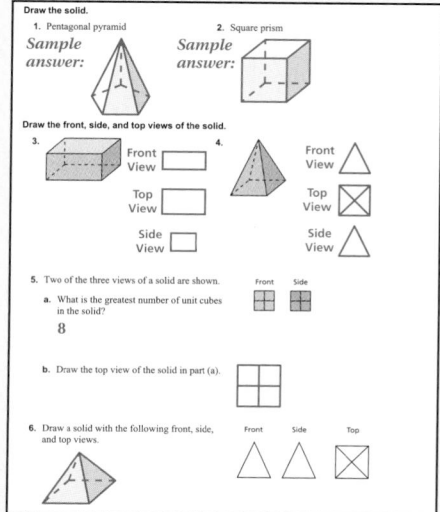

1. Prisms and cylinders both have two parallel, identical bases. The bases of a cylinder are circles. The bases of a prism are polygons. A prism has lateral faces that are parallelograms or rectangles. A cylinder has one smooth, round lateral surface.

2. Pyramids and cones both have one base and one vertex not at the base. The base of a cone is a circle. The base of a pyramid is a polygon. A pyramid has lateral faces that are triangles. A cone has one smooth lateral surface.

3–6. See Additional Answers.

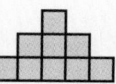 **Practice and Problem Solving**

7. front:

   side:

   top:

   surface area: 34 units^2
   volume: 10 units^3

8. front:

   side:

   top:

   surface area: 34 units^2
   volume: 9 units^3

9–21. See Additional Answers.

Practice and Problem Solving

22–25. See Additional Answers.

26. *Answer should include, but is not limited to*: an original drawing of a house; a description of any solids that make up any part of the house

27. See *Taking Math Deeper*.

28. See Additional Answers.

Fair Game Review

29. 28 m^2 **30.** 12 cm^2

31. 15 ft^2 **32.** B

Mini-Assessment

You and a friend attend a birthday party. Draw the front, side, and top views of the solid.

1.

   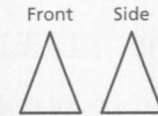

Front   Side   Top

2.

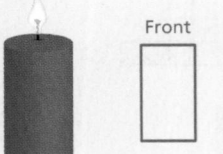

Front   Side   Top

3.

Front   Side   Top

Taking Math Deeper

Exercise 27

This type of problem begs to be touched, felt, and seen. Give students 9 cubes and ask them to construct different solids that have the given top and side views.

1 Help me see it.
Build different solids out of 9 cubes.

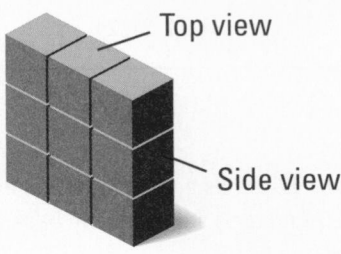

Top view

Side view

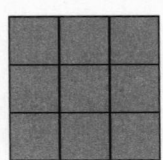

c. Front view

a. The greatest number of cubes is 9.

2 Here is a different one.

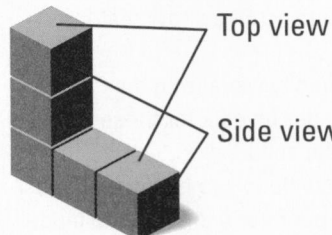

Top view

Side view

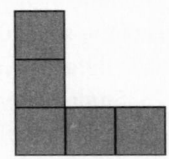

c. Front view

b. The least number of cubes is 5.

3 Here are the other possibilities (does not include reflections).

Lots of solids

Reteaching and Enrichment Strategies

If students need help. . .	If students got it. . .
Resources by Chapter • Practice A and Practice B • Puzzle Time Record and Practice Journal Practice Differentiating the Lesson Lesson Tutorials Skills Review Handbook	Resources by Chapter • Enrichment and Extension Start the next section

22. PYRAMID ARENA The Pyramid of Caius Cestius in Rome is in the shape of a square pyramid. Draw a sketch of the pyramid.

23. RESEARCH Use the Internet to find a picture of the Washington Monument. Describe its shape.

Draw a solid with the following front, side, and top views.

24.

front   side   top

25.

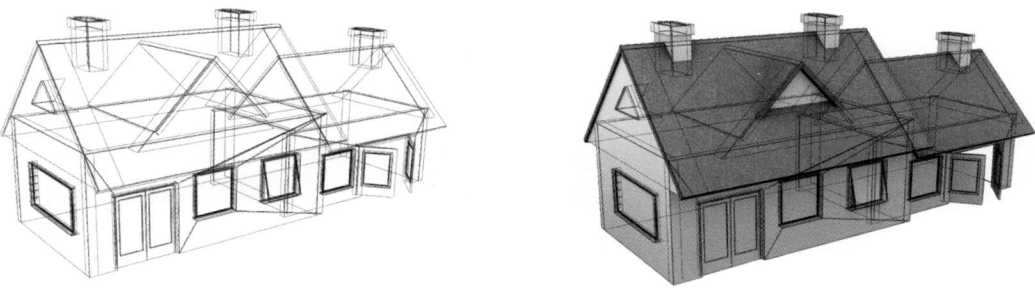

front   side   top

26. PROJECT Design and draw a house. Name the different solids that can be used to make a model of the house.

27. REASONING Two of the three views of a solid are shown.

  a. What is the greatest number of unit cubes in the solid?

  b. What is the least number of unit cubes in the solid?

  c. Draw the front views of both solids in parts (a) and (b).

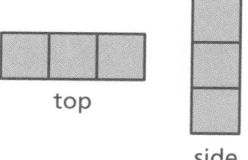

top

side

28. **Reasoning** Draw two different solids with five faces.

  a. Write the number of vertices and edges for each solid.

  b. Explain how knowing the numbers of edges and vertices helps you draw a three-dimensional figure.

Fair Game Review What you learned in previous grades & lessons

Find the area. *(Skills Review Handbook)*

29.

4 m

7 m

30.

3 cm

8 cm

31.

6 ft

3 ft

4 ft

32. MULTIPLE CHOICE You borrow $200 and agree to repay $240 at the end of 2 years. What is the simple interest rate per year? *(Section 4.4)*

  Ⓐ 5%   **Ⓑ** 10%   **Ⓒ** 15%   **Ⓓ** 20%

6.2 Surface Areas of Prisms

Essential Question How can you use a net to find the surface area of a prism?

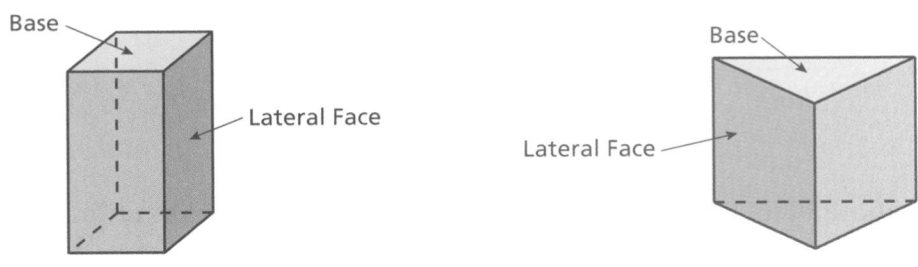

Rectangular Prism   **Triangular Prism**

The **surface area** of a prism is the sum of the areas of all its faces.
A two-dimensional representation of a solid is called a **net**.

1 ACTIVITY: Surface Area of a Right Rectangular Prism

Work with a partner.

a. Use the net for the rectangular prism to find its surface area.

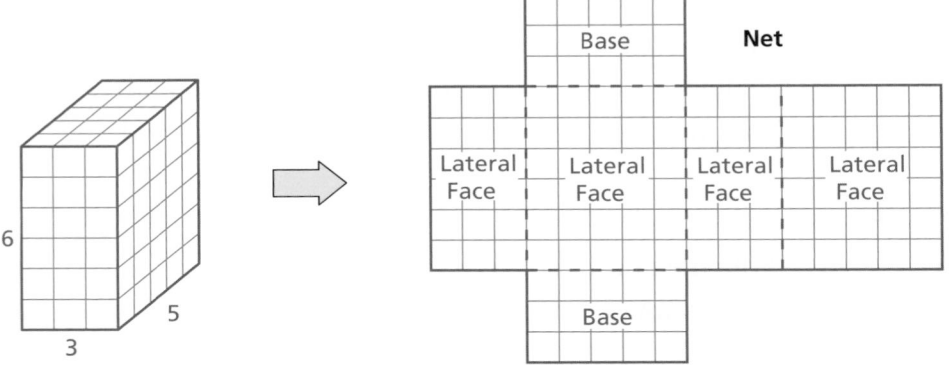

b. Copy the net for a rectangular prism. Label each side as h, w, or ℓ. Then use your drawing to write a formula for the surface area of a rectangular prism.

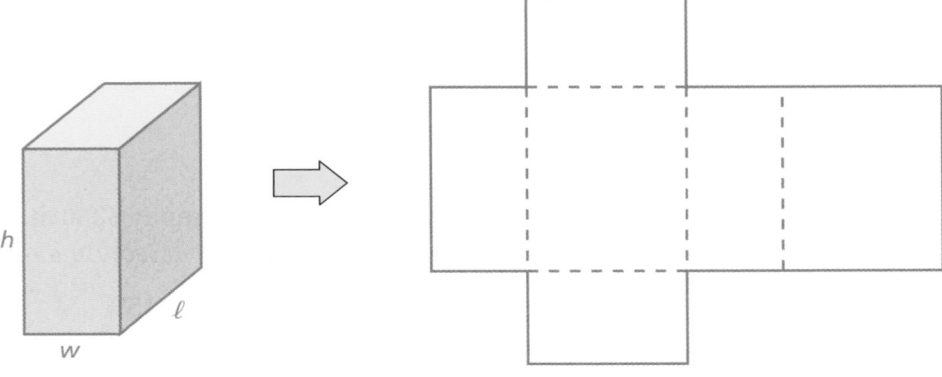

Laurie's Notes

Introduction

For the Teacher

- **Goal:** Students will develop an intuition about how to find the surface area of a prism.
- **Big Idea:** In this section and the next, you want students to see the connection between the prism and cylinder—*structurally they are the same*! The prism and cylinder each have two congruent bases and a lateral portion.
- **Teaching Tip:** It is very helpful to use snap together polygon frames to make the polyhedra. The solids can be folded and unfolded into the nets very quickly and easily.

Motivate

- Have available 2 cardboard boxes that have been folded. These are commonly used for donuts, pizza, and similar items.
- Holding the assembled boxes, ask students to visualize and then describe the cardboard net that would result if the boxes were "unfolded."
- Unfold the box and cut one of the sides so that students can see the net of the box.
- Explain the connection between the cardboard net and the surface area of the prism.

Activity Notes

Activity 1

- **?** "How many faces does a rectangular prism have and what shapes are they?" 6 faces; All are rectangles. Some students may also note that there are 3 pairs of congruent faces.
- **?** "How do you find the area of a rectangle?" length times width
- **?** "How many faces make up the lateral portion of the prism?" 4
- **?** "When the pieces of the lateral portion are unfolded, what shape do they form?" rectangle
- **?** "What are the dimensions of the lateral faces?" 6 by 16 or in the second net, h by $2\ell + 2w$
- At this point, some students may recognize that the dimensions of the rectangle that forms the lateral portion of the prism are the height of the prism and the perimeter of the base of the prism. If they do observe this, acknowledge that they are correct. Otherwise, they should see that the lateral portion is composed of two pairs of rectangles.
- **Extension:** A different, but related problem for students to investigate is to find how many different nets there are for a cube. These can be drawn on standard grid paper or investigated electronically at *illuminations.nctm.org*. There are 11 possible nets.

Previous Learning

Students should know the vocabulary of prisms.

Activity Materials	
Introduction	**Textbook**
• cardboard boxes	• polygon frames • colored pencils

Start Thinking! and Warm Up

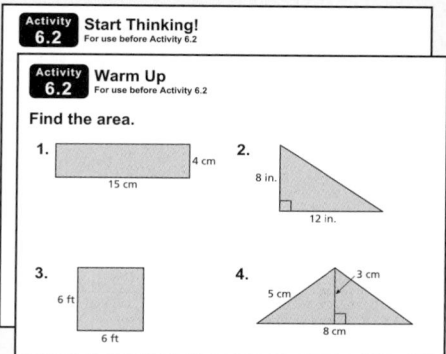

6.2 Record and Practice Journal

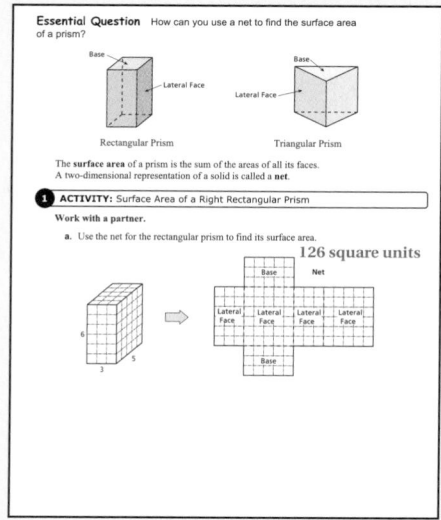

English Language Learners

Vocabulary

English learners may be familiar with the word *net* in everyday context. In finance, the net profit describes the bottom line of a financial transaction. In fishing, a net is a collection of knotted strings used to catch fish. In mathematical context, *net* or *geometric net* is used to mean the two-dimensional representation of a solid object.

Activity 2

- This activity is the same as the first activity, except the prism has a different base.
- **Teaching Tip:** Use a colored pencil to lightly color the two bases. The lateral portion that remains uncolored is a rectangle.

What Is Your Answer?

- Ask for volunteers to explain their reasoning.

Closure

- Draw the net for a pizza box. Label with approximate dimensions and find the surface area.

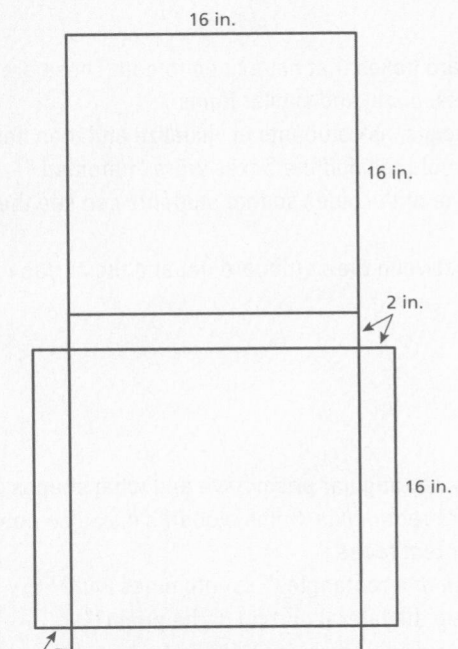

16 in.

16 in.

2 in.

16 in.

2 in.

$S = 640$ in.^{2}

6.2 Record and Practice Journal

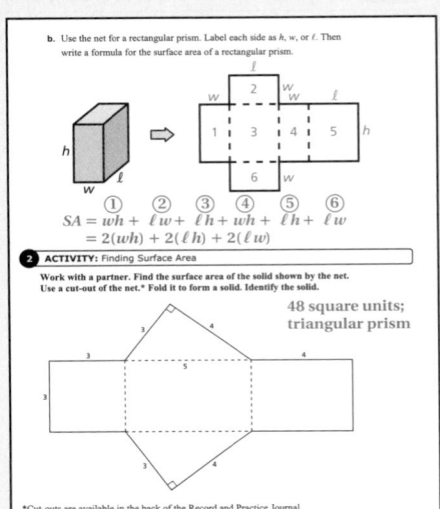

b. Use the net for a rectangular prism. Label each side as *h*, *w*, or *ℓ*. Then write a formula for the surface area of a rectangular prism.

$$SA = wh + \ell w + \ell h + wh + \ell h + \ell w$$
$$= 2(wh) + 2(\ell h) + 2(\ell w)$$

2 ACTIVITY: Finding Surface Area

Work with a partner. Find the surface area of the solid shown by the net. Use a cut-out of the net.* Fold it to form a solid. Identify the solid.

48 square units; triangular prism

*Cut-outs are available in the back of the Record and Practice Journal.

What Is Your Answer?

3. **IN YOUR OWN WORDS** How can you use a net to find the surface area of a prism? Draw a net, cut it out, and fold it to form a prism.

Find the sum of the areas of the faces shown by the net.

4. The greater the surface area of an ice block, the faster it will melt. Which will melt faster, the bigger block or the three smaller blocks? Explain your reasoning.

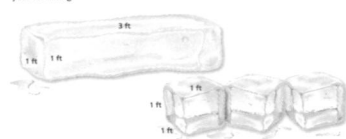

three smaller blocks

ACTIVITY: Finding Surface Area

Work with a partner. Find the surface area of the solid shown by the net. Copy the net, cut it out, and fold it to form a solid. Identify the solid.

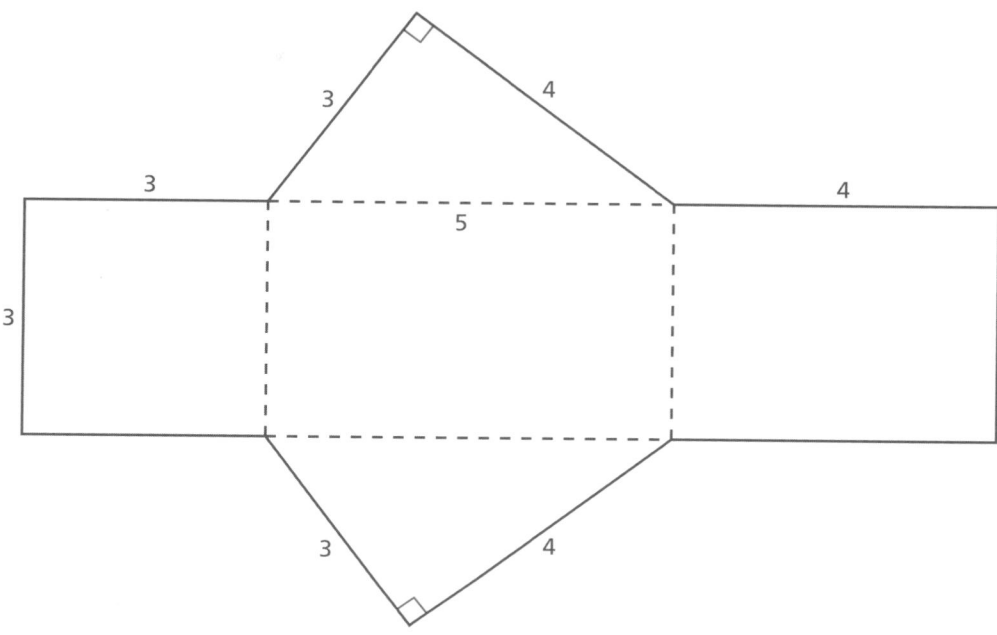

What Is Your Answer?

3. **IN YOUR OWN WORDS** How can you use a net to find the surface area of a prism? Draw a net, cut it out, and fold it to form a prism.

4. The greater the surface area of an ice block, the faster it will melt. Which will melt faster, the bigger block or the three smaller blocks? Explain your reasoning.

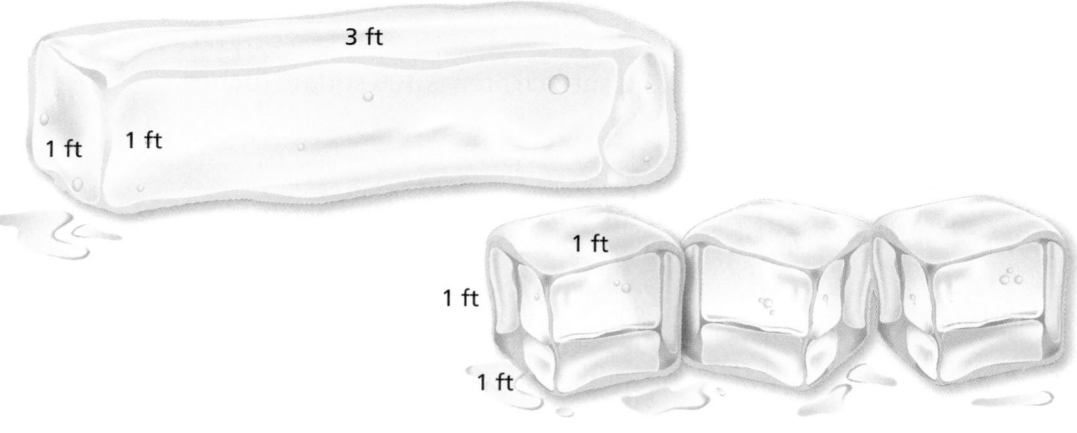

Practice Use what you learned about the surface area of a prism to complete Exercises 6–8 on page 260.

Check It Out
Lesson Tutorials
BigIdeasMath √com

Key Vocabulary 🔊
surface area, *p. 256*
net, *p. 256*

🔑 Key Idea

Surface Area of a Rectangular Prism

Words The surface area S of a rectangular prism is the sum of the areas of the bases and the lateral faces.

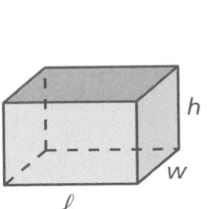

Algebra $S = 2\ell w + 2\ell h + 2wh$

Area of bases   Area of lateral faces

EXAMPLE 1 Finding the Surface Area of a Rectangular Prism

Find the surface area of the prism.

Draw a net.

$$S = 2\ell w + 2\ell h + 2wh$$

$$= 2(3)(5) + 2(3)(6) + 2(5)(6)$$

$$= 30 + 36 + 60$$

$$= 126$$

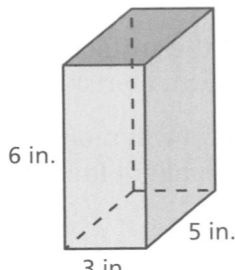

6 in.
5 in.
3 in.

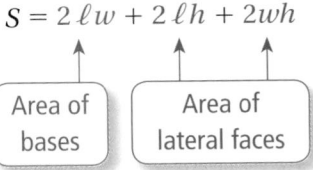

3 in.
5 in.
5 in.   5 in.   3 in.
6 in.

∴ The surface area is 126 square inches.

On Your Own

Find the surface area of the prism.

Now You're Ready
Exercises 9–11

1.

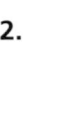

3 ft
2 ft

2.

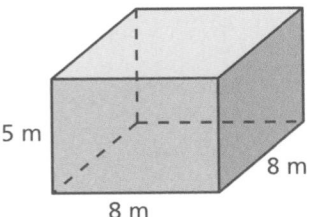

5 m
8 m
8 m
8 m

Laurie's Notes

Introduction

Connect
- **Yesterday:** Students explored surface area using the nets of two prisms.
- **Today:** Students will work with a formula for the surface area of a prism.

Motivate
- Hold a prism, perhaps a box that has a wrapping on it. It could be wrapping paper or simply clear shrink wrapping.
- **?** "Why would I need to know the surface area of this prism?" Listen for the concept of wrapping the box in some type of paper.
- Explain to students that when items are mass produced, someone needs to calculate the amount of wrapping needed to cover the prism. The wrapping is the surface area.

Lesson Notes

Key Idea
- Have a net available as a visual, whether it is a cardboard box or a rectangular prism made from the snap together polygon frames.
- **FYI:** Remind students that any two opposite faces can be called the bases. Once the bases are identified, the remaining 4 faces form the lateral portion.
- **?** "Can you explain why there are three parts to finding the surface area of the rectangular prism?" Students should recognize that the faces come in pairs, and there are 3 pairs.

Example 1
- The challenge for students is trying not to get bogged down in symbols. Students need to remember that there are 3 pairs of congruent faces. They need to make sure that they are calculating the area of each one of the 3 different faces, then doubling their answer to account for the pair.
- **Common Error:** When students multiply (2)(3)(5), they sometimes multiply $2 \times 3 \times 2 \times 5$, similar to using the Distributive Property. Remind them that multiplication is both commutative and associative, and they can multiply in order ($2 \times 3 \times 5$) or in a different order ($2 \times 5 \times 3$).
- **Teaching Tip:** Write the equation for surface area as: $S =$ bases + sides + (front and back). Students follow the words and find the area of each pair without thinking about the variables.

On Your Own
- Students need to record their work neatly so they, and you, can look back and see what corrections are needed.
- **Think-Pair-Share:** Students should read each question independently and then work with a partner to answer the questions. When they have answered the questions, the pair should compare their answers with another group and discuss any discrepancies.

Lesson Materials	
Introduction	**Textbook**
• wrapped box	• polygon frames

Start Thinking! and Warm Up

> **Lesson 6.2** Warm Up
> For use before Lesson 6.2
>
> **Lesson 6.2** Start Thinking!
> For use before Lesson 6.2
>
> Your uncle wants to build a shed that is 17 feet long by 13 feet wide by 8 feet high. Explain to your uncle how to find the surface area of the sides, front, and back so he knows how much paint to buy to paint the outside of the shed.

Extra Example 1

Find the surface area of a rectangular prism with a length of 6 yards, a width of 4 yards, and a height of 9 yards. 228 yd^2

On Your Own

1. 52 ft^2
2. 288 m^2

Extra Example 2

Find the surface area of a triangular prism.

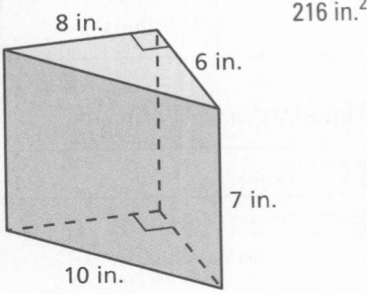

216 in.²

On Your Own

3. 150 m²

4. 60 cm²

Differentiated Instruction

Visual

Use a rectangular box to demonstrate three ways of finding surface area. The first method is to find the area of each face and then add areas. The second method is to open the box into a net and find the area of the net. The third method is to use the formula for finding surface area. Students should see that the three methods have the same result.

Key Idea

- This is the general formula for a prism without variables. Most students are comfortable with this form.
- **?** "How many faces are there that make up the bases?" two
- **?** "How many faces are there that make up the lateral faces?" It depends on how many sides the bases have.

Example 2

- Note that the net is a visual reminder of each face whose area must be found. Color coding the faces should help students keep track of their work.
- Encourage students to write the formula in words for each new problem: S = area of bases + areas of lateral faces.

On Your Own

- Give students sufficient time to do their work before asking volunteers to share their work *and* sketch at the board.
- Students having difficulty with Question 4 may want to redraw the triangular prism with the base on the bottom.

Closure

- **Writing:** Hold the prism from the Motivate section of today's lesson and ask students to write about how they would find the surface area of the prism. Look for: finding the areas of the bases and lateral faces and adding those areas together, *or* unwrapping the prism and finding the area of the wrapping paper.

The Dynamic Planning Tool
Editable Teacher's Resources at *BigIdeasMath.com*

 **Key Idea**

Surface Area of a Prism

The surface area S of a prism is the sum of the areas of the bases and the lateral faces.

$$S = \text{areas of bases} + \text{areas of lateral faces}$$

EXAMPLE ② **Finding the Surface Area of a Triangular Prism**

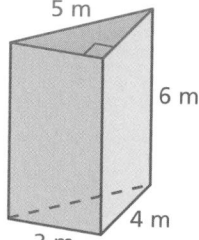

Find the surface area of the prism.

Draw a net.

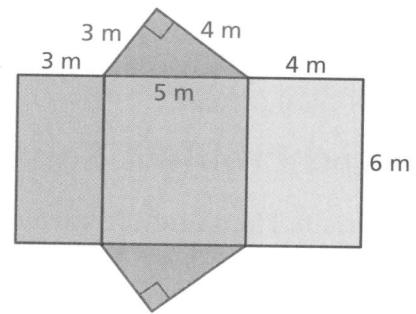

Remember

The area A of a triangle with base b and height h is $A = \frac{1}{2}bh$.

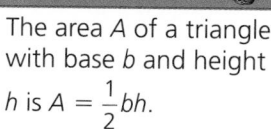

Area of a base

Red base: $\frac{1}{2} \cdot 3 \cdot 4 = 6$

Areas of lateral faces

Green lateral face: $3 \cdot 6 = 18$

Purple lateral face: $5 \cdot 6 = 30$

Blue lateral face: $4 \cdot 6 = 24$

Add the areas of the bases and the lateral faces.

$S = \text{areas of bases} + \text{areas of lateral faces}$

$= \underbrace{6 + 6} + 18 + 30 + 24$

> There are two identical bases. Count the area twice.

$= 84$

∴ The surface area is 84 square meters.

⬤ **On Your Own**

Now You're Ready
Exercises 12–14

Find the surface area of the prism.

3.
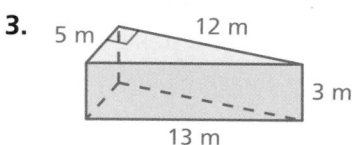

4.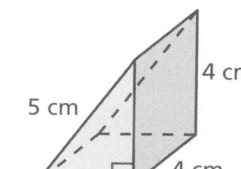

Vocabulary and Concept Check

1. **OPEN-ENDED** Describe a real-world situation in which you would want to find the surface area of a prism.

Find the indicated area for the rectangular prism.

2. Area of Face *A*

3. Area of Face *B*

4. Area of Face *C*

5. Surface area of the prism

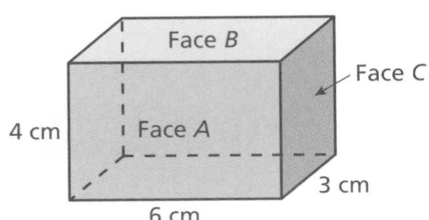

Practice and Problem Solving

Draw a net for the prism. Then find the surface area.

6.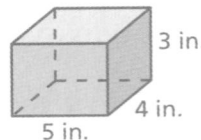

7.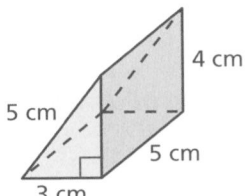

8.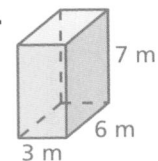

Find the surface area of the prism.

1️⃣ 9.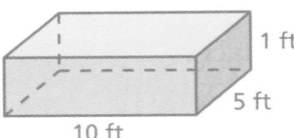

10.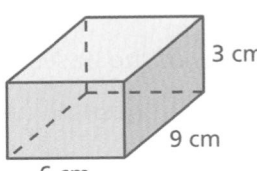

11.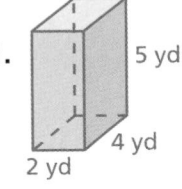

2️⃣ 12.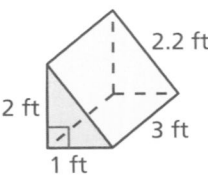

13.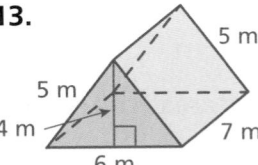

14.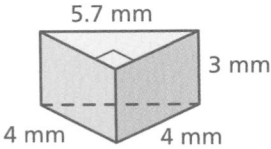

15. **GIFT BOX** What is the least amount of wrapping paper needed to wrap a gift box that measures 8 inches by 8 inches by 10 inches? Explain.

16. **TENT** What is the least amount of fabric needed to make the tent?

Assignment Guide and Homework Check

Level	Day 1 Activity Assignment	Day 2 Lesson Assignment	Homework Check
Basic	6–8, 24–27	1–5, 9–15 odd, 16, 19	9, 13, 16, 19
Average	6–8, 24–27	1–5, 11–13, 16, 17, 19, 20	12, 16, 17, 19
Advanced	6–8, 24–27	1–5, 14, 16–18, 20–23	14, 16, 18, 22

For Your Information

- **Exercises 15 and 16** Students should ignore any overlap in these two exercises.

Common Errors

- **Exercises 9–11** Students may find the area of only three of the faces instead of all six. Remind them that each face is paired with another. Show students the net of a rectangular solid to remind them of the six faces.
- **Exercises 9–11** Some students may multiply length by width by height to find the surface area. Show them that the surface area is the sum of the areas of all six faces, so they must multiply and add to find the solution.
- **Exercises 12–14** Students may try to use the formula for a rectangular prism to find the surface area of a triangular prism. Show them that this will not work by focusing on the area of the triangular base. For students who are struggling to identify all the faces, draw a net of the prism and tell them to label the length, width, and height of each part before finding the surface area.

6.2 Record and Practice Journal

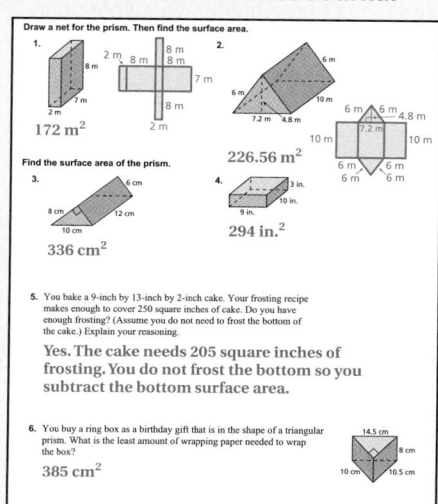

Technology
For the **T**eacher
Answer Presentation Tool
QuizShow

Vocabulary and Concept Check

1. *Sample answer:* You want to paint a large toy chest in the form of a rectangular prism, and in order to know how much paint to buy, you need to know the surface area.

2. 24 cm^2 3. 18 cm^2

4. 12 cm^2 5. 108 cm^2

Practice and Problem Solving

6.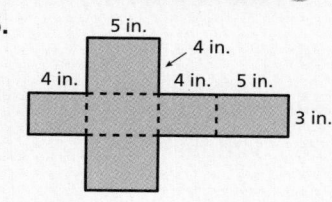

94 in.^2

7.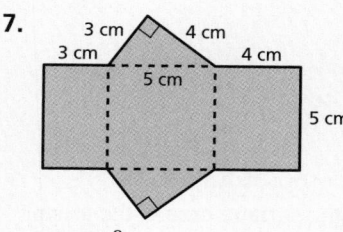

72 cm^2

8.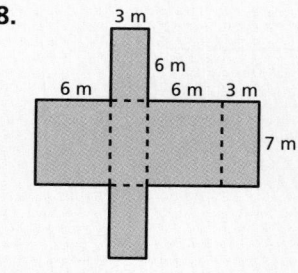

162 m^2

9. 130 ft^2 10. 198 cm^2

11. 76 yd^2 12. 17.6 ft^2

13. 136 m^2 14. 57.1 mm^2

15. 448 in.^2; The surface area of the box is 448 square inches, so that is the least amount of paper needed to cover the box.

16. 136 ft^2

Practice and Problem Solving

17. 156 in.^{2}

18. 68 m^2

19. 83 ft^2

20. See *Taking Math Deeper*.

21. 2 qt

22. $x = 4$ in.

23. $S = 2B + Ph$

Fair Game Review

24. 25 units

25. 48 units

26. 54 units

27. C

Mini-Assessment

Find the surface area of the prism.

1.

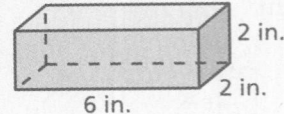

56 in.^{2}

2.

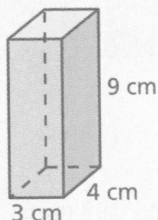

9 cm

4 cm

3 cm

150 cm^2

3. Find the least amount of fabric needed to make the tent. 152 ft^2

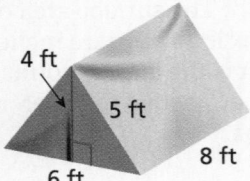

4 ft

5 ft

8 ft

6 ft

Taking Math Deeper

Exercise 20

This is a classic type of problem in manufacturing. For a given volume, what is the least amount of material I can use? The general answer is that the more cube-like, the more efficient the use of material. For instance, a cube-like tissue box is much more cost effective than a cereal box.

1 Help me see it.
Each storage box has a volume of 480 cubic inches. However, the shapes are quite different.

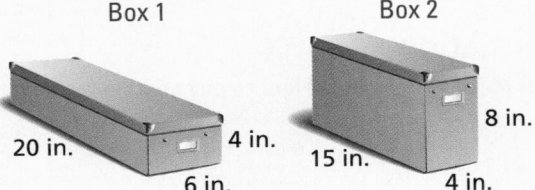

Box 1   Box 2

20 in.   4 in.   8 in.

6 in.   15 in.   4 in.

2 Find the surface area of each—in square feet.
Box 1
$2(20 \cdot 4) + 2(20 \cdot 6) + 2(4 \cdot 6) = 448$ in.$^2 \approx 3.11$ ft^2
Box 2
$2(15 \cdot 8) + 2(15 \cdot 4) + 2(8 \cdot 4) = 424$ in.$^2 \approx 2.94$ ft^2

Divide by 144 to get square feet.

3 Find the cost of each type and answer the question.

Rounding error

Box 1 Cost: $50(3.11 \text{ ft}^2)\left(1.25 \dfrac{\$}{\text{ft}^2}\right) \approx \$194.38$

Box 2 Cost: $50(2.94 \text{ ft}^2)\left(1.25 \dfrac{\$}{\text{ft}^2}\right) = \$183.75$

A company saves $10.63 by using Box 2.

A more exact answer of $10.42 can be found if the surface areas of the boxes are left in fraction form.

Project

Design a box that would have a volume of 480 cubic inches using the least possible amount of cardboard.

Reteaching and Enrichment Strategies

If students need help. . .	If students got it. . .
Resources by Chapter • Practice A and Practice B • Puzzle Time Record and Practice Journal Practice Differentiating the Lesson Lesson Tutorials Skills Review Handbook	Resources by Chapter • Enrichment and Extension Start the next section

Find the surface area of the prism.

17.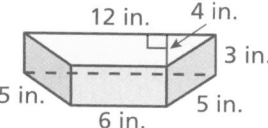
12 in. 4 in.
3 in.
5 in.   5 in.
6 in.

18.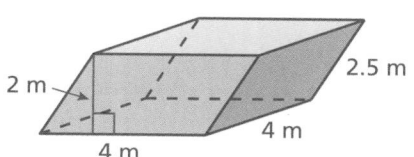
2 m
2.5 m
4 m
4 m

19. AQUARIUM A public library has an aquarium in the shape of a rectangular prism. The base is 6 feet by 2.5 feet. The height is 4 feet. How many square feet of glass were used to build the aquarium? (The top of the aquarium is open.)

20. STORAGE BOX The material used to make a storage box costs $1.25 per square foot. The boxes have the same volume. How much does a company save by choosing to make 50 of Box 2 instead of 50 of Box 1?

	Length	Width	Height
Box 1	20 in.	6 in.	4 in.
Box 2	15 in.	4 in.	8 in.

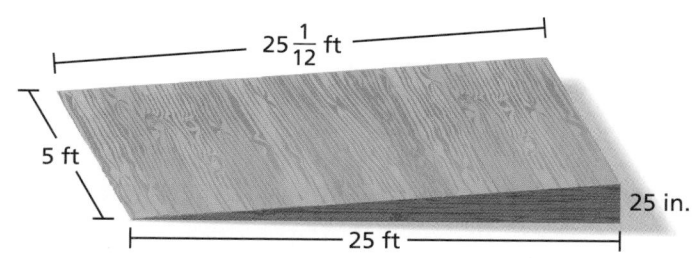
25 1/12 ft
5 ft
25 ft
25 in.

21. RAMP A quart of stain covers 100 square feet. How many quarts should you buy to stain the wheelchair ramp? (Assume you do not have to stain the bottom of the ramp.)

22. LABEL A label that wraps around a box of golf balls covers 75% of its lateral surface area. What is the value of *x*?

23. *Critical Thinking* Write a formula for the surface area of a rectangular prism using the height *h*, the perimeter *P* of a base, and the area *B* of a base.

3 in.
SUPER
Golf Balls
SUPER
Golf Balls
because YOU are a super golfer
2 in.
2 in.
x

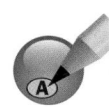
Fair Game Review *What you learned in previous grades & lessons*

Find the perimeter. *(Skills Review Handbook)*

24.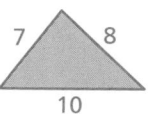
7   8
10

25.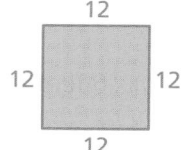
12
12   12
12

26.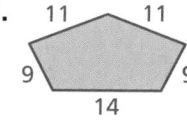
11   11
9   9
14

27. MULTIPLE CHOICE The class size increased 25% to 40 students. What was the original class size? *(Section 4.2)*

Ⓐ 10   Ⓑ 30   Ⓒ 32   Ⓓ 50

6.2b Circles

A **circle** is the set of all points in a plane that are the same distance from a point called the **center**.

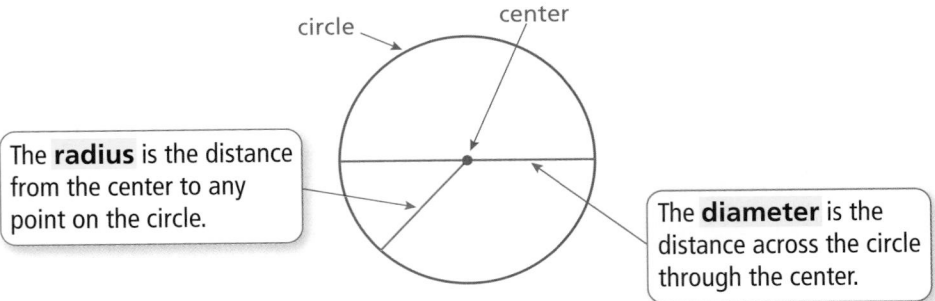

The **radius** is the distance from the center to any point on the circle.

The **diameter** is the distance across the circle through the center.

 **Key Idea**

Radius and Diameter

Words The diameter d of a circle is twice the radius r. The radius r of a circle is one-half the diameter d.

Algebra **Diameter:** $d = 2r$ **Radius:** $r = \dfrac{d}{2}$

EXAMPLE ① **Finding a Radius and a Diameter**

a. The diameter of a circle is 12 feet. Find the radius.

b. The radius of a circle is 8 meters. Find the diameter.

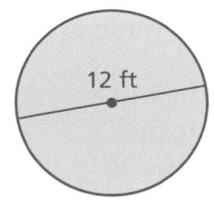

12 ft

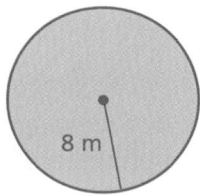

8 m

$r = \dfrac{d}{2}$ Radius of a circle

$= \dfrac{12}{2}$ Substitute 12 for d.

$= 6$ Divide.

∴ The radius is 6 feet.

$d = 2r$ Diameter of a circle

$= 2(8)$ Substitute 8 for r.

$= 16$ Multiply.

∴ The diameter is 16 meters.

● **Practice**

1. **DIAMETER** The radius of a dartboard is 9 inches. Find the diameter.

2. **RADIUS** The diameter of a clock is 1 foot. Find the radius.

◀)) Multi-Language Glossary at BigIdeasMath✓com.

Laurie's Notes

Introduction

Connect

- **Yesterday:** Students found surface areas of prisms.
- **Today:** Students will find circumferences and areas of circles. This is necessary for finding surface areas of cylinders in the following lesson.

Motivate

- Ask your students to get out a piece of scrap paper and a pencil. Tell them that they will have 1 minute to write a list of objects in the room that have a special characteristic.
- Announce that they need to list objects that are circular or have a circle on them. Use a stopwatch or clock with a second hand to time them.
- Items will vary but expect items such as: clock and/or watch face, bottom of coffee cup, pencil eraser, pupils of your eyes, a student's glasses, metal feet on a chair, etc.

Lesson Notes

Key Idea

- Draw a circle on the board. Label and discuss the center, a radius, and a diameter.
- Write the Key Idea.
- Discuss with students that if you know either the diameter or radius of a circle, you can find the other. There is a 2 : 1 relationship between **diameter** and **radius**.

Example 1

- Work through both parts of the example.
- Remind students to label answers with the appropriate units.

Practice

- **Common Error:** Students may incorrectly double the diameter or halve the radius.

Goal Today's lesson is finding **circumferences** and **areas** of **circles**.

Lesson Materials
Introduction
• stopwatch

Warm Up

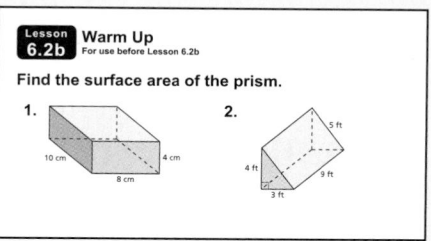

Extra Example 1

a. The diameter of a circle is 8.2 yards. Find the radius. 4.1 yd

b. The radius of a circle is 5 millimeters. Find the diameter. 10 mm

 Practice

1. 18 in.

2. 0.5 ft

Record and Practice Journal Practice
See Additional Answers.

Extra Example 2

Find (a) the circumference and

(b) the area of the circle. Use $\frac{22}{7}$ for π.

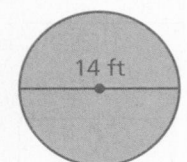
14 ft

a. 44 ft
b. 154 ft²

Practice

3. $C \approx 440$ cm;
$A \approx 15,400$ cm²

4. $C \approx 75.36$ in.;
$A \approx 452.16$ in.²

5. $C \approx 31.4$ in.;
$A \approx 78.5$ in.²

6. a. about 81.64 in.
b. about 14 rotations

Mini-Assessment

Find the radius or diameter of the circle.

1.
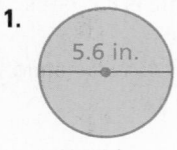
5.6 in.

2.
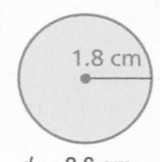
1.8 cm

$r = 2.8$ in. $d = 3.6$ cm

Find the circumference and area of the circle. Use 3.14 or $\frac{22}{7}$ for π.

3.
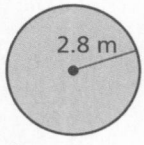
2.8 m

4.
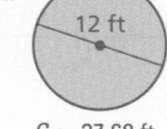
12 ft

$C \approx 17.6$ m
$A \approx 24.64$ m²

$C \approx 37.68$ ft
$A \approx 113.04$ ft²

5. A radio signal transmits in a circle from a tower. Radio signal A has a radius of 4 miles and radio signal B has a radius of 5 miles. How much more area does radio signal B cover than radio signal A? about 28.26 mi²

Laurie's Notes

Key Idea

- **Common Misconception:** Students may know that pi is a number; however, when they see it in a formula, they can become confused. Students may ask if π is a variable. Pi is a constant whose value is approximately 3.14 or $\frac{22}{7}$. In each formula, remind students that πd means the product of π and the diameter, $2\pi r$ (or $\pi 2r$) means the product of π and twice the radius, and πr^2 means the product of π and the square of the radius.
- **Big Idea:** The ratio of the circumference to the diameter is pi. Describe how to get the formula for circumference: multiply both sides of the equation by the diameter.

$$\frac{\text{circumference}}{\text{diameter}} = \text{pi}$$

$$\frac{C}{d} = \pi$$

$$C = \pi d$$

- **Common Error:** Students may square the product of π and r when finding area, instead of just the radius.

Example 2

- Work through each part of the example.
- ❓ "In step 2, why is the equal sign replaced by the approximately equal ($\approx$) sign?" 3.14 is an approximation for π.
- **Common Error:** Students may write $3^2 = 6$ instead of $3^2 = 9$.
- Remind students to label their answers with the correct units. Circumference is a linear measurement (cm) and area is a square measurement (cm²).
- **Common Error:** If students use an abbreviation for square centimeters, check to see that they write 28.26 sq. cm or 28.26 cm², and *not* 28.26² cm.

Practice

- **FYI:** Exercise 6 is a common application. Use a circular object to show students that one rotation is equal to the circumference.

Closure

- **Exit Ticket:** The diameter of a quarter (coin) is 24.26 millimeters.
  a. What is the radius of a quarter? 12.13 mm
  b. What is the circumference of a quarter? about 76.18 mm
  c. What is the area of the face of a quarter? about 462.01 mm²

Technology For the Teacher

Dynamic Classroom

The Dynamic Planning Tool
Editable Teacher's Resources at *BigIdeasMath.com*

The distance around a circle is called the **circumference**. The ratio $\frac{\text{circumference}}{\text{diameter}}$ is the same for *every* circle and is represented by the Greek letter π, called **pi**. The value of π can be approximated as 3.14 or $\frac{22}{7}$.

Key Ideas

Circumference of a Circle

Words   The circumference C of a circle is equal to the product of π and the diameter d or the product of π and twice the radius r.

Algebra $C = \pi d$ or $C = 2\pi r$

Area of a Circle

Words   The area A of a circle is the product of π and the square of the radius r.

Algebra $A = \pi r^2$

EXAMPLE 2   **Finding the Circumference and Area of a Circle**

Find (a) the circumference and (b) the area of the sticker. Use 3.14 for π.

3 cm

a. $C = 2\pi r$

$\approx 2 \cdot 3.14 \cdot 3$ Substitute.

$= 6.28 \cdot 3$ Simplify.

$= 18.84$ Simplify.

The circumference is about 18.84 centimeters.

b. $A = \pi r^2$

$\approx 3.14 \cdot (3)^2$

$= 3.14 \cdot 9$

$= 28.26$

The area is about 28.26 square centimeters.

Practice

Find the circumference and area of the object. Use 3.14 or $\frac{22}{7}$ for π.

3.
70 cm

4.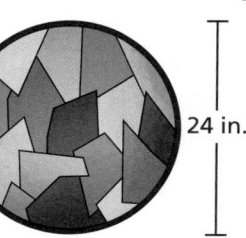
24 in.

5.
5 in.

6. TIRE The diameter of a bicycle tire is 26 inches.

   a. Find the circumference of the tire. Use 3.14 for π.

   b. How many rotations does the tire make to travel 95 feet? Explain your reasoning.

Essential Question How can you find the surface area of a cylinder?

1 ACTIVITY: Finding Area

Work with a partner. Use a cardboard cylinder.

- **Talk about how you can find the area of the outside of the roll.**

- **Use a ruler to estimate the area of the outside of the roll.**

- **Cut the roll and press it out flat. Then find the area of the flattened cardboard. How close is your estimate to the actual area?**

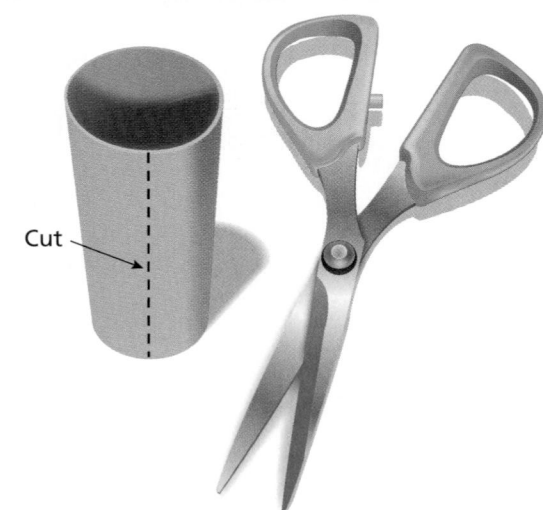

Cut

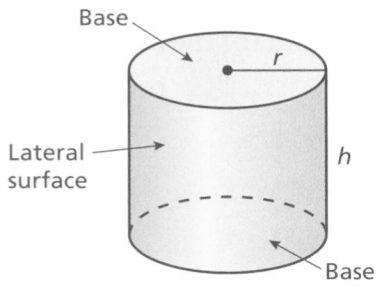

Base

r

Lateral surface

h

Base

The surface area of a cylinder is the sum of the areas of the bases and the lateral surface.

2 ACTIVITY: Finding Surface Area

Work with a partner.

son Farms

Mild

I PEPPERS

12 oz.

- **Trace the top and bottom of a can on paper. Cut out the two shapes.**

- **Cut out a long paper rectangle. Make the width the same as the height of the can. Wrap the rectangle around the can. Cut off the excess paper so the edges just meet.**

- **Make a net for the can. Name the shapes in the net.**

- **How are the dimensions of the rectangle related to the dimensions of the can?**

- **Explain how to use the net to find the surface area of the can.**

Laurie's Notes

Introduction

For the Teacher

- **Goal:** Students will develop an intuition about how to find the surface area of a cylinder.
- **Big Idea:** Recall the connection between a prism and a cylinder—*structurally they are the same*. Unlike the prism, you don't have cardboard models to unfold. A good model to use is a roll of paper towels.

Motivate

- Use two different cans (cylinders), where the taller can has a lesser radius. A tuna can and a 6-ounce vegetable can work well.
- ? Hold both cans. "Which can required more metal to make?" Answers will vary depending on can sizes.
- This question focuses attention on the surface areas of the cans, the need to consider their components, and how they were made.

Activity Notes

Activity 1

- Use cardboard rolls from paper towels or toilet paper, or make rolls from strips of file folder paper.
- Students should try to estimate the area before they cut. This requires that they use their spatial skills to think about the area of a surface that is curved.
- ? "What shape do you have when the roll is flattened out?" rectangle
- ? "What units do you use to measure the dimensions of your rectangle?" depending upon the ruler, centimeters or inches
- ? "What units do you use to label your answer?" cm^2 or in.^{2}

Activity 2

- Each pair of students will need scrap paper, tape, and scissors.
- **Teaching Tip:** To avoid having lots of scraps of paper on the floor, recycle plastic bags and tape them to 4 or 5 desks around the room. There should be a bag close enough to each group so that there is no excuse for a group leaving a mess.
- Discuss results when students have finished. Students should be able to describe the component parts of a cylinder (two circular bases and a rectangular lateral portion), how to find the area of each component, and how the dimensions of each component relate to the dimensions of the cylinder.
- Be sure to *wrap* the lateral portion around the two bases so that students see the relationship between the circumference of the base and the length of the rectangular portion.
- This hands-on experience of making the cylinder and calculating the surface area will help students remember the process and understand the formula.

Previous Learning

Students should know the vocabulary of cylinders.

Activity Materials	
Introduction	**Textbook**
• 2 cans	• cardboard rolls
	• scissors
	• scrap paper
	• tape
	• rulers
	• various cylinders

Start Thinking! and Warm Up

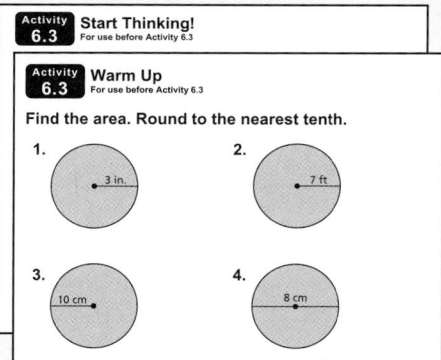

Activity 6.3 Start Thinking! For use before Activity 6.3

Activity 6.3 Warm Up For use before Activity 6.3

Find the area. Round to the nearest tenth.

1. 3 in.
2. 7 ft
3. 10 cm
4. 8 cm

6.3 Record and Practice Journal

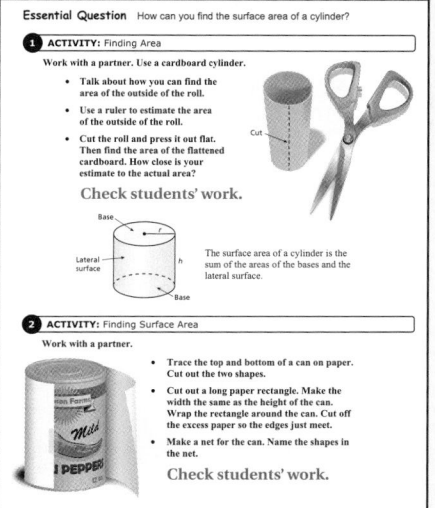

Essential Question How can you find the surface area of a cylinder?

1 ACTIVITY: Finding Area

Work with a partner. Use a cardboard cylinder.

- Talk about how you can find the area of the outside of the roll.
- Use a ruler to estimate the area of the outside of the can.
- Cut the roll and press it out flat. Then find the area of the flattened cardboard. How close is your estimate to the actual area?

Check students' work.

The surface area of a cylinder is the sum of the areas of the bases and the lateral surface.

2 ACTIVITY: Finding Surface Area

Work with a partner.

- Trace the top and bottom of a can on paper. Cut out the two shapes.
- Cut out a long paper rectangle. Make the width the same as the height of the can. Wrap the rectangle around the can. Cut off the excess paper so the edges just meet.
- Make a net for the can. Name the shapes in the net.

Check students' work.

English Language Learners

Vocabulary

Have students work in pairs, one English learner and one English speaker. Have each pair write a problem involving the surface area of a cylinder. On a separate piece of paper, students should solve their own problem. Then have students exchange their problem with another pair of students. Students solve the new problem. After solving the problem, the four students discuss the problems and solutions.

Laurie's Notes

Activity 3

- When estimating the dimensions of the common cylinders, encourage students to use their hands to visualize the size of the cylinder.
- If time permits, set up stations in the room with a different cylinder at each. Provide rulers. In small groups, students move from one station to the next. Make sure you have a good variety of common cylinders: soup can, soft drink can, tuna can, AA battery, etc.

? "What dimensions did you measure for each cylinder?" Students will often say diameter and height.

? "Could the radius be measured?" yes; Take $\frac{1}{2}$ of the diameter to find the radius.

- **Extension:** Gather the results of the activity, and then find the mean for several of the cylinders.

What Is Your Answer?

- Have students work in pairs. Review answers as a class.

Closure

- Hold the two cans from the Motivate section and ask, "Which of the two cans from the beginning of today's lesson required more metal to make?" Answers will vary depending on can sizes.

6.3 Record and Practice Journal

- How are the dimensions of the rectangle related to the dimensions of the can?

  Rectangle length = can circumference
  Rectangle width = can height

- Explain how to use the net to find the surface area of the can.

  Find the area of the 2 circles and the rectangle.

3 ACTIVITY: Estimation

Work with a partner. From memory, estimate the dimensions of the real-life items in parts (a)–(d) in inches. Then use the dimensions to estimate the surface area of each item in square inches.

Sample answers are given.

a.

$r \approx 1.5$ in.
$h \approx 4$ in.
$SA \approx 52$ in.^{2}

b.

$r \approx 1.25$ in.
$h \approx 4.5$ in.
$SA \approx 45$ in.^{2}

c.
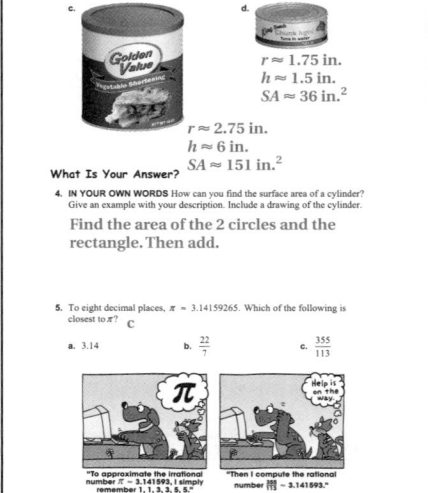

$r \approx 2.75$ in.
$h \approx 6$ in.
$SA \approx 151$ in.^{2}

d.

$r \approx 1.75$ in.
$h \approx 1.5$ in.
$SA \approx 36$ in.^{2}

What Is Your Answer?

4. **IN YOUR OWN WORDS** How can you find the surface area of a cylinder? Give an example with your description. Include a drawing of the cylinder.

  Find the area of the 2 circles and the rectangle. Then add.

5. To eight decimal places, $\pi \approx 3.14159265$. Which of the following is closest to π? **c**

  a. 3.14   b. $\frac{22}{7}$ c. $\frac{355}{113}$

Technology For the Teacher

Dynamic Classroom

The Dynamic Planning Tool
Editable Teacher's Resources at *BigIdeasMath.com*

ACTIVITY: Estimation

Work with a partner. From memory, estimate the dimensions of the real-life item in inches. Then use the dimensions to estimate the surface area of the item in square inches.

a.

b.

c.

d.

What Is Your Answer?

4. **IN YOUR OWN WORDS** How can you find the surface area of a cylinder? Give an example with your description. Include a drawing of the cylinder.

5. To eight decimal places, $\pi \approx 3.14159265$. Which of the following is closest to π?

   a. 3.14

   b. $\dfrac{22}{7}$

   c. $\dfrac{355}{113}$

"To approximate the irrational number $\pi \approx 3.141593$, I simply remember 1, 1, 3, 3, 5, 5."

"Then I compute the rational number $\dfrac{355}{113} \approx 3.141593$."

Practice

Use what you learned about the surface area of a cylinder to complete Exercises 5–7 on page 266.

Check It Out
Lesson Tutorials
BigIdeasMath ✓com

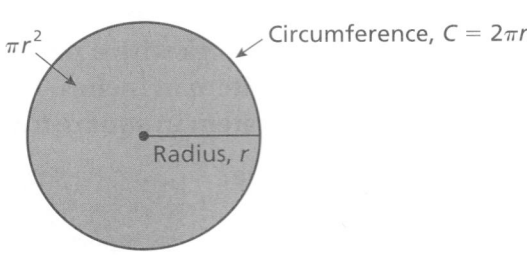

Area, $A = \pi r^2$ Circumference, $C = 2\pi r$

Radius, r

The diagram reviews some important facts for circles.

Key Idea

Surface Area of a Cylinder

Words The surface area S of a cylinder is the sum of the areas of the bases and the lateral surface.

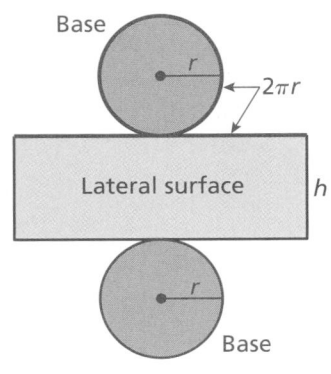

Base

r

$2\pi r$

Lateral surface h

r

Base

Remember

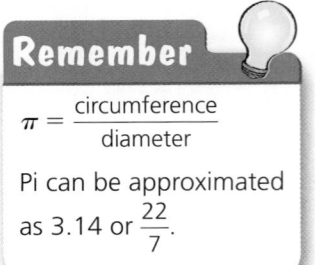

$\pi = \dfrac{\text{circumference}}{\text{diameter}}$

Pi can be approximated as 3.14 or $\dfrac{22}{7}$.

Algebra $S = 2\pi r^2 + 2\pi rh$

Area of bases

Area of lateral surface

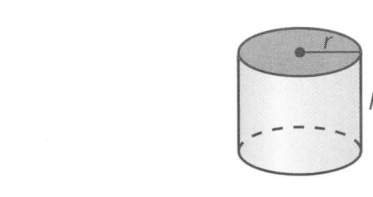

EXAMPLE ① **Finding the Surface Area of a Cylinder**

Find the surface area of the cylinder. Round your answer to the nearest tenth.

Draw a net.

$S = 2\pi r^2 + 2\pi rh$

$\quad = 2\pi(4)^2 + 2\pi(4)(3)$

$\quad = 32\pi + 24\pi$

$\quad = 56\pi \approx 175.8$

∴ The surface area is about 175.8 square millimeters.

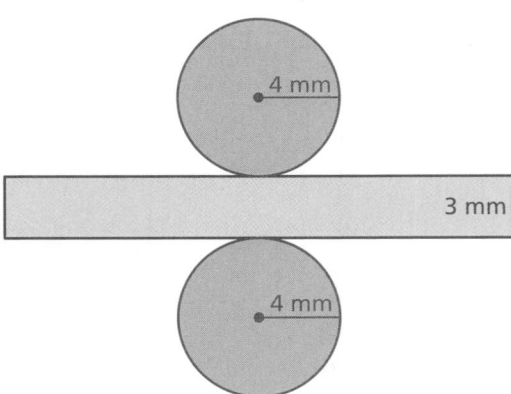

4 mm

3 mm

4 mm

On Your Own

Now You're Ready
Exercises 8–10

1. A cylinder has a radius of 2 meters and a height of 5 meters. Find the surface area of the cylinder. Round your answer to the nearest tenth.

Laurie's Notes

Introduction

Connect
- **Yesterday:** Students discovered how to find the surface area of a cylinder by examining the net that makes up a cylinder.
- **Today:** Students will work with a formula for the surface area of a cylinder.

Motivate
- Find two cans (cylinders) that have volumes in the ratio of 1 : 2 (i.e., 10 fl oz and 20 fl oz).
- ❓ "The larger can has twice the volume of the smaller can. Do you think the surface area is twice as much?" Answers may differ, but most students believe this is true. Return to this question at the end of the lesson.

Lesson Notes

Key Idea
- ❓ "How are cylinders and prisms alike?" Both have 2 congruent bases and a lateral portion. "Different?" Cylinders have circular bases, while the prism has a polygonal base.
- Refer to the diagram with the radius marked. Review the formulas for area and circumference.
- Write the formula in words first. Before writing the formula in symbols, ask direct questions to help students make the connection between the words and the symbols.
- ❓ "How do you find the sum of the areas of the bases?" Find the area of one base, πr^2, and then multiply by 2.
- ❓ "How do you find the area of the lateral portion?" The lateral portion is a rectangle whose dimensions are the height of the cylinder and a width that is the circumference of the base, so $2\pi rh$.
- Write the formula in symbols with each part identified (area of bases + lateral surface area).

Example 1
- Write the formula first to model good, problem-solving techniques.
- Notice that the values of the variables are substituted, with each term being left in terms of π.
- **Common Misconception:** Students are unsure of how to perform the multiplication with π in the middle of the term. Remind students that π is a number, a factor in this case, just like the other numbers. Because of the Commutative and Associative Properties, the whole numbers can be multiplied first. Then the two like terms, 32π and 24π, are combined. The last step is to substitute 3.14 for π.
- ❓ Review ≈. "What does this symbol mean? Why is it used?" approximately equal to; π is an irrational number and an estimate for pi is used in the calculation.

On Your Own
- Ask volunteers to share their answers.

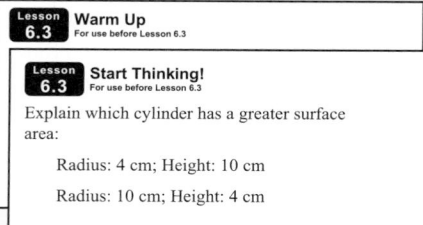

Extra Example 2

Find the lateral surface area of a cylinder with a radius of 2 inches and a height of 6 inches. Round your answer to the nearest tenth. $24\pi \approx 75.4$ in.^{2}

Extra Example 3

You earn $0.07 for recycling a can with a radius of 3 inches and a height of 4 inches (Extra Example 1). How much can you expect to earn for recycling a can with a radius of 2 inches and a height of 6 inches (Extra Example 2)? Assume that the recycle value is proportional to the surface area. $0.05

On Your Own

2. **a.** yes

   b. no; Only the lateral surface area doubled. Because the surface area of the can does not double, the recycle value does not double.

Differentiated Instruction

Visual

Encourage students to estimate their answers for reasonableness. For the surface area of a cylinder, a common error is using the diameter in the formula instead of the radius. Have students imagine the cylinder inside of a rectangular prism. By calculating the surface area of the prism, the student has an overestimate of the surface area of the cylinder.

Laurie's Notes

Example 2

- Note that only the lateral surface area is asked for in this example.
- If you have a small can with a label on it, use it as a model for this problem.
- Note again that the answer is left in terms of π until the last step.

Example 3

- Ask a volunteer to read the problem. Check to see if students understand what is being asked in this problem.
- **?** "How does the surface area of the can relate to the recycling value?" The value of the recycled can is the amount of metal used to make the can (which is the surface area).
- Have students compute the surface area of each can.
- **?** "Approximately how much more metal is there in the larger can?" $24\pi \approx 75$ in.^{2}
- This is a good review of proportions. Calculators are helpful.

On Your Own

- Give students sufficient time to do their work before asking volunteers to share their work at the board.

Closure

- Hold the two cans from the Motivate section of today's lesson and ask students to find the surface area of each. Answers will vary depending on can sizes.

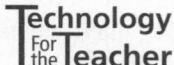

EXAMPLE **2** **Finding Surface Area**

How much paper is used for the label on the can of peas?

1 in.

2 in.

Find the *lateral* surface area of the cylinder.

$S = 2\pi rh$ ← Do not include the area of the bases in the formula.

$\quad = 2\pi(1)(2)$ \qquad Substitute.

$\quad = 4\pi \approx 12.56$ \qquad Multiply.

∴ About 12.56 square inches of paper is used for the label.

EXAMPLE **3** **Real-Life Application**

2 in.

5.5 in.

You earn \$0.01 for recycling the can in Example 2. How much can you expect to earn for recycling the tomato can? Assume that the recycle value is proportional to the surface area.

Find the surface area of each can.

Tomatoes

$S = 2\pi r^2 + 2\pi rh$

$\quad = 2\pi(2)^2 + 2\pi(2)(5.5)$

$\quad = 8\pi + 22\pi$

$\quad = 30\pi$

Peas

$S = 2\pi r^2 + 2\pi rh$

$\quad = 2\pi(1)^2 + 2\pi(1)(2)$

$\quad = 2\pi + 4\pi$

$\quad = 6\pi$

Use a proportion to find the recycle value x of the tomato can.

$$\frac{30\pi \text{ in.}^2}{x} = \frac{6\pi \text{ in.}^2}{\$0.01}$$

← surface area

← recycle value

$30\pi \cdot 0.01 = x \cdot 6\pi$ \qquad Use Cross Products Property.

$5 \cdot 0.01 = x$ \qquad Divide each side by 6π.

$0.05 = x$ \qquad Simplify.

∴ You can expect to earn \$0.05 for recycling the tomato can.

● **On Your Own**

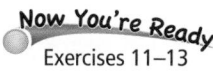

Now You're Ready
Exercises 11–13

2. **WHAT IF?** In Example 3, the height of the can of peas is doubled.

   a. Does the amount of paper used in the label double?

   b. Does the recycle value double? Explain.

6.3 Exercises

✓ Vocabulary and Concept Check

1. **CRITICAL THINKING** Which part of the formula $S = 2\pi r^2 + 2\pi rh$ represents the lateral surface area of a cylinder?

2. **CRITICAL THINKING** Given the height and the circumference of the base of a cylinder, describe how to find the surface area of the entire cylinder.

Find the indicated area of the cylinder.

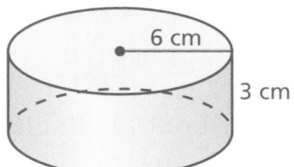

3. Area of a base

4. Surface area

 Practice and Problem Solving

Make a net for the cylinder. Then find the surface area of the cylinder. Round your answer to the nearest tenth.

5.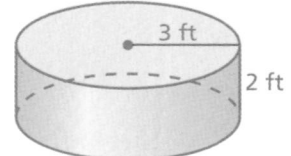

6.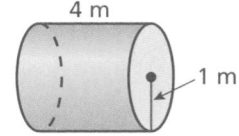

7.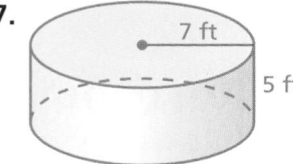

Find the surface area of the cylinder. Round your answer to the nearest tenth.

① 8.

9.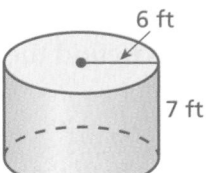

10.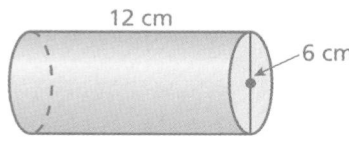

Find the lateral surface area of the cylinder. Round your answer to the nearest tenth.

② 11.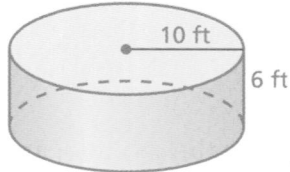

12.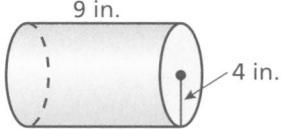

13.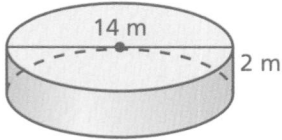

14. **TANKER** The truck's tank is a stainless steel cylinder. Find the surface area of the tank.

radius = 4 ft

Assignment Guide and Homework Check

Level	Day 1 Activity Assignment	Day 2 Lesson Assignment	Homework Check
Basic	5–7, 21–24	1–4, 8–15	2, 8, 12, 14
Average	5–7, 21–24	1–4, 9–15 odd, 16–18	2, 9, 11, 16
Advanced	5–7, 21–24	1–4, 10, 12, 15–20	2, 10, 12, 16

Common Errors

- **Exercises 8–10** Students may add the area of only one base. Remind them of the net for a cylinder and that there are two circles as bases.
- **Exercises 8–10** Students may double the radius instead of squaring it. Remind them of the area of a circle and also the order of operations.
- **Exercise 10** Students may use the diameter instead of the radius. Remind them that the radius is in the formula, so they should find the radius before finding the surface area.
- **Exercises 11–13** Students may multiply the height by the area of the circle instead of the circumference. Review with them how the lateral surface is created to show that the length of the rectangle is the circumference of the circular bases.

6.3 Record and Practice Journal

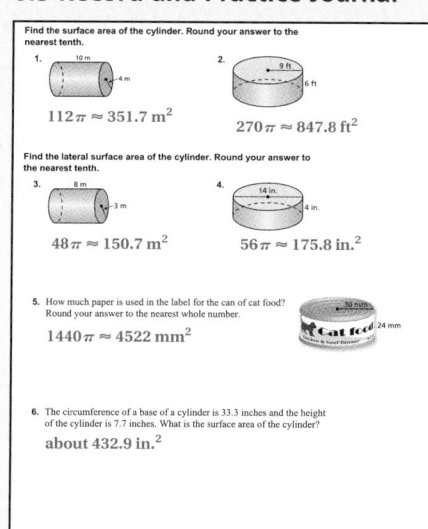

Vocabulary and Concept Check

1. $2\pi rh$

2. Use the given circumference to find the radius by solving $C = 2\pi r$ for r. Then use the formula for the surface area of a cylinder.

3. $36\pi \approx 113.0$ cm^2

4. $108\pi \approx 339.1$ cm^2

Practice and Problem Solving

5.

$30\pi \approx 94.2$ ft^2

6.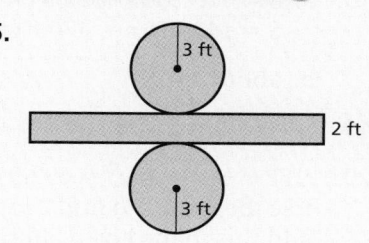

$10\pi \approx 31.4$ m^2

7.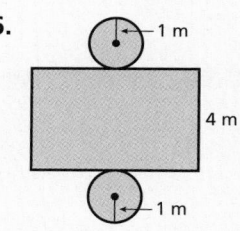

$168\pi \approx 527.5$ ft^2

8. $28\pi \approx 87.9$ mm^2

9. $156\pi \approx 489.8$ ft^2

10. $90\pi \approx 282.6$ cm^2

11. $120\pi \approx 376.8$ ft^2

12. $72\pi \approx 226.1$ in.^{2}

13. $28\pi \approx 87.9$ m^2

14. $432\pi \approx 1356.48$ ft^2

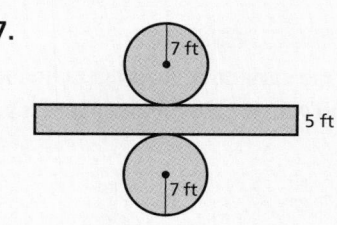

15. The error is that only the lateral surface area is found. The areas of the bases should be added;
$$S = 2\pi r^2 + 2\pi rh$$
$$= 2\pi (6)^2 + 2\pi (6)(11)$$
$$= 72\pi + 132\pi$$
$$= 204\pi \text{ ft}^2$$

16. about 36.4%

17. The surface area of the cylinder with the height of 8.5 inches is greater than the surface area of the cylinder with the height of 11 inches.

18. a. $S = 41.125\pi \approx 129.2 \text{ cm}^2$,
$S = 149.875\pi \approx 470.6 \text{ cm}^2$

b. about 4.0 lb

19. See *Taking Math Deeper*.

20. $(162\pi + 184) \text{ cm}^2 \approx 692.9 \text{ cm}^2$
Use the radius to find $2\pi r^2$. Add this to the given lateral surface area.

21. 117   **22.** 20.6

23. 56.52   **24.** B

Mini-Assessment

Find the surface area of the cylinder. Round your answer to the nearest tenth.

1.
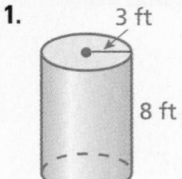
3 ft
8 ft

$66\pi \approx 207.2 \text{ ft}^2$

2.
2 in.
6 in.

$14\pi \approx 44.0 \text{ in.}^2$

3. Find the surface area of the roll of paper towels.

$67.5\pi \approx 212.0 \text{ in.}^2$

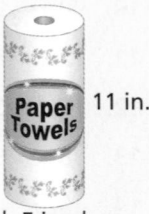
Paper Towels   11 in.

$\vdash$ 5 in. $\dashv$

Taking Math Deeper

Exercise 19

This is a real-life problem. That is, when you leave cheese in the refrigerator without covering it, the amount that dries out is proportional to the surface area.

1   Find the surface area of the uncut cheese.

$$S = 2\pi r^2 + 2\pi rh$$
$$= 2\pi \cdot 3^2 + 2\pi \cdot 3 \cdot 1$$
$$= 24\pi$$
a. $\approx 75.36 \text{ in.}^2$

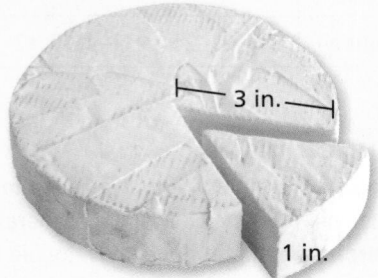
3 in.
1 in.

2   Find the surface area of the remaining cheese. One-eighth of the surface area is removed. But, two 3-by-1 rectangular regions are added.

$$S = \frac{7}{8} \cdot 24\pi + 2(3 \cdot 1)$$
$$= 21\pi + 6$$
b. $\approx 71.94 \text{ in.}^2$

1/8 removed

3 in.
1 in.

3   Answer the question.
b. The surface area decreased.

Reteaching and Enrichment Strategies

If students need help. . .	If students got it. . .
Resources by Chapter • Practice A and Practice B • Puzzle Time Record and Practice Journal Practice Differentiating the Lesson Lesson Tutorials Skills Review Handbook	Resources by Chapter • Enrichment and Extension • School-to-Work Start the next section

15. ERROR ANALYSIS Describe and correct the error in finding the surface area of the cylinder.

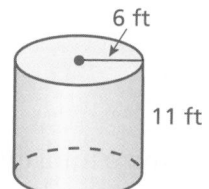

6 ft

11 ft

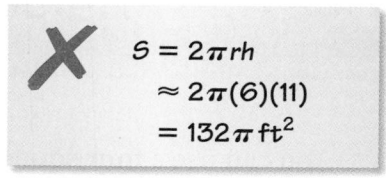

$S = 2\pi rh$
$\approx 2\pi(6)(11)$
$= 132\pi \text{ ft}^2$

├─ 16 in. ─┤

6 in.

8 in.

16. OTTOMAN What percent of the surface area of the ottoman is green (not including the bottom)?

17. REASONING You make two cylinders using 8.5-inch by 11-inch pieces of paper. One has a height of 8.5 inches and the other has a height of 11 inches. Without calculating, compare the surface areas of the cylinders.

18. INSTRUMENT A ganza is a percussion instrument used in samba music.

    a. Find the surface area of each of the two labeled ganzas.

    b. The weight of the smaller ganza is 1.1 pounds. Assume that the surface area is proportional to the weight. What is the weight of the larger ganza?

10 cm

24.5 cm

3.5 cm

5.5 cm

19. BRIE CHEESE The cut wedge represents one-eighth of the cheese.

    a. Find the surface area of the cheese before it is cut.

    b. Find the surface area of the remaining cheese after the wedge is removed. Did the surface area increase, decrease, or remain the same?

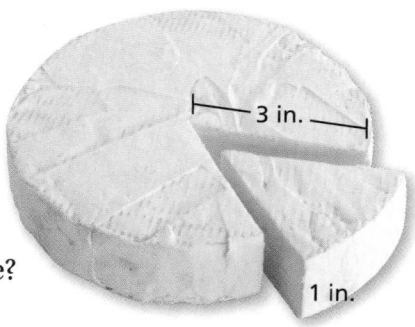

├─ 3 in. ─┤

1 in.

20. 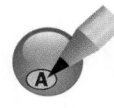 The lateral surface area of a cylinder is 184 square centimeters. The radius is 9 centimeters. What is the surface area of the cylinder? Explain how you found your answer.

Fair Game Review What you learned in previous grades & lessons

Evaluate the expression. *(Skills Review Handbook)*

21. $\frac{1}{2}(26)(9)$

22. $\frac{1}{2}(8.24)(3) + 8.24$

23. $\frac{1}{2}(18.84)(3) + 28.26$

24. MULTIPLE CHOICE A store pays $15 for a basketball. The percent of markup is 30%. What is the selling price? *(Section 4.3)*

   Ⓐ $10.50   Ⓑ $19.50   Ⓒ $30   Ⓓ $34.50

You can use a **four square** to organize information about a topic. Each of the four squares can be a category, such as *definition*, *vocabulary*, *example*, *non-example*, *words*, *algebra*, *table*, *numbers*, *visual*, *graph*, or *equation*. Here is an example of a four square for a three-dimensional figure.

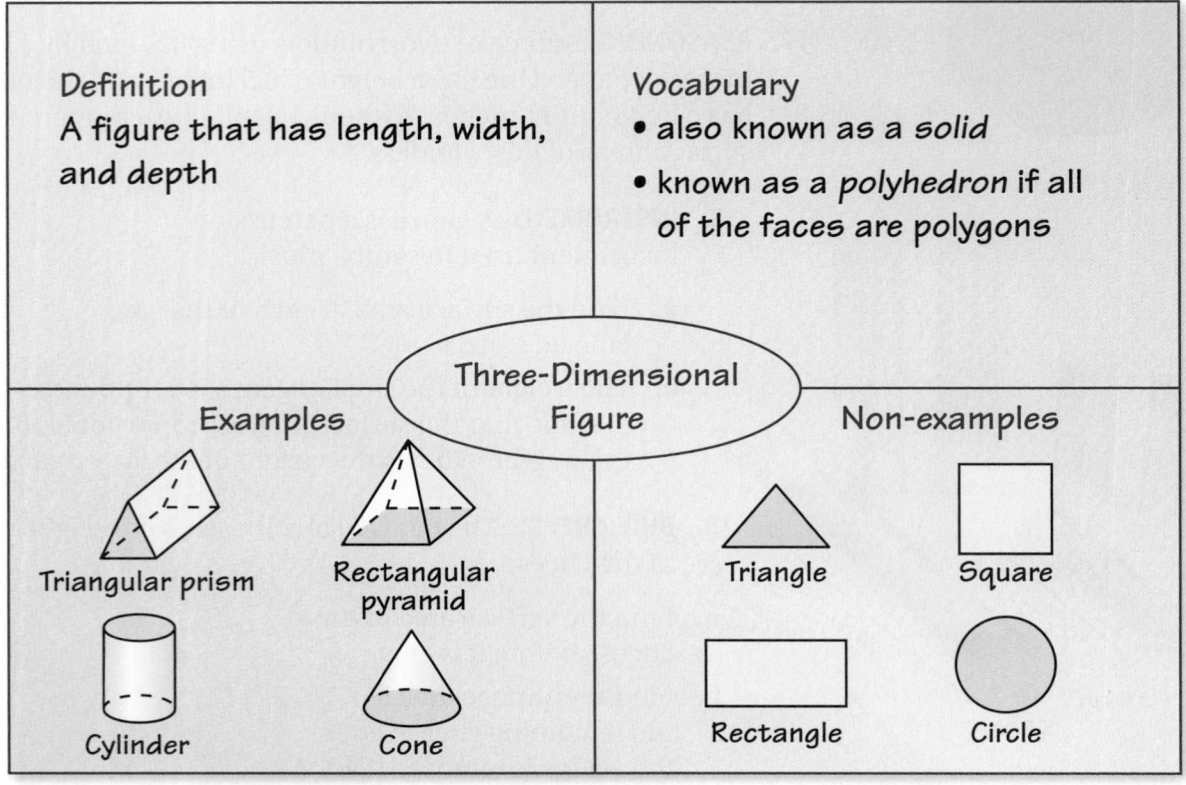

On Your Own

Make a four square to help you study these topics.

1. polyhedron
2. prism
3. pyramid
4. cylinder
5. cone
6. drawing a solid
7. surface area
   a. of a prism
   b. of a cylinder

After you complete this chapter, make four squares for the following topics.

8. surface area
   a. of a pyramid
   b. of a cone
   c. of a composite solid

"My four square shows that my new red skateboard is faster than my old blue skateboard."

Sample Answers

1.

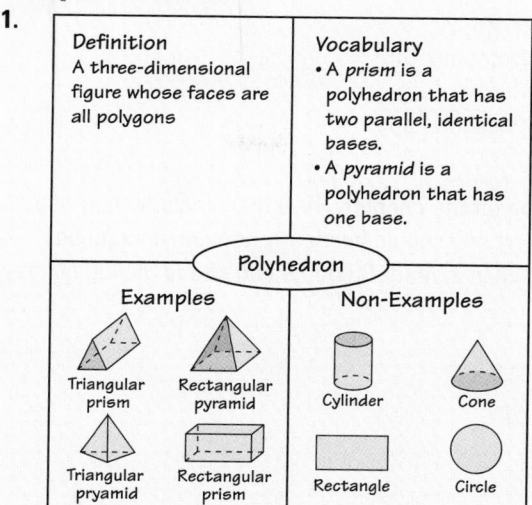

Definition	Vocabulary
A three-dimensional figure whose faces are all polygons	• A prism is a polyhedron that has two parallel, identical bases. • A pyramid is a polyhedron that has one base.

Polyhedron

Examples: Triangular prism, Rectangular pyramid, Triangular pryamid, Rectangular prism

Non-Examples: Cylinder, Cone, Rectangle, Circle

2.

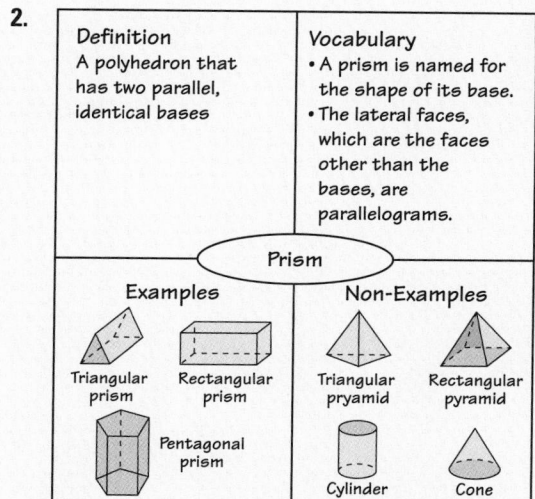

Definition	Vocabulary
A polyhedron that has two parallel, identical bases	• A prism is named for the shape of its base. • The lateral faces, which are the faces other than the bases, are parallelograms.

Prism

Examples: Triangular prism, Rectangular prism, Pentagonal prism

Non-Examples: Triangular pryamid, Rectangular pyramid, Cylinder, Cone

3.

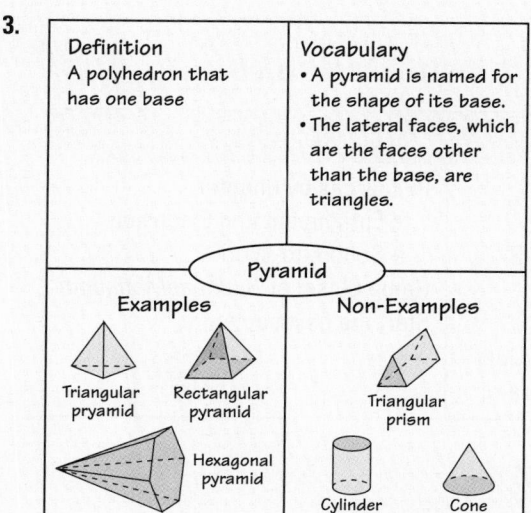

Definition	Vocabulary
A polyhedron that has one base	• A pyramid is named for the shape of its base. • The lateral faces, which are the faces other than the base, are triangles.

Pyramid

Examples: Triangular pryamid, Rectangular pyramid, Hexagonal pyramid

Non-Examples: Triangular prism, Cylinder, Cone

4–7. Available at *BigIdeasMath.com*.

List of Organizers
Available at *BigIdeasMath.com*

Comparison Chart
Concept Circle
Example and Non-Example Chart
Formula Triangle
Four Square
Idea (Definition) and Examples Chart
Information Frame
Information Wheel
Notetaking Organizer
Process Diagram
Summary Triangle
Word Magnet
Y Chart

About this Organizer

A **Four Square** can be used to organize information about a topic. Students write the topic in the "bubble" in the middle of the four square. Then students write concepts related to the topic in the four squares surrounding the bubble. Any concept related to the topic can be used. Encourage students to include concepts that will help them learn the topic. Students can place their four squares on note cards to use as a quick study reference.

Technology For the Teacher
Vocabulary Puzzle Builder

Answers

1. Front: Side:

Top:

2. Front: Side:

Top:

3. Front: Side:

Top:

4. $78\pi \approx 244.9 \text{ ft}^2$

5. $110\pi \approx 345.4 \text{ m}^2$

6. 132 cm^2 **7.** 100 mm^2

8. $108\pi \approx 339.12 \text{ in.}^2$ of material

9. $n - 2$

10. $7.5\pi \approx 23.6 \text{ in.}^2$ more paper

11. 4032 in.^2

Assessment Book

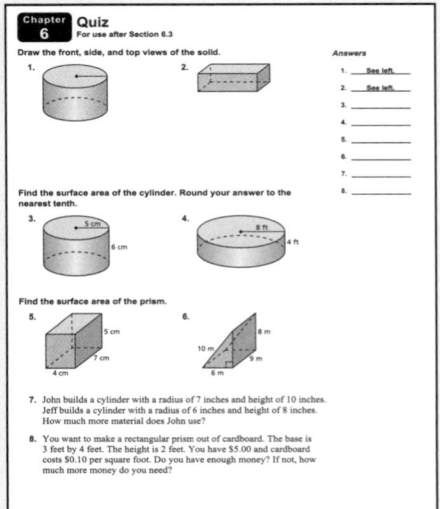

Alternative Quiz Ideas

100% Quiz **Math Log**
Error Notebook Notebook Quiz
Group Quiz Partner Quiz
Homework Quiz Pass the Paper

Math Log

Ask students to keep a math log for the chapter. Have them include diagrams, definitions, and examples. Everything should be clearly labeled. It might be helpful if they put the information in a chart. Students can add to the log as they are introduced to new topics.

Reteaching and Enrichment Strategies

If students need help. . .	If students got it. . .
Resources by Chapter • Study Help • Practice A and Practice B • Puzzle Time Lesson Tutorials *BigIdeasMath.com* Practice Quiz Practice from the Test Generator	Resources by Chapter • Enrichment and Extension • School-to-Work Game Closet at *BigIdeasMath.com* Start the next section

Technology For the Teacher

Answer Presentation Tool
Big Ideas Test Generator

Draw the front, side, and top views of the solid.   *(Section 6.1)*

1.

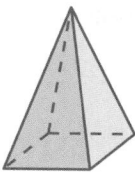

2.

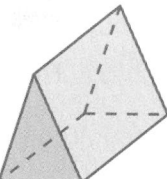

3.

Find the surface area of the cylinder. Round your answer to the nearest tenth.   *(Section 6.3)*

4.

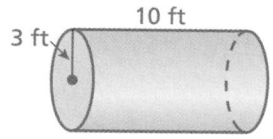

10 ft
3 ft

5.

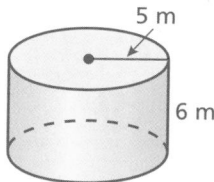

5 m
6 m

Find the surface area of the prism.   *(Section 6.2)*

6.

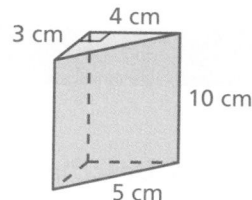

3 cm   4 cm
10 cm
5 cm

7.

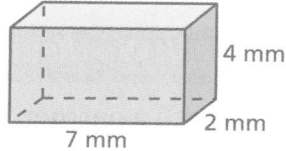

4 mm
2 mm
7 mm

8. MAILING TUBE What is the least amount of material needed to make the mailing tube?   *(Section 6.3)*

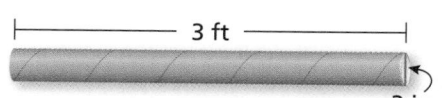

3 ft
3 in.

9. GEOMETRY Consider a prism that has n faces. Write an expression that represents the number of lateral faces.   *(Section 6.2)*

1.5 in.
1 in.
3 in.
4.5 in.

10. TOMATO PASTE How much more paper is used for the label of the large can of tomato paste than for the label of the small can?   *(Section 6.3)*

11. WOODEN CHEST All the faces of the wooden chest will be painted except for the bottom. Find the area to be painted, in *square inches*.   *(Section 6.2)*

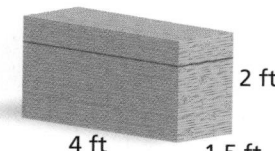

2 ft
4 ft
1.5 ft

6.4 Surface Areas of Pyramids

Essential Question How can you find the surface area of a pyramid?

Even though many well-known **pyramids** have square bases, the base of a pyramid can be any polygon.

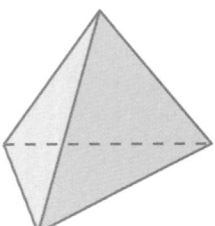

Triangular Base

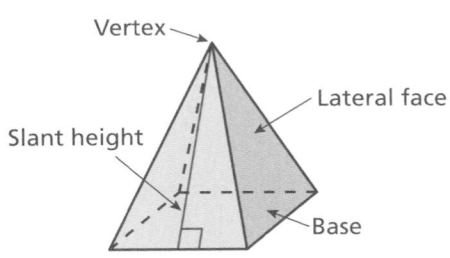

Vertex

Lateral face

Slant height

Base

Square Base

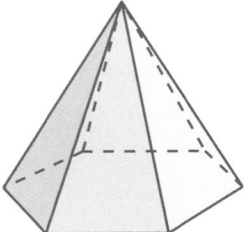

Hexagonal Base

---

1   ACTIVITY: Making a Scale Model

Work with a partner. Each pyramid has a square base.

- **Draw a net for a scale model of one of the pyramids. Describe your scale.**
- **Cut out the net and fold it to form a pyramid.**
- **Find the lateral surface area of the real-life pyramid.**

a. Cheops Pyramid in Egypt

Side = 230 m, Slant height ≈ 186 m

b. Muttart Conservatory in Edmonton

Side = 26 m, Slant height ≈ 27 m

c. Louvre Pyramid in Paris

Side = 35 m, Slant height ≈ 28 m

d. Pyramid of Caius Cestius in Rome

Side = 22 m, Slant height ≈ 29 m

Laurie's Notes

Introduction

For the Teacher

- **Goal:** Students will develop an intuitive understanding about how to find the surface area of a pyramid.
- **Big Idea:** In this section and the next, you want students to see the connection between the pyramid and cone—*structurally they are the same!* The pyramid and cone each have a base and a lateral portion.
- Discuss the vocabulary of pyramids and how they are named according to the base. Make a distinction between the slant height and the height.

Motivate

- Share information about the Great Pyramid of Egypt, also known as Cheops Pyramid.
- The Great Pyramid is the largest of the original *Seven Wonders of the World.* It was built in the 5th century B.C. and is estimated to have taken 100,000 men over 20 years to build it.
- The Great Pyramid is a square pyramid. It covers an area of 13 acres. The original height of the Great Pyramid was 485 feet, but due to erosion its height has declined to 450 feet. Each side of the square base is 755.5 feet in length (about 2.5 football field lengths).
- The Great Pyramid consists of approximately 2.5 million blocks that weigh from 2 tons to over 70 tons. The stones are cut so precisely that a credit card cannot fit between them.

Activity Notes

Activity 1

- This activity connects scale drawings with the study of pyramids.
- **?** "To make a net for a square pyramid, how many pieces will you need to make? Explain." 5 pieces; a square base and 4 congruent triangles
- To ensure a variety, assign one pyramid to each pair of students and make sure about $\frac{1}{4}$ of the class makes each pyramid.
- Students will need to decide on the scale they will use.
  Example: To make a scale model for pyramid A, assume the scale selected is 1 cm = 20 m.

  $$\frac{1 \text{ cm}}{20 \text{ m}} = \frac{x \text{ cm}}{230 \text{ m}} \rightarrow x = 11.5 \qquad \frac{1 \text{ cm}}{20 \text{ m}} = \frac{x \text{ cm}}{186 \text{ m}} \rightarrow x = 9.3$$

- Students will use their eyesight and knowledge of squares and isosceles triangles to construct the square and four isosceles triangles.
- When groups have finished, have several groups explain what scale they used and how they found the lateral surface area. Multiply the area of one triangular lateral face by 4.
- This hands-on experience of making the pyramid and finding the lateral surface area will help students remember the process and understand the formula.

Previous Learning

Students should know how to find the area of a triangle and should know the general properties of squares and isosceles triangles.

Activity Materials	
Introduction	**Textbook**
• models of pyramids	• scissors • tape • scrap paper • rulers • polygon frames

Start Thinking! and Warm Up

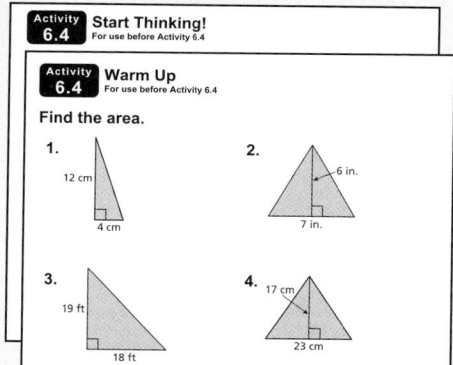

6.4 Record and Practice Journal

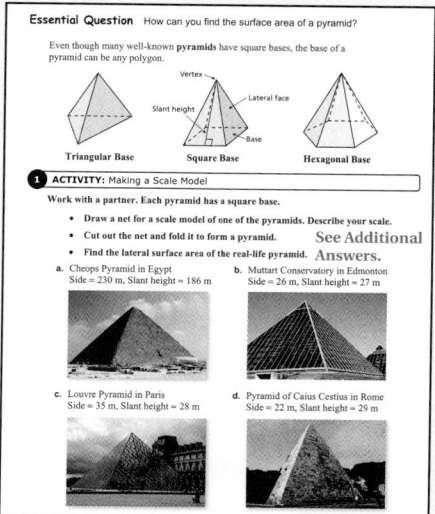

English Language Learners

Vocabulary

English learners may struggle with understanding the *slant height* of a pyramid. Use a skateboard ramp as an example. Ask students to find the length of the ramp. Most likely students will find the length of the slanted portion of the ramp. Compare this length to the slant height of a pyramid.

6.4 Record and Practice Journal

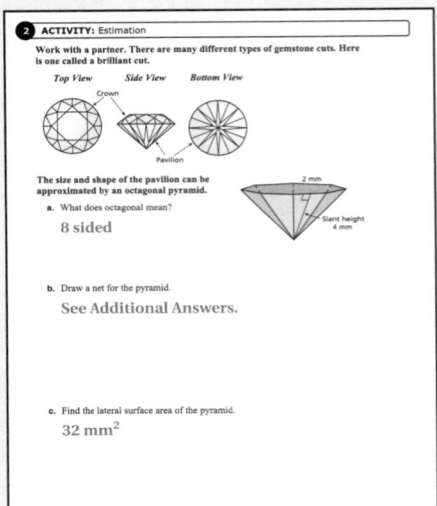

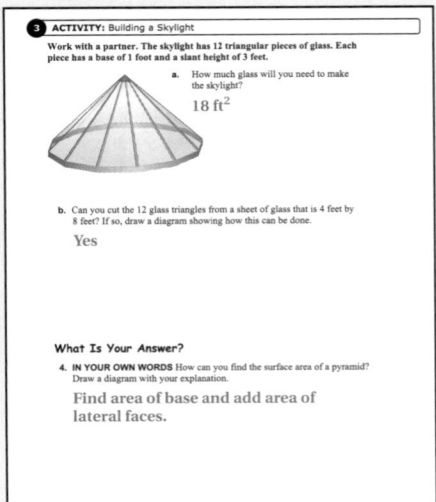

Activity 2

- Note that students are only asked to find the lateral surface area. This means that they are finding the surface area of 8 congruent isosceles triangles.
- **FYI:** The prefix octa- means eight. An octopus has 8 arms; when October was named in the Roman calendar, it was the 8th month; an octave on the piano has 8 notes; an octad is a group of 8 things.
- **?** "What common road sign is an octagon?" Stop sign
- The net includes the octagonal base, but the surface area of the base is not needed for this problem.
- Ask a volunteer to sketch his or her net at the board. If you have the appropriate snap together polygon frames, make the net.

Activity 3

- Again, only the lateral surface area is needed. Once students have found the area of one triangle, they need to multiply by 12.

What Is Your Answer?

- Have students work in pairs to answer the question.

Closure

- **Exit Ticket:** Sketch a net for a hexagonal pyramid and describe how to find the lateral surface area. Find the area of one of the lateral faces and multiply by 6.

Technology For the Teacher

Dynamic Classroom

The Dynamic Planning Tool
Editable Teacher's Resources at *BigIdeasMath.com*

ACTIVITY: Estimation

Work with a partner. There are many
different types of gemstone cuts. Here is
one called a brilliant cut.

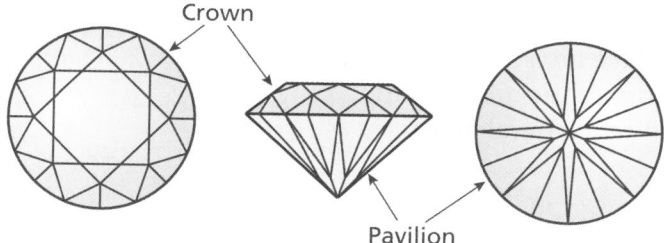

Top View   *Side View*   *Bottom View*

Crown

Pavilion

The size and shape of the pavilion can be
approximated by an octagonal pyramid.

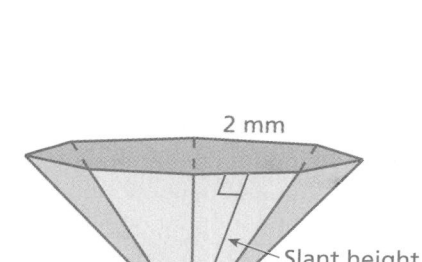

2 mm

Slant height
4 mm

a. What does octagonal mean?

b. Draw a net for the pyramid.

c. Find the lateral surface area of
the pyramid.

3 **ACTIVITY: Building a Skylight**

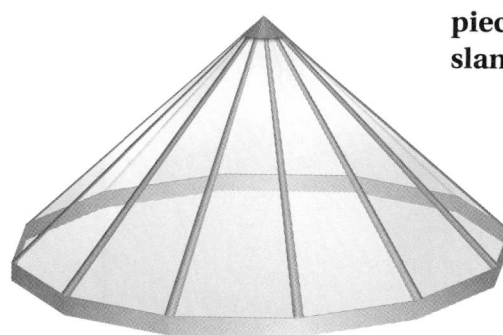

Work with a partner. The skylight has 12 triangular
pieces of glass. Each piece has a base of 1 foot and a
slant height of 3 feet.

a. How much glass will you need to make the skylight?

b. Can you cut the 12 glass triangles from a sheet of
glass that is 4 feet by 8 feet? If so, draw a diagram
showing how this can be done.

What Is Your Answer?

4. IN YOUR OWN WORDS How can you find the surface area of a pyramid?
Draw a diagram with your explanation.

Practice Use what you learned about the surface area of a pyramid to
complete Exercises 4–6 on page 274.

Check It Out
Lesson Tutorials
BigIdeasMath com

Key Vocabulary
regular pyramid, p. 272
slant height, p. 272

A **regular pyramid** is a pyramid whose base is a regular polygon. The lateral faces are triangles. The height of each triangle is the **slant height** of the pyramid.

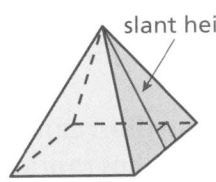 **Key Idea**

Surface Area of a Pyramid
The surface area S of a pyramid is the sum of the areas of the base and the lateral faces.

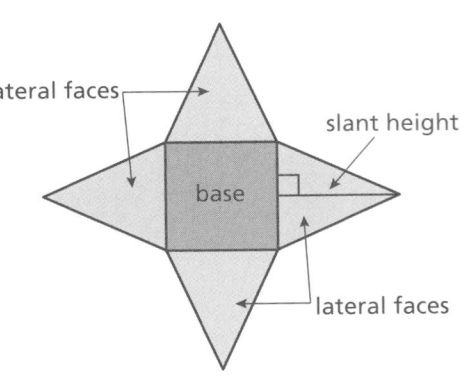

$S = $ area of base $+$ areas of lateral faces

Remember

In a regular polygon, all of the sides have the same length and all of the angles have the same measure.

EXAMPLE ① **Finding the Surface Area of a Square Pyramid**

Find the surface area of the regular pyramid.

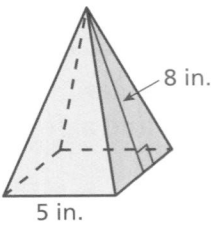

Draw a net.

Area of base   ***Area of a lateral face***

$5 \cdot 5 = 25$ $\dfrac{1}{2} \cdot 5 \cdot 8 = 20$

Find the sum of the areas of the base and the lateral faces.

$S = $ area of base $+$ areas of lateral faces

$= 25 + 20 + 20 + 20 + 20$

$= 105$

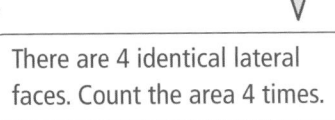
There are 4 identical lateral faces. Count the area 4 times.

∴ The surface area is 105 square inches.

On Your Own

1. What is the surface area of a square pyramid with a base side length of 9 centimeters and a slant height of 7 centimeters?

🔊 Multi-Language Glossary at BigIdeasMath✓com.

Laurie's Notes

Introduction

Connect

- **Yesterday:** Students discovered how to find the surface area of a pyramid by examining the net that makes up a pyramid.
- **Today:** Students will work with a formula for the surface area of a pyramid.

Motivate

- Ask students where they have heard about pyramids or have seen them before. Give groups of students 3–4 minutes to brainstorm a list. They may mention the pyramid on the back of U.S. dollar bills, camping tents, roof designs, tetrahedral dice, and of course, Egyptian pyramids.

Lesson Notes

Key Idea

- Introduce the vocabulary: regular pyramid, regular polygon, slant height.
- **?** "What information does the type of base give you about the lateral faces?" number of sides in the base = number of congruent isosceles triangles for the lateral surface area
- **?** "If you know the length of each side of the base, what else do you know?" the length of the base of the triangular lateral faces

Example 1

- Draw the net and label the known information. This should remind students of the work they did yesterday making a scale model of a pyramid.
- Write the formula in words first to model good, problem-solving techniques.
- Continue to ask questions as you find the total surface area: "How do you find the area of the base? How many lateral faces are there? What is the area of just one lateral face? How do you find the area of a triangle?"
- **Common Error:** In using the area formula for a triangle, the $\frac{1}{2}$ often produces a computation mistake. In this instance, students must multiply $\frac{1}{2} \times 5 \times 8$. Remind students that it's okay to change the order of the factors (Commutative Property). Rewriting the problem as $\frac{1}{2} \times 8 \times 5$ means that you can work with whole numbers: $\frac{1}{2} \times 8 \times 5 = 4 \times 5 = 20$.

On Your Own

- Encourage students to sketch a three-dimensional model of the pyramid and the net for the pyramid. Label the net with the known information.
- Ask a volunteer to share his or her work at the board.

Goal Today's lesson is finding the surface area of a pyramid using a formula.

Start Thinking! and Warm Up

| Lesson 6.4 | Warm Up |
| For use before Lesson 6.4 |

| Lesson 6.4 | Start Thinking! |
| For use before Lesson 6.4 |

Your neighbor needs to put a new roof on his gazebo. The roof is an octagonal pyramid. Why would knowing the surface area of the roof be useful information?

Extra Example 1

What is the surface area of a square pyramid with a base side length of 3 meters and a slant height of 6 meters? 45 m^2

On Your Own

1. 207 cm^2

Extra Example 2

Find the surface area of the regular pyramid.

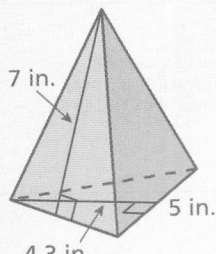

7 in.

5 in.

4.3 in.

63.25 in.^{2}

Extra Example 3

The slant height of the roof in Example 3 is 13 feet. One bundle of shingles covers 30 square feet. How many bundles of shingles should you buy to cover the roof? **16 bundles of shingles**

 On Your Own

2. 105.6 ft^2

3. 17 bundles

Differentiated Instruction

Kinesthetic

Photocopy nets of solids for students to cut out and assemble. Then have students draw their own nets to cut out and assemble.

Laurie's Notes

Example 2

- Remind students of the definition of a regular pyramid. This is important because the base, as drawn, doesn't look like an equilateral triangle. This is the challenge of representing a 3-dimensional figure on a flat 2-dimensional sheet of paper.
- Drawing the net is an important step. It allows the key dimensions to be labeled in a way that can be seen.
- Encourage mental math when multiplying $\frac{1}{2} \times 10 \times 8.7$ and $\frac{1}{2} \times 10 \times 14$. Ask students to share their strategies with other students.

Example 3

? "How does the lateral surface area of the roof relate to the bundles of shingles needed?" Lateral surface area divided by the area covered per bundle gives the number of bundles needed.

- Have students compute the lateral surface area. Some students may need to draw the triangular lateral face first before performing the computation.
- **FYI:** When shingles are placed on a roof, they need to overlap the shingle below. The coverage given per bundle takes into account the overlap.
- **Extension:** If a bundle of shingles sells for $34.75, what will the total cost be for the shingles? $764.50

On Your Own

- Give students sufficient time to do their work for each problem before asking volunteers to share their work at the board.

Closure

- **Exit Ticket:** Sketch a square pyramid with a slant height of 4 centimeters and a base side length of 3 centimeters. Sketch the net and find the surface area. 33 cm²

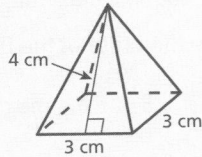

4 cm

3 cm

3 cm

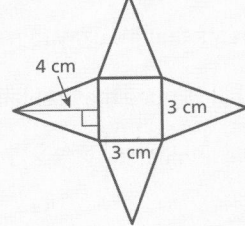

4 cm

3 cm

3 cm

Technology
For the **Teacher**

Dynamic Classroom

The Dynamic Planning Tool
Editable Teacher's Resources at *BigIdeasMath.com*

EXAMPLE 2 **Finding the Surface Area of a Triangular Pyramid**

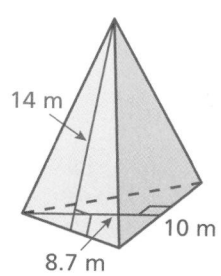

14 m

10 m

8.7 m

Find the surface area of the regular pyramid.

Draw a net.

Area of base

$\frac{1}{2} \cdot 10 \cdot 8.7 = 43.5$

Area of a lateral face

$\frac{1}{2} \cdot 10 \cdot 14 = 70$

10 m

8.7 m

14 m

Find the sum of the areas of the base and the lateral faces.

$S =$ area of base + areas of lateral faces

$= 43.5 + \underbrace{70 + 70 + 70}$

$= 253.5$

> There are 3 identical lateral faces. Count the area 3 times.

∴ The surface area is 253.5 square meters.

EXAMPLE 3 **Real-Life Application**

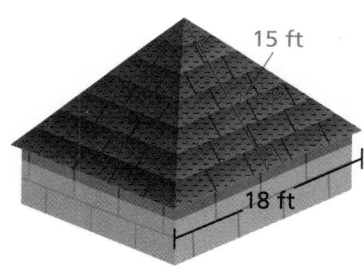

15 ft

18 ft

A roof is shaped like a square pyramid. One bundle of shingles covers 25 square feet. How many bundles should you buy to cover the roof?

The base of the roof does not need shingles. So, find the sum of the areas of the lateral faces of the pyramid.

Area of a lateral face

$\frac{1}{2} \cdot 18 \cdot 15 = 135$

There are four identical lateral faces. So, the sum of the areas of the lateral faces is

$135 + 135 + 135 + 135 = 540.$

Because one bundle of shingles covers 25 square feet, it will take $540 \div 25 = 21.6$ bundles to cover the roof.

∴ So, you should buy 22 bundles of shingles.

On Your Own

Now You're Ready
Exercises 4–12

2. What is the surface area of the pyramid at the right?

3. **WHAT IF?** In Example 3, one bundle of shingles covers 32 square feet. How many bundles should you buy to cover the roof?

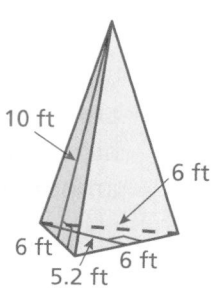

10 ft

6 ft

6 ft

6 ft

5.2 ft

Check It Out
Help with Homework
BigIdeasMath.com

 Vocabulary and Concept Check

1. **VOCABULARY** Which of the polygons could be the base for a regular pyramid?

 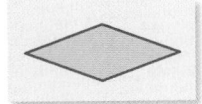 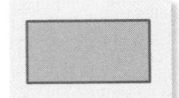 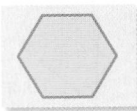

2. **VOCABULARY** Can a pyramid have rectangles as lateral faces? Explain.

3. **CRITICAL THINKING** Why is it helpful to know the slant height of a pyramid to find its surface area?

 **Practice and Problem Solving**

Use the net to find the surface area of the regular pyramid.

4.
3 in.
4 in.

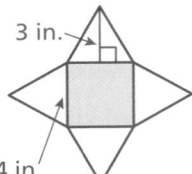

5.
9 mm
10 mm
Area of base
is 43.3 mm².

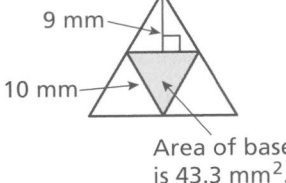

6.
6 m
6 m
Area of base
is 61.9 m².

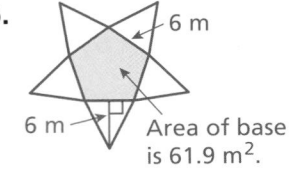

In Exercises 7–11, find the surface area of the regular pyramid.

 7.
9 ft
6 ft

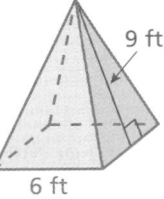

8.
6 cm
4 cm

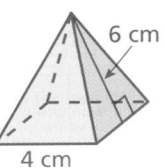

9.
10 yd
9 yd
7.8 yd

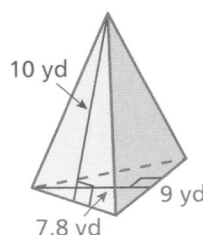

10.
10 in.
15 in.
13 in.

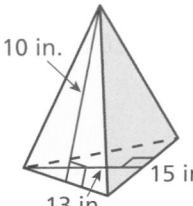

11.
20 mm
16 mm
Area of base
is 440.4 mm².

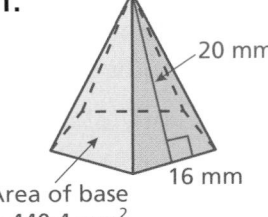

10 in.

12. **LAMPSHADE** The base of the lampshade is a regular hexagon with a side length of 8 inches. Estimate the amount of glass needed to make the lampshade.

13. **GEOMETRY** The surface area of a square pyramid is 85 square meters. The base length is 5 meters. What is the slant height?

Assignment Guide and Homework Check

Level	Day 1 Activity Assignment	Day 2 Lesson Assignment	Homework Check
Basic	4–6, 19–22	1–3, 7–13 odd, 12, 14	2, 9, 13, 14
Average	4–6, 19–22	1–3, 7–15 odd, 14	2, 9, 13, 14
Advanced	4–6, 19–22	1–3, 10, 11, 13, 15–18	10, 13, 16, 17

Common Errors

- **Exercises 7–11** Students may forget to add on the area of the base when finding the surface area. Remind them that when asked to find the surface area, the base is included.
- **Exercises 7–11** Students may add the wrong number of lateral face areas to the area of the base. Examine several different pyramids with different bases and ask if they can find a relationship between the number of sides of the base and the number of lateral faces. (They are the same.) Remind students that the number of sides on the base determines how many triangles make up the lateral surface area.
- **Exercise 12** Students may think that there is not enough information to solve the problem because it is not all labeled in the picture. Tell them to use the information in the word problem to finish labeling the picture. Also ask students to identify how many lateral faces are part of the lamp before they find the area of one face.

Vocabulary and Concept Check

1. the triangle and the hexagon

2. no; The lateral faces of a pyramid are triangles.

3. Knowing the slant height helps because it represents the height of the triangle that makes up each lateral face. So, the slant height helps you to find the area of each lateral face.

Practice and Problem Solving

4. 40 in.^{2}

5. 178.3 mm^2

6. 151.9 m^2

7. 144 ft^2

8. 64 cm^2

9. 170.1 yd^2

10. 322.5 in.^{2}

11. 1240.4 mm^2

12. 240 in.^{2}

13. 6 m

6.4 Record and Practice Journal

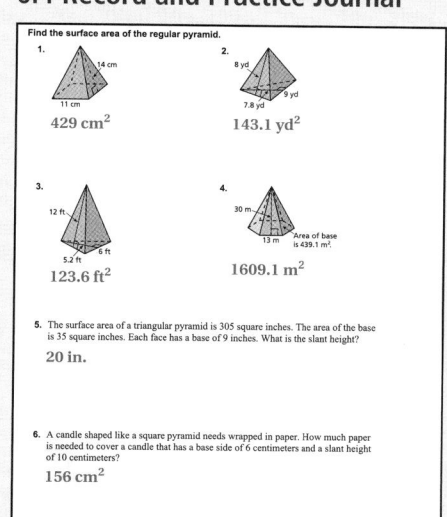

Find the surface area of the regular pyramid.

1. 429 cm^2
2. 143.1 yd^2
3. 123.6 ft^2
4. 1609.1 m^2

5. The surface area of a triangular pyramid is 305 square inches. The area of the base is 35 square inches. Each face has a base of 9 inches. What is the slant height?
   20 in.

6. A candle shaped like a square pyramid needs wrapped in paper. How much paper is needed to cover a candle that has a base side of 6 centimeters and a slant height of 10 centimeters?
   156 cm^2

Practice and Problem Solving

14. 34 ft²

15. See *Taking Math Deeper*.

16. The slant height is greater. The height is the distance between the top and the point on the base directly beneath it. The distance from the top to any other point on the base is greater than the height.

17. 124 cm²

18. greater than; It would take more material to make the lateral faces than the base.

Fair Game Review

19. $A \approx 452.16$ units²; $C \approx 75.36$ units

20. $A \approx 200.96$ units²; $C \approx 50.24$ units

21. $A \approx 572.265$ units²; $C \approx 84.78$ units

22. B

Mini-Assessment

Find the surface area of the regular pyramid.

1.

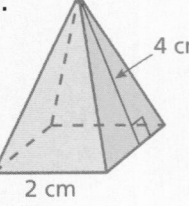

4 cm

2 cm

20 cm²

2.

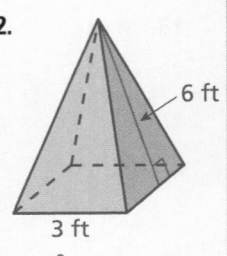

6 ft

3 ft

45 ft²

3. Find the surface area of the roof of the doll house. 480 in.²

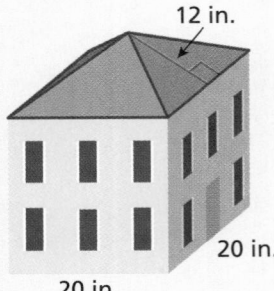

12 in.

20 in.

20 in.

T-275

Taking Math Deeper

Exercise 15

If you have ever sewn clothing from a pattern, you know that *on the bias* means that you are cutting against the weave of the fabric. Most patterns, like this one, don't allow cutting on the bias. The pieces must be cut with the weave.

① **a.** Find the area of the 8 pieces.

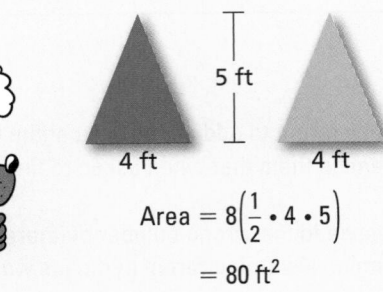

Four of each

5 ft

4 ft   4 ft

$$\text{Area} = 8\left(\frac{1}{2} \cdot 4 \cdot 5\right)$$
$$= 80 \text{ ft}^2$$

② **b.** Draw a diagram.

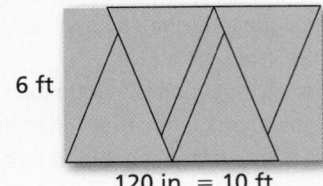

6 ft

120 in. = 10 ft

③ Answer the question.

For each color, you cut the four pieces from fabric that is 72 inches (6 ft) wide and 120 inches (10 ft) long.

Fabric Area = $2(6 \cdot 10) = 120$ ft²

Area of 8 Pieces = 80 ft²

c. Area of Waste = $120 - 80 = 40$ ft²

Project

Use construction paper and a pencil to create an "umbrella" using the least possible amount of paper. The umbrella should be similar to the one in the exercise, using a scale of 1 inch to 1 foot.

Reteaching and Enrichment Strategies

If students need help...	If students got it...
Resources by Chapter • Practice A and Practice B • Puzzle Time Record and Practice Journal Practice Differentiating the Lesson Lesson Tutorials Skills Review Handbook	Resources by Chapter • Enrichment and Extension • School-to-Work Start the next section

14. BMX You are building a bike ramp that is shaped like a square pyramid. You use two 4-foot by 8-foot sheets of plywood. How much plywood do you have left over?

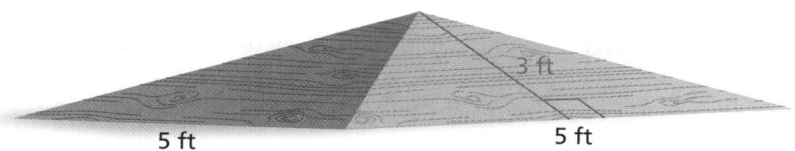

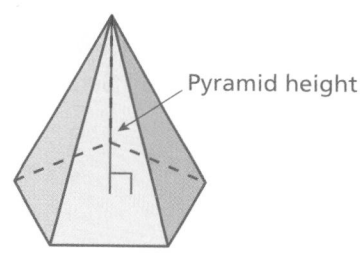

15. UMBRELLA You are making an umbrella that is shaped like a regular octagonal pyramid.

   a. Estimate the amount of fabric that is needed to make the umbrella.

   b. The fabric comes in rolls that are 72 inches wide. You don't want to cut the fabric "on the bias". Find out what this means. Then, draw a diagram of how you can cut the fabric most efficiently.

   c. How much fabric is wasted?

16. REASONING The *height* of a pyramid is the distance between the base and the top of the pyramid. Which is greater, the height of a pyramid or the slant height? Explain your reasoning.

17. TETRAHEDRON A tetrahedron is a triangular pyramid whose four faces are identical equilateral triangles. The total lateral surface area is 93 square centimeters. Find the surface area of the tetrahedron.

18. ⟨Reasoning⟩ Is the total area of the lateral faces of a pyramid *greater than*, *less than*, or *equal* to the area of the base? Explain.

Fair Game Review What you learned in previous grades & lessons

Find the area and circumference of the circle. Use 3.14 for π.   *(Skills Review Handbook)*

19.
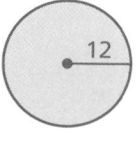
12

20.
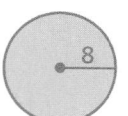
8

21.
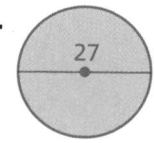
27

22. MULTIPLE CHOICE A youth baseball diamond is similar to a professional baseball diamond. The ratio of the perimeters is 2 : 3. The distance between bases on a youth diamond is 60 feet. What is the distance between bases on a professional diamond?   *(Section 5.3)*

   (A)   40 ft   (B)   90 ft   (C)   120 ft   (D)   180 ft

Essential Question How can you find the surface area of a cone?

A cone is a solid with one circular base and one vertex.

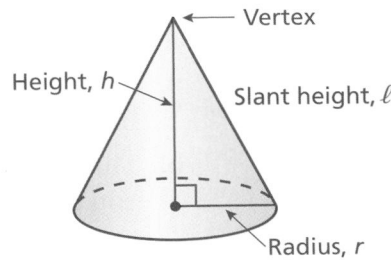

Vertex

Height, h

Slant height, ℓ

Radius, r

---

1   ACTIVITY: Finding the Surface Area of a Cone

Work with a partner.

- **Draw a circle with a radius of 3 inches.**
- **Mark the circumference of the circle into six equal parts.**
- **The circumference of the circle is $2(\pi)(3) = 6\pi$. So each of the six parts on the circle has a length of π. Label each part.**
- **Cut out one part as shown. Then, make a cone.**

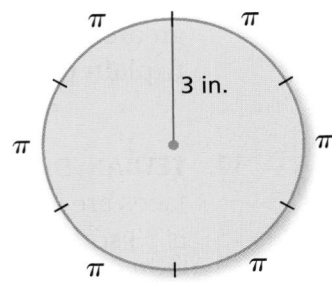

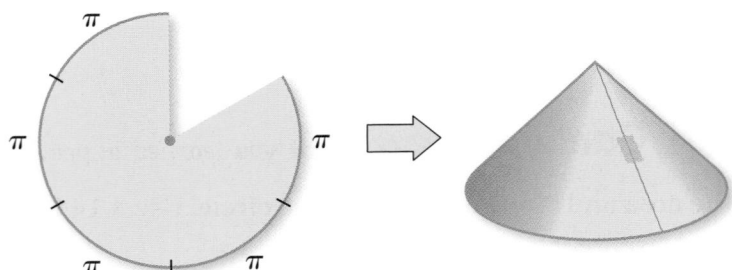

a. The base of the cone should be a circle. Explain why the circumference of the base is 5π.

b. Find the radius of the base.

c. What is the area of the original circle?

d. What is the area of the circle with one part missing?

e. Describe the surface area of the cone. Use your description to find the surface area, including the base.

Laurie's Notes

Introduction

For the Teacher

- **Goal:** Students will develop an intuitive understanding about how to find the surface area of a cone.
- **Big Idea:** Recall the connection between the pyramid and cone—*structurally they are the same.*
- Today's investigation will likely be a real surprise to students for several reasons. First, it's not obvious what the lateral portion of a cone is when it is opened up and placed flat. Secondly, it is very unusual to try to develop the formula for the surface area of a cone. Often it is simply stated and students are told to accept it as being true.
- There will be many computations. Caution students to be patient, write neatly, and keep their work organized.

Motivate

- Hold a paper cone (wrapper of an ice cream cone or homemade).
- Discuss with students the nets that they have seen in this chapter.
- **?** "What do you think the net for a cone is?" Students generally guess (incorrectly) that it's a triangle of some sort.

Activity Notes

Activity 1

- Discuss the vocabulary of cones. Make a distinction between the slant height and the height of the cone.
- **Common Error:** Students think height and slant height have the same length. In the diagram shown at the top of the page, the right triangle may help to explain the difference in length even though the Pythagorean Theorem is taught in future courses.
- **?** "How do you find the circumference of a circle? What is the circumference for this circle?" $C = 2\pi r = 2\pi(3) = 6\pi$
- It should seem reasonable to students that because the circumference is 6π and there are six equal pieces, each piece has a length of π inches.
- Students could use a ruler to approximate the radius of the base, *or*, substitute the circumference of the base (5π) in the circumference formula and solve for r.

$$5\pi = 2\pi r$$
$$5 = 2r \qquad \text{Divide both sides by } \pi.$$
$$\frac{5}{2} = r \qquad \text{Divide both sides by 2.}$$

- Have students talk through their work for finding the surface area. They should recognize the two components of finding the surface area of a cone. There is a circular base, so use the area of a circle formula, and there is the lateral surface that is a circle with a sector missing. The slant height of the cone is the radius of the flattened circle with a sector removed.

Previous Learning

Students should know how to find the area and circumference of a circle.

Activity Materials	
Introduction	**Textbook**
• paper cone	• scissors
	• tape
	• scrap paper
	• rulers

Start Thinking! and Warm Up

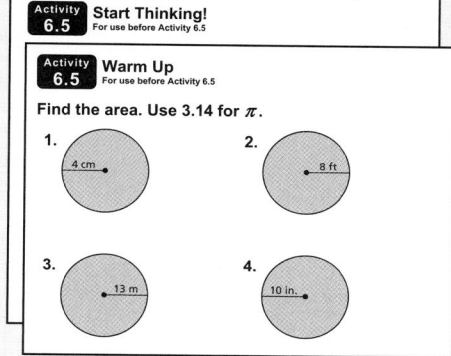

6.5 Record and Practice Journal

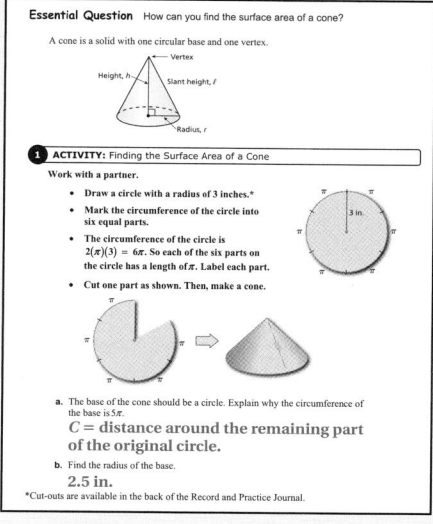

Differentiated Instruction

Kinesthetic

Place models of prisms, cylinders, pyramids, and cones around the room. Encourage students to sketch the objects from different points of view, such as from the floor or from above the object. Ask for volunteers to show their sketches of the objects. Discuss how the sketches differ depending on the position and the perspective of the drawer.

6.5 Record and Practice Journal

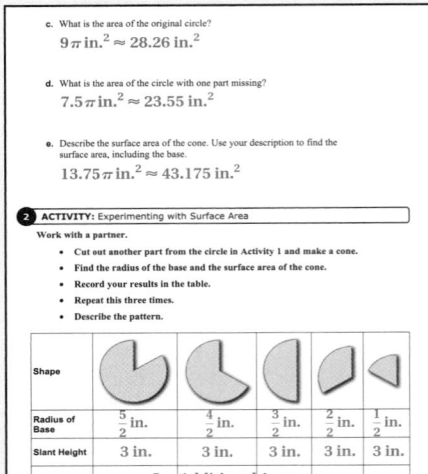

c. What is the area of the original circle?
9π in.$^2 \approx 28.26$ in.^{2}

d. What is the area of the circle with one part missing?
7.5π in.$^2 \approx 23.55$ in.^{2}

e. Describe the surface area of the cone. Use your description to find the surface area, including the base.
13.75π in.$^2 \approx 43.175$ in.^{2}

2 ACTIVITY: Experimenting with Surface Area

Work with a partner.
- Cut out another part from the circle in Activity 1 and make a cone.
- Find the radius of the base and the surface area of the cone.
- Record your results in the table.
- Repeat this three times.
- Describe the pattern.

Shape					
Radius of Base	$\frac{5}{2}$ in.	$\frac{4}{2}$ in.	$\frac{3}{2}$ in.	$\frac{2}{2}$ in.	$\frac{1}{2}$ in.
Slant Height	3 in.	3 in.	3 in.	3 in.	3 in.
Surface Area	See Additional Answers.				

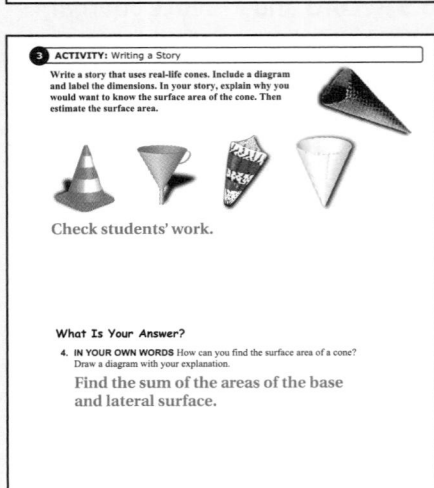

3 ACTIVITY: Writing a Story

Write a story that uses real-life cones. Include a diagram and label the dimensions. In your story, explain why you would want to know the surface area of the cone. Then estimate the surface area of the cone.

Check students' work.

What Is Your Answer?

4. IN YOUR OWN WORDS How can you find the surface area of a cone? Draw a diagram with your explanation.

Find the sum of the areas of the base and lateral surface.

Laurie's Notes

Activity 2

- **Big Idea:** To understand what happens when additional sectors of the circle are cut out, think of surface area of a cone in the following way:

$$\text{Surface Area} = \text{Area of Base} + \text{Lateral Area}$$
$$= \pi r^2 + \text{a fraction of the original circle}$$

The radius of the base decreases each time. The lateral area decreases by one-sixth of the original circle each time. The original area was $\pi r^2 = 9\pi$. Review the pattern in the answer key.

- Students may need assistance in recording their work for each subsequent sector that is removed. They should record their answers in terms of π and leave fractions as improper fractions.
- The radius of the base can be measured (approximated) or calculated from knowing the circumference.
- The slant height stays fixed. It is the original 3-inch radius.
- The surface area has two parts, the base and the lateral portion.
- **?** "What patterns do you observe in the table?" This may be difficult depending upon how the students record their work. Encourage students to leave their answers in terms of π.
- **?** "Each time a sector is removed, what happens to the area of the base?" decreases "What happens to the height of the cone?" It increases.
- Make sure students understand that you are asking about the *height* of the cone and not the *slant height* in the previous question.

Activity 3

- **Writing:** This activity allows students to display their creative writing skills.

What Is Your Answer?

- **Think-Pair-Share:** Students should read each question independently and then work with a partner to answer the questions. When they have answered the questions, the pair should compare their answers with another group and discuss any discrepancies.

Closure

- **Exit Ticket:** Sketch a net for a cone. What are the components of the net? circular base and a portion of a circle for the lateral surface

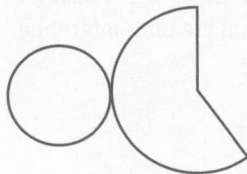

Technology For the Teacher

Dynamic Classroom

The Dynamic Planning Tool
Editable Teacher's Resources at *BigIdeasMath.com*

2 ACTIVITY: Experimenting with Surface Area

Work with a partner.

- Cut out another part from the circle in Activity 1 and make a cone.
- Find the radius of the base and the surface area of the cone.
- Record your results in the table.
- Repeat this three times.
- Describe the pattern.

Shape			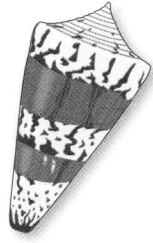		
Radius of Base					
Slant Height					
Surface Area					

3 ACTIVITY: Writing a Story

Write a story that uses real-life cones. Include a diagram and label the dimensions. In your story, explain why you would want to know the surface area of the cone. Then, estimate the surface area.

What Is Your Answer?

4. **IN YOUR OWN WORDS** How can you find the surface area of a cone? Draw a diagram with your explanation.

Use what you learned about the surface area of a cone to complete Exercises 4–6 on page 280.

Key Vocabulary
slant height, p. 278

The distance from the vertex of a cone to any point on the edge of its base is called the **slant height** of the cone.

Key Idea

Surface Area of a Cone

Words The surface area S of a cone is the sum of the areas of the base and the lateral surface.

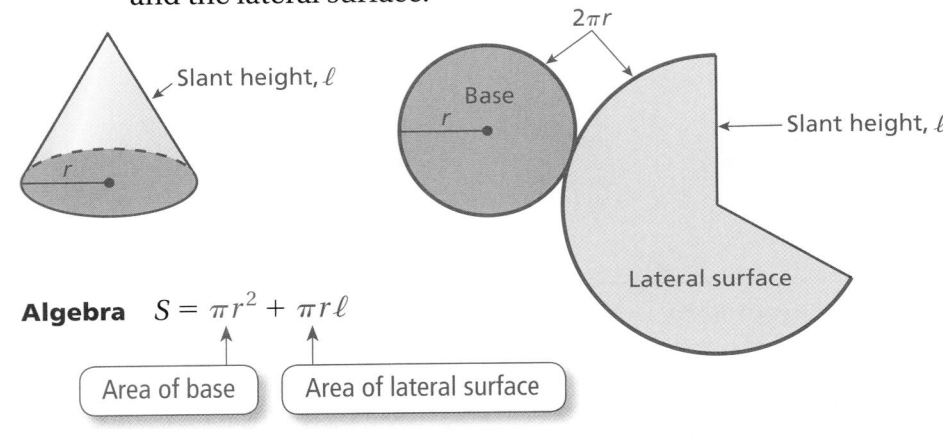

Algebra $S = \pi r^2 + \pi r \ell$

Area of base   Area of lateral surface

EXAMPLE (1)   **Finding the Surface Area of a Cone**

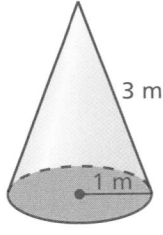

Find the surface area of the cone. Round your answer to the nearest tenth.

Draw a net.

$$S = \pi r^2 + \pi r \ell$$
$$= \pi(1)^2 + \pi(1)(3)$$
$$= \pi + 3\pi$$
$$= 4\pi \approx 12.6$$

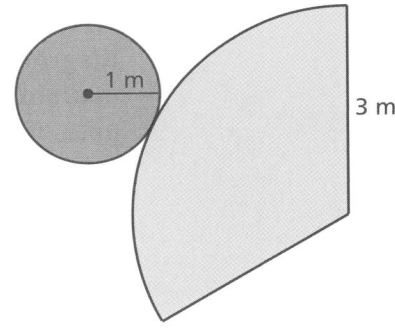

∴ The surface area is about 12.6 square meters.

On Your Own

Now You're Ready
Exercises 4–9

Find the surface area of the cone. Round your answer to the nearest tenth.

1.

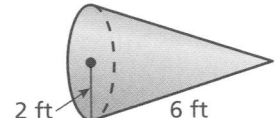

2.

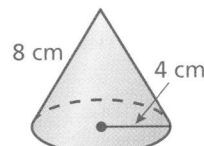

Laurie's Notes

Introduction

Connect

- **Yesterday:** Students discovered how to find the surface area of a cone by constructing a series of cones from a circle with a sector removed.
- **Today:** Students will work with a formula for the surface area of a cone.

Motivate

- Wear a party hat (or a dunce cap) to class.
- Tell a story as to why you might be wearing the hat.
- **?** "How much paper was used to make my hat?" Students should distinguish between the base (which is missing) and the lateral surface area.
- **?** "What information would you need in order to find the amount of paper used?" This is open ended, as the formula hasn't been given yet.

Lesson Notes

- Introduce the vocabulary: slant height.

Key Idea

- Write the formula in words. Draw the cone and net, labeled with the known information. Notice that the circumference of the base equals the arc length of the lateral surface.
- Write the symbolic formula. The area of the base is πr^2, which is no surprise. The area of the lateral surface may not be obvious to students. If your class has good numeric skills, you can show that the formula works for each case in the table they made yesterday. Otherwise, the manipulation is beyond the scope of most middle school students.
- Be sure that students recognize that in finding the lateral surface area, the r in the formula is the radius of the base.

Example 1

- Draw the net first and label the known information.
- **?** Continue to ask questions as you find the total surface area. "How do you find the area of the base? What is the slant height? What is the area of the lateral surface?"
- Notice that the work is done in terms of π. It is not until the last step that 3.14 is substituted for π.
- **Representation:** Encourage students to use parentheses to represent multiplication. Using the $\times$ symbol would make the expression confusing.
- **?** "Why can π be added to 3π?" like terms Students forget that the coefficient of π is 1.
- **Common Error:** Students may say $\pi + 3\pi = 3\pi^2$.

On Your Own

- Encourage students to sketch a three-dimensional model of the cone, and the net for the cone. Label the net with the known information.

Goal Today's lesson is finding the surface area of a cone using a formula.

Lesson Materials
Introduction
• party hat (cone shaped)

Start Thinking! and Warm Up

Lesson 6.5 Warm Up For use before Lesson 6.5

Lesson 6.5 Start Thinking! For use before Lesson 6.5

Use what you learned from the activity to explain to a partner how to find the lateral surface area of a cone.

Extra Example 1

Find the surface area of a cone with a radius of 6 inches and a slant height of 8 inches. Round your answer to the nearest tenth. $84\pi \approx 263.8$ in.^{2}

On Your Own

1. $16\pi \approx 50.2$ ft^2
2. $48\pi \approx 150.7$ cm^2

Laurie's Notes

Extra Example 2

The surface area of a cone is 48π square feet. The radius of the cone is 4 feet. What is the slant height ℓ of the cone? 8 ft

Extra Example 3

In Example 3, suppose the slant height of the party hat is 7 inches. How much paper do you need to make the hat? $24.5\pi \approx 77$ in.^{2}

On Your Own

3. 5 m

4. yes; The surface area changes from $\pi r \ell$ to $\pi r(2\ell) = 2(\pi r \ell)$, so the amount of paper doubles.

English Language Learners

Auditory

Have students work in groups and give each group a model of a prism, cylinder, pyramid, or cone. Ask students questions about the solids. "How many faces, edges, and vertices does a pyramid have?" "Which solid rolls?" "How would you describe a cone?" "How are a cone and cylinder alike?" "How are they different?"

Example 2

- Work through the problem with your students, annotating the steps as you go, as shown in the book.
- ❓ "What information is known to solve the problem?" Listen for not only the dimensions, but also the formula for the surface area of a cone.
- Substitute for the known variables, then give students time to work through the problem on their own. This will help you determine their comfort with manipulating an expression involving π.
- **Representation:** π and 5 are both factors in each expression, 75π and $5\pi\ell$. Sometimes it is helpful to represent the equation in the following way before dividing.

$$75\pi = 5\pi\ell$$
$$(15)(5\pi) = (5\pi)(\ell)$$

Now divide both sides by 5π.

Example 3

- If you have a party hat, use it to help students visualize the problem.

On Your Own

- Give students sufficient time to do their work for each problem before asking volunteers to share their work at the board.

Closure

- **Exit Ticket**: Have students find the amount of paper used to make your party hat. Answers will vary depending upon the party hat.

The Dynamic Planning Tool
Editable Teacher's Resources at *BigIdeasMath.com*

EXAMPLE 2 **Finding the Slant Height of a Cone**

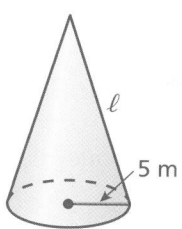

The surface area of the cone is 100π square meters. What is the slant height ℓ of the cone?

$$S = \pi r^2 + \pi r \ell$$ Write formula.

$$100\pi = \pi(5)^2 + \pi(5)(\ell)$$ Substitute.

$$100\pi = 25\pi + 5\pi\ell$$ Simplify.

$$75\pi = 5\pi\ell$$ Subtract 25π from each side.

$$15 = \ell$$ Divide each side by 5π.

⋮• The slant height is 15 meters.

EXAMPLE 3 **Real-Life Application**

You design a party hat. You attach a piece of elastic along a diameter. (a) How long is the elastic? (b) How much paper do you need to make the hat?

a. To find the length of the elastic, find the diameter of the base.

$$C = \pi d$$ Write formula.

$$22 \approx (3.14)d$$ Substitute.

$$7.0 \approx d$$ Solve for d.

⋮• The elastic is about 7 inches long.

5 in.

$C = 22$ in.

b. To find how much paper you need, find the lateral surface area.

$$S = \pi r \ell$$ ← Do not include the area of the base in the formula.

$$= \pi(3.5)(5)$$ Substitute.

$$= 17.5\pi \approx 55$$ Multiply.

⋮• You need about 55 square inches of paper to make the hat.

Remember

The diameter d of a circle is two times the radius r.

$$d = 2r$$

On Your Own

Now You're Ready
Exercises 10–14

3. WHAT IF? In Example 2, the surface area is 50π square meters. What is the slant height of the cone?

4. WHAT IF? In Example 3, the slant height of the party hat is doubled. Does the amount of paper used double? Explain.

Vocabulary and Concept Check

1. **VOCABULARY** Is the base of a cone a polygon? Explain.

2. **CRITICAL THINKING** In the formula for the surface area of a cone, what does $\pi r \ell$ represent? What does πr^2 represent?

3. **REASONING** Write an inequality comparing the slant height ℓ and the radius r of a cone.

Practice and Problem Solving

Find the surface area of the cone. Round your answer to the nearest tenth.

① **4.**
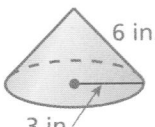
6 in.
3 in.

5.
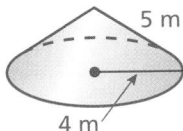
5 m
4 m

6.
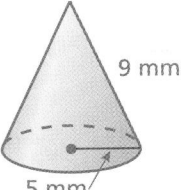
9 mm
5 mm

7.
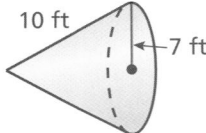
10 ft
7 ft

8.
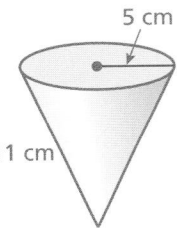
5 cm
11 cm

9.
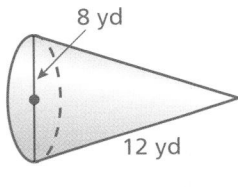
8 yd
12 yd

Find the slant height ℓ of the cone.

② **10.** $S = 33\pi \text{ in.}^2$
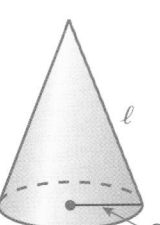
ℓ
3 in.

11. $S = 126\pi \text{ cm}^2$
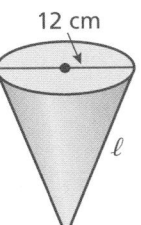
12 cm
ℓ

12. $S = 60\pi \text{ ft}^2$
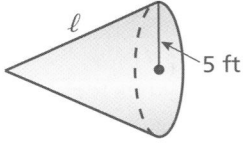
ℓ
5 ft

③ **13. NÓN LÁ** How much material is needed to make the Nón Lá Vietnamese leaf hat?

14. **PAPER CUP** A paper cup shaped like a cone has a diameter of 6 centimeters and a slant height of 7.5 centimeters. How much paper is needed to make the cup?

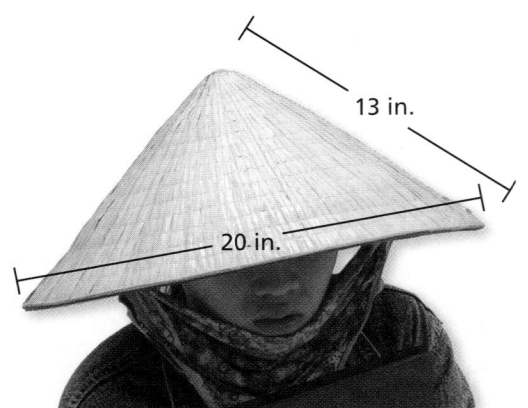
13 in.
20 in.

Assignment Guide and Homework Check

Level	Day 1 Activity Assignment	Day 2 Lesson Assignment	Homework Check
Basic	4–6, 23–26	1–3, 7–11, 13–16	8, 10, 14, 16
Average	4–6, 23–26	1–3, 7–11, 16–19	8, 10, 16, 18
Advanced	4–6, 23–26	1–3, 8, 10, 12, 16, 18–22	8, 10, 16, 20

Common Errors

- **Exercises 4–9** Students may forget to add the area of the base to the area of the lateral surface. Remind them of the net and the different parts for which they need to find the areas.
- **Exercises 4–12** Students may square the radius when finding the lateral surface area.
- **Exercises 9, 11, and 13–17** Students may use the diameter to find the surface area instead of the radius. Remind them to make sure they know the radius before finding the surface area.
- **Exercise 13** Students may include the surface area of the base when finding the amount of material needed for the hat. Ask them to describe how they would make the hat. Lead this discussion toward making the point that there is no base for the hat.
- **Exercises 15–17** Students may forget to convert the dimensions to the same unit of measure. Remind them to convert one dimension to the other unit of measure before finding the surface area.

6.5 Record and Practice Journal

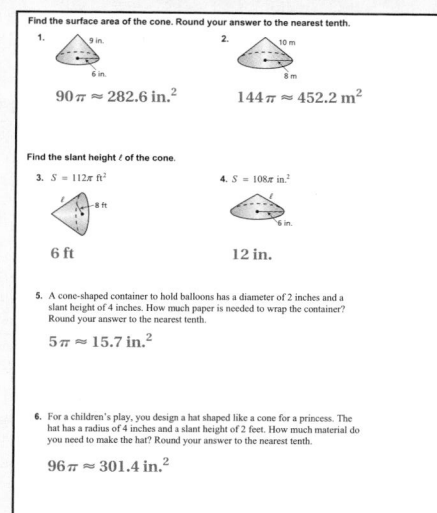

Find the surface area of the cone. Round your answer to the nearest tenth.

1. $90\pi \approx 282.6$ in.^{2}
2. $144\pi \approx 452.2$ m^2

Find the slant height ℓ of the cone.

3. $S = 112\pi$ ft^2; **6 ft**
4. $S = 108\pi$ in.^{2}; **12 in.**

5. A cone-shaped container to hold balloons has a diameter of 2 inches and a slant height of 4 inches. How much paper is needed to wrap the container? Round your answer to the nearest tenth.
$5\pi \approx 15.7$ in.^{2}

6. For a children's play, you design a hat shaped like a cone for a princess. The hat has a radius of 4 inches and a slant height of 2 feet. How much material do you need to make the hat? Round your answer to the nearest tenth.
$96\pi \approx 301.4$ in.^{2}

Vocabulary and Concept Check

1. no; The base of a cone is a circle. A circle is not a polygon.

2. $\pi r \ell$ is the lateral surface area and πr^2 is the area of the base.

3. $\ell > r$

Practice and Problem Solving

4. $27\pi \approx 84.8$ in.^{2}

5. $36\pi \approx 113.0$ m^2

6. $70\pi \approx 219.8$ mm^2

7. $119\pi \approx 373.7$ ft^2

8. $80\pi \approx 251.2$ cm^2

9. $64\pi \approx 201.0$ yd^2

10. 8 in.

11. 15 cm

12. 7 ft

13. $130\pi \approx 408.2$ in.^{2}

14. $22.5\pi \approx 70.65$ cm^2

15. $360\pi \approx 1130.4$ in.^{2}; $2.5\pi \approx 7.85$ ft^2

16. $8700\pi \approx 27{,}318$ mm^2; $87\pi \approx 273.18$ cm^2

17. $96\pi \approx 301.44$ ft^2; $\frac{32}{3}\pi \approx 33.49\overline{3}$ yd^2

Practice and Problem Solving

18. See *Taking Math Deeper*.

19. 12%

20. The slant height is greater. The height is the shortest distance from the vertex to the point on the base directly beneath the vertex. So, the distance from the vertex to any other point on the base is greater than the height.

21. the lateral surface area

22. the pyramid; The pyramid's surface area is $x^2 + 2xy$. The cone's surface area is $\frac{\pi}{4}x^2 + \frac{\pi}{2}xy$. Because $x^2 > \frac{\pi}{4}x^2$ and $2xy > \frac{\pi}{2}xy$, the pyramid's surface area is greater.

Fair Game Review

23. 45 in.^{2}

24. about 28.345 m^2

25. 16 ft^2

26. B

Mini-Assessment

Find the surface area of the cone. Round your answer to the nearest tenth.

1.
5 in.
2 in.
$14\pi \approx 44.0$ in.^{2}

2.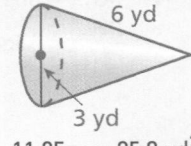
6 yd
3 yd
$11.25\pi \approx 35.3$ yd^2

3. How much paper was used to make the party hat? Round your answer to the nearest square inch.

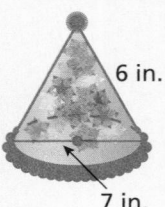
$21\pi \approx 66$ in.^{2}
6 in.
7 in.

Taking Math Deeper

Exercise 18

The terminology for the shingle packaging varies a little. It usually depends on the country in which the shingles are manufactured. The standard is usually that there are 3 bundles in a "square" and a square of shingles covers 100 square feet. (For some manufacturers, there are 4 bundles in a square.)

A bundle of asphalt shingles weighs about 70 pounds.

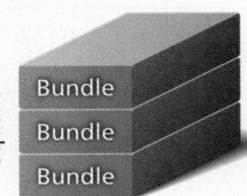
Square
Bundle
Bundle
Bundle

① Find the surface area of the roof.

$$S = \pi r \ell$$
$$= \pi \cdot 6 \cdot 13$$
$$= 78\pi$$
$$\approx 245 \text{ ft}^2$$

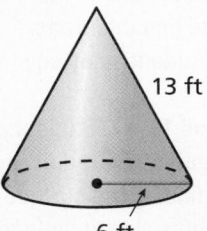
13 ft
6 ft

② Do the math.

$$\frac{245 \text{ ft}^2}{32 \text{ ft}^2 \text{ per bundle}} \approx 7.66 \text{ bundles}$$

③ Answer the question (in a real-life context).

Because you can't buy parts of bundles, you should buy 8 bundles of shingles to cover the roof.

Maybe 9 to be safe

Reteaching and Enrichment Strategies

If students need help...	If students got it...
Resources by Chapter • Practice A and Practice B • Puzzle Time Record and Practice Journal Practice Differentiating the Lesson Lesson Tutorials Skills Review Handbook	Resources by Chapter • Enrichment and Extension • School-to-Work • Financial Literacy Start the next section

Find the surface area of the cone with diameter d and slant height ℓ.

15. $d = 2$ ft
  $\ell = 18$ in.

16. $d = 12$ cm
  $\ell = 85$ mm

17. $d = 4$ yd
  $\ell = 10$ ft

18. ROOF A roof is shaped like a cone with a diameter of 12 feet. One bundle of shingles covers 32 square feet. How many bundles should you buy to cover the roof?

19. MEGAPHONE Two stickers are placed on opposite sides of the megaphone. Estimate the percent of the surface area of the megaphone covered by the stickers. Round your answer to the nearest percent.

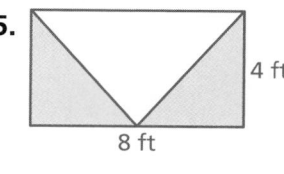

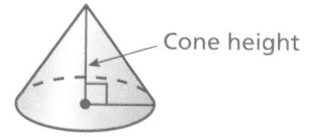

20. REASONING The height of a cone is the distance between the base and the vertex. Which is greater, the height of a cone or the slant height? Explain your reasoning.

21. GEOMETRY The surface area of a cone is also given as $S = \frac{1}{2}C\ell + B$, where C is the circumference and ℓ is the slant height. What does $\frac{1}{2}C\ell$ represent?

22. **Critical Thinking** A cone has a diameter of x millimeters and a slant height of y millimeters. A square pyramid has a base side length of x millimeters and a slant height of y millimeters. Which has the greater surface area? Explain.

Fair Game Review What you learned in previous grades & lessons

Find the area of the shaded region. Use 3.14 for π. *(Skills Review Handbook)*

23.

6 in.
4 in.
15 in.

24.

3 m
5 m

25.

4 ft
8 ft

26. MULTIPLE CHOICE Which best describes a translation? *(Section 5.5)*

    A a flip

    B a slide

    C a turn

    D an enlargement

Essential Question How can you find the surface area of a composite solid?

Share Your Work at...
My.BigIdeasMath.com

1 ACTIVITY: Finding a Surface Area

Work with a partner. You are manufacturing scale models of old houses.

a. Name the four basic solids of this composite figure.

b. Determine a strategy for finding the surface area of this model. Would you use a scale drawing? Would you use a net? Explain.

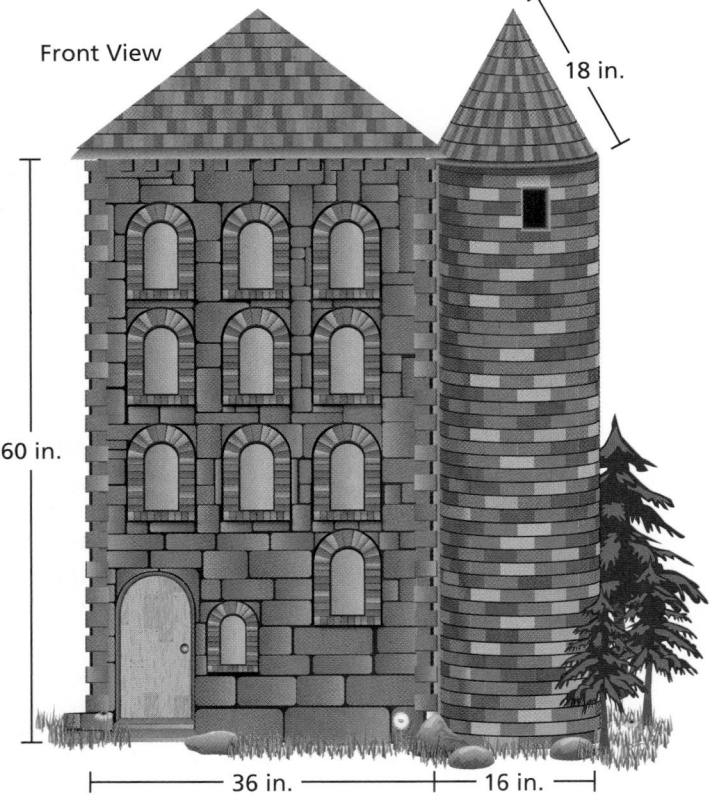

Front View

18 in.

60 in.

36 in.   16 in.

Many castles have cylindrical towers with conical roofs. These are called turrets.

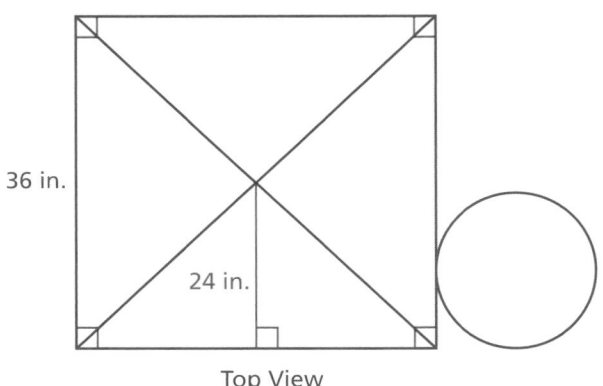

36 in.

24 in.

Top View

Laurie's Notes

Introduction

For the Teacher

- **Goal:** Students will explore strategies for finding the surface area of a composite figure.
- Look through the lesson and select one or more of the activities to focus on, depending upon the class time available and the depth to which one or more of the activities are explored.

Motivate

- **Story Time:** Coral Castle is a stone structure created by the Latvian-American eccentric Edward Leedskalnin. The structure is comprised of numerous megalithic stones (mostly coral), each weighing several tons. Edward Leedskalnin, out of love for a woman, built this coral monument. The question remains, how could he have carved and moved over 1100 tons of rock without any human assistance? Coral Castle has baffled scientists, engineers, and scholars since its opening in 1923.

Activity Notes

Activity 1

- This activity is very open-ended in terms of what strategy is used (i.e., using a net or a scale drawing) and any extensions you might pursue.
- Discuss with students models of buildings (replicas) that they have seen. These replicas are made of metal, wood, plastic, or lightweight cardboard.
- If you have examples of building replicas, bring them in to share with the class.
- The context of the activity requires the student to determine building expenses. One expense is the material for constructing the model, therefore they need to know the surface area involved.
- ❓ "What are the four basic solids in this composite figure?" square prism, square pyramid, cylinder, and cone
- Be sure that students observe that the base of the pyramid fits on (is congruent to) one face of the prism. The base of the cone fits on (is congruent to) the base of the cylinder.
- When students have computed the surface area of each portion and have the total surface area, ask a few questions.
- ❓ "How many surfaces of each solid did you find?" prism: 4 lateral faces; pyramid: 4 lateral faces; cylinder: only the lateral portion; cone: only the lateral portion
- ❓ "How did the scale drawing or net help you in finding the surface area?" Answers will vary.
- ❓ "Was it more difficult to find the surface area of one figure than another?" Answers will vary.
- ❓ "Can you think of alternative ways to find the surface area of the composite figure?" Answers will vary.

Previous Learning

Students should know how to find surface areas of prisms, cylinders, pyramids, and cones.

Start Thinking! and Warm Up

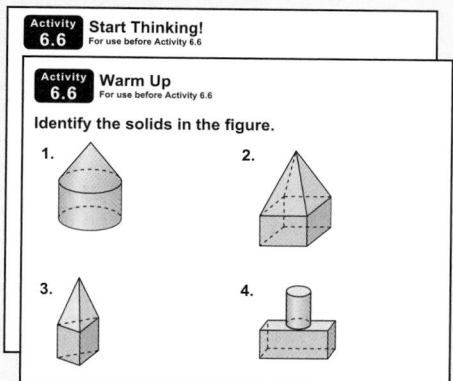

6.6 Record and Practice Journal

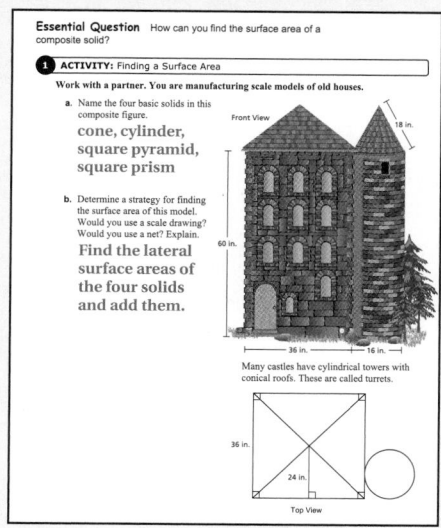

Differentiated Instruction

Kinesthetic

Provide blocks for students to use. Students can build their own composite figures and then calculate the surface areas.

6.6 Record and Practice Journal

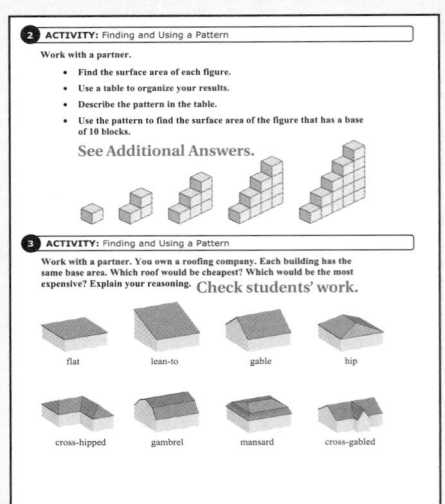

Laurie's Notes

Activity 2

- Encourage students to use a table to organize their findings. Have students gather data on each of the three views. For example:

Figure	Top	Front	Side	Total
1	1	1	1	6
2	2	2	3	14
3	3	3	6	24
4	4	4	10	36
5	5	5	15	50

- The top and front views are increasing by 1. The side view is increasing by consecutive whole numbers (2, 3, 4, 5, …). The total is increasing by consecutive even numbers (8, 10, 12, 14, …). Students often have difficulty describing the pattern when the amount of change is not a constant number.
- **Extension:** Have students add a column to the table and look at the number of cubes used in each figure (volume).

Activity 3

- Ask a student to read the problem.
- **?** "What does it mean that *each building has the same base area*?" same area but not necessarily the same dimensions
- This is a difficult concept for students to understand. Two bases with the same area might be 8×10 and 5×16.
- Students need to think about the design and the areas of various faces. They need to reason and explain. For instance, depending on the slopes of the lean-to and gable, it's reasonable to think that each of these could have the same amount of roofing. Students will need to talk about slope (pitch) of the roofs.

What Is Your Answer?

- Have students work in pairs to answer the questions.

Closure

- **Exit Ticket:** Identify at least 3 composite solids in the classroom. Answers will vary.

Technology **F**or the **T**eacher

Dynamic Classroom

The Dynamic Planning Tool
Editable Teacher's Resources at *BigIdeasMath.com*

2 ACTIVITY: Finding and Using a Pattern

Work with a partner.

- Find the surface area of each figure.
- Use a table to organize your results.
- Describe the pattern in the table.
- Use the pattern to find the surface area of the figure that has a base of 10 blocks.

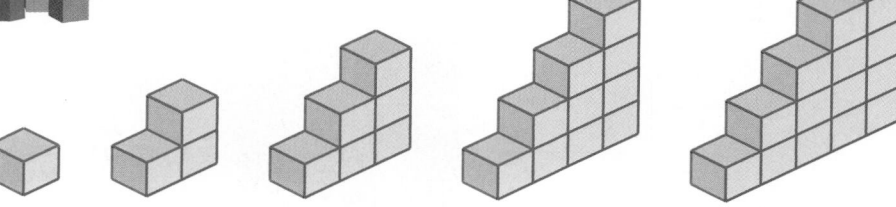

3 ACTIVITY: Finding and Using a Pattern

Work with a partner. You own a roofing company. Each building has the same base area. Which roof would be cheapest? Which would be the most expensive? Explain your reasoning.

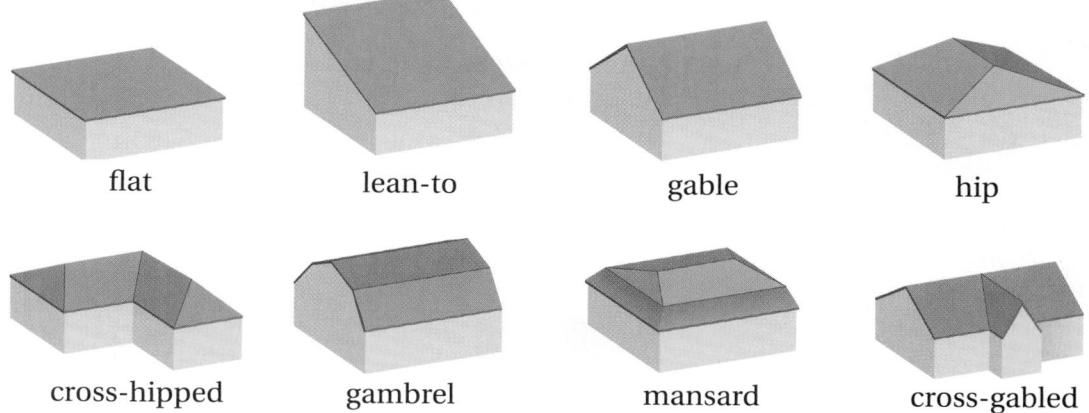

flat   lean-to   gable   hip

cross-hipped   gambrel   mansard   cross-gabled

What Is Your Answer?

4. **IN YOUR OWN WORDS** How can you find the surface area of a composite solid?

5. Design a building that has a turret and also has a mansard roof. Find the surface area of the roof.

Use what you learned about the surface area of a composite solid to complete Exercises 6–8 on page 286.

6.6 Lesson

Key Vocabulary ◀))
composite solid,
  p. 284

A **composite solid** is a figure that is made up of more than one solid.

composite solid

cylinder   cone

EXAMPLE (1) **Identifying Solids**

Identify the solids that make up Fort Matanzas.

Rectangular prism

Cylinder

Approximately a rectangular prism

EXAMPLE (2) **Standardized Test Practice**

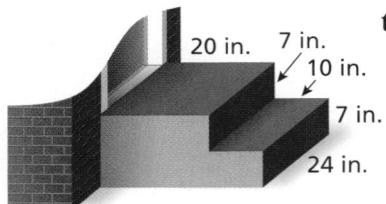

20 in.   7 in.
10 in.
7 in.
24 in.

You painted the steps to an apartment green. What is the surface area that you painted?

(A) 210 in.^{2} (B) 408 in.^{2} (C) 648 in.^{2} (D) 1056 in.^{2}

Find the area of each green face.

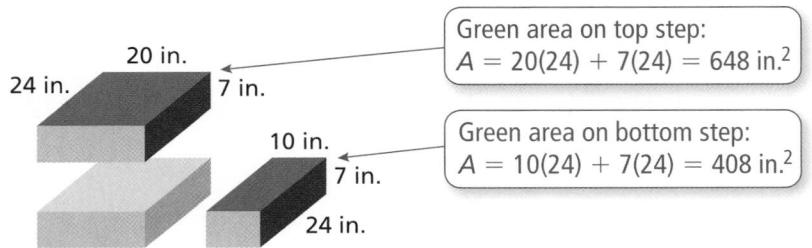

20 in.
24 in.
7 in.
10 in.
7 in.
24 in.

Green area on top step:
$A = 20(24) + 7(24) = 648$ in.^{2}

Green area on bottom step:
$A = 10(24) + 7(24) = 408$ in.^{2}

You painted $648 + 408 = 1056$ square inches.

∴ The correct answer is (D).

● **On Your Own**

Now You're Ready
Exercises 3–5

1. **WHAT IF?** In Example 2, you also painted the sides of the steps green. What is the surface area that you painted?

◀) Multi-Language Glossary at BigIdeasMath✓.com.

Laurie's Notes

Introduction

Connect
- **Yesterday:** Students explored the surface area of composite figures.
- **Today:** Students will work with formulas for the surface area of prisms, cylinders, pyramids, and cones to find the surface area of composite figures.

Motivate
- Share some tall building facts with students.
- The tallest building in England and the United Kingdom is the Canary Wharf Tower at 235 meters (771 feet) above ground level. The pyramid roof is 40 meters tall and is 30 meters square at the base.
- The Sears Tower in Chicago is a series of rectangular prisms with an antenna at the top rising to a total height of 1725 feet (526 meters). The Sears Tower is the tallest building in North America and is about twice as tall as the Canary Wharf Tower.

Lesson Notes

Example 1
- Ask if any students have visited a fortress.
- Discuss the purpose of the cylindrical portion, and why the rectangular prism is added to the height of the fortress, instead of the height being uniform. In other words, discuss form and function.
- **Big Idea:** Students should understand that when composite solids are made, certain portions of the original surface area may be covered up, meaning the area is no longer exposed. Students need to examine the solid and decide what surfaces are actually exposed.

Example 2
- ❓ How many faces will you paint?" four
- ❓ "Do you know the dimensions needed for each face?" yes
- Work through the problem as shown by computing the area of all four faces.
- ❓ "Is there another way you can find the area of the four steps that is more efficient?" Perhaps students notice that the length of each step is 24 inches so adding the widths together first and then multiplying by 24 is more efficient. $24(20 + 7 + 10 + 7) = 24(44)$
- ❓ "What property did we use?" Distributive Property

On Your Own
- Did students find the area of a 20×14 rectangle and then subtract a 7×10 rectangle or did they divide the gray region into two rectangles?

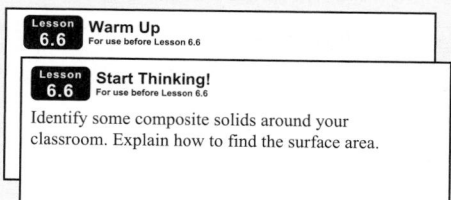
Extra Example 1
Identify the solids in the figure.

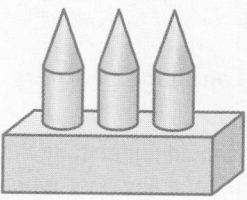

rectangular prism, cylinder, and cone

Extra Example 2
You painted the steps to your apartment yellow. What is the surface area that you painted?

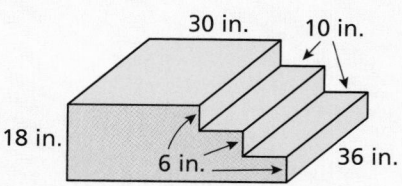

30 in. 10 in. 18 in. 6 in. 36 in.

2448 in.²

On Your Own
1. 1756 in.²

Laurie's Notes

Extra Example 3

Find the surface area of the composite solid. Round your answer to the nearest tenth.

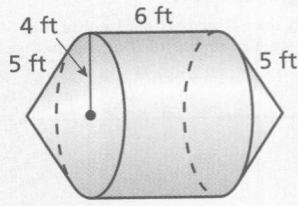

$88\pi \approx 276.3 \text{ ft}^2$

On Your Own

2. cone, cylinder;
$54\pi \approx 169.6 \text{ yd}^2$

3. rectangular prism,
triangular prism; 120 cm^2

English Language Learners

Vocabulary

Students should understand the meaning of face and base when dealing with solids. You may need to help them understand the concept of overlapping faces. Use models to help explain this concept.

Example 3

- Discuss the problem, and note the fact that faces overlap. If the face is not exposed, it is not included in the surface area.
- **?** "Why does the surface area formula for the prism begin with ℓw instead of $2\ell w$?" One of the bases is covered by the pyramid.
- **Common Error:** Students may only find the area of 4 faces instead of 5 for the prism. They think that it looks like a house and a house would be sitting on the ground with the bottom face not exposed. Make it clear that there is no context for this problem. It was not stated to be a house so consider all faces not covered by another solid when finding the surface area.

On Your Own

- **Neighbor Check:** Have students work independently and then have their neighbor check their work. Have students discuss any discrepancies.

Closure

- **Exit Ticket:** Draw a sketch of a house that is a square prism with a square pyramid sitting on top of it. The edge length of the square is 10 meters, the height is 10 meters, and the slant height of the pyramid is 8 meters. Find the surface area of the house and roof. (Exclude the base of the house.) 560 m^2

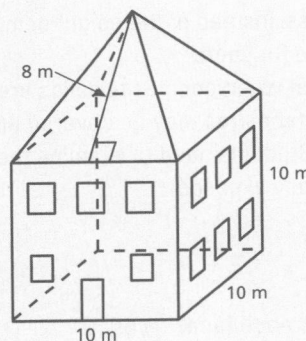

Technology For the Teacher

The Dynamic Planning Tool
Editable Teacher's Resources at *BigIdeasMath.com*

EXAMPLE ③ **Finding the Surface Area of a Composite Solid**

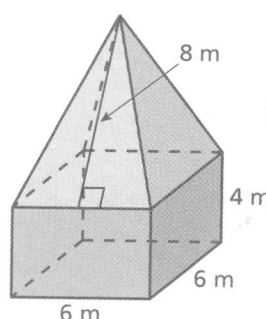

Find the surface area of the composite solid.

The solid is made up of a square prism and a square pyramid. Use the surface area formulas for a prism and a pyramid, but do not include the areas of the sides that overlap.

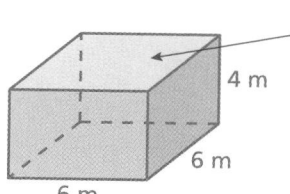

Do not include the top base of the prism in the surface area.

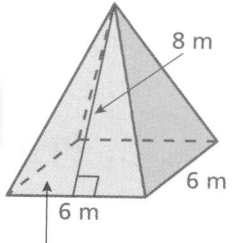

Do not include the base of the pyramid in the surface area.

Square prism

$$S = \ell w + 2\ell h + 2wh \qquad \text{Write formula.}$$
$$= 6(6) + 2(6)(4) + 2(6)(4) \qquad \text{Substitute.}$$
$$= 36 + 48 + 48 \qquad \text{Multiply.}$$
$$= 132 \qquad \text{Add.}$$

Square pyramid

$$S = \text{areas of lateral faces} \qquad \text{Write formula.}$$
$$= 4\left(\frac{1}{2} \cdot 6 \cdot 8\right) \qquad \text{Substitute.}$$
$$= 96 \qquad \text{Multiply.}$$

Find the sum of the surface areas: $132 + 96 = 228$.

⋮⋮ The surface area is 228 square meters.

🔵 **On Your Own**

Now You're Ready
Exercises 6–11

Identify the solids that make up the composite solid. Then find the surface area. Round your answer to the nearest tenth.

2.

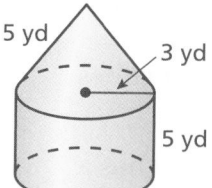

3.
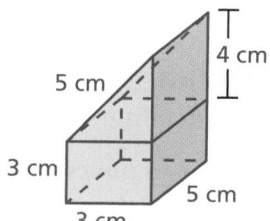

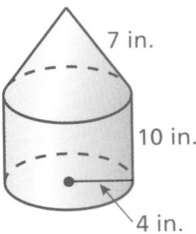

✓ Vocabulary and Concept Check

1. **OPEN-ENDED** Draw a composite solid formed by a triangular prism and a cone.

2. **REASONING** Explain how to find the surface area of the composite solid.

7 in.
10 in.
4 in.

Practice and Problem Solving

Identify the solids that form the composite solid.

① **3.**

4.

5.

Identify the solids that form the composite solid. Then find the surface area. Round your answer to the nearest tenth.

② ③ **6.** 4 ft 3 ft 6 ft 6 ft

7. 8 m 8 m 4 m 10 m

8. 4 in. 4 in. 5 in. 5 in. 5 in.

9. 2 cm 2.5 cm 2.5 cm 2 cm 3 cm 6 cm 5 cm

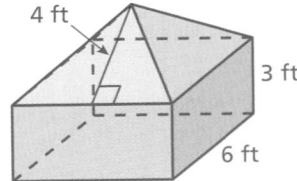

10. 7 in. 8 in. 8 in. 10 in. 6.9 in. 8 in.

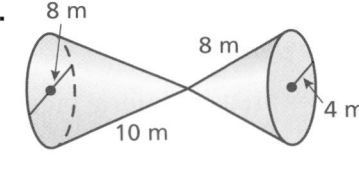

11. 8 ft 2 ft 4 ft 12 ft 5 ft

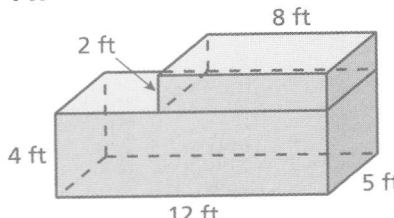

12. **OPEN-ENDED** The solid is made using eight cubes with side lengths of 1 centimeter.

a. Draw a new solid using eight cubes that has a surface area less than that of the original solid.

b. Draw a new solid using eight cubes that has a surface area greater than that of the original solid.

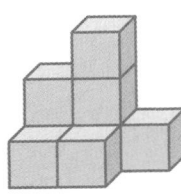

Assignment Guide and Homework Check

Level	Day 1 Activity Assignment	Day 2 Lesson Assignment	Homework Check
Basic	6–8, 19–22	1, 2, 3, 5, 9, 11, 12, 13	5, 9, 12, 13
Average	6–8, 19–22	1, 2, 9–15 odd, 12, 14	9, 12, 13, 14
Advanced	6–8, 19–22	1, 2, 12–18	12, 13, 14, 16

For Your Information

- **Exercise 18** Ask students to determine how many of the 27 cubes have three faces painted (8), have two faces painted (12), one face painted (6), and no faces painted (1).

Common Errors

- **Exercises 3–5** Students may have difficulty recognizing which solids are put together to form the shape. Tell them to trace the object and then draw lines that will break the object into parts of solids that are familiar to them.
- **Exercises 6–11** When finding the surface area of composite solids, students may include the parts that are no longer surface areas. Tell them to draw the solids without those parts and then find the area.
- **Exercises 8 and 11** Students will struggle finding the surface area of the portions that intersect. First, have them find the surface area that they know how to find. Next, draw pictures of what is remaining to be found and ask questions about how to find the area.

6.6 Record and Practice Journal

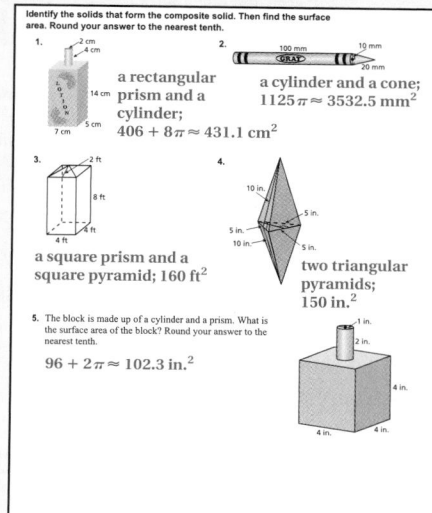

Identify the solids that form the composite solid. Then find the surface area. Round your answer to the nearest tenth.

1. a rectangular prism and a cylinder; $406 + 8\pi \approx 431.1$ cm^2

2. a cylinder and a cone; $1125\pi \approx 3532.5$ mm^2

3. a square prism and a square pyramid; 160 ft^2

4. two triangular pyramids; 150 in.^{2}

5. The block is made up of a cylinder and a prism. What is the surface area of the block? Round your answer to the nearest tenth. $96 + 2\pi \approx 102.3$ in.^{2}

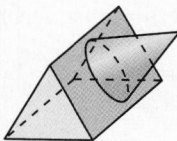

2. Find the lateral surface area and the area of one base of the cylinder, and then find the area of the lateral surface of the cone. Add up all of these areas.

Practice and Problem Solving

3. three cylinders

4. rectangular prism, triangular prism

5. rectangular prism, half of a cylinder

6. square prism, square pyramid; 156 ft^2

7. cones; $104\pi \approx 326.6$ m^2

8. cylinder, rectangular prism; $150 + 16\pi \approx 200.2$ in.^{2}

9. trapezoidal prism, rectangular prism; 152 cm^2

10. triangular prism, triangular pyramid; 351.6 in.^{2}

11. two rectangular prisms; 308 ft^2

12. **a.** *Sample answer:*

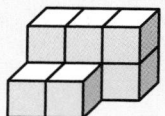

b. *Sample answer:*

Practice and Problem Solving

13. 63.4%

14. See *Taking Math Deeper*.

15. $144\pi \approx 452.2$ in.^{2}

16. 226 ft^2

17. $806\pi \approx 2530.8$ mm^2

18. less than; Removing the purple cubes reduces the surface area by the area of 2 cube faces. Removing the green cubes does not change the total surface area.

Fair Game Review

19. 10 ft^2

20. 16 cm^2

21. 47.5 in.^{2}

22. A

Mini-Assessment

Find the surface area of the composite solid. Round your answer to the nearest tenth.

1.

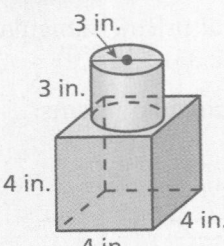

3 in.

3 in.

4 in.

4 in.

4 in.

$96 + 9\pi \approx 124.3$ in.^{2}

2.

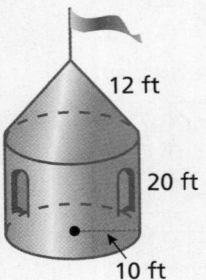

12 ft

20 ft

10 ft

$520\pi \approx 1632.8$ ft^2

Taking Math Deeper

Exercise 14

The barbell is made up of two hexagonal prisms and one cylinder. This problem has a lot of decimal calculations. Part of the skill is to learn to keep each part organized.

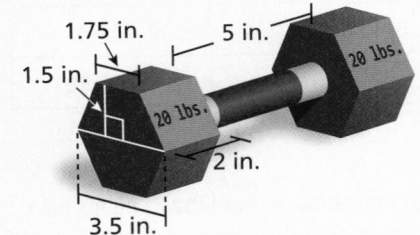

1.75 in.   5 in.

1.5 in.

20 lbs.

20 lbs.

2 in.

3.5 in.

① Find the area of each type of surface.

Figure	How many?	Area
2 □ 1.75	12	$12(1.75 \cdot 2) = 42.0$
1.75 ▱ 1.5 3.5	8	$8\left[\frac{1}{2}(3.5 + 1.75)(1.5)\right] = 31.5$
1 ●	2	$2(\pi \cdot 0.5^2) \approx 1.57$ Subtract 1" circular handle.
π ▮ 5	1	$5\pi \approx 15.7$

Total: 87.63 in.^{2}

② Find the total.

③ Help me see it. Draw a net for each hexagonal end of the barbell.

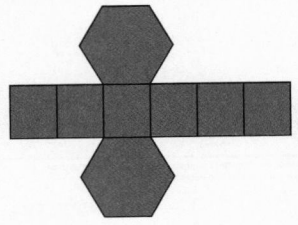

I see it.

Reteaching and Enrichment Strategies

If students need help. . .	If students got it. . .
Resources by Chapter • Practice A and Practice B • Puzzle Time Record and Practice Journal Practice Differentiating the Lesson Lesson Tutorials Skills Review Handbook	Resources by Chapter • Enrichment and Extension • School-to-Work • Financial Literacy • Technology Connection Start the next section

13. **BATTERIES** What is the percent increase in the surface area of the AAA battery to the AA battery? Round your answer to the nearest tenth of a percent.

AAA battery **AA battery**

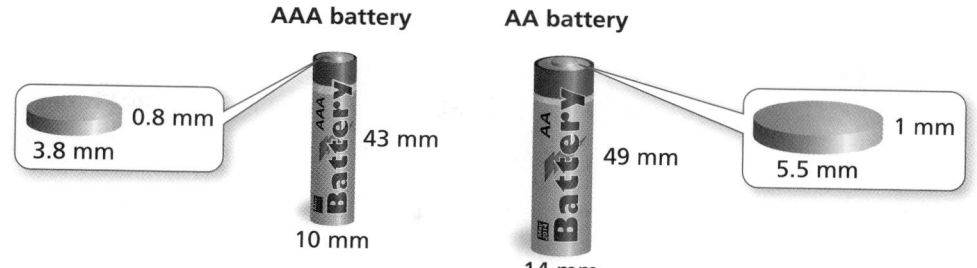

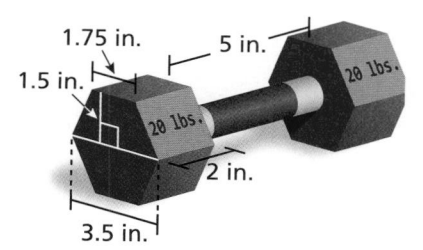

14. **BARBELL** The diameter of the handle of a barbell is 1 inch. The hexagonal weights are identical. What is the surface area of the barbell?

REASONING Find the surface area of the solid. Round your answer to the nearest tenth.

15.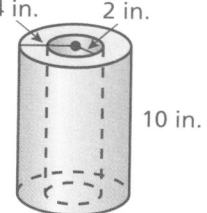

16.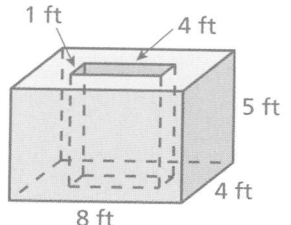

17.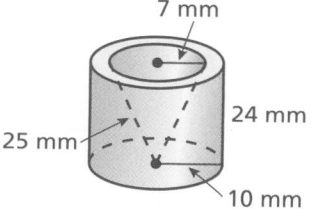

18. **Critical Thinking** The cube is made with 27 identical cubes. All cubes that cannot be seen are orange. Is the surface area of the solid formed without the purple cubes *greater than*, *less than*, or *equal to* the surface area of the solid formed without the green cubes? Explain your reasoning.

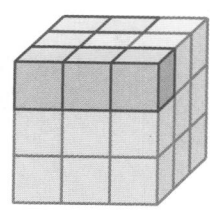

Fair Game Review What you learned in previous grades & lessons

Find the area. *(Skills Review Handbook)*

19.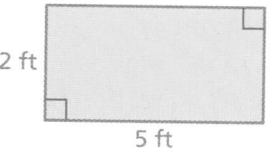

20.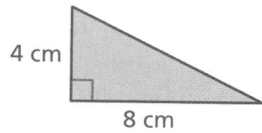

21.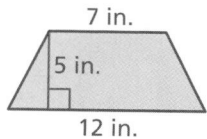

22. **MULTIPLE CHOICE** A cliff swallow nest is 86 meters above a canyon floor. The elevation of the nest is −56 meters. What is the elevation of the canyon floor? *(Section 2.4)*

Ⓐ −142   Ⓑ −30   Ⓒ 30   Ⓓ 142

Identify the solids that form the composite solid.   *(Section 6.6)*

1.

2.

3.

Find the surface area of the regular pyramid.   *(Section 6.4)*

4.
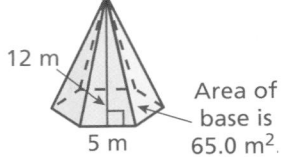

12 m

Area of
base is
65.0 m².

5 m

5.
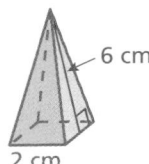

6 cm

2 cm

Find the surface area of the cone. Round your answer to the nearest tenth.
(Section 6.5)

6.
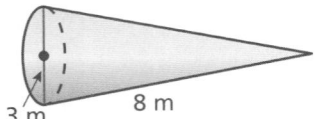

3 m   8 m

7.
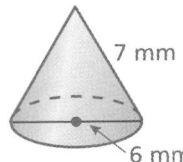

7 mm

6 mm

Find the surface area of the composite solid. Round your answer to the nearest tenth.   *(Section 6.6)*

8.
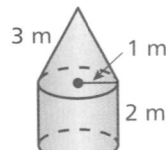

3 m   1 m

2 m

9.
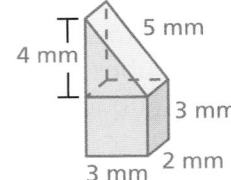

4 mm   5 mm

3 mm

3 mm   2 mm

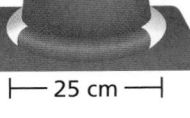

99 cm

12 cm

12 cm

25 cm

10. TRAFFIC CONE A square reflective sticker is placed on a
traffic cone to make it more visible at night. Estimate
the percent of the surface area of the traffic cone covered by
the sticker to the nearest percent.   *(Section 6.5)*

11. GEOMETRY The surface area of a cone is
150π square inches. The radius of the base is
10 inches. What is the slant height?   *(Section 6.5)*

12. TOOLBOX Find the surface area
of the toolbox.   *(Section 6.6)*

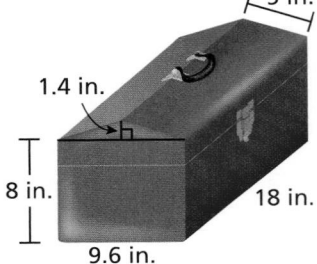

5 in.

1.4 in.

8 in.

18 in.

9.6 in.

Alternative Assessment Options

Math Chat
Structured Interview

Student Reflective Focus Question
Writing Prompt

Math Chat

- Have students work in pairs. Assign Quiz Exercises 10–12 to each pair. Each student works through all three problems. After the students have worked through the problems, they take turns talking through the processes that they used to get each answer. Students analyze and evaluate the mathematical thinking and strategies used.
- The teacher should walk around the classroom listening to the pairs and ask questions to ensure understanding.

Study Help Sample Answers

Remind students to complete Graphic Organizers for the rest of the chapter.

8a.

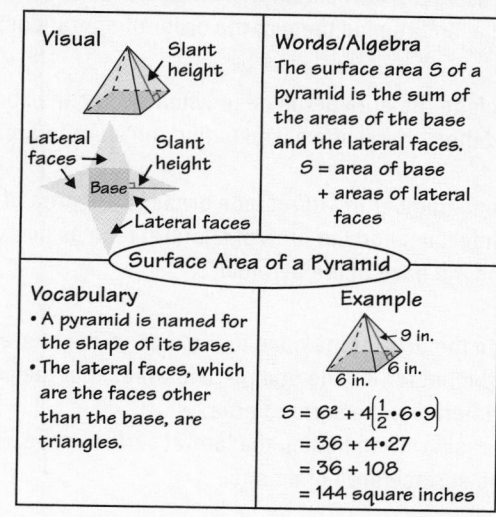

8b, c. Available at *BigIdeasMath.com*

Reteaching and Enrichment Strategies

If students need help...	If students got it...
Resources by Chapter • Study Help • Practice A and Practice B • Puzzle Time Lesson Tutorials *BigIdeasMath.com* Practice Quiz Practice from the Test Generator	Resources by Chapter • Enrichment and Extension • School-to-Work Game Closet at *BigIdeasMath.com* Start the Chapter Review

Technology For the Teacher
Answer Presentation Tool

Answers

1. cylinder, cone

2. rectangular prism, triangular prism

3. three rectangular prisms

4. 245 m^2 5. 28 cm^2

6. $14.25\pi \approx 44.7$ m^2

7. $30\pi \approx 94.2$ mm^2

8. $8\pi \approx 25.1$ m^2

9. 66 mm^2

10. 4%

11. about 5 in.

12. 807.84 in.^{2}

Assessment Book

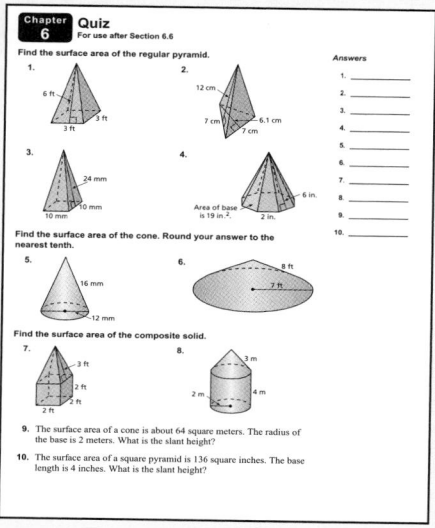

Answers

1.
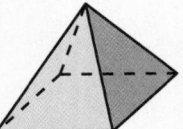

2.
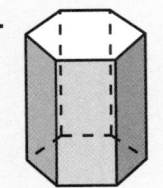

3.
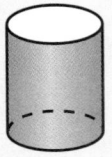

4. 100 in.^{2}

5. 400 cm^2

6. 108 m^2

Review of Common Errors

Exercises 1–3

- Students may mix up the different types of solids. Remind them of the definition of each solid and give them a few real-life examples of each.

Exercises 4–6

- Students may sum the areas of only 3 of the faces of the rectangular prism instead of all 6. Remind them that a rectangular prism has 6 faces.
- Students may try to use the formula for the surface area of a rectangular prism to find the surface area of a triangular prism. Show them that this will not work by comparing the nets of the two types of prisms.

Exercises 7–9

- Students may add the area of only one base. Remind them that a cylinder has *two* bases.
- Students may double the radius instead of squaring it, or forget the correct order of operations when using the formula for the surface area of a cylinder. Remind them of the formula, and remind them of the order of operations.

Exercises 10–12

- Students may forget to include the area of the base when finding the surface area of a pyramid. Remind them that when asked to find the surface area, the base is included.
- Students may add the wrong number of lateral face areas to the area of the base. Remind them that they must add the area of a lateral face as many times as there are sides on the base of the pyramid.

Exercises 13–15

- Students may forget to add the area of the base to the area of the lateral surface. Remind them of the net for a cone and the two areas that they must find and add together to determine the surface area.
- Students may square the radius when finding the lateral surface area. Remind them of the formula for the surface area of a cone.

Exercises 16–18

- When finding the surface area of a composite solid, students may include areas of the sides that overlap. Remind them that the overlapping sides are not part of the surface area.

Review Key Vocabulary

three-dimensional figure, *p. 252*
polyhedron, *p. 252*
lateral face, *p. 252*

surface area, *p. 256*
net, *p. 256*
regular pyramid, *p. 272*

slant height, *pp. 272, 278*
composite solid, *p. 284*

Review Examples and Exercises

6.1 Drawing 3-Dimensional Figures (*pp. 250–255*)

Draw a triangular prism.

Draw identical
triangular bases.

Connect corresponding
vertices.

Change any *hidden*
lines to dashed lines.

Exercises

Draw the solid.

1. Square pyramid

2. Hexagonal prism

3. Cylinder

6.2 Surface Areas of Prisms (*pp. 256–261*)

Find the surface area of the prism.

Draw a net.

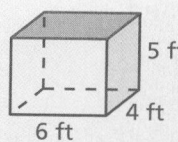

$$S = 2\ell w + 2\ell h + 2wh$$
$$= 2(6)(4) + 2(6)(5) + 2(4)(5)$$
$$= 48 + 60 + 40$$
$$= 148$$

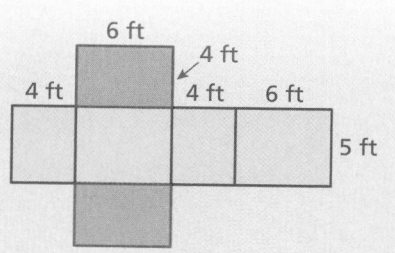

∴ The surface area is 148 square feet.

Exercises

Find the surface area of the prism.

4.

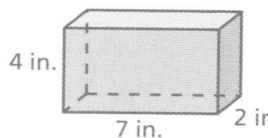

4 in.
7 in.
2 in.

5.

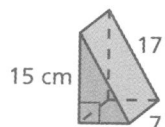

17 cm
15 cm
8 cm
7 cm

6.

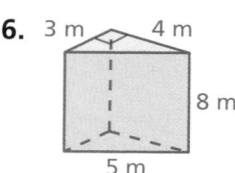

3 m   4 m
8 m
5 m

6.3 **Surface Areas of Cylinders** *(pp. 262–267)*

Find the surface area of the cylinder. Round your answer to the nearest tenth.

Draw a net.

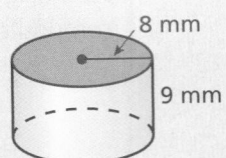

$$S = 2\pi r^2 + 2\pi r h$$
$$= 2\pi(8)^2 + 2\pi(8)(9)$$
$$= 128\pi + 144\pi$$
$$= 272\pi \approx 854.1$$

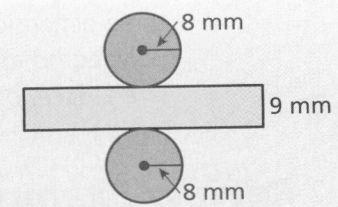

The surface area is about 854.1 square millimeters.

Exercises

Find the surface area of the cylinder. Round your answer to the nearest tenth.

7.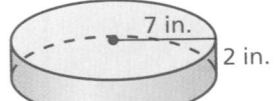

8.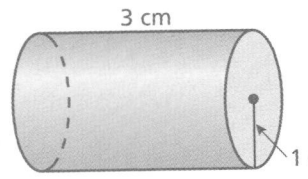

9.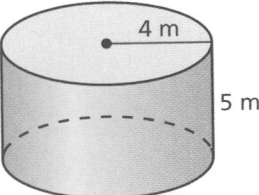

6.4 **Surface Areas of Pyramids** *(pp. 270–275)*

Find the surface area of the regular pyramid.

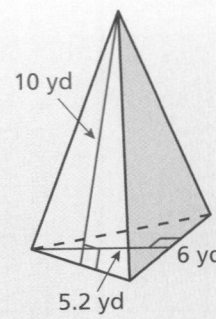

Draw a net.

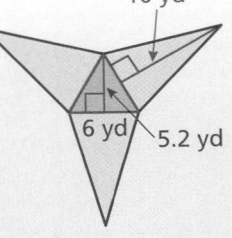

Area of base	*Area of a lateral face*
$\frac{1}{2} \cdot 6 \cdot 5.2 = 15.6$	$\frac{1}{2} \cdot 6 \cdot 10 = 30$

Find the sum of the areas of the base and all 3 lateral faces.

$$S = 15.6 + 30 + 30 + 30 = 105.6$$

The surface area is 105.6 square yards.

Exercises

Find the surface area of the regular pyramid.

10.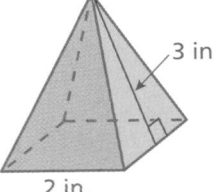

11.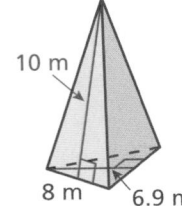

12.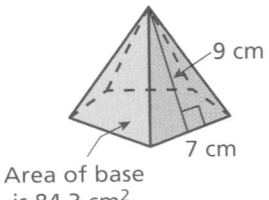

Area of base
is 84.3 cm².

Review Game

3-Dingo

Big Ideas
Game Closet

Materials per Group:
- one 3-Dingo card*
- objects to cover the 3-Dingo card squares, such as bingo chips

Directions:
This activity is played like bingo. Divide the class into groups. Call out a three-dimensional figure and all of the dimensions necessary to calculate the surface area. The group calculates the surface area and if that value is in a square on their card under the correct figure, they cover it. Keep calling out figures and their dimensions until a group wins.

Who Wins?
Just like bingo, the first group to get a row, column, or diagonal of covered squares wins. The winning team yells 3-Dingo!

*A 3-Dingo card has 5 columns of 5 squares and a three-dimensional figure (prism, cylinder, pyramid, cone, and composite solid) with variable dimensions shown at the top of each column of squares. Different values for the surface area of each figure are shown in each of the 5 squares below the figure. Different cards show the values in different orders, so no two 3-Dingo cards in a set are the same. A 3-Dingo card set is available at *BigIdeasMath.com*.

For the Student
Additional Practice
- Lesson Tutorials
- Study Help (textbook)
- Student Website
  Multi-Language Glossary
  Practice Assessments

Answers

7. $126\pi \approx 395.6$ in.^{2}
8. $8\pi \approx 25.1$ cm^2
9. $72\pi \approx 226.1$ m^2
10. 16 in.^{2}
11. 147.6 m^2
12. 241.8 cm^2
13. $5\pi \approx 15.7$ in.^{2}
14. $21\pi \approx 65.9$ cm^2
15. $48\pi \approx 150.7$ m^2
16. 228 ft^2
17. 116 yd^2
18. $45\pi \approx 141.3$ m^2

My Thoughts on the Chapter

What worked. . .

What did not work. . .

What I would do differently. . .

Surface Areas of Cones *(pp. 276–281)*

Find the surface area of the cone. Round your answer to the nearest tenth.

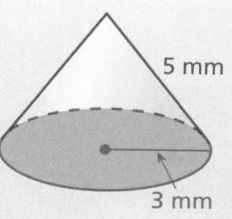

Draw a net.

$$S = \pi r^2 + \pi r \ell$$
$$= \pi(3)^2 + \pi(3)(5)$$
$$= 9\pi + 15\pi$$
$$= 24\pi \approx 75.4$$

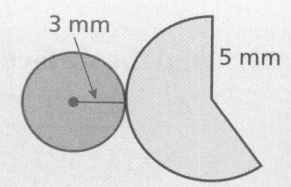

∴ The surface area is about 75.4 square millimeters.

Exercises

Find the surface area of the cone. Round your answer to the nearest tenth.

13.
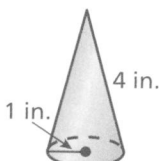

14.
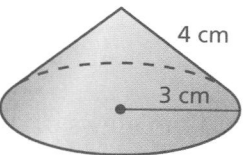

15.
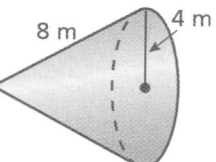

6.6 **Surface Areas of Composite Solids** *(pp. 282–287)*

Find the surface area of the composite solid. Round your answer to the nearest tenth.

The solid is made of a cone and a cylinder. Use the surface area formulas. Do not include the areas of the bases that overlap.

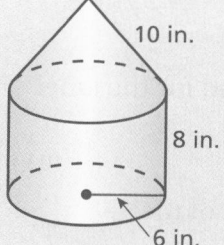

Cone

$$S = \pi r \ell$$
$$= \pi(6)(10)$$
$$= 60\pi \approx 188.4$$

Cylinder

$$S = \pi r^2 + 2\pi r h$$
$$= \pi(6)^2 + 2\pi(6)(8)$$
$$= 36\pi + 96\pi$$
$$= 132\pi \approx 414.5$$

∴ The surface area is about 188.4 + 414.5 = 602.9 square inches.

Exercises

Find the surface area of the composite solid. Round your answer to the nearest tenth.

16.
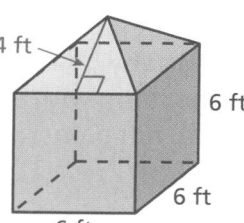

17.
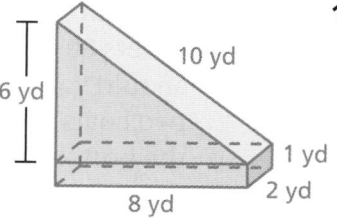

18.

Draw the solid.

1. Square prism

2. Pentagonal pyramid

3. Cone

Find the surface area of the prism or regular pyramid.

4.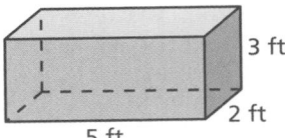

3 ft
2 ft
5 ft

5.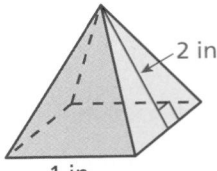

2 in.
1 in.

6.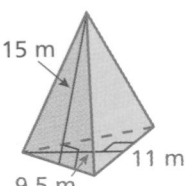

15 m
11 m
9.5 m

Find the surface area of the cylinder or cone. Round your answer to the nearest tenth.

7.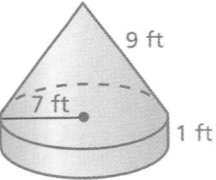

8 m
2 m

8.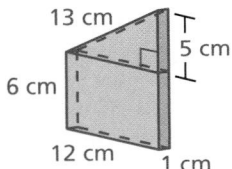

10 in.
7 in.

9. Draw the front, side, and top views of the solid in Exercise 8.

Identify the solids that form the composite solid. Then find the surface area. Round your answer to the nearest tenth.

10.

9 ft
7 ft
1 ft

11.

13 cm
5 cm
6 cm
12 cm
1 cm

12. CORN MEAL How much paper is used for the label of the corn meal container?

4 in.
5 in.

13. CRACKER BOX Find the surface area of the cracker box.

10 in.
8 in.
3 in.

14. COSTUME The cone-shaped hat will be part of a costume for a school play. What is the least amount of material needed to make this hat?

11 in.
6 in.

15. SKATEBOARD RAMP A quart of paint covers 80 square feet. How many quarts should you buy to paint the ramp with two coats? (Assume you will not paint the bottom of the ramp.)

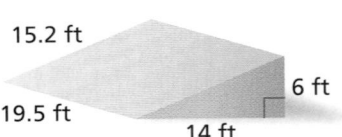

15.2 ft
19.5 ft
6 ft
14 ft

Test Item References

Chapter Test Questions	Section to Review
1–3, 9	6.1
4, 13, 15	6.2
7, 12	6.3
5, 6	6.4
8, 14	6.5
10, 11	6.6

Test-Taking Strategies

Remind students to quickly look over the entire test before they start so that they can budget their time. This test is very visual and requires that students remember many terms. It might be helpful for them to jot down some of the terms on the back of their test before they start.

Common Assessment Errors

- **Exercises 4 and 13** Students may sum the areas of only 3 of the faces or find the product of the length, width, and height to find the surface area of a rectangular prism. Showing them a net of the prism will help them see that it has 6 faces, and that they must multiply and add to find the surface area.
- **Exercises 5 and 6** Students may forget to include the area of the base or add the wrong number of lateral face areas when finding the surface area of a pyramid. Remind them that the base is included as part of the surface area, and that they must add the area of each lateral face.
- **Exercises 7 and 8** Students may use the diameter instead of the radius, or double the radius instead of squaring it. Remind them to halve the given diameter to find the radius and how to properly evaluate an exponent.
- **Exercises 10 and 11** When finding the surface area of a composite solid, students may include areas of the sides that overlap. Remind them that the overlapping sides are not part of the surface area.
- **Exercises 12 and 14** Students may include the area of the base(s) when calculating the amount of material. Point out that the container's label is only on the lateral surface and that the hat does not have a base.

Reteaching and Enrichment Strategies

If students need help. . .	If students got it. . .
Resources by Chapter • Practice A and Practice B • Puzzle Time Record and Practice Journal Practice Differentiating the Lesson Lesson Tutorials Practice from the Test Generator Skills Review Handbook	Resources by Chapter • Enrichment and Extension • School-to-Work • Financial Literacy Game Closet at *BigIdeasMath.com* Start Standardized Test Practice

Answers

1. 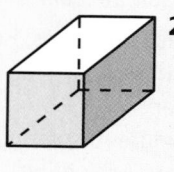   2.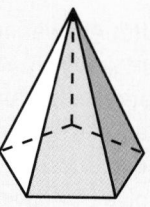

3. 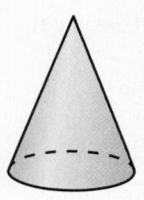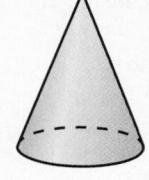   4. 62 ft^2

5. 5 in.^{2} 6. 299.8 m^2

7. 48π m$^2 \approx 150.7$ m^2

8. 60π in.$^2 \approx 188.4$ in.^{2}

9. Front: Side: Top:

10. cylinder, cone;
126π ft$^2 \approx 395.6$ ft^2

11. Rectangular prism,
Triangular prism; 246 cm^2

12. 20π in.$^2 \approx 62.8$ in.^{2}

13. 268 in.^{2}

14. 33π in.$^2 \approx 103.62$ in.^{2}

15. 13 quarts of paint

Assessment Book

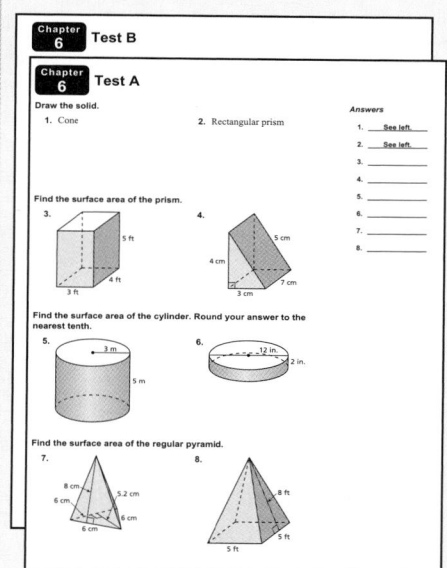

Available at *BigIdeasMath.com*

After Answering Easy Questions, Relax

Answer Easy Questions First

Estimate the Answer

Read All Choices before Answering

Read Question before Answering

Solve Directly or Eliminate Choices

Solve Problem before Looking at Choices

Use Intelligent Guessing

Work Backwards

About this Strategy

When taking a multiple choice test, be sure to read each question carefully and thoroughly. When taking a timed test, it is often best to skim the test and answer the easy questions first. Be careful that you record your answer in the correct position on the answer sheet.

Answers

1. C
2. F
3. 190 in.^{2}

Item Analysis

1. **A.** The student does not correctly match corresponding side lengths, instead using the proportion $\dfrac{PQ}{QR} = \dfrac{US}{ST}$.

   B. The student does not correctly match corresponding side lengths, instead using the proportion $\dfrac{PQ}{QR} = \dfrac{TU}{ST}$.

   C. Correct answer

   D. The student does not correctly match corresponding side lengths, instead using the proportion $\dfrac{PQ}{QR} = \dfrac{ST}{US}$.

2. **F.** Correct answer

   G. The student thinks that half of the entire rectangle is shaded and finds half the entire rectangle's area.

   H. The student finds the area of the part of the figure that is not shaded.

   I. The student finds the perimeter of the entire rectangle.

3. **Gridded Response:** Correct answer: 190 in.^{2}

   Common Error: The student finds the areas of only one face of each size, getting an answer of 95 square inches.

4. **A.** The student uses the correct numbers, but inverts one of the ratios.

   B. Correct answer

   C. The student finds the number of 2 minute intervals in 1 hour, but still keeps the 2 in the proportion. The student also inverts one of the ratios.

   D. The student finds the number of 2 minute intervals in 1 hour, but still keeps the 2 in the proportion.

5. **F.** The student includes the area of only one triangular face.

   G. The student includes the area of only three triangular faces.

   H. Correct answer

   I. The student does not multiply by $\dfrac{1}{2}$ when determining the area of a triangular face.

Technology For the Teacher

Big Ideas Test Generator

1. In the figure below, $\triangle PQR \sim \triangle STU$.

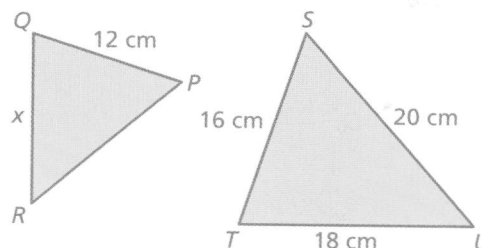

   What is the value of x?

   A. 9.6 cm

   B. $10\frac{2}{3}$ cm

   C. 13.5 cm

   D. 15 cm

2. The rectangle below is divided into six regions.

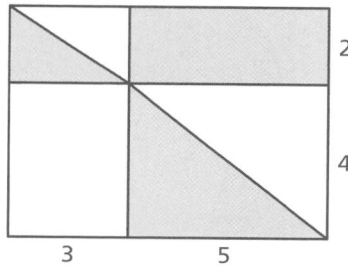

   What is the area of the part of the figure that is shaded?

   F. 23 units²

   G. 24 units²

   H. 25 units²

   I. 28 units²

3. A right rectangular prism and its dimensions are shown below.

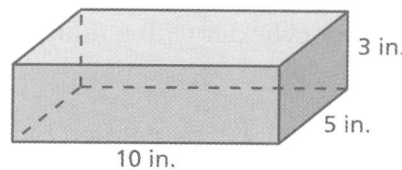

   What is the total surface area, in square inches, of the right rectangular prism?

4. You rode your bicycle 0.8 mile in 2 minutes. You want to know how many miles you could ride in 1 hour, if you ride at the same rate. Which proportion could you use to get your answer?

A. $\dfrac{0.8}{2} = \dfrac{60}{x}$

C. $\dfrac{0.8}{2} = \dfrac{30}{x}$

B. $\dfrac{0.8}{2} = \dfrac{x}{60}$

D. $\dfrac{0.8}{2} = \dfrac{x}{30}$

5. A right square pyramid is shown below.

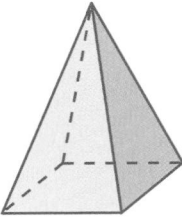

The square base and one of the triangular faces of the right square pyramid are shown below with their dimensions.

3 in.

Square Base

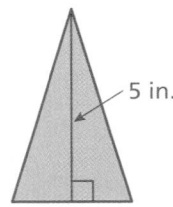

5 in.

3 in.

A Triangular Face

What is the total surface area of the right square pyramid?

F. 16.5 in.²

H. 39 in.²

G. 31.5 in.²

I. 69 in.²

6. A right circular cylinder with a radius of 3 centimeters and a height of 7 centimeters will be carved out of wood.

Part A   Draw and label a right circular cylinder with a radius of 3 centimeters and a height of 7 centimeters.

The two bases of the right circular cylinder will be painted blue. The rest of the cylinder will be painted red.

Part B   What is the surface area, in square centimeters, that will be painted blue? Show your work and explain your reasoning. (Use 3.14 for π.)

Part C   What is the surface area, in square centimeters, that will be painted red? Show your work and explain your reasoning. (Use 3.14 for π.)

Item Analysis (continued)

6. **4 points** The student demonstrates a thorough understanding of drawing and working with the surface area of right circular cylinders. In Part A, the student correctly draws and labels a right circular cylinder. In Part B, the student correctly find the area, in square centimeters, of the two bases, getting an answer of 56.52. In Part C, the student correctly finds the lateral area, in square centimeters, of the cylinder, getting an answer of 131.88.

   3 points The student demonstrates an understanding of drawing and working with the surface area of right circular cylinders, but the student's work demonstrates an essential but less than thorough understanding.

   2 points The student demonstrates a partial understanding of drawing and working with the surface area of right circular cylinders. The student's work demonstrates a lack of essential understanding.

   1 point The student demonstrates a limited understanding of drawing and working with the surface area of right circular cylinders. The student's response is incomplete and exhibits many flaws.

   0 points The student provided no response, a completely incorrect or incomprehensible response, or a response that demonstrates insufficient understanding of drawing and working with the surface area of right circular cylinders.

7. **A.** The student thinks that the greater factor determines the sign of a product.

   B. The student uses a division algorithm instead of multiplication.

   C. Correct answer

   D. The student does not take the reciprocal of the divisor and also incorrectly changes its sign.

8. **F.** The student divides by 2 in the equation in answer choice H, getting a solution of 3, thereby making answer choice F the equation with the greatest solution.

   G. The student divides by 2 in the equation in answer choice H, getting a solution of 3. The student makes a sign error in the equation in answer choice F, getting a solution of −8. The student makes a sign error in the equation in answer choice G, getting a solution of 8.

   H. Correct answer

   I. The student makes a sign error in the equation in answer choice H, getting a solution of −12. The student makes a sign error in the equation in answer choice I, getting a solution of 12.

9. **Gridded Response:** Correct answer: 10 in.

   Common Error: The student finds the area of each face, getting an answer of 100 inches.

4. B

5. H

6. *Part A*

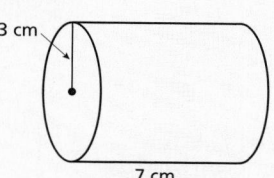

3 cm

7 cm

Part B 56.52 cm^2

Part C 131.88 cm^2

7. C

8. H

9. 10 in.

10. B

Answers for Extra Examples

1. **A.** Correct answer

   B. The student subtracts 36 from 75.

   C. The student finds 75% of 36.

   D. The student finds 75% of 36 and subtracts this result from 36.

2. **F.** The student reflects the figure in the *x*-axis.

   G. Correct answer

   H. The student translates the figure 7 units right.

   I. The student translates the figure 6 units up.

3. **A.** The student multiplies 0.04 by −0.2 instead of −0.02.

   B. Correct answer

   C. The student multiplies 0.04 by 0.02 instead of −0.02.

   D. The student multiplies 0.04 by 0.2 instead of −0.02.

Item Analysis (continued)

10. **A.** The student finds the *y*-coordinate when the *x*-coordinate is −2.

    B. Correct answer

    C. Because −1 is 1 less than 0, the student uses the ordered pair (0, −5) and subtracts 1 from −5.

    D. The student finds the *y*-coordinate when the *x*-coordinate is 1.

Extra Examples for Standardized Test Practice

1. Of the people at a party, 75% brought food to share. The host thanked all 36 people who brought food to share. How many people were at the party?

   A. 48

   B. 39

   C. 27

   D. 9

2. In the coordinate plane below, △*XYZ* is plotted and its vertices are labeled.

   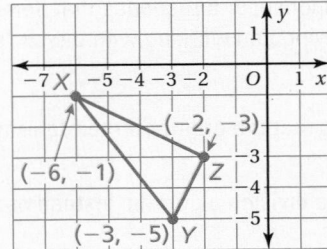

   Which of the following shows △*X'Y'Z'*, the image of △*XYZ* after it is reflected in the *y*-axis?

   F.

   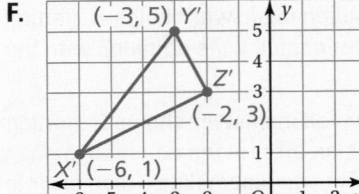

   H.

   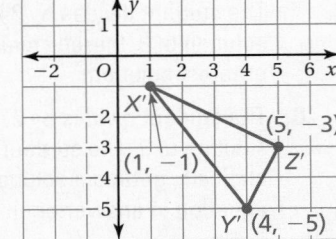

   G.

   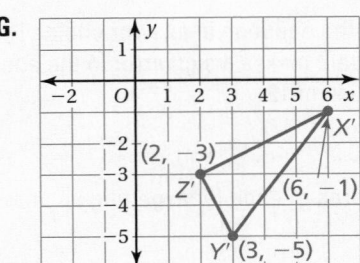

   I.

   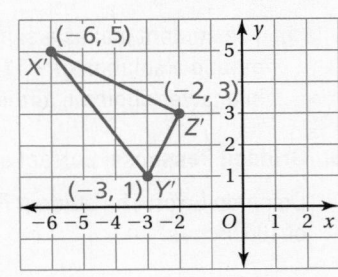

3. What is the next term in the sequence below?

   $$100, -2, 0.04$$

   A. −0.008

   B. −0.0008

   C. 0.0008

   D. 0.008

7. Anna was simplifying the expression in the box below.

$$-\frac{3}{8} \cdot \left[\frac{2}{5} \div (-4)\right] = -\frac{3}{8} \cdot \left[\frac{2}{5} \cdot \left(-\frac{1}{4}\right)\right]$$

$$= -\frac{3}{8} \cdot \left(-\frac{1}{10}\right)$$

$$= -\frac{3}{80}$$

What should Anna do to correct the error that she made?

A. Make the product inside the brackets positive.

B. Multiply by -10 instead of $-\frac{1}{10}$.

C. Make the final product positive.

D. Multiply by 4 instead of $-\frac{1}{4}$.

8. Which equation has the greatest solution?

F. $-3x + 9 = -15$

G. $12 = 2x + 28$

H. $\frac{x}{2} - 13 = -7$

I. $6 = \frac{x}{3} + 10$

9. A cube has a total surface area of 600 square inches. What is the length, in inches, of each edge of the cube?

10. A line contains the two points plotted in the coordinate plane below.

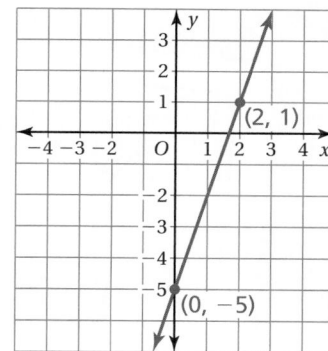

Another point on this line can be represented by the ordered pair $(-1, y)$. What is the value of y?

A. -11

B. -8

C. -6

D. -2

7 Volumes of Solids

"I petitioned my owner for a dog house with greater volume."

Add a bunk bed and we can have a sleep over.

"And this is what he built for me."

This setup is too good to be true.

"Do you know why the volume of a cone is one-third the volume of the cylinder with the same height and base?"

Connections to Previous Learning

- Distinguish between two-dimensional figures and three-dimensional solids, including the number of edges, faces, and vertices.
- Identify and build three dimensional solids from two dimensional figures, and vice versa.
- Find the volume of prisms without formulas.

- Find the volume of a prism.

- Justify and apply formulas for the volume of pyramids, prisms, cylinders, and cones.
- Use formulas to find the volume of three dimensional composite solids.
- Determine how changes in dimensions affect surface area and volume of similar solids, and apply these relationships to solve problems.

Math in History

For over 2500 years, mathematicians have been using the properties of similar triangles to indirectly measure things that were too difficult to measure directly.

★ Thales of Miletus (about 580 B.C.) was visiting the Great Pyramid of Egypt. When he saw the pyramid, he compared the length of its shadow to the length of his own shadow and was able to calculate the height of the pyramid.

★ In ancient China, properties of similar triangles were used in surveying, including the surveying that was used to build the Great Wall of China.

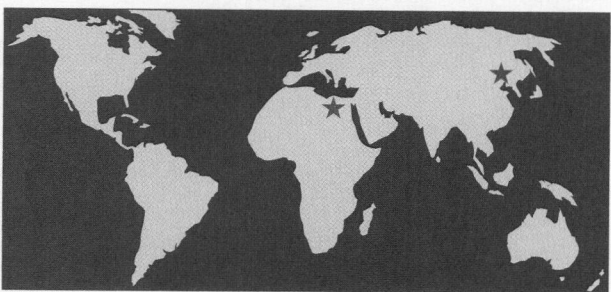

Pacing Guide for Chapter 7

Chapter Opener	1 Day
Section 1 Activity Lesson	1 Day 1 Day
Section 2 Activity Lesson	1 Day 1 Day
Section 3 Activity Lesson	1 Day 1 Day
Section 4 Activity Lesson	1 Day 1 Day
Study Help / Quiz	1 Day
Section 5 Activity Lesson	1 Day 1 Day
Section 6 Activity Lesson	1 Day 2 Days
Quiz / Chapter Review	1 Day
Chapter Test	1 Day
Standardized Test Practice	1 Day
Total Chapter 7	18 Days
Year-to-Date	131 Days

Check Your Resources

- Record and Practice Journal
- Resources by Chapter
- Skills Review Handbook
- Assessment Book
- Worked-Out Solutions

Technology
For
the Teacher

The Dynamic Planning Tool
Editable Teacher's Resources at
BigIdeasMath.com

- Simplifying fractions
- Naming similar figures
- Perimeter and area of figures
- Finding volumes of figures without formulas

Try It Yourself

1. *Answer should include, but is not limited to:* drawings of two rectangles with length-to-width ratios of 9 to 4

2. 35 cm

Record and Practice Journal

1. yes; $\dfrac{10}{5} = \dfrac{6}{3}$

2. no; $\dfrac{7}{4} \neq \dfrac{10}{7}$

3. no; $\dfrac{24}{12} \neq \dfrac{7}{5}$

4. yes; $\dfrac{2}{3} = \dfrac{6}{9}$

5. yes; $\dfrac{5}{10} = \dfrac{6}{12}$

6. yes; $\dfrac{12}{9} = \dfrac{20}{15}$

7. no; $\dfrac{3}{1} \neq \dfrac{2}{0.5}$

8. 4

9. 2.4

10. 12

11. 10

12. 20

13. 25

14. $2\dfrac{5}{8}$ in.

Math Background Notes

Vocabulary Review

- Similar
- Proportional
- Cross Products Property
- Ratio
- Corresponding Parts

Identifying Similar Figures

- Students should be able to identify similar figures.
- Remind students that they should check the orientation of the figures before writing ratios. The figures should be oriented in the same way so that corresponding sides are in the same place.
- **Teaching Tip:** Allow students to color code corresponding parts of figures. This will help visual learners to correctly orient and identify similar figures.
- **Teaching Tip:** Tactile learners may struggle with the concept of similar figures. Cut out and laminate several of the same type of polygons with the side measures marked. First, challenge students to orient all the figures in the same direction. Then, ask them to make groups of the polygons that they believe to be similar. Ask them to justify their choices by writing the appropriate ratios and checking the proportionality of the polygons.
- **Common Error:** Students may attempt to write ratios comparing Rectangle B to Rectangle C. Remind them that the goal of the example is to decide which rectangle is similar to Rectangle A, so one measure of Rectangle A should appear in the ratio.

Finding Measures in Similar Figures

- Students should be able to find missing measures in similar figures.
- You may wish to review the corresponding parts of similar figures with students. In Example 2, students need to realize that the figures are not oriented the same. Encourage them to re-draw and re-label one of the figures so that it is oriented in the same fashion as the other.
- **Common Error:** Students may try to write a proportion using the longest side and the base of each triangle, only to realize there are two variables in the proportion. Encourage them to identify the sides of the figure containing the most given information, and then try to write one ratio using this information first.

Reteaching and Enrichment Strategies

If students need help. . .	If students got it. . .
Record and Practice Journal • Fair Game Review Skills Review Handbook Lesson Tutorials	Game Closet at *BigIdeasMath.com* Start the next section

What You Learned Before

Number one on America's list of 10 worst ideas.

"I just figured out how to find your volume. We'll immerse you in a barrel of water and measure the water that overflows."

• Identifying Similar Figures

Example 1 Which rectangle is similar to Rectangle A?

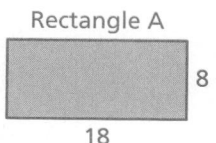

Rectangle A
18, 8

Rectangle B
27, 8

Rectangle C
27, 12

Rectangle A and Rectangle B

$$\frac{\text{Length of A}}{\text{Length of B}} = \frac{18}{27} = \frac{2}{3} \qquad \frac{\text{Width of A}}{\text{Width of B}} = \frac{8}{8} = 1 \qquad \text{Not proportional}$$

Rectangle A and Rectangle C

$$\frac{\text{Length of A}}{\text{Length of C}} = \frac{18}{27} = \frac{2}{3} \qquad \frac{\text{Width of A}}{\text{Width of C}} = \frac{8}{12} = \frac{2}{3} \qquad \text{Proportional}$$

∴ So, Rectangle C is similar to Rectangle A.

• Finding Measures in Similar Figures

Example 2 The two triangles are similar. Find the value of *x*.

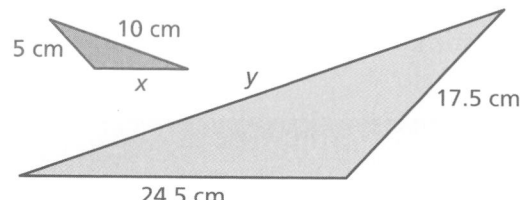

10 cm
5 cm
x
y
17.5 cm
24.5 cm

$$\frac{5}{17.5} = \frac{x}{24.5} \qquad \text{Write a proportion.}$$

$$122.5 = 17.5x \qquad \text{Use Cross Products Property.}$$

$$7 = x \qquad \text{Divide each side by 17.5.}$$

∴ So, *x* is 7 centimeters.

Try It Yourself

1. Construct two more rectangles that are similar to Rectangle A in Example 1.

2. Find the value of *y* in Example 2.

7.1 Volumes of Prisms

Essential Question How can you find the volume of a prism?

1 ACTIVITY: Pearls in a Treasure Chest

Work with a partner. A treasure chest is filled with valuable pearls. Each pearl is about 1 centimeter in diameter and is worth about $80.

Use the diagrams below to describe two ways that you can estimate the number of pearls in the treasure chest.

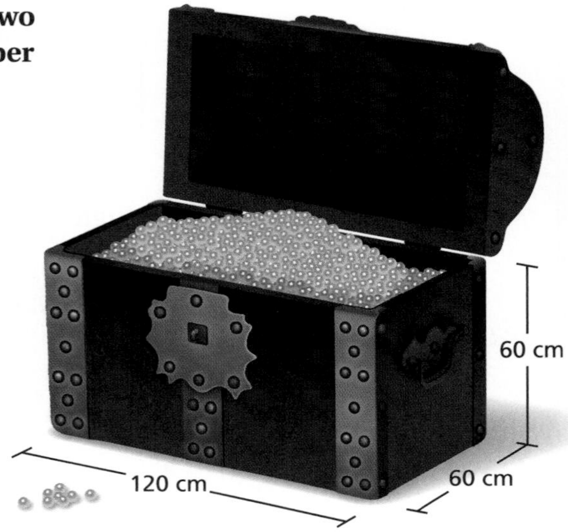

60 cm
120 cm
60 cm

a.

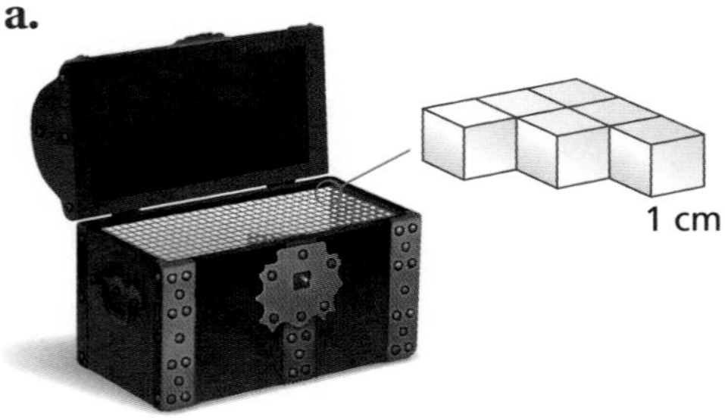

1 cm

b.

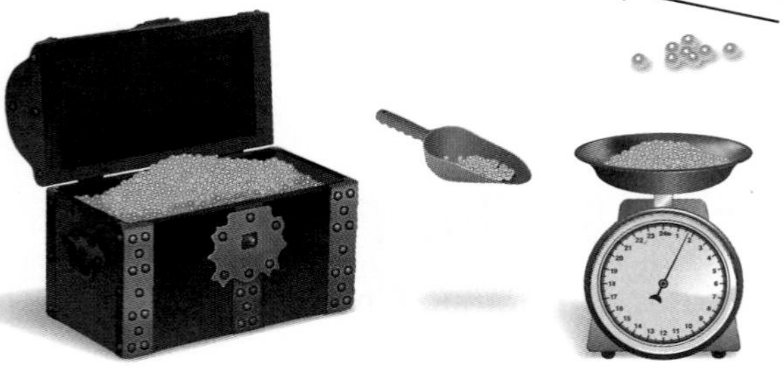

c. Use the method in part (a) to estimate the value of the pearls in the chest.

2 ACTIVITY: Finding a Formula for Volume

Work with a partner. You know that the formula for the volume of a rectangular prism is $V = \ell wh$.

a. Find a new formula that gives the volume in terms of the area of the base B and the height h.

b. Use both formulas to find the volume of each prism. Do both formulas give you the same volumes?

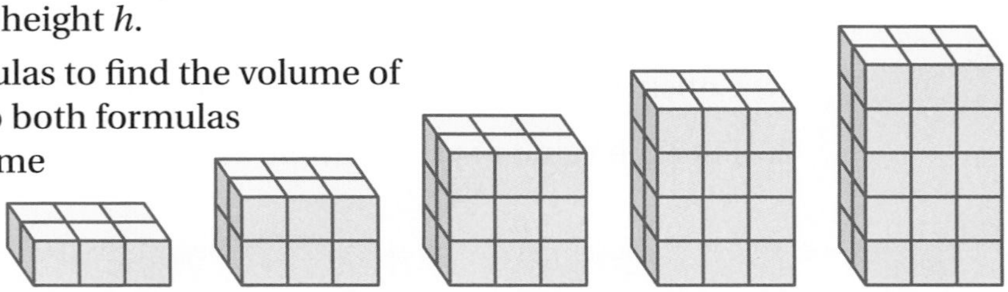

Laurie's Notes

Introduction

For the Teacher

- **Goal:** Students will develop an intuitive understanding of how to measure the volume of a prism.
- **Connection:** Students have explored volume and surface area of a prism. They should have a sense that volume is a *filling process* and that it can be found by stacking equal layers on top of one another.

Motivate

- Hold up a variety of common containers and ask what is commonly found inside. Examples: egg carton (12 eggs); playing cards box (52 cards); crayon box (8 crayons)
- Discuss with students these examples of volume. Each container is filled with objects of the same size. How many eggs fit in the egg carton, or how many crayons fit in the crayon box? Because the units are different (eggs, cards, crayons), you can't compare the volumes.

Activity Notes

Activity 1

- If you have beads (marbles), use them to model this activity. "I've filled this box with beads. How would you estimate the number of beads in the box?"
- **?** "How big is the treasure chest? Compare it to an object in this room." Students should recognize that 120 centimeters is more than 3 feet long.
- **?** "Do you think there is a thousand dollars worth of pearls in the treasure chest? a million dollars? a billion dollars?" Students will likely have only a wild guess about the value of the pearls in the chest at this point.
- In part (a) of this problem, watch for students to make a layer of centimeter cubes on the bottom and then put a total of 60 layers in the chest.
- **?** "How did you estimate the number of pearls using the method in part (a)?" The bottom layer holds about 7200 pearls. Times 60 layers is 432,000 pearls.
- **?** Explain how the method in part (b) could be used to estimate the number of pearls in the chest. *Sample answer:* Weigh the chest full, and then empty to find the weight of the pearls. Then weigh 10 pearls and use this information to estimate the total number of pearls.

Activity 2

- From the graphic, students should see that the bottom layer has 6 cubes, the second layer has 6 cubes, the third layer has 6 cubes, and so on.
- If students are not thinking about layers (height), suggest that writing the volume of each prism would be helpful: 6, 12, 18, 24, 30.
- **Big Idea:** The area of the base (denoted *B*) is 6. The height (denoted *h*) is how many layers?

Previous Learning

Students should know how to evaluate expressions and solve one-step equations.

Activity Materials	
Introduction	**Textbook**
• egg carton • crayon box • playing cards box	• marbles or beads • unit cubes

Start Thinking! and Warm Up

Activity 7.1 Start Thinking!
For use before Activity 7.1

Activity 7.1 Warm Up
For use before Activity 7.1

Multiply.

1. $7 \times 5 \times 8$ 2. $12 \times 7 \times 8$

3. $(13)(10)(7)$ 4. $11 \cdot 15 \cdot 3$

5. $(14)(20)(4)$ 6. $12 \cdot 16 \cdot 21$

7.1 Record and Practice Journal

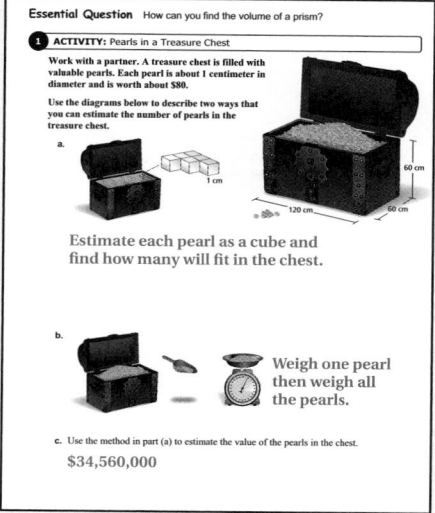

Essential Question How can you find the volume of a prism?

1 ACTIVITY: Pearls in a Treasure Chest

Work with a partner. A treasure chest is filled with valuable pearls. Each pearl is about 1 centimeter in diameter and is worth about $80.

Use the diagrams below to describe two ways that you can estimate the number of pearls in the treasure chest.

a.

Estimate each pearl as a cube and find how many will fit in the chest.

b.

Weigh one pearl then weigh all the pearls.

c. Use the method in part (a) to estimate the value of the pearls in the chest.
$34,560,000

Differentiated Instruction

Visual

Students may think that prisms with the same volume have the same surface area. Have them work together to find prisms with the same volume, but different dimensions. Then direct the students to find the surface areas of each of the prisms. Ask them to share their results.

7.1 Record and Practice Journal

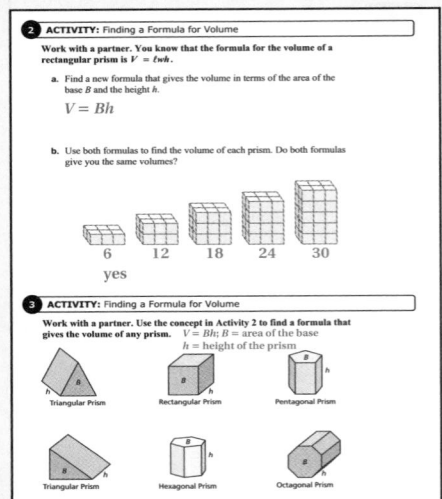

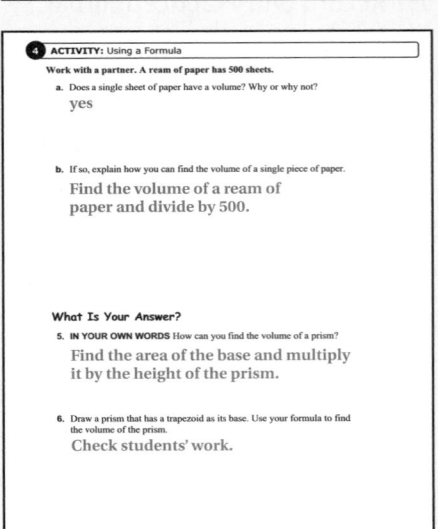

Activity 3

- The formula discovered in Activity 2 is now used in Activity 3.
- **Connection:** Students can memorize formulas and have little understanding of why the formula makes sense. It is important throughout this chapter that students see that the formulas are all similar. The volume is found by finding the area of the base (B) and then multiplying by the number of layers (h).
- Having models of these prisms is very helpful.
- **Common Misconception:** The height of a prism does not need to be the vertical direction. Demonstrate this by holding a rectangular prism (a tissue box is fine). Ask students to identify the base (a face of the prism) and the height (an edge). Chances are students will identify the (standard) bottom of the box as the base. Now, rotate the tissue box so that the base is vertical. Again ask students to identify the base and height. Students may stick with their first answers or may now switch to the "bottom face" as the base.
- A prism is named by its base. A triangular prism has 2 triangular bases.
- Give students time to discuss the solids in this activity. If you have physical models of each of these, ask six volunteers to describe how to find the volume of the solid. Expect the student volunteer to point to the base and the height as they give the formula for volume.

Activity 4

- **Common Misconception:** Students may believe a sheet of paper has no height and so, no volume, only area. It may be difficult to measure the height with tools available to us, but a sheet of paper does have a height.
- The ream of copy paper is a good visual model.

What Is Your Answer?

- **Think-Pair-Share:** Students should read each question independently and then work with a partner to answer the questions. When they have answered the questions, the pair should compare their answers with another group and discuss any discrepancies.

Closure

- **Writing Prompt:** To find the volume of a tissue box ...

Technology For the Teacher

Dynamic Classroom

The Dynamic Planning Tool
Editable Teacher's Resources at *BigIdeasMath.com*

3 **ACTIVITY: Finding a Formula for Volume**

Work with a partner. Use the concept in Activity 2 to find a formula that gives the volume of any prism.

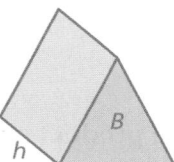

Triangular Prism

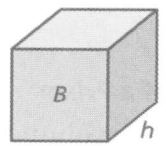

Rectangular Prism

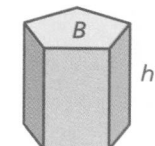

Pentagonal Prism

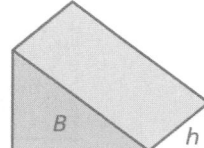

Triangular Prism

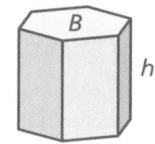

Hexagonal Prism

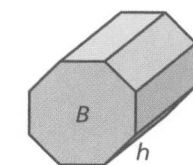

Octagonal Prism

4 **ACTIVITY: Using a Formula**

Work with a partner. A ream of paper has 500 sheets.

a. Does a single sheet of paper have a volume? Why or why not?

b. If so, explain how you can find the volume of a single sheet of paper.

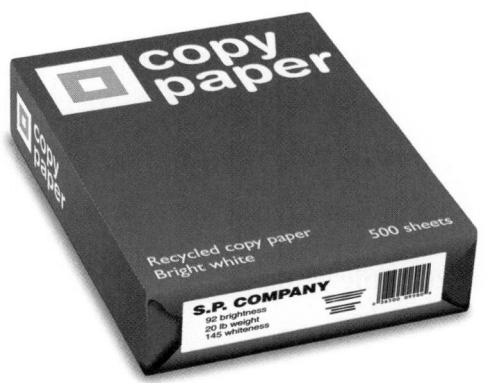

What Is Your Answer?

5. IN YOUR OWN WORDS How can you find the volume of a prism?

6. Draw a prism that has a trapezoid as its base. Use your formula to find the volume of the prism.

Use what you learned about the volumes of prisms to complete Exercises 4–6 on page 302.

Check It Out
Lesson Tutorials
BigIdeasMath com

The **volume** of a three-dimensional figure is a measure of the amount of space that it occupies. Volume is measured in cubic units.

Key Vocabulary
volume, *p. 300*

Key Idea

Volume of a Prism

Words The volume *V* of a prism is the product of the area of the base and the height of the prism.

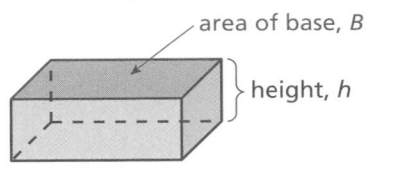
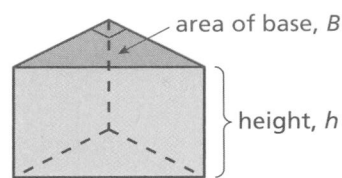

area of base, *B* — height, *h*

area of base, *B* — height, *h*

Algebra
$$V = Bh$$

Area of base ← → Height of prism

EXAMPLE 1 — Finding the Volume of a Prism

Study Tip

The area of the base of a rectangular prism is the product of the length ℓ and the width w.

You can use $V = \ell wh$ to find the volume of a rectangular prism.

Find the volume of the prism.

$V = Bh$	Write formula for volume.
$= 6(8) \cdot 15$	Substitute.
$= 48 \cdot 15$	Simplify.
$= 720$	Multiply.

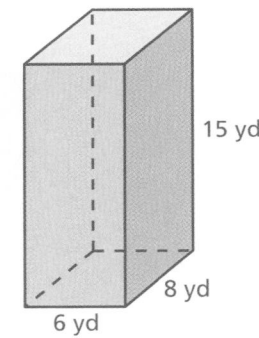

15 yd

8 yd

6 yd

∴ The volume is 720 cubic yards.

EXAMPLE 2 — Finding the Volume of a Prism

Find the volume of the prism.

$V = Bh$	Write formula for volume.
$= \dfrac{1}{2}(5.5)(2) \cdot 4$	Substitute.
$= 5.5 \cdot 4$	Simplify.
$= 22$	Multiply.

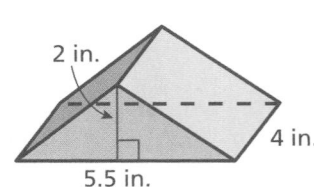

2 in.

4 in.

5.5 in.

∴ The volume is 22 cubic inches.

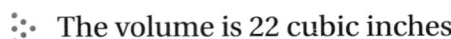 Multi-Language Glossary at BigIdeasMath✓com.

Laurie's Notes

Introduction

Connect
- **Yesterday:** Students explored how to find the volume of a prism.
- **Today:** Students will use the formula for the volume of a prism to solve problems.

Motivate
- **True Story:** Baseball legend Ken Griffey Jr. owed teammate Josh Fogg some money and paid him back in pennies. Griffey stacked 60 cartons, each holding $25 worth of pennies, in Fogg's locker.
- ❓ Ask the following questions.
  - "How does this story relate to the volume of a prism?" The volume of the carton is being measured in pennies.
  - "How big is a carton that can hold $25 worth of pennies?" open-ended
  - "How many pennies were in each carton?" 2500
  - "How much did Griffey owe Fogg?" $1500
  - "How much do you think each carton weighed?" open-ended

Lesson Notes

Key Idea
- ❓ "What is a prism?" three-dimensional solid with two congruent bases and lateral faces that are rectangles
- ❓ "What are cubic units? Give an example." Cubic units are cubes which fill a space completely without overlapping or leaving gaps. Cubic inches and cubic centimeters are common examples.
- Point out to students that the bases of the prisms have been shaded differently. The height will be perpendicular to the two congruent bases.
- **Teaching Tip:** Use words (area of base, height) and symbols (B, h) when writing the formula.
- **Review Vocabulary:** *Product* is the answer to a multiplication problem.

Example 1
- Discuss the *Study Tip* with students.
- ❓ "Could the face measuring 8 yards by 15 yards be the base?" Yes, the height would then be 6 yards.
- **Extension:** Point out to students that all of the measurements are in terms of yards. "What if the 6 yard edge had been labeled 18 feet. Now how would you find the volume?" Convert all 3 dimensions to yards or to feet.

Example 2
- Ask a volunteer to describe the base of this triangular prism.
- ❓ "What property is used to simplify the area of the base?" Commutative Property of Multiplication
- Caution students to distinguish between the height of the base and the height of the prism.

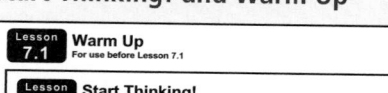

Goal Today's lesson is finding the **volumes** of prisms.

Start Thinking! and Warm Up

Lesson 7.1	Warm Up

For use before Lesson 7.1

Lesson 7.1	Start Thinking!

For use before Lesson 7.1

You are sending a gift to a friend. Explain how volume would be helpful in figuring out which size box to use to send the gift to your friend.

Extra Example 1
Find the volume of a rectangular prism with a length of 2 meters, a width of 6 meters, and a height of 3 meters. 36 m³

Extra Example 2
Find the volume of the prism.

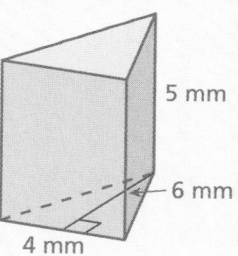

5 mm

6 mm

4 mm

60 mm³

On Your Own

1. 80 ft³

2. 270 m³

Extra Example 3

Two rectangular prisms each have a volume of 120 cubic centimeters. The base of Prism A is 2 centimeters by 4 centimeters. The base of Prism B is 4 centimeters by 6 centimeters.

a. Find the height of each prism.
    Prism A: 15 cm, Prism B: 5 cm

b. Which prism has the lesser surface area? Prism B

 ## On Your Own

3. yes; Because it has the same volume as the other two bags, but its surface area is 107.2 square inches which is less than both Bag A and Bag B.

English Language Learners

Vocabulary

Discuss the meaning of the words *volume* and *cubic units*. Have students add these words to their notebooks.

Laurie's Notes

On Your Own

*   Ask volunteers to share their work at the board.

Example 3

*   This example connects volume, surface area, and solving equations.
*   Ask a student to read the example.
*   ❓ "What type of measurement is 96 cubic inches?" volume
*   ❓ "What concept does part (b) refer to?" surface area
*   Work through part (a).
*   Before beginning part (b), ask students to review the formula for surface area of a rectangular prism. Note that only five of the six faces are considered.
*   Work through part (b).
*   ❓ "Both bags hold the same amount of popcorn. Are there any practical advantages of one bag over the other?" Bag A: can grip it in your hand more easily; Bag B: less likely to tip over and uses less paper

On Your Own

*   Given the discussion of the practical features of Bags A and B, students should have a sense of the problems in the design of Bag C.

Closure

*   Sketch a rectangular prism. Label the dimensions 4 centimeters, 6 centimeters, and 10 centimeters. Find the volume of the prism.
    $V = 4 \cdot 6 \cdot 10 = 240$ cm³

The Dynamic Planning Tool
Editable Teacher's Resources at *BigIdeasMath.com*

On Your Own

Find the volume of the prism.

1.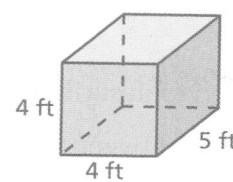

4 ft 4 ft 5 ft

2.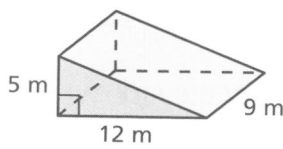

5 m 12 m 9 m

EXAMPLE 3 **Real-Life Application**

A movie theater designs two bags to hold 96 cubic inches of popcorn. (a) Find the height of each bag. (b) Which bag should the theater choose to reduce the amount of paper needed? Explain.

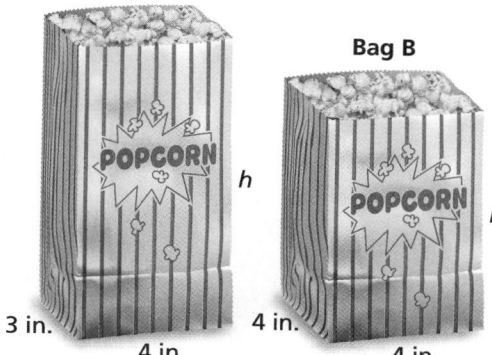

Bag A

Bag B

3 in. 4 in. 4 in. 4 in.

a. Find the height of each bag.

Bag A	Bag B
$V = Bh$	$V = Bh$
$96 = 4(3)(h)$	$96 = 4(4)(h)$
$96 = 12h$	$96 = 16h$
$8 = h$	$6 = h$

∴ The height is 8 inches.   ∴ The height is 6 inches.

b. To determine the amount of paper needed, find the surface area of each bag. Do not include the top base.

Bag A

$S = \ell w + 2\ell h + 2wh$

$= 4(3) + 2(4)(8) + 2(3)(8)$

$= 12 + 64 + 48$

$= 124$ in.^{2}

Bag B

$S = \ell w + 2\ell h + 2wh$

$= 4(4) + 2(4)(6) + 2(4)(6)$

$= 16 + 48 + 48$

$= 112$ in.^{2}

∴ The surface area of Bag B is less than the surface area of Bag A. So, the theater should choose Bag B.

On Your Own

3. You design Bag C that has a volume of 96 cubic inches. Should the theater in Example 3 choose your bag? Explain.

Bag C

4 in. 4.8 in. h

 Vocabulary and Concept Check

1. **VOCABULARY** What type of units are used to describe volume?

2. **CRITICAL THINKING** What is the difference between volume and surface area?

3. **CRITICAL THINKING** You are ordering packaging for a product. Should you be more concerned with volume or surface area? Explain.

 Practice and Problem Solving

Find the volume of the prism.

 4.

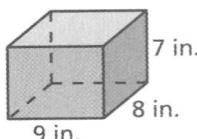

7 in.
8 in.
9 in.

5.

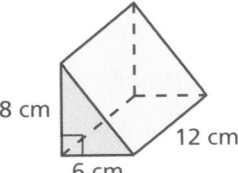

8 cm
12 cm
6 cm

6.

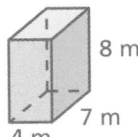

8 m
7 m
4 m

7.

5 yd
4 yd
8 yd

8.

 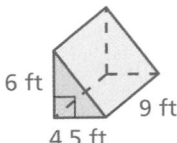

6 ft
9 ft
4.5 ft

9.

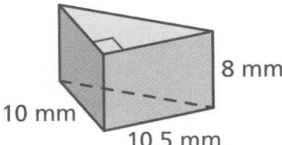

8 mm
10 mm
10.5 mm

10.

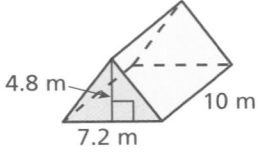

4.8 m
10 m
7.2 m

11.

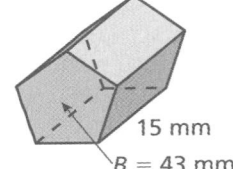

15 mm
$B = 43$ mm^2

12.

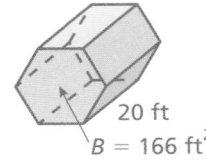

20 ft
$B = 166$ ft^2

13. **ERROR ANALYSIS** Describe and correct the error in finding the volume of the triangular prism.

7 cm
10 cm
5 cm

$$X \quad \begin{aligned} V &= Bh \\ &= 10(5)(7) \\ &= 50 \cdot 7 \\ &= 350 \text{ cm}^3 \end{aligned}$$

School Locker

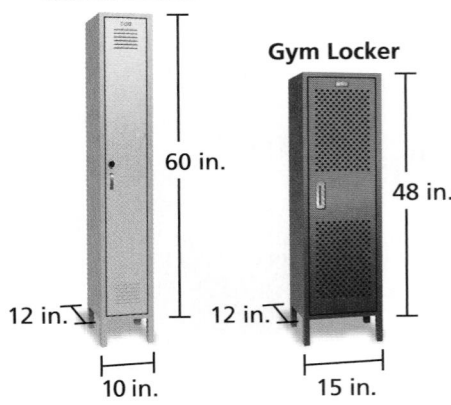

60 in.

12 in.

10 in.

Gym Locker

48 in.

12 in.

15 in.

14. **LOCKER** Each locker is shaped like a rectangular prism. Which has more storage space? Explain.

15. **CEREAL BOX** A cereal box is 9 inches by 2.5 inches by 10 inches. What is the volume of the box?

Assignment Guide and Homework Check

Level	Day 1 Activity Assignment	Day 2 Lesson Assignment	Homework Check
Basic	4–6, 25–28	1–3, 7–15, 17	7, 10, 14, 17
Average	4–6, 25–28	1–3, 7–15 odd, 16–20	7, 9, 16, 18
Advanced	4–6, 25–28	1–3, 13, 16–24	16, 18, 22, 23

Common Errors

- **Exercises 4–12** Students may write the units incorrectly, often writing square units instead of cubic units. Remind them that they are working in three dimensions, so the units are cubed. Give an example showing the formula for the base as three units multiplied together. For example, write the volume of Exercise 5 as $V = \frac{1}{2}(6 \text{ cm})(8 \text{ cm})(12 \text{ cm})$.

7.1 Record and Practice Journal

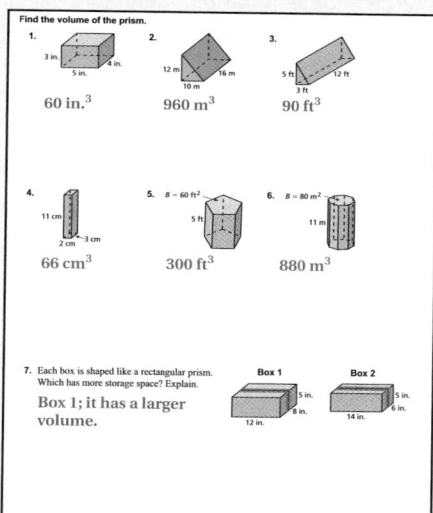

Vocabulary and Concept Check

1. cubic units

2. The volume of an object is the amount of space it occupies. The surface area of an object is the sum of the areas of all its faces.

3. *Sample answers:* Volume because you want to make sure the product will fit inside the package. Surface area because of the cost of packaging.

Practice and Problem Solving

4. 504 in.³   5. 288 cm³

6. 224 m³   7. 160 yd³

8. 121.5 ft³   9. 420 mm³

10. 172.8 m³   11. 645 mm³

12. 3320 ft³

13. The area of the base is wrong.
$$V = \frac{1}{2}(7)(5) \cdot 10$$
$$= 175 \text{ cm}^3$$

14. The gym locker has more storage space because it has a greater volume.

15. 225 in.³

16. 1440 in.³

17. 7200 ft³

18. sometimes; The prisms in Example 3 have different surface areas, but the same volume. Two prisms that are exactly the same will have the same surface area.

19. 1728 in.³

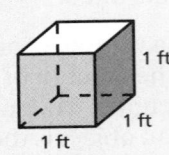

1 ft
1 ft
1 ft

$1 \times 1 \times 1 = 1 \text{ ft}^3$

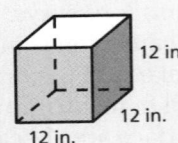

12 in.
12 in.
12 in.

$12 \times 12 \times 12 = 1728 \text{ in.}^3$

20. 48 packets

21. 20 cm

22. *Sample answer:* gas about $3 per gallon; $36

23. See *Taking Math Deeper.*

24. 240 cm³

 Fair Game Review

25. reflection

26. translation

27. rotation   **28.** D

Mini-Assessment

Find the volume of the prism.

1.

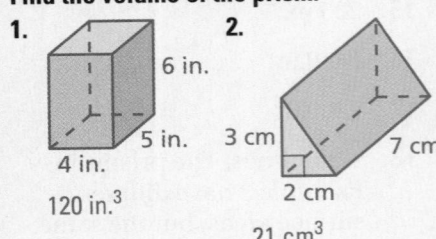

6 in.
5 in.
4 in.

120 in.³

2.

3 cm
7 cm
2 cm

21 cm³

3. Find the volume of the fish tank.

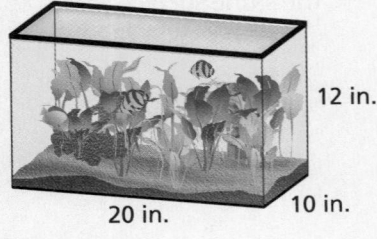

12 in.
20 in.
10 in.

2400 in.³

Taking Math Deeper

Exercise 23

This problem gives students a chance to relate dimensions of a solid with the volume of a solid. It also gives students an opportunity to work with prime factorization. Although aquariums are traditionally the shape of a rectangular prism, remember that other shapes are also possible.

① Find the volume of the aquarium in cubic inches.

$$\text{Volume} = (450 \text{ gal})\left(231 \frac{\text{in.}^3}{\text{gal}}\right) = 103{,}950 \text{ in.}^3$$

② You could choose two of the dimensions and solve for the third, or you can use prime factorization to find whole number solutions.

Find the prime factorization of 103,950.

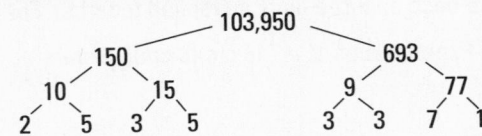

103,950

150   693

10   15   9   77

2 5 3 5   3 3 7 11

The prime factorization is $2 \times 3 \times 3 \times 3 \times 5 \times 5 \times 7 \times 11$.

③ Rearrange the factors to find one set of possible dimensions.

Length: $2 \times 3 \times 5 \times 5 = 150$ in.
Width: $3 \times 7 = 21$ in.
Height: $3 \times 11 = 33$ in.

A long tank

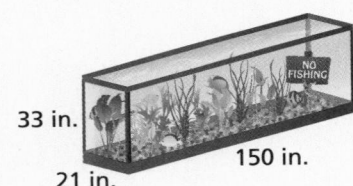

33 in.
150 in.
21 in.

Project

Research a local aquarium. Select an exhibit and draw a picture about the exhibit you selected. Include a short report about the details of the exhibit.

Reteaching and Enrichment Strategies

If students need help. . .	If students got it. . .
Resources by Chapter • Practice A and Practice B • Puzzle Time Record and Practice Journal Practice Differentiating the Lesson Lesson Tutorials Skills Review Handbook	Resources by Chapter • Enrichment and Extension Start the next section

Find the volume of the prism.

16.

12 in.

12 in.   10 in.

17.

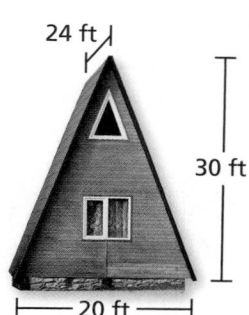

24 ft

30 ft

20 ft

18. REASONING Two prisms have the same volume. Do they *always*, *sometimes*, or *never* have the same surface area? Explain.

19. CUBIC UNITS How many cubic inches are in a cubic foot? Use a sketch to explain your reasoning.

20. CAPACITY As a gift, you fill the calendar with packets of chocolate candy. Each packet has a volume of 2 cubic inches. Find the maximum number of packets you can fit inside the calendar.

6 in.

8 in.   4 in.

21. HEIGHT Two liters of water are poured into an empty vase shaped like an octagonal prism. The base area is 100 square centimeters. What is the height of the water? (1 L = 1000 cm^3)

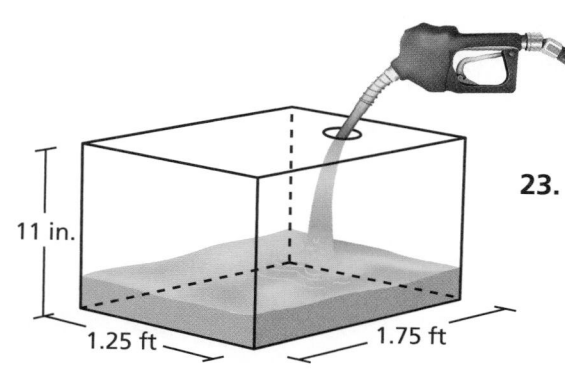

11 in.

1.25 ft   1.75 ft

22. GAS TANK The gas tank is 20% full. Use the current price of gas in your community to find the cost to fill the tank. (1 gal = 231 in.^{3})

23. OPEN-ENDED You visit an aquarium. One of the tanks at the aquarium holds 450 gallons of water. Draw a diagram to show one possible set of dimensions of the tank. (1 gal = 231 in.^{3})

24. *Critical Thinking* What is the volume of the rectangular prism?

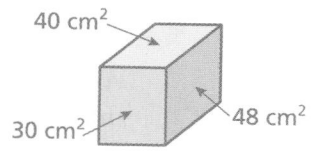

40 cm^2

30 cm^2 48 cm^2

 **Fair Game Review** *What you learned in previous grades & lessons*

Identify the transformation. *(Section 5.5, Section 5.6, and Section 5.7)*

25.

26.

27.

28. MULTIPLE CHOICE What is the approximate surface area of a cylinder with a radius of 3 inches and a height of 10 inches? *(Section 6.3)*

Ⓐ 30 in.^{2} Ⓑ 87 in.^{2} Ⓒ 217 in.^{2} Ⓓ 245 in.^{2}

7.2 Volumes of Cylinders

Essential Question How can you find the volume of a cylinder?

Share Your Work at... My.BigIdeasMath.com

1 ACTIVITY: Finding a Formula Experimentally

Work with a partner.

a. Find the area of the face of a coin.

b. Find the volume of a stack of a dozen coins.

c. Generalize your results to find the volume of a cylinder.

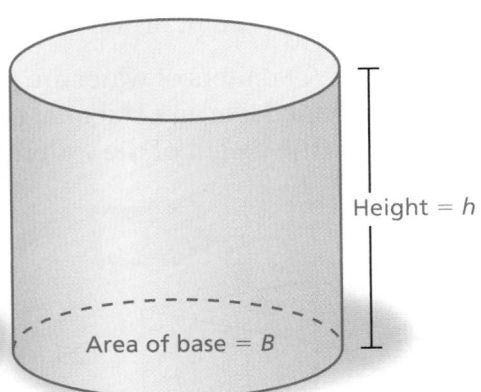

Height = *h*

Area of base = *B*

2 ACTIVITY: Making a Business Plan

Work with a partner. You are planning to make and sell 3 different sizes of cylindrical candles. You buy 1 cubic foot of candle wax for $20 to make 8 candles of each size.

a. Design the candles. What are the dimensions of each size?

b. You want to make a profit of $100. Decide on a price for each size.

c. Did you set the prices so that they are proportional to the volume of each size of candle? Why or why not?

Laurie's Notes

Introduction

For the Teacher

- **Goal:** Students will develop an understanding of the volume of a cylinder.
- **Big Idea:** Recall that a prism and a cylinder have two congruent bases and a lateral portion. Students will develop the formula for the volume of a cylinder in the same way as the volume of a prism using layers of the base.

Motivate

- Use round crackers, wafer candies, coins, or circular metal washers to model layers of very thin cylinders being stacked to make a cylinder.
- Display the models for students to see.
- **?** "What do these items have in common?" cylinder made up of thin layers
- Explain that today they will explore the volume of a cylinder using an approach similar to the crackers, wafers, coins, or washers.

Activity Notes

Activity 1

- **?** "How will you find the area of the base?" Area of a circle $= \pi r^2$; Students will likely measure the diameter to the nearest tenth centimeter.
- Discuss with students how they found the volume of the stack of 12 coins.
- **Big Idea:** Area is measured in square units. Volume is measured in cubic units.

Activity 2

- This is an open-ended activity that you can adjust to the skill level of your students.
- You can make the following assumptions: all of the candles have the same base with different heights: 2 inches, 3 inches, and 5 inches. Because there are 8 of each, the total height is 80 inches.
- **?** There are 12^3 or 1728 cubic inches of wax available. "What does the radius of the candle need to be to use up the wax?"
- At this stage, students need to think through how they can find the volume of a cylinder. They need to find the area of the base (πr^2) and multiply by 80, the total height of the 24 candles. Through trial and error, and with a calculator, they will find that a 2.5-inch radius would use 1570 cubic inches.
- Pricing the candles involves proportions. The 3-inch tall candle should cost 50% more than the 2-inch tall candle. The 5-inch tall candle should cost 150% more than the 2-inch tall candle.

Previous Learning

Students should know that cylinders are composed of 2 circular bases and a rectangle.

Start Thinking! and Warm Up

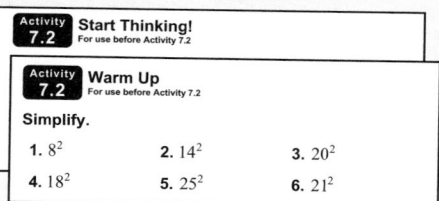

| Activity **7.2** | **Start Thinking!** For use before Activity 7.2 |

| Activity **7.2** | **Warm Up** For use before Activity 7.2 |

Simplify.

1. 8^2 2. 14^2 3. 20^2

4. 18^2 5. 25^2 6. 21^2

7.2 Record and Practice Journal

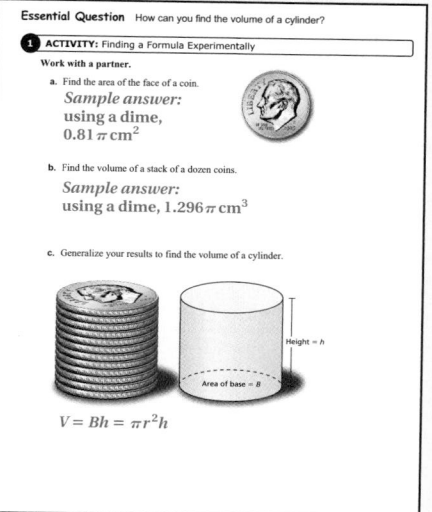

Essential Question How can you find the volume of a cylinder?

1 ACTIVITY: Finding a Formula Experimentally

Work with a partner.

a. Find the area of the face of a coin.
   Sample answer:
   using a dime,
   $0.81\pi\,cm^2$

b. Find the volume of a stack of a dozen coins.
   Sample answer:
   using a dime, $1.296\pi\,cm^3$

c. Generalize your results to find the volume of a cylinder.

$V = Bh = \pi r^2 h$

Differentiated Instruction

Money

Have students bring in coins of different denominations from their country of origin. Repeat Activity 1 using the measurements of the coins.

Activity 3

- Perhaps students have found volume by displacement in a science class.
- **Note:** This way of measuring volume is similar to the story of Archimedes who yelled "Eureka" when getting into a bathtub of water. He realized that he had displaced a volume of water equal to the volume of his body.
- If possible, have graduated cylinders and a large stone available to model this problem.

Activity 4

- Encourage students to take a guess even if their reasoning is no more than "it looks like it would hold a lot more."
- **?** "What is the area of each base (leave answers in terms of π)?" 9π and 4π
- **?** "So, which is greater, 4 layers of 9π or 9 layers of 4π?" They're the same.

What Is Your Answer?

- **Think-Pair-Share:** Students should read each question independently and then work with a partner to answer the questions. When they have answered the questions, the pair should compare their answers with another group and discuss any discrepancies.

Closure

- Refer to one of the cylinders used to motivate the activity and ask students to describe how they would find the volume of the cylinder.

7.2 Record and Practice Journal

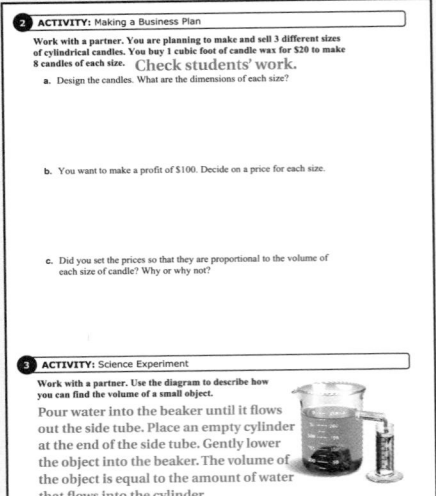

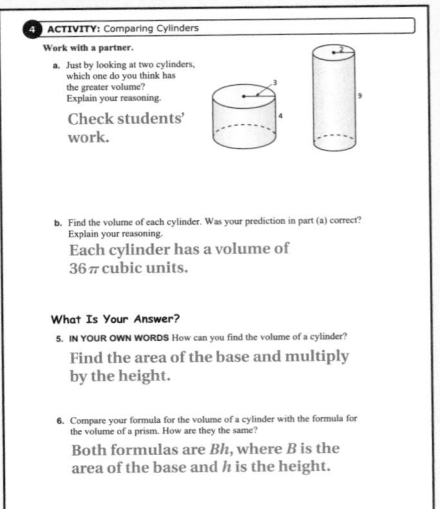

2 ACTIVITY: Making a Business Plan

Work with a partner. You are planning to make and sell 3 different sizes of cylindrical candles. You buy 1 cubic foot of candle wax for $20 to make 8 candles of each size. **Check students' work.**

a. Design the candles. What are the dimensions of each size?

b. You want to make a profit of $100. Decide on a price for each size.

c. Did you set the prices so that they are proportional to the volume of each size of candle? Why or why not?

3 ACTIVITY: Science Experiment

Work with a partner. Use the diagram to describe how you can find the volume of a small object.

Pour water into the beaker until it flows out the side tube. Place an empty cylinder at the end of the side tube. Gently lower the object into the beaker. The volume of the object is equal to the amount of water that flows into the cylinder.

4 ACTIVITY: Comparing Cylinders

Work with a partner.

a. Just by looking at two cylinders, which one do you think has the greater volume? Explain your reasoning. **Check students' work.**

b. Find the volume of each cylinder. Was your prediction in part (a) correct? Explain your reasoning. **Each cylinder has a volume of 36π cubic units.**

What Is Your Answer?

5. IN YOUR OWN WORDS How can you find the volume of a cylinder? **Find the area of the base and multiply by the height.**

6. Compare your formula for the volume of a cylinder with the formula for the volume of a prism. How are they the same? **Both formulas are Bh, where B is the area of the base and h is the height.**

Technology For the Teacher

Dynamic Classroom

The Dynamic Planning Tool
Editable Teacher's Resources at *BigIdeasMath.com*

3 ACTIVITY: Science Experiment

Work with a partner. Use the diagram to describe how you can find the volume of a small object.

4 ACTIVITY: Comparing Cylinders

Work with a partner.

a. Just by looking at the two cylinders, which one do you think has the greater volume? Explain your reasoning.

b. Find the volume of each cylinder. Was your prediction in part (a) correct? Explain your reasoning.

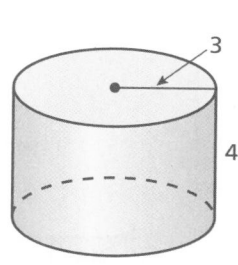
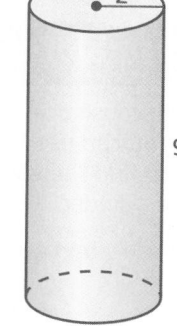

What Is Your Answer?

5. IN YOUR OWN WORDS How can you find the volume of a cylinder?

6. Compare your formula for the volume of a cylinder with the formula for the volume of a prism. How are they the same?

"Here's how I remember how to find the volume of <u>any</u> prism or cylinder."

"Base times tall, will fill 'em all."

Practice

Use what you learned about the volumes of cylinders to complete Exercises 3–5 on page 308.

Check It Out
Lesson Tutorials
BigIdeasMath com

Key Idea

Volume of a Cylinder

Words The volume V of a cylinder is the product of the area of the base and the height of the cylinder.

area of base, B

height, h

Algebra $V = Bh$

Area of base ⟍ ⟍ Height of cylinder

EXAMPLE 1 Finding the Volume of a Cylinder

Find the volume of the cylinder. Round your answer to the nearest tenth.

$$V = Bh \qquad \text{Write formula for volume.}$$
$$= \pi(3)^2(6) \qquad \text{Substitute.}$$
$$= 54\pi \approx 169.6 \qquad \text{Simplify.}$$

The volume is about 169.6 cubic meters.

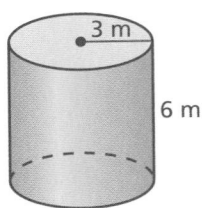

3 m

6 m

Study Tip

Because $B = \pi r^2$, you can use $V = \pi r^2 h$ to find the volume of a cylinder.

EXAMPLE 2 Finding the Height of a Cylinder

Find the height of the cylinder. Round your answer to the nearest whole number.

The diameter is 10 inches. So, the radius is 5 inches.

$$V = Bh \qquad \text{Write formula for volume.}$$
$$314 = \pi(5)^2(h) \qquad \text{Substitute.}$$
$$314 = 25\pi h \qquad \text{Simplify.}$$
$$4 \approx h \qquad \text{Divide each side by } 25\pi.$$

The height is about 4 inches.

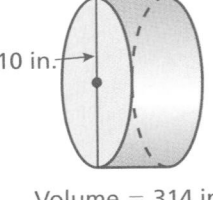

h

10 in.

Volume = 314 in.^{3}

On Your Own

Now You're Ready
Exercises 3–11
and 13–15

Find the volume V or height h of the cylinder. Round your answer to the nearest tenth.

1.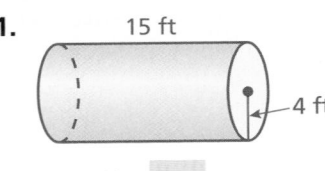

15 ft

4 ft

$V \approx$

2.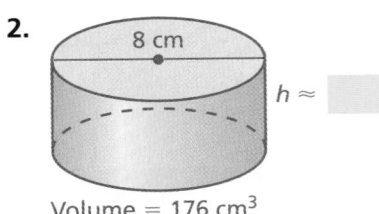

8 cm

$h \approx$

Volume = 176 cm^3

Laurie's Notes

Introduction

Connect

- **Yesterday:** Students discovered how to find the volume of a cylinder by considering the layers that make up a cylinder.
- **Today:** Students will work with a formula for the volume of a cylinder.

Motivate

- Hold two cans for the class to see. I often use cans of whole tomatoes—not only are the dimensions what I want but the contents are the same, so we can do a little cost analysis at the end of class.
- **?** "How do the volumes of these two cans compare?" The purpose here is to get students thinking about dimensions, not to do computations.

Lesson Notes

Key Idea

- **?** "How are cylinders and prisms alike?" two congruent bases and a lateral portion
- **?** "How are cylinders and prisms different?" Cylinders have circular bases while prisms have polygonal bases.
- Write the formula in words.
- Before writing the formula in symbols, ask how to find the area of the base.
- Note the use of color to identify the base in the formula and the diagram.

Example 1

- Model good problem solving by writing the formula first.
- Notice that the values of the variables are substituted, simplified, and left in terms of π. The last step is to substitute 3.14 for π.
- **?** Write "$\approx$" on the board. "What does this symbol mean?" approximately equal to "Why do you use this symbol?" π is an irrational number.
- **Extension:** Discuss how big this cylinder is. The diameter and the height are 6 meters. This is wider than my classroom and more than twice the height.
- **Big Idea:** Volume $= Bh$ is the general formula for both prisms and cylinders. The base of a cylinder is a circle, so the general formula can be rewritten as the specific formula $V = \pi r^2 h$.

Example 2

- Students will be unsure of how to divide 314 by 25π. One way is to find the product 25π, and then divide 314 by the product. The other way is to divide 314 by π, 3.14, and then divide that answer by 25.

On Your Own

- Students should work the problems alone and then check with a neighbor.

Start Thinking! and Warm Up

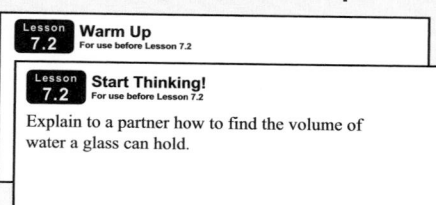

Lesson 7.2 **Warm Up** For use before Lesson 7.2

Lesson 7.2 **Start Thinking!** For use before Lesson 7.2

Explain to a partner how to find the volume of water a glass can hold.

Extra Example 1

Find the volume of a cylinder with a radius of 6 feet and a height of 3 feet. Round your answer to the nearest tenth. $108\pi \approx 339.1$ ft³

Extra Example 2

Find the height of a cylinder with a diameter of 4 yards and a volume of 88 cubic yards. Round your answer to the nearest whole number. $\frac{22}{\pi} \approx 7$ yd

On Your Own

1. $240\pi \approx 753.6$ ft³
2. $\frac{11}{\pi} \approx 3.5$ cm

Laurie's Notes

Extra Example 3

A jelly jar has a radius of 3 centimeters and a height of 8 centimeters. The jelly remaining in the jar has a height of 3 centimeters. How much jelly is missing from the jar? $45\pi \approx 141.3 \text{ cm}^3$

Extra Example 4

About how many gallons of water does the water cooler bottle in Example 4 contain, if the bottle is 1.25 feet tall?
about 7.4 gal

On Your Own

3. $125\pi \approx 392.5 \text{ cm}^3$

4. about 233,145 gal

English Language Learners

Vocabulary

Discuss the meaning of the words *volume* and *cubic units*. Have students add these words to their notebooks.

Example 3

? "What percent of the salsa is missing and what percent remains?"
60%; 40%

- Work through the problem.

- **Extension:** Find the original volume without using the volume formula. *Hint:* You could use the percent equation or set up a proportion.

Example 4

- It is helpful to have 3 rulers to model what a cubic foot looks like. Hold the 3 rulers so they form 3 edges of a cube that meet at a vertex.

- ? "About how many gallons do you think would fill a cubic foot?" There will be a range of answers.

- Work through the problem, finding the volume of the water cooler in cubic feet.

- The second part of the problem involves dimensional analysis, a technique used earlier in the text.

- Estimate first. If 1 ft$^3 \approx 7.5$ gal, how many gallons would 1.3345 cubic feet be? An estimate of 10 gallons is reasonable and that is choice B.

On Your Own

- Give students sufficient time to do their work before asking volunteers to share their work at the board.

Closure

- Hold the two cans used to motivate the lesson and ask students to find the volume of each. If the contents are the same (or pretend that they are the same), how should the prices compare?

Technology
For the Teacher

Dynamic Classroom

The Dynamic Planning Tool
Editable Teacher's Resources at *BigIdeasMath.com*

EXAMPLE **3**   **Real-Life Application**

How much salsa is missing from the jar?

The missing salsa fills a cylinder with a
height of $10 - 4 = 6$ centimeters and a
radius of 5 centimeters.

$$V = Bh \qquad \text{Write formula for volume.}$$
$$= \pi(5)^2(6) \qquad \text{Substitute.}$$
$$= 150\pi \approx 471 \qquad \text{Simplify.}$$

⁙ About 471 cubic centimeters of salsa are missing from the jar.

EXAMPLE **4**   **Standardized Test Practice**

**About how many gallons of water does the water cooler bottle
contain? ($1 \text{ ft}^3 \approx 7.5$ gal)**

(A) 5.3 gal   **(B)** 10 gal   **(C)** 17 gal   **(D)** 40 gal

Find the volume of the cylinder. The diameter is 1 foot. So, the radius is
0.5 foot.

$$V = Bh \qquad \text{Write formula for volume.}$$
$$= \pi(0.5)^2(1.7) \qquad \text{Substitute.}$$
$$= 0.425\pi \approx 1.3345 \qquad \text{Simplify.}$$

So, the cylinder contains about 1.3345 cubic feet of water. To find the
number of gallons it contains, multiply by $\dfrac{7.5 \text{ gal}}{1 \text{ ft}^3}$.

$$1.3345 \text{ ft}^3 \times \frac{7.5 \text{ gal}}{1 \text{ ft}^3} \approx 10 \text{ gal}$$

⁙ The water cooler bottle contains about 10 gallons of water. The
correct answer is **(B)**.

On Your Own

Now You're Ready
Exercise 12

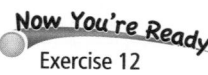

3. **WHAT IF?** In Example 3, the height of the salsa in the jar is
5 centimeters. How much salsa is missing from the jar?

4. A cylindrical water tower has a diameter of 15 meters and a
height of 5 meters. About how many gallons of water can the
tower contain? ($1 \text{ m}^3 \approx 264$ gal)

7.2 Exercises

✓ Vocabulary and Concept Check

1. **DIFFERENT WORDS, SAME QUESTION** Which is different? Find "both" answers.

   How much does it take to fill the cylinder?

   What is the capacity of the cylinder?

   How much does it take to cover the cylinder?

   How much does the cylinder contain?

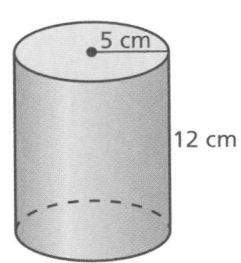

2. **REASONING** Without calculating, which of the solids has the greater volume? Explain.

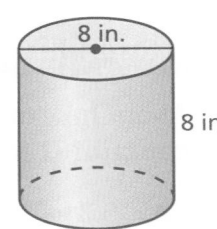
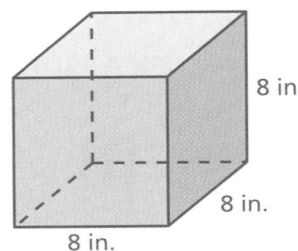

Practice and Problem Solving

Find the volume of the cylinder. Round your answer to the nearest tenth.

3.
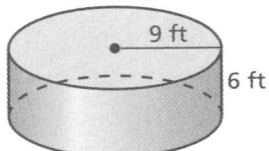

4.
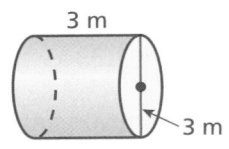

5.
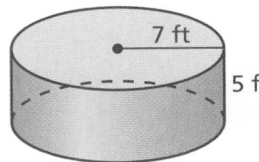

6.
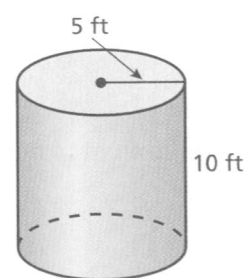

7.
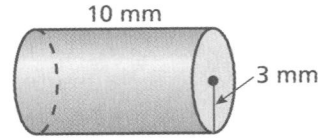

8.
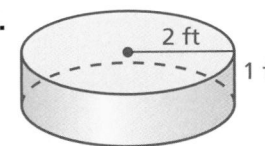

9.
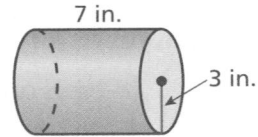

10.
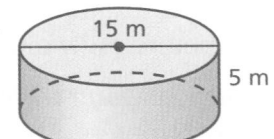

11.
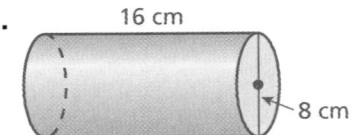

12. SWIMMING POOL A cylindrical swimming pool has a diameter of 16 feet and a height of 4 feet. About how many gallons of water can the pool contain? Round your answer to the nearest whole number. (1 ft³ ≈ 7.5 gal)

Assignment Guide and Homework Check

Level	Day 1 Activity Assignment	Day 2 Lesson Assignment	Homework Check
Basic	3–5, 21–23	1, 2, 6–12, 13, 15, 17	2, 6, 12, 13
Average	3–5, 21–23	1, 2, 6–11, 13, 15, 17, 18	2, 6, 13, 18
Advanced	3–5, 21–23	1, 2, 9–11, 14–20	2, 10, 14, 16

Common Errors

- **Exercises 3–11** Students may forget to square the radius when finding the area of the base. Remind them of the formula for the area of a circle.
- **Exercises 4, 10–15** Students may use the diameter in the formula for the area of a circle instead of finding the radius. Encourage them to write the dimensions that they are given before attempting to find the volume. For example, a student would write: diameter = 3 m, height = 3 m.
- **Exercise 12** Students may find the volume of the pool, but forget to find how many gallons of water that the pool contains. Encourage them to write down the information that they know about the problem and also what they are trying to find. This should help them answer each part of the question.

Vocabulary and Concept Check

1. How much does it take to cover the cylinder?; $170\pi \approx 533.8$ cm^2; $300\pi \approx 942$ cm^3

2. The cube has a greater volume because the cylinder could fit inside the cube and there is still room in the corners of the cube that are not in the cylinder.

Practice and Problem Solving

3. $486\pi \approx 1526.0$ ft^3

4. $\dfrac{27}{4}\pi \approx 21.2$ m^3

5. $245\pi \approx 769.3$ ft^3

6. $250\pi \approx 785.0$ ft^3

7. $90\pi \approx 282.6$ mm^3

8. $4\pi \approx 12.6$ ft^3

9. $63\pi \approx 197.8$ in.^{3}

10. $\dfrac{1125}{4}\pi \approx 883.1$ m^3

11. $256\pi \approx 803.8$ cm^3

12. about 6029 gal

7.2 Record and Practice Journal

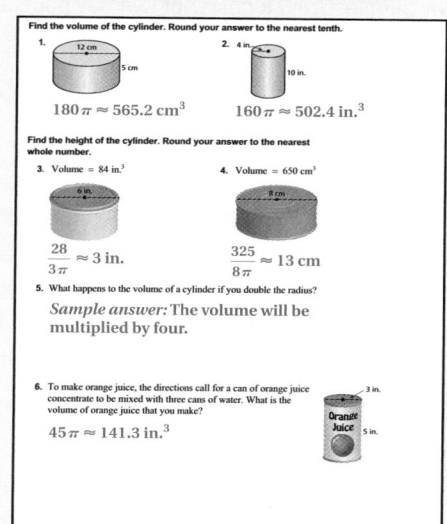

Find the volume of the cylinder. Round your answer to the nearest tenth.

1. 12 cm, 5 cm
$180\pi \approx 565.2$ cm^3

2. 4 in., 10 in.
$160\pi \approx 502.4$ in.^{3}

Find the height of the cylinder. Round your answer to the nearest whole number.

3. Volume = 84 in.^{3} 6 in.
$\dfrac{28}{3\pi} \approx 3$ in.

4. Volume = 650 cm^3 8 cm
$\dfrac{325}{8\pi} \approx 13$ cm

5. What happens to the volume of a cylinder if you double the radius?
Sample answer: The volume will be multiplied by four.

6. To make orange juice, the directions call for a can of orange juice concentrate to be mixed with three cans of water. What is the volume of orange juice that you make?
3 in., 5 in.
$45\pi \approx 141.3$ in.^{3}

13. $\frac{125}{8\pi} \approx 5$ ft

14. $\frac{125}{\pi} \approx 40$ in.

15. $\frac{240}{\pi} \approx 76$ cm

16. The volume is $\frac{1}{4}$ of the original volume. Because the diameter is halved, the radius is also halved.

So, $V = \pi\left(\frac{r}{2}\right)^2 h = \frac{1}{4}\pi r^2 h$.

17. See *Taking Math Deeper.*

18. 4710 lb

19. $8325 - 729\pi \approx 6036$ m³

20. **a.** $384\pi \approx 1205.76$ in.³

b. about 14.22 in.

c. about 19 min

Fair Game Review

21. $a = 0.5 \cdot 200$; 100

22. $a = 0.8 \cdot 400$; 320

23. D

Mini-Assessment

Find the volume of the cylinder. Round your answer to the nearest tenth.

1. 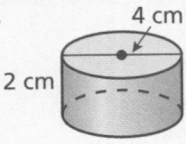 2.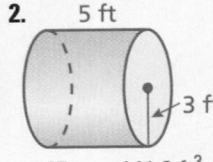

4 cm
2 cm

5 ft
3 ft

$8\pi \approx 25.1$ cm³ $45\pi \approx 141.3$ ft³

3. Find the volume of the can of beans. Round your answer to the nearest whole number.

3 in.
4.5 in.

OLD COUNTRY STYLE
BAKED BEANS
in tomato sauce

$\frac{81}{8}\pi \approx 32$ in.³

Taking Math Deeper

Exercise 17

If you have ever driven through farm country, you have probably seen hay and straw fields with large round bales. You seldom see the smaller *square bales*, but they are still used. The reason you don't see them is that they are usually moved to a storage shed as soon as they are baled.

① Find the volume of the round bale.

$V = \pi r^2 h$
$= \pi \cdot 2^2 \cdot 5$
≈ 62.8 ft³

Find Volume

② Find the volume of the square bale.

$V = \ell w h$
$= 2 \cdot 2 \cdot 4$
$= 16$ ft³

③ Find the number of square bales in a round bale.

$\frac{62.8 \text{ ft}^3}{16 \text{ ft}^3} = 3.925$

There are about 4 square bales in a round bale.

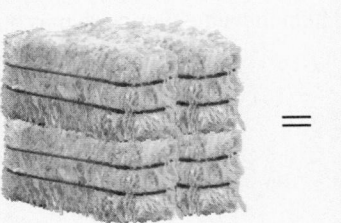 =

Both square and round bales vary in size. Suppose the square bale is only 18 in. by 18 in. by 3 ft. Its volume would be 6.75 cubic feet and there would be about 9.3 square bales in a round bale.

Project

In Germany, they use large rolls of hay to make Mr. and Mrs. Hay people—like snow people. Draw a poster of two hay people. Assuming the rolls are the size of those in Exercise 17, how much hay would be needed to make your *people*?

Reteaching and Enrichment Strategies

If students need help. . .	If students got it. . .
Resources by Chapter • Practice A and Practice B • Puzzle Time Record and Practice Journal Practice Differentiating the Lesson Lesson Tutorials Skills Review Handbook	Resources by Chapter • Enrichment and Extension Start the next section

Find the height of the cylinder. Round your answer to the nearest whole number.

13. Volume = 250 ft³

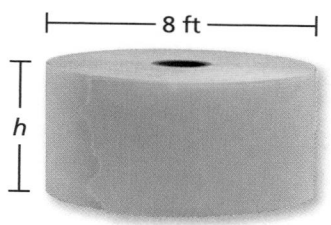

8 ft

h

14. Volume = 32,000 in.³

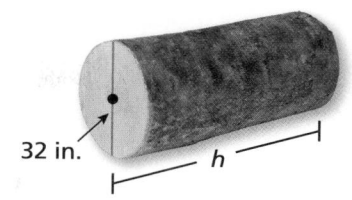

32 in.

h

15. Volume = 600,000 cm³

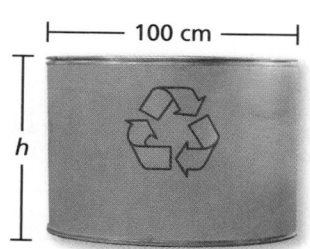

100 cm

h

16. CRITICAL THINKING How does the volume of a cylinder change when its diameter is halved? Explain.

5 ft

4 ft

Round Hay Bale

17. HAY BALES A traditional "square" bale of hay is actually in the shape of a rectangular prism. Its dimensions are 2 feet by 2 feet by 4 feet. How many "square" bales contain the same amount of hay as one large "round" bale?

18. ROAD ROLLER A tank on the road roller is filled with water to make the roller heavy. The tank is a cylinder that has a height of 6 feet and a radius of 2 feet. One cubic foot of water weighs 62.5 pounds. Find the weight of the water in the tank.

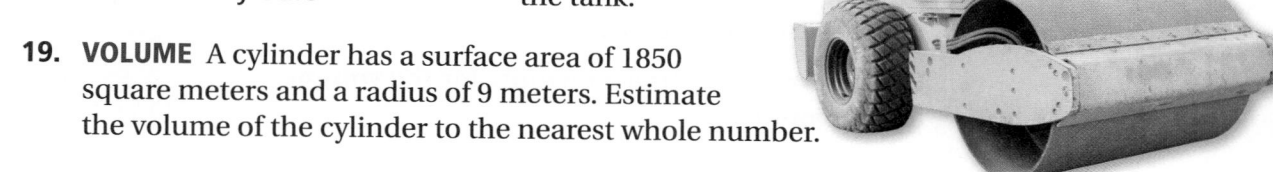

19. VOLUME A cylinder has a surface area of 1850 square meters and a radius of 9 meters. Estimate the volume of the cylinder to the nearest whole number.

20. *Critical Thinking* Water flows at 2 feet per second through a pipe with a diameter of 8 inches. A cylindrical tank with a diameter of 15 feet and a height of 6 feet collects the water.

  a. What is the volume, in cubic inches, of water flowing out of the pipe every second?
  b. What is the height, in inches, of the water in the tank after 5 minutes?
  c. How many minutes will it take to fill 75% of the tank?

Fair Game Review *What you learned in previous grades & lessons*

Write and solve an equation to answer the question. *(Section 4.1)*

21. 50% of 200 is what number?

22. 80% of 400 is what number?

23. MULTIPLE CHOICE The variables x and y vary directly. When x is 18, y is 24. Which equation relates x and y? *(Section 3.7)*

  Ⓐ $y = \dfrac{3}{4}x$ **Ⓑ** $y = 2x - 12$ **Ⓒ** $y = 4x - 3$ **Ⓓ** $y = \dfrac{4}{3}x$

1 2 3 4 5 6 7 8 9 0

Essential Question How can you find the volume of a pyramid?

1   ACTIVITY: Finding a Formula Experimentally

Work with a partner.

● **Draw the two nets on cardboard and cut them out.**

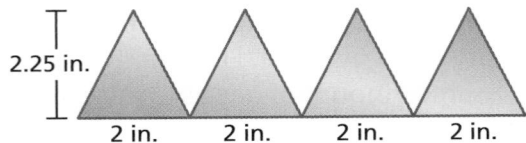

2.25 in.

2 in.   2 in.   2 in.   2 in.

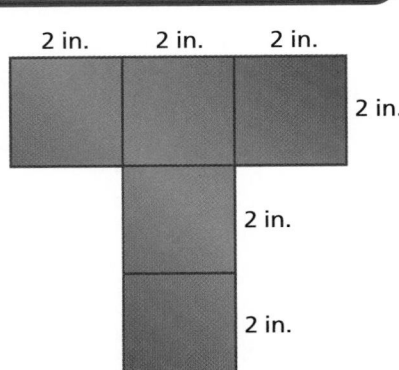

2 in.   2 in.   2 in.

2 in.

2 in.

2 in.

2 in.

● **Fold and tape the nets to form an open square box and an open pyramid.**

● **Both figures should have the same size square base and the same height.**

● **Fill the pyramid with pebbles. Then pour the pebbles into the box. Repeat this until the box is full. How many pyramids does it take to fill the box?**

● **Use your result to find a formula for the volume of a pyramid.**

2   ACTIVITY: Comparing Volumes

Work with a partner. You are an archeologist studying two ancient pyramids. What factors would affect how long it took to build each pyramid? Given similar conditions, which pyramid took longer to build? Explain your reasoning.

Cholula Pyramid in Mexico
Height: about 217 ft
Base: about 1476 ft by 1476 ft

Cheops Pyramid in Egypt
Height: about 480 ft
Base: about 755 ft by 755 ft

Laurie's Notes

Introduction

For the Teacher

- **Goal:** Students will develop an intuition of how to find the volume of a pyramid.

Motivate

- Share information about two well known pyramid-shaped buildings.
- The Luxor Resort and Casino in Las Vegas reaches 350 feet into the sky and is 36 stories tall. Luxor has 4407 rooms, making it the third largest hotel in the world.
- The Rock and Roll Hall of Fame in Cleveland was designed by architect I. M. Pei. The design of the glass-faced main building uses a pyramid shape to invoke the image of a guitar neck rising to the sky.

Activity Notes

Activity 1

- Have students cut out the nets and make the shapes.
- ? "How are the two shapes alike?" same base, same height "How are they different?" One is a square pyramid. The other is a square prism.
- ? "How do you think their volumes compare?" Most students guess that the prism has twice the volume as the pyramid.
- ? "How can you test your hunch about the volumes?" If students have looked at the activity, they'll want to fill the pyramid.
- After the first pour, students should start to suspect that their guess might be off.
- After the second pour, students are pretty sure the relationship is 3 to 1.
- This hands-on experience of making and filling the prism will help students remember the factor of $\frac{1}{3}$. The formula should now make sense to them. The volume of a pyramid should be $\frac{1}{3}$ the volume of a prism with the same base and height as the pyramid.

Activity 2

- ? "What do you know about the pyramids from looking only at the pictures?" They look like square pyramids.
- ? "What do you know about the pyramids from looking at their dimensions?" Cholula has the larger base. Cheops is taller.
- Give time for students to calculate the volume. From the first activity, they should feel comfortable finding the area of the base and multiplying by the height (this would be the prism's volume), and then taking $\frac{1}{3}$ of this answer.
- ? "Which pyramid has the greater volume, and by how much?" Cholula is about 66 million cubic feet greater in volume.
- **FYI:** Cholula is about 5 football fields long on each edge.

Activity Materials

Textbook

- nets of pyramid and prism
- scissors
- tape
- uncooked popcorn, rice, dried beans, mini uncooked pasta

Start Thinking! and Warm Up

Activity 7.3 Start Thinking!
For use before Activity 7.3

Activity 7.3 Warm Up
For use before Activity 7.3

Multiply.

1. $\frac{2}{3} \times 15$ 2. $\frac{3}{4} \times 8$ 3. $\frac{7}{10} \times 6$

4. $\frac{1}{3} \times 18$ 5. $\frac{5}{9} \times 30$ 6. $\frac{4}{13} \times 72$

7.3 Record and Practice Journal

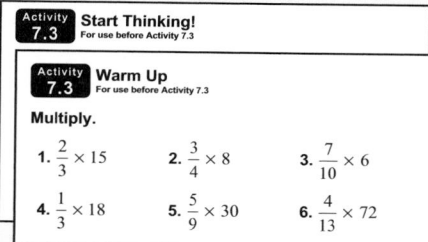

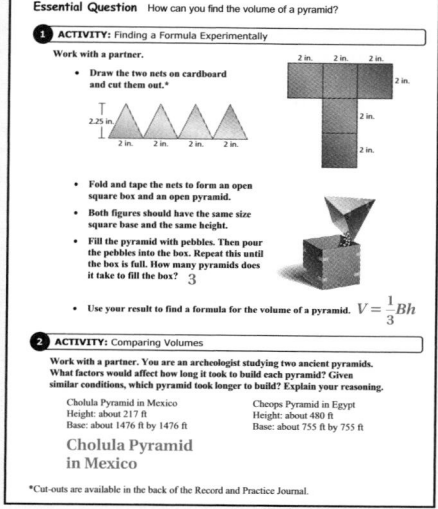

Essential Question How can you find the volume of a pyramid?

1 ACTIVITY: Finding a Formula Experimentally

Work with a partner.

- Draw the two nets on cardboard and cut them out.*
- Fold and tape the nets to form an open square box and an open pyramid.
- Both figures should have the same size square base and the same height.
- Fill the pyramid with pebbles. Then pour the pebbles into the box. Repeat this until the box is full. How many pyramids does it take to fill the box? 3
- Use your result to find a formula for the volume of a pyramid. $V = \frac{1}{3}Bh$

2 ACTIVITY: Comparing Volumes

Work with a partner. You are an archeologist studying two ancient pyramids. What factors would affect how long it took to build each pyramid? Given similar conditions, which pyramid took longer to build? Explain your reasoning.

Cholula Pyramid in Mexico
Height: about 217 ft
Base: about 1476 ft by 1476 ft

Cholula Pyramid in Mexico

Cheops Pyramid in Egypt
Height: about 480 ft
Base: about 755 ft by 755 ft

*Cut-outs are available in the back of the Record and Practice Journal.

Differentiated Instruction

Visual

Students may confuse pyramids and triangular prisms. Show the students models and point out the following characteristics. A pyramid has one base, which can be any polygon. The remaining faces are triangles. A triangular prism has two bases, which are triangles. The remaining faces are rectangles.

7.3 Record and Practice Journal

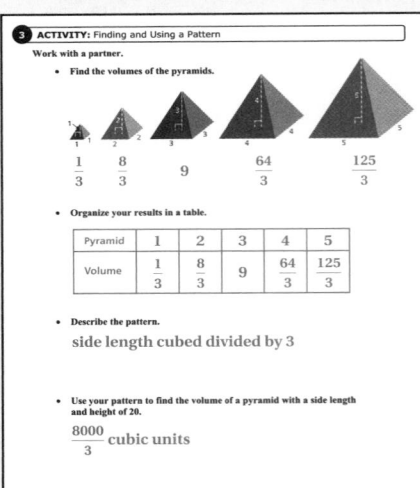

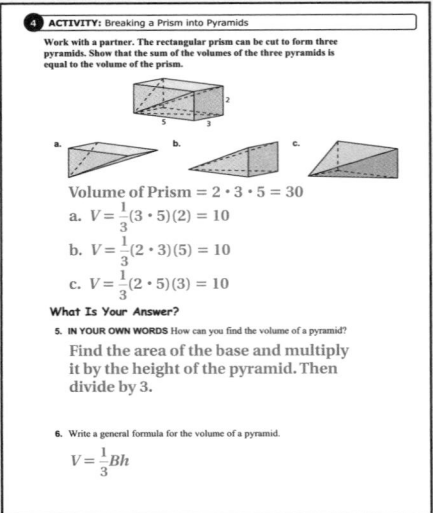

<h1>Laurie's Notes</h1>

Activity 3

- Ask a student to describe the five pyramids shown. You want to make sure that students recognize that the bases are all squares, and the height of the pyramid is the same as the length of the base edge.
- Reinforce good problem solving by having students organize their data in a table.
- Allow time for students to record the volume of each pyramid.
- **?** "What was the volume of the smallest pyramid and how did you find it?"

$$V = \frac{1}{3} \cdot 1^3 = \frac{1}{3}$$

- **?** "What was the volume of the next pyramid and how did you find it?"

$$V = \frac{1}{3} \cdot 2^3 = \frac{8}{3}$$

- Repeat this for several pyramids and ask a student to summarize the pattern. $V = \frac{1}{3} \cdot s^3$, where s is the side length and the height.

- **?** "Does the height of a pyramid have to equal the side length of the base?" No. They are equal in this problem, but they do not need to be.

Activity 4

- Discuss with students how to follow the color coding so that correct dimensions can be matched up.

What Is Your Answer?

- **Neighbor Check:** Have students work independently and then have their neighbor check their work. Have students discuss any discrepancies.

Closure

- Does the volume formula you wrote for Question 5 need to have a square base? Explain your thinking. No; Students should try to sketch or make pyramids with other polygonal bases.

Technology For the Teacher

Dynamic Classroom

The Dynamic Planning Tool
Editable Teacher's Resources at *BigIdeasMath.com*

3 ACTIVITY: Finding and Using a Pattern

Work with a partner.

- Find the volumes of the pyramids.
- Organize your results in a table.
- Describe the pattern.
- Use your pattern to find the volume of a pyramid with a side length and height of 20.

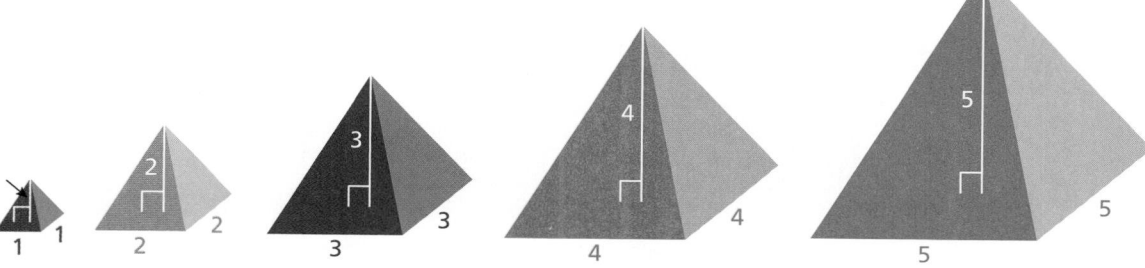

4 ACTIVITY: Breaking a Prism into Pyramids

Work with a partner. The rectangular prism can be cut to form three pyramids. Show that the sum of the volumes of the three pyramids is equal to the volume of the prism.

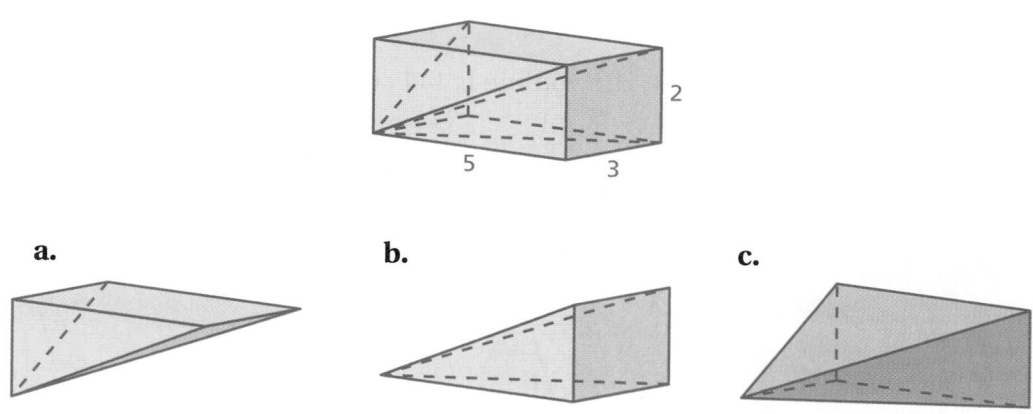

a.   b.   c.

What Is Your Answer?

5. **IN YOUR OWN WORDS** How can you find the volume of a pyramid?

6. Write a general formula for the volume of a pyramid.

Practice ➤ Use what you learned about the volumes of pyramids to complete Exercises 4–6 on page 314.

Key Idea

Volume of a Pyramid

Words   The volume *V* of a pyramid is one-third
the product of the area of the base and
the height of the pyramid.

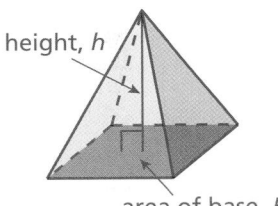

height, *h*

area of base, *B*

Area of base

Algebra $V = \dfrac{1}{3}Bh$

Height of pyramid

EXAMPLE ❶   **Finding the Volume of a Pyramid**

Find the volume of the pyramid.

$V = \dfrac{1}{3}Bh$ Write formula for volume.

$= \dfrac{1}{3}(48)(9)$ Substitute.

$= 144$ Multiply.

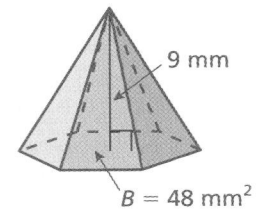

9 mm

$B = 48 \text{ mm}^2$

∴ The volume is 144 cubic millimeters.

EXAMPLE ❷   **Finding the Volume of a Pyramid**

Find the volume of the pyramid.

Study Tip

The area of the base of
a rectangular pyramid
is the product of the
length *ℓ* and the
width *w*.

You can use $V = \dfrac{1}{3}\ell wh$
to find the volume of a
rectangular pyramid.

a.

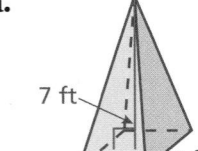

7 ft

4 ft

3 ft

$V = \dfrac{1}{3}Bh$

$= \dfrac{1}{3}(3)(4)(7)$

$= 28$

∴ The volume is
28 cubic feet.

b.

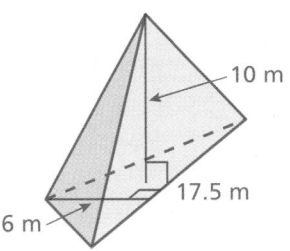

10 m

17.5 m

6 m

$V = \dfrac{1}{3}Bh$

$= \dfrac{1}{3}\left(\dfrac{1}{2}\right)(17.5)(6)(10)$

$= 175$

∴ The volume is
175 cubic meters.

Laurie's Notes

Introduction

Connect

- **Yesterday:** Students discovered how to find the volume of a pyramid by comparing it to the volume of a prism with the same base area and same height.
- **Today:** Students will work with a formula for the volume of a pyramid.

Lesson Notes

Key Idea

- Write the Key Idea.
- Write the formula in words.
- Write the formula in symbols.
- **?** "How will you find the area of the base?" It depends upon what type of polygon the base is.

Example 1

- Work through the example.
- Model good problem solving. Write the formula in words. Write the symbols underneath the words. Substitute the values for the symbols.
- **Common Error:** In using this volume formula, the $\frac{1}{3}$ often produces a computational mistake. In this problem, students must multiply $\frac{1}{3} \times 48 \times 9$. A common error is to take $\frac{1}{3}$ of both numbers $\left(\frac{1}{3} \text{ of } 48 \text{ and } \frac{1}{3} \text{ of } 9 \right)$ as if the Distributive Property were at work. I explain to students that if it had been a whole number, such as 2, they wouldn't think to multiply 2×48 and 2×9. They should only use $\frac{1}{3}$ as a factor once.

Example 2

- **?** "Describe the base of each pyramid." part (a): rectangular base; part (b): triangular base
- **?** "How do you find the area of the base in part (b)?" $\frac{1}{2} \times 17.5 \times 6$
- **FYI:** Be careful in using language, such as "one-half base times height." You may know that in this context, *base* refers to an edge of the base of a triangle and not the area of the base of the pyramid. This can be confusing to students.
- Students may need help with multiplying the fractions in each problem.
  In part (a), students should recognize that $\frac{1}{3} \times 3 = 1$. They are multiplying reciprocals. In part (b), $\frac{1}{3} \times \frac{1}{2} \times 6 = 1$. The Commutative Property allows the order of the factors to be rearranged.

Goal

Today's lesson is finding the volumes of pyramids.

Start Thinking! and Warm Up

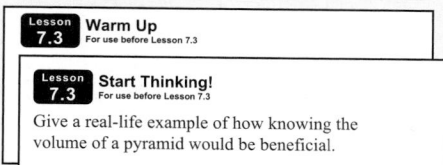

Extra Example 1

Find the volume of a pentagonal pyramid with a base area of 24 square feet and a height of 8 feet. 64 ft^3

Extra Example 2

a. Find the volume of a rectangular pyramid with a base of 2 meters by 6 meters, and a height of 3 meters. 12 m^3

b. Find the volume of a triangular pyramid with a height of 8 inches, and where the triangular base has a width of 4 inches and an altitude of 9 inches. 48 in.^{3}

On Your Own

1. 42 ft^3

2. $186\frac{2}{3} \text{ in.}^3$

3. 231 cm^3

Extra Example 3

a. The volume of lotion in Bottle B is how many times the volume in Bottle A? $1\frac{2}{3}$

b. Which is the better buy? Bottle B

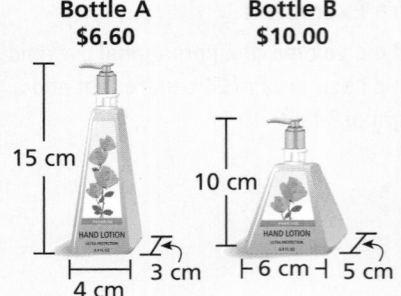

| Bottle A | Bottle B |
| $6.60 | $10.00 |

15 cm   10 cm

4 cm   3 cm   ⊢6 cm⊣ 5 cm

On Your Own

4. yes; Bottles B and C have the same volume, but Bottle C has a unit cost of $2.20.

English Language Learners

Forming Answers

Encourage English learners to form complete sentences in their responses. Students can use the question to help them form the answer.

Question: If you know the area of the base of a pyramid, what else do you need to know to find the volume?

Response: If you know the area of the base of a pyramid, you need to know the height of the pyramid to find the volume.

Laurie's Notes

On Your Own

- Have students name each pyramid and describe what they know about each base. Note that for the pentagonal pyramid, the area of the base has already been computed.
- In Question 2, none of the dimensions contain factors of 3. In computing the volume, $V = \frac{1}{3} \times 10 \times 8 \times 7$, suggest to students that they multiply the whole numbers for a product of 560 and then multiply by $\frac{1}{3}$. Remind students how to rewrite the improper fraction $\frac{560}{3}$ as a mixed number, $186\frac{2}{3}$.

Example 3

- If you have any lotion or shampoo that is in a pyramidal bottle, use it as a model.
- Work through the computation of volume for each bottle.
- Explain different approaches to multiplying the factors in Bottle A: (1) multiply in order left to right or (2) use the Commutative Property to multiply the whole numbers, and then multiply by $\frac{1}{3}$.
- Discuss the phrase "how many times." Because the volume of Bottle B is not a multiple of the volume of Bottle A, students are uncertain how to compare the volumes.
- ? "How do you decide which bottle is the better buy?" Students will try to describe how to find the cost for one cubic inch. This is the unit price.
- Use the language, cost per volume or cost per cubic inch.

On Your Own

- Give students time to complete this problem. Ask volunteers to share their work at the board.

Closure

- **Exit Ticket:** Sketch a rectangular pyramid with base 3 units by 4 units, and a height of 5 units. What is the volume? 20 units^3

Technology For the Teacher

The Dynamic Planning Tool
Editable Teacher's Resources at *BigIdeasMath.com*

 On Your Own

Now You're Ready
Exercises 4–12

Find the volume of the pyramid.

1.

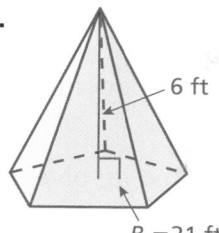

6 ft

$B = 21 \text{ ft}^2$

2.

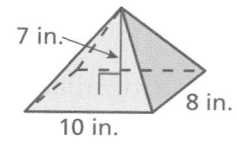

7 in.

8 in.

10 in.

3.

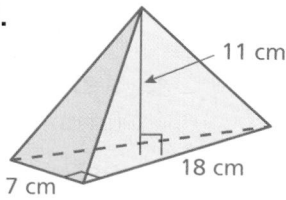

11 cm

7 cm

18 cm

EXAMPLE ③ **Real-Life Application**

Bottle A
$9.96

6 in.

1 in.

2 in.

Bottle B
$14.40

4 in.

1.5 in.

3 in.

(a) The volume of sunscreen in Bottle B is how many times the volume in Bottle A?

(b) Which is the better buy?

a. Use the formula for the volume of a pyramid to estimate the amount of sunscreen in each bottle.

Bottle A	**Bottle B**
$V = \dfrac{1}{3}Bh$	$V = \dfrac{1}{3}Bh$
$= \dfrac{1}{3}(2)(1)(6)$	$= \dfrac{1}{3}(3)(1.5)(4)$
$= 4 \text{ in.}^3$	$= 6 \text{ in.}^3$

∴ So, the volume of sunscreen in Bottle B is $\dfrac{6}{4}$, or 1.5 times the volume in Bottle A.

b. Find the unit cost for each bottle.

Bottle A	**Bottle B**
$\dfrac{\text{cost}}{\text{volume}} = \dfrac{\$9.96}{4 \text{ in.}^3}$	$\dfrac{\text{cost}}{\text{volume}} = \dfrac{\$14.40}{6 \text{ in.}^3}$
$= \dfrac{\$2.49}{1 \text{ in.}^3}$	$= \dfrac{\$2.40}{1 \text{ in.}^3}$

∴ The unit cost of Bottle B is less than the unit cost of Bottle A. So, Bottle B is the better buy.

 On Your Own

Now You're Ready
Exercise 18

4. Bottle C is on sale for $13.20. Is Bottle C a better buy than Bottle B in Example 3? Explain.

Bottle C

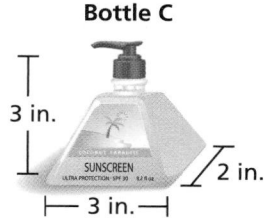

3 in.

2 in.

3 in.

Check It Out
Help with Homework
BigIdeasMath.com

✓ Vocabulary and Concept Check

1. **WRITING** How is the formula for the volume of a pyramid different from the formula for the volume of a prism?

2. **OPEN-ENDED** Describe a real-life situation that involves finding the volume of a pyramid.

3. **REASONING** A triangular pyramid and a triangular prism have the same base and height. The volume of the prism is how many times the volume of the pyramid?

Practice and Problem Solving

Find the volume of the pyramid.

4.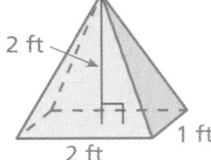
2 ft
2 ft
1 ft

5.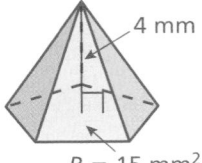
4 mm
$B = 15$ mm^2

6.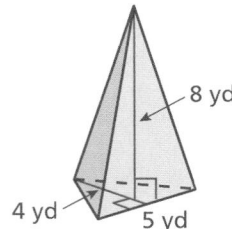
8 yd
4 yd
5 yd

7.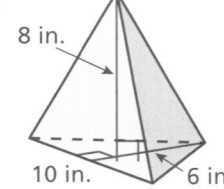
8 in.
10 in.
6 in.

8.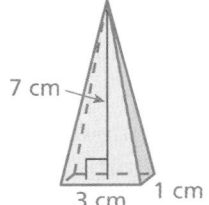
7 cm
3 cm
1 cm

9.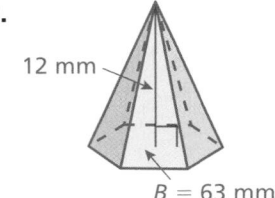
12 mm
$B = 63$ mm^2

10.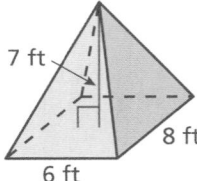
7 ft
8 ft
6 ft

11.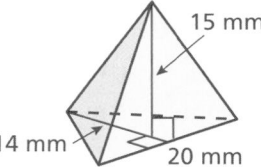
15 mm
14 mm
20 mm

12. **PARACHUTE** In 1483, Leonardo da Vinci designed a parachute. It is believed that this was the first parachute ever designed. In a notebook, he wrote "If a man is provided with a length of gummed linen cloth with a length of 12 yards on each side and 12 yards high, he can jump from any great height whatsoever without injury." Find the volume of air inside Leonardo's parachute.

Not drawn to scale

Assignment Guide and Homework Check

Level	Day 1 Activity Assignment	Day 2 Lesson Assignment	Homework Check
Basic	4–6, 21–25	1–3, 7–12, 17	3, 7, 10, 12
Average	4–6, 21–25	1–3, 7–17 odd, 18	3, 7, 13, 17
Advanced	4–6, 21–25	1–3, 9–11, 14–20 even, 19	10, 14, 16, 18

For Your Information

- **Exercise 12** Skydiver Adrian Nicholas tested the design, jumping from a hot-air balloon at 3000 meters. The parachute weighed over 90 kilograms.

Common Errors

- **Exercises 4–11** Students may write the units incorrectly, often writing square units instead of cubic units. This is especially true when the area of the base is given. Remind them that the units are cubed because there are three dimensions.

- **Exercises 4–11** Students may forget to multiply by one of the measurements, especially when finding the area of the base. Encourage them to find the area of the base separately and then substitute it into the equation. Using colored pencils for each part can also assist students. Tell them to write the formula using different colors for the base and height, as in the lesson. When they substitute values into the equation for volume, they will be able to clearly see that they have accounted for all of the dimensions.

Technology
For
the **T**eacher
Answer Presentation Tool
QuizShow

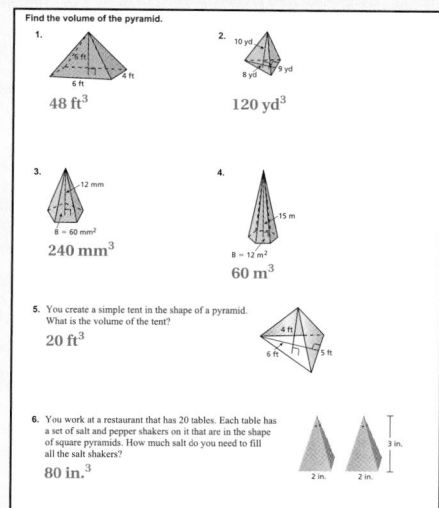

Vocabulary and Concept Check

1. The volume of a pyramid is $\frac{1}{3}$ times the area of the base times the height. The volume of a prism is the area of the base times the height.

2. *Sample answer:* You are comparing the sizes of two tents and want to know which one has more space inside of it.

3. 3 times

Practice and Problem Solving

4. $1\frac{1}{3}$ ft^3

5. 20 mm^3

6. $26\frac{2}{3}$ yd^3

7. 80 in.^{3}

8. 7 cm^3

9. 252 mm^3

10. 112 ft^3

11. 700 mm^3

12. 576 yd^3

13. 30 in.^{2}

14. 9 cm

15. 7.5 ft

16. See *Taking Math Deeper*.

17. 12,000 in.^{3}; The volume of one paperweight is 12 cubic inches. So, 12 cubic inches of glass is needed to make one paperweight. So, it takes $12 \times 1000 = 12,000$ cubic inches to make 1000 paperweights.

18. Spire B; 4 in.^{3}

19. *Sample answer:* 5 ft by 4 ft

20. yes; Prism: $V = xyz$
    Pyramid: $V = \frac{1}{3}(xy)(3z) = xyz$

Fair Game Review

21. 28

22. 72

23. 60

24. 20

25. B

Mini-Assessment

Find the volume of the pyramid.

1.
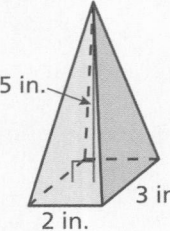
5 in.
3 in.
2 in.
10 in.^{3}

2.
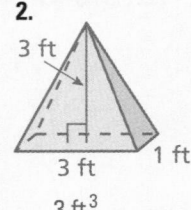
3 ft
3 ft
1 ft
3 ft^3

3. Find the volume of the paper weight.
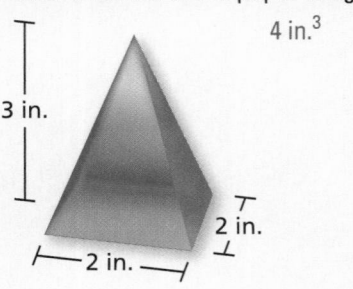
4 in.^{3}
3 in.
2 in.
2 in.

Taking Math Deeper

Exercise 16

Students have to think a bit about this question. At first it seems like you can't tell the shape of the base. However, you can count the number of support sticks to find the shape of the base.

1. Count the supports.
   a. There are 12. So, the base is a dodecagon (a 12-sided polygon).

2. Using a ruler, the base of the teepee appears to be about the same as its height. So, estimate the width of the base to be 10 feet.

3. Use a 10 by 10 grid to estimate the area of the base. It appears to have an area of about 80 square feet.

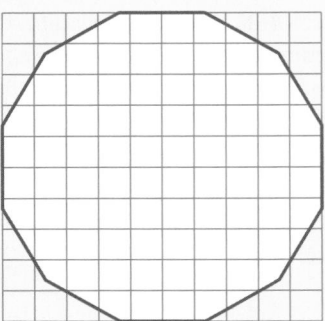

   b. $V = \frac{1}{3}Bh$
   $= \frac{1}{3} \cdot 80 \cdot 10$
   ≈ 267 ft^3

   You need to give some leeway in the answers. Anything from 250 cubic feet to 300 cubic feet is a reasonable answer.

Project

Write a report about what you think were the pros and cons of living in a teepee.

Reteaching and Enrichment Strategies

If students need help. . .	If students got it. . .
Resources by Chapter • Practice A and Practice B • Puzzle Time Record and Practice Journal Practice Differentiating the Lesson Lesson Tutorials Skills Review Handbook	Resources by Chapter • Enrichment and Extension • School-to-Work Start the next section

Copy and complete the table to find the area of the base *B* or the height *h* of the pyramid.

	Volume, *V*	Area of Base, *B*	Height, *h*
13.	60 in.³		6 in.
14.	144 cm³	48 cm²	
15.	135 ft³	54 ft²	

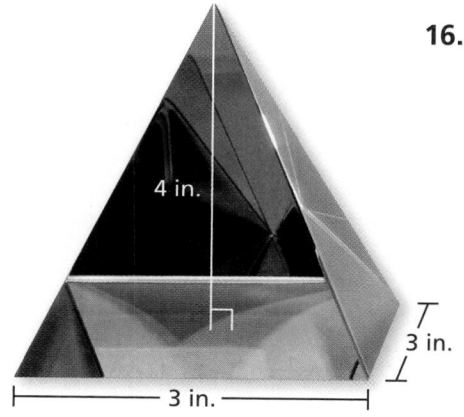

Paperweight

4 in.

3 in.

3 in.

16. TEEPEE Use the photo of the teepee.

   a. What is the shape of the base? How can you tell?

   b. The teepee's height is about 10 feet. Estimate the volume of the teepee.

17. PAPERWEIGHT How much glass is needed to manufacture 1000 paperweights? Explain your reasoning.

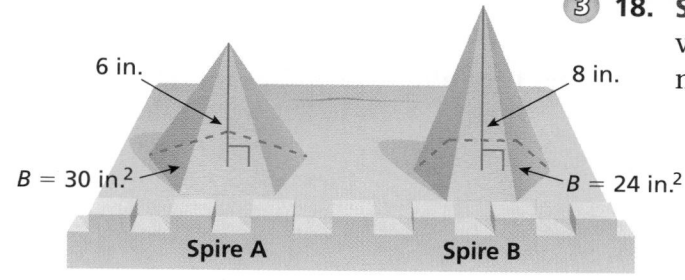

6 in.

B = 30 in.²

Spire A

8 in.

B = 24 in.²

Spire B

③ 18. SPIRE Which sandcastle spire has a greater volume? How much more sand is required to make the spire with the greater volume?

19. OPEN-ENDED A pyramid has a volume of 40 cubic feet and a height of 6 feet. Find one possible set of dimensions of the rectangular base.

20. Reasoning Do the two solids have the same volume? Explain.

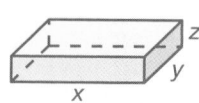

Simplify the expression. *(Skills Review Handbook)*

21. $\frac{1}{3} \times 12 \times 7$

22. $\frac{1}{3} \times 8 \times 27$

23. $\frac{1}{3} \times 6^2 \times 5$

24. $\frac{1}{3} \times 2^2 \times 15$

25. MULTIPLE CHOICE You spend 25% of your money on a shirt. Then you spend $\frac{1}{6}$ of the remainder on lunch. Lunch costs $8. What percent of your money is spent on lunch? *(Section 4.1)*

  Ⓐ 4.2%   **Ⓑ** 12.5%   **Ⓒ** 16.7%   **Ⓓ** 32%

Essential Question How can you remember the formulas for surface area and volume?

You discovered that the volume of a pyramid is one-third the volume of a prism that has the same base and same height. You can use a similar activity to discover that the volume of a cone is one-third the volume of a cylinder that has the same base and height.

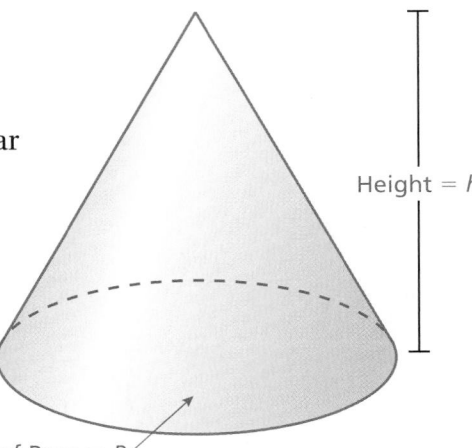

Height = h

Area of Base = B

Volume of a Cone = $\frac{1}{3}$(Area of Base) × (Height)

1 ACTIVITY: Summarizing Volume Formulas

Work with a partner. You can remember the volume formulas for all of the solids shown with just two concepts.

Volumes of Prisms and Cylinders

Volume = (Area of Base) × (Height)

Volumes of Pyramids and Cones

Volume = $\frac{1}{3}$ (Volume of Prism or Cylinder with same base and height)

Make a list of all the formulas you need to remember to find the area of a base. Talk about strategies for remembering these formulas.

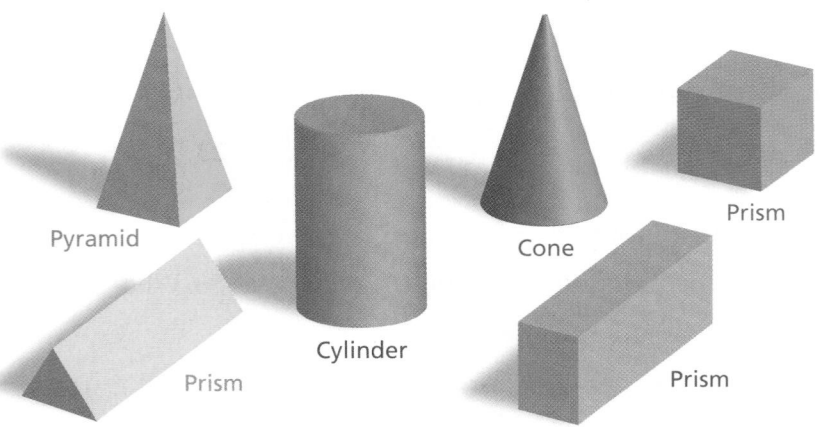

Pyramid

Prism

Cylinder

Cone

Prism

Prism

Laurie's Notes

Introduction

For the Teacher

- **Goal:** Students will develop a strategy to summarize volume and surface area formulas.
- **Big Idea:** It is not a big stretch for students to accept that the volume relationship between the cone and cylinder is the same as the volume relationship between the pyramid and prism.

  Volume of a Cylinder = (Area of Base) × (Height)

  Volume of a Cone = $\frac{1}{3}$ × (Area of Base) × (Height)

- **Common Error:** Students think height and slant height have the same measure. The slant height is used in finding the surface area of a cone, while the height is used in finding volume. In the diagram, the height is shown.

Motivate

- Give pairs of students two minutes to look around the classroom and make a list of all the geometric solids they see that are prisms, cylinders, pyramids, or cones.
- Make a column on the board for each type of solid. Ask one pair of students to list an item in each column. Continue to have pairs of students add to the lists, but only items that are not in the lists already.
- Was every group able to list 4 new items? If your classroom is like most, there are fewer pyramids and cones than prisms and cylinders.

Activity Notes

Activity 1

- **Connection:** Recall the connection between the pyramid and the cone—*structurally they are the same*. Each has a single base and a lateral portion that contains a vertex.
- **FYI:** You do not want students to think that it is necessary to memorize a lot of formulas. Instead, students need to consider the structure of the shape. Prisms and cylinders have the same structure (two congruent bases and a lateral portion) and pyramids and cones have the same structure (one base and a lateral portion that contains a vertex). There is one general formula for each pair of solids. Moreover, the two general formulas have a 1 : 3 relationship.
- **Teaching Tip:** Make a poster of each of the two volume formulas for your classroom, or ask a student to make them.
- **?** "How many general volume formulas are there?" *two*
- **?** "In each formula, you need to find the area of a base. What types of bases have you studied?" *most were squares, rectangles, triangles, or circles*
- Have pairs of students share their lists and strategies. Collect information at the board.

Previous Learning

Students should know how to find the surface area of a cone.

Activity Materials
Textbook
• models of cones • deck of cards or stack of scrap paper

Start Thinking! and Warm Up

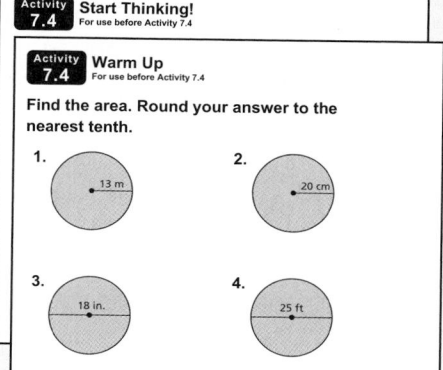

7.4 Record and Practice Journal

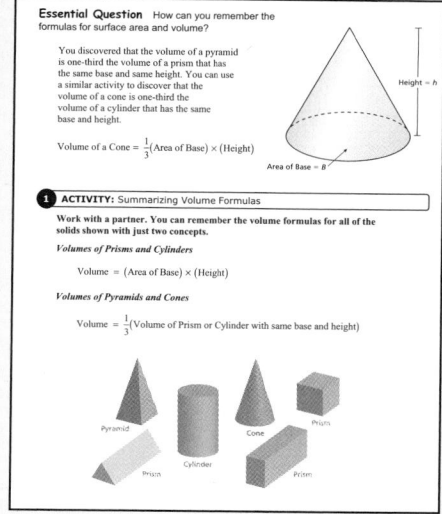

English Language Learners

Visual

Show students models of a pyramid and a prism. Then show them models of a cone and a cylinder. Have students note that the relationship between the volume of a cone and the volume of a cylinder with the same base and height is the same as the relationship between the volume of a pyramid and the volume of a prism with the same base and height.

7.4 Record and Practice Journal

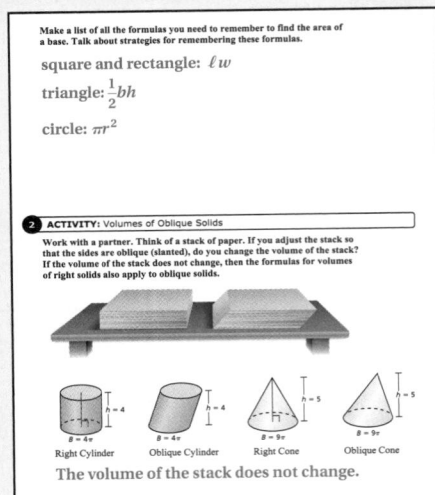

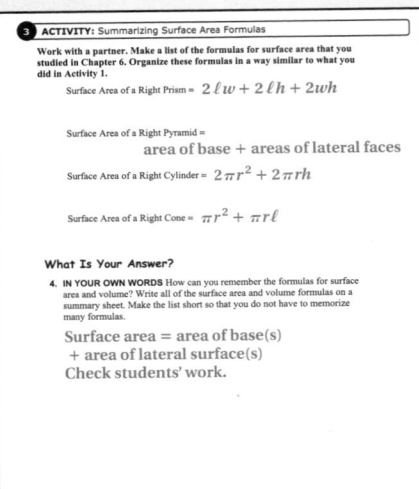

Activity 2

- It is helpful to have a stack of scrap paper or a deck of cards to model this activity.
- Give students time to discuss their thinking and then have them share their thoughts with the rest of the class.
- **Common Misconception:** Students may believe that the height decreases as the stack of paper is slid to one side. The thickness doesn't change, so the height remains the same. The volume remains the same because no sheets are removed.
- **?** What is changing in this problem? Some students may recognize that the surface area is changing because more area is being exposed. Model this with the cards or scrap paper.

Activity 3

- **FYI:** Remember, prisms and cylinders have the same structure (two congruent bases and a lateral portion) and pyramids and cones have the same structure (one base and a lateral portion that contains a vertex). There is one general formula for each pair of solids.
- **Teaching Tip:** Make a poster of each of the two surface area formulas for your classroom, or ask a student to make them.
- Students should refer back in their notes and book as needed to complete this activity.
- Have pairs of students share their lists. Collect information at the board.

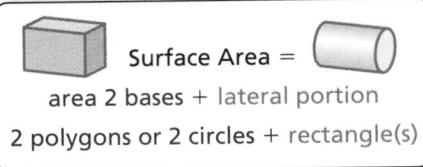

Surface Area =
area 2 bases + lateral portion
2 polygons or 2 circles + rectangle(s)

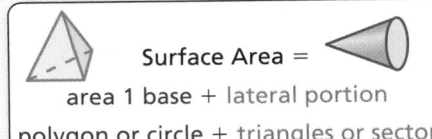

Surface Area =
area 1 base + lateral portion
polygon or circle + triangles or sector

What Is Your Answer?

- Have students work in pairs.

Closure

- Sketch a cube with edge length 2 centimeters, and a cylinder with height and diameter each 2 centimeters. Compare the volumes of the cube and cylinder.
- Sketch a cylinder and a cone, each with height and diameter of 3 centimeters. Compare the volumes of the cylinder and cone.

Technology For the Teacher

Dynamic Classroom

The Dynamic Planning Tool
Editable Teacher's Resources at *BigIdeasMath.com*

2 ACTIVITY: Volumes of Oblique Solids

Work with a partner. Think of a stack of paper. If you adjust the stack so that the sides are oblique (slanted), do you change the volume of the stack? If the volume of the stack does not change, then the formulas for volumes of right solids also apply to oblique solids.

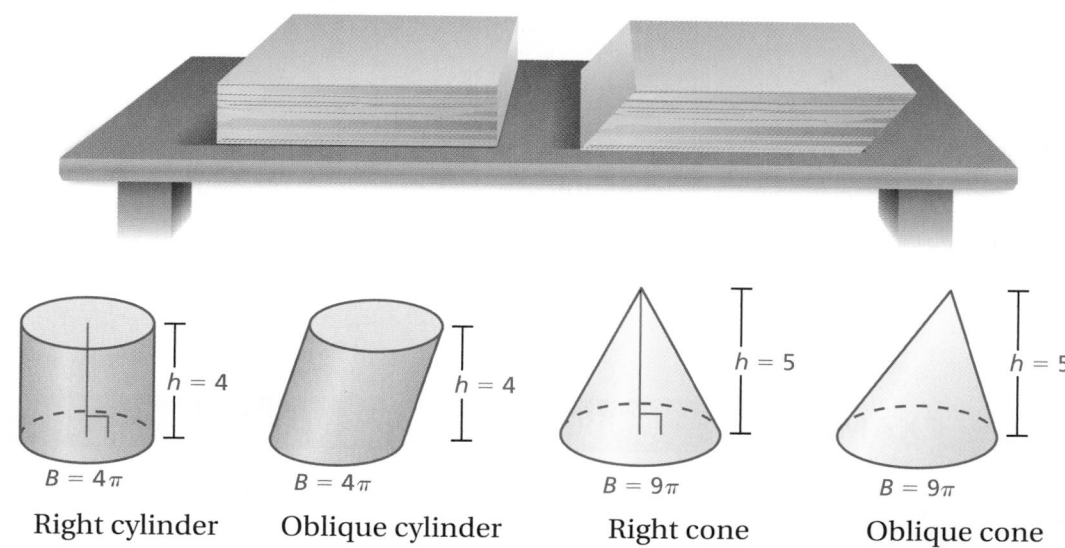

$B = 4\pi$ $B = 4\pi$ $B = 9\pi$ $B = 9\pi$

Right cylinder   Oblique cylinder   Right cone   Oblique cone

3 ACTIVITY: Summarizing Surface Area Formulas

Work with a partner. Make a list of the formulas for surface area that you studied in Chapter 6. Organize these formulas in a way similar to what you did in Activity 1.

Surface Area of a Right Prism =

Surface Area of a Right Pyramid =

Surface Area of a Right Cylinder =

Surface Area of a Right Cone =

What Is Your Answer?

4. **IN YOUR OWN WORDS** How can you remember the formulas for surface area and volume? Write all of the surface area and volume formulas on a summary sheet. Make the list short so that you do not have to memorize many formulas.

Practice

Use what you learned about the volumes of cones to complete Exercises 4–6 on page 320.

Key Idea

Volume of a Cone

Words The volume V of a cone is one-third the product of the area of the base and the height of the cone.

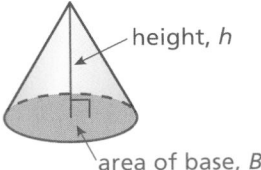

height, h

area of base, B

Algebra $V = \dfrac{1}{3}Bh$

Area of base

Height of cone

EXAMPLE **1** **Finding the Volume of a Cone**

Find the volume of the cone. Round your answer to the nearest tenth.

The diameter is 4 meters. So, the radius is 2 meters.

> **Study Tip**
>
> Because $B = \pi r^2$, you can use $V = \dfrac{1}{3}\pi r^2 h$ to find the volume of a cone.

$$V = \frac{1}{3}Bh \qquad \text{Write formula.}$$

$$= \frac{1}{3}\pi(2)^2(6) \qquad \text{Substitute.}$$

$$= 8\pi \approx 25.1 \qquad \text{Simplify.}$$

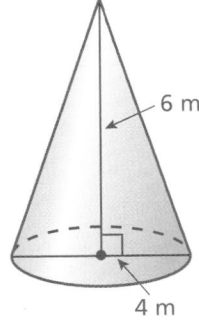

6 m

4 m

∴ The volume is about 25.1 cubic meters.

EXAMPLE **2** **Finding the Height of a Cone**

Find the height of the cone. Round your answer to the nearest tenth.

$$V = \frac{1}{3}Bh \qquad \text{Write formula.}$$

$$956 = \frac{1}{3}\pi(9)^2(h) \qquad \text{Substitute.}$$

$$956 = 27\pi h \qquad \text{Simplify.}$$

$$11.3 \approx h \qquad \text{Divide each side by } 27\pi.$$

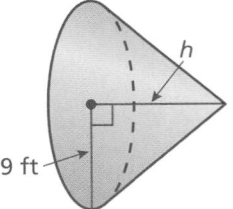

h

9 ft

Volume = 956 ft³

∴ The height is about 11.3 feet.

Laurie's Notes

Introduction

Connect
- **Yesterday:** Students developed a strategy to summarize volume and surface area formulas.
- **Today:** Students will work with the formula for the volume of a cone.

Motivate
- Bring an ice cream cone and ice cream scoop to class.
- **FYI:** An ice cream scoop with a radius of 1 inch will make a round (sphere) scoop of ice cream that is a little more than 4 cubic inches. Share this information with students.
- **?** "If I place a scoop of ice cream (with a 1 inch radius) on this cone, and the ice cream melts (because I received a phone call from a parent), will the ice cream overflow the cone?" This question is posed only to get students thinking about the volume of a cone. The volume of the cone will be found at the end of the lesson.

Lesson Notes

Key Idea
- Write the Key Idea.
- Write the formula in words. Draw the cone with the dimensions labeled.
- Write the symbolic formula.
- **?** "What is the base?" a circle. "How do you find its area?" $A = \pi r^2$

Example 1
- Notice that the work is done in terms of π. It is not until the last step that 3.14 is substituted for π.
- **Representation:** Encourage students to use the parentheses to represent multiplication. Using the $\times$ symbol would make the expression confusing.
- **Common Misconception:** Remind students that π is a number and because multiplication is commutative and associative, this expression could be rewritten as $\frac{1}{3}(6)(2)^2\pi$, making the computation less confusing.
- **?** "What is being squared in this expression?" only the 2

Example 2
- This example requires students to solve an equation for a variable.
- Work through the problem, annotating the steps as shown in the book.
- **?** "How does $\frac{1}{3}\pi(9)^2h$ equal $27\pi h$?" Only the 9 is being squared, which is 81. One-third of 81 is 27. The order of the factors doesn't matter.
- Students may have difficulty with the last step, dividing by 27π. It can be done in two steps—divide by 27 then divide by 3.14. Or, divide 956 by the product 27π, which is about 84.78.

Goal
Today's lesson is finding the volume of a cone.

Lesson Materials	
Introduction	**Textbook**
• ice cream cone • ice cream scoop	• sand timer
Closure	
• ice cream cone	

Start Thinking! and Warm Up

Lesson 7.4 Warm Up
For use before Lesson 7.4

Lesson 7.4 Start Thinking!
For use before Lesson 7.4

Explain which sugar cone can hold more ice cream:

Sugar cone 1: Radius 3 cm; Height 14 cm

Sugar cone 2: Diameter 7 cm; Height 13 cm

Extra Example 1
Find the volume of a cone with a diameter of 6 feet and a height of 3 feet. Round your answer to the nearest tenth. $9\pi \approx 28.3 \text{ ft}^3$

Extra Example 2
Find the height of a cone with a radius of 6 yards and a volume of 75 cubic yards. Round your answer to the nearest whole number. $\frac{75}{12\pi} \approx 2 \text{ yd}$

Laurie's Notes

On Your Own

1. $180\pi \approx 565.2$ cm^3

2. $\dfrac{96}{\pi} \approx 30.6$ yd

Extra Example 3

In Example 3, the height of the sand is 36 millimeters and the radius is 15 millimeters. The sand falls at a rate of 150 cubic millimeters per second. How much time do you have to answer the question? about 57 sec

On Your Own

3. about 42 sec

4. about 6 sec

Differentiated Instruction

Organization

Some students might benefit from finding the area of the base B of the cone first. Then they can substitute this value into the formula, $V = \dfrac{1}{3}Bh$.

On Your Own

- Ask volunteers to share their work at the board.

Example 3

- If you have a timer of this type, use it as a model.
- Ask a student volunteer to read the problem. Ask for ideas as to how the problem can be solved.
- **?** "How long is 30 millimeters?" 30 millimeters is equal to 3 centimeters, which is a little more than 1 inch. This helps students form a visual image of the actual size of the sand timer.
- **Teaching Tip:** Again, explain that $\dfrac{1}{3} \times 24$ is a whole number. Then multiply $8 \times 10^2 = 800$.
- Be sure to use units in labeling answers. The dimensional analysis technique shows that the answer will have units of seconds.
- **Extension:** I have a sand timer in my classroom. Students calculate the volume, measure the amount of time it takes to fall to the bottom, and use this information to calculate the rate at which the sand is falling.

On Your Own

- **Extension:** Question 4 is a preview of an upcoming lesson. The height and radius have each been decreased by a factor of 2. (They are $\dfrac{1}{2}$ the original dimensions). What happens to the volume? It is decreased by a factor of 8, or 2^3.

Closure

- **Exit Ticket:** Have students find the volume of the ice cream cone used to motivate the lesson.

Technology For the Teacher

Dynamic Classroom

The Dynamic Planning Tool
Editable Teacher's Resources at *BigIdeasMath.com*

Now You're Ready
Exercises 4–17

On Your Own

Find the volume *V* or height *h* of the cone. Round your answer to the nearest tenth.

1.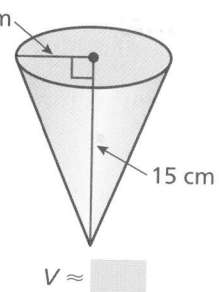

6 cm

15 cm

$V \approx$

2.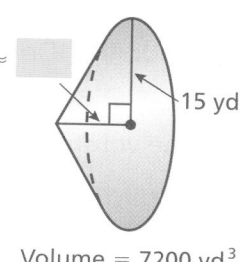

$h \approx$

15 yd

Volume = 7200 yd³

EXAMPLE **3** **Real-Life Application**

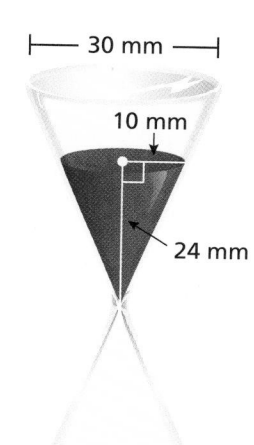

├── 30 mm ──┤

10 mm

24 mm

You must answer a trivia question before the sand in the timer falls to the bottom. The sand falls at a rate of 50 cubic millimeters per second. How much time do you have to answer the question?

Use the formula for the volume of a cone to find the volume of the sand in the timer.

$$V = \frac{1}{3}Bh$$ Write formula.

$$= \frac{1}{3}\pi(10)^2(24)$$ Substitute.

$$= 800\pi \approx 2512$$ Simplify.

The volume of the sand is about 2512 cubic millimeters. To find the amount of time you have to answer the question, multiply the volume by the rate at which the sand falls.

$$2512 \text{ mm}^3 \times \frac{1 \text{ sec}}{50 \text{ mm}^3} = 50.24 \text{ sec}$$

∴ You have about 50 seconds to answer the question.

On Your Own

3. WHAT IF? In Example 3, the sand falls at a rate of 60 cubic millimeters per second. How much time do you have to answer the question?

4. WHAT IF? In Example 3, the height of the sand in the timer is 12 millimeters and the radius is 5 millimeters. How much time do you have to answer the question?

 Vocabulary and Concept Check

1. **VOCABULARY** Describe the height of a cone.

2. **WRITING** Compare and contrast the formulas for the volume of a pyramid and the volume of a cone.

3. **REASONING** You know the volume of a cylinder. How can you find the volume of a cone with the same base and height?

 **Practice and Problem Solving**

Find the volume of the cone. Round your answer to the nearest tenth.

1 **4.**

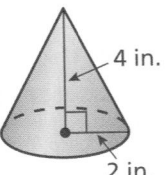

4 in.

2 in.

5.

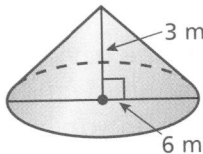

3 m

6 m

6.

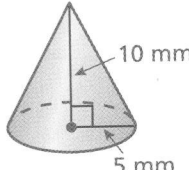

10 mm

5 mm

7.

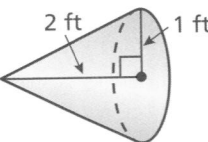

2 ft   1 ft

8.

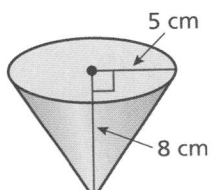

5 cm

8 cm

9.

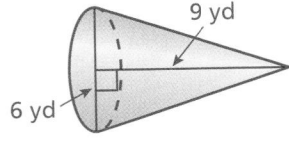

9 yd

6 yd

10.

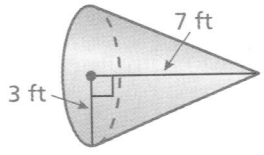

7 ft

3 ft

11.

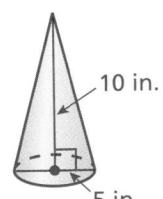

10 in.

5 in.

12.

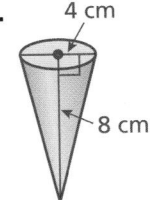

4 cm

8 cm

13. **ERROR ANALYSIS** Describe and correct the error in finding the volume of the cone.

8 m

6 m

$$V = \frac{1}{3}Bh$$

$$= \frac{1}{3}(\pi)(6)^2(8)$$

$$= 96\pi \, m^3$$

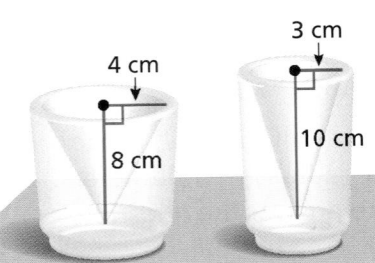

4 cm

8 cm

3 cm

10 cm

Glass A   Glass B

14. **GLASS** The inside of each glass is shaped like a cone. Which glass can hold more liquid? How much more?

Assignment Guide and Homework Check

Level	Day 1 Activity Assignment	Day 2 Lesson Assignment	Homework Check
Basic	4–6, 23–26	1–3, 7–17 odd, 14	2, 7, 14, 15
Average	4–6, 23–26	1–3, 7–19 odd, 20	2, 7, 15, 19
Advanced	4–6, 23–26	1–3, 13, 16–22	2, 16, 18, 21

For Your Information
- **Exercise 15** Because the volume is given in terms of π, students should not substitute 3.14 for π.

Common Errors
- **Exercises 4–12** Students may forget to cube the dimensions for the volume. Remind them that the volume of the cone is in three dimensions.
- **Exercises 4–12** Students may forget to square the radius when finding the area of the base. Remind them of the formula for the area of a circle. Encourage them to color-code the formula for volume as they solve for each part so that they do not forget to include one of the dimensions.
- **Exercises 5, 9, 11, 12** Students may use the diameter instead of the radius to find the area of the base. Remind them that the area of a circle is found using the radius.
- **Exercises 15–17** Students may try to use the Distributive Property before solving for h. For example, a student may incorrectly write $225 = \frac{1}{3}\pi(5^2) \cdot \frac{1}{3}h$. Remind them that factors are multiplied.

7.4 Record and Practice Journal

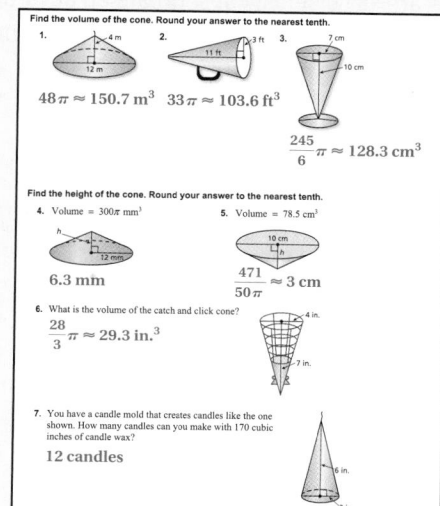

Vocabulary and Concept Check

1. The height of a cone is the distance from the vertex to the center of the base.

2. Both formulas are $\frac{1}{3}Bh$, but the base of a cone is always a circle.

3. Divide by 3.

Practice and Problem Solving

4. $\frac{16\pi}{3} \approx 16.7$ in.^{3}

5. $9\pi \approx 28.3$ m^3

6. $\frac{250\pi}{3} \approx 261.7$ mm^3

7. $\frac{2\pi}{3} \approx 2.1$ ft^3

8. $\frac{200\pi}{3} \approx 209.3$ cm^3

9. $27\pi \approx 84.8$ yd^3

10. $21\pi \approx 65.9$ ft^3

11. $\frac{125\pi}{6} \approx 65.4$ in.^{3}

12. $\frac{32\pi}{3} \approx 33.5$ cm^3

13. The diameter was used instead of the radius.
$$V = \frac{1}{3}(\pi)(3)^2(8)$$
$$= 24\pi \text{ m}^3$$

14. Glass A; $\frac{38\pi}{3} \approx 39.8$ cm^3

15. 1.5 ft

16. $\frac{27}{\pi} \approx 8.6$ cm

17. $\frac{40}{3\pi} \approx 4.2$ in.

18. 60π m^3

19. 24.1 min

20. See *Taking Math Deeper*.

21. $3y$

22. $4:1$

Fair Game Review

23. 315 m^3

24. 400 cm^3

25. $152\pi \approx 477.28$ ft^3

26. D

Mini-Assessment

Find the volume of the cone. Round your answer to the nearest tenth.

1.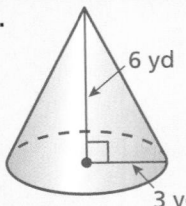

6 yd

3 yd

$18\pi \approx 56.5$ yd^3

2.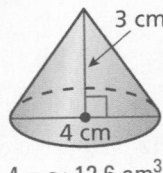

3 cm

4 cm

$4\pi \approx 12.6$ cm^3

3. The volume of the ice cream cone is 4.71 cubic inches. Find the height of the cone.

├─2 in.─┤ $\frac{14.13}{\pi} \approx 4.5$ in.

Taking Math Deeper

Exercise 20

This is a great type of problem to help students understand the importance of *planning ahead*. Also, in planning, remind students that you can't plan *exactly* how many cups will be used, nor can you plan how full each cup will be. So, the answers to the questions are just "ball park" figures.

① How many paper cups will you need?

$$\text{Volume of Cup} = \frac{1}{3}\pi(4)^2(11)$$
$$\approx 184.2 \text{ cm}^3$$

$$\text{Amount of Lemonade} = (10 \text{ gal})\left(3785 \, \frac{\text{cm}^3}{\text{gal}}\right)$$
$$= 37{,}850 \text{ cm}^3$$

$$\text{Number of Cups} \approx \frac{37{,}850}{184.2} \approx 205.5$$

a. You need about 206 cups.

├─ 8 cm ─┤

11 cm

Think Outside the Box

├─ $\frac{9}{11} \cdot 8$ ─┤

9 cm

② How many packs of 50 cups?

b. You should order 5 packs of 50 cups. This will give you 250 cups.

Suppose each cup is not filled to the brim, but only to a height of 9 centimeters. This could mean each cup has a volume of 101 cubic centimeters, which would imply that you would use about 375 cups... so 5 packs would *not* be enough.

③ How many cups are left over if you sell only 80% of the lemonade? 80% of 37,850 = 30,280 cm^3

$$\text{Number of Cups} \approx \frac{30{,}280}{184.2} \approx 164.4$$

c. You would have about $250 - 165 = 85$ cups left over.

Sell 80%

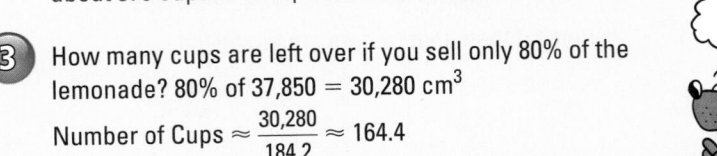

Project

You open a lemonade stand. The lemonade costs you $5.00 per gallon and cups are $6.00 per 50 cups. Create an advertisement including the price of your lemonade. How did you determine the price to charge customers?

Reteaching and Enrichment Strategies

If students need help. . .	If students got it. . .
Resources by Chapter • Practice A and Practice B • Puzzle Time Record and Practice Journal Practice Differentiating the Lesson Lesson Tutorials Skills Review Handbook	Resources by Chapter • Enrichment and Extension • School-to-Work Start the next section

Find the height of the cone. Round your answer to the nearest tenth.

② **15.** Volume $= \frac{1}{18}\pi$ ft³

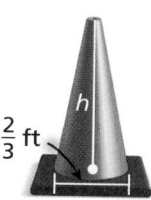

$\frac{2}{3}$ ft

16. Volume $= 225$ cm³

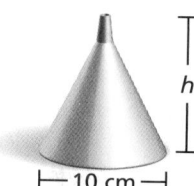

10 cm

17. Volume $= 3.6$ in.³

1.8 in.

18. REASONING The volume of a cone is 20π cubic meters. What is the volume of a cylinder having the same base and same height?

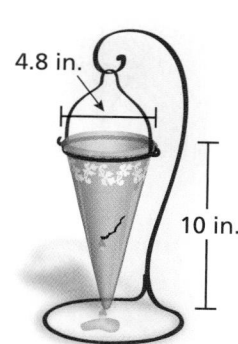

4.8 in.

10 in.

19. VASE Water leaks from a crack in a vase at a rate of 0.5 cubic inch per minute. How long does it take for 20% of the water to leak from a full vase?

20. LEMONADE STAND You have 10 gallons of lemonade to sell. (1 gal $\approx$ 3785 cm³)

a. Each customer uses one paper cup. How many paper cups will you need?

b. The cups are sold in packages of 50. How many packages should you buy?

c. How many cups will be left over if you sell 80% of the lemonade?

8 cm

11 cm

21. REASONING The cylinder and the cone have the same volume. What is the height of the cone?

22. *Critical Thinking* Cone A has the same height but twice the radius of Cone B. What is the ratio of the volume of Cone A to the volume of Cone B?

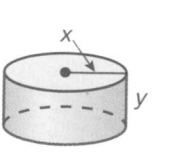

x

y

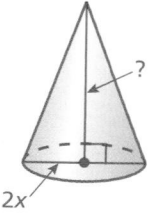

?

2x

Ⓐ **Fair Game Review** *What you learned in previous grades & lessons*

Find the volume of the solid. *(Section 7.1, Section 7.2, and Section 7.3)*

23.

9 m

7 m

5 m

24.

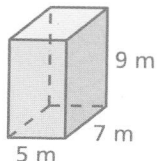

15 cm

10 cm

8 cm

25. 4 ft

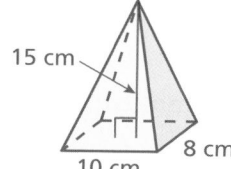

9.5 ft

26. MULTIPLE CHOICE Which scale has a scale factor of 3 : 1? *(Section 5.4)*

Ⓐ 1 in. : 2 ft   Ⓑ 3 cm : 1 mm   Ⓒ 5 ft : 15 yd   Ⓓ 0.5 ft : 2 in.

You can use a **formula triangle** to arrange variables and operations of a formula. Here is an example of a formula triangle for volume of a prism.

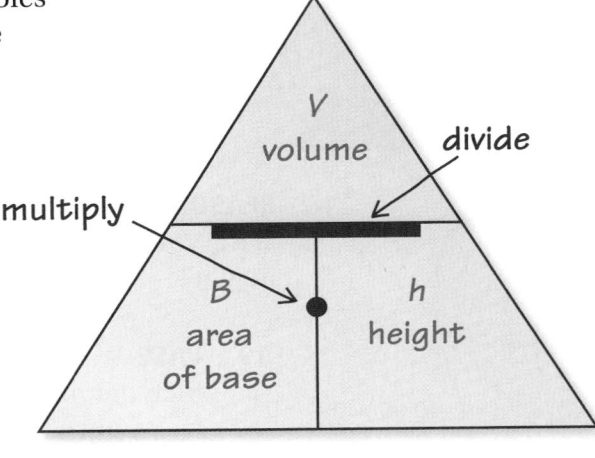

To find an unknown variable, use the other variables and the operation between them. For example, to find the area B of the base, cover up the B. Then you can see that you divide the volume V by the height h.

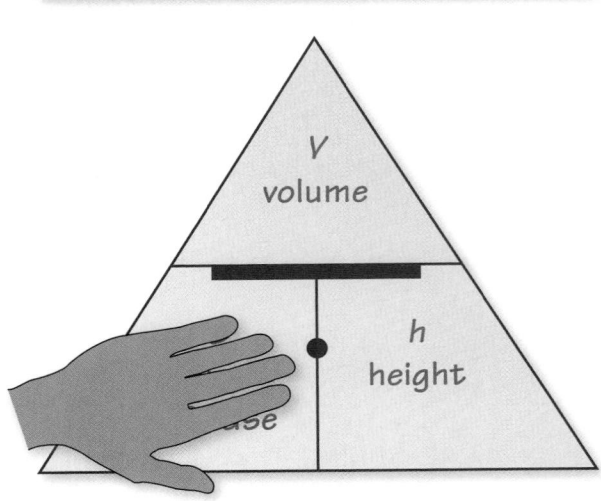

On Your Own

Make a formula triangle to help you study these topics. *Hint:* **Your formula triangles may have a different form than what is shown in the example.**

1. volume of a cylinder

2. volume of a pyramid

3. volume of a cone

After you complete this chapter, make formula triangles for the following topics.

4. volume of a composite solid

5. surface areas of similar solids

6. volumes of similar solids

"See how a formula triangle works? Cover any variable and you get its formula."

Sample Answers

1. Volume of a cylinder

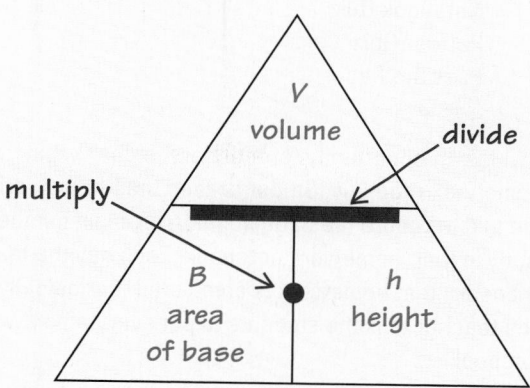

2. Volume of a pyramid

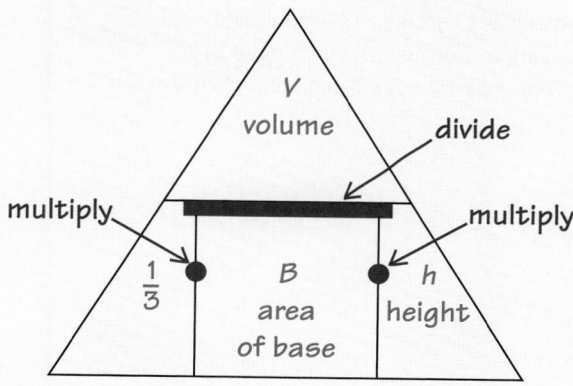

3. Volume of a cone

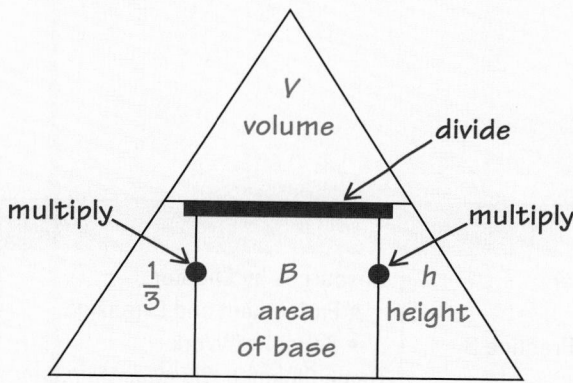

About this Organizer

A **Formula Triangle** can be used to arrange variables and operations of a formula. Students divide a triangle into the same number of parts as there are variables and factors in a formula. Then students write the variables and factors in the parts of the triangle and place either a multiplication or a division symbol, as appropriate, between the parts. This type of organizer can help students learn the formulas as well as see how the variables in the formulas are related. Students can place their formula triangles on note cards to use as a quick study reference.

Technology
For the Teacher
Vocabulary Puzzle Builder

Answers

1. 168 in.^{3}

2. 360 ft^3

3. 4925 mm^3

4. $14\pi \approx 44$ yd^3

5. 5 m^3

6. 664 ft^3

7. $50\pi \approx 157$ cm^3

8. $\dfrac{28.26}{\pi} \approx 9$ cm

9. 10,666.7 ft^3

10. 27 ft^3

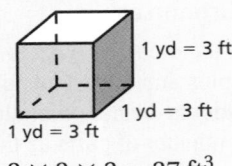

1 yd = 3 ft

1 yd = 3 ft

1 yd = 3 ft

$3 \times 3 \times 3 = 27$ ft^3

11. 13.5 in.

Assessment Book

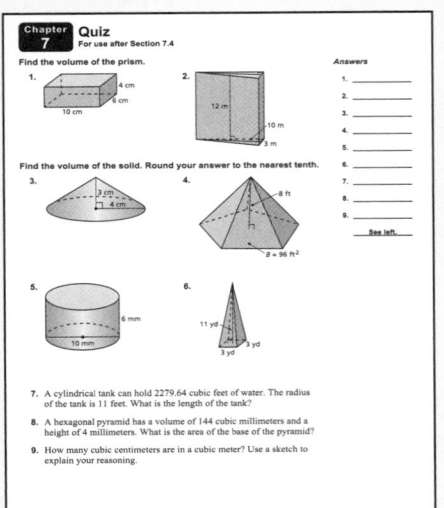

Alternative Quiz Ideas

100% Quiz	Math Log
Error Notebook	Notebook Quiz
Group Quiz	Partner Quiz
Homework Quiz	Pass the Paper

Homework Quiz

A homework notebook provides an opportunity for teachers to check that students are doing their homework regularly. Students keep their homework in a notebook. They should be told to record the page number, problem number, and copy the problem exactly in their homework notebook. Each day the teacher walks around and visually checks that homework is completed. Periodically, without advance notice, the teacher tells the students to put everything away except their homework notebook.

Questions are from students' homework.

1. What are the answers to Exercises 4–6 on page 302?
2. What are the answers to Exercises 13–15 on page 309?
3. What are the answers to Exercises 4–6 on page 314?
4. What are the answers to Exercises 7–9 on page 320?

Reteaching and Enrichment Strategies

If students need help. . .	If students got it. . .
Resources by Chapter • Study Help • Practice A and Practice B • Puzzle Time Lesson Tutorials *BigIdeasMath.com* Practice Quiz Practice from the Test Generator	Resources by Chapter • Enrichment and Extension • School-to-Work Game Closet at *BigIdeasMath.com* Start the next section

Technology For the Teacher

Answer Presentation Tool
Big Ideas Test Generator

Find the volume of the prism. *(Section 7.1)*

1.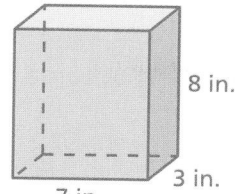

8 in.

3 in.

7 in.

2.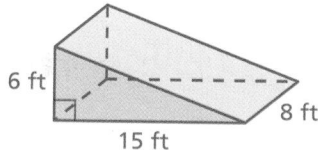

6 ft

15 ft

8 ft

3.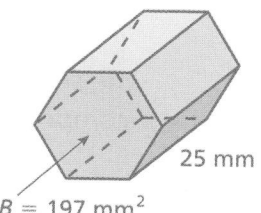

25 mm

$B = 197$ mm^2

Find the volume of the solid. Round your answer to the nearest tenth.
(Section 7.2, Section 7.3, and Section 7.4)

4.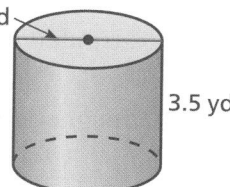

4 yd

3.5 yd

5.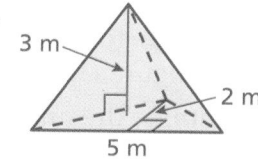

3 m

2 m

5 m

6.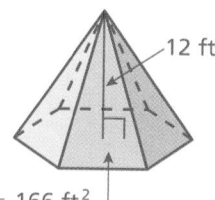

12 ft

$B = 166$ ft^2

7.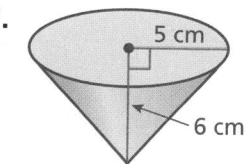

5 cm

6 cm

8. PAPER CONE The paper cone can hold 84.78 cubic centimeters of water. What is the height of the cone? *(Section 7.4)*

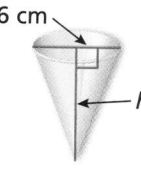

6 cm

h

9. ROOF A pyramid hip roof is a good choice for a house in a hurricane area. What is the volume of the roof to the nearest tenth? *(Section 7.3)*

20 ft

40 ft

40 ft

10. CUBIC UNITS How many cubic feet are there in a cubic yard? Use a sketch to explain your reasoning. *(Section 7.1)*

11. JUICE CAN You are buying two cylindrical cans of juice. Each can holds the same amount of juice. What is the height of Can B? *(Section 7.2)*

4 in.

6 in.

6 in.

h

Can A

Can B

Essential Question How can you estimate the volume of a composite solid?

1 ACTIVITY: Estimating Volume

Work with a partner. You work for a toy company and need to estimate the volume of a Minifigure that will be molded out of plastic.

a. Estimate the number of cubic inches of plastic that is needed to mold the Minifigure's head. Show your work.

b. Estimate the number of cubic inches of plastic that is needed to mold one of the Minifigure's legs. Show your work.

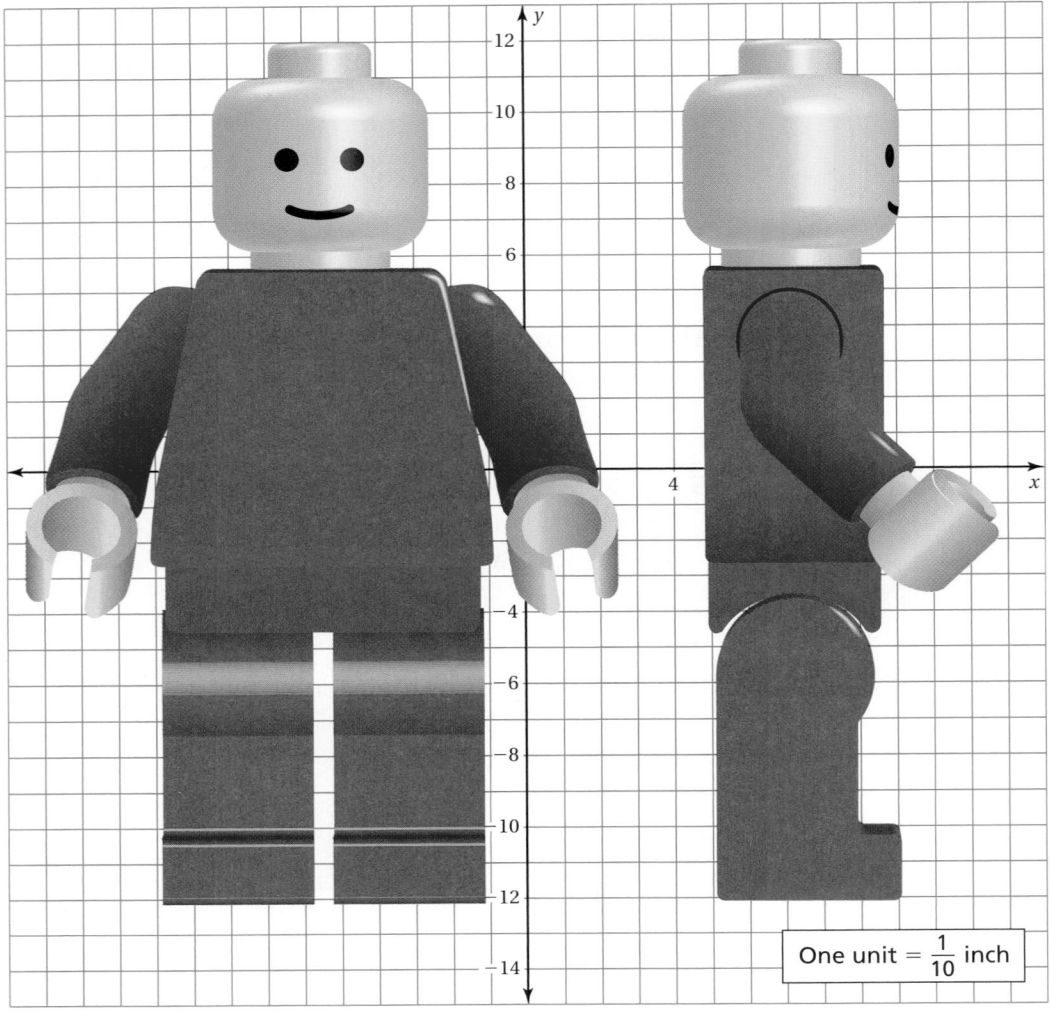

One unit = $\frac{1}{10}$ inch

Laurie's Notes

Introduction

For the Teacher

- **Goal:** Students will explore strategies for finding the volumes of composite figures.
- Each activity can be expanded enough to require the entire class period to complete. Therefore, decide in advance how far you want students to take each activity, or if you want students to investigate each activity.

Motivate

- Start looking at home for items that are composites of at least two solids, such as a jar of balsamic vinegar that is a square prism and a cylinder. Ask students to identify the various solids that form each item.

Activity Notes

Activity 1

- This activity is very open-ended. Students are asked to estimate the volume of the Minifigure's head (two cylinders) and one leg (approximately 2 prisms and 1 cylinder). Part of the challenge today will be keeping their work neat and organized. The answers provided are just one of many ways students can approach the activity.
- As the lead graphic engineer of a toy manufacturer, you have been asked to estimate the amount of plastic needed for the Minifigure. These *deluxe* pieces are solid plastic!
- ❓ "What are the basic solids in this composite figure?" square prism, rectangular prism, and cylinder
- Once students understand the context, it is time for them to work with a partner and decide how to proceed. The figure is drawn on a coordinate grid with a scale provided. Students worked with scale drawings previously.
- After students have computed the volume of the head and a leg, ask a few questions.
- ❓ "What did you use for the dimensions of the head?" *Sample answer:* a cylinder with an approximate height of 0.5 inch and a radius of 0.3 inch, and a cylinder with a height of 0.1 inch and a radius of 0.15 inch; The volume of the head $= \pi(0.3)^2(0.5) + \pi(0.15)^2(0.1) \approx 0.15$ in.^{3}
- ❓ "What solids did you work with to find the volume of one leg?" foot: rectangular prism; leg portion: square prism; hip portion: cylinder
- ❓ "What dimensions did you find for each solid?" *Sample answer:*

rectangular prism	square prism	cylinder
$V = (0.5)(0.4)(0.2)$	$V = (0.4)(0.4)(0.3)$	$V = \pi(0.2)^2(0.4)$
$V = 0.04$ in.^{3}	$V = 0.048$ in.^{3}	$V \approx 0.05$ in.^{3}

  The sum of the 3 composite parts of one leg ≈ 0.138 in.^{3}
- ❓ "Can you think of alternative ways to find the volume of the composite figure?" Students may mention displacement of water.
- ❓ "Did any group work in fractions instead of decimals?" open-ended
- **Extension:** Compute the volume of other body parts.

Previous Learning

Students should know how to find volumes of prisms, cylinders, pyramids, and cones.

Activity Materials	
Introduction	**Closure**
• composite solids from home	• composite solids

Start Thinking! and Warm Up

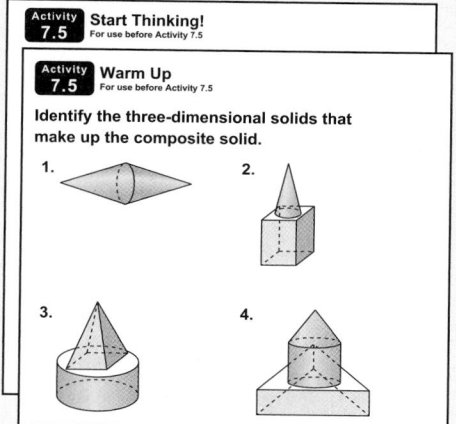

7.5 Record and Practice Journal

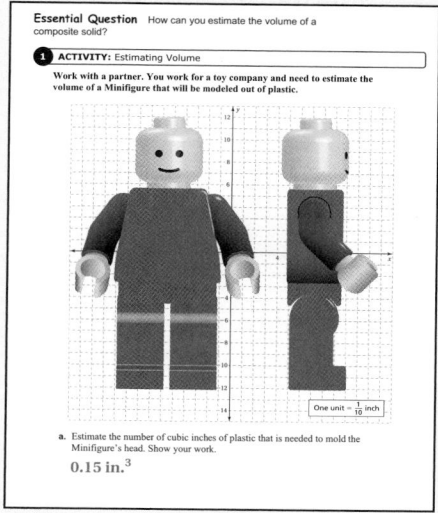

Discuss with students the limits of measuring tools. What if you only had a 12-inch ruler? How could you measure the height of a ceiling or the length of a hall? For the ceiling, you might have a long stick to use as the measuring device and then measure the stick. For the hallway, you could measure the length of your foot and then walk the distance. A curved object could be measured using a piece of string. Have students measure different objects or distances and compare their results.

7.5 Record and Practice Journal

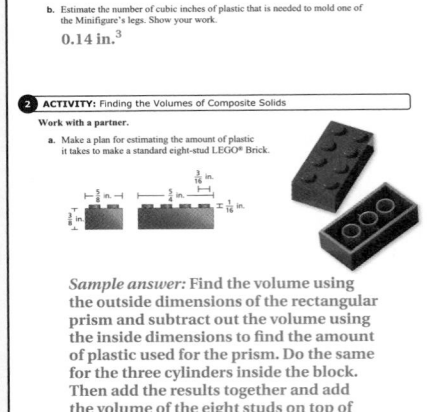

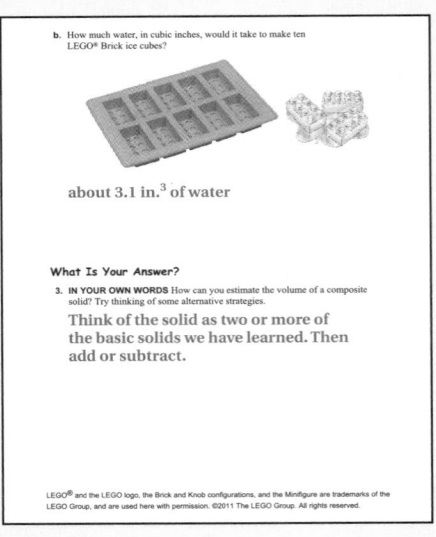

Laurie's Notes

Activity 2

- Part (a) asks students to make a plan for how much plastic would be needed. It does not ask students to do any computation, although they could, given a few assumptions.
- From the photos, it is known that the pieces are not solid. The thickness of plastic forming the cylinders and prisms varies. One would have to decide the thickness of the plastic, knowing that the outer walls appear to be thicker than the cylindrical walls.
- You might want to work through finding the volume of one top stud so that fractions can be discussed and reviewed.
- This is not a trivial problem. Precision is required so that when the LEGO® Bricks are stacked, the 3 bottom cylinders are centered among the 8 top cylinders (studs).
- **Alternate Approach:** You can fill a LEGO® Brick with water and measure how much water is contained. Then subtract the volume of water from the estimated volume of the brick to find the amount of plastic needed.
- Part (b) asks about a solid composite figure.
- There are 10 rectangular prisms measuring $\frac{5}{4}$ in. $\times \frac{5}{8}$ in. $\times \frac{3}{8}$ in.
  $V = 2\frac{119}{128}$ in.^{3} = 2.9296875 in.^{3}
- There are 80 cylinders with a height of $\frac{1}{16}$ in. and a radius of $\frac{3}{32}$ in.
  $V \approx 0.138$ in.^{3}
- The volume could also be found by filling the ice cube tray with water and measuring the volume of the water!

What Is Your Answer?

- **Think-Pair-Share:** Students should read the question independently and then work with a partner to answer the question. When they have answered the question, the pair should compare their answer with another group and discuss any discrepancies.

Closure

- Use one of the composite solids from home (or distribute items to different pairs of students) and have students measure and compute the volume of the item.

Technology
For the Teacher

Dynamic Classroom

The Dynamic Planning Tool
Editable Teacher's Resources at *BigIdeasMath.com*

Work with a partner.

a. Make a plan for estimating the amount of plastic it takes to make a standard eight-stud LEGO® Brick.

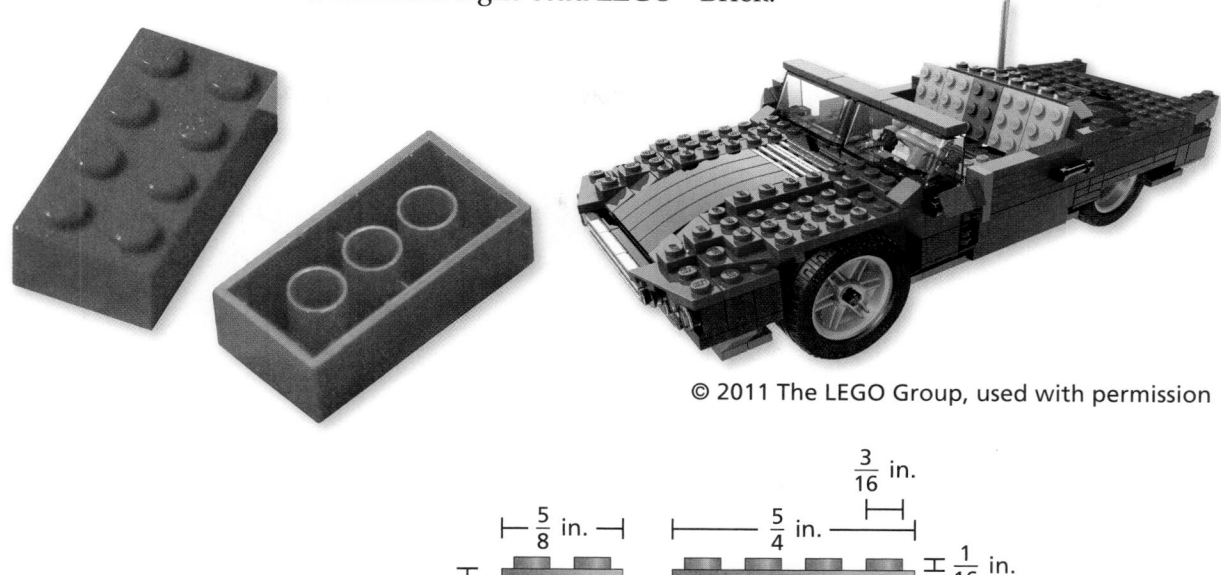

© 2011 The LEGO Group, used with permission

b. How much water, in cubic inches, would it take to make ten LEGO® Brick ice cubes?

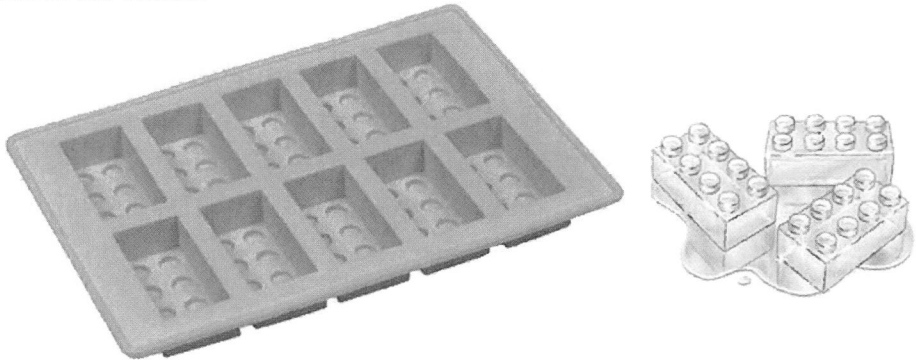

What Is Your Answer?

3. IN YOUR OWN WORDS How can you estimate the volume of a composite solid? Try thinking of some alternative strategies.

Practice

Use what you learned about the volumes of composite solids to complete Exercises 4–6 on page 328.

EXAMPLE ① Finding the Volume of a Composite Solid

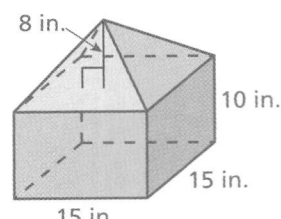

8 in.
10 in.
15 in.
15 in.

Find the volume of the composite solid.

The solid is made up of a square prism and a square pyramid. Find each volume.

Square prism

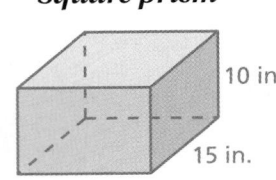

10 in.
15 in.
15 in.

Square pyramid

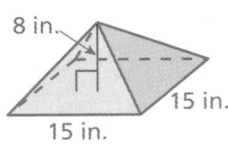

8 in.
15 in.
15 in.

$$V = Bh$$
$$= 15(15)(10)$$
$$= 2250$$

$$V = \frac{1}{3}Bh$$
$$= \frac{1}{3}(15)(15)(8)$$
$$= 600$$

Find the sum: $2250 + 600 = 2850$ in.³

:· The volume of the composite solid is 2850 cubic inches.

EXAMPLE ② Finding the Volume of a Composite Solid

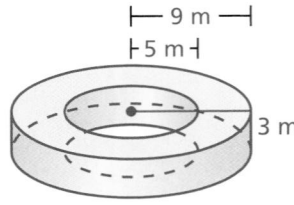

├─ 9 m ─┤
├5 m┤
3 m

Find the volume of the composite solid. Round your answer to the nearest tenth.

The solid is a cylinder with a cylinder-shaped hole. Find each volume.

Entire Cylinder

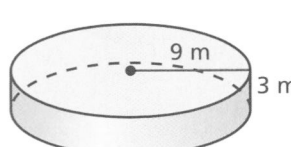

9 m
3 m

Cylinder-Shaped Hole

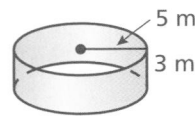

5 m
3 m

$$V = Bh$$
$$= \pi(9)^2(3)$$
$$= 243\pi$$

$$V = Bh$$
$$= \pi(5)^2(3)$$
$$= 75\pi$$

Find the difference: $243\pi - 75\pi = 168\pi \approx 527.5$ m³.

:· The volume of the composite solid is about 527.5 cubic meters.

Laurie's Notes

Introduction

Connect
- **Yesterday:** Students explored the volumes of composite figures.
- **Today:** Students will work with formulas to find the volume of a composite solid.

Motivate
- Discuss examples of composite solids. They could be items that are in your classroom, in your school, some portion of your school building, or a particular building in your town/city. Buildings are often made up of more than one solid. For the item(s) selected, discuss a strategy for how the volume would be found.

Lesson Notes

Example 1
- Work through the example, discussing the components of the solid.
- Model good problem solving techniques by writing the formula first, and then showing the substitution of the known variables.
- Remind students to show all of their work neatly and to label their answer with the correct units.
- **Connection:** Unlike working with the surface areas of composite figures, the composite volume is simply the sum of the individual volumes. In working with the surface areas of composite figures, recall that the areas of certain faces are in the interior of the solid and therefore the areas are subtracted or simply not computed.

Example 2
- ❓ "How is this solid made?" Students may describe it as a donut. A cylinder has been removed from the center of a cylinder.
- Work through the problem as shown by computing the volume of each cylinder and then subtracting. Note that each answer is left in terms of π and in the last step π is replaced by 3.14.
- ❓ "Is there another way to find the volume? Is it okay to subtract the two radii (9 m − 5 m) and find the volume of a cylinder with radius 4 meters? Explain." No. The answer is different. A cylinder with radius 4 and height 3 has a volume of 48π, which is quite different from 168π. This is actually a common student error.

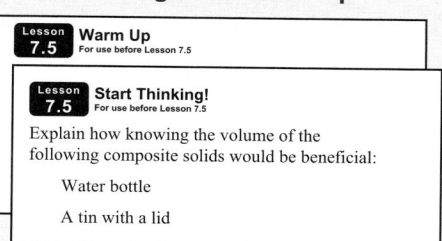
Extra Example 1

Find the volume of the composite solid. Round your answer to the nearest tenth.

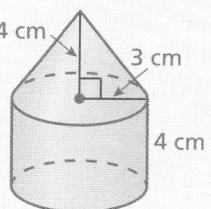

$48\pi \approx 150.7 \text{ cm}^3$

Extra Example 2

Find the volume of the composite solid. Round your answer to the nearest tenth.

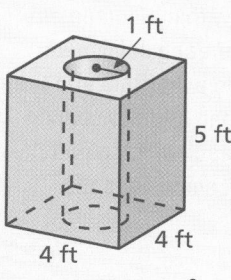

$80 - 5\pi \approx 64.3 \text{ ft}^3$

Laurie's Notes

On Your Own

1. 4.7 ft^3

2. $1680 - 48\pi \approx 1529.3 \text{ in.}^3$

Extra Example 3

Find the volume of the metal washer. Round your answer to the nearest tenth.

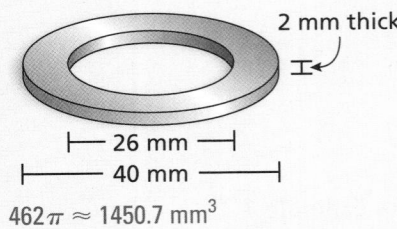

2 mm thick

|← 26 mm →|

|← 40 mm →|

$462\pi \approx 1450.7 \text{ mm}^3$

On Your Own

3. The volume of the silver ring would be half the volume of the silver ring in Example 3.

4. $\dfrac{27}{4}\pi \approx 21.2 \text{ ft}^3$

English Language Learners

Visual

Have students make a table to organize the different solids within a composite solid. Have a row for each solid, with columns that include a diagram and the volume. They can then add the column with the volume of each solid to find the total volume of the composite solid.

On Your Own

- Discuss each solid before students begin so they understand the dimensions. The dotted lines are needed to convey the three-dimensional structure, however it may confuse some students.
- Encourage students to use words to label each part of the problem so their work is readable. Example: top pyramid; rectangular prism; cylindrical hole
- If enough board space is available, have two different students share their work for each problem.

Example 3

? "Have you solved a problem similar to this before?" Yes, Example 2

? Walk students through the example by asking the following questions.
- "What is the height of the coin?" 2.2 mm
- "What is the radius of the coin?" 11.5 mm
- "What is the radius of the inner portion?" 8.5 mm
- "What is the width of the silver ring portion?" 3 mm
- "Is this number used in computing the volume?" no

- Work through the problem as shown. Students will need to remember how to multiply decimals.

On Your Own

- Note that in Question 3, only the thickness is decreased. The radius stays the same.

Closure

- **Exit ticket:** Draw a sketch of a farm silo composed of a cylinder with a cone on top. The cylinder is 24 feet in height and 16 feet in diameter. The cone is also 16 feet in diameter and 9 feet in height. Find the volume of the silo. $1728\pi \approx 5425.92 \text{ ft}^3$

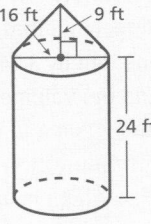

16 ft   9 ft

24 ft

Technology For the Teacher

Dynamic Classroom

The Dynamic Planning Tool
Editable Teacher's Resources at *BigIdeasMath.com*

On Your Own

Now You're Ready
Exercises 4–11

Find the volume of the composite solid. Round your answer to the nearest tenth.

1.

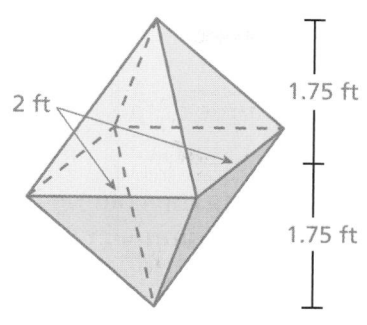

2 ft

1.75 ft

1.75 ft

2.

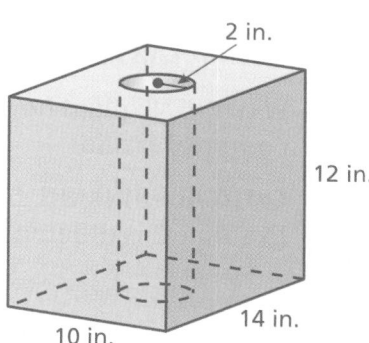

2 in.

12 in.

10 in.

14 in.

EXAMPLE 3 Real-Life Application

8.5 mm

11.5 mm

2.2 mm

What is the volume of the silver ring in an Argentine peso? Round your answer to the nearest tenth.

The coin is a cylinder. The silver ring is the portion remaining when the inner cylinder is removed. Find the volume of each cylinder.

Entire cylinder

11.5 mm

2.2 mm

Inner cylinder

8.5 mm

2.2 mm

$$V = Bh$$

$$= \pi(11.5)^2(2.2)$$

$$= 290.95\pi$$

$$V = Bh$$

$$= \pi(8.5)^2(2.2)$$

$$= 158.95\pi$$

Subtract the volume of the inner cylinder from the volume of the entire cylinder: $290.95\pi - 158.95\pi = 132\pi \approx 414.5$ mm³.

∴ The volume of the silver ring is about 414.5 cubic millimeters.

On Your Own

3. WHAT IF? In Example 3, how would the volume of the silver ring change if the coin were only half as thick?

4. Find the volume of the composite solid. Round your answer to the nearest tenth.

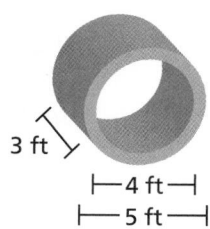

3 ft

4 ft

5 ft

Check It Out
Help with Homework
BigIdeasMath ✓com

Vocabulary and Concept Check

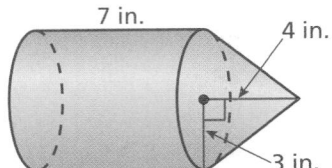

1. **VOCABULARY** What is a composite solid?

2. **WRITING** Explain how to find the volume of the composite solid.

3. **CRITICAL THINKING** Explain how finding the volume in Example 2 is different from finding the volume in Example 1.

Practice and Problem Solving

Find the volume of the composite solid. Round your answer to the nearest tenth.

4.

5.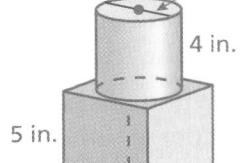

6.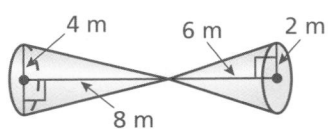

7.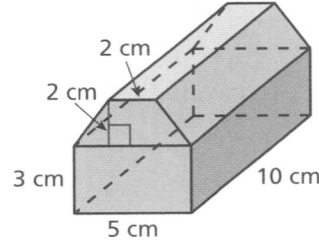

8.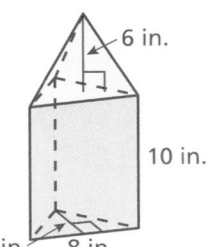

9.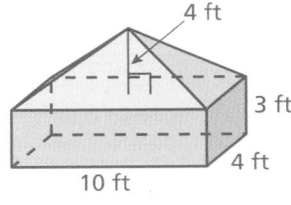

10.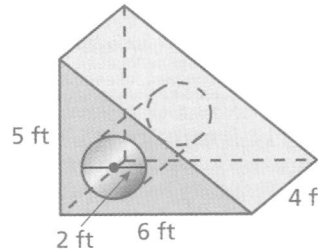

11.

12. **BIRD FEEDER** The cedar waxwing measures about 6 inches from head to tail. The green hexagonal part of the bird feeder has a base area of 18 square inches. Estimate how much bird seed the bird feeder will hold. Explain how you found your estimate.

Assignment Guide and Homework Check

Level	Day 1 Activity Assignment	Day 2 Lesson Assignment	Homework Check
Basic	4–6, 17–19	1–3, 7–11, 13	2, 7, 10, 13
Average	4–6, 17–19	1–3, 7–11, 13, 14	2, 7, 10, 13
Advanced	4–6, 17–19	1–3, 8–16 even, 15	2, 10, 12, 14

Common Errors

- **Exercises 4–11** Students may find the volumes of the two figures and choose one as the final volume instead of adding (or subtracting) the volumes. Remind them that finding the volumes of both solids is only the first step. They must then determine whether to add or subtract the volumes for the final volume of the composite solid.
- **Exercises 4–11** Some students may continue to make the same mistakes as in previous sections when finding the volumes of different solids. Remind them of the mistakes they have made before and work through the corrections so that they understand what they did wrong.
- **Exercise 12** Students may think that there is not enough information to solve the problem. Ask them what the height of the bird is and how the bird's height relates to the volume of the birdfeeder.

Technology For the Teacher

Answer Presentation Tool
QuizShow

7.5 Record and Practice Journal

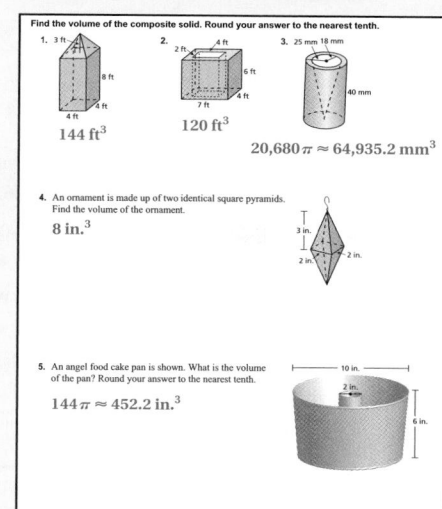

Find the volume of the composite solid. Round your answer to the nearest tenth.

1. 3 ft, 8 ft, 4 ft, 4 ft
144 ft³

2. 2 ft, 4 ft, 6 ft, 4 ft, 7 ft
120 ft³

3. 25 mm 18 mm, 40 mm
$20,680\pi \approx 64,935.2$ mm³

4. An ornament is made up of two identical square pyramids. Find the volume of the ornament.
8 in.³
3 in., 2 in., 2 in.

5. An angel food cake pan is shown. What is the volume of the pan? Round your answer to the nearest tenth.
$144\pi \approx 452.2$ in.³
10 in., 2 in., 6 in.

Vocabulary and Concept Check

1. A composite solid is a solid that is made up of more than one solid.

2. Find the volume of the cylinder and the cone. Then add them together.

3. In Example 2, you had to subtract the volume of the cylinder-shaped hole from the volume of the entire cylinder. In Example 1, you had to find the volumes of the square prism and the square pyramid and add them together.

Practice and Problem Solving

4. 16 ft³

5. $125 + 16\pi \approx 175.2$ in.³

6. $\frac{56}{3}\pi \approx 58.6$ m³

7. 220 cm³

8. 288 in.³

9. 173.3 ft³

10. $60 - 4\pi \approx 47.4$ ft³

11. $216 - 24\pi \approx 140.6$ m³

12. 126 in.³; The height of the green hexagonal part is about 6 inches. The height of the hexagonal pyramid is about 3 inches. Then find the volume of each solid and add them together.

13. a. *Sample answer:* 80%

    b. *Sample answer:*
$100\pi \approx 314$ in.^{3}

14. a. $159 + 18.00625\pi$
≈ 215.5 in.^{3}

    b. 1.5 lb

15. 13.875 in.^{3}; The volume
of the hexagonal prism is
$10.5(0.75)$ and the volume of
the hexagonal pyramid is
$\frac{1}{3}(6)(3)$.

16. See *Taking Math Deeper*.

 Fair Game Review

17. $\dfrac{25}{9}$ **18.** $\dfrac{81}{49}$

19. B

Mini-Assessment

**Find the volume of the composite solid.
Round your answer to the nearest tenth.**

1.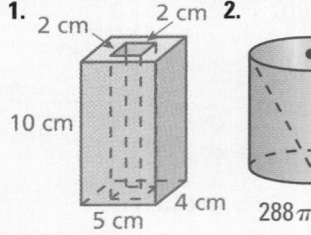

2 cm   2 cm   **2.**   6 in.

10 cm

5 cm   4 cm

160 cm^3 12 in.

$288\pi \approx 904.3$ in.^{3}

3. Find the volume of the wedding cake.
Round your answer to the nearest
tenth.

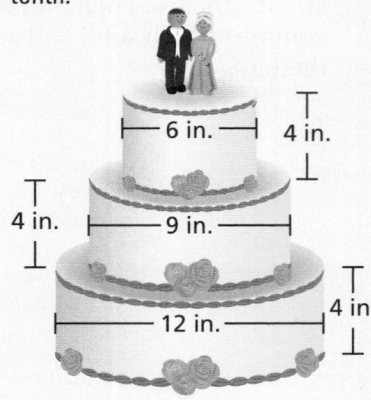

6 in.   4 in.

4 in.   9 in.

12 in.   4 in.

$261\pi \approx 819.5$ in.^{3}

Taking Math Deeper

Exercise 16

This is a logic puzzle. It gives students a chance to use deductive reasoning to
determine the volume of each type of solid.

1   This one   than this
is 8 more   one.

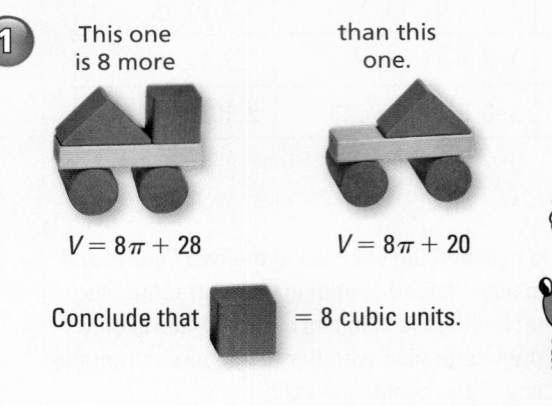

$V = 8\pi + 28$ $V = 8\pi + 20$

Conclude that   = 8 cubic units.

8 more

2 Note that 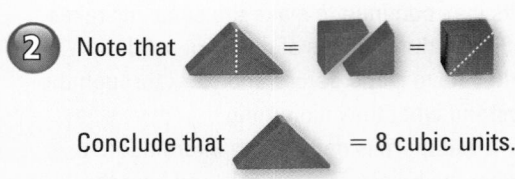 =   =

Conclude that   = 8 cubic units.

3

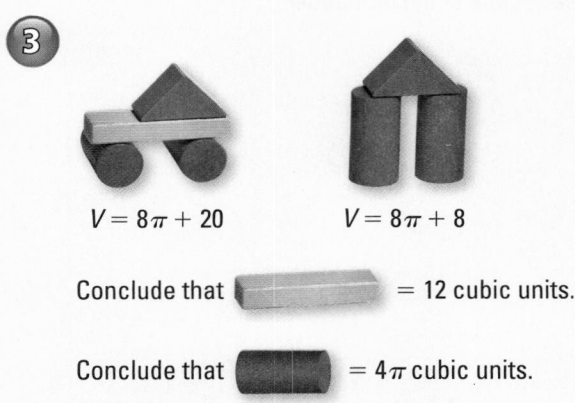

$V = 8\pi + 20$ $V = 8\pi + 8$

Conclude that   = 12 cubic units.

Conclude that   = 4π cubic units.

Reteaching and Enrichment Strategies

If students need help...	If students got it...

Resources by Chapter
  • Practice A and Practice B
  • Puzzle Time
Record and Practice Journal Practice
Differentiating the Lesson
Lesson Tutorials
Skills Review Handbook
	Resources by Chapter
	  • Enrichment and Extension
	  • School-to-Work
	  • Financial Literacy
	Start the next section

13. **CAKE** The raspberry layer cake has a diameter of 10 inches and a height of 5 inches.

   a. About what percent of the cake is remaining?

   b. Estimate the volume of the remaining cake.

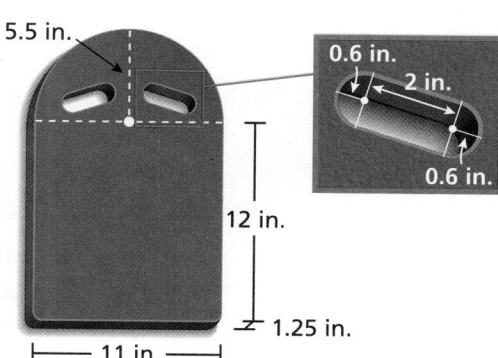

14. **KICKBOARD** A foam kickboard used for swimming has two identical hand grips.

   a. Find the volume of the kickboard.

   b. One cubic inch of foam weighs about 0.007 pound. How much does the kickboard weigh?

15. **PAPERWEIGHT** Estimate the amount of glass in the paperweight. Explain how you found your estimate.

16. **Puzzle** The volume of each group of solids is given. Find the volume of each of the four types of blocks.

$V = 8\pi + 8$

$V = 8\pi + 28$

$V = 8\pi + 20$

Fair Game Review What you learned in previous grades & lessons

The two figures are similar. Find the ratio (red to blue) of the areas. *(Section 5.2)*

17.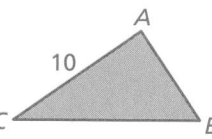

18.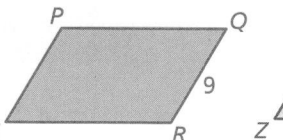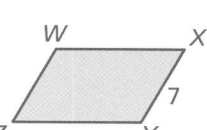

19. **MULTIPLE CHOICE** A fire hydrant releases 1200 gallons of water in 4 minutes. What is the rate of release in gallons per second? *(Section 3.1)*

   (A) 3 gal/sec   **(B)** 5 gal/sec   **(C)** 30 gal/sec   **(D)** 300 gal/sec

Essential Question When the dimensions of a solid increase by a factor of k, how does the surface area change? How does the volume change?

1 ACTIVITY: Comparing Volumes and Surface Areas

Work with a partner. Copy and complete the table. Describe the pattern. Are the solids similar? Explain your reasoning.

a.

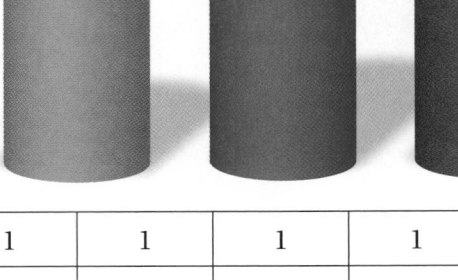

Radius	1	1	1	1	1
Height	1	2	3	4	5
Surface Area					
Volume					

b.

Radius	1	2	3	4	5
Height	1	2	3	4	5
Surface Area					
Volume					

Laurie's Notes

Introduction

For the Teacher

- **Goal:** Students will explore what happens to the surface areas and volumes of solids whose dimensions are increased by a factor of k.
- In this lesson, students will investigate similar figures in space. You may want to begin with a few review questions about similar figures.

Motivate

- **Story Time:** Retell a portion of the story of *Goldilocks and the 3 Bears*. Focus on the 3 sizes of porridge bowls, chairs, and beds.
- Share with students that Papa Bear's mattress was twice as long, twice as wide, and twice as high as Baby Bear's mattress. So, are there twice as many feathers in Papa Bear's feather bed mattress?
- This question will be answered at the end of the class.

Activity Notes

Activity 1

- Remind students to leave their answers in terms of π.
- ? To help students see the pattern in the first table, ask the following questions.
  - "Describe the changes in the dimensions." radius same, height increases by 1
  - "How does the surface area change?" increases by 2π
  - "How does the volume change?" increases by π
  - "Compare each figure's height to the original figure's height. Do the same for surface areas and volumes. What do you notice?" The volumes are multiplied by the same number as the heights.
- ? To help students see the pattern in the second table, ask the following questions.
  - "Describe the changes in the dimensions." radius and height each increase by 1
  - "Compare each figure's height to the original figure's height. Do the same for radii, surface areas, and volumes. What do you notice?" The heights and radii are multiplied by the same number. The surface areas are multiplied by the square of this number and the volumes are multiplied by the cube of this number.
- ? "Are the cylinders similar in part (a)? Explain." No, only the height increased, not the radius.
- ? "Are the cylinders similar in part (b)? Explain." Yes, both the radius and height are increasing by the same factor.

Previous Learning

Students should be familiar with similar figures, surface area formulas, and volume formulas.

Start Thinking! and Warm Up

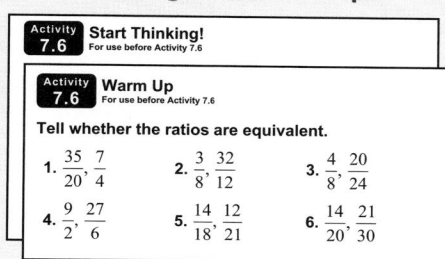

Activity 7.6 Start Thinking! For use before Activity 7.6

Activity 7.6 Warm Up For use before Activity 7.6

Tell whether the ratios are equivalent.

1. $\dfrac{35}{20}, \dfrac{7}{4}$ 2. $\dfrac{3}{8}, \dfrac{32}{12}$ 3. $\dfrac{4}{8}, \dfrac{20}{24}$

4. $\dfrac{9}{2}, \dfrac{27}{6}$ 5. $\dfrac{14}{18}, \dfrac{12}{21}$ 6. $\dfrac{14}{20}, \dfrac{21}{30}$

7.6 Record and Practice Journal

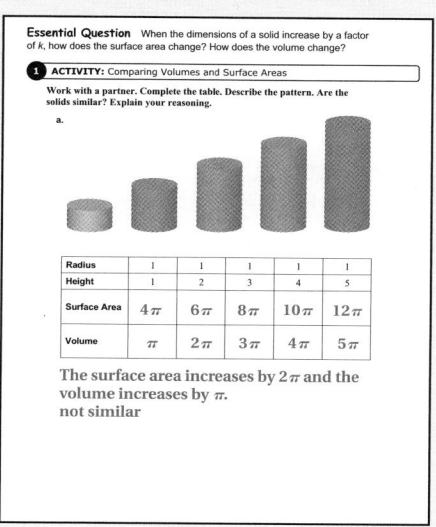

Essential Question When the dimensions of a solid increase by a factor of k, how does the surface area change? How does the volume change?

1 ACTIVITY: Comparing Volumes and Surface Areas

Work with a partner. Complete the table. Describe the pattern. Are the solids similar? Explain your reasoning.

a.

Radius	1	1	1	1	1
Height	1	2	3	4	5
Surface Area	4π	6π	8π	10π	12π
Volume	π	2π	3π	4π	5π

The surface area increases by 2π and the volume increases by π.
not similar

Differentiated Instruction

Differentiated Instruction

In the examples, check to be sure that students are correctly identifying corresponding sides. Remind them that they have to identify corresponding linear measures to write proportions before solving them.

7.6 Record and Practice Journal

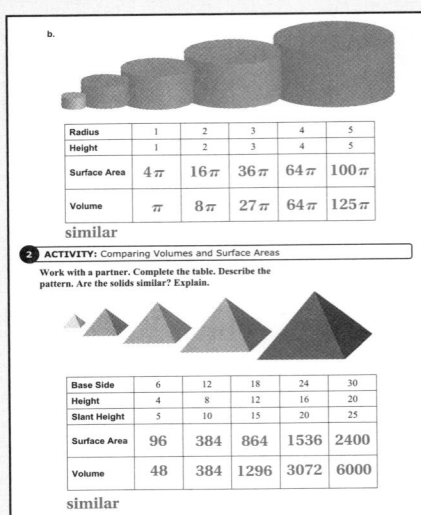

b.

Radius	1	2	3	4	5
Height	1	2	3	4	5
Surface Area	4π	16π	36π	64π	100π
Volume	π	8π	27π	64π	125π

similar

2 ACTIVITY: Comparing Volumes and Surface Areas

Work with a partner. Complete the table. Describe the pattern. Are the solids similar? Explain.

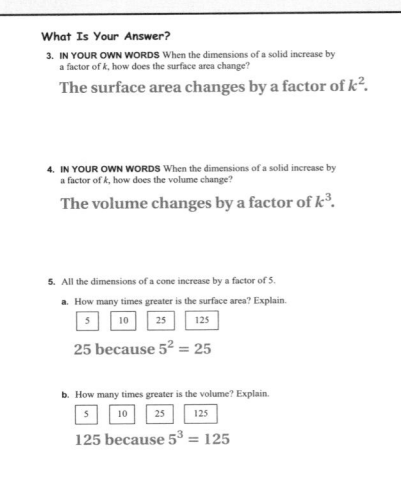

Base Side	6	12	18	24	30
Height	4	8	12	16	20
Slant Height	5	10	15	20	25
Surface Area	96	384	864	1536	2400
Volume	48	384	1296	3072	6000

similar

What Is Your Answer?

3. IN YOUR OWN WORDS When the dimensions of a solid increase by a factor of k, how does the surface area change?

The surface area changes by a factor of k^2.

4. IN YOUR OWN WORDS When the dimensions of a solid increase by a factor of k, how does the volume change?

The volume changes by a factor of k^3.

5. All the dimensions of a cone increase by a factor of 5.

a. How many times greater is the surface area? Explain.

| 5 | 10 | 25 | 125 |

25 because $5^2 = 25$

b. How many times greater is the volume? Explain.

| 5 | 10 | 25 | 125 |

125 because $5^3 = 125$

Activity 2

- You want students to see a pattern. It may be helpful for students to use a calculator.
- Ask students to describe patterns they see. Remind them to think about factors (multiplication) versus addition. The first pyramid should be referred to as the original pyramid. Describe any patterns in terms of the original pyramid.
- **?** "Are the pyramids similar? Explain." yes; The three dimensions are all changing by factors of 2, 3, 4, and 5 times the original dimensions.
- To help students see the factor by which the surface areas and volumes are multiplied, they should divide the new surface area (or volume) by the original surface area (or volume).

Example for Pyramid 4

Base Side	$24 \div 6 = 4$	Multiplied by a scale factor of 4
Height	$16 \div 4 = 4$	Multiplied by a scale factor of 4
Slant Height	$20 \div 5 = 4$	Multiplied by a scale factor of 4
Surface Area	$1536 \div 96 = 16$	Multiplied by a scale factor of 4^2
Volume	$3072 \div 48 = 64$	Multiplied by a scale factor of 4^3

- **Big Idea:** When the dimensions of a solid are all multiplied by a scale factor of k, the surface area is multiplied by a scale factor of k^2 and the volume is multiplied by a scale factor of k^3.

What Is Your Answer?

- **Think-Pair-Share:** Students should read each question independently and then work with a partner to answer the questions. When they have answered the questions, the pair should compare their answers with another group and discuss any discrepancies.

Closure

- Refer to Papa Bear's feather bed mattress. If the dimensions are all double Baby Bear's mattress, how many times more feathers are there? 8 times more feathers

Technology For the Teacher

Dynamic Classroom

The Dynamic Planning Tool
Editable Teacher's Resources at *BigIdeasMath.com*

ACTIVITY: Comparing Volumes and Surface Areas

Work with a partner. Copy and complete the
table. Describe the pattern. Are the solids
similar? Explain.

Base Side	6	12	18	24	30
Height	4	8	12	16	20
Slant Height	5	10	15	20	25
Surface Area					
Volume					

What Is Your Answer?

3. **IN YOUR OWN WORDS** When the dimensions of a solid increase by
   a factor of k, how does the surface area change?

4. **IN YOUR OWN WORDS** When the dimensions of a solid increase by
   a factor of k, how does the volume change?

5. All the dimensions of a cone increase by a factor of 5.

   a. How many times greater is the surface area? Explain.

   | 5 | 10 | 25 | 125 |

   b. How many times greater is the volume? Explain.

   | 5 | 10 | 25 | 125 |

Practice

Use what you learned about the surface areas and volumes of
similar solids to complete Exercises 4–6 on page 335.

Check It Out
Lesson Tutorials
BigIdeasMathcom

Key Vocabulary
similar solids, *p. 332*

Solids of the same type that have proportional corresponding linear measures are **similar solids**.

EXAMPLE 1 Identifying Similar Solids

Which cylinder is similar to Cylinder A?

Cylinder B

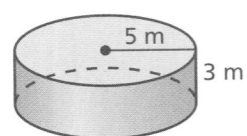

Check to see if corresponding linear measures are proportional.

Cylinder A

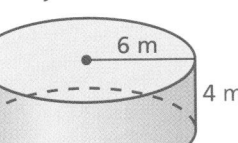

Cylinder A and Cylinder B

$$\frac{\text{Height of A}}{\text{Height of B}} = \frac{4}{3} \qquad \frac{\text{Radius of A}}{\text{Radius of B}} = \frac{6}{5}$$

Not proportional

Cylinder C

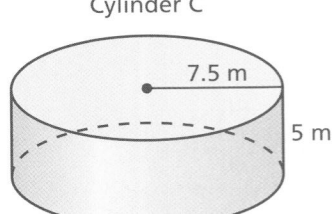

Cylinder A and Cylinder C

$$\frac{\text{Height of A}}{\text{Height of C}} = \frac{4}{5} \qquad \frac{\text{Radius of A}}{\text{Radius of C}} = \frac{6}{7.5} = \frac{4}{5}$$

Proportional

So, Cylinder C is similar to Cylinder A.

EXAMPLE 2 Finding Missing Measures in Similar Solids

The cones are similar. Find the missing slant height ℓ.

$$\frac{\text{Radius of X}}{\text{Radius of Y}} = \frac{\text{Slant height of X}}{\text{Slant height of Y}}$$

Cone X

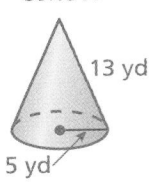

Cone Y

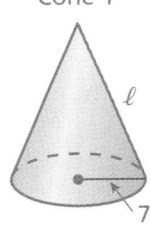

$$\frac{5}{7} = \frac{13}{\ell} \qquad \text{Substitute.}$$

$$5\ell = 91 \qquad \text{Use Cross Products Property.}$$

$$\ell = 18.2 \qquad \text{Divide each side by 5.}$$

The slant height is 18.2 yards.

On Your Own

Now You're Ready
Exercises 4–9

1. Cylinder D has a radius of 7.5 meters and a height of 4.5 meters. Which cylinder in Example 1 is similar to Cylinder D?

2. The prisms are similar. Find the missing width and length.

Prism A

Prism B

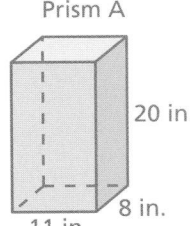

Multi-Language Glossary at BigIdeasMath✓com.

Laurie's Notes

Introduction

Connect
- **Yesterday:** Students explored what happens to the surface areas and volumes of solids when the dimensions are multiplied by a factor of *k*.
- **Today:** Students will use properties of similar solids to solve problems.

Motivate
- **Movie Time:** Hold an object that is miniature in size (model car, doll house item, statue, and so on). Tell students that this is a prop from a movie set. The movie plot is about giants. In order to make people look large, all of the props have been shrunk proportionally.
- Spend some time talking about movie making. Creating props larger than normal will make real people appear smaller than normal, and vice versa.

Lesson Notes

Example 1
- Note that the definition simply states that the corresponding linear measures must be in the same proportion. This means that solids can increase or decrease proportionally.
- **?** "What is a proportion?" an equation of two equal ratios
- **?** "How do you know if two ratios are equal?" Students might say by eyesight; by simple arithmetic, like $\frac{1}{2} = \frac{2}{4}$; that the ratios simplify to the same ratio. Students should recall the Cross Products Property.
- Work through the example.
- Be sure to write the words and the numbers. Use language such as "The ratio of the height of A to the height of B is 4 to 3."
- **?** "How do you know $\frac{6}{7.5} = \frac{4}{5}$?" Answers may vary depending upon students' number sense. By the Cross Products Property $6 \times 5 = 7.5 \times 4$.

Example 2
- **?** "By the definition of similar solids, what can you determine about two similar cones?" Corresponding linear measures are proportional.
- Set up the proportion and solve for the missing slant height.

On Your Own
- **Think-Pair-Share:** Students should read each question independently and then work with a partner to answer the questions. When they have answered the questions, the pair should compare their answers with another group and discuss any discrepancies.
- Ask volunteers to put their work on the board.

Start Thinking! and Warm Up

Extra Example 1
Which prism is similar to Prism A?

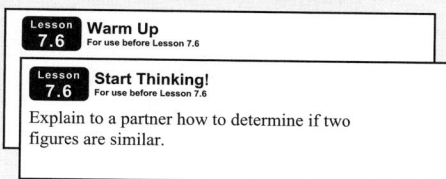

Prism A   Prism B   Prism C

Prism B

Extra Example 2
The square pyramids are similar. Find the length of the base of Pyramid E.

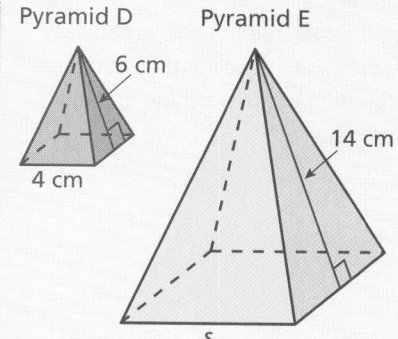

Pyramid D   Pyramid E

On Your Own
1. Cylinder B
2. $w = 3.2$ in.
   $\ell = 4.4$ in.

Laurie's Notes

Key Ideas

- Write the Key Ideas.
- **Example:** If the linear dimensions of B are double A, dimensions would be in the ratio of $\frac{1}{2}$, and the surface areas would be in the ratio of $\left(\frac{1}{2}\right)^2$ or $\frac{1}{4}$.
- Refer to yesterday's activity to confirm that this relationship was found in Activity 1, part (b) and Activity 2.

Example 3

 "Do you have enough information to solve this problem? Explain." Yes; The heights are in the ratio of $\frac{6}{10}$, so the surface areas are in the ratio of $\left(\frac{6}{10}\right)^2$.

- Set up the problem and solve.
- **FYI:** Notice that the problem is solved using the Multiplication Property of Equality. It could also be solved using the Cross Products Property.
- **Connection:** The ratio:

$$\frac{\text{original dimension}}{\text{new dimension}}$$

is the scale factor. The square of the scale factor is used to find the new surface area.

On Your Own

- Students should first identify the ratio of the corresponding linear measurements. Question 3: $\frac{5}{8}$; Question 4: $\frac{5}{4}$
- Ask student volunteers to share their work at the board.

Extra Example 3

The cones are similar. What is the surface area of Cone G? Round your answer to the nearest tenth.

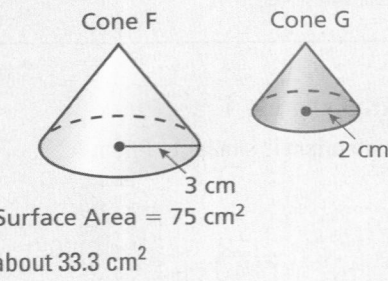

Cone F   Cone G

2 cm

3 cm

Surface Area = 75 cm^2

about 33.3 cm^2

On Your Own

3. 237.5 m^2

4. 171.9 cm^2

English Language Learners

Vocabulary

Have students add the key vocabulary *similar solids* to their notebooks with a description of the meaning in their own words.

Key Ideas

Linear Measures

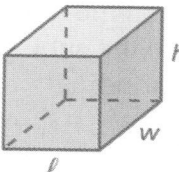
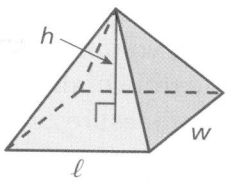
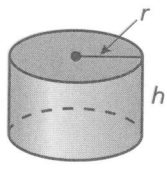
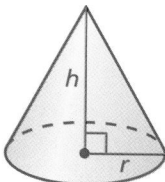

Surface Areas of Similar Solids

If two solids are similar, then the ratio of their surface areas is equal to the square of the ratio of their corresponding linear measures.

Solid A

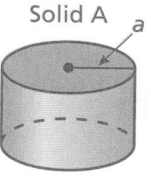

Solid B

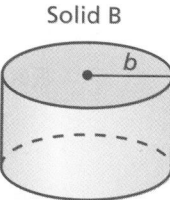

$$\frac{\text{Surface Area of A}}{\text{Surface Area of B}} = \left(\frac{a}{b}\right)^2$$

EXAMPLE ③ **Finding Surface Area**

Pyramid A

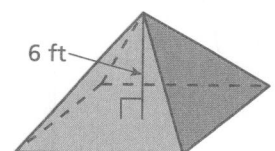

6 ft

Pyramid B

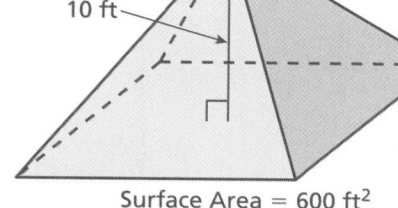

10 ft

Surface Area = 600 ft²

The pyramids are similar. What is the surface area of Pyramid A?

$$\frac{\text{Surface Area of A}}{\text{Surface Area of B}} = \left(\frac{\text{Height of A}}{\text{Height of B}}\right)^2$$

$\dfrac{S}{600} = \left(\dfrac{6}{10}\right)^2$ Substitute.

$\dfrac{S}{600} = \dfrac{36}{100}$ Evaluate power.

$\dfrac{S}{600} \cdot 600 = \dfrac{36}{100} \cdot 600$ Multiply each side by 600.

$S = 216$ Simplify.

∴ The surface area of Pyramid A is 216 square feet.

On Your Own

The solids are similar. Find the surface area of the red solid. Round your answer to the nearest tenth.

3.

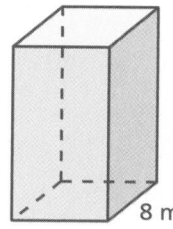

8 m

Surface Area = 608 m²

5 m

4.

5 cm

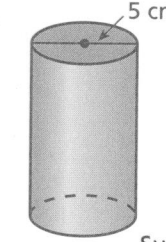

4 cm

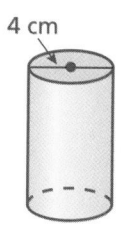

Surface Area = 110 cm²

 **Key Idea**

Volumes of Similar Solids

If two solids are similar, then the ratio of their volumes is equal to the cube of the ratio of their corresponding linear measures.

$$\frac{\text{Volume of A}}{\text{Volume of B}} = \left(\frac{a}{b}\right)^3$$

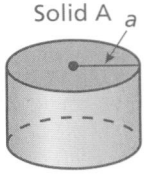
Solid A *a*

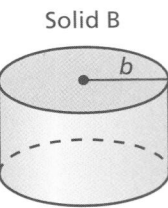
Solid B
b

EXAMPLE **4** **Standardized Test Practice**

Original Tank

Volume = 2000 ft³

The dimensions of the touch tank at an aquarium are doubled. What is the volume of the new touch tank?

Ⓐ   150 ft³   Ⓑ   4000 ft³

Ⓒ   8000 ft³   Ⓓ   16,000 ft³

The dimensions are doubled, so the ratio of the dimensions in the original tank to the dimensions in the new tank is 1 : 2.

$$\frac{\text{Original volume}}{\text{New volume}} = \left(\frac{\text{Original dimension}}{\text{New dimension}}\right)^3$$

$$\frac{2000}{V} = \left(\frac{1}{2}\right)^3 \qquad \text{Substitute.}$$

$$\frac{2000}{V} = \frac{1}{8} \qquad \text{Evaluate power.}$$

$$16,000 = V \qquad \text{Use Cross Products Property.}$$

> **Study Tip**
>
> When the dimensions of a solid are multiplied by k, the surface area is multiplied by k^2 and the volume is multiplied by k^3.

∴ The volume of the new tank is 16,000 cubic feet. The correct answer is Ⓓ.

● **On Your Own**

Now You're Ready
Exercises 10–13

The solids are similar. Find the volume of the red solid. Round your answer to the nearest tenth.

5.

5 cm

12 cm

Volume = 288 cm³

6.

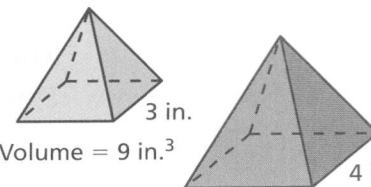

3 in.
Volume = 9 in.³

4 in.

Laurie's Notes

Key Idea

- Write the Key Idea.
- **Example:** If the linear dimensions of B are double A, dimensions would be in the ratio of $\frac{1}{2}$, and the volumes would be in the ratio of $\left(\frac{1}{2}\right)^3$ or $\frac{1}{8}$.
- Refer to yesterday's activity to confirm that this relationship was found in Activity 1, part (b) and Activity 2.

Example 4

- Write the problem in words to help students recognize how the numbers are being substituted.
- **Common Misconception:** Many students think that when you double the dimensions, the surface area and volume also double. This Big Idea takes time for students to fully understand.
- **Connection:** The ratio:

$$\frac{\text{original dimension}}{\text{new dimension}}$$

is the scale factor. The cube of the scale factor is used to find the new volume.

On Your Own

- Students should first identify the ratio of the corresponding linear measurements. Question 5: $\frac{5}{12}$; Question 6: $\frac{4}{3}$
- Ask student volunteers to share their work at the board.

Closure

- Use one of the miniature items used to motivate the lesson and ask a question related to surface area or volume. Some miniature items have a scale printed on the item.

The Dynamic Planning Tool
Editable Teacher's Resources at *BigIdeasMath.com*

Extra Example 4

The cylinders are similar. Find the volume of Cylinder J. Round your answer to the nearest tenth.

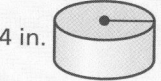

Cylinder H   Cylinder J

4 in.   6 in.

Volume = 314 in.³

1059.8 in.³

On Your Own

5. 20.8 cm³

6. 21.3 in.³

Vocabulary and Concept Check

1. Similar solids are solids of the same type that have proportional corresponding linear measures.

2. *Sample answer:*

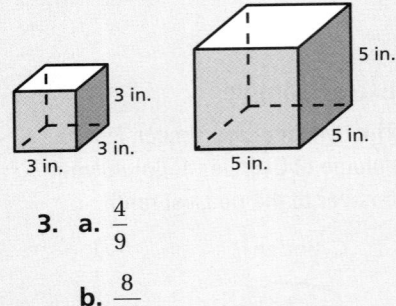

3. a. $\dfrac{4}{9}$

   b. $\dfrac{8}{27}$

Practice and Problem Solving

4. yes

5. no

6. yes

7. no

8. 25 in.

9. $b = 18$ m
   $c = 19.5$ m
   $h = 9$ m

Assignment Guide and Homework Check

Level	Day 1 Activity Assignment	Day 2 Lesson Assignment	Homework Check
Basic	4–6, 20–23	1–3, 7–16	7, 8, 10, 16
Average	4–6, 20–23	1–3, 7–14, 16, 18	7, 8, 10, 16
Advanced	4–6, 20–23	1–3, 7, 8–14 even, 16–19	7, 8, 10, 16

Common Errors

- **Exercises 4–7** Students may only compare two sets of measurements instead of all three. The bases of two figures may be similar, but the heights may not be proportional to the length or width. Remind them to check all of the corresponding linear measures when determining if two solids are similar. Ask them how many sets of ratios they need to write for each type of solid.

- **Exercises 8 and 9** Students may write the proportion incorrectly. For example, in Exercise 8, a student may write $\dfrac{10}{4} = \dfrac{10}{d}$ which gives $d = 4$ instead of $d = 25$. Remind them to write the proportion with the dimensions of one solid in the numerator and the dimensions of the other solid in the denominator.

7.6 Record and Practice Journal

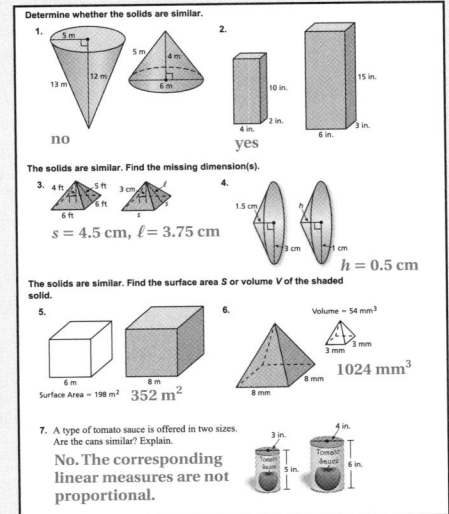

Check It Out
Help with Homework
BigIdeasMath √com

Vocabulary and Concept Check

1. **VOCABULARY** What are similar solids?

2. **OPEN-ENDED** Draw two similar solids and label their corresponding linear measures.

3. **REASONING** The ratio of the corresponding linear measures of Cube A to Cube B is $\frac{2}{3}$.

   a. Find the ratio of the area of one face of Cube A to the area of one face of Cube B.

   b. Find the ratio of the volume of Cube A to the volume of Cube B.

Practice and Problem Solving

Determine whether the solids are similar.

① 4.
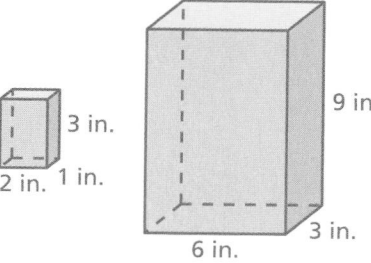

5.
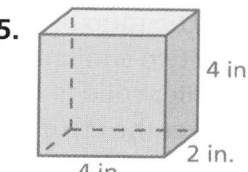

6.
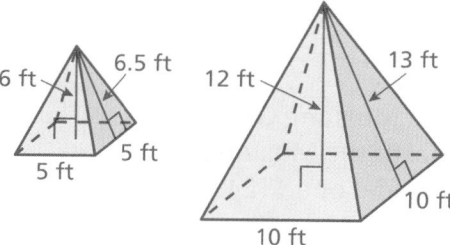

7.
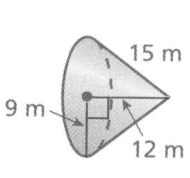
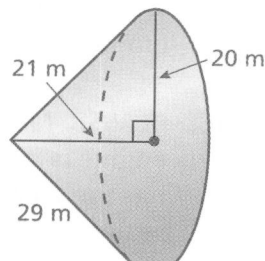

The solids are similar. Find the missing dimension(s).

② 8.
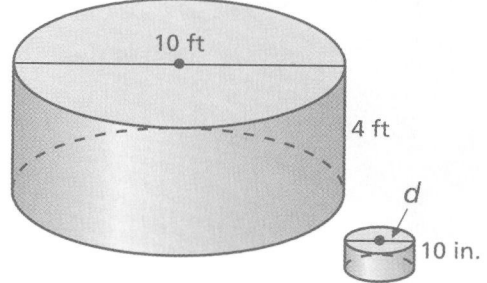

9.
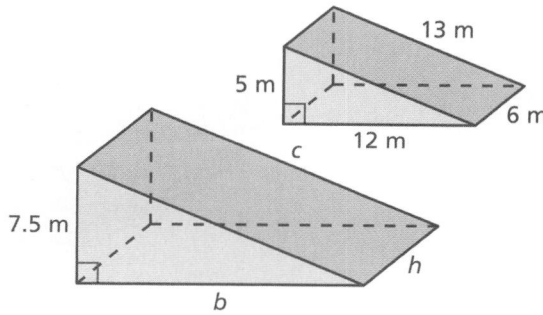

**The solids are similar. Find the surface area *S* or volume *V* of the red solid.
Round your answer to the nearest tenth.**

③ ④ 10.

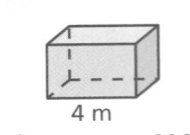 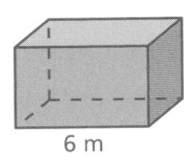

4 m

6 m

Surface Area = 336 m²

11.

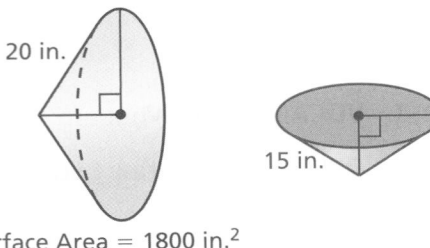

20 in.

15 in.

Surface Area = 1800 in.²

12.

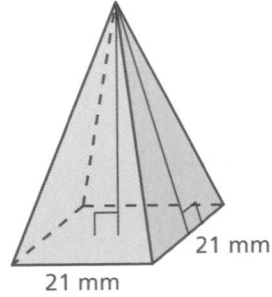 

21 mm

21 mm

7 mm

7 mm

Volume = 5292 mm³

13.

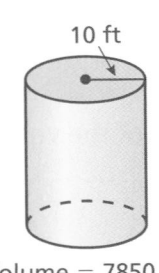 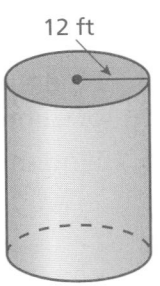

10 ft

12 ft

Volume = 7850 ft³

14. ERROR ANALYSIS The ratio of the corresponding
linear measures of two similar solids is 3 : 5. The
volume of the smaller solid is 108 cubic inches.
Describe and correct the error in finding the
volume of the larger solid.

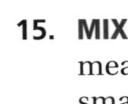

$$\frac{108}{V} = \left(\frac{3}{5}\right)^2$$

$$\frac{108}{V} = \frac{9}{25}$$

$$300 = V$$

The volume of the larger
solid is 300 cubic inches.

15. MIXED FRUIT The ratio of the corresponding linear
measures of two similar cans of fruit is 4 to 7. The
smaller can has a surface area of 220 square centimeters.
Find the surface area of the larger can.

16. CLASSIC MUSTANG The volume of a 1968 Ford Mustang GT
engine is 390 cubic inches. Which scale model of the Mustang
has the greater engine volume, a 1 : 18 scale model or a 1 : 24
scale model? How much greater?

Common Errors

- **Exercises 10–13** Students may forget to square or cube the ratio of the corresponding linear measures when finding the surface area or volume of the red solid. Remind them of the Key Ideas.
- **Exercises 10–13** Students may cube the ratio of corresponding linear measures when finding surface area, or square the ratio of corresponding linear measures when finding volume. Remind them of the Key Ideas. The ratio of corresponding linear measures is squared for surface area and cubed for volume.
- **Exercise 15** Students may invert the ratio of the surface area when writing the proportion. When they look at the ratio of corresponding linear measures as a fraction, ask whether the numerator or denominator corresponds to the smaller figure. This will help them know where to place the surface area in the proportion.

 Practice and Problem Solving

10. 756 m^2

11. 1012.5 in.^2

12. 196 mm^3

13. $13{,}564.8 \text{ ft}^3$

14. The ratio of the volumes of two similar solids is equal to the cube of the ratio of their corresponding linear measures.
$$\frac{108}{V} = \left(\frac{3}{5}\right)^3$$
$$\frac{108}{V} = \frac{27}{125}$$
$$V = 500 \text{ in.}^3$$

15. 673.75 cm^2

16. 1 : 18 scale model; about 0.04 in.^3

Practice and Problem Solving

17. **a.** yes; Because all circles are similar, the slant height and the circumference of the base of the cones are proportional.

b. no; because the ratio of the volumes of similar solids is equal to the cube of the ratio of their corresponding linear measures

18. See Additional Answers.

19. See *Taking Math Deeper*.

Fair Game Review

20. 39   **21.** 1

22. 0   **23.** C

Mini-Assessment

The solids are similar. Find the surface area S of the red solid. Round your answer to the nearest tenth.

1. 288 m²

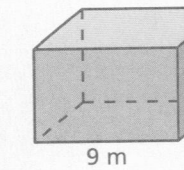

3 m   9 m

Surface Area = 32 m²

2. 37.8 in.²

8 in.

4 in.

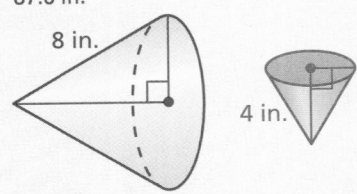

Surface Area = 151 in.²

3. The candles are similar. Find the volume of the larger candle. 600 in.³

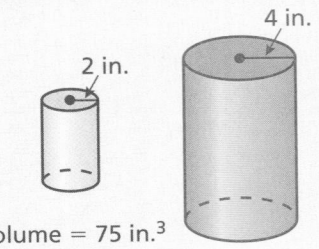

2 in.   4 in.

Volume = 75 in.³

Taking Math Deeper

Exercise 19

This problem is a straightforward application of the two main concepts of the lesson. That is, with similar solids, surface area is proportional to the square of the scale factor and volume is proportional to the cube of the scale factor. Even so, students have trouble with this problem because they don't see that enough information is given to create the table. The thing to learn here is that it is possible to compare two surfaces without knowing either one of them.

1 Make a table. Include the height of each doll.

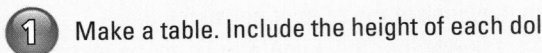

Heights go up by 1

Height	1	2	3	4	5	6	7
Surface Area	S	$4S$	$9S$	$16S$	$25S$	$36S$	$49S$
Volume	V	$8V$	$27V$	$64V$	$125V$	$216V$	$343V$

2 Compare the surface areas of the dolls.

3 Compare the volumes of the dolls.

Matryoshka dolls, or Russian nested dolls, are also called stacking dolls. A set of matryoshkas consists of a wooden figure which can be pulled apart to reveal another figure of the same sort inside. It has, in turn, another figure inside, and so on. The number of nested figures is usually five or more.

Reteaching and Enrichment Strategies

If students need help. . .	If students got it. . .
Resources by Chapter • Practice A and Practice B • Puzzle Time Record and Practice Journal Practice Differentiating the Lesson Lesson Tutorials Skills Review Handbook	Resources by Chapter • Enrichment and Extension • School-to-Work • Financial Literacy • Technology Connection Start the next section

17.  **Critical Thinking** You and a friend make paper cones to collect beach glass. You cut out the largest possible three-fourths circle from each piece of paper.

a. Are the cones similar? Explain your reasoning.

b. Your friend says that because your sheet of paper is twice as large, your cone will hold exactly twice the volume of beach glass. Is this true? Explain your reasoning.

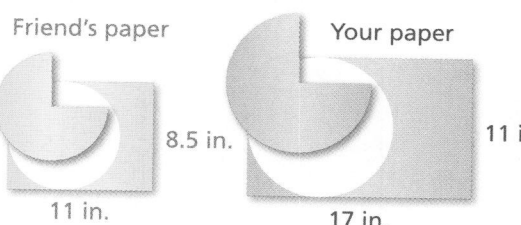

Friend's paper   Your paper

8.5 in.   11 in.

11 in.   17 in.

18. MARBLE STATUE You have a small marble statue of Wolfgang Mozart that is 10 inches tall and weighs 16 pounds. The original statue in Vienna is 7 feet tall.

a. Estimate the weight of the original statue. Explain your reasoning.

b. If the original statue were 20 feet tall, how much would it weigh?

Wolfgang Mozart

19. RUSSIAN DOLLS The largest doll is 7 inches tall. Each of the other dolls is 1 inch shorter than the next larger doll. Make a table that compares the surface areas and volumes of the seven dolls.

 **Fair Game Review** *What you learned in previous grades & lessons*

Add. *(Section 1.2)*

20. $69 + (-31) + 7 + (-6)$ **21.** $-2 + (-5) + (-12) + 20$ **22.** $10 + (-6) + (-5) + 1$

23. MULTIPLE CHOICE What is the mean of the numbers below? *(Skills Review Handbook)*

14, 6, 21, 8, 14, 19, 30

    Ⓐ 6   Ⓑ 15   Ⓒ 16   Ⓓ 56

1. Determine whether the solids are similar. *(Section 7.6)*

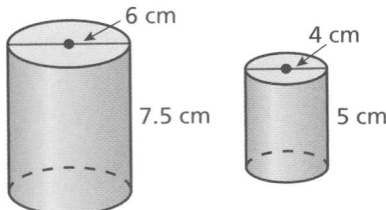

2. The prisms are similar. Find the missing width and height. *(Section 7.6)*

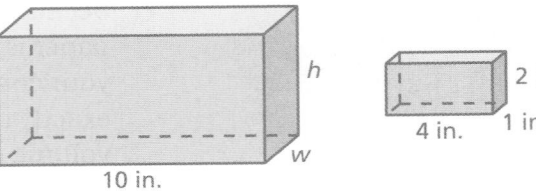

Find the volume of the composite solid. Round your answer to the nearest tenth.
(Section 7.5)

3.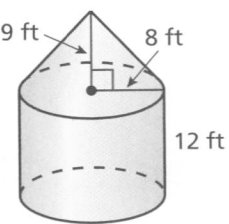

4.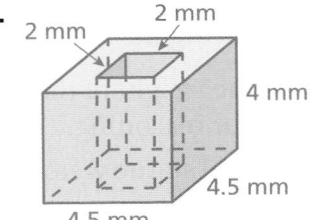

5. The solids are similar. Find the surface area of the red solid. *(Section 7.6)*

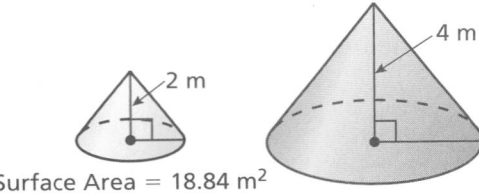

Surface Area = 18.84 m²

6. **ARCADE** You win a token after playing an arcade game. What is the volume of the gold ring? Round your answer to the nearest tenth. *(Section 7.5)*

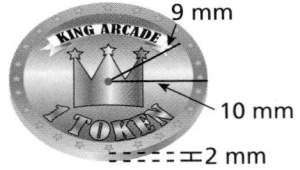

7. **SHED** What is the volume of the storage shed? *(Section 7.5)*

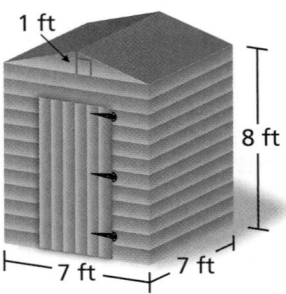

8. **JEWELRY BOXES** The ratio of the corresponding linear measures of two similar jewelry boxes is 2 to 3. The larger box has a volume of 162 cubic inches. Find the volume of the smaller jewelry box. *(Section 7.6)*

9. **GELATIN** You make a dessert with lemon gelatin and lime gelatin. What percent of the dessert is lime-flavored? Explain. *(Section 7.5)*

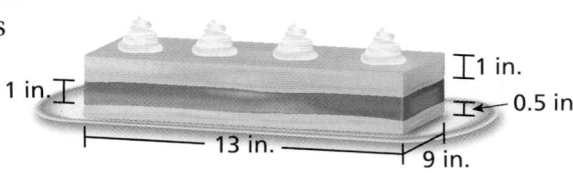

Alternative Assessment Options

Math Chat   Student Reflective Focus Question
Structured Interview   Writing Prompt

Math Chat

Ask students to use their own words to summarize what they know about finding volumes of composite solids and similar solids. Be sure that they include examples. Select students at random to present their summary to the class.

Study Help Sample Answers

Remind students to complete Graphic Organizers for the rest of the chapter.

4. Volume of a composite solid

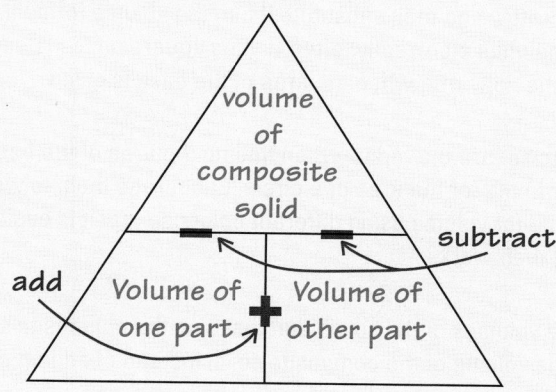

5. Surface areas of similar solids A and B

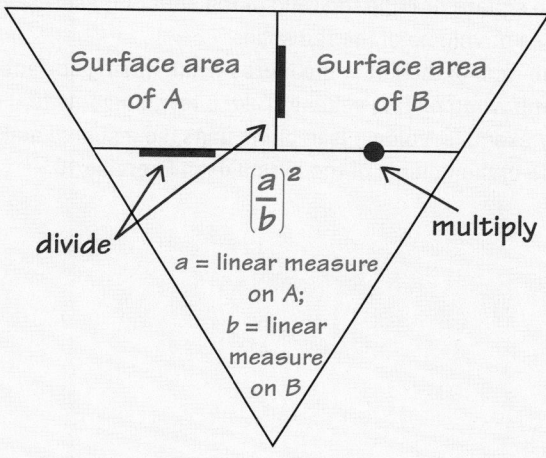

6. Available at *BigIdeasMath.com*.

Reteaching and Enrichment Strategies

If students need help. . .	If students got it. . .
Resources by Chapter • Study Help • Practice A and Practice B • Puzzle Time Lesson Tutorials *BigIdeasMath.com* Practice Quiz Practice from the Test Generator	Resources by Chapter • Enrichment and Extension • School-to-Work Game Closet at *BigIdeasMath.com* Start the Chapter Review

Technology
For the Teacher
Answer Presentation Tool

Assessment Book

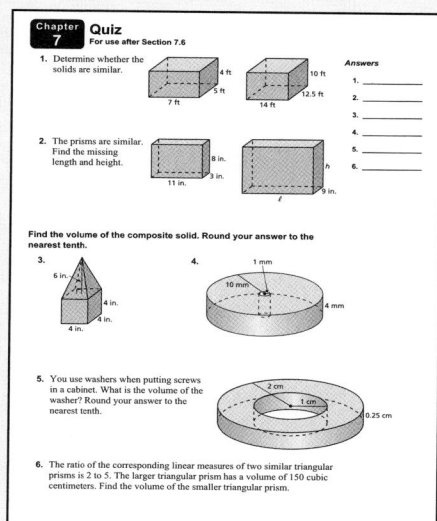

Answers

1. 96 in.^{3}

2. 120 m^3

3. 607.5 mm^3

4. $40\pi \approx 125.6$ in.^{3}

5. $\dfrac{75}{8\pi} \approx 3.0$ m

6. $393.75\pi \approx 1236.4$ ft^3

Review of Common Errors

Exercises 1–3

- Students may write linear or square units for volume rather than cubic units. Remind them that volume is measured in cubic units.

Exercises 4–6

- Students may forget to square the radius when finding the area of the base. Remind them of the formula for area of a circle.
- Students may use the diameter instead of the radius.

Exercises 7–9

- Students may forget to multiply the area of the base by the height and/or the factor of $\dfrac{1}{3}$ when finding the volume of a pyramid. Encourage them to find the area of the base separately and then substitute it into the volume formula.
- Students may write the units incorrectly, often writing square units instead of cubic units. This is especially true when the area of the base is given.

Exercises 10–12

- Students may forget to square the radius when finding the area of the base. Remind them of the formula for the area of a circle. Encourage them to write the parts of the formula for volume using different colors so that it is easier for them to keep track of their work.

Exercises 13–15

- Students may find the volumes of the solids that make up the composite solid and choose one as the volume of the composite solid, instead of adding or subtracting volumes as appropriate.

Exercises 16 and 17

- Students may forget to square or cube the ratio of the linear measures when finding the surface area or volume of the red solid.
- Students may cube the ratio of the linear measures when finding surface area, or square the ratio when finding volume. Take a few moments to examine the units. For example, volume is in cubic units, so it should make sense to cube the ratio of the heights of the pyramids in Exercise 16.

Review Key Vocabulary

volume, *p. 300*   similar solids, *p. 332*

Review Examples and Exercises

7.1 Volumes of Prisms *(pp. 298–303)*

Find the volume of the prism.

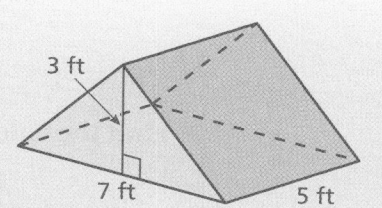

$V = Bh$ Write formula for volume.

$= \dfrac{1}{2}(7)(3) \cdot 5$ Substitute.

$= 52.5$ Multiply.

∴ The volume is 52.5 cubic feet.

Exercises

Find the volume of the prism.

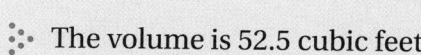

7.2 Volumes of Cylinders *(pp. 304–309)*

Find the height of the cylinder. Round your answer to the nearest whole number.

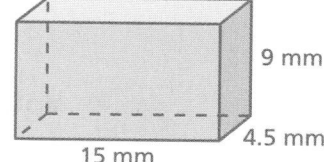

$V = Bh$ Write formula for volume.

$565 = \pi(6)^2(h)$ Substitute.

$565 = 36\pi h$ Simplify.

$5 \approx h$ Divide each side by 36π.

∴ The height is about 5 centimeters.

Volume = 565 cm³

Exercises

Find the volume *V* or height *h* of the cylinder. Round your answer to the nearest tenth.

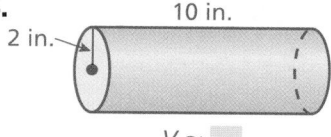

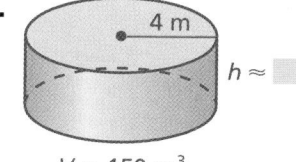

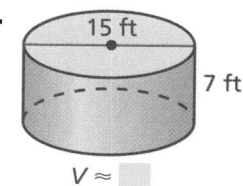

7.3 Volumes of Pyramids (pp. 310–315)

Find the volume of the pyramid.

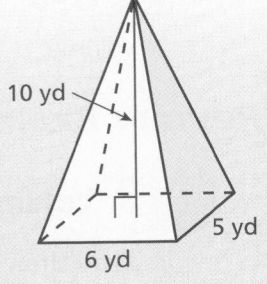

$$V = \frac{1}{3}Bh \qquad \text{Write formula for volume.}$$

$$= \frac{1}{3}(6)(5)(10) \qquad \text{Substitute.}$$

$$= 100 \qquad \text{Multiply.}$$

⋮ The volume is 100 cubic yards.

Exercises

Find the volume of the pyramid.

7.

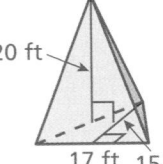

20 ft

17 ft 15 ft

8.

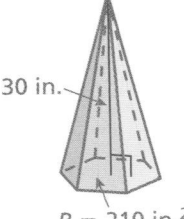

30 in.

$B = 210 \text{ in.}^2$

9.

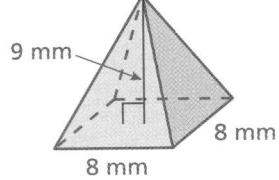

9 mm

8 mm

8 mm

7.4 Volumes of Cones (pp. 316–321)

Find the height of the cone. Round your answer to the nearest tenth.

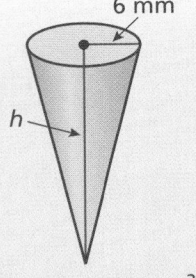

6 mm

h

Volume = 900 mm³

$$V = \frac{1}{3}Bh \qquad \text{Write formula for volume.}$$

$$900 = \frac{1}{3}\pi(6)^2(h) \qquad \text{Substitute.}$$

$$900 = 12\pi h \qquad \text{Simplify.}$$

$$23.9 \approx h \qquad \text{Divide each side by } 12\pi.$$

⋮ The height is about 23.9 millimeters.

Exercises

Find the volume V or height h of the cone. Round your answer to the nearest tenth.

10.

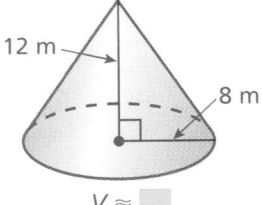

12 m

8 m

$V \approx$ ▢

11.

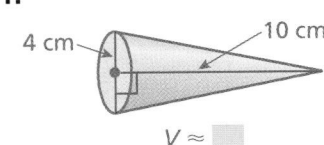

4 cm

10 cm

$V \approx$ ▢

12.

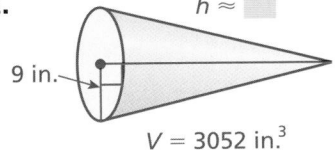

$h \approx$ ▢

9 in.

$V = 3052 \text{ in.}^3$

Review Game

Volume

Big Ideas
Game Closet

Materials

- a variety of containers of different shapes and sizes
- liquid measuring devices
- water

Directions

At the start of the chapter, have students bring in containers of different shapes and sizes. The containers should be shaped the same as the solids being studied in the chapter, they should be able to hold water, and they should be small enough so that they do not require a lot of water to fill. Collect containers until you have a sufficient variety of shapes and enough for the number of groups you want to have.

Work in groups. Give each group a container. Each group calculates the volume of the container and passes it to another group. Continue until all groups have calculated the volumes of the containers. When calculations are completed, each group measures the volume of a container using water. Measured volumes are shared with the class and compared to the calculated volumes.

Who Wins?

The group whose calculated volume is closest to the correct measured volume receives 1 point. The group with the most points wins.

Answers

7. 850 ft^3

8. 2100 in.^3

9. 192 mm^3

10. $256\pi \approx 803.8 \text{ m}^3$

11. $\dfrac{40}{3}\pi \approx 41.9 \text{ cm}^3$

12. $\dfrac{3052}{27\pi} \approx 36.0 \text{ in.}$

13. $360\pi \approx 1130.4 \text{ m}^3$

14. 132 ft^3

15. $12\pi \approx 37.7 \text{ cm}^3$

16. 576 m^3

17. 86.625 yd^2

My Thoughts on the Chapter

What worked. . .

What did not work. . .

What I would do differently. . .

7.5 Volumes of Composite Solids *(pp. 324–329)*

Find the volume of the composite solid. Round your answer to the nearest tenth.

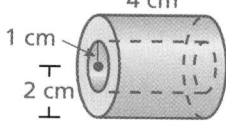

Square Prism

$$V = Bh$$
$$= (12)(12)(9)$$
$$= 1296$$

Cylinder

$$V = Bh$$
$$= \pi(5)^2(9)$$
$$= 225\pi \approx 706.5$$

Find the difference: $1296 - 706.5 = 589.5$.

∴ The volume of the composite solid is about 589.5 cubic feet.

Exercises

Find the volume of the composite solid. Round your answer to the nearest tenth.

13.

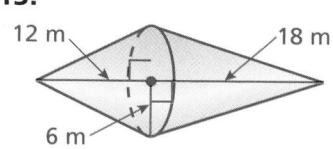

14.

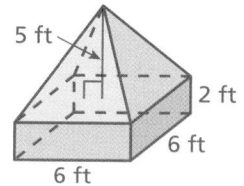

15.

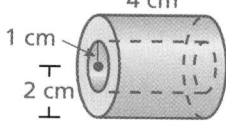

7.6 Surface Areas and Volumes of Similar Solids *(pp. 330–337)*

The cones are similar. What is the volume of the red cone? Round your answer to the nearest tenth.

$$\frac{\text{Volume of A}}{\text{Volume of B}} = \left(\frac{\text{Height of A}}{\text{Height of B}}\right)^3$$

$$\frac{V}{157} = \left(\frac{4}{6}\right)^3 \qquad \text{Substitute.}$$

$$\frac{V}{157} = \frac{64}{216} \qquad \text{Evaluate power.}$$

$$V \approx 46.5 \qquad \begin{array}{l}\text{Multiply each side}\\\text{by 157.}\end{array}$$

∴ The volume is about 46.5 cubic inches.

Exercises

The solids are similar. Find the surface area S or volume V of the red solid.

16.

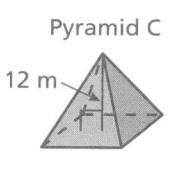
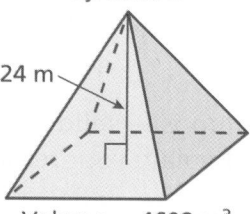

Volume = 4608 m³

17.

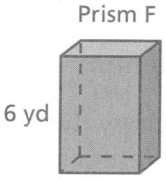
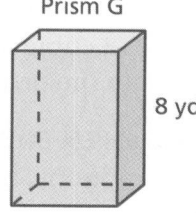

Surface Area = 154 yd²

Find the volume of the solid. Round your answer to the nearest tenth.

1.

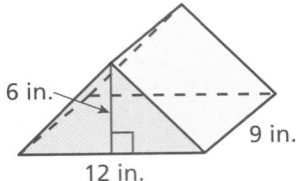

6 in.
9 in.
12 in.

2.

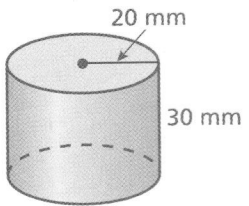

20 mm
30 mm

3.

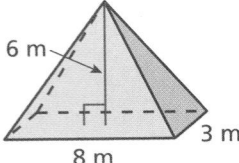

6 m
3 m
8 m

4.

6 cm
3 cm

5.

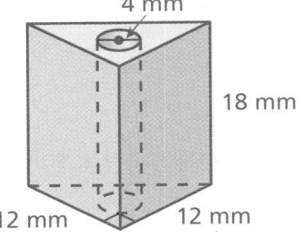

4 mm
18 mm
12 mm   12 mm

6.

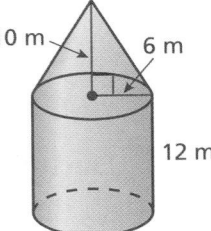

10 m   6 m
12 m

7. The pyramids are similar.

   a. Find the missing dimension.

   b. Find the surface area of the red pyramid.

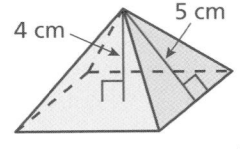

4 cm   5 cm

Surface Area = 96 cm²

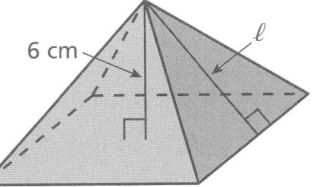

6 cm   ℓ

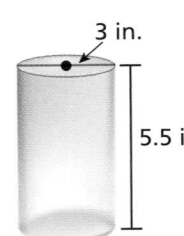

5 in.   3 in.
5 in.   5.5 in.

8. SMOOTHIES You are making smoothies. You will use either the cone-shaped glass or the cylindrical glass. Which glass holds more? About how much more?

9. CAPACITY A baseball team uses a heated tub to treat injuries. What is the capacity of the tub in liters? $(1 \text{ L} = 1000 \text{ cm}^3)$

80 cm
150 cm   150 cm

10. WAFFLE CONES The ratio of the corresponding linear measures of two similar waffle cones is 3 to 4. The smaller cone has a volume of about 18 cubic inches. Find the volume of the larger cone. Round your answer to the nearest tenth.

11. OPEN-ENDED Draw two different composite solids that have the same volume, but different surface areas. Explain your reasoning.

Test Item References

Chapter Test Questions	Section to Review
1, 9	7.1
2, 8	7.2
3	7.3
4, 8	7.4
5, 6, 11	7.5
7, 10	7.6

Test-Taking Strategies

Remind students to quickly look over the entire test before they start so that they can budget their time. This test is very visual and requires that students remember many terms. It might be helpful for them to jot down some of the terms on the back of their test before they start. Students should make sketches and diagrams to help them.

Common Assessment Errors

- **Exercises 1–6, 8–11** Students may write linear or square units for volume rather than cubic units.
- **Exercises 1–6, 8–11** Students may forget to multiply the area of the base by the height and/or the factor of $\frac{1}{3}$ (if appropriate) when finding volume. Encourage them to find the area of the base separately and then substitute it into the volume formula.
- **Exercises 2, 4, 5, 6, 8** When finding the area of the circular base of a cylinder or a cone, students may not square the radius, or they may use the diameter instead of the radius.
- **Exercises 5 and 6** Students may forget to add or subtract the volumes as appropriate to find the volume of the composite solid.
- **Exercises 7 and 10** Students may raise the ratio of the linear measures to the wrong power, or forget to square or cube the ratio altogether. Discuss why squaring or cubing the ratio makes sense.

Reteaching and Enrichment Strategies

If students need help. . .	If students got it. . .
Resources by Chapter • Practice A and Practice B • Puzzle Time Record and Practice Journal Practice Differentiating the Lesson Lesson Tutorials Practice from the Test Generator Skills Review Handbook	Resources by Chapter • Enrichment and Extension • School-to-Work • Financial Literacy Game Closet at *BigIdeasMath.com* Start Standardized Test Practice

Answers

1. 324 in.^{3}
2. $12{,}000\pi \approx 37{,}680$ mm^3
3. 48 m^3
4. $4.5\pi \approx 14.1$ cm^3
5. $1296 - 72\pi \approx 1069.9$ mm^3
6. $552\pi \approx 1733.3$ m^3
7. **a.** $\ell = 7.5$ cm
   b. 216 cm^2
8. cylindrical glass; about 6.2 in.^{3}
9. 1800 L
10. 42.7 in.^{3}
11. See Additional Answers.

Assessment Book

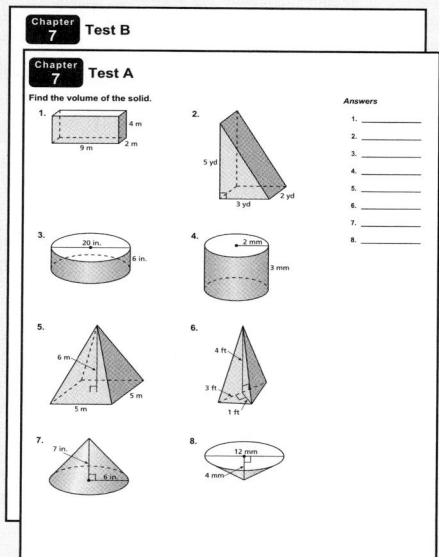

After Answering Easy Questions, Relax

Answer Easy Questions First

Estimate the Answer

Read All Choices before Answering

Read Question before Answering

Solve Directly or Eliminate Choices

Solve Problem before Looking at Choices

Use Intelligent Guessing

Work Backwards

About this Strategy

When taking a multiple choice test, be sure to read each question carefully and thoroughly. After skimming the test and answering the easy questions, stop for a few seconds, take a deep breath, and relax. Work through the remaining questions carefully, using your knowledge and test-taking strategies. Remember, you already completed many of the questions on the test!

Answers

1. A

2. H

3. 480 in.³

Item Analysis

1. **A.** Correct answer

   B. The student thinks that $3 \div \left(-\frac{1}{6}\right) = -\frac{1}{2}$ and finds $\frac{1}{6} + \left(-\frac{1}{2}\right)$.

   C. The student thinks that $\frac{1}{2} \div 3 = \frac{3}{2}$ and that $3 \div \left(-\frac{1}{6}\right) = -\frac{1}{2}$ and finds $\frac{3}{2} + \left(-\frac{1}{2}\right)$.

   D. The student makes a sign error in an otherwise correct sum.

2. **F.** The student incorrectly uses half the radius.

   G. The student incorrectly uses half the radius and also uses the formula for a right circular cylinder, neglecting to multiply by $\frac{1}{3}$.

   H. Correct answer

   I. The student uses the formula for a right circular cylinder, neglecting to multiply by $\frac{1}{3}$.

3. **Gridded Response:** Correct answer: 480 in.³

   Common Error: The student finds the base area to be the product of 8 and 6, not half the product, and gets an answer of 960 cubic inches.

4. **A.** The student chooses a graph that represents wind speeds that are less than or equal to 39 miles per hour or greater than 74 miles per hour.

   B. The student chooses a graph that represents wind speeds that are less than 39 miles per hour or greater than or equal to 74 miles per hour.

   C. Correct answer

   D. The student chooses a graph that represents wind speeds that are greater than 39 miles per hour and at most 74 miles per hour.

5. **F.** Correct answer

   G. The student does not correctly apply the concept of slope and does not determine that answer choice F does not lie on the line.

   H. The student does not correctly apply the concept of slope and does not determine that answer choice F does not lie on the line.

   I. The student does not correctly apply the concept of slope and does not determine that answer choice F does not lie on the line.

Technology
For the
Teacher

Big Ideas Test Generator

1. What is the value of the expression below when $h = \frac{1}{2}$ and $k = -\frac{1}{6}$?

$$h \div 3 + 3 \div k$$

A. $-17\frac{5}{6}$

B. $-\frac{1}{3}$

C. 1

D. $17\frac{5}{6}$

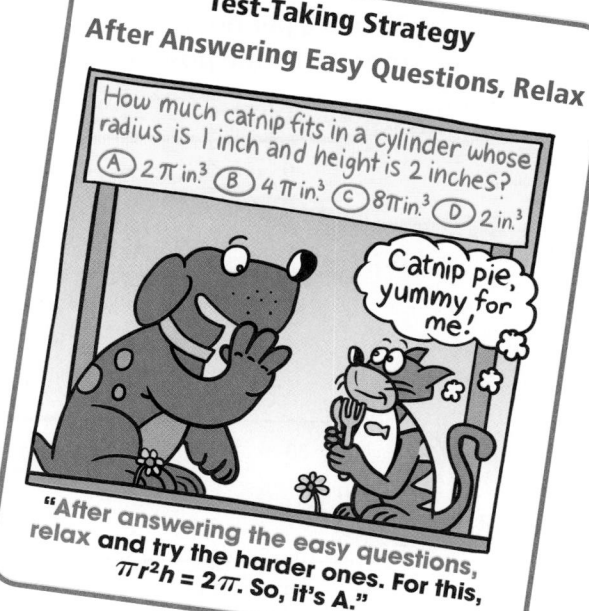
2. A right circular cone and its dimensions are shown below.

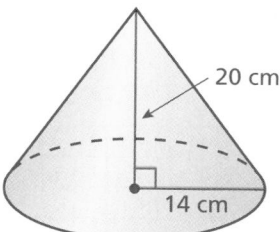

20 cm

14 cm

What is the volume of the right circular cone? $\left(\text{Use } \frac{22}{7} \text{ for } \pi. \right)$

F. $1,026\frac{2}{3}$ cm³

G. $3,080$ cm³

H. $4,106\frac{2}{3}$ cm³

I. $12,320$ cm³

3. A right triangular prism and its dimensions are shown below.

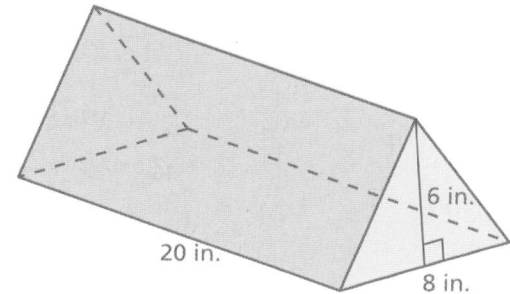

6 in.

20 in.

8 in.

What is the volume of the right triangular prism?

4. A tropical storm has maximum sustained surface winds of at least 39 miles per hour but less than 74 miles per hour. Which graph correctly represents the possible wind speeds of a tropical storm?

A.

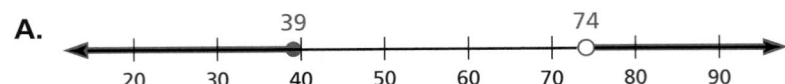

B.

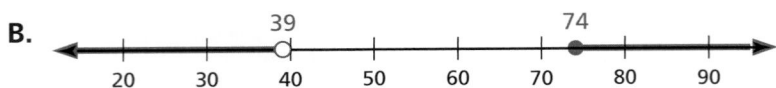

C.

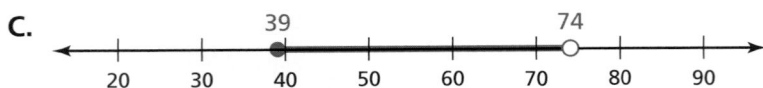

D.

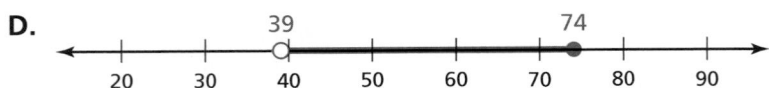

5. Use the coordinate plane to answer the question below.

Which point does *not* lie on the same line as the other three?

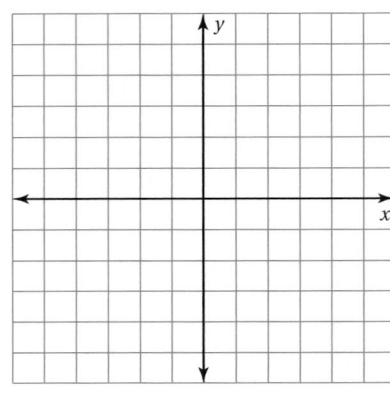

F. $(-5, 3)$

H. $(-1, -1)$

G. $(-3, 2)$

I. $(1, -4)$

6. Olga was solving an equation in the box shown.

What should Olga do to correct the error that she made?

A. Multiply both sides by $-\dfrac{5}{2}$ instead of $-\dfrac{2}{5}$.

B. Multiply both sides by $\dfrac{2}{5}$ instead of $-\dfrac{2}{5}$.

C. Distribute $-\dfrac{2}{5}$ to get $-4x - 6$.

D. Add 15 to -30.

$$-\frac{2}{5}(10x - 15) = -30$$

$$10x - 15 = -30\left(-\frac{2}{5}\right)$$

$$10x - 15 = 12$$

$$10x - 15 + 15 = 12 + 15$$

$$10x = 27$$

$$\frac{10x}{10} = \frac{27}{10}$$

$$x = \frac{27}{10}$$

Item Analysis (continued)

6. **A.** Correct answer

   B. The student thinks that multiplying by $\frac{2}{5}$ is the inverse operation of multiplying by $-\frac{2}{5}$.

   C. The student does not distribute the negative sign to the second term.

   D. The student makes an order of operations error by not first distributing the multiplication.

7. **F.** The student divides the number of inches per hour by the number of hours.

   G. Correct answer

   H. The student divides the number of hours by the number of inches per hour and makes a place value error.

   I. The student thinks that $2\frac{1}{2}$ is equivalent to 0.25 and adds this number to 0.08.

8. **Gridded Response:** Correct answer: 9 in.^{3}

   Common Error: The student divides the volume by r instead of r^2, getting an answer of 108 cubic inches.

9. **A.** The student finds $\frac{6}{9}$ of 240 and subtracts this result from 240.

   B. The student uses direct variation and finds $\frac{6}{9}$ of 240.

   C. Correct answer

   D. The student finds $\frac{6}{9}$ of 240 and adds this result to 240.

10. **F.** The student thinks that 12 students are equivalent to 12%.

    G. Correct answer

    H. The student finds what percent 12 is of 20.

    I. The student finds the percent of students who prefer to write with a pen.

Answers

4. C

5. F

6. A

Answers

7. G

8. 9 in.^{3}

9. C

10. G

11. 113.04 in.^{3}

Answers for Extra Examples

1. A. The student chooses the same rectangle with the vertices labeled with different points.

   B. Correct answer

   C. The student reflects the rectangle in the *y*-axis.

   D. The student rotates the rectangle 180°.

2. F. Correct answer

   G. The student subtracts $\frac{3}{8}$ from $\frac{2}{3}$ and then multiplies this difference by 72.

   H. The student finds $\frac{2}{3}$ of 72.

   I. The student finds the number of singers who are neither women nor sopranos.

Item Analysis (continued)

11. 2 points The student demonstrates a thorough understanding of finding the composite volume of a right circular cylinder and a right circular cone. The student correctly finds a composite volume of 113.04 cubic inches. The student provides clear and complete work and explanations.

1 point The student demonstrates a partial understanding of finding the composite volume of a right circular cylinder and a right circular cone. The student provides some correct work and explanation toward finding the composite volume.

0 points The student demonstrates insufficient understanding of finding the composite volume of a right circular cylinder and a right circular cone. The student is unable to make any meaningful progress toward finding the composite volume.

Extra Examples for Standardized Test Practice

1. On the grid below, rectangle *EFGH* is plotted and its vertices are labeled.

Which of the following shows rectangle *E′F′G′H′*, the image of rectangle *EFGH* after it is reflected in the *x*-axis?

A.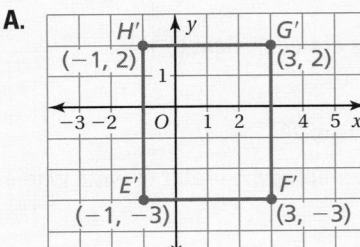

C.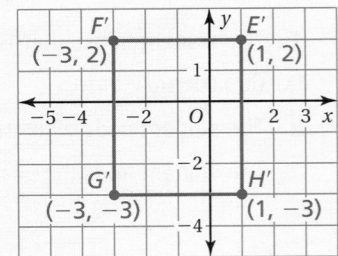

B.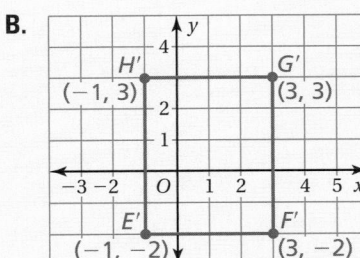

D.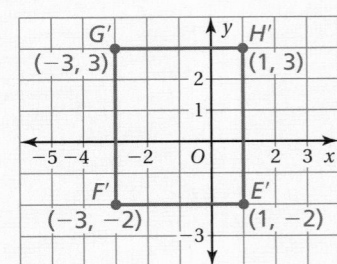

2. There are 72 singers in a choir. The conductor determined that $\frac{2}{3}$ of the singers are women and that $\frac{3}{8}$ of the women are sopranos. How many women in the choir are sopranos?

   F. 18

   G. 21

   H. 48

   I. 54

7. It has been raining at a rate of 0.08 inch per hour. At this rate, how much rain will fall in $2\frac{1}{2}$ hours?

    F. 0.032 in.   **H.** 0.3125 in.

    G. 0.2 in.   **I.** 0.33 in.

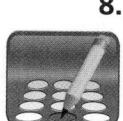

8. A right circular cylinder has a volume of 1296 cubic inches. If the radius of the cylinder is divided by 12, what would be the volume, in cubic inches, of the smaller cylinder?

9. If 9 friends share equally a large box of baseball cards, each friend gets 240 cards. If 6 friends share equally the same box of cards, how many cards does each friend get?

    A. 80   **C.** 360

    B. 160   **D.** 400

10. All students in a class were surveyed to find out their preferences for writing instruments. The survey found that 12 students prefer to write with a pencil and 20 students prefer to write with a pen. What percent of students in the class prefer to write with a pencil?

    F. 12%   **H.** 60%

    G. 37.5%   **I.** 62.5%

11. The figure below is a diagram for making a tin lantern.

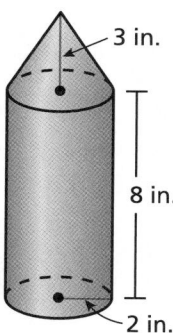

3 in.

8 in.

2 in.

The figure consists of a right circular cylinder without its top base and a right circular cone without its base. What is the volume, in cubic inches, of the entire lantern? Show your work and explain your reasoning. (Use 3.14 for π.)

8 Data Analysis and Samples

"I took a survey of pet owners on how many times per day you should treat your dog to a biscuit."

"What do you think?"

"I've completed a circle graph analyzing what you do each day."

Connections to Previous Learning

- Construct and analyze line graphs and double bar graphs.
- Differentiate between continuous and discrete data, and choose ways to represent it.

- Determine measures of central tendency including mean, median, mode, and range (variability).
- Select appropriate measures of central tendency to describe a data set.

- Evaluate the reasonableness of a sample to determine the appropriateness of generalizations made about the population.
- Construct and analyze histograms, stem-and-leaf plots, and circle graphs.

Math in History

The word "mathematics" comes from the Greek *máthema*, which means learning, study, and science. The plural form in English, like the French plural form *les mathématiques*, goes back to the Latin plural *mathematica* which is based on the Greek plural *mathematiká* used by Aristotle, and meaning roughly "all things mathematical."

Do you call it "math" or "maths"?

★ In American English, mathematics is treated as singular. For instance, in American English it would be proper to say "Mathematics *is* my favorite topic." So, when the word was shortened, it was shortened to "math" and the sentence would become "Math is my favorite topic."

★ In British English, mathematics is treated as plural. So, in British English, it would be proper to say "Mathematics *are* my favorite topics." Because of this, the abbreviated version of the word became plural and the sentence would become "Maths are my favorite topics."

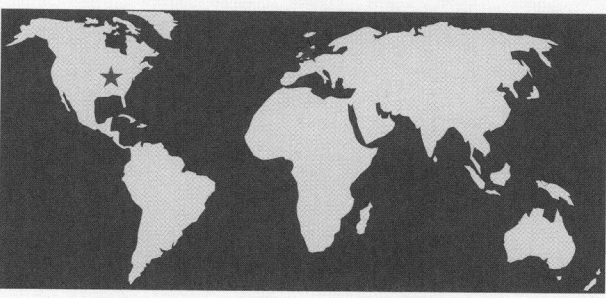

Pacing Guide for Chapter 8

Chapter Opener	1 Day
Section 1 Activity Lesson	 1 Day 1 Day
Section 2 Activity Lesson	 1 Day 1 Day
Study Help / Quiz	1 Day
Section 3 Activity Lesson	 1 Day 1 Day
Section 4 Activity Lesson Lesson b	 1 Day 1 Day 1 Day
Quiz / Chapter Review	1 Day
Chapter Test	1 Day
Standardized Test Practice	1 Day
Total Chapter 8	14 Days
Year-to-Date	145 Days

Check Your Resources

- Record and Practice Journal
- Resources by Chapter
- Skills Review Handbook
- Assessment Book
- Worked-Out Solutions

The Dynamic Planning Tool
Editable Teacher's Resources at
BigIdeasMath.com

Math Background Notes

Additional Topics for Review

- Bar graphs
- Common degrees in a circle
- Place value

Try It Yourself

1. 4.69
2. 5.7
3. 4.55, 4.65, 4.75
4. 0.6

Record and Practice Journal

1. a. 5   b. 4.5
   c. 3   d. 5
2. a. 825   b. 550
   c. none   d. 1800
3. a. 18   b. 19
   c. 19   d. 11
4. a. 22.5
   b. 20
   c. 10 and 20
   d. 40
5. a. 8.4   b. 8
   c. none   d. 5
6. a. $2\frac{3}{7}$ b. 3
   c. 3   d. 5
7. mean: 103; median: 105.5; mode: none; range: 23
8. a. mean: 83; median: 82.5; modes: 84 and 79; range: 29
   b. mean or median
   c. *Sample answer:* If both of the students have low scores, then it will decrease the mean and median. If both of the students have high scores, then it will increase the mean and median. If the scores are split or average, the answers could stay the same.

Vocabulary Review

- Mean
- Median
- Mode
- Range

Finding Mean, Median, Mode, and Range

- Students should know how to find the mean, median, mode, and range of data.
- Remind students that the mean is what most people refer to as the average of the data.
- Remind students that finding the median of a data set with an odd number of data entries is different than finding the median of a data set with an even number of data values. To find the median of an odd number of data values, order the data and select the value in the middle. To find the median of an even number of data values, order the data and find the mean (average) of the middle two values.
- **Common Error:** In Example 2, many students will forget to order the data before finding the median. Remind them that ordering the data is essential to finding the median. Ordered data will also make identifying the greatest and least values easier when it comes time to calculate the range. In this chapter, students will learn that ordering the data is also a helpful way to identify outliers.
- The mode of a data set is the value that occurs most often. It is possible for a data set to have more than one mode.
- If a data set has two values that occur with equal frequency, the data set is called bimodal.

Reteaching and Enrichment Strategies

If students need help. . .	If students got it. . .
Record and Practice Journal • Fair Game Review Skills Review Handbook Lesson Tutorials	Game Closet at *BigIdeasMath.com* Start the next section

What You Learned Before

"Mom, my owner, and Fluffy have agreed to participate in my random survey. Will you be my fourth participant?"

Finding Mean, Median, Mode, and Range

The table shows the top ten Olympic pole vault heights for men and women.

Example 1   What is the mean of the men's data?

$$\text{mean} = \frac{\text{sum of data}}{\text{number of data values}}$$

$$= \frac{56.71}{10} \approx 5.67$$

:: So, the mean height is about 5.67 meters.

Example 2   What is the median of the women's data?

4.45, 4.55, 4.55, 4.65, $\underbrace{4.65, 4.70}$, 4.75, 4.75, 4.80, 5.05

$$\frac{4.65 + 4.70}{2} \approx 4.68$$

:: So, the median is about 4.68 meters.

Example 3   What is the mode of the men's data?

:: Because it occurs most often, the mode is 5.70 meters.

Example 4   What is the range of the men's data?

range = greatest data value − least data value

$$= 5.96 - 5.45 = 0.51$$

:: So, the range is 0.51 meter.

Olympic Pole Vault Heights (meters)	
Men	**Women**
5.45	4.55
5.60	5.05
5.96	4.75
5.70	4.65
5.45	4.65
5.70	4.55
5.60	4.70
5.70	4.75
5.85	4.80
5.70	4.45

Try It Yourself

Use the table to answer the question. Round your answer to the nearest hundredth.

1. What is the mean of the women's data?

2. What is the median of the men's data?

3. What is the mode(s) of the women's data?

4. What is the range of the women's data?

Essential Question How can you use a stem-and-leaf plot to organize a set of numbers?

1 ACTIVITY: Decoding a Graph

Work with a partner. You intercept a secret message that contains two different types of plots. You suspect that each plot represents the same data. The graph with the dots indicates only ranges for the numbers.

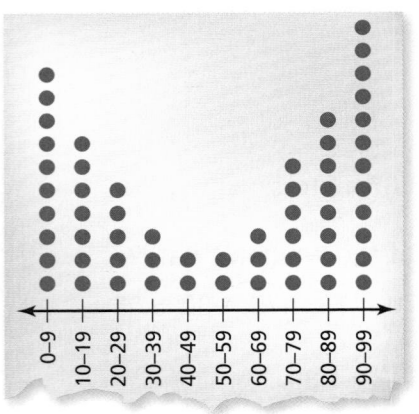

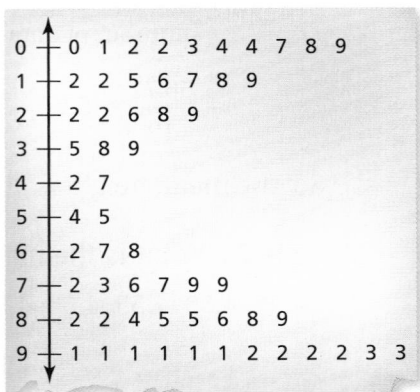

a. How many numbers are in the data set? How can you tell?

b. How many numbers are greater than or equal to 90? How can you tell?

c. Is 91 in the data set? If so, how many times is it in the set? How can you tell?

d. Make a list of all of the numbers in the data set.

e. You intercept a new secret message. Use the secret code shown below to decode the message.

Secret Code

A = 29	F = 31	K = 18	P = 4	U = 19
B = 33	G = 8	L = 26	Q = 10	V = 17
C = 7	H = 16	M = 22	R = 21	W = 12
D = 20	I = 5	N = 3	S = 2	X = 25
E = 15	J = 11	O = 9	T = 32	Y = 13
				Z = 1

‾‾ ‾‾ ‾‾ ‾‾ ‾‾ ‾‾ ‾‾ ‾‾ ‾‾ ‾‾ ‾‾ ‾‾ ‾‾ ‾‾ ‾‾ ‾‾ ‾‾ ‾‾ ‾‾
32 16 15   2 32 15 22   2 16 9 12 2   32 16 15   32 15 3 2

‾‾ ‾‾ ‾‾ ‾‾ ‾‾ ‾‾ ‾‾ ‾‾ ‾‾ ‾‾ ‾‾ ‾‾ ‾‾ ‾‾ ‾‾ ‾‾ ‾‾ ‾‾ ‾‾ ‾‾
32 16 15   26 15 29 17 15 2   2 16 9 12   32 16 15   9 3 15 2

Laurie's Notes

Introduction

For the Teacher

- **Goal:** Students will explore the features of a stem-and-leaf plot.
- As you prepare for this chapter, consider the knowledge and skills related to data analysis that your students have learned previously.
- In this chapter, students will learn about new types of graphical displays.

Motivate

- Make a stem-and-leaf plot using sticky notes. As students enter, hand each one a sticky note. Draw a vertical line on the board and use stems of 0, 1, 2, and 3. Ask students to write down the day of the month they were born, making their number large enough and dark enough to be seen from a distance.
- Ask those students born before the 10th of the month to come forward and put their sticky notes in a row adjacent to the 0 stem.
- Repeat for each of the stems; 10th–19th, 20th–29th, and 30th–31st.
- **?** "What observations can you make about the data?" most likely there are fewer in the 30s; would expect randomness and balanced distribution over other 3 stems, but it might not be!
- Do not sort—this will be done at the end of the class. Explain that today they will explore a new type of plot similar to what they have made with their sticky notes.

Activity Notes

Activity 1

- Although the names of these plots are not given, it is reasonable to assume that students will still know how to read information from each plot.
- **?** "What type of graph does the plot on the left remind you of? Explain." bar graph; If a rectangle enclosed the column of dots it would resemble a vertical bar graph.
- **?** "What do you think the two dots in the middle over 40–49 mean?" *Sample answer:* maybe two people between the ages of 40–49, or two weeks where there were between 40–49 sales, etc.
- **?** "If the two plots represent the same data, where would you find those same two blue dots?" Students should recognize the similarity in the shape of the graphs and guess that the 4 on the left is probably the tens digit.
- **?** "The same data *are* represented in each plot. How are the two plots the same and how do they differ?" They have the same shape, like a U. You don't know the actual numbers in the dot plot, they are just in a range.
- **Big Idea**: When you make a dot plot, you lose the specific numbers. A stem-and-leaf plot shows the distribution *and* retains the actual numbers. In this example, both plots show that the distribution is bimodal.
- Explain that the plot on the right is called a stem-and-leaf plot. The secret message spells out how to read the plot.

Previous Learning

Students have constructed and interpreted bar graphs and pictographs. Students have learned about mean, median, and distribution.

Activity Materials	
Introduction	**Textbook**
• sticky notes	• number cubes

Start Thinking! and Warm Up

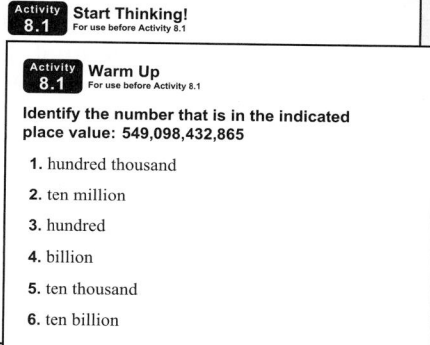

Activity 8.1 Start Thinking!
For use before Activity 8.1

Activity 8.1 Warm Up
For use before Activity 8.1

Identify the number that is in the indicated place value: 549,098,432,865

1. hundred thousand
2. ten million
3. hundred
4. billion
5. ten thousand
6. ten billion

8.1 Record and Practice Journal

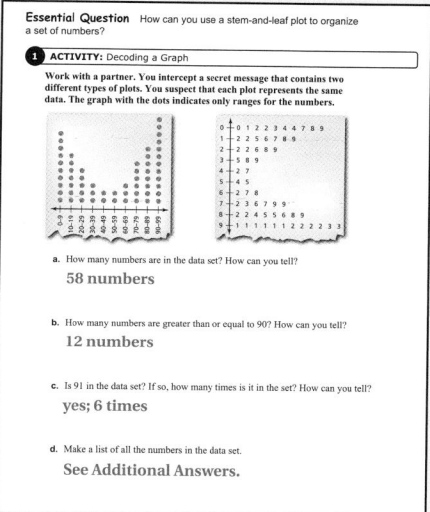

Essential Question How can you use a stem-and-leaf plot to organize a set of numbers?

1 ACTIVITY: Decoding a Graph

Work with a partner. You intercept a secret message that contains two different types of plots. You suspect that each plot represents the same data. The graph with the dots indicates only ranges for the numbers.

a. How many numbers are in the data set? How can you tell?
58 numbers

b. How many numbers are greater than or equal to 90? How can you tell?
12 numbers

c. Is 91 in the data set? If so, how many times is it in the set? How can you tell?
yes; 6 times

d. Make a list of all the numbers in the data set.
See Additional Answers.

Differentiated Instruction

Kinesthetic

To make the stem-and-leaf plot in Activity 2, have students copy the data onto individual pieces of paper. Then sort the data into piles based on the stem value. For each pile, cut one of the pieces to separate the stem and leaf, then tape the stem to the board. Cut the remaining leaves with the same stem and tape them to the board in order from least to greatest. Repeat the process for each of the piles to complete the stem-and-leaf plot.

8.1 Record and Practice Journal

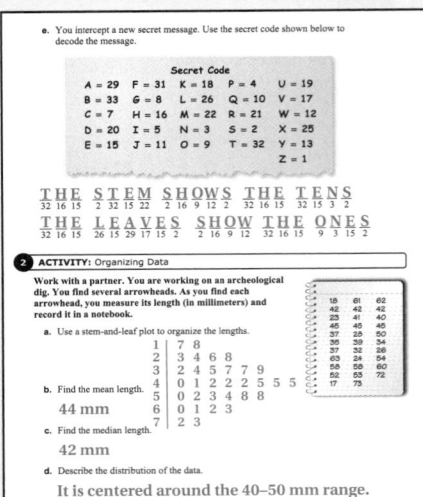

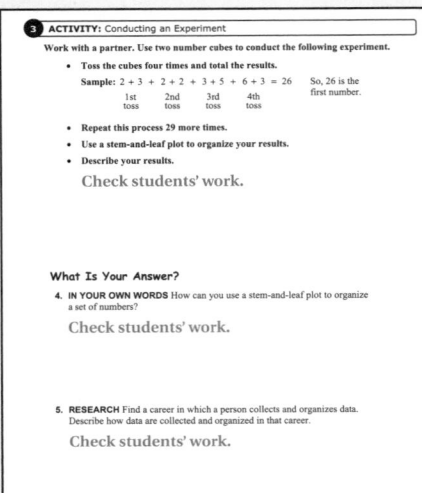

Activity 2

- Refer to the stem-and-leaf plot from Activity 1 to explain how to make a stem-and-leaf plot for this activity.
- **?** Ask questions to start the stem-and-leaf plot.
  - "How long is a millimeter? What would 18 millimeters look like?" A millimeter is about the thickness of a fingernail. 18 millimeters might be about the diameter of a nickel.
  - "What is the range of this data set?" $73 - 17 = 56$
  - "What will you need for stems in this plot?" 1 through 7
- Give sufficient time for students to complete the plot and find the mean and median.
- **?** "How does the stem-and-leaf plot help you to organize the data?" It sorts the data into sections of width 10, and then within each section the data is sorted in ascending order.
- **?** "Circle the median. How did the plot help you find the median?" Listen for an understanding that the data are sorted so they could count to the middle of the set.

Activity 3

- If time is running short, you may want to limit the number of trials. Each pair of students could generate 3–5 totals and then combine data with the rest of the class. Record the results on the board or the overhead.
- **?** "What is the least number you can toss?" 2 "Greatest?" 12
- **?** "What is the least sum you could have after 4 tosses?" 8 "Greatest?" 48
- When students have finished, ask a few students to draw their plots on the board (unless there is class data).
- Discuss the results.

What Is Your Answer?

- Have students work in pairs.

Closure

- Return to the sticky note plot made when motivating today's lesson. Ask a volunteer to sort the data. Now ask students to describe the plot, along with finding the median, mode, and range.

Technology For the Teacher

Dynamic Classroom

The Dynamic Planning Tool
Editable Teacher's Resources at *BigIdeasMath.com*

2 ACTIVITY: Organizing Data

Work with a partner. You are working on an archeological dig. You find several arrowheads.

18	61	62
42	42	42
23	41	40
45	45	45
37	28	50
35	39	34
37	32	26
63	24	54
58	58	60
52	53	72
17	73	

50 mm

As you find each arrowhead, you measure its length (in millimeters) and record it in a notebook.

   a. Use a stem-and-leaf plot to organize the lengths.

   b. Find the mean length.

   c. Find the median length.

   d. Describe the distribution of the data.

3 ACTIVITY: Conducting an Experiment

Work with a partner. Use two number cubes to conduct the following experiment.

● **Toss the cubes four times and total the results.**

   Sample: $2 + 3 \;+\; 2 + 2 \;+\; 3 + 5 \;+\; 6 + 3 \;= 26$

   1st toss   2nd toss   3rd toss   4th toss

   So, 26 is the first number.

● **Repeat this process 29 more times.**

● **Use a stem-and-leaf plot to organize your results.**

● **Describe your results.**

What Is Your Answer?

4. IN YOUR OWN WORDS How can you use a stem-and-leaf plot to organize a set of numbers?

5. RESEARCH Find a career in which a person collects and organizes data. Describe how data are collected and organized in that career.

Practice

Use what you learned about stem-and-leaf plots to complete Exercises 4–7 on page 352.

Check It Out
Lesson Tutorials
BigIdeasMath com

Key Vocabulary
stem-and-leaf plot,
   p. 350
stem, *p. 350*
leaf, *p. 350*

 **Key Idea**

Stem-and-Leaf Plots

A **stem-and-leaf plot** uses the digits of data values to organize a data set. Each data value is broken into a **stem** (digit or digits on the left) and a **leaf** (digit or digits on the right).

A stem-and-leaf plot shows how data are distributed.

Stem	Leaf
2	0 0 1 2 5 7
3	1 4 8
4	2
5	8 9

Key: $2|0 = 20$

The *key* explains what the stems and leaves represent.

EXAMPLE 1 **Making a Stem-and-Leaf Plot**

	A	B
1	DATE	MINUTES
2	JULY 9	55
3	JULY 9	3
4	JULY 9	6
5	JULY 10	14
6	JULY 10	18
7	JULY 10	5
8	JULY 10	23
9	JULY 11	30
10	JULY 11	23
11	JULY 11	10
12	JULY 11	2
13	JULY 11	36

Make a stem-and-leaf plot of the length of the 12 cell phone calls.

Step 1: Order the data.

2, 3, 5, 6, 10, 14, 18, 23, 23, 30, 36, 55

Step 2: Choose the stems and leaves. Because the data values range from 2 to 55, use the *tens* digits for the stems and the *ones* digits for the leaves.

Step 3: Write the stems to the *left* of the vertical line.

Step 4: Write the leaves for each stem to the *right* of the vertical line.

Phone call lengths

Order the stems vertically. The stem for data values less than 10 is 0.

Include stems without leaves.

Stem	Leaf
0	2 3 5 6
1	0 4 8
2	3 3
3	0 6
4	
5	5

Write the leaves horizontally.

Key: $1|4 = 14$ **minutes**

On Your Own

Now You're Ready
Exercises 8–11

1. Make a stem-and-leaf plot of the hair lengths.

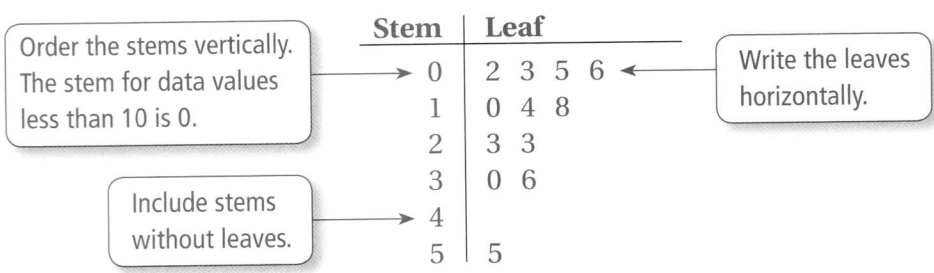

Hair Length (centimeters)									
5	1	20	12	27	2	30	5	7	38
40	47	1	2	1	32	4	44	33	23

 Multi-Language Glossary at BigIdeasMath com.

Laurie's Notes

Introduction

Connect
- **Yesterday:** Students explored the features of a stem-and-leaf plot.
- **Today:** Students will construct and interpret stem-and-leaf plots.

Motivate
- Share a little trivia about cell phones, the context for Example 1.
  - 1973—Dr. Martin Cooper is considered the inventor of the first portable handset. Dr. Cooper, former general manager for the systems division at Motorola, was the first person to make a call on a cell phone.
  - 1977—Cell phones go public. The first trials of cell phone testing began in the city of Chicago with 2000 customers. Japan began testing cellular phone service in 1979.

Lesson Notes

Key Idea
- A common question that students ask is "what if the data are not two-digit numbers?"
  - If there are 3-digit numbers within a small range (i.e., 431–476), the stem can be two digits (43, 44, 45, 46, 47).
  - If there are decimals (i.e., 3.4), the stem can be the whole number portion and the leaf can be the decimal portion.

Example 1
- By sorting the data first, the leaves are arranged in order when written to the right of the stem.
- Discuss with students the need to have a key that describes how to read the data in the plot.
- **Common Error:** When a data value repeats, remind students that the leaf must be listed again. The number of leaves must equal the number of data values in the set.
- **?** "Describe the stem-and-leaf plot. What does the plot tell you about the data?" Data is skewed towards the lower data. It tapers at the upper end. There is a gap from 36 to 55.
- **Big Idea:** Because the stem-and-leaf plot shows how data are distributed, a stem must be included in the plot even if there are no data values in that interval. In this example, the stem of "4" is still included to show the gap in the data.

On Your Own
- **Think-Pair-Share:** Students should read each question independently and then work with a partner to answer the questions. When they have answered the questions, the pair should compare their answers with another group and discuss any discrepancies.
- **?** "Describe what the plot tells you about the data." There are some lesser data values and greater data values with few in the middle.
- **?** "Explain why this might be the case." perhaps girls versus boys

Goal Today's lesson is constructing and interpreting stem-and-leaf plots.

Start Thinking! and Warm Up

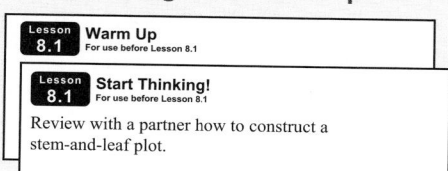

Lesson 8.1 **Warm Up** For use before Lesson 8.1

Lesson 8.1 **Start Thinking!** For use before Lesson 8.1
Review with a partner how to construct a stem-and-leaf plot.

Extra Example 1
Make a stem-and-leaf plot of the length of the eleven fish, in inches.
7, 12, 20, 14, 20, 25, 8, 18, 16, 20, 14

Fish Length

Stem	Leaf
0	7 8
1	2 4 4 6 8
2	0 0 0 5

Key: 1 | 8 = 18 inches

On Your Own

1. **Hair Length**

Stem	Leaf
0	1 1 1 2 2 4 5 5 7
1	2
2	0 3 7
3	0 2 3 8
4	0 4 7

Key: 1 | 2 = 12 cm

Extra Example 2

Use the stem-and-leaf plot of student test scores in Example 2.

a. How many students scored more than 85 points? 8

b. How many students scored at most 75 points? 3

c. What is the median test score? 84

 On Your Own

    2. **a.** 9 students

          b. 4 students

Extra Example 3

Use the stem-and-leaf plot in Example 2. Which statement is *not* true? B

A. The range of test scores is 34.

B. The mean test score is 85.

C. The test scores that occur most often are 81, 84, and 88.

D. Most of the test scores are over 80.

 On Your Own

    3. yes; From the stem-and-leaf plot, there are plants that are 22 inches, 25 inches, and 36 inches tall.

English Language Learners

Pair Activity

Form small groups of English learners and English speakers. Have them consider different types of data such as golf scores, the number of text messages each person in class sends in a week, the number of sixth, seventh, and eighth graders that ride the school bus, the numbers rolled on a number cube, and the number of hours that students in a class slept last night. Instruct students to discuss whether a stem-and-leaf plot would be an appropriate way to display the data.

Laurie's Notes

Example 2

- **Big Idea:** The leaves in the stem-and-leaf plot are sorted from least to greatest. The range can be found quickly, along with the median. To find the median, alternate counting in from each end of the data (the least and the greatest).

? "How many students are represented in the plot?" 18

? "What is the median? Explain how you decided." 84; There are 18 numbers so the middle will be the average of the 9th and 10th numbers, which are both 84.

On Your Own

- **Extension:** Ask students to describe the plot.

Example 3

- This example reviews data analysis concepts. You may need to remind students how to find the mean, median, mode, and range.

On Your Own

- **Neighbor Check:** Have students work independently and then have their neighbor check their work. Have students discuss any discrepancies.

Closure

- Explain how a stem-and-leaf plot is similar to a bar graph and how it differs from a bar graph. *Sample answer:* A stem-and-leaf plot shows the distribution of the data and the frequency of different intervals for the data, similar to a bar graph. However, all of the actual data values can be read in a stem-and-leaf plot, but not in a bar graph.

Technology For the Teacher

The Dynamic Planning Tool
Editable Teacher's Resources at *BigIdeasMath.com*

EXAMPLE 2 **Interpreting a Stem-and-Leaf Plot**

Test Scores

Stem	Leaf
6	6
7	0 5 7 8
8	1 1 3 4 4 6 8 8 9
9	0 2 9
10	0

Key: 9 | 2 = 92 points

The stem-and-leaf plot shows student test scores. (a) How many students scored less than 80 points? (b) How many students scored at least 90 points? (c) How are the data distributed?

a. There are five scores less than 80 points: 66, 70, 75, 77, and 78.

> Five students scored less than 80 points.

b. There are four scores of at least 90 points: 90, 92, 99, and 100.

> Four students scored at least 90 points.

c. There are few low test scores and few high test scores. So, most of the scores are in the middle.

On Your Own

Now You're Ready
Exercises 16–19

2. Use the grading scale at the right.

a. How many students received a B on the test?

b. How many students received a C on the test?

A:	90–100
B:	80–89
C:	70–79
D:	60–69
F:	59 and below

EXAMPLE 3 **Standardized Test Practice**

Which statement is *not* true?

Ⓐ Most of the plants are less than 20 inches tall.

Ⓑ The median plant height is 11 inches.

Ⓒ The range of the plant heights is 35 inches.

Ⓓ The plant height that occurs most often is 11 inches.

Plant Heights

Stem	Leaf
0	1 2 4 5 6 8 9
1	0 1 1 5 7
2	2 5
3	6

Key: 1 | 5 = 15 inches

There are 15 plant heights. So, the median is the eighth data value, 10 inches.

> The correct answer is Ⓑ.

On Your Own

3. You are told that three plants are taller than 20 inches. Is the statement true? Explain.

 ## Vocabulary and Concept Check

1. **VOCABULARY** The key for a stem-and-leaf plot is $3\,|\,4 = 34$. Which number is the stem? the leaf?

2. **WRITING** Describe how to make a stem-and-leaf plot for the data values 14, 22, 9, 13, 30, 8, 25, and 29.

3. **WRITING** How does a stem-and-leaf plot show the distribution of data?

 ## Practice and Problem Solving

Use the stem-and-leaf plot at the right.

4. How many data values are in the set?

5. What is the least value? greatest value?

6. What is the median? range?

7. Is the value 32 in the set? Explain.

Stem	Leaf
0	4 6 8
1	0
2	3 4
3	0 6 6 9
4	2

Key: $3\,|\,6 = 36$

Make a stem-and-leaf plot of the data.

 8.

Books Read

26	15	20	9
31	25	29	32
17	26	19	40

9.

Hours Online

8	12	21	14
18	6	15	24
12	17	2	0

10.

Test Scores (%)

87	82	95	91	69
88	68	87	65	81
97	85	80	90	62

11.

Points Scored

58	50	42	71	75
45	51	43	38	71
42	70	56	58	43

12. **ERROR ANALYSIS** Describe and correct the error in making a stem-and-leaf plot of the data.

51, 25, 47, 42, 55, 26, 50, 44, 55

Stem	Leaf
2	5 6
4	2 4 7
5	0 1 5 5

Key: $4\,|\,2 = 42$

13. **PUPPIES** The weights (in pounds) of eight puppies at a pet store are 12, 24, 17, 8, 18, 31, 24, and 15. Make a stem-and-leaf plot of the data. Describe the distribution of the data.

Assignment Guide and Homework Check

Level	Day 1 Activity Assignment	Day 2 Lesson Assignment	Homework Check
Basic	4–7, 23–27	1–3, 8–13, 15	8, 12, 13, 15
Average	4–7, 23–27	1–3, 9–15 odd, 12, 16–20	9, 12, 15, 16
Advanced	4–7, 23–27	1–3, 12, 14, 16–22	12, 14, 16, 21

Common Errors

- **Exercises 8–11** Students may forget to include the numbers that have zeros in the ones place in the leaf part of the plot. Remind them that they should be able to read the numbers in the data set by reading stem *and* leaf.
- **Exercises 8–11** Students may not include repeats of numbers. Remind them that the plot represents all of the data values, so they should be able to count the values in the leaf part and have all of the data accounted for.
- **Exercises 8–11** Students may forget to include stems that have no data values. Remind them of Example 1. It is necessary to include the stems with no data to help answer questions about the data set.
- **Exercises 16–19** Students may need to be reminded of the definitions of some of the terms so that they can answer the questions. Give an example of each term and how to find it using a stem-and-leaf plot.

8.1 Record and Practice Journal

Make a stem-and-leaf plot of the data.

1.

Class Sizes			
12	10	21	28
9	16	19	16
25	32	14	21

```
0 | 9
1 | 0 2 4 6 6 9
2 | 1 1 5 8
3 | 2
```

2.

Minutes Spent on Homework			
75	82	91	68
92	86	79	76
75	81	88	60

```
6 | 0 8
7 | 5 5 6 9
8 | 1 2 6 8
9 | 1 2
```

3. The number of text messages from eight phones are 8, 11, 14, 22, 5, 15, 7, and 20. Make a stem-and-leaf plot of the data. Describe the distribution of the data.

```
0 | 5 7 8
1 | 1 4 5
2 | 0 2
```
The data is evenly distributed.

4. The number of minutes seven members spent at band practice are 57, 49, 55, 62, 78, 72, and 75. Make a stem-and-leaf plot of the data. Describe the distribution of the data.

```
4 | 9
5 | 5 7
6 | 2
7 | 2 5 8
```
The data shows that few students do less than 50 minutes of practice.

5. The stem-and-leaf plot shows the number of miles students travel to get to school.

a. How many students travel more than 15 miles?

6 students

b. Find the mean, median, mode, and range of the data.

```
Stem | Leaf
0    | 5 7
1    | 2 4 8
2    | 0 1 5 7
3    | 3
```
Key: 1 | 4 = 14 miles

mean: 18.2
median: 19
mode: none
range: 28

Technology For the Teacher
Answer Presentation Tool
QuizShow

Vocabulary and Concept Check

1. 3 is the stem; 4 is the leaf

2. Write the data set in order from least to greatest.
8, 9, 13, 14, 22, 25, 29, 30
Use the tens digits for the stems and the ones digits for the leaves. Write the stems to the left of the vertical line. Then write the leaves to the right of the vertical line.

```
0 | 8 9
1 | 3 4
2 | 2 5 9
3 | 0
```
Key: 1 | 3 = 13

3. From the leaves, you can see where most of the data lies and whether there are many values that are low or high.

Practice and Problem Solving

4. 11

5. 4; 42

6. 24; 38

7. no; There is no 2 as a leaf for the stem 3.

8. **Books Read**

Stem	Leaf
0	9
1	5 7 9
2	0 5 6 6 9
3	1 2
4	0

Key: 1 | 5 = 15 books

9–13. See Additional Answers.

Practice and Problem Solving

14–21. See Additional Answers.

22. See *Taking Math Deeper*.

 Fair Game Review

23.

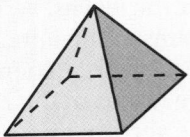

24.

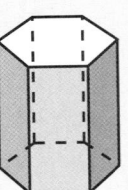

25.

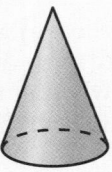

26. See Additional Answers.

27. B

Mini-Assessment

The table shows the number of hours 15 students were online this week.

Hours Online

21	14	8	13	17
18	9	12	7	21
15	12	21	15	7

1. Make a stem-and-leaf plot of the data.

Stem	Leaf
0	7 7 8 9
1	2 2 3 4 5 5 7 8
2	1 1 1

Key: 1 | 8 = 18 hours

2. Find the mean, median, mode, and range of the data. mean = 14, median = 14, mode = 21, range = 14

Taking Math Deeper

Exercise 22

With most problems involving data analysis, it helps to start by ordering the data.

1 Order the data.

85, 107, 108, 112, 115, 118, 119, 120, 122, 125, 127, 131, 136, 136, 140, 142, 152, 156

2 Make a frequency table.

85–99	I
100–114	III
115–129	IIII II
130–144	IIII
145–159	II

OR

80–99	I
90–99	
100–109	II
110–119	IIII
120–129	IIII
130–139	III
140–149	II
150–159	II

Make a stem-and-leaf plot.

8	5
9	
10	7 8
11	2 5 8 9
12	0 2 5 7
13	1 6 6
14	0 2
15	2 6

5 or 8 intervals

Key: 8 | 5 = 85 points

3 The stem-and-leaf plot gives the exact bowling scores. The frequency table only shows the distribution of the scores. The first frequency table is better for showing the distribution of a small data set.

Reteaching and Enrichment Strategies

If students need help. . .	If students got it. . .
Resources by Chapter • Practice A and Practice B • Puzzle Time Record and Practice Journal Practice Differentiating the Lesson Lesson Tutorials Skills Review Handbook	Resources by Chapter • Enrichment and Extension Start the next section

Make a stem-and-leaf plot of the data.

14.

Bikes Sold			
78	112	105	99
86	96	115	100
79	81	99	108

15.

Minutes in Line			
4.0	2.6	1.9	3.1
3.6	2.2	2.7	3.8
1.6	2.0	3.1	2.9

VOLLEYBALL The stem-and-leaf plot shows the number of digs for the top 15 players at a volleyball tournament.

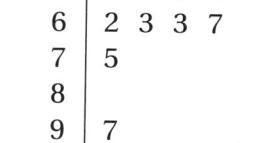

Stem	Leaf
4	1 1 3 3 5
5	0 2 3 4
6	2 3 3 7
7	5
8	
9	7

Key: 5 | 0 = 50 digs

② **16.** How many players had more than 60 digs?

17. Find the mean, median, mode, and range of the data.

18. Describe the distribution of the data.

19. Which data value is the outlier? Describe how the outlier affects the mean.

20. RESEARCH Use the Internet to find the heights of the players on your favorite professional sports team.

    a. Make a stem-and-leaf plot of the data.

    b. Analyze the stem-and-leaf plot and make two conclusions about the heights.

21. OPEN-ENDED Describe a real-life situation with eight data values that has a median of 33. Make a stem-and-leaf plot of the data.

22. *Critical Thinking* Make a frequency table and a stem-and-leaf plot of the bowling scores in the table. Compare and contrast the two data displays. Which display is better for showing how the data are distributed? Explain.

Bowling Scores					
131	108	115	140	152	122
120	118	156	142	112	107
136	85	127	119	136	125

Fair Game Review   *What you learned in previous grades & lessons*

Draw the solid.   *(Section 6.1)*

23. Square pyramid

24. Hexagonal prism

25. Cone

26. Cylinder

27. MULTIPLE CHOICE In a bar graph, what determines the length of each bar? *(Skills Review Handbook)*

    Ⓐ Frequency   Ⓑ Data value   Ⓒ Leaf   Ⓓ Change in data

8.2 Histograms

Essential Question How do histograms show the differences in distributions of data?

1 ACTIVITY: Analyzing Distributions

Work with a partner. The graphs (histograms) show four different types of distributions.

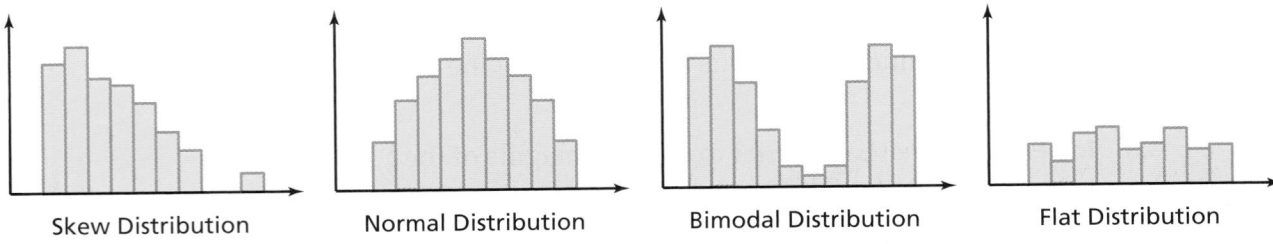

Skew Distribution   Normal Distribution   Bimodal Distribution   Flat Distribution

a. Describe a real-life example of each distribution.

b. Describe the mean, median, and mode of each distribution.

c. In which distributions are the mean and median about equal? Explain your reasoning.

d. How did each type of distribution get its name?

2 ACTIVITY: Analyzing Distributions

Work with a partner. A survey asked 100 adult men and 100 adult women to answer the following questions.

Question 1: What is your ideal weight?
Question 2: What is your ideal age?

Match the histogram to the question.

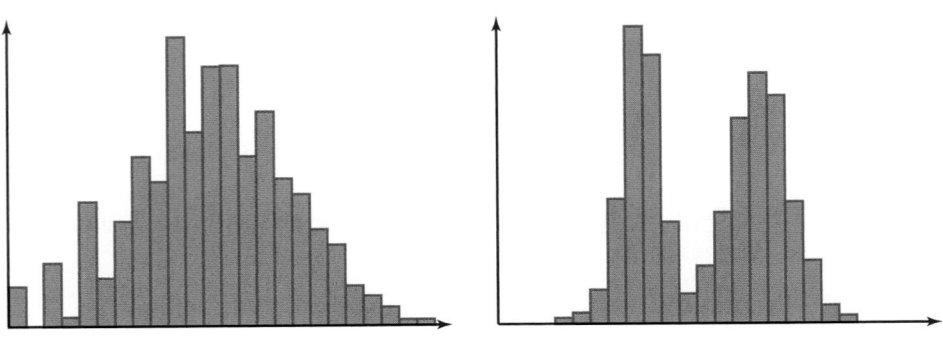

Laurie's Notes

Introduction

For the Teacher

- **Goal:** Students will develop an intuitive understanding of how a histogram shows the distribution of data.
- The histograms explored today will resemble bar graphs. The big idea of the day is to focus on the distribution of the data.

Motivate

- Have students construct a human histogram to show the length of their names (number of letters in first, middle, and last).
- Students line up in order from shortest to longest name length. There will be duplicate name lengths and those students should be adjacent to one another.
- Decide on the number of intervals and the width of each interval.
- Have students move to form their vertical bar.
- Ask which bar (interval) has the most students.

Activity Notes

Activity 1

- This is a very open-ended problem, with no single correct answer.
- **?** "What does the word *distribution* mean when talking about data and graphs?" listen for students to use ideas such as spread, gap, cluster
- You may want to give a few real-life examples to get them started. For each, suggest a *reasonable* mean, median, and mode.
- Students will be uncomfortable with the lack of numbers or scaling. They have to decide what the numbers are, and more importantly, the mean, median, and mode. The context determines the range, and the distribution helps students estimate a reasonable mean, median, and mode.
- Give multiple groups time to share their ideas about the distributions. Be sure to ask students to justify why they believe their context makes sense.

Activity 2

- Students need to analyze the distribution of the data in order to make a decision about how to match the histogram to the question. Students need to have some general knowledge about how the two genders might differ on the issue of ideal weight and ideal age.
- **?** "If you think the first graph matches Question 1, how would you label and scale the axes?" Horizontal axis would be labeled "weights" and scaled from 100 to 200 pounds; the vertical axis would be labeled "number of people" and because 200 people were surveyed, the scale might be from 0 to 25.
- Ask a similar question about Question 2, just to be sure that students understand the context and have a sense about what a reasonable range of numbers would be.
- The importance of this question is that students can give a reasonable explanation as to why they matched the graphs as they did.

Previous Learning

Students have constructed and interpreted bar graphs and pictographs. Students have students learned about mean, median, mode, and distribution.

Activity Materials
Textbook
• number cubes

Start Thinking! and Warm Up

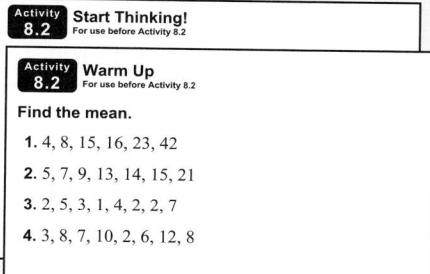

Activity 8.2	**Start Thinking!** For use before Activity 8.2

Activity 8.2	**Warm Up** For use before Activity 8.2

Find the mean.

1. 4, 8, 15, 16, 23, 42

2. 5, 7, 9, 13, 14, 15, 21

3. 2, 5, 3, 1, 4, 2, 2, 7

4. 3, 8, 7, 10, 2, 6, 12, 8

8.2 Record and Practice Journal

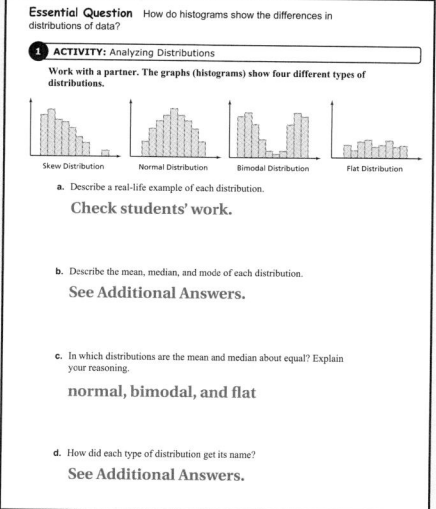

Essential Question How do histograms show the differences in distributions of data?

1 ACTIVITY: Analyzing Distributions

Work with a partner. The graphs (histograms) show four different types of distributions.

Skew Distribution   Normal Distribution   Bimodal Distribution   Flat Distribution

a. Describe a real-life example of each distribution.
   Check students' work.

b. Describe the mean, median, and mode of each distribution.
   See Additional Answers.

c. In which distributions are the mean and median about equal? Explain your reasoning.
   normal, bimodal, and flat

d. How did each type of distribution get its name?
   See Additional Answers.

English Language Learners

Visual

Histograms and bar graphs have a similar appearance. Students may confuse the two. Compare and contrast the two types of data displays. Both use the lengths of bars to represent data. The bars of a histogram represent numerical intervals and are touching. The bars of a bar graph represent categories and are separated.

8.2 Record and Practice Journal

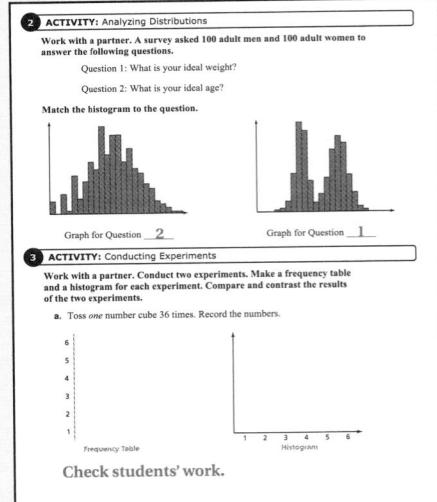

Check students' work.

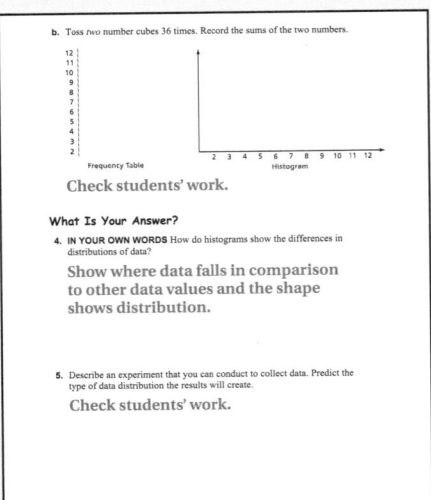

Check students' work.

What Is Your Answer?

4. IN YOUR OWN WORDS How do histograms show the differences in distributions of data?

Show where data falls in comparison to other data values and the shape shows distribution.

5. Describe an experiment that you can conduct to collect data. Predict the type of data distribution the results will create.

Check students' work.

Activity 3

? "How do you record information in a frequency table?" Make tally marks, and then total and record each outcome.

- **Teaching Tip:** If you have students with a tendency to roll the number cubes so far that they go off the desk, have them work in a corner of the room on the floor or roll the number cubes inside a small box.

? "What did you observe about the first experiment and the results you obtained?" It should have a flat distribution because each number is equally likely. This may not have occurred for every pair of students, but if class data are collated it would be very noticeable.

? "Compare the results of the first and second experiments and explain any differences." In the second experiment, there should be a normal distribution as 6, 7, and 8 are more likely compared to 2 and 12, which are the least likely sums to occur.

- **Connection:** Students may make what appears to be a standard bar graph and they may refer to it as a bar graph. Generally a histogram, as defined in the next lesson, is a bar graph that shows the frequency of data values *in intervals* of the same size. In today's investigation, the focus has been on the distribution and not on grouping the data into intervals.

What Is Your Answer?

- Ask volunteers to share their answers.

Closure

- Ask students to describe the type of distribution for their human histogram. Answers will vary depending on the students in each class.

Technology For the Teacher

The Dynamic Planning Tool
Editable Teacher's Resources at *BigIdeasMath.com*

3 ACTIVITY: Conducting Experiments

Work with a partner. Conduct two experiments. Make a frequency table and a histogram for each experiment. Compare and contrast the results of the two experiments.

a. Toss one number cube 36 times. Record the numbers.

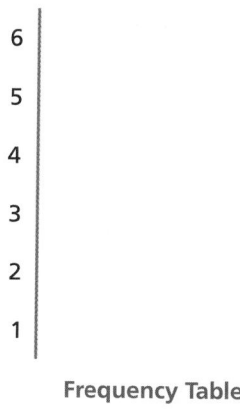

6
5
4
3
2
1

Frequency Table

1 2 3 4 5 6

Histogram

b. Toss two number cubes 36 times. Record the sums of the two numbers.

12
11
10
9
8
7
6
5
4
3
2

Frequency Table

2 3 4 5 6 7 8 9 10 11 12

Histogram

What Is Your Answer?

4. **IN YOUR OWN WORDS** How do histograms show the differences in distributions of data?

5. Describe an experiment that you can conduct to collect data. Predict the type of data distribution the results will create.

Practice

Use what you learned about histograms to complete Exercises 4 and 5 on page 358.

Key Vocabulary
histogram, *p. 356*

🔑 Key Idea

Histograms

A **histogram** is a bar graph that shows the frequency of data values in intervals of the same size.

The height of a bar represents the frequency of the values in the interval.

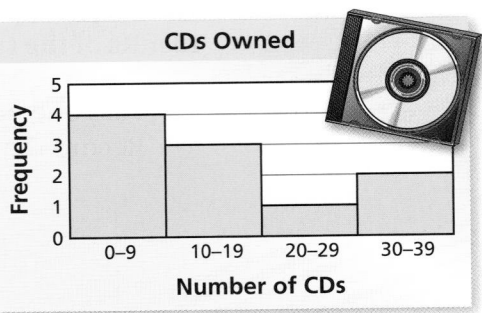

EXAMPLE (1) **Making a Histogram**

Remember

A *frequency table* groups data values into intervals. The *frequency* is the number of data values in an interval.

The frequency table shows the number of pairs of shoes that each person in a class owns. Display the data in a histogram.

Pairs of Shoes	Frequency
1–3	11
4–6	4
7–9	0
10–12	3
13–15	6

Step 1: Draw and label the axes.

Step 2: Draw a bar to represent the frequency of each interval.

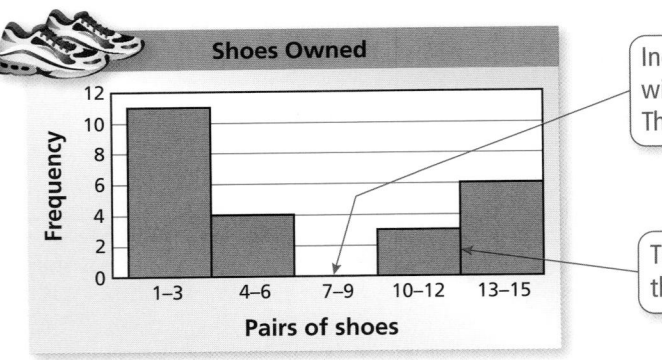

Include any interval with a frequency of 0. The bar height is 0.

There is no space between the bars of a histogram.

🔘 On Your Own

Now You're Ready
Exercises 6–8

1. The frequency table shows the ages of people riding a roller coaster. Display the data in a histogram.

Age	10–19	20–29	30–39	40–49	50–59
Frequency	16	11	5	2	4

🔊 Multi-Language Glossary at BigIdeasMath✓.com.

Laurie's Notes

Introduction

Connect
- **Yesterday:** Students explored how histograms show the different distributions of data.
- **Today:** Students will construct and interpret a histogram.

Motivate
- ❓ "Do you think anyone owns more than 100 pairs of shoes? 200 pairs?" Continue until students say no to some number.
- **FYI:** Former Philippine First Lady Imelda Marcos owned 1200 pairs of shoes.
- ❓ "Do you think anyone would pay $1000 for a pair of shoes? $2000?" Continue until students say no to some number.
- **FYI:** Singer Alison Krauss wore a pair of shoes valued at $2 million at the Oscar Award ceremony.

Lesson Notes

Key Idea
- Students have constructed bar graphs. A histogram is a particular type of bar graph where the data is numeric, and the data is grouped into intervals of equal size. A bar graph includes categorical data (i.e., favorite vegetable) and numeric data.
- ❓ Use the sample shown to ask questions, checking students' understanding of how to read a histogram.
  - "How many people were polled?" 10
  - "How many people polled had 9 or fewer CDs?" 4
  - "How many people polled had 15 CDs?" cannot determine

Example 1
- The intervals for the histogram have been pre-determined. The size of each interval is 3. Students should notice the pattern in the right-end value of each interval: 3, 6, 9, 12, 15.
- Labeling the axes can often present a challenge for students. Students may write a number below the hash mark instead of writing an interval between the hash marks.
- Explain that no space is left between the bars because the intervals are continuous.
- Remind students that the axes are labeled and the histogram is given a title explaining what the data is about.
- Ask questions about the completed histogram.

On Your Own
- Students should recognize that the data is about people ages 10 to 59. It would not be correct to leave a gap for people ages 0 to 9.
- Ask a volunteer to share their histogram at the board.

Goal Today's lesson is constructing and interpreting a **histogram**.

Start Thinking! and Warm Up

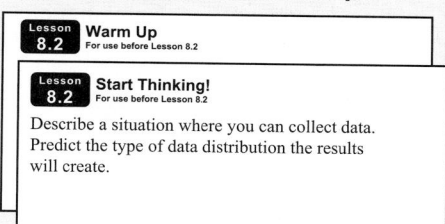

Describe a situation where you can collect data. Predict the type of data distribution the results will create.

Extra Example 1

The frequency table shows the number of T-shirts each person in a class owns. Display the data in a histogram.

T-shirts	Frequency
1–4	3
5–8	6
9–12	13
13–16	0
17–20	4

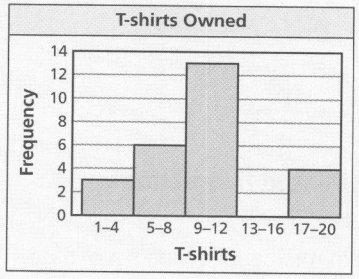

On Your Own

1.

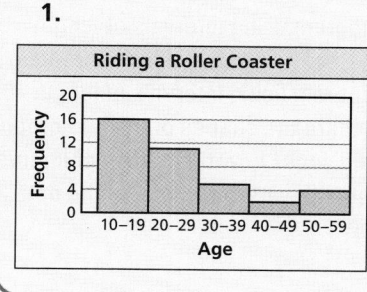

Extra Example 2

The histogram shows the number of different instruments each member of a jazz band can play.

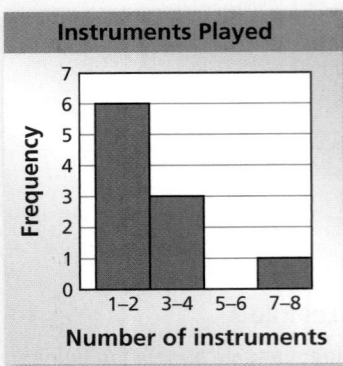

Instruments Played

Frequency

Number of instruments

a. Which interval contains the most data values? 1–2

b. How many band members can play more than 2 instruments? 4

c. How many band members can play at most 2 instruments? 6

On Your Own

2. **a.** 11 students

b. 18 students

Differentiated Instruction

Visual

Give students the following set of data.

1, 2, 2, 4, 5, 7, 7, 8, 9, 11, 13, 14, 14, 16, 17, 19, 20, 20, 21, 22, 24, 24, 25, 26, 26, 29

Instruct half of the class to create a histogram using intervals of length 5 and the other half of the class to create a histogram using intervals of length 10. Compare the shapes of the two graphs and discuss how the choice of intervals affects the interpretation of the data.

Example 2

- The focus of this example is to interpret information given in a histogram. Review the vocabulary *less than*, *at least*, *at most*, and *more than*.
- In addition to the questions posed, ask students to describe the distribution of the data. You could also ask about the number of races shown in the graph.

On Your Own

- This graph is almost symmetric. Ask students what it means for a graph to be symmetric.

Closure

- Explain how a histogram is similar to a bar graph and how it differs from a bar graph. *Sample answer:* Both have bars that represent the frequency. A histogram has intervals, but a bar graph has categories.

Technology For the Teacher

Dynamic Classroom

The Dynamic Planning Tool
Editable Teacher's Resources at *BigIdeasMath.com*

EXAMPLE 2 Using a Histogram

The histogram shows the winning speeds at the Daytona 500. (a) Which interval contains the most data values? (b) How many of the winning speeds are less than 140 miles per hour? (c) How many of the winning speeds are at least 160 miles per hour?

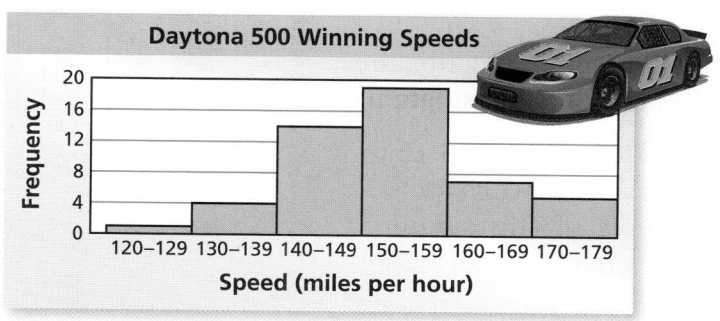

a. The interval with the tallest bar contains the most data values.

So, the 150–159 miles per hour interval contains the most data values.

b. One winning speed is in the 120–129 miles per hour interval and four winning speeds are in the 130–139 miles per hour interval.

So, $1 + 4 = 5$ winning speeds are less than 140 miles per hour.

c. Seven winning speeds are in the 160–169 miles per hour interval and five winning speeds are in the 170–179 miles per hour interval.

So, $7 + 5 = 12$ winning speeds are at least 160 miles per hour.

On Your Own

Now You're Ready
Exercises 9–11

2. The histogram shows the number of hours that students in a class slept last night.

a. How many students slept at least 8 hours?

b. How many students slept less than 12 hours?

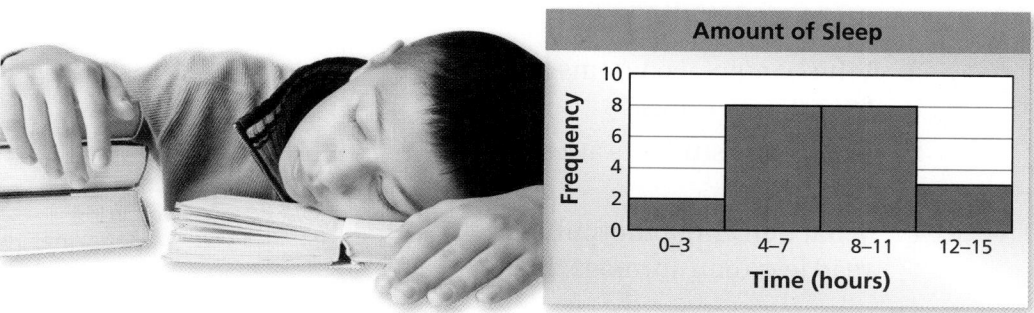

Check It Out
Help with Homework
BigIdeasMath ✓com

Vocabulary and Concept Check

1. **VOCABULARY** Which graph is a histogram? Explain your reasoning.

2. **REASONING** Describe the outliers in the histogram.

3. **CRITICAL THINKING** How can you tell when an interval of a histogram has a frequency of zero?

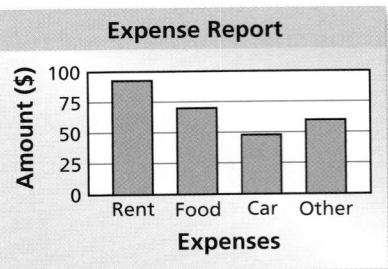

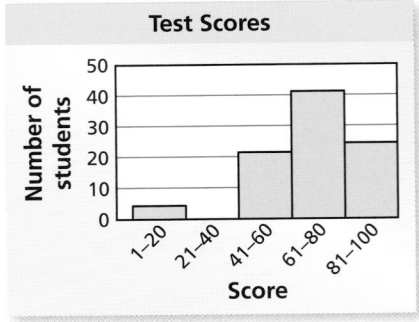

Practice and Problem Solving

Determine the type of distribution shown by the histogram.

4.

5.

Display the data in a histogram.

① 6.

States Visited	
States	Frequency
1–5	12
6–10	14
11–15	6
16–20	3

7.

Chess Team	
Wins	Frequency
10–13	3
14–17	4
18–21	4
22–25	2

8.

Movies Watched	
Movies	Frequency
0–1	5
2–3	11
4–5	8
6–7	1

② 9. **MAGAZINES** The histogram shows the number of magazines read last month by students in a class.

a. Which interval contains the fewest data values?

b. How many students are in the class?

c. What percent of the students read less than six magazines?

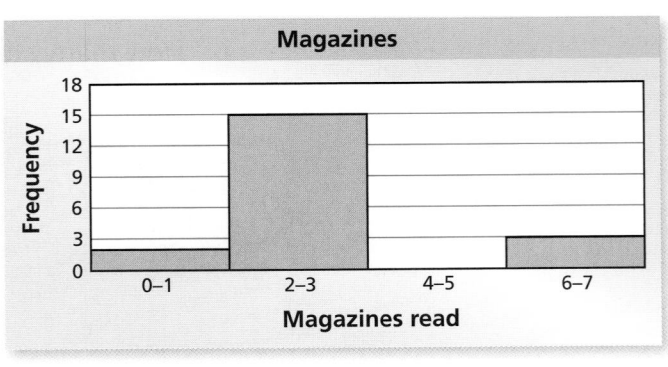

Assignment Guide and Homework Check

Level	Day 1 Activity Assignment	Day 2 Lesson Assignment	Homework Check
Basic	4, 5, 14–18	1–3, 6–9	2, 6, 8, 9
Average	4, 5, 14–18	1–3, 7, 9, 10, 11	2, 7, 9, 10
Advanced	4, 5, 14–18	1–3, 10–13	2, 10, 11, 12

For Your Information
- **Exercise 11** Pennsylvania has a total area of 46,055 square miles. Indiana has a total area of 36,418 square miles.

Common Errors
- **Exercises 6–8** Students may struggle with determining how to label the vertical axis of the histogram. Remind them to use consistent intervals and to base their decision on the frequency of the data. For example, if there is a high frequency, they should count by 5s or 10s, but if there is a low frequency, then they should count by 1s or 2s.
- **Exercise 9** Students may say that the interval 0–1 has the least number of students, but it is really 4–5. To help them answer this question, ask them to label above each interval the frequency of that interval. Writing the frequency themselves above each interval helps students to read the histogram.
- **Exercise 11** Students may only look at the frequency for each graph and say that Indiana has the greater area. Encourage them to look at the intervals as well. The two graphs are drawn using different intervals and cannot be compared strictly by the heights of the bars.

8.2 Record and Practice Journal

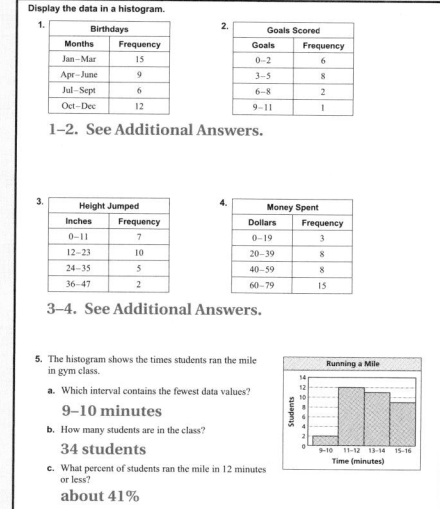

1–2. See Additional Answers.

3–4. See Additional Answers.

5. The histogram shows the times students ran the mile in gym class.
   a. Which interval contains the fewest data values?
      9–10 minutes
   b. How many students are in the class?
      34 students
   c. What percent of students ran the mile in 12 minutes or less?
      about 41%

Technology For the Teacher
Answer Presentation Tool
QuizShow

1. The *Test Scores* graph is a histogram because the number of students (frequency) achieving the test scores are shown in intervals of the same size (20).

2. The scores falling into the interval 1–20 are outliers because most of the scores are between 41 and 100.

3. No bar is shown on that interval.

Practice and Problem Solving

4. skew   5. flat

6.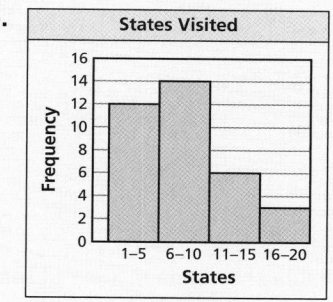

7.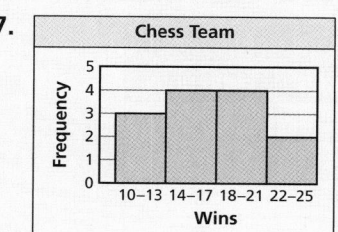

8.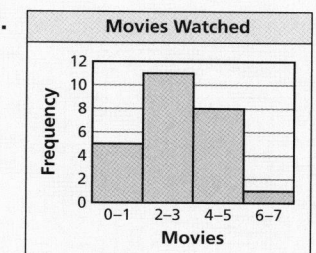

9. a. 4–5
   b. 20 students
   c. 85%

10. **a.** no; The histogram shows that only one state fell in the interval of 40–44.9%. This state did not necessarily have 40% of possible voters vote.

   b. yes; 36 states are between 50% and 64.9%.

11. Pennsylvania; You can see from the intervals and frequencies that Pennsylvania counties are greater in area, which makes up for it having fewer counties.

12. See Additional Answers.

13. See *Taking Math Deeper*.

Fair Game Review

14. 45   15. 27

16. 22.4   17. 51.2

18. D

Mini-Assessment

The table shows the number of songs downloaded last month by your friends.

Songs Downloaded	
Songs	Frequency
0–1	4
2–3	12
4–5	8
6–7	5

1. Display the data in a histogram.

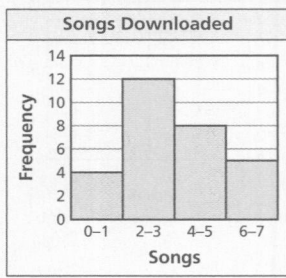

2. Which interval contains the most data values? The 2–3 songs interval contains the most data values.

Taking Math Deeper

Exercise 13

This exercise points out to students that making a histogram is as much an *art* as a *science*. The big question lies in "how many intervals should I use?"

With too many intervals, most distributions will appear flat. With too few intervals, most distributions will have only a few tall spikes. As with *Goldilocks and the 3 Bears,* the goal is to get the porridge "just right."

 Order the data.
   51, 54, 55, 57, 59, 67, 68, 70, 71, 73, 75, 76, 77, 77, 78, 78, 79, 79, 80, 80, 81, 81, 82, 82, 83, 83, 84, 85, 85, 88

 a. Make a frequency distribution and histogram.

51–55	III
56–60	II
61–65	
66–70	III
71–75	III
76–80	IIII IIII
81–85	IIII IIII
86–90	I

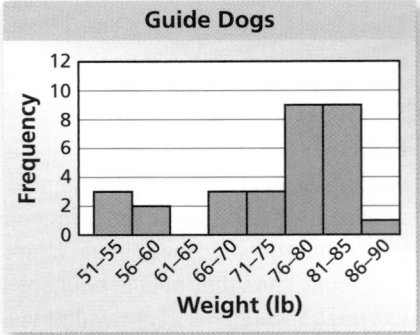

 b. Make a frequency distribution and histogram using a different sized interval.

51–60	IIII
61–70	III
71–80	IIII IIII II
81–90	IIII IIII

Too few.

c. The second histogram has only four intervals, and this does not convey as much information about the distribution as the first one.

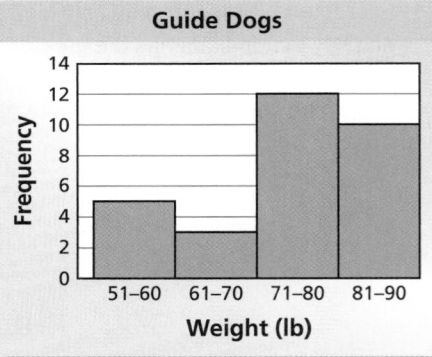

Reteaching and Enrichment Strategies

If students need help...	If students got it...
Resources by Chapter • Practice A and Practice B • Puzzle Time Record and Practice Journal Practice Differentiating the Lesson Lesson Tutorials Skills Review Handbook	Resources by Chapter • Enrichment and Extension Start the next section

10. **VOTING** The histogram shows the percent of the voting age population that voted in a recent presidential election. Explain whether each statement is supported by the graph.

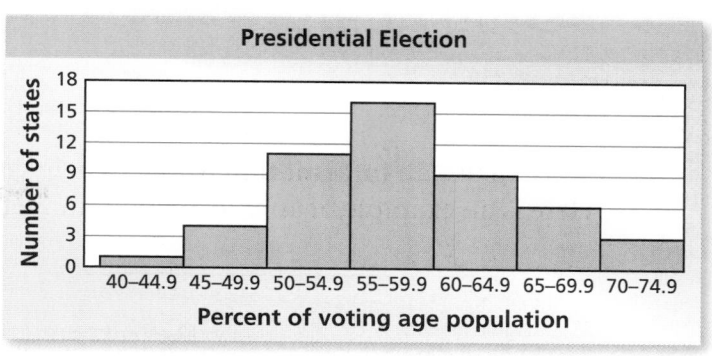

Presidential Election

   a. Only 40% of one state voted.

   b. Most states had between 50% and 64.9% that voted.

11. **AREA** The histograms show the areas of counties in Pennsylvania and Indiana. Which state do you think has the greater area? Explain.

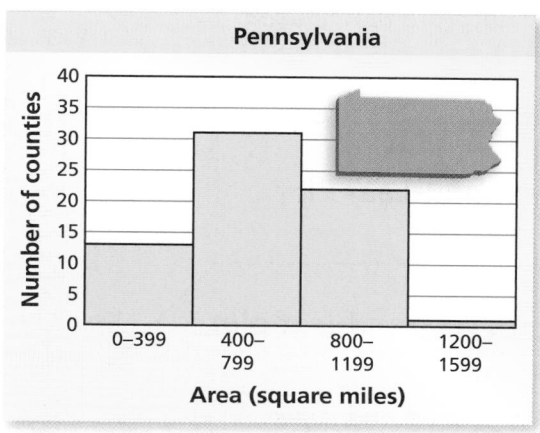

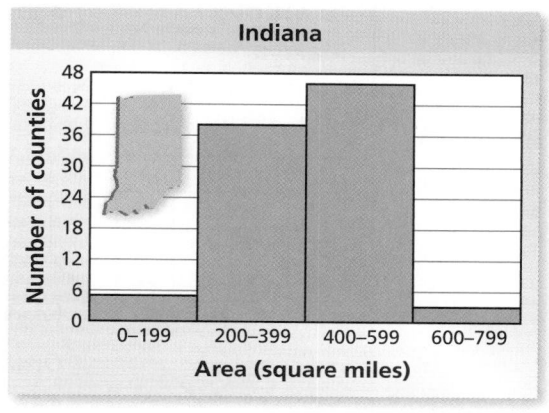

12. **REASONING** Can you find the mean, median, mode, and range of the data in Exercise 7? If so, find them. If not, explain why.

13.  The table shows the weights of guide dogs enrolled in a training program.

   a. Make a histogram of the data starting with the interval 51–55.

   b. Make another histogram of the data using a different sized interval.

   c. Compare and contrast the two histograms.

Weight (lb)					
81	88	57	82	70	85
71	51	82	77	79	77
83	80	54	80	81	73
59	84	75	76	68	78
83	78	55	67	85	79

Fair Game Review What you learned in previous grades & lessons

Find the percent of the number. *(Section 4.1)*

14. 25% of 180   **15.** 30% of 90   **16.** 16% of 140   **17.** 64% of 80

18. **MULTIPLE CHOICE** Two rectangles are similar. The smaller rectangle has a length of 8 feet. The larger rectangle has a length of 14 feet. What is the ratio of the area of the smaller rectangle to the area of the larger rectangle? *(Section 5.3)*

   (A) 7 : 4   (B) 4 : 7   (C) 9 : 16   (D) 16 : 49

You can use an **information frame** to help you organize and remember concepts. Here is an example of an information frame for a stem-and-leaf plot.

Definition
A *stem-and-leaf plot* is a data display that uses the digits of data values to organize the data. Stem-and-leaf plots show how data are distributed.

Example
Make a stem-and-leaf plot of the data.

Patient Ages			
32	21	5	43
28	27	16	9
42	13	16	24

Stem-and-Leaf Plot

Visual

Stem	Leaf
0	5 9
1	3 6 6
2	1 4 7 8
3	2
4	2 3

Key: 2 | 4 = 24

Making a stem-and-leaf plot

Order the data.

Choose the stems and leaves.

Write the stems to the *left* of the vertical line.

Write the leaves for each stem to the *right* of the vertical line.

On Your Own

Make an information frame to help you study the topic.

1. histogram

After you complete this chapter, make information frames for the following topics.

2. circle graph

3. making a prediction about a population

4. Pick three other topics that you studied earlier in this course. Make an information frame for each topic.

"I'm having trouble thinking of a good title for my **information frame**."

Sample Answer

1.

Definition

A histogram is a bar graph that shows the frequency of data values in intervals of the same size.

Example

Display the data in a histogram.

CDs owned	Frequency
0-9	4
10-19	3
20-29	1
30-39	2

Histogram

Visual

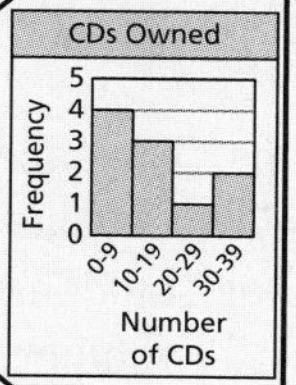

CDs Owned

Making a histogram

Draw and label the axes.

Draw a bar to represent the frequency of each interval in the frequency table.

List of Organizers

Available at *BigIdeasMath.com*

Comparison Chart
Concept Circle
Example and Non-Example Chart
Formula Triangle
Four Square
Idea (Definition) and Examples Chart
Information Frame
Information Wheel
Notetaking Organizer
Process Diagram
Summary Triangle
Word Magnet
Y Chart

About this Organizer

An **Information Frame** can be used to help students organize and remember concepts. Students write the topic in the middle rectangle. Then students write related concepts in the spaces around the rectangle. Related concepts can include *Words, Numbers, Algebra, Example, Definition, Non-Example, Visual, Procedure, Details*, and *Vocabulary*. Students can place their information frames on note cards to use as a quick study reference.

Answers

1–6. See Additional Answers.

7.

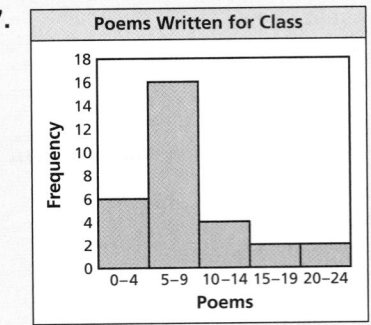

8. Weights of Packages

Stem	Leaf
0	6 7
1	2 3 5 6 8
2	2 5

Key: 0 | 6 = 6 ounces

Most of the data values are in the middle.

9. a. 2–3

   b. 12 games

   c. 25%

10. mean: 12;
median: 11;
mode: 8;
range: 14

Assessment Book

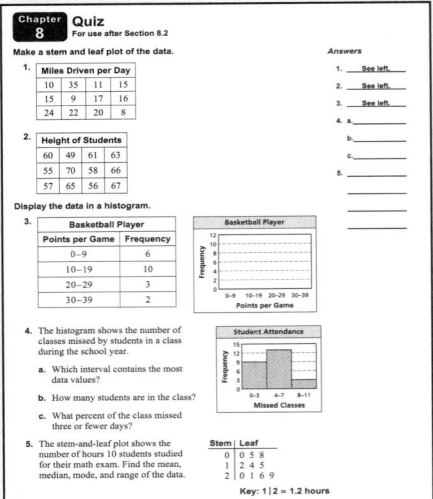

Alternative Quiz Ideas

100% Quiz	Math Log
Error Notebook	Notebook Quiz
Group Quiz	Partner Quiz
Homework Quiz	Pass the Paper

Group Quiz

Students work in groups. Give each group a large index card. Each group writes five questions that they feel evaluate the material they have been studying. On a separate piece of paper, students solve the problems. When they are finished, they exchange cards with another group. The new groups work through the questions on the card.

Reteaching and Enrichment Strategies

If students need help. . .	If students got it. . .
Resources by Chapter • Study Help • Practice A and Practice B • Puzzle Time Lesson Tutorials *BigIdeasMath.com* Practice Quiz Practice from the Test Generator	Resources by Chapter • Enrichment and Extension • School-to-Work Game Closet at *BigIdeasMath.com* Start the next section

Technology For the Teacher

Answer Presentation Tool
Big Ideas Test Generator

Make a stem-and-leaf plot of the data.   *(Section 8.1)*

1.

Cans Collected Each Month			
80	90	84	92
76	83	79	59
68	55	58	61

2.

Miles Driven Each Day				
21	18	12	16	10
16	9	15	20	28
35	50	37	20	11

3.

Ages of Tortoises			
86	99	100	124
92	85	110	130
115	129	83	104

4.

Kilometers Run Each Day				
6.0	5.6	6.2	3.0	2.5
3.5	2.0	5.0	3.9	3.1
6.2	3.1	4.5	3.8	6.1

Display the data in a histogram.   *(Section 8.2)*

5.

Soccer Team Goals	
Goals per Game	Frequency
0–1	5
2–3	4
4–5	0
6–7	1

6.

Minutes Practiced	
Minutes	Frequency
0–19	8
20–39	10
40–59	11
60–79	2

7.

Poems Written for Class	
Poems	Frequency
0–4	6
5–9	16
10–14	4
15–19	2
20–24	2

8. WEIGHTS The weights (in ounces) of nine packages are 7, 22, 16, 12, 6, 18, 15, 13, and 25. Make a stem-and-leaf plot of the data. Describe the distribution of the data.   *(Section 8.1)*

9. REBOUNDS The histogram shows the number of rebounds per game for a middle school basketball player this season.   *(Section 8.2)*

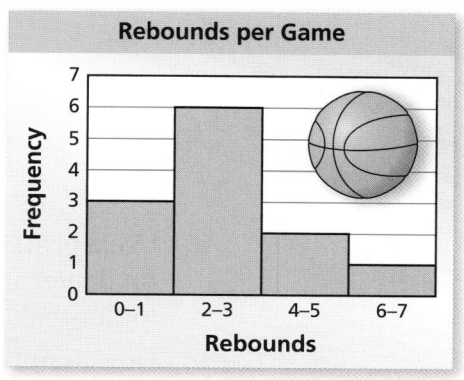

   a. Which interval contains the most data values?

   b. How many games were played by the player this season?

   c. What percent of the games did the player have 4 or more rebounds?

Stem	Leaf
0	6 8 8 9
1	0 1 2 3 7 8
2	0

Key: 0│9 = 9 hours

10. STAGE CREW The stem-and-leaf plot shows the number of hours 11 stage crew members spent building sets. Find the mean, median, mode, and range of the data.   *(Section 8.1)*

8.3 Circle Graphs

Essential Question How can you use a circle graph to show the results of a survey?

Share Your Work at...
My.BigIdeasMath.com

1 ACTIVITY: Reading a Circle Graph

Work with a partner. Six hundred middle school students were asked "What is your favorite sport?" The circle graph shows the results of the survey.

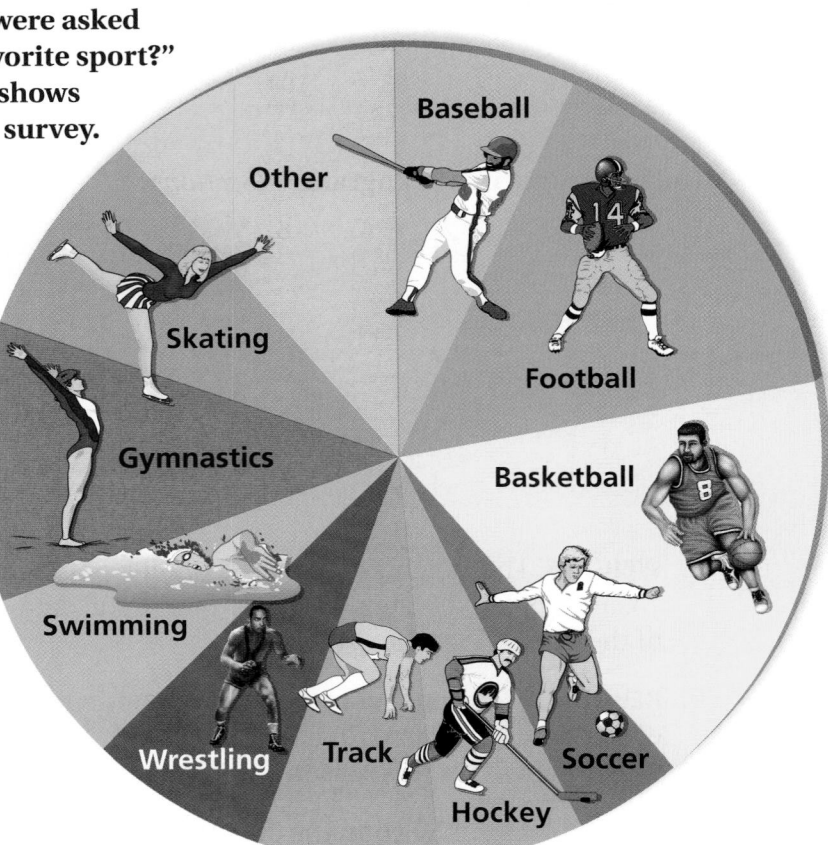

a. Use a protractor to find the angle measure (in degrees) of the section (pie piece) for football.

b. How many degrees are in a full circle?

c. Write and solve a proportion to determine the number of students who said that football is their favorite sport.

d. Repeat the process for the other sections of the circle graph.

Laurie's Notes

Introduction

For the Teacher

- **Goal:** Students will read the results of a survey displayed as a circle graph.
- A circle graph shows the relationship between the parts of a data set. The circle graph is also referred to as a pie chart.

Motivate

- Write three percent problems on the board and ask students to find the answers and place them in order from least to greatest.
  - 32% of 400 128   15% of 700 105   24% of 500 120
- Review finding a percent of a number.

Activity Notes

Activity 1

- Your students' familiarity with using a protractor and their facility with proportions will influence the amount of time needed for this activity. Use a calculator when solving the proportions to cut the work time.
- It is helpful to have students estimate the angle measure before they begin. Students should be able to state if the angle is acute or obtuse.
- **?** "How many students are represented in this circle graph?" 600
- **?** "If 10% of them had picked baseball as their favorite sport, how many students would that be? Explain." 60; 10% of 600 is 60.
- **?** "What would the angle measure be for the baseball sector if it was 10% of the whole?" 10% of 360° = 36°.
- **Teaching Tip:** Model how to measure an angle of any sector, say football. This will give you the opportunity to show how to line up the protractor on the vertex of the angle, and the 0° mark along a ray of the angle. angle measure ≈ 55°
- Model how to set up and solve a proportion to determine what percent 55° is of 360°.

  $$\frac{55}{360} = \frac{x}{100} \quad \rightarrow \quad x \approx 15; \text{ Football} \approx 15\% \text{ of the people surveyed.}$$

- **?** "If 15% of those surveyed picked football as their favorite sport, how many people would this be? Explain." find 15% of 600; 90 people
- Circulate around the room to ensure that angles are being measured accurately. All angle measures for this activity should be multiples of 5. Students may need to round each angle measure up or down for the sum of the angle measures to equal 360°. If you notice that an angle measure is clearly off, ask students to simply use their eyes to compare two angles.
- When students have finished, ask a few summary questions.
- **?** "The sectors of the circle graph are labeled with categories. What other information could be presented in the sectors?" percent of students who picked each sport; number of students who picked each sport

Previous Learning

Students should know how to measure with a protractor, and be able to set up and solve a proportion.

Activity Materials
Textbook
• protractors

Start Thinking! and Warm Up

Activity 8.3 Start Thinking!
For use before Activity 8.3

Activity 8.3 Warm Up
For use before Activity 8.3

Solve the proportion.

1. $\dfrac{s}{28} = \dfrac{3}{4}$ 2. $\dfrac{15}{x} = \dfrac{20}{8}$

3. $\dfrac{n}{10} = \dfrac{84}{20}$ 4. $\dfrac{3}{y} = \dfrac{9}{12}$

5. $\dfrac{36}{p} = \dfrac{54}{15}$ 6. $\dfrac{42}{h} = \dfrac{14}{20}$

8.3 Record and Practice Journal

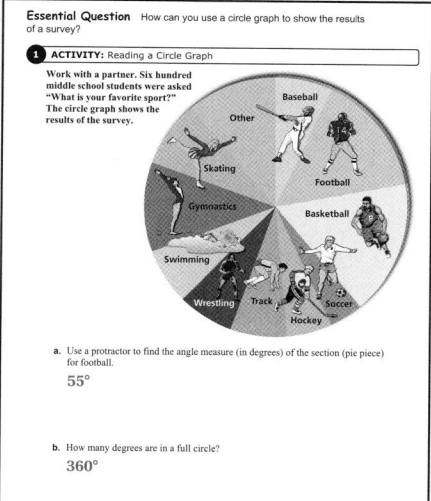

Essential Question How can you use a circle graph to show the results of a survey?

1 ACTIVITY: Reading a Circle Graph

Work with a partner. Six hundred middle school students were asked "What is your favorite sport?" The circle graph shows the results of the survey.

a. Use a protractor to find the angle measure (in degrees) of the section (pie piece) for football.
   55°

b. How many degrees are in a full circle?
   360°

Create a poster displaying the different types of data displays: frequency tables, bar graphs, pictographs, line plots, line graphs, double bar graphs, histograms, stem-and-leaf plots, and circle graphs. Labeling each of the displays will help English learners determine which type of graph is appropriate for a particular set of data.

Laurie's Notes

Activity 2

- **Whole Class Activity:** This is an activity that students *may* have time to complete. It can be done as a whole class activity at the front of the room.
- This reverses Activity 1. Students must take the raw data, decide what part of 100 it is, compute that percent of 360°, and draw the angle.
- **Alternate Approach:** Model how to work through the steps for one sport and leave the remainder for students to complete as homework.

What Is Your Answer?

- Ask a volunteer to share their circle graph from Question 4.

Closure

- "What information can be displayed in a circle graph?" Listen for the idea of one topic (such as favorite sport) that can be divided into several choices.

8.3 Record and Practice Journal

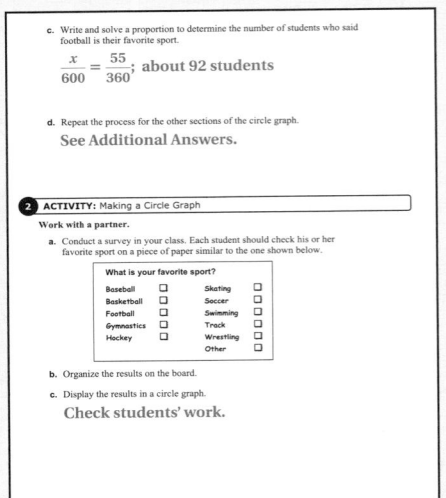

c. Write and solve a proportion to determine the number of students who said football is their favorite sport.

$\frac{x}{600} = \frac{55}{360}$; about 92 students

d. Repeat the process for the other sections of the circle graph.
See Additional Answers.

2 ACTIVITY: Making a Circle Graph

Work with a partner.

a. Conduct a survey in your class. Each student should check his or her favorite sport on a piece of paper similar to the one shown below.

What is your favorite sport?			
Baseball	☐	Skating	☐
Basketball	☐	Soccer	☐
Football	☐	Swimming	☐
Gymnastics	☐	Track	☐
Hockey	☐	Wrestling	☐
		Other	☐

b. Organize the results on the board.

c. Display the results in a circle graph.
Check students' work.

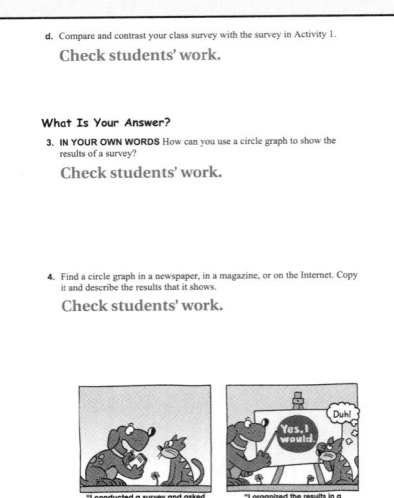

d. Compare and contrast your class survey with the survey in Activity 1.
Check students' work.

What Is Your Answer?

3. **IN YOUR OWN WORDS** How can you use a circle graph to show the results of a survey?
Check students' work.

4. Find a circle graph in a newspaper, in a magazine, or on the Internet. Copy it and describe the results that it shows.
Check students' work.

Technology **F**or the **T**eacher

Dynamic Classroom

The Dynamic Planning Tool
Editable Teacher's Resources at *BigIdeasMath.com*

2 **ACTIVITY: Making a Circle Graph**

Work with a partner.

a. Conduct a survey in your class. Each student should check his or her favorite sport on a piece of paper similar to the one shown below.

What is your favorite sport?

Baseball ❑   Skating ❑
Basketball ❑   Soccer ❑
Football ❑   Swimming ❑
Gymnastics ❑   Track ❑
Hockey ❑   Wrestling ❑
                        Other ❑

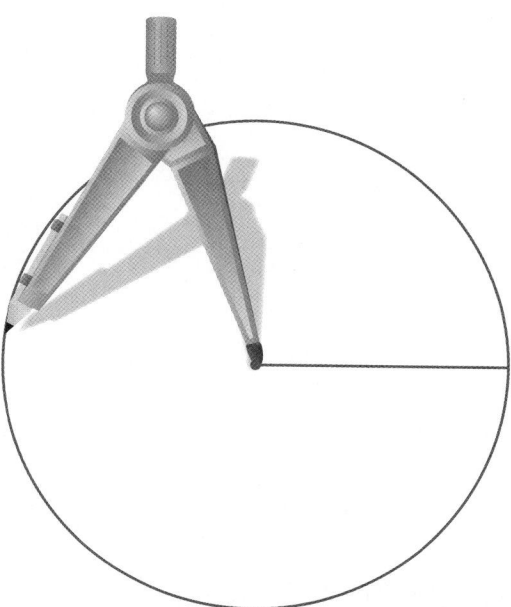

b. Organize the results on the board.

c. Display the results in a circle graph.

d. Compare and contrast your class survey with the survey in Activity 1.

What Is Your Answer?

3. **IN YOUR OWN WORDS** How can you use a circle graph to show the results of a survey?

4. Find a circle graph in a newspaper, in a magazine, or on the Internet. Copy it and describe the results that it shows.

"I conducted a survey and asked 30 people if they would like a million dollars."

"I organized the results in a circle graph."

Practice

Use what you learned about circle graphs to complete Exercises 5–7 on page 366.

Check It Out
Lesson Tutorials
BigIdeasMath.com

Key Vocabulary
circle graph, *p. 364*

Key Idea

Circle Graphs

A **circle graph** displays data as sections of a circle. The sum of the angle measures in a circle graph is 360°.

Favorite Fruit

The percents total 100%.

The circle represents all of the data.

40%

35%

25%

Each section represents part of the data.

When the data are given in percents, multiply the decimal form of each percent by 360° to find the angle measure for each section.

EXAMPLE **1** **Making a Circle Graph**

Favorite Amusement Park	People
Disney World	25
Busch Gardens	15
Universal Studios	12
Marineland	8

The table shows the results of a survey. Display the data in a circle graph.

Step 1: Find the total number of people.

$$25 + 15 + 12 + 8 = 60$$

Step 2: Find the angle measure for each section of the graph. Multiply the fraction of people that chose each park by 360°.

Disney World

$$\frac{25}{60} \cdot 360° = 150°$$

Busch Gardens

$$\frac{15}{60} \cdot 360° = 90°$$

Universal Studios

$$\frac{12}{60} \cdot 360° = 72°$$

Marineland

$$\frac{8}{60} \cdot 360° = 48°$$

Check

$$\frac{25}{60} + \frac{15}{60} + \frac{12}{60} + \frac{8}{60} = 1$$

$$150° + 90° + 72° + 48° = 360° \checkmark$$

Step 3: Use a protractor to draw the angle measures found in Step 2 on a circle. Then label the sections.

Favorite Amusement Park

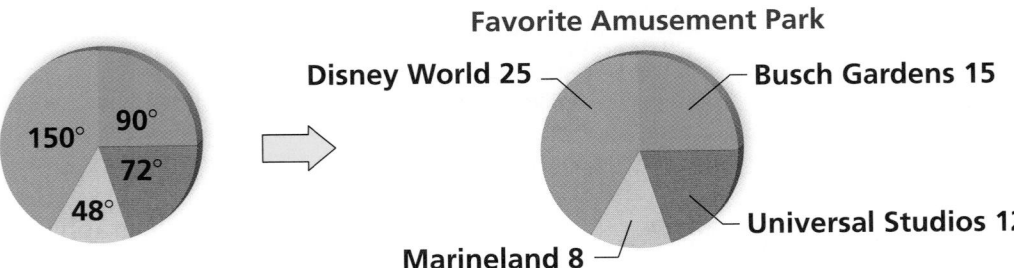

Multi-Language Glossary at BigIdeasMath.com.

Laurie's Notes

Introduction

Connect
- **Yesterday:** Students explored how information from a survey can be displayed in a circle graph.
- **Today:** Students will construct and interpret a circle graph.

Motivate
- Make a human circle graph with yarn.
- Ask students a simple survey question (i.e., favorite fruit). They line up according to their response (all of the apples together, bananas, and so on). Once they are sorted, have them form a circle without changing their order. This means the first and last in line come together to form the circle.
- Stand in the center of the circle. You hold one end of the yarn and different students hold the other ends to show the sectors in the circle graph.
- Ask questions based upon the sectors formed.

Lesson Notes

Key Idea
- Discuss what information is displayed in a circle graph.
- **Big Idea:** The data displayed in a circle graph must be univariate, meaning one variable, and the data must represent a whole. The sectors in the circle graph show how each part is related to the whole.

Example 1
- Constructing a circle graph from raw data involves a number of steps and computations. Work slowly through each step in the process, and give time for students to ask questions and think about the process.
- Ask students to look at the table of data for *Favorite Amusement Park*, the one variable involved.
- **?** "Did more than half the people pick Disney World? Explain." no; There are 60 people, and $\frac{30}{60}$ would be half.
- **?** "Which section of the circle graph will be the largest?" Disney World; "Smallest?" Marineland
- **?** "How will you know how big to make each angle?" Listen for using proportions to find the angle measure.
- In Step 2, point out to students that they are finding a fractional amount of 360°. Example: $\frac{15}{60} = \frac{1}{4} = 25\%$; so, you are finding 25% of 360°.
- Discuss how to check your answer by verifying that the fractions add to 1 whole and the angles add to 360°.
- **Common Error:** Students are often unclear as to how to begin. They need to draw a radius, which serves as the first side of the angle. From here, move around the circle making each sector adjacent to the next. It is common for students to leave gaps between each sector.

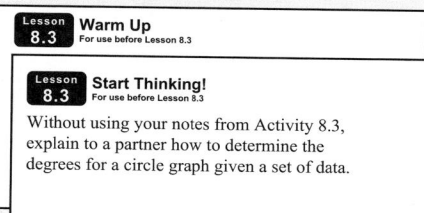

Lesson Materials	
Introduction	Textbook
• yarn	• protractors

Start Thinking! and Warm Up

Lesson 8.3	Warm Up For use before Lesson 8.3

Lesson 8.3	Start Thinking! For use before Lesson 8.3

Without using your notes from Activity 8.3, explain to a partner how to determine the degrees for a circle graph given a set of data.

Extra Example 1

The table shows the results of a survey. Display the data in a circle graph.

Favorite Pickle	People
Kosher Dill	24
Kosher Dill Spears	12
Sweet Gherkins	8
Bread & Butter Chips	4

Favorite Pickle

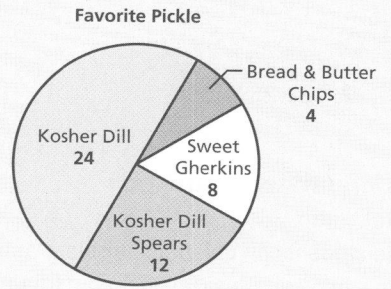

Laurie's Notes

On Your Own

1.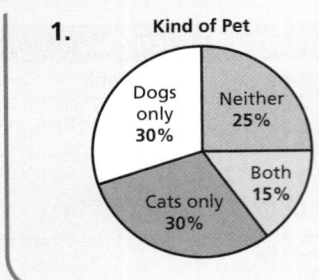

Kind of Pet

Extra Example 2

Students chose one of four books for their book report.

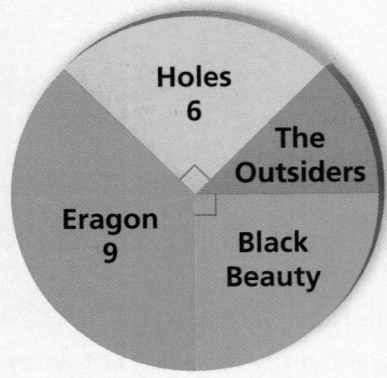

Book Reports

a. How many students are in the class? 24

b. What fraction of students chose *Eragon*? $\frac{3}{8}$

c. How many students chose *The Outsiders*? 3

On Your Own

2. 31.25%; 50%

Differentiated Instruction
Kinesthetic

Give students an unlabeled circle graph with 4 or 5 sections. Tell them the total number represented by the graph. Using a protractor, students are to determine the percent of the graph represented by each section, and then the number associated with each section.

On Your Own

 "How many teachers are part of this data?" unknown; It is common for students to think there were 100 teachers.

Example 2

- This graph has limited numerical data provided. But (from what is given), additional information can be determined.

? "What does the symbol in the blue and green regions mean?" The angles are right angles measuring 90°.

? "Which project was most popular?" Deforestation; "Least?" Plate Tectonics

? "How many students picked biotechnology?" 8 "Knowing that 8 represents 25% of the students, how many total students are there?" $4 \times 8 = 32$

On Your Own

- **Neighbor Check:** Have students work independently and then have their neighbor check their work. Have students discuss any discrepancies.

Closure

- Label the sectors of the circle graph in Example 2 with percents. Solar Energy 25%, Plate Tectonics 18.75%, Biotechnology 25%, Deforestation 31.25%

Technology For the Teacher

Dynamic Classroom

The Dynamic Planning Tool
Editable Teacher's Resources at *BigIdeasMath.com*

● On Your Own

1. The table shows the dog and cat ownership among teachers in a school. Display the data in a circle graph.

Kind of pet	Dogs only	Cats only	Both	Neither
Percent	30%	30%	15%	25%

EXAMPLE 2 Using a Circle Graph

Students chose one of four topics for their science projects. **(a)** What fraction of the students chose *Biotechnology*? **(b)** How many students are in the class? **(c)** How many students chose *Plate Tectonics*?

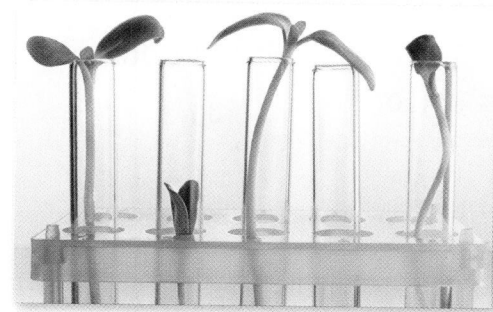

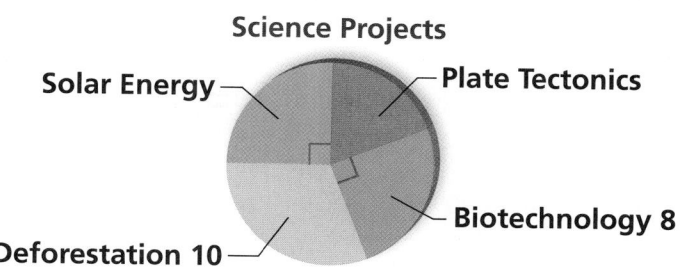

Science Projects

a. Because the *Biotechnology* section has a right angle, this section represents $\dfrac{90°}{360°} = \dfrac{1}{4}$ of the data.

⋮⋮ One-fourth of the students chose *Biotechnology*.

b. Let *x* be the number of students.

$\dfrac{1}{4}x = 8$ Use the circle graph and the results of part (a) to write an equation.

$x = 32$ Multiply each side by 4.

⋮⋮ There are 32 students in the class.

c. Because the *Solar Energy* and *Biotechnology* sections have the same angle measure, the same number of students chose each project. So, 8 students chose *Solar Energy*. Subtract to find the number of students who chose *Plate Tectonics*.

$32 - 8 - 8 - 10 = 6$

⋮⋮ Six students chose *Plate Tectonics*.

● On Your Own

2. What percent chose *Deforestation*? What percent chose either *Biotechnology* or *Solar Energy*?

Check It Out
Help with Homework
BigIdeasMath ✓com

 Vocabulary and Concept Check

1. **VOCABULARY** How do you make a circle graph when the data are given in percents?

2. **REASONING** Can one section of a circle graph be 110%? Explain.

3. **WHICH ONE DOESN'T BELONG?** Which one does *not* belong with the other three? Explain your reasoning.

$$360° \qquad 100\% \qquad 1 \qquad \frac{1}{2}$$

4. **DIFFERENT WORDS, SAME QUESTION** Which is different? Find "both" answers.

How many people chose drama?

What is 25% of 120?

What is 90% of 120?

What is $\frac{1}{4}$ of 120?

Favorite Movie

Total: 120

 Practice and Problem Solving

The circle graph shows the results of a survey on favorite fruit.

5. Which fruit is the most popular?

6. Compare the number of students who chose oranges with the number of students who chose apples.

7. The survey included 80 students. How many students chose bananas?

Favorite Fruit

Find the angle measure that corresponds to the percent of a circle.

 8. 20%   **9.** 15%   **10.** 70%   **11.** 3%

Display the data in a circle graph.

12.

Season	Rainfall (inches)
Spring	9
Summer	18
Fall	6
Winter	3

13.

Expense	Cost (dollars)
Play rights	400
Costume rental	650
Programs/tickets	300
Advertising	250
Other	400

Assignment Guide and Homework Check

Level	Day 1 Activity Assignment	Day 2 Lesson Assignment	Homework Check
Basic	5–7, 19–22	1–4, 8–14	2, 8, 12, 14
Average	5–7, 19–22	1–4, 10–16	2, 10, 12, 14
Advanced	5–7, 19–22	1–4, 8–14 even, 15–18	2, 8, 12, 16

Common Errors

- **Exercises 8–11** Students may multiply the percent by 360° without writing the percent as a fraction or decimal. Remind them that a percent is a part of a whole, so it needs to be in fraction or decimal form. Encourage them to estimate their answer. For example, 20% is close to 25%, and one-fourth of 360 is 90. So, the answer should be less than 90°.
- **Exercises 12 and 13** Students may make mistakes in creating the circle graph and end up with less than or more than 360° in the graph. Remind them that they need to determine the total number of data values. This number is the denominator in the fraction used to determine how many degrees to make each section. Encourage students to add all the degrees after they have been found to make sure that they add up to 360°.
- **Exercise 15** Students may immediately say that they can create a circle graph to represent the data because there are percents. Ask them to add up the percent values. Ask them to describe what the total percent means (some students may have given two answers).

Vocabulary and Concept Check

1. Multiply the decimal form of each percent by 360° to find the angle measure for each section.

2. no; The sum of all sections of a circle graph is 100%.

3. $\frac{1}{2}$ does not belong because it does not represent an entire circle.

4. What is 90% of 120?; 108; 30

Practice and Problem Solving

5. orange

6. The number of students who chose oranges is 4 times the number who chose apples.

7. 20 students

8. 72°   9. 54°

10. 252°   11. 10.8°

12.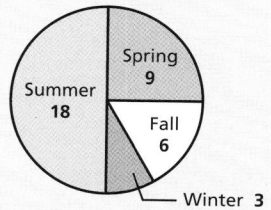

Rainfall (inches)

13.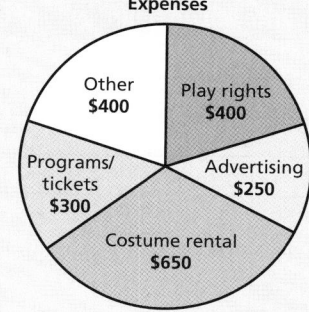

Expenses

8.3 Record and Practice Journal

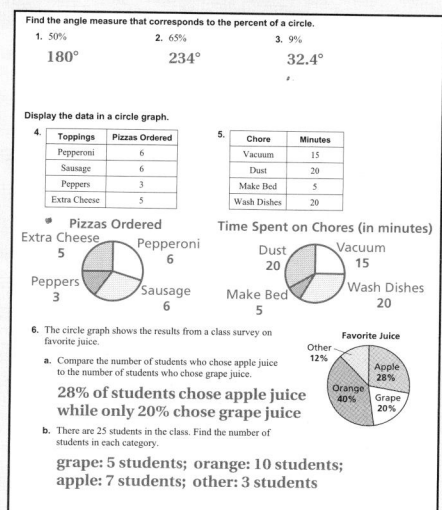

Find the angle measure that corresponds to the percent of a circle.

1. 50% 180°
2. 65% 234°
3. 9% 32.4°

Display the data in a circle graph.

4.
Toppings	Pizzas Ordered
Pepperoni	6
Sausage	6
Peppers	3
Extra Cheese	5

5.
Chore	Minutes
Vacuum	15
Dust	20
Make Bed	5
Wash Dishes	20

6. The circle graph shows the results from a class survey on favorite juice.

a. Compare the number of students who chose apple juice to the number of students who chose grape juice.
28% of students chose apple juice while only 20% chose grape juice

b. There are 25 students in the class. Find the number of students in each category.
grape: 5 students; orange: 10 students; apple: 7 students; other: 3 students

T-366

Practice and Problem Solving

14. See *Taking Math Deeper.*

15. no; The sum of the percents is greater than 100%. This would occur when students like more than one of these activities.

16. a. shirts: $2800
    pants: $1600
    shoes: $1300
    other: $1500

    b. 35°

17. *Sample answer:* Knowledge of percents, proportions, and degrees of a circle. How to convert from one form to another.

18. See Additional Answers.

Fair Game Review

19. $x = 40$

20. $n = 15$

21. $w = 1.5$

22. A

Mini-Assessment

The table shows the results of a survey. Display the results in a circle graph.

Favorite Movie Genre	People
Comedy	15
Action	25
Drama	10
Horror	10

Favorite Movie Genre

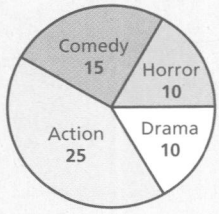

Taking Math Deeper

Exercise 14

This is a nice problem that helps students learn more about our neighbors in North America. It would be nice to start the problem with a map of North America. Show students a flat map and a globe. Discuss which type of map makes Canada look bigger.

 Help me see it.

c. Notice that the flat map makes Canada appear much larger than the U.S., but the globe makes it appear to be about the same size.

 a. Complete the table.

	U.S.	Mexico	Canada	Other
Area	3.72	0.76	3.85	1.12
Percent	39.4%	8.0%	40.7%	11.9%
Angle	140°	30°	145°	45°

b. Make a circle graph.

Round angles.

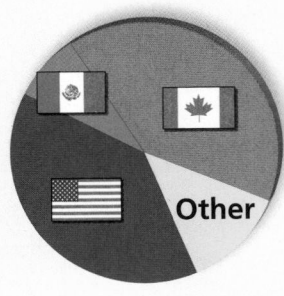

Other

Reteaching and Enrichment Strategies

If students need help. . .	If students got it. . .

Resources by Chapter
  • Practice A and Practice B
  • Puzzle Time
Record and Practice Journal Practice
Differentiating the Lesson
Lesson Tutorials
Skills Review Handbook
	Resources by Chapter
	  • Enrichment and Extension
	  • School-to-Work
	Start the next section

14. LAND AREA The table shows the land areas, in millions of square miles, of all the countries in North America.

Country	United States	Mexico	Canada	Other
Land area	3.72	0.76	3.85	1.12
Percent				
Angle in circle graph				

   a. Copy and complete the table. Round each angle to the nearest 5 degrees.
   b. Display the data in a circle graph.
   c. Find a map of North America. Do Canada and the United States appear to have the same area? Explain why or why not.

15. REASONING A survey asks a group of students what they like to do during summer vacation. The results show that 68% like to go to the beach, 45% like to go camping, 72% like to go to amusement parks, and 29% like to go to the mall. Can a circle graph be used to display these data? Explain your reasoning.

Department Store Sales

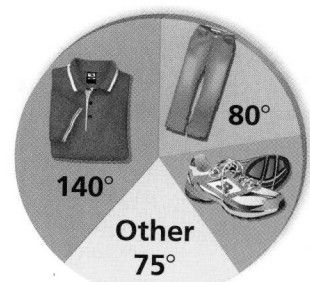

16. RETAIL A department store had $7200 in sales.

   a. Find the amount collected for each category.

   b. Long sleeve shirts were $\frac{1}{4}$ of the shirt sales. Find the angle measure of the section that would represent long sleeve shirts on the circle graph.

17. WRITING What math skills are needed to interpret data in a circle graph?

18. *Critical Thinking* Make a circle graph and a bar graph of the data in the table. Compare and contrast the two data displays. Which of the two better represents the data? Explain your reasoning.

Favorite Subject	Students
Art	12
English	56
Math	82
Music	28
Science	22

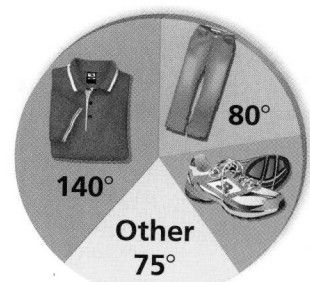

Fair Game Review What you learned in previous grades & lessons

Solve the proportion. *(Section 3.5)*

19. $\frac{5}{8} = \frac{x}{64}$

20. $\frac{6}{n} = \frac{51}{127.5}$

21. $\frac{138}{23} = \frac{9}{w}$

22. MULTIPLE CHOICE Which formula gives the surface area of a prism? *(Section 6.2)*

   Ⓐ $S = 2\ell w + 2\ell h + 2wh$

   Ⓑ $S = Ch + 2B$

   Ⓒ $S = 2\pi r^2 + 2\pi rh$

   Ⓓ $S = \frac{1}{2}C\ell + B$

8.4 Samples and Populations

Essential Question
How can you use a survey to make conclusions about the general population?

Share Your Work at...
My.BigIdeasMath.com

1 ACTIVITY: Interpreting a Survey

Work with a partner. Read the newspaper article. Analyze the survey by answering the following questions.

a. The article does not say how many "teens and young adults" were surveyed. How many do you think need to be surveyed so that the results can represent all teens and young adults in your state? in the United States? Explain your reasoning.

b. Outline the newspaper article. List all of the important points.

c. Write a questionnaire that could have been used for the survey. Do not include leading questions. For example, "Do you think your cell phone plan is restrictive?" is a leading question.

> ## The Daily Ti...
>
> VOL 01   No. 279   WEDNESDAY, OCTOBER 6, 2010
>
> ## TEXT MESSAGING SURVEY RESULTS
>
> A survey reports that almost one-third of teens and young adults believe that their text messaging plans are restrictive.
>
> About 40% say their plans lead to higher cell phone bills. According to those participating in the survey, the average number of text messages sent per day is between 6 and 7.
>
> The majority of survey participants say they would send more text messages if their cell phone plans were not as restrictive.

2 ACTIVITY: Conducting a Survey

Work with a partner. The newspaper article in Activity 1 states that the average number of text messages sent per day is between 6 and 7.

a. Does this statement seem correct to you? Explain your reasoning.

b. Plan a survey to check this statement. How will you conduct the survey?

c. Survey your classmates. Organize your data using one of the types of graphs you have studied in this chapter.

d. Write a newspaper article summarizing the results of your survey.

Laurie's Notes

Introduction

For the Teacher

- **Goal:** Students will develop an understanding of surveys and how they are used to describe a general population.

Motivate

- Conduct a quick survey of your class. Ask a couple of fun questions and then ask a math related question.
  - How many of you can roll your tongue?
  - Who likes spicy brown mustard better than yellow mustard?
  - Can you simplify a fraction?
- Discuss each of these questions, who would ask the question, and why they might be asked. Point out the following:
  - Tongue rolling is probably the most commonly used classroom example of a simple genetic trait in humans. The debate goes on in terms of whether tongue rolling is an inherited trait.
  - This question doesn't allow a person who doesn't like mustard at all to answer.
  - Teachers survey students all the time to help guide instruction.

Activity Notes

Activity 1

- **?** "Have any of you responded to a survey outside of school?" Encourage students to share the nature of the survey.
- **?** "Who conducts surveys and why?" individuals or groups (organizations) who want to find out information about a certain segment of the population
- Discuss the differences between a quick, informal survey of a local population and a formal survey of a very large population.
- **FYI:** The newspaper article was paraphrased, but the data is correct.
- Students are asked to think critically about the survey reported in the news article.
- Take time to discuss explanations offered for part (a).
- List the points generated by the class in part (b).
- Discuss questions students believe might have been asked in part (c). This is important to discuss before students try to write their own questionnaire. It is difficult for many students to think of questions that are not leading, yet are specific enough to elicit useful information.

Activity 2

- **Common Error:** Some students are very literal and may say that a person cannot send between 6 and 7 text messages a day. You can send 6 or 7, but you cannot send a partial text. This student does not understand how to interpret the results of a survey.
- Discuss the survey students would conduct in part (b).

Previous Learning

Students should know how to write and solve a proportion, and interpret a circle graph.

Start Thinking! and Warm Up

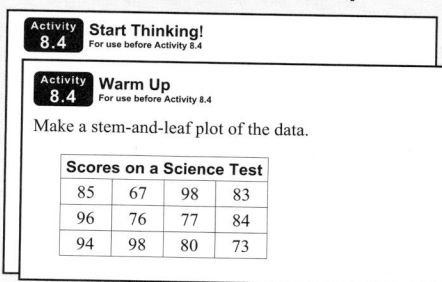

8.4 Record and Practice Journal

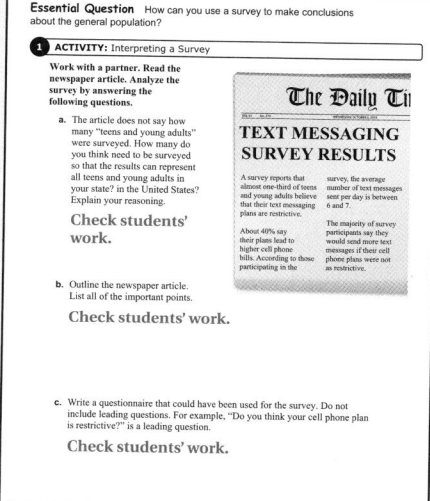

Visual

Form student groups and have each group create visual representations of the data in the "Text Messaging Survey Results." Ask students to discuss the pros and cons of showing survey results using visual representation. In Activity 2, have students predict the types of visuals they would use to interpret the questionnaire that they created.

Laurie's Notes

Activity 3

- Write "R U READY" on the board. All students, even those without a cell phone, will understand what you are asking. Ask them if they think their parents would understand, or their neighbors.
- Discuss texting shortcuts used in messaging on the computer as well as the cell phone.
- Depending upon time available, you may only be able to discuss possible questions for a survey. A subset of the class (volunteers) could write the survey and all the students could then conduct the survey.

What Is Your Answer?

- Ask volunteers to share their survey for Question 5.

Closure

- **Exit Ticket:** What survey question would you ask to find out what vegetable should be served more often for the hot lunch program at your school? Answers will vary.

8.4 Record and Practice Journal

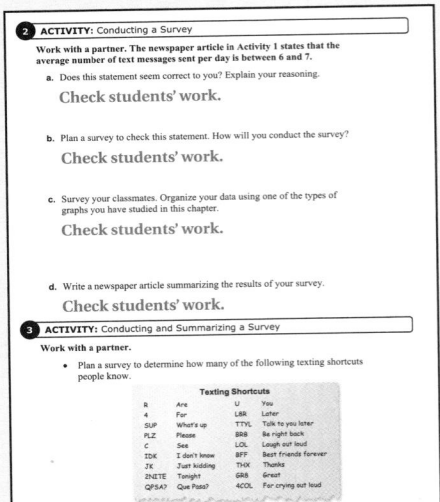

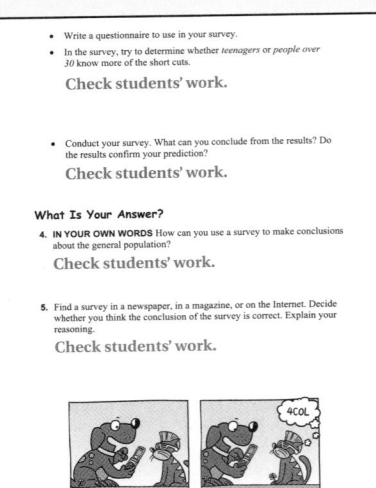

Technology
For the Teacher

Dynamic Classroom

The Dynamic Planning Tool
Editable Teacher's Resources at *BigIdeasMath.com*

Work with a partner.

- Plan a survey to determine how many of the following texting shortcuts people know.

- Write a questionnaire to use in your survey.

- In the survey, try to determine whether *teenagers* or *people over 30* know more of the shortcuts.

- Conduct your survey. What can you conclude from the results? Do the results confirm your prediction?

Texting Shortcuts

R	Are	U	You
4	For	L8R	Later
SUP	What's up	TTYL	Talk to you later
PLZ	Please	BRB	Be right back
C	See	LOL	Laugh out loud
IDK	I don't know	BFF	Best friends forever
JK	Just kidding	THX	Thanks
2NITE	Tonight	GR8	Great
QPSA?	Que Pasa?	4COL	For crying out loud

What Is Your Answer?

4. **IN YOUR OWN WORDS** How can you use a survey to make conclusions about the general population?

5. Find a survey in a newspaper, in a magazine, or on the Internet. Decide whether you think the conclusion of the survey is correct. Explain your reasoning.

"I'm sending my Mom a text message for Mother's Day."

"2 GR8 2 ME 2 EVR B 4GOT10. XX00"

Use what you learned about samples and populations to complete Exercises 3–5 on page 372.

Check It Out
Lesson Tutorials
BigIdeasMath.com

Key Vocabulary
population, *p. 370*
sample, *p. 370*

Key Idea

Samples and Population

A **population** is an entire group of people or objects. A **sample** is a part of the population.

A class is a part of an entire school.

All of the students in a school are a population.

All of the students in a class are a sample.

EXAMPLE 1 Identifying a Population and a Sample

Response	Residents
Favor road	533
Oppose road	267

An agency wants to know the opinions of county residents on the construction of a new road. The agency surveys 800 residents. Identify the population and the sample.

The population is all county residents. The sample consists of the 800 residents surveyed by the agency.

On Your Own

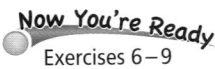
Now You're Ready
Exercises 6–9

1. You want to know how many students in your school are going to the volleyball game. You survey 50 students. Ten are going to the game. The rest are not going to the game. Identify the population and the sample.

Key Idea

Reasonable Samples

A reasonable sample is

- selected at random,
- representative of the population, and
- large enough to provide accurate data.

The results of a reasonable sample are proportional to the results of the population. So, reasonable samples can be used to make predictions about the population.

Multi-Language Glossary at BigIdeasMath.com.

Laurie's Notes

Introduction

Connect
- **Yesterday:** Students developed an understanding of surveys and how they are used to describe a general population.
- **Today:** Students will identify a reasonable sample for a population.

Motivate
- Tell students about something you read recently that reported, "Four out of five students who responded to the survey said they should have more homework!"
- After students quiet down, restate what you read, omitting a few key words. "Four out of five students said they should have more homework!"
- ? "How are the two claims different?" Students should recognize that you could survey 100 people, only 5 respond to the survey and 4 of the 5 answered one way. 95 people did not respond. In the other scenario, 80 of the 100 students answered one way.

Lesson Notes

Key Idea
- Discuss the difference between a sample and a population.
- ? "Can you give other examples of samples and populations?" 50 sheep, all the sheep in a pasture; every 10th light bulb, all the light bulbs produced

Example 1
- Have a volunteer read the problem.
- ? "What does the word *resident* mean in this problem?" Listen for student understanding of those who live in the county versus those who might be visiting or those who are part-time residents (snow birds).

On Your Own
- ? "The 50 students you surveyed are all in the school's drama program. Why might your sample be misleading?" If the drama program meets at the same time as the volleyball program, that would give misleading information about the whole school population.

Key Idea
- Discuss the three qualities of a reasonable sample. Relate the qualities to the previous question about the volleyball game.
- Discuss what it means for *the results of a reasonable sample being proportional to the results of the population.* For the volleyball problem, 10 out of 50, or 20% were going to the game. *If* this is a reasonable sample, you would *predict* 20% of the school population to attend the game.
- Remind students that a prediction is not a guarantee, only a prediction!

Extra Example 1

A company wants to know how many of its customers prefer the new improved product over the old product. The company surveys 1000 of its customers. Identify the sample and the population.
Sample: 1000 surveyed customers; Population: All of the company's customers

On Your Own

1. The population is all of the students in your school. The sample is the 50 students you surveyed.

Extra Example 2

You want to know the number of students in your classroom who do their homework right after school. You survey the first 10 students who arrive in the classroom.

a. What is the population of your survey? your class

b. What is the sample of your survey? the first 10 students that arrive

c. Is the sample reasonable? Explain. Yes. The sample is selected at random, representative of the population, and large enough to provide accurate data.

Extra Example 3

You ask 5 students to name their favorite sport. There are 600 students in the school.

Favorite Sport	Number of Students
Football	2
Soccer	2
Basketball	1

a. Predict the number *n* of students in the school who would name basketball as their favorite sport. 120

b. Is the prediction appropriate? Explain. No. The sample is not large enough to provide accurate data.

 **On Your Own**

2. C; 50 seniors at random

3. 384 students

English Language Learners
Vocabulary
English learners may easily grasp the concept of a sample, because it is similar to receiving a small amount of a product. However, the word *population* may be confusing. Relate both words to a sample of food. The sample is what the student tastes, and the population is all of the food.

Laurie's Notes

Example 2

• Work through the problem and discuss why the first three samples are not reasonable.

? "What other samples might not be reasonable?" Answers will vary.

Example 3

• Discuss the circle graph shown. Have students estimate the size of each angle shown. Green is a little more than 90°; Blue is about 110°; Red is about 150°.

? "How can you predict from the sample the number of students in the school who watch one movie a week?" Language will vary; listen for proportional to sample results.

• **Extension:** Predict the number of students in the school who watch no movies in one week.

On Your Own

• **Think-Pair-Share:** Students should read each question independently and then work with a partner to answer the questions. When they have answered the questions, the pair should compare their answers with another group and discuss any discrepancies.

Closure

• **Exit Ticket:** Describe a reasonable sample. Listen for students to mention the three qualities of a reasonable sample listed in the Key Idea.

Technology **F**or **T**eacher the **T**eacher

Dynamic Classroom

The Dynamic Planning Tool
Editable Teacher's Resources at *BigIdeasMath.com*

EXAMPLE **2** **Standardized Test Practice**

You want to estimate the number of students in a high school who ride the school bus. Which sample is best?

 Ⓐ 4 students in the hallway

 Ⓑ All students in the marching band

 Ⓒ 50 seniors at random

 Ⓓ 100 students at random during lunch

Choice A is not large enough to provide accurate data.

Choice B is not selected at random.

Choice C is not representative of the population because seniors are more likely to drive to school than other students.

 Choice Ⓓ is best. It is large and random.

EXAMPLE **3** **Making Predictions**

Movies per Week

You ask 75 randomly chosen students how many movies they watch each week. There are 1200 students in the school. **(a)** Predict the number n of students in the school who watch one movie each week. **(b)** Is the prediction appropriate? Explain.

a. Find the fraction of students in the sample who watch one movie.

$$\frac{\text{Students who watch one movie}}{\text{Number of students in sample}} = \frac{21}{75}$$

Multiply to find n.

$$n = \frac{21}{75}(1200) = 336$$

 About 336 students watch one movie each week.

b. The sample is selected at random, representative of the population, and large enough to provide accurate data.

 The sample is reasonable, so the prediction is appropriate.

On Your Own

Exercises 10–13

2. **WHAT IF?** In Example 2, you want to estimate the number of seniors in a high school who ride the school bus. Which sample should you use to make a prediction?

3. In Example 3, predict the number n of students in the school who watch two or more movies each week.

 ## Vocabulary and Concept Check

1. **VOCABULARY** Why would you survey a sample instead of a population?

2. **CRITICAL THINKING** What should you consider when conducting a survey?

 ## Practice and Problem Solving

The circle graph shows the results of a survey of 960 adults randomly chosen from different parts of the United States. In the survey, each adult was asked to name his or her favorite nut.

3. Do you think the results would be similar if the survey were conducted using middle school students? children in first grade? Explain your reasoning.

4. What other type of data display could be used to show the data?

5. Plan a survey to check the results of the survey. How could you conduct the survey so that the people surveyed would be chosen at random?

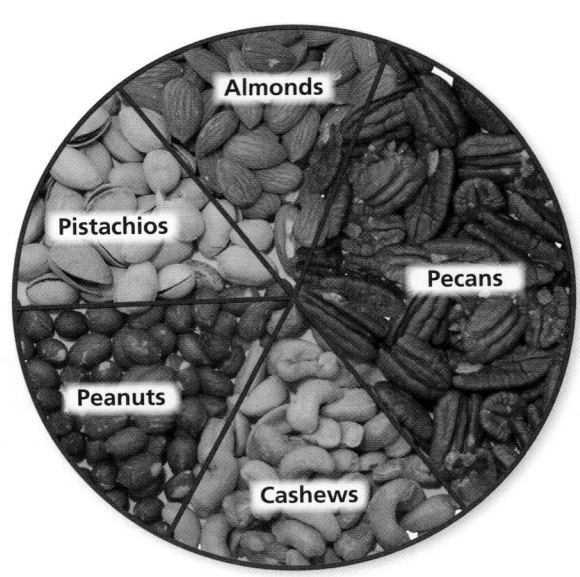

Almonds

Pistachios

Pecans

Peanuts

Cashews

Identify the population and the sample.

① 6.
Residents of New Jersey

Residents of Ocean County

7.

150 Quarters   All quarters in circulation

8.
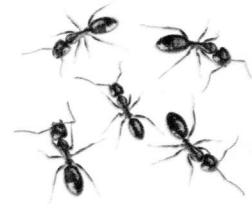

5 Ants   Colony of ants

9.

All books in library   10 library books

③ 10. **ERROR ANALYSIS** Consider the information given in Example 3. Describe and correct the error in predicting the number n of students in the school who watch zero movies each week.

$$ n = \frac{45}{75}(1200) $$

$$ n = 720 $$

Assignment Guide and Homework Check

Level	Day 1 Activity Assignment	Day 2 Lesson Assignment	Homework Check
Basic	3–5, 20–24	1, 2, 7–15 odd, 10, 14	2, 9, 11, 13, 14
Average	3–5, 20–24	1, 2, 7, 9, 10, 13–17	2, 9, 13, 14, 17
Advanced	3–5, 20–24	1, 2, 10, 12, 14–19	2, 12, 14, 18

Common Errors

- **Exercises 6–9** Students may get confused with the words population and sample. Encourage them to think about what it means to eat a sample of something, and then compare the whole to the population.
- **Exercise 11** Students may ignore the fact that the survey is conducted among students arriving at a band class. They may only see that it is a class. Ask them what kind of answers they would expect to get from different classes in the school (such as math), including band class. They should recognize that everyone going to band class will play an instrument, but not everyone in a math class will.
- **Exercises 14–16** Students may not understand why you would want to question the entire population for a survey. Ask them to estimate the population size for each survey, and then ask if it would be reasonable to ask everyone in that population for a response.

8.4 Record and Practice Journal

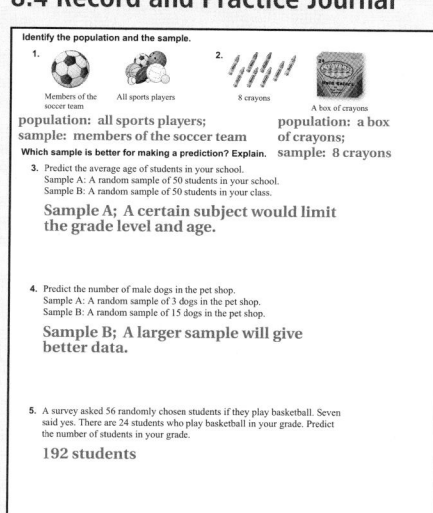

Vocabulary and Concept Check

1. Samples are easier to obtain.

2. You should make sure the people surveyed are selected at random and are representative of the population, as well as making sure your sample is large enough.

Practice and Problem Solving

3. *Sample answer:* The results may be similar for middle school students, but not for children in first grade. Children in first grade probably do not know all of these nuts.

4. bar graph

5. *Sample answer:* You could send a survey home with your classmates and have them ask one of their parents what their favorite nut is.

6. Population: Residents of New Jersey
   Sample: Residents of Ocean County

7. Population: All quarters in circulation
   Sample: 150 quarters

8. Population: Colony of ants
   Sample: 5 ants

9. Population: All books in library
   Sample: 10 books

10. There were 30, not 45, students who watched zero movies each week.
    $n = \dfrac{30}{75}(1200)$
    $n = 480$

11. **a.** Population: All students at your school
Sample: First 15 students at band class

   b. no; Your sample includes 15 students arriving at band class, and students who take band class play a musical instrument.

12. Sample B because it is a larger sample.

13. Sample A because it is representative of the population.

14–18. See Additional Answers.

19. See *Taking Math Deeper*.

 Fair Game Review

20. 62.5%

21. 31.25%

22. $77.\overline{7}\%$

23. $81.\overline{81}\%$

24. C

Mini-Assessment

Identify the population and the sample.

1. Employees attending a meeting sample; All employees of a company population

2. Four buffalo sample; A herd of buffalo population

3. You want to know the number of students in your school who play a sport. You survey the first 10 students who arrive for lunch.

   a. What is the population of your survey? All students in your school the sample? First 10 students who arrive for lunch

   b. Is the sample reasonable? Explain. No; The sample is not large enough to provide accurate data or may not be representative of the population.

Taking Math Deeper

Exercise 19

This problem may look straightforward. In reality, it is filled with questions that encompass the nature of statistical sampling.

 Straightforward Approach: 75% of the students in the sample said that they plan to go to college. Because 75% of 900 is 675, you can predict that 675 students in the high school plan to go to college.

2. Is the sample large enough to make an accurate prediction?

   One of the most surprising results in statistics is that relatively small sample sizes can produce accurate results. *IF* the sample of 60 students was random, then it is a large enough sample to provide results that are accurate.

3. Once you address the issues of sample size and randomness, there are still many other things to consider. Here are some of them:
   - How were the students asked the question? Were they asked in such a way that they felt free to tell the truth?
   - Do students in 9th grade really know whether they plan to go to college or not?
   - What time of year was the survey taken? If it was in the Spring, then the juniors and seniors should have a reasonable idea of whether they are going to college or not.

Project

Select a circle graph from the newspaper, a magazine, or online. Explain the graph. Explain another way that the data could be displayed.

Reteaching and Enrichment Strategies

If students need help. . .	If students got it. . .
Resources by Chapter • Practice A and Practice B • Puzzle Time Record and Practice Journal Practice Differentiating the Lesson Lesson Tutorials Skills Review Handbook	Resources by Chapter • Enrichment and Extension • School-to-Work Start the next section

② **11. INSTRUMENT** You want to know the number of students in your school who play a musical instrument. You survey the first 15 students who arrive at a band class.

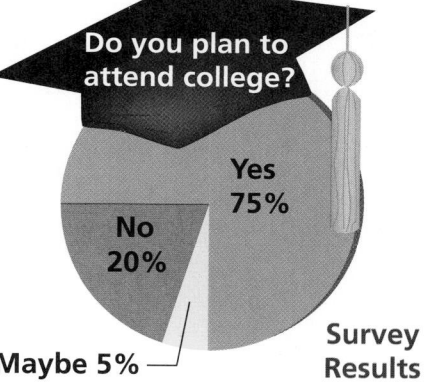

   a. What is the population of your survey? the sample?

   b. Is the sample reasonable? Explain.

Which sample is better for making a prediction? Explain.

12.

Predict the number of students in a school who like gym class.	
Sample A	A random sample of 8 students from the yearbook
Sample B	A random sample of 80 students from the yearbook

13.

Predict the number of defective pencils produced per day.	
Sample A	A random sample of 500 pencils from 20 machines
Sample B	A random sample of 500 pencils from 1 machine

Determine whether you would survey the population or a sample. Explain.

14. You want to know the average height of seventh-graders in the United States.

15. You want to know the favorite types of music of students in your homeroom.

16. You want to know the number of students in your state who have summer jobs.

Ticket Sales	
Adults	**Students**
522	210

17. THEATER A survey asked 72 randomly chosen students if they were going to attend the school play. Twelve said yes. Predict the number of students who attend the school.

18. CRITICAL THINKING Explain why 200 people with email addresses may not be a random sample.

19. ⟨Reasoning⟩ A guidance counselor surveys a random sample of 60 out of 900 high school students. Using the survey results, the counselor predicts that approximately 675 students plan to attend college. Do you agree with her prediction? Explain.

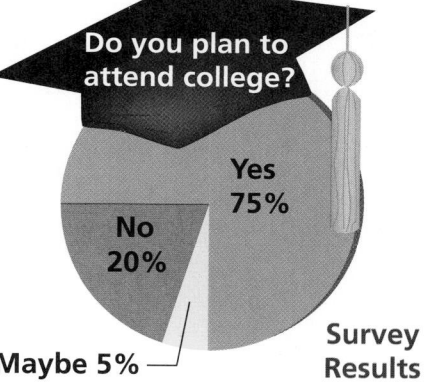

Do you plan to attend college?

Yes 75%

No 20%

Maybe 5%

Survey Results

 Fair Game Review *What you learned in previous grades & lessons*

Write the fraction as a percent. *(Skills Review Handbook)*

20. $\dfrac{5}{8}$ **21.** $\dfrac{5}{16}$ **22.** $\dfrac{21}{27}$ **23.** $\dfrac{36}{44}$

24. MULTIPLE CHOICE What is the volume of the cone? *(Section 7.4)*

   Ⓐ 16π cm^3 **Ⓑ** 108π cm^3

   Ⓒ 48π cm^3 **Ⓓ** 144π cm^3

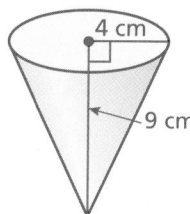

4 cm

9 cm

Check It Out
Lesson Tutorials
BigIdeasMath√com

EXAMPLE **1** **Making Predictions from Samples**

You want to know if students at a school are in favor of building a new soccer field. You conduct two surveys. For Survey 1, you randomly ask 50 students in the lunch room. For Survey 2, you randomly ask 50 student athletes. There are 1200 students in the school.

Response	Survey 1	Survey 2
Yes	17	32
No	25	11
Not Sure	8	7

a. **Use the results of each survey to predict the number n of students in the school that are in favor of building a new soccer field.**

Find the fraction of students in each sample that responded "yes."

Survey 1   **Survey 2**

$\dfrac{17}{50}$ $\dfrac{\text{Responded "yes"}}{\text{Number in sample}}$ $\dfrac{32}{50}$

Multiply each fraction by the total number of students in the school to find n for each survey.

Survey 1   **Survey 2**

$n = \dfrac{17}{50}(1200) = 408$ $n = \dfrac{32}{50}(1200) = 768$

∴ Using Survey 1, you can predict that 408 students are in favor of building a new soccer field. Using Survey 2, you can predict that 768 students are in favor of building a new soccer field.

b. **Which prediction is more reliable? Explain.**

The sample in Survey 1 is selected at random, representative of the population, and large enough to provide accurate data. This sample is reasonable.

The sample in Survey 2 is not representative of the population because student athletes are more likely to be in favor of building a soccer field. This sample is not reasonable.

∴ The prediction from Survey 1 is more reliable because it uses a reasonable sample.

Laurie's Notes

Introduction

Connect

- **Yesterday:** Students identified a reasonable sample for a population.
- **Today:** Students will use samples to make predictions and draw conclusions about two populations.

Motivate

- ❓ "Have any of you (or your parents) been asked to participate in a survey, perhaps at the mall or on the telephone?" Answers will vary.
- ❓ "If a candidate for governor wants to know how voters feel about new road construction in the southern part of the state, who might the candidate survey?" registered voters who drive in the southern part of the state
- Explain that today they will compare survey results from different samples.

Lesson Notes

Example 1

- Ask a volunteer to read the information stated.
- ❓ "How do you use results from a sample to make predictions?" Use proportional reasoning.
- Make sure students understand that Survey 1 and 2 both consist of 50 students, but Survey 2 only consists of athletes.
- Work through parts (a) and (b) of the example.
- ❓ "Would surveying 50 band students give reliable results? Explain." Uncertain; Band students may be more likely to favor building a new field if they would be performing on the field.

Warm Up

Warm Up
For use before Lesson 8.4b

Identify the population and the sample.

1. You want to know how many students in your school are buying a yearbook. You survey 100 students. Sixty are buying a yearbook. The rest are not buying a yearbook.

Extra Example 1

You want to know which field trip students in a school prefer. You conduct two surveys. For Survey 1, you randomly ask 25 students in art classes. For Survey 2, you randomly ask 25 students in the cafeteria. There are 600 students in the school.

Field Trip	Survey 1	Survey 2
Art Museum	14	8
Zoo	8	7
Planetarium	3	10

a. Use the results of each survey to predict the number n of students in the school that prefer a field trip to the art museum. Survey 1: 336; Survey 2: 192
b. Which prediction is more reliable? Explain. The prediction from Survey 2 is more reliable because it uses a reasonable sample.

Record and Practice Journal Practice

See Additional Answers.

T-373A

Extra Example 2

The double box-and-whisker plot shows average temperatures of a random sample of days in Boston and Las Vegas. Compare the average temperatures in Las Vegas to the average temperatures in Boston.

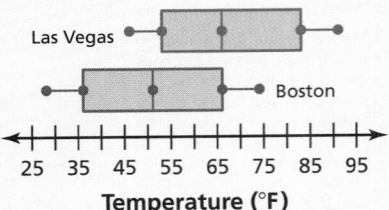

Temperature (°F)

In general, the average temperature in Las Vegas is greater than the average temperature in Boston.

Practice

1. See Additional Answers.

2. Check students' work.

3. In general, boys are taller than girls.

4. See Additional Answers.

Mini-Assessment

1. You want to know if students in a school will attend a classical music concert. You conduct two surveys. For Survey 1, you randomly ask 30 students in the lunchroom. For Survey 2, you randomly ask 30 students in the band and orchestra. There are 1500 students in the school.

Response	Survey 1	Survey 2
Yes	7	22
No	15	5
Not Sure	8	3

a. Use the results of each survey to predict the number n of students in the school that will attend the concert. Survey 1: 350; Survey 2: 1100

b. Which prediction is more reliable? Explain. The prediction from Survey 1 is more reliable because it uses a reasonable sample.

Laurie's Notes

Discuss

- Review box-and-whisker plots before beginning Example 2.
- A box-and-whisker plot uses 5 key values displayed on a number line. Quartiles divide the data into four parts, each representing 25% of the data.
- Box-and-whisker plots allow two data sets of different sizes to be compared visually.

Example 2

- Before discussing the problem, ask students if they have any observations about the two box-and-whisker plots.
- ❓ "What is the range of gas mileages for trucks?" 12 miles per gallon "For cars?" 16 miles per gallon
- ❓ "What is the greatest gas mileage for a truck?" 26 miles per gallon "For a car?" 36 miles per gallon
- **Note:** You cannot determine how many cars or trucks are represented in the box-and-whisker plot.
- Analyze the box-and-whisker plot as shown.
- ❓ "True or false: No truck had greater gas mileage than the upper 50% of the cars. Explain." True; The median of the cars is greater than the greatest value of the trucks.

Practice

- **Extension:** If there is a current issue at your school that students are interested in, have the students conduct a random sample and draw conclusions.
- **Common Error:** Exercise 4, students may forget to order the data before finding the key values.

Closure

- Draw the following box-and-whisker plot on the board. The double box-and-whisker plot shows the science mid-term test scores for 6th and 7th graders. Have students compare the Grade 6 test scores to the Grade 7 test scores.

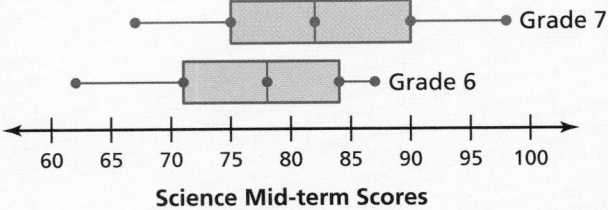

Science Mid-term Scores

Technology **F**or the **T**eacher

The Dynamic Planning Tool
Editable Teacher's Resources at *BigIdeasMath.com*

EXAMPLE **2** **Using Samples to Compare Populations**

The double box-and-whisker plot shows the gas mileages of random samples of cars and trucks. Compare the gas mileages of cars to the gas mileages of trucks.

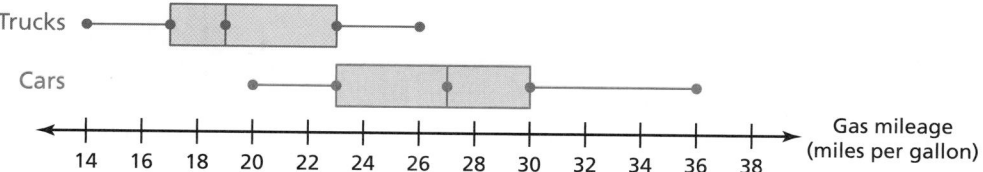

Trucks: The third quartile is 23 miles per gallon. So, 25% of the trucks get 23 miles per gallon or *more*.

Cars: The first quartile is 23 miles per gallon. So, 75% of the cars get 23 miles per gallon or *more*.

⋮ In general, cars get better gas mileage than trucks.

Practice

1. **SPORTS** You want to survey students in your grade about their favorite sport. Describe samples that are (a) reasonable and (b) not reasonable.

2. **PROJECT** Conduct the surveys in Exercise 1 at your school.

   a. Use the results of each survey to predict the favorite sport of the students in your grade.

   b. Do you think your predictions are reasonable? Explain.

3. **HEIGHT** The double box-and-whisker plot shows the heights of random samples of boys and girls in a school. Compare the heights of boys to the heights of girls.

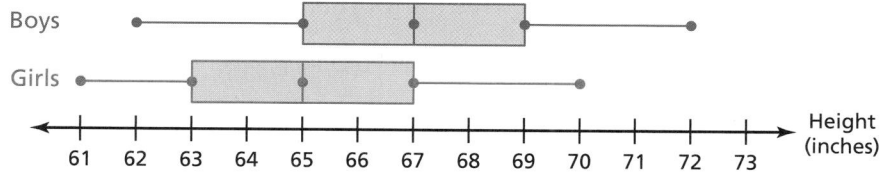

4. **GRADES** Each table shows a random sample of test grades in a class. Create a double box-and-whisker plot of the data. Compare the two data sets.

Grades in Mr. Smith's Class			
72	68	84	87
76	75	52	73
88	84	69	71
76	78	86	82

Grades in Mrs. Higsbee's Class			
85	88	93	78
76	65	71	86
90	96	85	88
79	90	82	94

Check It Out
Progress Check
BigIdeasMath.com

Identify the population and the sample. *(Section 8.4)*

1.

Passengers on a train

Passengers in the first train car

2.

DVDs in a video store   DVDs in the comedy section

3. Display the data in a circle graph. *(Section 8.3)*

Favorite Book Genre	Students
Fantasy	20
Historical Fiction	10
Mystery	15
Nonfiction	5

Which sample is better for making a prediction? Explain. *(Section 8.4)*

4.
Predict the number of people who plan to vote in this year's election.	
Sample A	A random sample of 5000 registered voters
Sample B	A random sample of 50,000 United States citizens

5.
Predict the number of defective light bulbs produced per day.	
Sample A	A sample of the last 1000 light bulbs produced during a day
Sample B	A random sample of 1000 light bulbs produced throughout a day

6. **EXERCISE EQUIPMENT** You want to know how many students in your school support the purchase of new exercise equipment for the gym this year. You survey the first 20 students who arrive for football team tryouts. *(Section 8.4)*

   a. What is the population for your survey? the sample?

   b. Is the sample reasonable? Explain.

7. **WORLD LANGUAGES** Every student in a middle school takes one world language class. *(Section 8.3)*

   a. What fraction of the students take French?

   b. How many students are in the school?

   c. How many students take Spanish?

World Language Classes

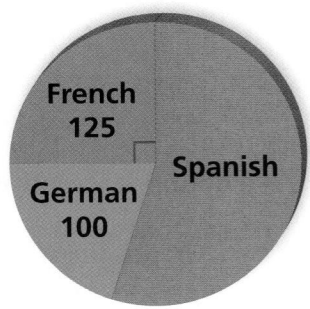

French 125

Spanish

German 100

8. **NEWSPAPERS** A survey asks 48 randomly chosen students if they plan to buy a school newspaper this week. Of the 48 surveyed, 32 plan to buy a school newspaper. Predict the number of students enrolled at the school. *(Section 8.4)*

Newspaper Sales	
Faculty	Students
18	360

Alternative Assessment Options

Math Chat   Student Reflective Focus Question

Structured Interview   Writing Prompt

Student Reflective Focus Question

Ask students to summarize the similarities and differences between samples and populations. Be sure that they include examples. Select students at random to present their summary to the class.

Study Help Sample Answers

Remind students to complete Graphic Organizers for the rest of the chapter.

2.
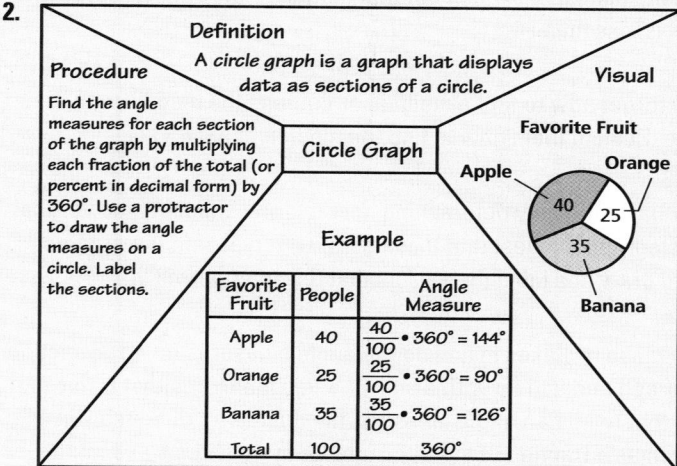

3.
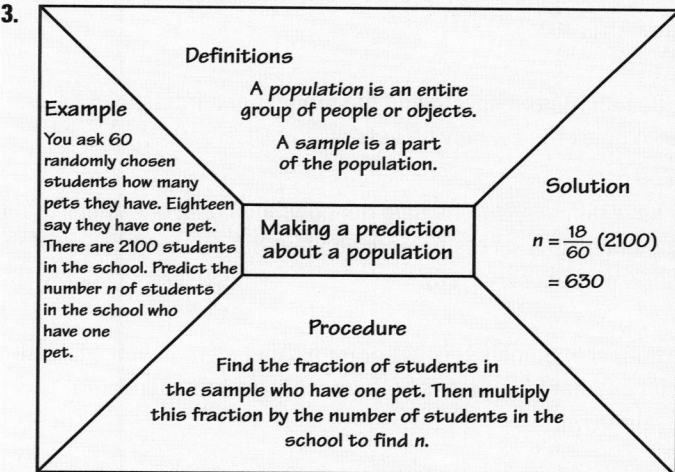

4. Available at *BigIdeasMath.com*

Reteaching and Enrichment Strategies

If students need help. . .	If students got it. . .

Resources by Chapter
  • Study Help
  • Practice A and Practice B
  • Puzzle Time
Lesson Tutorials
BigIdeasMath.com Practice Quiz
Practice from the Test Generator
	Resources by Chapter
	  • Enrichment and Extension
	  • School-to-Work
	Game Closet at *BigIdeasMath.com*
	Start the Chapter Review

Answers

1. population: passengers on a train
   sample: passengers in the first train car

2. population: DVDs in the video store
   sample: DVDs in the comedy section

3.
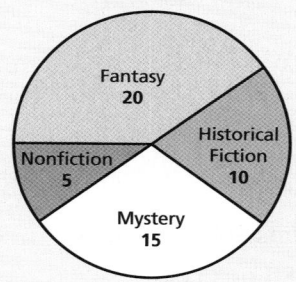

4. Sample A because the population you want is registered voters.

5. Sample B because the sample is random.

6–7. See Additional Answers.

8. 540 students

Technology For the Teacher

Answer Presentation Tool

Assessment Book

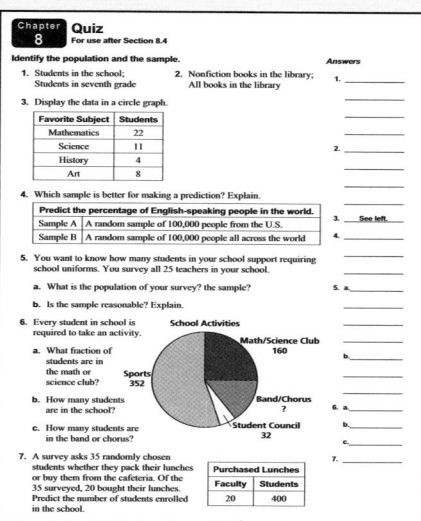

Additional Review Options
- **Quiz**Show
- Big Ideas Test Generator
- Game Closet at *BigIdeasMath.com*
- Vocabulary Puzzle Builder
- Resources by Chapter
  - Puzzle Time
  - Study Help

Answers

1. **Hats Sold Each Day**

Stem	Leaf
0	5 8 9
1	2 2 3 4 5 8
2	1 5
3	0

Key: $2 \mid 1 = 21$ hats

2. **Ages of Park Volunteers**

Stem	Leaf
1	3 3 5 7 9
2	0 1
3	
4	0 8
5	2 5
6	0

Key: $1 \mid 3 = 13$ years old

3. 6 tuna

4. 87 pounds

Review of Common Errors

Exercises 1 and 2
- Students may forget to include the zeros in the ones place as leaves, or forget to include repeats of numbers. Remind them that *all* of the numbers in a data set should be represented in a stem-and-leaf plot.
- Students may forget to include stems that do not have leaves. Remind them that a stem-and-leaf plot shows how data are distributed, so it is important to include all of the stems in a range of data.

Exercise 3
- Students may misinterpret the question and count the data values represented with a stem of 9. Point out that neither of the values represented with this stem is less than 90.

Exercise 4
- Students may forget how to find a median, or confuse the median with the mean or mode. Remind them how to find the median of a data set.

Exercises 5 and 6
- Students may have difficulty determining how to draw and label the vertical axis of the histogram. Suggest that they find the interval with the greatest frequency and draw and label the axis so that this interval will fit.

Exercises 7 and 8
- Students may make mistakes in finding the angle measures for each section of the circle graph and end up with a sum of angle measures that is not 360°. Remind them that to find an angle measure, the number of data items in a group is written as a fraction of the total number of data items and multiplied by 360°. Encourage students to check their work and make sure the sum of the angle measures is 360°.

Exercise 9
- Students may have forgotten how to find the product of a fraction and a whole number. A quick review may be helpful.

Exercise 10
- Students may have difficulty determining the population and the sample. Point out that the population is all of the parents and that the sample is a part of the population (the 50 parents surveyed).

Exercise 11
- Students may think that Sample B is better for making a prediction because it is larger. Point out that Sample A is better because it represents the population of your town.

8 Chapter Review

Review Key Vocabulary

stem-and-leaf plot, *p. 350*   histogram, *p. 356*   population, *p. 370*
stem, *p. 350*   circle graph, *p. 364*   sample, *p. 370*
leaf, *p. 350*

Review Examples and Exercises

8.1 Stem-and-Leaf Plots *(pp. 348–353)*

Make a stem-and-leaf plot of the number of DVDs rented each day at a store.

Step 1: Order the data. 19, 25, 28, 39, 50, 50, 53

Step 2: Choose the stems and leaves. Because the data range from 19 to 53, use the *tens* digits for the stems and the *ones* digits for the leaves.

Step 3: Write the stems to the *left* of the vertical line.

Step 4: Write the leaves for each stem to the *right* of the vertical line.

Day	DVDs Rented
Sun.	50
Mon.	19
Tue.	25
Wed.	28
Thu.	39
Fri.	53
Sat.	50

Order the stems vertically. The stem for data values less than 10 is 0.

Include stems without leaves.

Write the leaves horizontally.

DVDs Rented

Stem	Leaf
1	9
2	5 8
3	9
4	
5	0 0 3

Key: 2|5 = 25 DVDs

Exercises

Make a stem-and-leaf plot of the data.

1.

Hats Sold Each Day			
5	18	12	15
21	30	8	12
13	9	14	25

2.

Ages of Park Volunteers			
13	17	40	15
48	21	19	52
13	55	60	20

The stem-and-leaf plot shows the weights (in pounds) of yellowfin tuna caught during a fishing contest.

3. How many tuna weigh less than 90 pounds?

4. What is the median weight of the tuna?

Weights of Tuna

Stem	Leaf
7	6
8	0 2 5 7 9
9	5 6
10	2

Key: 8|5 = 85 pounds

8.2 Histograms (pp. 354–359)

The frequency table shows the number of crafts each member of the Craft Club made for a fundraiser. Display the data in a histogram.

Crafts	Frequency
0–2	10
3–5	8
6–8	5
9–11	0
12–14	2

Step 1: Draw and label the axes.

Step 2: Draw a bar to represent the frequency of each interval.

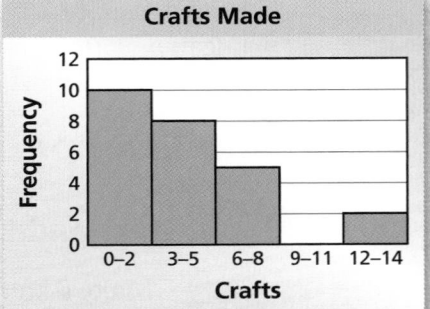

Crafts Made

Exercises

Display the data in a histogram.

5.

Heights of Gymnasts	
Heights (in.)	Frequency
50–54	1
55–59	8
60–64	5
65–69	2

6.

Minutes Studied	
Minutes	Frequency
0–19	5
20–39	9
40–59	12
60–79	3

8.3 Circle Graphs (pp. 362–367)

The table shows the results of a survey of 50 students. Display the data in a circle graph.

Favorite P.E. Activity	Students
Badminton	15
Volleyball	10
Kickball	25

Step 1: Find the angle measure for each section of the graph.

Multiply the fraction of students who chose each activity by 360°.

Badminton

$$\frac{15}{50} \cdot 360° = 108°$$

Volleyball

$$\frac{10}{50} \cdot 360° = 72°$$

Kickball

$$\frac{25}{50} \cdot 360° = 180°$$

Step 2: Use a protractor to draw the angle measures on a circle. Label the sections.

180°   108°   72°

Kickball 25   Badminton 15   Volleyball 10

Review Game
Sampling and Predictions

For the Student
Additional Practice
- Lesson Tutorials
- Study Help (textbook)
- Student Website
  Multi-Language Glossary
  Practice Assessments

Materials per group:
- pencils
- paper

Directions:

Divide the class into groups of 2 to 4 students. Each group comes up with a survey question. For instance, a group's survey question could be "How many pets does a student have?" Make sure each group has a different question.

Each group surveys 25% (or some other predetermined percentage) of the class and makes a display of the results, choosing the most appropriate data display. Groups then use their display to predict the results for the entire class.

Each group then surveys the entire class and makes a display of these results, using the same type of data display chosen to analyze the sample. Groups then write a short paragraph in which they compare the entire class actual results to the entire class predicted results.

The larger the sample size, the more accurate the predictions may be. If possible, arrange to involve more students, in addition to the students in your class, in the surveys.

Who Wins?

The group whose actual results are closest to the predicted results wins.

Answers

5.

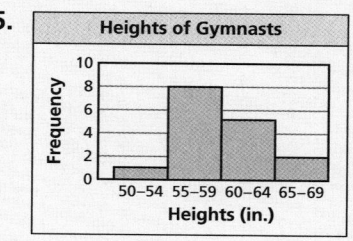

6.

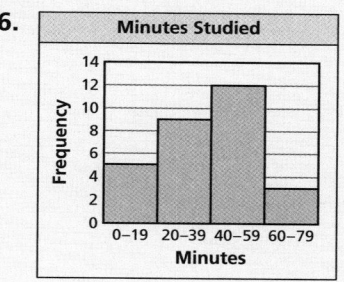

7.

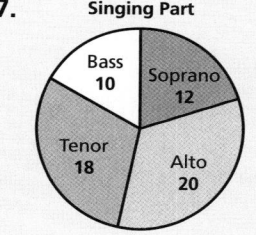

8.

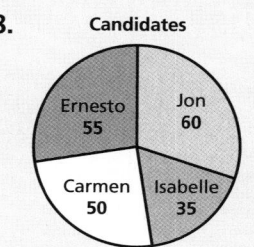

9. 180 students

10. population: all of the parents
sample: 50 parents

11. Sample A because Sample B is not representative of the people in your town.

My Thoughts on the Chapter

What worked. . .

What did not work. . .

What I would do differently. . .

Display the data in a circle graph.

7.

Singing Part	Students
Soprano	12
Alto	20
Tenor	18
Bass	10

8.

Candidate	Votes
Jon	60
Isabelle	35
Carmen	50
Ernesto	55

8.4 Samples and Populations *(pp. 368–373)*

You ask 80 randomly chosen students how many pets they have. There are 600 students in the school. (a) Predict the number _n_ of students in the school who have exactly one pet. (b) Is the prediction appropriate? Explain.

a. Find the fraction of students in the sample who have exactly one pet.

$$\frac{\text{Students who have exactly one pet}}{\text{Number of students in sample}} = \frac{42}{80}$$

Multiply to find _n_.

$$n = \frac{42}{80}(600) = 315$$

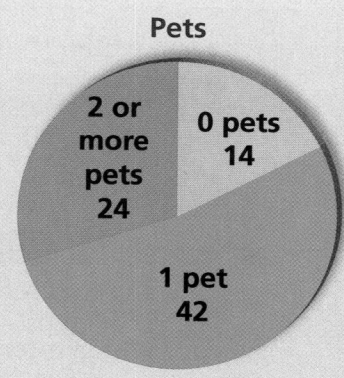

Pets

⋮ About 315 students in the school have exactly one pet.

b. The sample is selected at random, is representative of the population, and is large enough to provide accurate data.

⋮ The sample is reasonable, so the prediction is appropriate.

9. Use the information in the Example above. Predict the number _x_ of students in the school who have two or more pets.

10. Your principal wants to know how many parents plan to attend Back-to-School Night. The principal surveys 50 parents and finds that 40 plan to attend. Identify the population and the sample.

11. Which sample is better for making a prediction? Explain.

Predict the number of people in your town who support building a new library.	
Sample A	A random sample of 500 people in your town
Sample B	A random sample of 5000 people in your state

Check It Out
Test Practice
BigIdeasMath.com

Make a stem-and-leaf plot of the data.

1.

Quiz Scores (%)			
96	88	80	72
80	94	92	100
76	80	68	90

2.

CDs Sold Each Day				
45	31	29	38	38
67	40	62	45	60
40	39	60	43	48

3. Display the data in a histogram.

Television Watched Per Week	
Hours	**Frequency**
0–9	14
10–19	16
20–29	10
30–39	8

4. Display the data in a circle graph.

Category	Amount Spent ($)
Clothing	30
Entertainment	10
Food	5
Savings	15

5. Which sample is better for making a prediction? Explain.

Predict the number of students in your school who play at least one sport.	
Sample A	A random sample of 10 students from the school student roster
Sample B	A random sample of 80 students from the school student roster

6. WATER The histogram shows the number of glasses of water that the students in a class drink in one day.

    a. Which interval contains the fewest data values?

    b. How many students are in the class?

    c. Health experts recommend drinking at least 8 glasses of water per day. What percent of the students drink the recommended amount?

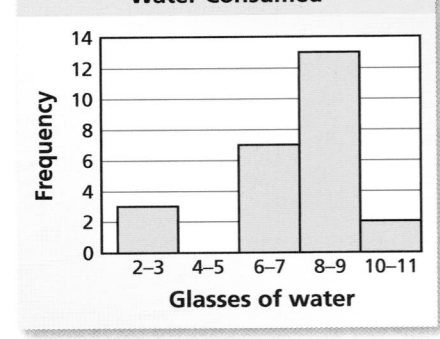

7. FIELD TRIP Of 60 randomly chosen students surveyed, 16 chose the aquarium as their favorite field trip. There are 720 students in the school. Predict the number of students in the school who would choose the aquarium as their favorite field trip.

8. MALL There are 240 stores in a mall.

    a. Find the number of stores in each category.

    b. Electronics stores make up $\frac{1}{5}$ of the "Other" category. Find the angle measure of the section that would represent electronics.

Test Item References

Chapter Test Questions	Section to Review
1, 2	8.1
3, 6	8.2
4, 8	8.3
5, 7	8.4

Test-Taking Strategies

Remind students to quickly look over the entire test before they start so that they can budget their time. On this test, students are asked to display as well as analyze data. It can be difficult for students to determine how to begin. So, remind students to use the **Stop** and **Think** strategy before they answer a question.

Common Assessment Errors

- **Exercises 1 and 2** Students may forget to include the zeros in the ones place as leaves, or forget to include repeats of numbers. Remind them that *all* of the numbers in a data set should be represented in a stem-and-leaf plot.
- **Exercise 2** Students may forget to include 5 as a stem because it does not have leaves. Remind them that a stem-and-leaf plot shows how data are distributed, so it is important to include all of the stems in a range of data.
- **Exercise 3** Students may have difficulty determining how to draw and label the vertical axis of the histogram. Suggest that they find the interval with the greatest frequency and draw and label the axis so that this interval will fit.
- **Exercise 4** Students might make mistakes in finding the angle measures for each section of the circle graph and end up with a sum of angle measures that is not 360°. Remind them that to find an angle measure, the number of data items in a group is written as a fraction of the total number of data items and multiplied by 360°. Encourage students to check their work and make sure the sum of the angle measures is 360°.
- **Exercise 7** Students might write the fraction of students in the sample as $\frac{60}{16}$ and predict that 2700 students in the school would choose the aquarium as their favorite field trip, which is more than the number of students in the school. Remind them that they should make sure their answer is reasonable.

Reteaching and Enrichment Strategies

If students need help...	If students got it...
Resources by Chapter • Practice A and Practice B • Puzzle Time Record and Practice Journal Practice Differentiating the Lesson Lesson Tutorials Practice from the Test Generator Skills Review Handbook	Resources by Chapter • Enrichment and Extension • School-to-Work • Financial Literacy Game Closet at *BigIdeasMath.com* Start Standardized Test Practice

Answers

1–3. See Additional Answers.

4.

Amount Spent

- Entertainment $10
- Food $5
- Clothing $30
- Savings $15

5. Sample B because the sample size is larger.

6. a. 4–5

   b. 25 students

   c. 60%

7. 192 students

8. a. Other: 70
   Fashion and Accessories: 80
   Specialty: 30
   Restaurants: 60

   b. 21°

Assessment Book

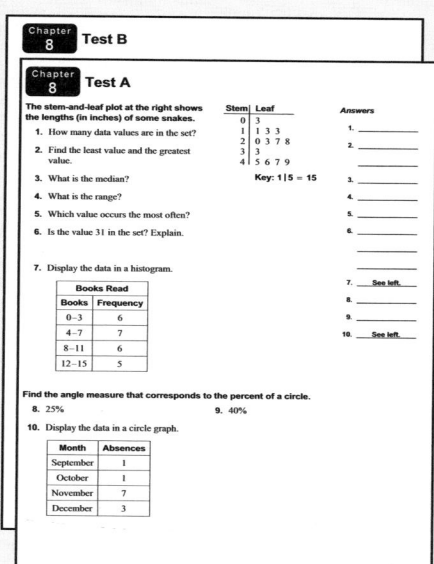

T-378

After Answering Easy Questions, Relax

Answer Easy Questions First

Estimate the Answer

Read All Choices before Answering

Read Question before Answering

Solve Directly or Eliminate Choices

Solve Problem before Looking at Choices

Use Intelligent Guessing

Work Backwards

About this Strategy

When taking a multiple choice test, be sure to read each question carefully and thoroughly. Sometimes you do not know the answer. So . . . guess intelligently! Look at the choices and choose the ones that are possible answers.

Answers

1. D
2. F
3. 7

Item Analysis

1. **A.** The student chooses the greatest deposit, but this will earn less interest than answer choice D.

   B. The student chooses an answer choice that has a longer term than answer choice C and a greater interest rate than answer choices A and D, but this will earn less interest than answer choices A and D.

   C. The student chooses the greatest interest rate, but this will earn the least interest of all the answer choices.

   D. Correct answer

2. **F.** Correct answer

   G. The student visually estimates the size of the trumpet section to be 3 times the rhythm section.

   H. The rhythm, saxophone, and trombone sections together comprise 56% of the graph. The student thinks that this means 56 students and estimates the number of students in the trumpet section to be about half of the 56 students.

   I. The student determines that 44% of the students are in the trumpet section and thinks that this means 44 students.

3. **Gridded Response**: Correct answer: 7

   Common Error: The student thinks that the missing leaf represents the ones digit of the actual median, getting an answer of 8.

4. **A.** The student divides before subtracting to isolate the variable term.

   B. Correct answer

   C. The student incorrectly thinks that $-6 - 4 = -2$.

   D. The student incorrectly combines unlike terms, thinking that $4 + y = 4y$.

5. **F.** The student finds the number of guests that could be served with 4-ounce servings, but does not subtract 96 from this answer.

   G. The student uses direct variation instead of inverse variation and incorrectly determines the number of guest that could be served with 4-ounce servings. The student also does not subtract this answer from 96.

   H. Correct answer

   I. The student uses direct variation instead of inverse variation and incorrectly determines the number of guest that could be served with 4-ounce servings.

Technology
For
the **T**eacher

Big Ideas Test Generator

1. Which deposit will earn the most simple interest for the given terms and annual interest rates?

   A. $3300 for 1 year at 4%

   B. $2000 for 1 year at 5.5%

   C. $2500 for 6 months at 8%

   D. $3000 for 18 months at 3%

2. The band instructor made the circle graph below to show the percent of students in each section of the Jazz Band.

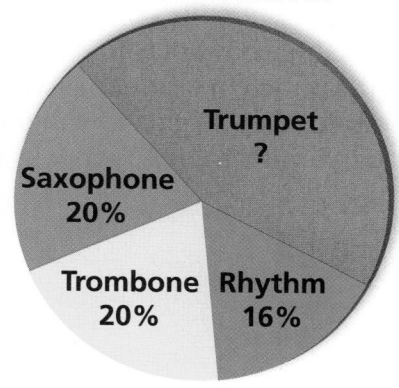

Jazz Band Sections

Trumpet ?
Saxophone 20%
Trombone 20%
Rhythm 16%

There are 4 students in the rhythm section. What is the number of students in the trumpet section?

   F. 11

   G. 12

   H. 28

   I. 44

3. One of the leaves is missing in the stem-and-leaf plot.

   The median of the data set represented by the stem-and-leaf plot is 38. What is the value of the missing leaf?

Stem	Leaf
1	3 4
2	
3	4 5 7 7 7 ? 9
4	0 1 1 4
5	0 2 3

Key: 1|4 = 14

Test-Taking Strategy
Use Intelligent Guessing

What is the mean length of these hyena fangs: 4 in., 3 in., 3 in., 4 in., 5 in., 5 in.?

(A) 6 in.   (B) $\frac{1}{3}$ ft   (C) 2 in.   (D) 5 in.

MEOW!

"The mean can't be 6 or 2 or 5 inches. So, you can use intelligent guessing to find that the answer is $\frac{1}{3}$ ft, or 4 in."

4. Mario was solving the equation in the box below.

$$3y + 4 + y = 5 - 11$$
$$3y + 4 = -6$$
$$3y = -10$$
$$y = -\frac{10}{3}$$

What should Mario do to correct the error that he made?

A. Divide -6 by 3 to get 2.

B. Add $3y$ and y to get $4y$.

C. Subtract 4 from -6 to get -2.

D. Simplify $3y + 4 + y$ to get $7y$.

5. The president of a service organization made a large bowl of fruit punch for a party. She needs to decide whether to serve the punch in 6-ounce servings or 4-ounce servings. She determined that she could serve 96 guests with 6-ounce servings. How many more guests could she serve with 4-ounce servings than with 6-ounce servings?

F. 144

G. 64

H. 48

I. 32

6. At the end of the school year, your teacher counted up the number of absences for each student. The results are shown in the histogram below.

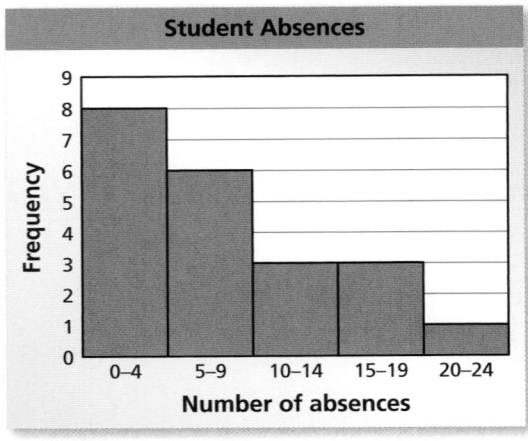

Based on the histogram, how many students had fewer than 10 absences?

Item Analysis (continued)

6. **Gridded Response:** Correct answer: 14 students

   Common Error: The student only considers the bar immediately to the left of 10, getting an answer of 6.

7. **A.** Correct answer

   B. The student thinks that $\big| 2 - (-6) \big| = 4$, or the student thinks that $\big| 2 - (-6) \big| = -4$ and $-\big| 4 \big| = 4.$

   C. The student thinks that $\big| 2 - (-6) \big| = -4$, or the student thinks that $\big| 2 - (-6) \big| = 4$ and $-\big| 4 \big| = 4.$

   D. The student thinks that $\big| 2 - (-6) \big| = -8$, or the student thinks that $-\big| 4 \big| = 4.$

8. **F.** The student finds the sum of the area of the base and only one of the lateral faces.

   G. The student finds only the sum of the areas of the lateral faces.

   H. Correct answer

   I. The student tries to find the volume of the pyramid, using the slant height instead of the height.

9. **A.** The student incorrectly thinks that doubling all the edge lengths doubles the volume.

   B. The student incorrectly thinks that doubling all the edge lengths multiplies the volume by 2^2.

   C. The student incorrectly thinks that doubling all the edge lengths multiplies the volume by $(2 + 2 + 2)$.

   D. Correct answer

Answers

4. B

5. H

6. 14 students

7. A

8. H

9. D

10. *Part A* 53

   Part B 57

   Part C 63.5

   Part D 69.05

1. **A.** The student chooses a mapping which reflects the point in the *y*-axis.

   B. The student chooses a mapping which reflects the point in the *y*-axis.

   C. Correct answer

   D. The student chooses a mapping which rotates the point 180°.

2. **F.** The student finds what percent $15.00 is of $6.00.

   G. Correct answer

   H. The student subtracts $6.00 from $15.00 to get $9.00 and thinks that this means 90%.

   I. The student finds what percent $6.00 is of $15.00.

Item Analysis (continued)

10. **4 points** The student demonstrates a thorough understanding of determining range, mode, median, and mean from a stem-and-leaf plot. In Part A, the student explains that a stem-and-leaf plot is ordered and that it is easy to determine the minimum of 45 and the maximum of 98 to get a range of 53. In Part B, the student explains that the leaf that is repeated the most in any one row indicates the mode, which in this case is 57. In Part C, the student explains that a stem-and-leaf plot is already ordered, which is a necessary step toward finding the median, which in this case is 63.5. In Part D, the student explains that a stem-and-leaf plot organizes parts of numbers by place value and can make finding the sum of the data easier. In this case, the sum of the data can be found by the following expression: 3(40) + 5(50) + 4(60) + 6(80) + 2(90) + (5 + 8 + 8 + 2 + 7 + 7 + 7 + 8 + 0 + 1 + 6 + 8 + 3 + 3 + 4 + 8 + 9 + 9 + 0 + 8) for a total of 1381, which divided by 20 gets a mean of 69.05.

3 points The student demonstrates an understanding of determining range, mode, median, and mean from a stem-and-leaf plot, but the student's work demonstrates an essential but less than thorough understanding.

2 points The student demonstrates a partial understanding of determining range, mode, median, and mean from a stem-and-leaf plot. The student's work demonstrates a lack of essential understanding.

1 point The student demonstrates a limited understanding of determining range, mode, median, and mean from a stem-and-leaf plot. The student's response is incomplete and exhibits many flaws.

0 points The student provides no response, a completely incorrect or incomprehensible response, or a response that demonstrates insufficient understanding of determining range, mode, median, and mean from a stem-and-leaf plot.

Extra Examples for Standardized Test Practice

1. Which point is mapped to its image after being reflected in the *x*-axis?

   A. $(-3, -2) \rightarrow (3, -2)$

   B. $(4, 5) \rightarrow (-4, 5)$

   C. $(-2, 8) \rightarrow (-2, -8)$

   D. $(-7, -1) \rightarrow (7, 1)$

2. The value of one of Kevin's baseball cards was $6.00 when he first got it. The value of this card is now $15.00. What was the percent increase in the value of the card?

   F. 250%   **H.** 90%

   G. 150%   **I.** 40%

7. What is the value of the expression below when $a = -6$ and $b = 4$?

$$\frac{|2 - a|}{-|b|}$$

A. -2

B. -1

C. 1

D. 2

8. A right square pyramid and its dimensions are shown below.

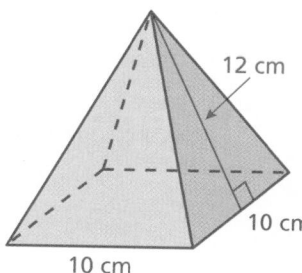

12 cm

10 cm

10 cm

What is the surface area of the pyramid?

F. 160 cm^2

G. 240 cm^2

H. 340 cm^2

I. 400 cm^2

9. Suppose that the volume of a right rectangular prism is V. The lengths of all the edges of the prism are then doubled. What is the volume of the new prism that is created?

A. $2V$

B. $4V$

C. $6V$

D. $8V$

10. A stem-and-leaf plot is shown.

For Parts A–D, explain how the stem-and-leaf plot can help to find each measure. Then find each measure using the stem-and-leaf plot. Show your work and explain your reasoning for each answer.

Stem	Leaf
4	5 8 8
5	2 7 7 7 8
6	0 1 6 8
7	
8	3 3 4 8 9 9
9	0 8

Key: 5|2 = 52

Part A range

Part B mode

Part C median

Part D mean

9 Probability

"If there are 7 cats in a sack and I draw one at random,..."

"... what is the probability that I will draw you?"

"I'm just about finished making my two number cubes."

"Now, here's how the game works. You toss the two cubes."

"If the sum is even I win. If it's odd, you win."

Connections to Previous Learning

- Write common fractions as percents, only include halves, fourths, tenths, and hundredths.

- Interpret and compare ratio and rates.
- Use reasoning about multiplication and division to solve ratio and rate problems.
- Use equivalent forms of fractions, decimals, and percents.

- Determine outcomes of experiments.
- Predict if outcomes are likely or unlikely.
- Predict if experiments are fair or unfair.
- Determine, compare, and make predictions based on experimental or theoretical probability of independent or dependent events.

Math in History

Ancient civilizations often had superstitions about different numbers.

★ The Greek mathematician Euclid (around 300 B.C.) believed that any number that was equal to the sum of its proper factors was a "perfect" number. The least perfect number is 6, which is equal to $1 + 2 + 3$.

★ In China, even numbers are considered lucky, especially the number 8. One exception is the number 4, which is considered very unlucky.

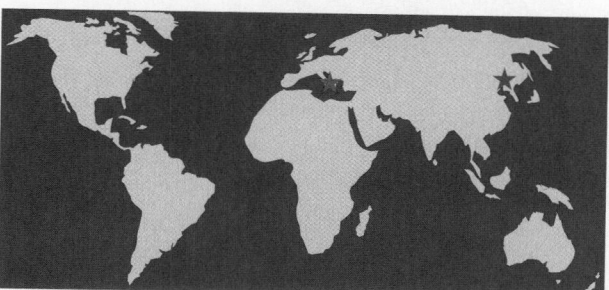

Pacing Guide for Chapter 9

Chapter Opener	1 Day
Section 1 Activity Lesson	 1 Day 1 Day
Section 2 Activity Lesson	 1 Day 1 Day
Study Help / Quiz	1 Day
Section 3 Activity Lesson	 1 Day 1 Day
Section 4 Activity Lesson	 1 Day 1 Day
Quiz / Chapter Review	1 Day
Chapter Test	1 Day
Standardized Test Practice	1 Day
Total Chapter 9	13 Days
Year-to-Date	158 Days

Check Your Resources

- Record and Practice Journal
- Resources by Chapter
- Skills Review Handbook
- Assessment Book
- Worked-Out Solutions

Technology For the Teacher

The Dynamic Planning Tool
Editable Teacher's Resources at
BigIdeasMath.com

Math Background Notes

Additional Topics for Review

- Greatest Common Factor
- Multiplying fractions
- Converting between fractions, decimals, and percents

Try It Yourself

1. $\frac{1}{3}$ 2. $\frac{3}{4}$

3. $\frac{1}{2}$ 4. $\frac{1}{3}$

Record and Practice Journal

1. $\frac{5}{6}$ 2. $\frac{1}{2}$

3. $\frac{1}{2}$ 4. $\frac{9}{13}$

5. $\frac{2}{3}$ 6. $\frac{5}{7}$

7. $\frac{60}{90}; \frac{2}{3}$ 8. $\frac{14}{56}; \frac{1}{4}$

9. $2:3$ 10. $5:3$

11. $1:2$ 12. $1:4$

13. $7:4$ 14. $3:2$

15. $3:20$

Vocabulary Review

- Greatest Common Factor
- Ratio
- Simplest Form

Simplifying Fractions

- Students should know how to simplify fractions.
- Remind students that the most efficient way to simplify (or reduce) fractions is to divide both the numerator and the denominator by the greatest common factor.
- Although dividing the numerator and denominator by the greatest common factor is the most efficient way to simplify a fraction, dividing out any common factor is the first step to simplifying a fraction.
- Remind students that they must divide the numerator and denominator by the same factor.
- **Teaching Tip:** Some students may find the mental math involved in finding the greatest common factor a bit challenging. Encourage these students to find the prime factorization of the numerator and denominator. Divide out any common factors and then multiply the left-over factors back together. The result is a simplified fraction.

Writing Ratios

- Students should know how to write ratios.
- Remind students that a ratio is a comparison between two quantities.
- Ratios can be written in three different ways. They can be expressed as a fraction, using a colon, or using the word "to."
  Example: $\frac{3}{4}$, $3:4$, 3 to 4
- **Common Error:** Order matters when writing ratios. A ratio of $3:4$ carries a different meaning than a ratio of $4:3$. Remind students to write the ratio in the same order that the problem asks for it.
- It is best to express the final ratio in simplest form. Writing a ratio in simplest form is similar to simplifying fractions.

Reteaching and Enrichment Strategies

If students need help...	If students got it...
Record and Practice Journal • Fair Game Review Skills Review Handbook Lesson Tutorials	Game Closet at *BigIdeasMath.com* Start the next section

What You Learned Before

M = Mouse
B = Dog Biscuit

Why do we always have to use this spinner?

"Let's spin to decide what we will have for lunch."

● Simplifying Fractions

Example 1 Simplify $\dfrac{12}{36}$.

$$\dfrac{12 \div 12}{36 \div 12} = \dfrac{1}{3}$$

Simplify fractions by using the Greatest Common Factor.

Example 2 Simplify $\dfrac{33}{60}$.

$$\dfrac{33 \div 3}{60 \div 3} = \dfrac{11}{20}$$

● Writing Ratios

Example 3

a. Write the ratio of girls to boys in Classroom A.

$$\dfrac{\text{Girls in Classroom A}}{\text{Boys in Classroom A}} = \dfrac{11}{14}$$

	Boys	Girls
Classroom A	14	11
Classroom B	12	8

⋮• So, the ratio of girls to boys in Classroom A is $\dfrac{11}{14}$.

b. Write the ratio of boys in Classroom B to the total number of students in both classes.

$$\dfrac{\text{Boys in Classroom B}}{\text{Total number of students}} = \dfrac{12}{14 + 11 + 12 + 8} = \dfrac{12}{45} = \dfrac{4}{15}$$

← Write in simplest form.

⋮• So, the ratio of boys in Classroom B to the total students is $\dfrac{4}{15}$.

Try It Yourself

Write the ratio in simplest form.

1. Baseballs to footballs

2. Footballs to total pieces of equipment

3. Sneakers to ballet slippers

4. Sneakers to total number of shoes

9.1 Introduction to Probability

Essential Question How can you predict the results of spinning a spinner?

1 ACTIVITY: Helicopter Flight

Play with a partner.

- You begin flying the helicopter at (0, 0) on the coordinate plane. Your goal is to reach the cabin at (20, 14).
- Spin any one of the spinners. Move one unit in the indicated direction.
- If the helicopter encounters any obstacles, you must start over.
- Record the number of moves it takes to land exactly on (20, 14).
- After you have played once, it is your partner's turn to play.
- The player who finishes in the fewest moves wins.

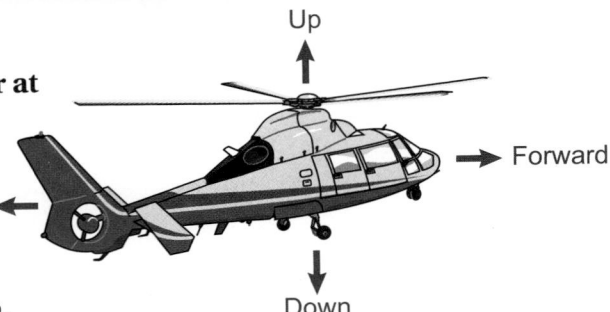

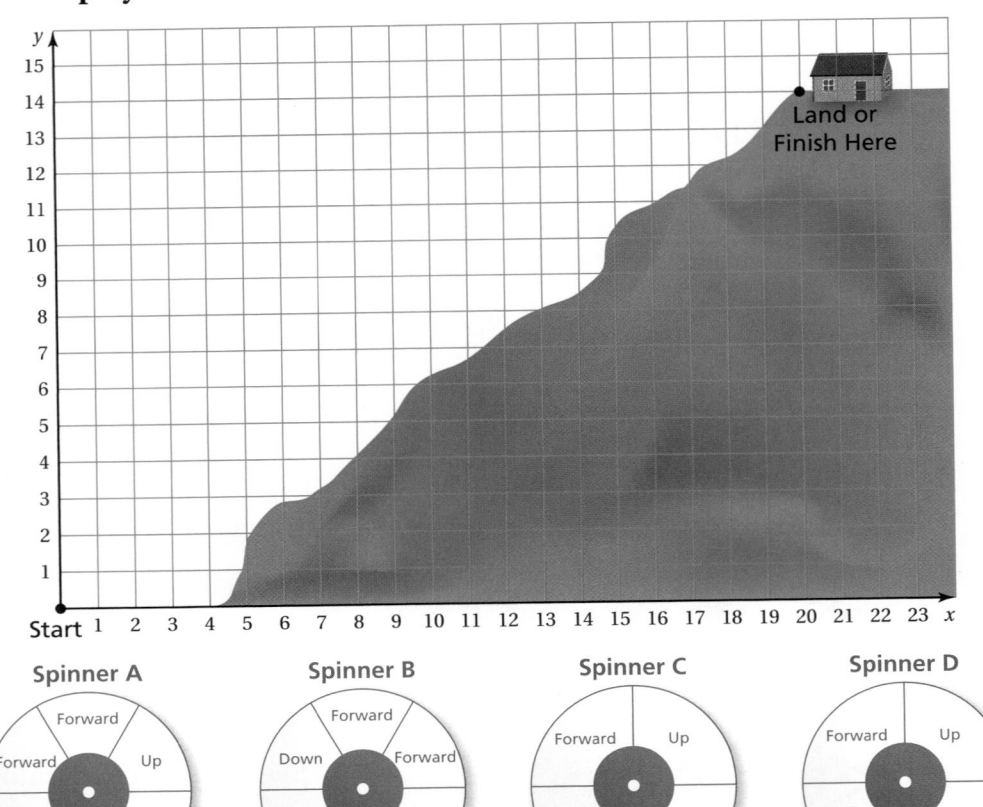

Spinner A

Spinner B

Spinner C

Spinner D

Laurie's Notes

Introduction

For the Teacher

- **Goal:** Students will develop an intuitive understanding of how to predict the results of a spinner.

Motivate

- Draw the following spinners on the board. (N = no homework and Y = homework)

- Tell students they will get one spin to decide if they have homework or not.
- ❓ "If you only have one spin, which spinner do you want to use? Why?" Any of the first three spinners can be selected.
- **Common Error:** Students sometimes believe that how you place the outcomes on the spinner influences the outcomes.

Activity Notes

Activity 1

- This activity is modeled after a problem from the *NCTM Illuminations* website.
- The word *probability* is not used in this investigation even though your students may use it when describing the spinners. The goal is to get a sense about what happens when the sections in a spinner are equal in size.
- **Model:** Show students how to use a paper clip and a pencil as a spinner. Place the spinner card on a flat surface. Place the paper clip (spinner) over the center of the card. Put the pencil point through the paper clip. Spin the paper clip around the pencil point.
- Be sure that students understand that they may select *any* spinner, and the spinner determines the direction to move.
- **Management Tip:** You may wish to change the rules to shorten the number of total turns to reach (20, 14). When you cannot move in the direction indicated by the spinner, stay in the same position for that turn.
- Find out who was able to reach (20, 14) in the fewest turns.
- ❓ "Is there a strategy for using the fewest number of turns or is it all luck?" Students should recognize that certain spinners were more helpful for particular directions of the helicopter.

Previous Learning

Students should know how to plot ordered pairs in a coordinate plane and how to make a bar graph and frequency table.

Activity Materials
Textbook
• paper clips

Start Thinking! and Warm Up

| Activity 9.1 | **Start Thinking!** For use before Activity 9.1 |

Activity 9.1 Warm Up For use before Activity 9.1

Write the decimal as a fraction in simplest form.

1. 0.4   2. 0.63   3. 0.82

4. 0.36   5. 1.55   6. 0.834

9.1 Record and Practice Journal

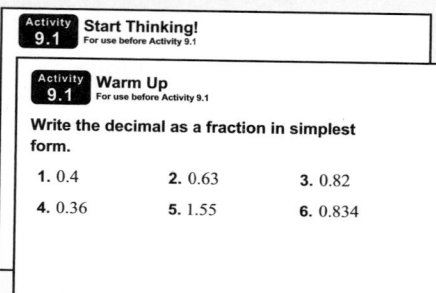

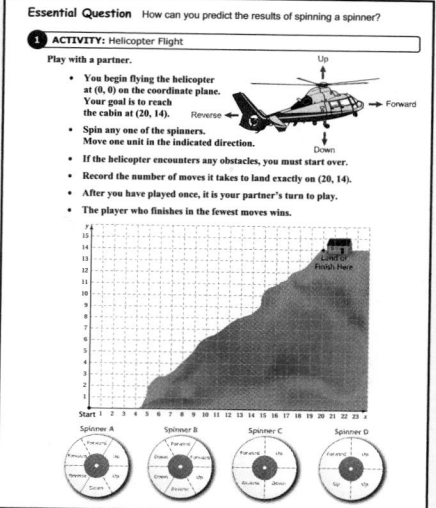

Essential Question How can you predict the results of spinning a spinner?

1 ACTIVITY: Helicopter Flight

Play with a partner.

- You begin flying the helicopter at (0, 0) on the coordinate plane. Your goal is to reach the cabin at (20, 14).
- Spin any one of the spinners. Move one unit in the indicated direction.
- If the helicopter encounters any obstacles, you must start over.
- Record the number of moves it takes to land exactly on (20, 14).
- After you have played once, it is your partner's turn to play.
- The player who finishes in the fewest moves wins.

Differentiated Instruction

Auditory

Ask students what it means when something happens by chance. Discuss with students situations in which the outcomes happen by chance, such as flipping a coin, rolling a number cube, or spinning a spinner. Ask students what it means for an event to be impossible.

9.1 Record and Practice Journal

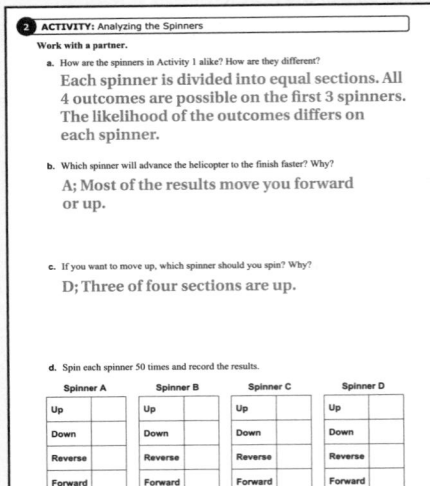

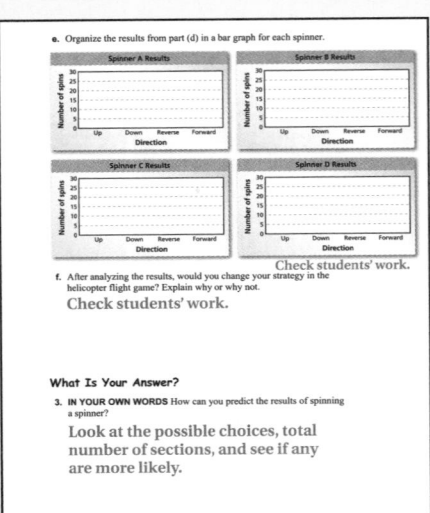

Laurie's Notes

Activity 2

- If students are not comfortable discussing the different spinners, they should begin with part (d) instead of part (a).

- **?** "How are the spinners in Activity 1 similar? How are they different?" *Sample answers:* Similar in that equal sections are used within a spinner, and all 4 outcomes are possible on the first 3 spinners. Different in that the likelihood of the outcomes differs on each spinner. Listen for students' understanding of how to measure *likelihood*.

- **?** "Which spinner will advance the helicopter to the finish the fastest? Why?" Spinner A, because $\frac{2}{3}$ of the results move you up or forward. These are the two directions you want to travel for the majority of the mission. If you were about to hit the hill, however, Spinner D would be most helpful in avoiding a crash.

- **Extension:** Collect class data for each spinner (50 spins × number of pairs of students).

- Listen to student analysis of part (f). Students should recognize that the spinner you choose determines what direction you will likely move.

- **Extension:** Ask students whether there is an advantage to traveling straight up and then forward, or moving on a diagonal, alternating between up and forward.

What Is Your Answer?

- **Neighbor Check:** Have students work independently and then have their neighbor check their work. Have students discuss any discrepancies.

Closure

- You might offer students an opportunity to spin for no homework if they can come to a consensus as to which of your spinners to use from the beginning of the lesson.

Technology For the Teacher

Dynamic Classroom

The Dynamic Planning Tool
Editable Teacher's Resources at *BigIdeasMath.com*

Work with a partner.

a. How are the spinners in Activity 1 alike? How are they different?

b. Which spinner will advance the helicopter to the finish fastest? Why?

c. If you want to move up, which spinner should you spin? Why?

d. Spin each spinner 50 times and record the results.

Spinner A	
Up	
Down	
Reverse	
Forward	

Spinner B	
Up	
Down	
Reverse	
Forward	

Spinner C	
Up	
Down	
Reverse	
Forward	

Spinner D	
Up	
Down	
Reverse	
Forward	

e. Organize the results from part (d) in a bar graph for each spinner.

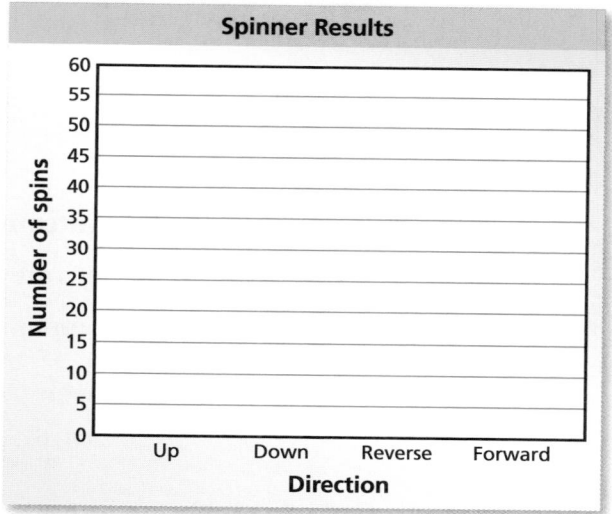

f. After analyzing the results, would you change your strategy in the helicopter flight game? Explain why or why not.

What Is Your Answer?

3. IN YOUR OWN WORDS How can you predict the results of spinning a spinner?

Use what you learned about probability and spinners to complete Exercises 4 and 5 on page 388.

Check It Out
Lesson Tutorials
BigIdeasMath com

Key Vocabulary
experiment, *p. 386*
outcomes, *p. 386*
event, *p. 386*
probability, *p. 387*

Key Ideas

Outcomes and Events

An **experiment** is an activity with varying results. The possible results of an experiment are called **outcomes**. A collection of one or more outcomes is an **event**. The outcomes of a specific event are called *favorable outcomes*.

For example, randomly selecting a marble from a group of marbles is an experiment. Each marble in the group is an outcome. Selecting a green marble from the group is an event.

Possible outcomes

Event: Choosing a green marble
Number of favorable outcomes: 2

EXAMPLE 1 **Identifying Outcomes**

You roll the number cube.

a. What are the possible outcomes?

The six possible outcomes are rolling a 1, 2, 3, 4, 5, and 6.

b. What are the favorable outcomes of rolling an even number?

even	*not* even
2, 4, 6	1, 3, 5

∴ The favorable outcomes of the event are rolling a 2, 4, and 6.

c. What are the favorable outcomes of rolling a number greater than 5?

greater than 5	*not* greater than 5
6	1, 2, 3, 4, 5

∴ The favorable outcome of the event is rolling a 6.

On Your Own

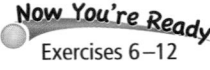
Exercises 6–12

1. You randomly choose a letter from a hat that contains the letters A through K. (a) What are the possible outcomes? (b) What are the favorable outcomes of choosing a vowel?

◀ Multi-Language Glossary at BigIdeasMath ✓com.

Laurie's Notes

Introduction

Connect

- **Yesterday:** Students developed an intuitive understanding of how to predict the results of a spinner.
- **Today:** Students will describe the outcomes of an experiment.

Motivate

- Give students another chance to have no homework. Hold a standard deck of cards and tell them one student is going to select a card from the deck. If it is a 7, there is no homework. If it is not a 7, there is homework.
- Students immediately say it is not fair. Act indignant, and suggest that if they do not draw a card, then there definitely will be homework. This will change their attitude.
- If you are willing to accept the outcome, let one student draw a card. The *odds* are in your favor!

Lesson Notes

Key Ideas

- Discuss the vocabulary words: experiment, outcomes, and event. You can relate the vocabulary to the helicopter rescue mission.
- **?** Ask students to identify the favorable outcomes for the events of choosing each color of marble. green (2), blue (1), red (1), yellow (1), purple (1)

Example 1

- Make sure that students understand that there can be more than one favorable outcome.
- **?** "What are some other examples of experiments and events? What are the favorable outcomes for these events?" *Sample answer:* An experiment is spinning a spinner with the numbers 1–12. An event is spinning a number greater than 10, with the favorable outcomes 11 and 12.

On Your Own

- **Think-Pair-Share:** Students should read the question independently and then work with a partner to answer the question. When they have answered the question, the pair should compare their answer with another group and discuss any discrepancies.

Goal Today's lesson is identifying the favorable outcomes of an event.

Lesson Materials
Introduction
• deck of cards

Start Thinking! and Warm Up

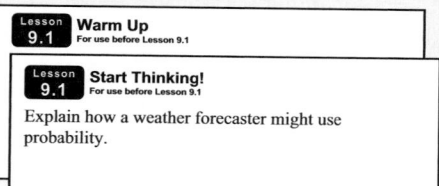

| Lesson 9.1 | **Warm Up** For use before Lesson 9.1 |

| Lesson 9.1 | **Start Thinking!** For use before Lesson 9.1 |

Explain how a weather forecaster might use probability.

Extra Example 1

You roll a number cube.

a. What are the possible outcomes? 1, 2, 3, 4, 5, 6

b. What are the favorable outcomes of rolling an odd number? 1, 3, 5

c. What are the favorable outcomes of rolling a number greater than 4? 5, 6

On Your Own

1. a. A, B, C, D, E, F, G, H, I, J, K

   b. A, E, I

Extra Example 2

You spin the spinner shown in Example 2.

a. How many ways can spinning blue occur? 1 way
b. How many ways can spinning *not* green occur? 5 ways
c. What are the favorable outcomes of spinning *not* green? red, red, red, purple, blue

 On Your Own

2. a. 8 outcomes

   b. 2 ways

   c. 5 ways; blue, blue, red, green, purple

English Language Learners

Vocabulary

Some English learners may confuse the words *outcome* and *event*. Help students see how the two words are related using Example 2 parts (a) and (b).

Possible Outcomes   *Favorable*
(red   red   red) ← *Outcomes*
purple blue   green   *of event*
                             choosing red

Example 2

- **Common Error:** When answering part (a), many students will say that there are only 4 possible outcomes, not 6, because there are only four colors in the spinner. Explain to students that they need to count every occurrence of a color as a possible outcome. Students may be able to understand this concept better after part (c).

Key Idea

- Write the Key Idea.
- Discuss possible events which have probabilities near each benchmark. Make them personal for your situation, if possible. Examples: the sun rising tomorrow = 1, math homework = 0.75, winning the softball game = 0.50, Laurie skipping her morning coffee = 0.25, a winter in Vermont with no snow = 0
- Avoid using potentially sensitive events, such as you will die, you will grow taller/get heavier in the next 5 years, you will eat dinner.
- Spend time discussing what *equally likely* means. Give a number of examples which are/are not equally likely.

Closure

- **Exit Ticket:** What are the favorable outcomes of drawing a face card from a deck of cards? There are three face cards (jack, queen, and king) and four suits, so there are 12 favorable outcomes.

Technology
For the **T**eacher

The Dynamic Planning Tool
Editable Teacher's Resources at *BigIdeasMath.com*

EXAMPLE 2 **Counting Outcomes**

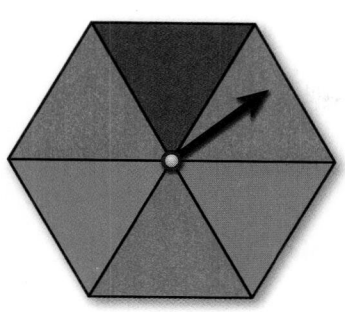

You spin the spinner.

a. How many possible outcomes are there?

The spinner has 6 sections. So, there are 6 possible outcomes.

b. In how many ways can spinning red occur?

The spinner has 3 red sections. So, spinning red can occur in 3 ways.

c. In how many ways can spinning *not* purple occur? What are the favorable outcomes of spinning *not* purple?

The spinner has 5 sections that are *not* purple. So, spinning *not* purple can occur in 5 ways.

purple	*not* purple
purple	red, red, red, green, blue

The favorable outcomes of the event are red, red, red, green, and blue.

 On Your Own

Now You're Ready
Exercises 13–18

2. You randomly choose a marble.

   a. How many possible outcomes are there?

   b. In how many ways can choosing blue occur?

   c. In how many ways can choosing *not* yellow occur? What are the favorable outcomes of choosing *not* yellow?

 **Key Idea**

Probability

The **probability** of an event is a number that measures the likelihood that the event will occur. Probabilities are between 0 and 1, including 0 and 1. The diagram relates likelihoods (above the diagram) and probabilities (below the diagram).

Study Tip

Probabilities can be written as fractions, decimals, or percents.

		Equally likely to		
Impossible		happen or not happen		Certain
	Unlikely		Likely	
0	$\frac{1}{4}$	$\frac{1}{2}$	$\frac{3}{4}$	1
0	0.25	0.5	0.75	1
0%	25%	50%	75%	100%

 Vocabulary and Concept Check

1. **VOCABULARY** Is rolling an even number on a number cube an *outcome* or an *event*? Explain.

2. **REASONING** Can the probability of an event be 1.5? Explain.

3. **OPEN-ENDED** Give a real-life example of an event that is impossible. Give a real-life example of an event that is certain.

 Practice and Problem Solving

Use the spinners shown.

4. You want to move down. Which spinner should you spin? Explain.

5. You want to move forward. Does it matter which spinner you spin? Explain.

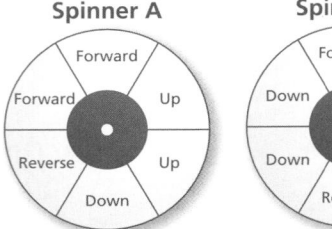

Spinner A   Spinner B

① 6. What are the possible outcomes of randomly choosing one of the tiles shown below?

You randomly choose one of the tiles shown above. Find the favorable outcomes of the event.

7. Choosing a 6

8. Choosing an odd number

9. Choosing a number greater than 5

10. Choosing an odd number less than 5

11. Choosing a number less than 3

12. Choosing a number divisible by 3

You randomly choose one marble from the bag. (a) Find the number of ways the event can occur. (b) Find the favorable outcomes of the event.

 13. Choosing blue

14. Choosing green

15. Choosing purple

16. Choosing yellow

17. Choosing *not* red

18. Choosing *not* blue

19. **ERROR ANALYSIS** Describe and correct the error in finding the number of ways that choosing *not* purple can occur.

purple	*not* purple
purple	red, blue, green, yellow

Choosing not purple can occur in 4 ways.

Color Distinguishment Help at BigIdeasMath✓com.

Assignment Guide and Homework Check

Level	Day 1 Activity Assignment	Day 2 Lesson Assignment	Homework Check
Basic	4, 5, 28–32	1–3, 6, 7–23 odd, 20	9, 13, 20, 21
Average	4, 5, 28–32	1–3, 7–25 odd, 20	9, 13, 20, 21
Advanced	4, 5, 28–32	1–3, 12, 16–22 even, 19, 24–27	12, 18, 24, 26

Common Errors

- **Exercises 7–12** Students may forget to include, or include too many, favorable outcomes. Encourage them to write out all of the possible outcomes and then circle the favorable outcomes for the given event.
- **Exercises 13–18** Students may forget to include the repeats of a color when describing the favorable outcomes and how many ways the event can occur. Ask students to describe how many times you could pull out a specific color from the bag if you happened to pull out the same color each time (without replacing). For example, if the event is choosing a red marble, you can pull out a red marble three times before there are no more red marbles in the bag. So, the event can occur three times.

9.1 Record and Practice Journal

A bag is filled with 4 red marbles, 3 blue marbles, 3 yellow marbles, and 2 green marbles. You randomly choose one marble from the bag. (a) Find the number of ways the event can occur. (b) Find the favorable outcomes of the event.

1. Choosing red
   a. 4
   b. red, red, red, red

2. Choosing green
   a. 2
   b. green, green

3. Choosing yellow
   a. 3
   b. yellow, yellow, yellow

4. Choosing *not* blue
   a. 9
   b. red, red, red, red, yellow, yellow, yellow, green, green

5. In order to figure out who will go first in a game, your friend asks you to pick a number between 1 and 25.
   a. What are the possible outcomes?
      1, 2, 3, 4, 5, 6, 7, 8, 9, 10, 11, 12, 13, 14, 15, 16, 17, 18, 19, 20, 21, 22, 23, 24, 25
   b. What are the favorable outcomes of choosing an even number?
      2, 4, 6, 8, 10, 12, 14, 16, 18, 20, 22, 24
   c. What are the favorable outcomes of choosing a number less than 20?
      1, 2, 3, 4, 5, 6, 7, 8, 9, 10, 11, 12, 13, 14, 15, 16, 17, 18, 19

Technology For the Teacher
Answer Presentation Tool
QuizShow

Vocabulary and Concept Check

1. event; It is a collection of several outcomes.

2. no; Probabilities are between 0 and 1.

3. *Sample answer:* flipping a coin and getting both heads and tails; rolling a number cube and getting a number between 1 and 6

Practice and Problem Solving

4. B; It has more spaces labeled down.

5. no; They both have the same number of forward outcomes.

6. 1, 2, 3, 4, 5, 6, 7, 8, 9

7. 6   **8.** 1, 3, 5, 7, 9

9. 6, 7, 8, 9   **10.** 1, 3

11. 1, 2   **12.** 3, 6, 9

13. **a.** 2 ways   **b.** blue, blue

14. **a.** 1 way   **b.** green

15. **a.** 2 ways
    b. purple, purple

16. **a.** 1 way   **b.** yellow

17. **a.** 6 ways
    b. yellow, green, blue, blue, purple, purple

18. **a.** 7 ways
    b. red, red, red, purple, purple, green, yellow

19. There are 7 marbles that are *not* purple, even though there are only 4 colors. Choosing *not* purple could be red, red, red, blue, blue, green, or yellow.

Practice and Problem Solving

20. 7 ways

21. false; five

22. true

23. false; red

24. true

25. no; More sections on a spinner does not necessarily mean you are more likely to spin red. It depends on the size of the sections of the spinner.

26. 30 classical CDs

27. See *Taking Math Deeper*.

 Fair Game Review

28. 1

29. 30

30. −20

31. $-3\frac{1}{2}$

32. A

Mini-Assessment

You randomly choose one number below. Find the favorable outcomes of the event.

10, 11, 12, 13, 14, 15, 16, 17, 18, 19

1. Choosing a 14 14

2. Choosing an even number 10, 12, 14, 16, 18

3. Choosing an odd number less than 15 11, 13

4. Choosing a number greater than 16 17, 18, 19

5. Choosing a number divisible by 2 10, 12, 14, 16, 18

Taking Math Deeper

Exercise 27

This problem previews the concept of dependent events.

① With all five cards available, the number of possible outcomes is 5.

Choose 1 at random.

② With only four cards left, the number of possible outcomes is reduced to 4.

Choose 1 at random.

③ Answer the question.

After the baker is chosen, the number of possible outcomes decreases.

Project

Create a game that uses picture cards and the changing probabilities indicated in the problem. Create the cards. Play the game with a partner.

Reteaching and Enrichment Strategies

If students need help. . .	If students got it. . .
Resources by Chapter • Practice A and Practice B • Puzzle Time Record and Practice Journal Practice Differentiating the Lesson Lesson Tutorials Skills Review Handbook	Resources by Chapter • Enrichment and Extension Start the next section

20. COINS You have 10 coins in your pocket. Five are Susan B. Anthony Dollars, two are Golden Dollars, and three are Presidential Dollars. You randomly choose a coin. In how many ways can choosing *not* a Presidential Dollar occur?

Golden Dollar

Susan B. Anthony Dollar

Presidential Dollar

Spinner A

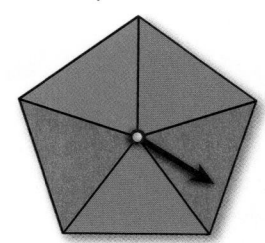

Tell whether the statement is *true* or *false*. If it is false, change the italicized word to make the statement true.

21. There are *three* possible outcomes of spinning Spinner A.

22. Spinning *red* can occur in four ways on Spinner B.

23. Spinning blue and spinning *green* are equally likely on Spinner A.

24. It is *impossible* to spin purple on Spinner B, so it is certain to spin not purple on spinner B.

25. LIKELIHOOD There are more red sections on Spinner B than on Spinner A. Does this mean that you are more likely to spin red on Spinner B? Explain.

Spinner B

Baker · Dancer · Pirate · Fireman · Bellhop

26. MUSIC A bargain bin contains classical and rock CDs. There are 60 CDs in the bin. You are equally likely to randomly choose a classical CD or a rock CD from the bin. How many of the CDs are classical CDs?

27. **Reasoning** You randomly choose one of the cards. Then, you randomly choose a second card. Describe how the number of possible outcomes changes after the first card is chosen.

 **Fair Game Review** What you learned in previous grades & lessons

Multiply. *(Section 2.3)*

28. $\frac{1}{2} \times 2$

29. $\frac{5}{6} \times 36$

30. $-\frac{4}{5} \times 25$

31. $\frac{1}{8} \times (-28)$

32. MULTIPLE CHOICE You are making half of a recipe that requires $\frac{3}{4}$ cup of sugar. How much sugar should you use? *(Section 2.3)*

Ⓐ $\frac{3}{8}$ cup

Ⓑ $\frac{5}{8}$ cup

Ⓒ $\frac{5}{4}$ cups

Ⓓ $\frac{3}{2}$ cups

Essential Question How can you find a theoretical probability?

1 ACTIVITY: Black and White Spinner Game

Work with a partner. You work for a game company. You need to create a game that uses the spinner below.

a. Write rules for a game that uses the spinner. Then play it.

b. After playing the game, do you want to revise the rules? Explain.

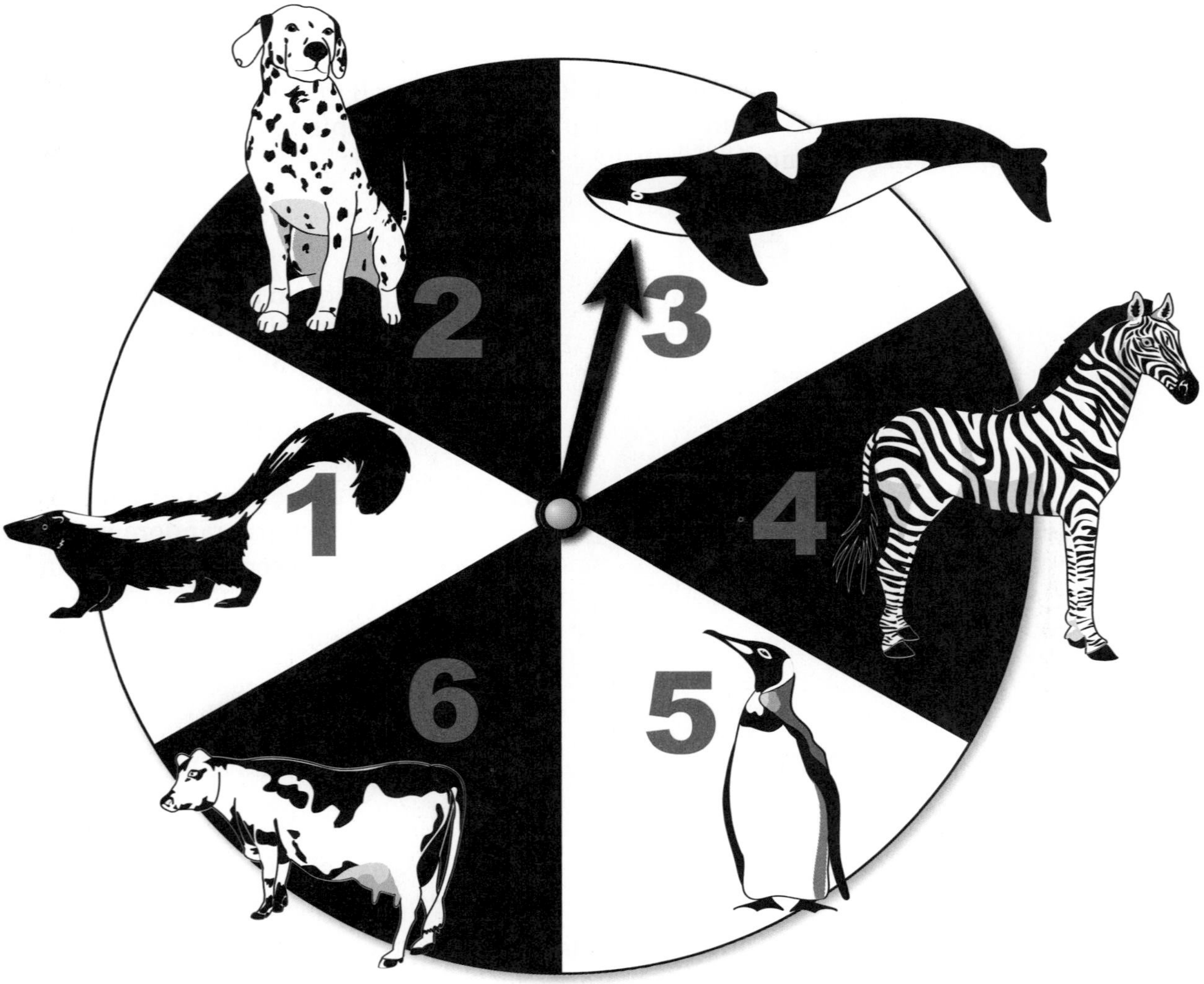

c. Each pie-shaped section of the spinner is the same size. What is the measure of the central angle of each section?

d. What is the probability that the spinner will land on 1? Explain.

Laurie's Notes

Introduction

For the Teacher

- **Goal:** Students will develop an understanding of theoretical probability.

Motivate

❓ Ask a few trivia questions about mammals.
- "What is the world's largest mammal?" blue whale; over 200 tons
- "What is the world's largest land mammal?" African elephant; about 16,500 pounds
- "What is the world's smallest mammal?" Thailand bumblebee bat; weighs less than a penny
- "What is the world's smallest mammal in terms of length?" Etruscan Pygmy Shrew; about 1.5 inches long for body and head

Activity Notes

Activity 1

- Students will have fun with this activity. The six mammals add interest and increase student creativity.
- ❓ "Name the mammals from 1 to 6." skunk, dog, whale, zebra, penguin, cow
- It is hard to predict what rules students might develop, and how simple or involved the rules might be. Here are a few things that students may focus on:
  - 5-letter mammals whale, zebra
  - mammals that spend most of their time in or around water whale, penguin
  - 4-legged mammals skunk, dog, zebra, cow
  - neighborhood animals dog, skunk, cow
  - ignore the mammals and look at prime numbers 2, 3, 5
- The rules are open-ended. Sometimes students write simple rules—you spin x and you win. Other times students will create a game board, or they might do a point accumulation, or have cards that they distribute. This activity can become a multi-day project if you have time.
- The opportunity to revise the rules is included in case students think of a new or necessary change while playing the game.
- ❓ "What is the measure of the central angle for section 1 (the skunk)?" 60°
- Discuss the students' answers for part (d).
- A *central angle* is an angle with its vertex at the center of a circle and whose sides are radii of the circle.

Previous Learning

Students should know how many degrees are in a circle and how to find the angle measure of a central angle.

Activity Materials
Textbook
- paper clip - index cards

Start Thinking! and Warm Up

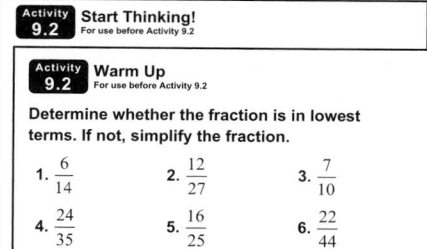

Activity 9.2 Start Thinking! For use before Activity 9.2

Activity 9.2 Warm Up For use before Activity 9.2

Determine whether the fraction is in lowest terms. If not, simplify the fraction.

1. $\dfrac{6}{14}$ 2. $\dfrac{12}{27}$ 3. $\dfrac{7}{10}$

4. $\dfrac{24}{35}$ 5. $\dfrac{16}{25}$ 6. $\dfrac{22}{44}$

9.2 Record and Practice Journal

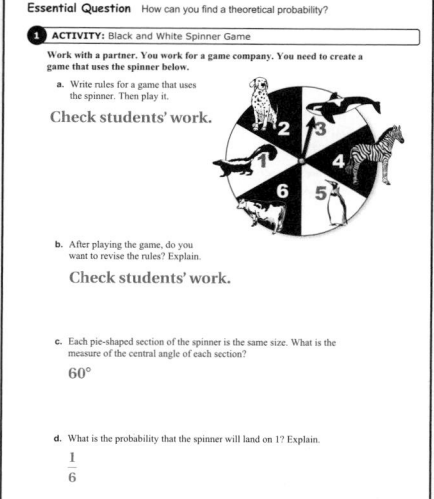

Essential Question How can you find a theoretical probability?

1 ACTIVITY: Black and White Spinner Game

Work with a partner. You work for a game company. You need to create a game that uses the spinner below.

a. Write rules for a game that uses the spinner. Then play it.

Check students' work.

b. After playing the game, do you want to revise the rules? Explain.

Check students' work.

c. Each pie-shaped section of the spinner is the same size. What is the measure of the central angle of each section?

60°

d. What is the probability that the spinner will land on 1? Explain.

$\dfrac{1}{6}$

Vocabulary

This chapter contains many new terms that may cause English learners to struggle. Students should write the key vocabulary in their notebooks along with definitions and diagrams so that they become familiar and comfortable with the vocabulary.

9.2 Record and Practice Journal

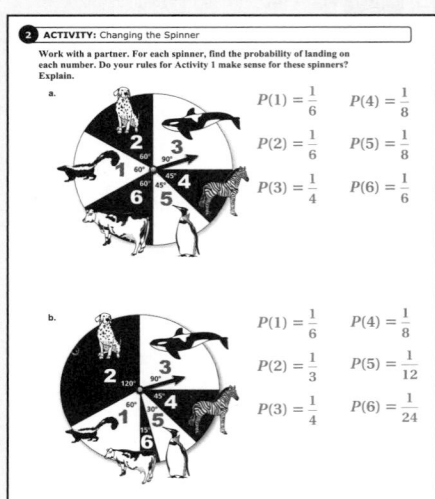

Laurie's Notes

Activity 2

- In this activity, students must consider spinners with different central angle measures. The Black and White Spinner Game students created for Activity 1 will be played using these new spinners. First, they need to calculate the probability of each section.
- **?** "How can you find the probability of each section?" Listen for the idea of estimating the central angle measure and writing the ratio of the central angle measure to 360°. You may need to give a few hints to get this idea across.
- **Big Idea:** To find the probability of each section, write and simplify the ratio of the measure of the central angle to the measure of the whole circle.

  Example: $P(\text{Whale}) = \dfrac{90°}{360°} = \dfrac{1}{4}$

- Discuss the results for each spinner. Note that for the second spinner, the probability of the black pieces and the probability of the white pieces each sum to $\dfrac{1}{2}$.

Activity 3

- We define *fair* for an experiment in the lesson. We mean the informal, English definition in this activity.
- **?** "What would make the game fair?" if each player has an equal chance of winning
- Give time for students to play 10 games on each spinner. Have each pair of students gather data in a table.
- Collect class data for wins and losses on each spinner.
- **?** "What does the data tell you about the fairness of the three spinners?" The data suggests that Spinners 1 and 2b are fair.

What Is Your Answer?

- For Question 6, have 12 index cards available for each pair of students.
- **Big Idea:** Even × Even = Even; Even × Odd = Odd × Even = Even; Odd × Odd = Odd. The probability of having an even product is $\dfrac{3}{4}$, so Player 2 is more likely to win.

Closure

- Find the following: $P(\text{no homework}) = \dfrac{1}{3}$

  $P(\text{homework}) = \dfrac{2}{3}$

Technology
For
the **T**eacher

The Dynamic Planning Tool
Editable Teacher's Resources at *BigIdeasMath.com*

2 ACTIVITY: Changing the Spinner

Work with a partner. For each spinner, find the probability of landing on each number. Do your rules from Activity 1 make sense for these spinners? Explain.

a.

b.

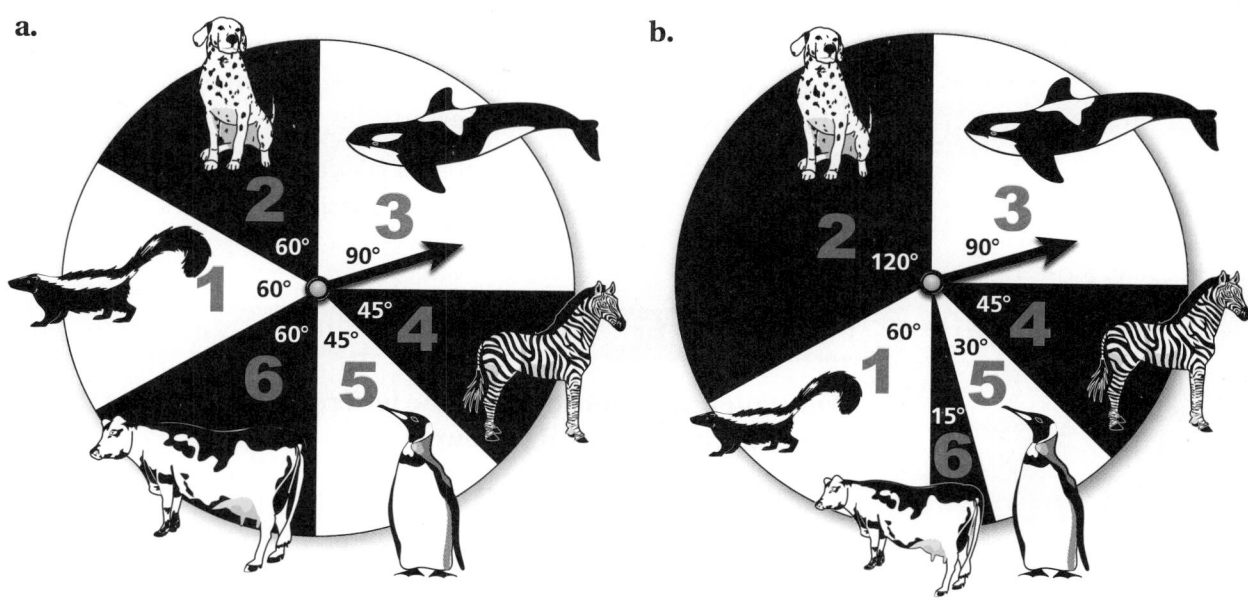

3 ACTIVITY: Is This Game Fair?

Work with a partner. Apply the following rules to each spinner in Activities 1 and 2. Is the game fair? If not, who has the better chance of winning?

- Take turns spinning the spinner.
- If the spinner lands on an odd number, Player 1 wins.
- If the spinner lands on an even number, Player 2 wins.

What Is Your Answer?

4. **IN YOUR OWN WORDS** How can you find a theoretical probability?

5. Find and describe a career in which probability is used. Explain why probability is used in that career.

6. Two people play the following game.

   Each player has 6 cards numbered 1, 2, 3, 4, 5, and 6. At the same time, each player holds up one card. If the product of the two numbers is odd, Player 1 wins. If the product is even, Player 2 wins. Continue until both players are out of cards. Which player is more likely to win? Why?

Practice

Use what you learned about theoretical probability to complete Exercises 4–7 on page 394.

Check It Out
Lesson Tutorials
BigIdeasMath ✓com

Key Vocabulary 🔊
theoretical probability,
  p. 392
fair experiment,
  p. 393

 **Key Idea**

Theoretical Probability

When all possible outcomes are equally likely, the **theoretical probability** of an event is the ratio of the number of favorable outcomes to the number of possible outcomes. The probability of an event is written as $P(\text{event})$.

$$P(\text{event}) = \frac{\text{number of favorable outcomes}}{\text{number of possible outcomes}}$$

EXAMPLE **1** **Finding a Theoretical Probability**

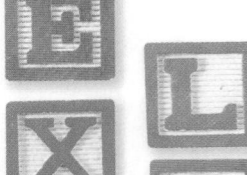   

You randomly choose one of the letters shown. What is the theoretical probability of choosing a vowel?

$$P(\text{event}) = \frac{\text{number of favorable outcomes}}{\text{number of possible outcomes}}$$

$$P(\text{vowel}) = \frac{3}{7}$$

There are 3 vowels.

There is a total of 7 letters.

∴ The probability of choosing a vowel is $\frac{3}{7}$ or about 43%.

EXAMPLE **2** **Using a Theoretical Probability**

The theoretical probability that you randomly choose a green marble from a bag is $\frac{3}{8}$. There are 40 marbles in the bag. How many are green?

$$P(\text{green}) = \frac{\text{number of green marbles}}{\text{total number of marbles}}$$

$$\frac{3}{8} = \frac{n}{40}$$ Substitute. Let n be the number of green marbles.

$$15 = n$$ Multiply each side by 40.

∴ There are 15 green marbles in the bag.

🔘 **On Your Own**

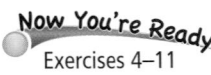
Now You're Ready
Exercises 4–11

1. In Example 1, what is the theoretical probability of choosing an X?

2. The theoretical probability that you spin an odd number on a spinner is 0.6. The spinner has 10 sections. How many sections have odd numbers?

🔊 Multi-Language Glossary at BigIdeasMath✓com.

Laurie's Notes

Introduction

Connect

- **Yesterday:** Students developed an understanding of theoretical probability.
- **Today:** Students will compute the theoretical probability of an event.

Motivate

- Shuffle a standard deck of cards. Explain that a deck of cards has 4 suits, with 13 different cards in each suit.
- Ask students to consider three different events: randomly choosing an ace, randomly choosing a spade, or randomly choosing the ace of spades.
- ❓ "Do all of these events have the same probability? Which is most likely? Least likely?" Students do not need to give exact answers yet, but they should be able to make a reasonable guess. Most likely—spade; Least likely—ace of spades
- ❓ "Would the probabilities change if the two jokers remained in the deck?" yes

Lesson Notes

Key Idea

- Write the Key Idea.
- Discuss theoretical probability and give several examples with which students would be familiar, such as cards, dice, and marbles in a bag. Stress that the outcomes must be equally likely.

Example 1

- Work through the example.
- ❓ "What is the probability of *not* choosing a vowel?" $\frac{4}{7}$

Example 2

- ❓ Read the problem and ask a few questions to help students understand.
  - "What does it mean that the theoretical probability is $\frac{3}{8}$?" Listen for students to suggest that for every 8 marbles, three of them are green.
  - "What proportion can be used to solve the problem?" $\frac{3}{8} = \frac{n}{40}$
  - "What property is used to solve for *n*?" Multiplication Property of Equality
- **Connection:** Students might also think about equivalent fractions $\frac{3}{8} = \frac{15}{40}$
- ❓ "How many marbles are *not* green?" 25

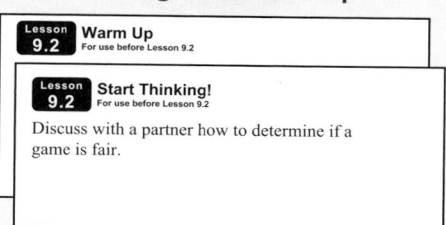

Start Thinking! and Warm Up

> **Lesson 9.2** Warm Up
> For use before Lesson 9.2

> **Lesson 9.2** Start Thinking!
> For use before Lesson 9.2
>
> Discuss with a partner how to determine if a game is fair.

Extra Example 1

The letters in the word JACKSON are placed in a hat. You randomly choose a letter from the hat. What is the theoretical probability of choosing a vowel? $\frac{2}{7}$

Extra Example 2

The theoretical probability that you randomly choose a red marble from a bag is $\frac{5}{8}$. There are 40 marbles in the bag. How many are red? 25

On Your Own

1. $\frac{1}{7}$ or about 14.3%

2. 6 sections

Extra Example 3

You and your friend play the game below.

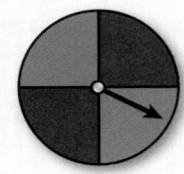

Scoring Rules:

- You get one point when the spinner lands on purple.
- Your friend gets one point when the spinner lands on green.
- The first person to 10 points wins.

a. Is the spinner fair? yes; The spinner is fair because it is equally likely to land on purple or green.

b. Is the game fair? yes; The game is fair because each player has the same probability of winning.

c. Predict the number of turns it will take you or your friend to win. 20 turns

On Your Own

3. yes; The probability of blue or green is the same as the probability of red or blue.

Differentiated Instruction

Visual

Use a circle graph to represent probability. If there are two possible outcomes and one event has a 25% chance of happening, color one-fourth of the circle graph blue and the remaining three-fourths red. Visual learners will see that the circle graph is filled (all possible outcomes equal one) and that the larger the region of the circle, the greater probability that the event will occur.

Laurie's Notes

Discuss

- Explain the vocabulary word *fair experiment*.
- Discuss the difference between a fair game (in the *Study Tip*) and a fair experiment. A fair game means that all players have an equal chance of winning, and a fair experiment means that all possible outcomes are equally likely. It is possible that an experiment is fair, but a game is not, as shown in Example 3.

Example 3

- Read the problem. Have a spinner available to model the scoring rules.
- **?** "What is the probability that the spinner lands on red?" $P(\text{red}) = \frac{2}{6} = \frac{1}{3}$
- **?** "Is the game fair?" no
- **Note:** Even though the game is not fair, the spinner is fair. All the possible outcomes (which color it will land on) are equally likely. It is the rules of the game that make it not fair.
- **Extension:** Describe two different ways the rules could be changed to make the game fair.
- **?** "In part (c), explain why the equation $\frac{2}{3}x = 10$ helps you determine how many turns you must take to win." you need 10 points to win; your probability of scoring is $\frac{2}{3}$, meaning *you score 2 out of 3 turns you take*

Closure

- Revisit the events from the beginning of the lesson and ask students to determine the probability of each.

  Randomly choosing an ace $\frac{4}{52} = \frac{1}{13}$

  Randomly choosing a spade $\frac{13}{52} = \frac{1}{4}$

  Randomly choosing the ace of spades $\frac{1}{52}$

Technology For the Teacher

Dynamic Classroom

The Dynamic Planning Tool
Editable Teacher's Resources at *BigIdeasMath.com*

An experiment is **fair** if all of its possible outcomes are equally likely.

The spinner is equally likely to land on 1 or 2. The spinner is fair.

The spinner is more likely to land on 1 than on either 2 or 3. The spinner is *not* fair.

EXAMPLE 3 Making a Prediction

Scoring Rules:

• You get one point when the spinner lands on blue or green.

• Your friend gets one point when the spinner lands on red.

• The first person to get 10 points wins.

You and your friend play the game. (a) Is the spinner fair? (b) Is the game fair? (c) Predict the number of turns it will take you to win.

a. Yes, the spinner is fair because it is equally likely to land on red, blue, or green.

b. Find and compare the theoretical probabilities of the events.

$$\textbf{You: } P(\text{blue or green}) = \frac{\text{number of blue or green sections}}{\text{total number of sections}}$$

$$= \frac{4}{6} = \frac{2}{3}$$

$$\textbf{Your friend: } P(\text{red}) = \frac{\text{number of red sections}}{\text{total number of sections}}$$

$$= \frac{2}{6} = \frac{1}{3}$$

∴ It is more likely that the spinner will land on blue or green than on red. Because your probability is greater, the game is *not* fair.

c. Write and solve an equation using $P(\text{blue or green})$ found in part (b). Let x be the number of turns it will take you to win.

$\frac{2}{3}x = 10$ Write equation.

$x = 15$ Multiply each side by $\frac{3}{2}$.

∴ So, you can predict that it will take you 15 turns to win.

On Your Own

Now You're Ready
Exercises 12–14

3. WHAT IF? In Example 3, you get one point when the spinner lands on blue or green. Your friend gets one point when the spinner lands on red or blue. The first person to get 5 points wins. Is the game fair? Explain.

 Vocabulary and Concept Check

1. **VOCABULARY** An event has a theoretical probability of 0.5. What does this mean?

2. **OPEN-ENDED** Describe an event that has a theoretical probability of $\frac{1}{4}$.

3. **WHICH ONE DOESN'T BELONG?** Which spinner does *not* belong with the other three? Explain your reasoning.

Spinner 1

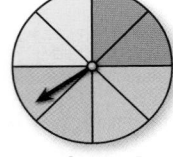
Spinner 2

Spinner 3

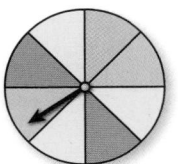
Spinner 4

 **Practice and Problem Solving**

Use the spinner to determine the theoretical probability of the event.

① 4. Spinning red

5. Spinning a 1

6. Spinning an odd number

7. Spinning a multiple of 2

8. Spinning a number less than 7

9. Spinning a 7

10. **LETTERS** Each letter of the alphabet is printed on an index card. What is the theoretical probability of randomly choosing any letter except Z?

② 11. **GAME SHOW** On a game show, a contestant randomly chooses a chip from a bag that contains numbers and strikes. The theoretical probability of choosing a strike is $\frac{3}{10}$. There are 30 chips in the bag. How many are strikes?

A number cube is rolled. Determine if the game is fair. If it is *not* fair, who has the greater probability of winning?

③ 12. You win if the number is odd. Your friend wins if the number is even.

13. You win if the number is less than 3. If it is not less than 3, your friend wins.

14. **SCORING POINTS** You get one point if a 1 or a 2 is rolled on the number cube. Your friend gets one point if a 5 or a 6 is rolled. The first person to 5 points wins.

a. Is the number cube fair? Is the game fair? Explain.

b. Predict the number of turns it will take you to win.

Assignment Guide and Homework Check

Level	Day 1 Activity Assignment	Day 2 Lesson Assignment	Homework Check
Basic	4–7, 21–25	1–3, 8–15	8, 10, 12, 14
Average	4–7, 21–25	1–3, 9–14, 16, 17	10, 14, 16, 17
Advanced	4–7, 21–25	1–3, 11, 13, 14, 16–20	11, 14, 16, 17

Common Errors

- **Exercises 4–9** Students may write a different probability than what is asked, or forget to include a favorable outcome. For example, in Exercise 4 a student may not realize that there are two red sections and will write the probability as $\frac{1}{6}$ instead of $\frac{1}{3}$. Remind them to read the event carefully and to write the favorable outcomes before finding the probability.

- **Exercise 11** Students may write an incorrect proportion when finding how many strikes are in the bag. Encourage them to write the proportion in words before substituting and solving.

- **Exercises 12 and 13** Students may look at the events and guess that the game is fair or not fair without finding the probability of each event. Encourage them to prove that their guess is true by finding the probability of each event.

9.2 Record and Practice Journal

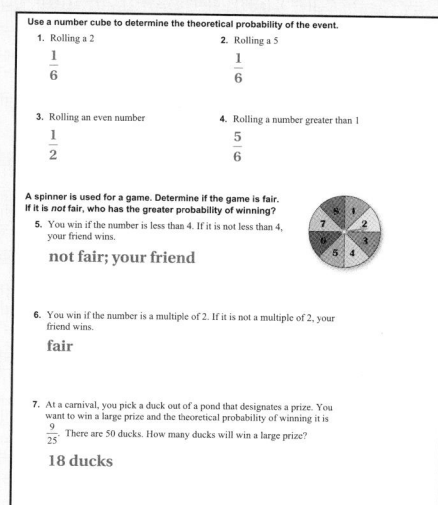

Use a number cube to determine the theoretical probability of the event.

1. Rolling a 2
   $\frac{1}{6}$

2. Rolling a 5
   $\frac{1}{6}$

3. Rolling an even number
   $\frac{1}{2}$

4. Rolling a number greater than 1
   $\frac{5}{6}$

A spinner is used for a game. Determine if the game is fair. If it is *not* fair, who has the greater probability of winning?

5. You win if the number is less than 4. If it is not less than 4, your friend wins.
   not fair; your friend

6. You win if the number is a multiple of 2. If it is not a multiple of 2, your friend wins.
   fair

7. At a carnival, you pick a duck out of a pond that designates a prize. You want to win a large prize and the theoretical probability of winning it is $\frac{9}{25}$. There are 50 ducks. How many ducks will win a large prize?
   18 ducks

 Vocabulary and Concept Check

 Vocabulary and Concept Check

1. There is a 50% chance you will get a favorable outcome.

2. *Sample answer:* picking a 1 out of 1, 2, 3, 4

3. Spinner 4; The other three spinners are fair.

 Practice and Problem Solving

4. $\frac{1}{3}$ or about 33.3%

5. $\frac{1}{6}$ or about 16.7%

6. $\frac{1}{2}$ or 50%

7. $\frac{1}{2}$ or 50%

8. 1 or 100%

9. 0 or 0%

10. $\frac{25}{26}$ or about 96.2%

11. 9 chips

12. fair

13. not fair, your friend

14. a. yes; yes; Each outcome is equally likely, and both players have equal chances of winning.

    b. 15 turns

15. $\frac{1}{44}$ or about 2.3%

16. See *Taking Math Deeper*.

17. a. $\frac{4}{9}$ or about 44.4%

b. 5 males

18.

	Mother's Genes	
	X	X
Father's Genes — X	XX	XX
Father's Genes — Y	XY	XY

19. There are 2 combinations for each.

20. a.

	Parent 1	
	C	s
Parent 2 — C	CC	Cs
Parent 2 — s	Cs	ss

$\frac{1}{4}$ or 25%

b. $\frac{3}{4}$ or 75%

Fair Game Review

21. $\frac{1}{4}$ **22.** $-\frac{1}{9}$

23. $-\frac{21}{40}$ **24.** $\frac{1}{45}$

25. C

Mini-Assessment

Use the spinner to determine the theoretical probability of the event.

1. $P(\text{purple})$ $\frac{1}{6}$

2. $P(3)$ $\frac{1}{6}$

3. $P(\text{even})$ $\frac{1}{2}$

4. $P(\text{multiple of 3})$ $\frac{1}{3}$

5. $P(\text{less than 5})$ $\frac{2}{3}$

Taking Math Deeper

Exercise 16

Begin by making a list that shows how many birthdays occurred in each month.

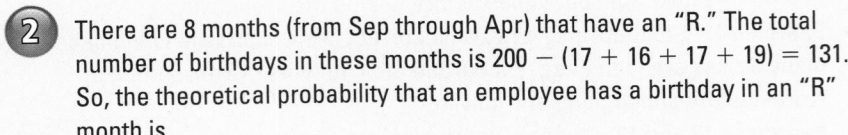

The broken vertical axis on the bar graph makes the bar heights appear more varied than they really are.

① Make a list.

Jan	16	May	17	Sep	18
Feb	15	Jun	16	Oct	18
Mar	17	Jul	17	Nov	15
Apr	16	Aug	19	Dec	16

Check your list by making sure that the numbers total 200.

$$16 + 15 + 17 + 16 + 17 + 16 + 17 + 19 + 18 + 18 + 15 + 16 = 200$$

② There are 8 months (from Sep through Apr) that have an "R." The total number of birthdays in these months is $200 - (17 + 16 + 17 + 19) = 131$. So, the theoretical probability that an employee has a birthday in an "R" month is

a. $P = \frac{131}{200} = 65.5\%$.

③ To find the probability that an employee's birthday is in the first half of the year, add the totals for January through June.

$$16 + 15 + 17 + 16 + 17 + 16 = 97$$

b. $P(\text{birthday Jan}-\text{Jun}) = \frac{97}{200} = 48.5\%$

Did you know? In the Pacific Northwest, it is only safe to pick and eat oysters during months that have an "R." During the non-R months, the water is too warm.

Project

Survey 50 people about the month of their birthday. Do the results match your conjecture from Exercise 16 part (b)?

Reteaching and Enrichment Strategies

If students need help. . .	If students got it. . .
Resources by Chapter • Practice A and Practice B • Puzzle Time Record and Practice Journal Practice Differentiating the Lesson Lesson Tutorials Skills Review Handbook	Resources by Chapter • Enrichment and Extension Start the next section

15. HISTORY You write a report about your favorite president. Your friend writes a report on a randomly chosen president. What is the theoretical probability that you write reports on the same president?

16. BIRTHDAYS The bar graph shows the birthday months of all 200 employees at a local business.

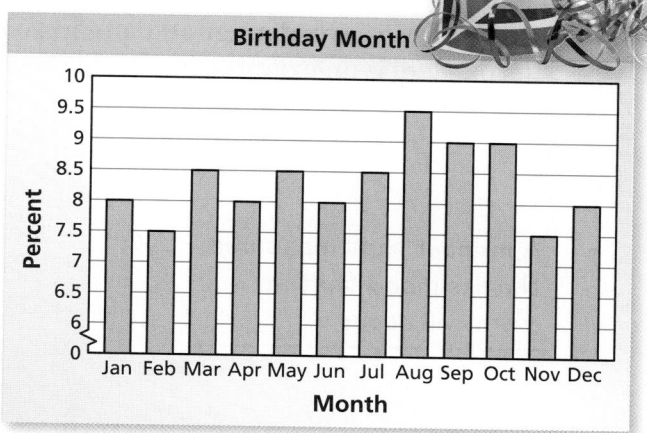

  a. What is the theoretical probability of randomly choosing a person at the business who was born in a month with an R in its name?

  b. What is the theoretical probability of randomly choosing a person at the business who has a birthday in the first half of the year?

17. SCHEDULING There are 16 females and 20 males in a class.

  a. What is the theoretical probability of randomly choosing a female from the class?

  b. One week later, there are 45 students in the class. The theoretical probability of randomly selecting a female is the same as last week. How many males joined the class?

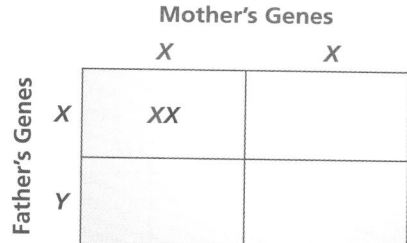

A Punnett square is a grid used to show possible gene combinations for the offspring of two parents. In the Punnett square shown, a boy is represented by *XY*. A girl is represented by *XX*.

18. Complete the Punnett square.

19. Explain why the probability of two parents having a boy or having a girl is equally likely.

20.  **Critical Thinking** Two parents each have the gene combination *Cs*. The gene *C* is for curly hair. The gene *s* is for straight hair.

  a. Make a Punnett square for the two parents. If all outcomes are equally likely, what is the probability of a child having the gene combination *CC*?

  b. Any gene combination that includes a *C* results in curly hair. If all outcomes are equally likely, what is the probability of a child having curly hair?

Fair Game Review What you learned in previous grades & lessons

Multiply. *(Section 2.3)*

21. $\dfrac{1}{2} \times \dfrac{1}{2}$

22. $-\dfrac{1}{6} \times \dfrac{2}{3}$

23. $-\dfrac{3}{5} \times \dfrac{7}{8}$

24. $\dfrac{4}{5} \times \dfrac{1}{36}$

25. MULTIPLE CHOICE What is the mean of the numbers 11, 6, 12, 22, 7, 8, and 4? *(Skills Review Handbook)*

  Ⓐ 4 　　　　Ⓑ 8 　　　　Ⓒ 10 　　　　Ⓓ 70

You can use a **word magnet** to organize information associated with a vocabulary word or term. Here is an example of a word magnet for probability.

Probability

A number that measures the likelihood that an event will occur

Can be written as a fraction, decimal, or percent.

Always between 0 and 1, inclusive

If *P*(event) = 0, the event is *impossible*.

If *P*(event) = 0.25, the event is *unlikely*.

If *P*(event) = 0.5, the event is *equally likely* to happen or not happen.

If *P*(event) = 0.75, the event is *likely*.

If *P*(event) = 1, the event is *certain*.

On Your Own

Make a word magnet to help you study these topics.

1. event

2. outcome

3. theoretical probability

After you complete this chapter, make word magnets for the following topics.

4. experimental probability

5. independent events

6. dependent events

7. Choose three other topics that you studied earlier in this course. Make a word magnet for each topic.

"I'm going to sell my word magnet poster at the Fraidy Cat Festival."

Sample Answers

1.

Event

A collection of one or more outcomes

Outcomes are the possible results of an experiment.

The *probability* of an event, *P*(event), is the likelihood that the event will occur.

An *experiment* is an activity with varying results.

Example of an event: rolling an even number when you roll a number cube

Example of an event: choosing a red or blue marble from a bag that contains red, blue, green, and yellow marbles

2.

Outcome

A possible result of an experiment

An *experiment* is an activity with varying results.

The outcomes for an event are called favorable outcomes. The other outcomes are unfavorable outcomes.

An *event* is a collection of one or more outcomes.

Example: You roll a number cube. The six possible outcomes are rolling a 1, 2, 3, 4, 5, and 6.

3.

Theoretical Probability

The ratio of the number of favorable outcomes to the number of possible outcomes, when all possible outcomes are equally likely

The probability of an event is written as *P*(event).

Example: You flip a coin. The theoretical probability of flipping heads and the theoretical probability of flipping tails is $P(\text{heads}) = P(\text{tails}) = \frac{1}{2}$.

Theoretical probability of an event:

$$P(\text{event}) = \frac{\text{number of favorable outcomes}}{\text{number of possible outcomes}}$$

List of Organizers
Available at *BigIdeasMath.com*

Comparison Chart
Concept Circle
Example and Non-Example Chart
Formula Triangle
Four Square
Idea (Definition) and Examples Chart
Information Frame
Information Wheel
Notetaking Organizer
Process Diagram
Summary Triangle
Word Magnet
Y Chart

About this Organizer

A **Word Magnet** can be used to organize information associated with a vocabulary word or term. As shown, students write the word or term inside the magnet. Students write associated information on the blank lines that "radiate" from the magnet. Associated information can include, but is not limited to: other vocabulary words or terms, definitions, formulas, procedures, examples, and visuals. This type of organizer serves as a good summary tool because any information related to a topic can be included.

Technology For the Teacher
Vocabulary Puzzle Builder

Answers

1. 2, 4, 6, 8

2. 4, 8

3. 3, 4, 5, 6, 7, 8, 9

4. 2

5. 0

6. 4

7. $\frac{1}{4}$ or 25%

8. $\frac{9}{10}$ or 90%

9. $\frac{1}{2}$ or 50%

10. fair

11. not fair; your friend

12. 19

13. 78

Assessment Book

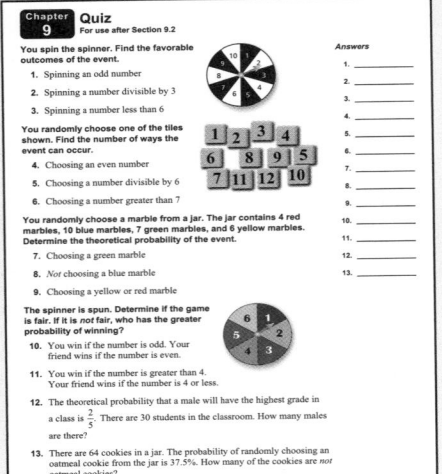

100% Quiz	Math Log
Error Notebook	Notebook Quiz
Group Quiz	Partner Quiz
Homework Quiz	Pass the Paper

100% Quiz

This is a quiz where students are given the answers and then they have to explain and justify each answer.

Reteaching and Enrichment Strategies

If students need help. . .	If students got it. . .
Resources by Chapter • Study Help • Practice A and Practice B • Puzzle Time Lesson Tutorials *BigIdeasMath.com* Practice Quiz Practice from the Test Generator	Resources by Chapter • Enrichment and Extension • School-to-Work Game Closet at *BigIdeasMath.com* Start the next section

Technology For the Teacher

Answer Presentation Tool
Big Ideas Test Generator

Check It Out
Progress Check
BigIdeasMath ✓com

You spin the spinner. Find the favorable outcomes of the event. *(Section 9.1)*

1. Spinning an even number

2. Spinning a number divisible by 4

3. Spinning a number greater than or equal to 3

You randomly choose one butterfly. Find the number of ways the event can occur. *(Section 9.1)*

4. Choosing red

5. Choosing brown

6. Choosing *not* blue

You randomly choose one push pin from the jar. Determine the theoretical probability of the event. *(Section 9.2)*

7. Choosing a yellow pin

8. *Not* choosing a blue pin

9. Choosing a green or red pin

12 Green
6 White
8 Red
4 Blue
10 Yellow

The spinner is spun. Determine if the game is fair. If it is *not* fair, who has the greater probability of winning? *(Section 9.2)*

10. You win if the number is even. Your friend wins if the number is odd.

11. You win if the number is less than 4. Your friend wins if the number is 4 or greater.

12. **TICKETS** The theoretical probability that your ticket will be drawn from a bucket to win a bicycle is $\frac{1}{35}$. There are 665 tickets in the bucket. How many tickets are yours? *(Section 9.2)*

13. **APPLES** There are 104 apples in a bushel. The probability of randomly choosing a Granny Smith apple from the bushel is 25%. How many of the apples are *not* Granny Smith apples? *(Section 9.2)*

Essential Question What is meant by experimental probability?

1 ACTIVITY: Throwing Sticks

Play with a partner. This game is based on an Apache game called "Throw Sticks."

- Take turns throwing three sticks into the center of the circle and moving around the circle according to the chart.

- If your opponent lands on or passes your playing piece, you must start over.

- The first player to pass his or her starting point wins.

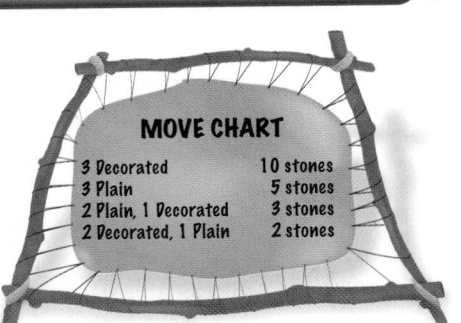

MOVE CHART

3 Decorated	10 stones
3 Plain	5 stones
2 Plain, 1 Decorated	3 stones
2 Decorated, 1 Plain	2 stones

Each stick has one plain side and one decorated side.

The game board has 40 stones arranged in a circle. The stones are placed in groups of 10.

Players start on opposite sides of the circle.

Player 1 Starting Point

Player 2 Starting Point

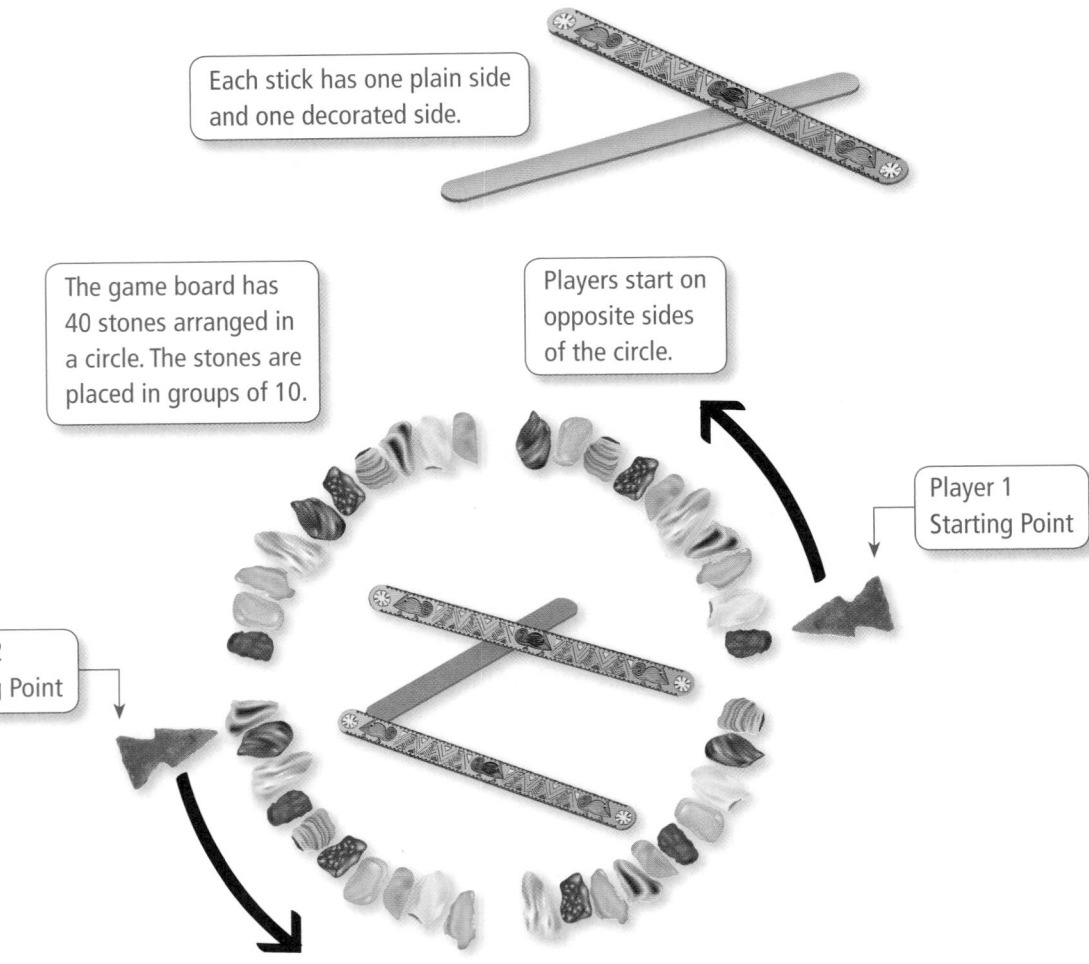

Laurie's Notes

Introduction

For the Teacher

- **Goal:** Students will develop an understanding of experimental probability.
- **Big Idea:** It is only through repeated trials that the experimental probability begins to approach the theoretical probability.

Motivate

- Ask ten students to stand at the front of the room. Hand each one a penny.
- **?** "If [student 1] tosses his or her penny, what is the probability it will land on heads?" $\frac{1}{2}$
- **?** "If [student 4] tosses his or her penny, what is the probability it will land on heads?" $\frac{1}{2}$
- **?** "If [student 9] tosses his or her penny, what is the probability it will land on heads? $\frac{1}{2}$
- **?** "If all of the students toss their pennies, what is the probability they will all land on heads?" Students may say $\frac{1}{2}$, and you will need to hold off on telling them they are incorrect.

Activity Notes

Activity 1

- **Preparation:** Make 3 sticks for each pair of students, where one side is decorated and the other side is plain. Each student will need a game token to represent moving around the circle.
- This activity is modeled after a game played by Native Americans.
- Have students read the directions. When students understand the directions, they should begin.
- **Teaching Tip:** A spot on the floor might be a better location to play the game than on top of a desk.
- The goal of the activity is to develop a sense about outcomes and likelihood.
- Allow the play to continue long enough for one of the players to win. If one pair of students finishes quickly, they could try to play a second game.
- **?** "About how many tosses did it take for someone to win?" Answers will vary.
- **?** "What is the fewest number of tosses necessary to win? What would you need to toss?" 5 tosses with 4 of the 5 showing all three decorated sticks

Start Thinking! and Warm Up

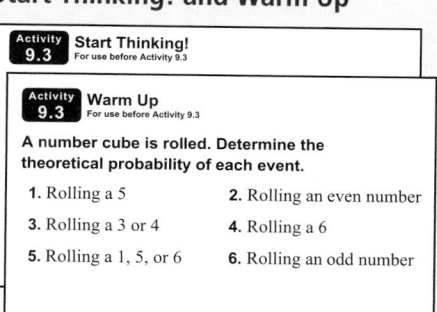

Activity 9.3 Start Thinking!
For use before Activity 9.3

Activity 9.3 Warm Up
For use before Activity 9.3

A number cube is rolled. Determine the theoretical probability of each event.

1. Rolling a 5
2. Rolling an even number
3. Rolling a 3 or 4
4. Rolling a 6
5. Rolling a 1, 5, or 6
6. Rolling an odd number

9.3 Record and Practice Journal

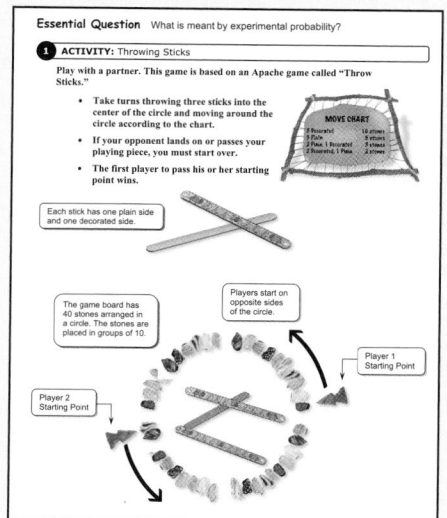

Essential Question What is meant by experimental probability?

1 ACTIVITY: Throwing Sticks

Play with a partner. This game is based on an Apache game called "Throw Sticks."

- Take turns throwing three sticks into the center of the circle and moving around the circle according to the chart.
- If your opponent lands on or passes your playing piece, you must start over.
- The first player to pass his or her starting point wins.

Differentiated Instruction

Kinesthetic

Set up 6 groups and assign each group a number from 1 to 6. Each group will predict whether a number less than, greater than, or equal to their assigned number will occur when rolling a number cube 20 times. Have each group track their results in a frequency table. Each group then determines what fraction of the results is less than, equal to, or greater than their assigned number and presents their findings to the class.

9.3 Record and Practice Journal

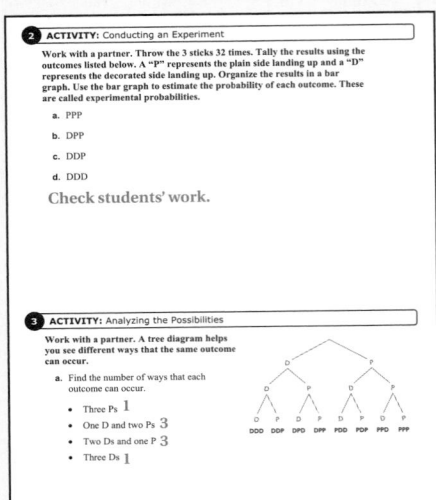

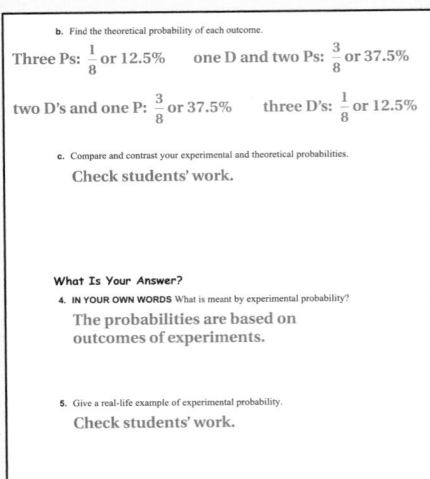

Laurie's Notes

Activity 2

- Have students organize the data in a frequency table.
- Give students grid paper to make the bar graph.
- The expected results are that 3 decorated or 3 plain are least likely, while 2 of one type and 1 of the other type is more common.
- **?** "Of the 32 trials, how many outcomes were all plain (PPP)? All decorated (DDD)?" poll several pairs of students
- **?** "Of the 32 trials, how many outcomes were two plain, one decorated (PPD)? Two decorated, one plain (DDP)?" poll several pairs of students
- Discuss experimental probability. For example, if 4 of the 32 trials were DDD, the experimental probability would be $\frac{4}{32}$ or $\frac{1}{8}$.

Activity 3

- Give time for students to examine the tree diagram to see if they can make sense of the information displayed.
- If necessary, help students interpret the tree diagram. The first stick can land one of two ways: D or P. Following down the D-branch; the second stick can land one of two ways: D or P. Following down the DD-branches, the last stick can land one of two ways: D or P. Notice there are 8 different paths from the top to the bottom. The bottom row summarizes the 8 paths.
- Discuss the theoretical probability of these events.
- **?** "Is DPD equivalent to DDP? Explain." yes; Two of the sticks are decorated and one is not. The order in which they are listed does not matter.
- Discuss results for part (c). Some pairs of students may have results where the experimental and theoretical probabilities are very close, while others may not. If time permits, collate the results of Activity 2 from the entire class. Then answer part (c).

Closure

- Referring to the 10 pennies question from the beginning of the lesson, what is the probability of 3 pennies being tossed and all of them landing heads?

  $\frac{1}{8}$; Using a tree diagram, there are 8 total possible outcomes and only one outcome is 3 heads.

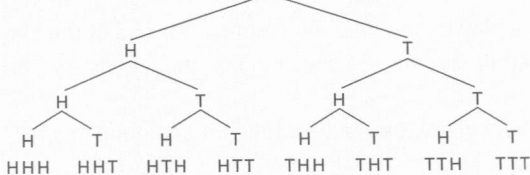

2 ACTIVITY: Conducting an Experiment

Work with a partner. Throw the 3 sticks 32 times. Tally the results using the outcomes listed below. Organize the results in a bar graph. Use the bar graph to estimate the probability of each outcome. These are called **experimental probabilities**.

a. PPP

b. DPP

c. DDP

d. DDD

3 ACTIVITY: Analyzing the Possibilities

Work with a partner. A tree diagram helps you see different ways that the same outcome can occur.

a. Find the number of ways that each outcome can occur.

- Three Ps
- One D and two Ps
- Two Ds and one P
- Three Ds

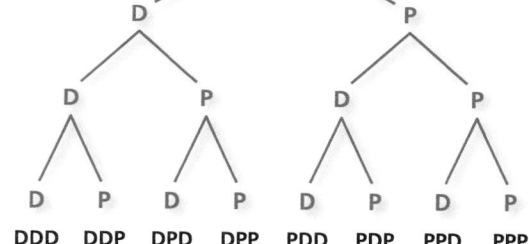

b. Find the theoretical probability of each outcome.

c. Compare and contrast your experimental and theoretical probabilities.

What Is Your Answer?

4. **IN YOUR OWN WORDS** What is meant by experimental probability?

5. Give a real-life example of experimental probability.

Practice

Use what you learned about experimental probability to complete Exercises 3–6 on page 402.

Check It Out
Lesson Tutorials
BigIdeasMath ✓com

Key Vocabulary 🔊
experimental
  probability, *p. 400*

 **Key Idea**

Experimental Probability

Probability that is based on repeated trials of an experiment is called **experimental probability**.

$$P(\text{event}) = \frac{\text{number of times the event occurs}}{\text{total number of trials}}$$

EXAMPLE ① **Standardized Test Practice**

Thirteen out of 20 emails in your inbox are junk emails. What is the experimental probability that your next email is junk?

Ⓐ 35%   Ⓑ 45%   Ⓒ 55%   Ⓓ 65%

$$P(\text{event}) = \frac{\text{number of times the event occurs}}{\text{total number of trials}}$$

$$P(\text{junk}) = \frac{13}{20}$$ ⟵ You have 13 emails that are junk.
                      ⟵ You have a total of 20 emails.

∴ The probability is $\frac{13}{20}$, 0.65, or 65%. The correct answer is Ⓓ.

EXAMPLE ② **Making a Prediction**

It rains 2 out of the last 12 days in March. If this trend continues, how many rainy days would you expect in April?

Find the experimental probability of a rainy day.

$$P(\text{event}) = \frac{\text{number of times the event occurs}}{\text{total number of trials}}$$

It rains 2 days.

$$P(\text{rain}) = \frac{2}{12} = \frac{1}{6}$$

There is a total of 12 days.

"April showers bring May flowers." Old Proverb, 1557

To make a prediction, multiply the probability of a rainy day by the number of days in April.

$$\frac{1}{6} \cdot 30 = 5$$

∴ You can predict that there will be 5 rainy days in April.

🔊 Multi-Language Glossary at BigIdeasMath ✓com.

Laurie's Notes

Introduction

Connect
- **Yesterday:** Students developed an understanding of experimental probability.
- **Today:** Students will compute the experimental probability of an event.

Motivate
- Play *Mystery Bag*. Before students arrive, place 10 cubes of the same shape and size in a paper bag; five of one color and five of a second color.
- Ask a volunteer to be the detective.
- **?** "There are 10 cubes in my bag. Can you guess what color they are?" not likely
- Let the student remove a cube and look at its color.
- **?** "Can you guess what color my cubes are?" not likely
- *Replace the cube.* Let the student pick again and see the color. Repeat your question.
- Try this 5–8 times until the student is ready to guess. The number of trials will depend upon the results and the student. You want students to see that they are collecting data and making a prediction.

Lesson Notes

Key Idea
- Discuss experimental probability and give several examples.
- **?** "What is the difference between theoretical and experimental probability? Listen for an understanding that theoretical is what should happen and experimental is what does happen in an experiment.

Example 1
- **?** "How do you change a fraction to a percent?" In this problem, write an equivalent fraction with a denominator of 100. $\frac{13}{20} = \frac{65}{100} = 65\%$
- **?** **Extension:** "What is $P(\text{not junk}) + P(\text{junk})$?" $\frac{7}{20} + \frac{13}{20} = 1$
- **Big Idea:** When an event has only two outcomes (i.e., heads or tails; decorated or plain), the sum of the experimental probabilities must equal 1.

Example 2
- Note the important phrase, *if this trend continues*. Knowing the weather for the last 12 days in March, you make a prediction about the weather in April.
- **?** "Does it seem reasonable to use information from late March to predict weather in April?" Some students may say the weather in April is different, which is why the problem was phrased *if this trend continues*.
- **Big Idea:** The experimental probability is used to make a prediction when you expect a trend to continue, or you believe the experiment reflects what might be true about a larger population.

Goal Today's lesson is finding the **experimental probability** of an event.

Lesson Materials	
Introduction	**Closure**
• bag • colored cubes	• bag • colored cubes

Start Thinking! and Warm Up

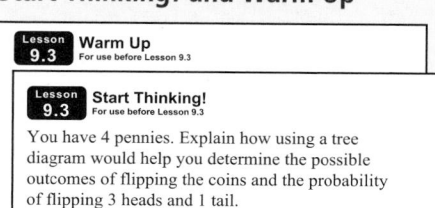

Lesson 9.3 Warm Up For use before Lesson 9.3

Lesson 9.3 Start Thinking! For use before Lesson 9.3

You have 4 pennies. Explain how using a tree diagram would help you determine the possible outcomes of flipping the coins and the probability of flipping 3 heads and 1 tail.

Extra Example 1
Fifteen out of 25 emails in your inbox are junk emails. What is the experimental probability that your next email is junk? 60%

Extra Example 2
It rains 3 out of the last 15 days in May. If this trend continues, how many rainy days would you expect in June? 6 days

JUNE						
SUN	MON	TUE	WED	THU	FRI	SAT
			1	2	3	4
5	6	7	8	9	10	11
12	13	14	15	16	17	18
19	20	21	22	23	24	25
26	27	28	29	30		

Laurie's Notes

On Your Own

1. $\frac{7}{20}$ or 35%

2. a. $\frac{1}{40}$ or 2.5%

   b. 125

Extra Example 3

Using the bar graph and results from Example 3, what is the experimental probability of rolling an even number? 48% How does this compare to the theoretical probability of rolling an even number? The theoretical probability is 50%. The experimental and theoretical probabilities are similar.

On Your Own

3. experimental probability:

   $\frac{4}{5} = 0.8 = 80\%$;

   theoretical probability:

   $\frac{5}{6} \approx 0.833 = 83.3\%$.

   The experimental and theoretical probabilities are similar.

English Language Learners

Group Activity

Set up groups of English learners and English speakers. Have each group predict the number of times a flipped coin will land on *heads* when flipped 30 times. Each group should track their results in a frequency table. Ask groups to present their predictions and results to the class. Then combine the results of all the groups and discuss how the combined results compare to the individual group results.

On Your Own

- **Neighbor Check:** Have students work independently and then have their neighbor check their work. Have students discuss any discrepancies.
- ? "What do you notice about P(junk) and P(not junk)?" sum to 1

Example 3

- Work through the example.
- If there is sufficient time, you might consider having students gather data for rolling a number cube 50 times or, if available, run a computer simulation.
- ? "Compare the experimental and theoretical probabilities." Listen for the fact that they are close to each other.
- Discuss with students the idea that the greater the number of trials used to compute an experimental probability, the closer that probability will be to the theoretical probability.

Closure

- Use the colored cubes in the Mystery Bag. Reveal the contents, or use a different color ratio if you wish, and ask the following:
  - "If the contents of the Mystery Bag came from a bag of 1000 colored cubes, how many of each color would you predict in the bag of 1000?" Answers will vary depending upon materials.

Technology For the Teacher

Dynamic Classroom

The Dynamic Planning Tool
Editable Teacher's Resources at *BigIdeasMath.com*

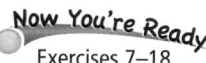

On Your Own

Now You're Ready
Exercises 7–18

1. In Example 1, what is the experimental probability that your next email is *not* junk?

2. At a clothing company, an inspector finds 5 defective pairs in a shipment of 200 jeans.

    a. What is the experimental probability of a pair of jeans being defective?

    b. About how many would you expect to be defective in a shipment of 5000 pairs of jeans?

EXAMPLE ③ **Comparing Experimental and Theoretical Probabilities**

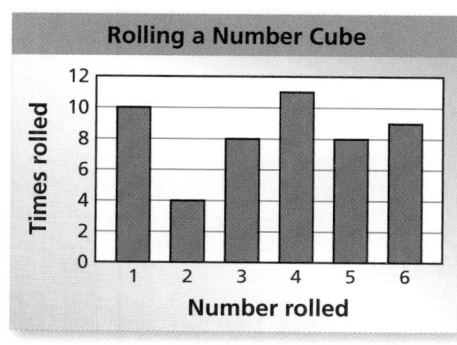

The bar graph shows the results of rolling a number cube 50 times. What is the experimental probability of rolling an odd number? How does this compare with the theoretical probability of rolling an odd number?

Find the experimental probability of rolling a 1, 3, or 5.

The bar graph shows 10 ones, 8 threes, and 8 fives. So, an odd number was rolled $10 + 8 + 8 = 26$ times in a total of 50 rolls.

Experimental Probability

$$P(\text{event}) = \frac{\text{number of times the event occurs}}{\text{total number of trials}}$$

$$P(\text{odd}) = \frac{26}{50}$$

    An odd number was rolled 26 times.

    There was a total of 50 rolls.

$$= \frac{13}{25}$$

Theoretical Probability

$$P(\text{event}) = \frac{\text{number of favorable outcomes}}{\text{number of possible outcomes}}$$

$$P(\text{odd}) = \frac{3}{6}$$

    There are 3 odd numbers.

    There is a total of 6 numbers.

$$= \frac{1}{2}$$

:: The experimental probability is $\frac{13}{25} = 0.52 = 52\%$. The theoretical probability is $\frac{1}{2} = 0.5 = 50\%$. The experimental and theoretical probabilities are similar.

On Your Own

Now You're Ready
Exercise 19

3. In Example 3, what is the experimental probability of rolling a number greater than 1? How does this compare with the theoretical probability of rolling a number greater than 1?

 Vocabulary and Concept Check

1. **VOCABULARY** Describe how to find the experimental probability of an event.

2. **REASONING** You flip a coin 10 times and find the experimental probability of flipping tails to be 0.7. Does this seem reasonable? Explain.

 Practice and Problem Solving

You have three sticks. Each stick has one red side and one blue side. You throw the sticks 10 times and record the results. Use the table to find the experimental probability of the event.

Outcome	Frequency
3 red	4
3 blue	0
2 blue, 1 red	2
2 red, 1 blue	4

3. Tossing 3 red   4. Tossing 2 blue, 1 red

5. Tossing 2 red, 1 blue   6. *Not* tossing all red

Use the bar graph to find the experimental probability of the event.

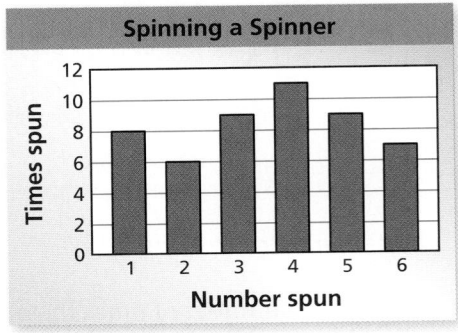

Spinning a Spinner

7. Spinning a 6   8. Spinning an even number

9. *Not* spinning a 1   10. Spinning a number less than 3

11. Spinning a 1 or a 3   12. Spinning a 7

13. **ERROR ANALYSIS** Describe and correct the error in finding $P(4)$ using the bar graph.

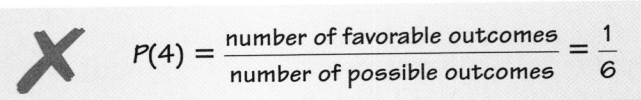

$$P(4) = \frac{\text{number of favorable outcomes}}{\text{number of possible outcomes}} = \frac{1}{6}$$

14. **EGGS** You check 20 cartons of eggs. Three of the cartons have at least one cracked egg. What is the experimental probability that a carton of eggs has at least one cracked egg?

15. **BOARD GAME** There are 105 lettered tiles in a board game. You choose the tiles shown. How many of the 105 tiles would you expect to be vowels?

16. **CARDS** You have a package of 20 assorted thank-you cards. You pick the four cards shown. How many of the 20 cards would you expect to have flowers on them?

Assignment Guide and Homework Check

Level	Day 1 Activity Assignment	Day 2 Lesson Assignment	Homework Check
Basic	3–6, 24–28	1, 2, 7–19 odd, 14, 16	2, 7, 14, 16, 19
Average	3–6, 24–28	1, 2, 11–15, 18–21	2, 12, 18, 19
Advanced	3–6, 24–28	1, 2, 13, 14, 16, 18–23	2, 16, 18, 22

Common Errors

- **Exercises 7–12** Students may forget to total all of the trials before writing the experimental probability. They may have an incorrect number of trials in the denominator. Remind them that they need to know the total number of trials when finding the probability.
- **Exercise 19** Students may try to use some of the information given for the experimental probability when finding the theoretical probability. For example, they may say that the total possible outcomes theoretically is 20 because that is the number of trials; however, it is only 2 because there are only two possibilities when you flip a coin. Remind them that the theoretical probability does not use the information from the experiment.

9.3 Record and Practice Journal

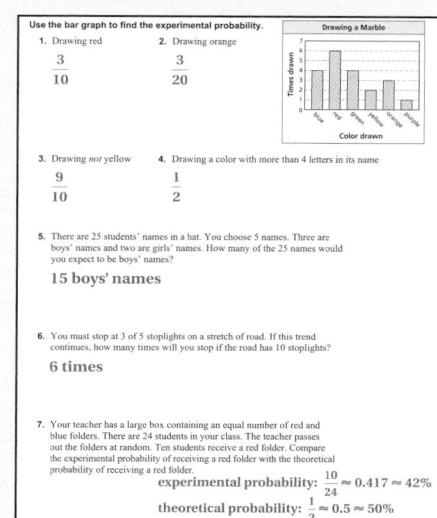

Technology For the Teacher

Answer Presentation Tool
QuizShow

1. Perform an experiment several times. Count how often the event occurs and divide by the number of trials.

2. yes; You could flip tails 7 out of 10 times, but with more trials the probability of flipping tails should get closer to 0.5.

 Practice and Problem Solving

3. $\frac{2}{5}$ or 40%   4. $\frac{1}{5}$ or 20%

5. $\frac{2}{5}$ or 40%   6. $\frac{3}{5}$ or 60%

7. $\frac{7}{50}$ or 14%   8. $\frac{12}{25}$ or 48%

9. $\frac{21}{25}$ or 84%   10. $\frac{7}{25}$ or 28%

11. $\frac{17}{50}$ or 34%   12. 0 or 0%

13. The theoretical probability was found, not the experimental probability.
$$P(4) = \frac{11}{50}$$

14. $\frac{3}{20}$ or 15%

15. 45 tiles

16. 5 cards

Practice and Problem Solving

17. 25

18. a. $\frac{5}{9}$ or about 55.6%

 b. 40 songs

19. The experimental probability of 60% is close to the theoretical probability of 50%.

20–22. See *Taking Math Deeper*.

23. *Sample answer:* Roll two number cubes 50 times and find each product. Record how many times the product is at least 12. Divide this number by 50 to find the experimental probability.

Fair Game Review

24. $x = 20$ **25.** $x = 5$

26. $x = -13$ **27.** $x = 24$

28. C

Mini-Assessment

You have three sticks. Each stick has one red side and one blue side. You throw the sticks 10 times and record the results. Use the table to find the experimental probability of the event.

Outcome	Frequency
3 red	3
3 blue	2
2 blue, 1 red	4
2 red, 1 blue	1

1. Tossing 3 blue $\frac{1}{5}$

2. Tossing 2 blue, 1 red $\frac{2}{5}$

3. *Not* tossing all blue $\frac{4}{5}$

4. Tossing 2 red, 1 blue $\frac{1}{10}$

5. Tossing 3 red $\frac{3}{10}$

Taking Math Deeper

Exercises 20–22

This is a good opportunity for students to compare *experimental probabilities* (results of trials), shown in the bar graph, with the *theoretical probabilities* (all possible outcomes), shown in the table.

① Make a list of the different sums and the number of times rolled from the bar graph. Check that the total is 60. Find the experimental probability of each sum.

2: 2/60	3: 4/60	4: 5/60	5: 6/60	6: 13/60	7: 10/60
8: 6/60	9: 8/60	10: 2/60	11: 3/60	12: 1/60	

20. The most likely sum is 6.

② Make a list of all possible ways to get each sum.

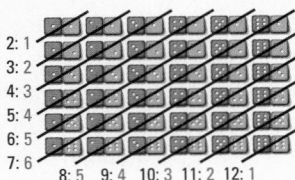

2: 1
3: 2
4: 3
5: 4
6: 5
7: 6
8: 5 9: 4 10: 3 11: 2 12: 1

7 is most likely.

Find the theoretical probability of each sum.

2: 1/36	3: 2/36	4: 3/36	5: 4/36	6: 5/36	7: 6/36
8: 5/36	9: 4/36	10: 3/36	11: 2/36	12: 1/36	

21. The most likely sum is 7.

③ **22.** Compare the two types of probabilities with a double bar graph.

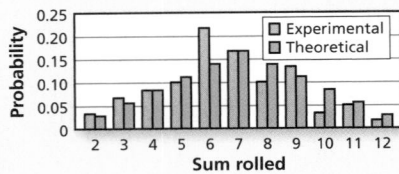

Reteaching and Enrichment Strategies

If students need help. . .	If students got it. . .
Resources by Chapter • Practice A and Practice B • Puzzle Time Record and Practice Journal Practice Differentiating the Lesson Lesson Tutorials Skills Review Handbook	Resources by Chapter • Enrichment and Extension • School-to-Work Start the next section

17. **QUALITY CONTROL** An inspector estimates that $\frac{1}{2}$% of MP3 players are defective. In a shipment of 5000 MP3 players, predict the number that are defective.

18. **MUSIC** During a 24-hour period, the ratio of pop songs played to rap songs played on a radio station is $60 : 75$.

   a. What is the experimental probability that the next song played is rap?

   b. Out of the next 90 songs, how many would you expect to be pop?

③ 19. **FLIPPING A COIN** You flip a coin 20 times. You flip heads 12 times. Compare your experimental probability of flipping heads with the theoretical probability of flipping heads.

You roll a pair of number cubes 60 times. You record your results in the bar graph shown.

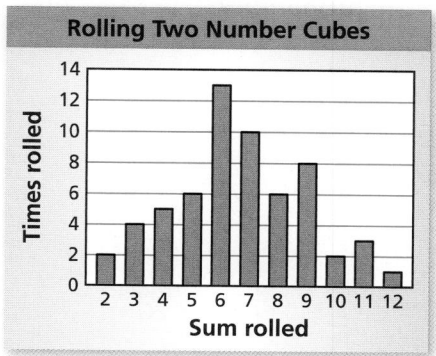

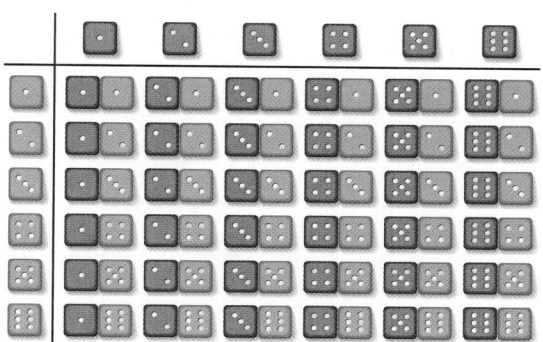

20. Use the bar graph to find the experimental probability of rolling each sum. Which sum is most likely?

21. Use the table to find the theoretical probability of rolling each sum. Which sum is most likely?

22. Compare the probabilities you found in Exercises 20 and 21.

23. **Critical Thinking** You roll two number cubes. Describe and perform an experiment to find the probability that the product of the two numbers rolled is at least 12. How many times did you roll the number cubes?

 **Fair Game Review** *What you learned in previous grades & lessons*

Solve the equation. *(Section 2.5)*

24. $5x = 100$ 25. $75 = 15x$ 26. $2x = -26$ 27. $-4x = -96$

28. **MULTIPLE CHOICE** What is the least common denominator of the fractions $\frac{1}{16}$, $\frac{2}{19}$, and $\frac{3}{76}$? *(Skills Review Handbook)*

   Ⓐ 16   Ⓑ 76   Ⓒ 304   Ⓓ 1216

Essential Question What is the difference between dependent and independent events?

1 ACTIVITY: Dependent Events

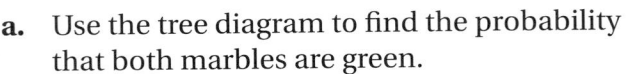

Work with a partner. You have three marbles in a bag. There are two green marbles and one purple marble. You randomly draw two marbles from the bag.

a. Use the tree diagram to find the probability that both marbles are green.

First Draw

Second Draw

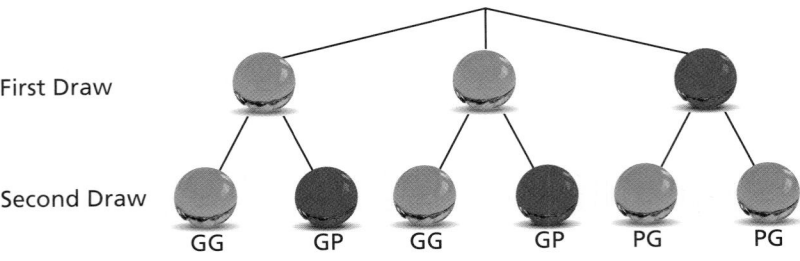

GG   GP   GG   GP   PG   PG

b. In the tree diagram, does the probability of getting a green marble on the second draw *depend* on the color of the first marble? Explain.

2 ACTIVITY: Independent Events

Work with a partner. Using the same marbles from Activity 1, randomly draw a marble from the bag. Then put the marble back in the bag and draw a second marble.

a. Use the tree diagram to find the probability that both marbles are green.

First Draw

Second Draw

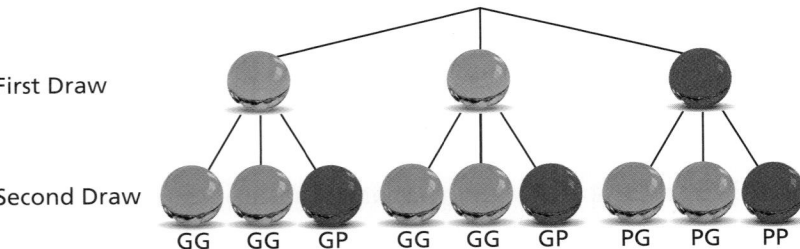

GG GG GP   GG GG GP   PG PG PP

b. In the tree diagram, does the probability of getting a green marble on the second draw *depend* on the color of the first marble? Explain.

Laurie's Notes

Introduction

For the Teacher

- **Goal:** Students will explore the differences between dependent and independent events.

Motivate

- Hand two volunteers a bag containing 2 red and 2 blue cubes (any similar objects will work). The goal is to draw twice and to end up with two of the same color. The only difference is that person A *replaces* their cube after the first draw and person B *does not replace* their cube after the first draw. So, Person A is always drawing 1 out of 4 items. Person B is drawing 1 out of 4 items, and then 1 out of 3 items.
- Have each person do 10 trials and gather data. This should help students start to see a pattern.
- **Big Idea:** The probability on the second draw *depends upon* what is drawn on the first draw when you do *not* replace the cube.
- Draw a tree diagram to show the difference between Person A and Person B. This will help students read the tree diagrams in the activity.

Activity Notes

Activity 1 and Activity 2

- The tree diagrams are used in the activities to help compute the theoretical probability.
- **Common Error:** Students read the tree diagram incorrectly and think a person is drawing six times in Activity 1 and nine times in Activity 2 because they think that each branch represents a draw. Remind students that the first horizontal row of circles displays the possible outcomes of the first draw and the second horizontal row of circles displays the possible outcomes of the second draw.
- **?** "On the first draw, you will either draw a green marble or a purple marble. Why does the tree diagram show more than just one of each?" because there are two different green marbles that could be drawn and only one purple
- **Teaching Tip:** Use the labels green marble 1 and green marble 2 to help students recognize why you need two green and one purple in the tree diagram for the first draw.
- **?** "When you *do not* replace the marble, are the probabilities on the second draw the same as on the first draw? Explain." no; There are only two marbles instead of three to draw from. The probability of the second draw depends upon the first marble drawn.
- **?** "When you *do* replace the marble, are the probabilities on the second draw the same as on the first draw? Explain." yes; There are the same marbles in the bag each time.
- **?** "For Activity 1, what is the probability of drawing 2 purple marbles? for Activity 2?" $0; \frac{1}{9}$

Activity Materials	
Introduction	**Closure**
• paper bags • colored cubes	• deck of cards

Start Thinking! and Warm Up

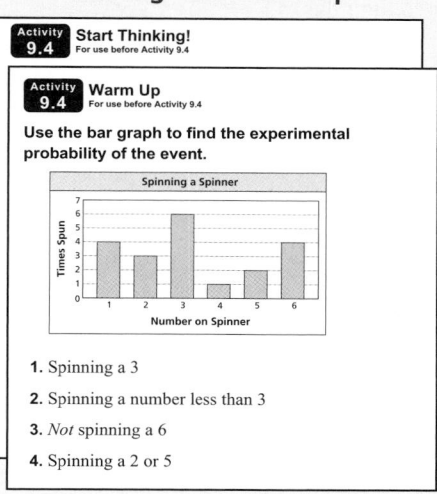

Activity 9.4 Start Thinking! For use before Activity 9.4

Activity 9.4 Warm Up For use before Activity 9.4

Use the bar graph to find the experimental probability of the event.

1. Spinning a 3
2. Spinning a number less than 3
3. *Not* spinning a 6
4. Spinning a 2 or 5

9.4 Record and Practice Journal

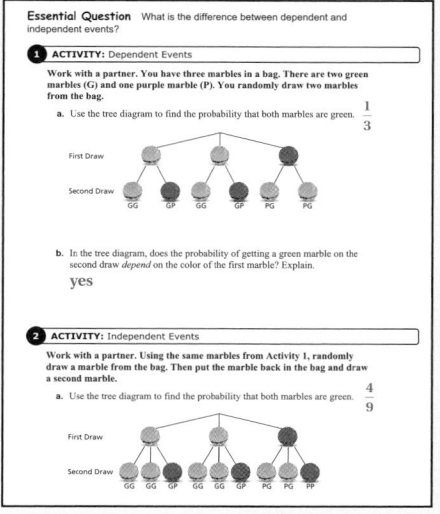

Differentiated Instruction

Kinesthetic

Before class, choose three books to put on your desk. Ask 3 students to choose a book by writing the title on a piece of paper. Then ask another 3 students to select a book, one at a time, and take it back to their seat. Discuss when the selection of the books was independent (The students wrote their choice.) and when the selection of books was dependent (The students took the books to their desks.).

9.4 Record and Practice Journal

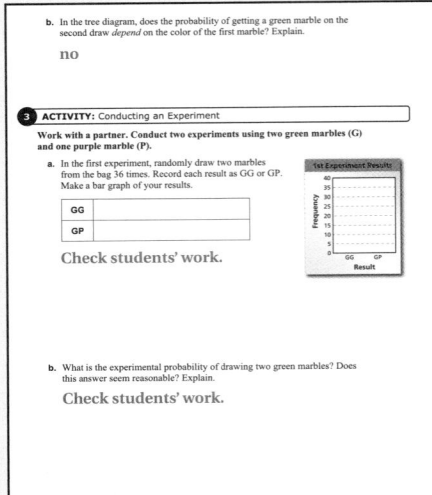

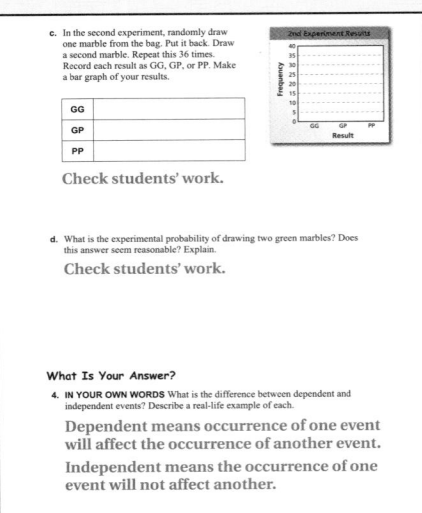

Activity 3

- Students will now collect the data for the same events described in Activities 1 and 2. The experimental probabilities should be close to the theoretical probabilities.

- Explain that in part (a), it does not matter if you draw one marble followed by another, or you draw two at one time. The result is the same. You will either have two green marbles in your hand (GG), or one of each color (GP).

- In part (c), you cannot draw two marbles at once. You draw one, replace it, and draw again. This results in a third possible outcome (PP).

- Have pairs of students hold up the results of each experiment so that bar graphs can be analyzed visually.

- **?** "For the first experiment, compare the lengths of the two bars. What do you notice? Interpret your answer." The GP bar should be about twice as long as the GG bar, so you are twice as likely to draw one of each color than you are to draw two greens.

- **?** "For the second experiment, compare the lengths of the three bars. What do you notice? Interpret your answer." The GG bar and the GP bar are about the same length. The PP bar is not very tall, maybe about a fourth as tall as either of the other bars. So, you are as likely to draw two greens as you are to draw one of each color, and you are 4 times more likely to draw two green or one of each color as you are to draw two purple.

What Is Your Answer?

- Student language may not be precise at this point and students may only be comfortable describing the difference in dependent and independent events in terms of purple and green marbles.

Closure

- You have a deck of playing cards (4 suits, 13 cards in each suit). Give an example of dependent events and independent events using the deck of cards.

  Sample answer:

  Dependent events—draw one card, set it aside, then draw another

  Independent events—draw one card, replace it, then draw another

Technology
For the Teacher

Dynamic Classroom

The Dynamic Planning Tool
Editable Teacher's Resources at *BigIdeasMath.com*

ACTIVITY: Conducting an Experiment

Work with a partner. Conduct two experiments.

a. In the first experiment, randomly draw two marbles from the bag 36 times. Record each result as GG or GP. Make a bar graph of your results.

b. What is the experimental probability of drawing two green marbles? Does this answer seem reasonable? Explain.

c. In the second experiment, randomly draw one marble from the bag. Put it back. Draw a second marble. Repeat this 36 times. Record each result as GG, GP, or PP. Make a bar graph of your results.

d. What is the experimental probability of drawing two green marbles? Does this answer seem reasonable? Explain.

1st Experiment

GG	
GP	

2nd Experiment

GG	
GP	
PP	

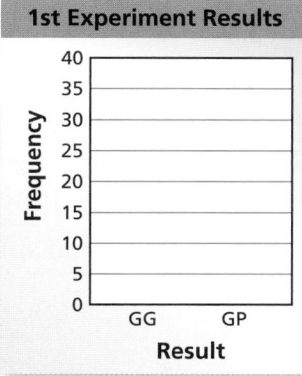

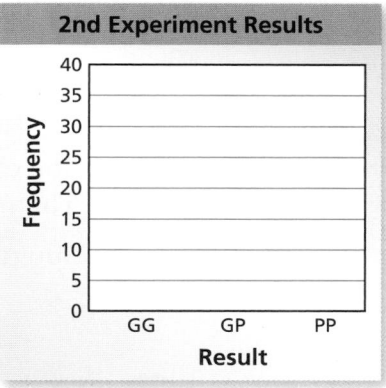

What Is Your Answer?

4. IN YOUR OWN WORDS What is the difference between dependent and independent events? Describe a real-life example of each.

Practice

Use what you learned about independent and dependent events to complete Exercises 5 and 6 on page 409.

Key Vocabulary
independent events, p. 406
dependent events, p. 406

Two events are **independent events** if the occurrence of one event *does not* affect the likelihood that the other event will occur.

Two events are **dependent events** if the occurrence of one event *does* affect the likelihood that the other event will occur.

EXAMPLE 1 Identifying Independent and Dependent Events

Tell whether the events are *independent* or *dependent*. Explain.

a. You flip heads on one coin and tails on another coin.

The outcome of flipping one coin does not affect the outcome of flipping the other coin.

⁘ So, the events are independent.

b. Your teacher chooses one student to lead a group, and then chooses another student to lead another group.

The teacher cannot pick the same student to lead both groups. So, there are fewer students to choose from when the leader of the second group is chosen.

⁘ So, the events are dependent.

On Your Own

Now You're Ready
Exercises 5–9

Tell whether the events are *independent* or *dependent*. Explain.

1. You choose a blue marble from a bag and set it aside. Then you choose a green marble from the bag.

2. You roll a 5 on a number cube and spin blue on a spinner.

Key Idea

Probability of Independent Events

Words The probability of two independent events A and B is the probability of A times the probability of B.

Symbols $P(A \text{ and } B) = P(A) \cdot P(B)$

probability of both events

probability of first event

probability of second event

◀ Multi-Language Glossary at BigIdeasMath.com.

Laurie's Notes

Introduction

Connect
- **Yesterday:** Students developed an understanding of the difference between dependent and independent events.
- **Today:** Students will use a formal definition to compute theoretical probabilities of dependent and independent events.

Motivate
- ❓ "Playing a game with number cubes, do you roll two ones very often?" no
- ❓ "What is the probability of rolling two ones?" Students may quickly say $\frac{1}{6}$ before realizing that it is $\frac{1}{36}$.
- ❓ "Does the roll on the first number cube affect what you roll on the second number cube?" no
- ❓ "Are rolling two number cubes independent or dependent events?" independent
- Explain that today they will compute the probability of independent and dependent events.

Lesson Notes

Example 1
- Write the definitions of independent and dependent events. Relate the definitions to the previous investigations or other contexts familiar to your students.
- **Model:** For part (a), use two coins. For part (b), have 4 students stand at the front of the class. Select one person to lead the group on the left side of the room. Once that person moves to the left side of the room, it is visually clear that they cannot be selected to lead the right side of the room. So, the chance of being selected for the left side is $\frac{1}{4}$, but the right side is $\frac{1}{3}$.

On Your Own
- **Think-Pair-Share:** Students should read each question independently and then work with a partner to answer the questions. When they have answered the questions, the pair should compare their answers with another group and discuss any discrepancies.

Key Idea
- ❓ "What are independent events?" two events where the occurrence of the first event does not affect the likelihood that the other event will occur
- This can seem like a daunting formula to students because it is very symbolic. Be sure to use the words and the notation.

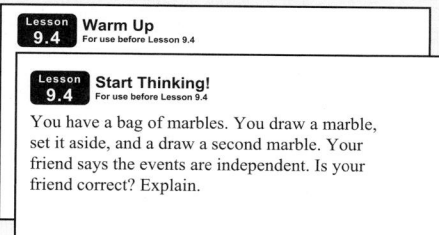
Extra Example 1

Tell whether the events are *independent* or *dependent*. Explain.

a. You choose a marble from a bag and set it aside. Then you choose another marble from the bag. dependent; There is one less marble on the second draw.

b. You choose a marble from a bag, record its color, and place it back into the bag. Then you choose another marble from the bag. independent; There are the same number of marbles for each draw.

On Your Own
1. dependent; There is one less marble on the second draw.
2. independent; The outcome of rolling a number cube does not affect the outcome of spinning a spinner.

Extra Example 2

You flip two quarters. What is the probability that you flip two tails? $\frac{1}{4}$

On Your Own

3. $\frac{1}{2}$

4. $\frac{1}{3}$

English Language Learners

Vocabulary

Make sure students understand the difference between *independent events* and *dependent events.* Have students give examples that are independent or dependent events and have them explain their reasoning.

Example 2

- Use two coins to model the problem.
- Students will ask if you can flip the coins at the same time or if you have to flip one after the other. It does not matter, but you need to identify one of the quarters as the first coin for recording purposes, so it is easier to flip one after the other.
- Use coins with different dates so that you can tell the two coins apart if you flip them at the same time.
- **Extension:** Ask students to find the probability of flipping two tails and the probability of flipping one head and one tail. $\frac{1}{4}; \frac{1}{2}$
- Notice that $P(2 \text{ heads}) + P(2 \text{ tails}) + P(1 \text{ head and 1 tail}) = 1$

On Your Own

- In Question 3, point out to students that this problem is easier using a tree diagram rather than a formula.
- For Question 4, provide a coin and a number cube to each pair of students. Encourage students to draw a tree diagram for this problem.
- Sample tree diagram for Question 4:

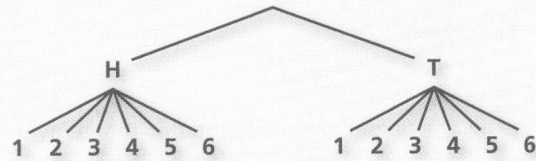

- **Think-Pair-Share:** Students should read each question independently and then work with a partner to answer the questions. When they have answered the questions, the pair should compare their answers with another group and discuss any discrepancies.

EXAMPLE **2** **Finding the Probability of Independent Events**

You flip two quarters. What is the probability that you flip two heads?

Method 1: Use a tree diagram to find the probability.

Let H = Heads and T = Tails.

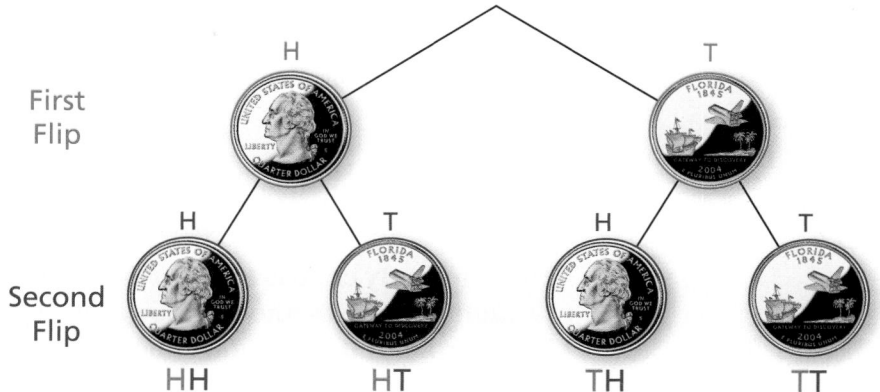

First
Flip

Second
Flip

HH   HT   TH   TT

$$P(\text{two heads}) = \frac{\text{number of times two heads occur}}{\text{total number of outcomes}}$$

$$= \frac{1}{4}$$

∴ The probability that you flip two heads is $\frac{1}{4}$.

Method 2: Use the formula for independent events.

$$P(A \text{ and } B) = P(A) \cdot P(B)$$

$$P(\text{heads and heads}) = P(\text{heads}) \cdot P(\text{heads})$$

$$= \frac{1}{2} \cdot \frac{1}{2} \qquad \text{Substitute.}$$

$$= \frac{1}{4} \qquad \text{Multiply.}$$

∴ The probability that you flip two heads is $\frac{1}{4}$.

On Your Own

Now You're Ready
Exercises 10–18

3. You flip two coins. What is the probability that you flip one heads and one tails?

4. You flip a coin and roll a number cube. What is the probability that you flip tails and roll a number less than 5?

 **Key Idea**

Probability of Dependent Events

Words The probability of two dependent events
probability of *A* times the probability of *B* after *A* occurs.

Symbols $P(A \text{ and } B) = P(A) \cdot P(B \text{ after } A)$

probability of both events

probability of first event

probability of second event after first event occurs

EXAMPLE **3** **Finding the Probability of Dependent Events**

You randomly choose a flower from the vase to take home. Your friend randomly chooses another flower from the vase to take home. What is the probability that you choose a purple flower and your friend chooses a yellow flower?

Purple: 7
Yellow: 9
Pink: 12

Choosing a flower changes the number of flowers left in the vase. So, the events are dependent.

$$P(\text{first is purple}) = \frac{7}{28} = \frac{1}{4}$$

There are 7 purple flowers.

There is a total of 28 flowers.

$$P(\text{second is yellow}) = \frac{9}{27} = \frac{1}{3}$$

There are 9 yellow flowers.

There is a total of 27 flowers left.

Use the formula to find the probability.

$$P(A \text{ and } B) = P(A) \cdot P(B \text{ after } A)$$

$$P(\text{purple and yellow}) = P(\text{purple}) \cdot P(\text{yellow after purple})$$

$$= \frac{1}{4} \cdot \frac{1}{3} \qquad \text{Substitute.}$$

$$= \frac{1}{12} \qquad \text{Simplify.}$$

∴ The probability of choosing a purple flower and then a yellow flower is $\frac{1}{12}$, or about 8%.

On Your Own

 Now You're Ready
Exercises 19–25

5. WHAT IF? In Example 3, what is the probability that both flowers are purple?

Laurie's Notes

Key Idea

- **?** "What are dependent events?" When one event affects another event.
- Write the Key Idea.
- This can seem like a daunting formula to students because it is very symbolic. When explaining the formula, be sure to use the words and the notation.

Example 3

- Work through the example.
- **Model:** Model this example using 28 pencils, seven of which are similar in some way. It could also be modeled using 28 students, seven of which are similar in some way (gender, wearing a blue shirt, and so on).
- Explain that this formula is very similar to the formula for the probability of independent events, except that the probability of the second event has been affected by the occurrence of the first event.

Closure

- Use a number cube and a coin.

  Find the probability of rolling a four and landing on heads. $\frac{1}{12}$

  Find the probability of rolling an even number and landing on tails. $\frac{1}{4}$

The Dynamic Planning Tool
Editable Teacher's Resources at *BigIdeasMath.com*

Extra Example 3

In Example 3, what is the probability that you choose a pink flower and your friend chooses a yellow flower? $\frac{1}{7}$

On Your Own

5. $\frac{1}{18}$ or about 5.6%

Differentiated Instruction

Visual

Use a table to help students understand why $P(A \text{ and } B) = P(A) \cdot P(B)$ for independent events A and B. Let A be the event of choosing jeans or shorts. Let B be the event of choosing a tank top, a T-shirt, or a button-down shirt.

	Jeans (J)	Shorts(S)
Tank Top (T)	JT	ST
T-shirt (R)	JR	SR
Button-down Shirt (B)	JB	SB

From the table, the probability of randomly choosing shorts and a tank top is $\frac{1}{6}$. The probability of choosing shorts is $P(S) = \frac{3}{6} = \frac{1}{2}$. The probability of choosing a tank top is $P(T) = \frac{1}{3}$. So, the probability of choosing shorts and a tank top is $P(S \text{ and } T) = \frac{1}{2} \cdot \frac{1}{3} = \frac{1}{6}$.

Vocabulary and Concept Check

1. Draw a tree diagram or multiply $P(A)$ by $P(B)$.

2.
| Flip 1 | Flip 2 | Flip 3 | Outcome |
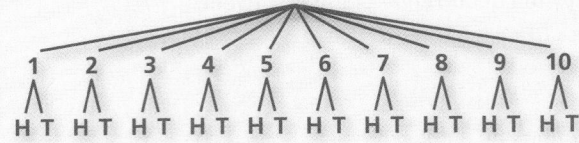

Flip 1, Flip 2, Flip 3, Outcome:
HHH, HHT, HTH, HTT, THH, THT, TTH, TTT

3. *Sample answer:* independent events: a traffic jam and a sunny day; dependent events: temperatures below freezing and ice

4. Find the probability of rolling an odd number and then an even number.; $\frac{1}{4}$; $\frac{1}{12}$

Practice and Problem Solving

5. independent; The outcome of the first roll does not affect the outcome of the second roll.

6. independent; The outcome of the first coin flip does not affect the outcome of the second coin flip.

7. independent; You replace the marble, so the probability doesn't change.

8. dependent; There is one less marble to choose from on the second draw.

9. dependent; There is one less person to choose from on the second draw.

Assignment Guide and Homework Check

Level	Day 1 Activity Assignment	Day 2 Lesson Assignment	Homework Check
Basic	5, 6, 36–39	1–4, 7–23 odd, 26–29	7, 11, 17, 19, 26, 28
Average	5, 6, 36–39	1–4, 7–25 odd, 28–30	7, 11, 17, 19, 28, 30
Advanced	5, 6, 36–39	1–4, 14, 18, 23–25, 29–35	14, 18, 24, 30, 34

For Your Information

- **Exercises 15–18** Provide this tree diagram for students who are struggling.

1 2 3 4 5 6 7 8 9 10
H T H T H T H T H T H T H T H T H T H T

- **Exercise 25** You may need to help students see that the letters spell *Orlando*.

9.4 Record and Practice Journal

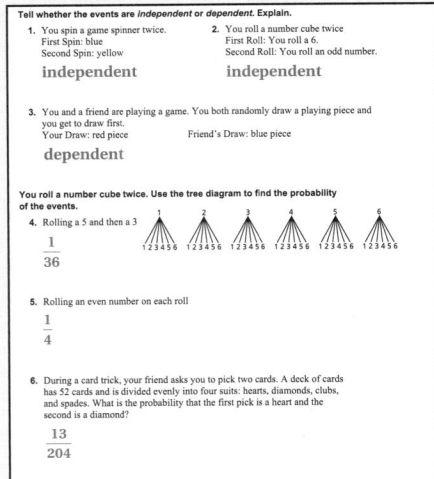

9.4 Exercises

✓ Vocabulary and Concept Check

1. **VOCABULARY** Events A and B are independent. Describe two ways to find $P(A \text{ and } B)$.

2. **FILL IN THE BLANKS** Copy and complete the tree diagram to find the possible outcomes for flipping a coin three times.

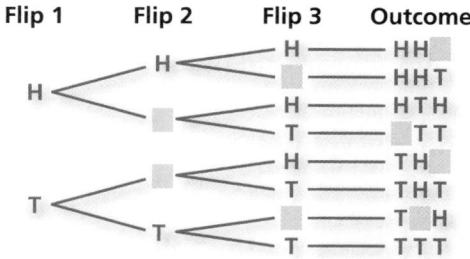

3. **OPEN-ENDED** Describe a real-life example of two independent events. Describe a real-life example of two dependent events.

4. **DIFFERENT WORDS, SAME QUESTION** Which is different? Find "both" answers.

Find the probability of rolling a 1 and then a 2, 4, or 6.

Find the probability of rolling a 1 and then an even number.

Find the probability of rolling an odd number and then an even number.

Find the probability of rolling a number less than 2 and then an even number.

✏️ Practice and Problem Solving

Tell whether the events are *independent* or *dependent*. Explain.

① 5. You roll a number cube twice.

First Roll: You roll a 4.
Second Roll: You roll an even number.

6. You flip a coin twice.

First Flip: Heads
Second Flip: Heads

7. You randomly draw a marble from a bag containing 2 red marbles and 5 green marbles. You put the marble back and then draw a second marble.

First Draw: Green   Second Draw: Red

8. You randomly draw a marble from a bag containing 2 red marbles and 5 green marbles. You keep the marble and then draw a second marble.

First Draw: Green   Second Draw: Red

9. You and your friend are in a drawing for two door prizes. You can win only one prize.

First Draw: Your name is drawn.   Second Draw: Your friend's name is drawn.

A spinner has three equal sections numbered 1, 2, and 3. You spin it twice. Use the tree diagram to find the probability of the events.

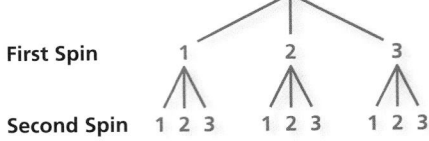

First Spin

Second Spin

2 **10.** Spinning a 1 and then a 3

11. Spinning an odd number and then a 2

12. Spinning a 3 and then an even number

13. Spinning an even number and then an odd number

14. Spinning an odd number on each spin

You spin the spinner and flip a coin. Find the probability of the events.

15. Spinning a 4 and flipping heads

16. Spinning an even number and flipping tails

17. Spinning a multiple of 3 and flipping heads

18. Spinning white and *not* flipping tails

You randomly choose one of the lettered tiles. Without replacing the first tile, you choose a second tile. Find the probability of choosing the first tile, then the second tile.

3 **19.** R and N

20. A and L

21. D and O

22. N and yellow

23. O and *not* yellow

24. *Not* O and O

25. If you randomly choose all seven tiles in order, what is the probability that you will spell the name of a popular vacation destination in Florida?

26. EARRINGS A jewelry box contains two gold hoop earrings and two silver hoop earrings. You randomly choose two earrings. What is the probability that both are silver hoop earrings?

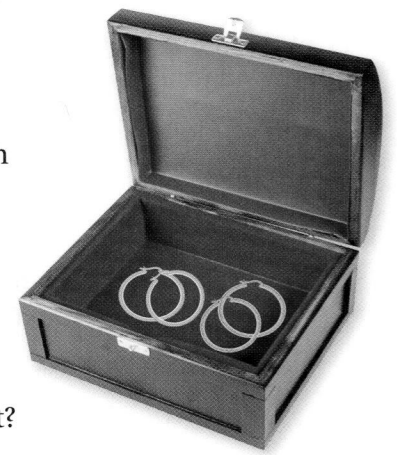

27. PASSWORD You forgot the last two digits of your password for a website.

    a. You choose a two-digit number at random. What is the probability that your choice is correct?

    b. Suppose you remember that both digits are even numbers. How does this change the probability that your choice is correct?

Common Errors

- **Exercise 2** Students may struggle filling in the missing parts of the tree diagram. Encourage them to look at the diagram one level at a time, starting from the left.
- **Exercises 5–9** Students may mix up independent and dependent events or may have difficulty determining which type of event it is. Remind them that independent events are similar to doing two different things, or an event where you *start over* before the next trial. Dependent events have at least one less possible outcome after the first draw, roll, or flip.
- **Exercises 10–14** If students use the formula, they may find the probability of the second spin as if the experiment was dependent. Remind them that these are independent events.
- **Exercises 19–24** Students may forget to subtract one from the total number of possible outcomes when finding the probability of choosing the second tile. Remind them that the second draw has one less possible outcome because you have removed one of the tiles.

 ## Practice and Problem Solving

10. $\frac{1}{9}$　　　**11.** $\frac{2}{9}$

12. $\frac{1}{9}$　　　**13.** $\frac{2}{9}$

14. $\frac{4}{9}$　　　**15.** $\frac{1}{20}$ or 5%

16. $\frac{1}{4}$ or 25%　　**17.** $\frac{3}{20}$ or 15%

18. $\frac{1}{4}$ or 25%

19. $\frac{1}{42}$ or about 2.4%

20. $\frac{1}{42}$ or about 2.4%

21. $\frac{1}{21}$ or about 4.8%

22. $\frac{1}{14}$ or about 7.1%

23. $\frac{4}{21}$ or about 19%

24. $\frac{5}{21}$ or about 23.8%

25. $\frac{1}{2520}$ or about 0.04%

26. $\frac{1}{6}$ or about 16.7%

27. a. $\frac{1}{100}$ or 1%

   b. It increases the probability that your choice is correct to $\frac{1}{25}$ or 4%, because each digit could be 0, 2, 4, 6, or 8.

28. $\frac{1}{5}$ or 20%

29. a. $\frac{1}{9}$ or about 11.1%

   b. It increases the probability that your guesses are correct to $\frac{1}{4}$ or 25%, because you are only choosing between 2 choices for each question.

30. a. $\frac{1}{5}$ or 20%

   b. 2

31. $\frac{16}{25}$ or 64%

32. See *Taking Math Deeper*.

33. 1 : 5; 5 : 1

34. 2 : 1; 1 : 2

35. 1 : 35; 35 : 1

Fair Game Review

36. $x = -3.3$

37. $n = -10.8$

38. $p = 4$

39. B

Mini-Assessment

Tell whether the events are *independent* or *dependent*. Explain.

1. You randomly draw a marble from a bag containing 3 red marbles and 4 green marbles. You put the marble back and then draw a second marble. independent; The outcome of the first event does not affect the outcome of the second event because the marble was replaced.

2. You randomly draw a marble from a bag containing 3 red marbles and 4 green marbles. You set the marble aside and then draw a second marble. dependent; The outcome of the first event does affect the outcome of the second event because there are fewer marbles from which to choose.

Taking Math Deeper

Exercise 32

A good way to list the possible outcomes of an experiment is to use a tree diagram.

 Draw the first column of a tree diagram.

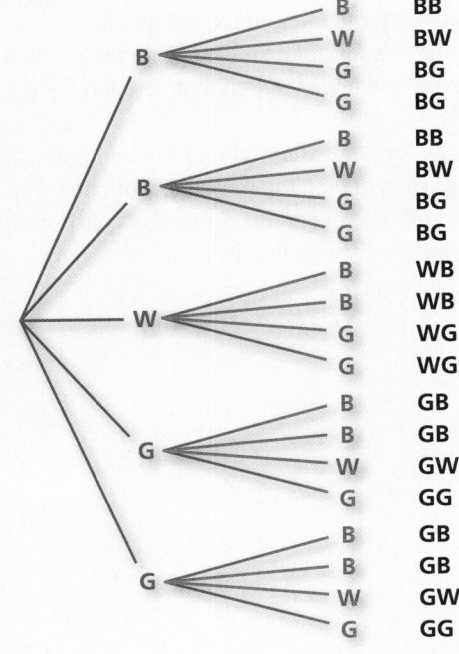

20 outcomes

 Add the second column of the tree diagram.

 Use the tree diagram to list the possible outcomes.

Project

Check out the sunglasses display at your favorite store. How many different types of sunglasses are displayed? How many of each type do they have? Suppose you and your friend make separate trips to the store to buy sunglasses. What is the probability that you both select the same type of sunglasses? Explain.

Reteaching and Enrichment Strategies

If students need help. . .	If students got it. . .
Resources by Chapter • Practice A and Practice B • Puzzle Time Record and Practice Journal Practice Differentiating the Lesson Lesson Tutorials Skills Review Handbook	Resources by Chapter • Enrichment and Extension • School-to-Work Start the next section

28. FISH You randomly choose two fish from the bowl. What is the probability that the first is red and the second is gold?

29. TAKING A TEST You are guessing at two questions on a multiple choice test. Each question has three choices: A, B, and C.

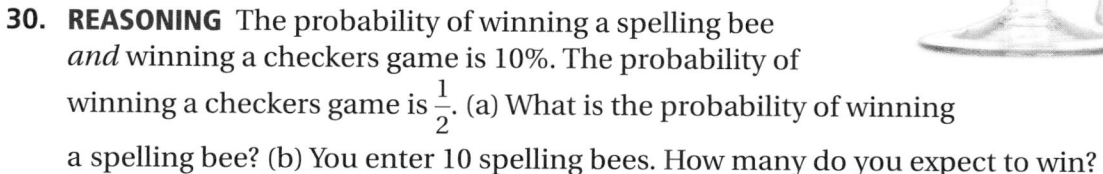

   a. What is the probability that you guess the correct answers to both questions?

   b. Suppose you can eliminate one of the choices for each question. How does this change the probability that your guesses are correct?

30. REASONING The probability of winning a spelling bee *and* winning a checkers game is 10%. The probability of winning a checkers game is $\frac{1}{2}$. (a) What is the probability of winning a spelling bee? (b) You enter 10 spelling bees. How many do you expect to win?

31. SHOES Twenty percent of the shoes manufactured by a company are black. One shoe is chosen and replaced. Then a second shoe is chosen. What is the probability that *neither* shoe is black?

32. *Critical Thinking* You randomly choose a pair of sunglasses from the shelf below. Then you randomly choose a second pair of sunglasses without replacing the first pair. List all of the possible outcomes.

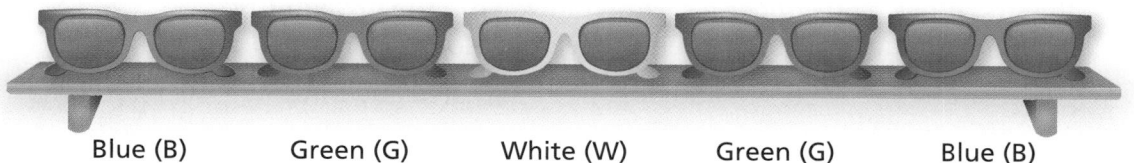

     Blue (B)   Green (G)   White (W)   Green (G)   Blue (B)

ODDS The *odds in favor of* an event is the ratio of the number of favorable outcomes to the number of unfavorable outcomes. The *odds against* an event is the ratio of the number of unfavorable outcomes to the number of favorable outcomes. Find the *odds in favor of* and the *odds against* the event when rolling a number cube.

33. Rolling a 6

34. Rolling a number less than 5

35. Rolling a 6, then rolling a 3

Fair Game Review What you learned in previous grades & lessons

Solve the equation. *(Section 2.4, Section 2.5, and Section 2.6)*

36. $6 = 9.3 + x$

37. $\frac{n}{2} = -5.4$

38. $-4p + 6 = -10$

39. MULTIPLE CHOICE Which intervals can be used to make a histogram? *(Section 8.2)*

   A 16–18, 19–21, 22–26, 27–32

   B 91–110, 111–130, 131–150

   C 11–20, 21–40, 41–50, 51–70

   D 50–60, 60–70, 70–80, 80–90

Use the bar graph to find the experimental probability of the event. *(Section 9.3)*

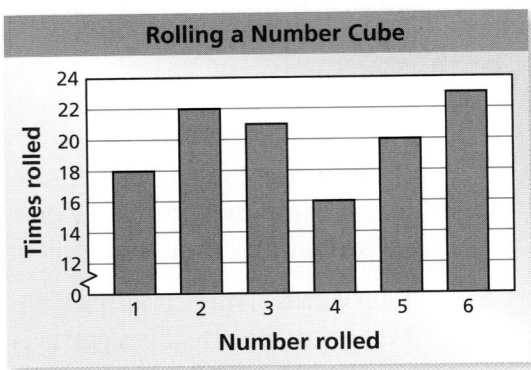

1. Rolling a 4

2. Rolling a multiple of 3

3. Rolling a 2 or a 3

4. Rolling a number less than 7

You randomly choose a playing piece and flip a coin. Find the probability of the events. *(Section 9.4)*

5. Choosing red and flipping tails

6. Choosing black and flipping heads

7. *Not* choosing red and *not* flipping heads

You randomly choose one of the letter blocks. Without replacing the first block, you choose a second block. Find the probability of choosing the first block, then the second block. *(Section 9.4)*

8. E and M   **9.** M and *not* E   **10.** Red and blue

11. PENS There are 30 pens in a box. You choose the five pens shown. How many of the 30 pens would you expect to have red ink? *(Section 9.3)*

12. SWEATERS A drawer contains three tan sweaters and two black sweaters. You randomly choose two sweaters. What is the probability that both sweaters are black? *(Section 9.4)*

13. SPINNER You spin the spinner 40 times. It lands on red 32 times. Compare the experimental probability of the spinner landing on red with the theoretical probability of the spinner landing on red. *(Section 9.3)*

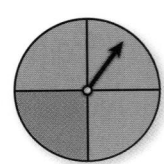

14. FISH You randomly choose one fish from each bowl for your aquarium. What is the probability of choosing two gold fish? *(Section 9.4)*

Alternative Assessment Options

Math Chat   Student Reflective Focus Question

Structured Interview   Writing Prompt

Math Chat

- Have individual students work problems from the quiz on the board. The student explains the process used and justifies each step. Students in the class ask questions of the student presenting.
- The teacher probes the thought process of the student presenting, but does not teach or ask leading questions.

Study Help Sample Answers

Remind students to complete Graphic Organizers for the rest of the chapter.

4.

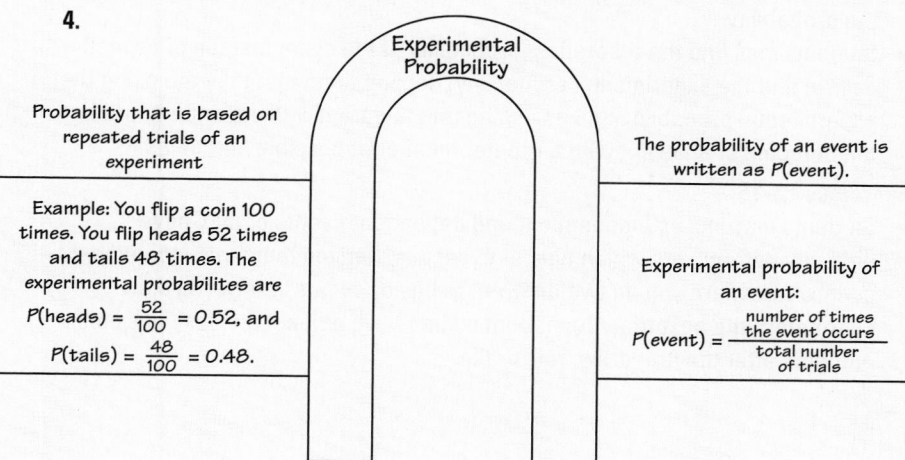

Experimental Probability

Probability that is based on repeated trials of an experiment

Example: You flip a coin 100 times. You flip heads 52 times and tails 48 times. The experimental probabilites are $P(\text{heads}) = \frac{52}{100} = 0.52$, and $P(\text{tails}) = \frac{48}{100} = 0.48$.

The probability of an event is written as $P(\text{event})$.

Experimental probability of an event:

$$P(\text{event}) = \frac{\text{number of times the event occurs}}{\text{total number of trials}}$$

5–7. Available at *BigIdeasMath.com*.

Answers

1. $\frac{2}{15}$ or about 13.3%

2. $\frac{11}{30}$ or about 36.7%

3. $\frac{43}{120}$ or about 35.8%

4. 1 or 100%

5. $\frac{1}{12}$ or about 8.3%

6. $\frac{5}{12}$ or about 41.7%

7. $\frac{5}{12}$ or about 41.7%

8. $\frac{1}{14}$ or about 7.1%

9. $\frac{5}{28}$ or about 17.9%

10. $\frac{1}{14}$ or about 7.1%

11. 6

12. $\frac{1}{10}$ or 10%

13. The experimental probability of 80% is close to the theoretical probability of 75%.

14. $\frac{1}{3}$

Reteaching and Enrichment Strategies

If students need help. . .	If students got it. . .
Resources by Chapter • Study Help • Practice A and Practice B • Puzzle Time Lesson Tutorials *BigIdeasMath.com* Practice Quiz Practice from the Test Generator	Resources by Chapter • Enrichment and Extension • School-to-Work Game Closet at *BigIdeasMath.com* Start the Chapter Review

Technology For the Teacher

Answer Presentation Tool

Assessment Book

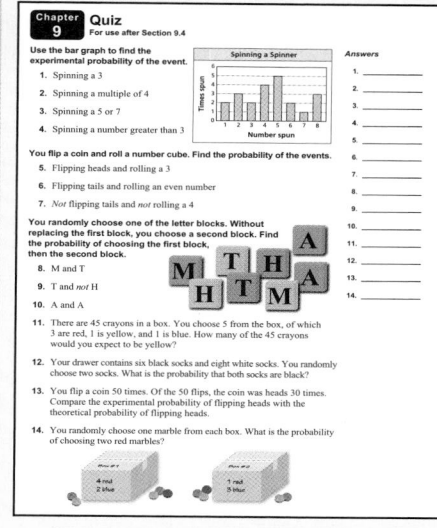

For the Teacher
Additional Review Options

- **Quiz***Show*
- Big Ideas Test Generator
- Game Closet at *BigIdeasMath.com*
- Vocabulary Puzzle Builder
- Resources by Chapter
  - Puzzle Time
  - Study Help

Answers

1. 2
2. 3
3. 5
4. 8
5. 1, 1, 2, 2, 2

Review of Common Errors

Exercises 1–5

- Students may forget to include, or include too many, favorable outcomes. Encourage them to write out all of the possible outcomes and then circle the favorable outcomes for the given event.

Exercise 6

- Students may look at the events and guess that the game is fair or not fair without finding the probability of each event. Encourage them to prove that their guess is true by finding the probability of each event.

Exercises 8–11

- Students may forget to total all of the trials before writing the experimental probability. They may have an incorrect number of trials in the denominator. Remind them that they need to know the total number of trials when finding the probability.
- Students may find the theoretical probability of the event instead of using the data to find the experimental probability. Remind them that they are using the experimental probability and assuming that this trend will continue to predict the outcome of an event with a greater number of possible outcomes.

Exercises 12–15

- Students may mix up independent and dependent events or may have difficulty determining which type of event it is. Remind them that independent events are where you do two different things or events where you start over before the next trial. Dependent events have at least one less possible outcome after the first draw, roll, or flip.

Review Key Vocabulary

experiment, *p. 386*
outcomes, *p. 386*
event, *p. 386*
probability, *p. 387*
theoretical probability, *p. 392*

fair experiment, *p. 393*
experimental
    probability, *p. 400*
independent events, *p. 406*
dependent events, *p. 406*

Review Examples and Exercises

9.1 Introduction to Probability *(pp. 384–389)*

You randomly choose one toy racecar.

a. In how many ways can choosing a green car occur?

b. In how many ways can choosing a car that is *not* green occur? What are the favorable outcomes of choosing a car that is *not* green?

a. There are 5 green cars. So, choosing a green car can occur in 5 ways.

b. There are 2 cars that are *not* green. So, choosing a car that is *not* green can occur in 2 ways.

green	*not* green
green, green, green, green, green	blue, red

∴ The favorable outcomes of the event are blue and red.

Exercises

You spin the spinner. Find the number of ways the event can occur.

1. Spinning a 1

2. Spinning a 3

3. Spinning an odd number

4. Spinning a number greater than 0

5. On the spinner, what are the favorable outcomes of spinning a number less than 3?

9.2 Theoretical Probability (pp. 390–395)

The theoretical probability that you choose a purple grape from a bag is $\frac{2}{9}$. There are 36 grapes in the bag. How many are purple?

$$P(\text{purple}) = \frac{\text{number of purple grapes}}{\text{total number of grapes}}$$

$\frac{2}{9} = \frac{n}{36}$ Substitute. Let n be the number of purple grapes.

$8 = n$ Multiply each side by 36.

⋮ There are 8 purple grapes in the bag.

Exercises

6. You get one point when the spinner at the right lands on an odd number. Your friend gets one point when it lands on an even number. The first person to get 5 points wins. Is the game fair? If it is not fair, who has the greater probability of winning?

7. The probability that you spin an even number on a spinner is $\frac{2}{3}$. The spinner has 12 sections. How many sections have even numbers?

9.3 Experimental Probability (pp. 398–403)

The bar graph shows the results of spinning the spinner 70 times. What is the experimental probability of spinning a 2?

The bar graph shows 14 ones, 12 twos, 16 threes, 15 fours, and 13 fives. So, a two was spun 12 times in 70 spins.

$$P(\text{event}) = \frac{\text{number of times the event occurs}}{\text{total number of trials}}$$

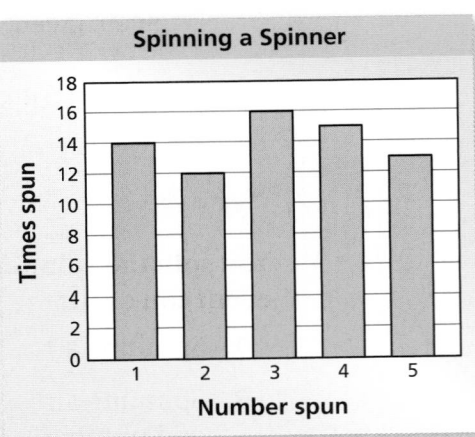

$P(2) = \frac{12}{70}$ ←   A 2 was spun 12 times.

       ←   There was a total of 70 spins.

$ = \frac{6}{35}$

⋮ The experimental probability is $\frac{6}{35}$, or about 17%.

Review Game

Making Predictions

Big Ideas
Game Closet

For the Student
Additional Practice

- Lesson Tutorials
- Study Help (textbook)
- Student Website
  Multi-Language Glossary
  Practice Assessments

Materials per team:

- deck of cards
- paper
- pencil
- calculator

Directions:

Each team shuffles their deck of 52 cards. The first 39 cards are flipped over and placed in a discard pile. Teams are to keep track of what cards are discarded and what cards are left in the deck. The 13 cards remaining in the deck are shuffled by the teacher.

With the whole class, teams take turns predicting the next card to be flipped over from their deck. Predictions can include black card, red card, face card, single number card (if they are brave), etc. The team calculates the probability of the prediction on the board for the class to see.

Each team will do this for 5 cards. For each correct prediction, the team gets a point. The teacher holds onto each team's remaining cards until the game is over, in case of a tie.

Who Wins?

The team with the most points wins. If there is a tie, a one card draw will be the tie breaker. With the remaining cards in the team's deck, each team will make a prediction. The team whose prediction is correct wins.

Answers

6. not fair; your friend

7. 8

8. $\frac{8}{35}$ or about 22.9%

9. $\frac{43}{70}$ or about 61.4%

10. $\frac{57}{70}$ or about 81.4%

11. $\frac{2}{5}$ or 40%

12. $\frac{1}{5}$ or 20%

13. $\frac{1}{10}$ or 10%

14. $\frac{3}{10}$ or 30%

15. $\frac{1}{5}$ or 20%

My Thoughts on the Chapter

What worked. . .

Teacher Tip

Not allowed to write in your teaching edition? Use sticky notes to record your thoughts.

What did not work. . .

What I would do differently. . .

Exercises

Use the bar graph on page 414 to find the experimental probability of the event.

8. Spinning a 3

9. Spinning an odd number

10. *Not* spinning a 5

11. Spinning a number greater than 3

9.4 **Independent and Dependent Events** *(pp. 404–411)*

You randomly choose a marble without replacing it. Your friend then chooses another marble. What is the probability that you choose a red marble and your friend chooses a blue marble?

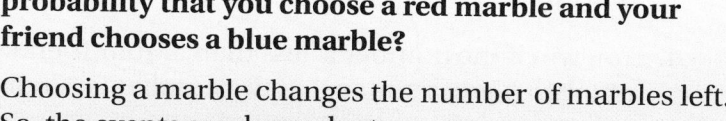

Choosing a marble changes the number of marbles left. So, the events are dependent.

$$P(\text{first is red}) = \frac{5}{12}$$

> There are 5 red marbles.
>
> There is a total of 12 marbles.

$$P(\text{second is blue}) = \frac{3}{11}$$

> There are 3 blue marbles.
>
> There is a total of 11 marbles left.

Use the formula to find the probability.

$$P(\text{red and blue}) = P(\text{red}) \cdot P(\text{blue after red})$$

$$= \frac{5}{12} \cdot \frac{3}{11} \qquad \text{Substitute.}$$

$$= \frac{5}{44} \qquad \text{Simplify.}$$

∴ The probability of choosing a red marble followed by a blue marble is $\frac{5}{44}$, or about 11%.

Exercises

You randomly choose one of the lettered tiles. Without replacing the first tile, you choose a second tile. Find the probability of choosing the first tile, then the second tile.

12. R and A

13. A and A

14. R and *not* D

15. You choose one of the lettered tiles and flip a coin. What is the probability of choosing an A and flipping heads?

Check It Out
Test Practice
BigIdeasMath com

You randomly choose one game piece. (a) Find the number of ways the event can occur. (b) Find the favorable outcomes of the event.

1. Choosing green

2. Choosing *not* yellow

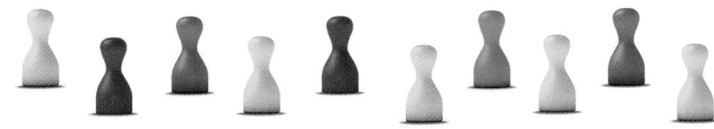

The spinner is spun. Determine if the game is fair. If it is *not* fair, who has the greater probability of winning?

3. You win if the number is odd. Your friend wins if the number is even.

4. You win if the number is less than 5. Your friend wins if the number is greater than 5. If the number is 5, nobody wins.

Use the bar graph to find the experimental probability of the event.

5. Rolling a 1 or a 2

6. Rolling an odd number

7. *Not* rolling a 5

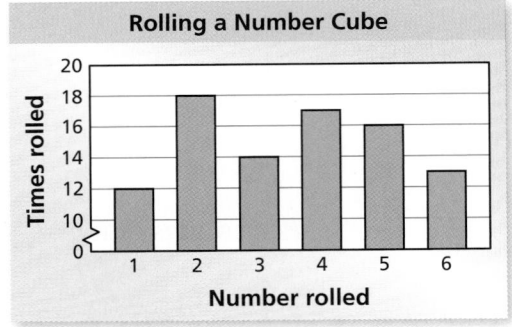

You randomly choose one chess piece. Without replacing the first piece, you choose a second piece. Find the probability of choosing the first piece, then the second piece.

8. Bishop and bishop

9. King and queen

10. King and pawn

11. King and *not* pawn

12. **MINTS** You have a bag of 60 assorted mints. You randomly choose six mints. Two of the mints you choose are peppermints. How many of the 60 mints would you expect to be peppermints?

13. **NAMES** The names of 49 middle school students are placed in a hat. The probability of randomly drawing the name of a seventh-grade student is $\frac{3}{7}$. How many seventh-grade students' names are in the hat?

14. **BEADS** Thirty percent of the beads in a bag are blue. One bead is randomly chosen and replaced. Then a second bead is chosen. What is the probability that *neither* bead is blue?

Test Item References

Chapter Test Questions	Section to Review
1, 2	9.1
3, 4, 13	9.2
5–7, 12	9.3
8–11, 14	9.4

Test-Taking Strategies

Remind students to quickly look over the entire test before they start so that they can budget their time. There is a lot of vocabulary in this chapter, so students should have been making flash cards as they worked through the chapter. Words that get mixed up should be jotted on the back of the test before they start. Students need to use the **Stop** and **Think** strategy before they answer a question.

Common Assessment Errors

- **Exercises 1 and 2** Students may forget to include, or include too many, favorable outcomes. Encourage them to write out all of the possible outcomes and then circle the favorable outcomes for the given event.
- **Exercises 3 and 4** Students may look at the events and guess that the game is fair or not fair without finding the probability of each event. Encourage them to prove that their guess is true by finding the probability of each event.
- **Exercises 5–7** Students may forget to total all of the trials before writing the experimental probability. Remind them that they need to know the total number of trials when finding the probability.
- **Exercises 5–7** Students may find the theoretical probability of the event instead of using the data to find the experimental probability. Remind them they are using the experimental probability to predict the outcome of an event with a greater number of possible outcomes.
- **Exercises 8–10** Students may mix up independent and dependent events or may have difficulty determining which type of event it is. Remind them that independent events are two different events where you start over before the next trial. Dependent events have at least one less possible outcome after the first draw, roll, or flip.

Reteaching and Enrichment Strategies

If students need help. . .	If students got it. . .
Resources by Chapter • Practice A and Practice B • Puzzle Time Record and Practice Journal Practice Differentiating the Lesson Lesson Tutorials Practice from the Test Generator Skills Review Handbook	Resources by Chapter • Enrichment and Extension • School-to-Work Game Closet at *BigIdeasMath.com* Start Standardized Test Practice

Answers

1. **a.** 1

   b. green

2. **a.** 5

   b. red, blue, red, green, blue

3. not fair; you

4. fair

5. $\frac{1}{3}$ or about 33.3%

6. $\frac{7}{15}$ or about 46.7%

7. $\frac{37}{45}$ or about 82.2%

8. $\frac{1}{120}$ or about 0.8%

9. $\frac{1}{240}$ or about 0.4%

10. $\frac{1}{30}$ or about 3.3%

11. $\frac{7}{240}$ or about 2.9%

12. 20

13. 21

14. $\frac{49}{100}$ or 49%

Assessment Book

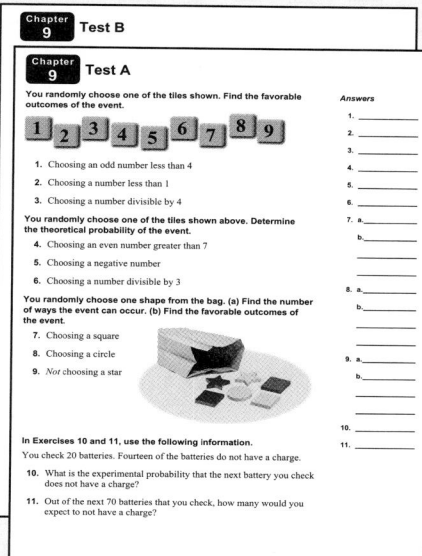

Test-Taking Strategies

Available at *BigIdeasMath.com*

After Answering Easy Questions, Relax
Answer Easy Questions First
Estimate the Answer
Read All Choices before Answering
Read Question before Answering
Solve Directly or Eliminate Choices
Solve Problem before Looking at
  Choices
Use Intelligent Guessing
Work Backwards

About this Strategy

When taking a multiple choice test, be sure to read each question carefully and thoroughly. Sometimes you don't know the answer. So . . . guess intelligently! Look at the choices and choose the ones that are possible answers.

Answers

1. C
2. F
3. $\frac{1}{5}$ or 0.2

Item Analysis

1. **A.** The student does not understand the concepts of certainty and likelihood.

   B. The student does not understand the difference between likely and unlikely.

   C. Correct answer

   D. The student does not understand that even a highly unlikely event is not impossible.

2. **F.** Correct answer

   G. The student reflects the figure in the y-axis.

   H. The student rotates the figure 180°.

   I. The student rotates the figure 90° counterclockwise.

3. **Gridded Response:** Correct answer: $\frac{1}{5}$ or 0.2

   Common Error: The student only considers that Sunday is one day of the week and gets an answer of $\frac{1}{7}$.

4. **A.** The student incorrectly cancels the two 9s when multiplying $\frac{9}{5}$ and -9, and thinks that the product is -5. The student also thinks that the sum of -5 and 32 is -27.

   B. The student thinks that the sum of -16 and 32 is -16.

   C. Correct answer

   D. The student incorrectly cancels the two 9s when multiplying $\frac{9}{5}$ and -9, and thinks that the product is -5.

5. **F.** The student thinks that because there are three categories, each category gets $\frac{1}{3}$ of the circle.

   G. Correct answer

   H. The student thinks that because both fat and protein are the same, each gets $\frac{1}{4}$ of the circle, and that because carbohydrates is the greatest amount, it gets $\frac{1}{2}$ of the circle.

   I. The student thinks that 240 means 240°.

Technology For the Teacher

Big Ideas Test Generator

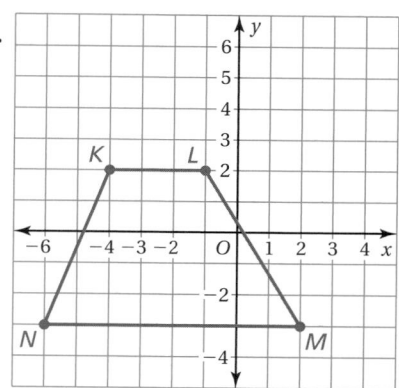

What's the probability of drawing 1 hyena out of a bag with 2 hyenas and 3 mice?

Ⓐ -10% Ⓑ 40% Ⓒ 60% Ⓓ 500%

40% < 60% I'm hoping 40%.

"You know it can't be -10% or 500%. So, you can intelligently guess between 40% and 60%."

1. A school athletic director asked each athletic team member to name his or her favorite professional sports team. The results are below:

   - Columbus Crew: 3
   - Detroit Shock: 4
   - Florida Marlins: 20
   - Florida Panthers: 8
   - Jacksonville Jaguars: 26
   - Miami Dolphins: 22
   - Miami Heat: 15
   - New York Buzz: 5
   - Orlando Magic: 18
   - Pensacola Power: 7
   - Tampa Bay Buccaneers: 17
   - Tampa Bay Lightning: 12
   - Tampa Bay Rays: 28
   - Other: 6

   One athletic team member is picked at random. What is the likelihood that this team member's favorite professional sports team is *not* located in Florida?

   A. certain

   B. likely, but not certain

   C. unlikely, but not impossible

   D. impossible

2. Trapezoid *KLMN* is graphed in the coordinate plane shown.

   Rotate trapezoid *KLMN* 90° clockwise about the origin. What are the coordinates of point *M′*, the image of point *M* after the rotation?

   F. $(-3, -2)$

   G. $(-2, -3)$

   H. $(-2, 3)$

   I. $(3, 2)$

3. Each student in your class voted for his or her favorite day of the week. Their votes are shown below:

   - Sunday: 6
   - Saturday: 10
   - Friday: 8
   - Other day: 6

   A student from your class is picked at random. What is the probability that this student's favorite day of the week is Sunday?

4. A formula for converting a temperature in degrees Celsius C to a temperature in degrees Fahrenheit F is shown below.

$$F = \frac{9}{5}C + 32$$

When the temperature in degrees Celsius is $-9°$, what is the temperature, to the nearest degree, in degrees Fahrenheit?

A. $-27°$

B. $-16°$

C. $16°$

D. $27°$

5. A nutritionist calculated the calories in a meal that came from three sources, as shown below.

- Calories from carbohydrates: 240
- Calories from fat: 180
- Calories from protein: 180

The nutritionist wants to make a circle graph to display this data. What should be the angle measure for the section labeled "carbohydrates"?

F. $120°$

G. $144°$

H. $180°$

I. $240°$

6. A right circular cone has a diameter of 10 centimeters and a slant height of 13 centimeters. Stan was computing its surface area in the box below.

$$\pi r^2 + \pi r\ell = 3.14 \cdot 5^2 + 3.14 \cdot 5 \cdot 13$$
$$= 15.7^2 + 15.7 \cdot 13$$
$$= 246.49 + 204.1$$
$$= 450.59 \ cm^2$$

What should Stan do to correct the error that he made?

A. Use the formula $\frac{1}{3}\pi r^2 h$.

B. Label the answer with the unit cm^3.

C. Square the 5 before multiplying by 3.14.

D. Distribute the 3.14 to get $3.14 \cdot 5 + 3.14 \cdot 13$.

7. Which expression is *not* equal to the other three?

F. 6

G. -6

H. $|6|$

I. $|-6|$

Item Analysis (continued)

Answers

4. C
5. G
6. C
7. G

6. **A.** The student confuses surface area with volume.

   B. The student confuses the unit for surface area with the unit for volume.

   C. Correct answer

   D. The student distributes incorrectly by distributing over multiplication.

7. **F.** The student thinks that quantities in absolute value bars are always negative.

   G. Correct answer

   H. The student does not realize that answer choices H and I are both equal to 6.

   I. The student does not realize that answer choices H and I are both equal to 6.

8. **Gridded Response:** Correct answer: $\frac{1}{16}$ or 0.0625

   Common Error: The student does not realize the compound nature of the event and gets an answer of $\frac{1}{4}$ or 0.25 or equivalent.

9. **A.** The student divides the $300 by the 2 additional cousins.

   B. Correct answer

   C. The student thinks that the cost per cousin does not change.

   D. The student uses direct variation instead of inverse variation.

10. **F.** The student divides 16 by 10 and makes that answer a percent.

   G. The student thinks that 10 umbrellas mean 10%.

   H. The student realizes that 10 out of 16 is $\frac{5}{8}$, but then uses the digits of the fraction to make a percent.

   I. Correct answer

8. $\frac{1}{16}$ or 0.0625

9. B

10. I

11. *Part A* 60%

   Part B 300

Answers for Extra Examples

1. **A.** The student finds the height for a prism, not a pyramid, with the given dimensions.

   B. The student divides the volume by the perimeter of the base and does not multiply each side of the equation by 3.

   C. Correct answer

   D. The student divides the volume by the length of one side of the base and does not multiply each side of the equation by 3.

2. **F.** The student finds 95% of 228, rounded to the nearest machine.

   G. The student subtracts 95 from 100 and adds this difference to 228.

   H. The student finds 5% of 228, rounded to the nearest machine, and adds this amount to 228.

   I. Correct answer

Item Analysis (continued)

11. **2 points** The student demonstrates a thorough understanding of determining and applying experimental probability. In Part A, the student gets an answer of 60% and provides appropriate work or explanation. In Part B, the student gets an answer of 300 and provides an appropriate explanation.

   1 point The student demonstrates a partial understanding of determining and applying experimental probability. The student gets a correct answer for Part A, or a correct answer for Part B based on the student's answer to Part A. The student's work or explanation provides evidence of partial understanding.

   0 points The student demonstrates insufficient understanding of determining and applying experimental probability. The student does not make any meaningful progress towards an answer to Part A or Part B.

Extra Examples for Standardized Test Practice

1. A right square pyramid has a volume of 360 cubic inches. Each side of its base has a length of 5 inches. What is the height, in inches, of the pyramid?

   A. 14.4   **C.** 43.2

   B. 18   **D.** 72

2. An inspector determined that 95% of the machines she tested functioned properly. There were 228 machines that functioned properly. How many machines did the inspector test?

   F. 217   **H.** 239

   G. 233   **I.** 240

8.  A spinner is divided into 8 congruent sections, as shown below.

You spin the spinner twice. What is the probability that the arrow will stop in a yellow section both times?

9. For a vacation trip, 4 cousins decided to rent a minivan and share the cost equally. They determined that each cousin would have to pay $300.

The minivan had 2 unused seats, so 2 additional cousins joined them. If the larger group of cousins now share the cost equally, what would each cousin have to pay?

- **A.** $150
- **B.** $200
- **C.** $300
- **D.** $450

10. A travel store has 16 umbrellas in stock. Ten of the umbrellas are black. What percent of the umbrellas are black?

- **F.** 1.6%
- **G.** 10%
- **H.** 58%
- **I.** 62.5%

11.  A professional golfer hit 1125 tee shots, of which 675 landed in the fairway. A statistic called "driving accuracy" is found by using the formula below.

$$\text{driving accuracy} = \frac{\text{number of fairways hit}}{\text{number of tee shots}}$$

Part A What was the golfer's driving accuracy, expressed as a percent? Show your work and explain your reasoning.

Part B Based only on your answer to Part A, how many of the golfer's next 500 tee shots would you expect to land in the fairway? Explain your reasoning.

Additional Topics

Topic 1   Angles

Topic 2   Geometry

"Move 4 of the lines to make 3 equilateral triangles."

"Well done, Descartes!"

"I'm at 3rd base. You are running to 1st base and Fluffy is running to 2nd base."

"Should I throw the ball to 2nd to get Fluffy out or throw it to 1st to get you out?"

Connections to Previous Learning

- Measure angles in whole number degrees using a protractor. Sketch angles of specified measure.
- Draw points, lines, line segments, rays, angles (right, acute, obtuse), and perpendicular and parallel lines. Identify these in two-dimensional figures.

- Solve real-world and mathematical problems by writing and solving equations of the form $x + p = q$ and $px = q$ for cases in which p, q, and x are all nonnegative rational numbers.

- Use facts about supplementary, complementary, vertical, and adjacent angles in a multi-step problem to write and solve simple equations for an unknown angle in a figure.
- Draw geometric shapes with given conditions. Focus on constructing triangles from three measures of angles or sides, noticing when the conditions determine a unique triangle, more than one triangle, or no triangle.
- Describe the two-dimensional figures that result from slicing three-dimensional figures, as in plane sections of right rectangular prisms and right rectangular pyramids.

Pacing Guide for Additional Topics

Opener	1 Day
Topic 1	1 Day
Topic 2	2 Days
Total Additional Topics	4 Days
Year-to-Date	162 Days

Math in History

Until the invention of the Global Positioning System (GPS), triangulation was the primary method that surveyors used to determine accurate locations of objects. When the horizontal distance between two objects is known, a surveyor can use triangulation to find the height, distance, and angular position of any third object within sight of the original objects.

★ In *Eratosthenes Batavus* (or "*The Dutch Eratosthenes*"), Willebrord Snellius (1580–1626), a Dutch astronomer and mathematician, described the method of triangulation and the results of his studies.

★ In 1783, General William Roy directed the Principal Triangulation of Great Britain. This survey included all of Great Britain and took about 70 years to complete.

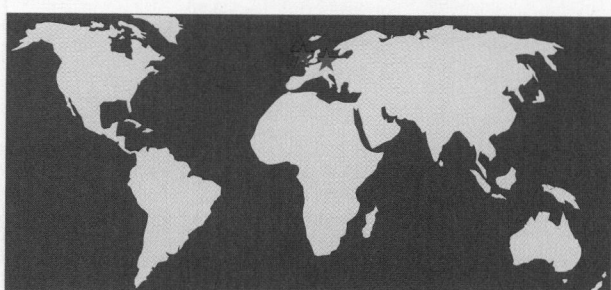

Technology For the Teacher

The Dynamic Planning Tool
Editable Teacher's Resources at
BigIdeasMath.com

- Classifying two-dimensional shapes
- Identifying solids

Try It Yourself

1. 70°; acute

2. 90°; right

3. 115°; obtuse

4.
55°

5.
160°

6.
85°

7.
180°

Record and Practice Journal

1. 60°; acute   2. 90°; right

3. 120°; obtuse

4. 65°; acute

5. 180°; straight

6. 165°; obtuse

7.
80°

8.
35°

9.
100°

10.
175°

11.
87°

12.
122°

Math Background Notes

Vocabulary Review

- Protractor
- Acute
- Obtuse
- Right
- Straight
- Vertex
- Endpoint
- Ray

Measuring Angles

- Students should know how to measure angles using a protractor.
- Remind students that an angle can be classified by its measure. A right angle is 90°, an acute angle is less than 90°, an obtuse angle is between 90° and 180°, and a straight angle is 180°.
- **Common Error:** Students may use the wrong set of angles on a protractor. Encourage them to decide which set to use by comparing the angle measure to 90°.
- **Teaching Tip:** Ask students to find real-life examples of angles, such as the hands of a clock.
- **Representation:** Show students how to measure an angle when one of the rays does not pass through the 0° mark on a protractor. Suppose in Example 1(a), the center of the protractor is aligned at the angle's vertex but the rays pass through the 40° mark and the 60° mark. Students can find the angle measure by subtracting 40 from 60.

Drawing Angles

- Students should know how to draw angles of specified measure.
- **Common Error:** Again, students may use the wrong set of angles on a protractor. Encourage them to decide which set to use by comparing the angle measure to 90°.

Reteaching and Enrichment Strategies

If students need help. . .	If students got it. . .
Record and Practice Journal • Fair Game Review Skills Review Handbook Lesson Tutorials	Game Closet at *BigIdeasMath.com* Start the next section

What You Learned Before

"Look at this baby crocodile! Isn't it cute?"

Yes, it's very acute.

Measuring Angles

Example 1 Use a protractor to find the measure of each angle. Then classify the angle as *acute*, *obtuse*, *right*, or *straight*.

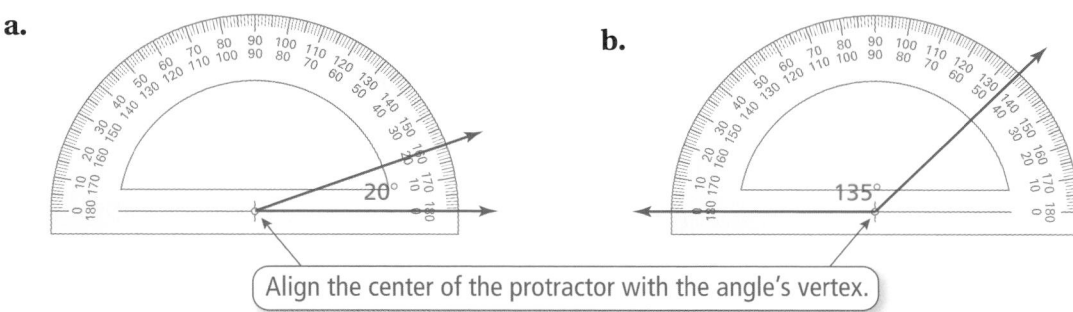

a. 20°

b. 135°

Align the center of the protractor with the angle's vertex.

⁝• The angle measure is 20°. So, the angle is acute.

⁝• The angle measure is 135°. So, the angle is obtuse.

Drawing Angles

Example 2 Use a protractor to draw a 45° angle.

Draw a ray. Place the center of the protractor on the endpoint of the ray and align the protractor so the ray passes through the 0° mark. Make a mark at 45°. Then draw a ray from the endpoint at the center of the protractor through the mark at 45°.

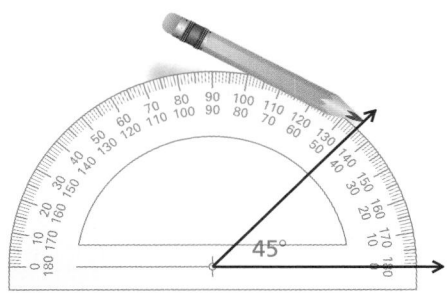

45°

Try It Yourself

Use a protractor to find the measure of the angle. Then classify the angle as *acute*, *obtuse*, *right*, or *straight*.

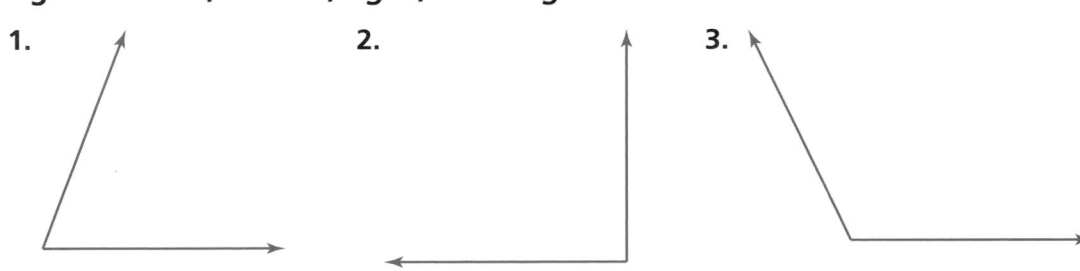

1.

2.

3.

Use a protractor to draw an angle with the given measure.

4. 55°   **5.** 160°   **6.** 85°   **7.** 180°

Topic 1   Angles

 Key Ideas

Complementary Angles

Words   Two angles are **complementary angles** if the sum of their measures is 90°.

Examples

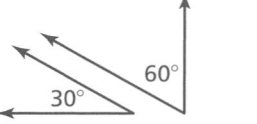   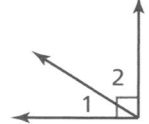

∠1 and ∠2 are complementary angles.

Supplementary Angles

Words   Two angles are **supplementary angles** if the sum of their measures is 180°.

Examples

135°   45°   3 4

∠3 and ∠4 are supplementary angles.

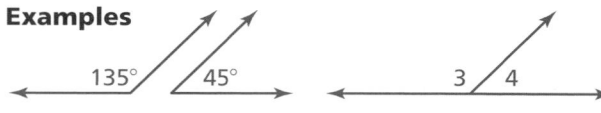 **EXAMPLE   1   Classifying Pairs of Angles**

Tell whether the angles are *complementary*, *supplementary*, or *neither*.

a.

60°   120°

60° + 120° = 180°

⋮∙   So, the angles are supplementary.

b.

51°
39°

39° + 51° = 90°

⋮∙   So, the angles are complementary.

c.

78° 112°

112° + 78° = 190°

⋮∙   So, the angles are *neither* complementary nor supplementary.

● **Practice**

Tell whether the angles are *complementary*, *supplementary*, or *neither*.

1.
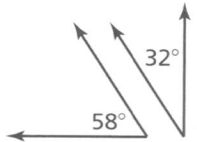
32°
58°

2.
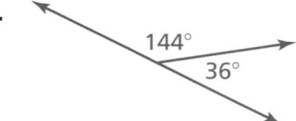
144°
36°

3.
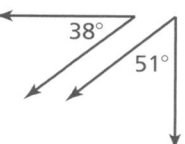
38°
51°

🔊 Multi-Language Glossary at BigIdeasMath ✓com.

Laurie's Notes

Introduction

Connect

- **Previously:** Students measured and drew angles.
- **Today:** Students will classify pairs of angles.

Motivate

- Share with students the Greek origin of the word geometry. It comes from "geo" meaning "Earth" and "metron" meaning "measurement."
- Have students get in groups of 4 and give them a few minutes to find other words that begin with "geo." You may want to provide dictionaries.
- Two examples students may be familiar with are geology and geography.

Lesson Notes

Key Ideas

- Write the *Key Ideas*.
- **Common Misconception:** Students sometimes believe that complementary or supplementary angles must be adjacent to (touching) each other because this is the way they are often drawn. Discuss that the angles do not need to have this orientation. For example, $\angle A$ and $\angle B$ are complementary, however, it is not immediately obvious because of their orientation.

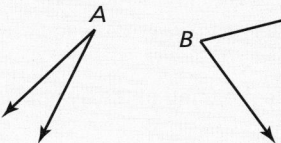

Example 1

- In this example, students find the sum of the angle measures and determine if it is 90°, 180°, or neither. Make sure students do not rely on their eyesight. They should actually add the angle measures.

Practice

- **Common Error:** Students may mix up the terms *complementary* and *supplementary*.

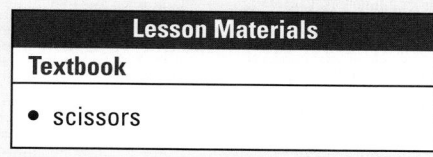

Lesson Materials
Textbook
• scissors

Warm Up

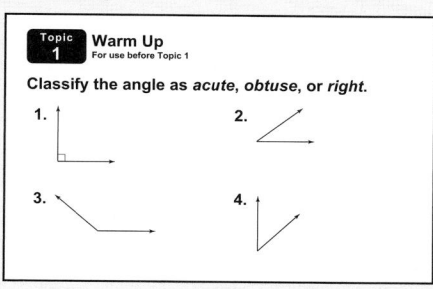

Extra Example 1

Tell whether the angles are *complementary*, *supplementary*, or *neither*.

a.
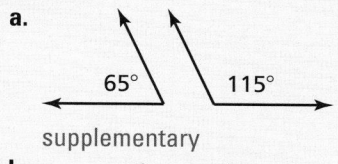
supplementary

b.
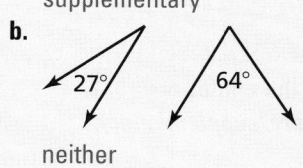
neither

c.
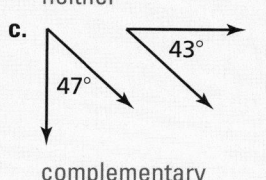
complementary

Practice

1. complementary
2. supplementary
3. neither

Record and Practice Journal Practice

See Additional Answers.

Extra Example 2

Tell whether the angles are *adjacent* or *vertical*. Then find the value of *x*.

a.

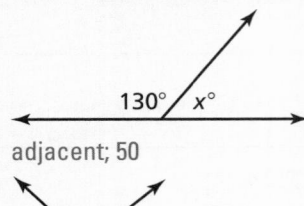

130°   x°

adjacent; 50

b.

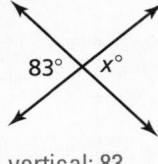

83°   x°

vertical; 83

● Practice

4. adjacent; 112
5. vertical; 90
6. adjacent; 18
7. 76

Mini-Assessment

Tell whether the angles are *complementary*, *supplementary*, or *neither*.

1.

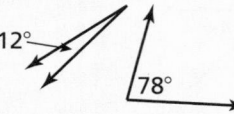

12°   78°

complementary

2.

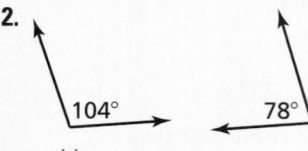

104°   78°

neither

3. Tell whether the angles are *adjacent* or *vertical*. Then find the value of *x*.

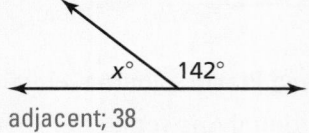

x°   142°

adjacent; 38

Laurie's Notes

Key Ideas

? "What does the word *adjacent* mean?" side-by-side Point out two students in the class that are seated adjacent to one another and two students that are not adjacent to one another.

● Write the Key Ideas. Define and sketch adjacent angles and vertical angles.

● **Model:** When two lines intersect, two pairs of vertical angles are formed. Vertical angles have the same measure. Demonstrate this with a pair of scissors that have straight blades. To make a small cut, you do not open your hands very wide because you want the vertical angle to be small. If you want to make a larger cut, you open your hands wide so that the vertical angle will be greater.

● In the example of vertical angles, explain that angles with the same number of red arcs have the same measure.

Example 2

● It is helpful for students to use a familiar context to visualize this concept. Find out if there is a familiar pair of streets or bike paths in your town that intersect at non-right angles as shown. Then discuss these examples first.

● Work through each part as shown.

? **Extension:** In part (a), "What is the measure of the two remaining angles? How do you know?" 65°; the two remaining angles are supplementary with 115°.

● Remind students of the meaning of the symbol used to mark right angles.

Practice

● **Extension:** Describe other attributes of each angle pair. In Exercises 4 and 5, the angles are supplementary. In Exercise 6, the angles are complementary.

● **Common Error:** Students may mix up the terms *adjacent* and *vertical*.

● Closure

● True or False?
  1. Adjacent angles are always acute. false
  2. Complementary angles could have the same measure. true
  3. Vertical angles have the same measure. true
  4. The sum of the measures of a pair of supplementary angles is 90°. false

 **Key Ideas**

Adjacent Angles

Words Two angles are **adjacent angles** if they share a common side and have the same vertex.

Examples

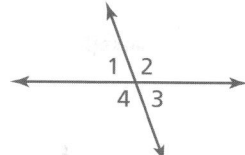

∠1 and ∠2 are adjacent.

∠2 and ∠4 are not adjacent.

Vertical Angles

Words Two angles are **vertical angles** if they are opposite angles formed by the intersection of two lines. Vertical angles have the same measure.

Examples

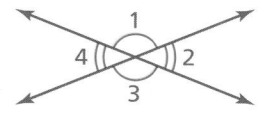

∠1 and ∠3 are vertical angles.

∠2 and ∠4 are vertical angles.

EXAMPLE ② **Finding Angle Measures**

Tell whether the angles are *adjacent* or *vertical*. Then find the value of *x*.

a.

The angles are vertical angles.
Vertical angles have the same measure.

⋮ So, *x* is 115.

b.

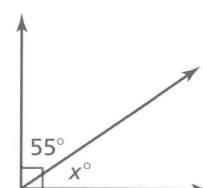

The angles are adjacent angles. Because the angles are complementary, the sum of their measures is 90°.

$$x + 55 = 90$$
$$x = 35$$

⋮ So, *x* is 35.

Practice

Tell whether the angles are *adjacent* or *vertical*. Then find the value of *x*.

4.

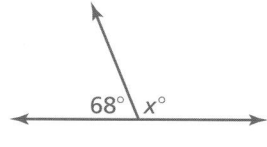

5.

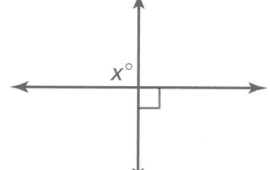

6.

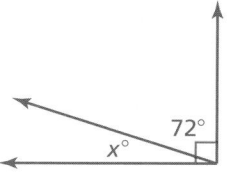

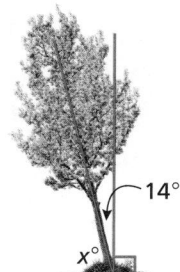

7. LANDSCAPING The tree is tilted 14°. Find the value of *x*.

EXAMPLE 1 **Constructing a Triangle with Given Side Lengths**

Construct a triangle with side lengths of 3 centimeters, 4 centimeters, and 5 centimeters.

Step 1: Cut different colored straws to the appropriate lengths.

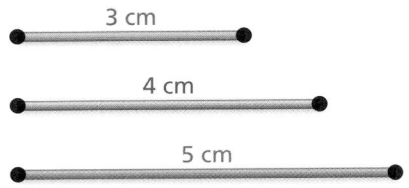

Step 2: Arrange the straws side by side.

Step 3: Form the triangle by connecting the ends of the red and green straws.

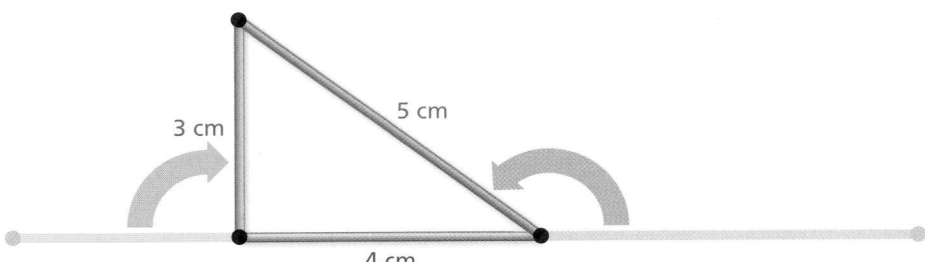

Practice

Construct a triangle with the given side lengths.

1. 6 cm, 5 cm, 3 cm
2. 5 cm, 12 cm, 13 cm
3. 2 in., 3 in., 2 in.

4. **REASONING** Repeat Exercises 1–3, but place the sides in a different order when arranging them side by side.

   a. Did you construct any triangles that look different from the original triangles? Do the triangles have the same shape and size?

   b. Given three side lengths that form a triangle, how many different triangles do you think you can construct? Explain.

Laurie's Notes

Introduction

Connect

- **Yesterday:** Students classified angles.
- **Today:** Students will construct triangles given side lengths or angle measures. They will also describe the intersection of a plane and a solid.
- **Management Tip:** To help facilitate and manage the materials, place them in sandwich bags. You might consider pre-cutting the straws or provide scissors for students to cut their own. To help keep the room clean, cluster 4–6 desks together in a circle and tape a recycled paper or plastic bag to the front edge of one of the desks. Students are expected to put excess straws pieces (if they are cutting their own) in the bag when they are finished with the activity.
- **Alternative:** Example 1 can be done using a compass and straight edge. While this may be appropriate for some students, most students will benefit from the tactile experience of using the straws.

Motivate

- Play a quick game that will help students remember vocabulary relating to triangles. Divide the class into two groups. Give a vocabulary word and each group must write the definition on a piece of paper and hand it to you. The first team with a correct definition gets a point. The team with the most points at the end wins.
- Some examples: obtuse angle, acute angle, right angle, scalene triangle, isosceles triangle, right triangle, equilateral triangle, equiangular triangle

Lesson Notes

Example 1

- Tell students to treat the endpoints of the sides like pivot points, or hinges. They need to rotate the two outer sides around the pivot points so that they meet to form a triangle.
- **?** "Can you make a different triangle using these three side lengths?" No. There is one unique triangle.
- This activity can be modeled using technology. Consider using an applet or software to simulate the triangle construction.

Practice

- Students may have difficulty with Exercise 4. The orientations of the triangles may be different, but when three side lengths form a triangle, it is a unique triangle.
- **Connection:** Exercise 4 hints at the Side-Side-Side (SSS) Congruence Postulate.
- If you choose to answer these questions in class, have students observe other students' triangles when answering Exercise 4.

Warm Up

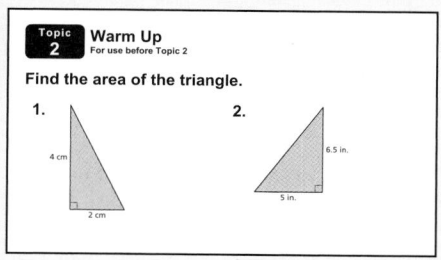

Extra Example 1

Construct a triangle with side lengths of 2 centimeters, 3 centimeters, and 4 centimeters.

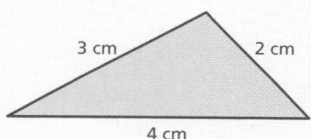

Practice

1–3. See Additional Answers.

4. a. The triangles look different because they are rotated, but they are the same size and shape.

b. one; The triangles formed may have different orientations but they are the same.

Record and Practice Journal Practice
See Additional Answers.

Extra Example 2

Construct a triangle with side lengths of 4.5 centimeters, 2.5 centimeters, and 1 centimeter, if possible. not possible

 **Practice**

5. not possible

6. See Additional Answers.

7. not possible

8. See Additional Answers.

9. See Additional Answers.

10. not possible

11. See Additional Answers;
The sum of any two side lengths must be greater than the remaining side length.

Example 2

- Example 2 is similar to Example 1 but the side lengths do not form a triangle.
- Consider using technology to model this problem after students have tried to make the triangle.

Practice

- **Common Error:** Students may look at the side lengths and guess instead of trying to construct the triangle.
- For Exercise 11, you may need to guide students to the conclusion that the sum of any two side lengths must be greater than the remaining side length.
- **Connection:** Draw a picture of two streets or paths that form a right angle. If you are walking, it is shorter to cut across from point A to B instead of staying on the path. This is because in $\triangle ABC$, $AC + CB > AB$.

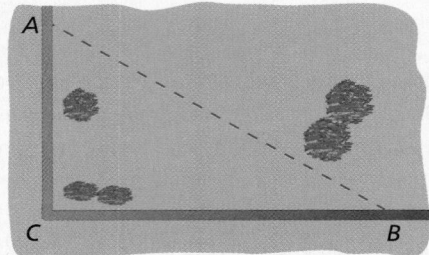

EXAMPLE 2 **Constructing a Triangle with Given Side Lengths**

Construct a triangle with side lengths of 1 inch, 2.5 inches, and 1 inch, if possible.

Step 1: Cut different colored straws to the appropriate lengths.

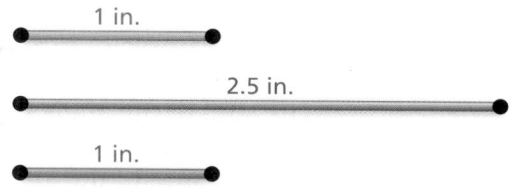

Step 2: Arrange the straws side by side.

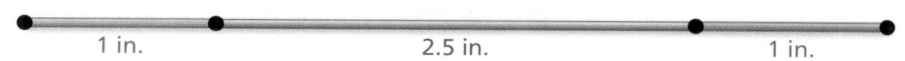

Step 3: The red and green straws are not long enough for their ends to connect. So, a triangle cannot be constructed.

Practice

Construct a triangle with the given side lengths, if possible.

5. 2 cm, 4 cm, 1 cm

6. 6 cm, 8 cm, 10 cm

7. 1 in., 2 in., 1 in.

8. 5 cm, 7 cm, 4 cm

9. 2 in., 2 in., 2 in.

10. 1 in., 5 in., 3 in.

11. **REASONING** Complete the table below for each set of side lengths in Exercises 5–10. Write a rule that compares the sum of any two side lengths to the third side length.

Side Length			
Sum of Other Two Side Lengths			

The sum of the angle measures of any triangle is 180°. You can use a protractor to construct a triangle given three angle measures.

EXAMPLE 3 **Constructing a Triangle with Given Angle Measures**

Construct a triangle with the given angle measures, if possible.

a. 35°, 25°, 100°

The sum of the angle measures is 35° + 25° + 100° = 160°.

∴ The sum of the angle measures is not 180°. So, you cannot construct a triangle with these angle measures.

b. 30°, 60°, 90°

The sum of the angle measures is 30° + 60° + 90° = 180°. Use a protractor to construct a triangle with these angle measures.

Step 1: Use a protractor to draw the 30° angle.

Step 2: Use a protractor to draw the 60° angle.

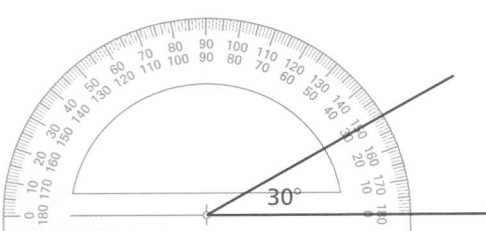

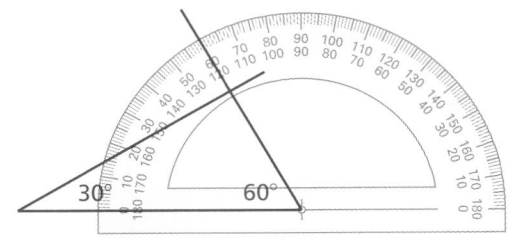

Step 3: The protractor shows that the measure of the remaining angle is 90°.

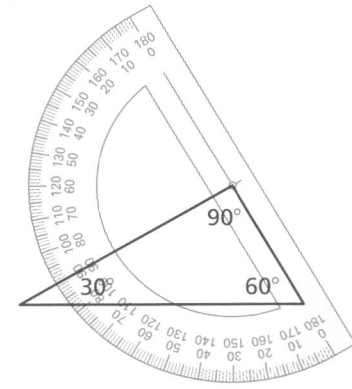

Practice

Construct a triangle with the given angle measures, if possible.

12. 50°, 30°, 110°

13. 60°, 60°, 60°

14. 60°, 40°, 80°

15. 45°, 75°, 100°

16. 20°, 20°, 120°

17. 70°, 70°, 40°

18. REASONING Construct a triangle with angle measures 90°, 45°, and 45°. Could you have drawn the sides longer? How many triangles can you construct given three angle measures whose sum is 180°?

Laurie's Notes

Discuss

? "What attributes does a triangle have besides the three sides?" three angles

? "What do you know about the angles of a triangle?" The sum of the angle measures of a triangle is 180°.

? "How do you measure the angles of a triangle?" protractor

- Review with students how to use a protractor.

- **Common Error:** Students may read the wrong scale when measuring angles, meaning they use the inside scale instead of the outside scale, or vice versa. Encourage students to think about whether the angle they are measuring is acute (less than 90°) or obtuse (between 90° and 180°).

Example 3

- In part (a), show students what a construction would look like for angle measures that do not form a triangle. Draw the first two angles and extend the lines to create a triangle. Show students that the third angle formed does not match the third given angle. Point out that changing the third angle would in turn change the other two angles as well.

- Work through both parts of the example.

Practice

- Exercise 18 is connected to the idea of similar triangles. The angles of a triangle do not determine the side lengths.

- **Common Error:** Students may forget to check the measure of the third angle and incorrectly think that the angle measures form a triangle regardless of their sum.

Extra Example 3

Construct a triangle with the given angle measures, if possible.

a. 110°, 20°, 50°

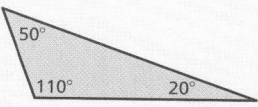

b. 65°, 25°, 80°
not possible

● Practice

12. not possible

13.

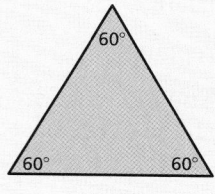

14.

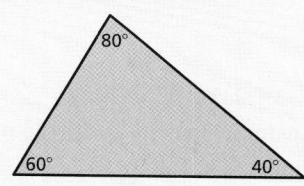

15. not possible

16. not possible

17.

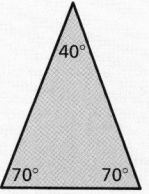

18.

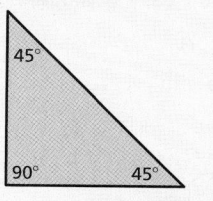

Yes; infinitely many

Extra Example 4

Describe the intersection of the plane and the solid.

a.

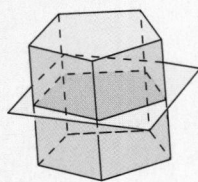

pentagon

b.

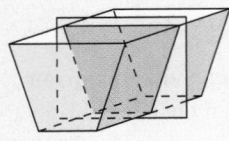

trapezoid

Practice

19. triangle

20. triangle

21. rectangle

22. rectangle

23. triangle

24. rectangle

25. The intersection is the shape of the base.

Mini-Assessment

Construct a triangle with the given side lengths, if possible.

1. 1.5 in., 1 in., 1 in.

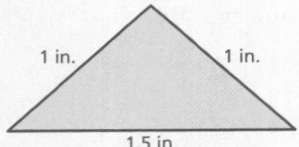

2. 3 cm, 5 cm, 9 cm not possible

Construct a triangle with the given angle measures, if possible.

3. 45°, 55°, 70° not possible

4. 100°, 35°, 45°

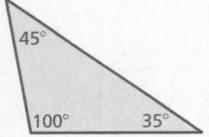

T-427

Laurie's Notes

Example 4

- It is helpful for students to see and hold models of solids when thinking about how a plane intersects a solid. If you have a set of transparent plastic models, a rubber band stretched around it will help students visualize the intersection.
- Consider using technology that displays the animation of a plane cutting through a solid.

Practice

- **Common Error:** Students may incorrectly describe the intersection because of the orientation of the drawing.

Closure

- Construct a triangle using the following information, if possible.

  a. side lengths: 4 cm, 6 cm, 8 cm

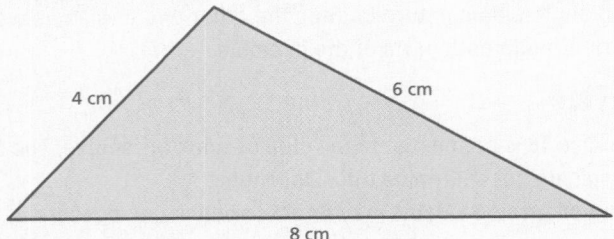

  b. angle measures: 30°, 30°, 120°

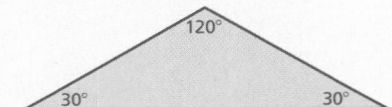

Technology For the Teacher

Dynamic Classroom

The Dynamic Planning Tool
Editable Teacher's Resources at *BigIdeasMath.com*

Consider a plane "slicing" through a solid. The intersection of the plane and the solid is a two-dimensional shape. For example, the diagram shows that the intersection of the plane and the rectangular prism is a rectangle.

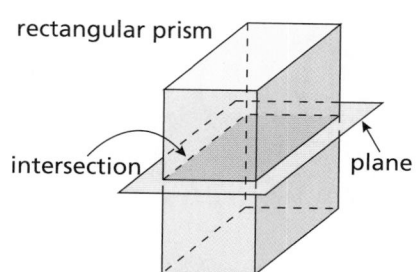

rectangular prism

intersection

plane

EXAMPLE **4** **Describing the Intersection of a Plane and a Solid**

Describe the intersection of the plane and the solid.

a.

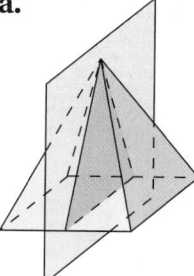

b.

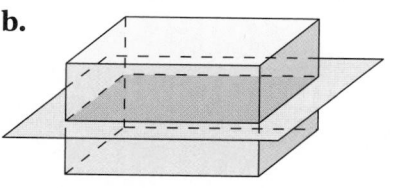

c.

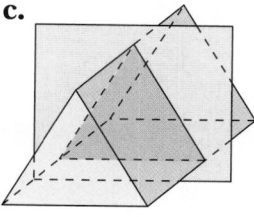

a. The intersection is a triangle.

b. The intersection is a rectangle.

c. The intersection is a triangle.

Practice

Describe the intersection of the plane and the solid.

19.

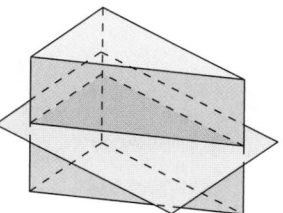

20.

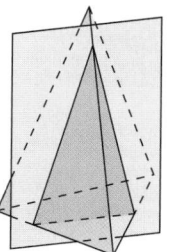

21.

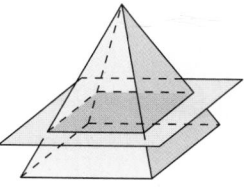

22.

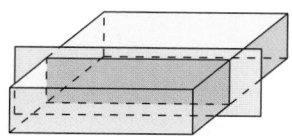

23.

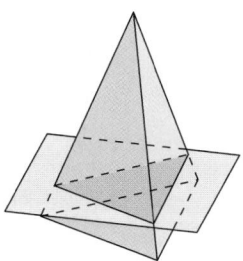

24.

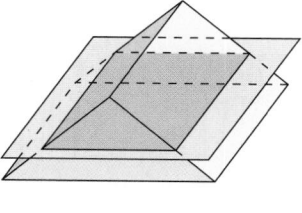

25. **REASONING** A plane that intersects a prism is parallel to the bases of the prism. Describe the intersection of the plane and the prism.

Appendix A
My Big Ideas Projects

A.1 Literature Project
The Mathematics of Jules Verne

A.2 History Project
Mathematics in Ancient Greece

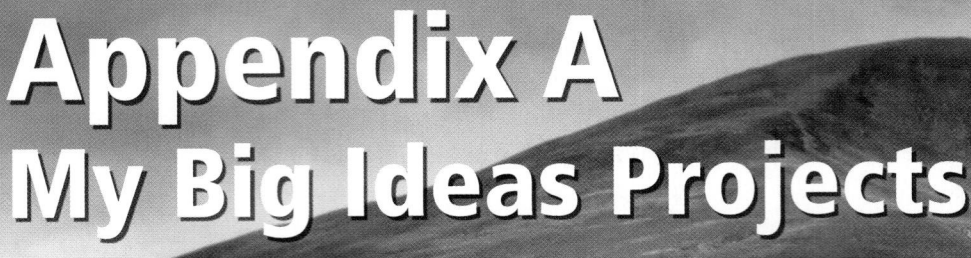

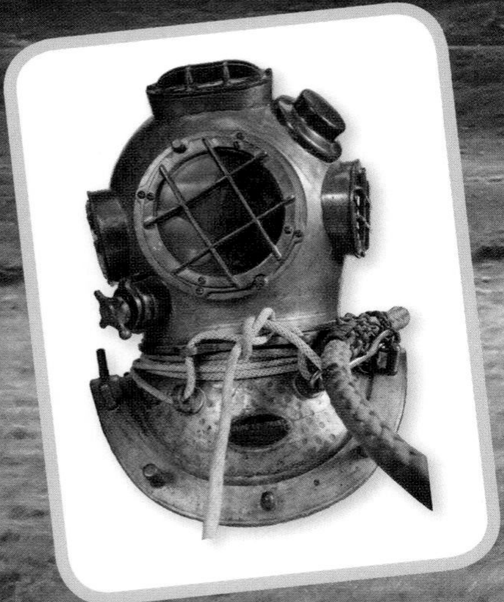

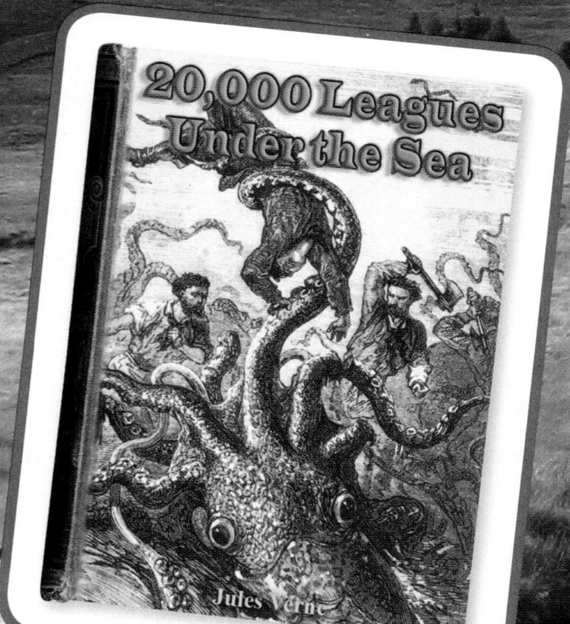

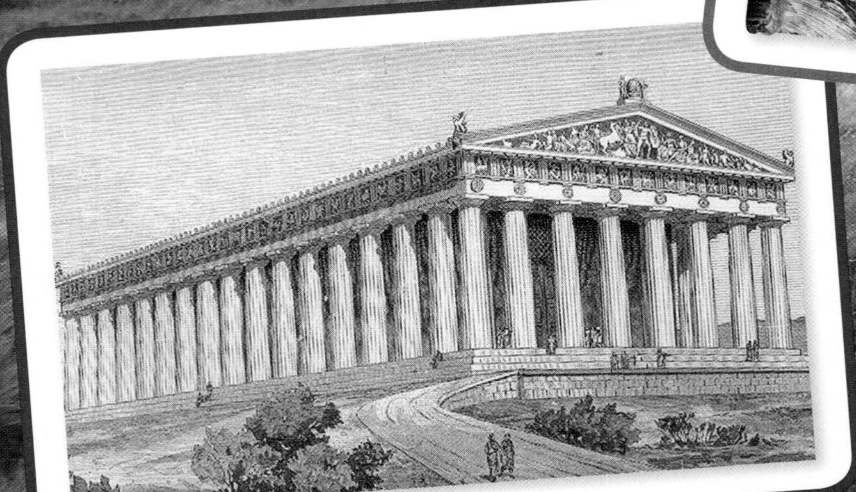

About the Appendix

- The interdisciplinary projects can be used anytime throughout the year.
- The projects offer students an opportunity to build on prior knowledge, to take mathematics to a deeper level, and to develop organizational skills.
- Students will use the Essential Questions to help them form "need to knows" to focus their research.

Essential Question

- **Literature Project**

  How does the knowledge of mathematics influence science fiction writing?
- **History Project**

  How do you use mathematical knowledge that was originally discovered by the Greeks?

Additional Resources

BigIdeasMath.com

Essential Question

- **Art Project**
  How does the knowledge of mathematics help you create a kaleidoscope?
- **Science Project**
  How does the classification of living organisms help you understand the similarities and differences of animals?

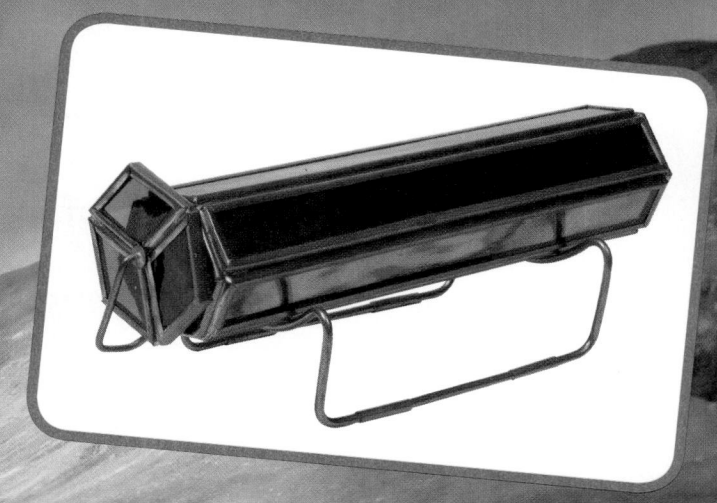

My Big Ideas Projects

A.1 Literature Project

The Mathematics of Jules Verne

Share Your Work at...
My.BigIdeasMath.com

Share Your Work at... My.BigIdeasMath.com

1 Project Overview

Jules Verne (1828–1905) was a famous French science fiction writer. He wrote about space, air, and underwater travel before aircraft and submarines were commonplace, and before any means of space travel had been devised.

For example, in his 1865 novel *From the Earth to the Moon*, he wrote about three astronauts who were launched from Florida and recovered through a splash landing. The first actual moon landing wasn't until 1969.

Essential Question How does the knowledge of mathematics influence science fiction writing?

Read one of Jules Verne's science fiction novels. Then write a book report about some of the mathematics used in the novel.

Sample: A league is an old measure of distance. It is approximately equal to 4 kilometers. You can convert 20,000 leagues to miles as follows.

$$20{,}000 \text{ leagues} \cdot \frac{4 \text{ km}}{1 \text{ league}} \cdot \frac{1 \text{ mile}}{1.6 \text{ km}} = 50{,}000 \text{ miles}$$

Project Notes

Introduction

For the Teacher

- **Goal:** Students will read one of Jules Verne's science-fiction novels and write a book report about some of the mathematics used in the novel. A sample for *Twenty Thousand Leagues Under the Sea* is included below. Samples for *Journey to the Center of the Earth* and *Around the World in Eighty Days* can be found at *BigIdeasMath.com*.
- **Management Tip:** You may want to have students work together in groups.

Essential Question

- How does the knowledge of mathematics influence science fiction writing?

References
Go to *BigIdeasMath.com* to access links related to this project.

Things to Think About

Summary of *Twenty Thousand Leagues Under the Sea*

- In 1866, several ships around the world had contact with a strange sea creature. The United States sent the Abraham Lincoln, a frigate capable of great speed, to pursue the monster. Monsieur Pierre Aronnax, Professor of Natural History at the Museum of Paris, was asked to join the expedition. Aronnax was accompanied by his devoted servant Conseil, who followed his master anywhere. Ned Land, a Canadian whaler, also joined the expedition.

- They searched for months for the monster but could not find it. Just as they were about to give up their search, it was spotted. After giving chase, there was a collision, and the three main characters fell into the sea. As it turns out, the "monster" was actually the Nautilus, a spectacular submarine that was captained by Nemo. Aronnax, Conseil, and Land were rescued by Captain Nemo, but became his prisoners. They had free run of the submarine, were well fed, and were able to take part in exciting activities, but they were not allowed to leave the submarine permanently.

- The submarine was powered by electricity. It had moving covers on the side windows, which allowed everyone to see what was happening in the ocean. The crew used deep sea diving suits to walk on the ocean floor in pursuit of pearls, volcanoes, and sea creatures. They traveled to the South Pole and fought off gigantic octopi.

- In less than 10 months, the crew traveled twenty thousand leagues. All the while, Ned Land planned their escape. Finally, the Nautilus got caught in "the maelstrom," a whirlpool off the coast of Norway from which no ship had ever escaped. At the same time, the three prisoners were trying to get away. They were able to escape, but they had no knowledge of the fate of the Nautilus.

Meet with a reading or language arts teacher and review curriculum maps to identify whether students have read *20,000 Leagues Under the Sea*, *Journey to the Center of the Earth*, or *Around the World in Eighty Days*. If the books have already been read, you may want to discuss the work students have completed and then review the books with them. If the books have not been read, perhaps you can both work simultaneously and share notes. Or, you may want to explore activities that the reading or language arts teacher has done in the past to support student learning in this particular area.

Project Notes

Mathematics Used in the Story

- 20,000 leagues was the distance traveled during the 10 months, not the depth below the surface. One league is about 4 kilometers, so the journey was about 50,000 miles.
- They interchanged the units of distance frequently; using inches, feet, yards, meters, fathoms (1 fathom = 6 feet), miles, and of course, leagues.
- Other units of measure used include knots (1 nautical mile per hour, approximately 6076 feet per hour), miles per hour, revolutions per second, pounds per square inch, atmospheres, horsepower, hundredweight, degrees, and cubic yards.
- One of the first ships to collide with the "monster" ended up with a hole that was in the shape of an isosceles triangle.
- One of the weapons was described as being able to "throw with ease a conical projectile of nine pounds a mean distance of ten miles."
- It was mentioned that water covers $\frac{7}{10}$ of the Earth's surface.
- Proportions were used frequently. For example, the common narwhal, or unicorn of the sea, was compared to the "monster." "Increase its size fivefold or tenfold, give it strength proportionate to its size, lengthen its destructive weapons, and you obtain the animal required."

Scientific Predictions that Have Happened

- submarines
- deep sea diving gear with metal helmets receiving air by use of pumps and regulators
- traveling underwater to the South Pole
- depth gauges
- electric reflectors to light the sea
- compressed air guns
- electric lanterns
- handrails made of metal and charged with electricity

Journey to the Center of the Earth and *Around the World in Eighty Days*

Go to *BigIdeasMath.com* for sample reports.

Closure

- **Rubric** An editable rubric for this project is available at *BigIdeasMath.com*.
- Students may present their reports to the class or school as a television report or public information broadcast.

2 Things to Include

- Describe the major events in the plot.

- Write a brief paragraph describing the setting of the story.

- List and identify the main characters. Explain the contribution of each character to the story.

- Explain the major conflict in the story.

- Describe at least four examples of mathematics used in the story.

- Which of Jules Verne's scientific predictions have come true since he wrote the novel?

Jules Verne (1828–1905)

3 Things to Remember

- You can download one of Jules Verne's novels at *BigIdeasMath.com*.

- Add your own illustrations to your project.

- Organize your report in a folder, and think of a title for your report.

Mathematics in Ancient Greece

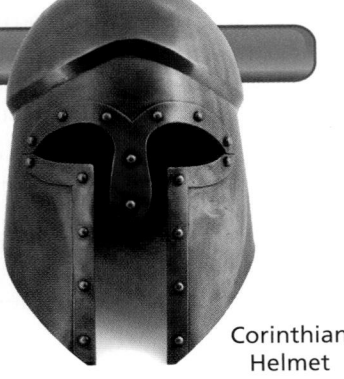

Share Your Work at...
My.BigIdeasMath.com

1 Getting Started

The ancient Greek period began around 1100 B.C. and lasted until the Roman conquest of Greece in 146 B.C.

The civilization of the ancient Greeks influenced the languages, politics, educational systems, philosophy, science, mathematics, and arts of Western Civilization. It was a primary force in the birth of the Renaissance in Europe between the 14th and 17th centuries.

Corinthian Helmet

Essential Question How do you use mathematical knowledge that was originally discovered by the Greeks?

Sample: Ancient Greek symbols for the numbers from 1 through 10 are shown in the table.

I	II	III	IIII	Γ	ΓI	ΓII	ΓIII	ΓIIII	△
1	2	3	4	5	6	7	8	9	10

These same symbols were used to write the numbers between 11 and 39. Here are some examples.

$$\triangle \Gamma \text{III} = 18 \qquad \triangle \triangle \triangle \Gamma = 35 \qquad \triangle \triangle \text{IIII} = 24$$

Alexander the Great

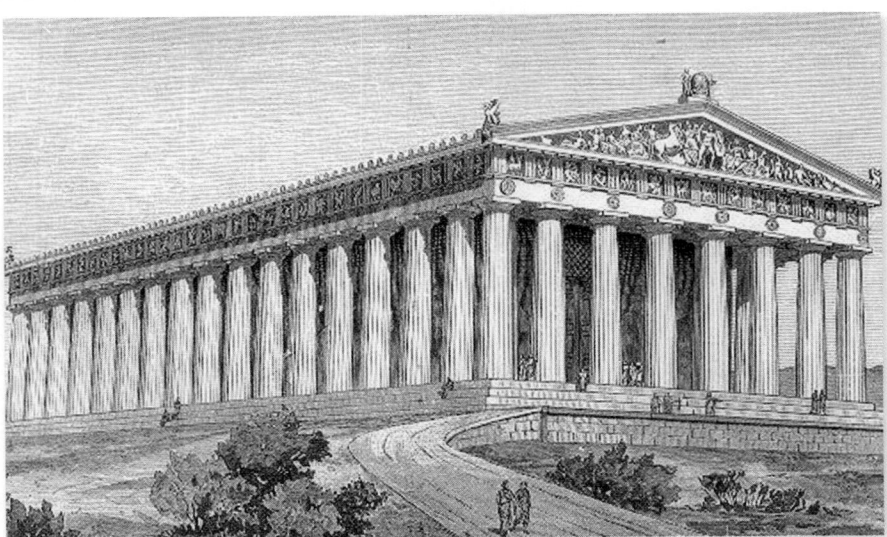

Parthenon

Project Notes

Introduction

For the Teacher

- **Goal:** Students will discover how ancient Greeks used and applied mathematics in many areas of life.
- **Management Tip:** Students can work in groups to create a report about the mathematics used in ancient Greece.

Essential Question

- How do you use mathematical knowledge that was originally discovered by the Greeks?

Things to Think About

? How have the following Greeks contributed to the field of mathematics?

- **Pythagoras (c. 570 B.C.– 490 B.C.):** Most students will recognize the name Pythagoras as being the person after whom the Pythagorean Theorem was named. The theorem states that the sum of the squares of the lengths of the legs of a right triangle equals the square of the length of the hypotenuse. The Babylonians discovered this theorem 1000 years before Pythagoras, but he is credited with proving the theorem which bears his name. He also founded the idea of irrational numbers.

- **Aristotle (c. 384 B.C.– 322 B.C.):** Aristotle is famous for being a philosopher, an astronomer, and a scientist. He is also responsible for developing the study of logic, which he called analytics. He believed that all mathematical ideas needed to be proven, and so he created ways to do this, which were later called proofs. Mathematicians still study his logic and syllogism today.

- **Euclid (c. 300 B.C.):** Euclid is known for his discovery of geometry, which he called *The Elements*. *The Elements* consisted of plane geometry, number theory, irrational numbers, and solid geometry. Within the chapters of number theory, Euclid included a way to find the greatest common divisor (GCD) of two integers, without factoring the integers. Mathematicians often refer to this as Euclid's algorithm.

- **Archimedes (c. 287 B.C.– 212 B.C.):** One of the most famous stories about Archimedes was his discovery that an object placed in water becomes lighter by the amount equal to the weight of the water it displaces. He supposedly made this discovery when he got into his bathtub, and then shouted "Eureka!" as he ran through the town. Archimedes is responsible for calculating an approximation of pi and creating a way to find the areas and volumes of solid figures.

- **Eratosthenes (c. 276 B.C.– 194 B.C.):** In addition to being a mathematician, Eratosthenes was a geographer and an astronomer. His most famous mathematical discovery is called the Sieve of Eratosthenes. It is used to find prime numbers. He also discovered an accurate method to measure the circumference of Earth, the distance from Earth to the sun, and the distance from Earth to the moon.

References

Go to *BigIdeasMath.com* to access links related to this project.

Cross-Curricular Instruction

Meet with a history teacher and review curriculum maps to identify whether students have covered the ancient Greeks. If the topic has been covered, you may want to discuss the work students have completed and then review prior knowledge with them. If the history teacher has not discussed these concepts, perhaps you can both work simultaneously on these concepts and share notes. Or, you may want to explore activities that the history teacher has done in the past to support student learning in this particular area.

Project Notes

❓ Who taught Alexander the Great?

- Alexander the Great attended the School of Royal Pages when he was thirteen. The school hired Aristotle to be a tutor for him and about 50 other boys his age. Aristotle taught them philosophy, politics, ethics, geography, and marine biology.

❓ How did the ancient Greeks represent fractions?

- The ancient Greeks did not use the fraction bar to separate the numerator from the denominator. Instead, they used a ′ (which is called a diacritical mark) to show the denominator.

$$\frac{1}{2} = \beta'$$

- In more complex fractions the numerator was written with an overbar.

$$\frac{17}{28} = \overline{\iota\varsigma}\kappa\eta$$

(**Note:** By the Alexandrian Age, the Greek Attic system of enumeration, shown on page A4 in the pupil's edition, was being replaced by the Ionian Greek system of enumeration as shown above.)

❓ How did the ancient Greeks use mathematics?

- Unlike the ancient Egyptians, who used mathematics to measure the depth of the Nile River or to build the great pyramids, the ancient Greeks used mathematics to expand their knowledge and philosophy.
- They used arithmetic (then called logistic) in business and in war. A leader needed to know how to use numbers in order to line up his troops for battle.
- Philosophers used number theory, which was called arithmetic, to prove what was previously accepted without proof.
- Some specific mathematical uses came from discoveries by Pythagoras and Archimedes. Pythagoras is famous for proving the Pythagorean Theorem. Archimedes is famous for approximating the value of pi so that the circumferences of circles could be computed. In addition to Archimedes' discovery of the law of displacement, he created a system of levers and pulleys to move heavy objects, such as ships. He also invented an object, which was later called the Archimedes Screw, that helped farmers irrigate their fields by raising the level of the water in the rivers.

Closure

- **Rubric** An editable rubric for this project is available at *BigIdeasMath.com*.
- You may hold a class debate where students can compare, defend, and discuss their findings with another student or group of students.

- Describe at least one contribution that each of the following people made to mathematics.

  Pythagoras (c. 570 B.C.–c. 490 B.C.)

  Aristotle (c. 384 B.C.–c. 322 B.C.)

  Euclid (c. 300 B.C.)

  Archimedes (c. 287 B.C.–c. 212 B.C.)

  Eratosthenes (c. 276 B.C.–c. 194 B.C.)

- Which of the people listed above was the teacher of Alexander the Great? What subjects did Alexander the Great study when he was in school?

- How did the ancient Greeks represent fractions?

- Describe how the ancient Greeks used mathematics. How does this compare with the ways in which mathematics is used today?

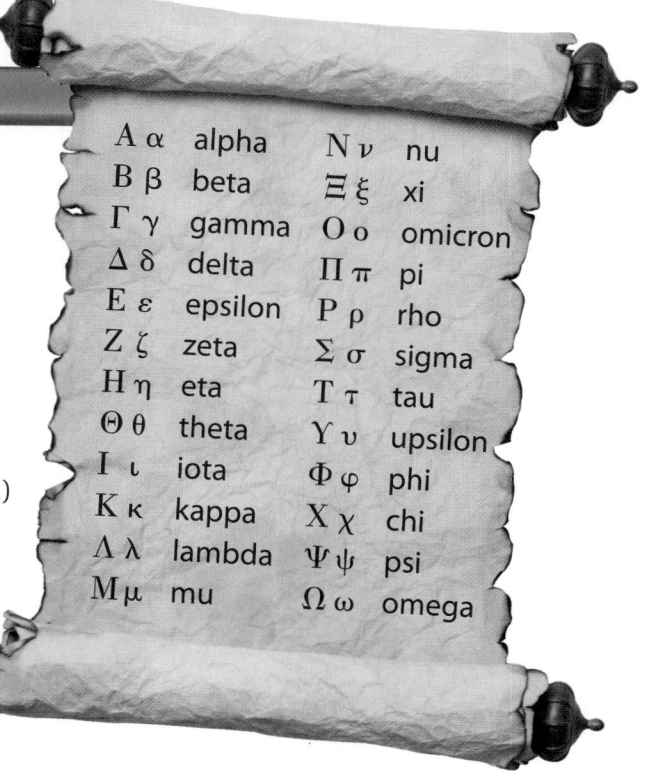

A α	alpha	N ν	nu
B β	beta	Ξ ξ	xi
Γ γ	gamma	O o	omicron
Δ δ	delta	Π π	pi
E ε	epsilon	P ρ	rho
Z ζ	zeta	Σ σ	sigma
H η	eta	T τ	tau
Θ θ	theta	Υ υ	upsilon
I ι	iota	Φ φ	phi
K κ	kappa	X χ	chi
Λ λ	lambda	Ψ ψ	psi
M μ	mu	Ω ω	omega

3 **Things to Remember**

- Add your own illustrations to your project.

- Try to include as many different math concepts as possible. Your goal is to include at least one concept from each of the chapters you studied this year.

- Organize your report in a folder, and think of a title for your report.

Greek Pottery

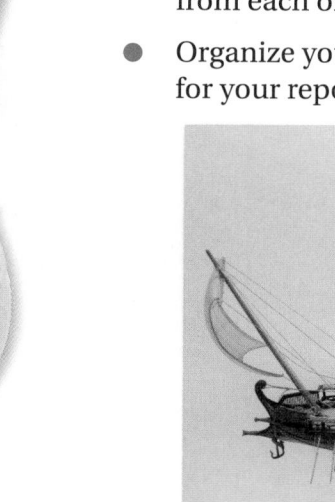

Trireme Greek Warship

Building a Kaleidoscope

Share Your Work at... My.BigIdeasMath.com

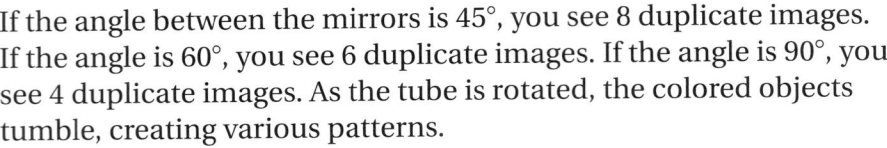

1   Getting Started

A kaleidoscope is a tube of mirrors containing loose colored beads, pebbles, or other small colored objects. You look in one end and light enters the other end, reflecting off the mirrors.

Mirrors set at 60°

Essential Question How does the knowledge of mathematics help you create a kaleidoscope?

If the angle between the mirrors is 45°, you see 8 duplicate images. If the angle is 60°, you see 6 duplicate images. If the angle is 90°, you see 4 duplicate images. As the tube is rotated, the colored objects tumble, creating various patterns.

Write a report about kaleidoscopes. Discuss the mathematics you need to know in order to build a kaleidoscope.

Sample: A kaleidoscope whose mirrors meet at 60° angles has reflective symmetry and rotational symmetry.

Reflect

Rotate 120°

Antique Kaleidoscope

Project Notes

Introduction

For the Teacher

- **Goal:** Students will discover how a kaleidoscope functions.
- **Management Tip:** Students may wish to search online for a virtual kaleidoscope. If time permits, students can visit *BigIdeasMath.com* for links to websites containing instructions on how to make a kaleidoscope. Students can create a kaleidoscope on their own or in groups.

Essential Question

- How does the knowledge of mathematics help you create a kaleidoscope?

Things to Think About

? What is a kaleidoscope?

- A kaleidoscope has two, three, or four mirrors inside of a tube. It also has a collection of glass pieces, beads, pebbles, water, or other substances that reflect light.
- When the viewer looks through one end of the tube, light is reflected through the other end. The results are multiple reflections of the materials, making beautiful patterns. As the tube is rotated, the patterns change. Whatever the pattern, the reflections and symmetry are determined by the angle(s) of the mirrors. The intersection of the lines of symmetry becomes the center of the rotational symmetry.

? Who invented the kaleidoscope?

- The ancient Greeks contemplated the use of reflections from mirrors. The ancient Egyptians placed two highly polished pieces of limestone together at different angles to see different reflections.
- Sir David Brewster of Scotland (the inventor of the modern lighthouse) is credited with the invention of the kaleidoscope. In 1816, he was working with optical tools, including prisms, and noticed the beautiful reflections and symmetrical patterns that were formed. In 1817, he named his invention the kaleidoscope, which means 'beautiful form to see.' Brewster filed for a patent of the kaleidoscope, but an error in the paperwork cost Brewster the financial rewards he deserved. His initial design was a tube with pairs of mirrors at one end, pairs of translucent disks at the other end, and beads between the two ends. Initially intended as a science tool, the kaleidoscope was quickly copied as a toy.
- In America, Charles Bush popularized the kaleidoscope. Although the early kaleidoscopes sold for around $2.00, today they often sell for $2000.00.

Materials

- List of materials needed to construct a kaleidoscope is available at *BigIdeasMath.com*

References

Go to *BigIdeasMath.com* to access links related to this project.

Cross-Curricular Instruction

Meet with an art teacher and review curriculum maps to identify whether students have covered kaleidoscopes. If the topic has already been covered, you may want to discuss the work students have completed and then review prior knowledge with them. If the art teacher has not discussed these concepts, perhaps you can both work simultaneously on these concepts and share notes. Or, you may want to explore activities that the art teacher has done in the past to support student learning in this particular area.

? How does a kaleidoscope work?

- Kaleidoscopes can be made of two, three, or four mirrors. The angle between the mirrors determines the number of reflections. The measures of the angles between the mirrors must divide evenly into 360°. The number of reflections is determined by dividing the number of degrees into 360°.

- If two mirrors are used, they are placed in a V formation. The table shows the number of reflections in a kaleidoscope based upon the angle formed by the two mirrors.

Degrees	Number of reflections
90	$360 \div 90 = 4$
60	$360 \div 60 = 6$
45	$360 \div 45 = 8$
36	$360 \div 36 = 10$
30	$360 \div 30 = 12$
22.5	$360 \div 22.5 = 16$
15	$360 \div 15 = 24$
10	$360 \div 10 = 36$
1	$360 \div 1 = 360$

- If three mirrors are used, they are placed in a triangular formation. This system works similarly to the two-mirror system except the third mirror replaces the side of the V formation that does not have a mirror. This results in a continuation of reflections throughout the field of view. With the triangular formation, it is important that the measure of each angle divides into 360°, but the sum of the angles must be 180°. Only three combinations work: 60°-60°-60°, 45°-45°-90°, and 30°-60°-90°.

Closure

- **Rubric** An editable rubric for this project is available at *BigIdeasMath.com*.
- Students may present their reports to a parent panel or community members.

2 Things to Include

- How does the angle at which the mirrors meet affect the number of duplicate images that you see?

- What angles can you use other than 45°, 60°, and 90°? Explain your reasoning.

- Research the history of kaleidoscopes. Can you find examples of kaleidoscopes being used before they were patented by David Brewster in 1816?

- Make your own kaleidoscope.

- Describe the mathematics you used to create your kaleidoscope.

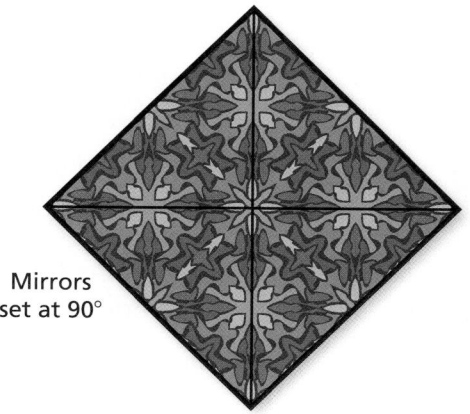

Mirrors
set at 90°

Mirrors
set at 60°

3 Things to Think About

- Add your own drawings and pattern creations to your project.

- Organize your report in a folder, and think of a title for your report.

Giant Kaleidoscope, San Diego harbor

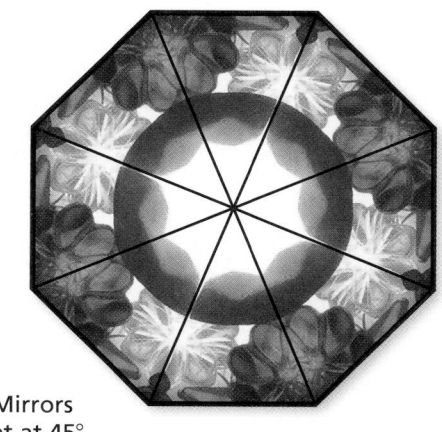

Mirrors
set at 45°

Classifying Animals

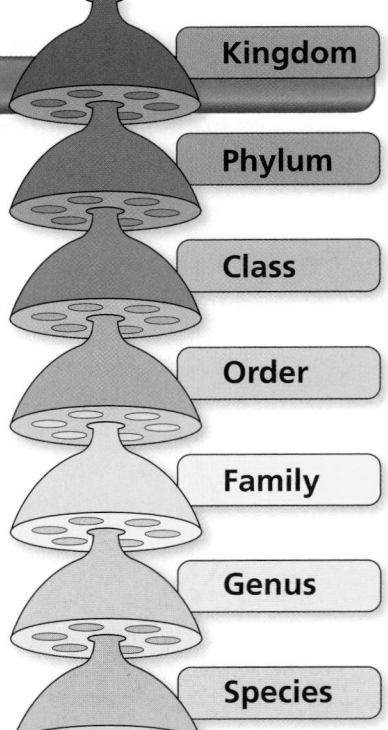

Kingdom

Phylum

Class

Order

Family

Genus

Species

 1 Getting Started

Biologists classify animals by placing them in phylums, or groups, with similar characteristics. Latin names, such as Chordata (having a spinal cord) or Arthropoda (having jointed limbs and rigid bodies) are used to describe these groups.

Biological classification is difficult, and scientists are still developing a complete system. There are seven main ranks of life on Earth; kingdom, phylum, class, order, family, genus, and species. However, scientists usually use more than these seven ranks to classify organisms.

Essential Question How does the classification of living organisms help you understand the similarities and differences of animals?

Write a report about how animals are classified. Choose several different animals and list the phylum, class, and order of each animal.

Wasp
Phylum: Arthropoda
Class: Insecta
Order: Hymenoptera
　　　(membranous wing)

Sample: A bat is classified as an animal in the phylum Chordata, class Mammalia, and order Chiroptera. *Chiroptera* is a Greek word meaning "hand-wing."

Bat
Phylum: Chordata
Class: Mammalia
Order: Chiroptera
　　　(hand-wing)

Monkey
Phylum: Chordata
Class: Mammalia
Order: Primate (large brain)

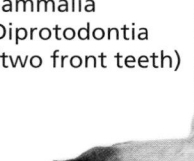

Kangaroo
Phylum: Chordata
Class: Mammalia
Order: Diprotodontia
　　　(two front teeth)

Project Notes

Introduction

For the Teacher
- **Goal:** Students will discover how to classify animals.
- **Management Tip:** You may want students to work together in groups.

Essential Question
- How does the classification of living organisms help you understand the similarities and differences of animals?

Things to Think About

? How are living organisms classified?
- All living things are classified into groups based on their similarities and differences. Here is a mnemonic that may help: King Phillip Came Over From Great Spain – Kingdom, Phylum, Class, Order, Family, Genus, and Species.

? How many kingdoms are there?
- The kingdom that categorizes animals is called Animalia. The other kingdoms categorize plants, fungi, bacteria, and protists.

? What are some examples of phyla in the animal kingdom?
- There are many phyla in the Animalia Kingdom, a few of them are: Chordata (all animals with backbones such as fish, birds, mammals, reptiles, and amphibians), Arthropoda (insects, spiders, and crustaceans), Mollusca (snails, squid, and clams), Annelida (segmented worms), and Echinodermata (starfish and sea urchins).

? What are some examples of classes in the animal phyla?
- The phylum Chordata is broken into several classes including Mammalia (mammals), Aves (birds), Reptilia (reptiles), and Amphibia (amphibians).

? What are some examples of orders in the animal classes?
- The class Mammalia is broken into many orders. A few of them are: Rodentia (rats and mice), Primates (Old- and New-World monkeys), Chiroptera (bats), Carnivora (dogs and cats), Perissodactyla (horses and zebras), and Proboscidea (elephants).

? What are some examples of families in the animal orders?
- A few of the families for the order Carnivora are: Canidae (dogs), Felidae (cats), Ursidae (bears), Hyaenidae (hyaena and aardwolves), and Mustelidae (weasels and wolverines).

? What are some examples of genera in the animal families?
- The Felidae family is divided into: Acinonyx (cheetah), Panthera (lion and tiger), Neofelis (clouded leopard), and Felis (domestic cats).

? What are some examples of species in the animal genera?
- When naming the species, scientists place the genus in front of the species, so that Panthera is divided into Panthera leo (lion) and Panthera tigris (tiger).

References

Go to *BigIdeasMath.com* to access links related to this project.

Meet with a science teacher and review curriculum maps to identify whether students have covered animal classification. If the topic has already been covered, you may want to discuss the work students have completed and then review prior knowledge with them. If the science teacher has not discussed these concepts, perhaps you can both work simultaneously on these concepts and share notes. Or, you may want to explore activities that the science teacher has done in the past to support student learning in this particular area.

Project Notes

? How can you use a graphic organizer to help classify animals?

- You can use many different graphic organizers. Examples are shown.
- If you are trying to remember pieces of information about a particular animal, you may want to use a four square shown below.

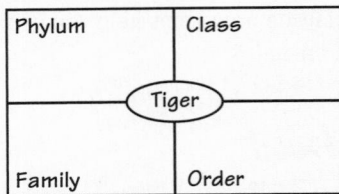

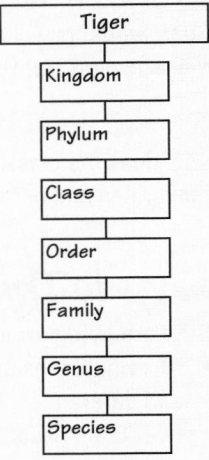

? Can animals be divided into vertebrates and invertebrates?

- All vertebrates are from the phylum Chordata. Invertebrates may be from different phyla.

Invertebrates	Vertebrates
Protozoa	Fish
Echinoderms (starfish)	Amphibians (frogs)
Annelids (earthworms)	Reptiles (alligators)
Mollusks (octopus)	Birds
Arthropods (crabs)	Mammals
Crustaceans (crabs)	Marsupials (kangaroos)
Arachnids (spiders)	Primates (gorillas)
Insects	Rodents (mice)
	Cetaceans (whales, dolphins)
	Animals such as seals

? What percent of the animal species are invertebrates?

- 97% of the animal species in the world are invertebrates. Even though vertebrates represent a much smaller portion of the animal kingdom, they dominate their environments because of their size and mobility.

Closure

- **Rubric** An editable rubric for this project is available at *BigIdeasMath.com*.
- Students may present their reports to the class or compare their report with other students' reports.

2 Things to Include

- List the different classes of phylum Chordata. Have you seen a member of each class?

- List the different classes of phylum Arthropoda. Have you seen a member of each class?

- Show how you can use graphic organizers to help classify animals. Which types of graphic organizers seem to be most helpful? Explain your reasoning.

- Summarize the number of species in each phlyum in an organized way. Be sure to include fractions, decimals, and percents.

Parrot
Phylum: Chordata
Class: Aves
Order: Psittaciformes
 (strong, curved bill)

Frog
Phylum: Chordata
Class: Amphibia
Order: Anura (no tail)

Spider
Phylum: Arthropoda
Class: Arachnida
Order: Araneae (spider)

3 Things to Remember

- Organize your report in a folder, and think of a title for your report.

Lobster
Phylum: Arthropoda
Class: Malacostraca
Order: Decopada (ten footed)

Crocodile
Phylum: Chordata
Class: Reptilia
Order: Crocodilia
 (pebble-worm)

Cougar
Phylum: Chordata
Class: Mammalia
Order: Carnivora (meat eater)

Selected Answers

Integers and Absolute Value
(pages 6 and 7)

1. 9, −1, 15

3. −6; All of the other expressions are equal to 6.

5. 6   **7.** 10   **9.** 13   **11.** 12   **13.** 8   **15.** 18

17. 45   **19.** 125   **21.** $|-4| < 7$ **23.** $|-4| > -6$ **25.** $|5| = |-5|$

27. Because $|-5| = 5$, the statement is incorrect. $|-5| > 4$

29. −8, 5   **31.** $-7, -6, |5|, |-6|, 8$ **33.** $-17, |-11|, |20|, 21, |-34|$

35. −4

37. a. MATE    **b.** TEAM

39. $n \geq 0$ **41.** The number closer to 0 is the greater integer.

43. a. Player 3   **b.** Player 2   **c.** Player 1

45. false; The absolute value of zero is zero, which is neither positive nor negative.

47. 144   **49.** 3170

Adding Integers
(pages 12 and 13)

1. Change the sign of the integer.

3. positive; 20 has the greater absolute value and is positive.

5. negative; The common sign is a negative sign.

7. false; A positive integer and its absolute value are equal, not opposites.

9. −10   **11.** 7   **13.** 0   **15.** 10

17. −7   **19.** −11   **21.** −4   **23.** −34

25. −10 and −10 are not opposites. $-10 + (-10) = -20$

27. $48   **29.** −27   **31.** 21   **33.** −85

35. Use the Associate Property to add 13 and −13 first. −8

37. *Sample answer:* Use the Commutative Property to switch the last two terms. −12

39. *Sample answer:* Use the Commutative Property to switch the last two terms. 11

41. −13   **43.** *Sample answer:* $-26 + 1; -12 + (-13)$

45. $b = 2$ **47.** $6 + (-3) + 8$

49. Find the number in each row or column that already has two numbers in it before guessing.

51. 8   **53.** 183

Section 1.3 — Subtracting Integers
(pages 18 and 19)

1. You add the integer's opposite.

3. What is 3 less than -2?; -5; 5

5. C

7. B

9. 13

11. -5

13. -10

15. 3

17. 17

19. 1

21. -22

23. -20

25. $-3 - 9$

27. 6

29. 9

31. 7

33. $m = 14$

35. $c = 15$

37. 2

39. 3

41. *Sample answer:* $x = -2, y = -1$; $x = -3, y = -2$

43. sometimes; It's positive only if the first integer is greater.

45. always; It's always positive because the first integer is always greater.

47. all values of a and b

49. when a and b have the same sign and $|a| > |b|$ or $|a| = |b|$, or $b = 0$

51. -45

53. 468

55. 2378

Section 1.4 — Multiplying Integers
(pages 26 and 27)

1. a. They are the same. **b.** They are different.

3. negative; different signs

5. negative; different signs

7. false; The product of the first two negative integers is positive. The product of the positive result and the third negative integer is negative.

9. -21

11. 12

13. 27

15. 12

17. 0

19. -30

21. 78

23. 121

25. $-240,000$

27. 54

29. -105

31. 0

33. -1

35. -36

37. 54

39. The answer should be negative. $-10^2 = -(10 \cdot 10) = -100$

41. 32

43. $-7500, 37,500$

45. -12

47. a.

Month	Price of Skates	
June	165	$= \$165$
July	$165 + (-12)$	$= \$153$
August	$165 + 2(-12)$	$= \$141$
September	$165 + 3(-12)$	$= \$129$

b. The price drops $12 every month.

c. no; yes; In August you have $135 but the cost is $141. In September you have $153 and the cost is only $129.

49. 3

51. 14

53. D

Section 1.5 — Dividing Integers
(pages 32 and 33)

1. They have the same sign. They have different signs. The dividend is zero.

3. *Sample answer:* $-4, 2$ **5.** negative **7.** negative

9. -3 **11.** 3 **13.** 0 **15.** -6 **17.** 7 **19.** -10

21. undefined **23.** 12

25. The quotient should be 0. $0 \div (-5) = 0$ **27.** 15 pages

29. -8 **31.** 65 **33.** 5

35. 4 **37.** -400 ft/min **39.** 5

41. *Sample answer:* $-20, -15, -10, -5, 0$; Start with -10, then pair -15 with -5 and -20 with 0.

43.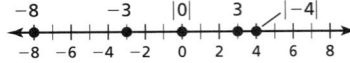

45. B

Section 1.6 — The Coordinate Plane
(pages 38 and 39)

1. 4

3. $(2, -2)$ is in Quadrant IV, $(-2, 2)$ is in Quadrant II.

5. $(3, 1)$ **7.** $(-2, 4)$ **9.** $(2, -2)$ **11.** $(-4, 2)$ **13.** $(4, 0)$

15–25. See graph below.

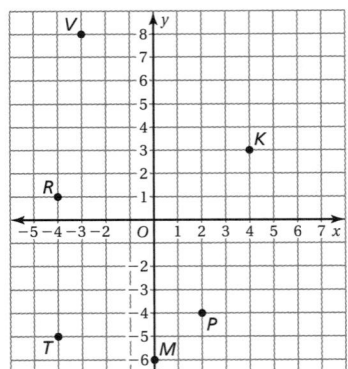

15. Quadrant I

17. y-axis

19. Quadrant IV

21. Quadrant II

23. Quadrant III

25. Quadrant II

27. The numbers are reversed. To plot $(4, 5)$, start at $(0, 0)$ and move 4 units right and 5 units up.

29. $(-2, 1)$ **31.** sometimes; It is true only for $(0, 0)$.

33. always; The x-coordinate of a point in Quadrant II is negative, and so is the y-coordinate of a point in Quadrant IV.

35. Flamingo Café

37. Because the rainforest is in Quadrant IV, the x-coordinate of the point will be positive and the y-coordinate of the point will be negative.

39. $(2, 2)$

41–43. See graph below.

41. Quadrant II

43. x-axis

45. $-\dfrac{16}{2} < -\dfrac{12}{3}$

47. $3.45 > 3\dfrac{3}{8}$

Section 2.1 — Rational Numbers
(pages 54 and 55)

1. A number is rational if it can be written as $\dfrac{a}{b}$ where a and b are integers and $b \neq 0$.

3. rational numbers, integers

5. rational numbers, integers, whole numbers

7. repeating

9. terminating

11. 0.875

13. $-0.\overline{7}$

15. $1.8\overline{3}$

17. $-5.58\overline{3}$

19. The bar should be over both digits to the right of the decimal point. $-\dfrac{7}{11} = -0.\overline{63}$

21. $\dfrac{9}{20}$

23. $-\dfrac{39}{125}$

25. $-1\dfrac{16}{25}$

27. $-12\dfrac{81}{200}$

29. $-2.5, -1.1, -\dfrac{4}{5}, 0.8, \dfrac{9}{5}$

31. $-\dfrac{9}{4}, -0.75, -\dfrac{6}{10}, \dfrac{5}{3}, 2.1$

33. $-2.4, -2.25, -\dfrac{11}{5}, \dfrac{15}{10}, 1.6$

35. spotted turtle

37. $-1.82 < -1.81$

39. $-4\dfrac{6}{10} > -4.65$

41. $-2\dfrac{13}{16} < -2\dfrac{11}{14}$

43. Michelle

45. No; The base of the skating pool is at -10 feet, which is deeper than $-9\dfrac{5}{6}$ feet.

47. **a.** when a is negative
   b. when a and b have the same sign, $a \neq 0 \neq b$

49. $\dfrac{7}{30}$

50. 21.15

Section 2.2 — Adding and Subtracting Rational Numbers
(pages 60 and 61)

1. Because $|-8.46| > |5.31|$, subtract $|5.31|$ from $|-8.46|$ and the sign is negative.

3. What is 3.9 less than -4.8?; -8.7; -0.9

5. $-\dfrac{5}{14}$

7. $2\dfrac{3}{10}$

9. -0.9

11. 1.844

13. $1\dfrac{1}{2}$

15. $\dfrac{1}{18}$

17. $-18\dfrac{13}{24}$

19. -2.6

21. 14.963

23. $\dfrac{3}{8} - \dfrac{5}{6} = -\dfrac{11}{24}$

25. $\dfrac{1}{18}$

27. $-3\dfrac{9}{10}$

29. No, the cook needs $\dfrac{1}{12}$ cup more.

31–33. Subtract the least number from the greatest number.

35. $-\dfrac{n}{4}$

37. $-\dfrac{b}{24}$

39. 35.88

41. $8\dfrac{2}{3}$

43. C

Section 2.3 — Multiplying and Dividing Rational Numbers
(pages 66 and 67)

1. The same rules for signs of integers are applied to rational numbers.

3. $-\dfrac{1}{3}$

5. $-\dfrac{3}{7}$

7. negative

9. positive

11. $-\dfrac{2}{3}$

13. $-\dfrac{1}{100}$

15. $2\dfrac{5}{14}$

17. $3.\overline{63}$

19. -6

21. -2.5875

23. $\dfrac{1}{3}$

25. $2\dfrac{1}{2}$

27. $-4\dfrac{17}{27}$

29. 0.025

31. 47.43

33. -0.064

35. The wrong fraction was inverted.

$$-\frac{1}{4} \div \frac{3}{2} = -\frac{1}{4} \times \frac{2}{3}$$
$$= -\frac{2}{12}$$
$$= -\frac{1}{6}$$

37. 8 packages

39. 1.3

41. $-4\dfrac{14}{15}$

43. $-1\dfrac{11}{36}$

45. $191\dfrac{11}{12}$ yd

47. How many spaces are between the boards?

49. **a.** $-2,\ 4,\ -8,\ 16,\ -32,\ 64$

   b. When -2 is raised to an odd power, the product is negative. When -2 is raised to an even power, the product is positive.

   c. negative

51. -5.4

53. $-8\dfrac{5}{18}$

Lesson 2.3b

Number Properties
(pages 67A and 67B)

1.
$$2 + 3 + (-2) = 2 + (-2) + 3 \qquad \text{Comm. Prop. of Add.}$$
$$= 0 + 3 \qquad \text{Additive Inverse Property}$$
$$= 3 \qquad \text{Addition Prop. of Zero}$$

3.
$$4 \cdot 19 \cdot \frac{1}{2} = 4 \cdot \frac{1}{2} \cdot 19 \qquad \text{Comm. Prop. of Mult.}$$
$$= 2 \cdot 19 \qquad \text{Multiply 4 and } \frac{1}{2}.$$
$$= 38 \qquad \text{Multiply 2 and 19.}$$

5.
$$5\left(\frac{7}{8} \cdot \frac{2}{5}\right) = 5\left(\frac{2}{5} \cdot \frac{7}{8}\right) \qquad \text{Comm. Prop. of Mult.}$$
$$= \left(5 \cdot \frac{2}{5}\right) \cdot \frac{7}{8} \qquad \text{Assoc. Prop. of Mult.}$$
$$= 2 \cdot \frac{7}{8} \qquad \text{Multiply 5 and } \frac{2}{5}.$$
$$= \frac{7}{4} \qquad \text{Multiply 2 and } \frac{7}{8}.$$

7. 0

9. *Sample answer:* Find a map, Lose a compass, Lose a compass

Section 2.4

Solving Equations Using Addition or Subtraction *(pages 74 and 75)*

1. Subtraction Property of Equality

3. No, $m = -8$ not -2 in the first equation.

5. $a = 19$

7. $k = -20$

9. $c = 3.6$

11. $q = -\dfrac{1}{6}$

13. $g = -10$

15. $y = -2.08$

17. $q = -\dfrac{7}{18}$

19. $w = -1\dfrac{13}{24}$

21. The 8 should have been subtracted rather than added.

$$x + 8 = 10$$
$$\underline{-8 \quad -8}$$
$$x = 2$$

23. $c + 10 = 3;\ c = -7$

25. $p - 6 = -14;\ p = -8$

27. $P + 2.54 = 1.38;\ -\$1.16 \text{ million}$

29. $x + 8 = 12;\ 4 \text{ cm}$

31. $x + 22.7 = 34.6;\ 11.9 \text{ ft}$

33. Because your first jump is higher, your second jump went a farther distance than your first jump.

35. $m + 30.3 + 40.8 = 180;\ 108.9°$

37. -9

39. $6, -6$

41. -56

43. -9

45. B

Section 2.5 — Solving Equations Using Multiplication or Division (pages 80 and 81)

1. Multiplication is the inverse operation of division, so it can undo division.

3. dividing by 5

5. multiplying by -8

7. $h = 5$

9. $n = -14$

11. $m = -2$

13. $x = -8$

15. $p = -8$

17. $n = 8$

19. $g = -16$

21. $f = 6\frac{3}{4}$

23. They should divide by -4.2.

$$-4.2x = 21$$
$$\frac{-4.2x}{-4.2} = \frac{21}{-4.2}$$
$$x = -5$$

25. $\frac{2}{5}x = \frac{3}{20}$; $x = \frac{3}{8}$

27. $\frac{x}{-1.5} = 21$; $x = -31.5$

29. $\frac{x}{30} = 12\frac{3}{5}$; 378 ft

31–33. Sample answers are given.

31. a. $-2x = 4.4$ **b.** $\frac{x}{1.1} = -2$

33. a. $4x = -5$ **b.** $\frac{x}{5} = -\frac{1}{4}$

35. $-1.26n = -10.08$; 8 days

37. -50 ft

39. $-5, 5$

41. -7

43. 12

45. B

Lesson 2.5b — Algebraic Expressions (pages 81A and 81B)

1. Terms: $y, 10, -\frac{3}{2}y$

Like terms: y and $-\frac{3}{2}y$

3. Terms: $7, 4p, -5, p, 2q$

Like terms: 7 and -5, $4p$ and p

5. $-\frac{3}{8}b$

7. $3q + 2$

9. $7g + 3$

Section 2.6 — Solving Two-Step Equations (pages 86 and 87)

1. Eliminate the constants on the side with the variable. Then solve for the variable using either division or multiplication.

3. D

5. A

7. $b = -3$

9. $t = -4$

11. $g = 4.22$

13. $p = 3\frac{1}{2}$

15. $h = -3.5$

17. $y = -6.4$

19. Each side should be divided by -3, not 3.

$$-3x + 2 = -7$$
$$-3x = -9$$
$$\frac{-3x}{-3} = \frac{-9}{-3}$$
$$x = 3$$

21. $a = 1\frac{1}{3}$

23. $b = 13\frac{1}{2}$

25. $v = -\frac{1}{30}$

27. $2.5 + 2.25x = 9.25$; 3 games

29. $v = -5$

31. $d = -12$

33. $m = -9$

35. *Sample answer:* You travel halfway up a ladder. Then you climb down two feet and are 8 feet above the ground. How long is the ladder? $x = 20$

37. the initial fee

39. Find the number of insects remaining and then find the number of insects you caught.

41. decrease the length by 10 cm; $2(25 + x) + 2(12) = 54$

43. $-6\frac{2}{3}$

45. 6.2

Lesson 2.6b — Solving Inequalities
(pages 87A–87D)

1. $x < 3$;

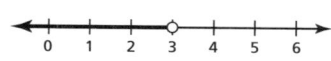

3. $r > 0.7$;

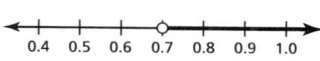

5. $z \leq \frac{1}{5}$;

7. $b \geq -40$;

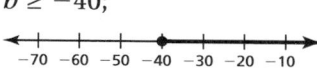

9. $m \leq \frac{4}{3}$;

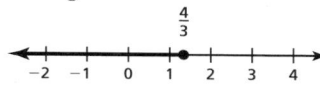

11. $x \geq 25$;

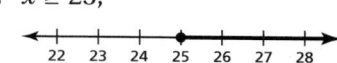

13. $j > -10.5$;

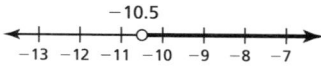

15. $k < 12$;

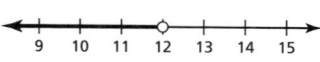

17. $z \leq -20$;

19. $n < 3$;

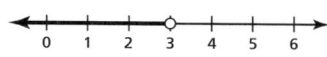

21. $c \geq 18$;

Section 3.1 — Ratios and Rates
(pages 102 and 103)

1. It has a denominator of 1.

3. *Sample answer:* A basketball player runs 10 ft down the court in 2 sec.

5. $0.10 per fl oz

7. $72

9. 840 MB

11. $\frac{5}{9}$

13. $\frac{7}{3}$

15. $\frac{4}{3}$

17. 60 mi/h

19. $2.40 per lb

21. 54 words per min

23. 90 calories per serving

25. 4.5 servings per package

27. 4.8 MB per min

29. a. It costs $122 for 4 tickets.

   b. $30.50 per ticket

   c. $305

31. The 9-pack is the best buy at $2.55 per container.

33. Try searching for "fire hydrant colors."

35–37.

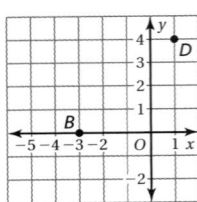

Section 3.2

Slope
(pages 108 and 109)

1. yes; Slope is the rate of change of a line.

3. 5; A ramp with a slope of 5 increases 5 units vertically for every 1 unit horizontally. A ramp with a slope of $\frac{1}{5}$ increases 1 unit vertically for every 5 units horizontally.

5. $\frac{3}{2}$

7. 1

9. $\frac{4}{5}$

11.

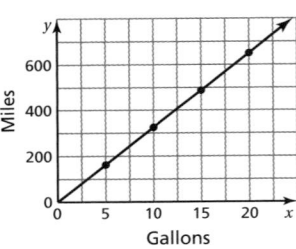

slope $= 32.5$

13.

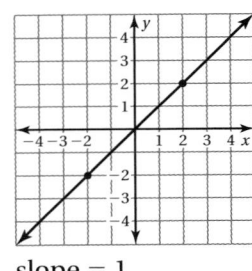

slope $= 1$

15. The change in y should be in the numerator. The change in x should be in the denominator.

Slope $= \frac{5}{4}$

17. a.

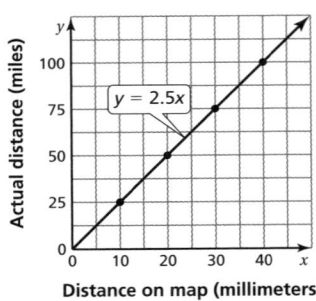

   b. 2.5; Every millimeter represents 2.5 miles.

   c. 120 mi

   d. 90 mm

19. $y = 6$

21. $<$

23. $-\frac{4}{5}$

25. 3

Proportions
(pages 114 and 115)

1. Both ratios are equal.

3. *Sample answer:* $\frac{6}{10}, \frac{12}{20}$

5. yes

7. no

9. yes

11. no

13. yes

15. yes

17. yes

19. no

21. yes

23. yes; Both can do 45 sit-ups per minute.

25. yes

27. yes

29. yes; They are both $\frac{4}{5}$.

31. **a.** Pitcher 3   **b.** Pitcher 2 and Pitcher 4

33. **a.** no

   b. *Sample answer:* If the collection has 50 quarters and 30 dimes, when 10 of each coin are added, the new ratio of quarters to dimes is $3:2$.

35. -13

37. -18

39. D

Writing Proportions
(pages 120 and 121)

1. You can use the columns or the rows of the table to write a proportion.

3. *Sample answer:* $\frac{x}{12} = \frac{5}{6}$; $x = 10$

5. $\frac{x}{50} = \frac{78}{100}$

7. $\frac{x}{150} = \frac{96}{100}$

9. $\frac{n \text{ winners}}{85 \text{ entries}} = \frac{34 \text{ winners}}{170 \text{ entries}}$

11. $\frac{100 \text{ meters}}{x \text{ seconds}} = \frac{200 \text{ meters}}{22.4 \text{ seconds}}$

13. $\frac{\$24}{3 \text{ shirts}} = \frac{c}{7 \text{ shirts}}$

15. $\frac{5 \text{ 6th grade swimmers}}{16 \text{ swimmers}} = \frac{s \text{ 6th grade swimmers}}{80 \text{ swimmers}}$

17. $y = 16$

19. $c = 24$

21. $g = 14$

23. $\frac{1}{200} = \frac{19.5}{x}$; Dimensions for the model are in the numerators and the corresponding dimensions for the actual space shuttle are in the denominators.

25. Draw a diagram of the given information.

Hint

27. $x = 9$

29. $x = 140$

Solving Proportions
(pages 126 and 127)

1. mental math; Multiplication Property of Equality; Cross Products Property

3. yes; Both cross products give the equation $3x = 60$.

5. $h = 80$

7. $n = 15$

9. $y = 7\frac{1}{3}$

11. $k = 5.6$

13. $n = 10$

15. $d = 5.76$

17. $m = 20$

19. $d = 15$

21. $k = 5.4$

23. 108 pens

25. $x = 1.5$

27. $k = 4$

29. $769.50

31. **a.** 16 mo   **b.** 40 mo

33. Make a table to solve the problem.

35. 2; $\dfrac{1/2}{1/4} = \dfrac{1}{2} \times \dfrac{4}{1} = 2$

37. 6400

39. 7920

Converting Measures Between Systems
(pages 134 and 135)

1. To convert between measurements, multiply by the ratio of the given relationship such that the desired unit is in the numerator, or set up and solve a proportion using the given relationship as one of the ratios.

3. Find the number of inches in 5 cm; 5 cm ≈ 1.97 in.; 5 in. ≈ 12.7 cm

5. >

7. >

9. <

11. 9.5

13. 21.08

15. 3.13

17. 121.92

19. 64.96

21. 2.38

23. about 584 km

25. >

27. >

29. <

31. 72

33. 4.72

35. 77,400 kg

37. A kilometer is shorter than a mile. So, the given speed when converted should be greater than 110.

39. about 3.7 gal

41.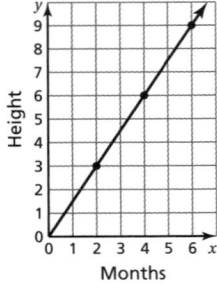

slope $= \dfrac{3}{2}$

43. C

Section 3.7 — Direct Variation
(pages 140 and 141)

1. As one quantity increases, the other quantity increases.

3. the second graph; The points do not lie on a line.

5. no; The line does not pass through the origin.

7. yes; The points lie on a line that passes through the origin.

9. no; The line does not pass through the origin.

11. yes; The line passes through the origin.

13. yes; The line passes through the origin.

15. yes; The equation can be written as $y = kx$.

17. no; The equation cannot be written as $y = kx$.

19. yes; The equation can be written as $y = kx$.

21. no; The equation cannot be written as $y = kx$.

23. yes

25. $y = 5x$

27. $y = 24x$

29. $y = \dfrac{9}{8}x$

31. You can draw the ramp on a coordinate plane and write a direct variation equation.

Hint

33. no

35. Every graph of direct variation is a line; however, not all lines show direct variation because the line must pass through the origin.

37. $y = -60$

39. $d = -59\dfrac{1}{2}$

Lesson 3.7b — Proportional Relationships
(pages 141A and 141B)

1. $(0, 0)$: You earn \$0 for working 0 hours.

$(1, 15)$: You earn \$15 for working 1 hour; unit rate: $\dfrac{\$15}{1\,h}$

$(4, 60)$: You earn \$60 for working 4 hours; unit rate: $\dfrac{\$60}{4\,h} = \dfrac{\$15}{1\,h}$

3. $y = 1.5$

Section 3.8 — Inverse Variation
(pages 146 and 147)

1. As x increases, y decreases.

3. *Sample answer:* The wingspan of a bird varies inversely with its wing beat frequency.

5. inverse variation; The equation can be written as $y = \dfrac{k}{x}$.

7. direct variation; The equation can be written as $y = kx$.

9. neither; The equation cannot be written as $y = kx$ or $y = \dfrac{k}{x}$.

Section 3.8

Inverse Variation *(continued)*
(pages 146 and 147)

11. direct variation; The equation can be written as $y = kx$.

13. inverse variation; The equation can be written as $y = \dfrac{k}{x}$.

15. neither; The equation cannot be written as $y = kx$ or $y = \dfrac{k}{x}$.

17.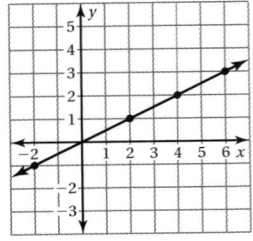

direct variation

19.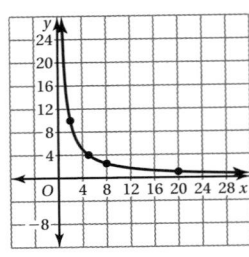

inverse variation

21. inverse variation; The equation can be written as $y = \dfrac{k}{x}$.

23. $y = \dfrac{4}{x}$

25. a. yes; $t = \dfrac{12}{s}$ **b.** 3 h

27. decreases **29.** 88 **31.** 63 **33.** yes **35.** yes **37.** B

Section 4.1

The Percent Equation
(pages 162 and 163)

1. A part of the whole is equal to a percent times the whole.

3. 55 is 20% of what number?; 275; 11

5. 37.5% **7.** 84 **9.** 64

11. $45 = p \cdot 60$; 75%

13. $a = 0.32 \cdot 25$; 8

15. $12 = 0.005 \cdot w$; 2400

17. $102 = 1.2 \cdot w$; 85

19. 30 represents the part of the whole.

$30 = 0.6 \cdot w$

$50 = w$

21. $5400 **23.** 26 years old **25.** 56 signers

27. If the percent is less than 100%, the percent of a number is less than the number. If the percent is equal to 100%, the percent of a number will equal the number. If the percent is greater than 100%, the percent of a number is greater than the number.

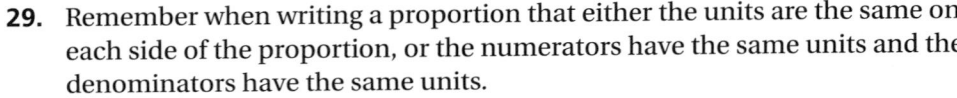

29. Remember when writing a proportion that either the units are the same on each side of the proportion, or the numerators have the same units and the denominators have the same units.

31. 92% **33.** 0.88 **35.** 0.36

Percents of Increase and Decrease
(pages 168 and 169)

1. If the original amount decreases, the percent of change is a percent of decrease. If the original amount increases, the percent of change is a percent of increase.

3. The new amount is now 0.

5. decrease; 66.7% 7. increase; 225% 9. decrease; 12.5%

11. decrease; 37.5% 13. 10 m 15. 37 points

17. 153 students 19. 42.16 kg

21. They should have subtracted 10 in the last step because 25 is decreased by 40%.
   40% of 25 = 0.4 • 25 = 10
   So, 25 − 10 = 15.

23. increase; 100% 25. increase; 133.3%

27. Increasing 20 to 40 is the same as increasing 20 by 20. So, it is a 100% increase. Decreasing 40 to 20 is the same as decreasing 40 by one-half of 40. So, it is a 50% decrease.

29. **a.** 100% increase **b.** 300% increase

31. less than; *Sample answer:* Let x represent the number. A 10% increase is equal to $x + 0.1x$, or $1.1x$. A 10% decrease of this new number is equal to $1.1x − 0.1(1.1x)$, or $0.99x$. Because $0.99x < x$, the result is less than the original number.

33. 10 girls 35. 35% 37. 56.25

Discounts and Markups
(pages 176 and 177)

1. *Sample answer:* Multiply the original price by 100% − 25% = 75% to find the sale price.

3. **a.** 6% tax on a discounted price; The discounted price is less, so the tax is less.

   b. 30% markup on a $30 shirt; 30% of $30 is less than $30.

5. $35.70 7. $76.16 9. $53.33

11. $450 13. $172.40 15. 20%

17. no; Only the amount of markup should be in the numerator, $\frac{105 − 60}{60} = 0.75$.
   So, the markup is 75%.

19. $36

21. "Multiply $45.85 by 0.1" and "Multiply $45.85 by 0.9, then subtract from $45.85." Both will give the sale price of $4.59. The first method is easier because it is only one step.

23. no; $31.08 25. $30 27. 180 29. C

Section 4.4 — Simple Interest
(pages 182 and 183)

1. I = simple interest, P = principal, r = annual interest rate (in decimal form), t = time (in years)

3. You have to change 6% to a decimal and 8 months to years.

5. **a.** $300 **b.** $1800

7. **a.** $292.50 **b.** $2092.50

9. **a.** $308.20 **b.** $1983.20

11. **a.** $1722.24 **b.** $6922.24

13. 3%

15. 4%

17. 2 yr

19. 1.5 yr

21. $1440

23. 2 yr

25. $2720

27. $6700.80

29. $8500

31. 5.25%

33. 4 yr

35. 12.5 yr; Substitute $2000 for P and I, 0.08 for r, and solve for t.

37. Year 1 = $520; Year 2 = $540.80; Year 3 = $562.43

39. $n = 5$

41. $z = 9$

Section 5.1 — Identifying Similar Figures
(pages 198 and 199)

1. They have the same measure.

3. *Sample answer:* A photograph of size 3 in. × 5 in. and another photograph of size 6 in. × 10 in.

5. $\angle A$ and $\angle W$, $\angle B$ and $\angle X$, $\angle C$ and $\angle Y$, $\angle D$ and $\angle Z$;
   Side AB and Side WX, Side BC and Side XY, Side CD and Side YZ, Side AD and Side WZ

7.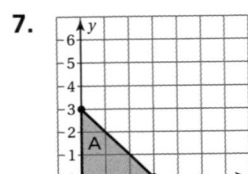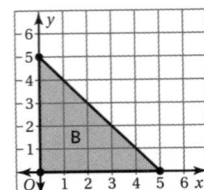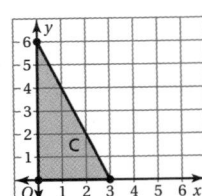
   A and B; Corresponding side lengths are proportional and corresponding angles have the same measure.

9. similar; Corresponding angles have the same measure. Because $\frac{4}{6} = \frac{6}{9} = \frac{8}{12}$, the corresponding side lengths are proportional.

11. no

13. 48°

15. 42°

17. Simplify the ratios of length to width for each photo to see if any of the photos are similar.

19. yes; One could be a trapezoid and the other could be a parallelogram.

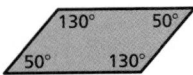 

21. **a.** yes
    b. yes; This is true for all similar triangles because the height of a triangle is a dimension of the triangle like the side lengths.

23. $\frac{16}{81}$

25. $\frac{49}{16}$

27. B

Section 5.2 — Perimeters and Areas of Similar Figures
(pages 204 and 205)

1. The ratio of the perimeters is equal to the ratio of the corresponding side lengths.

3. 120 in.^{2}; Because the ratio of the corresponding side lengths is $\frac{1}{2}$, the ratio of the areas is equal to $\left(\frac{1}{2}\right)^2$. To find the area, solve the proportion $\frac{30}{x} = \frac{1}{4}$.

5. $\frac{5}{8}; \frac{25}{64}$

7. $\frac{14}{9}; \frac{196}{81}$

9. perimeter triples

11. Area is 16 times larger.

13. 45 in.

15. false; $\dfrac{\text{Area of } \triangle ABC}{\text{Area of } \triangle DEF} = \left(\dfrac{AB}{DE}\right)^2$

17. 39,900%; The ratio of the corresponding lengths is $\frac{6 \text{ in.}}{120 \text{ in.}} = \frac{1}{20}$. So, the ratio of the areas is $\frac{1}{400}$ and the area of the actual merry-go-round is 180,000 square inches. The percent of increase is $\frac{180,000 - 450}{450} = 399 = 39,900\%$.

19. $\frac{3}{4}$

21. 25% increase

23. 42.7% decrease

Section 5.3 — Finding Unknown Measures in Similar Figures
(pages 210 and 211)

1. You can set up a proportion and solve for the unknown measure.

3. 15

5. 14.4

7. 8.4

9. 35 ft

11. 108 yd

13. 3 times

15. 12.5 bottles

17. 31.75

19. 3.88

21. 41.63

Section 5.4 — Scale Drawings
(pages 216 and 217)

1. A scale is the ratio that compares the measurements of the drawing or model with the actual measurements. A scale factor is a scale without any units.

3. Convert one of the lengths into the same units as the other length. Then, form the scale and simplify.

5. 10 ft by 10 ft

7. 112.5%

9. 50 mi

11. 110 mi

13. 15 in.

15. 21.6 yd

17. The 5 cm should be in the numerator.
$$\frac{1 \text{ cm}}{20 \text{ m}} = \frac{5 \text{ cm}}{x \text{ m}}$$
$$x = 100 \text{ m}$$

19. 2.4 cm; 1 cm : 10 mm

Section 5.4   Scale Drawings *(continued)*
(pages 216 and 217)

21. a. *Answer should include, but is not limited to:* Make sure words and picture match the product.

   b. Answers will vary.

23. Find the size of the object that would represent the model of the Sun.

25–27.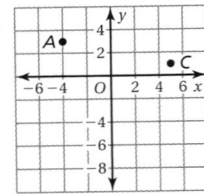

29. C

Lesson 5.4b   Scale Drawings
(pages 217A and 217B)

1. 15 ft^2

3. 3 ft^2

5.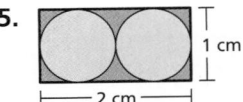

Section 5.5   Translations
(pages 224 and 225)

1. A

3. yes; Translate the letters T and O to the end.

5. no

7. yes

9. no

11. $A'(-3, 0)$, $B'(0, -1)$, $C'(1, -4)$, $D'(-3, -5)$

13.

15.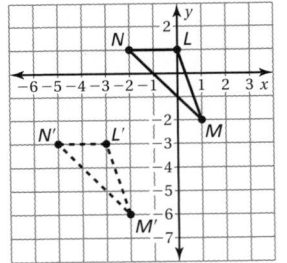

17. 2 units left and 2 units up

19. 6 units right and 3 units down

21. a. 5 units right and 1 unit up

   b. no; It would hit the island.

   c. 4 units right and 4 units up

23. If you are doing more than 10 moves and have not moved the knight to g5, you might want to start over.

25. no

27. yes

Section 5.6

Reflections
(pages 230 and 231)

1. The third one because it is not a reflection.

3. Quadrant IV

5. yes

7. no

9. no

11. $M'(-2, -1), N'(0, -3), P'(2, -2)$

13. $D'(-2, 1), E'(0, 2), F'(1, 5), G'(-1, 4)$

15. $T'(-1, -1), U'(-4, 2), V'(-6, -2)$

17. $J'(-2, 2), K'(-7, 4), L'(-9, -2), M'(-3, -1)$

19. x-axis

21. y-axis

23.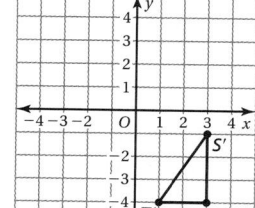

25. the first one; The left side of the face is a mirror image of the right side.

27.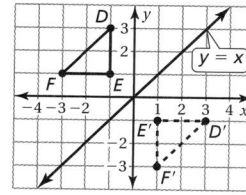
The x-coordinate and y-coordinate for each point are switched in the image.

29. straight

31. acute

Section 5.7

Rotations
(pages 236 and 237)

1. a. reflection   **b.** rotation   **c.** translation

3. Quadrant I

5. Quadrant III

7. No

9. yes; 180° clockwise or counterclockwise

11. It only needs to rotate 90° to produce an identical image.

13. $A'(-1, -4), B'(-4, -3),$
$C'(-4, -1), D'(-1, -2)$

15. $A'(0, 1), B'(-1, -2),$
$C'(-3, -2), D'(-2, 1)$

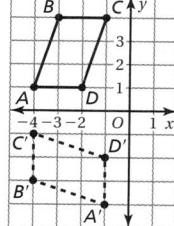

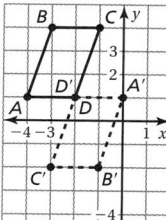

17. because both ways will produce the same image

19. Use Guess, Check, and Revise to solve this problem.

21. triangular prism

23. C

1. Prisms and cylinders both have two parallel, identical bases. The bases of a cylinder are circles. The bases of a prism are polygons. A prism has lateral faces that are parallelograms or rectangles. A cylinder has one smooth, round lateral surface.

3. *Sample answer:* Prisms: A cereal box is a rectangular prism. A pup tent with parallel triangular bases at the front and back is a triangular prism.

   Pyramids: The Egyptian pyramids are rectangular pyramids. A house roof forms a pyramid if it has lateral faces that are triangles that meet at a common vertex.

   Cylinders: Some examples of cylinders are a soup can, a tuna fish can, and a new, unsharpened, round pencil.

   Cones: some examples of cones are a traffic cone, an ice cream sugar cone, a party hat, and the sharpened end of a pencil.

5. base: circle; solid: cylinder

7. front: side: top: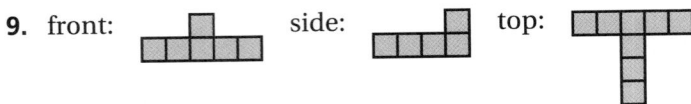

   surface area: 34 units^2; volume: 10 units^3

9. front: side: top:

   surface area: 38 units^2; volume: 9 units^3

11. 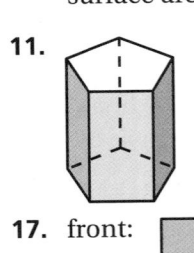   13.   15.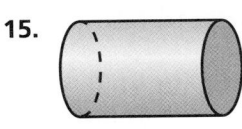

17. front:

    side:

    top: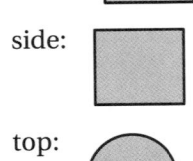

19. front:

    side:

    top: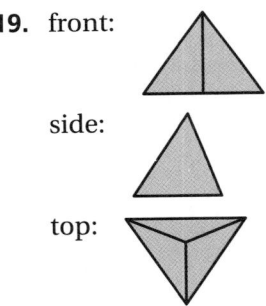

21. front:

    side:

    top: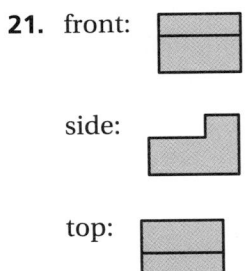

23. The Washington Monument is an *obelisk*. It consists of a pyramid sitting on top of a solid that tapers as it rises.

    Hint

25. 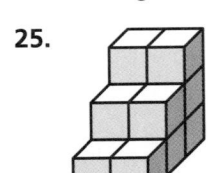   27. Use cubes to create solids that are possible.

    29. 28 m^2 31. 15 ft^2

Section 6.2 — Surface Areas of Prisms
(pages 260 and 261)

1. *Sample answer:* You want to paint a large toy chest in the form of a rectangular prism, and in order to know how much paint to buy, you need to know the surface area.

3. 18 cm^2 **5.** 108 cm^2

7.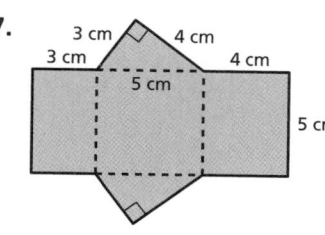

72 cm^2

9. 130 ft^2 **11.** 76 yd^2 **13.** 136 m^2

15. 448 in.^{2}; The surface area of the box is 448 square inches, so that is the least amount of paper needed to cover the box.

17. 156 in.^{2} **19.** 83 ft^2 **21.** 2 qt

23. $S = 2B + Ph$ **25.** 48 units **27.** C

Lesson 6.2b — Circles
(pages 261A and 261B)

1. 18 in.

3. $C \approx 440$ cm; $A \approx 15,400$ cm^2 **5.** $C \approx 31.4$ in.; $A \approx 78.5$ in.^{2}

Section 6.3 — Surface Areas of Cylinders
(pages 266 and 267)

1. $2\pi rh$

3. $36\pi \approx 113.0$ cm^2

5.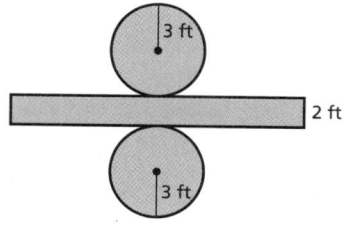

$30\pi \approx 94.2$ ft^2

7.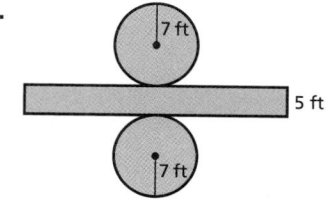

$168\pi \approx 527.5$ ft^2

9. $156\pi \approx 489.8$ ft^2 **11.** $120\pi \approx 376.8$ ft^2 **13.** $28\pi \approx 87.9$ m^2

15. The error is that only the lateral surface area is found. The areas of the bases should be added;
$$S = 2\pi r^2 + 2\pi rh$$
$$= 2\pi (6)^2 + 2\pi (6)(11)$$
$$= 72\pi + 132\pi$$
$$= 204\pi \text{ ft}^2$$

17. The surface area of the cylinder with the height of 8.5 inches is greater than the surface area of the cylinder with the height of 11 inches.

19. After removing the wedge, is there any new surface area added?

Hmmm.

21. 117 **23.** 56.52

Section 6.4 — Surface Areas of Pyramids
(pages 274 and 275)

1. the triangle and the hexagon

3. Knowing the slant height helps because it represents the height of the triangle that makes up each lateral face. So, the slant height helps you to find the area of each lateral face.

5. 178.3 mm^2 7. 144 ft^2 9. 170.1 yd^2

11. 1240.4 mm^2 13. 6 m

15. Determine how long the fabric needs to be so you can cut the fabric most efficiently.

17. 124 cm^2

19. $A \approx 452.16$ units^2; $C \approx 75.36$ units

21. $A \approx 572.265$ units^2; $C \approx 84.78$ units

Section 6.5 — Surface Areas of Cones
(pages 280 and 281)

1. no; The base of a cone is a circle. A circle is not a polygon.

3. $\ell > r$ 5. $36\pi \approx 113.0$ m^2 7. $119\pi \approx 373.7$ ft^2

9. $64\pi \approx 201.0$ yd^2 11. 15 cm   13. $130\pi \approx 408.2$ in.^{2}

15. $360\pi \approx 1130.4$ in.^{2}; $2.5\pi \approx 7.85$ ft^2 17. $96\pi \approx 301.44$ ft^2; $\frac{32}{3}\pi \approx 33.49\overline{3}$ yd^2

19. 12%   21. the lateral surface area

23. 45 in.^{2} 25. 16 ft^2

Section 6.6 — Surface Areas of Composite Solids
(pages 286 and 287)

1. *Sample answer:*

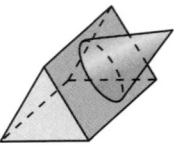

3. three cylinders

5. rectangular prism, half of a cylinder

7. cones; $104\pi \approx 326.6$ m^2

9. trapezoidal prism, rectangular prism; 152 cm^2 11. two rectangular prisms; 308 ft^2

13. 63.4%   15. $144\pi \approx 452.2$ in.^{2} 17. $806\pi \approx 2530.84$ mm^2

19. 10 ft^2 21. 47.5 in.^{2}

1. cubic units

3. *Sample answers:* Volume because you want to make sure the product will fit inside the package. Surface area because of the cost of packaging.

5. 288 cm³   7. 160 yd³   9. 420 mm³   11. 645 mm³

13. The area of the base is wrong.

$$V = \frac{1}{2}(7)(5) \cdot 10 = 175 \text{ cm}^3$$

15. 225 in.³   17. 7200 ft³

19. 1728 in.³

21. 20 cm

23. You can write the volume in cubic inches and use prime factorization to find the dimensions.

Hint

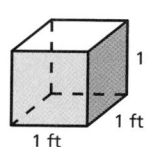   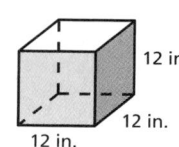

$1 \times 1 \times 1 = 1 \text{ ft}^3$ $12 \times 12 \times 12 = 1728 \text{ in.}^3$

25. reflection   27. rotation

Section 7.2

Volumes of Cylinders
(pages 308 and 309)

1. How much does it take to cover the cylinder?; $170\pi \approx 533.8 \text{ cm}^2$; $300\pi \approx 942 \text{ cm}^3$

3. $486\pi \approx 1526.0 \text{ ft}^3$ 5. $245\pi \approx 769.3 \text{ ft}^3$ 7. $90\pi \approx 282.6 \text{ mm}^3$ 9. $63\pi \approx 197.8 \text{ in.}^3$

11. $256\pi \approx 803.8 \text{ cm}^3$ 13. $\frac{125}{8\pi} \approx 5 \text{ ft}$ 15. $\frac{240}{\pi} \approx 76 \text{ cm}$

Hint

17. Divide the volume of one round bale by the volume of one square bale.

19. $8325 - 729\pi \approx 6036 \text{ m}^3$ 21. $a = 0.5 \cdot 200$; 100

23. D

Section 7.3

Volumes of Pyramids
(pages 314 and 315)

1. The volume of a pyramid is $\frac{1}{3}$ times the area of the base times the height. The volume of a prism is the area of the base times the height.

3. 3 times greater   5. 20 mm³   7. 80 in.³   9. 252 mm³

11. 700 mm³   13. 30 in.²   15. 7.5 ft

17. 12,000 in.³; The volume of one paperweight is 12 cubic inches. So, 12 cubic inches of glass is needed to make one paperweight. So it takes $12 \times 1000 = 12{,}000$ cubic inches to make 1000 paperweights.

19. *Sample answer:*
5 ft by 4 ft   21. 28   23. 60   25. B

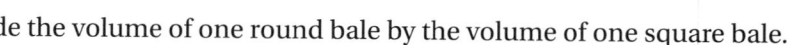

Volumes of Cones
(pages 320 and 321)

1. The height of a cone is the distance from the vertex to the center of the base.

3. Divide by 3.

5. $9\pi \approx 28.3 \text{ m}^3$

7. $\dfrac{2\pi}{3} \approx 2.1 \text{ ft}^3$

9. $27\pi \approx 84.8 \text{ yd}^3$

11. $\dfrac{125\pi}{6} \approx 65.4 \text{ in.}^3$

13. The diameter was used instead of the radius.

$$V = \frac{1}{3}(\pi)(3)^2(8) = 24\pi \text{ m}^3$$

15. 1.5 ft

17. $\dfrac{40}{3\pi} \approx 4.2 \text{ in.}$

19. 24.1 min

21. $3y$

23. 315 m^3

25. $152\pi \approx 477.28 \text{ ft}^3$

Volumes of Composite Solids
(pages 328 and 329)

1. A composite solid is a solid that is made up of more than one solid.

3. In Example 2, you had to subtract the volume of the cylinder-shaped hole from the volume of the entire cylinder. In Example 1, you had to find the volumes of the square prism and the square pyramid and add them together.

5. $125 + 16\pi \approx 175.2 \text{ in.}^3$

7. 220 cm^3

9. 173.3 ft^3

11. $216 - 24\pi \approx 140.6 \text{ m}^3$

13. **a.** *Sample answer:* 80%   **b.** *Sample answer:* $100\pi \approx 314 \text{ in.}^3$

15. 13.875 in.^3; The volume of the hexagonal prism is $10.5(0.75)$ and the volume of the hexagonal pyramid is $\dfrac{1}{3}(6)(3)$.

17. $\dfrac{25}{9}$

19. B

Surface Areas and Volumes of Similar Solids
(pages 335–337)

1. Similar solids are solids of the same type that have proportional corresponding linear measures.

3. **a.** $\dfrac{4}{9}$ **b.** $\dfrac{8}{27}$

5. no

7. no

9. $b = 18 \text{ m}; c = 19.5 \text{ m}; h = 9 \text{ m}$

11. 1012.5 in.^2

13. $13{,}564.8 \text{ ft}^3$

15. 673.75 cm^2

17. **a.** yes; Because all circles are similar, the slant height and the circumference of the base of the cones are proportional.

   b. no; because the ratio of the volumes of similar solids is equal to the cube of the ratio of their corresponding linear measures

19. Choose two variables, one to represent the surface area of the smallest doll and one to represent the volume of the smallest doll. Use these variables to find the surface areas and volumes of the other dolls.

21. 1   **23.** C

Section 8.1   Stem-and-Leaf Plots
(pages 352 and 353)

1. 3 is the stem; 4 is the leaf

3. From the leaves, you can see where most of the data lies and whether there are many values that are low or high.

5. 4; 42   **7.** no; There is no 2 as a leaf for the stem 3.

9. **Hours Online**

Stem	Leaf
0	0 2 6 8
1	2 2 4 5 7 8
2	1 4

Key: 2 | 1 = 21 hours

11. **Points Scored**

Stem	Leaf
3	8
4	2 2 3 3 5
5	0 1 6 8 8
6	
7	0 1 1 5

Key: 3 | 8 = 38 points

13. **Weights**

Stem	Leaf
0	8
1	2 5 7 8
2	4 4
3	1

Key: 2 | 4 = 24 pounds

Most of the weights are in the middle.

15. **Minutes in Line**

Stem	Leaf
1	6 9
2	0 2 6 7 9
3	1 1 6 8
4	0

Key: 4 | 0 = 4.0 minutes

17. mean: 56.6; median: 53; modes: 41, 43, 63; range: 56

19. 97; It increases the mean.

21. *Sample answer:* Points by a basketball player in his first 8 games

Points

Stem	Leaf
2	1 3 4
3	2 4
4	0 1 5

Key: 3 | 2 = 32 points

23.

25.

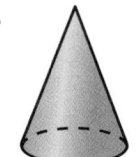

27. B

Section 8.2

Histograms
(pages 358 and 359)

1. The *Test Scores* graph is a histogram because the number of students (frequency) achieving the test scores are shown in intervals of the same size (20).

3. No bar is shown on that interval.

5. flat

7.

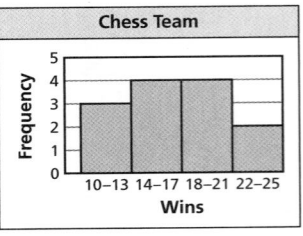

9. a. 4–5

    b. 20 students

    c. 85%

11. Pennsylvania; You can see from the intervals and frequencies that Pennsylvania counties are greater in area, which makes up for it having fewer counties.

13. Don't use a smaller interval because the distribution will appear flat.

15. 27

17. 51.2

Section 8.3

Circle Graphs
(pages 366 and 367)

1. Multiply the decimal form of each percent by 360° to find the angle measure for each section.

3. $\frac{1}{2}$ does not belong because it does not represent an entire circle.

5. orange

7. 20 students

9. 54°

11. 10.8°

13.

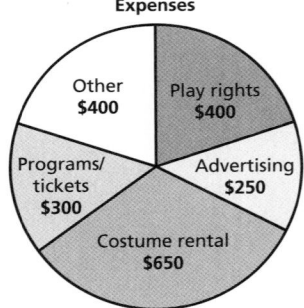

15. no; The sum of the percentages is greater than 100%. This would occur when students like more than one of these activities.

17. *Sample answer:* Knowledge of percentages, proportions, and degrees of a circle. How to convert from one form to another.

19. $x = 40$

21. $w = 1.5$

Section 8.4

Samples and Populations
(pages 372 and 373)

1. Samples are easier to obtain.

3. *Sample answer:* The results may be similar for middle school students, but not for children in first grade. Children in first grade probably do not know all of these nuts.

5. *Sample answer:* You could send a survey home with your classmates and have them ask one of their parents what their favorite nut is.

7. Population: All quarters in circulation
   Sample: 150 quarters

9. Population: All books in library
   Sample: 10 books

11. **a.** Population: All students at your school
   Sample: First 15 students at band class

   b. no; Your sample includes 15 students arriving at band class, and students who take band class play a musical instrument.

13. Sample A because it is representative of the population.

15. A population because there are few enough students in your homeroom to not make the surveying difficult.

17. 1260 students

19. Use the survey results to find the number of students in the school that plan to attend college.

21. 31.25%

23. 81.$\overline{81}$%

Lesson 8.4b

Comparing Populations
(pages 373A and 373B)

1. **a.** *Sample answer:* randomly asking students at lunch
   b. *Sample answer:* randomly asking students on the football team

3. In general, boys are taller than girls.

Section 9.1   Introduction to Probability
(pages 388 and 389)

1. event; It is a collection of several outcomes.

3. *Sample answer:* flipping a coin and getting both heads and tails; rolling a number cube and getting a number between 1 and 6

5. no; They both have the same number of forward outcomes.

7. 6

9. 6, 7, 8, 9

11. 1, 2

13. a. 2 ways   **b.** blue, blue

15. a. 2 ways   **b.** purple, purple

17. a. 6 ways   **b.** yellow, green, blue, blue, purple, purple

19. There are 7 marbles that are *not* purple, even though there are only 4 colors. Choosing *not* purple could be red, red, red, blue, blue, green, or yellow.

21. false; five

23. false; red

25. no; More sections on a spinner does not necessarily mean you are more likely to spin red. It depends on the size of the sections of the spinner.

27. Do the number of outcomes increase, decrease, or stay the same?

29. 30

31. $-3\frac{1}{2}$

Section 9.2   Theoretical Probability
(pages 394 and 395)

1. There is a 50% chance you will get a favorable outcome.

3. Spinner 4; The other three spinners are fair.

5. $\frac{1}{6}$ or about 16.7%

7. $\frac{1}{2}$ or 50%

9. 0 or 0%

11. 9 chips

13. not fair, your friend

15. $\frac{1}{44}$ or about 2.3%

17. a. $\frac{4}{9}$ or about 44.4%   **b.** 5 males

19. There are 2 combinations for each.

21. $\frac{1}{4}$

23. $-\frac{21}{40}$

25. C

Section 9.3   Experimental Probability
(pages 402 and 403)

1. Perform an experiment several times. Count how often the event occurs and divide by the number of trials.

3. $\frac{2}{5}$ or 40%

5. $\frac{2}{5}$ or 40%

7. $\frac{7}{50}$ or 14%

9. $\frac{21}{25}$ or 84%

11. $\frac{17}{50}$ or 34%

13. The theoretical probability was found, not the experimental probability. $P(4) = \dfrac{11}{50}$

15. 45 tiles   **17.** 25

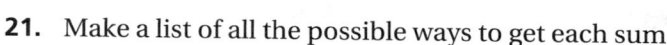

19. The experimental probability of 60% is close to the theoretical probability of 50%.

21. Make a list of all the possible ways to get each sum.

23. *Sample answer:* Roll two number cubes 50 times and find each product. Record how many times the product is at least 12. Divide this number by 50 to find the experimental probability.

25. $x = 5$ **27.** $x = 24$

Section 9.4   Independent and Dependent Events
(pages 409–411)

1. Draw a tree diagram or multiply $P(A)$ by $P(B)$.

3. *Sample answer:* independent events: a traffic jam and a sunny day; dependent events: temperatures below freezing and ice

5. independent; The outcome of the first roll does not affect the outcome of the second roll.

7. independent; You replace the marble, so the probability doesn't change.

9. dependent; There is one less person to choose from on the second draw.

11. $\dfrac{2}{9}$ **13.** $\dfrac{2}{9}$ **15.** $\dfrac{1}{20}$ or 5%   **17.** $\dfrac{3}{20}$ or 15%

19. $\dfrac{1}{42}$ or about 2.4%   **21.** $\dfrac{1}{21}$ or about 4.8%   **23.** $\dfrac{4}{21}$ or about 19%   **25.** $\dfrac{1}{2520}$ or about 0.04%

27. a. $\dfrac{1}{100}$ or 1%

     b. It increases the probability that your choice is correct to $\dfrac{1}{25}$ or 4%, because each digit could be 0, 2, 4, 6, or 8.

29. a. $\dfrac{1}{9}$ or about 11.1%

     b. It increases the probability that your guesses are correct to $\dfrac{1}{4}$ or 25%, because you are only choosing between 2 choices for each question.

31. $\dfrac{16}{25}$ or 64%   **33.** $1 : 5; 5 : 1$ **35.** $1 : 35; 35 : 1$

37. $n = -10.8$ **39.** B

Topic 1   Angles
(pages 422 and 423)

1. complementary   **3.** neither   **5.** vertical; 90   **7.** 76

1.

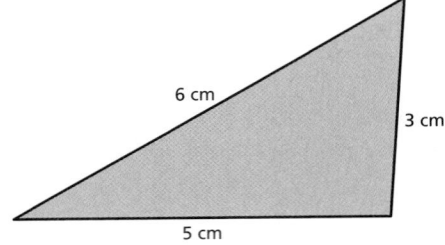

3.

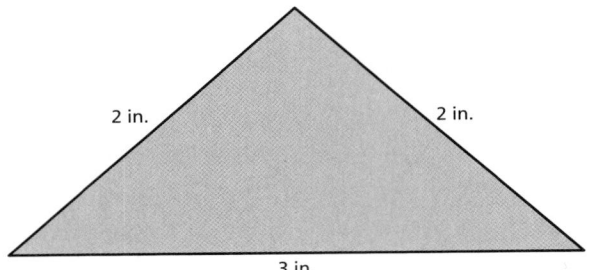

5. not possible

7. not possible

9.

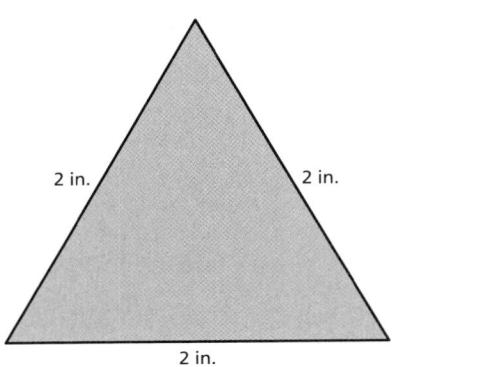

11. Side lengths in Exercise 5:

Side length	2 cm	4 cm	1 cm
Sum of other two side lengths	5 cm	3 cm	6 cm

Side lengths in Exercise 7:

Side length	1 in.	2 in.	1 in.
Sum of other two side lengths	3 in.	2 in.	3 in.

Side lengths in Exercise 9:

Side length	2 in.	2 in.	2 in.
Sum of other two side lengths	4 in.	4 in.	4 in.

Side lengths in Exercise 6:

Side length	6 cm	8 cm	10 cm
Sum of other two side lengths	18 cm	16 cm	14 cm

Side lengths in Exercise 8:

Side length	5 cm	7 cm	4 cm
Sum of other two side lengths	11 cm	9 cm	12 cm

Side lengths in Exercise 10:

Side length	1 in.	5 in.	3 in.
Sum of other two side lengths	8 in.	4 in.	6 in.

The sum of any two side lengths must be greater than the remaining side length.

13.

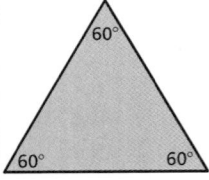

15. not possible

17.

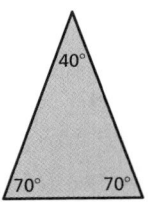

19. triangle

21. rectangle

23. triangle

25. The intersection is the shape of the base.

Key Vocabulary Index

Mathematical terms are best understood when you see them used and defined *in context*. This index lists where you will find key vocabulary. A full glossary is available in your Record and Practice Journal and at *BigIdeasMath.com*.

Student Index

This student-friendly index will help you find vocabulary, key ideas, and concepts. It is easily accessible and designed to be a reference for you whether you are looking for a definition, real-life application, or help with avoiding common errors.

a net, 270
a prism, 252
a pyramid, 253
scale, 212–217, 217A–217B
similar figures, 206
a solid, 251
    views of, 253
three-dimensional figures,
    250–255

E

Equality
    Addition Property of, 72
    Division Property of, 78
    Multiplication Property of, 78
    Subtraction Property of, 72
Equation(s)
    equivalent
        defined, 72
    modeling, 70–71, 76, 82–83
    percent, 158–163
        error analysis, 162
        real-life application, 161
        reasoning, 162–163
    proportion, 110, 112
    solving
        by addition, 70–75
        by division, 76–81
        error analysis, 74, 80, 86
        by multiplication, 76–81
        practice problems, 74, 80, 86
        real-life applications, 73, 79
        reasoning, 87
        by subtraction, 70–75
        two-step, 86
        using reciprocals, 79
    two-step
        error analysis, 86
        modeling, 82–83
        practice problems, 86
        real-life application, 85
        solving, 82–87
        writing, 87
    writing, 87
Equivalent equations, defined, 72
Error analysis
    algebra, 74, 80
    converting units of measure, 134
    coordinate planes, 38
    direct variation, 140
    division
        of integers, 32
    equations
        percent, 162

solving, 74, 80, 86
two-step, 86
integers
    absolute value of, 6
    adding, 12
    dividing, 32
    multiplying, 26
    subtracting, 18
interest, simple, 182
inverse variation, 146
percents, 162
    of change, 168
    of decrease, 168
prisms
    volume of, 302
probability, 388
    outcomes, 402
proportions, 120, 126
rational numbers
    as decimals, 54
    dividing, 66
    multiplying, 66
    subtracting, 60
samples and populations, 372
similar solids, 336
slope, 108
stem-and-leaf plots, 352
surface area of a cylinder, 267
volume
    of a cone, 320
    of a prism, 302
    of similar solids, 336
Event(s), *See also* Probability
    defined, 386
    dependent, 404–411
        defined, 406
        practice problems, 409
    independent, 404–411
        defined, 406
        practice problems, 409
Example and non-example chart,
    218
Experiment(s), *See also* Probability
    defined, 386
Experimental probability, *See also*
        Probability
    defined, 400
Exponent(s)
    using, 25
Expression(s)
    algebraic
        like terms, 81A
        practice problems, 81A–81B

real-life application, 81B
    simplest form, 81A
    simplifying, 81A–81B
equivalent
    defined, 72
error analysis, 26
evaluating, 31
simplifying, 81A–81B
    practice problems, 81A–81B
    real-life application, 81B

F

Fair experiment, defined, 393
Financial literacy, 57, 179
Formula(s)
    area
        of a circle, 261B, 264
        of a lateral surface, 265
        of a triangle, 259
    circumference, 261B
    diameter, 261A, 279
    interest, simple, 178
    pi, 261B
    radius, 261A
    surface area
        of a square prism, 285
        of a square pyramid, 285
    for surface area of
        a cone, 279
        a cylinder, 264
        a rectangular prism, 258
        a regular pyramid, 272
    volume
        of a composite solid, 326
        of a cone, 316, 318
        of a cylinder, 306
        of a prism, 298–299
        of a pyramid, 312
        of a square prism, 326
        of a square pyramid, 326
Formula triangle, 422
Four square, 268
Fraction(s)
    adding
        practice problems, 60
    decimals as, 53
    dividing
        practice problems, 66
    multiplying
        practice problems, 66
    simplifying
        practice problems, 60, 66

Student Index

Additional Answers

Chapter 1

Section 1.2

Practice and Problem Solving

43. *Sample answer:*
$-26 + 1;\ -12 + (-13)$

44. $d = -10$ **45.** $b = 2$

46. $m = -7$ **47.** $6 + (-3) + 8$

48. a. point C; E is $15 + (-13) = 2$ higher than C, so C is deeper.

    b. point B; D is $-18 + 15 = -3$ from B, so D is 3 units lower than B.

Section 1.5

Fair Game Review

42.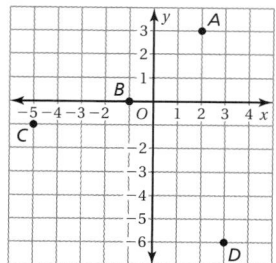

43.

44.

Section 1.6

On Your Own

5–8.

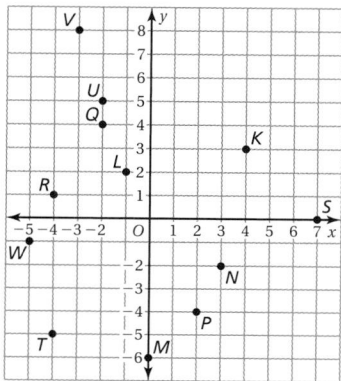

Practice and Problem Solving

15–26. See graph below.

32. never; All points in Quadrant III have negative y-coordinates.

33. always; The x-coordinate of a point in Quadrant II is negative, and so is the y-coordinate of a point in Quadrant IV.

39. $(2, 2)$

40–43. See graph below.

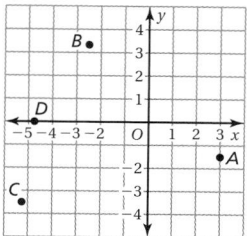

40. Quadrant IV   **41.** Quadrant II

42. Quadrant III   **43.** x-axis

44. a. $(7, -3)$

    b. school; $(2, -1)$ is closer to $(0, 0)$ than $(7, -3)$.

Chapter 2

Section 2.1

Practice and Problem Solving

40. $-5\frac{3}{11} < -5.\overline{2}$ **41.** $-2\frac{13}{16} < -2\frac{11}{14}$

42. *Sample answer:* $-0.4,\ -0.\overline{45}$

43. Michelle   **44.** math quiz

45. No; The base of the skating pool is at -10 feet, which is deeper than $-9\frac{5}{6}$ feet.

47. a. when a is negative

    b. when a and b have the same sign, $a \neq 0 \neq b$

Lesson 2.3b

Practice

3. $4 \cdot 19 \cdot \frac{1}{2} = 4 \cdot \frac{1}{2} \cdot 19$ Comm. Prop. of Mult.

    $= 2 \cdot 19$ Multiply 4 and $\frac{1}{2}$.

    $= 38$ Multiply 2 and 19.

4. $\frac{1}{3} \cdot 2 \cdot \frac{1}{2} = \frac{1}{3} \cdot \left(2 \cdot \frac{1}{2}\right)$ Assoc. Prop. of Mult.

    $= \frac{1}{3} \cdot 1$ Multiply 2 and $\frac{1}{2}$.

    $= \frac{1}{3}$ Mult. Prop. of One

5. $5\left(\dfrac{7}{8}\cdot\dfrac{2}{5}\right)=5\left(\dfrac{2}{5}\cdot\dfrac{7}{8}\right)$ Comm. Prop. of Mult.

$\qquad = \left(5\cdot\dfrac{2}{5}\right)\cdot\dfrac{7}{8}$ Assoc. Prop. of Mult.

$\qquad = 2\cdot\dfrac{7}{8}$ Multiply 5 and $\dfrac{2}{5}$.

$\qquad = \dfrac{7}{4}$ Multiply 2 and $\dfrac{7}{8}$.

6. $-1.45+(-8.55+2.7)=(-1.45+(-8.55))+2.7$
$\qquad\qquad\qquad\qquad$ Assoc. Prop. of Add.
$\qquad\qquad = -10+2.7$
$\qquad\qquad\qquad$ Add -1.45 and -8.55.
$\qquad\qquad = -7.3$
$\qquad\qquad\qquad$ Add -10 and 2.7.

Record and Practice Journal Practice

1. 5   **2.** 12.5   **3.** -4 **4.** 8   **5.** 4

6. $\dfrac{3}{2}$ **7.** 0

8. *Sample answer:* Win a bid of 6, Win a bid of 4, Lose a bid of 10

Lesson 2.5b
Record and Practice Journal Practice

1. Terms: $3x, 4, -7x, -6$;
Like terms: $3x$ and $-7x$, 4 and -6

2. Terms: $-9, 2.5y, -0.7y, 6.4y$;
Like terms: $2.5y, -0.7y,$ and $6.4y$

3. $3a+9$ **4.** $-\dfrac{1}{8}y+7$ **5.** $-3m+\dfrac{2}{3}$

6. $-0.7w+1.1$ **7.** $7d-5$ **8.** $-p-8$

9. $2\ell+12$ **10.** $15x+4$ **11.** $16x+12$

12. $20w+9m$ **13.** $10.2x$

Section 2.6
Practice and Problem Solving

40. -21 ft

41. decrease the length by 10 cm;
$2(25+x)+2(12)=54$

Lesson 2.6b
Practice

22. a. $20w+128\geq265;\ w\geq6.85$

b.

No; Yes; After 6 weeks you will only have $248, which is not enough. After 8 weeks you will have $288, which is enough.

Record and Practice Journal Practice

1. $b<12$;

2. $-7.5\leq z$;

3. $c\geq-2$;

4. $8\dfrac{1}{4}>y$;

5. $x\leq-72$;

6. $t\geq-26.4$;

7. $n<15$;

8. $f>-12$;

9. $m>-7.2$;

10. $k\geq1$;

11. $a<4$;

12. $p\leq-2.2$;

13. $x\leq14$;

14. $p<20$;

15. a. $w+16\leq20;\ w\leq4$

    b. The solution $w\leq4$ means that your dog drunk at most 4 quarts of water.

16. a. $4x+24\geq60;\ x\geq9$

    b.

no; yes; You have to answer 9 or more questions to win the game.

Chapter 3

Section 3.1

Record and Practice Journal

1. Numerical rates are sample answers.

Description	Verbal Rate	Numerical Rate
Your pay rate for washing cars	dollars per hour	$\dfrac{\$5}{h}; \dfrac{\$50}{h}$
The average rainfall rate in a rain forest	inches per year	$\dfrac{100 \text{ in.}}{\text{yr}}; \dfrac{5 \text{ in.}}{\text{yr}}$
Your average driving rate along an interstate	miles per hour	$\dfrac{60 \text{ mi}}{h}; \dfrac{600 \text{ mi}}{h}$
The growth rate for the length of a baby alligator	inches per month	$\dfrac{0.5 \text{ in.}}{\text{mo}}; \dfrac{10 \text{ in.}}{\text{mo}}$
Your running rate in a 100-meter dash	meters per second	$\dfrac{8 \text{ m}}{\text{sec}}; \dfrac{80 \text{ m}}{\text{sec}}$
The population growth rate of a large city	people per year	$\dfrac{25{,}000 \text{ people}}{\text{yr}}; \dfrac{10 \text{ people}}{\text{yr}}$
The average pay rate for a professional athlete	dollars per year	$\dfrac{\$3{,}000{,}000}{\text{yr}}; \dfrac{\$3000}{\text{yr}}$
The fertilization rate for an apple orchard	pounds per acre	$\dfrac{150 \text{ lb}}{\text{acre}}; \dfrac{1 \text{ lb}}{\text{acre}}$

Section 3.2

Record and Practice Journal

1. c. mi/h to ft/sec: Multiply by 5280 to convert miles to feet and divide by 3600 to convert hours to seconds.
   ft/sec to mi/h: Divide by 5280 to convert feet to miles and multiply by 3600 to convert seconds to hours.

Practice and Problem Solving

12.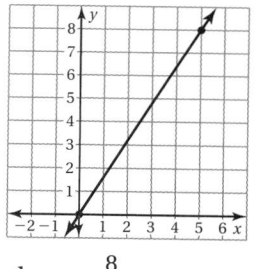

    slope $= \dfrac{8}{5}$

13.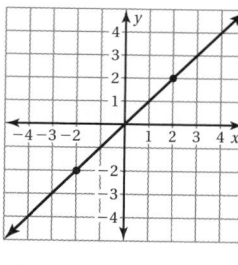

    slope $= 1$

14.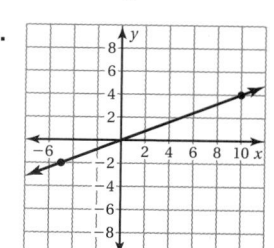

    slope $= \dfrac{2}{5}$

17. a.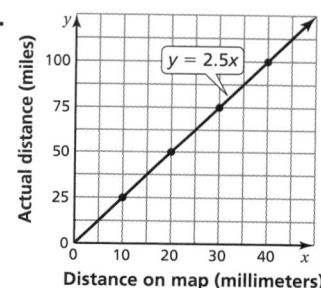

    b. 2.5; Every millimeter represents 2.5 miles.

    c. 120 mi

    d. 90 mm

Section 3.7

Practice and Problem Solving

32. a.

Location	Earth	Jupiter	Moon
Weight (lb)	100	214	16.67
Weight (lb)	120	256.8	20

   b. *Sample answer:* The amount of gravity changes.

35. Every graph of direct variation is a line; however, not all lines show direct variation because the line must pass through the origin.

Lesson 3.7b

Practice

1. $(0, 0)$: You earn \$0 for working 0 hours.

    $(1, 15)$: You earn \$15 for working 1 hour;
    unit rate: $\dfrac{\$15}{1\,\text{h}}$

    $(4, 60)$: You earn \$60 for working 4 hours;
    unit rate: $\dfrac{\$60}{4\,\text{h}} = \dfrac{\$15}{1\,\text{h}}$

2. $(0, 0)$: The balloon rises 0 feet in 0 seconds.

    $(1, 5)$: The balloon rises 5 feet in 1 second;
    unit rate: $\dfrac{5\,\text{ft}}{1\,\text{sec}}$

    $(6, 30)$: The balloon rises 30 feet in 6 seconds;
    unit rate: $\dfrac{30\,\text{ft}}{6\,\text{sec}} = \dfrac{5\,\text{ft}}{1\,\text{sec}}$

3. $y = 1.5$

Record and Practice Journal Practice

1. $(0, 0)$: The car travels 0 miles in 0 hours.

    $(1, 60)$: The car travels 60 miles in 1 hour.

    $(2, 120)$: The car travels 120 miles in 2 hours.

2. $(0, 0)$: 0 pounds of shrimp costs \$0.

    $(4, 40)$: 4 pounds of shrimp costs \$40.

    $(7, 70)$: 7 pounds of shrimp costs \$70.

3. $(0, 0)$: You receive 0 emails in 0 days.

    $(3, 45)$: You receive 45 emails in 3 days.

    $(4, 60)$: You receive 60 emails in 4 days.

4. $(0, 0)$: There are 0 cups of blueberries in 0 pies.

    $(2, 12)$: There are 12 cups of blueberries in 2 pies.

    $(4, 24)$: There are 24 cups of blueberries in 4 pies.

5. **a.** Waiter A: 20%; Waiter B: 15%

    b. Waiter A: 1000 cents, or \$10;
    Waiter B: 750 cents, or \$7.50

6. **a.** Salesman A: 5%; Salesman B: 7.5%

    b. \$250

    c. \$125 less

Section 3.8

Practice and Problem Solving

17.
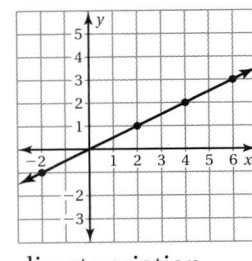
direct variation

18.
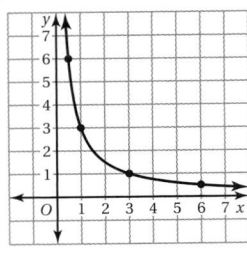
inverse variation

19.
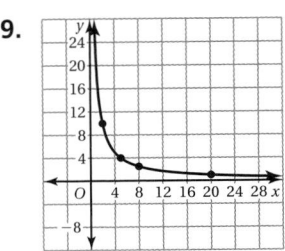
inverse variation

20.
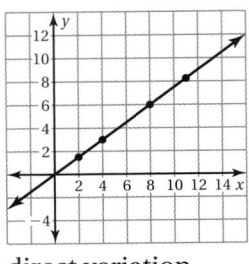
direct variation

21. inverse variation; The equation can be written as $y = \dfrac{k}{x}$.

22. direct variation; The equation can be written as $y = kx$.

Chapter 3 Test

5.
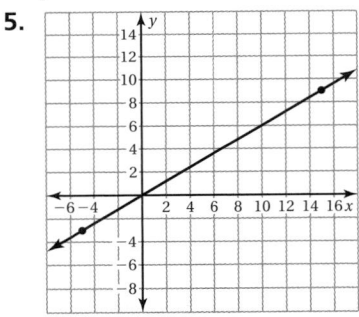
slope $= \dfrac{3}{5}$

6.
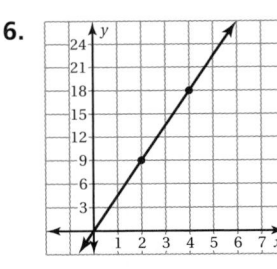
slope $= \dfrac{9}{2}$

Chapter 4

Section 4.2

Practice and Problem Solving

31. less than; *Sample answer:* Let x represent the number. A 10% increase is equal to $x + 0.1x$, or $1.1x$. A 10% decrease of this new number is equal to $1.1x - 0.1(1.1x)$, or $0.99x$. Because $0.99x < x$, the result is less than the original number.

Chapter 5

Section 5.1

Practice and Problem Solving

7.

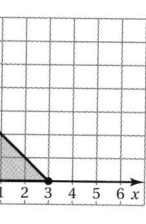

A and B; Corresponding side lengths are proportional and corresponding angles have the same measure.

8.

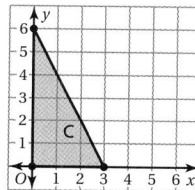
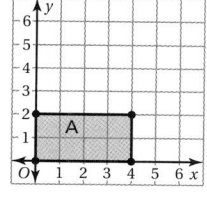
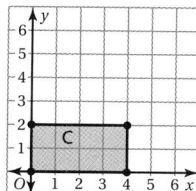

A, B and C; Corresponding side lengths are proportional and corresponding angles have the same measure.

16. $\angle F$ and $\angle Q$, $\angle G$ and $\angle R$, $\angle H$ and $\angle T$; Side FG and Side QR, Side GH and Side RT, Side FH and Side QT

18. a. sometimes; They are similar only when corresponding side lengths are proportional and corresponding angles have the same measure.

   b. always; All angles have the same measure and all sides are proportional.

   c. sometimes; Corresponding angles always have the same measure, but corresponding side lengths are not always proportional.

   d. never; They do not have the same shape.

19. yes; One could be a trapezoid and the other could be a parallelogram.

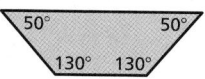

20. a. yes   **b.** no

21. a. yes

   b. yes; This is true for all similar triangles because the height of a triangle is a dimension of the triangle like the side lengths.

22. yes;

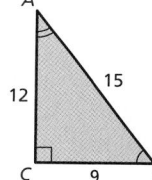 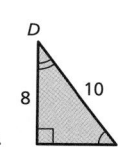

Because corresponding angles have the same measure and corresponding side lengths are proportional, $\triangle ABC$ is similar to $\triangle JKL$.

Record and Practice Journal Practice

1. $\angle A$ and $\angle E$, $\angle B$ and $\angle F$, $\angle C$ and $\angle G$, $\angle D$ and $\angle H$; Side AB and Side EF, Side BC and Side FG, Side CD and Side GH, Side DA and Side HE

2. $\angle X$ and $\angle L$, $\angle Y$ and $\angle M$, $\angle Z$ and $\angle N$; Side XY and Side LM, Side YZ and Side MN, Side ZX and Side NL

Section 5.2

Practice and Problem Solving

18. a. $\dfrac{1}{4}; \dfrac{1}{4}; \dfrac{1}{16}$

   b. The ratio of the circumferences is equal to the ratio of the radii. The ratio of the square of the radii is equal to the ratio of the areas. These are the same proportions that are used for similar figures.

Section 5.3

Record and Practice Journal

1. b.

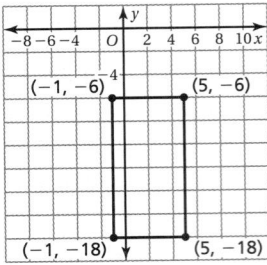

$$\frac{\text{Shaded Length}}{\text{Unshaded Length}} \overset{?}{=} \frac{\text{Shaded Width}}{\text{Unshaded Width}}$$

$$\frac{6}{12} \overset{?}{=} \frac{3}{6}$$

$$\frac{1}{2} = \frac{1}{2}$$

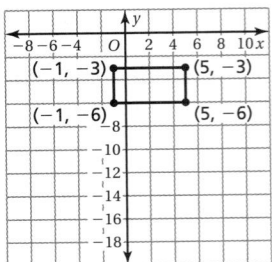

$$\frac{\text{Shaded Length}}{\text{Unshaded Length}} \overset{?}{=} \frac{\text{Shaded Width}}{\text{Unshaded Width}}$$

$$\frac{6}{6} \overset{?}{=} \frac{3}{3}$$

$$1 = 1$$

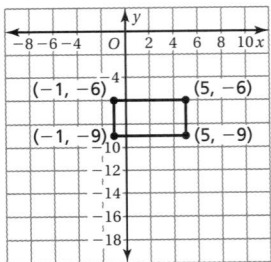

$$\frac{\text{Shaded Length}}{\text{Unshaded Length}} \overset{?}{=} \frac{\text{Shaded Width}}{\text{Unshaded Width}}$$

$$\frac{6}{6} \overset{?}{=} \frac{3}{3}$$

$$1 = 1$$

Lesson 5.4b

Practice

4.

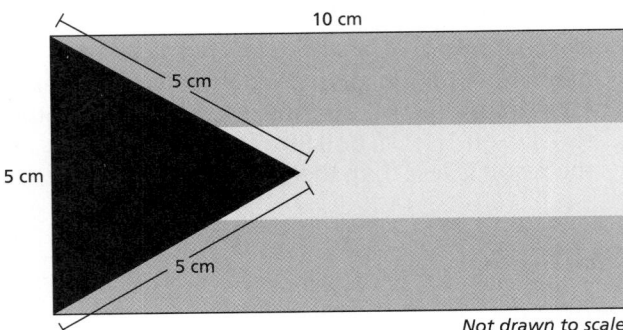

Not drawn to scale

5.

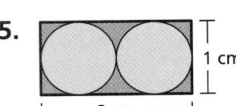

Record and Practice Journal Practice

1. 150,000 yd^2 **2.** 12 ft^2

3. Check students' drawings. Dimensions of flag are 4 inches by 7 inches.

4. Check students' drawings. Dimensions of flag are 1 inch by 1.75 inches.

Section 5.5

Practice and Problem Solving

15.

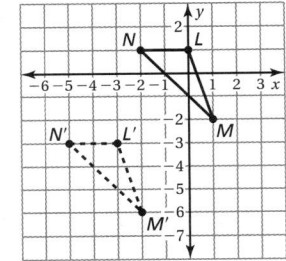

21. **a.** 5 units right and 1 unit up

   b. no; It would hit the island.

   c. 4 units right and 4 units up

22. 2 units right and 6 units down

Section 5.6

On Your Own

4. a.

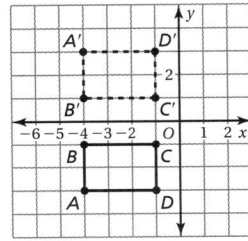

b.

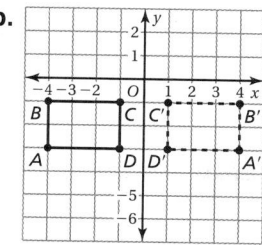

c. yes; They are all rectangles of the same size and shape.

Practice and Problem Solving

27.

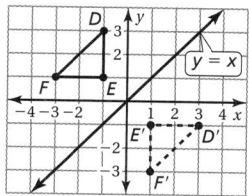

The x-coordinate and y-coordinate for each point are switched in the image.

Section 5.7

On Your Own

3. a.

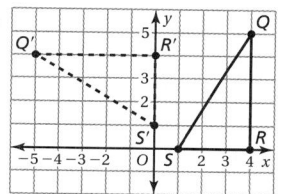

b.

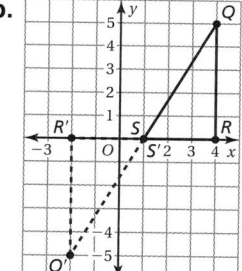

c. yes; The size and shape do not change in a rotation.

Practice and Problem Solving

15.

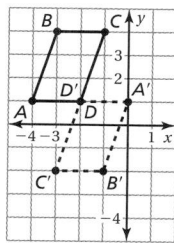

$A'(0, 1), B'(-1, -2), C'(-3, -2), D'(-2, 1)$

16.

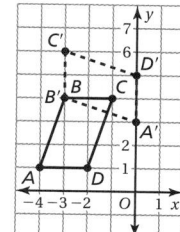

$A'(0, 3), B'(-3, 4), C'(-3, 6), D'(0, 5)$

18. a. *Sample answer:* The original figure is a 2×3 rectangle. The image is a 4×6 rectangle.

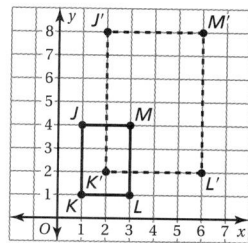

b. no; yes; The rectangles are similar because the ratios of the corresponding sides are both $\frac{2}{1}$.

c. A dilation changes the size of the figure. The other transformations maintain the original size and shape.

Chapter 5 Test

7.

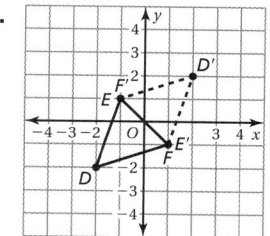

$D'(2, 2), E'(1, -1), F'(-1, 1)$

8.

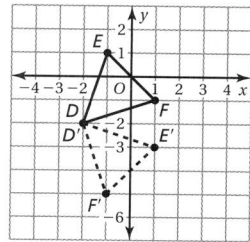

$D'(-2, -2), E'(1, -3), F'(-1, -5)$

Chapter 6

Section 6.1

Record and Practice Journal

1. b.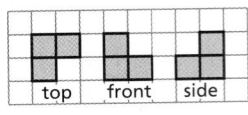

   c.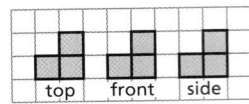

   d.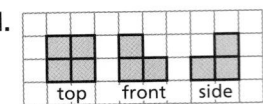

   e.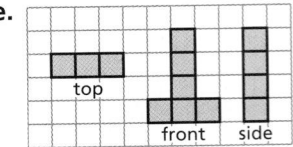

   f.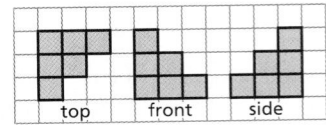

   g.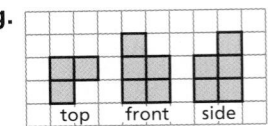

2. a.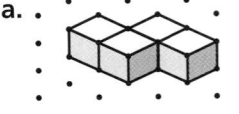

   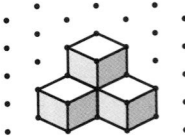

On Your Own

3. front:

   side:

   top: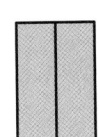

4. front:

   side:

   top:

5. front: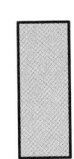

   side: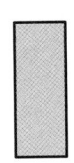

   top:

Vocabulary and Concept Check

3. *Sample answer:* Prisms: A cereal box is a rectangular prism. A pup tent with parallel triangular bases at the front and back is a triangular prism.

   Pyramids: The Egyptian pyramids are rectangular pyramids. A house roof forms a pyramid if it has lateral faces that are triangles that meet at a common vertex.

   Cylinders: Some examples of cylinders are a soup can, a tuna fish can, and a new, unsharpened, round pencil.

   Cones: Some examples of cones are a traffic cone, an ice cream sugar cone, a party hat, and the sharpened end of a pencil.

4. base: hexagon
   solid: hexagonal pyramid

5. base: circle
   solid: cylinder

6. base: pentagon
   solid: pentagonal prism

Practice and Problem Solving

9. front:

side:

top:

surface area: 38 units²
volume: 9 units³

10.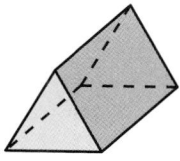

11.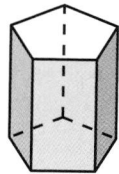

12.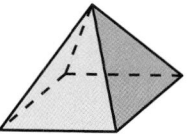

13.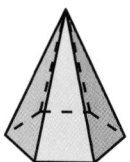

14.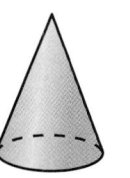

15.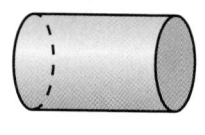

16. front:

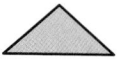

side:

top:

17. front:

side:

top:

18. front:

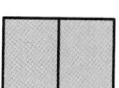

side:

top:

19. front:

side:

top:

20. front:

side:

top:

21. front:

side:

top:

22.

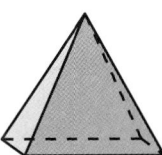

23. The Washington Monument is an *obelisk*. It consists of a pyramid sitting on top of a solid that tapers as it rises.

24.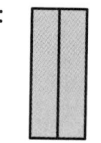

25.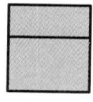

26. *Answer should include, but is not limited to*: an original drawing of a house; a description of any solids that make up any part of the house

28. *Sample answer:*

a.

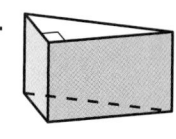   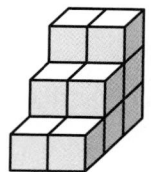

Triangular prism
6 vertices
9 edges

Square pyramid
5 vertices
8 edges

b. More than one solid can have the same number of faces, so knowing the number of edges and vertices can help you to draw the intended solid.

Lesson 6.2b

Record and Practice Journal Practice

1. 3 ft   **2.** 5.4 cm   **3.** $2\frac{1}{4}$ in.   **4.** 5 mm

5. 8 yd   **6.** 22 m   **7.** 44 m   **8.** 14.444 ft

9. 47.1 in.   **10.** 154 yd²

11. 28.26 cm²   **12.** 314 mm²

13. 339.12 ft²; The garden area is $\frac{3}{4}$ of a circle with a 12-foot radius.

Section 6.4
Record and Practice Journal
1. **a.** $SA = 85{,}560 \text{ m}^2$

   b. $SA = 1404 \text{ m}^2$

   c. $SA = 1960 \text{ m}^2$

   d. $SA = 1276 \text{ m}^2$

2. **b.**

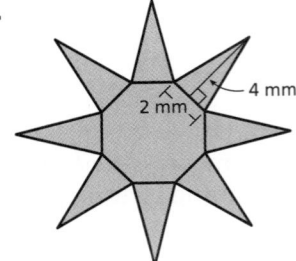

Section 6.5
Record and Practice Journal
2. Surface Area:

   $3\pi\left(\dfrac{5}{2}\right) + \left(\dfrac{5}{2}\right)^2 \pi \text{ in.}^2$

   $3\pi\left(\dfrac{4}{2}\right) + \left(\dfrac{4}{2}\right)^2 \pi \text{ in.}^2$

   $3\pi\left(\dfrac{3}{2}\right) + \left(\dfrac{3}{2}\right)^2 \pi \text{ in.}^2$

   $3\pi\left(\dfrac{2}{2}\right) + \left(\dfrac{2}{2}\right)^2 \pi \text{ in.}^2$

   $3\pi\left(\dfrac{1}{2}\right) + \left(\dfrac{1}{2}\right)^2 \pi \text{ in.}^2$

Section 6.6
Record and Practice Journal
2.

Base of *n* Blocks	1	2	3	4	5
Surface Area	6	14	24	36	50

For each 1 unit increase of *n*, the increase in surface area is two square units greater than the last increase.

For 10 blocks, $S = 150$.

Chapter 7

Section 7.6
Practice and Problem Solving
18. **a.** 9483 pounds; The ratio of the height of the original statue to the height of the small statue is 8.4 : 1. So, the ratio of the weights, or volumes is $\left(\dfrac{8.4}{1}\right)^3$.

   b. 221,184 lb

Chapter 7 Test
11. *Sample answer:*

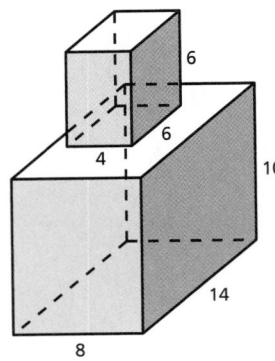

$V = 1264$ cubic units
$S = 784$ square units

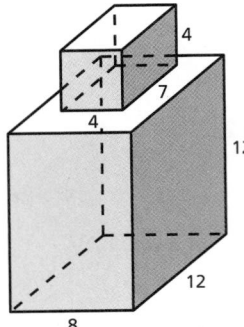

$V = 1264$ cubic units
$S = 760$ square units

Chapter 8

Section 8.1
Record and Practice Journal
1. **d.** 0, 1, 2, 2, 3, 4, 4, 7, 8, 9, 12, 12, 15, 16, 17, 18, 19, 22, 22, 26, 28, 29, 35, 38, 39, 42, 47, 54, 55, 62, 67, 68, 72, 73, 76, 77, 79, 79, 82, 82, 84, 85, 85, 86, 88, 89, 91, 91, 91, 91, 91, 91, 92, 92, 92, 92, 93, 93

Practice and Problem Solving
9. **Hours Online**

Stem	Leaf
0	0 2 6 8
1	2 2 4 5 7 8
2	1 4

Key: 2 | 1 = 21 hours

10. Test Scores

Stem	Leaf
6	2 5 8 9
7	
8	0 1 2 5 7 7 8
9	0 1 5 7

Key: 8 | 1 = 81%

11. Points Scored

Stem	Leaf
3	8
4	2 2 3 3 5
5	0 1 6 8 8
6	
7	0 1 1 5

Key: 3 | 8 = 38 points

12. Stems that fall in the range should still be shown even if there are no leaves for that stem.

Stem	Leaf
2	5 6
3	
4	2 4 7
5	0 1 5 5

Key: 4 | 2 = 42

13. Weights

Stem	Leaf
0	8
1	2 5 7 8
2	4 4
3	1

Key: 2 | 4 = 24 pounds

Most of the weights are in the middle.

14. Bikes Sold

Stem	Leaf
7	8 9
8	1 6
9	6 9 9
10	0 5 8
11	2 5

Key: 11 | 2 = 112 bikes

15. Minutes in Line

Stem	Leaf
1	6 9
2	0 2 6 7 9
3	1 1 6 8
4	0

Key: 4 | 0 = 4.0 minutes

16. 6 players

17. mean: 56.6; median: 53; modes: 41, 43, 63; range: 56

18. Most of the data are in the 40s, 50s, and 60s.

19. 97; It increases the mean.

20. a. *Sample answer:*

Heights

Stem	Leaf
7	3 4 4 6 7 9 9 9
8	0 0 1 1 1 3 4

Key: 7 | 3 = 73 inches

b. The heights are balanced. The range of heights is 11 inches. All players are taller than 6 feet.

21. *Sample answer:* Points by a basketball player in his first 8 games

Points

Stem	Leaf
2	1 3 4
3	2 4
4	0 1 5

Key: 3 | 2 = 32 points

Fair Game Review

26.

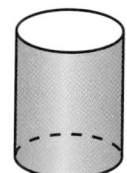

Section 8.2

Record and Practice Journal

1. **b.** skew: The mean and median are towards the left. The mode could be anywhere.

    normal: The mean and median are towards the middle. The mode could be anywhere.

    bimodal: The mean and median are towards the middle. The mode is probably near one of the high points.

    flat: The mean and median are in the middle. The mode could be anywhere.

   d. skew: The distribution is leaning or "skewed" to one side.

    normal: The average value occurs the most often, which is what you would expect (i.e., what is normal).

    bimodal: The distribution looks like it has two modes..

    flat: The distribution has a flat top, as any value is just as likely to occur as any other.

Practice and Problem Solving

12. no; You only know what interval each of the data values falls into, not the specific data values.

Record and Practice Journal Practice

1.
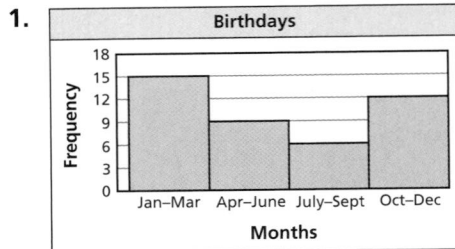

2.
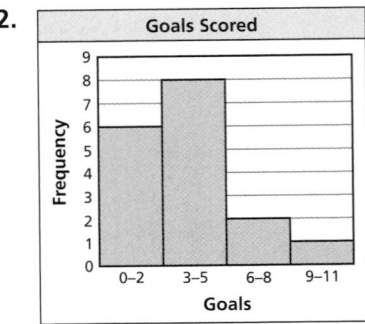

3.
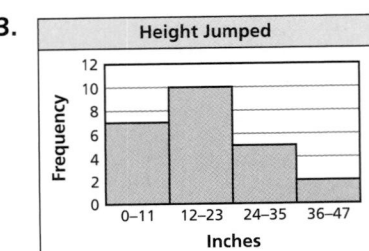

4.
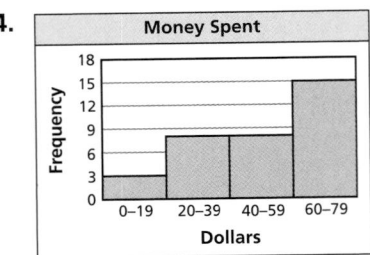

8.1–8.2 Quiz

1. **Cans Collected Each Month**

Stem	Leaf
5	5 8 9
6	1 8
7	6 9
8	0 3 4
9	0 2

   Key: 6 | 1 = 61 cans

2. **Miles Driven Each Day**

Stem	Leaf
0	9
1	0 1 2 5 6 6 8
2	0 0 1 8
3	5 7
4	
5	0

   Key: 3 | 5 = 35 miles

3. **Age of Tortoises**

Stem	Leaf
8	3 5 6
9	2 9
10	0 4
11	0 5
12	4 9
13	0

   Key: 9 | 2 = 92 years

4. Kilometers Run Each Day

Stem	Leaf
2	0 5
3	0 1 1 5 8 9
4	5
5	0 6
6	0 1 2 2

Key: 2 | 0 = 2.0 kilometers

5.

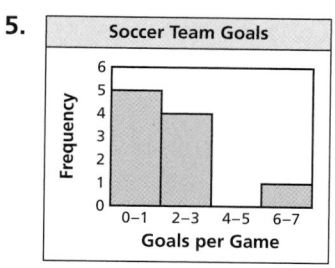

6.

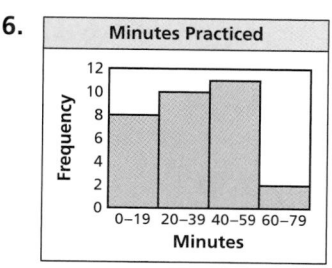

Section 8.3

Record and Practice Journal

1. d. baseball: 25°; about 42 students
basketball: 60°; 100 students
soccer: 15°; 25 students
hockey: 20°; about 33 students
track: 25°; about 42 students
wrestling: 25°; about 42 students
swimming: 25°; about 42 students
gymnastics: 40°; about 67 students
skating: 30°; 50 students
other: 40°; about 67 students

Practice and Problem Solving

18.

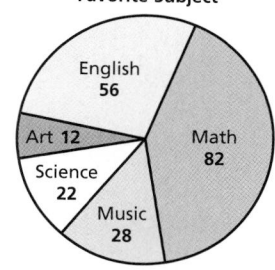

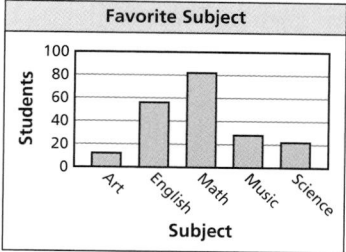

Sample answer: The circle graph does a better job of representing the data because the size of each section is based on a specific scale (100% or 360°).

Section 8.4

Practice and Problem Solving

14. A sample because it is much easier for which to collect data.

15. A population because there are few enough students in your homeroom to not make the surveying difficult.

16. A sample because it is much easier for which to collect data.

17. 1260 students

18. Not everyone has an email address, so the sample may not be representative of the entire population.

Lesson 8.4b

Practice

1. a. *Sample answer:* randomly asking students at lunch

b. *Sample answer:* randomly asking students on the football team

4.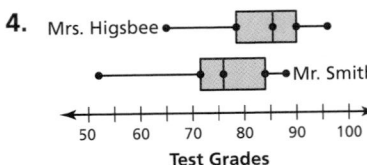

In general, Mrs. Higsbee's class has better test grades than Mr. Smith's class.

Record and Practice Journal Practice

1. a. Survey 1: 420 students; Survey 2: 680 students

   b. The prediction from Survey 1 is more reasonable because it uses a reasonable sample.

2. In general, boys have a larger shoe size than girls.

3.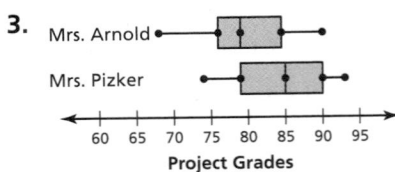

In general, Mrs. Pizker's class received higher grades on the project.

8.3–8.4 Quiz

6. a. population: all students in your school
   sample: 20 students arriving at football tryouts

   b. no; The sample is not random or representative of the entire student body because most football players use the gym.

7. a. $\dfrac{1}{4}$

   b. 500 students

   c. 275 students

Chapter 8 Test

1. Quiz Scores

Stem	Leaf
6	8
7	2 6
8	0 0 0 8
9	0 2 4 6
10	0

Key: 7 | 2 = 72%

2. CDs sold each day

Stem	Leaf
2	9
3	1 8 8 9
4	0 0 3 5 5 8
5	
6	0 0 2 7

Key: 4 | 0 = 40 CDs

3.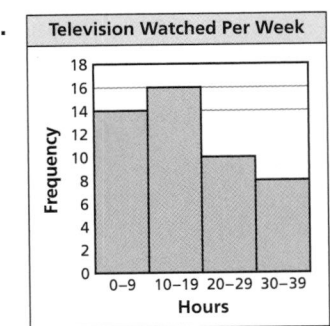

Additional Topics

Topic 1

Record and Practice Journal Practice

1. neither

2. complementary

3. complementary

4. supplementary

5. supplementary

6. neither

7. 50

8. adjacent; 52

9. vertical; 126

10. vertical; 36

11. adjacent; 10

12. $m\angle 1 = 128°$; $m\angle 2 = 52°$; $m\angle 3 = 128°$

Topic 2

Practice

1.

6 cm

3 cm

5 cm

2.

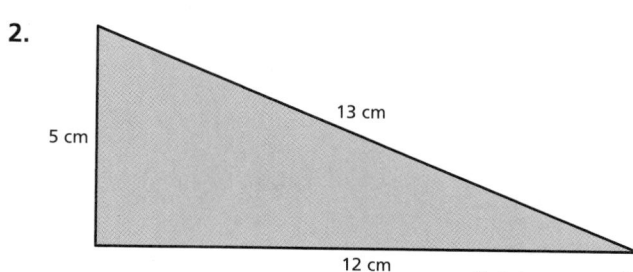

13 cm
5 cm
12 cm
Not drawn to scale

3.

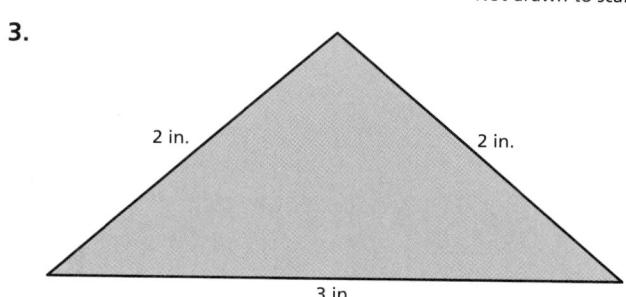

2 in.   2 in.
3 in.

6.

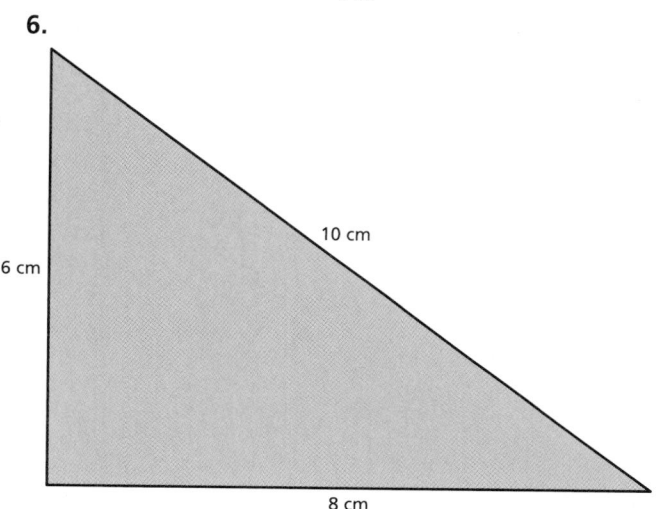

10 cm
6 cm
8 cm

8.

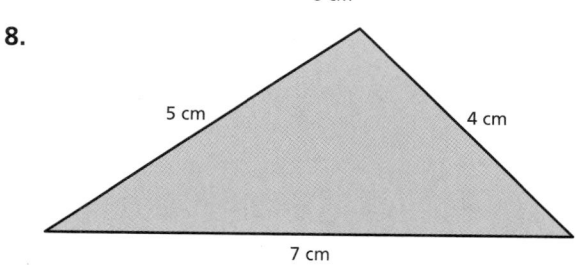

5 cm   4 cm
7 cm

9.

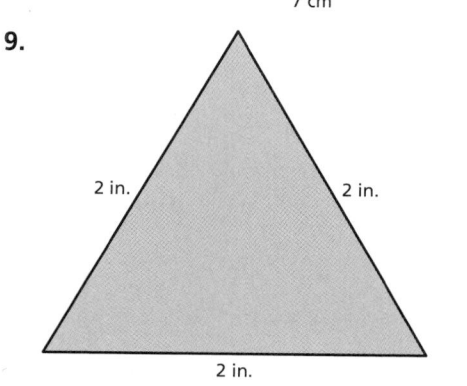

2 in.   2 in.
2 in.

11. Side lengths in Exercise 5:

Side length	2 cm	4 cm	1 cm
Sum of other two side lengths	5 cm	3 cm	6 cm

Side lengths in Exercise 6:

Side length	6 cm	8 cm	10 cm
Sum of other two side lengths	18 cm	16 cm	14 cm

Side lengths in Exercise 7:

Side length	1 in.	2 in.	1 in.
Sum of other two side lengths	3 in.	2 in.	3 in.

Side lengths in Exercise 8:

Side length	5 cm	7 cm	4 cm
Sum of other two side lengths	11 cm	9 cm	12 cm

Side lengths in Exercise 9:

Side length	2 in.	2 in.	2 in.
Sum of other two side lengths	4 in.	4 in.	4 in.

Side lengths in Exercise 10:

Side length	1 in.	5 in.	3 in.
Sum of other two side lengths	8 in.	4 in.	6 in.

Record and Practice Journal Practice

1.

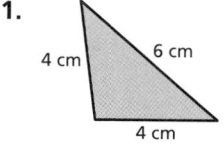

4 cm   6 cm
4 cm

2. not possible

3.

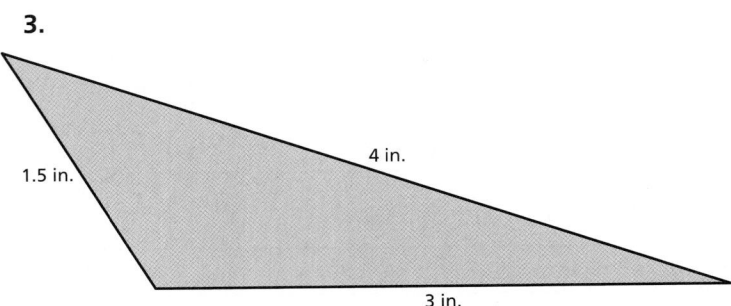

4.

5. not possible

6. triangle

7. rectangle

8. rectangle

Photo Credits

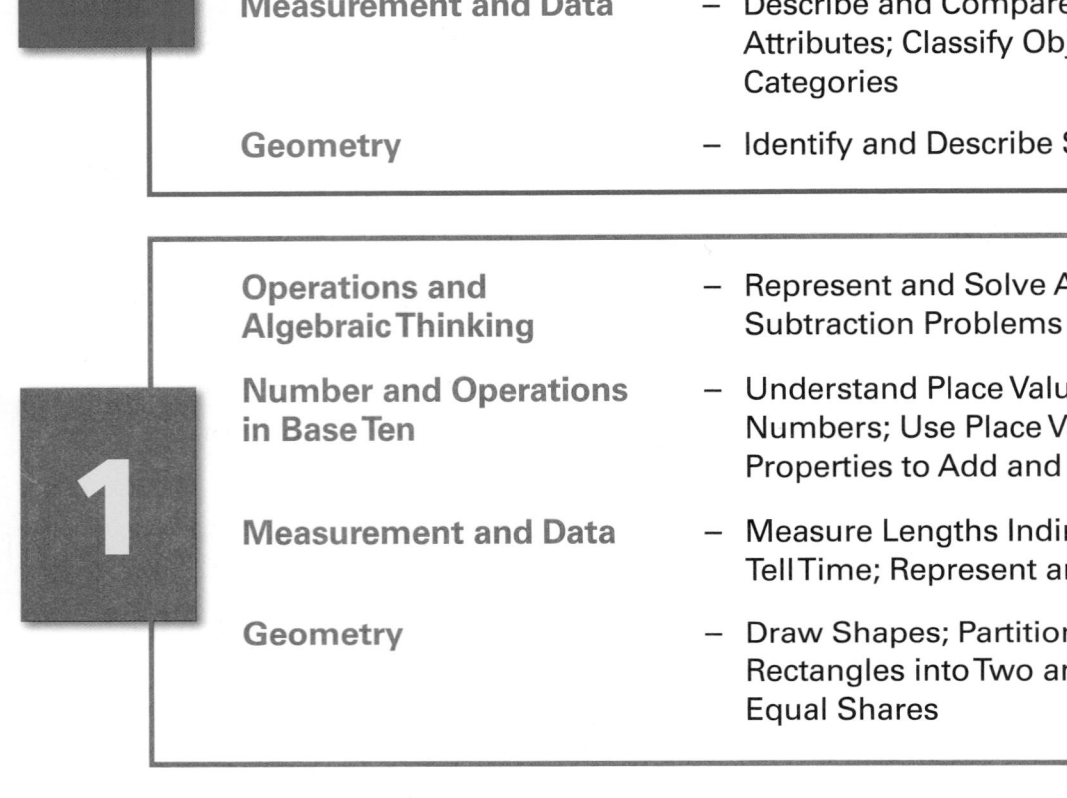

K

Counting and Cardinality	– Count to 100 by Ones and Tens; Compare Numbers
Operations and Algebraic Thinking	– Understand and Model Addition and Subtraction
Number and Operations in Base Ten	– Work with Numbers 11–19 to Gain Foundations for Place Value
Measurement and Data	– Describe and Compare Measurable Attributes; Classify Objects into Categories
Geometry	– Identify and Describe Shapes

1

Operations and Algebraic Thinking	– Represent and Solve Addition and Subtraction Problems
Number and Operations in Base Ten	– Understand Place Value for Two-Digit Numbers; Use Place Value and Properties to Add and Subtract
Measurement and Data	– Measure Lengths Indirectly; Write and Tell Time; Represent and Interpret Data
Geometry	– Draw Shapes; Partition Circles and Rectangles into Two and Four Equal Shares

2

Operations and Algebraic Thinking	– Solving One- and Two-Step Problems Involving Addition and Subtraction; Build a Foundation for Multiplication
Number and Operations in Base Ten	– Understand Place Value for Three-Digit Numbers; Use Place Value and Properties to Add and Subtract
Measurement and Data	– Measure and Estimate Lengths in Standard Units; Work with Time and Money
Geometry	– Draw and Identify Shapes; Partition Circles and Rectangles into Two, Three, and Four Equal Shares

3

Operations and Algebraic Thinking — Represent and Solve Problems Involving Multiplication and Division; Solve Two-Step Problems Involving Four Operations

Number and Operations in Base Ten — Round Whole Numbers; Add, Subtract, and Multiply Multi-Digit Whole Numbers

Number and Operations — Fractions — Understand Fractions as Numbers

Measurement and Data — Solve Time, Liquid Volume, and Mass Problems; Understand Perimeter and Area

Geometry — Reason with Shapes and Their Attributes

4

Operations and Algebraic Thinking — Use the Four Operations with Whole Numbers to Solve Problems; Understand Factors and Multiples

Number and Operations in Base Ten — Generalize Place Value Understanding; Perform Multi-Digit Arithmetic

Number and Operations — Fractions — Build Fractions from Unit Fractions; Understand Decimal Notation for Fractions

Measurement and Data — Convert Measurements; Understand and Measure Angles

Geometry — Draw and Identify Lines and Angles; Classify Shapes

5

Operations and Algebraic Thinking — Write and Interpret Numerical Expressions

Number and Operations in Base Ten — Perform Operations with Multi-Digit Numbers and Decimals to Hundredths

Number and Operations — Fractions — Add, Subtract, Multiply, and Divide Fractions

Measurement and Data — Convert Measurements within a Measurement System, Understand Volume

Geometry — Graph Points in the First Quadrant of the Coordinate Plane; Classify Two-Dimensional Figures

Mathematics Reference Sheet

Conversions

U.S. Customary
1 foot = 12 inches
1 yard = 3 feet
1 mile = 5280 feet
1 acre ≈ 43,560 square feet
1 cup = 8 fluid ounces
1 pint = 2 cups
1 quart = 2 pints
1 gallon = 4 quarts
1 gallon = 231 cubic inches
1 pound = 16 ounces
1 ton = 2000 pounds
1 cubic foot ≈ 7.5 gallons

U.S. Customary to Metric
1 inch ≈ 2.54 centimeters
1 foot ≈ 0.3 meter
1 mile ≈ 1.6 kilometers
1 quart ≈ 0.95 liter
1 gallon ≈ 3.79 liters
1 cup ≈ 237 milliliters
1 pound ≈ 0.45 kilogram
1 ounce ≈ 28.3 grams
1 gallon ≈ 3785 cubic centimeters

Time
1 minute = 60 seconds
1 hour = 60 minutes
1 hour = 3600 seconds
1 year = 52 weeks

Temperature
$$C = \frac{5}{9}(F - 32)$$

$$F = \frac{9}{5}C + 32$$

Metric
1 centimeter = 10 millimeters
1 meter = 100 centimeters
1 kilometer = 1000 meters
1 liter = 1000 milliliters
1 kiloliter = 1000 liters
1 milliliter = 1 cubic centimeter
1 liter = 1000 cubic centimeters
1 cubic millimeter = 0.001 milliliter
1 gram = 1000 milligrams
1 kilogram = 1000 grams

Metric to U.S. Customary
1 centimeter ≈ 0.39 inch
1 meter ≈ 3.28 feet
1 kilometer ≈ 0.6 mile
1 liter ≈ 1.06 quarts
1 liter ≈ 0.26 gallon
1 kilogram ≈ 2.2 pounds
1 gram ≈ 0.035 ounce
1 cubic meter ≈ 264 gallon

Circumference and Area of a Circle

$C = \pi d$ or $C = 2\pi r$

$A = \pi r^2$

$\pi \approx \dfrac{22}{7}$, or 3.14

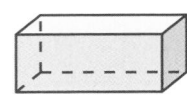

The Coordinate Plane

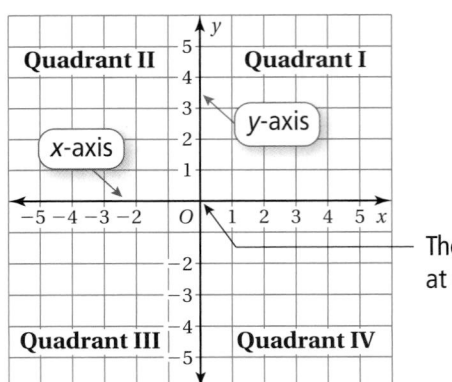

The origin is at (0, 0).

Surface Area and Volume

Prism

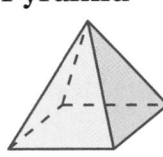

S = areas of bases
      + areas of lateral faces

$V = Bh$

Pyramid

S = area of base
      + areas of lateral faces

$V = \dfrac{1}{3}Bh$

Cylinder

$S = 2\pi r^2 + 2\pi rh$

$V = Bh$

Cone

$S = \pi r^2 + \pi r\ell$

$V = \dfrac{1}{3}Bh$